MW01204060

Small and Decentralized Wastewater Management Systems

McGraw-Hill Series in Water Resources and Environmental Engineering

CONSULTING EDITOR

George Tchobanoglous, *University of California, Davis*

Bailey and Ollis: *Biochemical Engineering Fundamentals*
Bouwer: *Groundwater Hydrology*
Canter: *Environmental Impact Assessment*
Chanlett: *Environmental Protection*
Chapra: *Surface Water-Quality Modeling*
Chow, Maidment, and Mays: *Applied Hydrology*
Crites and Tchobanoglous: *Small and Decentralized Wastewater Management Systems*
Davis and Cornwell: *Introduction to Environmental Engineering*
deNevers: *Air Pollution Control Engineering*
Eckenfelder: *Industrial Water Pollution Control*
Eweis, Ergas, Chang, and Schroeder: *Bioremediation Principles*
LaGrega, Buckingham, and Evans: *Hazardous Waste Management*
Linsley, Franzini, Freyberg, and Tchobanoglous: *Water Resources and Engineering*
McGhee: *Water Supply and Sewerage*
Mays and Tung: *Hydrosystems Engineering and Management*
Metcalf & Eddy, Inc.: *Wastewater Engineering: Collection and Pumping of Wastewater*
Metcalf & Eddy, Inc.: *Wastewater Engineering: Treatment, Disposal, Reuse*
Peavy, Rowe, and Tchobanoglous: *Environmental Engineering*
Sawyer and McCarty: *Chemistry for Environmental Engineering*
Tchobanoglous, Theisen, and Vigil: *Integrated Solid Waste Management: Engineering Principles and Management Issues*
Wentz: *Hazardous Waste Management*
Wentz: *Safety, Health, and Environmental Protection*

Small and Decentralized Wastewater Management Systems

Ron Crites

Managing Engineer
Brown and Caldwell

Formerly
Director—Water Resources
Nolte and Associates

George Tchobanoglous

Professor Emeritus
Department of Civil and Environmental Engineering
University of California, Davis

Consultant
Nolte and Associates

The preparation of this textbook was sponsored by
Nolte and Associates

Boston Burr Ridge, IL Dubuque, IA Madison, WI
New York San Francisco St. Louis
Bangkok Bogotá Caracas Lisbon London Madrid Mexico City
Milan New Delhi Seoul Singapore Sydney Taipei Toronto

About the Cover Photograph

Wastewater management in the community of Stinson Beach, located north of San Francisco on the coast, is accomplished entirely with individual onsite systems. Stinson Beach was one of the first communities in the United States to develop and implement an Onsite Wastewater Management District.

Photographs

All of the photographs were taken and printed by George Tchobanoglous, unless otherwise noted.

WCB/McGraw-Hill

A Division of The ***McGraw-Hill*** *Companies*

SMALL AND DECENTRALIZED WASTEWATER MANAGEMENT SYSTEMS

This book is printed on acid-free paper.

2 3 4 5 6 7 8 9 0 DOC/DOC 1 0 9 8

ISBN 0-07-289087-8

Vice president and editorial director: *Kevin T. Kane*
Publisher: *Tom Casson*
Executive editor: *Eric Munson*
Editorial assistant: *George Haag*
Marketing manager: *John T. Wannemacher*
Project manager: *Karen J. Nelson*
Production supervisor: *Lori Koetters*
Designer: *Gino Cieslik*
Compositor: *Publication Services, Inc.*
Typeface: *10/12 Times Roman*
Printer: *R. R. Donnelley & Sons Company*

Library of Congress Cataloging-in-Publication Data

Crites, Ron.
Small and decentralized wastewater management systems / Ron Crites, George Tchobanoglous.
p. cm.
Includes index.
ISBN 0-07-289087-8
1. Sewage disposal plants. 2. Appropriate technology.
I. Tchobanoglous, George. II. Title.
TD746.C75 1998
628.3–dc21 97-43819

www.mhhe.com

We dedicate this book to our wives, Pam and Rosemary,
for their patience, understanding, and encouragement.

ABOUT THE AUTHORS

RONALD W. CRITES is a managing engineer with Brown and Caldwell in Sacramento, California. During the preparation of this book he was director—water resources with Nolte and Associates in Sacramento, California. He received his B.S. degree in Civil Engineering from California State University, Chico, and his M.S. and his engineer's degree in sanitary engineering from Stanford University. His primary interests and experience are in natural systems for wastewater treatment, onsite systems, water reuse, and biosolids management. He has 30 years of experience in wastewater engineering consulting. He has authored or coauthored over 110 technical publications, including four textbooks. He has conducted seminars and workshops on constructed wetlands, land treatment, and water reuse both nationally and internationally. He is a member of ASCE, AWWA, WEF, IAWQ, ASA, and WateReuse. He has contributed to seven Water Environment Federation Manuals of Practice. He is a registered civil engineer in California, Hawaii, Massachusetts, and Oregon.

GEORGE TCHOBANOGLOUS is an emeritus professor of environmental engineering in the Department of Civil and Environmental Engineering at the University of California at Davis. He received his B.S. in civil engineering from the University of the Pacific, his M.S. degree in sanitary engineering from the University of California at Berkeley, and his Ph.D. in environmental engineering from Stanford University. His principal research interests are in the areas of wastewater treatment and reuse, wastewater filtration, UV disinfection, aquatic wastewater management systems, wastewater management for small and decentralized systems, and solid waste management. He has authored over 300 technical publications including 12 textbooks and two reference works. The textbooks are used in more than 200 colleges and universities throughout the United States, and they are also used extensively by practicing engineers in the United States and abroad. Professor Tchobanoglous serves nationally and internationally as a consultant to both governmental agencies and private companies. An active member of numerous professional societies, he is a past president of the American Association of Environmental Engineering Professors. He is a registered civil engineer in California.

PREFACE

Since the passage of the Clean Water Act in 1972, the primary focus of wastewater management activities in the United States has been on community point source wastewater discharges. Yet, more than 60 million people in the United States live in homes that are served by decentralized collection and treatment systems. It is now recognized that complete sewerage of the country may never be possible or desirable, for geographical, economic, and sustainability reasons. Given the fact that complete sewerage is unlikely for many residents, decentralized wastewater management becomes of great importance to the future management of the environment. Clearly, there is a need for a text that puts the engineering and scientific details of small and decentralized wastewater management systems into a textbook format. This textbook is a response to that need. Both the student and the practitioner will find in this book the engineering principles, the data, the engineering and scientific formulas, and examples of the day-to-day issues associated with the management of wastewater from both small and individual wastewater flows.

ORGANIZATION

The scope of the book is comprehensive and includes the design of alternative collection systems, both conventional and innovative systems for wastewater treatment, and reuse or disposal methods for the treated effluent. The constituents in wastewater and their fate in the environment are discussed in Chaps. 2 and 3. Process analysis and design of wastewater treatment systems are discussed in Chaps. 4, 5, and 7 through 11. Alternative wastewater collection systems are discussed in Chap. 6. Effluent repurification and reuse are discussed in Chap. 12; effluent disposal from small decentralized wastewater systems is described in Chap. 13. Biosolids and septage management is discussed in Chap. 14 and management of decentralized wastewater systems is discussed in Chap. 15.

IMPORTANT FEATURES OF THIS BOOK

To aid in the planning, analysis, and design of wastewater management systems, design data and information are summarized and presented in more than 300 tables. To illustrate the principles and facilities involved in the field of wastewater management, over 570 illustrations, graphs, and diagrams are included. To help the reader understand the material presented in this textbook, detailed solved examples are presented in Chaps. 2 through 14. Whenever possible, spreadsheet solutions are presented. To help the readers of this textbook hone their analytical skills, a series of discussion topics and problems are included at the end of each chapter. Selected references are also included at the end of each chapter.

To further increase the utility of this textbook, a series of appendixes have been included. Conversion factors from U.S. Customary Units to the International System (SI) of Units are presented in App. A. Conversion factors commonly used for the analysis and design of wastewater management systems are presented in App. B. Physical characteristics of water and selected gases are presented in Apps. C and D, respectively. Dissolved-oxygen concentrations in water as a function of temperature are presented in App. E. Carbonate equilibrium is considered in App. F. Tables of most probable numbers (MPN) are presented in App. G.

USE OF THIS BOOK

Enough material is presented in this textbook to support up to three quarter-length or two semester-length courses at either the undergraduate or graduate level. The first four chapters along with Chaps. 7 and 12 compose a basic introduction to the field of wastewater management. The material presented in this text can be used in a number of different courses. Suggested outlines are presented below for courses dealing with (1) conventional wastewater treatment with an emphasis on smaller treatment systems (less than 1 to 5 Mgal/d), (2) decentralized wastewater management, and (3) natural systems for wastewater management. A suggested outline for an introductory course in wastewater treatment is presented below.

Topic	Chapter	Sections
Introduction	1	All
Wastewater characteristics	2	All
Fate of constituents	3	All
Introduction to process analysis	4	All
Wastewater pretreatment	5	All
Biological treatment of wastewater	7	All
Land treatment systems	10	10-1 to 10-2
Effluent disposal and reuse	12	12-1 to 12-2
Biosolids management	14	All

A suggested outline for an undergraduate course dealing with decentralized wastewater management systems is presented below.

Topic	Chapter	Sections
Introduction	1	All
Wastewater characteristics	2	All
Fate of constituents	3	3-5 and 3-7
Introduction to process analysis	4	4-1 through 4-3
Wastewater pretreatment	5	5-1, 5-10 through 5-14
Alternative wastewater collection systems	6	All
Treatment of septic tank effluent	7, 11	7-6 through 7-9, 11 (all)
Effluent disposal for decentralized wastewater systems	13	All
Septage management	14	14-1 and 14-3
Management of decentralized systems	15	All

The following outline is appropriate for a course dealing with natural systems for wastewater management.

Topic	Chapter	Sections
Introduction	1	All
Wastewater characteristics	2	All
Fate of constituents	3	All
Introduction to process analysis	4	All
Wastewater pretreatment	5	5-1, 5-4, 5-14
Biological treatment of wastewater	7	7-1, 7-5, 7-9
Lagoon treatment systems	8	All
Wetlands and aquatic treatment systems	9	All
Land treatment systems	10	All
Intermittent and recirculating packed-bed filters	11	All
Effluent reuse	12	12-1, 12-7 through 12-11
Effluent disposal for decentralized wastewater systems	13	All
Biosolids and septage management	14	All

In an undertaking of the magnitude of this textbook, it is impossible to avoid errors. Any corrections, criticisms, or suggestions for improvements will be appreciated by the authors. Additional information and data are also welcomed.

Ronald Crites
Davis, California

George Tchobanoglous
Davis, California

ACKNOWLEDGMENTS

This textbook could not have been written without the valuable help of a number of people. The help and support of the following individuals are acknowledged gratefully: Harold Ball of Orenco Systems for his reviews and input into Chaps. 5, 6, and 11 and for making available the services of Chris Jordan who prepared selected drawings in Chaps. 5, 6, and 11; Terry Bounds of Orenco Systems for input and review of Chaps. 4, 5 and 6; Adrian Carolan of Schreiber Corp. for providing written material on the extended aeration process in Chap. 7; Robert Emerick for reviewing the section on UV disinfection in Chap. 12; Prof. Robert Gearheart for his review of Chap. 9; Jim Kreissl of EPA for his review of Chap. 9; Prof. Robert Lang for reviewing the entire manuscript; Kara Nelson for reviewing Chap. 9; Frank Loge for reviewing the section on UV disinfection in Chap. 12; Mike Parker of i.e. Engineering for his review and input into Chaps. 6 and 11 and the use of selected STEP system details; Sherwood Reed for multiple reviews and suggestions for Chap. 9; Dr. Chet Rock, a reviewer for McGraw-Hill, for providing useful suggestions for improving the text; Doreen Brown Salazar for reviewing Chaps. 2 and 7; Dr. Ed Schroeder for his helpful suggestions during the preparation of the manuscript; Dr. Joann Silverstein, a McGraw-Hill reviewer, for useful comments and vision; Dr. Richard Stowell for reviewing Chap. 8; Steve Wert for reviewing Chaps. 11 and 13; and Dr. Mac Wesner for reviewing Sec. 12-4 on membranes.

We thank George S. Nolte, Jr., for sponsoring the preparation of the manuscript. The following people from Nolte and Associates contributed significantly to the development of the text: Glenn Dombeck for reviewing Chap. 9, Buster Ide for preparing many of the graphics, Tom Mingee for reviewing Chap. 12, and Dave Richard for reviewing Chaps. 5 and 12. Finally, to Michael B. Anderson, who was tireless in his attention to detail and prepared many of the examples and the solutions manual, our special thanks.

Janet Williams created the original hand-drawn artwork in Chaps. 2, 7, 10, 11, and 12. Esther Sandoval provided word processing.

We thank our series editor, Eric Munson, for his unfailing support and encouragement; Karen Nelson, our production coordinator, for her organizational skills and help in making this text more user friendly; and George Watson, our technical editor, for his thoroughness and attention to detail.

CONTENTS

FOREWORD

Nolte and Associates, George Tchobanoglous, and Ron Crites have developed a comprehensive text and reference work of small and decentralized wastewater management systems that addresses changing technological and community needs. Engineers and environmental companies are continuously searching for systems that are appropriate in scale, cost, and flexibility, and in harmony with the emerging awareness of the principles of sustainability. This book is timely, as the needs of small communities are changing dramatically. Funding for large-scale infrastructure systems is difficult to obtain. Elected community leaders are increasingly influenced by persons who have grown up under the environmental movement and who question the relevance of projects that are dominated by high energy consumption, large pipes, concrete, steel, and other structural impacts that require environmental mitigation.

I have known George and Ron for many years and admire their quest to expand the knowledge, tools, and techniques available to engineers to solve problems, a force that drives us all. They have organized the text material in an effective manner, from setting the stage for the significant potential of decentralized wastewater systems, to defining the whole system and its component parts. They have identified the fate of constituents in the environment and have illustrated process analysis and the design of wastewater treatment facilities, including fascinating new work on alternative technologies—filter systems, pond treatment, land application, wetland systems, etc. This book will provide an excellent resource for all engineering students and practicing wastewater and environmental engineers seeking answers to questions and problems and guidance in dealing with small and decentralized wastewater management systems.

It has taken quite some time to produce this text and reference work. The field experiences of George, Ron, and many other design team professionals redirected its content numerous times. In short, it bears little resemblance to what it started out to be, and it is a tribute to their commitment to engineering excellence. As our collective experiences continue to add to the body of knowledge in this area, it will be a challenge to keep the pace. Nolte and Associates is pleased to have helped make this encompassing work a reality for both students and experienced civil and environmental engineers. We are committed to further the work we have supported by dedicating a portion of our website, *Nolte.com,* to communications on decentralized wastewater systems. Please visit, enjoy, and contribute!

George S. Nolte, Jr., President
Nolte and Associates
Sacramento, California
October 1997

CHAPTER 1

Small and Decentralized Wastewater Management Systems: An Overview

The focus of the textbooks currently used in undergraduate and graduate wastewater treatment courses is directed primarily toward the design of large treatment systems, but the reality is that, in the United States, almost all of the large treatment plants have been built. The projected need for small treatment plants is far greater than that for large treatment plants. Construction costs for new and improved small and decentralized systems will run into the millions of dollars. For these reasons, the focus of this text is on treatment plants with flows of about 1.0 Mgal/d (3785 m^3/d) down to individual home systems, with which most new engineers will deal in their professional careers. It should be noted that the wastewater analyses and designs for treatment plants serving significantly larger flows are based on the fundamental principles and processes described in this book; only the size is larger and a few additional technologies are used.

The objectives of small and decentralized wastewater management (DWM) systems are (1) protecting public health, (2) protecting the receiving environment from degradation or contamination, and (3) reducing costs of treatment by retaining water and solids near their point of origin through reuse. To answer the question of what level of wastewater management is required, it is necessary to have knowledge of the constituents of concern in wastewater, the impacts of these constituents when discharged to the environment, the transformation and long-term fate of these constituents in treatment processes and in the environment, and the treatment methods that can be used to remove or modify the constituents found in wastewater. These topics are the subject matter of this textbook. The constituents in wastewater and their fate in the environment are discussed in Chaps. 2 and 3. Process analysis and design for treatment systems are discussed in Chaps. 4, 5, and 7 through 11. Alternative collection systems are discussed in Chap. 6. Effluent repurification and reuse is discussed in Chap. 12, with effluent disposal from small DWM systems described in Chap. 13. Residual biosolids and septage management is discussed in Chap. 14, and management of decentralized wastewater systems is discussed in Chap. 15.

To introduce the subject of small and decentralized systems, the topics considered in this chapter include: (1) terminology, (2) an introduction to decentralized systems, (3) the role of technology—old and new, (4) the need for management of decentralized systems, and (5) the challenges in the implementation of small and decentralized wastewater management. Because the concept and facilities associated with decentralized systems may be new to the reader, the primary focus of the discussion in this chapter is on decentralized wastewater management.

1-1 TERMINOLOGY

Decentralized wastewater management (DWM) may be defined as the collection, treatment, and disposal/reuse of wastewater from individual homes, clusters of homes, isolated communities, industries, or institutional facilities, as well as from portions of existing communities at or near the point of waste generation (Tchobanoglous, 1995). Where treatment plants have been built to serve portions of a community, they have often been identified as satellite treatment plants, but are classified as decentralized plants in this text. Centralized wastewater management, on the other hand, consists of conventional or alternative wastewater collection systems (sewers), centralized treatment plants, and disposal/reuse of the treated effluent, usually far from the point of origin. Decentralized systems maintain both the solid and liquid fractions of the wastewater near their point of origin, although the liquid portion and any residual solids can be transported to a centralized point for further treatment and reuse (Tchobanoglous, 1996).

In the literature and in government regulations, a variety of terms have been used to refer to individual constituents in wastewater that are of concern in wastewater collection, treatment, reuse, or disposal, including contaminants, impurities, pollutants, and characteristics. The terms *contaminants, impurities,* and *pollutants* are often used interchangeably. The terminology used commonly for key concepts and terms in the field of wastewater management is summarized in Table 1-1. In some cases, confusion arises with the use of these terms when applied in different settings. For example, a contaminant in one setting may not be a contaminant in another setting. To avoid confusion, the term *constituent* will be used in this text to refer to an individual compound or element, such as ammonia nitrogen. The term *characteristic* is used to refer to a group of constituents, such as physical or biological characteristics.

1-2 DECENTRALIZED WASTEWATER MANAGEMENT

The purpose of this section is to further introduce the concept of DWM by considering: (1) the significance of DWM, (2) the applications of DWM, and (3) the elements of DWM.

TABLE 1-1
Terminology commonly used in the field of wastewater management

Term	Definition
Biosolids	The material that remains after sludge and septage are stabilized biologically or chemically
Characteristics (wastewater)	General classes of wastewater constituents such as physical, chemical, biological, and biochemical
Composition	The makeup of wastewater, including the physical, chemical, and biological constituents
Constituents*	Individual components, elements, or biological entities such as suspended solids or ammonia nitrogen
Contaminants	Constituents added to the water supply through use
Decentralized wastewater management	Collection, treatment, and reuse of wastewater at or near its source of generation
Effluent	The liquid discharged from a processing step
Impurities	Constituents added to the water supply through use
Parameter	A measurable factor such as temperature
Pollutants	Constituents added to the water supply through use
Reclamation	Treatment of wastewater for subsequent reuse application
Recycled water	Water suitable for reuse, replaces reclaimed water
Repurification	Treatment of wastewater so that it can be used for a variety of applications including indirect or direct potable reuse
Reuse	Beneficial use of reclaimed or repurified wastewater
Septage	The semiliquid material that is pumped out of septic (or interceptor) tanks, consisting of liquid, scum, and sludge
Sludge	The material that settles out of wastewater in Imhoff tanks, clarifiers (primary and secondary), lagoons, and aquatic and land treatment systems

*To avoid confusion the term *constituents* will be used in this text in place of contaminants, impurities, and pollutants.

Significance of Decentralization

At the present time, more than 60 million people in the United States live in homes that are served by decentralized collection and treatment systems. In the early 1970s, with the passage of the Clean Water Act, it was announced that it was only a matter of time before centralized sewerage facilities would be available to almost all residents. Now, more than 25 years later, it is recognized that complete sewerage of the country may never be possible or desirable, for both geographical and economic reasons. Given the fact that complete sewerage is unlikely for many residents, it is clear that decentralized wastewater management is of great importance to the future management of the environment. The concept of decentralized management of wastewater, therefore, deserves the kind of attention that has, heretofore, been reserved for conventional centralized wastewater management systems (Tchobanoglous, 1996). Recently, the U.S. Environmental Protection Agency (EPA) evaluated the barriers to implementation of decentralized wastewater treatment systems (U.S. EPA, 1997).

Typical situations in which decentralized wastewater management should be considered or selected include:

1. Where the operation and management of existing onsite systems must be improved.
2. Where individual onsite systems are failing and the community cannot afford the cost of a conventional wastewater management system.
3. Where the community or facility is remote from existing sewers.
4. Where localized water reuse opportunities are available.
5. Where fresh water for domestic supply is in short supply.
6. Where existing wastewater treatment plant capacity is limited and financing is not available for expansion.
7. Where, for environmental reasons, the quantity of effluent discharged to the environment must be limited.
8. Where the expansion of the existing wastewater collection and treatment facilities would involve unnecessary disruption of the community.
9. Where the site or environmental conditions that require further wastewater treatment or exportation of wastewater are isolated to certain areas.
10. Where residential density is sparse.
11. Where regionalization would require political annexation that would be unacceptable to the community.
12. Where specific wastewater constituents are treated or altered more appropriately at the point of generation.

Four examples that involve the use of decentralized wastewater management systems are presented to illustrate the range of possible applications. In the first case, because a planned community was remote from existing sewers, the county asked a developer to provide an independent wastewater management facility. The developer proposed a decentralized wastewater management system, which was acceptable to both the county and the permitting agency of the state (see Fig. 1-1). In the system that was constructed, effluent from septic tanks at individual residences is collected and is treated further using a recirculating pea gravel filter (see Chap. 11). After ultraviolet (UV) light disinfection, the effluent is used for landscape irrigation in the summer (see Chap. 12) and disposed of by subsurface soil absorption in the winter.

In the second case, an unsewered community was having problems with failing leachfields. Alternatives were evaluated that included conventional gravity sewers, septic tank effluent pump (STEP) systems, septic tank effluent gravity (STEG) systems, and vacuum sewers. The selected system involved replacing all existing septic tanks with new watertight septic tanks, a STEP collection system, a recirculating pea gravel filter, and effluent irrigation of trees in the summer and effluent storage in the winter.

FIGURE 1-1
View of Stonehurst housing development near Martinez, California, served with a decentralized wastewater management system (Crites et al., 1997).

In the third case, an unsewered community has varying residential densities, and varying concentrations of nitrate-nitrogen in the underlying groundwater. Proposed solutions include an onsite wastewater management district for the lowest housing density [less than 4 units/ac (10 units/ha)], improved treatment for nitrogen removal of existing septic tank/leachfield systems for the moderately dense residential areas, and satellite treatment plants for the collected septic tank effluent from the densest residential areas. Sewering of the entire area was expensive and unacceptable politically, because of the need for annexation.

In the fourth case, retention of wastewater solids in septic tanks has been proposed as part of a decentralized wastewater management system for a large city. The cost savings in wastewater solids treatment and the avoidance of disruptions that would be caused by the construction of large regional collection and treatment facilities were strong reasons in favor of the decentralized approach.

Applications of Decentralized Wastewater Management

To protect the environment, discharge requirements for treated wastewater are becoming increasingly strict for both large and small discharges. The challenge is to be able to provide the required level of treatment in decentralized systems, subject to serious economic constraints. Alternative wastewater collection and treatment options are summarized in Table 1-2 for: (1) individual residences, (2) clusters of homes, (3) public facilities, (4) commercial establishments, (5) industrial parks, (6) small communities, and (7) small portions of large communities.

TABLE 1-2
Typical wastewater treatment and containment options for small and decentralized systems

Type of treatment	Examples	Type of system* S	Type of system* D	Chapter
Wastewater collection	Pressure sewers without grinder pumps	✓	✓	6
	Pressure sewers with grinder pumps	✓	✓	6
	Small diameter variable slope sewers	✓	✓	6
	Vacuum sewers	✓	✓	6
Preliminary	Coarse screens	✓		5
	Fine screens	✓		5
	Grit removal	✓		5
	Oil and grease removal	✓	✓	5
Primary	Septic tanks	✓	✓	5
	Imhoff tanks	✓	✓	5
	Rotary disk filter	✓		5
Advanced primary	Septic tank with effluent filter vault	✓	✓	5
	Septic tank with attached growth reactor element		✓	13
Secondary	Aerobic units	✓	✓	7
	Aerobic/anaerobic	✓	✓	7
	Intermittent sand filter	✓	✓	8,11
	Recirculating gravel filter	✓	✓	11
	Peat filter		✓	11
	Lagoons	✓		8
	Constructed wetlands	✓	✓	9,13
	Aquatic treatment	✓		9
Advanced	Land treatment	✓	✓	10
	Intermittent and recirculating packed-bed filters	✓	✓	11
	Filtration, rapid	✓		12
	Constructed wetlands	✓	✓	9
	Disinfection, chlorine, UV radiation	✓	✓	12
	Repurification (including the use of membranes and carbon adsorption)	✓	✓	12
	Recycle treatment systems			
	Toilet flushing		✓	12
	Landscape watering and toilet flushing		✓	12
Containment	Holding tanks		✓	5
	Privy		✓	

*S = small centralized and D = decentralized.

Individual residences. Wastewater from individual dwellings and other community facilities in unsewered locations is usually managed by onsite treatment and disposal systems. Although a variety of onsite systems have been used, the most common system consists of a septic tank for the partial treatment of the wastewater and long-term storage of the solids and a subsurface disposal field for final treatment and disposal of the septic tank effluent. Although the blackwater (principally toilet and kitchen wastes) and graywater (principally shower and clothes washing water) are usually combined, they have been separated in some systems. Alternative treatment systems for individual residences include intermittent and recirculating packed-bed (usually sand) filters and various aerobic treatment systems.

Cluster systems. Groups or clusters of individual residences can combine their wastewater and have it treated and reused in decentralized wastewater management systems. Typically, large septic tanks or a series of smaller tanks are used. Imhoff tanks, commonly used in the past, are making a comeback in modified forms.

Public facilities. Public facilities such as schools, highway rest areas, prisons, campgrounds, and recreational areas often are isolated from centralized wastewater management systems and are good candidates for DWM. Remote developments, such as the one described briefly in "Individual residences" above, are well served by DWM systems.

Commercial establishments. Wastewater from restaurants, for example, requires further treatment beyond primary settling to remove oils and grease. Recirculating granular-medium filters are used in conjunction with septic tanks where a higher level of treatment is required. Lagoon systems, aquatic treatment, and land treatment systems become alternatives to consider as the flows increase.

Industrial parks. Office buildings and isolated industrial facilities can be served by DWM systems. Water recycling can be achieved in these cases by using a variety of technologies ranging from recirculating packed-bed filters to activated sludge combined with membrane technology (see Chaps. 11 and 12).

Community systems. Decentralized wastewater management in community systems can involve septic tanks for solids retention and the use of small-diameter pipelines to convey the clarified effluent. Pre-engineered and -constructed "package" plants, and individually designed plants, are used where the flows are higher and operating staff are available and affordable. In some cases, such as in "Public facilities" above, it may be possible and desirable to develop a combination of an onsite wastewater management district for the low-density residential area of a community with a sewered area of higher density development. The methods used for effluent disposal will also vary with the size of the system and the local reuse opportunities.

Elements of Decentralized Wastewater Management

The elements that DWM systems comprise include: (1) wastewater pretreatment, (2) wastewater collection, (3) wastewater treatment, (4) effluent reuse or disposal, and (5) biosolids and septage management. Although the components are the same as for large centralized systems, the difference is in the application of technology. It should also be noted that not every DWM system will incorporate all of the above elements.

Wastewater pretreatment. The objective of wastewater pretreatment is to remove solids, oil and grease, and other floatable or settleable materials so that the remaining wastewater can be treated effectively and reused or disposed of safely. For example, the use of individual septic tanks at the point of origin can be considered an integral part of DWM because it manages the solids separately from the septic tank effluent.

Wastewater collection. Where the density of residential development has increased to the point that continued use of individual onsite systems for effluent treatment and disposal is no longer feasible, some form of wastewater collection is often needed. Although the use of conventional gravity-flow sewers for the collection of wastewater continues to be the accepted norm for sewerage practice in the United States, alternative collection systems that are consistent with DWM are becoming increasingly popular. In some areas the use of conventional gravity sewers is becoming counterproductive because the use of water conservation devices continues to increase. The minimum flows required for gravity-flow sewers to operate make them problematic where development occurs slowly in a large development or where water conservation reduces the wastewater flows significantly. In many cases, the water used to flush conventional gravity-flow collection systems for the removal of accumulated solids far exceeds the water saved through water conservation measures.

Wastewater treatment. Representative wastewater treatment facilities that have been used for small and decentralized systems are presented in Table 1-2. In the past, removal of biochemical oxygen demand (BOD), suspended solids, and pathogens was the focus of treatment. Today, nutrient removal, removal of toxics, and beneficial reuse are of increasing importance. Detailed discussions of wastewater treatment are provided in Chaps. 5 and 7 through 12.

Reuse or disposal. The methods of wastewater reuse and/or disposal are presented in Table 1-3. As the level of treatment increases, the potential for beneficial reuse of the treated water also increases. As described in Chap. 12, reuse of treated effluent requires that water quality criteria are met rigorously. For rural DWM systems, agricultural and landscape irrigation will be the most likely form of reuse. In humid areas, land treatment and groundwater recharge will be more common.

In urban areas, a number of self-contained recycle systems have been developed to take sanitary wastewater from buildings, treat it, and return the bulk of the treated effluent for reuse as toilet and urinal flushing. One such unit involves three treatment steps: (1) the solids in the wastewater are collected and treated aerobically, (2) the effluent from the biological treatment unit is then passed through a self-cleaning ultrafiltration step where residual organics, microorganisms, and suspended solids are removed, and (3) the effluent is then passed through an activated carbon column for polishing (see Chap. 12). The material removed in the ultrafiltration step is returned to the first processing step for further treatment. The effluent from the carbon filters is disinfected with ozone or UV light before it is reused for toilet-flushing water. Although such processes are expensive, they have been used for office buildings located in unsewered areas, and where water for domestic use is in short supply.

TABLE 1-3
Typical wastewater reuse and disposal options for small and decentralized systems

Option	Examples
Constructed wetlands	Free water surface Subsurface flow
Discharge to water bodies	Streams, lakes, ponds, reservoirs, bays, ditches, rivers, oceans
Evaporation systems	Evapotranspiration beds Evaporation ponds
Land application	Surface application Spray application Drip application
Reuse applications	Agricultural irrigation Landscape irrigation Groundwater recharge Habitat wetlands Nonpotable supply Industrial supply Recreational lakes Water supply augmentation

Subsurface soil disposal	Soil absorption systems
	  Conventional leachfields
	  Shallow trench pressure dosed leachfields
	  Shallow sand-filled pressure dosed leachfields
	  Drip irrigation (integral or external emitters)
	Seepage beds
	Mound systems
	Fill systems
	At-grade systems

Biosolids and septage management. The solids removed from wastewater require stabilization followed by disposal or reuse. Septage, the material pumped out of septic tanks, also requires further stabilization prior to disposal or reuse. Treatment and beneficial reuse of septage and biosolids, generally by composting and land application, are described in Chap. 14.

1-3 THE ROLE OF TECHNOLOGY—OLD AND NEW

Perhaps the most significant change that has occurred in the past 15 years in the implementation of small and decentralized wastewater management systems is the development of new technology and hardware and the reapplication of old technology using new equipment. Some of these technologies are highlighted in the following discussion, for both small and decentralized systems. Additional details on these and other technologies are presented in Chaps. 5 and 7 through 13.

Technologies for Small Systems

A number of new technologies have been introduced for small treatment systems that have made it possible to produce an effluent of the same quality, or even better, as compared to large treatment plants. Important examples include the use of: (1) alternative wastewater collection technologies, (2) rotary disk screens, (3) cyclic activated sludge processes, (4) aquatic treatment systems, (5) constructed wetlands, and (6) land treatment systems.

Alternative collection systems. In many areas that are now being developed, the use of conventional gravity-flow sewers may not be economically feasible for reasons of topography, high water table, structurally unstable soils, and rocky conditions. Further, in small unsewered communities, the cost of installing conventional gravity-flow sewers is prohibitive, especially where the density of development is low. To overcome these difficulties, (1) small-diameter, variable-grade effluent sewers, (2) pressure sewers, and (3) vacuum sewers have been developed as alternatives. Because infiltration/inflow is, for all practical purposes, eliminated when alternative sewers are used, the size of the alternative sewers can be kept to a more economical minimum. More economical effluent collection and lower costs for solids management are driving forces behind the considerations for DWM. Alternative wastewater collection systems are discussed in Chap. 6.

Rotary disk screens. Fine screens, typically with openings of about 0.01 in (0.25 mm) have been developed as a replacement for primary sedimentation facilities. A rotary disk screen has been used in the City of San Diego 1.0 Mgal/d (3785 m^3/d) aquaculture facility for more than 10 years (see Fig. 1-2). It should be noted that fine screens were used extensively in the 1920s, but were abandoned because of the accumulation of grease.

Cyclic activated sludge processes. In the recent past, most activated sludge processes were of the complete-mix design. However, as our understanding of the mechanisms of biological nitrogen and phosphorus removal have advanced, a number of cyclic activated sludge processes have been developed. By cycling the activated sludge treatment process through aerobic (in the presence of oxygen) and anaerobic (in the absence of oxygen) periods, it is now possible to remove both nitrogen and phosphorus biologically. It is interesting to note the original activated sludge process developed by Ardern and Lockett in 1914 operated on a fill and draw basis (Ardern and Lockett, 1914). Representative examples of currently used cyclic processes include: the oxidation ditch, the Schreiber countercurrent aeration process, the Biolac™ process, the intermittent decanted extended aeration process, and the sequencing batch reactor (see Chap. 7). Overhead views of the oxidation ditch and the Schreiber process are shown in Fig. 1-3.

Aquatic treatment systems. The use of floating aquatic plants was pioneered in the 1970s at the National Aeronautics and Space Administration (NASA) space center as a potential wastewater treatment system for space travel. Despite

FIGURE 1-2
View of rotary disk screen used as a replacement for primary sedimentation. The size of the openings in the screen material is 0.25 mm. The solids removed on the screen are scraped and washed off with high-pressure water jets.

(*a*)

(*b*)

FIGURE 1-3
An overhead view of extended aeration activated sludge processes: (*a*) oxidation ditch at Patterson, California, and (*b*) countercurrent aeration activated sludge process (Courtesy of the Schreiber Corp.).

FIGURE 1-4
View of aquatic treatment system employing water hyacinths used for secondary treatment at San Pasqual treatment facility in San Diego County, California.

several failures, the technology has evolved and has been integrated with aerated lagoons, extended aeration, and constructed wetlands to offer several new wastewater treatment flow diagrams (see Chap. 9 and Fig. 1-4).

Constructed wetlands. Constructed wetlands for wastewater treatment have evolved from research in Germany with emergent plants to be a significant wastewater treatment technology for septic tank effluent, pond effluent, and biological secondary effluent. Reeds, rushes, and cattails serve as a matrix for attached biological growth. Constructed wetlands can be free water surface or subsurface flow (through gravel) as described in Chap. 9. Wetlands can also function as water reuse and as wildlife habitat (see Fig. 1-5).

Land treatment systems. Like intermittent sand filters, land treatment systems were developed in the nineteenth century and were subsequently forgotten until the 1960s. The effectiveness of land treatment was established in the 1860s in England and was used in the 1870s from Paris to Moscow (Jewell and Seabrook, 1979; Rafter, 1897). Land treatment systems are economical for many rural locations and have evolved into significant alternatives for centralized wastewater management. A typical land treatment system for small flows is shown in Fig. 1-6. For DWM, land treatment is even more attractive because it is a passive, natural technology that achieves high levels of nutrient removal without significant need for operational labor, energy, or chemicals.

Biosolids and septage management. Biosolids management is implemented predominantly by land application or by composting and distribution as a soil amendment or conditioner (see Chap. 14). Traditional management of septage (the

FIGURE 1-5
View of constructed wetland used for the treatment of wastewater from a resort in Crete.

contents pumped periodically from septic tanks) is either to discharge it to the nearest centralized wastewater treatment plant or to land application. Separate septage treatment is rare, except on Cape Cod, Massachusetts, where several septage-only treatment systems have been constructed. Alternative flow diagrams for treatment are presented in Chap. 14.

FIGURE 1-6
View of overland flow land treatment system at Davis, California, used for secondary treatment.

Technologies for Decentralized Systems

Recognizing that funds for operation and maintenance are often limited, the development of technology has been focused on DWM systems that feature low energy, low labor, and low maintenance requirements.

Septic tanks and effluent screens. New developments in pretreatment include watertight septic tanks made of concrete, fiberglass, and plastic, and the septic tank effluent filter screen. Watertight septic tanks are important to minimize extraneous flows into effluent gravity collection systems and to sustain biological growth within the septic tank. The effluent filter screen (Orenco, 1996) increases the longevity and reliability of downstream processes and piping systems by retaining solids in the septic tank more consistently (see Fig. 1-7). Operationally, septic tank effluent flows into the vault through inlet holes located in the center (clear zone) of the tank. Before passing into the center of the vault, the effluent must pass through a series of fine screens located on the inside of the vault. When needed, the screen element can be removed and cleaned. An advantage of the effluent screen is that it can be installed in both new and existing septic tanks. The development of the effluent vault is also significant because it has made feasible the use of small, lightweight, high-head [300-ft (100-m)] multistage well pumps (see Fig. 1-8) that are used for STEP systems or pressure dosing of soil absorption systems.

FIGURE 1-7
Septic tank equipped with effluent filter vault for limiting the discharge of solids.

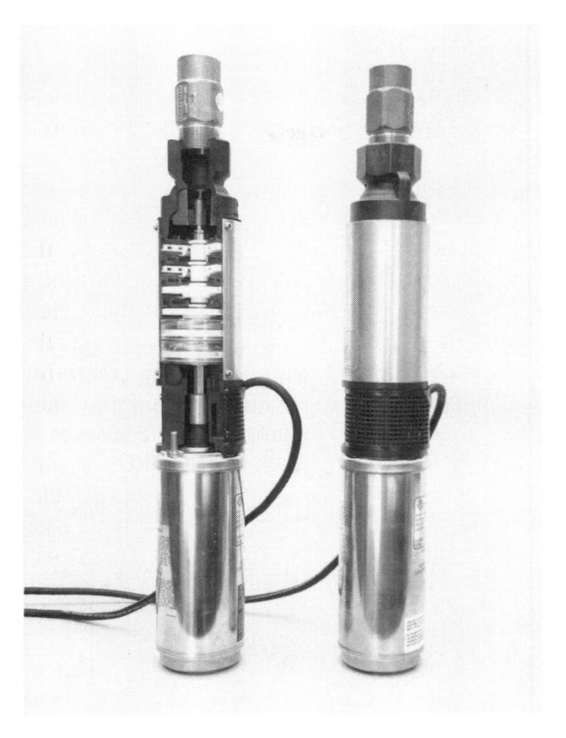

FIGURE 1-8
Typical high-head multistage turbine pump used in conjunction with effluent filter vault.

Septic tanks with recirculating trickling filters. Another development in pretreatment is the use of a recirculating trickling filter as an integral part of the septic tank for enhanced nitrogen removal. Septic tank effluent is recirculated over plastic medium that supports attached growth nitrifying bacteria. The resultant nitrates are converted into nitrogen gas by denitrification in the septic tank. More details on performance of the recirculating trickling filters (RTFs) are presented in Chap. 13. Other pretreatment techniques are presented in Chap. 5.

Intermittent and recirculating packed-bed filters. Sand and fine gravel have been used effectively for many years in intermittent and recirculating filters. The intermittent sand filter (ISF) was used extensively for DWM and for small communities in the 1880s through the 1920s in Massachusetts (Mancl and Peeples, 1991). The technology was rediscovered in the 1970s for oxidation lagoon upgrades (see Chap. 8), and has become an integral part of DWM for small systems (see Chap. 11). A typical intermittent sand filter used for an individual home is shown in Fig. 1-9. Essentially the same design, dating from about 1915, is shown in Fig. 1-10.

Shallow pressure-dosed soil absorption systems. Final treatment and disposal of the effluent from a septic tank or other individual treatment unit is accomplished currently and most commonly by means of subsurface soil absorption. A soil absorption system, commonly known as a *leachfield* or *drainfield,* consists of a series

FIGURE 1-9
Typical intermittent sand filter used for an individual residence before the orifice shields and final layer of gravel are added.

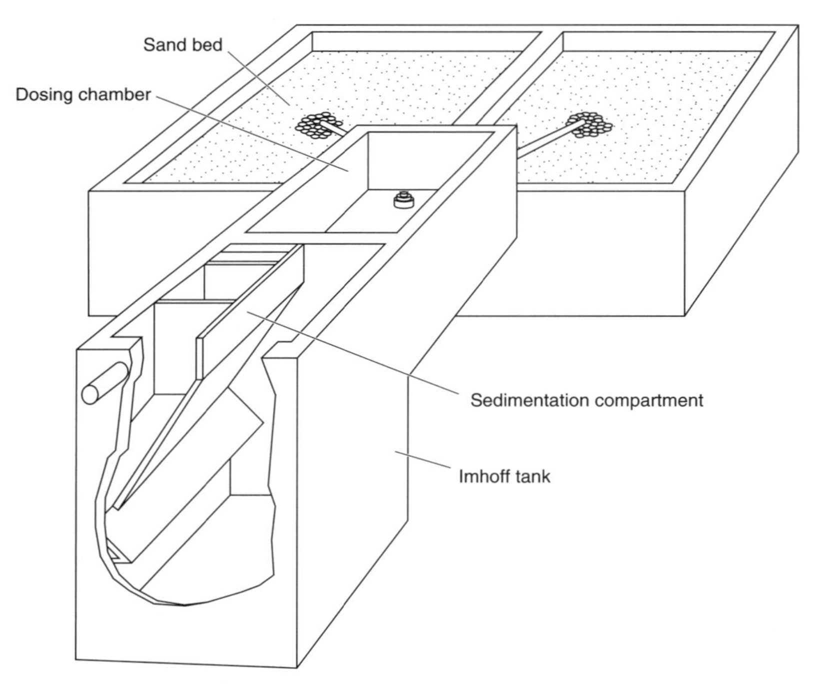

FIGURE 1-10
Typical intermittent sand filter recommended for a family of five persons by the public health service in 1915 (Frank and Rhynus, 1920).

of relatively deep 3- to 8-ft (0.9- to 2.4-m) trenches filled with gravel. The problem with deep trenches is that, in addition to being more expensive to construct, they fail to take advantage of the treatment capabilities of the upper soil mantle because they are typically located below the region of maximum bacterial activity in the soil. The trend in new trench designs is to use shallow trenches [1 ft (0.3 m)] without gravel or other porous medium (see Fig. 1-11*a*). The use of shallow trenches enhances biological and chemical treatment of the effluent because of the bacterial activity in the shallow soil and the increased opportunity for adsorption of phosphorus, metals, and viruses. Higher removals of BOD, total suspended solids (TSS), and nitrogen are also likely. Shallow trenches were actually used in the early 1900s (see Fig. 1-11*b*), as recommended by the Public Health Service in 1915 (Lumsden et al., 1915).

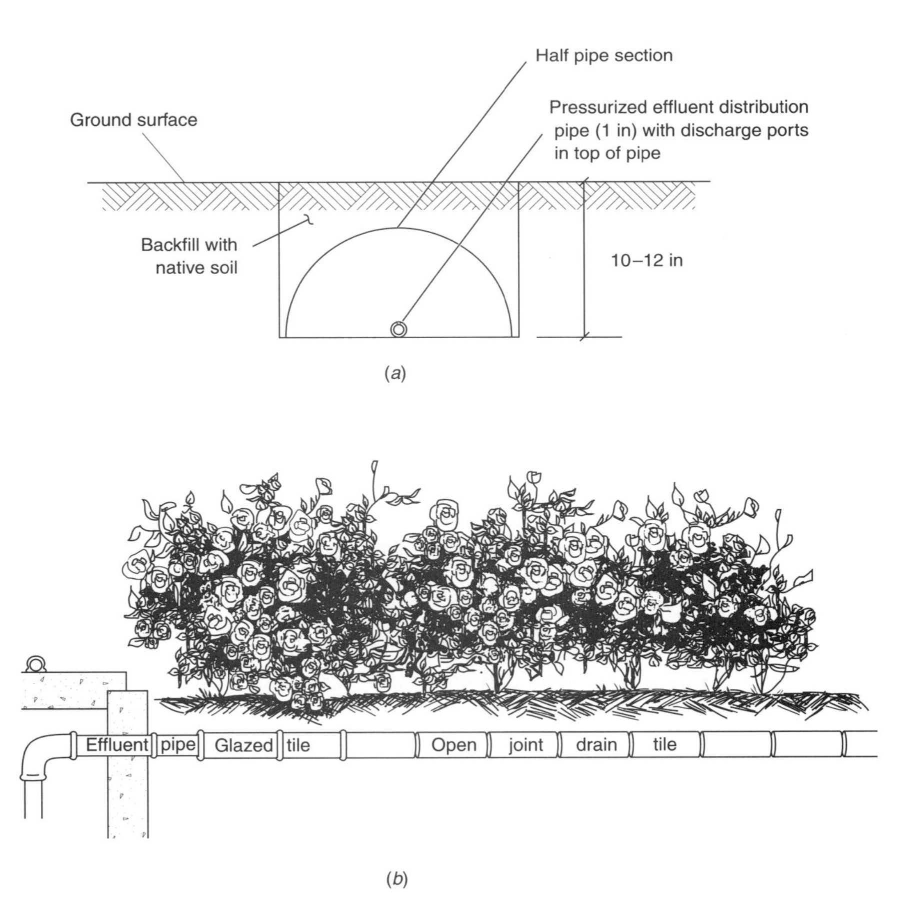

FIGURE 1-11
Views of shallow soil absorption system: (*a*) 1990s version and (*b*) 1895 version (Lumsden et al., 1915).

1-4 MANAGEMENT OF DECENTRALIZED SYSTEMS

Although most of the treatment units used in decentralized wastewater management systems require very little maintenance, they rarely receive any. As a result, many system failures have occurred. With onsite systems the principal mode of failure has been a premature clogging of the infiltrative capacity of the disposal field, below the required capacity for managing the daily flow. In most cases where premature failure has occurred, it has been found that the disposal fields have been improperly designed, constructed, or operated, and either overloaded by solids from unmanaged septic tanks or overloaded hydraulically by leaky septic tanks.

When onsite systems are used on large lots, the failure of an individual system may result in a localized environmental problem. However, as the density of development increases and lot sizes become smaller, the failure of one or more onsite systems can pose a nuisance problem and, in some cases, a public health problem. To ensure that individual decentralized systems will function properly, especially in densely developed areas, it is usually necessary to organize a maintenance district or to contract with a public or private operating agency to conduct periodic inspections and any necessary maintenance. Large-scale decentralized wastewater management systems should be allowed only if a responsible management agency has been designated. Management of decentralized systems is described in Chap. 15.

Without wastewater management oversight, onsite systems must be designed and operated conservatively. With management, onsite systems can be designed to operate at significantly higher rates and the size of the physical facilities can be reduced. Systems can be monitored and should a system component fail, it can be restored as needed. With management a small, remote DWM system can be as environmentally safe and responsible as a centralized wastewater management system. In planning new developments, careful attention is given to environmentally responsible design concepts. Water reuse and solids recycling can be incorporated into a DWM system with assurance that water and environmental quality will be protected.

1-5 CHALLENGES IN THE IMPLEMENTATION OF DECENTRALIZED WASTEWATER MANAGEMENT

In many cases, small communities have limited economic resources and expertise to manage DWM systems (Nelson and Dow, 1994). Problems are often experienced in design, contracting, inadequate construction supervision, project management, billing, accounting, budgeting, operations, and maintenance. Overcoming these problems makes the implementation of decentralized wastewater management systems a challenging undertaking.

While the implementation of DWM systems is formidable, from an economic and social point of view, the engineering involved is equally challenging. To implement DWM systems, the designer must not only be knowledgeable about the elements involved in the design of conventional centralized wastewater management systems, but must have additional information about such items as: septic tanks and Imhoff tanks, used for the pretreatment of household wastes; alternative wastewater collection systems, including the use of pressure and small-diameter, variable-slope sewers; intermittent and recirculating sand filters; and soil absorption systems, including shallow pressure-dosed systems. Clearly, both students and practicing engineers will find the field of DWM challenging.

PROBLEMS AND DISCUSSION TOPICS

1-1. What factors contributed to the use of centralized wastewater collection, treatment, and disposal?

1-2. What types of wastewater management are used in the community in which you grew up, and what is the current monthly homeowner charge for wastewater management?

1-3. In the community in which you grew up, what are the services that are provided for the service charge?

1-4. With respect to wastewater reclamation and reuse, what are the disadvantages of centralized versus decentralized systems?

1-5. How can the use of DWM systems reduce the impact of growth on public water supplies?

1-6. What savings have been made in water use from low flush toilets? If older conventional toilets use 4.5 gal/flush and low-flush toilets require 1 gal/flush, estimate the potential annual water savings that could be made, per 1000 persons, if the conventional toilets were replaced.

1-7. Will the switch to low-flush toilets and water conservation result in smaller wastewater treatment plants?

1-8. How will future technological innovations impact the field of decentralized wastewater management?

1-9. How can the management of decentralized wastewater systems impact the choice of wastewater technology?

1-10. Based on your current understanding, which type of wastewater management system (centralized or decentralized) do you favor? List your reasons.

1-11. How can the use of DWM systems contribute to the maintenance of a sustainable environment?

1-12. Why are compounds of nitrogen and phosphorus that are found in wastewater of concern in DWM?

REFERENCES

Ardern, E., and W. T. Lockett (1914) Experiments on the Oxidation of Sewage without the Aid of Filters, *J. Soc. Chem. Ind.*, Vol. 33, pp. 523, 1122.

Crites, R., C. Lekven, S. Wert, and G. Tchobanoglous (1997) Decentralized Wastewater System for a Small Residential Development in California, *Small Flow Journal,* Vol. 3, Issue 1, Morgantown, WV.

Frank, L. C., and C. P. Rhynus (1920) The Treatment of Sewage from Single Houses and Small Communities, *Public Health Bulletin No. 101,* U.S. Public Health Service, Washington, DC.

Jewell, W. J., and B. L. Seabrook (1979) History of Land Application as a Treatment Alternative, EPA 430/9-79-012, U.S. Environmental Protection Agency, Washington, DC.

Lumsden, L. L., C. W. Stiles, and A. W. Freeman (1915) Safe Disposal of Human Excreta at Unsewered Homes, *Public Health Bulletin No. 68,* U.S. Public Health Service, Government Printing Office, Washington, DC.

Mancl, K. M., and J. A. Peeples (1991) One Hundred Years Later: Reviewing the Work of the Massachusetts State Board of Health on the Intermittent Sand Filtration of Wastewater from Small Communities. Proceedings of the Sixth National Symposium on Individual and Small Community Sewage Systems. American Society of Agricultural Engineers, pp. 22–30, Chicago, IL.

Nelson, V. I., and D. B. Dow (1994) National Consortium for Decentralized Wastewater Technology and Management, Onsite Wastewater Treatment, Proceedings of the Seventh International Symposium on Individual and Small Community Sewage Systems, pp. 11–15, Atlanta, GA.

Orenco Systems, Inc. (1996) Equipment catalog.

Rafter, G. W. (1897) Sewage Irrigation, USGS Water Supply and Irrigation Paper No. 3, U.S. Department of the Interior, Washington, DC.

Tchobanoglous, G. (1995) Decentralized Systems for Wastewater Management. Presented at the Water Environment Association of Ontario Annual Conference, Toronto, Canada.

Tchobanoglous, G. (1996) Appropriate Technologies for Wastewater Treatment and Reuse, Australian Water & Wastewater Association, *Water Journal,* Vol. 23, No. 4.

U.S. EPA (1997) Response to Congress on Use of Decentralized Wastewater Treatment Systems. EPA 832-R-97-001b. Environmental Protection Agency Office of Wastewater Management, Washington, DC.

CHAPTER 2

Constituents in Wastewater

Of fundamental importance in the implementation of wastewater management facilities, regardless of size, is (1) knowledge of the constituents found in wastewater and (2) knowledge of the fate of these constituents when released to the environment. These topics are considered in Chaps. 2 and 3. The purpose of this chapter is, therefore, to provide the necessary background information and data on the constituents found in wastewater that will be required for the analysis and design of wastewater management facilities and systems. Topics to be considered include: (1) wastewater constituents, (2) sampling and analytical procedures, (3) the physical characteristics of wastewater, (4) the inorganic chemical characteristics of wastewater, (5) aggregate organic chemical characteristics, (6) characterization of individual organic compounds, (7) the biological characteristics of wastewater, and (8) toxicity testing. Typical concentration values for the constituents found in wastewater and septic tank effluent are presented and discussed in Chap. 4.

2-1 WASTEWATER CONSTITUENTS

The constituents found in wastewater can be classified as physical, chemical, and biological. The analyses commonly used to quantify the constituents found in wastewater are reported in Table 2-1. The size range of the constituents found in wastewater is summarized in Fig. 2-1. Constituents of concern in wastewater, and the reasons for concern, are summarized in Table 2-2. Of the constituents listed in Table 2-2, suspended solids, biodegradable organics, and pathogenic organisms are of major importance, and most wastewater management facilities are designed to accomplish their removal. Although the other constituents are also of concern, the need for their removal must be considered on a case-to-case basis. Before considering the physical, chemical, and biological characteristics of wastewater, it is appropriate to consider briefly the analytical procedures used to obtain wastewater characterization data.

TABLE 2-1
Common analyses used to assess the constituents found in wastewater*

Test†	Abbreviation/definition	Use or significance of test results
Physical characteristics		
Total solids	TS	To determine the most suitable type of operations and processes for its treatment.
Total volatile solids	TVS	
Total fixed solids	TFS	
Total suspended solids	TSS	
Volatile suspended solids	VSS	
Fixed suspended solids	FSS	
Total dissolved solids	TDS (TS − TSS)	To assess reuse potential of wastewater.
Volatile dissolved solids	VDS	
Total fixed dissolved solids	FDS	
Settleable solids		To determine those solids that will settle by gravity in a specified time period.
Particle size distribution	PSD	To assess the performance of treatment processes.
Turbidity	NTU	Used to assess the quality of treated wastewater.
Color	Light brown, gray, black	To assess the condition of wastewater (fresh or septic).
Transmittance	%T	Used to assess the suitability of treated effluent for UV disinfection.
Odor	TON	To determine if odors will be a problem.
Temperature	°C or °F	Important in the design and operation of biological processes in treatment facilities.
Density	ρ	
Conductivity	EC	Used to assess the suitability of treated effluent for agricultural applications.
Inorganic chemical characteristics		
Free ammonia	NH_4^+	Used as a measure of the nutrients present and the degree of decomposition in the wastewater; the oxidized forms can be taken as a measure of the degree of oxidation. Used as a measure of the nutrients present.
Organic nitrogen	Org N	
Total Kjeldahl nitrogen	TKN (org N + NH_4^+)	
Nitrites	NO_2	
Nitrates	NO_3	
Inorganic phosphorus	Inorg P	
Total phosphorus	TP	
Organic phosphorus	Org P	
pH	pH = log 1/[H+]	A measure of the acidity or basicity of an aqueous solution.

TABLE 2-1
(Continued)

Test[†]	Abbreviation/definition	Use or significance of test results
Alkalinity	$\sum HCO_3^- + CO_3^{-2} + OH^- - H^+$	A measure of the buffering capacity of the wastewater.
Chloride	Cl^-	To assess the suitability of wastewater for agricultural reuse.
Sulfate	SO_4^{-2}	To assess the potential for the formation of odors and to assess the treatability of the waste sludge.
Metals	As, Cd, Ca, Cr, Co, Cu, Pb, Mg, Hg, Mo, Ni, Se, Na, Zn	To assess the suitability of the wastewater for reuse and for toxicity effects in treatment. Trace amounts of metals are important in biological treatment.
Specific inorganic elements and compounds		To assess presence or absence of a specific constituent.
Various gases	O_2, CO_2, NH_3, H_2S, CH_4	The presence or absence of specific gases.
	Organic chemical characteristics	
Five-day carbonaceous biochemical oxygen demand	$CBOD_5$	A measure of the amount of oxygen required to stabilize a waste biologically.
Ultimate carbonaceous biochemical oxygen demand	UBOD (also BOD_u, L)	A measure of the amount of oxygen required to stabilize a waste biologically.
Nitrogenous oxygen demand	NOD	A measure of the amount of oxygen required to oxidize biologically the ammonia nitrogen in the wastewater to nitrate.
Chemical oxygen demand	COD	Often used as a substitute for the BOD test.
Total organic carbon	TOC	Often used as a substitute for the BOD test.
Specific organic compounds and classes of compounds		To determine presence of specific organic compounds and to assess whether special design measures will be needed for removal.
	Biological characteristics	
Coliform organisms	MPN (most probable number)	To assess presence of pathogenic bacteria and effectiveness of disinfection process.
Specific microorganisms	Bacteria, protozoa, helminths, viruses	To assess presence of specific organisms in connection with plant operation and for reuse.
Toxicity	TU_A and TU_C	Toxic unit acute, toxic unit chronic.

*Adapted, in part, from Tchobanoglous and Schroeder (1985).
[†] Details on the various tests may be found in Standard Methods (1995).

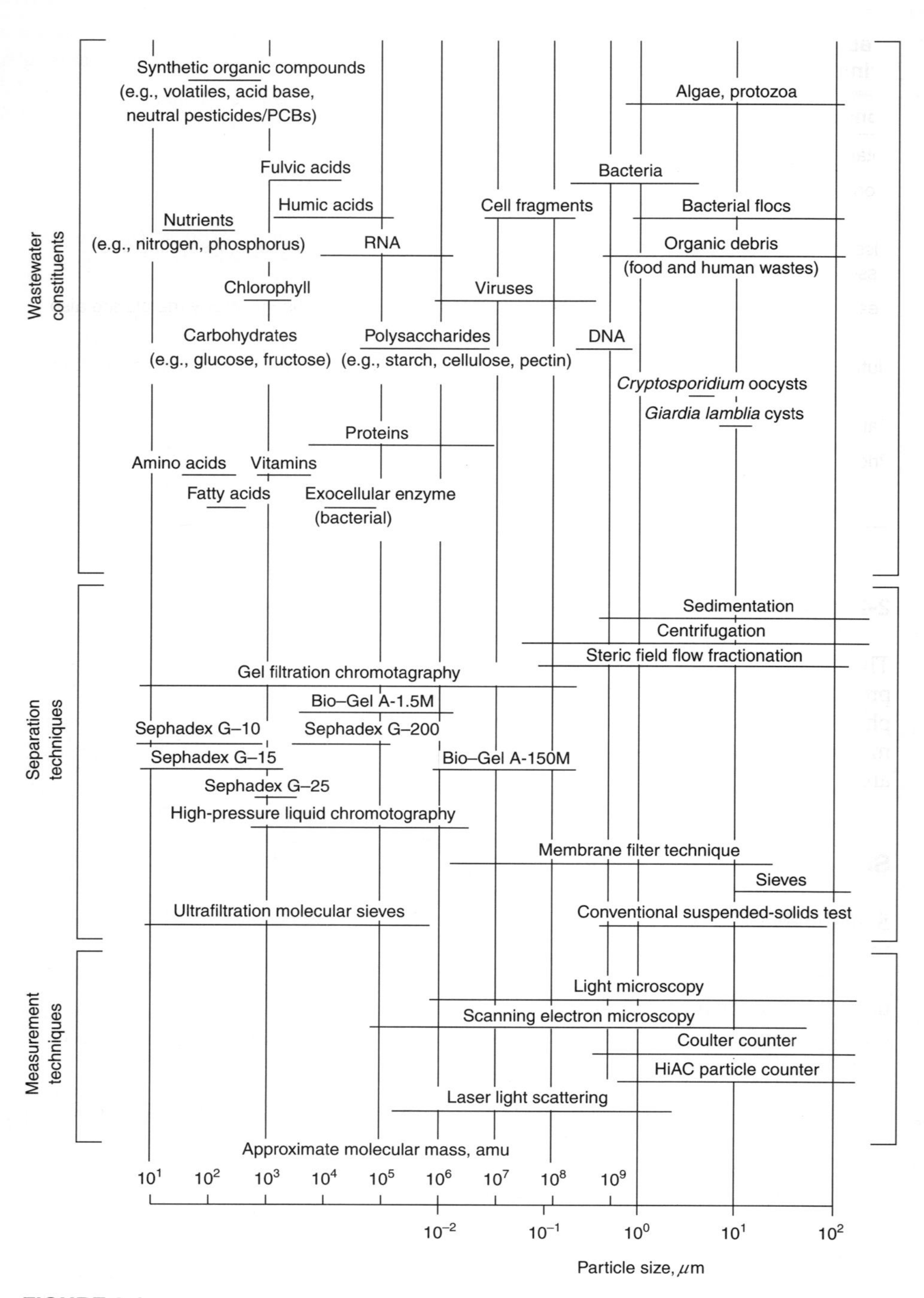

FIGURE 2-1
Size ranges of constituents found in wastewater, excluding oversize debris such as balls, cigarette butts, pieces of wood, and rags (adapted from Levine et al., 1985).

TABLE 2-2
Principal constituents of concern in wastewater treatment

Constituents	Reason for concern
Total suspended solids	Sludge deposits and anaerobic conditions.
Biodegradable organics	Depletion of natural oxygen resources and the development of septic conditions.
Dissolved inorganics (e.g., total dissolved solids)	Inorganic constituents added by usage. Recycling and reuse applications.
Heavy metals	Metallic constituents added by usage. Many metals are also classified as priority pollutants.
Nutrients	Excessive growth of undesirable aquatic life, eutrophication, nitrate contamination of drinking water.
Pathogens	Communicable diseases.
Priority organic pollutants	Suspected carcinogenicity, mutagenicity, teratogenicity, or high acute toxicity. Many priority pollutants resist conventional treatment methods (known as refractory organics).

2-2 SAMPLING AND ANALYTICAL PROCEDURES

The sampling techniques and the analyses used to characterize wastewater vary from precise quantitative chemical determinations to the more qualitative biological and physical determinations. Sampling techniques, the methods of analysis, the units of measurement for chemical constituents, and some useful concepts from chemistry are considered below.

Sampling

Sampling programs are undertaken for a variety of reasons such as to obtain: (1) routine operating data on overall plant performance, (2) data that can be used to document the performance of a given treatment operation or process, (3) data that can be used to implement proposed new programs, and (4) data needed for reporting regulatory compliance. To meet the goals of the sampling program, the data collected must be:

1. *Representative.* The data must represent the wastewater or environment being sampled.
2. *Reproducible.* The data obtained must be reproducible by others following the same sampling and analytical protocols.
3. *Defensible.* Documentation must be available to validate the sampling plan. The data must have a known degree of accuracy and precision.
4. *Useful.* The data can be used to meet the objectives of the monitoring plan (Pepper et al., 1996).

(a)

(b)

FIGURE 2-2
Collection of samples for analysis: (*a*) septic tank effluent and (*b*) cartridge filter for concentrating viruses for testing (filter being emptied).

Because the data from the analysis of the samples will ultimately serve as a basis for implementing wastewater management facilities and programs, the techniques used in a wastewater sampling program must be such that representative samples are obtained. There are no universal procedures for sampling; sampling programs must be tailored individually to fit each situation (see Fig. 2-2). Special procedures are necessary to handle sampling problems that arise when wastes vary considerably in composition.

Before a sampling program is undertaken, a detailed sampling protocol must be developed, along with a quality assurance project plan (QAPP) [known previously as quality assurance/quality control (QA/QC)]. As a minimum, the following items must be specified in the QAPP (Pepper et al., 1996). Additional details on the subject of sampling may be found in Standard Methods, 1995.

1. *Sampling plan.* Number of sampling locations, number and type of samples, time intervals (e.g., real-time and/or time-delayed samples).
2. *Sample types and size.* Catch or grab samples, composite samples, or integrated samples; size of samples.
3. *Sample labeling and chain of custody.* Sample labels, sample seals, field log book, chain of custody record, sample analysis request sheets, sample delivery to the laboratory, receipt and logging of sample, and assignment of sample for analysis.
4. *Sampling methods.* Specific techniques and equipment to be used (e.g., manual or automatic sampling).

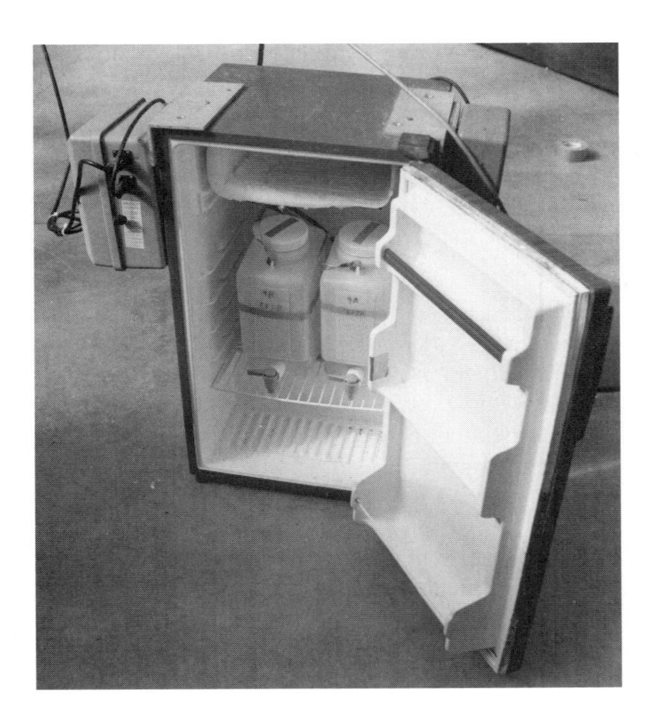

FIGURE 2-3
Composite sampler used to collect and refrigerate samples over a 24-h period.

5. *Sampling storage and preservation.* Type of containers (e.g., glass or plastic), preservation methods, maximum allowable holding times.
6. *Sample constituents.* A list of the parameters to be measured.
7. *Analytical methods.* A list of the field and laboratory test methods and procedures to be used, and the detection limits for the individual methods.

If the physical, chemical, and biological integrity of the samples is not maintained during interim periods between sample collection and sample analysis, a carefully performed sampling program will become worthless. Considerable research on the problem of sample preservation has failed to perfect a universal treatment or method, or to formulate a set of fixed rules applicable to samples of all types. Prompt analysis is undoubtedly the most positive assurance against error due to sample deterioration. When analytical and testing conditions dictate a lag between collection and analysis, such as when a 24-hour composite sample is collected, provisions must be made for preserving samples (see Fig. 2-3). Current methods of sample preservation for the analysis of properties subject to deterioration must be used (Standard Methods, 1995). Probable errors due to deterioration of the sample should be noted in reporting analytical data.

Methods of Analysis

The quantitative methods of analysis are either gravimetric, volumetric, or physicochemical. In the physicochemical methods, properties other than mass or volume are measured. Instrumental methods of analysis such as turbidimetry, colorimetry, potentiometry, polarography, adsorption spectrometry, fluorometry, spectroscopy, and nuclear radiation are representative of the physicochemical analyses. Details concerning the various analyses may be found in Standard Methods (1995), the accepted reference that details the conduct of water and wastewater analyses.

Regardless of the method of analysis used, the detection level must be specified. Several detection limits are defined and are listed below in order of increasing levels (Standard Methods, 1995).

1. *Instrumental detection level (IDL).* Constituent concentration that produces a signal greater than 5 times the signal/noise ratio of the instrument. The IDL is, in many respects, similar to *critical level* and *criterion of detection.* The last level is stated as 1.645 times the standard deviation s of blank analyses.
2. *Lower level of detection (LLD).* Constituent concentration in reagent water that produces a signal $2(1.645s)$ above the mean of blank analyses. This signal sets both Type I and Type II errors at 5 percent. Other names for this level are *detection level* and *level of detection* (LOD).
3. *Method detection level (MDL).* Constituent concentration that, when processed through the complete method, produces a signal with a 99 percent probability that it is different from the blank. For seven replicates of the sample, the mean must be $3.14s$ above the blank, where s is the standard deviation of the seven replicates. The MDL is computed from replicate measurements 1 to 5 times the actual MDL. The MDL will be larger than the LLD because of the few replications and the sample processing steps and may vary with constituent and matrix.
4. *Level of quantification (LOQ).* Constituent concentration that produces a signal sufficiently greater than the blank that can be detected within specified levels by good laboratories during routine operating conditions. Typically it is the concentration that produces a signal $10s$ above the reagent blank signal.

The approximate relationship between the detection limits is

$$\text{IDL:LLD:MDL:LOQ} = 1:2:4:10$$

Units of Measurement for Physical and Chemical Parameters

The results of the analysis of wastewater samples are expressed in terms of physical and chemical units of measurement. The most common units are reported in Table 2-3. Measurements of chemical parameters are usually expressed in the physical unit of milligrams per liter (mg/L) or grams per cubic meter (g/m^3). The concentration of trace constituents is usually expressed as micrograms per liter (μg/L). As noted in Table 2-3, the concentration can also be expressed as parts per million (ppm), which is a mass-to-mass ratio. The relationship between mg/L and ppm is

$$\text{ppm} = \frac{\text{mg/L}}{\text{specific gravity of fluid}} \tag{2-1}$$

For dilute systems, such as those encountered in natural waters and wastewater, in which one liter of sample weighs approximately one kilogram, the units of mg/L

TABLE 2-3
Units commonly used to express analytical results

Basis	Application	Unit
	Physical analyses:	
Density	$\frac{\text{Mass of solution}}{\text{Unit volume}}$	$\frac{\text{kg}}{\text{m}^3}$
Percent by volume	$\frac{\text{Volume of solute} \times 100}{\text{Total volume of solution}}$	% by vol.
Percent by mass	$\frac{\text{Mass of solute} \times 100}{\text{Combined mass of solute + solvent}}$	% by mass
Volume ratio	$\frac{\text{Milliliters}}{\text{Liter}}$	$\frac{\text{mL}}{\text{L}}$
Mass per unit volume	$\frac{\text{Picograms}}{\text{Liter of solution}}$	$\frac{\text{pg}}{\text{L}}$
	$\frac{\text{Nanograms}}{\text{Liter of solution}}$	$\frac{\text{ng}}{\text{L}}$
	$\frac{\text{Micrograms}}{\text{Liter of solution}}$	$\frac{\mu\text{g}}{\text{L}}$
	$\frac{\text{Milligrams}}{\text{Liter of solution}}$	$\frac{\text{mg}}{\text{L}}$
	$\frac{\text{Grams}}{\text{Cubic meter of solution}}$	$\frac{\text{g}}{\text{m}^3}$
Mass ratio	$\frac{\text{Milligrams}}{10^6 \text{ milligrams}}$	ppm
	Chemical analyses:	
Molality	$\frac{\text{Moles of solute}}{\text{1000 grams of solvent}}$	$\frac{\text{mol}}{\text{kg}}$
Molarity	$\frac{\text{Moles of solute}}{\text{Liter of solution}}$	$\frac{\text{mol}}{\text{L}}$
Normality	$\frac{\text{Equivalents of solute}}{\text{Liter of solution}}$	$\frac{\text{equiv}}{\text{L}}$
	$\frac{\text{Milliequivalents of solute}}{\text{Liter of solution}}$	$\frac{\text{meq}}{\text{L}}$

Note: 10^{12} pg = 10^9 ng = 10^6 μg = 10^3 mg = 1 gm
mg/L = g/m^3

or g/m^3 are interchangeable with ppm. Dissolved gases, considered to be chemical constituents, are measured in units of ppm (volume/volume basis), μg/m^3, or mg/L. Gases that evolve as by-products of wastewater treatment, such as carbon dioxide and methane (anaerobic decomposition), are measured in terms of ft^3 (m^3 or L). Conversion of gas concentrations between ppm and μg/m^3 is given by Eq. (2-2).

$$\mu\text{g/m}^3 = \frac{(\text{concentration, ppm})(\text{molecular weight, g/mole of gas})(10^6\ \mu\text{g/g})}{(22.414 \times 10^{-3}\ \text{m}^3/\text{mole of gas})} \tag{2-2}$$

In Eq. (2-2) the volume occupied by a gas at standard conditions (32°F and 14.7 lb/in^2 or 0°C and 101.325 kPa) is 22.414 L (1 L = 10^{-3} m^3). Parameters such as temperature, odor, hydrogen ion, and biological organisms, are expressed in other units, as explained in Example 2-1.

EXAMPLE 2-1. CONVERSION OF GAS CONCENTRATION UNITS. The off-gas from a small-diameter pressure sewer was found to contain 9 ppm$_v$ (by volume) of hydrogen sulfide (H_2S). Determine the concentration in μg/m^3 and in mg/L at standard conditions (0 °C, 1 atm).

Solution

1. Compute the concentration in μg/m^3 using Eq. (2-2). The molecular weight of H_2S = 34.08[2(1.01) + 32.06].

$$9\ \text{ppm}_v = \left(\frac{9\ \text{m}^3}{10^6\ \text{m}^3}\right)\left(\frac{34.08\ \text{g/mole H}_2\text{S}}{22.4 \times 10^{-3}\ \text{m}^3/\text{mole of H}_2\text{S}}\right)\left(\frac{10^6\ \mu\text{g}}{\text{g}}\right) = 13{,}693\ \mu\text{g/m}^3$$

2. The concentration in mg/L is

$$13{,}693\ \mu\text{g/m}^3 = \left(\frac{13{,}693\ \mu\text{g}}{\text{m}^3}\right)\left(\frac{\text{mg}}{10^3\ \mu\text{g}}\right)\left(\frac{\text{m}^3}{10^3\ \text{L}}\right) = 0.014\ \text{mg/L}$$

Comment. If gas measurements, expressed in μg/L, are made at other than standard conditions, the concentration must be corrected to standard conditions, using the ideal gas law, before converting to ppm.

Useful Chemical Relationships

Other useful relationships from elementary chemistry used in the analysis of wastewater test results include mole fraction, electroneutrality, chemical equilibrium, and solubility product.

Mole fraction. The ratio of the number of moles of a given solute to the total number of moles of all components in solution is defined as the *mole fraction*. In equation form,

$$x_B = \frac{n_B}{n_A + n_B + n_C + \cdots + n_N} \tag{2-3}$$

where x_B = mole fraction of solute B
n_B = number of moles of solute B
n_A = number of moles of solute A
n_C = number of moles of solute C
n_N = number of moles of solute N

The application of Eq. (2-3) is illustrated in Example 2-2.

EXAMPLE 2-2. DETERMINATION OF MOLE FRACTION. Determine the mole fraction of oxygen in water if the concentration of dissolved oxygen is 10.0 mg/L.

Solution. Determine the mole fraction of oxygen using Eq. (2-3) written as follows:

$$x_{O_2} = \frac{n_{O_2}}{n_{O_2} + n_W}$$

1. Determine the moles of oxygen:

$$n_{O_2} = \frac{10 \text{ mg/L}}{32 \times 10^3 \text{ mg/mole } O_2} = 3.125 \times 10^{-4} \text{ mole/L}$$

2. Determine the moles of water:

$$n_W = \frac{1000 \text{ g/L}}{18 \text{ g/mole of water}} = 55.556 \text{ mole/L}$$

3. The mole fraction of oxygen is

$$x_{O_2} = \frac{3.125 \times 10^{-4}}{3.125 \times 10^{-4} + 55.556} = 5.62 \times 10^{-6}$$

Electroneutrality. The principle of *electroneutrality* requires that the sum of the positive ions (cations) must equal the sum of negative ions (anions) in solution; thus

$$\sum \text{cations} = \sum \text{anions} \tag{2-4}$$

where cations = positively charged species in solution, eq/L or meq/L
anions = negatively charged species in solution, eq/L or meq/L

Equation (2-4) can be used to check the accuracy of chemical analyses by taking into account the percentage difference defined as follows (Standard Methods, 1995):

$$\% \text{ difference} = 100 \times \left(\frac{\sum \text{cations} - \sum \text{anions}}{\sum \text{cations} + \sum \text{anions}} \right) \tag{2-5}$$

The acceptance criteria are as given below.

$\sum$ anions, meq/L	Acceptable difference, %
0–3.0	±0.2
3.0–10.0	±2
10–800	±5

The application of Eqs. (2-4) and (2-5) is illustrated in Example 2-3.

EXAMPLE 2-3. CHECKING THE ACCURACY OF ANALYTICAL MEASUREMENTS. The following analysis has been completed on a filtered effluent, from an extended aeration wastewater treatment plant, that is to be used for landscape watering. Check the accuracy of the analysis to determine if it is sufficiently accurate, using the criteria given above.

Cation	Conc., mg/L	Anion	Conc., mg/L
Ca^{+2}	82.2	HCO_3^-	220
Mg^{+2}	17.9	SO_4^{-2}	98.3
Na^+	46.4	Cl^-	78.0
K^+	15.5	NO_3^-	25.6

Solution

1. Prepare a cation-anion balance:

	Concentration				Concentration		
Cation	**mg/L**	**mg/meq**	**meq/L**	**Anion**	**mg/L**	**mg/meq**	**meq/L**
Ca^{+2}	82.2	20.04	4.10	HCO_3^-	220	61.02	3.61
Mg^{+2}	17.9	12.15	1.47	SO_4^{-2}	98.3	48.03	2.05
Na^+	46.4	23.00	2.02	Cl^-	78.0	35.45	2.20
K^+	15.5	39.10	0.40	NO_3^-	25.6	62.01	0.41
		$\sum$ cations	7.99			$\sum$ anions	8.27

2. Check the accuracy of the cation-anion balance using Eq. (2-5).

$$\% \text{ difference} = 100 \times \left(\frac{\sum \text{cations} - \sum \text{anions}}{\sum \text{cations} + \sum \text{anions}}\right)$$

$$= 100 \times \left(\frac{7.99 - 8.27}{7.99 + 8.27}\right) = -1.72$$

For a total anion concentration between 3 and 10 meq/L, the acceptable difference must be equal to or less than 2 percent (see table given above); thus, the analysis is of sufficient accuracy.

Comment. If the cation-anion balance is not of sufficient accuracy, the problem may be analytical or a constituent may be missing.

Chemical equilibrium. A reversible chemical reaction in which reactants A and B combine to yield products C and D may be written as

$$aA + bB \leftrightarrow cC + dD \tag{2-6}$$

where the coefficients a, b, c, and d correspond to the number of moles of constituents A, B, C, and D, respectively. When the chemical species come to a state of equilibrium, as governed by the law of mass action, the numerical value of the ratio of the products over the reactants is known as the *equilibrium constant* K and is written as

$$\frac{[C]^c[D]^d}{[A]^a[B]^b} = K \tag{2-7}$$

Brackets are used in Eq. (2-7) to denote molar concentrations. For a given reaction, the value of the equilibrium constant will change with temperature and the ionic strength of the solution.

Solubility product. The equilibrium constant for a reaction involving a precipitate and its constituent ions is known as the *solubility product*. For example, the reaction for calcium carbonate ($CaCO_3$) is

$$CaCO_3 \leftrightarrow Ca^{+2} + CO_3^{-2} \tag{2-8}$$

Because the activity of the solid phase is usually taken as 1, the solubility product is written as

$$[Ca^{+2}][CO_3^{-2}] = K_{sp} \tag{2-9}$$

where K_{sp} = solubility product constant.

2-3 PHYSICAL CHARACTERISTICS

The principal physical characteristics of a wastewater, as reported in Table 2-1, are its solids content, particle size distribution, turbidity, color, transmittance/absorption, odor, temperature, density, and conductivity.

Solids

Wastewater contains a variety of solid materials varying from rags to colloidal material. In the characterization of wastewater, coarse materials are usually removed before the sample is analyzed for solids. The various solids classifications are identified in Table 2-4. The interrelationship between the various solids fractions found in wastewater is illustrated graphically in Fig. 2-4. As shown in Fig. 2-4, a filtration step is used to separate the total suspended solids (TSS) from the total solids (TS). The apparatus used to determine TSS is shown in Fig. 2-5.

The analysis of laboratory data is illustrated in Example 2-4. In Table 2-4 volatile solids (VS) are presumed to be organic matter, although some organic matter will not burn and some inorganic solids break down at high temperatures. Thus, both TS and TSS are composed of fixed solids and volatile solids. Similarly, total dissolved solids (TDS) is also composed of both fixed and volatile solids. The standard test for settleable solids consists of placing a wastewater sample in a 1-L Imhoff cone (see Fig. 2-6) and noting the volume of solids in millimeters that settle after a specified time period (1 h). Typically, about 60 percent of the suspended solids in a municipal wastewater are settleable.

Although TSS test results are used commonly as a measure of performance of treatment processes, and for regulatory control purposes, it is important to note that the test itself has no fundamental significance. The principal reasons that the test lacks a fundamental basis are as follows: (1) The measured values of TSS are dependent on the type of filter used in their determination (see Fig. 2-7). More TSS will be measured if the pore size of the filter used is reduced, as discussed subsequently under serial filtration. (2) Depending on the sample size used for the determination of TSS, auto filtration, where the suspended solids that have been intercepted by the filter also serve as filter, can occur. Auto filtration will cause an apparent increase in the measured TSS value over the actual value. (3) TSS is a lumped parameter;

TABLE 2-4
Definitions for solids found in wastewater*

Test	Description
Total solids (TS)	The residue remaining after a wastewater sample has been evaporated and dried at a specified temperature (103 to 105°C).
Total volatile solids (TVS)	Those solids that can be volatilized and burned off when the TS are ignited (500 ± 50°C).
Total fixed solids (TFS)	The residue that remains after TS are ignited (500 ± 50°C).
Total suspended solids (TSS)	Portion of the TS retained on a filter (see Fig. 2-4) with a specified pore size, measured after being dried at specified temperature. The filter most commonly used for the determination of TSS is the Whatman glass fiber filter, which has a nominal pore size of about 1.58 μm.
Volatile suspended solids (VSS)	Those solids that can be volatilized and burned off when the TSS are ignited (500 ± 50°C).
Fixed suspended solids (FSS)	The residue that remains after TSS are ignited (500 ± 50°C).
Total dissolved solids (TDS) (TS − TSS)	Those solids that pass through the filter, and are then evaporated and dried at specified temperature. It should be noted that what is measured as TDS comprises colloidal and dissolved solids. Colloids are typically in the size range from 0.001 to 1 μm.
Volatile dissolved solids (VDS) (TVS − TSS)	Those solids that can be volatilized and burned off when the TDS are ignited (500 ± 50°C).
Fixed dissolved solids (FDS)	The residue that remains after TDS are ignited (500 ± 50°C).
Settleable solids	Suspended solids, expressed as milliliters per liter, that will settle out of suspension within a specified period of time.

*Adapted from Standard Methods (1995).

because the number and size distribution of the particles that compose the measured value is unknown.

Particle Size Distribution

As noted above, TSS is a lumped parameter. In an effort to understand more about the nature of the particles that compose the TSS in wastewater, measurement of particle size is undertaken and an analysis of the distribution of particle sizes is conducted (Tchobanoglous, 1995). Information on particle size is of importance in assessing the effectiveness of treatment processes (e.g., secondary sedimentation, effluent filtration, and effluent disinfection). Because the effectiveness of both chlorine and UV disinfection is dependent on particle size, the determination of particle size has become more important, especially with the move toward greater effluent reuse in the western United States.

Information on the size of the biodegradable organic particles is significant from a treatment standpoint, as the biological conversion rate of these particles is

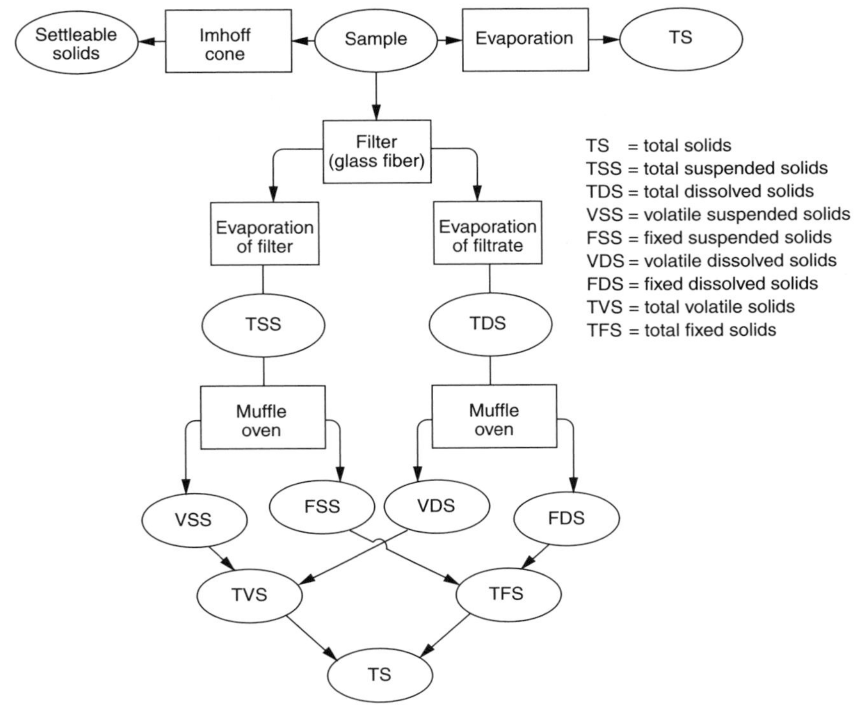

FIGURE 2-4
Interrelationships of the various solids fractions found in wastewater.

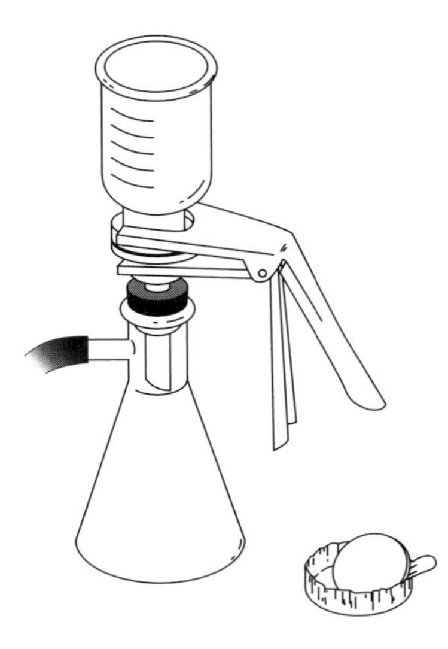

FIGURE 2-5
Apparatus used to determine TSS. After a wastewater sample has been filtered, the preweighed filter paper is placed in an aluminum dish for drying before weighing.

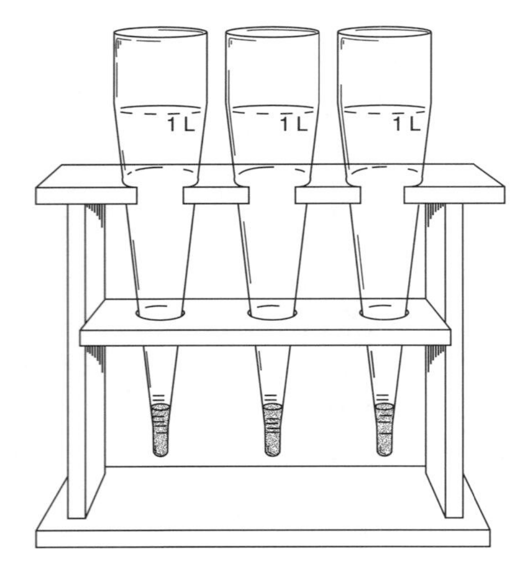

FIGURE 2-6
Imhoff cone used to determine settleable solids. The settleable solids that accumulate in the bottom of the cone after a total of 60 min are reported as mL/L.

(*a*)

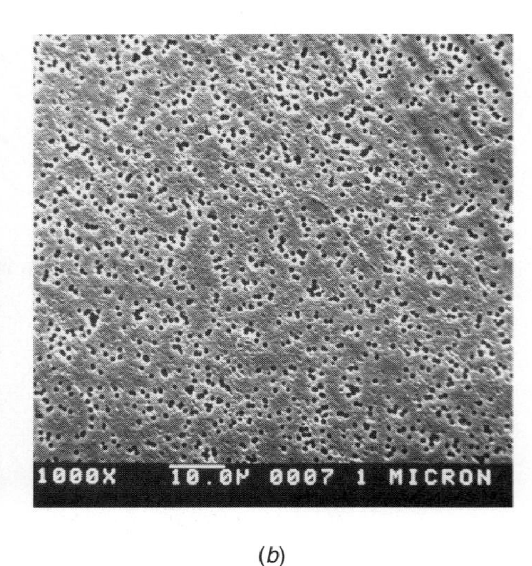

(*b*)

FIGURE 2-7
Micrographs of two laboratory filters for the analysis of TSS: (*a*) glass fiber filter with a nominal pore size of 1.2 μm and (*b*) polycarbonate membrane filter with a nominal pore size of 1.0 μm.

dependent on size (see discussion in Sec. 2-5, which deals with biochemical oxygen demand). Methods that have been used to determine particle size are summarized in Table 2-5. As reported in Table 2-5, the methods can be divided into two general categories: (1) methods based on observation and measurement and (2) methods based on separation and analysis techniques. The methods used most commonly to study and quantify the particles in wastewater are (1) serial filtration, (2) electronic particle counting, and (3) direct microscopic observation.

EXAMPLE 2-4. ANALYSIS OF SOLIDS DATA FOR A WASTEWATER SAMPLE. The following test results were obtained for an effluent wastewater sample taken from a septic tank without an effluent filter vault (see Sec. 5-10, Chap. 5). All of the tests were performed using a sample size of 100 mL. Determine the concentration TS, total volatile solids (TVS), TSS, volatile suspended solids (VSS), TDS, and volatile dissolved solids (VDS). The samples used in the solids analyses were all either dried or dried and ignited in accordance with Standard Methods (1995).

TS and TVS using evaporation dish with unfiltered sample

Tare mass of evaporating dish = 62.6775 g

Mass of evaporating dish plus residue after evaporation = 62.7264 g

Mass of evaporating dish plus residue after ignition = 62.6971 g

TSS and VSS by filtration

Tare mass of filter = 1.6623 g

Mass of filter and residue on filter after evaporation = 1.6728 g

Mass of filter and residue on filter after ignition = 1.6645 g

Solution

1. Determine the total solids:

$$\text{TS} = \frac{\left[\begin{pmatrix}\text{mass of evaporation}\\ \text{dish plus residue, g}\end{pmatrix} - \begin{pmatrix}\text{mass of}\\ \text{evaporation dish, g}\end{pmatrix}\right]}{\text{sample size, L}} \times 1000 \text{ mg/g}$$

$$= \frac{(62.7264 - 62.6775) \times 1000 \text{ mg/g}}{0.10 \text{ L}} = 489 \text{ mg/L}$$

2. Determine the total volatile solids:

$$\text{TVS} = \frac{(62.7264 - 62.6971) \times 1000 \text{ mg/g}}{0.10 \text{ L}} = 293 \text{ mg/L}$$

3. Determine the total suspended solids:

$$\text{TSS} = \frac{(1.6728 - 1.6623) \times 1000 \text{ mg/g}}{0.10 \text{ L}} = 105 \text{ mg/L}$$

4. Determine the volatile suspended solids:

$$\text{VSS} = \frac{(1.6728 - 1.6645) \times 1000 \text{ mg/g}}{0.10 \text{ L}} = 83 \text{ mg/L}$$

5. Determine the total dissolved solids:

$$\text{TDS} = \text{TS} - \text{TSS} = 489 - 105 = 384 \text{ mg/L}$$

6. Determine the volatile dissolved solids:

$$\text{VDS} = \text{TVS} - \text{VSS} = 293 - 83 = 210 \text{ mg/L}$$

TABLE 2-5

Analytical techniques applicable to particle size analysis of wastewater contaminants*

Technique	Typical size range, μm
Observation and measurement	
Microscopy	
Light	0.2 to >100
Transmission electron	0.2 to >100
Scanning electron	0.002 to 50
Image analysis	0.2 to >100
Particle counters	
Conductivity difference	0.2 to >100
Equivalent light scattering	0.005 to >100
Light blockage	0.2 to >100
Separation and analysis	
Centrifugation	0.08 to >100
Field flow fractionation	0.09 to >100
Gel filtration chromatography	<0.0001 to >100
Sedimentation	0.05 to >100
Membrane filtration (see Chap. 12)	0.0001 to 1

*Adapted from Levine et al. (1985).

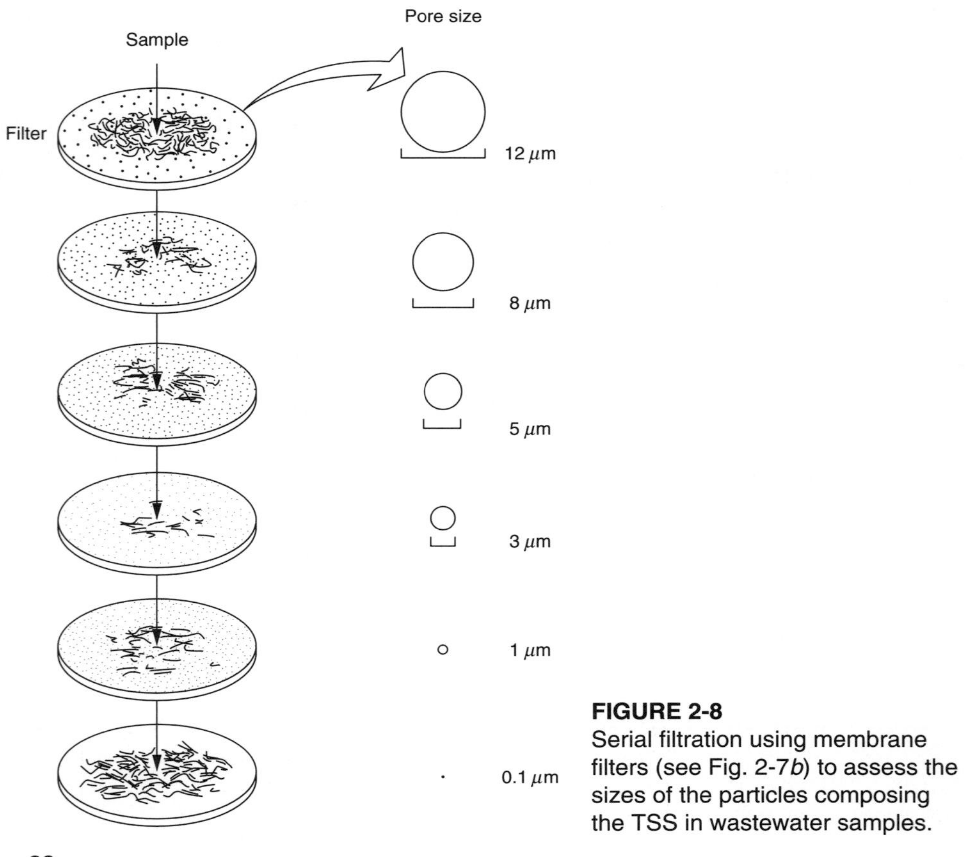

FIGURE 2-8
Serial filtration using membrane filters (see Fig. 2-7*b*) to assess the sizes of the particles composing the TSS in wastewater samples.

TABLE 2-6
Typical data on the distribution of filterable solids and filtrate turbidity in treated (effluent) wastewater obtained by serial filtration

			Percent of mass retained in indicated size range and turbidity of filtrate passing through filter with the smallest pore size					
Sample (date, time)	Total, TSS[a]	Initial sample, turbidity	>0.1 <1.0	>1.0 <3.0	>3.0 <5.0	>5.0 <8.0	>8.0 <12.0	>12.0
Monterey[b] (12/08/93)								
TSS, mg/L	40.4		22.2	3.3	1.8	2.4	1.3	8.9
Turbidity, NTU		9.0	0.56	3.9	5.2	5.7	6.2	6.7
Monterey[b] (01/24/94)								
TSS, mg/L	38.3		21.3	8.7	4.3	0.8	0.4	2.8
Turbidity, NTU		10.0	1.4	4.6	6.4	9.8	9.8	9.8
UCD[c] (8/17/97)								
TSS, mg/L	4.67		0.55	0.53	0.24	0.20	0.24	2.91
Turbidity, NTU		0.58	0.14	0.25	0.35	0.41	0.42	0.50
UCD[d] (8/17/97)								
TSS, mg/L	1.47		0.45	0.33	0.30	0.14	0.10	0.15
Turbidity, NTU		0.45	0.14	0.31	0.33	0.38	0.40	0.41

[a] Total TSS retained on a polycarbonate membrane filter with a pore size of 0.1 μm.
[b] Secondary effluent, Monterey, CA (Courtesy Jaques, 1994).
[c] Secondary effluent, University of California, Davis, CA.
[d] Tertiary filtered secondary effluent, University of California, Davis, CA.

Serial filtration. In the serial filtration method, a wastewater sample is passed sequentially through a series of membrane filters (see Fig. 2-8) with circular openings of known diameter (typically 12, 8, 5, 3, 1, and 0.1 μm), and the amount of suspended solids retained in each filter is measured. Typical results from such a measurement are reported in Table 2-6, and graphically in Fig. 2-9. What is interesting to note in Table 2-6 is the amount of TSS found between 0.1 and 1.0 μm. If a 0.1-μm filter had been used to determine TSS for the treated effluent at Monterey instead of a filter with a nominal pore size equal to or greater than 1.0 μm, as specified in Standard Methods for the TSS test, more than 20 mg/L of additional TSS would have been measured. Although some information is gained on the size and distribution of the particles in the wastewater sample, little information is gained on the nature of the individual particles. This method is useful in assessing the effectiveness of treatment methods (e.g., microfiltration) for the removal of residual TSS.

Electronic particle size counting. In electronic particle size counting, particles in wastewater are counted by diluting a sample and then passing the diluted sample through a calibrated orifice or past laser beams. As the particles pass through the orifice, the conductivity of the fluid changes as a result of the presence of the particle. The change in conductivity is correlated to the size of an equivalent sphere. In a similar fashion, as a particle passes by a laser beam, it reduces the intensity of the laser because of particle scattering. The reduced intensity is correlated to the

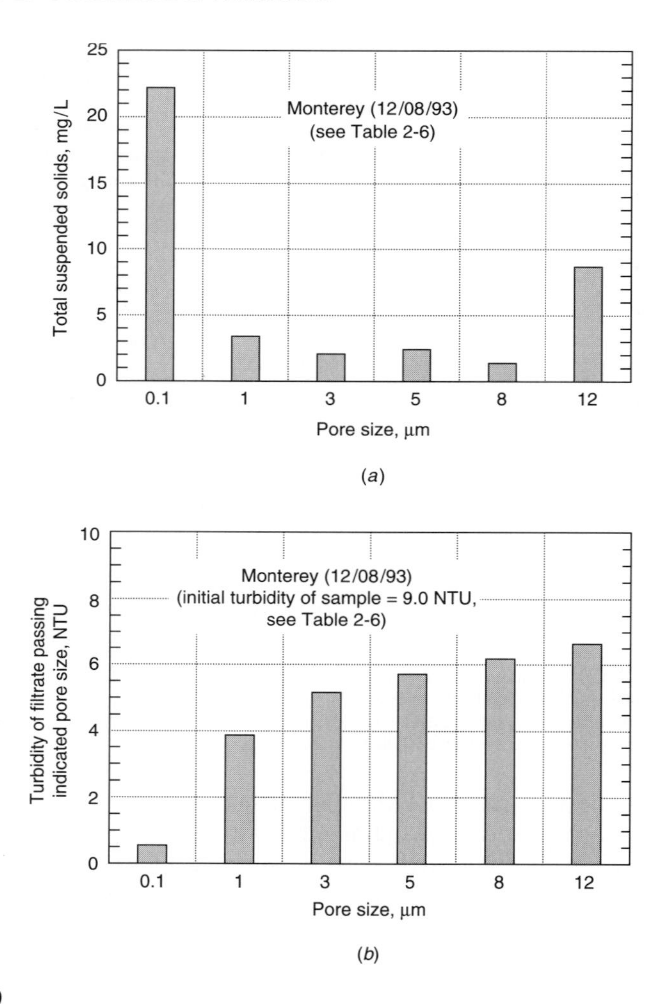

FIGURE 2-9
Typical data on TSS and turbidity obtained by serial filtration: (*a*) TSS retained on filter with indicated pore size after passing through filter with the next largest pore size (see Fig. 2-8) and (*b*) turbidity of filtrate passing through filter with indicated pore size.

diameter of the particle. The particles that are counted are grouped into particle size ranges (e.g., 0.5 to 2 , 2 to 5, 5 to 20 μm, etc.). In turn, the volume fraction corresponding to each particle size range can be computed.

Typical effluent volume fraction data from two activated sludge treatment plants are reported in Fig. 2-10. As shown, the particle size data for small particles are the same for both treatment plants. However the particle size data for the large particles is quite different, primarily because of the type of activated sludge process used and the design and operation of the secondary clarifiers (see discussion in Chap. 7). Particle size information, such as that shown in Fig. 2-10, is useful in assessing the performance of secondary sedimentation facilities, effluent filtration, and chlorine and UV irradiation disinfection.

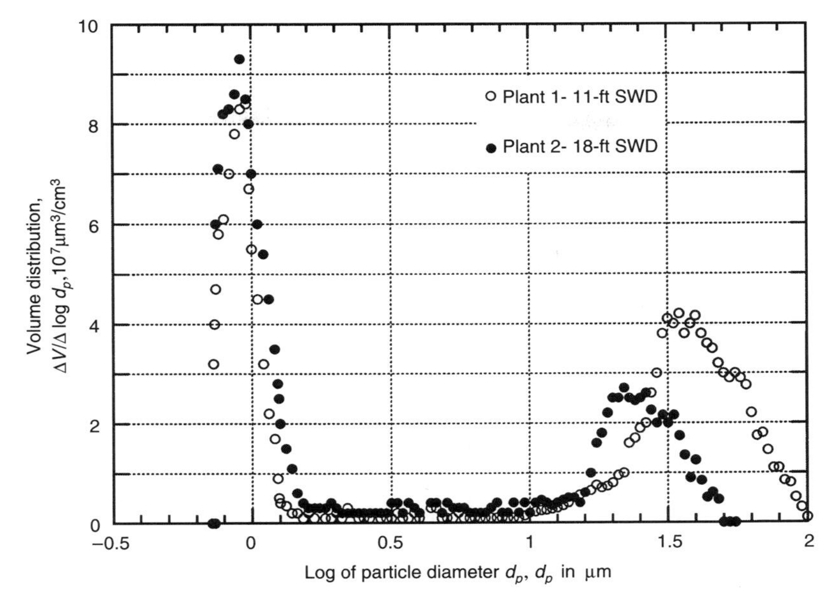

FIGURE 2-10
Volume fraction of particle sizes found in the effluent from two activated sludge plants, with clarifiers having different side water depths (SWD).

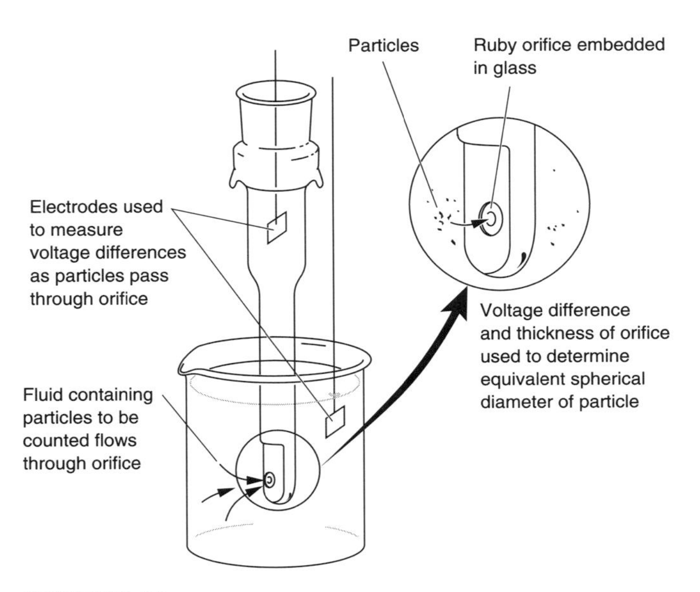

FIGURE 2-11
Typical particle-counting chamber used to enumerate the particles in a wastewater sample.

Microscopic observation. Particles in wastewater can also be enumerated microscopically by placing a small sample in a particle-counting chamber. To aid in differentiating different types of particles, various types of stains can be used. In general, microscopic counting of particles is impractical on a routine basis, given the number of particles per milliliter of wastewater. Nevertheless, this method can be used to qualitatively assess the nature and size of the particles in wastewater.

A quantitative assessment of wastewater particles can be obtained with a microscope by means of a process called *optical imaging*. A small sample of wastewater is placed on a microscope slide. The images of the wastewater particles are collected with a video camera attached to a microscope and transmitted to a computer where various measurements of the wastewater particles can be assessed. The types of measurements that can be obtained are dependent on the computer software, but typically include the mean, minimum, and maximum diameter, the aspect ratio (length to width ratio), the circumference, the surface area, the volume, and the centroid of various particles. Particle imaging greatly reduces the time required to measure various characteristics of wastewater particles, but the cost of the software and equipment is often prohibitive for many small laboratories.

Turbidity

Turbidity, a measure of the light-scattering properties of water, is another test used to indicate the quality of treated effluents and natural waters with respect to colloidal and residual suspended matter. The measurement of turbidity is based on a comparison of the intensity of light scattered by a sample as compared to the light scattered by a reference suspension under the same conditions (Standard Methods, 1995). Formazin suspensions are used as the primary reference standard. The results of turbidity measurements are reported as nephelometric turbidity units (NTU).

Colloidal matter will scatter or absorb light and thus prevent its transmission. The effect of particle size on turbidity is illustrated in Table 2-6 and in Fig. 2-9*b*. As shown, most of the turbidity is associated with particle sizes below 3 μm, with the greatest contribution being from the particles between 0.1 and 1.0 μm. In general, there is no clearly defined relationship between turbidity and the concentration of suspended solids in untreated wastewater. There is, however, a reasonable relationship between turbidity and suspended solids for treated secondary effluent from the activated sludge process. The general form of the relationship is as follows:

$$\text{TSS, mg/L} \approx (\text{fTSS})(T) \tag{2-10}$$

where fTSS = factor varying from 2.0 to 2.7 for converting turbidity readings to TSS, (mg/L TSS)/NTU, and T = turbidity, NTU.

Color

Color in wastewater is caused by suspended solids, colloidal material, and dissolved substances. Color caused by suspended solids is known as *apparent color* whereas color caused by colloidal and dissolved substances is known as *true color*. True color is obtained by filtering a sample. Because the measured value will depend on the pore size of the filter, the type of filter and pore size must be specified. The color of a wastewater sample is determined by comparing the color of the sample to the color produced by various solutions of potassium chloroplatinate (K_2PtCl_6). One color unit corresponds to the color produced by 1.0 mg/L of platinum. The sources of color in wastewater include infiltration/inflow, industrial discharges, and the decomposition of organic compounds in wastewater. Depending on the time of year, infiltration/inflow will contain a variety of humic substances (e.g., tannins, humic acid, and humates). Derived principally from the decomposition of the lignin found in leaves and other organic plant materials, humic substances typically impart a yellow color to the water. Industrial discharges may contain organic dyes as well as metallic compounds which can impart a variety of colors to the wastewater.

In a qualitative way, color can be used to assess the general condition of wastewater. If light brown in color, the wastewater is usually less than 6 hours old. A light-to-medium gray color is characteristic of wastewaters that have undergone some decomposition or that have been in the collection system for some time. If the color is dark gray or black, the wastewater typically is septic, having undergone extensive bacterial decomposition under anaerobic (in the absence of oxygen) conditions. The blackening of wastewater is often due to the formation of various sulfides, particularly ferrous sulfide (FeS). The formation of sulfides occurs when hydrogen sulfide, produced from the reduction of sulfate under anaerobic conditions, combines with a divalent metal, such as iron, which may be present.

Transmittance/Absorption

Transmittance, defined as the ability of a liquid to transmit light of a specified wavelength through a known depth of solution, is computed from the following relationship:

$$\%\ \text{transmittance}\ T = \left(\frac{I}{I_o}\right)100 \qquad (2\text{-}11)$$

where I = final intensity of transmitted light (radiation) after passing through a solution of known depth and I_o = initial intensity of incident light (radiation).

Transmittance is measured with a spectrophotometer using a specified wavelength. Percent transmittance is affected by all substances in wastewater that can absorb or scatter light. Unfiltered and filtered transmittance are measured in wastewater in connection with the evaluation and design of UV disinfection systems (see Chap. 12). Absorbance, the loss of radiant energy as light passes through a fluid, is defined as follows:

$$\%\ a = \left(\frac{I_o - I}{I_o}\right)100 \qquad (2\text{-}12)$$

The principal wastewater characteristics that affect the percent transmission include selected inorganic compounds (e.g., copper and iron), organic compounds (e.g., organic dyes, humic substances, and conjugated ring compounds such as benzene and toluene), and TSS. Of the inorganic compounds which affect transmittance, iron is considered to be the most important with respect to UV absorbance because dissolved iron can absorb UV light directly and because iron will adsorb onto suspended solids, bacterial clumps, and other organic compounds. The sorbed iron can prevent the UV light from penetrating the particle and inactivating organisms that may be embedded within the particle. Organic compounds, identified as being absorbers of UV light, are compounds with six conjugated carbons or a five- or six-member conjugated ring. The reduction in transmittance observed during storm events is often ascribed to the presence of humic substances from stormwater flows.

Odor

The determination of odor has become increasingly important as the public has become more concerned with the proper operation of wastewater treatment facilities. The odor of fresh wastewater is usually not offensive, but a variety of malodorous compounds are released when wastewater is decomposed biologically under anaerobic conditions. The principal malodorous compound is hydrogen sulfide (the smell of rotten eggs). Other compounds such as indole, skatole, and mercaptans, formed under anaerobic conditions, may cause odors that are more offensive than that of hydrogen sulfide (WEF, 1995). Detection thresholds for various malodorous compounds found in wastewater are reported in Table 2-7. Because of public concern, special care is called for in the design of wastewater treatment facilities to avoid conditions that will allow the development of odors (see Sec. 5-14, Chap. 5, for a discussion of odor control).

Odors can be measured by sensory and instrumental methods. The sensory measurement of odors by the human olfactory system can provide meaningful information at very low detection levels. As a result, the sensory method is often used to measure odors from treatment plants.

TABLE 2-7
Odor thresholds of odorous compounds associated with untreated wastewater

Odorous compound	Chemical formula	Molecular weight	Odor threshold, ppm_v*	Characteristic odor
Ammonia	NH_3	17.0	46.8	Ammoniacal
Chlorine	Cl_2	71.0	0.314	
Crotyl mercaptan	$CH_3{-}CH{=}CH{-}CH_2{-}SH$	90.19	0.000029	Skunklike
Dimethyl sulfide	$CH_3{-}S{-}CH_3$	62	0.0001	Decayed vegetables
Diphenyl sulfide	$(C_6H_5)_2S$	186	0.0047	
Ethyl mercaptan	$CH_3CH_2{-}SH$	62	0.00019	Decayed cabbage
Hydrogen sulfide	H_2S	34	0.00047	Rotten eggs
Indole	C_8H_6NH	117	0.0001	
Methyl amine	CH_3NH_2	31	21.0	
Methyl mercaptan	CH_3SH	48	0.0021	Decayed cabbage
Skatole	C_9H_9NH	132	0.019	Fecal matter
Sulfur dioxide	SO_2	64.07	0.009	
Thiocresol	$CH_3{-}C_6H_4{-}SH$	124	0.000062	Skunk, rancid

*Parts per million by volume.

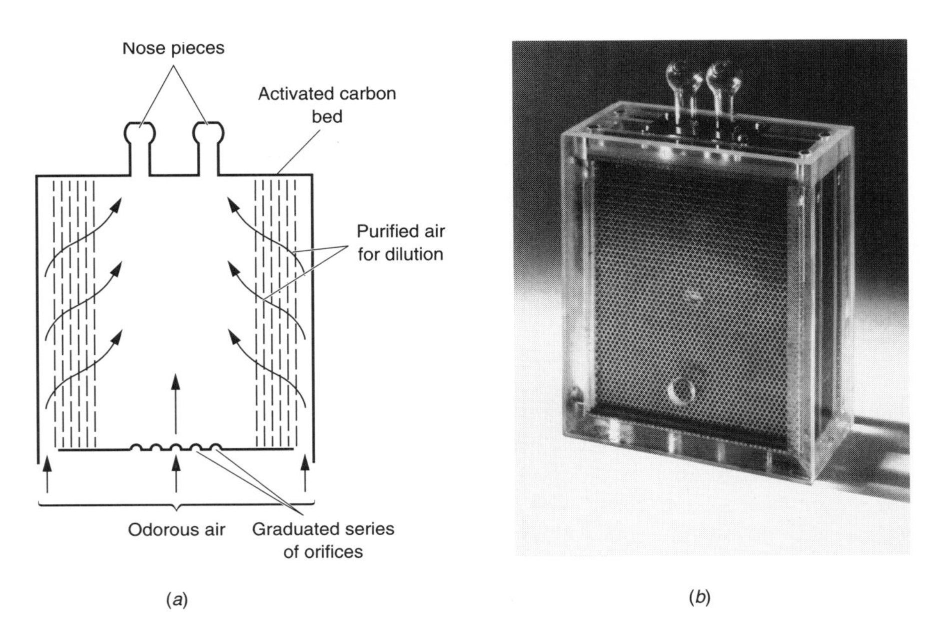

FIGURE 2-12
Typical device used for the detection of odors in the field: (*a*) schematic and (*b*) photographic view.

Specific odorant concentrations can also be measured by instrumental devices. Direct measurements of hydrogen sulfide can be made in the field to as low as 1 part per billion (ppb) with a hand-held meter. A typical device used for the detection of odors is shown in Fig. 2-12.

The threshold odor of a water or wastewater sample is determined by diluting the sample with odor-free water. The *threshold odor number* (TON) corresponds to the greatest dilution of the sample with odor-free water at which an odor is just perceptible. The recommended sample size for odor measurements is 200 mL. The numerical value of the TON is determined as follows:

$$\text{TON} = \frac{A + B}{A} \tag{2-13}$$

where A = mL of sample and B = mL of odor-free water.

The odor emanating from the liquid sample is determined as discussed above with human subjects (often a panel of subjects). Details for this procedure which was approved by the Standard Methods Committee in 1985 may be found in Standard Methods (1995). Odor detection procedures given by ASTM should also be consulted (ASTM, 1979). The application of Eq. (2-13) is illustrated in Example 2-5.

EXAMPLE 2-5. DETERMINATION OF THRESHOLD ODOR NUMBER. A 25-mL sample of treated wastewater requires 175 mL of distilled water to reduce the odor to a level that is just perceptible. What is the threshold odor number (TON)?

Solution. Determine the TON using Eq. (2-13).

$$\text{TON} = \frac{A + B}{A} = \frac{25 \text{ mL} + 175 \text{ mL}}{25 \text{ mL}} = 8.0$$

Comment. The value of the threshold number will depend on the nature of the odor compound. Also, depending on the odor compound, different persons will respond differently in assessing when an odor is just perceptible.

Temperature

The temperature of wastewater is commonly higher than that of the water supply because of the addition of warm water from domestic use. The measurement of temperature is important because most wastewater treatment systems include biological processes that are temperature-dependent. The temperature of wastewater will vary from season to season and also with geographic location. In cold regions, the temperature will vary from about 45 to 65°F (7 to 18°C) while in warmer regions the temperature will vary from 55 to 86°F (13 to 30°C). The variation of wastewater temperature throughout a year for four locations is shown in Fig. 2-13.

The temperature of water is a very important parameter because of its effect on chemical reactions and reaction rates, aquatic life, and the suitability of the water for beneficial uses. Increased temperature, for example, can cause a change in the species of fish that can exist in the receiving water body. Industrial establishments

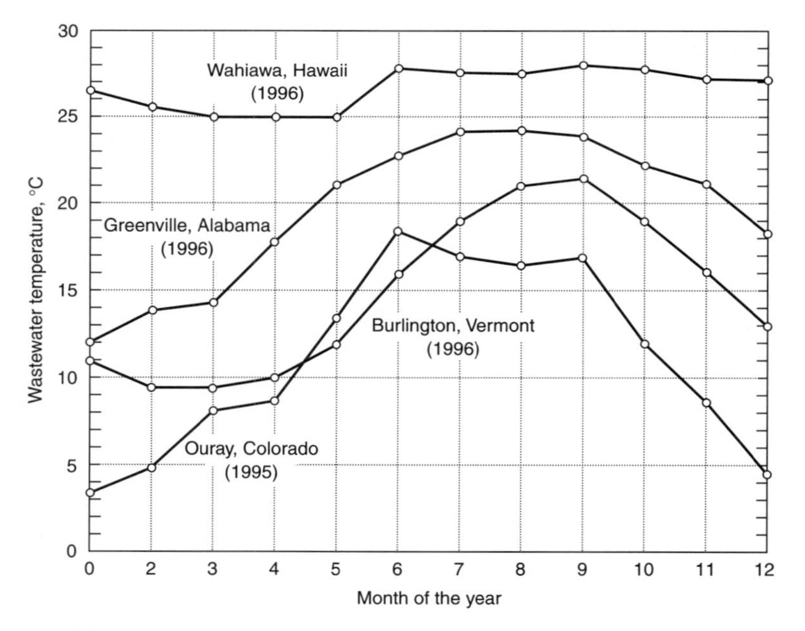

FIGURE 2-13
Typical variation in the temperature of wastewater at various locations.

that use surface water for cooling-water purposes are particularly concerned with the temperature of the intake water. In addition, oxygen is less soluble in warm water than in cold water. The increase in the rate of biochemical reactions that accompanies an increase in temperature, combined with the decrease in the quantity of oxygen present in surface waters, can often cause serious depletions in dissolved oxygen concentrations during the summer months.

Optimum temperatures for bacterial activity are in the range from about 77 to 95°F (25 to 35°C). Aerobic digestion and nitrification stop when the temperature rises to 122°F (50°C). When the temperature drops to about 15°C, methane-producing bacteria become quite inactive, and at about 41°F (5°C), the autotrophic-nitrifying bacteria practically cease functioning. At 36°F (2°C), even the chemoheterotrophic bacteria acting on carbonaceous material become essentially dormant. The effects of temperature on the performance of biological treatment processes are considered in greater detail in Chap. 9.

Density, Specific Gravity, and Specific Weight

The density of wastewater, ρ_w, is defined as its mass per unit volume expressed as slug/ft^3 in U.S. customary units and g/L or kg/m^3 in Système International (SI) units. Density is an important physical characteristic of wastewater because of the potential for the formation of density currents in sedimentation tanks, constructed wetlands, and in other treatment units. The density of domestic wastewater which does not contain significant amounts of industrial waste is essentially the same as that of water at the same temperature.

In some cases the specific gravity of the wastewater, s_w, is used in place of the density. The specific gravity is defined as

$$s_w = \frac{\rho_w}{\rho_o} \tag{2-14}$$

where ρ_w = density of wastewater and ρ_o = density of water. Both the density and specific gravity of wastewater are temperature-dependent and will vary with the concentration of total solids in the wastewater.

The specific weight of a fluid, γ, is its weight per unit volume. In U.S. customary units, it is expressed in lb$_f$/ft^3. The relationship between γ, ρ, and the acceleration due to gravity, g, is $\gamma = \rho g$. At normal temperatures γ is about 62.4 lb$_f$/ft^3 (9.81 kN/m^3). Values for both density and specific weight as a function of temperature are given in App. C.

Conductivity

The electrical conductivity (EC) of a water is a measure of the ability of a solution to conduct an electrical current. Because the electrical current is transported by the ions in solution, the conductivity increases as the concentration of ions increases. In effect, the measured EC value is used as a surrogate measure of total dissolved solids (TDS) concentration. At present, the EC of a water is one of the most important parameters used to determine the suitability of a water for irrigation. The salinity of treated wastewater to be used for irrigation is determined by measuring its electrical conductivity.

The electrical conductivity in U.S. customary units is expressed as micromhos per centimeter (μmho/cm) and in SI units as millisiemens per meter (mS/m). It should be noted that 10 μmho/cm is equivalent to 1 mS/m. Equation (2-15) can be used to estimate the TDS of a water sample based on the measured EC value (Standard Methods, 1995).

$$\text{TDS (mg/L)} \approx \text{EC } (\mu\text{mho/cm or dS/m}) \times (550 - 700) \tag{2-15}$$

The above relationship does not necessarily apply to raw wastewater or high-strength industrial wastewater. The above relationship can also be used to check the acceptability of chemical analyses (see Standard Methods, 1995).

2-4 INORGANIC CHEMICAL CHARACTERISTICS

The chemical constituents of wastewater are typically classified as inorganic and organic. Inorganic constituents in wastewater include: (1) individual elements such as calcium (Ca), chloride (Cl), iron (Fe), chromium (Cr), and zinc (Zn) and (2) a wide variety of compounds such as nitrate (NO_3) and sulfate (SO_4). The organic constituents, of most interest in wastewater, are classified as aggregate and individual. Aggregate organic constituents comprise a number of individual constituents that cannot be distinguished separately, but are of great significance in the treatment, disposal, and reuse of wastewater. Specific organic constituents are also of importance in the treatment, disposal, and reuse of wastewater. Inorganic chemical constituents are considered in this section. Aggregate organic characteristics and individual organic constituents are considered in Secs. 2-5 and 2-6, respectively.

Inorganic chemical constituents of concern include nutrients, nonmetallic constituents, metals, and gases. Inorganic chemical nutrients include free ammonia, organic nitrogen (determined as ammonia following sample digestion), nitrites, nitrates, organic phosphorus, and inorganic phosphorus. Nitrogen and phosphorus are important because these two nutrients have been identified most commonly as being responsible for the growth of undesirable aquatic plants. Other tests, such as pH, alkalinity, chloride, and sulfate are performed to assess the suitability of reusing treated wastewater and in controlling the various treatment processes. Tests for metals, as well as other constituents, are used to assess the suitability of digested biosolids and compost for land application. Because the concentration of the species of nitrogen and phosphorus are dependent on the hydrogen ion concentration in solution, pH is considered first in the following discussion.

pH

The usual means of expressing the hydrogen-ion concentration of a solution is in terms of pH, which is defined as the negative logarithm of the hydrogen-ion con-

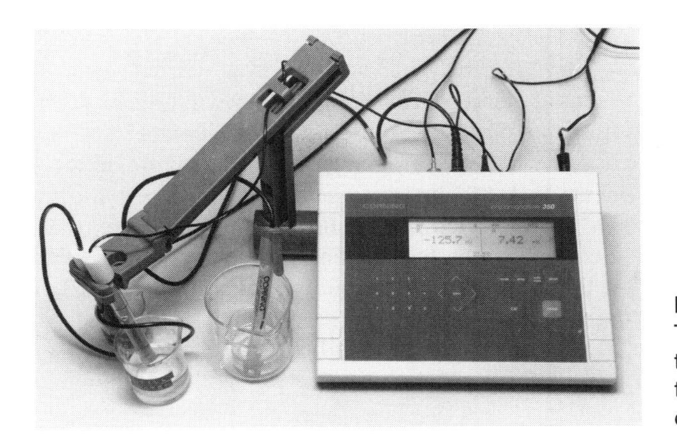

FIGURE 2-14
Typical meter used for the measurement of the pH and specific ion concentrations.

centration:

$$\mathrm{pH} = -\log_{10}[\mathrm{H}^+] \tag{2-16}$$

The hydrogen-ion concentration is usually measured instrumentally by using a pH meter (see Fig. 2-14). Various pH papers and indicator solutions that change color at various pH values are also used.

The hydrogen-ion concentration in water is closely connected with the extent to which water molecules dissociate. Water will dissociate into hydrogen and hydroxyl ions as follows:

$$\mathrm{H_2O} \leftrightarrow \mathrm{H^+} + \mathrm{OH^-} \tag{2-17}$$

Applying the law of mass action [Eq. (2-7)] to Eq. (2-17) yields,

$$\frac{[\mathrm{H^+}][\mathrm{OH^-}]}{\mathrm{H_2O}} = K \tag{2-18}$$

Because the concentration of water in a dilute aqueous system is essentially constant, the concentration can be incorporated into the equilibrium constant K to give

$$[\mathrm{H^+}][\mathrm{OH^-}] = K_w \tag{2-19}$$

K_w is known as the ionization constant, or ion product, of water and is approximately equal to 1×10^{-14} at a temperature of 25°C. Equation (2-20) can be used to calculate the hydroxyl-ion concentration when the hydrogen-ion concentration is known, and vice versa.

If the negative logarithm of the hydroxyl-ion concentration is defined as pOH, it can be seen from Eq. (2-19) that, for water at 25°C,

$$\mathrm{pH} + \mathrm{pOH} = 14 \tag{2-20}$$

The hydrogen-ion concentration range suitable for the existence of most biological life is relatively narrow, typically between pH 5 and 9. Wastewaters with pH values below 5 or greater than 9 are often difficult to treat by biological means. If the pH value is not adjusted before discharge, the wastewater effluent may alter the pH of the receiving water. Most wastewater treatment plant effluent discharges must be within specified pH limits.

Nitrogen

Nitrogen and phosphorus are essential elements for biological growth and, as such, are known as nutrients or biostimulants. Trace quantities of other elements, such as iron, are also needed for biological growth, but nitrogen and phosphorus are, in most cases, the major nutrients of importance. Because nitrogen is an essential building block in the synthesis of protein, nitrogen data are required to evaluate the treatability of wastewater by biological processes. Insufficient nitrogen can necessitate the addition of nitrogen to make the waste treatable. Total nitrogen is comprised of ammonia, nitrite, nitrate, and organic nitrogen.

Ammonia nitrogen exists in aqueous solution as either the ammonium ion or ammonia gas, depending on the pH of the solution, in accordance with the following equilibrium reaction:

$$NH_4^+ + OH^- \Leftrightarrow NH_3 + H_2O \qquad (2\text{-}21)$$

At pH levels above 9.3, the equilibrium is displaced to the right; at levels below pH 9.3, the ammonium ion is predominant, as shown in Fig. 2-15*a*. In wastewater samples containing small amounts of suspended solids, ammonia can be measured colorimetrically by adding Nessler's reagent and observing the intensity of the yellow-brown colloid formed. To avoid interference with solids, in more concentrated samples, the ammonia can be boiled off before Nesslerization, as described below. Ammonia can also be determined titrimetrically or with specific-ion electrodes.

Nitrite nitrogen, determined colorimetrically, is relatively unstable and is easily oxidized to the nitrate form. It is an indicator of past pollution in the process of stabilization and seldom exceeds 1 mg/L in wastewater or 0.1 mg/L in surface waters or groundwaters. Although present in low concentrations, nitrite can be very important in wastewater or water-pollution studies because it is extremely toxic to most fish and other aquatic species. Nitrites present in wastewater effluents are oxidized by chlorine, and thus increase the chlorine dosage requirements and the cost of disinfection.

Nitrate nitrogen, the most highly oxidized form of nitrogen found in wastewaters, is usually determined by colorimetric methods. Where secondary effluent is to be reclaimed for groundwater recharge, the nitrate concentration is important. The U.S. (EPA) drinking-water standards limit it to 45 mg/L as NO_3^- (10 mg/L as $NO_3{-}N$) because of its serious and occasionally fatal effects on infants. In the stomachs of young infants nitrate can be reduced to nitrite which can then bond with hemoglobin and reduce the transfer of oxygen to the body's cells resulting in a bluish skin color. Commonly known as the *blue baby syndrome*, the medical term for the condition is *methemoglobinemia*. Nitrates in wastewater effluents may vary in concentration from 2 to 30 mg/L as N, depending on the degree of nitrification and denitrification.

Organic nitrogen is determined by the Kjeldahl method in which an aqueous sample is first boiled to drive off the ammonia and then digested by boiling in sulfuric acid. During the digestion the organic nitrogen present in the sample is converted to ammonia, which is distilled and measured by Nesslerization. Total Kjeldahl nitrogen

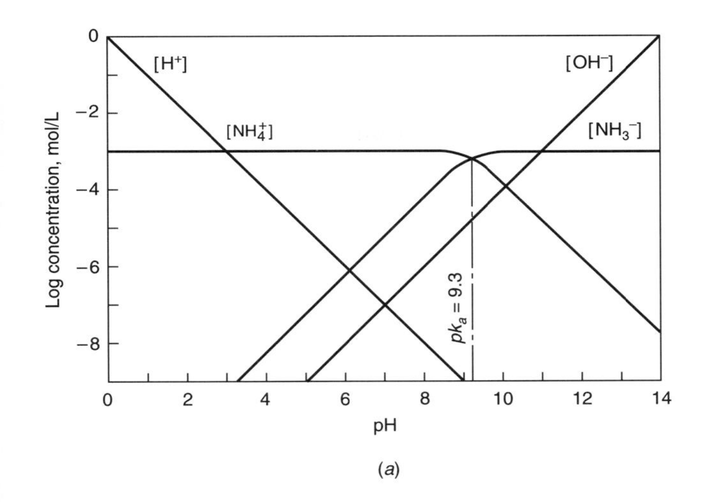

(*a*)

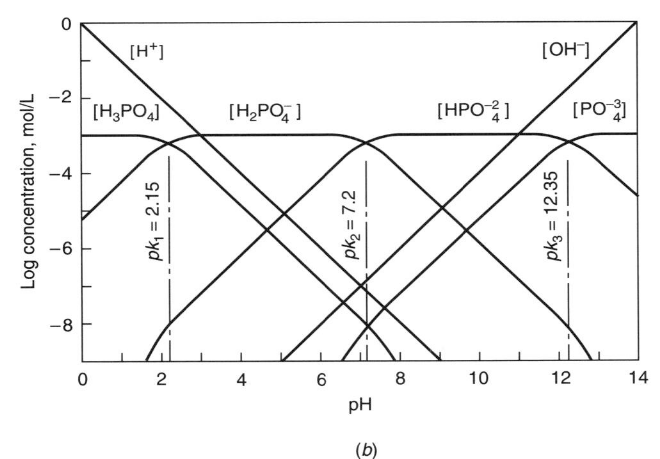

(*b*)

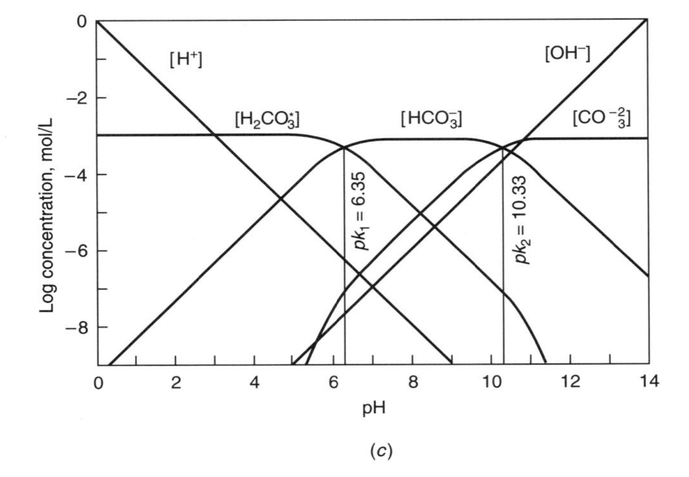

(*c*)

FIGURE 2-15
Log concentration versus pH diagrams for: (*a*) ammonia, (*b*) phosphate, and (*c*) carbonate. By sliding the constituent curves up or down, pH values can be obtained at any concentration value.

is determined in the same manner as organic nitrogen, except that the ammonia is not driven off before the digestion step. Total Kjeldahl nitrogen is, therefore, the total of the organic and ammonia nitrogen.

Phosphorus

Phosphorus is also essential to the growth of algae and other biological organisms. Because of noxious algal blooms that occur in surface waters, efforts have been made to control the amount of phosphorus compounds that enter surface waters in domestic and industrial waste discharges and natural runoff. Municipal wastewaters, for example, may contain from 4 to 12 mg/L of phosphorus as P. The usual forms of phosphorus that are found in aqueous solutions include the orthophosphate, polyphosphate, and organic phosphate. The orthophosphates (e.g., PO_4^{-3}, HPO_4^{-2}, $H_2PO_4^-$, H_3PO_4, HPO_4^{-2} complexes) are available for biological metabolism without further breakdown. The distribution of the several phosphate species as a function of pH is shown in Fig. 2-15*b*.

The polyphosphates include those molecules with two or more phosphorus atoms, oxygen atoms, and in some cases, hydrogen atoms combined in a complex molecule (e.g., $P_2O_7^{-4}$, $HP_2O_7^{-3}$, $H_2P_2O_7^{-2}$, $H_3P_2O_7^-$, $H_4P_2O_7$, $HP_2O_7^{-3}$ complexes and $P_3O_{10}^{-5}$, $HP_3O_{10}^{-4}$, $H_2P_3O_{10}^{-3}$, $H_3P_3O_{10}^{-2}$, $HP_3O_{10}^{-4}$ complexes). Polyphosphates undergo hydrolysis in aqueous solutions and revert to the orthophosphate forms; however, the hydrolysis process is usually quite slow. The organically bound phosphorus is usually of minor importance in most domestic wastes, but it can be an important constituent of industrial wastes and wastewater sludges. Analytically, orthophosphates can be determined using gravimetric, volumetric, and physicochemical methods. Polyphosphates and organic phosphorus must first be converted to the orthophosphate form before they can be analyzed.

Alkalinity

The alkalinity of a water is its acid-neutralizing capacity (Standard Methods, 1995). In wastewater, alkalinity results from the presence of the hydroxides [OH^-], carbonates [CO_3^{-2}], and bicarbonates [HCO_3^-] of elements such as calcium, magnesium, sodium, potassium, or ammonia. Of these, calcium and magnesium bicarbonates are most common. Borates, silicates, phosphates, and similar compounds can also contribute to the alkalinity; however, they are seldom of significance, except perhaps in some agricultural and industrial wastewaters. The alkalinity in wastewater helps to resist changes in pH caused by the addition of acids. Wastewater is normally alkaline, receiving its alkalinity from the water supply, the groundwater, and the materials added during domestic use. The distribution of the carbonate species as a function of pH is shown in Fig. 2-15*c*. Because of the importance of the carbonate equilibrium in wastewater treatment, the fundamental relationships between the carbonate species are considered further in App. F.

Alkalinity is determined by titrating against a standard acid; the results are expressed in terms of calcium carbonate, $CaCO_3$. For most practical purposes alkalinity can be defined in terms of molar quantities as

$$[\text{Alk}],\ \text{mole/L} = [\text{HCO}_3^-] + [\text{CO}_3^{-2}] + [\text{OH}^-] - [\text{H}^+] \tag{2-22}$$

The corresponding expression in terms of equivalents is

$$[\text{Alk}],\ \text{eq/m}^3 = \text{meq/L} = [\text{HCO}_3^-] + 2[\text{CO}_3^{-2}] + [\text{OH}^-] - [\text{H}^+] \tag{2-23}$$

In practice, alkalinity is expressed in terms of calcium carbonate. To convert from meq/L to mg/L as $CaCO_3$ it is helpful to remember that

$$\text{Milliequivalent mass of CaCO}_3 = \frac{100\ \text{mg/mmol}}{2\ \text{meq/mmol}} \tag{2-24}$$

$$= 50\ \text{mg/meq}$$

Thus 3 meq/L of alkalinity would be expressed as 150 mg/L as $CaCO_3$:

$$\text{Alkalinity } A \text{ as CaCO}_3 = \frac{3.0\ \text{meq}}{\text{L}} \times \frac{50\ \text{mg CaCO}_3}{\text{meq CaCO}_3}$$

$$= 150\ \text{mg/L as CaCO}_3$$

Chlorides

The chloride concentration in wastewater is an important parameter with respect to wastewater reuse applications. Chlorides in natural water result from the leaching of chloride-containing rocks and soils with which the water comes in contact. In coastal areas chloride concentrations can be caused by brackish and saltwater intrusion. In addition, agricultural, industrial, and domestic wastewaters discharged to surface waters are potential sources of chlorides. In wastewater, chlorides are added through usage. For example, human excreta contains about 6 g of chlorides per person per day. In areas where the hardness of water is high, use of regeneration-type water softeners will also add large quantities of chlorides. Because conventional methods of waste treatment do not remove chloride to any significant extent, higher than usual chloride concentrations can be taken as an indication that the body of water is being used for waste disposal.

Sulfur

The sulfate ion occurs naturally in most water supplies and is present in wastewater as well. Sulfur is required in the synthesis of proteins and is released in their degradation. Sulfate is reduced biologically under anaerobic conditions to sulfide which, in turn, can combine with hydrogen to form hydrogen sulfide (H_2S). The following generalized reactions are typical:

$$\text{Organic matter} + \text{SO}_4^{-2} \xrightarrow{\text{bacteria}} \text{S}^{-2} + \text{H}_2\text{O} + \text{CO}_2 \tag{2-25}$$

$$\text{S}^{-2} + 2\text{H}^+ \rightarrow \text{H}_2\text{S} \tag{2-26}$$

If lactic acid is used as the precursor organic compound, the reduction of sulfate to sulfide occurs as follows:

$$\underset{\text{Lactic acid}}{2CH_3CH(OH)COOH} + \underset{\text{sulfate}}{SO_4^{-2}} \xrightarrow{\text{bacteria}} \underset{\text{acetate}}{2CH_3COOH} + \underset{\text{sulfide ion}}{S^{-2}} + 2H_2O + 2CO_2 \quad (2\text{-}27)$$

Hydrogen sulfide gas, which will diffuse into the atmosphere above the wastewater in sewers that are not flowing full, tends to collect at the crown of the pipe. The accumulated H_2S can then be oxidized biologically to sulfuric acid, which is corrosive to concrete sewer pipes. This corrosive effect, known as "crown rot," can seriously threaten the structural integrity of the sewer pipe (ASCE, 1989; U.S. EPA, 1985e).

Sulfates are reduced to sulfides in sludge digesters and may upset the biological process if the sulfide concentration exceeds 200 mg/L. Fortunately, such concentrations are rare. The presence of H_2S in digester gas is corrosive to the gas piping and, if combusted along with digester gas in dual fuel engines at less than optimal temperatures, the products of combustion can damage the engine and severely corrode exhaust-gas heat-recovery equipment, especially if allowed to cool below the dew point.

Other Inorganic Nonmetallic Constituents

In addition to the constituents described above, other inorganic nonmetallic constituents that are of importance in wastewater treatment, recycling, and disposal include boron, chlorine, and silica.

Metals

Metals of importance in the treatment, reuse, and disposal of treated effluents and sludge are summarized in Table 2-8. All living organisms require varying amounts (macro and micro) of metallic elements, such as iron, chromium, copper, zinc, and cobalt, for proper growth. Although macro and micro amounts of metals are required for proper growth, the same metals can be toxic when present in elevated concentrations. As more use is made of treated wastewater effluent for irrigation and landscape watering, a variety of metals must be determined to assess any adverse affects that may occur. Calcium, magnesium, and sodium are of importance in determining the sodium adsorption ratio (SAR), used to assess the suitability of treated effluent for agricultural use (see Chap. 10). Where composted sludge is applied in agricultural applications, arsenic, cadmium, copper, lead, mercury, molybdenum, nickel, selenium, and zinc must be determined.

Metals are determined by atomic absorption, inductively coupled plasma, or colorimetrically (with less precision). Various classes of metals are defined (Standard Methods, 1995):

1. *Dissolved metals* are those metals present in unacidified samples that pass through a 0.45-μm membrane filter.

TABLE 2-8
Metals of importance in wastewater management

Metal	Symbol	Nutrients necessary for biological growth		Concentration threshold of inhibitory effect on heterotrophic organisms, mg/L	Used to determine SAR for land application of effluent	Used to determine if sludge is suitable for land application
		Macro	Micro*			
Arsenic	As			0.05		√
Cadmium	Cd			1.0		√
Calcium	Ca	√			√	
Chromium	Cr		√	10†, 1‡		
Cobalt	Co		√			
Copper	Cu		√	1.0		√
Iron	Fe	√				
Lead	Pb		√	0.1		√
Magnesium	Mg	√	√		√	
Manganese	Mn		√			
Mercury	Hg			0.1		√
Molybdenum	Mo		√			√
Nickel	Ni		√	1.0		√
Potassium	K	√				
Selenium	Se		√			√
Sodium	Na	√			√	
Tungsten	W		√			
Vanadium	V		√			
Zinc	Zn		√	1.0		√

*Often identified as trace elements needed for biological growth.
†Total chromium.
‡Hexavalent chromium.

2. *Suspended metals* are those metals present in unacidified samples that are retained on a 0.45-μm membrane filter.
3. *Total metals* are the total of the dissolved and suspended metals or the concentration of metals determined on an unfiltered sample after digestion.
4. *Acid-extractable metals* are those metals in solution after an unfiltered sample is treated with a hot dilute mineral acid.

Gases

Measurements of dissolved gases, such as ammonia, carbon dioxide, hydrogen sulfide, methane, and oxygen, are made to help in the operation of wastewater treatment systems. Measurements of dissolved oxygen and ammonia are made to monitor and control aerobic biological treatment processes. The presence of hydrogen sulfide is determined not only because it is an odorous and toxic gas, but also because its formation can lead to the corrosion of concrete sewers (U.S. EPA, 1985e). Methane, carbon dioxide, and ammonia measurements are used in connection with the operation of anaerobic digesters. Data on the density and specific weight of selected gases and on composition of air may be found in App. D.

TABLE 2-9
Henry's law constants for several gases that are slightly soluble in water*

	H, 10^{-4}, atm/mol fraction							
T, °C	Air	CO_2	CO	H_2	H_2S	CH_4	N_2	O_2
0	4.32	0.0728	3.52	5.79	2.68	2.24	5.29	2.55
10	5.49	0.104	4.42	6.36	3.67	2.97	6.68	3.27
20	6.64	0.142	5.36	6.83	4.83	3.76	8.04	4.01
30	7.71	0.186	6.20	7.29	6.09	4.49	9.24	4.75
40	8.70	0.233	6.96	7.51	7.45	5.20	10.4	5.35
50	9.46	0.283	7.61	7.65	8.84	5.77	11.3	5.88
60	10.1	0.341	8.21	7.65	10.30	6.26	12.0	6.29

*Adapted from Perry et al. (1984).

Henry's law. The equilibrium or saturation concentration of gas dissolved in a liquid is a function of the type of gas and the partial pressure of the gas adjacent to the liquid. The relationship between the partial pressure of the gas in the atmosphere above the liquid and the concentration of the gas in the liquid is given by Henry's law:

$$P_g = Hx_g \tag{2-28}$$

where P_g = partial pressure of gas, atm
H = Henry's law constant, atm/mole fraction
x_g = equilibrium mole fraction of dissolved gas

$$= \frac{\text{mol gas } (n_g)}{\text{mol gas } (n_n) + \text{mol water } (n_w)}$$

Henry's law constant is a function of the type of gas, temperature, and nature of the liquid. Values of H for water for various gases are listed in Table 2-9. Use of the data in Table 2-9 is illustrated in Example 2-6.

EXAMPLE 2-6. SATURATION CONCENTRATION OF OXYGEN IN WATER. What is the saturation of oxygen in water in contact with dry air at 1 atm and 20°C?

Solution

1. Dry air contains about 21 percent oxygen by volume (see App. D). Therefore P_g = 0.21.
2. From Table 2-9, at 20°C, $H = 4.01 \times 10^4$ atm, and

$$x_g = \frac{P_g}{H} = \frac{0.21 \text{ atm}}{4.01 \times 10^4 \text{ atm}}$$

$$= 5.24 \times 10^{-6}$$

3. One liter of water contains 1000/18 = 55.6 g-mol, thus

$$\frac{n_g}{n_g + n_w} = 5.24 \times 10^{-6}$$

$$\frac{n_g}{n_g + 55.6} = 5.24 \times 10^{-6}$$

Because the number of moles of dissolved gas in a liter of water is much less than the number of moles of water,

$$n_g + 55.6 \approx 55.6$$

and

$$n_g \approx (55.6)\, 5.24 \times 10^{-6}$$
$$\approx 2.91 \times 10^{-4}$$

4. Determine the saturation concentration of oxygen:

$$C_s \approx \frac{2.9 \times 10^{-4}}{\mathrm{L}} \left(\frac{32\ \mathrm{g}}{\mathrm{mol}}\right)\left(\frac{10^3\ \mathrm{mg}}{\mathrm{g}}\right)$$
$$\approx 9.31\ \mathrm{mg/L}$$

Unitless form of Henry's law. In the literature, the unitless form of Henry's law is often used to compute the solubility of trace gases in water or wastewater. The unitless form is usually written as

$$\frac{C_g}{C_s} = H_c \tag{2-29}$$

where C_g = concentration of constituent in gas phase, μg/m^3
C_s = saturation concentration of constituent in liquid, μg/m^3
H_c = Henry's law constant, unitless

Assuming atmospheric conditions prevail, the following equation is used to convert the values of Henry's constant given in terms of atm·m^3/g-mol to the unitless form of Henry's law used in Eq. (2-29):

$$H_c = \frac{H}{RT} \tag{2-30}$$

where H_c = Henry's law constant, unitless as used in Eq. (2-29)
H = Henry's law constant values expressed in atm·m^3/g-mol
R = universal gas law constant, 0.000082057 atm·m^3/g-mol·°K
T = temperature, °K (273 + °C)

2-5 AGGREGATE ORGANIC CHEMICAL CHARACTERISTICS

The organic matter in wastewater typically consists of proteins (40 to 60 percent), carbohydrates (25 to 50 percent), and oils and fats (8 to 12 percent). Urea, the major constituent of urine, is another important organic compound contributing to fresh wastewater. Because urea decomposes rapidly, urea is seldom found in other than very fresh wastewater. Along with the proteins, carbohydrates, fats and oils, and urea, wastewater typically contains small quantities of a large number of different synthetic organic molecules, with structures ranging from simple to extremely complex. Over the years, a number of different analyses have been developed to determine the organic content of wastewaters. In general, the analyses may be classified into those used to measure an aggregate amount of organic matter comprising organic constituents with similar characteristics and those analyses used to quantify individual organic compounds (Standard Methods, 1995). Individual organic compounds are considered in the following section.

Characterization of Aggregate Organic Matter in Wastewater

Analyses of aggregate organics are made to characterize untreated and treated wastewaters, to assess the performance of treatment processes, and to study receiving waters. Laboratory methods commonly used today to measure gross amounts of organic matter (typically greater than 1 mg/L) in wastewater include (1) 5-day biochemical oxygen demand (BOD_5), (2) chemical oxygen demand (COD), and (3) total organic carbon (TOC). Oil and grease and surfactants, because of their special importance in the design and operation of wastewater treatment plants, are considered separately in this section.

Biochemical Oxygen Demand (BOD)

The BOD test is the most common test used in the field of wastewater treatment. If sufficient oxygen is available, the aerobic biological decomposition of an organic waste will continue until all of the waste is consumed. Three more or less distinct activities occur. First, a portion of the waste is oxidized to end products to obtain energy for cell maintenance and the synthesis of new cell tissue. Simultaneously, some of the waste is converted into new cell tissue using part of the energy released during oxidation. Finally, when the organic matter is used up, the new cells begin to consume their own cell tissue to obtain energy for cell maintenance. This third process is called endogenous respiration. Using the term COHNS (which represents the elements carbon, oxygen, hydrogen, nitrogen, and sulfur) to represent the organic waste and the term $C_5H_7NO_2$ to represent cell tissue, the three processes are defined by the following generalized chemical reactions (see Fig. 2-16):

Oxidation

$$\text{COHNS} + O_2 + \text{bacteria} \rightarrow CO_2 + H_2O + NH_3 + \text{other end products} + \text{energy} \tag{2-31}$$

Synthesis

$$\text{COHNS} + O_2 + \text{bacteria} + \text{energy} \rightarrow \underset{\text{New cell tissue}}{C_5H_7NO_2} \tag{2-32}$$

Endogenous respiration

$$C_5H_7NO_2 + 5O_2 \rightarrow 5CO_2 + NH_3 + 2H_2O \tag{2-33}$$

If only the oxidation of the organic carbon that is present in the waste is considered, the ultimate BOD is the oxygen required to complete the three reactions given above. This oxygen demand is known as the *ultimate carbonaceous* or *first-stage* BOD, and

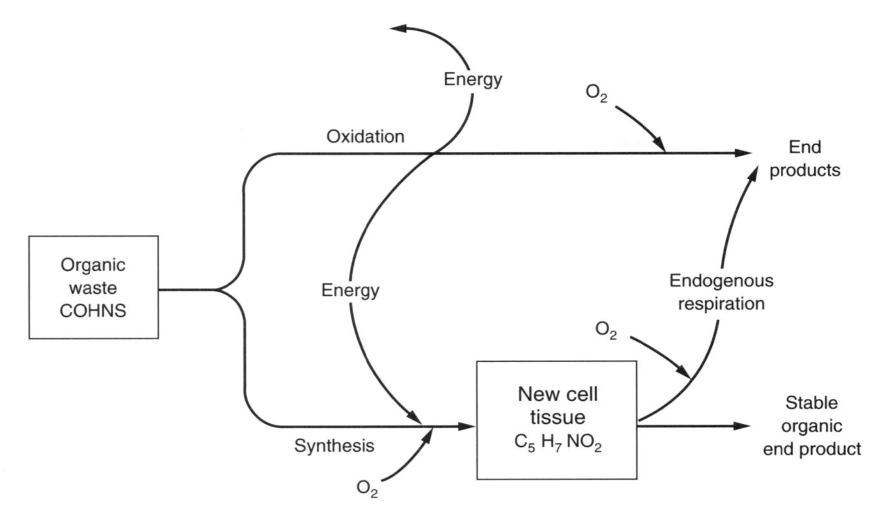

FIGURE 2-16
Schematic diagram illustrating the conversion of an organic waste to end products and residual cell tissue.

is usually denoted as UBOD. In the standard test for BOD (see Fig. 2-17*a*), a small sample of the wastewater to be tested is placed in a BOD bottle (vol. 300 mL). The bottle is then filled with dilution water saturated in oxygen and containing the nutrients required for biological growth. Before stoppering the bottle, the oxygen concentration in the bottle is measured. After incubating the bottle for 5 days at 20°C, the dissolved-oxygen concentration is measured again. The BOD of the sample is the difference in the dissolved-oxygen concentration values, expressed in milligrams per liter, divided by the decimal fraction of sample used. The computed BOD value is known as the 5-day, 20°C biochemical oxygen demand. When testing waters with low concentrations of microorganisms, a seeded BOD test is conducted (see Fig. 2-17*b*). The organisms contained in the effluent from primary sedimentation facilities are used commonly as the seed for the BOD test. Seed organisms can also be obtained commercially.

The standard incubation period is usually 5 days at 20°C, but other lengths of time and temperatures can be used. Longer time periods (typically 7 days) which correspond to work schedules are often used, especially in small plants where the laboratory staff is not available on weekends. The temperature , however, should be constant throughout the test. After incubation, the dissolved oxygen of the sample is measured and the BOD is calculated using Eq. (2-34) or (2-35).

When the dilution water is not seeded:

$$\text{BOD, mg/L} = \frac{D_1 - D_2}{P} \tag{2-34}$$

When the dilution water is seeded:

$$\text{BOD, mg/L} = \frac{(D_1 - D_2) - (B_1 - B_2)f}{P} \tag{2-35}$$

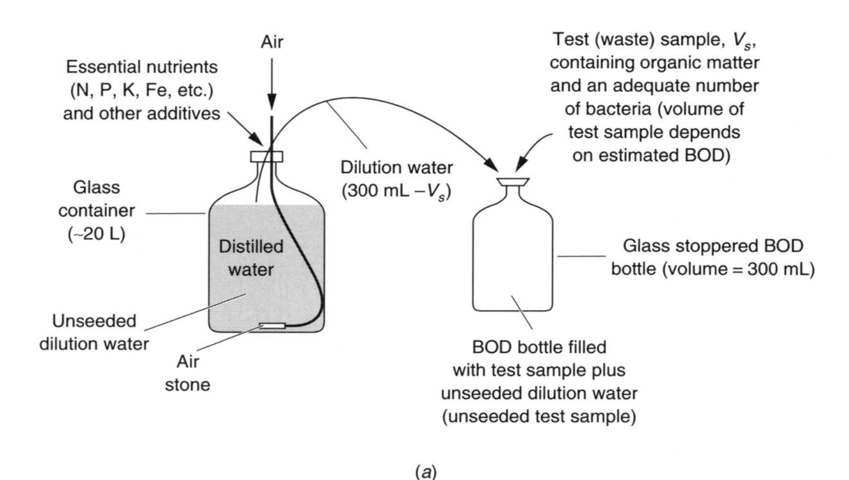

(*a*)

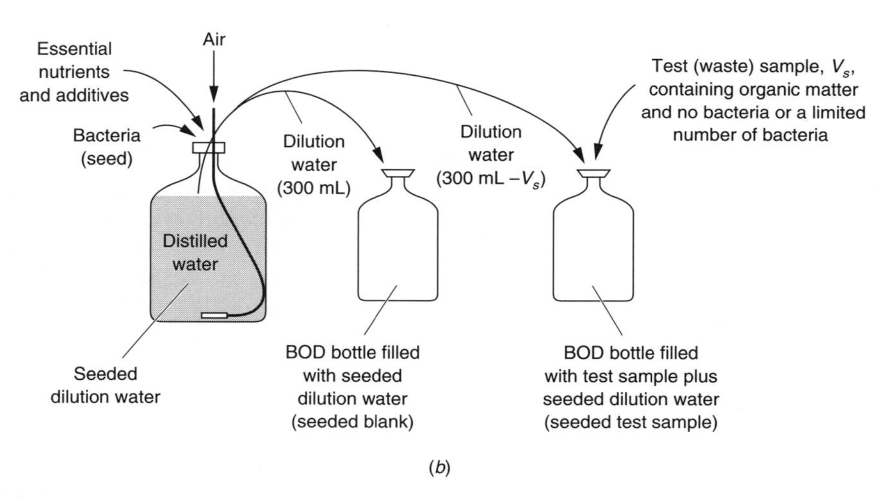

(*b*)

FIGURE 2-17
Procedure for setting up BOD test bottles: (*a*) with unseeded dilution water and (*b*) with seeded dilution water (from Tchobanoglous and Schroeder, 1985).

where D_1 = dissolved oxygen of diluted sample immediately after preparation, mg/L
D_2 = dissolved oxygen of diluted sample after 5 days incubation at 20°C, mg/L
B_1 = dissolved oxygen of seed control before incubation, mg/L
B_2 = dissolved oxygen of seed control after incubation, mg/L
f = fraction of seeded dilution water volume in sample to volume of seeded dilution water in seed control
P = fraction of wastewater sample volume to total combined volume

The application of Eq. (2-35) is illustrated in Example 2-7.

EXAMPLE 2-7. DETERMINATION OF BOD FROM LABORATORY DATA. The following information is available for a seeded 5-day BOD test conducted on a wastewater sample. A volume of 15 mL of the waste sample was added directly into a 300-mL BOD incubation bottle. The initial dissolved oxygen (DO) of the diluted sample was 8.8 mg/L and the final DO after 5 days was 1.9 mg/L. The corresponding initial and final DOs of the seeded dilution water were 9.1 and 7.9, respectively. What is the 5-day BOD (BOD_5) of the wastewater sample?

Solution. Determine the 5-day BOD using Eq. (2-35):

$$\text{BOD, mg/L} = \frac{(D_1 - D_2) - (B_1 - B_2)f}{P}$$

$$f = [(300 - 15)/300] = 0.95$$

$$P = 15/300 = 0.05$$

$$\text{BOD}_5\text{, mg/L} = \frac{(8.8 - 1.9) - (9.1 - 7.9)0.95}{0.05} = 115.2 \text{ mg/L}$$

Modeling of BOD reaction. The rate of BOD exertion is modeled on the assumption that the amount of organic material remaining at any time t is governed by a first-order function, as given below:

$$\text{BOD}_r = \text{UBOD}(e^{-k_1 t}) \tag{2-36}$$

where BOD_r = amount of waste remaining at time t (days) expressed in oxygen equivalents, mg/L

UBOD = the total or ultimate carbonaceous BOD, mg/L

k_1 = first-order reaction rate constant, d^{-1}

t = time, d

Thus the BOD exerted up to time t is given by

$$\text{BOD}_t = \text{UBOD} - \text{BOD}_r = \text{UBOD} - \text{UBOD}(e^{-k_1 t}) = \text{UBOD}(1 - e^{-k_1 t}) \tag{2-37}$$

Equation (2-37) is the standard expression used to define the BOD for wastewater. The basis for this equation is discussed in Sec. 2-3 in conjunction with the analysis of a batch reactor. It should be noted that in the literature dealing with the characterization of wastewater, the terms BOD_u or L are often used to denote ultimate carbonaceous BOD (UBOD).

The value of k_1 for untreated wastewater is generally about 0.12 to 0.46 d^{-1} (base e), with a typical value of about 0.23 d^{-1}. The range of k_1 values for effluents from biological treatment processes is from 0.12 to 0.23 d^{-1}. For a given wastewater, the value of k_1 at 20°C can be determined experimentally by observing the variation with time of the dissolved oxygen in a series of incubated samples. If k_1 at 20°C is equal to 0.23 d^{-1}, the 5-day oxygen demand is about 68 percent of the ultimate first-stage demand. Occasionally, the first-order reaction rate constant will be expressed in log (base 10) units. The relationship between k_1 (base e) and K_1 (base 10) is as follows:

$$K_1 \text{ (base 10)} = \frac{k_1 \text{ (base } e)}{2.303} \tag{2-38}$$

It has been found that k_1 varies with temperature as

$$k_{1_T} = k_{1_{(20)}}(1.047)^{T-20} \tag{2-39}$$

Equation (2-39), along with Eq. (2-37), makes it possible to convert test results from different time periods and temperatures to the standard 5-day 20°C test as illustrated in Example 2-8. The basis for Eq. (2-39) is presented in the following section in the discussion dealing with reaction rates.

EXAMPLE 2-8. CALCULATION OF BOD. Determine the 1-day BOD and ultimate first-stage BOD for a wastewater whose 5-day, 20°C BOD is 200 mg/L. The reaction constant k (base e) $= 0.23\ \text{d}^{-1}$. What would have been the 5-day BOD if the test had been conducted at 25°C?

Solution

1. Determine the ultimate carbonaceous BOD:

$$\text{BOD}_5 = \text{UBOD} - \text{BOD}_t = \text{UBOD}(1 - e^{-k_1 t})$$

$$200 = \text{UBOD}(1 - e^{-5\times 0.23}) = \text{UBOD}(1 - 0.316)$$

$$\text{UBOD} = 293 \text{ mg/L}$$

2. Determine the 1-day BOD:

$$\text{BOD}_t = \text{UBOD}(1 - e^{-k_1 t})$$

$$\text{BOD}_1 = 293(e^{-0.23\times 1}) = 293(1 - 0.795) = 60.1 \text{ mg/L}$$

3. Determine the 5-day BOD at 25°C:

$$k_{1_T} = k_{1_{(20)}}(1.047)^{T-20}$$

$$k_{1_{(25)}} = 0.23(1.047)^{25-20} = 0.29\ \text{d}^{-1}$$

$$\text{BOD}_5 = \text{UBOD}(1 - e^{-k_1 t}) = 293(1 - e^{-0.29\times 5}) = 224 \text{ mg/L}$$

Respirometric determination of BOD. Respirometric methods involve the direct measurement of the amount of oxygen consumed during the conversion of the organic material in the sample, under controlled conditions of constant temperature and pressure and mixing. Four different types of respirometers are available: manometric, electrolytic, volumetric, and direct-input. The difference between the respirometers is the manner in which the need for oxygen is sensed and the method of providing the needed oxygen. In the manometric respirometer (e.g., the Gilson and Warburg respirometers, the forerunners of the modern respirometers) the oxygen uptake is related to the change in pressure caused by the consumption of oxygen at constant volume. In electrolytic respirometers, the oxygen pressure over the sample is maintained constant by continuously replacing the oxygen used by the micro-

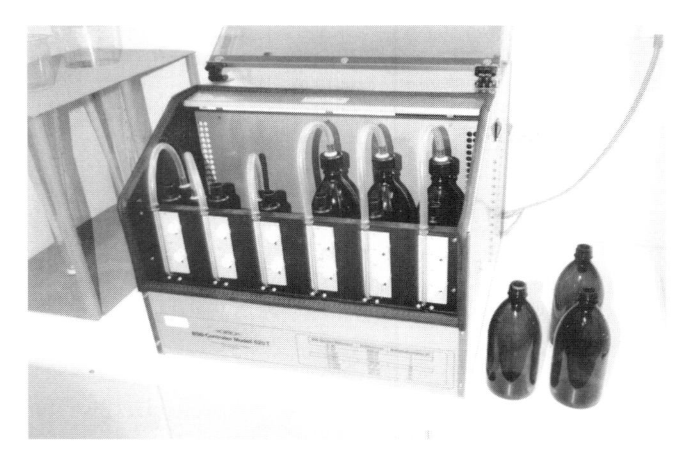

FIGURE 2-18
Typical example of a respirometric setup used to determine continuous BOD values.

organisms. Oxygen replacement is accomplished by means of an electrolysis reaction in which oxygen is produced in response to changes in the pressure. In volumetric respirometers, oxygen uptake is related to changes in volume due to oxygen uptake at constant pressure. In direct-input respirometers, pure oxygen is supplied in response to pressure changes (Standard Methods, 1995).

The results of the respirometric tests can be correlated to BOD_5 test results obtained by the conventional bottle method. The respirometric method can also be used to measure the UBOD. Another important use of respirometric techniques is as a screening tool to assess the biotreatability of specific compounds, to identify the presence of toxic constituents in the waste stream, and to determine the effects of toxic compounds on biotreatability and of chemical additions on oxygen uptake rates and biotreatability. A typical respirometer is illustrated in Fig. 2-18.

Analysis of BOD data. To determine the ultimate carbonaceous UBOD value, the value of k_1 must be known. Because both values (i.e., UBOD and k_1) are unknown in a conventional BOD test, the usual procedure is to determine these values from a series of BOD measurements. A number of methods are available for determining the values of UBOD and k_1. These methods are considered in Sec. 3-2, Chap. 3, in the discussion dealing with the determination of reaction rate coefficients for transformation processes.

Effect of particle size on BOD reaction rates. If a separation and analysis technique such as membrane filtration (see Figs. 2-8 and 2-9) is used to quantify the size distribution of the solids in the influent wastewater, the various size fractions can be correlated to observed oxygen (BOD) uptake rates, determined by using a respirometer. As reported in Table 2-10, the observed BOD reaction rate coefficients are affected significantly by the size of the particles in wastewater. From the data given in Table 2-10, it is clear that the treatment

TABLE 2-10
Effect of the size of the biodegradable particles found in wastewater on observed BOD reaction rates*

Fraction	Size range, μm	K (base 10), d^{-1}
Settleable	>100	0.08
Supracolloidal	1–100	0.09
Colloidal	0.1–1.0	0.22
Soluble	<0.1	0.39

* Adapted from Balmat (1957).

of a wastewater can be affected by modifying the particle size distribution. Further, wastewaters with significantly different particle size distributions will respond differently, depending on the method of treatment (e.g., in constructed wetlands).

Limitations in the BOD test. Although the BOD_5 test is commonly used, it suffers from several serious deficiencies. The most serious one is that the test has no stoichiometric validity. That is, the arbitrary 5-day period usually does not correspond to the point where all of the waste is consumed. Thus, it is not known where the 5-day BOD value falls along the curve (see Fig. 2-19). The 5-day value is used because the test was developed in England, where the maximum time of flow of most rivers from headwaters to the ocean is about 4.8 days. Other limitations of the test include the need for acclimated seed, the potential for nitrification to occur, and the general limits on the precision of the test. For example, if oxygen depletion in a test bottle is too great, the test is invalid. The BOD test is not highly reproducible, and values below 2 mg/L or beyond two significant figures are suspect.

From an analytical standpoint, BOD is a poor parameter because, like the TSS test, BOD is a lumped (aggregate) parameter. The individual waste constituents that make up the measured BOD value are unknown. Furthermore, the size distribution of the particles found in different wastewaters and their contributions to the measured BOD values are unknown. Because BOD is a lumped parameter, the development of sophisticated models to describe the transformations that are used is not warranted. Despite its limitations, the use of BOD as a regulatory parameter is acceptable because the test represents the potential oxygen depletion effect the wastewater may have on the receiving water body and the degree of treatment which the wastewater has undergone.

Nitrification. An oxygen demand can also result from the biological oxidation of ammonia (see Fig. 2-20). The simplified reactions that define the nitrification process are as follows:

Conversion of ammonia to nitrite (as typified by *Nitrosomonas*):

$$NH_3 + \frac{3}{2}O_2 \rightarrow HNO_2 + H_2O \tag{2-40}$$

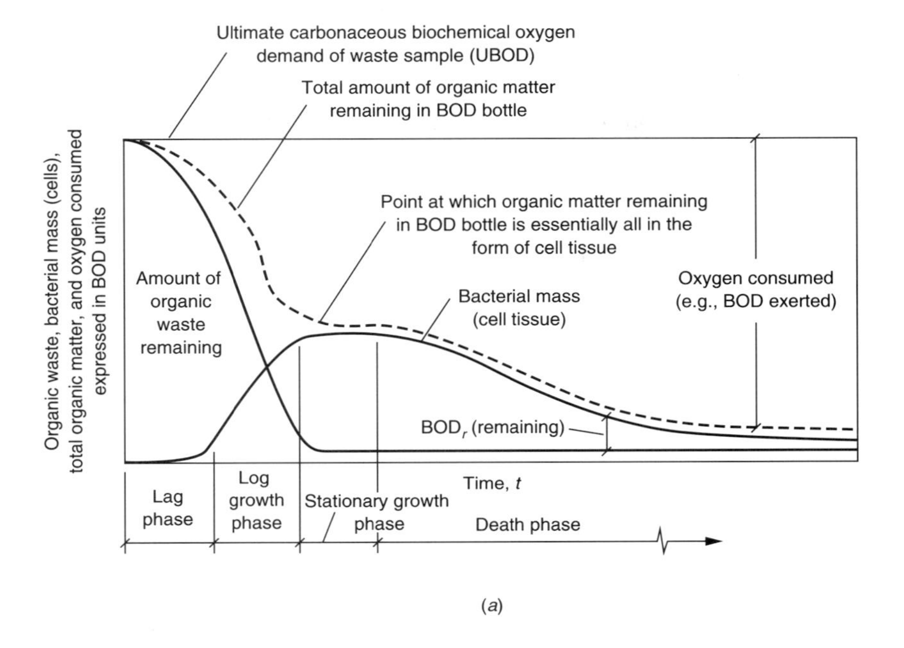

(*a*)

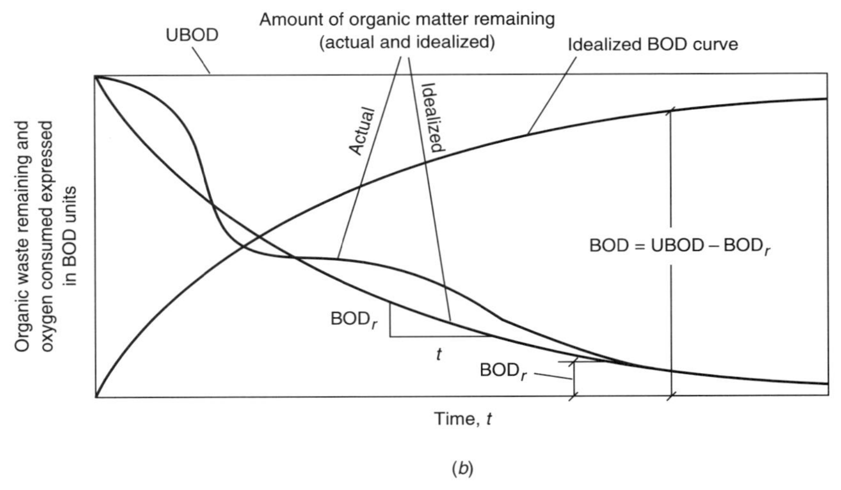

(*b*)

FIGURE 2-19
Definition sketch for the exertion of BOD with time: (*a*) interrelationship between the amount of organic waste remaining, the bacterial mass (cell tissue), and oxygen consumed; (*b*) the idealized representation of the test results.

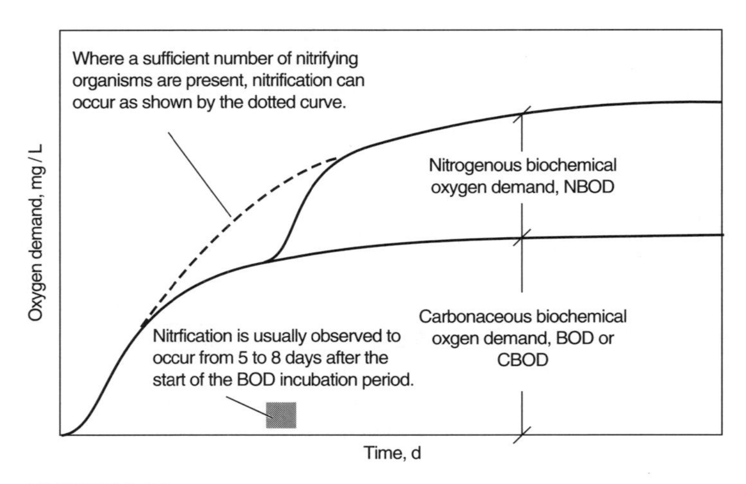

FIGURE 2-20
Definition sketch for the exertion of carbonaceous and nitrogenous biochemical oxygen demand in a waste sample (from Tchobanoglous and Schroeder, 1985).

Conversion of nitrite to nitrate by *Nitrobacter:*

$$HNO_2 + \frac{1}{2}O_2 \rightarrow HNO_3 \tag{2-41}$$

Overall conversion of ammonia to nitrate:

$$NH_3 + 2O_2 \rightarrow HNO_3 + H_2O \tag{2-42}$$

The oxygen required for the conversion of ammonia to nitrate is known as the NOD (nitrogenous oxygen demand). Typically, the oxygen demand due to nitrification will occur from 5 to 8 days after the start of a conventional BOD test. However, as shown in Fig 2-20, nitrification can occur at the outset of the test, if sufficient nitrifying organisms are present initially. The oxygen and alkalinity requirements to complete the reaction given by Eq. (2-42) are considered in Example 2-9.

EXAMPLE 2-9. OXYGEN AND ALKALINITY REQUIREMENTS FOR NITRIFICATION. Determine the amount of oxygen and alkalinity required to complete the overall nitrification reaction as given by Eq. (2-42).

Solution

1. Determine the amount of oxygen required:

$$\underset{14}{NH_3} + \underset{2(32)}{2O_2} \rightarrow HNO_3 + H_2O$$

$$O_2 \text{ required} = 64/14 = 4.57 \text{ mg } O_2/\text{mg N}$$

2. Determine the amount of alkalinity required to complete the nitrification reaction. The alkalinity required can be estimated by writing Eq. (2-42) as given below where the alkalinity needed to carry out the reaction is given as bicarbonate (HCO_3^-).

$$\underset{14}{NH_4^+} + \underset{2(50)}{2HCO_3^-} + 2O_2 \rightarrow NO_3^- + 2CO_2 + 3H_2O$$

$$\text{Alkalinity } Alk \text{ required} = 100/14 = 7.14 \text{ mg } Alk \text{ as } CaCO_3/\text{mg N}$$

Comment. Because of the high ammonia concentration in septic tank effluent (see Sec. 4-2, Chap. 4), it may not be possible to nitrify the septic tank effluent completely because of insufficient alkalinity. The problem of insufficient alkalinity is especially serious in areas with very soft (i.e., low-hardness) water.

Carbonaceous Biochemical Oxygen Demand (CBOD)

When nitrification occurs, the measured BOD value will be higher than the true value because of the oxidation of nitrogenous material (see Fig. 2-20). If a given percentage of carbonaceous biochemical oxygen demand removal must be achieved to meet regulatory permit limits, early nitrification can pose a serious problem. The effects of nitrification can be overcome either by using various chemicals to suppress the nitrification reactions or by treating the sample to eliminate the nitrifying organisms (Young, 1973). Pasteurization and chlorination/dechlorination are two methods that have also been used to suppress the nitrifying organisms. When the nitrification reaction is suppressed, the resulting BOD is known as the *carbonaceous biochemical oxygen demand* (CBOD). In effect, the CBOD is a measure of the oxygen demand exerted by the oxidizable carbon in the sample. The CBOD test, in which the nitrification reaction is suppressed chemically, should be used only on samples that contain small amounts of organic carbon (e.g., treated effluent). Large errors will occur in the measured BOD values (up to 20 percent) when the CBOD test is used on wastewater containing significant amounts of organic matter, such as untreated wastewater (Albertson, 1995).

Chemical Oxygen Demand (COD)

The COD test is used to measure the oxygen equivalent of the organic material in wastewater that can be oxidized chemically by using dichromate in an acid solution, as illustrated in the following equation, when the organic nitrogen is in the reduced state (oxidation number $= -3$) (Sawyer et al., 1994).

$$C_nH_aO_bN_c + dCr_2O_7^{-2} + (8d + c)H^+ \rightarrow nCO_2 + \frac{a + 8d - 3c}{2}H_2O + cNH_4^+ + 2dCr^{+3} \tag{2-43}$$

where

$$d = \frac{2n}{3} + \frac{a}{6} + \frac{b}{3} - \frac{c}{2}$$

Although it would be expected that the value of the ultimate carbonaceous BOD would be as high as the COD, this is seldom the case. Some of the reasons for the observed differences are as follows:

1. Many organic substances which are difficult to oxidize biologically, such as lignin, can be oxidized chemically.
2. Inorganic substances that are oxidized by the dichromate increase the apparent organic content of the sample.
3. Certain organic substances may be toxic to the microorganisms used in the BOD test.
4. High COD values may occur because of the presence of inorganic substances with which the dichromate can react.

From an operational standpoint, one of the main advantages of the COD test is that it can be completed in about 2.5 hours (compared to 5 or more days for the BOD test). To further reduce the time, a rapid COD test, which takes only about 15 min, has been developed.

Total Organic Carbon (TOC)

The TOC test, done instrumentally, is used to determine the total organic carbon in an aqueous sample. The test methods for TOC utilize heat and oxygen, ultraviolet radiation, chemical oxidants, or some combination of these methods to convert organic carbon to carbon dioxide which is measured with an infrared analyzer or by other means. The TOC of a wastewater can be used as a measure of its pollutional characteristics and in some cases it has been possible to relate TOC to BOD and COD values. The TOC test is also gaining in favor because it takes only 5 to 10 min to complete. If a valid relationship can be established between results obtained with the TOC test and the results of the BOD test for a given wastewater, use of the TOC test for process control is recommended.

More recently, a continuous online TOC analyzer has been developed, in conjunction with the space program, that can be used to detect TOC concentrations in the parts per billion (ppb) range (see Fig. 2-21). Two of these instruments are currently (1997) being used to detect the residual TOC in the treated effluent from microfiltration and reverse osmosis (RO) treatment units at Aqua 2000 in San Diego. Continuous TOC measurements may be used to monitor the performance of the full-scale RO units, to be used in conjunction with the repurification project in which repurified effluent is proposed to be blended with other waters in a surface water storage reservoir. After a suitable detention time, the blended water will be treated and used as a potable water supply (see Chap. 12 for a more detailed description of this project).

Interrelationships of BOD, COD, and TOC

Typical values for the ratio of BOD_5/COD for untreated municipal wastewater are in the range from 0.3 to 0.8 (see Table 2-11). If the BOD_5/COD ratio for untreated wastewater is 0.5 or greater, the waste is considered to be easily treatable by

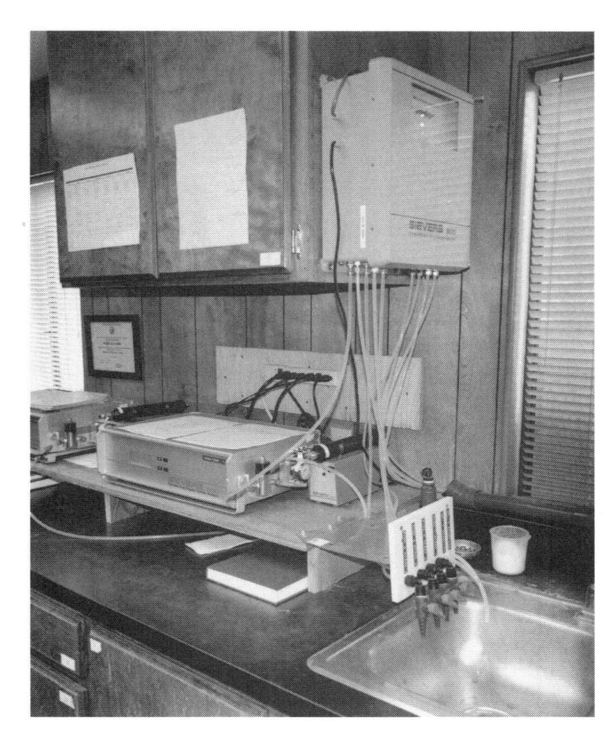

FIGURE 2-21
TOC analyzer developed to detect TOC concentrations in the ppb range.

TABLE 2-11
Comparison of ratios of various parameters used to characterize wastewater

Type of wastewater	BOD_5/COD	BOD_5/TOC
Untreated	0.3–0.8	1.2–2.0
After primary settling	0.4–0.6	0.8–1.2
Final effluent	0.1–0.3*	0.2–0.5†

*$CBOD_5/COD$.
†$CBOD_5/TOC$.

biological means. If the ratio is below about 0.3, either the waste may have some toxic components or acclimated microorganisms may be required in its stabilization. The corresponding BOD_5/TOC ratio for untreated wastewater varies from 1.2 to 2.0. In using these ratios, it is important to remember that they will change significantly with the degree of treatment the waste has undergone, as reported in Table 2-11. The theoretical basis for these ratios is explored in Example 2-10.

EXAMPLE 2-10. DETERMINATION OF BOD_5/COD, BOD_5/TOC, AND TOC/COD RATIOS. Determine the theoretical BOD_5/COD, BOD_5/TOC, and TOC/BOD ratios for the following compound $C_5H_7NO_2$. Assume the value of the BOD_5 first-order reaction rate constant is $0.23\ d^{-1}$ (base e) [$0.10\ d^{-1}$ (base 10)].

Solution

1. Determine the COD of the compound using Eq. (2-31):

$$\underset{113}{C_5H_7NO_2} + \underset{160}{5O_2} \rightarrow 5CO_2 + NH_3 + 2H_2O$$

$$COD = 160/113 = 1.42 \text{ mg } O_2/\text{mg } C_5H_7NO_2$$

2. Determine the BOD_5 of the compound:

$$\frac{BOD_5}{UBOD} = (1 - e^{-k_1 t}) = (1 - e^{-0.23 \times 5}) = 1 - 0.32 = 0.68$$

$$BOD_5 = 0.68 \times 1.42 \text{ mg } O_2/\text{mg } C_5H_7NO_2 = 0.97 \text{ mg } BOD_5/\text{mg } C_5H_7NO_2$$

3. Determine the TOC of the compound:

$$TOC = (5 \times 12)/113 = 0.53 \text{ mg TOC/mg } C_5H_7NO_2$$

4. Determine BOD_5/COD, BOD_5/TOC, and TOC/BOD ratios:

$$\frac{BOD_5}{COD} = \frac{0.68 \times 1.42}{1.42} = 0.68$$

$$\frac{BOD_5}{TOC} = \frac{0.68 \times 1.42}{0.53} = 1.82$$

$$\frac{TOC}{BOD} = \frac{0.53}{1.42} = 0.37$$

Oil and Grease

The term *oil and grease,* as commonly used, includes the fats, oils, waxes, and other related constituents found in wastewater. The term *fats, oil, and grease* (FOG) used previously in the literature has been replaced by the term oil and grease. The oil and grease content of a wastewater is determined by extraction of the waste sample with trichlorotrifluoroethane (oil and grease are soluble in trichlorotrifluoroethane). Other extractable substances includes mineral oils, such as kerosene and lubricating and road oils. Oil and grease are quite similar chemically; they are compounds (esters) of alcohol or glycerol (glycerin) with fatty acids. The glycerides of fatty acids that are liquid at ordinary temperatures are called oils, and those that are solids are called grease (or fats).

Because of their properties, the presence of oil and grease in wastewater can cause many problems in septic tanks (see Fig. 2-22), wastewater collection systems, and in the treatment of wastewater. The scum layer that builds up in a septic tank must be removed periodically, so that the clear space between the scum and sludge layer is not reduced to the point where solids are carried over to the second compartment or are discharged to the leachfield, where they can lead to the clogging of the leachfield. The discharge of oils and grease in septic tank effluent is especially troublesome where an intermittent sand filter is used for further treatment. Oils and grease that accumulate within the filter limit the transfer of oxygen and can ultimately lead to the failure of the filter. If oil and grease are not removed in waste-

FIGURE 2-22
Accumulation of oil and grease on the surface of a septic tank. The L-shaped scum probe is resting on the surface of the scum.

water pretreatment processes, they tend to accumulate in downstream processes. If grease is not removed before discharge of treated wastewater, it can interfere with the biological life in the surface waters and create unsightly films. The thickness of oil required to form a translucent film on the surface of a water body is about 0.0000120 in, as given below.

Appearance	Film thickness, in	Quantity spread, gal/mi^2
Barely visible	0.0000015	25
Silvery sheen	0.0000030	50
First trace of color	0.0000060	100
Bright bands of color	0.0000120	200
Colors begin to dull	0.0000400	666
Colors are much darker	0.0000800	1332

Source: Eldridge (1942).

Surfactants

Surfactants, or surface-active agents, are large organic molecules that are composed of a strongly hydrophobic (insoluble in water) and a strongly hydrophilic (water soluble) group. Their presence in wastewater is usually from the discharge of household detergents, industrial laundering, and other cleaning operations. Surfactants tend to collect at the air-water interface, and can cause foaming in wastewater-treatment plants and in the surface waters into which treated effluent is discharged. During aeration of wastewater, these compounds collect on the surface of the air bubbles and thus create a very stable foam. The determination of surfactants is accomplished by measuring the color change in a standard solution of methylene blue dye. Another name for surfactant is methylene blue active substance (MBAS). Before 1965, the type of surfactant present in synthetic detergents, called alkylbenzene-sulfonate (ABS), was especially troublesome because it resisted breakdown by biological means. As a result of legislation in 1965, ABS has been replaced in detergents by linear-alkyl-sulfonate (LAS), which is biodegradable.

2-6 CHARACTERIZATION OF INDIVIDUAL ORGANIC COMPOUNDS

Individual organic compounds are determined to assess the presence of priority pollutants identified by the U.S. Environmental Protection Agency. Priority pollutants (both inorganic and organic) have been and are continuing to be selected on the basis of their known or suspected carcinogenicity, mutagenicity, teratogenicity, or high acute toxicity. Many of the organic priority pollutants are also classified as *volatile organic compounds* (VOCs), as discussed below.

Analysis of Individual Organic Compounds

The analytical methods used to determine individual organic compounds require the use of sophisticated instrumentation capable of measuring trace concentrations in the range of 10^{-12} to 10^{-3} mg/L. Gas chromatographic (GC) and high-performance liquid chromatographic (HPLC) methods are most commonly used to detect individual organic compounds. Different types of detectors are used with each method, depending on the nature of the compound being analyzed. Typical detectors used in conjunction with gas chromatography include electrolytic conductivity, electron capture (ECD), flame ionization (FID), photoionization (PID) and GC mass spectrometer (GCMS). Typical detectors for high-performance liquid chromatography include photodiode array (PDAD) and post column reactor (PCR). It should also be noted that many of the individual organic constituents can be determined by two or more of the above methods (Standard Methods, 1995).

Characteristics of Volatile Organic Compounds (VOCs)

Volatile organic compounds are of great concern because (1) once such compounds are in the vapor state they are much more mobile, and therefore more likely to be released to the environment; (2) the presence of some of these compounds in confined work areas and in the atmosphere may pose a significant public health risk; and (3) they contribute to a general increase in reactive hydrocarbons in the atmosphere, which can lead to the formation of photochemical oxidants. Organic compounds that have boiling point less than 100°C and/or vapor pressure greater than 1 mm Hg at 25°C are generally considered to be VOCs. The release of these compounds in wastewater collection and at treatment plants, especially at the headworks, is of particular concern with respect to the health of collection system and wastewater treatment plant workers.

Individual Organic Compounds

There are over 180 individual organic compounds that can be determined by using one or more of the methods cited above. Representative organic compounds, by category, that can be analyzed are reported in Table 2-12. In the listing given in Table 2-12, it will be noted that some of the compounds appear under a number of classifications. For example, benzene is classified as a volatile organic compound as

TABLE 2-12
Typical classes of organic compounds and representative examples determined as individual compounds

Class of organic compound	Representative examples
Volatile organic compounds (VOCs)	Benzene Bromoform Chlorobenzene Chloromethane Toluene Tetrachloroethene Trichloroethene Vinyl chloride
Methane	Methane
Volatile aromatic organic compounds	Benzene Chlorobenzene Tetrachloroethene Toluene Trichloroethene
Volatile halocarbons	Bromoform Carbon tetrachloride Methyl chloride Tetrachloroethene Trichloroethene Vinyl chloride
Trihalomethanes and chlorinated organic solvents	Bromoform Chloroform Tetrachloroethane
Disinfection by-products	Aldehydes (various) Haloacetic acids (various) Trichlorophenol
Extractable base/neutrals and acids	1,3-Dichlorbenzene Naphthalene Fluorene Heptachlor
Polychlorinated aromatic hydrocarbons (PAHs)	Anthracene Fluorene Naphthalene
Pesticides	Various
Herbicides	Various

well as a volatile aromatic organic compound. As instrumental methods of analysis have improved, the detection limits for these compounds have become increasingly small, typically below 10 ng/L. The specific organic compounds that are analyzed for will depend on the application. For example, for indirect reuse applications scans of disinfection by-products may be required where chlorine is used for disinfection.

2-7 BIOLOGICAL CHARACTERISTICS

The biological characteristics of wastewater are of fundamental importance in the control of diseases caused by pathogenic organisms of human origin, and because of the extensive and fundamental role played by bacteria and other microorganisms in the decomposition and stabilization of organic matter, both in nature and in wastewater treatment plants. The purpose of this section is to introduce: (1) the microorganisms found in surface waters and wastewater, (2) the methods and techniques used for the enumeration of bacteria, (3) the method of enumerating viruses, (4) the pathogenic microorganisms associated with human disease, and (5) the use of indicator organisms. The organisms responsible for the treatment of wastewater are considered further in Chap. 7.

Microorganisms Found in Surface Waters and Wastewater

The principal groups of organisms found in surface water and wastewater include bacteria, fungi, algae, protozoa, plants and animals, and viruses. The general classification of these organisms, their growth and metabolic requirements, their oxygen requirements, and a general description of the organisms found in wastewater are considered in the following discussion.

General classification. The microorganisms in wastewater can be classified as eukaryotes, eubacteria, and archaea (see Table 2-13). As reported in Table 2-13, two general types of cell structure have been identified: eukaryotic and prokaryotic (see Fig. 2-23). In eukaryotic cells (see Fig. 2-23*a*) the nucleus, which contains several DNA molecules, is enclosed by a nuclear membrane. In addition, eukaryotic cells typically contain other membrane-enclosed structures such as mitochondria and chloroplasts. Algae, fungi, and protozoa are classified as eukaryotic microorganisms. All plants, including seed plants, ferns, and mosses, and animals, including invertebrates and vertebrates, are classified as multicellular eukaryotic macroorganisms. In prokaryotic cells (see Fig. 2-23*b*) the nucleoid region contains a single DNA molecule and is not surrounded by a nuclear membrane. Bacteria and archaea are the only prokaryotes (Madigan et al., 1997; Roberts and Janovy, 1996).

Viruses are *intracellular parasites* that require the machinery of a host cell to support their growth. Although viruses contain the genetic information (either DNA or RNA) needed to replicate themselves, they are unable to reproduce outside of a host cell. Viruses are composed of a nucleic acid core (RNA or DNA) surrounded by an outer coat of protein and glycoprotein. Viruses are classified separately according

TABLE 2-13
General classification of microorganisms*

Group	Cell structure	Characterization	Typical size[†]	Representative members
Eukaryotes	Eukaryotic[‡]	Multicellular with extensive differentiation of cells and tissue	10–100 μm in diameter	Plants (seed plants, ferns, mosses); animals (vertebrates, invertebrates)
		Unicellular or coenocytic or mycelial; little or no tissue differentiation	10–100 μm in diameter	Algae, fungi, protozoa
Eubacteria	Prokaryotic[§]	Cell chemistry similar to eukaryotes	0.2–2.0 μm in diameter	Most bacteria
Archaea	Prokaryotic[§]	Distinctive cell chemistry	0.2–2.0 μm in diameter	Methanogens, halophiles, thermacidophiles

* Adapted from Ingraham and Ingraham (1995), Madigan et al. (1997), and Stanier et al. (1986).
[†] For additional size information see Table 2-19.
[‡] Contain true nucleus.
[§] Contain no nuclear membrane.

to the host infected. Bacteriophages, as the name implies, are viruses that infect bacteria as the host.

Carbon and energy requirements. To reproduce and function properly, microorganisms must have a source of carbon, inorganic elements (nutrients), and energy for the synthesis of new cellular material (see Table 2-14). Microorganisms that require organic carbon as the carbon source for the formation of cell tissue are called *heterotrophs*. Organisms that derive cell carbon solely from carbon dioxide are called *autotrophs*. Organisms (e.g., protozoa, fungi, and most bacteria) that derive their energy from the oxidation of organic compounds are known as *chemoorganotrophs*. Organisms (e.g., nitrifying bacteria) that derive their energy from the oxidation of reduced inorganic compounds, such as ammonia, nitrite, and sulfide, are known as *chemolithotrophs*. Organisms that are able to use light as an energy source are called *phototrophs*. Phototrophic organisms may be either photoheterotrophs (certain sulfur bacteria) or photoautotrophs (algae and photosynthetic bacteria).

Nutritional requirements. In addition to carbon, a number of inorganic and organic nutrients are required for proper cell synthesis and growth. The principal inorganic nutrients needed by microorganisms are N, S, P, K, Mg, Ca, Fe, Na, and Cl. Minor nutrients of importance include Zn, Mn, Mo, Se, Co, Cu, Ni, and W (Madigan et al., 1997). Information on the distribution of the inorganic nutrients in the dry mass of a typical bacterial cell is presented in Table 2-15. Even though the quantities of some of the nutrients are small, the absence of any of these substances would limit and, in some cases, alter growth.

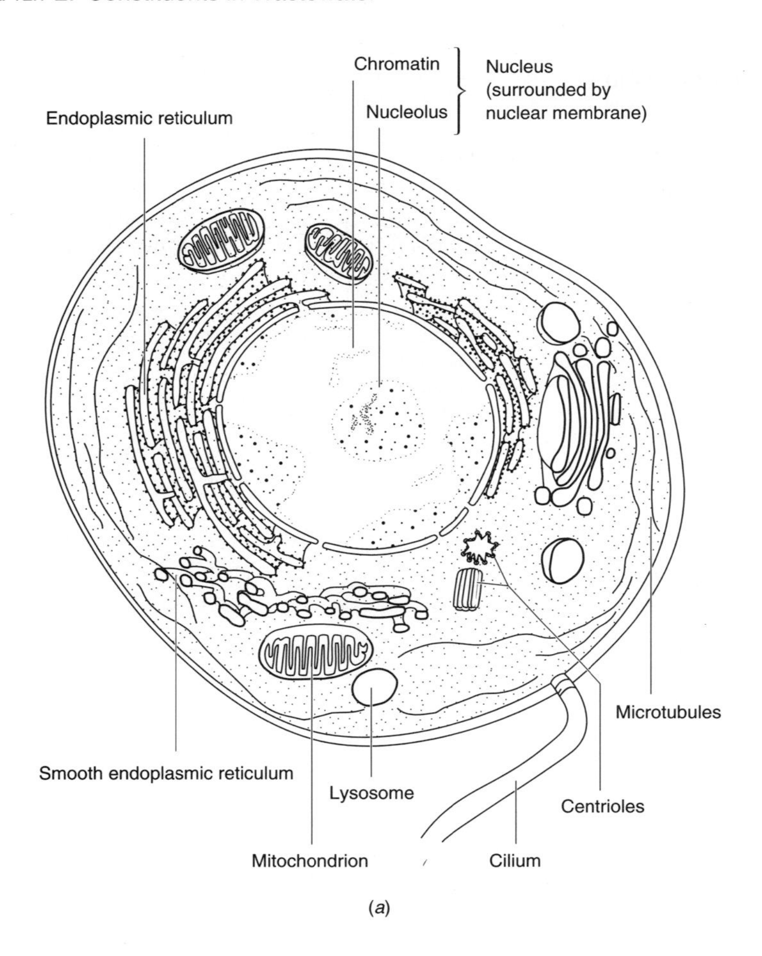

(*a*)

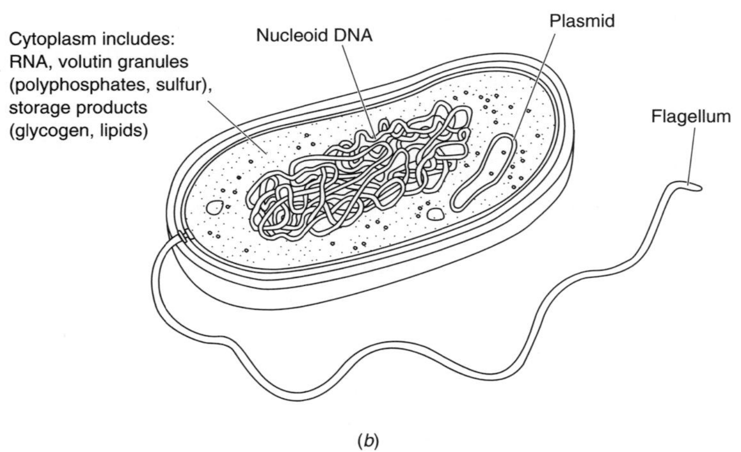

(*b*)

FIGURE 2-23
Typical structure of microorgansim cells: (*a*) eukaryotic and (*b*) prokaryotic.

TABLE 2-14
Classification of microorganisms by sources of carbon and energy*

Classification	Carbon source	Energy source	Representative organisms
Heterotrophic			
Chemoorganotrophs†	Organic carbon	Organic oxidation-reduction reaction	Bacteria, fungi, protozoa, animals
Photoheterotrophs	Organic carbon	Light	Photosynthetic bacteria
Autotrophic			
Chemolithotrophs‡	CO_2	Inorganic oxidation-reduction reaction	Bacteria
Photoautotrophs	CO_2	Light	Higher plants, algae, photosynthetic bacteria

*From Madigan et al. (1997), Stanier et al. (1986).
† Also known as chemoheterotrophs.
‡ Also known as chemoautotrophs.

Organic nutrients, known as *growth factors*, which cannot be synthesized by all organisms from other carbon sources, may also be needed by some organisms as precursors or constituents of organic cell material. Although growth factor requirements differ from one organism to another, the major growth factors fall into the following three classes: (1) amino acids, (2) purines and pyrimidines, and (3) vitamins (Madigan et al., 1997).

TABLE 2-15
Typical composition of bacterial cells*

	Percentage of dry mass	
Element	Range	Typical
Carbon	45–60	50
Oxygen	16–30	24
Nitrogen	12–16	12
Hydrogen	5–8	6
Phosphorus	2–5	3
Sulfur	0.8–1.5	1
Potassium	0.8–1.5	1
Sodium	0.5–2.0	1
Calcium	0.4–0.7	0.5
Magnesium	0.4–0.7	0.5
Chlorine	0.4–0.7	0.5
Iron	0.1–0.4	0.2
All others	0.2–0.5	0.3

*Adapted from Grady and Lim (1980), Stanier et al. (1986), and Tchobanoglous and Schroeder (1985).

TABLE 2-16
Oxygen requirements for various microorganisms*

Classification	Relationship to oxygen	Type of respiration
Aerobic		
Obligate	Required	Aerobic respiration
Facultative	Not required, but growth better with O_2	Aerobic, anaerobic respiration, fermentation
Microaerophilic	Required, but at levels lower than atmospheric	Aerobic respiration
Anaerobic		
Aerotolerant	Not required, and growth no better when O_2 present	Inorganic oxidation-reduction reaction
Obligate	Harmful or lethal	Fermentation or anaerobic respiration

*Adapted from Madigan et al. (1997), Stanier et al. (1986).

Oxygen requirements. Microorganisms are also classified metabolically depending on their ability to grow in either the presence or absence of molecular oxygen (see Table 2-16). The most common classifications are *obligately aerobic* (can exist only in the presence of oxygen), *obligately anaerobic* (can exist only in an environment that is devoid of oxygen), *facultative anaerobes* (can exist in either the presence or absence of molecular oxygen), and *aerotolerant anaerobes* (insensitive to the presence of molecular oxygen).

Environmental requirements. Environmental conditions of temperature and pH have an important effect on the survival and growth of bacteria. In general, optimal growth occurs within a fairly narrow range of temperature and pH, although the bacteria may be able to survive within much broader limits. Temperatures below the optimum typically have a more significant effect on growth rate than temperatures above the optimum; it has been observed that growth rates double with approximately every 10°C increase in temperature until the optimum temperature is reached. According to the temperature range in which they function best, bacteria may be classified as *psychrophilic, mesophilic, thermophilic,* or *hyperthermophilic.* Typical temperature ranges for bacteria in each of these categories are presented in Table 2-17. The pH of the environment is also a key factor in the growth of

TABLE 2-17
Typical temperature ranges for various bacteria

	Temperature, °C	
Type	Range	Optimum
Psychrophilic*	−10–20	12–14
Mesophilic	10–50	32–42
Thermophilic	40–70	55–65
Hyperthermophilic	70–95	80–90

*Also called *cryophilic*.
Note: 1.8(°C) + 32 = °F.

TABLE 2-18
Typical descriptions of the microorganisms found in natural waters and wastewater

Organism	Description
Bacteria	Bacteria are single-cell prokaryotic organisms. The interior of the cell contains a colloidal suspension of proteins, carbohydrates, and other complex organic compounds, called the *cytoplasm*. The cytoplasmic area contains ribonucleic acid (RNA), whose major role is in the synthesis of proteins. Also within the cytoplasm is the area of the nucleus, which is rich in deoxyribonucleic acid (DNA). DNA contains all the information necessary for the reproduction of all the cell components and may be considered to be the blueprint of the cell. Their usual mode of reproduction is by binary fission, although some species reproduce sexually or by budding.
Fungi	Fungi are multicellular, nonphotosynthetic, heterotrophic eukaryotes. Fungi are strict aerobes which reproduce sexually or asexually, by fission, budding, or spore formation. Molds or "true fungi" produce microscopic units (hyphae), which collectively form a filamentous mass called the *mycelium*. Yeasts are fungi that cannot form a mycelium and are therefore unicellular. Fungi have the ability to grow under low-moisture, low-nitrogen conditions and can tolerate an environment with a relatively low pH. The ability of the fungi to survive under low-pH and nitrogen-limiting conditions, coupled with their ability to degrade cellulose, makes them very important in the composting of sludge.
Protozoa	Protozoa are motile, microscopic eukaryotes that are usually single cells. The majority of protozoa are aerobic heterotrophs, some are aerotolerant anaerobes, and a few are anaerobic. Protozoa are generally an order of magnitude larger than bacteria and often consume bacteria as an energy source. In effect, the protozoa act as polishers of the effluents from biological waste-treatment processes by consuming bacteria and particulate organic matter.
Rotifers	Rotifers are aerobic heterotrophic multicellular animal eukaryotes. The name is derived from the fact that they have two sets of rotating cilia on their heads which are used for motility and capturing food. Rotifers are very effective in consuming dispersed and flocculated bacteria and small particles of organic matter. Their presence in an effluent indicates a highly efficient aerobic biological purification process.
Algae	Algae are unicellular or multicellular, autotrophic, photosynthetic eukaryotes. They are of importance in biological treatment processes. In wastewater treatment lagoons, the ability of algae to produce oxygen by photosynthesis is vital to the ecology of the water environment.
Viruses	Viruses are composed of a nucleic acid core (either DNA or RNA) surrounded by an outer shell of protein called a *capsid*. Viruses are obligate intracellular parasites that multiply only within a host cell, where they redirect the cell's biochemical system to reproduce themselves. Viruses can also exist in an extracellular state in which the virus particle (known as a *virion*) is metabolically inert. Bacteriophages are viruses that infect bacteria as the host; they have not been implicated in human infections.

organisms. Most bacteria cannot tolerate pH levels above 9.5 or below 4.0. Generally, the optimum pH for bacterial growth lies between 6.5 and 7.5.

General description. A general description of the microorganisms found in wastewater is given in Table 2-18 in the terminology introduced in the previous paragraphs. The term *parasite* used in the description of protozoa and virus can be defined as a relationship in which one organism lives at the expense of another. Parasites that live on the surface of a host organism are *ectoparasites*. Parasites that live internally within the host are known as *endoparasites* (Roberts and Janovy, 1996).

Another important feature of microorganisms is their ability to form resistant forms. For example, selected species of bacteria can form endospores (formed within the cell), the structure of which is extremely complex. The endospore, which contains all of the information necessary for reproduction, is coated with several layers of proteins. Endospores are extremely resistant to heat and disinfecting chemicals. It has been speculated that endospores may remain dormant for decades and perhaps even centuries. A spore can become viable in a suitable environment in a three-step process: activation, germination, and outgrowth (Madigan et al., 1997). The resistant forms in protozoans are known as cysts or oocysts. Resistant forms in helminths are eggs and oocysts.

Data on the shape, resistant form, and size of the microorganisms found in wastewater are presented in Table 2-19. Information on the size of the microorganisms, especially the resistant form, is needed to determine the type of treatment that will be required to treat and/or remove them.

Pathogenic Organisms

Pathogenic organisms found in wastewater may be discharged by human beings who are infected with disease or who are carriers of a particular disease. The pathogenic organisms found in wastewater can be classified into three broad categories: bacteria, parasites (protozoa and helminths), and viruses. The principal pathogenic organisms found in untreated wastewater are reported in Table 2-20, along with the diseases and disease symptoms associated with each pathogen. Bacterial pathogenic organisms of human origin typically cause diseases of the gastrointestinal tract, such as typhoid and paratyphoid fever, dysentery, diarrhea, and cholera. Because these organisms are highly infectious, they are responsible for many thousands of deaths each year in areas with poor sanitation, especially in the tropics. It has been estimated that up to 4.5 billion people are or have been infected with some parasite (Madigan et al., 1997). Typical data on the quantity of selected pathogenic organisms found in wastewater and the corresponding concentration needed for an infectious dose are reported in Table 2-21.

Bacteria. Many types of harmless bacteria colonize the human intestinal tract and are routinely shed in the feces. Because pathogenic bacteria are present in the feces of infected individuals, domestic wastewater contains a wide variety and concentration range of nonpathogenic and pathogenic bacteria. One of the most common bacterial pathogens found in domestic wastewater is the genus *Salmonella.* The *Salmonella* group contains a wide variety of species that can cause disease in humans and animals. Typhoid fever, caused by *Salmonella typhi,* is the most severe and serious. The most common disease associated with *Salmonella* is food poisoning, identified as *salmonellosis. Shigella,* a less common genus of bacteria, is responsible for an intestinal disease known as *bacillary dysentery* or *shigellosis.* Waterborne outbreaks of shigellosis have been reported from recreational swimming areas and where wastewater has contaminated wells used for drinking water (Crook, 1998).

TABLE 2-19

Typical data on the shape, size, and resistant forms of classes of microorganisms and selected species found in wastewater*

Microorganism	Shape	Size, μm[†]	Resistant form
Bacteria			
Bacilli	Rod	0.3–1.5 $D \times$ 1–10 L	Endospores or dormant cells
Cocci	Spherical	0.5–4	Endospores or dormant cells
Escherichia coli	Rod	0.6–1.2 $D \times$ 2–3 L	Endospores or dormant cells
Spirilla	Spiral	0.6–2 $D \times$ 20–50 L	Endospores or dormant cells
Vibrio	Rod, curved	0.4–2 $D \times$ 1–10 L	Endospores or dormant cells
Protozoa			
Cryptosporidium[‡]			
Oocysts	Spherical	3–6	Oocysts
Sporozite	Teardrop	1–3 $W \times$ 6–8 L	
Entamoeba histolytica			
Cysts	Spherical	10–15 D	Cyst
Trophozite	Semispherical	10–20	
Giardia lamblia[§]			
Cysts	Oval	6–8 $W \times$ 8–14 L	Cysts
Trophozite	Pear or kite	6–8 $W \times$ 12–16 L	
Helminths			
Ancylostoma duodenale (hookworm) eggs	Elliptical or egg	36–40 $W \times$ 55–70 L	Embryonated egg
Ascaris lumbricoides (roundworm) eggs	Lemon or egg	35–50 $W \times$ 45–70 L	Embryonated egg
Trichuris trichiura (whipworm) eggs	Elliptical or egg	20–24 $W \times$ 50–55 L	Embryonated egg
Viruses			
MS2	Spherical	0.022–0.026	Virion
Enterovirus	Spherical	0.020–0.030	Virion
Norwalk	Spherical	0.020–0.035	Virion
Polio	Spherical	0.025–0.030	Virion
Rotavirus	Spherical	0.070–0.080	Virion

*Compiled from various sources.

[†] D = diameter, L = length, and W = width.

[‡] Member of the Phylum Apicomplexa.

[§] Member of the Phylum Sarcomastigophora, Order Diplomonadida.

TABLE 2-20
Infectious agents potentially present in untreated domestic wastewater*

Organism	Disease	Remarks/symptoms
Bacteria		
Campylobacter jejuni	Gastroenteritis	Diarrhea
Escherichia coli (enteropathogenic)	Gastroenteritis	Diarrhea
Legionella pneumophila	Legionnaires' disease	Malaise, myalgia, fever, headache, respiratory illness
Leptospira (spp.)	Leptospirosis	Jaundice, fever (Weil's disease)
Salmonella typhi	Typhoid fever	High fever, diarrhea, ulceration of small intestine
Salmonella (~2100 serotypes)	Salmonellosis	Food poisoning
Shigella (4 spp.)	Shigellosis	Bacillary dysentery
Vibrio cholerae	Cholera	Extremely heavy diarrhea, dehydration
Yersinia enterocolitica	Yersinosis	Diarrhea
Protozoa		
Balantidium coli	Balantidiasis	Diarrhea, dysentery
Cryptosporidium parvum	Cryptosporidiosis	Diarrhea
Cyclospora	Cyclosporasis	Severe diarrhea, stomach cramps, nausea, and vomiting lasting for extended periods
Entamoeba histolytica	Amebiasis (amoebic dysentery)	Prolonged diarrhea with bleeding, abscesses of the liver and small intestine
Giardia lamblia	Giardiasis	Mild to severe diarrhea, nausea, indigestion
Helminths†		
Ascaris lumbricoides	Ascariasis	Roundworm infestation
Enterobius vermicularis	Enterobiasis	Pinworm
Fasciola hepatica	Fascioliasis	Sheep liver fluke
Hymenolepis nana	Hymenolepiasis	Dwarf tapeworm
Taenia saginata	Taeniasis	Beef tapeworm
T. solium	Taeniasis	Pork tapeworm
Trichuris trichiura	Trichuriasis	Whipworm
Viruses		
Adenovirus (31 types)	Respiratory disease	
Enteroviruses (72 types, e. g., polio, echo, and coxsackie viruses)	Gastroenteritis, heart anomalies, meningitis	
Hepatitis A virus	Infectious hepatitis	Jaundice, fever
Norwalk agent	Gastroenteritis	Vomiting
Parvovirus (3 types)	Gastroenteritis	
Rotavirus	Gastroenteritis	

*Adapted from Feachem et al. (1983), Madigan et al. (1997), and Crook (1997).
† The helminths listed are those with a worldwide distribution.

TABLE 2-21
Microorganism concentration found in septic tank effluent and untreated wastewater and the corresponding infectious dose*

Organism	Concentration in septic tank effluent and raw wastewater, MPN/100 mL†	Infectious dose, number
Bacteria		
Coliform, total	10^7–10^9	
Coliform, fecal	10^6–10^8	10^6–10^{10}‡
Clostridium perfringens	10^3–10^5	1–10^{10}
Enterococci	10^4–10^5	
Fecal streptococci	10^4–10^6	
Pseudomonas aeruginosa	10^3–10^4	
Shigella	10^0–10^3 (?)	10–20
Salmonella	10^2–10^4	
Protozoa		
Cryptosporidium parvum oocysts	10^1–10^4	1–10
Entamoeba histolytica cysts	10^{-1}–10^3	10–20
Giardia lamblia cysts	10^3–10^4	$<$20
Helminths		
Ova	10^1–10^3	
Ascaris lumbricoides		1–10
Viruses		
Enteric virus	10^3–10^4	1–10
Coliphage	10^3–10^4	

*Adapted in part from Crook (1998) and Feachem et al. (1983).
† Most probable number per 100 mL, a statistical estimate of concentration.
‡ *Escherichia coli* (enteropathogenic).

Other bacteria isolated from raw wastewater include *Vibrio, Mycobacterium, Clostridium, Leptospira,* and *Yersinia* species. *Vibrio cholerae* is the disease agent for cholera, which is not common in the United States but is still prevalent in other parts of the world. Humans are the only known hosts, and the most frequent mode of transmission is through water. *Mycobacterium tuberculosis* has been found in municipal wastewater, and outbreaks have been reported among persons swimming in water contaminated with wastewater (Crook, 1998).

Waterborne gastroenteritis of unknown cause is frequently reported, with the suspected agent being bacterial. One potential source of this disease is certain gram-negative bacteria normally considered to be nonpathogenic. These include the enteropathogenic *Escherichia coli* and certain strains of *Pseudomonas,* which may affect the newborn, and have been implicated in gastrointestinal disease outbreaks. *Campylobacter jejuni* has been identified as the cause of a form of bacterial diarrhea in humans. While it has been well established that this organism causes disease in animals, it has also been implicated as the etiologic agent in human waterborne disease outbreaks (Crook, 1998).

Protozoa. Of the disease-causing organisms reported in Table 2-20, the protozoans *Cryptosporidium parvum, Cyclospora,* and *Giardia lamblia* (see Fig. 2-24) are of great concern because of their significant impact on individuals with compromised immune systems, including very young children, the elderly, persons with cancer, and individuals with acquired immunodeficiency syndrome (AIDS). The life cycle of *Cryptosporidium parvum* and *Giardia lamblia* is illustrated in Fig. 2-25. As shown, infection is caused by the ingestion of water contaminated with oocysts and cysts. It is also important to note that nonhuman sources of *Cryptosporidium parvum* and *Giardia lamblia* are present in the environment.

Pathogenic protozoan disease outbreaks have been significant, highlighted by the 1993 outbreak of cryptosporidiosis in Milwaukee in which 400,000 persons became ill and cyclosporiasis which caused disease outbreaks in 10 states. As noted in Table 2-20, these protozoan organisms may cause symptoms which can include severe diarrhea, stomach cramps, nausea, and vomiting lasting for extended periods. With respect to wastewater these organisms are of concern because they are found in wastewater and because conventional disinfection techniques using chlorine and UV irradiation have not proven to be effective in their inactivation or destruction. The most resistant forms are the oocysts of *Cryptosporidium parvum* and the cysts of *Giardia lamblia* (see Table 2-19).

Helminths. The most important helminthic parasites that may be found in wastewater are intestinal worms, including the stomach worm *Ascaris lumbricoides,* the tapeworms *Taenia saginata* and *Taenia solium,* the whipworm *Trichuris trichiura,* the hookworms *Ancylostoma duodenale* and *Necator americanus,* and the threadworm *Strongyloides stercoralis.* The infective stage of some helminths is either the adult organism or larvae, while the eggs or ova of other helminths constitute the infective stage. The free-living nematode larvae stages are not pathogenic to human beings. The eggs and larvae, which range in size from about 10 μm to more than 100 μm, are resistant to environmental stresses and may survive usual wastewater disinfection procedures, although eggs can be removed by commonly used wastewater treatment processes such as sedimentation, filtration, and stabilization ponds.

Viruses. More than 100 different types of enteric viruses capable of producing infection or disease are excreted by humans. Enteric viruses multiply in the intestinal tract and are released in the fecal matter of infected persons. From the standpoint of health, the most important human enteric viruses are the enteroviruses (polio, echo, and coxsackie), Norwalk viruses, rotaviruses, reoviruses, caliciviruses, adenoviruses, and hepatitis A virus. Of the viruses that cause diarrheal disease, only the Norwalk virus and rotavirus have been shown to be major waterborne pathogens. The reoviruses and adenoviruses, known to cause respiratory illness, gastroenteritis, and eye infections, have been isolated from wastewater. There is no evidence that the human immunodeficiency virus (HIV), the pathogen that eventually causes the acquired immunodeficiency syndrome (AIDS), can be transmitted via the waterborne route (Crook, 1998; Madigan et al., 1997; Rose and Gerba, 1991). The biology of viruses is delineated in Voyles (1993).

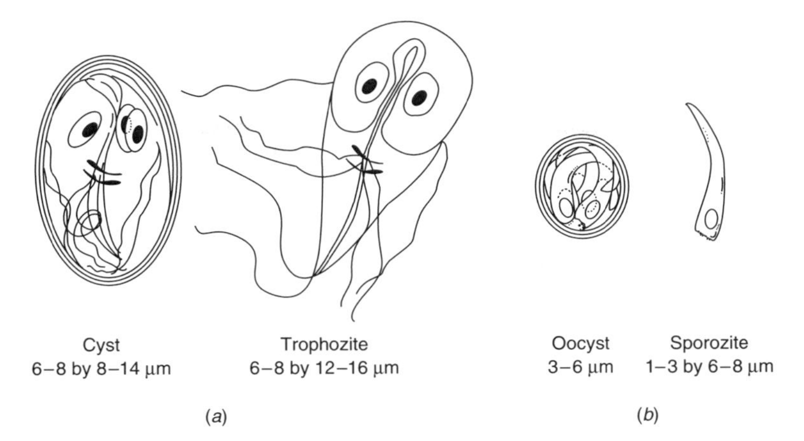

FIGURE 2-24
Definition sketch for (*a*) *Giardia lamblia* cyst and trophozite and (*b*) *Cryptosporidium parvum* oocyst and sporozite.

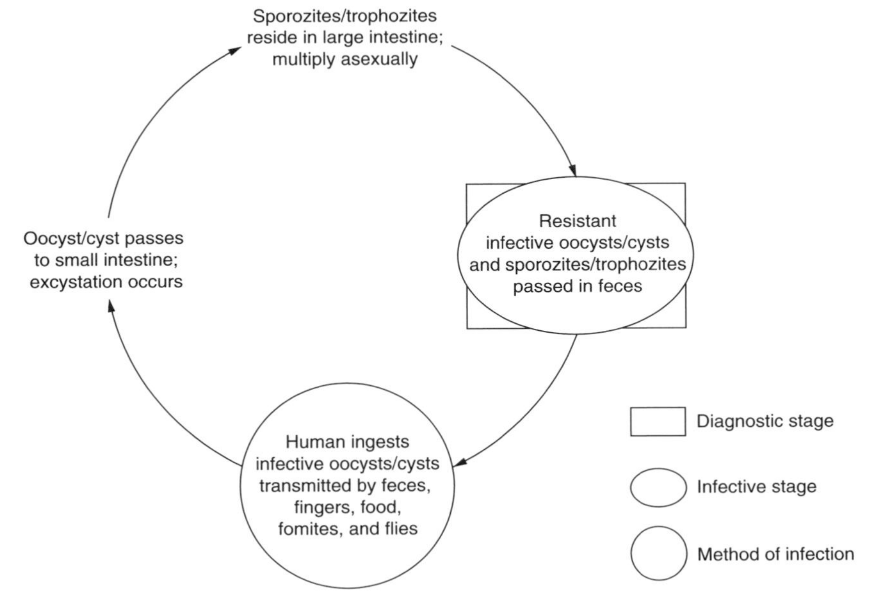

FIGURE 2-25
Life cycle of *Cryptosporidium parvum* and *Giardia lamblia*.

Survival of pathogenic organisms. Of great concern in the management of disease-causing organisms is the survival of these organisms in the environment. Typical data on the survival of microorganisms in the environment are presented in Table 2-22. Although the data given in Table 2-22 can be used as a guide, numerous exceptions have been reported in the literature. Additional data on the effect of temperature on the survival of microorganisms is given in Sec. 14-5, Chap. 14.

TABLE 2-22
Typical pathogen survival times at 20 to 30°C in various environments[*]

Pathogen	Survival time,[†] d		
	Fresh water and wastewater	Crops	Soil
Bacteria			
Fecal coliforms[‡]	<60 but usually <30	<30 but usually <15	<120 but usually <50
Salmonella (spp.)[‡]	<60 but usually <30	<30 but usually <15	<120 but usually <50
Shigella[‡]	<30 but usually <10	<10 but usually <5	<120 but usually <50
Vibrio cholerae[§]	<30 but usually <10	<5 but usually <2	<120 but usually <50
Protozoa			
E. histolytica cysts	<30 but usually <15	<10 but usually <2	<20 but usually <10
Helminths			
A. lumbricoides eggs	Many months	<60 but usually <30	<Many months
Viruses[‡]			
Enteroviruses[†]	<120 but usually <50	<60 but usually <15	<100 but usually <20

[*]Adapted from Feachem et al. (1983).
[†]Includes polio, echo, and coxsackie viruses.
[‡]In seawater, viral survival is less, and bacterial survival is very much less than in fresh water.
[§]*V. cholerae* survival in aqueous environments is a subject of current uncertainty.

Use of Indicator Organisms

Because the numbers of pathogenic organisms present in wastes and polluted waters are few and difficult to isolate and identify, the coliform organism, which is more numerous and more easily tested for, is commonly used as an indicator organism. The intestinal tract of humans contains a large population of rod-shaped bacteria known collectively as coliform bacteria. Each person discharges from 100 to 400 billion coliform bacteria per day, in addition to other kinds of bacteria. Thus, the presence of coliform bacteria is taken as an indication that pathogenic organisms may also be present. The absence of coliform bacteria is taken as an indication that the water is free from disease-producing organisms.

While coliform and fecal coliform organisms may be present, it has not been demonstrated that they are in fact indicators of the presence of enteric viruses and protozoa. Further, concern over newly emerging pathogenic organisms that may arise from nonhuman reservoirs (e.g., *Cryptosporidium parvum* and *Giardia lamblia*) has led to the questioning of the use of indicators that arise primarily from fecal inputs. *Cryptosporidium* oocysts and *Giardia* cysts are not as readily inactivated by chlorine and UV disinfection as are the bacterial surrogates now in use. In a recently completed study, it was concluded that coliform bacteria are adequate indicators for the potential presence of pathogenic bacteria and viruses, but are inadequate as an indicator of waterborne protozoa. It was also found that waterborne disease outbreaks have occurred in water systems that have not violated their water quality standards (Craun et al., 1997).

Given the limitations in using coliform organisms as indicators of potential contamination by wastewater, attention has now focused on the use of bacteriophages as an indicator organisms and more specifically as indicators of enteric viruses.

Bacteriophages are viruses that can infect prokaryotic cells (typically bacteria). There are six major families of bacteriophages, five of which are DNA-based and one of which is RNA-based. RNA bacteriophages that infect *E. coli* are known as *coliphages*. Coliphages that attach directly to the cell wall are known as *somatic*. Coliphages that infect only male strains of *E. coli* (which possess pilli) are known as male-specific (F+) coliphages. More correctly, they should be identified as F+ specific RNA bacteriophage (Voyles, 1993). Within the male-specific family there appear to be four serotypes. Groups II and III predominate in human feces and wastewater whereas groups I and IV predominate in animal feces, with the exception of pigs, which may harbor group II. Group III appears to be found exclusively in human feces (Hsu et al., 1996). The isohedral virus MS2, used in wastewater disinfection studies, is an example of an F+ specific RNA bacteriophage. Interest in using F+ specific RNA coliphages as indicators of enteroviruses is based on the fact that specific coliphages, such as MS2, are approximately the same size (i.e., 0.022–0.026 μm) and general shape as pathogenic viruses of interest such as polio (see Table 2-19) because they are of fecal origin, because they are more resistant in the environment and to treatment processes than bacterial indicators, and because their enumeration by plaque assay (described subsequently) is simple, rapid, and inexpensive (Hsu et al., 1996).

Enumeration and Identification of Bacteria

Individual bacteria are typically enumerated by one of four methods: (1) direct count, (2) plate culture, (3) membrane filtration, and (4) multiple-tube fermentation. Presence-absence (P-A) tests have been developed to assess water quality qualitatively. Colonies of bacteria are often identified by using the heterotrophic plate count (HPC) method. In addition, a number of staining and fluorescent methods have been developed for the identification of specific bacteria. These tests are considered in the following discussion.

Direct counts. Direct counts can be obtained by microscopically using a Petroff-Hauser counting chamber (see Fig. 2-26). Counting cells, using a cell counter such as shown in Fig. 2-26, are designed so that each square in the counting chamber corresponds to a given volume (depth is known). Because it is impossible to differentiate between live and dead cells, the measured counts are total counts. Another technique for obtaining direct counts is with an electronic particle counter in which a sample containing bacteria is passed through an orifice (see Fig. 2-11). As each bacterium passes through the orifice, the electrical conductivity of the fluid in the orifice decreases. The number of times the conductivity is reduced and the values to which it is reduced are correlated to the number of bacteria. Unfortunately the electronic particle counter cannot differentiate between bacteria (whether live or dead) and inert particles, both of which are counted as particles.

Plate culture. The pour-plate and spread-plate count methods are used to culture and identify and to enumerate bacteria. In the pour-plate method (see Fig. 2-27*a*), a sample of wastewater to be tested is diluted serially. A small amount

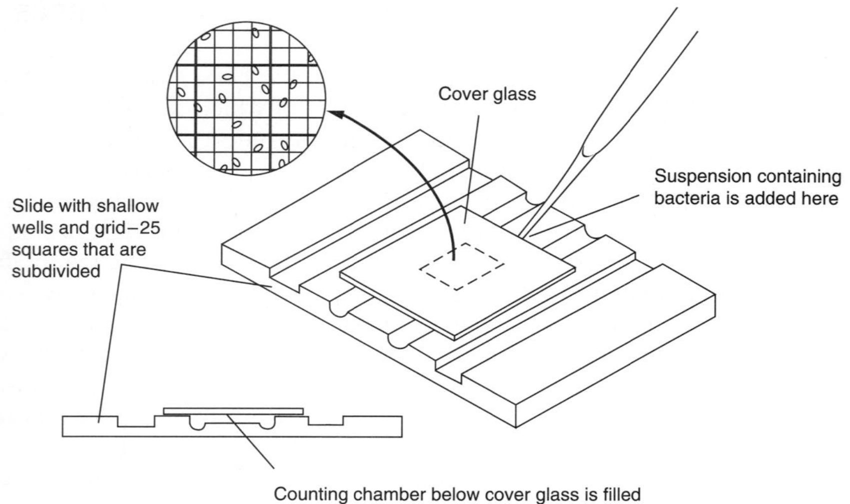

FIGURE 2-26
Schematic of Petroff-Hauser counting chamber for bacteria.

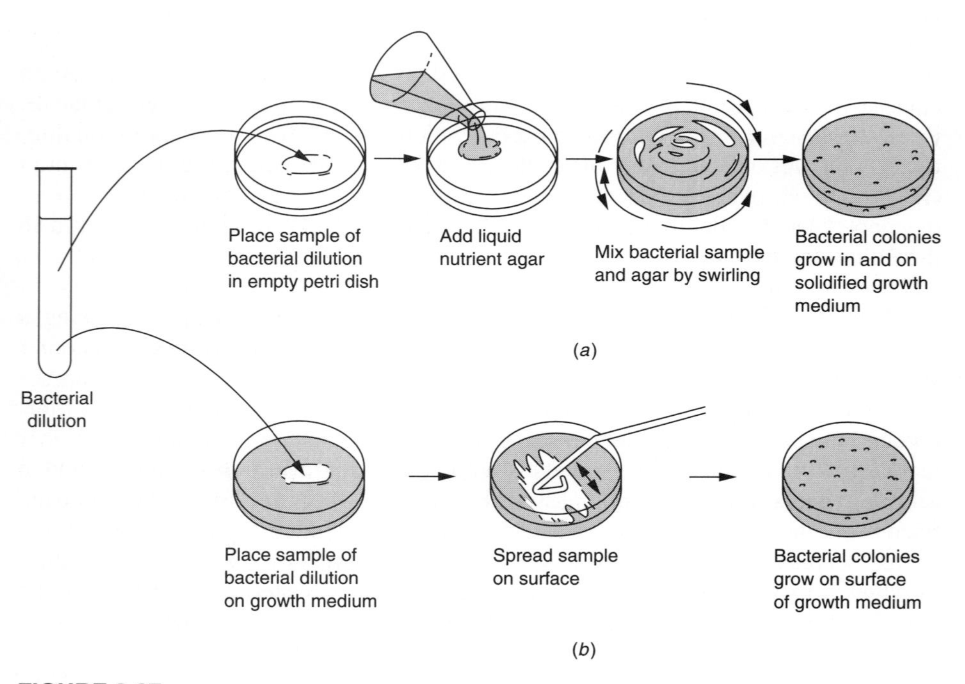

FIGURE 2-27
Schematic of plate culture methods used for the enumeration of bacteria: (*a*) pour plate and (*b*) spread plate.

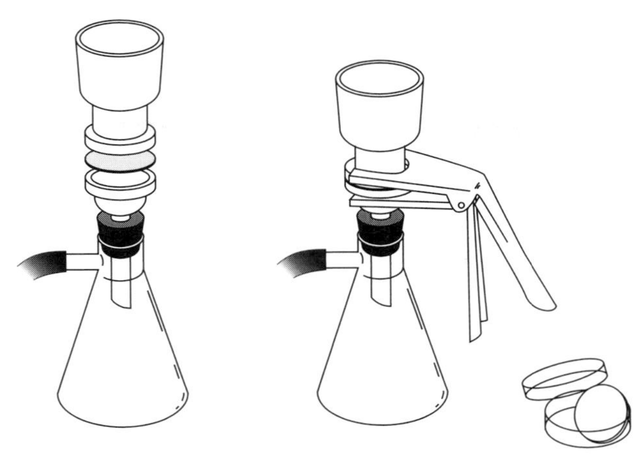

FIGURE 2-28
Schematic of the membrane-filter technique used for the enumeration of bacteria.

of each dilution is then mixed with a warmed liquid culture medium, poured into a culture dish, and incubated under controlled conditions. The colonies on culture plates, with separate distinct bacterial colonies, are counted, assuming that each colony developed from a single bacterium. The total number of bacteria is determined by using the appropriate dilutions. In the spread-plate method (see Fig. 2-27*b*), a small amount of the diluted wastewater is placed and spread on the surface of a prepared culture dish containing a suitable solid medium. Both the pour- and spread-plate methods are extremely sensitive, because individual cells can be counted.

Membrane-filter technique. In the membrane-filter technique (see Fig. 2-28), a known volume of water sample is passed through a membrane filter that has a very small pore size. Bacteria are retained on the filter because they are larger than the size of the pores of the membrane filter. The membrane filter containing the bacteria is then placed in contact with agar that contains nutrients necessary for the growth of the bacteria. After incubation, the coliform colonies can be counted and the concentration in the original water sample determined. The membrane-filter technique has the advantage of being faster than the MPN procedure (described in "Multiple-tube fermentation," below) and of giving a direct count of the number of coliforms. Both methods are subject to limitations in interpretation (Standard Methods, 1995).

Multiple-tube fermentation. The multiple-tube fermentation technique is based on the principle of dilution to extinction as illustrated in Fig. 2-29. Concentrations of total coliform bacteria are most often reported as the *most probable number per 100 mL* (MPN/100 mL). The MPN is based on the application of the Poisson distribution for extreme values to the analysis of the number of positive and negative results obtained in testing multiple portions of equal volume and in portions constituting a geometric series. It is emphasized that the MPN is not the absolute

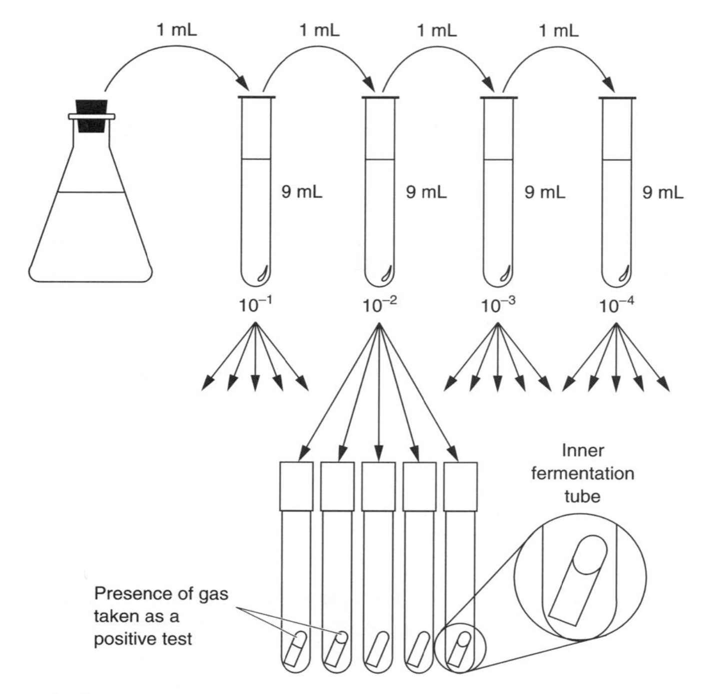

FIGURE 2-29
Schematic of the multiple-tube fermentation technique used for the enumeration of bacteria.

concentration of organisms that are present, but only a statistical estimate of that concentration. The complete multiple-tube fermentation procedure for total coliform involves three test phases identified as the presumptive, confirmed, and completed test. A similar procedure is available for the fecal coliform group as well as for other bacterial groups (Standard Methods, 1995).

The MPN can be determined by using the Poisson distribution directly, MPN tables derived from the Poisson distribution, or the Thomas equation. The joint probability (based on the Poisson distribution) of obtaining a given result from a series of three dilutions is given by Eq. (2-44). It should be noted that Eq. (2-44) can be expanded to account for any number of serial dilutions.

$$y = \frac{1}{a}[(1 - e^{-n_1\lambda})^{p_1}(e^{-n_1\lambda})^{q_1}][(1 - e^{-n_2\lambda})^{p_2}(e^{-n_2\lambda})^{q_2}][(1 - e^{-n_3\lambda})^{p_3}(e^{-n_3\lambda})^{q_3}] \tag{2-44}$$

where
y = probability of occurrence of a given result
a = constant for a given set of conditions
n_1, n_2, n_3 = sample size in each dilution, mL
λ = coliform density, number/mL
p_1, p_2, p_3 = number of positive tubes in each sample dilution
q_1, q_2, q_3 = number of negative tubes in each sample dilution

When the Poisson equation or MPN tables are not available, the Thomas equation (Thomas, 1942) can be used to estimate the MPN:

$$\text{MPN/100 mL} = \frac{\text{Number of positive tubes} \times 100}{\sqrt{\begin{pmatrix}\text{mL of sample in}\\ \text{negative tubes}\end{pmatrix} \times \begin{pmatrix}\text{mL of sample in}\\ \text{all tubes}\end{pmatrix}}} \tag{2-45}$$

In applying the Thomas equation to situations in which some of the dilutions have all five tubes positive, the count of positive tubes should begin with the highest dilution in which at least one negative result has occurred. The application of the Thomas equation is illustrated in Example 2-11.

EXAMPLE 2-11. CALCULATION OF MPN USING MULTIPLE-TUBE FERMENTATION TEST RESULTS. The results of a coliform analysis using the multiple-tube fermentation test for the effluent from an intermittent sand filter (see Chap. 11) are as given below. Using these data, determine the coliform density (MPN/100 mL) using the Poisson equation, the Thomas equation, and the MPN tables given in App. G.

Size of portion, mL	Number positive	Number negative
1.0	4	1
0.1	3	2
0.01	2	3
0.001	0	5

Solution

1. Determine the MPN using the Poisson equation [Eq. (2-44)]. Substitute the appropriate values for n, p, and q and solve the Poisson equation by successive trials.

$$\begin{array}{lll} n_1 = 1.0 & p_1 = 4 & q_1 = 1 \\ n_2 = 0.1 & p_2 = 3 & q_2 = 2 \\ n_3 = 0.01 & p_3 = 2 & q_3 = 3 \\ n_4 = 0.001 & p_4 = 0 & q_4 = 5 \end{array}$$

a. Substitute the coefficient values in Eq. (2-44) and determine ya values for selected values of λ.

$$y = \frac{1}{a}[(1 - e^{-1.0\lambda})^4(e^{-1.0\lambda})^1][(1 - e^{-0.1\lambda})^3(e^{-0.1\lambda})^2][(1 - e^{-0.01\lambda})^2(e^{-0.01\lambda})^3][(1 - e^{-0.001\lambda})^0(e^{-0.001\lambda})^5]$$

λ	ya
3.80	3.6754×10^{-7}
3.84	3.6773×10^{-7}
3.85	3.6774×10^{-7}
3.86	3.6773×10^{-7}
3.90	3.6755×10^{-7}

b. The maximum value of *ya* occurs for a λ value of 3.85 organisms per milliliter. Thus the MPN/100 mL is

$$\text{MPN/100 mL} = 100 \times 3.85 = 385$$

2. Determine the MPN using the Thomas equation (Eq. 2-45):
   a. Number of positive tubes $(4 + 3 + 2) = 9$
   b. Milliliters of sample in negative tubes $= [(1 \times 1.0) + (2 \times 0.1) + (3 \times 0.01) + (5 \times 0.001)] = 1.235$.
   c. Milliliters of sample in all tubes $= [(5 \times 1.0) + (5 \times 0.1) + (5 \times 0.01) + (5 \times 0.001)] = 5.555$.

$$\text{MPN/100 mL} = \frac{9 \times 100}{\sqrt{(1.235) \times (5.555)}} = 344/100 \text{ mL}$$

3. From App. G, eliminating the portion with no positive tubes, as outlined, the MPN/100 mL is 390.

Comment. It should be noted that MPN tables were developed for use before the advent of the small hand-held scientific calculator as a means of computing the results from a multiple-tube fermentation test. With the use of the scientific calculator, the results from all of the serial dilutions can be considered.

Presence-absence test. The presence-absence (P-A) test for coliform organisms is a modification of the multiple-tube fermentation technique described above. Instead of using multiple dilutions, a single 100-mL sample is tested for the P-A of coliform organisms by using lauryl sulfate tryptose lactose broth, as in the MPN test. Coliform organisms are present if a distinct yellow color forms, indicating that lactate fermentation has occurred in the sample.

In addition to the modified MPN test, several commercial enzymatic assays have been developed that can be used to detect both total coliform bacteria as well as *E. coli.* In the enzymatic assays, wastewater samples are added to bottles or MPN tubes containing powdered ingredients composed of salts and specific enzyme substrates that serve as the sole carbon source. Samples containing coliform organisms turn yellow, and samples containing *E. coli* will fluoresce when exposed to longwave UV illumination. Other tests are also available for detecting *E. coli* (Standard Methods, 1995).

Heterotrophic plate count. The heterotrophic plate count (HPC) is a procedure for estimating the number of live heterotrophic bacteria (see Table 2-14) in wastewater samples. The HPC is often used to evaluate the performance of treatment processes and regrowth in effluent distribution systems in reuse applications. The HPC can be determined using the (1) pour-plate method, (2) spread-plate method, or (3) membrane-filter method as described above. In the HPC test, colonies of bacteria, which may be derived from pairs, chains, clusters, or single cells, are measured. The results are reported as colony-forming units per milliliter (CFU/mL). Test details may be found in Standard Methods (1995).

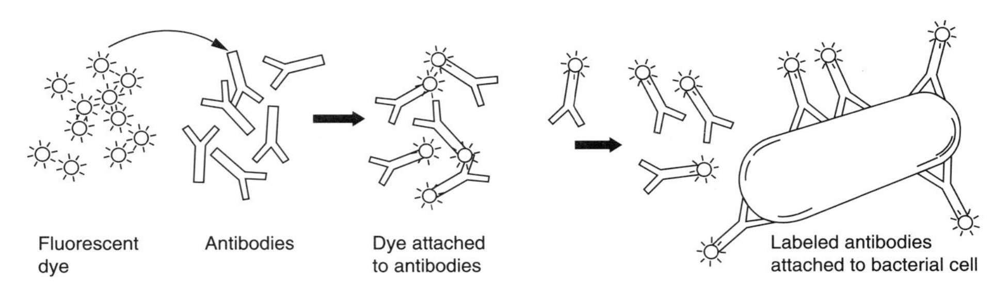

FIGURE 2-30
Schematic of the fluorescence method used for the enumeration of bacteria.

Identification of specific bacteria and protozoa. Over the years, a variety of techniques have been developed for the identification of specific bacteria, including growth-dependent methods, the use of fluorescent antibodies, and the use of nucleic acid probes (BioVir, 1997; Madigan et al., 1997). Growth-dependent methods have been considered previously. A brief description of the last two methods is given because of their use in the study of biological treatment and disinfection processes. The availability of the techniques described below has made it possible to study specific reactions and organisms, and will ultimately help to further our understanding of the microbial interactions which occur in natural systems.

In the fluorescence method, known as the immunofluorescent-antibody (IFA) procedure, a monoclonal antibody (a soluble protein, also known as immunoglobulin, produced by a single B cell clone) is tagged with a fluorescent dye such as rhodamine B or fluorescein. The covalent attachment of the dye to the antibody does not affect the specificity of the antibody. Once the tagged antibody has found the organism in question and becomes attached to the surface (see Fig. 2-30), the sample can be examined by fluorescence microscopy. Organisms to which the antibodies attach will glow when exposed to the fluorescent light of the microscope. This method of analysis is commonly used for bacteria, and is also the method of choice for the identification of *Cryptosporidium parvum* and *Giardia lamblia*.

A nucleic acid probe, as used in microbial studies, is a molecule having a strong interaction only with a genomic sequence unique to the targeted organisms in question, and possessing a means for detection once a probe-target interaction has been achieved (Keller and Manak, 1989). In the nucleic acid probe method, a nucleic acid probe is synthesized that is complementary to either a DNA or RNA sequence of the targeted organism or organisms (see Fig. 2-31). The probe is then labeled with either a radioisotope, a fluorescent dye, or an enzyme. The use of fluorescent or enzyme labels has greatly enhanced the use of probes, and has reduced the use of radioactive probes and handling problems associated with the management of radioactive materials. The next step in the nucleic acid probe technique is to hybridize the probe to the nucleic acid in a bacterial cell. With a nucleic acid probe specific to DNA, the hybridization step generally involves filtering a solution of lysed organisms onto a polycarbonate filter and then soaking the filter in a liquid medium that contains the labeled nucleic acid probe. This process is commonly referred to as *dot-blot hybridization.*

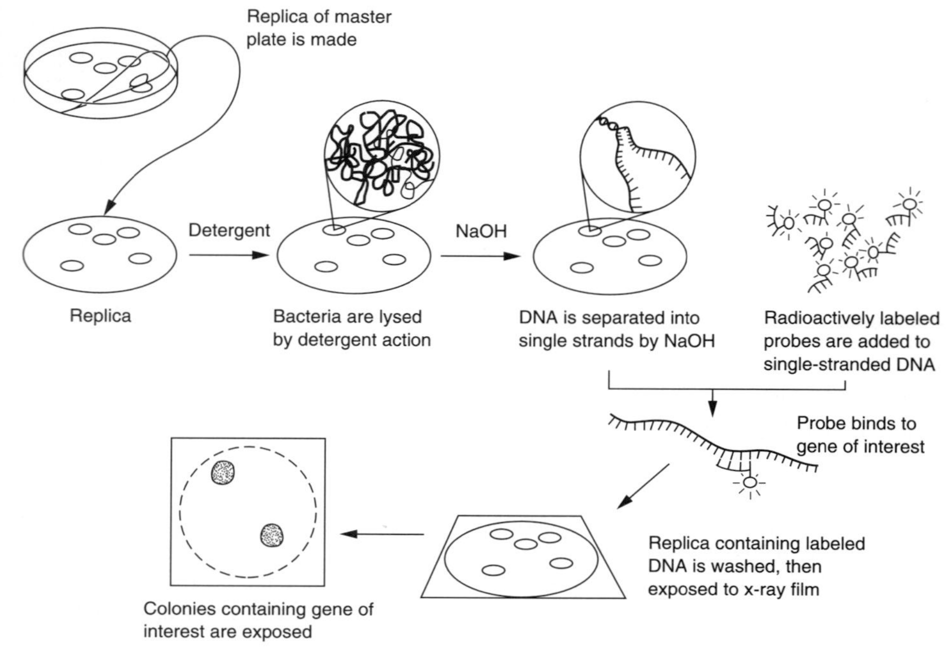

FIGURE 2-31
Schematic of the DNA probe method used for the identification of bacteria.

With a nucleic acid probe specific to RNA, either dot-blot hybridization or a method called *in-situ hybridization* can be used. In-situ hybridization involves adding the nucleic acid probe to a solution of cells with permeablized cell walls (see Fig. 2-32). The final step in the nucleic acid probe approach is to detect the hybridized probe which is dependent on the method used to hybridize the probe. When dot-blot hybridization is used, the hybridized probe is typically detected by placing the filter on a sheet of film sensitive to the type of label put on the nucleic acid probe. When in-situ hybridization is used, cells containing the hybridized probe are generally detected by fluorescent microscopy. In-situ hybridization has the advantage over dot-blot hybridization of allowing the location of cells in the native habitat to be visualized (Loge, 1997).

Nucleic acid probes, with their ability to identify specific nucleic acid sequences, have become very important in the identification of pathogenic organisms. Nucleic acid probes have also been used to study biological treatment processes and the disinfection of wastewater following biological wastewater treatment. With the accelerated development of nucleic acid probes, it is certain that greater use will be made of this technique in developing a better understanding of the biological processes used to treat wastewater. Additional details may be found in Madigan et al. (1997).

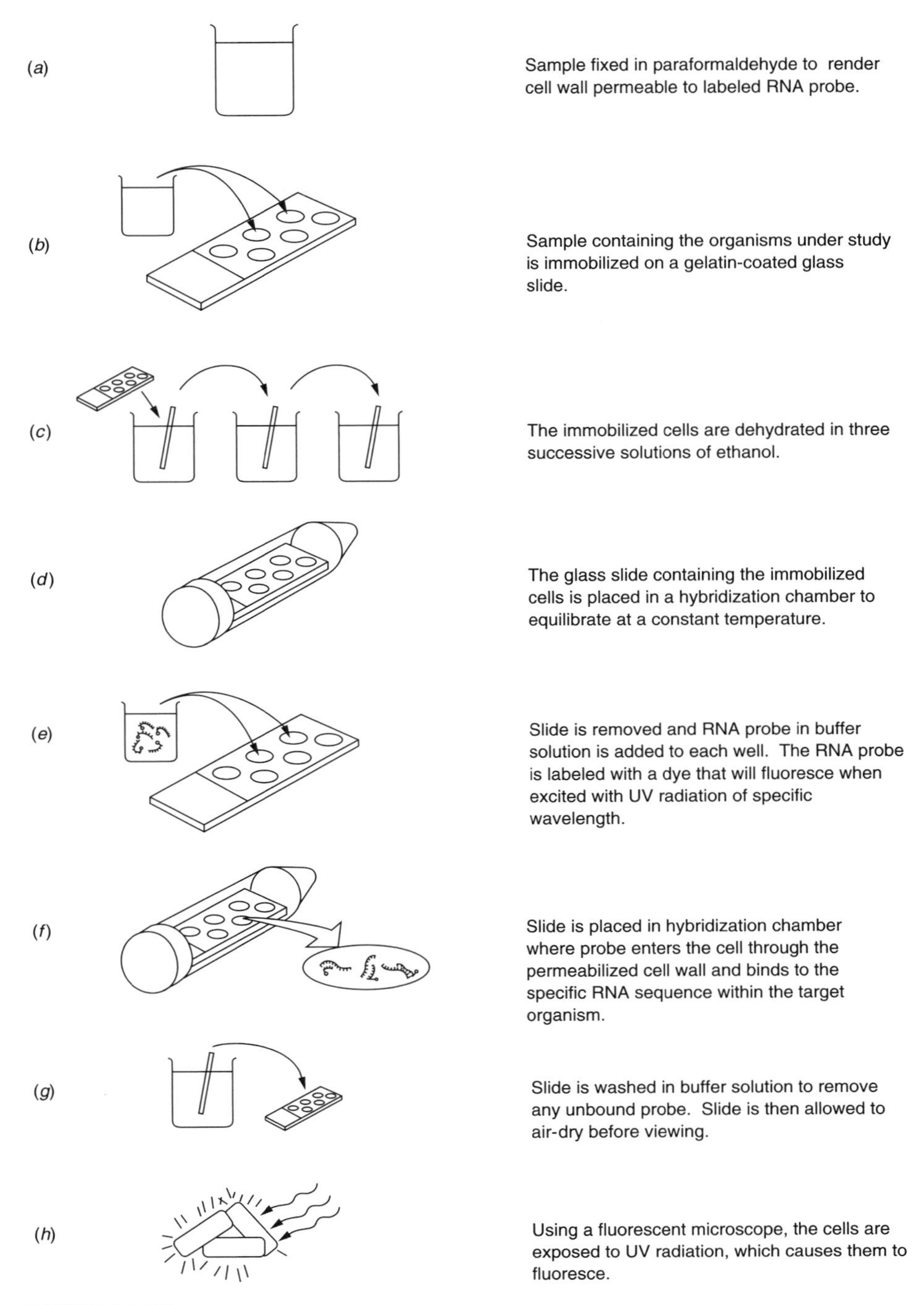

FIGURE 2-32
Schematic of the RNA probe method used for the in-situ identification of bacteria (after D. Thompson, 1997). The RNA probe method is effective because each bacterium contains from 10^4 to 10^6 RNA molecules per cell, allowing multiple probe attachments and resulting in a bright image.

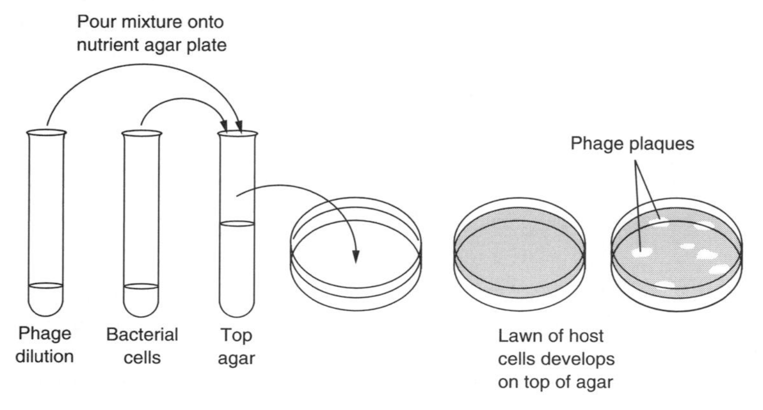

FIGURE 2-33
Schematic of the technique used for the enumeration of viruses.

Enumeration and Identification of Viruses

Virus particles, which vary in size from 0.01 to 0.10 μm (see Table 2-19), are too small to see with a light microscope. While an electron scanning microscope can be used, the process of preparing the sample for examination is costly and time-consuming, and most commercial laboratories do not have an electron scanning microscope. To assess the presence of viruses in wastewater samples, the plaque assay technique is typically used. In this method the effect of viruses on host cells is measured. To conduct a virus assay (see Fig. 2-33), a dilute suspension containing virus is mixed with a small amount of agar and the sensitive host cells. The mixture is then poured onto the surface of a nutrient agar plate. Within a short period of time, the bacteria which were distributed over the agar plate grow to form a uniform growth over the entire surface of the nutrient agar plate. The uniform growth is sometimes referred to as a *lawn*. Each virus particle which has become attached to a bacterial cell will begin to destroy the host cell. The destroyed cell will appear as a hole or plaque on the agar plate. The number of plaques is counted and their number is reported as the number of plaque-forming units (PFUs) per volume of sample used. The total number of viruses per milliliter of sample is determined, taking into account the dilution that was used initially, if any.

New Microorganisms

Within the past 5 years there has been a disturbing increase in the number of disease outbreaks in the United States, especially in light of the fact that it was thought that most endemic contagious diseases had been controlled or eliminated (Levins et al., 1994). The bacteria *Legionella pneumophila,* the causative agent in Legionnaires' disease, found in wastewater and reclaimed wastewater, is an example of a new disease-causing organism that has only recently been identified (Levins et al.,

1994). The significance of the disease outbreaks and the identification of new disease organisms is that the concern for public health must remain the primary objective of wastewater management.

2-8 TOXICITY TESTS

Toxicity tests are used to:

1. Assess the suitability of environmental conditions for aquatic life.
2. Establish acceptable receiving water concentrations for conventional parameters (such as DO, pH, temperature, salinity, and turbidity).
3. Study the effects of water quality parameters on wastewater toxicity.
4. Assess the toxicity of wastewater to one or more freshwater, estuarine, or marine test organisms.
5. Establish the relative sensitivity of a group of standard aquatic organisms to effluent as well as standard toxicants.
6. Assess the degree of wastewater treatment needed to meet water pollution control requirements.
7. Determine the effectiveness of wastewater treatment methods.
8. Establish permissible effluent discharge rates.
9. Determine compliance with federal and state water quality standards and water quality criteria associated with NPDES permits (Standard Methods, 1995).

Such tests provide results that are useful in protecting human health, aquatic biota, and the environment from impacts caused by the release of constituents found in wastewater into surface waters.

Toxicity Terminology

Terms commonly encountered in considering the conduct of toxicity tests and the analysis, interpretation, and application of test results are summarized in Table 2-23. Because the terms reported in Table 2-23 are subject to change as new and improved methods of toxicity testing are developed, it is imperative that the latest version of Standard Methods and related U.S. EPA protocols be reviewed before any toxicity testing is undertaken.

Toxicity Testing

Toxicity tests are classified according to (1) duration: short-term, intermediate, and/or long-term; (2) method of adding test solutions: static, recirculation, renewal, or flow-through; and (3) purpose: NPDES permit requirements, mixing zone determinations, etc. Toxicity testing has been widely validated in recent years. Even though organisms vary in sensitivity to effluent toxicity, the EPA has documented

TABLE 2-23
Terms used in evaluating the effects of contaminants on living organisms*

Term	Description
Acute toxicity	Exposure that will result in significant response shortly after exposure (typically a response is observed within 48 or 96 h).
Chronic toxicity	Exposure that will result in sublethal response over a long term, often one-tenth of the life span or more.
Chronic value (ChV)	Geometric mean of the NOEC and LOEC from partial and full-cycle tests and early-life-stages tests.
Cumulative toxicity	Effects on an organism caused by successive exposures.
Dose	Amount of a constituent that enters the test organism.
Effective concentration (EC)	Constituent concentration estimated to cause a specified effect in a specified time period (e.g., 96-h EC50).
Exposure time	Time period during which a test organism is exposed to a test constituent.
Inhibiting concentration (IC)	Constituent concentration estimated to cause a specified percentage inhibition or impairment in a qualitative function.
Lethal concentration (LC)	Constituent concentration estimated to produce death in a specified number of test organisms in a specified time period (e.g., 96-h LC50).
Lowest-observed-effect concentration (LOEC)	Lowest constituent concentration in which the measured values are statistically different from the control.
Maximum allowable toxicant concentration (MATC)	Constituent concentration that may be present in receiving water without causing significant harm to productivity or other uses.
Median tolerance limit (TLm)	An older term used to denote the constituent concentration at which at least 50 percent of the test organisms survive for a specified period of time. Use of the term *median tolerance limit* has been superseded by the terms *median lethal concentration* (LC50) and *median effective concentration* (EC50).
No-observed-effect concentration (NOEC)	Highest constituent concentration at which the measured effects are no different from the control.
Sublethal toxicity	Exposure that will damage organism, but not cause death.
Toxicity	Potential for a test constituent to cause adverse effects on living organisms.

*Adapted from Standard Methods (1995).

that: (1) toxicity of effluents correlates well with toxicity measurements in the receiving waters when effluent dilution was measured and (2) predictions of impacts from both effluent and receiving water toxicity tests compare favorably with ecological community responses in the receiving waters. The EPA has conducted nationwide tests with freshwater, estuarine and marine ecosystems. Methods include both acute as well as chronic exposures. Typical short-term chronic toxicity test methods

TABLE 2-24
Typical examples of short-term chronic toxicity test methods using various freshwater and marine/estuarine aquatic species*

Species/common name	Test duration	Test end points
	Freshwater species	
Cladoceran *Ceriodaphnia dubia*	Approximately 7 d (until 60% of control have 3 broods)	Survival, reproduction
Fathead minnow *Pimephales promelas*	7 d	Larval growth, survival
	9 d	Embryo-larval survival, percent hatch, percent abnormality
Freshwater algae *Selenastrum capricomutum*	4 d	Growth
	Marine/estuarine species	
Sea urchin *Arbacia punctulata*	1.5 h	Fertilization
Red macroalgae *Champia parvula*	7–9 d	Cystocarp production (fertilization)
Mysid *Mysidopsis bahia*	7 d	Growth, survival, fecundity
Sheepshead minnow *Caprinodon variegatus*	7 d	Larval growth, survival
	7–9 d	Embryo-larval survival, percent hatch, percent abnormality
Inland silverside *Menidia beryijina*	7 d	Larval growth, survival

*From U.S. EPA (1988, 1989).

are reported in Table 2-24. Detailed contemporary testing and analysis protocols are summarized in Standard Methods (1995) and in U.S. EPA publications (1985a, b, c, d).

Analysis of Toxicity Data

Methods used to analyze both short-term (acute) and long-term (chronic) toxicity data are considered in the following discussion.

Acute toxicity data. The median lethal concentration (LC50) when mortality is the test end point, or median effective concentration (EC50) when a sublethal effect (e.g., immobilization, fatigue in swimming, "avoidance") is the end point, are typically used to define acute toxicity (Stephen, 1982). A typical bioassay setup

FIGURE 2-34
Typical setup used to conduct fish bioassays where mortality is the test end point.

using fish where mortality is the test end point is shown in Fig. 2-34. A fish swimming chamber is used to assess sublethal effects. A fish is placed in a chamber where the flow-through velocity can be increased until the fish is swept out of the chamber. The washout velocity for fish exposed to a specific compound can be compared to the washout velocity for the control fish.

Because the LC50 value is the median value, it is important to provide some information on the variability of the test population. The LC50 values can be determined graphically or analytically using the Spearman Karber, moving average, binomial, and probit methods. The 95 percent confidence limits are usually specified. Most standard statistical packages available for desktop computers include a probit analysis program. Determination of LC50 values, both graphically and by means of probit analysis, is illustrated in Example 2-12. Typically, LC50 values are computed for survival at both 48- and 96-hour exposures.

EXAMPLE 2-12. ANALYSIS OF ACUTE TOXICITY DATA. Determine graphically and by probit analysis the 48- and 96-h LC50 values in percent by volume for the following toxicity test data obtained using flathead minnows.

Concentration of waste, % by volume	No. of test animals	No. of test animals dead after*	
		48 h	96 h
60	20	16 (80)	20 (100)
40	20	12 (60)	18 (90)
20	20	8 (40)	16 (80)
10	20	4 (20)	12 (60)
5	20	0 (0)	6 (30)
2	20	0 (0)	2 (10)

*Percentage values are given in parentheses.

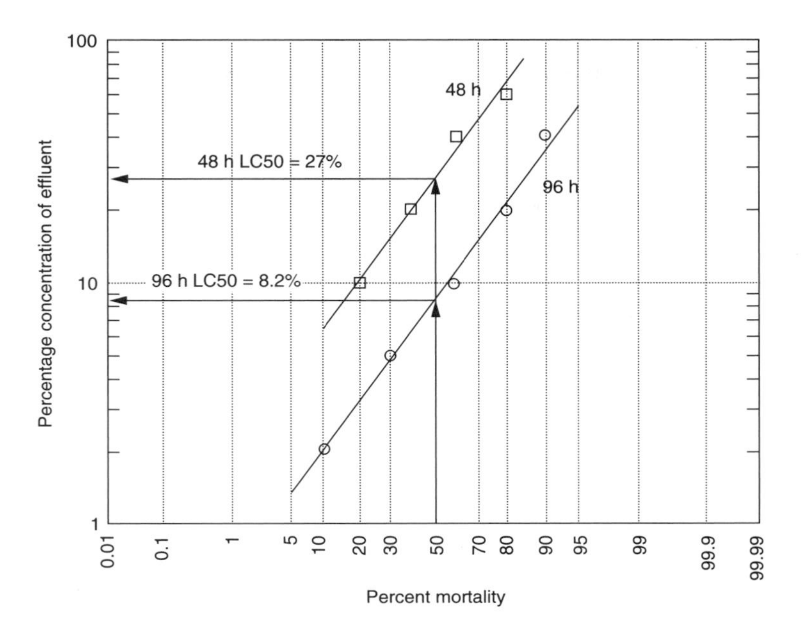

Solution

1. Plot the concentration of wastewater in percent by volume (log scale) against test animals surviving in percent (probability scale). The required plot is shown in the above figure.

2. Fit a line to the data points by eye, giving most consideration to the points lying between 16 and 84 percent mortality, which corresponds to approximately 1 standard deviation.

3. Find the wastewater concentration causing 50 percent mortality. The estimated LC50 values are
   a. 48-h LC50 = 27.0%
   b. 96-h LC50 = 8.2%

4. Compare the results obtained with a probit analysis to the values determined in step 3. The probit analysis results are as follows:
   a. 48-h LC50 = 27.6%
   95% confidence limits 21.0 and 37.8%
   b. 96-h LC50 = 8.1%
   95% confidence limits 5.8 and 10.9%

Comment. Although the LC50 values obtained by the graphical analysis approach are approximate, they are quite close to the values obtained by the probit analysis approach and serve as a good check. To obtain confidence limits, a probit or similar analysis must be performed.

Chronic toxicity data. Results of chronic toxicity tests often are analyzed statistically to determine the lowest-observed-effect concentration (LOEC), the no-observed-effect concentration (NOEC), and the chronic value (ChV). Statistical significance generally is assumed to mean significantly different at P = 0.05. The chronic value is calculated as the geometric mean of the LOEC and the NOEC.

Chronic toxicity limits may be specified with either NOEC or ChV as the end point. The term *maximum acceptable toxicant concentration* (MATC) often is used interchangeably with the chronic value. Like acute toxicity data, lethal concentration (LC) or effective concentration (EC) values can be used with chronic toxicity data to describe chronic toxicity tolerance levels. Recently, the concept of the inhibiting concentration (IC) has been introduced to characterize effects in chronic tests. A variety of nonparametric and parametric statistical methods is available to determine NOECs, LOECs, LCs, ECs, and ICs (Standard Methods, 1995).

Application of Toxicity Data

In applying acute and chronic toxicity test results, the toxic units (TU) approach has been adopted by a number of federal and state agencies. In the toxic units approach (U.S. EPA, 1985a), a TU concentration is established for the protection of aquatic life.

Toxic unit acute (TU_a). The TU_a is defined as the reciprocal of the wastewater concentration that caused the acute effect by the end of the exposure period:

$$TU_a = 100/LC50 \tag{2-46}$$

Toxic unit chronic (TU_c). The TU_c is defined as the reciprocal of the effluent concentration at which the measured effects, by the end of the chronic exposure period, are no different from the control:

$$TU_c = 100/NOEC \tag{2-47}$$

where NOEC is the no-observed-effect concentration.

Depending on the use to be made of the toxicity test results, a variety of different numerical values have been used for TU_a and TU_c as a basis for assessing the suitability of a given effluent for discharge to the environment. For example, to protect against acute toxicity it has been suggested that the MATC should be less than $0.3 \times TU_c$. Because the limiting values vary from location to location, current regulatory standards must be reviewed in applying toxicity results.

PROBLEMS AND DISCUSSION TOPICS

2-1. Check the accuracy of the following laboratory results (A, B, C, or D, to be assigned by instructor). What is the percent difference for the cation-anion balance? On the basis of the criteria provided in Sec. 2-2, are the results acceptable? If electroneutrality is not achieved, what common ions may be unaccounted for?

Constituent	Concentration, mg/L			
	A	B	C	D
Ca^{+2}	28.6	83.9	39.1	64.4
Mg^{+2}	40.2	66.2	29.6	43.5
Na^{+}	56.4	35.2	57.0	62.7
K^{+}	21.1	37.0	22.2	46.2
HCO_3^-	255	286	198	263
SO_4^{-2}	76.0	179	78.7	152
Cl^-	55.3	95.0	48.8	62.6
NO_3^-	18.2	26.5	39.7	39.8

2-2. If a water sample contains a calcium ion concentration of 3.5 mg/L and a carbonate ion concentration of 6.4 mg/L, is the water oversaturated or undersaturated with respect to $CaCO_3$? The solubility product K_{sp} for $CaCO_3$ at 25°C is 5.0×10^{-9}.

2-3. Consider the reaction $HCO_3^- \rightarrow H^+ + CO_3^{-2}$. If the equilibrium constant K_{eq} for the reaction is 5.0×10^{-11}, what is the proportion of bicarbonate to carbonate when the pH is 6.0? What is the proportion of bicarbonate to carbonate when the pH is 8.0?

2-4. Use the following test results (A, B, C, or D, to be assigned by instructor) from a wastewater sample to determine the following parameters: total solids, total volatile solids, total fixed solids, total suspended solids, volatile suspended solids, and total dissolved solids. Report the results in mg/L.

Measurement	Units	A	B	C	D
Sample size	mL	100	200	150	50
Tare mass of evaporation dish	g	55.3218	59.2514	65.9783	60.3774
Mass of dish and residue after evaporation	g	55.3735	59.3546	66.0524	60.4026
Mass of dish and residue after ignition	g	55.3436	59.2942	66.0086	60.3872
Tare mass of glass filter	g	1.4365	1.4738	1.5784	1.6783
Mass of filter and residue after evaporation	g	1.4504	1.4950	1.5967	1.6839
Mass of filter and residue after ignition	g	1.4402	1.4796	1.5834	1.6793

2-5. Determine the saturation value for oxygen in water at 14°C (see Table 2-9).

2-6. The following results were obtained when the effluent from a septic tank was analyzed for BOD. Due to the abundance of microorganisms in the effluent, seeding of the sample was not required. Determine the BOD_5 for sample A, B, C, or D (to be assigned by instructor).

Parameter	Unit	A	B	C	D
Volume of sample used	mL	15	10	15	10
Initial DO	mg/L	9.4	8.9	9.5	9.3
Five-day DO	mg/L	1.2	1.7	1.1	2.2

2-7. If the effluent in Prob. 2-6 was from a source that required seeded dilution water, determine the BOD_5 for sample A, B, C, or D (to be assigned by instructor). Assume that the decrease in the dissolved oxygen concentration of the seeded dilution water over the 5-day test period was 1.1 mg/L.

2-8. The 5-day, 20°C BOD of a wastewater is 210 mg/L. What will be the ultimate and 10-day BOD? If the sample had been incubated at 30°C, what would the 5-day BOD have been ($k_1 = 0.23\ d^{-1}$)?

2-9. Determine the theoretical COD value for a solution containing 200 mg/L of glycine [$CH_2(NH_2)COOH$]. Also estimate the BOD, the nitrogenous oxygen demand, and the carbon to nitrogen ratio.

2-10. What are the advantages and disadvantages of the COD test with respect to the UBOD test? Cite two references in your response.

2-11. If the ultimate BOD of a restaurant septic tank effluent is 360 mg/L and the 5-day BOD is 250 mg/L, determine the BOD reaction rate constant.

2-12. Compare the surface area to volume ratio of three spherical bacteria with diameters of 1, 2, and 4 μm. What are the implications of the difference in the ratios, if any?

2-13. Use the Poisson formula to compute the most probable number (MPN) of coliform organisms per 100 mL for the following test results (choose one column only). Positive test results are indicated by (+), and negative results by (−).

Sample volume, mL	A		B		C		D	
	(+)	(−)	(+)	(−)	(+)	(−)	(+)	(−)
100	5	0	5	0			5	0
10	5	0	5	0			5	0
1	4	1	4	1	5	0	4	1
0.1	3	2	2	3	2	3	4	1
0.01	1	4			0	5	0	5

2-14. Use the MPN table and other information given in App. G to determine the MPN/100 mL for the sample data in the preceding problem. Compare the value obtained from the MPN table to the MPN value computed using the Poisson formula. If the values differ, explain the variation.

2-15. You are traveling in a foreign land where the water supply is unsafe. You are offered bottled water. From a public health standpoint is there any difference between bottled water with or without gas?

2-16. What important characteristics should be analyzed for in the wastewater from individual residences using a septic tank, eating establishments, and roadside rest stops? Why?

REFERENCES

Albertson, O. E. (1995) Is $CBOD_5$ Test Viable for Raw and Settled Wastewater? *Journal of Environmental Engineering,* Vol. 121, No. 7, pp. 515–520.

ASCE (1989) *Sulfide in Wastewater Collection and Treatment Systems,* Manual of Practice No. 69, American Society of Civil Engineers, New York.

American Society for Testing and Materials (1979) *Standard Practice for the Determination of Odor and Taste Thresholds by the Forced-Choice Ascending Concentration Series Method of Limits,* E679, ASTM, Philadelphia.

Balmat, J. L. (1957) Biochemical Oxidation of Various Particulate Fractions of Sewage, *Sewage and Industrial Wastes,* Vol. 29, No. 7, 1957.

BioVir (1997) Literature on Enteric Virus, *Cryptosporidium,* and *Giardia,* BioVir Laboratories, Benicia, CA.

Craun, G. F., P. S. Berger, and R. L. Calderon (1997) Coliform Bacteria and Waterborne Disease Outbreaks, *Journal American Water Works Association,* Vol. 89, Issue 3.

Crook, J. (1998) Water Reclamation and Reuse Criteria, Chap. 7 in T. Asano (ed.), *Wastewater Reclamation and Reuse,* Technomic Publishing, Lancaster, PA.

Eldridge, E. F. (1942) *Industrial Waste Treatment Practice,* McGraw-Hill, New York.

Feachem, R. G., D. J. Bradley, H. Garelick, and D. D. Mara (1983) *Sanitation and Disease: Health Aspects of Excreta and Wastewater Management,* published for the World Bank by John Wiley & Sons, New York.

Grady, C. P. L., Jr., and H. C. Lim (1980) *Biological Wastewater Treatment: Theory and Application,* Marcel Dekker, New York.

Hsu, F-C, H. Chung, A. Amante, Y. S. Carol Shieh, D. Wait, and M. D. Sobsey (1996) Distinguishing Human from Animal Fecal Contamination in Water by Typing Male-Specific RNA Coliphages, Proceedings 1996 Water Quality Technology Conference, American Water Works Association, Denver, CO.

Ingraham, J. L., and C. A. Ingraham (1995) *Introduction to Microbiology,* Wadsworth Publishing, Belmont, CA.

Jaques, R. S. (1994) Personal communication, Monterey Regional Water Pollution Control Agency, Monterey, CA.

Keller, G. H., and M. L. Manak (1989) *DNA Probes,* Stockton Press, New York.

Levine, A. D., G. Tchobanoglous, and T. Asano (1985) Characterization of the Size Distribution of Contaminants in Wastewater: Treatment and Reuse Implications, *Journal Water Pollution Control Federation,* Vol. 57, No. 7, pp. 205–216.

Levine, A. D., G. Tchobanoglous, and T. Asano (1991) Size Distributions of Particulate Contaminants in Wastewater and Their Impact on Treatability, *Water Research,* Vol. 25, No. 8.

Levins, R., et al. (1994) The Emergence of New Diseases: Lessons learned from the emergence of new diseases and the resurgence of old ones may help us prepare for future epidemics, *American Scientist,* Vol. 82, pp. 53–60.

Loge, F. (1997) Personal communication, Department of Civil and Environmental Engineering, University of California at Davis.

Madigan, M. T., J. M. Martinko, and J. Parker (1997) *Brock Biology of Microorganisms,* 8th ed., Prentice Hall, Upper Saddle River, NJ.

Pepper, I. L., C. P. Gerba, and M. L. Brusseau (eds.) (1996) *Pollution Science,* Academic Press, San Diego, CA.

Perry, R. H., D. W. Green, and J. O. Maloney (1984) *Perry's Chemical Engineers' Handbook,* 6th ed., McGraw-Hill, New York.

Roberts, L. S., and J. Janovy, Jr. (1996) *Foundations of Parasitology,* 5th ed., Wm. C. Brown Publishers, Dubuque, IA.

Rose, J. B., and C. P. Gerba (1991) Assessing Potential Health Risks from Viruses and Parasites in Reclaimed Water in Arizona and Florida, U.S.A., *Water Science Technology,* Vol. 23, pp. 2091–2098.

Sawyer, C. N., P. L. McCarty, and G. F. Parkin (1994) *Chemistry for Environmental Engineering,* 4th ed., McGraw-Hill, New York.

Standard Methods (1995) *Standard Methods for the Examination of Water and Waste Water,* 19th ed., American Public Health Association, Washington, DC.

Stanier, R. Y., J. L. Ingraham, M. L. Wheelis, and P. R. Painter (1986) *The Microbial World,* 5th ed., Prentice-Hall, Englewood Cliffs, NJ.

Stephen, C. E. (1982) Methods for Calculating an LC 50, in F. L. Mayer and J. L. Hamelink (eds.), *Aquatic Toxicology and Hazard Evaluation,* pp. 65–84, ASTM STP 634, American Society for Testing and Materials, Philadelphia, PA.

Tchobanoglous, G. (1995) Particle-Size Characterization: The Next Frontier, *Journal of Environmental Engineering,* ASCE, Vol. 121, No. 12.

Tchobanoglous, G., and E. D. Schroeder (1985) *Water Quality: Characteristics, Modeling, Modification,* Addison-Wesley, Reading, MA.

Thomas, H. A., Jr. (1942) Bacterial Densities from Fermentation Tube Tests, *Journal American Water Works Association,* Vol. 34, No. 4, p. 572.

Thompson, D. (1997) Personal communication, Department of Civil and Environmental Engineering, University of California at Davis.

U.S. EPA (1985a) *Methods for Measuring the Acute Toxicity of Effluents to Freshwater and Marine Organisms,* EPA-600/4-85/013, U.S. EPA Environmental Monitoring and Support Laboratory, U.S. Environmental Protection Agency, Cincinnati, OH.

U.S. EPA (1985b) *Technical Support Document for Water Quality-Based Toxics Control,* U.S. EPA Office of Water, EPA-440/4-85/032, U.S. Environmental Protection Agency, Washington, DC.

U.S. EPA (1985c) *Short Term Methods for Estimating Chronic Toxicity of Effluents and Receiving Waters to Freshwater Organisms,* EPA-660/4-85/014, U.S. Environmental Protection Agency, Washington, DC.

U.S. EPA (1985d) *User's Guide to the Conduct and Interpretation of Complex Effluent Toxicity Tests at Estuarine/Marine Sites,* EPA-600/X-86/224, U.S. Environmental Protection Agency, Washington, DC.

U.S. EPA (1985e) *Odor Control and Corrosion Control in Sanitary Sewerage Systems and Treatment Plants,* Design Manual, EPA-625/1-85-018, U.S. Environmental Protection Agency, Washington, DC.

U.S. EPA (1988) *Short Term Methods for Estimating the Chronic Toxicity of Effluents and Receiving Waters to Marine and Estuarine Organisms,* EPA-600/4-88/028, U.S. Environmental Protection Agency, Washington, DC.

U.S. EPA (1989) *Short Term Methods for Estimating Chronic Toxicity of Effluents and Receiving Waters to Freshwater Organisms,* EPA-660/2nd ed., U.S. Environmental Protection Agency, Washington, DC.

Voyles, B. A. (1993) *The Biology of Viruses,* Mosby, St. Louis.
WEF (1995) *Odor Control in Wastewater Treatment Plants,* WEF Manual of Practice No. 22, ASCE Manuals and Reports on Engineering Practice No. 82, Water Environment Federation, Alexandria, VA.
Young, J. C. (1973) Chemical Methods for Nitrification Control, *Journal Water Pollution Control Federation,* Vol. 45, No. 4.

CHAPTER 3

Fate of Wastewater Constituents in the Environment

The purpose of this chapter is to introduce the methods and techniques used to assess the fate and impact of the constituents found in wastewater during wastewater treatment and when released to the environment. The approach used most commonly is to develop mathematical models that can be used to describe the observed phenomena. In turn, the development of mathematical models requires that the mechanisms involved in the transformation of constituents released to the environment be understood. The purpose of this chapter is sevenfold: (1) to introduce the mass balance principle, the basis for all mathematical models; (2) to consider the types of reactions, reaction rates, and reaction kinetics that are used in modeling; (3) to consider the types of reactors commonly used, their hydraulic characteristics, and their applications; (4) to review reactor treatment kinetics; (5) to provide an overview of the fate of constituents released to the environment; (6) to introduce the modeling of the fate of constituents released to water bodies (lakes, reservoirs, and rivers); and (7) to assess the impact of wastewater effluent discharge standards.

3-1 MASS BALANCE PRINCIPLE

The conservation of mass is the fundamental basis for all modeling analyses. The conservation of mass equation is applicable whether the discharge is to a lake, reservoir, stream, estuary, coastal area, or groundwater aquifer. However, the physical characteristics of the different environments require different approaches and approximations to solve for the constituent concentrations. The conservation of mass principle involves the accounting of the mass of any water quality constituent in a stationary volume of fixed dimensions called a *control volume*. To illustrate the basic concepts involved, a mass balance analysis will be performed on the contents of the container shown schematically in Fig. 3-1. The first step involved in preparing a mass balance is to define the system boundary so that all the flows of mass into

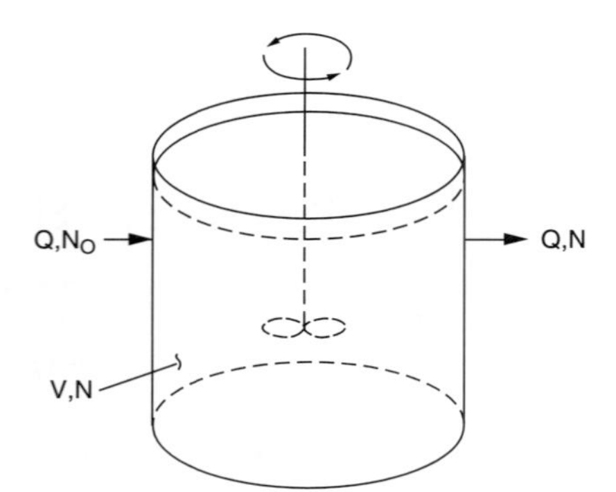

FIGURE 3-1
Definition sketch for mass balance analysis.

and out of the system boundary can be identified. In Fig. 3-1, the system boundary is shown by a dashed line. Proper selection of the system boundary is extremely important in simplifying the mass balance computations.

To apply a mass balance analysis to the liquid contents of the container shown in Fig. 3-1, it will be assumed that (1) the volumetric flowrate into and out of the container is constant, (2) the liquid within the container is not subject to evaporation (constant volume), (3) the liquid within the container is mixed completely, and (4) a chemical reaction involving a reactant A is occurring within the container. For the stated assumptions, the materials mass balance can be formulated as follows:

1. General word statement:

$$\begin{bmatrix}\text{Rate of}\\ \text{accumulation of}\\ \text{reactant within}\\ \text{the system}\\ \text{boundary}\end{bmatrix} = \begin{bmatrix}\text{rate of flow}\\ \text{of reactant}\\ \text{into the}\\ \text{system}\\ \text{boundary}\end{bmatrix} - \begin{bmatrix}\text{rate of flow}\\ \text{of reactant}\\ \text{out of the}\\ \text{system boundary}\end{bmatrix} + \begin{bmatrix}\text{rate of}\\ \text{generation of}\\ \text{reactant within}\\ \text{the system}\\ \text{boundary}\end{bmatrix} \tag{3-1}$$

2. Simplified word statement:

$$\text{Accumulation} = \text{inflow} - \text{outflow} + \text{generation} \tag{3-2}$$

3. Symbolic representation (refer to Fig. 3-1):

$$\frac{d[N]}{dt} = \frac{[N_o]}{V}Q - \frac{[N]}{V}Q + r_N V \tag{3-3}$$

where $d[N]/dt$ = rate of change of moles of reactant within the control volume, MT^{-1}
N_o = moles of reactant entering the control volume, M
V = volume contained within control volume, L^3
Q = volumetric flowrate into and out of control volume, L^3T^{-1}
N = moles of reactant leaving the control volume, M
r_N = rate of reaction within control volume, $ML^{-3}T^{-1}$

In Eq. (3-3), a positive sign is used for the rate-of-generation term because the necessary negative sign is part of the rate expression. Before attempting to solve any mass balance expression, a unit check should always be made to assure that units of the individual quantities are consistent. For Eq. (3-3), as given above, the resulting units for each term are M/T. To solve Eq. (3-3), the form of the rate expression must be known.

If there is no inflow or outflow, the container is known as a *batch reactor,* and Eq. (3-3) becomes

$$r_N = \frac{1}{V}\frac{d[N]}{dt} = \frac{\text{moles}}{\text{(liquid volume)(time)}} \tag{3-4}$$

If N is replaced by the term VC_A, where V is the volume and C_A is the concentration of constituent A, Eq. (3-4) becomes

$$r_A = \frac{1}{V}\frac{d(VC_A)}{dt} = \frac{1}{V}\frac{(VdC_A + C_A dV)}{dt} \tag{3-5}$$

If the volume remains constant, Eq. (3-5) reduces to

$$r_A = \frac{dC_A}{dt} \tag{3-6}$$

Reaction rates are considered further in the following discussion.

3-2 TYPES OF REACTIONS, REACTION RATES, AND REACTION KINETICS

Selection of reaction-rate expressions for the fate processes is based on (1) information obtained from the literature, (2) experience with the design and operation of similar systems, or (3) data derived from pilot plant studies. In cases where significantly different wastewater characteristics occur or new applications of existing technology or new processes are being considered, pilot plant testing is recommended. The types of reactions, the rates of reaction, the types of rate expressions, the effects of temperature on reaction rate coefficients, and determination of reaction rate coefficients are considered in the following discussion.

Types of Reactions

The two principal types of reactions that occur in wastewater treatment are classified as homogeneous and heterogeneous (nonhomogeneous). In *homogeneous reactions,* the reactants are distributed uniformly throughout the fluid so that the potential for reaction at any point within the fluid is the same. Homogeneous reactions may be either irreversible or reversible.

Examples of irreversible reactions are

$$\text{A} \rightarrow \text{B}$$

$$\text{A} + \text{A} \rightarrow \text{P}$$

$$a\text{A} + b\text{B} \rightarrow \text{P}$$

Examples of reversible reactions are

$$A \Leftrightarrow B$$

$$A + B \Leftrightarrow C + D$$

Heterogeneous reactions occur between one or more constituents that can be identified with specific sites, such as those on an ion-exchange resin and on the detritus in constructed wetlands. Reactions that require the presence of a solid-phase catalyst are also classified as heterogeneous.

Rates of Reaction

The term used to describe the change (decrease or increase) in the number of moles of a reactive substance per unit time per unit volume (for homogeneous reactions), or per unit surface area or mass (for heterogeneous reactions), is the *rate of reaction* (Denbigh and Turner, 1984). The rate at which a reaction proceeds is an important consideration in wastewater treatment. For example, treatment processes may be designed on the basis of the rate at which the reaction proceeds rather than the equilibrium position of the reaction, because the reaction takes too long to go to completion.

For homogeneous reactions, the rate of reaction r is given by Eq. (3-7).

$$r = \frac{1}{V}\frac{d[N]}{dt} = \frac{\text{moles}}{(\text{volume})(\text{time})} \tag{3-7}$$

For heterogeneous reactions where S is the surface area, the corresponding expression is

$$r = \frac{1}{S}\frac{d[N]}{dt} = \frac{\text{moles}}{(\text{area})(\text{time})} \tag{3-8}$$

For reactions involving two or more reactants with unequal stoichiometric coefficients, the rate expressed in terms of one reactant will not be the same as the rate for the other reactants. For example, for the reaction

$$a\text{A} + b\text{B} \rightarrow c\text{C} + d\text{D}$$

the concentration changes for the various reactants are given by

$$-\frac{1}{a}\frac{d[\text{A}]}{dt} = -\frac{1}{b}\frac{d[\text{B}]}{dt} = \frac{1}{c}\frac{d[\text{C}]}{dt} = \frac{1}{d}\frac{d[\text{D}]}{dt} \tag{3-9}$$

Thus, for reactions in which the stoichiometric coefficients are not equal, the rate of reaction is given by

$$r = \frac{1}{c_i}\frac{d[C_i]}{dt} \tag{3-10}$$

where the coefficient term $(1/c_i)$ is negative for reactants and positive for products.

Types of Rate Expressions

Typical rate expressions that have been used to describe the conversion of waste constituents in treatment processes and the fate of constituents released in the environment include the following:

$$r = \pm k \quad \text{(zero order)} \tag{3-11}$$

$$r = \pm kC \quad \text{(first order)} \tag{3-12}$$

$$r = \pm k(C - C_s) \quad \text{(first order)} \tag{3-13}$$

$$r = \pm kC^2 \quad \text{(second order)} \tag{3-14}$$

$$r = \pm kC_{\mathrm{A}}C_{\mathrm{B}} \quad \text{(second order)} \tag{3-15}$$

$$r = \pm \frac{kC}{K + C} \quad \text{(saturation type)} \tag{3-16}$$

$$r = \pm \frac{kC}{(1 + Rt)^n} \quad \text{(first order retarded)} \tag{3-17}$$

The sum of the exponent to which the concentration is raised is known as the order of the reaction. For example, the first-order reaction ($r_C = -kC$) is used to model the exertion of BOD and bacterial decay. Although Eq. (3-15) is second-order overall, it is first-order with respect to C_A and C_B individually. Equation (3-16) is known as a *saturation type* of equation. When C is large, the rate of reaction is zero-order.

The rate expression given by Eq. (3-17) is known as a *retarded first-order* rate expression, because the rate constant changes with distance or time. The term R in the denominator is the retardation factor. In wastewater treatment applications, the exponent n in Eq. (3-17) is related to the particle size distribution. For example, if all of the particles are the same size and composition, the value of the exponent n is equal to one and the retardation factor R is equal to zero. Application of the retarded first-order rate expression is considered further in Sec. 3-4.

Effects of Temperature on Reaction-Rate Coefficients

The temperature dependence of the specific reaction-rate constants is important because of the need to use constants that are determined at one temperature for systems of another temperature. For example, the reaction-rate constant determined for the BOD reaction at 20°C must often be used for systems at temperatures other than 20°C. The temperature dependence of the rate constant is given by the van't Hoff-Arrhenius equation:

$$\frac{d(\ln k)}{dT} = \frac{E}{RT^2} \tag{3-18}$$

where k = reaction-rate constant
T = temperature, K
E = a constant characteristic of reaction called *activation energy*
R = ideal gas constant, 1543 ft·lb/lb-mol·°R (8.314 J/mol·K)

Integration of Eq. (3-18) between the limits T_1 and T_2 gives

$$\ln \frac{k_2}{k_1} = \frac{E(T_2 - T_1)}{RT_1T_2} \tag{3-19}$$

With k_1 known for a given temperature and with E known, k_2 can be calculated from this equation. The activation energy E can be calculated by determining the k at two different temperatures and by using Eq. (3-19). Common values of E for wastewater treatment processes are in the range of 8400 to 84,000 J/mol (2000 to 20,000 cal/mol).

Because most wastewater treatment operations and processes are carried out at or near the ambient temperature, the quantity $E/(RT_1T_2)$ in Eq. (3-19) may be assumed to be a constant for all practical purposes. If the value of the quantity is designated by C, then Eq. (3-19) can be rewritten as

$$\ln \frac{k_2}{k_1} = C(T_2 - T_1) \tag{3-20}$$

$$\frac{k_2}{k_1} = e^{C(T_2 - T_1)} \tag{3-21}$$

Replacing e^C in Eq. (3-21) with a temperature coefficient θ yields

$$\frac{k_2}{k_1} = \theta^{(T_2 - T_1)} \tag{3-22}$$

which is commonly used in the environmental engineering field to adjust the value of the operative rate constant to reflect the effect of temperature. An alternative form of the temperature-correction equation may be obtained by expanding Eq. (3-22) as a series and dropping all but the first two terms. It should be noted, however, that although the value of θ is assumed to be constant, it will often vary considerably with temperature. Therefore, caution must be used in selecting appropriate values for θ for different temperature ranges. Typical values for various operations and processes for different temperature ranges are given, where available, in the sections in which the individual topics are discussed.

Determination of Reaction-Rate Coefficients

Typically, reaction-rate coefficients for the fate processes are determined from the results obtained from batch experiments (no inflow or outflow), from continuous-flow experiments, and from pilot- and field-scale experiments. Using the data from batch experiments, the coefficients can be determined by a variety of methods including (1) the method of integration and (2) the differential method. As summarized in Table 3-1, the method of integration involves the substitution of experimental data on the amount of reactant remaining at various times into the integrated form of the rate expression. In the differential method, where the order of the reaction is unknown, the concentrations remaining at two different times are used to solve the differential form of rate expression for the order of the reaction. Once the reaction order is known, the reaction-rate coefficient is determined by substitution using the test data. The application of these two methods is illustrated in Example 3-1.

TABLE 3-1
Integration and differential methods used to determine reaction rate coefficients

Rate expression	Integrated form	Method used to determine the reaction rate coefficient
	Integration method	
Zero-order reaction:		
$r_C = \dfrac{d[C]}{dt} = -k$	$[C] = [C_o] - kt$	Graphically, by plotting $[C]$ versus t
First-order reaction:		
$r_C = \dfrac{d[C]}{dt} = -k[C]$	$\ln\left[\dfrac{C}{C_o}\right] = -kt$	Graphically, by plotting $-\log[C/C_o]$ versus t
Second-order reaction:		
$r_C = \dfrac{d[C]}{dt} = -k[C]^2$	$\dfrac{1}{[C]} - \dfrac{1}{[C_o]} = kt$	Graphically, by plotting $1/[C]$ versus t
	Differential method	
$r_C = \dfrac{d[C]}{dt} = -k[C]^n$		$n = \dfrac{\log(-d[C_1]/dt) - \log(-d[C_2]/dt)}{\log[C_1] - \log[C_2]}$ Once the order of the reaction is known, the reaction rate coefficient can be determined by substitution

EXAMPLE 3-1. DETERMINATION OF THE REACTION ORDER AND THE REACTION RATE COEFFICIENT. Given the following set of data obtained using a batch reactor, determine the order of the reaction and the reaction rate coefficient using the integration and differential methods.

Time, d	Concentration $[C]$, mol/L
0	100.0
1	71.2
2	51.6
3	37.0
4	25.6
5	19.9
6	13.1
7	9.5
8	6.9

Solution—Part 1

1. Determine the reaction order and the reaction rate constant using the integration method. Develop the data needed to plot the experimental data functionally, assuming the reaction is either first- or second-order.

Time, d	$[C]$, mol/L	$-\log[C/C_o]$	$1/[C]$
0	100.0	0.000	0.010
1	71.2	0.148	0.014
2	51.6	0.287	0.019
3	37.0	0.432	0.027
4	25.6	0.592	0.039
5	19.9	0.701	0.050
6	13.1	0.883	0.076
7	9.5	1.022	0.105
8	6.9	1.161	0.145

2. To determine whether the reaction is first- or second-order, plot $-\log[C/C_o]$ and $1/[C]$ versus t as shown below. Because the plot of $-\log[C/C_o]$ versus t is a straight line on the log plot, the reaction is first-order with respect to the concentration C.

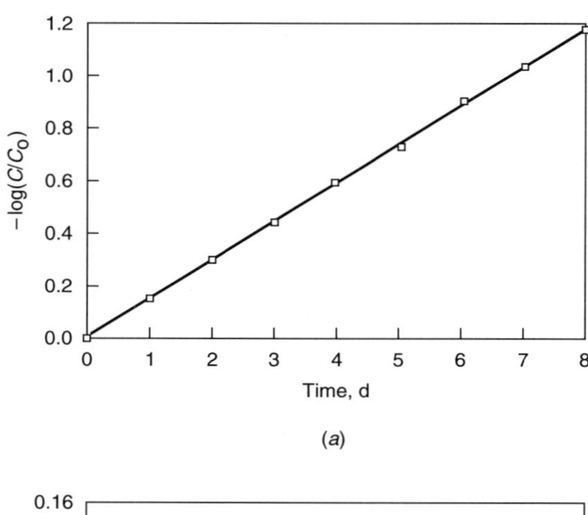

(a)

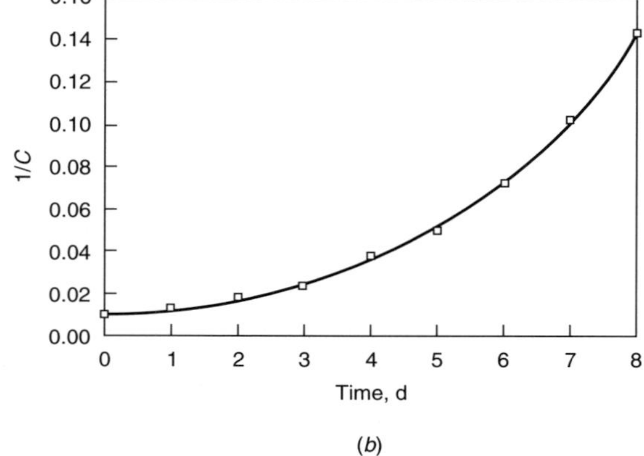

(b)

3. Determine the reaction-rate coefficient.

$$\text{Slope} = k/2.303$$

$$\text{Slope from the plot} = 1.161/8\ d = 0.145\text{d}^{-1}$$

$$k = 2.303 \times 0.144/\text{min} = 0.332\text{d}^{-1}$$

Solution—Part 2. Determine the reaction order and the reaction-rate constant using the differential method:

$$n = \frac{\log(-d[C_1]/dt) - \log(-d[C_2]/dt)}{\log[C_1] - \log[C_2]}$$

1. Use the experimental data obtained at 2 and 5 d:

Time, d	$[C]$, mol/L	$\left(\frac{[C_{t+1}] - [C_{t-1}]}{2}\right) \approx \frac{d[C_t]}{dt}$
2	51.6	[(37.0 − 71.2)/2] ≈ −17.10
5	19.9	[(13.1 − 25.6)/2] ≈ −6.25

2. Substitute and solve for n:

$$n = \frac{\log(17.10) - \log(6.25)}{\log(51.6) - \log(19.9)}$$

$$= \frac{1.23 - 0.80}{1.71 - 1.30} = \frac{0.43}{0.41} = 1.05$$

3. The reaction is first order, use $n = 1$.
4. The reaction rate constant is

$$\ln\left[\frac{C}{C_o}\right] = -kt$$

$$\ln\left[\frac{51.6}{100}\right] = -kt$$

$$k = 0.33\text{d}^{-1}$$

In the applications described above, the initial concentration of a constituent is generally known. However, as noted in Sec. 2-5 in Chap. 2, in the conventional BOD test both UBOD and k_1 are unknown. To determine these values, the usual procedure is to run a series of BOD measurements with time. With these measurements, the UBOD and k_1 values can be determined by a number of methods including the method of least-squares, the method of moments (Moore et al., 1950), the daily-difference method (Tsivoglou, 1958), the rapid-ratio method (Sheehy, 1960), the Thomas method (Thomas, 1950), and the Fujimoto method (Fujimoto, 1961). In the Fujimoto method, which is perhaps the most direct, an arithmetic plot is prepared of BOD_{t+1} versus BOD_t. The value at the intersection of the plot with a line of slope 1 corresponds to the ultimate BOD. After the UBOD has been determined, the rate constant is determined from the standard BOD equation. The application of the Fujimoto method is illustrated in Example 3-2.

EXAMPLE 3-2. DETERMINATION OF UBOD AND k_1 FROM BOD TEST RESULTS. Assuming the difference between the initial concentration at time zero ($t = 0$) and the concentration value remaining at each time step for the data given in Example 3-1 corresponds to the BOD, determine the UBOD and k using the Fujimoto method.

Solution

1. Prepare an arithmetic plot of BOD_{t+1} versus BOD_t, using the data provided expressed as BOD_t (see figure below). Plot a line of slope 1 through the origin. The value at the intersection of the two lines (100 mg/L) corresponds to the ultimate BOD (UBOD).

Time, d	BOD_r mg/L	BOD_t mg/L
0	100.0	0.000
1	71.2	28.8
2	51.6	48.4
3	37.0	63.0
4	25.6	74.4
5	19.9	80.1
6	13.1	86.9
7	9.5	90.5
8	6.9	93.1

2. Determine the reaction-rate constant using Eq. (2-35) for $t = 2$ d:

$$BOD_t = UBOD(1 - e^{-kt})$$

$$48.4 = 100(1 - e^{-kt})$$

$$-0.516 = -e^{-kt}$$

$$k = 0.662/2 = 0.33\text{d}^{-1}$$

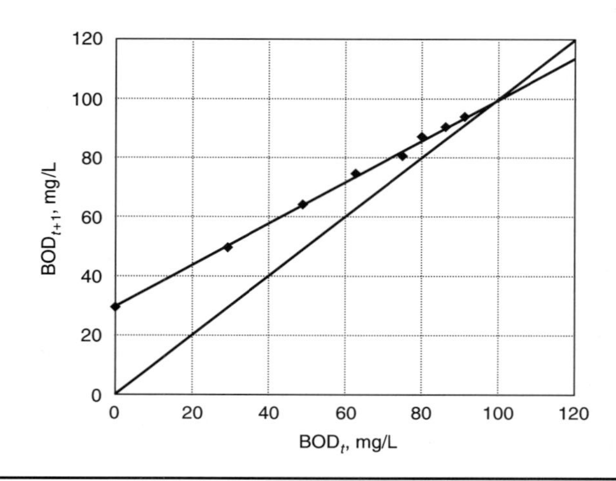

3-3 REACTORS: TYPES, HYDRAULIC CHARACTERISTICS, AND APPLICATIONS

The analysis of the chemical and biological reactions that occur in the environment and in treatment systems is complex, because of the many interactions which tend to confound the results statistically. To understand the nature of the mechanisms that may be operative, models of the constituent transformation and treatment processes are developed. The models are based on a consideration of the fluid system in which the reactions are occurring. In what follows, the types of reactors used for wastewater treatment are described, their hydraulic characteristics are defined, and typical applications for the various reactors are described. The reactions that occur in the various types of reactors are considered in the following section. Operational factors that must be considered in the selection of the type of reactor or reactors to be used for wastewater treatment are considered in Chap. 4.

Types of Reactors

Controlled chemical and biological reactions used for the treatment of wastewater are carried out in containers or tanks commonly known as *reactors*. The principal types of reactors now used include: (1) the batch reactor; (2) the plug-flow reactor, also known as a tubular flow reactor; (3) the complete-mix reactor, also known as a continuous-flow stirred-tank reactor; (4) complete-mix reactors in series; (5) the packed-bed reactor; (6) the fluidized bed reactor; and (7) the upflow sludge-blanket reactor. Descriptions of these reactors are presented in Table 3-2. The classification of the first four reactors is based on their hydraulic characteristics. Homogeneous reactions are usually carried out in such reactors. Heterogeneous reactions are usually carried out in the latter three types of reactors.

Hydraulic Characteristics of Reactors

The hydraulic characteristics of complete-mix, plug-flow, plug-flow with axial dispersion, and complete-mix reactors in series subject to continuous and slug inputs of tracer are considered in the following discussion.

Complete mix. If a continuous flow of a conservative (nonreactive) tracer (e.g., a dye such as rhodamine B) at concentration C_o were injected into the inlet of a complete-mix reactor, initially filled with clear water, the appearance of the tracer at the outlet would be as shown in Fig. 3-2*a*. Analytically, following the approach given in Sec. 3-1, the effluent tracer concentration as a function of time can be determined by writing a mass balance around the reactor.

1. General word statement:

$$\begin{matrix}\text{Rate of accumulation} \\ \text{of tracer within} \\ \text{the reactor}\end{matrix} = \begin{matrix}\text{rate of flow} \\ \text{of tracer into} \\ \text{the reactor}\end{matrix} - \begin{matrix}\text{rate of flow} \\ \text{of tracer out of} \\ \text{the reactor}\end{matrix} \tag{3-23}$$

TABLE 3-2
Principal types of reactors used for the treatment of wastewater

Type of reactor	Identification sketch	Description and/or application
Batch		Flow is neither entering nor leaving the reactor. The liquid contents are mixed completely. For example, the BOD test discussed in Chap. 2 is carried out in a bottle batch reactor.
Plug-flow, also known as tubular-flow		Fluid particles pass through the tank and are discharged in the same sequence in which they enter. The particles retain their identity and remain in the tank for a time equal to the theoretical detention time. This type of flow is approximated in long tanks with a high length-to-width ratio in which longitudinal dispersion is minimal or absent.
Complete-mix, also known as continuous-flow stirred-tank		Complete mixing occurs when the particles entering the tank are dispersed immediately throughout the tank. The particles leave the tank in proportion to their statistical population. Complete mixing can be accomplished in round or square tanks if the contents of the tank are uniformly and continuously distributed.
Complete-mix reactors in series	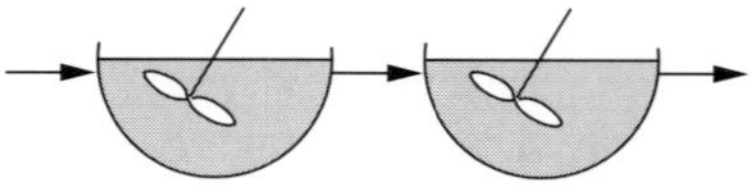	The series of complete-mix reactors is used to model the flow regime that exists between the hydraulic flow patterns corresponding to the complete-mix and plug-flow reactors. If the series is composed of one reactor, the complete-mix regime prevails. If the series consists of an infinite number of reactors in series, the plug-flow regime prevails.
Packed-bed	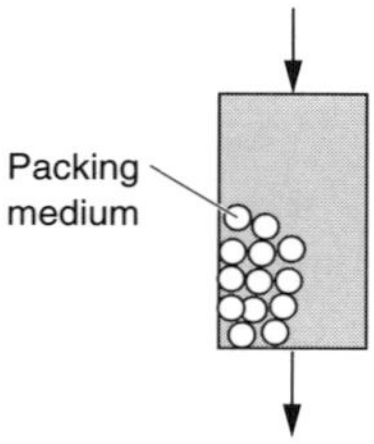 Packing medium	Packed-bed reactors are filled with some type of packing medium, such as rock, slag, ceramic, or plastic. With respect to flow, they can be completely filled (anaerobic filter) or intermittently dosed (trickling filter).
Fluidized-bed	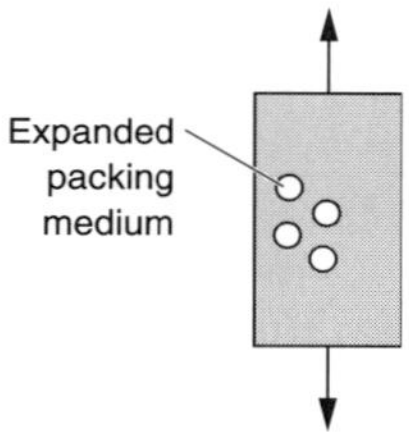 Expanded packing medium	The fluidized-bed reactor is similar to the packed-bed reactor in many respects, but the packing medium is expanded by the upward movement of fluid (air or water) through the bed. The porosity of the packing can be varied by controlling the flowrate of the fluid.

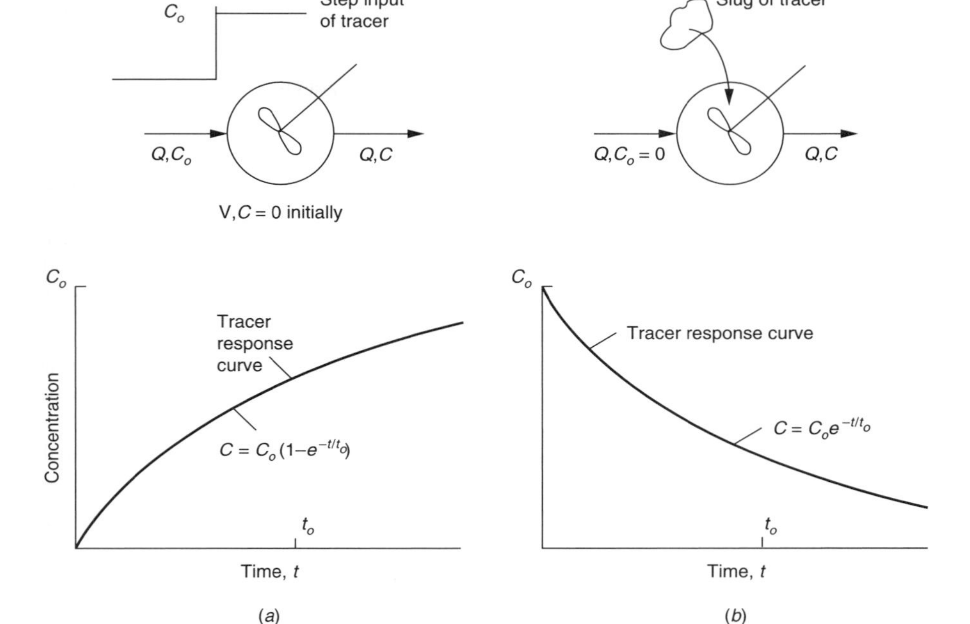

FIGURE 3-2
Concentration of tracer at outlet of complete-mix reactor: (*a*) subject to a constant input tracer at concentration C_o and (*b*) subject to a slug input of tracer.

2. Simplified word statement:

$$\text{Accumulation} = \text{inflow} - \text{outflow} \tag{3-24}$$

3. Symbolic representation (refer to Fig. 3-2):

$$\frac{dC}{dt}V = QC_o - QC \tag{3-25}$$

Rewriting Eq. (3-25) and simplifying yields

$$\frac{dC}{dt} = \frac{Q}{V}(C_o - C) \tag{3-26}$$

Integrating between the limits of $C = C_o$ to $C = C$ and $t = 0$ to $t = t$ yields

$$\int_{C=C_o}^{C=C} \frac{dC}{C_o - C} = \frac{Q}{V}\int_{t=0}^{t=t} dt \tag{3-27}$$

The resulting expression after integration is

$$C = C_o(1 - e^{-t(Q/V)}) = C_o(1 - e^{-t/t_o}) = C_o(1 - e^{-\theta}) \tag{3-28}$$

where t_o = the theoretical detention time, V/Q, and θ = the normalized detention time, t/t_o. It will be noted that Eq. (3-28) has the same form as the BOD equation given previously in Chap. 2 [Eq. (2-37)].

The corresponding response to a slug input of tracer which is mixed instantaneously and is then purged with clear water (see Fig. 3-2*b*) is given by

$$C = C_o e^{-t(Q/V)} = C_o e^{-t/t_o} = C_o e^{-\theta} \tag{3-29}$$

where C_o = the initial concentration of the tracer in the reactor.

Plug flow. The hydraulic characteristics of an ideal plug-flow reactor are illustrated in Fig. 3-3. In the case of the plug-flow reactor, the reactor is initially filled with clear water before being subjected to a continuous step input of tracer. If an observer were positioned at the outlet of the reactor, the appearance of the tracer in the effluent would occur as shown in Fig. 3-3*a* (t equals the actual time, and t_o equals the theoretical detention time V/Q). Under ideal plug-flow conditions, t equals t_o. The effect of an impulse disturbance, such as that caused by a slug injection of tracer uniformly distributed across the reactor, is also shown in Fig. 3-3*b*. To verify the form of the plot given in Fig. 3-3*b*, it will be instructive to prepare a materials balance for an ideal plug-flow reactor (no axial dispersion) in which the concentration C of a nonreactant tracer is distributed uniformly across the cross-sectional area of the control volume. The materials balance for a nonreactive tracer for the differential volume element shown in Fig. 3-4 can be written as follows:

$$\underset{\text{Accum}}{\frac{\partial C}{\partial t}\Delta V} = \underset{\text{inflow}}{QC\big|_x} - \underset{\text{outflow}}{QC\big|_{x+\Delta x}} \tag{3-30}$$

where C = concentration of constituent C, g/m^3
ΔV = differential volume element, m^3
Q = volumetric flow rate, m^3/s
r_c = reaction rate for constituent C, g/m$^3\cdot$s

Substituting the differential form for the term $QC\big|_{x+\Delta x}$ in Eq. (3-30) results in

$$\frac{\partial C}{\partial t}\Delta V = QC - Q\left(C + \frac{\Delta C}{\Delta x}\Delta x\right) \tag{3-31}$$

Substituting $A\,\Delta x$ for ΔV and simplifying yields

$$\frac{\partial C}{\partial t}A\,\Delta x = -Q\frac{\Delta C}{\Delta x}\Delta x \tag{3-32}$$

Dividing by A and Δx yields

$$\frac{\partial C}{\partial t} = -\frac{Q}{A}\frac{\Delta C}{\Delta x} \tag{3-33}$$

Taking the limit as Δx approaches zero yields

$$\frac{\partial C}{\partial t} = -\frac{Q}{A}\frac{\partial C}{\partial x} = -v\frac{\partial C}{\partial x} \tag{3-34}$$

where v = the velocity of flow.

Because both sides of the equation are the same (note $\partial t = \partial x/V$), except for the minus sign, the only way that the equation can be satisfied is if the change in

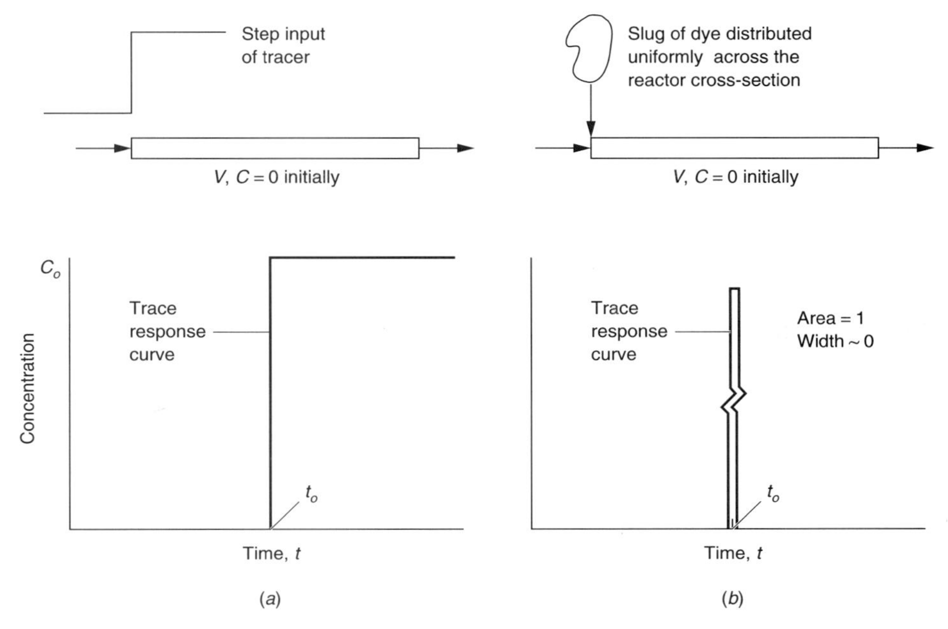

FIGURE 3-3
Concentration of tracer at outlet of plug-flow reactor: (a) subject to step input of tracer and (b) subject to a slug input of tracer.

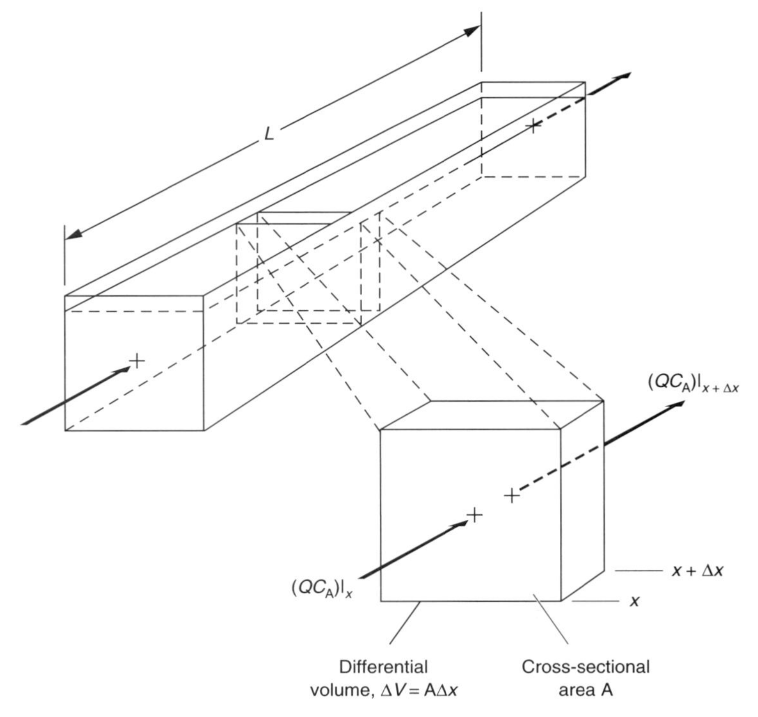

FIGURE 3-4
Definition sketch for the hydraulic analysis of a plug-flow reactor.

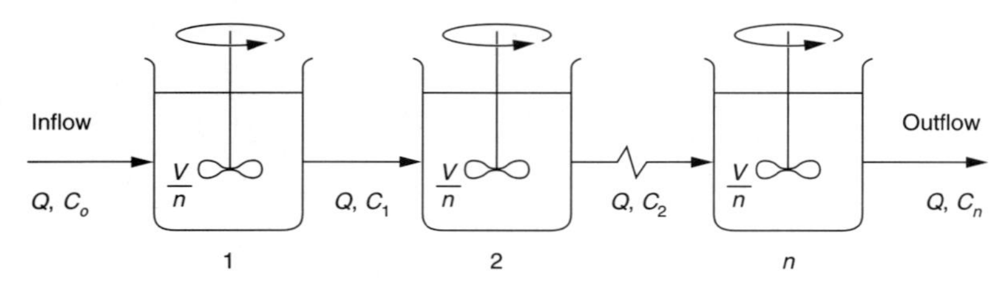

FIGURE 3-5
Definition sketch for the hydraulic analysis of complete-mix reactors in series.

concentration with distance is equal to zero. Thus, the influent concentration must be equal to the effluent concentration, which is consistent with the depiction in Fig. 3-3.

Complete-mix reactors in series. In some situations, the use of a series of complete-mix reactors may have certain advantages with respect to treatment. To understand the hydraulic characteristics of reactors in series (see Fig. 3-5), assume that a slug of tracer is placed into the first reactor of a series of equally sized reactors so that the resulting instantaneous concentration of tracer in the first reactor is C_o. The total volume of all the reactors is V and the volume of an individual reactor is V/n. From Eq. (3-29), the effluent concentration from the first reactor is given by

$$C_1 = C_o e^{-n(Q/V)t} = C_o e^{-n(t/t_o)} = C_o e^{-n\theta} \qquad (3\text{-}35)$$

Writing a materials balance for the second reactor results in the following:

$$\underset{\text{Accum}}{\frac{V}{n}\frac{dC_2}{dt}} = \underset{\text{inflow}}{QC_1} - \underset{\text{outflow}}{QC_2} \qquad \text{or} \qquad \frac{dC_2}{dt} + \frac{nQ}{V}C_2 = \frac{nQ}{V}C_1 \qquad (3\text{-}36)$$

Substituting for C_1 in Eq. (3-36) results in

$$\frac{dC_2}{dt} + \frac{nQ}{V}C_2 = \frac{nQC_o}{V}e^{-n(Q/V)t} \qquad (3\text{-}37)$$

Equation (3-37) can be solved as follows. To solve, gather terms and write Eq. (3-37) in differential form:

$$C_2' + \beta C_2 = \beta C_o e^{-\beta t} \qquad (3\text{-}38)$$

where $C_2' = dC_2/dt$
$\beta = nQ/V$

The most straightforward solution procedure for an ordinary linear first-order differential equation is to use an integrating factor of the form $e^{\beta t}$. Multiplying both sides of Eq. (3-38) by the integrating factor yields

$$e^{\beta t}(C_2' + \beta C_2) = \beta C_o$$

The left-hand side of the above expression can be written as

$$(e^{\beta t}C_2)' = \beta C_o$$

The differential sign is removed by integrating the above expression

$$e^{\beta t}C_2 = \beta C_o \int dt = \beta C_o t + K$$

Dividing by $e^{\beta t}$ yields

$$C_2 = \beta C_o t e^{-\beta t} + K e^{-\beta t}$$

But when $t = 0$, $C_2 = 0$, so the constant of integration $K = 0$; thus

$$C_2 = C_o(nQt/V)e^{-nQt/V} = C_o n\theta e^{-n\theta}$$

Similarly, for reactor 3

$$C_3 = \frac{C_o}{2!}(nQt/V)^2 e^{-(nQt/V)} = \frac{C_o}{2!}(n\theta)^2 e^{-(n\theta)}$$

The generalized expression for the effluent concentration for the ith reactor is

$$C_i = \frac{C_o}{(i-1)!}(nQt/V)^{i-1} e^{-(nQt/V)} = \frac{C_o}{(i-1)!}(n\theta)^{i-1} e^{-(n\theta)} \tag{3-39}$$

The effluent-concentration curves that are obtained by using Eq. (3-39) for 1, 2, 4, 6, and 75 reactors in series are shown in Fig. 3-6. It is interesting to note that a

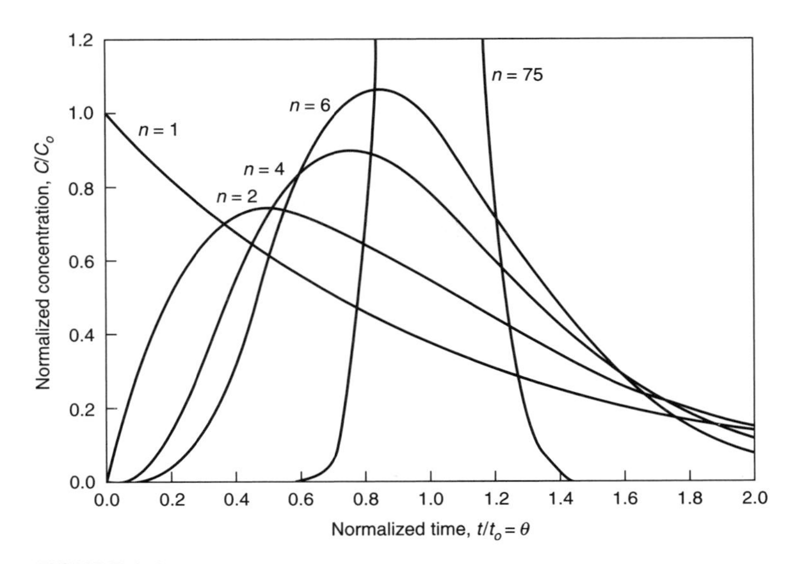

FIGURE 3-6
Effluent tracer concentration curves for reactors in series, subject to a slug input of tracer into the first reactor of the series. Concentration values greater than 1 occur because the same amount of tracer is placed in the first reactor in each series of reactors.

model comprising four complete-mix reactors in series can be used to describe the hydraulic characteristics of constructed wetlands. In Fig. 3-6, the concentration in the first reactor increases because the same amount of tracer is used regardless of the number of reactors in series.

An alternative approach that is often used is to plot the fraction of tracer remaining in the system at any time t. The fraction F of tracer remaining in the system, at any time t, is equal to

$$F = \frac{(V/n)C_1 + (V/n)C_2 + \cdots + (V/n)C_n}{(V/n)C_o}$$

$$= \frac{C_1 + C_2 + \cdots + C_n}{C_o} \tag{3-40}$$

If Eq. (3-39) is used to obtain the individual effluent concentrations, Eq. (3-40) becomes, for four equal-sized reactors in series,

$$F_{4R} = \frac{C_o e^{-4\theta} + C_o(4\theta)e^{-4\theta} + (C_o/2)(4\theta)^2 e^{-4\theta} + (C_o/6)(4\theta)^3 e^{-4\theta}}{C_o}$$

$$= \left[1 + 4\theta + \frac{(4\theta)^2}{2} + \frac{(4\theta)^3}{6}\right] e^{-4\theta} \tag{3-41}$$

The fraction of a tracer remaining in a series of 1, 2, 4, 6, and 75 complete-mix reactors in series is given in Fig. 3-7. The use of such curves is illustrated in Example 3-3.

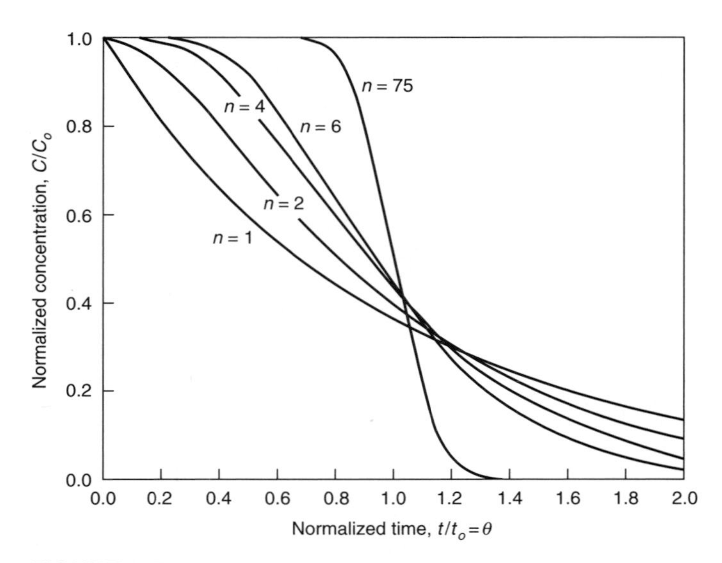

FIGURE 3-7
Fraction of tracer remaining in a system comprising reactors in series.

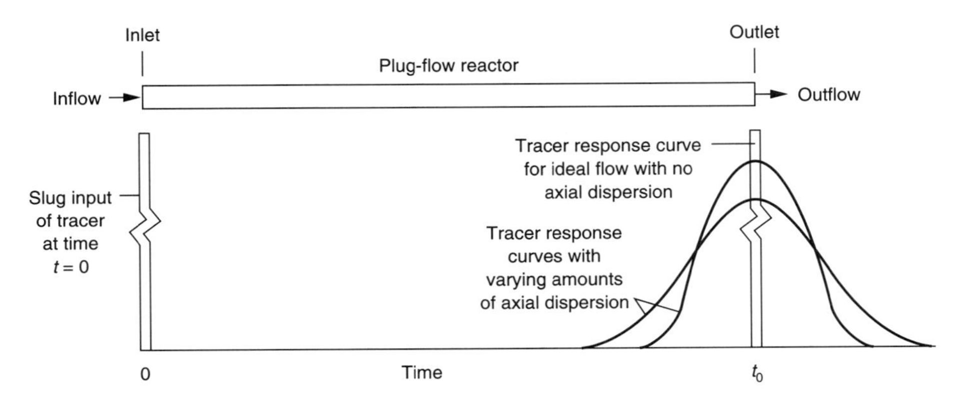

FIGURE 3-8
Axial dispersion in a plug-flow reactor.

Plug flow with axial dispersion. In practice, the flow in plug-flow reactors is seldom ideal. What typically occurs is that a portion of the tracer arrives at the outlet before the bulk of the tracer. The forward movement of the tracer is due to advection and diffusion as described in Sec. 3-4. In a tubular plug-flow reactor (e.g., a pipeline), the arrival of the tracer at the outlet can be reasoned by remembering that the velocity distribution in the pipeline will be parabolic. Depending on the degree of axial dispersion, the effluent distribution will appear as shown in Fig. 3-8. When the dispersion factor becomes infinite, the plug-flow reactor with axial dispersion is equivalent to a complete-mix reactor.

EXAMPLE 3-3. ANALYSIS OF REACTOR HYDRAULIC CHARACTERISTICS. According to the curve given in Fig. 3-7, if a given reactor volume is divided into 75 individual reactors connected in series, 80 percent of the water ($F = 0.8$) will be retained for 90 percent of the theoretical detention time ($\theta = t/t_o = 0.9$). How many reactors in series are required to retain 80 percent of the flow for 70 percent of the theoretical detention time?

Solution

1. Referring to Fig. 3-7, when six reactors in series are used, 80 percent of the flow will be retained for 65 percent of the time. Thus, the number of reactors in series must be greater than six.

2. Try seven or eight reactors in series.
   a. The fraction remaining for seven and eight reactors in series is given by the following two expressions, respectively.

$$F_{7R} = \left[1 + 7\theta + \frac{(7\theta)^2}{2} + \frac{(7\theta)^3}{6} + \frac{(7\theta)^4}{24} + \frac{(7\theta)^5}{120} + \frac{(7\theta)^6}{720}\right] e^{-7\theta}$$

$$F_{8R} = \left[1 + 8\theta + \frac{(8\theta)^2}{2} + \frac{(8\theta)^3}{6} + \frac{(8\theta)^4}{24} + \frac{(8\theta)^5}{120} + \frac{(8\theta)^6}{720} + \frac{(8\theta)^7}{5040}\right] e^{-8\theta}$$

   b. Prepare a computation table of the fraction remaining versus the normalized detention time.

θ	F_{7R}	F_{8R}
0.0	1.00	1.00
0.1	1.00	1.00
0.2	1.00	1.00
0.3	0.99	1.00
0.4	0.98	0.98
0.5	0.93	0.95
0.6	0.87	0.89
0.7	0.78	0.80
0.8	0.67	0.69
0.9	0.56	0.57
1.0	0.45	0.45

c. On the basis of values given in the above table, eight reactors in series will retain 80 percent of the flow for 70 percent of the theoretical detention time (see figure below).

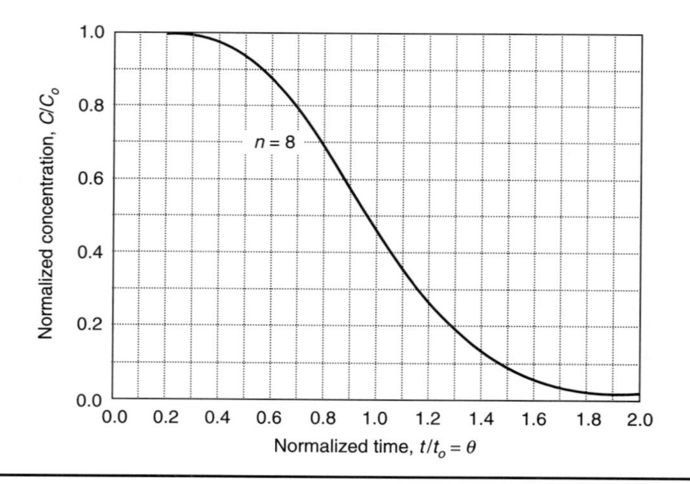

When varying amounts of axial dispersion are encountered, the flow is sometimes identified as "arbitrary flow." This type of flow, which is more difficult to describe mathematically, is often encountered in septic tanks, aeration tanks, and constructed wetlands. The output from a plug-flow reactor with axial dispersion (arbitrary flow) is often modeled as a number of complete-mix reactors in series, as outlined above.

Application of Reactors for Wastewater Treatment

The principal applications of various types of reactors used for wastewater treatment are summarized in Table 3-3 and in Fig. 3-9. Of the reactor types listed in Table 3-3, the complete-mix and plug-flow reactors are used most commonly.

TABLE 3-3
Principal applications of reactor types used for wastewater treatment

Type of reactor	Application in wastewater treatment
Batch	Sequencing batch reactor. Mixing of concentrated solutions into working solutions.
Plug flow	Chlorine contact basin. Activated-sludge biological treatment. Aquatic treatment systems.
Complete mix	Activated-sludge biological treatment.
Complete mix reactors in series	Constructed wetlands. Lagoon treatment systems. Used to simulate nonideal performance of plug-flow reactors.
Packed bed	Nonsubmerged and submerged trickling filters. Land treatment. Intermittent and recirculating sand filters.
Fluidized bed	Fluidized-bed reactors for biological treatment.

(*a*)

(*b*)

FIGURE 3-9
Typical reactor applications: (*a*) plug-flow used for aquatic treatment system (photograph taken shortly after planting) and (*b*) complete-mix used for the activated-sludge process.

3-4 REACTOR TREATMENT KINETICS

In wastewater treatment, the chemical and biological reactions that are needed to bring about the treatment of wastewater are carried out in the reactors described in the previous section. In this section, the focus is on modeling the reactions that occur in the reactors used for wastewater treatment. The reactors considered include: (1) batch, (2) complete mix, (3) complete-mix reactors in series, (4) ideal plug flow, (5) ideal plug flow with retarded reaction rate, and (6) plug flow with axial dispersion.

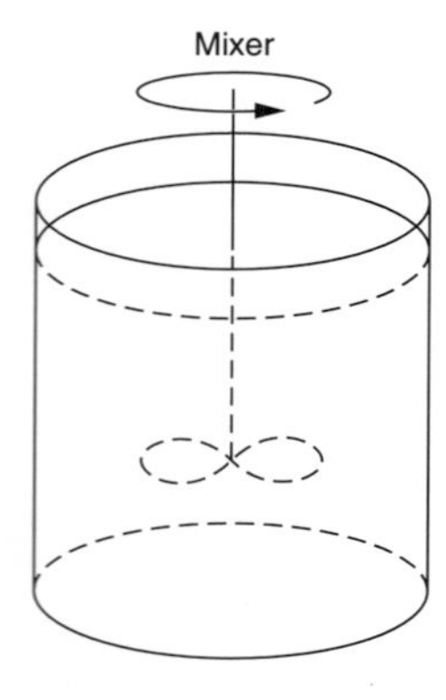

FIGURE 3-10
Schematic of batch reactor. Note there is no inflow or outflow in a batch reactor.

Batch Reactor with Reaction

The derivation of the materials balance equation for a batch reactor, in which the liquid contents are mixed completely, can be illustrated by considering the reactor shown in Fig. 3-10. A materials balance on a reactive constituent is written as follows:

$$\underset{\text{Accum}}{\frac{dC}{dt}V} = \underset{\text{inflow}}{QC_o} - \underset{\text{outflow}}{QC} + \underset{\text{generation}}{r_C V} \tag{3-42}$$

Because $Q = 0$ the resulting equation for a batch reactor is

$$\frac{dC}{dt} = r_C \tag{3-43}$$

Before proceeding further, it will be instructive to explore the difference between the rate-of-change term that appears as part of the accumulation term and the rate-of-generation or decay term. In general, these terms are not equal, except in the special case of a batch reactor, where there is no inflow or outflow. The key point to remember is that when flow is not occurring, the concentration per unit volume is changing according to the applicable rate expression. On the other hand, when flow is occurring, the concentration in the reactor is also being modified by the inflow or outflow from the reactor (Tchobanoglous and Schroeder, 1985).

If the rate of reaction is defined as first order ($r_c = -kC$), integrating between the limits $C = C_o$ and $C = C$ and $t = 0$ and $t = t$ yields

$$\int_{C=C_o}^{C=C} \frac{dC}{kC} = -\int_{t=0}^{t=t} dt = t \tag{3-44}$$

The resulting expression is

$$\frac{C}{C_o} = e^{-kt} \tag{3-45}$$

Equation (3-45) is the same as the BOD equation [Eq. (2-36)] considered previously in Chap. 2.

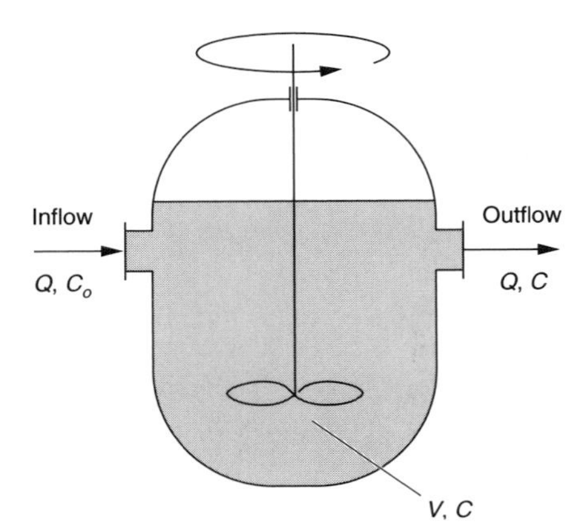

FIGURE 3-11
Schematic of complete-mix reactor with inflow and outflow.

Complete-Mix Reactor with Reaction

The general form of the mass balance equation for a complete-mix reactor as shown in Fig. 3-11, in which the liquid in the reactor is mixed completely, is given below.

$$\frac{dC}{dt}V = QC_o - QC + r_C V \tag{3-46}$$

Assuming first-order removal kinetics ($r_c = -kC$), Eq. (3-46) can be rearranged and written as follows.

$$C' + \beta C = \frac{Q}{V}C_o \tag{3-47}$$

where $C' = dC/dt$
$\beta = k + Q/V$

The non-steady-state (time-variant) solution of Eq. (3-47) can be derived by using the integrating factor method illustrated previously in Sec. 3-3 for complete-mix reactors in series. The solution is given by

$$C = \frac{QC_o}{V\beta}(1 - e^{-\beta t}) + C_o e^{-\beta t} \tag{3-48}$$

where $C = C_o$ at $t = 0$. In most applications in the field of wastewater treatment, the solution of mass balance equations, such as the one given by Eq. (3-48), can be simplified by noting that the long-term (so-called steady-state) concentration is of principal concern.

If it is assumed that only the steady-state effluent concentration is desired, then Eq. (3-46) can be simplified by noting that, under steady-state conditions, the rate accumulation term is equal to zero ($dC/dt = 0$). Using this fact, Eq. (3-46) can be written as

$$0 = QC_o - QC - kCV \tag{3-49}$$

When solved for C, Eq. (3-49) yields the following expression:

$$C = \frac{C_o}{[1 + k(V/Q)]} \tag{3-50}$$

It should also be noted that when $t \rightarrow \infty$, Eq. (3-48) becomes the same as Eq. (3-50).

Complete-Mix Reactors in Series with Reaction

When complete-mix reactors are used in series, the steady-state solution is of concern, as it is used for design. The steady-state form of the mass balance for the second reactor of the two-reactor system (see Fig. 3-5) is given by

$$\frac{dC_2}{dt}\frac{V}{2} = 0 = QC_1 - QC_2 + r_C\frac{V}{2} \tag{3-51}$$

Assuming first-order removal kinetics ($r_c = -kC$), Eq. (3-51) can be rearranged and solved for C_2, yielding

$$C_2 = \frac{C_1}{[1 + (kV/2Q)]}$$

But from Eq. (3-50), the value of C_1 is equal to

$$C_1 = \frac{C_o}{[1 + (kV/2Q)]}$$

Combining the above two expressions yields

$$C_2 = \frac{C_o}{[1 + k(V/2Q)]^2}$$

For n reactors in series, the corresponding expression is

$$C_n = \frac{C_o}{[1 + (kV/nQ)]^n} \tag{3-52}$$

Solving Eq. (3-52) for the detention time yields

$$t_o = \frac{V}{Q}\left(\frac{1}{(C_n/C_o)^{1/n}} - 1\right)\left(\frac{n}{k}\right) \quad \text{or} \quad t_o = \left[\left(\frac{C_o}{C_n}\right)^{1/n} - 1\right]\left(\frac{n}{k}\right) \tag{3-53}$$

Ideal Plug-Flow Reactor with Reaction

The derivation of the materials balance equation for an ideal plug-flow reactor, in which the concentration C of the constituent is uniformly distributed across the cross-sectional area of the control volume, and there is no longitudinal dispersion, can be illustrated by considering the differential volume element shown in Fig. 3-4. For the differential volume element ΔV, the materials balance on a reactive constituent C is written as follows:

$$\frac{\partial C}{\partial t}\Delta V = QC\big|_x - QC\big|_{x+\Delta x} + r_C\,\Delta V \tag{3-54}$$

Accum inflow outflow generation

where $\partial C/\partial t$ = change in concentration with time, $g/m^3 \cdot s$
C = concentration of constituent C, g/m^3
ΔV = differential volume element, m^3
Q = volumetric flow rate, m^3/s
r_C = reaction rate for constituent C, $g/m^3 \cdot s$

Substituting the differential form for the term $QC\big|_{x+\Delta x}$ in Eq. (3-54) results in

$$\frac{\partial C}{\partial t}\Delta V = QC - Q\left(C + \frac{\Delta C}{\Delta x}\Delta x\right) + r_C\,\Delta V \tag{3-55}$$

Substituting $A\,\Delta x$ for ΔV and dividing by A and Δx yields

$$\frac{\partial C}{\partial t} = -\frac{Q}{A}\frac{\Delta C}{\Delta x} + r_C \tag{3-56}$$

Taking the limit as Δx approaches zero yields

$$\frac{\partial C}{\partial t} = -\frac{Q}{A}\frac{\partial C}{\partial x} + r_C \tag{3-57}$$

If steady-state conditions are assumed ($\partial C/\partial t = 0$) and the rate of reaction is defined as $r_c = -kC^n$, integrating between the limits $C = C_o$ and $C = C$ and $x = 0$ and $x = L$ yields:

$$\int_{C=C_o}^{C=C} \frac{dC}{kC^n} = -\frac{A}{Q}\int_0^L dx = -\frac{AL}{Q} = -\frac{V}{Q} = -t_o \tag{3-58}$$

where t_o is the hydraulic detention time. Equation (3-58) is the steady-state solution to the materials balance equation for a plug-flow reactor without dispersion. If it is assumed that n is equal to 1, Eq. (3-58) becomes

$$\frac{C}{C_o} = e^{-kt_o} \tag{3-59}$$

which is equivalent to Eq. (3-45), derived previously for the batch reactor.

Ideal Plug-Flow Reactor with Retarded Reaction

A retarded rate expression is used where the rate constant is changing with distance or time. To illustrate the concept, consider the removal of total suspended solids in a constructed wetland. As shown in Fig. 3-12, the removal rate constant for the original particle size distribution is k_1. If it is assumed that the largest particle size will be removed after the wastewater has passed a unit distance, then the new removal rate constant for the remaining particle size distribution will be k_2. From actual

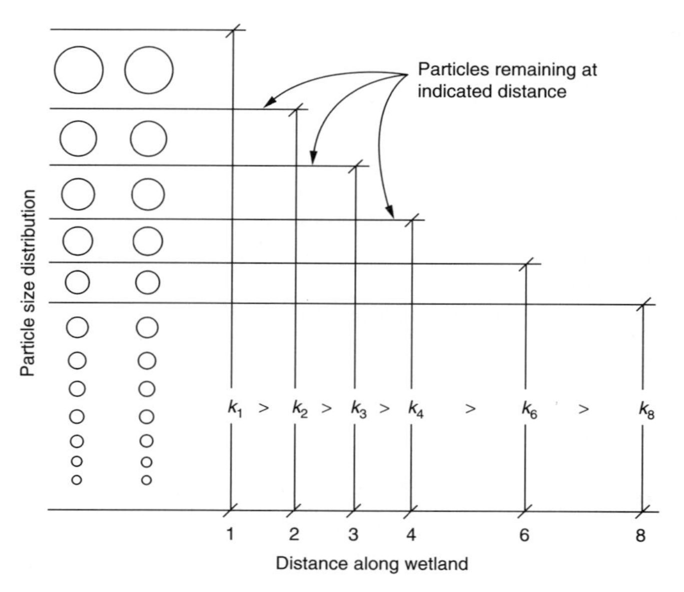

FIGURE 3-12
Definition sketch (idealized) to illustrate the change that can occur in the rate coefficient with treatment (Tchobanoglous, 1969).

observations, it has been found that k_1 is greater than k_2, and that k_2 will be greater than k_3, and so on.

The expression for a plug-flow reactor in which a retarded first-order reaction is occurring can be written as follows:

$$\int_{C=C_o}^{C=C} \frac{dC}{C} = -\int_{t=0}^{t=t} \frac{k}{(1+Rt)^n} dt \tag{3-60}$$

The integrated forms of Eq. (3-60) for $n = 1$ and $n \neq 1$ are given below.

$$C = C_o \exp\left[-\frac{k}{R}\ln(1+Rt)\right] \qquad \text{(for } n = 1\text{)} \tag{3-61}$$

$$C = C_o \exp\left\{-\frac{k}{R(n-1)}\left[1 - \frac{1}{(1+Rt)^{n-1}}\right]\right\} \qquad \text{(for } n \neq 1\text{)} \tag{3-62}$$

The use of Eq. (3-60) in analyzing the effects of retardation is illustrated in Example 3-4.

EXAMPLE 3-4. ANALYSIS OF IMPACT OF REACTION RATE RETARDATION. Derive an expression that can be used to compute the effluent concentration from a constructed wetland, designed as an ideal plug-flow reactor, assuming the removal of the constituent in question can be described by a retarded second-order equation. Assume the exponent n in the retardation term is equal to 1. If the value of the retardation coefficient is 0.5, compare the effluent concentration with and without retardation.

Solution

1. Starting with Eq. (3-60), derive an expression that can be used to compute the effluent concentration:

$$\int_{C=C_o}^{C=C} \frac{dC}{C^2} = -\int_{t=0}^{t=t} \frac{k}{(1+Rt)} dt$$

   a. Integrating the above expression yields

$$\frac{1}{C}\Bigg|_{C=C_o}^{C=C} = \frac{k}{R}\ln(1+Rt)\Bigg|_{t=0}^{t=t}$$

   b. Carrying out the above substitutions and solving for C yields

$$C = \frac{RC_o}{R + kC_o \ln(1+Rt)}$$

2. Determine the effect of retardation:

   a. The expression for the effluent concentration for second-order removal kinetics without retardation is

$$C = \frac{1}{\left(kt + \dfrac{1}{C_o}\right)}$$

   b. Compare effluent concentrations for the following conditions

$$C_o = 1.0$$
$$k = 0.3$$
$$R = 0.6$$
$$t = 1.0$$

$$C_{\text{eff(retarded)}} = \frac{0.6(1.0)}{0.6 + 0.3(1.0)\ln[1 + 0.6(1.0)]} = 0.81$$

$$C_{\text{eff(unretarded)}} = \frac{1}{\left[0.3(1.0) + \dfrac{1}{(1.0)}\right]} = 0.77$$

Comment. From the above computations it can be seen that the effect of retardation is not as significant for a second-order reaction. The impact is much greater for first-order reactions. The effect of retardation in the performance of constructed wetlands is considered in Chap. 9.

Plug-Flow Reactor with Axial Dispersion and Reaction

In most full-scale plug-flow reactors, the flow usually is nonideal because of entrance and exit flow disturbances and axial dispersion. Depending on the magnitude of these effects, the ideal effluent-tracer curves may look like the curves shown in Fig. 3-13. Two approaches are used to model a plug-flow reactor with dispersion. In the first approach, an additional term is added to account for the transport due to axial dispersion. In the second approach, a nonideal plug-flow reactor is modeled as a series of complete mix reactors, as discussed below.

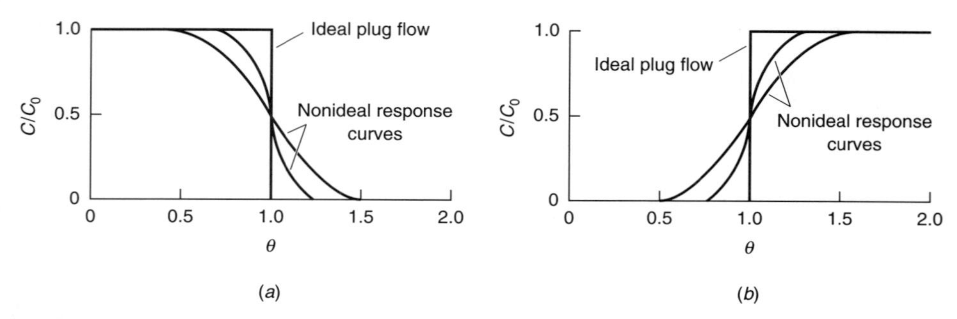

FIGURE 3-13
Ideal and nonideal effluent tracer curves for plug-flow reactor with axial dispersion: (*a*) based on continuous input of a tracer and (*b*) based on continuous step input of a tracer.

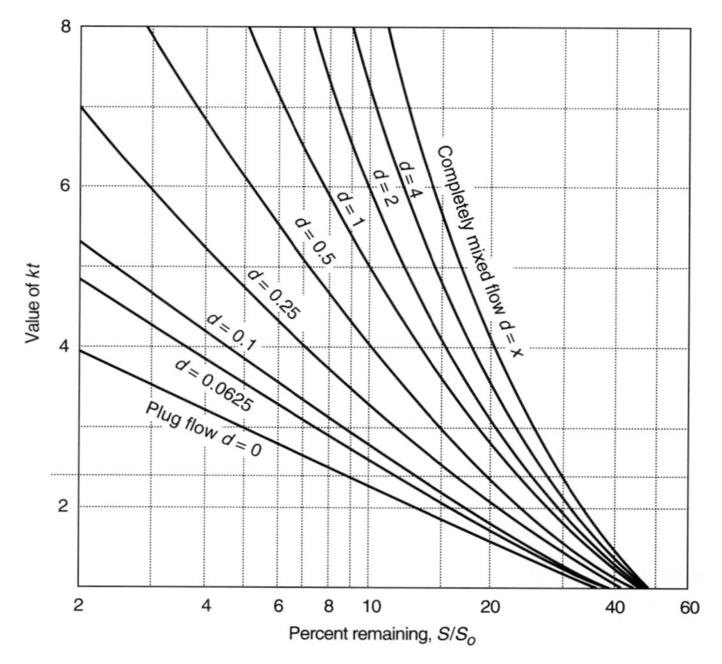

FIGURE 3-14
Graphical solution of Eq. (3-63) in terms of kt_o versus percent remaining for various dispersion factors (from Thirumurthi, 1969).

Using first-order removal kinetics, Wehner and Wilhelm (1958) have developed a solution for a plug-flow reactor with an arbitrary flow-through pattern, varying from complete mix to ideal plug flow. The equation developed by Wehner and Wilhelm is as follows:

$$\frac{C}{C_o} = \frac{4a \exp(1/2d)}{(1 + a)^2 \exp(a/2d) - (1 - a)^2 \exp(-a/2d)} \tag{3-63}$$

where C = effluent concentration
C_o = influent concentration
$a = \sqrt{1 + 4kt_o d}$
d = dispersion factor = D/uL
D = axial dispersion coefficient, ft^2/h (m^2/h)
u = fluid velocity, ft/h (m/h)
L = characteristic length, ft (m)
k = first-order reaction constant, 1/h
t_o = detention time, h

To facilitate the use of Eq. (3-63) for the design of treatment processes such as stabilization ponds and constructed wetlands, Thirumurthi (1969) developed Fig. 3-14, in which the term kt_o is plotted against C/C_o for dispersion factors varying from zero for an ideal plug-flow reactor to infinity for a complete-mix reactor. The application of Fig. 3-14 is illustrated in Example 3-5.

EXAMPLE 3-5. COMPARISON OF THE PERFORMANCE OF A CONSTRUCTED WETLAND DESIGNED AS A PLUG-FLOW REACTOR WITHOUT AND WITH AXIAL DISPERSION. A constructed wetland was designed as an ideal plug-flow reactor using a first-order BOD removal rate constant of $0.5\ d^{-1}$ at 20°C and a detention time of 5 d. Once in operation, a considerable amount of axial dispersion was observed in the wetland. What effect will the observed axial dispersion have on the performance of the wetland? The dispersion factor for the wetland, d, has been estimated to be about 0.5.

Solution

1. Estimate the percentage removal for an ideal plug-flow reactor using Eq. (3-59):

$$\frac{C}{C_o} = e^{-kt_o}$$

$$= e^{-0.5\times 5} = 0.082 = 8.2\%$$

Percentage removal $100 - 8.2 = 91.8\%$.

2. Determine the percentage removal for the constructed wetland using Fig. 3-14.
   a. The value of kt_o is

$$kt_o = (0.5/\text{d} \times 5\ \text{d}) = 2.5$$

   b. The percent remaining from Fig. 3-14 is equal to

$$C/C_o = 0.20 = 20\%$$

Percentage removal $100 - 20 = 80.0\%$.

Comment. Clearly, axial dispersion can affect the predicted design performance of a constructed wetland. Because of axial dispersion and temperature effects, the performance of constructed wetlands and waste stabilization ponds has often been less than expected. A design approach that can be used to take into account these effects is given in Sec. 4-8, Chap. 4.

3-5 FATE OF CONSTITUENTS RELEASED TO THE ENVIRONMENT

To understand what degree of wastewater treatment will be required to sustain the environment, it will be necessary to consider the transport and fate of constituents in the environment. Constituents released to the environment are subject to a variety of transport and transformation processes and operations that can alter their composition. Transport processes generally affect all water quality parameters in the same way, whereas fate and transformation processes are constituent-specific.

The physical, chemical, and biological processes that control the fate of the constituents discharged to water bodies are numerous and varied. The focus of this section is to introduce the important transformations that occur naturally in the environment. The fate of untreated and treated wastewater constituents released in the environment can be illustrated by considering: (1) the biochemical cycles and (2) the transformation and removal mechanisms that occur in nature. Because of their potentially deleterious effects, the long-term persistence of constituents released to

FIGURE 3-15
Aerobic carbon, nitrogen, and sulfur cycle in nature. The state of the carbon, nitrogen, and sulfur after each transformation is identified in the circles. The transformations that occur between states are represented by the connecting lines (adapted from Fair and Geyer, 1954).

the environment is also considered. Additional details on the fate of constituents released to the environment may be found in Hemond and Fechner (1994), Thibodeaux (1979), Schnoor (1996), and Thomann and Mueller (1987).

Biochemical Cycles in Nature

The biochemical cycles of carbon, nitrogen, phosphorus, and sulfur are of fundamental importance in sustaining life on earth. Understanding the biochemical transformations that occur in nature is necessary in the development of treatment processes in which similar transformations are utilized for the treatment of wastewater under controlled conditions.

The aerobic cycle. The growth and decay of organic material in nature occurs under aerobic and anaerobic conditions. In the aerobic cycle (see Fig. 3-15), oxygen is used for the oxidation of the organic material. In the anaerobic cycle (see Fig. 3-16), discussed below, the decay of the organic material occurs in the absence

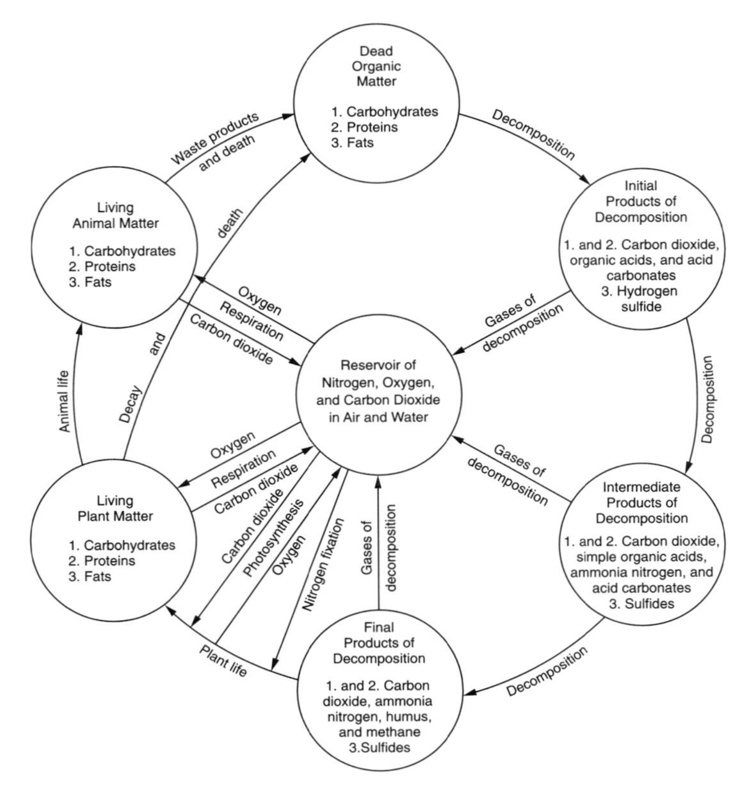

FIGURE 3-16
Anaerobic carbon, nitrogen, and sulfur cycle in nature. The state of the carbon, nitrogen, and sulfur after each transformation is identified in the circles. The transformations that occur between states are represented by the lines connecting the circles (adapted from Fair and Geyer, 1954).

of oxygen. The aerobic carbon cycle in its simplest form, as shown in Fig. 3-15, involves the conversion of carbon dioxide to organic carbon compounds in living plant matter and microorganisms and the subsequent conversion of the organic carbon back to carbon dioxide. Dead plant and animal organic matter is first broken down into initial and intermediate products, before the final stabilized products are obtained. Both chemorganotrophic and chemolithotrophic bacteria are involved in the many biodegradation processes required to obtain the final stabilized products. In the aerobic carbon cycle, the final products of degradation are oxidized more fully and hence are at a lower energy level than the final products of the anaerobic carbon cycle. This lower energy is explained by the fact that much more energy is released in aerobic than in anaerobic degradation. The portion of the cycle involved with the building or synthesis of the organic matter necessary for plant and animal life is the same in both the aerobic and anaerobic cycles. Also, as shown in Figs. 3-15 and 3-16, the elements of nitrogen and sulfur, considered separately below, are included because they are integral parts of the two cycles. It is important to remember that these elements are not the only ones necessary for the growth and maintenance of living organisms (Fair and Geyer, 1954).

The anaerobic cycle. The anaerobic carbon cycle is shown in Fig. 3-16. When the available oxygen is depleted, facultative and anaerobic organisms supplant aerobic organisms in the decomposition of the dead and decaying plant and animal matter. With the exception of the portion of the cycle dealing with the decomposition of dead plant and animal organic matter, the anaerobic cycle is the same as the aerobic cycle. As shown in Fig. 3-16, the initial decomposition products under anaerobic conditions are decomposition gases including ammonia, carbon dioxide, and hydrogen sulfide; organic acids; and acid carbonates. Intermediate products include decomposition gases, simpler organic acids, acid carbonates, and sulfides. Final decomposition products consist of decomposition gases including methane, humus material, and sulfides. The humus material remaining after anaerobic decomposition typically contains the inorganic elements necessary for life, allowing the cycle to continue (Fair and Geyer, 1954).

Nitrogen cycle. The various forms of nitrogen that are present in nature and the pathways by which the forms are changed are depicted in Figs. 3-15 and 3-16. As illustrated in the figures, nitrogen is cycled between organic and inorganic forms. In an aerobic environment, the decomposition of dead plant and animal protein by bacteria yields ammonia. In turn, ammonia is oxidized to nitrite and nitrate. Nitrate released to the environment along with ammonia fixed from the atmosphere is then assimilated into plant tissue containing organic nitrogen. In turn, plant tissue and microorganisms are consumed and converted to animal protein, thus completing the cycle. Although the group of microorganisms that can fix ammonia from nitrogen gas in the atmosphere is small, they are of critical importance, especially where nitrogen is limited.

The nitrogen present in fresh wastewater is contained primarily in proteinaceous matter and urea. Decomposition by bacteria readily changes both forms to ammonia. Significant conversion of proteinaceous matter and urea occurs in long, flat collection systems, as evidenced by the increased ammonia concentrations measured at the point of treatment as compared to fresh wastewater. Further, because nitrogen in the form of nitrates can be used to make protein by algae and other plants, it may be necessary to remove or to reduce the nitrogen that is present to prevent these growths. For example, nitrate must be removed from repurified water that is to be blended with other waters in water supply storage reservoirs to limit the growth of algae.

Phosphorus cycle. Like nitrogen, phosphorus in the environment cycles between organic and inorganic forms. Organic compounds containing phosphorus are found in all living matter. Orthophosphate (PO_4^{-3}) is the only form of phosphorus that is used readily by most plants and microorganisms. The phosphorus cycle involves two major steps, each of which is bacterially mediated: (1) conversion of organic to inorganic phosphorus and (2) conversion of inorganic to organic phosphorus. Conversion of insoluble forms of phosphorus such as calcium phosphate [$Ca_2(HPO_4)_2$] into soluble forms, principally PO_4^{-3}, is also carried out by microorganisms. Referring to the nitrogen cycle shown in Fig. 3-16, the organic phosphorus in dead plant and animal tissue and animal waste products is converted bacterially in a series of steps to PO_4^{-3}. The PO_4^{-3} released to the environment is then incorporated into plant and animal tissue containing phosphorus, thus completing the cycle.

Because the concentration of phosphorus in many natural water environments such as inland lakes and reservoirs is low, algal growth is usually limited. However, when releases containing excessive amounts of phosphorus such as septic tank effluent, untreated and partially treated wastewater, and irrigation return waters are added, algal blooms have often resulted. Because the problem of controlling waste releases from individual onsite systems is difficult, especially in those locations without a management district, collection systems have been used to serve homes fronting on sensitive water bodies.

Sulfur cycle. Sulfur is a component of all living matter, principally as a component of some amino acids in proteins. The principal forms of sulfur that are of special significance in water quality management are organic sulfur, hydrogen sulfide (H_2S), elemental sulfur (S), and sulfate (SO_4^{-2}). The interrelationships of these forms of sulfur are depicted in Figs. 3-15 and 3-16. The sulfur cycle is similar to the nitrogen cycle in that sulfate, like nitrate, is taken up by plants and microorganisms for the production of cell tissue. In turn, plants and microorganisms are consumed by animals. The ability to produce hydrogen sulfide from waste products and dead or decaying proteins is common to many organisms. In the presence of free oxygen, hydrogen sulfide is oxidized rapidly to elemental sulfur. Under aerobic conditions, sulfide is oxidized to sulfur and then to sulfate by photosynthetic sulfur bacteria and colorless blue-green algae. An important part of the sulfur cycle is the reduction of sulfate to hydrogen sulfide. The reduction of sulfate to sulfide is carried out by a small group of anaerobic bacteria known collectively as *sulfate-reducing bacteria.* A few bacteria are also capable of reducing elemental sulfur to sulfide.

The sulfur cycle is especially important in wastewater management with respect to organism survival, odor production, and bacterial/chemical corrosion. Because

TABLE 3-4
Constituent transformation and removal processes in the environment

Process	Comments
Adsorption/desorption	Many chemical constituents tend to attach or sorb onto solids. The implication for wastewater discharges is that a substantial fraction of some toxic chemicals are associated with the suspended solids in the effluent. Adsorption combined with solids settling results in the removal from the water column of constituents that might not otherwise decay.
Algal synthesis	The synthesis of algal cell tissue using the nutrients found in wastewater.
Bacterial conversion	Bacterial conversion (both aerobic and anaerobic) is the most important process in the transformation of constituents released to the environment. The exertion of BOD and NOD are the most common examples of bacterial conversion encountered in water quality management. The depletion of oxygen in the aerobic conversion of organic wastes is also known as *deoxygenation.* Solids discharged with treated wastewater are partly organic. Upon settling to the bottom, they decompose bacterially either anaerobically or aerobically, depending on local conditions. The bacterial transformation of toxic organic compounds is also of great significance.
Chemical reactions	Important chemical reactions that occur in the environment include hydrolysis, photochemical, and oxidation-reduction reactions. Hydrolysis reactions occur between contaminants and water.
Filtration	Removal of suspended and colloidal solids by straining (mechanical and chance contact), sedimentation, interception, impaction, and adsorption.
Flocculation	Flocculation is the term used to describe the aggregation of smaller particles into larger particles that can be removed by sedimentation and filtration. Flocculation is brought about by differential velocity gradients which allow some particles to overtake other particles and form larger particles.
Gas absorption/desorption	The process whereby a gas is taken up by a liquid is known as *absorption.* For example, when the dissolved-oxygen concentration in a body of water with a free surface is below the saturation concentration in the water, a net transfer of oxygen occurs from the atmosphere to the water. The rate of transfer (mass per unit time per unit surface area) is proportional to the amount by which the dissolved oxygen is below saturation. The addition of oxygen to water is also known as *reaeration.* Desorption occurs when the concentration of the gas in the liquid exceeds the saturation value, and there is a transfer from the liquid to the atmosphere.
Natural decay	In nature, contaminants will decay for a variety of reasons, including mortality in the case of bacteria and photooxidation for certain organic constituents. Natural decay follows first-order kinetics.
Photochemical reactions	Solar radiation is known to trigger a number of chemical reactions. Radiation in the near-ultraviolet (UV) and visible range is known to cause the breakdown of a variety of organic compounds.
Photosynthesis/respiration	During the day, algal cells in water bodies will produce oxygen by means of photosynthesis. Dissolved oxygen concentrations as high as 30 to 40 mg/L have been measured. During the evening hours algal respiration will consume oxygen. Where heavy growths of algae are present, oxygen depletion has been observed during the evening hours.

TABLE 3-4
(Continued)

Process	Comments
Sedimentation	The suspended solids discharged with treated wastewater ultimately settle to the bottom of the receiving water body. This settling is enhanced by flocculation and hindered by ambient turbulence. In rivers and coastal areas, turbulence is often sufficient to distribute the suspended solids over the entire water depth.
Sediment oxygen demand	The residual solids discharged with treated wastewater will, in time, settle to the bottom of streams and rivers. Because the particles are partly organic, they can be decomposed anaerobically as well as aerobically, depending on conditions. Algae which settle to the bottom will also be decomposed, but much more slowly. The oxygen consumed in the aerobic decomposition represents another dissolved oxygen demand in the water body.
Volatilization	Volatilization is the process whereby liquids and solids vaporize and escape to the atmosphere. Organic compounds that readily volatilize are known as VOCs (volatile organic compounds). The physics of this phenomenon is very similar to gas absorption, except that the net flux is out of the water surface.

hydrogen sulfide is toxic to many organisms, the reduction of sulfate to sulfide, which can then accumulate in the environment, is undesirable. Hydrogen sulfide can also combine and precipitate heavy metals such as iron, zinc, and cobalt. Because these elements are required for proper bacterial growth, high levels of H_2S may also inhibit bacterial growth. When hydrogen sulfide is produced, the potential for the release of odors always exists. When hydrogen sulfide is produced in collection systems it can lead to their deterioration, especially if they are constructed of reinforced-concrete pipe (U.S. EPA, 1985).

Constituent Transformation and Removal Processes

Bacterially mediated transformations have been introduced and discussed in connection with the discussion of biochemical cycles (see Figs. 3-15 and 3-16). In addition to the bacterially mediated transformation processes, a number of other removal mechanisms are operative in most water bodies. The principal transformation and removal processes that affect constituents discharged to the environment are listed in Table 3-4. The constituents affected by the transformation and removal processes described in Table 3-4 are listed in Table 3-5. The relative importance of the individual transformation and removal processes will be site-specific, and will depend on the water quality parameter under evaluation. For example, deoxygenation brought about by bacterial activity, surface reaeration, sediment oxygen demand, and photosynthesis/respiration are of major importance in assessing the oxygen resources of a stream.

TABLE 3-5
Constituents in wastewater affected by the processes given in Table 3-4

Process	Constituents affected
Adsorption/desorption	Metals; trace organics; NH_4^+; PO_4^{-3}
Algal synthesis	NH_4^+, PO_4^{-3}, pH, etc.
Bacterial conversion—aerobic/anaerobic	BOD_5; nitrification; denitrification; sulfate reduction; anaerobic fermentation (in bottom sediments), conversion of priority organic pollutants, etc.
Chemical reactions (hydrolysis, ion exchange, oxidation/reduction, etc.)	Decomposition of organic compounds, specific ion exchange, element substitution
Filtration	TSS; colloidal particles
Flocculation	Suspended matter (SS); colloidal particles
Gas absorption/desorption	O_2, CO_2, CH_4, NH_3, H_2S
Natural decay	Plants; animals; protists (algae, fungi, protozoa); eubacteria (most bacteria); archaebacteria; viruses; radioactive substances; plant mass
Photochemical reactions	Oxidation of inorganic and organic compounds
Photosynthesis/respiration	Algae, duckweed; submerged macrophytes; NH_4^+; PO_4^{-3}; pH; etc.
Sedimentation	TSS
Sediment oxygen demand	O_2, particulate BOD
Volatilization	VOCs; NH_3; CH_4; H_2S, other gases

Along with the identification of the constituent transformation processes, it is important to understand the time frame over which these transformations occur. The rate at which transformations occur is related to the rate at which reactions occur in solution or the rate at which material is transferred across a surface. Typical rate expressions that have been used to study the transformation and removal processes reported in Table 3-4 are given in Table 3-6. A typical application of one of the rate expressions given in Table 3-6 is illustrated in Example 3-6.

In applying the reaction-rate expressions given in Table 3-6, it is important to note the distinction between volume- and area-based reaction-rate coefficients. As noted in Sec. 3-2, volume-based coefficients are developed on the assumption that the reaction is occurring uniformly throughout the volume. Area-based coefficients account for the fact that mass transfer is occurring across a surface. The distinction is especially important in the analysis of constructed wetlands (see discussion in Sec. 9-2, Chap. 9).

TABLE 3-6
Typical rate expressions for selected processes given in Table 3-4*

Process	Rate expression	Comments
Bacterial conversion	$r_c = -k\text{BOD}$	r_c = rate of conversion, M/L^3T k = first-order reaction-rate coefficient, 1/T BOD = amount of organic material remaining, M/L^3
Chemical reactions	$r_c = \pm kC^n$	r_c = rate of conversion, M/L^3T k = reaction-rate coefficient, $(M/L^3)^{n-1}/T$ C = concentration of constituent, $(M/L^3)^n$ n = reaction order (e.g., for second order $n = 2$)
Gas absorption/ desorption	$r_{ab} = k_{ab}\frac{A}{V}(C_s - C)$/ $r_{de} = -k_{de}\frac{A}{V}(C - C_s)$	r_{ab} = rate of absorption, M/L^3T r_{de} = rate of desorption, M/L^3T k_{ab} = coefficient of absorption, L/T k_{de} = coefficient of desorption, L/T A = area, L^2 V = volume, L^3 C_s = saturation concentration of constituent in liquid, M/L^3 (see Eq. 2-29) C = concentration of constituent in liquid, M/L^3
Natural decay	$r_d = -k_d N$	r_d = rate of decay, no./T k_d = first-order reaction rate coefficient, 1/T N = amount of organisms remaining, number
Sedimentation	$r_s = \frac{v_s}{H}(\text{SS})$	r_s = rate of sedimentation, 1/T v_s = settling velocity, L/T H = depth, L SS = settleable solids, L^3/L^3
Volatilization	$r_v = -k_v(C - C_s)$	r_v = rate of volatilization per unit time per unit volume, M/L^3T k_v = volatilization constant, 1/T C = concentration of constituent in liquid, M/L^3 C_s = saturation concentration of constituent in liquid, M/L^3 (see Eq. 2-29)

*Adapted in part from Ambrose et al. (1988), Tchobanoglous and Burton (1991).

EXAMPLE 3-6. VOLATILIZATION OF AN ORGANIC COMPOUND. Develop an expression that can be used to predict the rate of volatilization of an organic compound from the surface of a lake. Assume that the lake is well mixed vertically, and that the rate of volatilization can be described by the following relationship, as given in Table 3-6.

$$r_v = -k_v(C - C_s)$$

where r_v = rate of mass transfer (volatilization), $g/m^2 \cdot h$
k_v = coefficient of mass transfer, m/h
C = contaminant concentration, g/m^3
C_s = saturation concentration of contaminant, g/m^3

Solution

1. Write a mass balance for a volume element of a lake, assuming no inflow or outflow and that the liquid in the volume element is mixed completely.

$$\underset{\text{Accum}}{\frac{dC}{dt}(A_s h)} = \underset{\text{inflow}}{0} - \underset{\text{outflow}}{0} + \underset{\substack{\text{outflow due} \\ \text{to evaporation}}}{r_v A_s}$$

where A_s = surface area, m^2; h = depth, m; and other terms are as defined above.

2. Substituting for r_v in the mass balance and simplifying yields

$$\frac{dC}{dt} = -\frac{k_v}{h}(C - C_s)$$

3. The integrated form of the above expression is given by

$$\frac{C_t - C_s}{C_o - C_s} = e^{-(k_v/h)t}$$

where C_o = concentration of contaminant at time $t = 0$, g/m^3; C_t = concentration of contaminant at time t, g/m^3; and other terms are as defined previously.

Long-Term Persistence of Constituents in the Environment

The environmental persistence of potentially hazardous constituents is one of the critical issues in the long-term management of these constituents in wastewater. The half-life concept can be used to characterize and compare the relative environmental persistence of various constituents. At the relatively low concentrations encountered in wastewater, the decay of an individual constituent can be described adequately as a first-order function as follows.

$$\frac{dC}{dt} = -k_T C \tag{3-64}$$

where C = concentration at time t
t = time
k_T = first-order reaction rate constant

The integrated form of Eq. (3-64) is

$$\ln \frac{C_o}{C} = k_T t \tag{3-65}$$

where C_o = concentration at time zero.

When half of the initial material has decayed away, C_o/C is equal to 2; the corresponding time is given by the following expression.

$$t_{1/2} = \frac{\ln 2}{k_T} = \frac{0.693}{k_T} \tag{3-66}$$

The half-life time is often used to characterize the environmental persistence of resistant waste constituents. The application of the half-life concept is illustrated in Example 3-7.

EXAMPLE 3-7. EVALUATION OF CONTAMINANT PERSISTENCE IN THE ENVIRONMENT. Determine the time required for the concentrations of toluene and dieldrin spilled in a shallow wastewater treatment pond to be reduced to one-half their initial values. Assume the first-order removal constants for toluene and Dieldrin are 0.0665/h and 2.665×10^{-5}/h, respectively.

Solution. Using the data given in the problem statement and Eq. (3-66), determine the time required for the concentrations in the treatment pond to reach one-half their original values.

1. For toluene

$$t_{1/2} = \frac{0.693}{k_T} = \frac{0.693}{0.0665/\text{h}}$$

$$= 10.4 \text{ h}$$

2. For Dieldrin

$$t_{1/2} = \frac{0.693}{k_T} = \frac{0.693}{2.665 \times 10^{-5}/\text{h}}$$

$$= 26{,}000 \text{ h} \approx 3 \text{ yr}$$

Comment. The time required for the concentration of Dieldrin to reach one-half of the initial value can be used as an argument for the development and use of agricultural chemicals that are more readily broken down in the environment.

3-6 MODELING THE FATE OF CONSTITUENTS RELEASED TO WATER BODIES

As noted previously, the fate of constituents released to the environment is governed by transport and transformation processes. Transformation processes were considered in Sec. 3-5. Transport processes move constituents from location to location with the movement of the fluid. The use of simplified mathematical models to determine the fate of constituents released to lakes and reservoirs and streams is illustrated in the following paragraphs. The availability and application of more complex mathematical models is discussed at the end of this section.

Transport Processes

After initial dilution, contaminants discharged to a water body are transported by two basic processes: (1) advection and (2) dispersion (see Fig. 3-17). *Advection* is

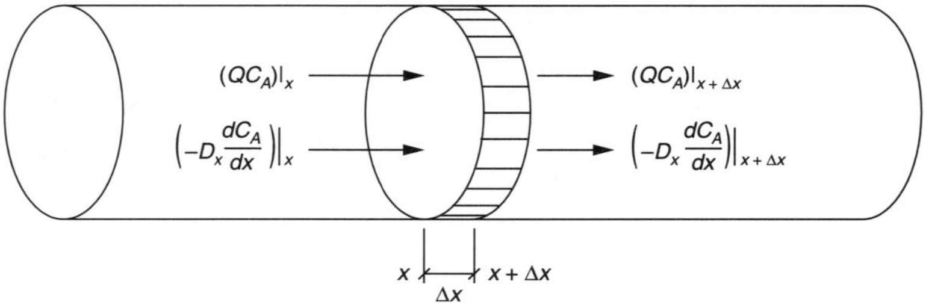

FIGURE 3-17
Schematic representation of advection and dispersion in one dimension.

the term applied to the transport of a constituent resulting from the flow of the water in which the constituent is dissolved or suspended. Turbulent velocity fluctuations, in conjunction with concentration gradients and molecular diffusion, lead to a mass transport phenomenon called *dispersion.*

Advection. Referring to Fig. 3-17, the transport of constituent A by advection in the x direction into and out of the control volume can be represented as:

$$\text{Rate of mass entering control volume due to advection} = QC\big|_x$$

$$\text{Rate of mass leaving control volume due to advection} = QC\big|_{x+\Delta x} = Q\left(C + \frac{\partial C}{\partial x}dx\right)$$

Diffusion. Turbulent velocity fluctuations, in conjunction with concentration gradients, lead to a mass transport phenomenon called diffusion, which can be described as local mixing by turbulent eddies. The rate of mass transport is proportional to the concentration gradient (or longitudinal rate of concentration variation). The mass transport due to diffusion is most commonly represented as follows:

$$M_{Ax} = -D_x\frac{\partial C_A}{\partial x} \tag{3-67}$$

where M_{Ax} = mass flux of material A in x direction, M/L^2T
D_x = diffusion coefficient (also called diffusivity) in the x direction, L^2/T
C_A = concentration of material A, M/L^3

Referring to Fig. 3-17, the transport of constituent A by diffusion in the x direction into and out of the control volume can be represented as

$$\text{Rate of mass entering control volume due to diffusion} = -A_xD_x\left.\frac{\partial C_A}{\partial x}\right|_x$$

$$\text{Rate of mass leaving control volume due to diffusion} = -A_xD_x\left.\frac{\partial C_A}{\partial x}\right|_{x+\Delta x}$$

$$= -\left[A_xD_x\frac{\partial C_A}{\partial x} + A_x\frac{\partial}{\partial x}\left(D_x\frac{\partial C_A}{\partial x}\right)\right]$$

where A_x = cross sectional area in the x direction, L^2.

Combined effects of advection and diffusion. The combined effects of advection and diffusion on the fate of a nonconservative constituent can be modeled by considering a one-dimensional mass balance for the control volume given in Fig. 3-17:

$$\frac{\partial C_A}{\partial t} A_x \, \Delta x = \left(u A_x C_x - A_x D_x \frac{\partial C_A}{\partial x} \right)\bigg|_x - \left(u A_x C_x - A_x D_x \frac{\partial C_A}{\partial x} \right)\bigg|_{x+\Delta x} + r_A A_x \, \Delta x \tag{3-68}$$

where u = average velocity in the x direction, L/T, and r_A = rate of reaction of material A, M/L^3T.

Taking the limit as Δx approaches zero yields the following expression:

$$\frac{\partial C_A}{\partial t} = D_x \frac{\partial^2 C_A}{\partial x^2} - u \frac{\partial C_A}{\partial x} + r_A \tag{3-69}$$

In the above discussion, a one-dimensional analysis was considered. In reality, diffusion occurs in three directions. Further, diffusion varies with direction and the degree of turbulence. Historically, one-dimensional models have been used to study the fate of constituents in rivers and streams with relatively constant cross-sectional areas. Clearly, the fate of constituents released to estuaries near ocean shorelines and in deeper ocean waters must be modeled by using three dimensional models (Thomann and Mueller, 1987). A number of researchers have developed approximations that can be used. If the diffusion term in Eq. (3-69) is neglected, as can often be done, the resulting equation is the same as Eq. (3-57) derived for plug flow.

Modeling the Fate of Constituents Released in Lakes and Reservoirs

In locations where streams are not available, it may be necessary to release treated wastewater into lakes or reservoirs. In some locations, the effluent from septic tanks and stormwater runoff, which may contain BOD, nutrients, and other constituents, has been released to lakes and reservoirs. The modeling of the fate of constituents in lakes and reservoirs depends on the water depth. Small and shallow lakes and reservoirs tend to remain well mixed because of wind-induced turbulence. Deeper lakes generally stratify during the summer, but for many of these, overturning occurs twice a year, mixing upper and lower strata.

If a lake, such as that shown in Fig. 3-18, can be assumed to behave like a complete mix reactor, then the following equation can be written to model the fate of a given constituent:

$$\underbrace{\frac{dC}{dt} V}_{\text{Accum}} = \underbrace{(Q_{in} C_{in} + Q_{sr} C_{sr} + Q_{gw} C_{gw} + Q_{rf} C_{rf} + Q_{ww} C_{ww})}_{\text{inflow}} - \underbrace{(Q_{ev} C_{ev} + Q_{ws} C + Q_{out} C)}_{\text{outflow}} + \underbrace{\sum_{i=1}^{n} r_{ci} V}_{\text{generation}} \tag{3-70}$$

where r_{ci} = rate expression for generation and/or decay reactions, all other terms are as used previously, and the following subscripts apply:

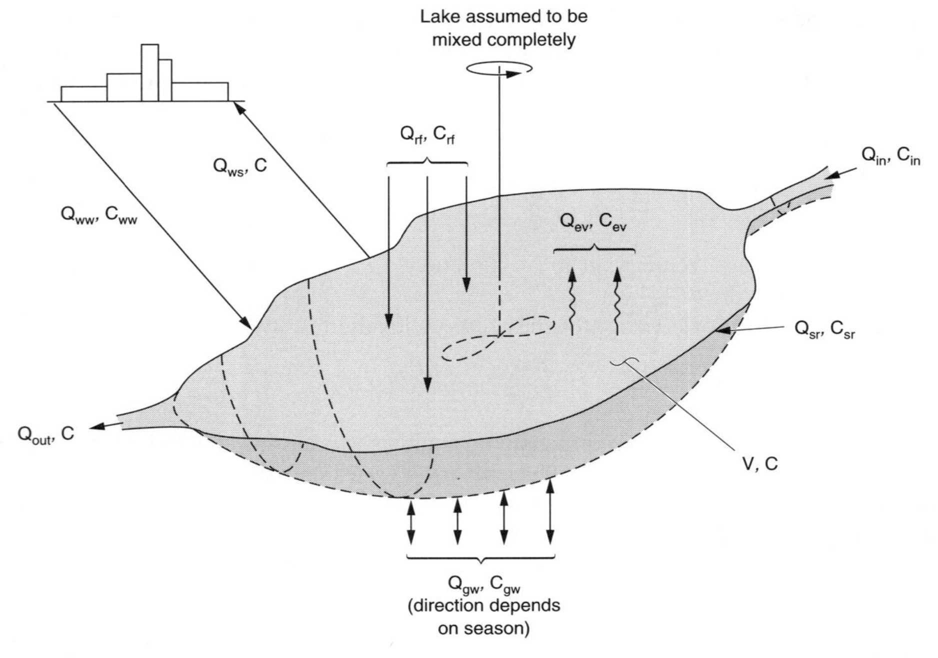

FIGURE 3-18
Schematic diagram of a lake used for the analysis of water quality changes due to external inputs.

in = inflow (stream or river)
sr = surface runoff
gw = groundwater
rf = rainfall
ww = wastewater discharge
ev = evaporation
ws = water supply
out = outflow (stream or river)

In the above equation, the groundwater term can be either an inflow or an outflow term, depending on the local groundwater level. Because constituent concentrations in vapor are typically small, the term containing C_{ev} is often omitted. The generation term is the summation of all of the factors that may influence a given constituent (see Table 3-4).

Rearranging Eq. (3-70) and assuming the groundwater term is an inflow term, C_{ev} is equal to zero, and the generation term can be described as a single first-order reaction, the following expression is obtained:

$$\frac{dC}{dt} + \frac{C(Q_{ws} + Q_{out} + kV)}{v} = \frac{Q_{in}C_{in} + Q_{sr}C_{sr} + Q_{gw}C_{gw} + Q_{rf}C_{rf} + Q_{ww}C_{ww}}{V} \tag{3-71}$$

For the steady-state case (where $dC/dt = 0$), the concentration is given by

$$C = \frac{Q_{\text{in}}C_{\text{in}} + Q_{\text{sr}}C_{\text{sr}} + Q_{\text{gw}}C_{\text{gw}} + Q_{\text{rf}}C_{\text{rf}} + Q_{\text{ww}}C_{\text{ww}}}{Q_{\text{ws}} + Q_{\text{out}} + kV} \tag{3-72}$$

For the non-steady-state case let

$$\beta = \frac{Q_{\text{ws}} + Q_{\text{out}} + kV}{V} \tag{3-73}$$

and

$$M = \frac{Q_{\text{in}}C_{\text{in}} + Q_{\text{sr}}C_{\text{sr}} + Q_{\text{gw}}C_{\text{gw}} + Q_{\text{rf}}C_{\text{rf}} + Q_{\text{ww}}C_{\text{ww}}}{V} \tag{3-74}$$

Using the above expressions for β and M, Eq. (3-71) can be written as

$$\frac{dC}{dt} + \beta C = M \tag{3-75}$$

The integrated form of the above expression can be derived by using the integrating factor method introduced in Sec. (3-3). The integrated equation is

$$C = \frac{M}{\beta}(1 - e^{-\beta t}) + C_o e^{-\beta t} \tag{3-76}$$

where C_o is the concentration at time $t = 0$. As t goes to infinity, Eq. (3-76) approaches the steady-state form [compare to Eq. (3-72)]:

$$C = \frac{M}{\beta} \tag{3-77}$$

If the volume of the lake does not remain constant, then the following term must be added to Eq. (3-70) to account for the changes in volume:

$$\frac{dV}{dt}C \tag{3-78}$$

The application of Eq. (3-77) is illustrated in Example 3-8.

EXAMPLE 3-8. ANALYSIS OF LAKE PHOSPHORUS LOADING. Consider a lake of surface area 1.2 mi^2, with an average depth of 15 ft. Assume the lake is located in an area where average annual rainfall is 27.6 in, and where average annual evaporation is 33.8 in. The lake is fed by a stream whose average flow is 5.2 ft^3/s. The surface runoff to the lake is assumed to be negligible. Lake outflow is managed so that a fairly constant lake surface elevation is maintained. Currently, there is no development around the lake. If the phosphorus concentrations for the stream and the rainwater are 0.08 mg/L and 0.01 mg/L, respectively, calculate the steady-state phosphorus concentration for the lake. Assume that the groundwater flow term is equal to zero, that the phosphorus precipitation rate is first order, and that the precipitation rate constant is 0.005 d^{-1}.

A residential development of 100 single-family dwellings is proposed for the above lake. The homes are to be located around the perimeter of the lake. The lake will serve as the water supply for the development. Water withdrawn from the lake is to be filtered and distributed throughout the development. Wastewater generated by the residents will be treated by individual septic tank/leachfield systems, and will enter the lake indirectly. If the average phosphorus concentration in the leachfield effluent reaching the lake is

6.0 mg/L, determine the steady-state phosphorus concentration in the lake. What is the percent increase in phosphorus concentration, based on the predevelopment phosphorus concentration? Assume that the average number of persons per residence is 3.0 and that the water usage rate is 75 gal/capita·d. Also assume that of the water withdrawn, 50 gal/capita·d will be discharged as wastewater and will be returned indirectly to the lake from septic tank leachfields.

Solution—Part 1 Predevelopment phosphorus concentration

1. Convert all flows to ft^3/s and calculate outflow.
   a. Rainfall:

$$\left(\frac{27.6\text{ in}}{\text{yr}}\right)\left(\frac{\text{ft}}{12\text{ in}}\right)\left(\frac{1\text{ yr}}{365\text{ d}}\right)\left(\frac{1\text{ d}}{86{,}400\text{ s}}\right)(1.2)(5280\text{ ft})^2 = 2.44\text{ ft}^3/\text{s}$$

   b. Evaporation:

$$\left(\frac{33.8\text{ in}}{\text{yr}}\right)\left(\frac{\text{ft}}{12\text{ in}}\right)\left(\frac{1\text{ yr}}{365\text{ d}}\right)\left(\frac{1\text{ d}}{86{,}400\text{ s}}\right)(1.2)(5280\text{ ft})^2 = 2.99\text{ ft}^3/\text{s}$$

   c. Calculate outflow:

$$\text{Outflow} = \text{inflow} + \text{rainfall} - \text{evaporation}$$
$$= 5.20 + 2.44 - 2.99 = 4.65\text{ ft}^3/\text{s}$$

2. Calculate M:
   From Eq. (3-74),

$$M = \frac{(Q_{\text{in}}C_{\text{in}} + Q_{\text{rf}}C_{\text{rf}})}{V}$$

$$V = 1.2(5280\text{ ft})^2(15\text{ ft}) = 5.02 \times 10^8\text{ ft}^3$$

$$M = \frac{(5.2\text{ ft}^3/\text{s})(0.08\text{ mg/L})(2.44\text{ ft}^3/\text{s})(0.01\text{ mg/L})}{5.02 \times 10^8\text{ ft}^3}$$
$$= 8.78 \times 10^{-10}\text{ mg/(L·s)}$$

3. Calculate β:
   From Eq. (3-73),

$$\beta = \frac{(Q_{\text{out}} + kV)}{V}$$

$$= \frac{\left(4.65\text{ ft}^3/\text{s} + \left(\frac{0.005}{\text{d}}\right)\left(\frac{1\text{ d}}{86{,}400\text{ s}}\right)(5.02 \times 10^8\text{ ft}^3)\right)}{5.02 \times 10^8\text{ ft}^3}$$

$$= 6.71 \times 10^{-8}\text{ s}^{-1}$$

4. Calculate the lake phosphorus concentration:
   From Eq. (3-77),

$$C = \frac{M}{\beta} = \left(\frac{8.78 \times 10^{-10}\text{ mg·s}}{6.71 \times 10^{-8}\text{ s·L}}\right) = 0.013\text{ mg/L}$$

5. Prepare a mass balance for phosphorus:
   Once the lake phosphorus concentration is known, both the rate of phosphorus outflow and the rate at which dissolved phosphorus is leaving the system as a result of chemical precipitation can be calculated. These values can be used with the inflow and rainfall phosphorus load rates to complete the mass balance.

a. Rate of phosphorus outflow in stream:

$$Q_{\text{out}}C_{\text{out}} = \left(\frac{4.65\ \text{ft}^3}{\text{s}}\right)\left(\frac{0.013\ \text{mg}}{\text{L}}\right)\left(\frac{28.32\ \text{L}}{\text{ft}^3}\right) = 1.72\ \text{mg/s}$$

b. Rate of phosphorus outflow due to precipitation:

$$kCV = \left(\frac{0.005}{\text{d}}\right)\left(\frac{0.013\ \text{mg}}{\text{L}}\right)(5.02 \times 10^8\ \text{ft}^3)\left(\frac{\text{d}}{24\ \text{h}} \times \frac{\text{h}}{3600\ \text{s}}\right)\left(\frac{28.32\ \text{L}}{\text{ft}^3}\right)$$

$$= 10.75\ \text{mg/s}$$

c. Rate of phosphorus inflow in stream:

$$Q_{\text{in}}C_{\text{in}} = \left(\frac{5.20\ \text{ft}^3}{\text{s}}\right)\left(\frac{0.08\ \text{mg}}{\text{L}}\right)\left(\frac{28.32\ \text{L}}{\text{ft}^3}\right) = 11.78\ \text{mg/s}$$

d. Rate of phosphorus inflow due to rainfall:

$$Q_{\text{rf}}C_{\text{rf}} = \left(\frac{2.44\ \text{ft}^3}{\text{s}}\right)\left(\frac{0.01\ \text{mg}}{\text{L}}\right)\left(\frac{28.32\ \text{L}}{\text{ft}^3}\right) = 0.69\ \text{mg/s}$$

6. Check phosphorus mass balance:
   If the phosphorus concentrations have been calculated correctly, the sum of phosphorus inputs should equal the sum of phosphorus outputs. In other words,

$$\begin{matrix}\text{Rate of} & & \text{Rate of} & & \text{Rate of} & & \text{Rate of}\\ \text{phosphorus} & + & \text{phosphorus} & = & \text{phosphorus} & + & \text{phosphorus}\\ \text{inflow in} & & \text{inflow in} & & \text{outflow in} & & \text{outflow due to}\\ \text{stream} & & \text{rain} & & \text{stream} & & \text{precipitation}\end{matrix}$$

$$\frac{11.78\ \text{mg}}{\text{s}} + \frac{0.69\ \text{mg}}{\text{s}} = \frac{1.72\ \text{mg}}{\text{s}} + \frac{10.75\ \text{mg}}{\text{s}}$$

It can be seen that the phosphorus load rate equation is balanced.

Solution—Part 2 Postdevelopment phosphorus concentration

1. Convert all flows to ft^3/s and calculate outflow.
   a. Water supply:

$$(100\ \text{homes})\left(\frac{3\ \text{persons}}{\text{home}}\right)\left(\frac{75\ \text{gal}}{\text{person·d}}\right)\left(\frac{\text{ft}^3}{7.48\ \text{gal}}\right)\left(\frac{\text{d}}{24\ \text{h}} \times \frac{\text{h}}{3600\ \text{s}}\right) = 0.035\ \text{ft}^3/\text{s}$$

   b. Wastewater:

$$(100\ \text{homes})\left(\frac{3\ \text{persons}}{\text{home}}\right)\left(\frac{50\ \text{gal}}{\text{person·d}}\right)\left(\frac{\text{ft}^3}{7.48\ \text{gal}}\right)\left(\frac{\text{d}}{24\ \text{h}} \times \frac{\text{h}}{3600\ \text{s}}\right) = 0.023\ \text{ft}^3/\text{s}$$

   c. Compute outflow:

$$\text{Outflow} = \text{inflow} + \text{rainfall} + \text{wastewater} - \text{evaporation} - \text{water supply}$$

$$= 5.20 + 2.44 + 0.023 - 2.99 - 0.035 = 4.64\ \text{ft}^3\ \text{s}^{-1}$$

2. Calculate M, β, and the phosphorus concentration:

$$M = \frac{(5.2\ \text{ft}^3/\text{s})(0.08\ \text{mg/L}) + (2.44\ \text{ft}^3/\text{s})(0.01\ \text{mg/L}) + (0.023\text{ft}^3/\text{s})(6.0\ \text{mg/L})}{5.02 \times 10^8\ \text{ft}^3}$$

$$= 1.16 \times 10^{-9}\ \text{mg/(L·s)}$$

$$\beta = \frac{\left(\dfrac{0.035\ \text{ft}^3 + 4.640\ \text{ft}^3}{\text{s}}\right) + \left(\dfrac{0.005}{\text{d}}\right)\left(\dfrac{\text{d}}{24\ \text{h}} \times \dfrac{\text{h}}{3600\ \text{s}}\right)(5.02 \times 10^8\ \text{ft}^3)}{5.02 \times 10^8\ \text{ft}^3}$$

$$= 6.72 \times 10^{-8}\ \text{s}^{-1}$$

$$C = \frac{M}{\beta} = \frac{1.16 \times 10^{-9}\ \text{mg/s}}{6.72 \times 10^{-8}\ \text{s}\cdot\text{L}} = 0.017\ \text{mg/L}$$

3. Determine the percent increase in phosphorus concentration due to development:

$$\text{Increase} = \frac{(0.017 - 0.013)}{0.013} 100\% = 31\%$$

Comment. As a result of the proposed residential development, it can be expected that the lake phosphorus concentration will rise 31 percent. Typically, before a project like this can be approved, the impact of additional nutrients such as nitrogen and potassium on aquatic plant life would have to be investigated. The increased potential impact to the lake if residents utilize commercial fertilizers for their lawns and gardens must also be considered.

Modeling of Dissolved Oxygen in Rivers and Streams

One of the earliest models developed to predict the effects of the discharge of biodegradable organic material on the oxygen resources in streams and rivers, was formulated by Streeter and Phelps in the early 1920s, based on their studies of the Ohio River (Streeter and Phelps, 1925; Phelps, 1944). The model proposed by Streeter and Phelps, commonly known as the *oxygen sag model,* has been applied in hundreds of studies throughout the United States. Because this model has been applied so extensively, it is important to understand its development and application.

If wastewater containing biodegradable organic matter is discharged into a stream with inadequate dissolved oxygen, the water downstream of the point of discharge (typically an outfall) will become anaerobic and will be turbid and dark. Settleable solids, if present, will be deposited on the stream bed and anaerobic decomposition will occur (see Fig. 3-19). Over the reach of stream where the dissolved-oxygen concentration is zero, a zone of putrefaction will occur with the production of hydrogen sulfide, ammonia, and other odorous gases. Because game fish require a minimum of 4 to 5 mg/L of dissolved oxygen, there will be no game fish in this portion of the stream. Farther downstream the water will become clearer and the dissolved-oxygen content will increase until the effects of pollution are negligible.

Where waste is discharged into a body of water, a reserve of dissolved oxygen is necessary if nuisance conditions are to be avoided. If a large supply of diluting water with adequate dissolved oxygen is available, the BOD of the waste can be satisfied without developing malodorous conditions. At high temperatures, when bacterial action is most rapid, the solubility of oxygen is reduced. Because the solubility of oxygen in water depends on temperature, conditions in a polluted stream usually are worse in warm weather, particularly when associated with low flows.

Initial dilution. When a waste with a constituent concentration C_w is discharged at a rate of flow Q_w into a stream containing the same constituent at

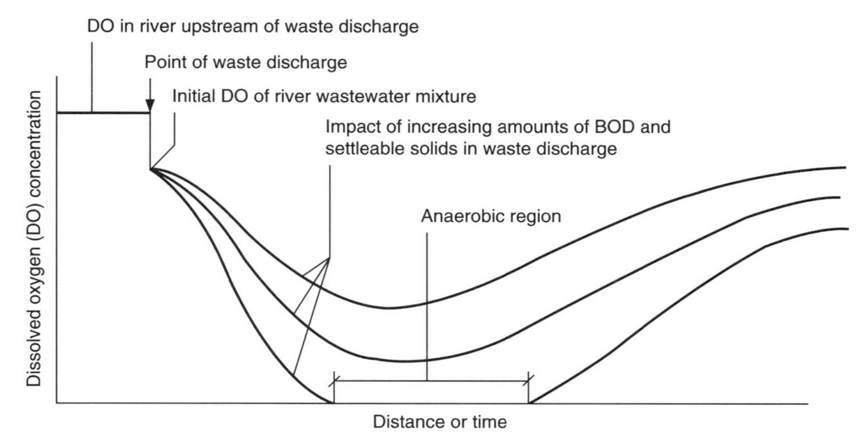

FIGURE 3-19
Impact of BOD and settleable solids on a stream with limited assimilation capacity.

concentration C_r, that is flowing at rate Q_r, the concentration C of the resulting mixture, assuming both streams are intermixed completely, is given by the following materials balance:

$$C_rQ_r + C_wQ_w = C(Q_r + Q_w)$$

$$C = \frac{C_rQ_r + C_wQ_w}{Q_r + Q_w} \tag{3-79}$$

It should be noted that conservation of mass and complete intermixing of the two streams is assumed in the development of this expression. It is applicable to oxygen content, BOD, suspended solids, heat, and other characteristic constituents of the waste. In practice, because complete intermixing seldom occurs, a mixing length is usually determined.

When water containing biodegradable constituents is exposed to air, oxygen is absorbed to replace the dissolved oxygen that is consumed in satisfying the BOD of the waste. The processes of deoxygenation and reoxygenation occur simultaneously. If the rate of deoxygenation is more rapid than the rate of reoxygenation, an increasing oxygen deficit results. If the dissolved-oxygen content becomes zero, aerobic conditions will no longer be maintained and putrefaction will set in. The oxygen deficit is defined as

$$D = C_s - C \tag{3-80}$$

where C_s is the saturation oxygen content and C is the actual oxygen content at the given temperature.

Deoxygenation and reoxygenation. The rates of deoxygenation and reoxygenation are formulated as follows:

Deoxygenation:

$$r_D = -k_1\text{BOD}_t \tag{3-81}$$

where r_D = rate of deoxygenation, mg/L·d
k_1 = deoxygenation rate constant, 1/d
BOD_t = carbonaceous BOD remaining at time t, mg/L

Reoxygenation:

$$r_R = k_2 D = k_2(C_s - C) \tag{3-82}$$

where r_R = rate of reoxygenation, mg/L·d
k_2 = reaeration rate constant, 1/d
D = dissolved-oxygen deficit, mg/L

Change in oxygen concentration. The rate at which the dissolved-oxygen concentration changes in a section of a river or stream is equal to

$$\underset{\text{Accum}}{\frac{\partial C}{\partial t}dV} = \underset{\text{inflow}}{QC} - \underset{\text{outflow}}{Q\left(C + \frac{\partial C}{\partial x}dx\right)} + \underset{\text{deoxygenation}}{r_D dV} + \underset{\text{reoxygenation}}{r_R dV} \tag{3-83}$$

Substituting for r_D and r_R yields

$$\frac{\partial C}{\partial t}dV = QC - Q\left(C + \frac{\partial C}{\partial x}dx\right) - k_1\mathrm{BOD}_t dV + k_2(C_s - C)dV \tag{3-84}$$

If steady-state conditions are assumed, $\partial C/\partial t = 0$ and the above expression can be simplified to

$$0 = -Q\frac{dC}{dx}dx - k_1\mathrm{BOD}_t dV + k_2(C_s - C)dV \tag{3-85}$$

Substituting Adx for dV, and dt for Adx/Q, in the above equation yields:

$$\frac{dC}{dt} = -k_1\mathrm{BOD}_t + k_2(C_s - C) \tag{3-86}$$

It should be noted that Eq. (3-86) can also be derived directly by perfoming a mass balance on a control volume that moves at the average velocity of flow v. Equation (3-86) can be written in terms of the dissolved-oxygen deficit by noting that if Eq. (3-80) is differentiated, the resulting expression is

$$\frac{dD}{dt} = -\frac{dC}{dt} \tag{3-87}$$

Substituting $-dD/dt$ for dC/dt and D for $(C_s - C)$ in Eq. (3-86) yields

$$\frac{dD}{dt} = k_1\mathrm{BOD}_t - k_2 D \tag{3-88}$$

The minus sign in Eq. (3-88) accounts for the fact that reoxygenation reduces the deficit. Substituting for BOD_t (see Eq. (2-36) in Chap. 2) and rearranging yields

$$\frac{dD}{dt} + k_2 D = k_1\mathrm{UBOD}e^{-k_1 t} \tag{3-89}$$

The amount of dissolved oxygen at any time can be determined using the integrated form of Eq. (3-89) if the rates of deoxygenation and reoxygenation are known. The following integrated expression is known as the *Streeter-Phelps equation* (Streeter and Phelps, 1925):

$$D_t = \frac{k_1 \text{UBOD}}{k_2 - k_1}(e^{-k_1 t} - e^{-k_2 t}) + D_o e^{-k_2 t} \tag{3-90}$$

where D_t = dissolved-oxygen deficit at time t, mg/L
UBOD = ultimate carbonaceous BOD at the point of discharge, mg/L
t = time in days
D_o = initial dissolved-oxygen deficit, mg/L
k_1, k_2 = as defined previously

When the wastewater is discharged into a stream, the values of D_t, UBOD, and D_o refer to the characteristics of the mixture. The deoxygenation coefficient k_1 can be determined either by laboratory or field tests. The reoxygenation coefficient k_2 can be estimated by knowing the characteristics of the stream and using one of the many empirical formulas that have been proposed. A generalized formula, proposed by O'Connor and Dobbins (1958) is

$$k_2 = \frac{1440(D_L U)^{1/2}}{H^{3/2}} \tag{3-91}$$

where k_2 = reaeration rate constant, 1/d
D_L = coefficient of molecular diffusion at 20°C, ft^2/h
U = mean stream velocity, ft/s
H = mean depth of flow, ft

If the coefficient of molecular diffusion of oxygen at 20°C is 8.1×10^{-5} ft^2/h, Eq. (3-91) can be written as follows (Thomann and Mueller, 1987):

$$k_2 = \frac{12.9 U^{1/2}}{H^{3/2}} \tag{3-92}$$

Typical k_2 values are given in Table 3-7. The values in the table can be corrected for the effects of temperature by using Eq. (3-22) with $\theta = 1.024$.

TABLE 3-7
Typical reaeration coefficient values in streams*

Water body	Ranges of k_2 at 20°C, d^{-1} (base e)†
Small ponds and backwaters	0.10–0.23
Sluggish streams and large lakes	0.23–0.35
Large streams of low velocity	0.35–0.46
Large steams of moderate velocity	0.46–0.69
Swift streams	0.69–1.15
Rapids and waterfalls	> 1.15

*From Report of the Engineering Board of Review (1925).
†For other temperatures, use $k_{2T} = k_{2(20)} 1.024^{T-20}$.

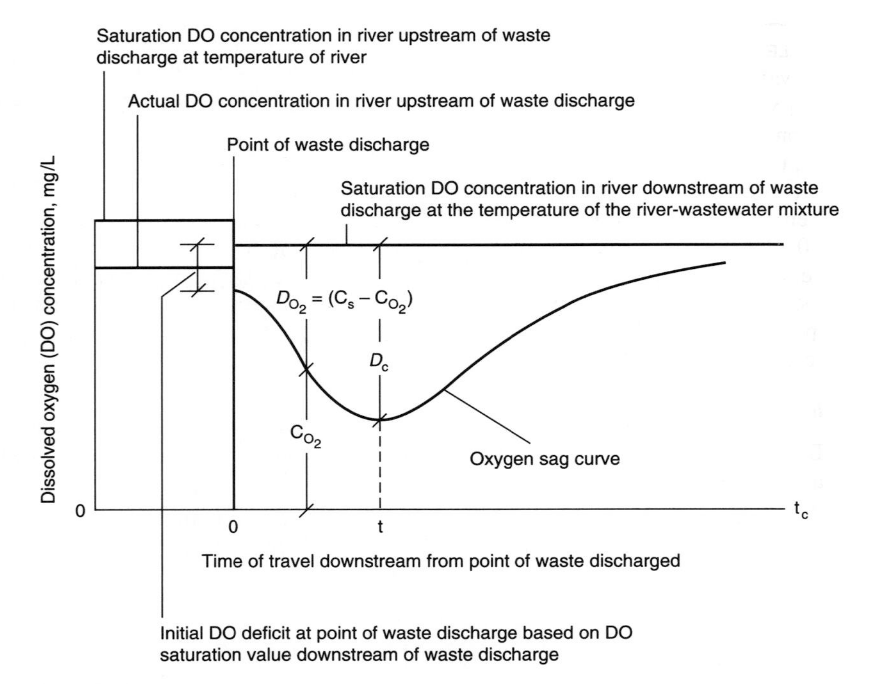

FIGURE 3-20
Definition sketch for a typical oxygen-sag analysis curve for a river.

If the dissolved-oxygen deficit is determined at several points downstream and the values are plotted, the resulting curve (see Fig. 3-20) is known as the *oxygen-sag* curve. The difference between the oxygen-sag and the deoxygenation curves represents the effect of reoxygenation.

The critical time t_c at which minimum dissolved oxygen occurs can be found by differentiating Eq. (3-90) and equating to zero:

$$t_c = \frac{1}{k_2 - k_1} \ln \left[\frac{k_2}{k_1} \left(1 - \frac{D_o(k_2 - k_1)}{k_1 \text{UBOD}} \right) \right] \tag{3-93}$$

The corresponding critical oxygen deficit is

$$D_c = \frac{k_1 \text{UBOD}}{k_2} e^{-k_1 t_c} \tag{3-94}$$

The oxygen-sag analysis is of practical value in predicting the oxygen content at any point along a stream into which wastewater is discharged in which significant photosynthesis/respiration effects are not expected. It permits an estimate of the degree of waste treatment required, or of the amount of dilution necessary to maintain a certain dissolved-oxygen content in the stream. The oxygen-sag analysis is also used to assess the assimilative capacity of a stream—the amount of waste that can be assimilated without injurious effects on water quality and the stream biota. The oxygen sag analysis is illustrated in Example 3-9.

EXAMPLE 3-9. OXYGEN-SAG ANALYSIS FOR A RIVER. A city discharges 65 Mgal/d of treated wastewater at 20°C into a nearby river. The average effluent 20°C BOD_5 is 30 mg/L. The k_1 value at 20°C is estimated to be 0.30 d^{-1} (base e). The effluent following post-aeration contains about 4.0 mg/L dissolved oxygen. The characteristics of the river water, upstream of the effluent discharge point, are as follows: minimum rate of flow is 100 ft^3/s, the average depth is 6 ft, the average velocity in the river is about 1.0 ft/s, the temperature of the river water is 16°C, the 20°C BOD_5 value in the river water is 2.0 mg/L, and the river water is at 92 percent of the dissolved-oxygen saturation value. Determine the maximum impact of this discharge and critical oxygen deficit and its location. Also estimate the 20°C BOD_5 of a sample taken at the critical point. Use temperature coefficients of 1.047 for k_1 and 1.024 for k_2. Also plot the dissolved-oxygen sag curve.

Solution

1. Define relationships that can be used to determine the flowrate, BOD, DO deficit, and temperature just downstream of the point of effluent discharge, assuming that the wastewater that is discharged is mixed completely with the river water.

Item	River upstream of discharge	Waste discharge	River downstream of discharge
Flowrate	Q_r	Q_w	$Q = Q_r + Q_w$
BOD	L_r	L_w	$L_o = (Q_r L_r + Q_w L_w)/Q$
DO	D_r	D_w	$D_o = (Q_r D_r + Q_w D_w)/Q$
Temperature	T_r	T_w	$T_o = (Q_r T_r + Q_w T_w)/Q$

2. Determine the dissolved oxygen in the stream before discharge:

$$\text{Saturation concentration, 16°C (see App. C)} = 9.86 \text{ mg/L}$$

$$\text{Dissolved oxygen in stream} = 0.92(9.86) = 9.07 \text{ mg/L}$$

3. Determine the temperature, dissolved oxygen, BOD_5, and UBOD of the mixture:

$$\text{Temperature of mixture} = \frac{65(1.55)(20) + 100(16)}{65(1.55) + 100} = 18.0°\text{C}$$

$$\text{Dissolved oxygen of mixture} = \frac{65(1.55)(4.0) + 100(9.07)}{65(1.55) + 100} = 6.53 \text{ mg/L}$$

$$\text{BOD}_5 \text{ of mixture} = \frac{65(1.55)(30) + 100(2.0)}{65(1.55) + 100} = 16.0 \text{ mg/L}$$

$$\text{UBOD of mixture} = \frac{16.0}{1 - e^{-0.3(5)}} = 20.6 \text{ mg/L}$$

4. Determine the value of k_2 using Eq. (3-92):

$$k_2 = \frac{12.9U^{1/2}}{H^{3/2}}$$

$$= \frac{12.9(1.0)^{1/2}}{6^{3/2}} = 0.88 \text{ d}^{-1}$$

5. Correct the rate constants to 18.0°C:

$$k_1 = 0.30(1.047)^{18.0-20} = 0.27\ \text{d}^{-1}$$

$$k_2 = 0.88(1.024)^{18.0-20} = 0.84\ \text{d}^{-1}$$

6. Determine t_c and x_c:

   a. Dissolved-oxygen saturation at 18.0°C is

$$\text{DO}_{\text{sat}} \text{ at } 18.0°\text{C (from App. C)} = 9.45\ \text{mg/L}$$

   b. The initial DO deficit is

$$D_o = 9.45 - 6.53 = 2.92\ \text{mg/L}$$

   c. The critical time t_c, from Eq. (3-93), is

$$t_c = \frac{1}{k_2 - k_1}\ln\left[\frac{k_2}{k_1}\left(1 - \frac{D_o(k_2 - k_1)}{k_1\text{UBOD}}\right)\right]$$

$$= \frac{1}{0.84 - 0.27}\ln\left[\frac{0.84}{0.27}\left(1 - \frac{2.92(0.84 - 0.27)}{0.27(20.6)}\right)\right] = 1.37\ \text{d}$$

   d. The critical distance x_c is

$$x_c = vt_c = (1.0\ \text{ft/s})(86{,}400\ \text{s/d})(1.37\ \text{d})/(5280\ \text{ft/mi}) = 22.4\ \text{mi}$$

7. Determine the critical deficit D_c using Eq. (3-94):

$$D_c = \frac{k_1\text{UBOD}}{k_2}e^{-k_1 t_c}$$

$$= \frac{0.27(20.6)}{0.84}e^{-0.27(1.37)} = 4.58\ \text{mg/L}$$

$$\text{DO}_c = 9.45 - 4.58 = 4.87\ \text{mg/L}$$

8. Determine the BOD_5 of a sample taken at x_c.

$$\text{BOD}_t = (20.6)e^{-0.27(1.37)} = 14.2\ \text{mg/L}$$

$$20°\text{C BOD}_5 = 14.2(1 - e^{-0.3(5)}) = 11.0\ \text{mg/L}$$

9. The dissolved oxygen-sag curve is plotted in the following figure.

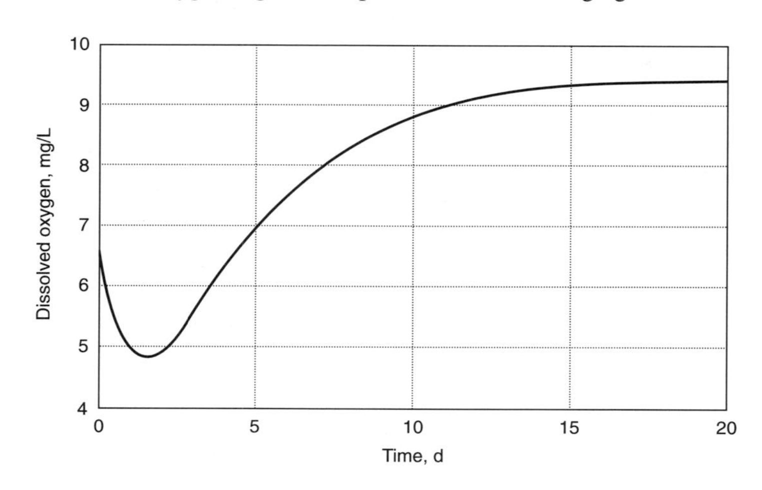

Comment. Although the oxygen-sag analysis suffers from a number of limitations, the exercise is nevertheless useful because it forces one to consider the transformations that are occurring in the stream and their interrelationships. To overcome the apparent limitations in the oxygen-sag analysis, segmented models in which a number of variables are considered have been developed. These and other models are considered in Thomann and Mueller (1987), and briefly in the following discussion.

Large-Scale Modeling of Water Quality in Water Bodies

With the advent of high-speed computers, the development of one-, two-, and three-dimensional mathematical models that can be used to determine the fate of constituents released to water bodies has developed at a rapid pace, in many cases far outstripping the availability of field data that can be used for general model verification. The principal types of mathematical models developed to study the impact of constituents released to water bodies are reported in Table 3-8.

TABLE 3-8
Typical modeling techniques used to assess the impacts of point and nonpoint sources of contaminants on various water bodies*

Water body	Method of analysis	Remarks
Lakes and reservoirs	Fully mixed	Small and shallow lakes and reservoirs.
	Stratified (one-, two-, and three-dimensional analysis)	Lakes with a depth greater than 15 ft are typically stratified.
Streams and rivers	One-dimensional without and with longitudinal dispersion (nonsegmented and segmented)	Historical method of analysis for streams that are many times longer than deep or wide (see oxygen-sag analysis).
	Instantaneous source	One-time release of a contaminant (e.g., an accidental spill).
	Continuous discharge	Typical point (e.g., wastewater treatment plant) and nonpoint discharges.
	Oxygen-sag analysis	Continuous discharge of carbonaceous and nitrogenous materials that can exert an oxygen demand. One of the first stream models developed in the 1920s.
	Two-dimensional (segmented)	Applied to rivers of varying cross section.
Estuaries	One-dimensional (segmented)	Can be applied to narrow well-mixed estuaries.
	Two-dimensional (segmented) Longitudinal-lateral Two-layer (stratified vertically)	Partially mixed estuaries, estuaries with embayments.
	Three-dimensional	Computationally intensive.
Ocean discharge	Initial mixing (stagnant and flow through water column)	Mixing controlled by the design and placement of the diffuser.
	Far field modeling	Used to model turbulent diffusion and far field regions.
	Transition region With spatially uniform current With complex current patterns	Region between near field and far field.

*Adapted from Brown and Barnwell (1987), Jorgensen and Gromiec (1989), Orlob (1983), and Schnoor (1996).

3-7 IMPACT OF WASTEWATER EFFLUENT DISCHARGE STANDARDS

To determine the degree of treatment that may be required, it will be necessary to review current standards and regulations regarding the discharge of contaminants to the environment.

Effluent Discharge Standards and Permits

Wastewater discharges are most commonly controlled through effluent standards and discharge permits. In the United States, the National Pollution Discharge Elimination System (NPDES), administered by the individual states with federal EPA oversight, is used for the control of wastewater discharges. Under this system, discharge permits are issued with limits on the quantity and quality of effluents. These limits are based on a case-by-case evaluation of potential environmental impacts and, in the case of multiple dischargers, on waste load allocation studies aimed at distributing the available assimilative capacity of the water body. Discharge permits are designed as an enforcement tool, with the ultimate goal of meeting ambient water quality standards.

Water quality standards are sets of qualitative and quantitative criteria designed to maintain or enhance the quality of receiving waters. In the United States, these standards are promulgated by the individual states. Receiving waters are divided into several classes depending on their uses, existing or intended, with different sets of criteria designed to protect uses such as drinking water supply, bathing, boating, and fishing for freshwater streams, and outdoor sports and shellfish harvesting for estuaries and coastal areas.

For toxic compounds, chemical-specific or whole-effluent toxicity studies are used to develop standards and criteria. In the chemical-specific approach, individual criteria are used for each of the toxic chemicals detected in the wastewater. Based on the results of laboratory studies, criteria can be developed to protect aquatic life against acute and chronic effects and to safeguard humans against deleterious health effects, including cancer. The chemical-specific approach, however, does not consider the possible additive, antagonistic, or synergistic effects of multiple chemicals. The biological availability of the compound, which depends on its form in the wastewater, is also not considered in this approach. The whole-effluent toxicity approach can be used to overcome the shortcomings of the chemical-specific approach. In the whole-effluent approach, toxicity or bioassay tests are used to determine the concentration at which the wastewater induces acute or chronic toxicity effects. In bioassay testing, selected organisms are exposed to effluent diluted in specified ratios with samples of receiving water (see Sec. 2-8 in Chap. 2). At various points during the test, the organisms impacted by various effects, such as lower reproduction rates, reduced growth, or death, are quantified. To protect aquatic life, discharge limits are established based on the results of the tests.

Future Regulations

Regulations are developed to implement legislation. For this reason, regulations are always subject to change as more information becomes available regarding the characteristics of wastewater, effectiveness of treatment processes, and environmental effects. It is anticipated that the focus of future regulations will be on the implementation of the Water Quality Act of 1987. Control of the pollutional effects of storm water and nonpoint sources, toxics in wastewater (priority pollutants), and, as noted above, the overall management of biosolids including the control of toxic substances will receive the most attention. Nutrient removal, the control of pathogenic organisms, and the removal of organic and inorganic substances such as VOCs and total dissolved solids will also continue to receive attention in specific applications.

PROBLEMS AND DISCUSSION TOPICS

3-1. Use the information provided in the table below (choose one column only) to create a mass balance for a complete-mix reactor. Assume that the constituent utilization reaction rate is first order. Calculate the reactor effluent constituent concentration for steady-state conditions.

Parameter	Units	A	B	C	D
Flow	gal/min	17	29	16	31
Influent concentration	mg/L	86	59	75	101
Rate coefficient k	d^{-1}	0.23	0.11	0.30	0.18
Reactor volume	ft^3	19,000	15,000	25,000	11,000

3-2. If the first-order rate coefficients in the previous problem were determined for water at a temperature of 15°C, calculate the steady-state effluent concentration if the water is at 25°C. Use a temperature coefficient θ of 1.048.

3-3. Determine the reaction order and the reaction-rate coefficient for the following concentrations, using both the differential and integration methods (choose one column only).

Time,	BOD, mg/L			
d	A	B	C	D
0	85	157	212	134
1	66	7.9	5.1	90
2	52	4.1	2.6	60
3	40	2.7	1.7	40
4	31	2.1	1.3	27
5	24	1.6	1.0	18
6	19	1.4	0.9	12
7	15	1.2	0.7	8

3-4. Given the following experimental data obtained from a constructed wetland, determine the temperature coefficient, θ.

Temperature, °C	Reaction-rate constant k, d^{-1}
15.0	0.53
20.5	0.99
25.5	1.37
31.5	2.80

3-5. Use the Fujimoto method to determine the ultimate BOD and the first-order reaction-rate coefficient k for the following BOD data (choose one column only). (*Note:* Using the ultimate BOD, calculate k for each day, and report the average of the seven k values.)

Time,	BOD, mg/L			
d	A	B	C	D
1	40	30	88	67
2	75	55	152	117
3	102	76	197	156
4	125	93	230	186
5	143	108	254	208
6	159	120	271	225
7	171	130	283	239

3-6. Use the data provided in the table below to plot concentration versus time for nonretarded and retarded first-order reactions occurring in batch reactors (choose one column only). At what time is the concentration within the retarded-rate batch reactor double the concentration within the nonretarded-rate batch reactor? Assume the exponent $n = 1$.

Parameter	Units	A	B	C	D
Initial concentration	mg/L	99	115	57	164
Coefficient k	d^{-1}	0.17	0.33	0.27	0.22
Retardation factor r	d^{-1}	2.5	1.8	2.8	3.6

3-7. Consider four complete-mix reactors in series. If a slug (i.e., nonsteady) concentration of inert tracer exists in the first reactor, calculate and plot the change in concentration versus time for all four reactors. Assume the initial concentration of the inert tracer in all but the first reactor is zero. Also assume that the tracer concentration in the flow entering the system is zero. The initial slug concentration, reactor volume (assume all are same size), and flow rate are provided in the columns below (choose one column).

Parameter	Units	A	B	C	D
Initial concentration*	mg/L	75	98	120	66
Reactor volume	ft^3	250,000	410,000	330,000	220,000
Flow rate	ft^3/s	1.3	2.1	1.9	0.9

*In first reactor.

3-8. Consider four complete-mix reactors in series. If a steady (i.e., constant) concentration of inert tracer enters the first reactor, calculate and plot the change in concentration versus time for all four reactors. Assume the initial concentration of the inert tracer in all four reactors is zero. The influent concentration, reactor volume (assume all are same size), and flow rate are provided in the columns below (choose one column).

Parameter	Units	A	B	C	D
Influent concentration	mg/L	80	124	92	101
Reactor volume	ft^3	420,000	290,000	150,000	360,000
Flow rate	ft^3/s	1.8	1.1	0.6	1.5

3-9. Consider five complete-mix reactors in series. Using the information provided in the table below, calculate the fraction of tracer remaining in the system after 1 day. Assume that the initial concentration of tracer is zero in all the reactors except the first one. Also, assume that inflow to the system contains no tracer.

Parameter	Units	A	B	C	D
Individual reactor volume	ft^3	120,000	95,000	64,000	88,000
Flow	gal/min	2,100	1,200	1,800	1,700

3-10. Three complete-mix reactors, connected in series, will be utilized to reduce the steady-state concentration of a biodegradable constituent (assume first-order decay). Using the information provided below (choose one column), determine the required volume of the reactors (assume all reactors are same size).

Parameter	Units	A	B	C	D
Influent concentration	mg/L	157	223	186	194
Effluent concentration	mg/L	30	25	20	35
Flow	gal/min	125	98	62	78
Decay constant	d^{-1}	0.35	0.41	0.22	0.18

3-11. Consider four complete-mix reactors in series. If a steady (i.e., constant) concentration of a biodegradable constituent enters the first reactor, calculate and plot the change in concentration versus time for all four reactors. Assume the initial concentration of the biodegradable constituent is zero in all reactors. The influent concentration, reactor volume (assume all are same size), flow rate, and first-order reaction rate coefficient are provided in the columns below (choose one column).

Parameter	Units	A	B	C	D
Influent concentration	mg/L	212	180	159	176
Reactor volume	ft^3	550,000	760,000	480,000	920,000
Flow rate	ft^3/s	1.2	1.8	1.1	2.5
Rate coefficient	d^{-1}	0.23	0.35	0.17	0.44

3-12. Compare the performance of three complete-mix reactors arranged in series (one after the other, all of the flow goes through all three reactors) to three reactors arranged in parallel (one-third of the flow goes to each reactor). Which configuration provides better treatment for the steady-state condition? Use the data (one column only) from the table below. Assume the utilization reaction rate is first order.

Parameter	Units	A	B	C	D
Flow	ft^3/s	1.2	0.8	2.3	1.7
Influent concentration	mg/L	85	101	67	92
Rate coefficient k	d^{-1}	0.27	0.44	0.35	0.21
Reactor volume (each)	ft^3	22,000	35,000	17,000	48,000

3-13. Plug-flow kinetics can be approximated by increasing the number of complete-mix reactors. Ideal plug flow occurs when the number of complete-mix reactors approaches infinity. The dispersion factor in the Wehner-Wilhelm equation can be used to simulate the effect of increasing the number of complete-mix reactors. A high dispersion factor behaves like one complete-mix reactor, while a low dispersion factor behaves like infinitely many complete-mix reactors, or plug flow. An intermediate dispersion factor value yields intermediate results, characteristic of arbitrary flow.

Calculate the effluent concentration using the data provided in the table below (choose one column only). Assuming steady-state conditions, first determine the effluent concentration for plug flow, then for a varying number of complete-mix reactors (1, 10, and 100). Next, calculate the effluent concentration using the Wehner-Wilhelm equation, using dispersion factor values of 0.001, 0.1, 10, and 1000. Comment on the effect of increasing the dispersion factor. Assume the utilization reaction rate is first order. Also, note that the volume given in the table is the total volume; therefore, for n reactors, the individual reactor volume is the given volume divided by n.

Parameter	Units	A	B	C	D
Influent concentration	mg/L	105	88	210	170
Flow	ft^3/s	2.8	1.7	1.3	2.2
Total reactor volume	ft^3	250,000	490,000	680,000	355,000
Reaction rate k	d^{-1}	0.55	0.49	0.38	0.60

3-14. A treatment wetland has been designed to operate as an ideal plug-flow reactor. However, the facility operator observes an appreciable amount of flow dispersion occurring in the wetland. Using the information provided in the table below, compare the predicted ideal plug-flow removal rate to the arbitrary flow removal rate (d is the Wehner-Wilhelm dispersion factor). What is the percent reduction in the removal rate? Assume the utilization reaction rate is first order.

Parameter	Units	A	B	C	D
Volume	ac-ft	7.1	15.2	10.2	8.1
Flow	Mgal/d	0.5	1.2	0.7	0.4
Rate coefficient k	d^{-1}	0.29	0.42	0.32	0.25
Dispersion factor d		0.52	0.37	0.61	0.46

3-15. Consider a lake subject to the inputs and outputs presented in the table below (choose one column only). If 200 residential units are to be constructed around the lake, calculate the steady-state phosphorus concentration in the lake before and after development, assuming each residential unit withdraws 250 gal/day of lake water, and returns 150 gal/day to the lake with a phosphorus concentration of 15.0 mg/L. Assume that the lake outflow is controlled so that a fairly constant lake level is achieved, and that groundwater inflow and outflow are negligible. Also, assume that the phosphorus utilization reaction rate is first order.

Parameter	Units	A	B	C	D
Lake area	mi^2	3.2	1.7	2.5	4.1
Average lake depth	ft	25	19	31	32
Inflow	ft^3/s	15	21	14	10
Inflow concentration	mg/L	0.09	0.05	0.08	0.07
Rainfall	in/yr	24.1	31.3	18.9	21.6
Rainfall concentration	mg/L	0.011	0.005	0.007	0.009
Evaporation	in/yr	45.8	53.0	69.2	49.1
Reaction rate k	d^{-1}	0.006	0.004	0.003	0.005

3-16. For the following conditions (choose one column only), use the Streeter-Phelps equation to find the dissolved oxygen (DO) concentration downstream from a waste discharge point as a function of time. Plot the results. What is the minimum DO concentration? When does it occur?

Parameter	Units	A	B	C	D
Initial DO concentration	mg/L	8.5	8.3	8.7	8.1
Saturation DO concentration	mg/L	9.0	9.2	9.5	9.0
Ultimate BOD concentration	mg/L	28	25	32	18
Deoxidation rate constant k_1	d^{-1}	0.22	0.34	0.15	0.18
Reaeration rate constant k_2	d^{-1}	0.35	0.55	0.35	0.29

REFERENCES

Ambrose, R. B., Jr., J. P. Connolly, E. Southerland, T. O. Barnwell, Jr., and J. L. Schnoor (1988) Waste Allocation Simulation Models, *Journal of Water Pollution Control Federation,* Vol. 60, No. 9, pp. 1646–1655.

Brown, L. C., and T. O. Barnwell (1987) The Enhanced Stream Water Quality Models QUAL2E and QUAL2E-UNCAS: Documentation and User's Manual, Report EPA/600/3-87/007, Environmental Protection Agency, Office of Research and Development, Washington, DC.

Denbigh, K. G., and J. C. R. Turner (1984) *Chemical Reactor Theory: An Introduction,* 3rd ed., Cambridge University Press, New York.

Fair, G. M., and J. C. Geyer (1954) *Water Supply and Waste-Water Disposal,* John Wiley & Sons, New York.

Fujimoto, Y. (1961) Graphical Use of First-Stage BOD Equation, *Journal of Water Pollution Control Federation,* Vol. 36, No. 1, p. 69.

Hemond, H. F., and E. J. Fechner (1994) *Chemical Fate and Transport in the Environment,* Academic Press, San Diego.

Jorgensen, S. E., and M. J. Gromiec (1989) *Mathematical Submodels in Water Quality Systems,* Elsevier, New York.

Moore, E. W., H. A. Thomas, and W. B. Snow (1950) Simplified Method for Analysis of BOD Data, *Sewage and Industrial Wastes,* Vol. 22, No. 10.

O' Connor, D. J., and W. E. Dobbins (1958) Mechanism of Reaeration in Natural Streams, *Transactions of the American Society of Civil Engineers,* Vol. 123, pp. 641–666.

Orlob, G. T. (ed.) (1983) *Mathematical Modeling of Water Quality: Streams, Lakes, and Reservoirs,* Wiley-Interscience, John Wiley & Sons, Chichester, England.

Phelps, E. B. (1944) *Stream Sanitation,* Wiley, New York.

Report of the Engineering Board of Review (1925) Sanitary District of Chicago, Part III, App. 1.

Schnoor, J. L. (1996) *Environmental Modeling,* Wiley-Interscience, John Wiley & Sons, New York.

Sheehy, J. P. (1960) Rapid Methods for Solving Monomolecular Equations, *Journal of Water Pollution Control Federation,* Vol. 32, No. 6.

Streeter, H. W., and E. B. Phelps (1925) A Study of the Pollution and Natural Purification of the Ohio River, Public Health Bulletin, Vol. 146, U.S. Public Health Service, Washington, DC.

Tchobanoglous, G. (1969) A Study of the Filtration of Treated Sewage Effluent, Ph.D. dissertation, Stanford University, Stanford, CA.

Tchobanoglous, G., and E. D. Schroeder (1985) *Water Quality: Characteristics, Modeling, Modification,* Addison-Wesley, Reading, MA.

Tchobanoglous, G., and F. L. Burton (1991) *Wastewater Engineering. Treatment, Disposal, Reuse,* 3rd ed., McGraw-Hill, New York.

Thibodeaux, L. J. (1979) *Chemodynamics: Environmental Movement of Chemicals in Air, Water, and Soil,* John Wiley & Sons, New York.

Thirumurthi, D. (1969) Design of Waste Stabilization Ponds, *Journal of Sanitation Engineering Division,* ASCE, Vol. 95, No. SA2.

Thomann, R. V., and J. A. Mueller (1987) *Principles of Surface Water Quality Modeling and Control,* Harper & Row, New York.

Thomas, H. A., Jr. (1950) Graphical Determination of BOD Curve Constants, *Water & Sewage Works,* Vol. 97, p. 123.

Tsivoglou, E. C. (1958) Oxygen Relationships in Streams, Robert A. Taft Sanitary Engineering Center, Technical Report W-58-2, Cincinnati, OH.

U.S. EPA (1985) *Odor Control and Corrosion Control in Sanitary Sewerage Systems and Treatment Plants,* Design Manual, EPA-625/1-85-018, U.S. Environmental Protection Agency, Washington, DC.

Wehner, J. F., and R. F. Wilhelm (1958) Boundary Conditions of Flow Reactor, *Chemical Engineering Science,* Vol. 6, p. 89.

CHAPTER 4

Introduction to Process Analysis and Design

The analysis and design of wastewater management facilities to reduce or eliminate the constituents found in wastewater involves consideration of those factors and issues that will affect the sizing, performance, and reliability of these facilities. The initial stages of a project, starting with the facilities plan and continuing through the conceptual and preliminary design phases, are considered to be critical to the success of the overall analysis and design process. It is during these initial stages that the design flowrates and constituent mass loadings are developed; process selection is made; the design criteria are developed, refined, and established; issues related to risk assessment and process reliability are examined; and facility layouts are prepared. At the completion of preliminary design, the project is fully defined so that preparation of the detailed plans and specifications can proceed expeditiously.

Important factors and issues, typical to most projects, addressed in this chapter include: (1) wastewater sources and flowrates; (2) wastewater constituent concentrations; (3) variations in wastewater flowrates and constituent concentrations; (4) statistical analysis of wastewater flowrates, constituent concentrations, and mass loading rates; (5) selection of design parameters for wastewater treatment plants; (6) selection of design parameters for septic tank effluent treatment systems; (7) risk assessment; (8) reliability considerations in process selection and design; and (9) process design considerations. The design and physical features of the various unit operations and processes that make up wastewater treatment systems are covered in the following chapters.

4-1 WASTEWATER SOURCES AND AVERAGE FLOWRATES

The sources of wastewater and the corresponding average flowrates in wastewater collection systems, and the flowrates from individual residences, are considered in this section. The effects of water conservation are also considered in this section. The variation observed in flowrates is considered in Section 4-3.

Sources of Domestic Wastewater Discharged to Collection Systems

The sources of wastewater discharged to collection systems are residential areas and commercial, institutional, recreational, and industrial facilities. For existing sources, direct measurement at various locations within the collection system is the preferred method of determining flowrates and variations. For new developments, the information and data given in this section can be used to estimate average flowrates.

Residential areas. For residential areas, the wastewater flowrate is primarily a function of the population. Residential sources and corresponding typical flows per unit are presented in Table 4-1. If the community has a water system but not a wastewater collection system, the average wastewater flowrate can be estimated by multiplying the water use by a factor of 60 to 80 percent, depending on the extent of the landscaping.

Commercial areas. Wastewater flowrates for commercial areas can be estimated using the information presented in Table 4-2. If available, actual flowrates from similar facilities should be used. For a mixed commercial area where the exact nature of the usage is unknown, the range of flowrates based on area is typically between 800 and 2000 gal/ac·d (7.5 and 19 m^3/ha·d) (Tchobanoglous and Burton, 1991).

TABLE 4-1
Typical wastewater flowrates from residential sources discharged to collection systems

		Flow, gal/unit·d		Flow, L/unit·d	
Facility	**Unit**	**Range**	**Typical**	**Range**	**Typical**
Apartment:					
High rise	Person	30–75	55	110–280	210
Low rise	Person	30–80	55	110–300	210
Hotel	Guest	30–50	40	110–190	150
Individual residence:					
Newer home	Person	40–100	70	150–360	270
Older home	Person	30–80	50	110–300	190
Summer cottage	Person	30–60	40	110–190	150
Motel:					
With kitchen	Unit	90–180	100	340–680	380
Without kitchen	Unit	75–150	95	280–570	360
Trailer park	Person	30–50	40	110–190	150

TABLE 4-2
Typical wastewater flowrates from commercial sources*

		Flow, gal/unit·d		Flow, L/unit·d	
Facility	**Unit**	**Range**	**Typical**	**Range**	**Typical**
Airport	Passenger	2–4	3	8–15	11
Apartment house	Person	40–80	50	150–300	190
Automobile service station	Vehicle served	8–15	12	30–57	45
	Employee	9–15	13	34–57	49
Bar	Customer	1–5	3	4–19	11
	Employee	10–16	13	38–61	49
Boardinghouse	Person	25–60	40	95–230	150
Department store	Toilet room	400–600	500	1500–2300	1900
	Employee	8–15	10	30–57	38
Hotel	Guest	40–60	50	150–230	190
	Employee	8–13	10	30–49	38
Industrial Building (sanitary waste only)	Employee	7–16	13	26–61	49
Laundry (self-service)	Machine	450–650	550	1700–2500	2100
	Wash	45–55	50	170–210	190
Office	Employee	7–16	13	26–61	49
Public lavatory	User	3–6	5	11–23	19
Restaurant (with toilet)	Meal	2–4	3	8–15	11
Conventional	Customer	8–10	9	30–38	34
Short order	Customer	3–8	6	11–30	23
Bar/cocktail lounge	Customer	2–4	3	8–15	11
Shopping center	Employee	7–13	10	26–49	38
	Parking space	1–3	2	4–11	8
Theater	Seat	2–4	3	8–15	11

*Adapted in part from Tchobanoglous and Burton (1991).

Institutional facilities. Typical flowrates for institutional facilities are presented in Table 4-3. As noted above, whenever possible, actual flowrates from similar facilities should be evaluated and adjusted to reflect local conditions.

Recreational facilities. Flowrates from recreational facilities are highly variable, and usually seasonal in nature. If the recreational area has a plentiful water supply, the values in Table 4-4 can be used. If the water supply is limited and water conservation is employed, actual water use records should be used for estimating wastewater flowrates.

TABLE 4-3
Typical wastewater flowrates from institutional sources*

Facility	Unit	Flow, gal/unit·d Range	Flow, gal/unit·d Typical	Flow, L/unit·d Range	Flow, L/unit·d Typical
Assembly hall	Seat	2–4	3	8–15	11
Hospital, medical	Bed	125–240	165	470–910	630
	Employee	5–15	10	19–57	38
Hospital, mental	Bed	75–140	100	280–530	380
	Employee	5–15	10	19–57	38
Prison	Inmate	80–150	120	300–570	450
	Employee	5–15	10	19–57	38
Rest home	Resident	50–120	90	190–450	340
	Employee	5–15	10	19–57	38
School, day					
With cafeteria, gym and showers	Student	15–30	25	57–110	95
With cafeteria only	Student	10–20	15	38–76	57
Without cafeteria and gym	Student	5–17	11	19–64	42
School, boarding	Student	50–100	75	190–380	280

*Adapted in part from Tchobanoglous and Burton (1991).

Nondomestic Sources of Wastewater Discharged to Collection Systems

Flows from industrial facilities vary with the type of facility and production levels. Because most industrial processes are continually being redesigned to minimize water use, values from similar industries should be obtained when estimating industrial flowrates. In many industrial facilities, process wastewater is treated separately. Typical wastewater flowrates for various industrial activities are presented in Table 4-5. Because there is a considerable range in the reported values, the values in Table 4-5 *should be used only as a guide* for preliminary planning.

Infiltration/Inflow Discharged to Collection Systems

Infiltration/inflow is the term used to describe the flow of extraneous water into wastewater collection systems. *Infiltration* refers to water which enters a collection system from the ground through building connections, defective pipes, pipe joints, connections, or access port (manhole) walls. *Inflow* comprises two components: *steady inflow* and *direct inflow*. Water discharged from cellar and foundation drains, cooling water discharges, and drains from springs and swampy areas is identified as *steady inflow.* Stormwater runoff which enters a collection system through roof leaders, yard and areaway drains, access port covers, cross connections from storm drains and catch basins, and combined sewers is known as *direct inflow.*

TABLE 4-4
Typical wastewater flowrates from recreational facilities*

		Flow, gal/unit·d		Flow, L/unit·d	
Facility	**Unit**	**Range**	**Typical**	**Range**	**Typical**
Apartment, resort	Person	50–70	60	190–260	230
Bowling alley	Alley	150–250	200	570–950	760
Cabin, resort	Person	8–50	40	30–190	150
Cafeteria	Customer	1–3	2	4–11	8
	Employee	8–12	10	30–45	38
Camp					
Pioneer type	Person	15-30	25	57–110	95
Children's with central toilet and bath	Person	35–50	45	130–190	170
Day with meals	Person	10–20	15	38–76	57
Day without meals	Person	10–15	13	38–57	49
Luxury, private bath	Person	75–100	90	280–380	340
Trailer	Trailer	75–150	125	280–570	470
Campground (developed)	Person	20–40	30	76–150	110
Cocktail lounge	Seat	12–25	20	45–95	76
Coffee shop	Customer	4–8	6	15–30	23
	Employee	8–12	10	30–45	38
Country club	Member present	60–130	100	230–490	380
	Employee	10–15	13	38–57	49
Dining hall	Meal served	4–10	7	15–38	26
Dormitory, bunkhouse	Person	20–50	40	76–190	150
Fairground	Visitor	1–2	2	4–8	8
Hotel, resort	Person	40–60	50	150–230	190
Picnic park with flush toilets	Visitor	5–10	8	19–38	30
Store, resort	Customer	1–4	3	4–15	11
	Employee	8–12	10	30–45	38
Swimming pool	Customer	5–12	10	19–45	38
	Employee	8–12	10	30–45	38
Theater	Seat	2–4	3	8–15	11
Visitor center	Visitor	4–8	5	15–30	19

*Adapted in part from Tchobanoglous and Burton (1991).

TABLE 4-5
Typical rates of water use for various industries*

Industry	Range of flow, gal/ton product	Range of flow, $m^3/10^3$ kg product
Cannery		
Green beans	10,000–16,000	42–67
Peaches and pears	3,000–4,000	13–17
Other fruits and vegetables	900–6,000	4–25
Chemical		
Ammonia	20,000–60,000	83–250
Carbon dioxide	12,000–18,000	50–75
Lactose	100,000–180,000	420–750
Sulfur	1,800–2,200	8–9
Food and beverage		
Beer	2,000–3,600	8–15
Bread	400–900	2–4
Meat packing	3,000–4,000†	13–17
Milk products	2,000–4,000	8–17
Whiskey	12,000–18,000	50–75
Pulp and paper		
Pulp	40,000–160,000	170–670
Paper	20,000–30,000	83–125
Textiie		
Bleaching	40,000–60,000‡	170–250
Dyeing	6,000–12,000‡	25–50

*Adapted in part from Tchobanoglous and Burton (1991).
† Live weight.
‡ Cotton.

Infiltration into the collection system, due to the presence of high groundwater, results in an increase in the quantity of wastewater that must be handled and a reduction in the wastewater constituent concentrations. The amount of flow that can enter a sewer from groundwater, or infiltration, may range from 100 to 10,000 gal/d·in-mi (0.0093 to 0.93 m^3/d·mm-km) or more, depending on the type and condition of the collection system and the maintenance the collection system has received. The number of inch-miles (millimeter-kilometers) in a wastewater collection system is the sum of the products of sewer diameters, in inches (millimeters), times the lengths, in miles (kilometers), of sewers of corresponding diameters. Another approach that has been used to define the quantity of infiltration is to use an area-based allowance. Typical values based on area will vary from 20 to 3000 gal/ac·d (0.19 to 28 m^3/ha·d) (Tchobanoglous and Burton, 1991). Because direct inflow can cause an almost immediate increase in flowrates in collection systems, it is important to determine the peak flowrates that must be handled.

Individual Residences Served by Septic Tanks

The flow from individual residences served by septic tanks is based on allocating the total water use to household and personal uses. Assuming that the household use consists of 10 gal for dishwashing, 25 gal for laundry, and 5 gal for miscellaneous uses, and that personal use consists of 2 gal for drinking and cooking, 3 gal for oral hygiene, 14 gal for bathing, and 16 gal for toilet flushing, the flow from a residence would be:

$$\text{Flow, gal/home}\cdot\text{d} = 40\ \text{gal/home}\cdot\text{d} + 35\ \text{gal/person}\cdot\text{d} \times (\text{number of persons/home}) \qquad (4\text{-}1)$$

Applying Eq. (4-1) to a home with 3.0 persons results in an average flow per resident of 48.3 gal, which correlates well with the values given in Table 4-1. Equation (4-1) can be revised to account for other household uses and the use of low-flush toilets and fixtures. For example, if 2.0 gal/flush toilets are used, the corresponding average flow for a residence with 3.0 occupants, based on five flushes per resident per day, would be 42.3 gal/capita·d.

Strategies for Reducing Water Use and Wastewater Flowrates

Because of the importance of conserving both resources and energy, various means for reducing wastewater flowrates and pollutant loadings from domestic sources are gaining increasing attention. The reduction of wastewater flowrates from domestic sources results directly from the reduction in interior water use. Therefore, the terms *interior water use* and *domestic wastewater flowrates* are used interchangeably. Representative water use rates for various external and internal activities are reported in Table 4-6. Information on the relative distribution of water use within a

TABLE 4-6
Typical rates of water use for various devices and appliances*

Industry device/appliance	Unit	Range of flow	Typical
Automatic home-type washing machine	gal/load	30–50	40
Automatic home-type dishwasher	gal/load	4–8	6
Bathtub	gal/use	23–30	26
Continuous-flowing drinking fountain	gal/min	1–2	1
Dishwashing machine, commercial			
Conveyor type, at 15 lb_f/in^2	gal/load	4–6	5
Stationary rack type, at 15 lb_f/in^2	gal/load	6–9	8
Fire hose, 1 ½ in, ½-in nozzle, 65-ft head	gal/load	35–40	38
Garbage disposal unit, home type	gal/load	1500–1900	1800
Garbage grinder, home type	gal/person·d	1–3	2
Garden hose, ⅝ in, 25-ft head	gal/min	3–5	4
Garden hose, ¾ in, 25-ft head	gal/min	4–5	5
Sprinkler	gal/min	2–4	3
Lawn sprinkler, 3000-ft^2 lawn, 1 in/wk	gal/wk	1500–1900	1800
Shower head, ⅝ in, 25-ft head	gal/min	23–28	25
Washbasin	gal/use	1–2	2
Toilet, flush valve type, 25 lb_f/in^2	gal/min	23–28	25
Toilet, tank type	gal/use	4–6	5

*Adapted in part from Tchobanoglous and Burton (1991).

TABLE 4-7
Typical distribution of residential interior water use*

Use	Percent of total: Range	Percent of total: Typical
Bathing (bath/shower)	15–25	20
Dishwashing	5–10	7
Clothes washing	15–25	20
Faucets	8–12	10
Kitchen food-waste grinder	2–5	3
Toilets	20–40	30
Miscellaneous (e.g., toilet leakage)	8–12	10
		100

*Without water-conserving fixtures.

TABLE 4-8
Flow-reduction devices and appliances

Device/appliance	Description and/or application
Faucet aerators	Increases the rinsing power of water by adding air and concentrating flow, thus reducing the amount of wash water used.
Flow-limiting shower heads	Restricts and concentrates water passage by means of orifices that limit and divert shower flow for optimum use by the bather.
Low-flush toilets	Reduces the discharge of water per flush.
Pressure-reducing valve	Maintains home water pressure at a lower level than that of the water distribution system. Decreases the probability of leaks and dripping faucets.
Pressurized shower	Water and compressed air are mixed together. Impact provides the sensation of conventional shower.
Retrofit kits for bathroom	Kits may consist of shower flow restrictors, toilet dams or fixture displacement bags, and toilet leak detector tablets.
Toilet dam	A partition in the toilet tank that reduces the amount of water per flush.
Toilet leak detectors	Tablets that dissolve in the toilet tank and release dye to indicate leakage of the flush valve.
Vacuum toilet	Vacuum along with a small amount of water is used to remove feces from toilet.
Water-efficient dishwasher	Reduces the amount of water used to wash dishes.
Water-efficient clothes washer	Reduces the amount of water used to wash clothes.

TABLE 4-9
Reductions achieved by flow-reduction devices and appliances*,†

	Flow reduction	
Device/appliance	**gal/capita·d or unit**	**L/capita·d or unit**
Faucet aerator	0.5	2
Limiting-flow shower heads		
3 gal/min	7	26
0.5 gal/min	14	53
Low-flush toilets		
3.4 gal/flush	8	30
0.5 gal/flush	20	76
Pressure-reducing valve, percent	3–6	3–6
Retrofit kits for bathroom fixtures	4–7	15–26
Toilet dam	4	15
Toilet leak detectors, gal/toilet·d	24	91
Water-efficient dishwasher	1	4
Water-efficient clothes washer (typically front-loaded)	1.5	6

*Adapted in part from Tchobanoglous and Burton (1991).
† As compared to conventional (non-water-conserving) devices or appliances.

residence is reported in Table 4-7. The principal devices and appliances that are used to reduce interior domestic water use and wastewater flows are described in Table 4-8. The actual flow reduction that is possible using these devices and appliances, as compared with the flows from the conventional devices, is reported in Table 4-9.

Another method of achieving flow reduction that has been adopted by a number of communities is to restrict the use of appliances, such as automatic dishwashers and kitchen food-waste grinders (i.e., garbage disposal units), that tend to increase water consumption. In some communities, the washing out of bottles and cans that are to be recycled is discouraged because of the additional organic load placed on wastewater treatment facilities. The use of one or more of the flow-reduction devices is now specified for all new residential dwellings in many communities; in others, the use of waste food grinders has been limited in new housing developments. Further, many individuals concerned about conservation have installed such devices on their own, as a means of reducing water consumption. A comparison of residential interior water use (and resulting per capita wastewater flows) is given in Table 4-10 for homes without and with water-conserving fixtures. In Sweden, to further reduce water usage, a toilet has been developed to isolate urine from feces. By isolating the two waste streams, water can be conserved and alternative methods of treatment can be implemented.

TABLE 4-10
Typical comparisons of interior water use without and with water conservation practices and devices

	Flow, gal/capita·d	
Use	Without water conservation	With water conservation
Bathing (bath/shower)	14	12
Dishwashing	4.9	3
Clothes washing	14	12
Faucets	7	5
Kitchen food-waste grinder	2.1	
Toilets	21	10
Miscellaneous (e.g., toilet leakage)	7.0	8
Total	70.0	50

4-2 WASTEWATER CONSTITUENT CONCENTRATIONS

The physical, chemical, and biological characteristics of wastewater vary throughout the day. An adequate determination of the waste characteristics will result only if the sample tested is representative. Typically, composite samples made up of portions of samples collected at regular intervals during a day are used (see Fig. 4-1). The amount of liquid used from each sample is proportional to the rate of flow at the time the sample was collected. Adequate characterization of wastewater is of fundamental importance in the design of treatment and disposal processes.

Quantity of Waste Discharged by Individuals

Typical data on the total quantities of waste discharged per person per day from individual residences are reported in Table 4-11. The data presented in Table 4-11 have been gathered from numerous sources (primarily in the United States) by the authors. The total number of pathogenic organisms discharged will depend on whether an individual is ill and is shedding pathogens. If one or more members of a family are ill and shedding pathogens, the number of measured organisms can increase by several orders of magnitude. The corresponding constituent concentrations, assuming the quantities of waste given in Table 4-11 were diluted in 50 and 120 gal of water, are reported in Table 4-12. As noted in Table 4-12, the typical mass amounts of waste discharged (column 3) are based on the assumption that 25 percent of the homes are equipped with kitchen food-waste grinders. The dilution values used correspond to the average flowrates for individual residences (50 gal/capita·d) and collection systems (120 gal/capita·d).

Composition of Wastewater from Industrial Activities

Typical constituent concentration data for two industrial operations are presented in Table 4-13. From the data in Table 4-13, it can be observed that flow values and

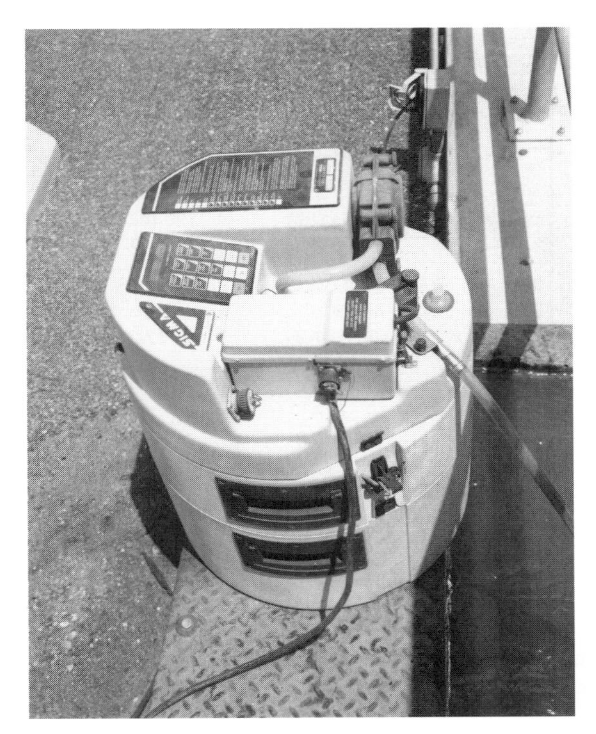

FIGURE 4-1
Composite sampler used to collect flow-weighted samples.

TABLE 4-11
Quantity of waste discharged by individuals on a dry weight basis*

	Value, lb/capita·d			Value, g/capita·d		
Constituent (1)	Range (2)	Typical without ground up kitchen waste (3)	Typical with ground up kitchen waste (4)	Range (5)	Typical without ground up kitchen waste (6)	Typical with ground up kitchen waste (7)
BOD_5	0.11–0.26	0.180	0.220	50–120	80	100
COD	0.30–0.65	0.420	0.480	110–295	190	220
TSS	0.13–0.33	0.200	0.250	60–150	90	110
NH_3 as N	0.011–0.026	0.017	0.019	5–12	7.6	8.4
Org. N as N	0.009–0.022	0.012	0.013	4–10	5.4	5.9
TKN† as N	0.020–0.048	0.029	0.032	9–21.7	13	14.3
Org. P as P	0.002–0.004	0.0026	0.0028	0.9–1.8	1.2	1.3
Inorg. P as P	0.004–0.006	0.0044	0.0048	1.8–2.7	2.0	2.2
Total P as P	0.006–0.010	0.0070	0.0076	2.7–4.5	3.2	3.5
Oil and grease	0.022–0.088	0.0661	0.075	10–40	30	34

*Developed from numerous sources. Data on the number of microorganisms present in septic tank effluent and untreated wastewater may be found in Table 2-21 in Chap. 2.
† TKN is total Kjeldahl nitrogen.

TABLE 4-12

Typical data on the unit loading factors and expected wastewater constituent concentrations from individual residences

			Concentration, mg/L	
			Volume, gal/capita·d (L/capita·d)	
Constituent	Unit	Typical value*	50 (189)	120 (454)
BOD_5	g/capita·d	85	450	187
COD	g/capita·d	198	1050	436
TSS	g/capita·d	95	503	209
NH_3 as N	g/capita·d	7.8	41.2	17.2
Org. N as N	g/capita·d	5.5	29.1	12.1
TKN as N	g/capita·d	13.3	70.4	29.3
Org. P as P	g/capita·d	1.23	6.5	2.7
Inorg. P as P	g/capita·d	2.05	10.8	4.5
Total P as P	g/capita·d	3.28	17.3	7.2
Oil and grease	g/capita·d	31	164	68

*Data from Table 4-11, Columns 6 and 7, assuming 25% of the homes have kitchen food-waste grinders.

TABLE 4-13

Typical range of effluent constituent concentrations resulting from two industrial activities

		Winery		Tomato cannery	
Constituent	Unit	Peak season*	Off season†	Peak season‡	Off season§
Flowrate	Mgal/d	0.30–0.46	0.12–0.27	1.1–5.9	0.3–1.7
pH		3.8–7.8	3.8–7.8	7.2–8.0	7.2–8.0
BOD	mg/L			460–1100	29–56
COD	mg/L	7500–44,000	750–13,000		
SS	mL/L	0–2	0–1	6–80	0.5–2.2
TSS	mg/L	57–3,950	12–400	270–760	69–120
TDS	mg/L	315–1,240	214–720	480–640	360–520
EC	μmho/cm			880–990	610–700
Nitrate	mg/L	0.63–362	0.23–53	0.4–5.6	2.2–0.1
Ammonia	mg/L	2.25			
Phosphorus	mg/L	0.25–2.75		1.5–7.4	0.3–3.9
Sulfate	mg/L	20–144	20–57	15–23	9.9–7.1
Sulfide	mg/L	0.05–2.3	0.05		
DO	mg/L	2.3–6.3	2.3–6.3	0.9–3.8	1.6–9.8
Temperature	°C			18–23	13–19

*Peak season runs from September through March.

† Off season runs from April to August.

‡ Peak season runs from early July to late September, when fresh-harvest tomatoes are canned. Treatment consists of screening and brief sedimentation.

§ Off season runs from November to June, when canned tomatoes are remanufactured into tomato paste, tomato sauce, and other tomato products (e.g., salsa, ketchup, spaghetti sauce). Treatment typically consists of screening, aeration, and sedimentation.

water quality measurements may vary over several orders of magnitude over a period of a year. Because of this variation, it is often difficult to define "typical operating conditions" for industrial activities. If industrial wastes are to be discharged into the collection system, it will be necessary to characterize them adequately before and after pretreatment to avoid plant upsets, especially with small systems.

Composition of Wastewater in Collection Systems

Typical data on the composition of untreated domestic wastewater as found in wastewater collection systems are reported in Table 4-14. The data presented in Table 4-14 are based on an average flow of 120 gal/capita·d and include constituents

TABLE 4-14
Typical composition of untreated domestic wastewater*

		Concentration	
Contaminants	**Unit**	**Range**	**Typical**†
Solids, total (TS)	mg/L	350–1200	700
Dissolved solids, total (TDS)	mg/L	280–850	500
Fixed	mg/L	145–525	300
Volatile	mg/L	105–325	200
Suspended solids, total (TSS)	mg/L	100–350	210
Fixed	mg/L	20–75	55
Volatile	mg/L	80–275	160
Settleable solids	mL/L	5–20	10
Biochemical oxygen demand, 5-d, 20°C (BOD_5, 20°C)	mg/L	110–400	210
Total organic carbon (TOC)	mg/L	80–290	160
Chemical oxygen demand (COD)	mg/L	250–1000	500
Nitrogen (total as N)	mg/L	20–85	35
Organic	mg/L	8–35	13
Free ammonia	mg/L	12–50	22
Nitrites	mg/L	0–0	0
Nitrates	mg/L	0–0	0
Phosphorus (total as P)	mg/L	4–15	7
Organic	mg/L	1–5	2
Inorganic	mg/L	3–10	5
Chlorides‡	mg/L	30–100	50
Sulfate‡	mg/L	20–50	30
Oil and grease	mg/L	50–150	90
Volatile organic compounds (VOCs)	mg/L	<100 to >400	100–400
Total coliform	no./100 mL	10^6–10^9	10^7–10^8
Fecal coliform	no./100 mL	10^3–10^7	10^4–10^5
Cryptosporidium oocysts	no./100 mL	10^{-1}–10^2	10^{-1}–10^1
Giardia lamblia cysts	no./100 mL	10^{-1}–10^3	10^{-1}–10^2

*Adapted from Tchobanoglous and Burton (1991).
† Based on a flow of 120 gal/capita·d. Additional data on the number of microorganisms present in septic tank effluent and untreated wastewater may be found in Table 2-21 in Chap. 2.
‡ Values should be increased by amount present in domestic water supply.

added by commercial, institutional, and industrial sources. As shown in Table 4-14, there is a significant range in the values reported for the individual constituents. Recognizing that there is no such thing as a typical wastewater, it must be emphasized that the typical data presented in Table 4-14 *should be used only as a guide.* The constituent concentrations presented in Table 4-12, developed from the waste amounts given in Table 4-11, can also be compared to the values given in Table 4-14. It is interesting to note that the values given in Table 4-12 correspond quite closely to the values given in Table 4-14 for typical wastewater.

Mineral Increase Resulting from Water Use

Data on the increase in the mineral content of wastewater resulting from water use, and the variation of the increase within a sewerage system, are especially important in evaluating the reuse potential of wastewater. Typical data on the incremental increase in mineral content that can be expected in municipal wastewater resulting from domestic use are reported in Table 4-15. Increases in the mineral content of wastewater result from domestic use, from the addition of highly mineralized water

TABLE 4-15
Typical mineral pickup from domestic water use*

	Increment range, mg/L†	
Constituent	**In septic tank effluent**	**In municipal wastewater**
Anions:		
Bicarbonate (HCO_3)	100–200	50–100
Carbonate (CO_3)	2–20	0–10
Chloride (Cl)	40–100	20–50‡
Sulfate (SO_4)	30–60	15–30
Cations:		
Calcium (Ca)	10–20	6–16
Magnesium (Mg)	8–16	4–10
Potassium (K)	10–20	7–15
Sodium (Na)	60–100§	40–70§
Other constituents:		
Aluminum (Al)	0.2–0.3	0.1–0.2
Boron (B)	0.1–0.4	0.1–0.4
Flouride (F)	0.2–0.4	0.2–0.4
Manganese (Mn)	0.2–0.4	0.2–0.4
Silica (SiO_2)	2–10	2–10
Total alkalinity (as $CaCO_3$)	60–120	60–120
Total dissolved solids (TDS)	200–400	150–380

*Adapted from Tchobanoglous and Burton (1991).
†Based on 50 and 120 gal/capita·d, respectively.
‡Reported values do not include commercial and industrial additions.
§Excluding the addition from domestic water softeners.

TABLE 4-16
Typical data on the expected effluent wastewater characteristics from a residential septic tank without and with an effluent filter vault*

		Concentration, mg/L					
		Without effluent filter			With effluent filter		
Constituent (1)	Typical complete mix value† mg/L (2)	Range (3)	Typical without ground up kitchen waste (4)	Typical with ground up kitchen waste (5)	Range (6)	Typical without ground up kitchen waste (7)	Typical with ground up kitchen waste (8)
BOD_5	450	150–250	180	190	100–140	130	140
COD	1050	250–500	345	400	160–300	250	300
TSS	503	40–140	80	85	20–55	30	30
NH_3 as N	41.2	30–50	40	44	30–50	40	44
Org. N as N	29.1	20–40	28	31	20–40	28	31
TKN as N	70.4	50–90	68	75	50–90	68	75
Org. P as P	6.5	4–8	6	6	4–8	6	6
Inorg. P as P	10.8	8–12	10	10	8–12	10	10
Total P as P	17.3	12–20	16	16	12–20	16	16
Oil and grease	164	20–50	25	30	10–20	15	20

*With assistance from Bounds (1997).
† Data from Table 4-12, column 4. Concentration if waste constituents were mixed completely.

from private wells and groundwater, and from industrial use. Domestic and industrial water softeners also contribute significantly to the increase in mineral content and, in some areas, may represent the major source. Occasionally, water added from private wells and groundwater infiltration will (because of its high quality) serve to dilute the mineral concentration in the wastewater.

Composition of Septic Tank Effluent

Typical data on the composition of septic tank effluent, based in part on the data given in Tables 4-11 and 4-12, are presented in Table 4-16 for septic tanks without and with effluent filter vaults (see Sec. 5-7, in Chap. 5) and with and without kitchen food-waste grinders. The beneficial effect of using an effluent filter vault, in terms of reduced constituent concentrations, is clearly evident by comparing columns 4 and 5 and 7 and 8. For the purpose of comparison, the constituent concentrations that would have been expected if the wastes discharged to the septic tank had been mixed completely are reported in column 3. The importance of the septic tank as a pretreatment process can be appreciated by comparing column 2 to columns 4 and 5, or to columns 7 and 8. Here again, because of the significant variations observed in the constituent concentrations in septic tank effluent, the values given in Table 4-16 *should be used only as a guide.*

4-3 VARIATIONS IN WASTEWATER FLOWRATES, CONSTITUENT CONCENTRATIONS, AND MASS LOADING RATES

The rated capacity of wastewater treatment plants is normally based on the average annual daily flowrate at the design year. As a practical matter, however, wastewater treatment plants have to be designed to meet a number of conditions that are influenced by flowrates, wastewater characteristics and constituent concentrations, and a combination of both (mass loading). Peaking conditions that must be considered include peak hydraulic flowrates and peak process constituent mass loading rates. The variation in wastewater flowrates, constituent concentrations, and mass loading rates in wastewater collection systems and from individual residences is considered in this section.

Definition of Terms

Before considering the variations in flowrates and constituent concentrations, it will be helpful to define some terminology that is used commonly to quantify the variations that are observed. The principal terms used to describe these observed variations are defined in Table 4-17. As will be discussed in Secs. 4-5 and 4-6, these terms are also of importance in the selection and sizing of individual unit treatment processes and operations.

Variation in Wastewater Flowrates

The variation in flowrates in wastewater collection systems and from individual homes is considered in the following discussion. The importance of flowrate variations in the design of wastewater treatment plants is identified in Table 4-18.

In wastewater collection systems. Typical flowrate variations that can be expected in collection systems are illustrated in Figs. 4-2 and 4-3. As shown in both figures, the flow pattern is bimodal with the first peak occurring about 10 A.M. to 12 noon and the second peak occurring between 7 and 9 P.M. In the past, it was quite common to have a single pronounced peak in the morning and a flatter peak in the early evening hours. However, as more couples both work, the bimodal flow pattern is becoming the norm. In many communities the evening flow peak is higher than the morning peak. It should be noted that instantaneous and peak day flowrates measured in collection systems may be considerably below the actual peak flows, because of limitations in the metering facilities.

From individual residences. The flowrate variations that can be expected from an individual residence are, as shown in Fig. 4-4, quite variable, going from no flow in the early morning hours (other than leakage) to a peak flowrate that may be 4 to 6 times as great as the average flowrate. While the flowrate variation from an individual home is quite variable and unpredictable, the flowrate variation for 60 or more homes is similar to that observed for small systems (see Fig. 4-2).

TABLE 4-17
Terminology used to quantify observed variations in flowrate and constituent concentrations

Item	Description
Average dry weather flow (ADWF)	The average of the daily flows sustained during dry weather periods with limited infiltration.
Average wet weather flow (AWWF)	The average of the daily flows sustained during wet weather periods when infiltration is a factor.
Average annual daily flow	The average flowrate occurring over a 24-h period based on annual flowrate data.
Instantaneous peak	Highest record flowrate occurring for a period consistent with the recording equipment. In many situations the recorded peak flow may be considerably below the actual peak flow because of metering and recording equipment limitations.
Minimum (average) hour	The average of the minimum flows sustained for the period of an hour in the record examined.
Minimum (average) day	The average of the minimum flows sustained for the period of a day in the record examined.
Minimum (average) week	The average of the minimum flows sustained for the period of a week in the record examined.
Minimum (average) month	The average of the minimum flows sustained for the period of a month in the record examined.
Peak (average) hour	The average of the peak flows sustained for a period of an hour in the record examined.
Peak (average) day	The average of the peak flows sustained for a period of a day in the record examined.
Peak (average) week	The average of the peak flows sustained for a period of a week in the record examined.
Peak (average) month	The average of the peak flows sustained for a period of a month in the record examined.
Sustained flow	The flowrate value sustained or exceeded for a given period of time (e.g., 1 hour, 1 day, 1 week, 1 month).

Peak, Minimum, and Sustained Wastewater Flowrates

Peak flows are flows of a given magnitude that are sustained for a specified period of time. Because it is difficult to compare numerical peak flow values from different wastewater treatment plants, peak flowrate values are normalized by dividing by the long-term average flowrate. The resultant ratio, known as a *peaking factor*, is defined as follows.

$$\text{Peaking factor, PF} = \frac{\text{Peak average flowrate}}{\text{Average flowrate}} \qquad (4\text{-}2)$$

TABLE 4-18
Typical flowrate and mass loading factors used for the design and operation of wastewater treatment plant facilities

Factor	Application
	Based on flowrate
Average daily flow	Used to develop flowrate ratios and to estimate pumping and chemical costs
Minimum (average) hour	Sizing turndown of pumping facilities and low range of plant flowmeter
Minimum (average) day	Sizing of influent channels to control solids deposition; sizing effluent recycle requirements for trickling filters
Minimum (average) month	Selection of minimum number of operating units required during low-flow periods
Peak (average) hour	Sizing of pumping facilities and conduits. Sizing of physical unit operations: grit chambers, sedimentation tanks, and filters; sizing chlorine contact tanks
Peak (average) day	Sizing of equalization basins, chlorine contact tanks, sludge pumping system
Peak (average) week	Record keeping and reporting
Peak (average) month	Record keeping and reporting; sizing of chemical storage facilities
	Based on mass loading
Minimum (average) month	Process turndown requirements
Minimum (average) day	Sizing of trickling filter recycle
Peak (average) day	Sizing of selected process units
Peak (average) month	Sizing of sludge storage facilities; sizing of composting requirements
Sustained (average) flows	Sizing of selected process units

The most common method of determining the peaking factor is from the analysis of flowrate data. Where flowrate records are available, at least 2 years of data should be analyzed to define the peak to average day peaking factor. The procedure used to define the average peaking factor is as follows. The first step is to determine the average flowrate for the period of record. The next step involves searching the record for the highest recorded flowrates that were sustained over a period of one day (see Fig. 4-5). The peak day average flowrate is obtained by integrating the area under the curve for the day. The peak factor is obtained by dividing the peak day average flow by the average flowrate. The minimum flow to average rate is obtained in a similar manner. A curve of sustained peak to average and minimum to average flows, such as shown in Fig. 4-6, is obtained by repeating the above process for two consecutive days, three consecutive days, etc., and plotting the measured ratios for the period of interest.

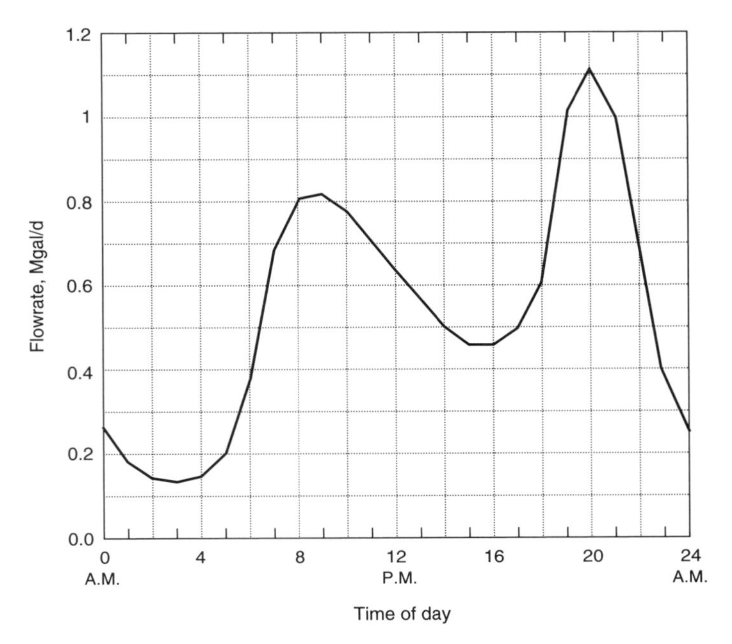

FIGURE 4-2
Typical hourly variation in domestic wastewater flowrates measured at the inlet to a small wastewater treatment plant.

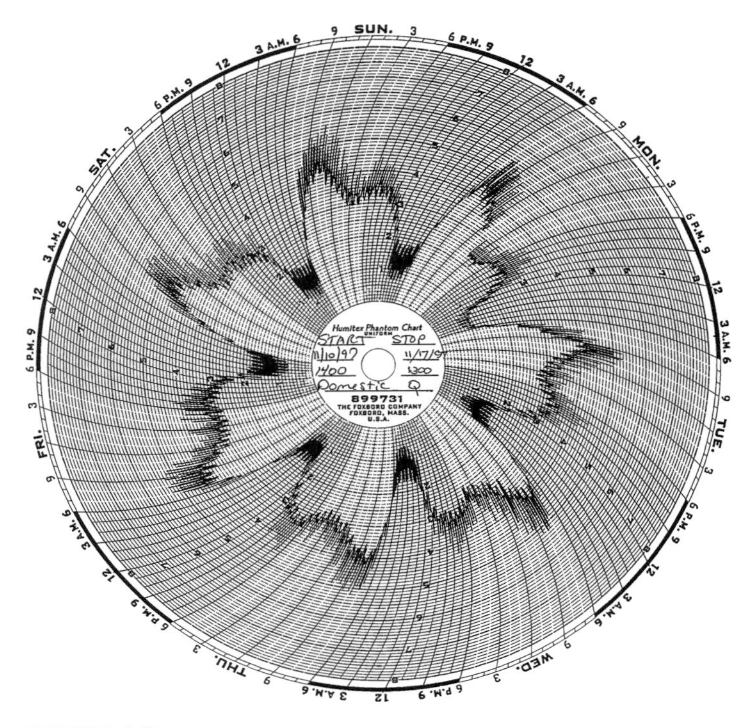

FIGURE 4-3
Typical weekly flowrate record measured at a pump station in San Diego. (Courtesy J. Swerlein, 1998)

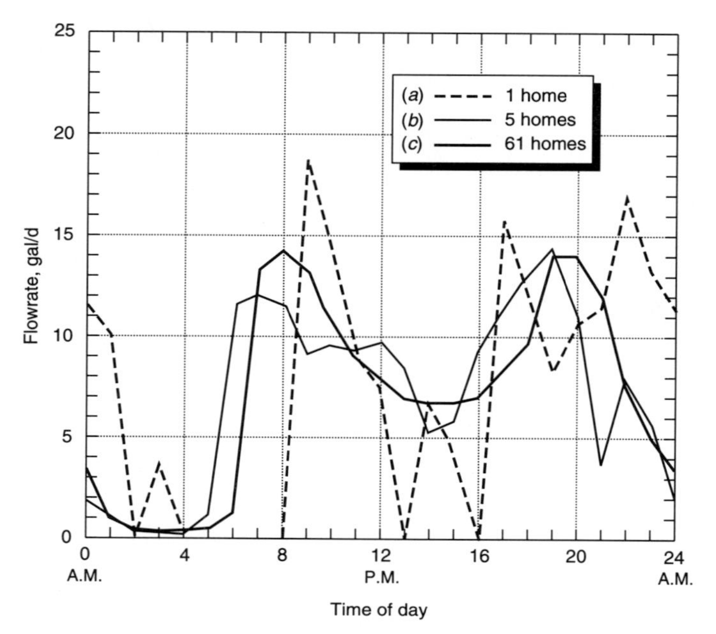

FIGURE 4-4
Typical daily flowrates for individual residences: (*a*) single home, (*b*) average of about five homes, (*c*) average of 61 homes (*b, c* courtesy Baker, 1990).

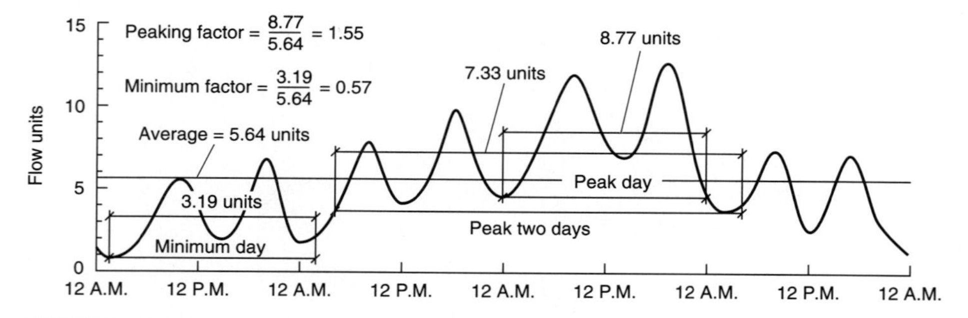

FIGURE 4-5
Definition sketch for determining sustained peak flows.

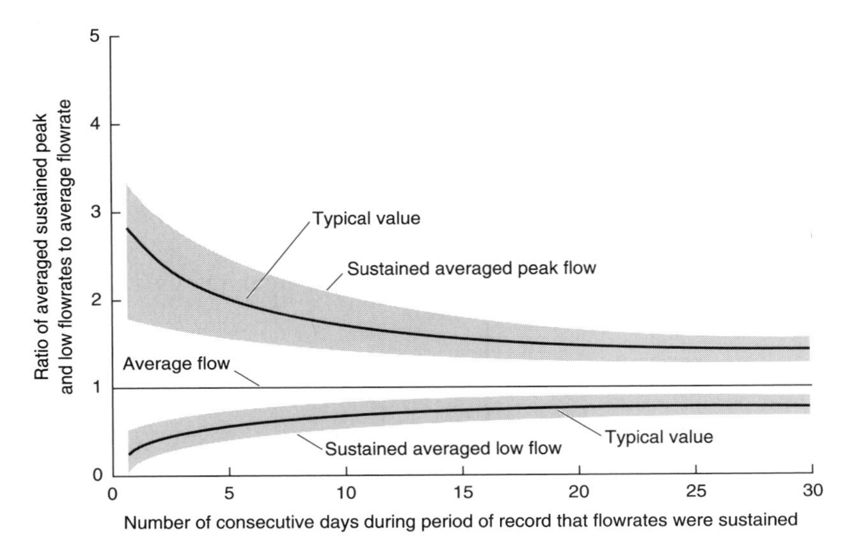

FIGURE 4-6
Typical ratios of average sustained peak and low daily flowrates to average annual daily flowrates for small treatment plants (0.1 to 1.0 Mgal/d).

In wastewater collection systems. Peak hydraulic flowrates are important so that the unit operations and processes and their interconnecting conduits can be sized appropriately to handle the applied flowrates (see Table 4-18). Peak hourly flowrates are important in the sizing of pumping stations and in sizing pumps. Typical sustained peak to average and minimum to average flowrates for small treatment plants are given graphically in Fig. 4-6. The flows considered in Fig. 4-6 include domestic and small amounts of commercial and industrial flows. The effects of infiltration/inflow would be superimposed on the values given in Fig. 4-6.

From individual residences. The peaking factor for individual residences in homes with septic tanks will vary with the time of the week and usage. Peak hourly rates are associated with the discharge of clothes washers or shower/bath water. In most cases, peak flows are attenuated as they flow through the household plumbing. Peak effluent flowrates from the septic tank are dampened further because of the volume of the tank and the nature of the outlet piping, which serves to limit the discharge of peak flows. The ratio of the peak day to the average day can vary from 1.5 to 5 or more, depending on the season of the year and usage (e.g., weekend guests at a summer cottage), with values of 2 to 3 being most common. Low to average flowrates can vary from zero to about 0.8, with a value of about 0.5 being typical.

Variations in Constituent Concentrations in Collection Systems

The principal factors responsible for variations in constituent concentrations include: (1) the established habits of community residents, which cause short-term (hourly, daily, and weekly) variations; (2) seasonal conditions, which usually cause both short- and longer-term variations; and (3) nondomestic discharges, which cause both short- and long-term variations. The presence of extraneous water such as infiltration/inflow in collection systems tends to decrease the concentrations of BOD and TSS, but the actual decrease observed depends on the characteristics of the water entering the collection system. In some cases, concentrations of some inorganic constituents may actually increase where the groundwater contains high levels of dissolved constituents. In both separate and combined conventional wastewater collection systems, seasonal variations in BOD and TSS are primarily a function of the amount of stormwater that enters the system.

Short-term variations. The typical variation observed in BOD and TSS over a 24-h period is illustrated in Fig. 4-7. The BOD variation follows the flow variation. The peak BOD (organic matter) concentration, measured at the inlet to wastewater treatment plants, typically occurs in the evening around 9 P.M. Wastewater from combined wastewater collection systems usually contains more inorganic matter than wastewater from sanitary sewer systems because of the larger quantities of storm drainage that enter the combined sewer system.

Seasonal variations. For domestic flow only, and neglecting the effects of infiltration, the unit (per capita) loadings and the strength of the wastewater from

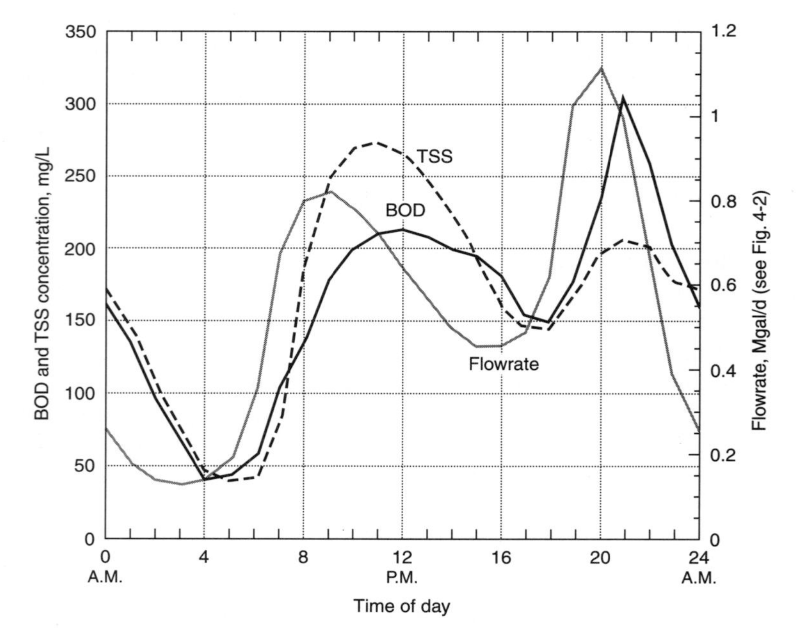

FIGURE 4-7
Daily variations in BOD and TSS concentrations measured at the inlet to a small wastewater treatment plant.

most seasonal sources, such as resorts, will remain about the same on a daily basis throughout the year, even though the total flowrate varies. The total mass of BOD and TSS of the wastewater, however, will increase directly with the population served. In general, BOD and TSS values are below average during the spring and other periods of high rainfall.

Nondomestic discharges. The concentrations of both BOD and TSS in nondomestic wastewater can vary significantly throughout the day. For example, the BOD and TSS contributed from a small vegetable processing facility during the noon washup period may far exceed values contributed during working hours. Problems with high short-term loadings most commonly occur in small treatment plants that have limited reserve capacity to handle these so-called shock loadings. Where industrial wastes are to be accommodated in small wastewater treatment facilities, special attention must be given to (1) developing an adequate characterization of the wastewater, (2) defining current and proposed wastewater flowrates, and (3) assessing of the technical and economic feasibility of flow equalization.

Variations in Constituent Concentrations in the Effluent from Septic Tanks

The concentrations of both BOD and TSS discharged from a septic tank serving an individual residence will vary somewhat with (1) the day of the week, (2) the time of day, and (3) the activities within the home. However, because the septic tank serves as an online equalization basin, the variations in the constituent concentrations are dampened considerably and will approach the typical values reported previously.

Variation in Mass Loading Rates for Wastewater Treatment Plants

As noted previously, knowledge of the expected flowrates, concentrations, and mass loading factors is of fundamental importance in the design and operation of wastewater treatment facilities. The importance of mass loading in the design of treatment facilities is identified in Table 4-18. The development of flow-weighted constituent concentration and mass loading rates is considered below.

Flow-weighted constituent concentrations. Flow-weighted constituent concentrations are obtained by multiplying the flow (typically hourly values over a 24-h period) by the corresponding constituent concentration, summing the results, and dividing by the summation of the flows as given by Eq. (4-3).

$$C_w = \frac{\sum_{i=1}^{n} C_i q_i}{\sum_{i=1}^{n} q_i} \tag{4-3}$$

Whenever possible, flow-weighted constituent concentrations should be used because they are a more accurate representation of the actual wastewater strength that must be treated. Determination of the simple arithmetic average and flow-weighted constituent concentrations is illustrated in Example 4-1.

EXAMPLE 4-1. CALCULATION OF FLOW-WEIGHTED BOD AND TSS CONCENTRATIONS. Compute the flow-weighted BOD and TSS values using the data provided in Figures 4-2 and 4-7 for a small community of about 5000 persons. Compare the flow-weighted values to the simple arithmetic averages. What is the significance of the difference?

Solution

1. Create a spreadsheet for calculating the flow-weighted values. Divide the BOD, TSS, and flow curves into 24 one-hour periods. Enter the time intervals (e.g., 12 to 1 A.M.) in column 1.
2. For each time period, calculate the average BOD value during the interval. For example, the average BOD value during the first interval (12 to 1 A.M.) is

Time interval (1)	BOD, mg/L (2)	TSS, mg/L (3)	Flowrate q, Mgal/d (4)	BOD × q (5) = (2) × (4)	TSS × q (6) = (3) × (4)
12–1 A.M.	146.5	157.5	0.220	32.2	34.7
1–2	112.5	124.0	0.160	18.0	19.8
2–3	78.5	91.0	0.135	10.6	12.3
3–4	52.5	62.0	0.135	7.1	8.4
4–5	43.0	43.5	0.165	7.1	7.2
5–6	52.0	41.0	0.280	14.6	11.5
6–7	83.5	63.5	0.525	43.8	33.3
7–8	123.5	140.5	0.745	92.0	104.7
8–9	159.5	223.5	0.815	130.0	182.2
9–10	191.0	260.5	0.800	152.8	208.4
10–11	206.5	272.0	0.750	154.9	204.0
11–12	212.0	270.0	0.680	144.2	183.6
12–1 P.M.	210.5	257.5	0.605	127.4	155.8
1–2	204.0	237.0	0.535	109.1	126.8
2–3	197.5	210.0	0.480	94.8	100.8
3–4	188.5	178.0	0.460	86.7	81.9
4–5	169.0	154.0	0.475	80.3	73.2
5–6	153.0	146.0	0.555	84.9	81.0
6–7	164.5	157.0	0.825	135.7	129.5
7–8	204.5	183.5	1.075	219.8	197.3
8–9	267.5	202.0	1.060	283.6	214.1
9–10	283.5	203.5	0.830	235.3	168.9
10–11	232.5	190.5	0.525	122.1	100.0
11–12	182.0	176.0	0.325	59.2	57.2
Totals	3918.0	4044.0	13.160	2446.1	2496.4
Average values	163.3	168.5	0.548		
Flow-weighted concentration values				185.9	189.7

$$\text{Value at beginning of interval} = 161$$

$$\text{Value at end of interval} = 132$$

$$\text{Average BOD} = \frac{161 + 132}{2} = 146.5$$

The average BOD values for each successive time interval are entered in column 2.

3. Enter the average values for TSS and flow in columns 3 and 4, respectively.
4. For each time period, multiply the average BOD value (column 2) by the average flowrate (column 4), and enter the results in column 5.
5. For each time period, multiply the average TSS (column 3) value by the average flowrate (column 4), and enter the results in column 6.
6. Calculate the sum and simple arithmetic average for columns 2 through 6.
7. Divide the sum of the values in columns 5 and 6 (BOD $\times$ flow and TSS $\times$ flow, respectively) by the sum of the values in column 4 (flow) to obtain the flow-weighted average for BOD and TSS. The resulting values are given in the last two lines of the spreadsheet.

Comment. When the computation of a simple average is compared to a flow-weighted value, the differences can be significant. In this example, if simple averages were used, the BOD loading would have been understated by 22.6 mg/L (12 percent), and the TSS loading by 21.2 mg/L (11 percent). If simple averages had been used in establishing process loading values in this case, the treatment facilities would be underdesigned by at least 11 percent.

Mass loading rates. Constituent mass loadings, usually expressed in pounds per day (kilograms per day), represent the mass of material (e.g., lb BOD) that must be treated. The mass loading can be computed by using Eq. (4-4) when the flowrate is expressed in million gallons per day, or Eq. (4-5) when the flowrate is expressed in cubic meters per day. Note that in SI units, the concentration expressed in milligrams per liter is equivalent to grams per cubic meter.

$$\text{Mass loading, lb/d} = (\text{concentration, mg/L})(\text{flowrate, Mgal/d})\left[\frac{8.34\text{ lb}}{\text{Mgal}\cdot(\text{mg/L})}\right] \tag{4-4}$$

$$\text{Mass loading, kg/d} = \frac{(\text{concentration, g/m}^3)(\text{flowrate, m}^3\text{/d})}{10^3\text{ g/kg}} \tag{4-5}$$

A typical daily variation in mass loading rates is illustrated in Fig. 4-7. The peak hourly BOD mass loading may vary as much as 3 to 4 times the minimum hourly BOD mass load in a 24-h period. For treatment plants to function properly, variations in the mass loading rates must be accounted for in the design of the biological treatment systems.

Sustained Wastewater Mass Loadings

If treatment processes are to function properly, data must be available during the design phase on the expected sustained peak mass loadings of constituents that are to be expected under varying loading conditions. When existing plant data are not

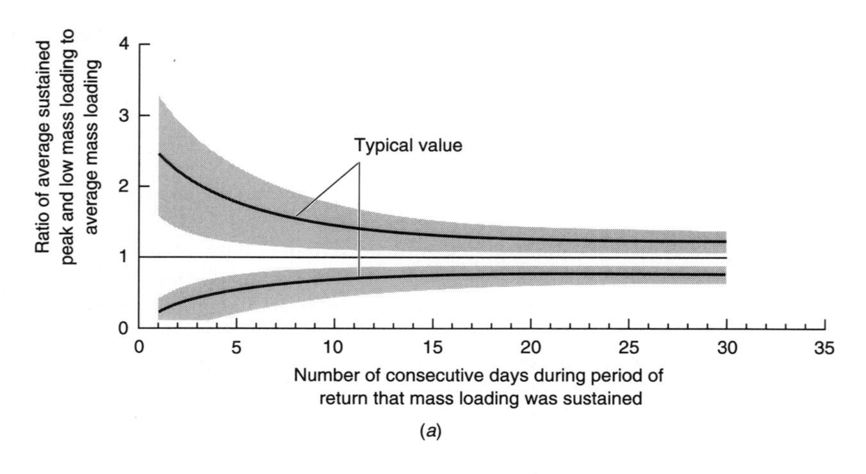

(a)

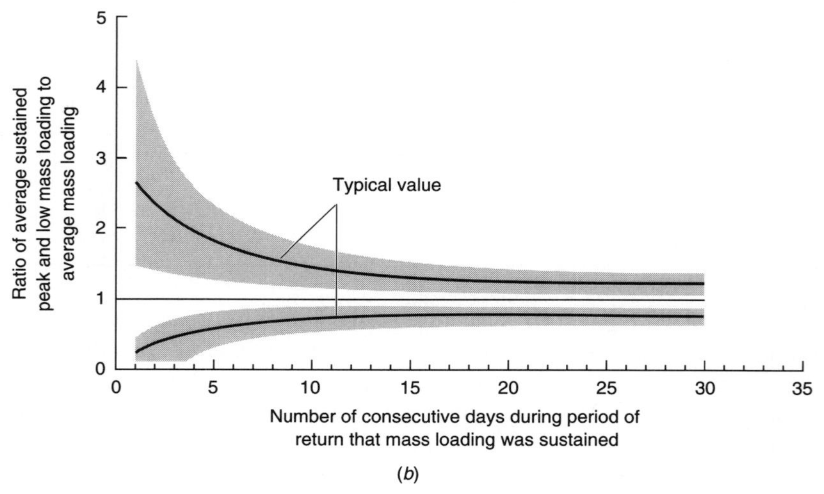

(b)

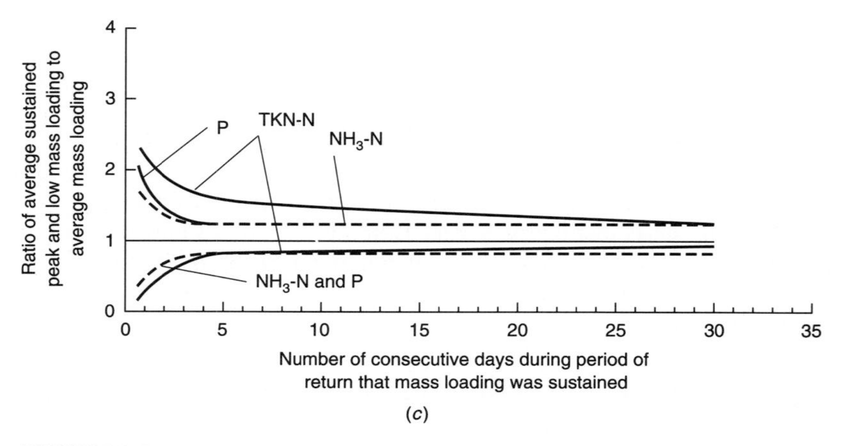

(c)

FIGURE 4-8

Typical curves for average sustained peak and low mass loadings to average mass loadings for: (*a*) BOD, (*b*) TSS, and (*c*) nitrogen and phosphorus (from Tchobanoglous and Burton, 1991).

available, the curves for BOD, TSS, TKN (total Kjeldahl nitrogen), NH_3 (ammonia), and phosphorus given in Fig. 4-8 can be used. The procedure used to develop the mass loading curves shown in Fig. 4-8 is the same as that described in the section dealing with sustained flowrates.

4-4 STATISTICAL ANALYSIS OF VARIATIONS IN FLOWRATE, CONSTITUENT CONCENTRATIONS, AND MASS LOADING RATES

The statistical analysis of wastewater flowrate and constituent concentration data involves the determination of statistical parameters used to quantify a series of measurements. Commonly used statistical parameters and graphical techniques for the analysis of wastewater management data are reviewed below.

Common Statistical Parameters

Commonly used statistical measures include the mean, median, mode, standard deviation, and coefficient of variation, based on the assumption that the data are distributed normally. Although the terms just cited are the most commonly used statistical measures, two additional statistical measures are needed to quantify the nature of a given distribution. The two additional measures are the coefficient of skewness and the coefficient of kurtosis. If a distribution is highly skewed, as determined by the coefficient of skewness, normal statistics cannot be used. For most wastewater data that are skewed, it has been found that the log of the value is normally distributed. Where the log of the values is normally distributed, the distribution is said to be *log normal.* The common statistical measures used for the analysis of wastewater management data [Eqs. (4-6) through (4-14)] are summarized in Table 4-19.

Graphical Analysis of Data

Graphical analysis of wastewater management data is used to determine the nature of the distribution. For most practical purposes, the type of the distribution can be determined by plotting the data on both arithmetic- and logarithmic-probability paper and noting whether the data can be fitted with a straight line. The three steps involved in the use of arithmetic and logarithmic-probability paper are as follows.

1. Arrange the measurements in a data set in order of increasing magnitude and assign a rank serial number.
2. Compute a corresponding plotting position for each data point using Eq. (4-15).

$$\text{Plotting position } (\%) = \left(\frac{m}{n+1}\right) \times 100 \tag{4-15}$$

where m = rank serial number and n = number of observations. The term $(n+1)$ is used to account for the fact that there may be an observation that is either larger or smaller than the largest or smallest in the data set. In effect, the plotting position

TABLE 4-19
Statistical parameters used for the analysis of wastewater management data

Parameter		Definition
Mean value $$\bar{x} = \frac{\sum f_i x_i}{n}$$	(4-6)	$\bar{x}$ = mean value f_i = frequency (for ungrouped data $f_i = 1$) x_i = the midpoint of the ith data range (for ungrouped data x_i = the ith observation)
Standard deviation $$s = \sqrt{\frac{\sum f_i (x_i - \bar{x})^2}{n-1}}$$	(4-7)	n = number of observations (Note $\sum f_i = n$) s = standard deviation
Coefficient of variation $$CV = \frac{100s}{\bar{x}}$$	(4-8)	CV = coefficient of variation, % a_3 = coefficient of skewness a_4 = coefficient of kurtosis
Coefficient of skewness $$\alpha_3 = \frac{\sum f_i (x_i - \bar{x})^3 / n - 1}{s^3}$$	(4-9)	M_g = geometric mean s_g = geometric standard deviation
Coefficient of kurtosis $$\alpha_4 = \frac{\sum f_i (x_i - \bar{x})^4 / n - 1}{s^4}$$	(4-10)	Median value: If a series of observations are arranged in order of increasing value, the middle-most observation, or the arithmetic mean of the two middle-most observations, in a series is known as the *median.*
Geometric mean $$\log M_g = \frac{\sum f_i (\log x_i)}{n}$$	(4-11)	Mode: The value occurring with the greatest frequency in a set of observations is known as the mode. If a continuous graph of the frequency distribution is drawn, the mode is the value of the high point, or hump, of the curve. In a symmetrical set of observations, the mean, median, and mode will be the same value.
Geometric standard deviation $$\log s_g = \sqrt{\frac{\sum f_i (\log^2 x_g)}{n-1}}$$	(4-12)	
Using probability paper $$s = P_{84.1} - \bar{x} \text{ or } \bar{x} - P_{15.9}$$	(4-13)	Coefficient of skewness: When a frequency distribution is asymmetrical, it is usually defined as being a skewed distribution.
$$s_g = \frac{P_{84.1}}{M_g} = \frac{M_g}{P_{15.9}}$$	(4-14)	Coefficient of kurtosis: Used to define the peakedness of the distribution. The value of the kurtosis for a normal distribution is 3. A peaked curve will have a value greater than 3, whereas a flatter curve will have a value less than 3.

represents the percent or frequency of observations that are equal to or less than the indicated value.

3. Plot the data on arithmetic- and logarithmic-probability paper. The probability scale is labeled "Percent of values equal to or less than the indicated value."

If the data, plotted on arithmetic probability paper, can be fit with a straight line, then the data are assumed to be normally distributed. Significant departure from a straight line can be taken as an indication of skewness. If the data are skewed, logarithmic probability paper can be used. The implication here is that the logarithm of the observed values is normally distributed. On logarithmic-probability paper, the straight line of best fit passes through the geometric mean, M_g, and through the intersection of $M_g \times s_g$ at a value of 84.1 percent and M_g/s_g at a value of 15.9 percent. The geometric standard deviation s_g can be determined by using Eq. (4-14), given in Table 4-19. The use of arithmetic- and logarithmic-probability paper is illustrated in Example 4-2.

EXAMPLE 4-2. STATISTICAL ANALYSIS OF WASTEWATER CONSTITUENT CONCENTRATION DATA. Determine the appropriate statistical parameters for the following set of effluent data from a recirculating gravel filter, collected over a 28-month period. The filter is used to treat the wastewater from a 37-home residential development (Crites et al., 1997).

	Value, mg/L			Value, mg/L			Value, mg/L	
Month	TSS	COD	Month	TSS	COD	Month	TSS	COD
1	3.75	1.50	11	8.40	9.00	21	2.25	6.25
2	7.80	1.00	12	9.50	3.33	22	3.00	5.60
3	8.75	2.50	13	15.00	18.00	23	5.00	1.50
4	2.75	2.50	14	9.25	8.75	24	2.80	3.00
5	3.25	3.75	15	6.75	10.00	25	2.00	3.00
6	3.25	4.75	16	7.00	5.00	26	6.75	1.50
7	3.80	5.00	17	6.00	3.75	27	3.80	10.00
8	4.75	5.00	18	2.25	3.75	28	4.00	13.25
9	7.00	6.25	19	5.40	8.40			
10	7.50	12.50	20	4.75	5.00			

Solution

1. Determine the nature of the distribution by plotting the data on arithmetic- and log-probability paper.
   a. Determine the plotting position using Eq. (4-15), where n equals 28:

$$\text{Plotting position (\%)} = \left(\frac{m}{n+1}\right) \times 100$$

		Value, mg/L				Value, mg/L	
Number	Plotting position	TSS	COD	Number	Plotting position	TSS	COD
1	3.5	2.00	1.00	15	51.7	5.00	5.00
2	6.9	2.25	1.50	16	55.2	5.40	5.00
3	10.3	2.25	1.50	17	58.6	6.00	5.00
4	13.8	2.75	1.50	18	62.1	6.75	5.60
5	17.2	2.80	2.50	19	65.5	6.75	6.25
6	20.7	3.00	2.50	20	69.0	7.00	6.25
7	24.1	3.25	3.00	21	72.4	7.00	8.40
8	27.6	3.25	3.00	22	75.9	7.50	8.75
9	31.0	3.75	3.33	23	79.3	7.80	9.00
10	34.5	3.80	3.75	24	82.8	8.40	10.00
11	37.9	3.80	3.75	25	86.2	8.75	10.00
12	41.4	4.00	3.75	26	89.7	9.25	12.50
13	44.8	4.75	4.75	27	93.1	9.50	13.25
14	48.3	4.75	5.00	28	96.6	15.0	18.00

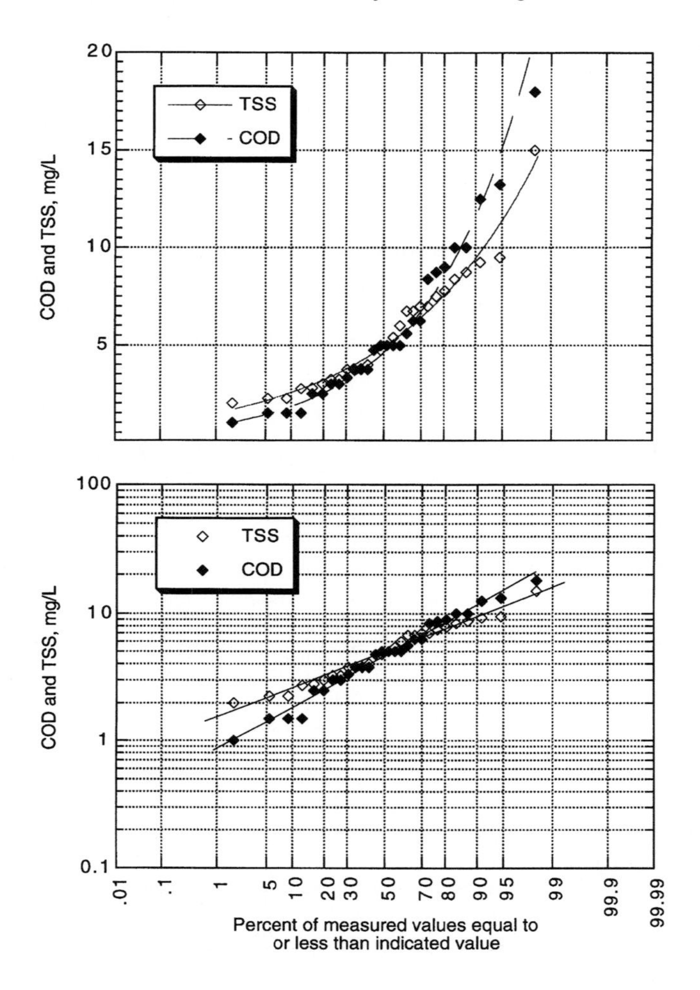

b. Plot the above data on both arithmetic- and log-probability paper. As shown in the above figures, both the TSS and COD data are log-normal.

2. Determine the geometric mean for TSS and COD and the corresponding geometric standard deviation using Eq. (4-14).

$$s_g = \frac{P_{84.1}}{M_g} = \frac{M_g}{P_{15.9}}$$

Constituent	M_g	s_g
TSS	5.0	1.6
COD	4.7	2.0

4-5 DESIGN PARAMETERS FOR WASTEWATER TREATMENT FACILITIES

Typical flowrate and constituent mass loading factors that are important in the design and operation of wastewater treatment facilities are described in Table 4-18. The selection of design flowrate, constituent concentrations, and constituent mass loading rates for the design of wastewater treatment facilities is considered in this section. The selection of the same parameters for the effluent from individual septic tanks is considered in the following section.

Design Flowrates

The rationale for selecting design flowrates is based on hydraulic and process considerations. Process units and hydraulic conduits have to be sized to accommodate the anticipated peak flowrates that will pass through the treatment plant. Many of the process units are designed on the basis of detention time or overflow rate (flowrate per unit of surface area) to achieve the desired removal rates of BOD and TSS. Because the performance of these units can be affected significantly by varying flowrate conditions and mass loadings, minimum and peak flowrates must be considered in design. Typical flowrate factors used for the design of wastewater treatment plants are summarized in Table 4-18.

Average flowrates. The development and forecasting of average flowrates is necessary to define the rated design capacity, as well as the hydraulic requirements of the treatment system. Average flowrates need to be developed both for the initial period of operation and for the design conditions. Average flowrates are often identified as the: (1) average annual daily flowrate; (2) average dry weather flowrate (ADWF), measured during periods when the amount of infiltration is limited; and (3) average wet weather flowrate (AWWF), measured during periods when the effects of infiltration are apparent. Because wastewater treatment plants must be able to process wet weather flows, the rated design capacity of a wastewater treatment plant is often based on the AWWF.

In defining the average flowrate, the nature of the distribution of the flowrate data must be considered. For example, if the flowrate data are distrbuted arithmetically, the average value is appropriate. If the data are distrbuted logarithmically, the geometric mean should be used. When the flowrate data are neither arithmetically or logarithmically distributed, it may be necessary to separate the flowrate data into dry and wet weather periods or to use some other transformation of the data (e.g., the square or cube root of the data may be distributed normally).

Where no information is available, typical ADWF values used for design can be estimated using the information given previously in Tables 4-12 and 4-18. Average design flowrates have also been established by local, state, and federal agencies. A total dry weather base flow of 120 gal/capita·d (460 L/capita·d) has been established by U.S. EPA as a historical average where infiltration/inflow is not excessive. The base flow includes 70 gal/capita·d (265 L/capita·d) for domestic flows,

10 gal/capita·d (38 L/capita·d) for commercial and small industrial flows, and 40 gal/capita·d (150 L/capita·d) for infiltration (Federal Register, 1989). The base flow should be adjusted according to local conditions and information. Domestic base flowrate values vary from about 65 to 80 gal/capita·d (245 to 300 L/capita·d). Commercial and small industrial flows vary from about 10 to 20 gal/capita·d (38 to 76 L/capita·d). The AWWF can be estimated by adding the average infiltration allowance during the wet weather period to the average dry weather residential and commercial and small industrial flow.

Peak, minimum, and sustained flowrates. Peak flows are typically associated with wet weather periods, and are often identified as peak wet weather flow (PWWF). As shown in Table 4-18, the peak hour value is used for the design of pumping facilities and pipelines and for sizing some treatment plant facilities. Other flows of interest include peak day, week, and month as well as minimum hour, day, and month. The flowrate peaking factors most frequently used in design are those for peak hour and peak day. As noted earlier, peaking factors should be determined from an analysis of existing flowrate records.

If no records are available, the recommended hourly peaking factor to be used for design of small treatment plants (0.10 to 1.0 Mgal/d) served by a collection system is 4 to 1. Although higher values have been recorded, the use of higher values has not proven to be cost-effective. The sustained flow curve given in Figure 4-6 can be used to obtain various peak and minimum to average ratios that can then be used to estimate peak and minimum flows without significant amounts of infiltration/inflow. Peak flow allowances for the design of pressure sewers are considered in Chap. 6.

In selecting peaking factors for PWWFs, the characteristics of the collection system serving the wastewater treatment plant must also be considered. Improvements to or rehabilitation of the collection system may also increase or decrease the peaking factors. Typical peaking factors for infiltration/inflow are in the range from 1.5 to 2.5. In a community with which the senior author was associated, the ADWF was about 1.0 Mgal/d whereas the peak wet weather flow was greater than 10 Mgal/d. Because such variations are commonplace, great care must be exercised in using published peaking factor data. Estimating design flowrates, including the use of peaking factors, is illustrated in Example 4-3.

EXAMPLE 4-3. ESTIMATION OF DESIGN WASTEWATER FLOWRATE. A small resort community, to be located in the foothills near popular skiing areas, is in the planning stage. Because of differences in topography, the town is to be served by a hybrid wastewater collection system. The downtown core area, located at an elevation above the proposed location of the community wastewater treatment facility, will be served by a conventional gravity wastewater collection system. In addition, two satellite housing subdivisions at elevations below the treatment plant are planned as part of the town. Wastewater from the two residential developments will be collected and conveyed by separate STEP (septic tank effluent pump) systems, as described in Chap. 6. Effluent from the two pressure sewer systems will be discharged to the gravity collection system at different points. Estimate the average and peak daily flowrates and the overall peaking factor.

The downtown core area will contain the following commercial and institutional establishments which will discharge directly into the gravity collection system:

1. Three motels without kitchens with 25, 50, and 90 rooms
2. Two gasoline service stations (2 employees each)
3. One laundromat (15 washing machines)
4. One post office (2 employees)
5. Three restaurants (1 conventional, 2 fast-food)
6. One shopping center (10 employees, 50 parking spaces)
7. One town hall/meeting place (seating for 90)
8. One elementary school, with cafeteria (120 students)

The following residential areas are planned:

1. *Downtown core area*—35 residential units that will discharge directly to the gravity sewer
2. *Ridgeview subdivision*—84 residential units that will discharge into the gravity collection system from a pressure sewer
3. *Crestmont subdivision*—71 residential units that will discharge into the gravity collection system from a pressure sewer

Both the downtown core residential area and the Ridgeview development are composed of four- and five-bedroom homes, occupied primarily by families. The Crestmont units, which are smaller, are intended for retired persons. Assume the following conditions and parameters apply:

1. During peak seasons (summer/winter), motel occupancy will be 90 percent.
2. During peak seasons, there will be a total of 120 vehicles served per day by the gasoline service stations.
3. Apply office wastewater generation rates for the post office.
4. During peak seasons, there will be a total of 500 restaurant customers per day (150 to the conventional restaurant, 350 to the fast-food restaurants).
5. Downtown and Ridgeview housing density will be 3.7 persons per residence.
6. Crestmont housing density will be 2.3 persons per residence.
7. The allowance for infiltration for the gravity collection system is 1000 gal/d·in-mi.
8. Use a daily peaking factor of 3.5 to 1 for the commercial and institutional facilities served by the gravity collection system.
9. The gravity collection system will comprise 1 mi of 6-in diameter VCP (vitrified clay pipe).
10. Use a peaking factor of 2.0 for infiltration into the gravity collection system.

Solution

1. Determine the commercial and institutional daily wastewater flowrate from the downtown core area, using the typical values provided in Tables 4-1, 4-2, and 4-3.

Source	Unit	Unit flow	Flowrate, gal/d
Motels	148.5 units	95 gal/unit·d	14,108
Gas stations	4 employees	13 gal/employee·d	52
	120 vehicles	12 gal/vehicle·d	1,440
Laundromat	15 machines	550 gal/machine·d	8,250
Post office	2 employees	13 gal/employee·d	26
Conventional restaurant	150 customers	9 gal/customer·d	1,350
Fast-food restaurant	350 customers	6 gal/customer·d	2,100
Shopping center	10 employees	10 gal/employee·d	100
	50 parking spaces	2 gal/parking space·d	100
Town hall	90 seats	3 gal/seat·d	270
Elementary school	120 students	15/gal/student·d	1,800
Total			29,596

2. Estimate the daily wastewater flow from the downtown core residential area using Eq. (4-1):

$$\text{Household flow} = \frac{40\text{ gal}}{\text{home}\cdot\text{d}} + \left(\frac{35\text{ gal}}{\text{person}\cdot\text{d}}\right)\left(\frac{\text{Number of persons}}{\text{home}}\right)$$

$$= \frac{40\text{ gal}}{\text{home}\cdot\text{d}} + \left(\frac{35\text{ gal}}{\text{person}\cdot\text{d}}\right)\left(\frac{3.7\text{ persons}}{\text{home}}\right) = \frac{170\text{ gal}}{\text{home}\cdot\text{d}}$$

The downtown residential area comprises 35 homes. Therefore, the residential flow from this area is

$$\text{Downtown residential flow} = \left(\frac{170\text{ gal}}{\text{home}\cdot\text{d}}\right)(35\text{ homes}) = \frac{5950\text{ gal}}{\text{d}}$$

3. Calculate the amount of infiltration and inflow (I/I) from the downtown area assuming an infiltration rate of 1000 gal/d·in-mi:

$$\text{I/I} = (1000\text{ gal/d}\cdot\text{in-mi})(6\text{ in})(1\text{ mile}) = 6000\text{ gal/d}$$

4. Estimate the daily wastewater flow from the Ridgeview and Crestmont developments using Eq. (4-1):

a. Ridgeview subdivision:

$$\text{Household flow} = \frac{40\text{ gal}}{\text{home}\cdot\text{d}} + \left(\frac{35\text{ gal}}{\text{person}\cdot\text{d}}\right)\left(\frac{3.7\text{ persons}}{\text{home}}\right) = \frac{170\text{ gal}}{\text{home}\cdot\text{d}}$$

$$\text{Total flow} = \left(\frac{170\text{ gal}}{\text{home}\cdot\text{d}}\right)(84\text{ homes}) = \frac{14{,}280\text{ gal}}{\text{d}}$$

b. Crestmont subdivision:

$$\text{Household flow} = \frac{40\text{ gal}}{\text{home}\cdot\text{d}} + \left(\frac{35\text{ gal}}{\text{person}\cdot\text{d}}\right)\left(\frac{2.3\text{ persons}}{\text{home}}\right) = \frac{121\text{ gal}}{\text{home}\cdot\text{d}}$$

$$\text{Total flow} = \left(\frac{121\text{ gal}}{\text{home}\cdot\text{d}}\right)(71\text{ homes}) = \frac{8591\text{ gal}}{\text{d}}$$

c. Because STEP collection systems consist of solvent-welded and/or rubber-gasketed plastic pipe under pressure, flow increases due to inflow and infiltration are negligible (see Chap. 6).

5. Determine the average and peak daily flowrates and the overall peaking factor.
   a. Sum the average flowrates from the downtown core area (commercial and institutional, residential, infiltration/inflow) and the satellite developments, and apply appropriate peaking factors from Table 4-18.

Source	Average flowrate, gal/d	Peaking factor	Peak flow, gal/d
Commercial and institutional	29,596	3.5	103,589
Downtown residential	5,950	2.5	14,875
Infiltration/inflow	6,000	2.0	12,000
Ridgeview subdivision (170 × 84)	14,280	2.5	35,700
Crestmont subdivision (121 × 71)	8,591	2.5	21,478
Total	64,417		187,639

   b. The average daily flowrate = 64,417 gal/d.
   c. The peak daily flowrate = 187,639 gal/d.
   d. The overall peaking factor is

$$\text{Overall peaking factor} = \frac{187{,}639 \text{ gal/d}}{64{,}417 \text{ gal/d}} = 2.9$$

Comment. The computed peaking factor is typical of what would be expected with a combined system employing a relatively short reach of gravity sewer and two STEP systems. In this example, the peaking factor is affected more by the fact that the community is a resort area.

Design Constituent Concentrations

As noted previously in Sec. 4-4, knowledge of the expected flowrates, concentrations, and mass loading factors is of fundamental importance in the design and operation of wastewater treatment facilities. In sizing facilities, flow-weighted constituent concentration values should be used. If only average constituent concentration values are available, the average values should be increased by about 10 percent. If no constituent concentration values are available, then the values given in Tables 4-12 and 4-13 should be used as a guide in developing typical expected values. The expected values should then be increased by 10 to 15 percent to more closely reflect expected flow-weighted constituent concentration values.

Design Mass Loadings

The importance of mass loading in the design of treatment facilities was identified previously in Table 4-18. For example, the sizing of the aeration facilities and the amount of sludge produced are directly related to the mass of BOD that must be processed. Further, the size of the sludge-processing facilities must be sized to deal with sustained organic loadings of at least a week. The development of a BOD mass loading curve is illustrated in Example 4-4.

EXAMPLE 4-4. DEVELOPMENT OF BOD SUSTAINED MASS LOADING VALUES. Develop sustained BOD peak mass loading values for a treatment plant with a design flowrate of 0.5 Mgal/d (1893 m^3/d). Assume that the long-term daily average BOD concentration is 220 mg/L.

Solution

1. Set up a computation table to determine sustained BOD mass loading values:

Length of sustained peak, d	Peaking factor	Peak BOD mass loading, lb/d
1	2.4	2200
2	2.1	1926
3	1.9	1742
4	1.8	1651
5	1.7	1559
10	1.4	1283
15	1.3	1192
20	1.25	1146
30	1.15	1055

2. Obtain peaking factors for the sustained peak BOD loading rate from Fig. 4-8*a*, and enter them in column 2.

3. Determine the sustained mass loading rates for various time periods using Eq. (4-4).

   a.

$$\text{Average daily BOD mass loading, lb/d} = (\text{concentration, mg/L})(\text{flowrate, Mgal/d})\left[\frac{8.34\ \text{lb}}{\text{Mgal}\cdot(\text{mg/L})}\right]$$

   b.

$$\begin{aligned}\text{Daily BOD mass loading, lb/d} &= (220\ \text{mg/L})(0.5\ \text{Mgal/d})[8.34\ \text{lb/Mgal}\cdot(\text{mg/L})] \\ &= 917\ \text{lb/d}\end{aligned}$$

   c. Determine the peak sustained mass loading by multiplying the daily BOD mass loading times the peaking factor, and enter the values in column 3.

Comment. If the sustained peak loading period were to last for 10 days, the total amount of BOD that would be received at a treatment facility during the 10-day period would be 12,830 lb.

4-6 DESIGN PARAMETERS FOR SEPTIC TANK EFFLUENT TREATMENT FACILITIES

Typical flowrate and constituent concentrations are important in the design and operation of facilities used to treat septic tank effluent. The selection of design and peak flowrates is considered in the following discussion.

Design Flowrates

The average and peak design flowrates for individual residences are of importance in the design and operation of intermittent and recirculating sand filters and drip irrigation systems.

Average flowrate. The average flowrate discharged to a septic tank can be estimated using Eq. (4-1), if the number of residents is known. Alternatively, an average value of 45 to 50 gal/capita·d can be used. In either case, the flowrate values are based on the assumption that a watertight septic tank is used (see Chap. 5).

Peak and minimum flowrates. Peak flows of short duration generally do not affect the design of most downstream processes because the effluent is usually pumped to the treatment units, and because the septic tank has adequate temporary storage capacity. What is of concern is the peak day flow, because it will impact the design of treatment facilities. As noted previously, the ratio of the peak day to the average day can vary from 2 to 5 or more, depending on the season of the year and usage (e.g., weekend guests at a summer cottage). Because peak flows can occur for a number of days, it is recommended that a peaking factor of 2.5 be used for the design of downstream treatment processes from septic tanks (see Table 4-20). Peaking factors for a small commercial establishment and communities are also given in Table 4-20.

TABLE 4-20
Peaking factors for wastewater flows from individual residences, small commercial establishments, and small communities

Peaking factor*	Individual residence		Small commercial establishment		Small community	
	Range	Typical†	Range	Typical	Range	Typical
Peak hour	4–10	4	6–10	4	3–6	4
Peak day	2–5	2.5	2–6	3.0	2–4	2.5
Peak week	1.25–4	2.0	2–6	2.5	1.5–3	1.75
Peak month	1.15–3	1.5	1.25–4	1.5	1.2–2	1.25

*Ratio of peak flow to average flow.

†Higher values are often reported, but the given values are suitable for sizing onsite wastewater management facilities.

TABLE 4-21
Comparison of design flows based on a per capita allowance times a peaking factor versus design flow based on a per bedroom allowance

Number of bedrooms	Number of persons	Flowrate,* gal/capita·d	Peaking factor	Design flow based on peak per capita flow, gal/d	Design flow based on per bedroom allowance, gal/d
1	2	55	2.5	275	150
2	3	48	2.5	360	300
3	4	45	2.5	450	450
4	5	42	2.5	525	600

*Computed using Eq. (4-1).

In many states it is quite common to use a flow allowance for design of 150 gal/d per bedroom, which in theory accounts for peak flow. Unfortunately, bedrooms do not generate wastewater, people do. Thus, it is recommended that a per capita design allowance, based on peak flow, be used for design. The relationship between the design flows based on a peak flow per capita, derived by using Eq. (4-1), and a per bedroom allowance are summarized in Table 4-21.

Design Constituent Concentrations

As noted earlier, because the septic tank serves as an online equalization basin, the variations in the constituent concentrations are dampened considerably. Therefore, the values given in Table 4-16 can be used as a guide in estimating the constituent concentrations from septic tanks serving individual residences.

4-7 RISK ANALYSIS CONSIDERATIONS IN DESIGN AND EFFLUENT REUSE

When health effects can occur as the result of an environmental action, risk analysis is used to quantify the corresponding risks. Typically, a complete risk analysis is divided into two parts: (1) risk assessment and (2) risk management. Risk assessment involves the study and analysis of the potential effect of certain hazards to human health. Using statistical information, risk assessment is intended to be a tool for making informed decisions. Risk management is the process of reducing risks that are determined to be unacceptable. Both risk assessment and risk management are introduced in the following discussion. Noncarcinogenic effects and ecological risk assessment are also considered briefly. Additional details may be found in Kolluru et al. (1996); Pepper (1996); Neely (1994); U.S. EPA (1986a, 1989, 1990).

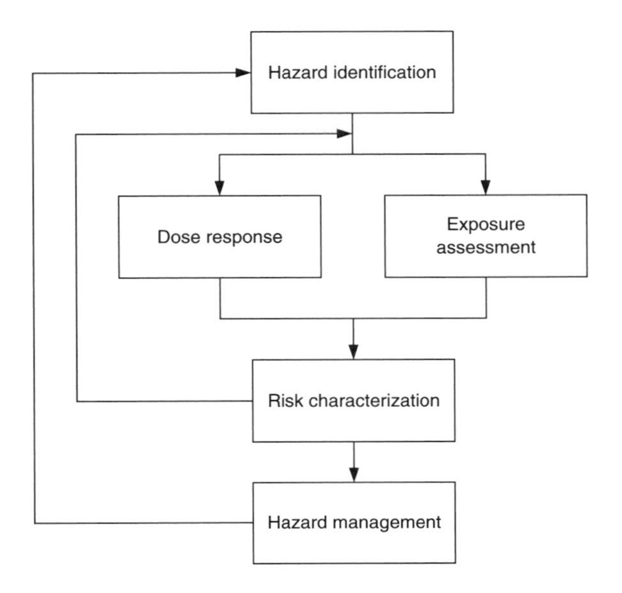

FIGURE 4-9
Definition sketch for the conduct of health effects risk assessment.

Risk Assessment

Health effects risk assessment takes place in four discrete steps, as diagrammed in Fig. 4-9.

1. Hazard identification
2. Exposure assessment
3. Dose response
4. Risk characterization

These four steps are described below, along with a brief discussion of the use of risk assessment in standard setting.

Hazard identification. This step involves weighing the available evidence and determining whether a substance or constituent exhibits a particular adverse health hazard. As part of hazard identification, evidence is gathered on the potential for a substance to cause adverse health effects in humans or unacceptable environmental impacts. For humans, the principal sources for this information are clinical studies, controlled epidemiological studies, experimental animal studies, and evidence gathered from accidents and natural disasters.

Exposure assessment. Exposure is the process by which an organism comes into contact with a hazard; exposure or access is what bridges the gap between a hazard and a risk (Kolluru, et al., 1996). For humans, exposure can occur through different pathways including inhalation of air, ingestion of water or food, absorption through the skin via dermal contact, or absorption through the skin via radiation. The key steps in exposure assessment are identification of a potential

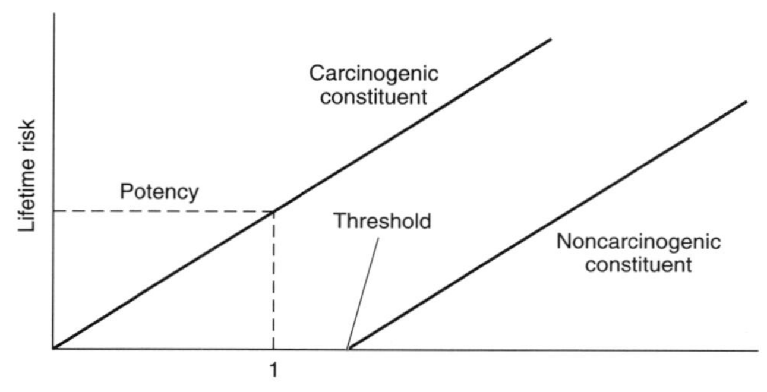

FIGURE 4-10
Definition sketch for dose-response curves for carcinogenic and noncarcinogenic constituents. As shown, it is assumed that the dose-response curve for a carcinogenic constituent has no threshold value.

receptor population, evaluation of exposure pathways and routes, and quantification of exposures (NRC, 1991). For example, an exposure scenario to assess the impacts of drinking groundwater which contains a known amount of reclaimed water which has been recharged would be as follows: an adult weighing 70 kg drinks 2.0 L of groundwater containing 50 μg/L of trichloroethylene (TCE) every day for 70 years. In this case, the adult is the receptor, drinking groundwater is the pathway, and drinking 2.0 L/d containing 50 μg/L of TCE for 70 years is the quantification of exposure.

Dose response assessment. The fundamental goal of a dose-response assessment is to define a relationship (typically mathematical) between the amount of a toxic constituent to which a human is exposed and the risk that there will be an unhealthy response to that dose in humans. Typical dose response relationships for carcinogenic and noncarcinogenic constituents are illustrated in Fig. 4-10. It should be noted that it is assumed that there is no threshold for potentially carcinogenic constituents. Although the dose response curve for a carcinogenic constituent is shown passing through the origin there are no data available at extremely low doses (see Fig. 4-11*a*). Therefore, mathematical models have been developed to define the dose response at low concentrations. Typical dose response models that have been proposed and used for human exposure include: (1) the single-hit model, (2) multistage model, (3) the linear multistage model, (4) the multihit model, and (5) the probit model. The characteristics of these models are summarized in Table 4-22. The mathematical relationship used to describe the relationship between risk and dose for the single-hit model is

$$P(d) = 1 - \exp[-(q_o - q_1 d)] \tag{4-16}$$

where $P(d)$ = lifetime risk (probability) of developing cancer
q_o and q_1 = empirical parameters picked to fit the data
d = dose

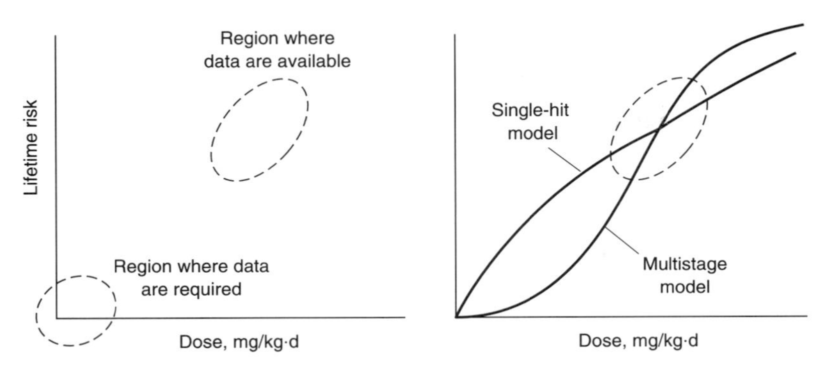

FIGURE 4-11
Definition sketch for dose-response curves: (*a*) illustration of where data are available and where data are required and (*b*) two different models used to define the dose-response relationship (adapted from Crump, 1984).

The mathematical formulation used to describe the relationship between risk and dose for the multistage model is

$$P(d) = 1 - \exp\left[-\sum_{i=0}^{n} q_i d^i\right] \tag{4-17}$$

where $P(d)$ = lifetime risk (probability) of developing cancer
q_i = positive empirical parameters picked to fit the data
d = dose

The relationship between these two models and the risk data given in Fig. 4-11*a* is shown in Fig. 4-11*b*. A variety of models have also been proposed to define the risk associated with low levels of microorganisms (Haas, 1983).

TABLE 4-22
Models used to assess nonthreshold effects of toxic constituents*

Model†	Description
One-hit	A single exposure can lead to the development of a tumor.
Multistage	The formation of a tumor is the result of a sequence of biological events.
Linear multistage	Modification of the multistage model. The model is linear at low doses with a constant of proportionality that statistically will produce less than 5% chance of underestimating risk.
Multihit	Several interactions are required before cell becomes transformed.
Probit	Tolerance of exposed population is assumed to follow a lognormal (probit) distribution.

*Adapted from Cockerham and Shane (1994) and Pepper et al. (1996).
†In all of the models, it is assumed that exposure to the toxic constituent will always produce an effect regardless of the dose.

The U.S. EPA has defined lifetime risk as follows:

$$\text{Lifetime risk} = \text{CDI} \times \text{PF} \tag{4-18}$$

where CDI = chronic daily intake over a 70-yr lifetime, mg/kg·d
PF = potency factor, $(\text{mg/kg·d})^{-1}$

The chronic daily intake (CDI) is computed as follows.

$$\text{CDI} = \frac{\text{total dose, mg/d}}{\text{body weight, kg}} \tag{4-19}$$

In its most general form the total dose is defined as

$$\text{Total dose} = \begin{pmatrix}\text{constituent}\\ \text{concentration}\end{pmatrix}\begin{pmatrix}\text{intake}\\ \text{rate}\end{pmatrix}\begin{pmatrix}\text{exposure}\\ \text{duration}\end{pmatrix}\begin{pmatrix}\text{absorption}\\ \text{factor}\end{pmatrix} \tag{4-20}$$

Recommended standard values for daily intake calculations have also been developed by the U.S. EPA. For example, the average body weights for an adult and child are 70 and 10 kg, respectively, and the corresponding rates of water ingestion are 2 and 1 L per day (U.S. EPA, 1986a, 1986b).

The potency factor PF, often identified as the slope factor, is the slope of the dose-response curve, at very low doses (see Fig 4-10). The U.S. EPA has selected the linear multistage model as the basis for assessing risk. In effect, the PF corresponds to the risk resulting from a lifetime average dose of 1.0 mg/kg·d. The U.S. EPA maintains an information database on toxic substances known as the Integrated Risk Information System (IRIS) (U.S. EPA, 1996). As an example, typical toxicity data for several compounds are reported in Table 4-23. The application of the data given in Table 4-23 is illustrated in Example 4-5.

TABLE 4-23
Toxicity data for selected potential carcinogenic constituents*

	Potency factor PF, $(\text{mg/kg·d})^{-1}$	
Chemical	**Oral route**	**Inhalation route**
Arsenic	1.75	50
Benzene	2.9×10^{-2}	2.9×10^{-2}
Chloroform	6.1×10^{-3}	8.1×10^{-2}
Heptachlor	3.4	
Methylene chloride	7.5×10^{-3}	1.4×10^{-2}
Tetrachloroethylene	5.1×10^{-2}	$1.0–3.3 \times 10^{-3}$
Trichloroethylene (TCE)	1.1×10^{-2}	1.3×10^{-2}
Vinyl chloride	2.3	0.295

*IRIS database (U.S. EPA, 1996).

EXAMPLE 4-5. RISK ASSESSMENT FOR THE DRINKING OF GROUNDWATER CONTAINING TRACE AMOUNTS OF TRICHLOROETHYLENE (TCE). Estimate the lifetime risk for an adult associated with drinking groundwater containing 50 μg/L of TCE.

Solution

1. Compute the CDI using Eq. (4-19).

$$\text{CDI} = \frac{\text{total dose, mg/d}}{\text{body weight, kg}}$$

$$= \frac{(50 \times 10^{-6}\ \text{g/L})(10^3\ \text{mg/g})(2\ \text{L/d})}{70\ \text{kg}} = 0.00143\ \text{mg/kg·d}$$

2. Compute the lifetime risk using Eq. (4-18) and data from Table 4-23.

$$\text{Lifetime risk} = \text{CDI} \times \text{PF}$$

The potency factor from Table 4-23 for the oral route is $1.1 \times 10^{-2}\ (\text{mg/kg·d})^{-1}$. Thus,

$$\text{Lifetime risk} = (0.00143\ \text{mg/kg·d})[1.1 \times 10^{-2}\ (\text{mg/kg·d})^{-1}]$$

$$= 1.6 \times 10^{-5}$$

From the results of this analysis, the estimated probability of developing additional cancers as a result of drinking the groundwater containing 50 μg/L of TCE is 1.6 per 100,000 persons.

Risk characterization. The final step in risk assessment is risk characterization, in which the questions of who is affected and what are the likely effects are defined to the extent they are known. Risk characterization involves the integration of exposure and dose response assessments to arrive at the quantitative probabilities that effects will occur in humans for a given set of exposure conditions. It must be recognized that the present state of knowledge concerning the impacts of specific constituents is incomplete. Thus, each step in risk assessment involves uncertainty. In hazard identification, most assessments depend on animal tests and yet animal biological systems are different from human ones. In dose-response, it is often unknown whether safe levels or thresholds exist for any toxic chemical. Exposure assessment usually involves modeling, with the attendant uncertainty as to substance releases, release characteristics, meteorology, and hydrology. Because of the uncertainties associated with any risk assessment, the results of such an analysis should only be used as guide in decision making (Haas, 1983).

Risk assessment in standard setting. Examples of risk assessment in wastewater management include health effects from the reuse of repurified water (Tanaka et al., 1998) (see Sec. 12-10) and health and environmental effects from land application of biosolids (see Sec. 14-6). A risk of 1 in 10,000 is often used in environmental risk assessment (U.S. EPA, 1986a). Risks of less than 1 in 10,000 are considered minimal.

Risk Management

Risk management involves the development of standards and guidelines and management strategies for specific constituents including both toxic constituents and infectious microorganisms. For example, if a toxic constituent or infectious microorganisms are present at higher than the maximum allowable concentration based on the risk assessment, risk management involves the determination of what management and/or technology is necessary to limit the risk to an acceptable value. Thus, the development and screening of alternatives; selection, design, and implementation; and monitoring and review are important elements of risk management.

Noncarcinogenic effects. In addition to the carcinogenic dose response information, the U.S. EPA has developed reference doses (RfD) for a number of constituents based on the assumption that thresholds exist for certain toxic effects (see Fig. 4-10), such as cellular necrosis, but may not exist for other toxic effects, such as carcinogenicity. In general, RfDs are established based on reported results from human epidemiological data, chronic animal studies, and other available toxicological information. The RfD values represent an estimate (with uncertainty spanning perhaps an order of magnitude) of a daily exposure to the human population (including sensitive subgroups) that is likely to be without an appreciable risk of deleterious effects during a lifetime (U.S. EPA, 1989; WCPH, 1997). RfD values are available in the IRIS database (U.S. EPA, 1996) and in the Health Effects Assessment Summary Tables (U.S. EPA, 1990, 1991).

The RfD is used as a reference point for gauging the potential effects of other doses. Usually, doses that are less than the RfD are not likely to be associated with health risks. As the frequency of exposure exceeds the RfD and the size of excess increases, the probability increases that adverse health effects may be observed in a human population. The RfD is derived from the following formula:

$$\text{RfD} = \frac{\text{NOAEL or LOAEL}}{(\text{UF}_1 \times \text{UF}_2 \cdots) \times \text{MF}} \tag{4-21}$$

where NOAEL = no-observable-adverse-effect level
LOAEL = lowest-observable-adverse-effect level
UF_1, UF_2 = uncertainty factors
MF = modifying factor

In the above equation, uncertainty factors are based on experimental species, effects, and duration of the study, while modifying factors represent professional assessments reflecting the confidence in the study. The LOAEL is only used when a suitable NOAEL is unavailable.

Ecological Risk Assessment

Ecological risk assessment is similar to health risk assessment for humans in that the ecological effects of exposure to one or more stressors is assessed. A *stressor* is defined as a substance, circumstance, or energy field that can cause an adverse effect

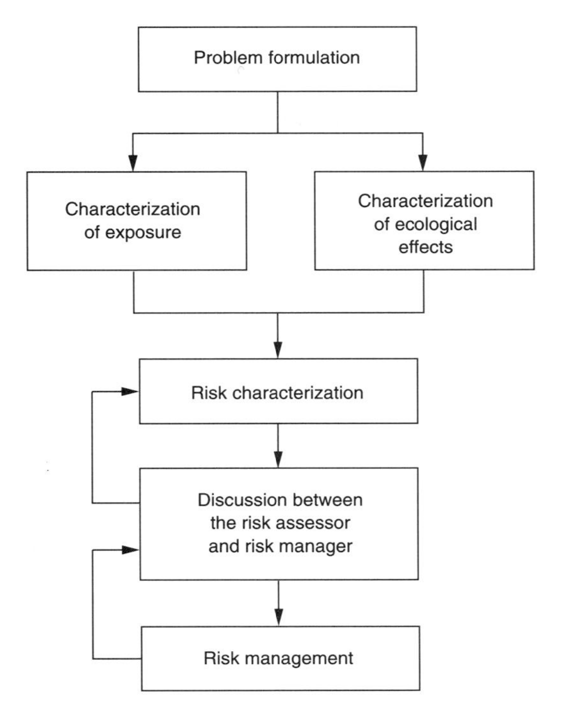

FIGURE 4-12
Definition sketch for the conduct of ecological risk assessment (adapted from U.S. EPA, 1992).

on a biological system. It should be noted that ecological risk assessments are undertaken for a variety of reasons such as to assess the potential impacts of the discharge of treated effluent to an existing wetland or biosolids on land. The framework for ecological risk assessment is illustrated in Fig. 4-12 involving: (1) problem formulation in which the characteristics of the stressor are identified, (2) identification and characterization of the ecosystem at risk and the exposure modes, (3) identification of likely ecological risks, and (4) risk characterization in which all of the information and data are integrated along with input from the risk manager (U.S. EPA, 1992). Because the field of ecological risk assessment is continually undergoing change, the latest reports and publications should be consulted.

4-8 RELIABILITY CONSIDERATIONS IN PROCESS SELECTION AND DESIGN

Important factors in process selection and design are treatment plant performance and reliability in meeting permit requirements. In most wastewater treatment plant discharge permits, effluent constituent requirements based on 7-day and 30-day average concentrations are specified. Because wastewater treatment effluent quality is variable for a number of reasons (varying organic loads, changing environmental conditions, etc.), it is necessary to ensure that the treatment system is designed to produce effluent concentrations equal to or less than the permit limits.

Two approaches in process selection and design are (1) the use of arbitrary safety factors and (2) statistical analysis of treatment plant performance to determine a functional relationship between effluent quality and the probable frequency of occurrence. The latter approach, termed the *reliability concept,* is preferred because it can be used to provide a consistent basis for analysis of uncertainty and a rational basis for the analysis of performance and reliability. Treatment plant reliability can be defined as the probability that a system can meet established performance criteria consistently over extended periods of time. Two components of reliability are the inherent reliability of the process and mechanical reliability. An approach to estimating the reliability of wastewater reclamation and reuse using enteric virus monitoring data is presented in Tanaka et al. (1998).

Inherent Process Reliability

Reliability of a system may be defined as the probability of adequate performance for at least a specified period of time under specified conditions, or, in terms of treatment plant performance, the percent of the time that effluent concentrations meet the permit requirements. For example, a treatment process with a reliability of 99 percent is expected to meet the performance requirements 99 percent of the time. For 1 percent of the time, or three to four times per year, the permit limits are expected to be exceeded. For each specific case where the reliability concept is to be employed, the levels of reliability must be evaluated, including the cost of the facilities required to achieve specified levels of reliability, associated operating and maintenance costs, and the cost of adverse environmental effects of a discharge violation.

Because of the variations in effluent quality, a treatment plant must be designed to produce an average effluent concentration below the permit requirements. The question is: What mean value guarantees that an effluent concentration is consistently less than a specified limit with a certain reliability? One approach involves the use of the coefficient of reliability (COR) method developed by Niku et al. (1979). In the COR method, the mean constituent values (design values) are related to the standards that must be achieved on a probability basis. The mean value, m_x, may be obtained by the relationship

$$m_x = (\text{COR})X_s \tag{4-22}$$

where m_x = mean constituent value
X_s = a fixed standard
COR = coefficient of reliability

The coefficient of reliability is determined by

$$\text{COR} = (V_x^2 + 1)^{1/2} \times \exp\{-Z_{1-\alpha}[\ln(V_x^2 + 1)]^{1/2}\} \tag{4-23}$$

where V_x = ratio of the standard deviation of existing distribution (σ_x) to the mean value of the existing distribution (m_x); V_x is termed the *coefficient of variation*
$Z_{1-\alpha}$ = number of standard deviations away from mean of a normal distribution
$1 - \alpha$ = cumulative probability of occurrence (reliability level)

TABLE 4-24
Values of standardized normal distribution*

Cumulative probability $1 - \alpha$	Percentile $Z_{1-\alpha}$
99.9	3.090
99	2.326
98	2.054
95	1.645
92	1.405
90	1.282
80	0.842
70	0.525
60	0.253
50	0

*Niku et al. (1979).

Values of $Z_{1-\alpha}$ for various cumulative probability levels, $1 - \alpha$, are given in Table 4-24. Values of COR for determining effluent concentrations for different coefficients of variation at different levels of reliability are reported in Table 4-25. Selection of an appropriate design value of V_x must be based on experience from operating facilities (actual or published data). The use of the reliability concept is illustrated in Example 4-6.

Another method of determining design conditions to meet effluent standards is the graphical probability method, similar to the method used in Example 4-6. Based on an analysis of the performance data from 37 activated sludge plants, it was concluded that the log-normal distribution for effluent BOD and TSS may be used to predict the effluent quality performance and the reliability of wastewater treatment plants (Niku et al., 1979). Values equal to or less than the indicated value

TABLE 4-25
Coefficient of reliability as a function of V_x and reliability*

	Reliability, %							
V_x	50	80	90	92	95	98	99	99.9
0.3	1.04	0.81	0.71	0.69	0.64	0.57	0.53	0.42
0.4	1.08	0.78	0.66	0.63	0.57	0.49	0.44	0.33
0.5	1.12	0.75	0.61	0.58	0.51	0.42	0.37	0.26
0.6	1.17	0.73	0.57	0.54	0.47	0.37	0.32	0.21
0.7	1.22	0.72	0.54	0.50	0.43	0.33	0.28	0.17
0.8	1.28	0.71	0.52	0.48	0.40	0.30	0.25	0.15
0.9	1.35	0.70	0.50	0.46	0.38	0.28	0.22	0.12
1.0	1.41	0.70	0.49	0.44	0.36	0.26	0.20	0.11
1.2	1.56	0.70	0.46	0.41	0.33	0.22	0.17	0.08
1.5	1.80	0.70	0.45	0.39	0.30	0.19	0.14	0.06

*Niku et al. (1979).

can be determined at the appropriate percentiles. For example, the peak day may be determined at the 99+ percentile based on occurring once every 365 days. These values can be compared to the values obtained from using the COR approach for selecting the appropriate mean effluent concentrations for design. The use of log-probability plots to determine design values is illustrated in Example 4-6.

EXAMPLE 4-6. ESTIMATING EFFLUENT DESIGN BOD AND TSS CONCENTRATIONS BASED ON RELIABILITY CONSIDERATIONS. An existing activated sludge plant is required to be expanded and upgraded to meet new permit requirements. The new effluent requirements are as given below. Determine the mean design effluent BOD and TSS concentrations required to meet 95 percent reliability level for the 30-day standard and 99 percent reliability for the 7-day standard using the COR method and the log-probability graphical method. Average monthly effluent BOD and TSS data for the existing facility for a period of 1 year are also given below.

Parameter	30-day mean	7-day mean
BOD_5, mg/L	20	45
TSS, mg/L	30	45

Month	BOD, mg/L	TSS, mg/L
Jan	34.0	15.0
Feb	27.1	18.0
Mar	29.0	17.5
Apr	25.0	22.5
May	25.1	22.0
Jun	22.0	24.9
Jul	21.7	28.0
Aug	20.5	25.1
Sep	17.0	19.5
Oct	18.5	20.0
Nov	23.1	20.1
Dec	24.0	21.5

Solution—Part 1: COR method

1. Determine the statistics for the given data using a standard statistical package:

	Value	
Parameter	BOD	TSS
Minimum	17	15
Maximum	34	28
Sum	287.0	254.1
Points	12	12
Mean	23.92	21.18
Median	23.55	20.80
RMS	24.33	21.46
Standard deviation	4.65	3.64
Variance	21.67	13.22
Standard error	1.34	1.05
Skewness	0.61	0.21
Kurtosis	0.05	−0.53

2. Determine the coefficient of reliability using Eq. (4-23) at a cumulative probability of 95 percent:

$$\text{COR} = (V_x^2 + 1)^{1/2} \times \exp\{-Z_{1-\alpha}[\ln(V_x^2 + 1)]^{1/2}\}$$

a. Determine the value of V_x using the results of the statistical analysis.
  i. For BOD:

$$V_x = \frac{\sigma_x}{m_x} = \frac{4.65}{23.91} = 0.194$$

  ii. For TSS:

$$V_x = \frac{\sigma_x}{m_x} = \frac{3.64}{21.18} = 0.172$$

b. The value of $Z_{1-\alpha}$ for a cumulative probability of 95 percent from Table 4-24 is 1.645.
c. Determine the coefficient of reliability.
  i. For BOD:

$$\text{COR} = (0.194^2 + 1)^{1/2} \times e^{\{-1.645[\ln(0.194^2+1)]^{1/2}\}} = 0.74$$

  ii. For TSS:

$$\text{COR} = (0.172^2 + 1)^{1/2} \times e^{\{-1.645[\ln(0.172^2+1)]^{1/2}\}} = 0.77$$

3. Determine the coefficient of reliability using Eq. (4-23) at a cumulative probability of 99 percent.
a. Determine the value of V_x using the results of the statistical analysis.
  i. For BOD:

$$V_x = 0.194$$

  ii. For TSS:

$$V_x = 0.172$$

b. The value of $Z_{1-\alpha}$ for a cumulative probability of 99 percent from Table 4-24 is 2.326.
c. Determine the coefficient of reliability:
  i. For BOD:

$$\text{COR} = (0.194^2 + 1)^{1/2} \times e^{\{-2.326[\ln(0.194^2+1)]^{1/2}\}} = 0.65$$

  ii. For TSS:

$$\text{COR} = (0.172^2 + 1)^{1/2} \times e^{\{-2.326[\ln(0.172^2+1)]^{1/2}\}} = 0.68$$

4. Determine the design effluent concentrations for 95 percent reliability for the 30-day standard.
a. Mean design BOD $= \text{COR} \times X_s$
$= 0.74 \times 20 \text{ mg/L} = 14.8 \text{ mg/L}$

b. Mean design TSS $= 0.77 \times 30 \text{ mg/L} = 23.1 \text{ mg/L}$

5. Determine the design effluent concentrations for 99 percent reliability for the 7-day standard.

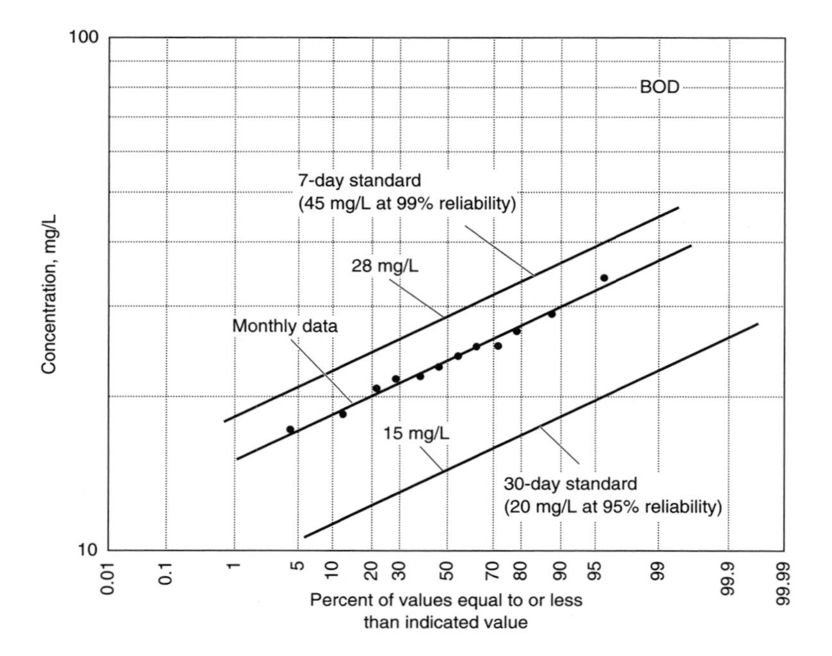
BOD
7-day standard
(45 mg/L at 99% reliability)
28 mg/L
Monthly data
15 mg/L
30-day standard
(20 mg/L at 95% reliability)
Concentration, mg/L
100
10
0.01
0.1
1
5
10
20
30
50
70
80
90
95
99
99.9
99.99
Percent of values equal to or less
than indicated value

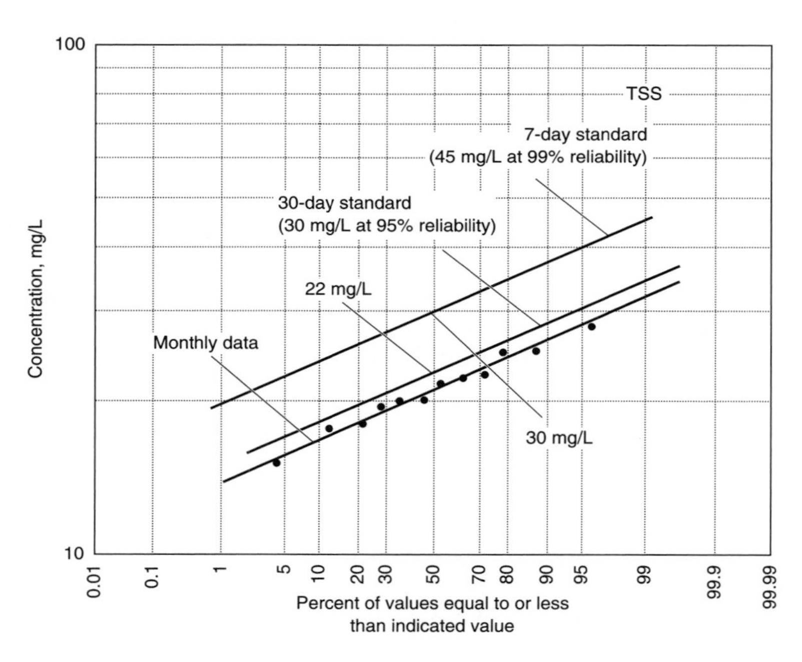
TSS
7-day standard
(45 mg/L at 99% reliability)
30-day standard
(30 mg/L at 95% reliability)
22 mg/L
Monthly data
30 mg/L
Concentration, mg/L
100
10
0.01
0.1
1
5
10
20
30
50
70
80
90
95
99
99.9
99.99
Percent of values equal to or less
than indicated value

a. Mean design BOD $= \text{COR} \times X_s$

$$= 0.65 \times 45 \text{ mg/L} = 29.3 \text{ mg/L}$$

b. Mean design TSS $= 0.68 \times 45 \text{ mg/L} = 30.6 \text{ mg/L}$

6. Use the most conservative values for design:

$$\text{BOD}_{\text{design}} = 14.8 \text{ mg/L}$$
$$\text{TSS}_{\text{design}} = 23.1 \text{ mg/L}$$

Solution—Part 2: Log-probability graphical method

1. Plot the monthly data for BOD and TSS on log-probability paper as illustrated in Example 4-2. The required plots for BOD and TSS are shown on the previous page.

2. Estimate the design effluent concentrations for BOD and TSS for (*a*) 95 percent reliability for the 30-day standard and (*b*) 99 percent reliability for the 7-day standard.

   a. Determine the design effluent concentrations for BOD. The BOD concentrations are determined by passing lines with the same slope as the data through the points at 20 mg/L and 95 percent and 45 mg/L and 99 percent and noting the corresponding values at 50 percent. The values so determined are

$$\text{BOD}_{\text{design}} \text{ at } 20 \text{ mg/L and } 95\% = 15.0 \text{ mg/L}$$
$$\text{BOD}_{\text{design}} \text{ at } 45 \text{ mg/L and } 99\% = 28.0 \text{ mg/L}$$

   b. Determine the design effluent concentrations for TSS. The TSS concentrations are determined by passing lines with the same slope as the data through the points at 30 mg/L and 95 percent and 45 mg/L and 99 percent and noting the corresponding values at 50 percent. The values so determined are

$$\text{TSS}_{\text{design}} \text{ at } 30 \text{ mg/L and } 95\% = 22 \text{ mg/L}$$
$$\text{TSS}_{\text{design}} \text{ at } 45 \text{ mg/L and } 99\% = 30 \text{ mg/L}$$

It is interesting to note that the values determined graphically are essentially the same as those determined analytically using the COR method.

Comment. When the concept of reliability is used, the mean effluent values selected for design will typically be significantly lower than permit requirements. In cases where the coefficient of variability is high and the reliability requirements are stringent, additional unit operations or processes such as filtration may have to be used to meet permit requirements consistently.

Mechanical Process Reliability

A number of approaches are available for analyzing mechanical reliability of a treatment plant and include:

1. Fault tree analysis
2. Event tree analysis
3. Failure modes and effects analysis
4. Critical component analysis

TABLE 4-26
Statistical measures used to assess equipment reliability*

Statistical measure	Description
Mean time before failure (MTBF)	A measure of the mechanical reliability of equipment, determined by the number of failures. The usual approach is to divide operating hours by the number of failures.
Expected time before failure (ETBF)	Similar to the MTBF, but with the actual elapsed time of 1 year used as the total time in service.
Inherent availability (AVI)	Fraction of calendar time that the component or unit was operating.
Operating availability (AVO)	Fraction of time the component or unit can be expected to be operational excluding preventative maintenance.

*Adapted from U.S. EPA (1982); WCPH (1996, 1997).

All four of these approaches are cited frequently in the literature and are used by a variety of industries. The critical component analysis (CCA) approach was developed by the U.S. EPA to determine the in-service reliability, maintainability, and operational availability of selected critical wastewater treatment components (U.S. EPA, 1982). The objective of the CCA is to determine which mechanical components in the wastewater treatment plant will have the most immediate impact on effluent quality should failure occur (WCPH, 1996, 1997). The statistical parameters most commonly used applying the CCA method are summarized in Table 4-26.

In a recently completed study of the San Pasqual Aqua III treatment system in San Diego (see also Chaps. 9 and 12), a complete process reliability analysis

TABLE 4-27
Summary statistics on the mechanical reliability for Aqua III*,†

	Statistical measure‡			
	MTBF, yr	90% CL MTBF, yr	AVO	AVI
Preliminary (headworks)	0.35	0.57	0.9953	0.9998
Primary	0.82	0.65	0.9967	0.9981
Secondary	2.12	1.75	0.9757	0.9953
Package plant	2.24	1.78	0.9994	0.9995
UV disinfection	0.58	0.25	0.9991	0.9984
Reverse osmosis	1.22	0.99	0.9900	0.9903
Aeration tower	1.16	0.50	0.7835	0.9995
Carbon tower	1.86	1.02	0.9963	0.9999
Product water	0.56	0.45	0.9771	0.9964

*Adapted from WCPH (1997).
†Aqua III data December 9, 1994 through September 30, 1995.
‡See Table 4-26.

was performed (WCPH, 1996, 1997). The Aqua III treatment system was designed to produce 1.0 Mgal/d of reclaimed water. The treatment facility includes preliminary treatment (coarse screening and grit removal), primary treatment (rotary drum and disk screens), secondary treatment (water hyacinth ponds), and tertiary treatment with a package plant consisting of coagulation, softening, sedimentation, and filtration. Advanced water treatment consists of ultraviolet disinfection, reverse osmosis, air stripping, and granular activated carbon adsorption. Sodium hypochlorite is used for plant effluent disinfection, with the required contact time taking place within the distribution system. The results of the process reliability analysis are presented in Table 4-27. As shown, the preliminary treatment process has the lowest MTBF. Typical problems experienced with the preliminary treatment works included: tripped breaker, packing leak, and gear box failure. With the exception of three treatment processes, the AVO for the remaining processes was greater than 0.99. The AVI was greater than 99 percent for all of the treatment processes (WCPH, 1996, 1997).

4-9 PROCESS SELECTION AND DESIGN CONSIDERATIONS

Starting with the permit requirements for finished water quality, the design engineer must select the most appropriate combination of processes to convert the initial wastewater characteristics to acceptable levels for reuse or discharge. The purpose of this section is to introduce the reader to issues involved in process selection and to the design considerations that are necessary for their implementation.

Selection of Treatment Methods

The selection of treatment methods and processes depends on the constituents to be removed and the degree to which the constituents must be removed. The classification of treatment methods, the levels of treatment, the applicability of the various methods, and important factors in their selection are considered in the following discussion.

Classification of Treatment Methods

The constituents in wastewater are removed by physical, chemical, and biological means. The individual methods usually are classified as physical unit operations, chemical unit processes, and biological unit processes. These operations and processes occur in a variety of combinations in treatment systems. Treatment methods in which the application of physical forces predominate are known as *physical unit operations.* Examples of physical unit operations include: flocculation, sedimentation, flotation, filtration, screening, mixing, and gas transfer. Treatment methods in which the removal or conversion of contaminants is brought about by the addition of chemicals or by other chemical reactions are known as *chemical unit processes.* Examples of chemical unit processes include precipitation, adsorption, and

TABLE 4-28
Levels of wastewater treatment

Treatment level	Description
Preliminary	Removal of wastewater constituents that may cause maintenance or operational problems with the treatment operations, processes, and ancillary systems.
Primary	Removal of a portion of the suspended solids and organic matter from the wastewater.
Advanced primary	Enhanced removal of suspended solids and organic matter from the wastewater. Typically accomplished by chemical addition or filtration.
Secondary	Removal of biodegradable organics and suspended solids. Disinfection is also typically included in the definition of conventional secondary treatment.
Secondary with nutrient removal	Removal of biodegradable organics, suspended solids, and nutrients (nitrogen, phosphorus, or both nitrogen and phosphorus).
Tertiary	Removal of residual suspended solids, usually by granular medium filtration. Disinfection is also typically a part of tertiary treatment. Nutrient removal is often included in this definition.
Advanced	Removal of dissolved and suspended materials remaining after normal biological treatment when required for water reuse or for the control of eutrophication in receiving waters.

disinfection. Treatment methods in which the removal of contaminants is brought about by biological activity are known as *biological unit processes*. Biological treatment is used primarily to remove the biodegradable organic constituents in wastewater. These substances are converted into gases that can escape to the atmosphere and into biological cell tissue that can be removed by settling. Biological treatment is also used to remove nutrients (nitrogen and phosphorus) in wastewater.

Levels of treatment. Terms that are often used in the literature to describe different levels of treatment are identified in Table 4-28. Although the terms identified in Table 4-28 are in common use, a more rational approach would be to establish the level of contaminant removal (treatment) required before the wastewater can be reused or discharged to the environment. The required unit operations and processes necessary to achieve that required degree of treatment can then be grouped together on the basis of fundamental considerations without regard to the level of treatment (Tchobanoglous and Burton, 1991).

Application of unit operations and processes. Unit operations and processes used for the treatment of the constituents of concern in wastewater are identified in Table 4-29 for wastewater and in Table 4-30 for sludge and septage.

Important factors in process selection. The most important factors that must be considered in the analysis and selection of unit operations and processes are summarized in Table 4-31. While all of the factors listed in Table 4-31 are important,

TABLE 4-29

Unit operations and processes used to remove major constituents found in wastewater

Constituent	Unit operation, unit process, or treatment system	
	Small systems	**Large systems**
Suspended solids	Sedimentation/flotation, effluent filter vaults Intermittent and recirculating packed-bed filters Natural processes (e.g., constructed wetlands, land treatment, etc.)	Screening and comminution Grit removal Sedimentation Filtration Flotation Chemical polymer addition Coagulation/sedimentation Natural processes
Biodegradable organics	Extended aeration activated sludge process variations Intermittent and recirculating packed-bed filters Lagoon processes Natural processes	Activated-sludge variations Fixed film reactor (e.g., trickling filters, rotating biological contactors, variations) Lagoon variations Physical-chemical systems Natural processes
Volatile organics	Natural processes	Air stripping Off-gas treatment Carbon adsorption Natural processes
Pathogens	Chlorination Hypochlorination UV radiation Natural processes	Chlorination Hypochlorination Bromine chloride Ozonation UV radiation Natural processes
Nitrogen	Nitrification/denitrification (e.g., packed-bed reactors) Natural processes	Nitrification/denitrification (e.g., suspended and attached growth processes and variations) Ammonia stripping Ion exchange Breakpoint chlorination Natural processes
Phosphorus	Biological phosphorus removal Natural processes	Metal salt addition Lime coagulation/sedimentation Biological phosphorus removal Biological-chemical phosphorus removal Natural systems
Refractory organics	Natural processes	Carbon adsorption Tertiary ozonation Natural processes
Heavy metals	Chemical precipitation Natural processes	Chemical precipitation Ion exchange Natural processes
Dissolved solids	Ion exchange Reverse osmosis	Ion exchange Reverse osmosis Electrodialysis

TABLE 4-30
Sludge processing and disposal methods

Processing/ disposal method	Unit operation, unit process, or treatment method	
	Small systems	**Large systems**
Preliminary operations	Sludge pumping Sludge grinding	Sludge pumping Sludge grinding Sludge blending and storage Sludge degritting
Thickening	Gravity thickening Gravity belt thickening Lagoons	Gravity thickening Flotation thickening Centrifugation Gravity belt thickening
Stabilization	Aerobic digestion Sludge storage basins Composting	Lime stabilization Heat treatment Anaerobic digestion Aerobic digestion Composting
Conditioning		Chemical conditioning Heat treatment
Disinfection	Composting Lime stabilization Long-term storage	Composting Pasteurization Lime stabilization Long-term storage
Dewatering	Belt filter press Sludge drying beds Lagoons Reed beds	Centrifuge Belt filter press Plate and frame filter press Sludge drying beds Lagoons
Composting	Aerated static pile Windrow	Aerated static pile Windrow In vessel
Heat drying		Dryer variations Multiple effect evaporation
Thermal reduction		Multiple hearth incineration Fluidized-bed incineration Coincineration with solid wastes Wet air oxidation Vertical deep-well reactor
Ultimate disposal	Land application Landfill	Land application Landfill

applicability is perhaps the most important, because it depends to a large extent on the experience and acumen of the engineer. In starting out their careers, most young engineers will typically select the wrong pump for a given application, because of a lack of experience. Therefore, it is recommended that factors given in Table 4-31 be used as a checklist in process selection until more experience is gained.

TABLE 4-31

Important factors that must be considered in evaluating and selecting unit operations and processes*

Factor	Comment
1. Process applicability	The applicability of a process is evaluated on the basis of past experience, data from full-scale plants, published data, and pilot plant studies. If new or unusual conditions are encountered, pilot plant studies are essential.
2. Applicable flow range	The process should be matched to the expected range of flowrates. For example, stabilization ponds are not suitable for extremely large flowrates.
3. Applicable flow variation	Most unit operations and processes have to be designed to operate over a wide range of flowrates. Most processes work best at a relatively constant flowrate. If the flow variation is too great, flow equalization may be necessary.
4. Influent wastewater characteristics	The characteristics of the influent wastewater affect the types of processes to be used (e.g., chemical or biological) and the requirements for their proper operation.
5. Inhibiting and unaffected constituents	What constituents are present and may be inhibitory to the treatment processes? What constituents are not affected during treatment?
6. Climatic constraints	Temperature affects the rate of reaction of most chemical and biological processes. Temperature may also affect the physical operation of the facilities. Warm temperatures may accelerate odor generation and also limit atmospheric dispersion.
7. Reaction kinetics and reactor selection	Reactor sizing is based on the governing reaction kinetics. Data for kinetic expressions usually are derived from experience, published literature, and the results of pilot plant studies.
8. Performance	Performance is usually measured in terms of effluent quality, which must be consistent with the effluent discharge requirements.
9. Treatment residuals	The types and amounts of solid, liquid, and gaseous residuals produced must be known or estimated. Often, pilot plant studies are used to identify and quantify residuals.
10. Sludge processing	Are there any constraints that would make sludge processing and disposal infeasible or expensive? How might recycle loads from sludge processing affect the liquid unit operations or processes? The selection of the sludge processing system should go hand-in-hand with the selection of the liquid treatment system.
11. Environmental constraints	Environmental factors, such as prevailing winds and wind directions and proximity to residential areas, may restrict or affect the use of certain processes, especially where odors may be produced. Noise and traffic may affect selection of a plant site. Receiving waters may have special limitations, requiring the removal of specific constituents such as nutrients.
12. Chemical requirements	What resources and what amounts must be committed for a long period of time for the successful operation of the unit operation or process? What effects might the addition of chemicals have on the characteristics of the treatment residuals and the cost of treatment?

(*continued*)

TABLE 4-31
(Continued)

Factor	Comment
13. Energy requirements	The energy requirements, as well as probable future energy cost, must be known if cost-effective treatment systems are to be designed.
14. Other resource requirements	What, if any, additional resources must be committed to the successful implementation of the proposed treatment system using the unit operation or process being considered?
15. Personnel requirements	How many people and what levels of skills are needed to operate the unit operation or process? Are these skills readily available? How much training will be required?
16. Operating and maintenance requirements	What special operating or maintenance requirements will need to be provided? What spare parts will be required and what will be their availability and cost?
17. Ancillary processes	What support processes are required? How do they affect the effluent quality, especially when they become inoperative?
18. Reliability	What is the long-term reliability of the unit operation or process being considered? Is the operation or process easily upset? Can it stand periodic shock loadings? If so, how do such occurrences affect the quality of the effluent?
19. Complexity	How complex is the process to operate under routine or emergency conditions? What levels of training must the operators have to operate the process?
20. Compatibility	Can the unit operation or process be used successfully with existing facilities? Can plant expansion be accomplished easily?
21. Land availability	Is there sufficient space to accommodate not only the facilities currently being considered but possible future expansion? How much of a buffer zone is available to provide landscaping to minimize visual and other impacts?

*From Tchobanoglous and Burton (1991).

Design Considerations for Wastewater Treatment Plants

The steps that are involved in process analysis and design for both small and large wastewater treatment plants include (1) conduct of flow and wastewater characterization studies, (2) preliminary process selection, (3) conduct of bench-scale and pilot-plant studies, (4) development of alternative treatment flow diagrams, (5) selection of design criteria, (6) layout of the physical facilities, (7) preparation of hydraulic profiles, (8) preparation of solids balances, (9) preparation of construction drawings, specifications, and bid documents, and (10) engineering cost estimates.

Influent flowrates and wastewater characterization. The wastewater flowrates and constituent quantities that require removal are critical to process design. The key design criteria and sizing factors for secondary treatment plant facilities are identified and described in Table 4-32. If adequate data on wastewater characteristics are not available to the designer, a wastewater characterization effort should be conducted.

Bench tests and pilot plant studies. Where the applicability of a process for a given situation is unknown, but the potential benefits of using the process are significant, bench-scale or pilot-scale tests must be conducted. Bench-scale tests are conducted in the laboratory with small quantities of the wastewater in question (Tchobanoglous and Burton, 1991). Pilot-scale tests are typically conducted with flows that are <1 to 10 percent of the design flows (see Fig. 4-13). The purpose of conducting pilot plant studies is to establish the suitability of the process in the treatment of a specific wastewater under specific environmental conditions and to obtain the necessary data on which to base a full-scale design. Factors that should be considered in planning pilot plant studies for wastewater treatment are presented in Table 4-33.

Development of alternative treatment process flow diagrams. A flow diagram is the grouping together of a number of unit operations and processes to achieve a specific treatment objective. Alternative flow diagrams are usually prepared to compare the cost-effectiveness of each treatment alternative. Process applicability is crucial in the development of process flow diagrams. An error in judgment in the assessment of the appropriateness or applicability of a particular process can be disastrous. Numerous resources exist, in addition to wastewater technology textbooks, including past experience with the process, other operating facilities, published information in technical journals, manuals of practice published by the Water Environment Federation, process design manuals published by U.S. EPA, and pilot plant study results. Examples of typical flow diagrams are presented in Figs. 4-14 through 4-16.

Selection of design criteria. Once alternative flow diagrams are prepared, the appropriate design criteria for each unit operation and process are assembled. Design criteria are selected on the basis of theory, published data in the literature, the results of bench-scale tests and pilot-scale studies, and the past experience of the designer. The criteria in this textbook are intended to be general in nature to allow the concept design of the various processes and operations described in Chaps. 5 through 14.

Plant layout of treatment facilities. Various plant layouts of treatment facilities are developed for the physical site. Each of the unit operations and processes is placed on a plan of the site and connected with the required piping. Facilities requiring similar operations are grouped together as much as possible.

Hydraulic profiles. Hydraulic profiles are prepared for average and peak flow conditions. Headlosses through each operation and process are computed to (1) determine if the hydraulic gradient is adequate for the wastewater to flow through the treatment facilities; (2) establish the head requirements for the pumps, where pumping is needed; and (3) allow optimization of the location of treatment units

TABLE 4-32

Effect of flowrates and constituent mass loadings on the selection and sizing of wastewater treatment plant facilities*

Unit operation or process	Critical design factors	Sizing criteria	Effects of design criteria on plant performance
Wastewater pumping and piping	Maximum hour flowrate	Flowrate	Wet well may flood, collection system may surcharge, or treatment units may overflow if peak rate is exceeded.
Screening	Maximum hour flowrate	Flowrate	Headlosses through bar rack and screens increase at high flowrates.
	Minimum hour flowrate	Channel approach velocity	Solids may deposit in approach channel at low flowrates.
Grit removal	Maximum hour flowrate	Overflow rate	At high flowrates, grit removal efficiency decreases in flow-through type grit chambers causing grit problems in other processes.
Primary sedimentation	Maximum hour flowrate	Overflow rate	Solids removal efficiency decreases at high overflow rates; increases loading on secondary treatment system.
	Minimum hour flowrate	Detention time	At low flowrates, long detention times may cause the wastewater to be septic.
Activated sludge	Maximum hour flowrate	Mean cell residence time	Solids washout at high flowrates; may need effluent recycle at low flowrates.
	Maximum organic load	Food/microorganism ratio	High oxygen demand may exceed aeration capacity and cause poor treatment performance.
Trickling filters	Maximum hour flowrate	Hydraulic loading	Solids washout at high flowrates may cause loss of process efficiency.
	Maximum hour flowrate	Hydraulic and organic loading	Increased recycle at low flowrates may be required to sustain process.
	Maximum organic load	Mass loading/medium volume	Inadequate oxygen during peak load may result in loss of process efficiency and cause odors.
Secondary sedimentation	Maximum hour flowrate†	Overflow rate or detention time	Reduced solids removal efficiency at high overflow rates or short detention times.
	Minimum hour flowrate	Detention time	Possible rising sludge at long detention time.
Chlorine contact tank	Maximum hour flowrate	Detention time	Reduced bacteria kill at reduced detention time.
UV disinfection	Maximum day or week flowrate	Dose	Increased dose at reduced flowrates.

*Adapted from Tchobanoglous and Burton (1991).

†Typically, the 99 percentile value is used.

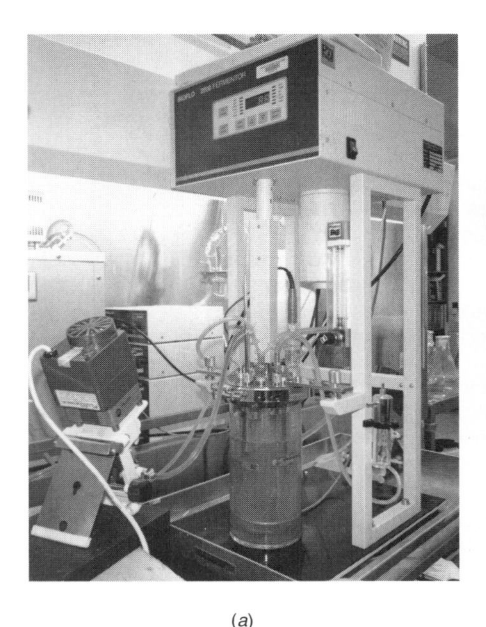

(a)

(b)

FIGURE 4-13
Views of bench and pilot test facilities: (*a*) bench-scale instrumented reactor used for wastewater treatment studies and (*b*) pilot-scale filter column used to study the filtration of algae from lagoon effluent.

on the site in terms of hydraulics. Typical headlosses through various wastewater treatment units are reported in Table 4-34. A typical hydraulic profile for the flow diagram depicted in Fig. 4-16*c* is shown in Fig. 4-17.

Solids balances. A solids balance should be conducted for the planned wastewater treatment plant to estimate the amount of sludge that will be produced, to aid in the planning of sludge processing and storage facilities. A typical solids balance analysis for the flow diagram depicted in Fig. 4-16*b* is shown in Fig. 4-18.

Construction documents. Construction documents consist of construction drawings, specifications, bid documents, and cost estimates. Construction drawings or plans consist of a combination of drawings organized according to discipline, i.e., civil, architectural, structural, mechanical, electrical, and instrumentation. Specifications are also organized by discipline and building components or materials. Bid documents include the plans, technical specifications, and instructions to bidders. Finally, the engineer's cost estimate is used as a guide in evaluating the bids submitted by various contractors.

Cost estimates. The development of a cost estimate typically includes projections of capital costs, annual operation and maintenance costs, and life cycle costs. Capital costs include equipment costs and facility construction and start-up costs.

TABLE 4.33
Considerations in setting up pilot plant testing programs

Item	Consideration
Reasons for conducting pilot testing	Test new process Simulation of another process Predict process performance Optimize system design Document process performance, for example: • Quantify effects of water quality parameters on UV performance • Assess effects of reactor hydraulics on UV performance • Assess effects of effluent filtration on UV performance • Investigate photoreactivation and impacts Satisfy regulatory agency requirements Satisfy legal requirements
Pilot plant size	Bench or laboratory scale Pilot scale Full scale (prototype)
Nonphysical design factors	Available time, money, and labor Degree of innovation and motivation involved Quality of water or wastewater Must seasonal effects be considered Location of facilities Complexity of process Similar testing experience Dependent and independent variables
Physical design factors	Scaleup factors Size of prototype Flow variations expected Facilities and equipment required and setup Materials of construction
Design of pilot testing program	Dependent variables including ranges Independent variables including ranges Time required Test facilities and appurtenances Time of year (seasonal effects) Test protocols Statistical design of data acquisition program

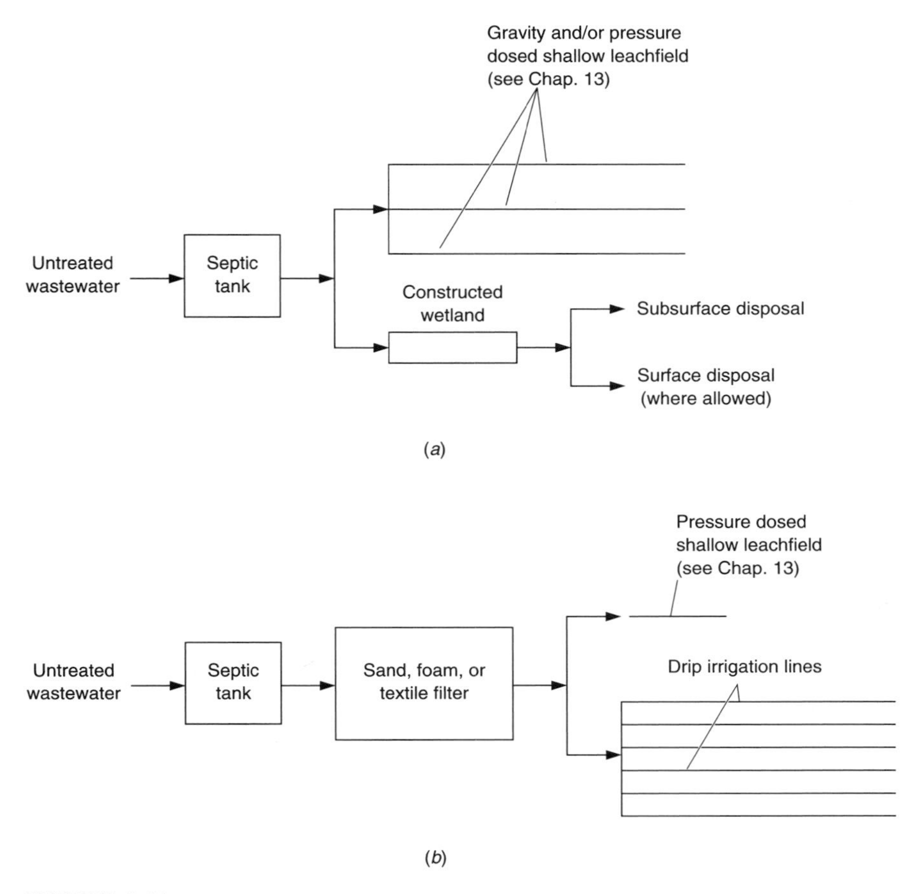

FIGURE 4-14
Flow diagrams for onsite wastewater management: (*a*) conventional system, employing a septic tank and gravity and pressure dosed leachfield or constructed wetland and subsurface or surface disposal, and (*b*) system with septic tank, intermittent packed-bed filter, and shallow leachfield and/or drip irrigation disposal.

Capital costs are based on recent construction contract bid results, contractor quotations, vendor information, and published material prices. Because construction costs vary over time, it is important that cost data be referenced to a standard cost index such as the Engineering News Record Construction Cost Index (ENRCCI) for construction costs. Operation and maintenance costs comprise personnel salaries and operating and maintenance costs (equipment repairs and replacements). Total life cycle costs are the total costs accrued during the project lifetime. They are computed by combining amortized capital costs with annual operation and maintenance costs. The cost ($/gal) of wastewater treated is computed by dividing the estimated life cycle costs ($/yr) by the annual quantity of wastewater (gal/yr). Typically, life cycle analyses are based on a 20-year facility life.

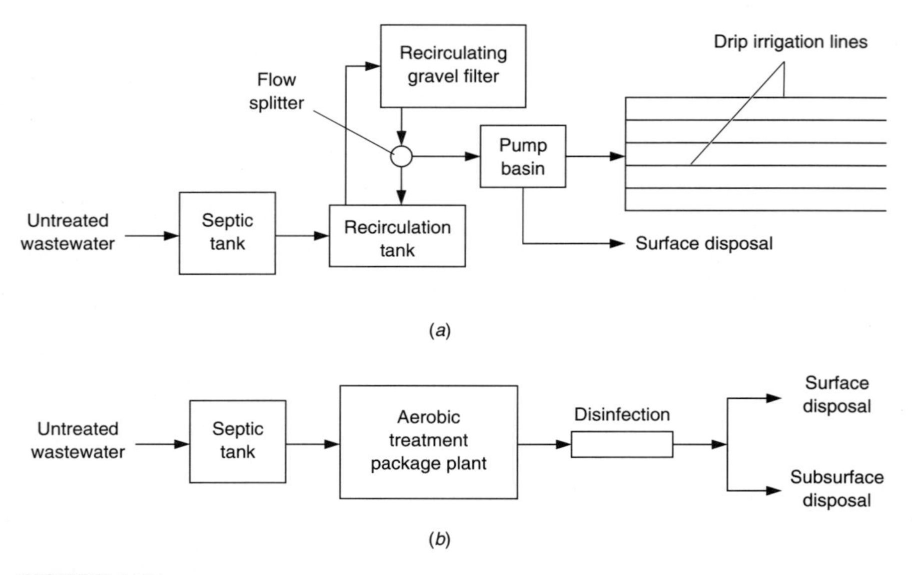

FIGURE 4-15
Flow diagrams for typical treatment systems receiving small flows (e.g., from clusters of homes, roadside rest stops, or isolated commercial or institutional facilities): (*a*) large septic tank with recirculating gravel filter and drip irrigation or surface disposal and (*b*) package treatment system, with effluent chlorine or UV disinfection and surface and/or subsurface discharge.

TABLE 4-34
Typical headlosses across various treatment units*

Treatment unit	Headloss range, ft
Bar screen	0.5–1.0
Grit chamber	
Aerated	1.5–4.0
Velocity controlled	1.5–3.0
Primary sedimentation	1.5–3.0
Aeration tank	0.7–2.0
Trickling filter	
Low rate	10.0–20.0
High rate, rock media	6.0–16.0
High rate, plastic media	16.0–40.0
Secondary sedimentation	1.5–3.0
Filtration	10.0–16.0
Carbon adsorption	10.0–20.0
Chlorine contact tank	0.7–6.0

*Adapted in part from Qasim (1985) and WPCF (1991).

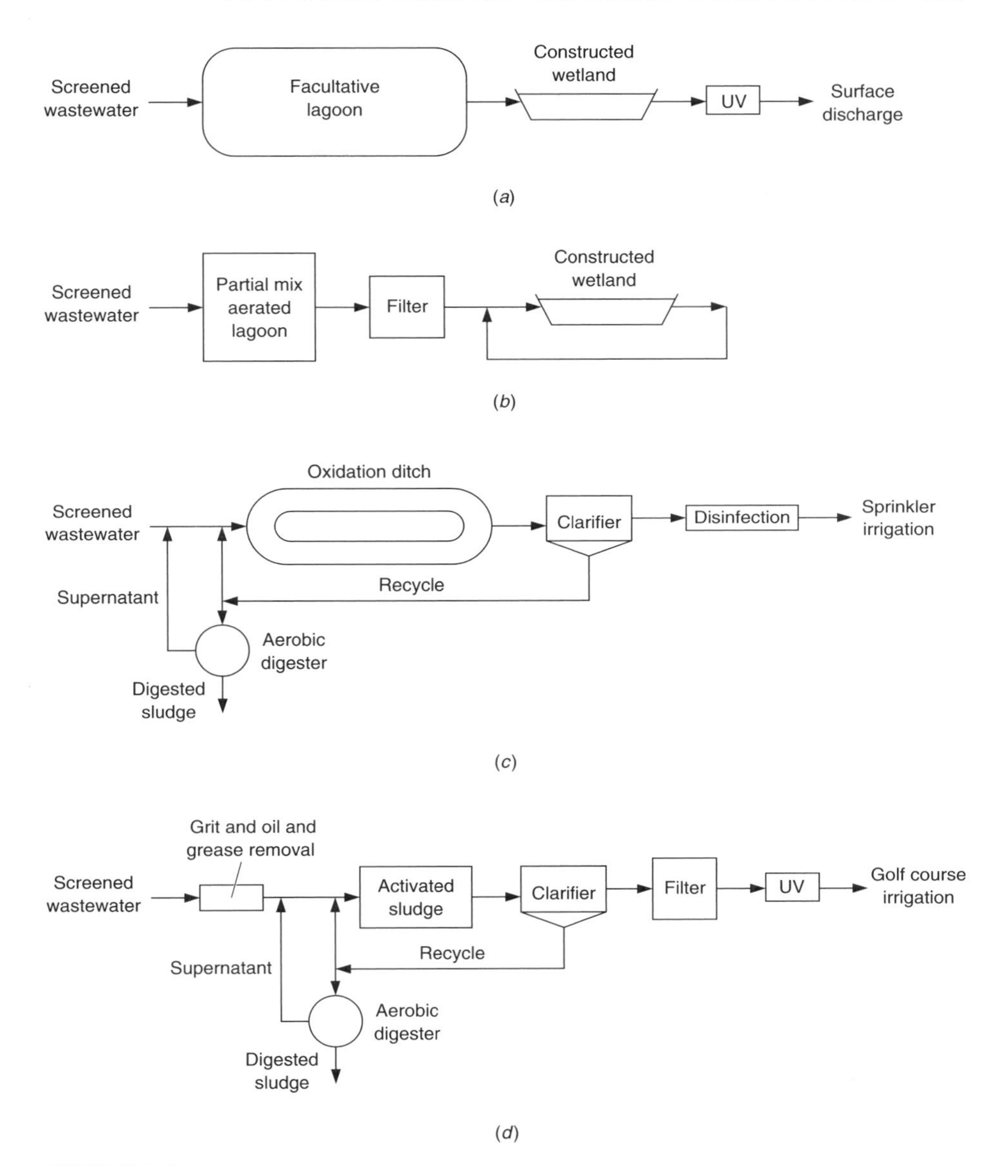

FIGURE 4-16
Flow diagrams for typical wastewater treatment systems receiving flows from small communities: (*a*) facultative lagoon followed by constructed wetland, UV disinfection, and surface discharge; (*b*) partial mix aerated lagoon with sand filter for effluent filtration and application to wetlands for reuse; (*c*) oxidation ditch without primary sedimentation, with aerobic digester for stabilization of solids, and chlorine or UV disinfection followed by irrigation; (*d*) extended aeration activated-sludge plant, with facilities for the removal of grit and oil and grease, and for effluent filtration and UV disinfection with effluent reuse for year-round golf course irrigation.

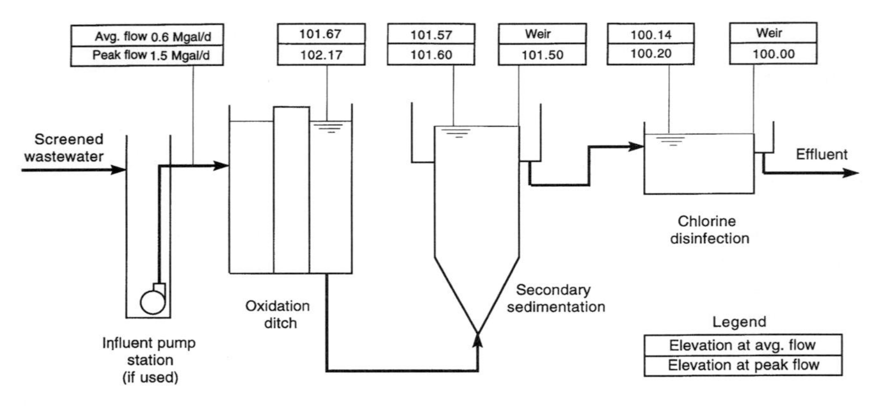

FIGURE 4-17
Typical example of a hydraulic profile. The hydraulic profile is for the wastewater treatment plant shown in Fig. 14-16*c*.

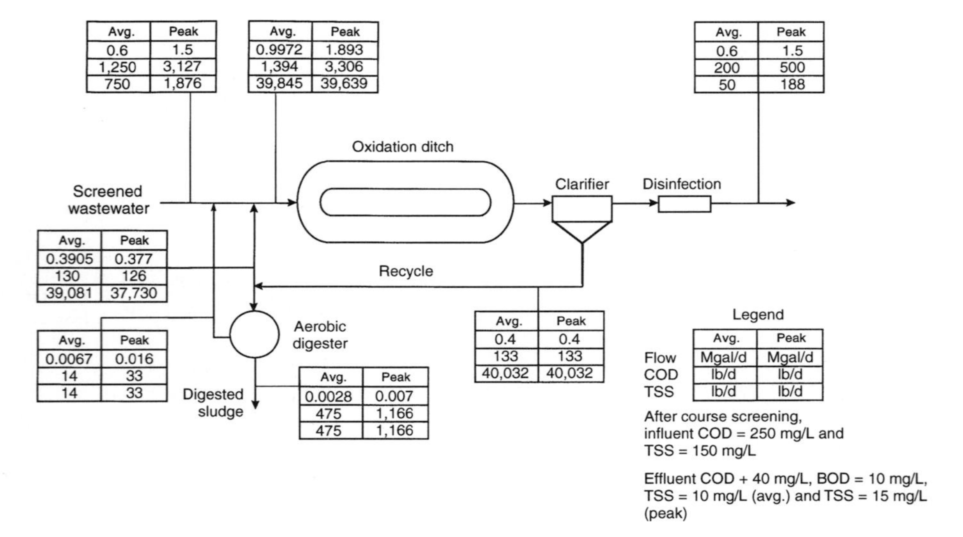

FIGURE 4-18
Typical example of a solids balance, used to determine the quantities of sludge that must be processed. The solids balance shown in this figure is for the wastewater treatment plant depicted in Fig. 4-16*c*.

PROBLEMS AND DISCUSSION TOPICS

4-1. A water conservation program has been proposed for a community to allow them to avoid expanding their wastewater treatment plant. If the current residential unit flowrate is 70 gal/capita·d and the proposed water conservation rate reduction is 50 percent, comment on the reasonableness of the proposal.

4-2. A rural community of 80 homes is considering the installation of a sewer system. Estimate the wastewater flowrate if 40 of the homes are newer, 30 are older, and 10 are summer cottages. What percent of the water supply does the wastewater flow represent if the water use is 8000 gal/d?

4-3. Estimate the wastewater flowrate that could be generated from a commercial area. The area consists of two conventional restaurants, a 200-seat theater, 3 department stores, and an 80-room hotel.

4-4. Estimate the typical flow and the maximum expected flow from a recreational area that consists of a resort hotel with 250 rooms, a resort store, a 120-person campground, and a visitor center. State all of your assumptions clearly.

4-5. A community has an average wastewater use of 80 gal/cap·d. Determine the equivalent dwelling unit (EDU) value if the average household has 2.8 people. Calculate the EDUs for the following commercial and institutional facilities: a 150-inmate prison, a 120-bed medical hospital, and a 160-employee office.

4-6. A hotel wishes to reduce its wastewater flows by using water conservation. The average wastewater flow limit is 50,000 gal/d and there are 240 rooms in the hotel. The flow reduction needed is 10,000 gal/d. Existing shower heads produce 3.5 gal/min and the toilets use 3.8 gal/flush. Develop a water conservation program that will accomplish the goal. What other elements of conservation could be employed?

4-7. A food processing industry generates 20,000 lb/d of BOD and a flow of 1.0 Mgal/d. What is the average BOD concentration? How many population equivalents does the industry represent in a community where the average per capita flowrate is 65 gal/d and the BOD generated per capita is found in Table 4-11?

4-8. Compute the flow-weighted BOD and TSS concentrations from the following data set.

Time	Flow, Mgal/d	BOD, mg/L	TSS, mg/L
0400	0.15	135	110
0800	0.32	205	220
1200	0.25	220	245
1600	0.20	200	220
2000	0.35	210	200
2400	0.21	175	160

4-9. The data in the table below consist of population values for a small community, and average monthly influent flow values to the community wastewater treatment plant, from 1993 through 1996. The community is located in an area of high groundwater. Using these data, answer the following questions.

a. What is the nature of the distribution of the monthly flowrate values? Use the plotting position method described in Sec. 4-4 [see Eq. (4-15)] to plot the monthly influent values versus the corresponding probabilities for each year on arithmetic- and log-probability paper, and check for linearity.

b. What are the average annual flow, ADWF, and AWWF for each year? If the data are arithmetically distributed, use the arithmetic mean; if the data are log-normally distributed, use the geometric mean (see Table 4-19). Assume the dry season occurs from May to October, and the wet season occurs from November to April.

c. What is the per capita flow contribution from commercial and light industrial activities? Assume that the residential contribution to the dry weather flow is 75 gal/capita, and that commercial and light industrial activities make up the remaining flow.

d. What is the per capita flow contribution from infiltration and inflow for each year? Assume that the difference between the wet and dry weather flows is due to infiltration and inflow.

Community population and average monthy influent flowrate data for the period from 1993 through 1996

Year:	1993	1994	1995	1996
Population:	8170	8890	10,440	11,620
Month	Influent flowrate, Mgal/d			
January	2.19	3.47	2.07	2.55
February	1.55	2.48	2.96	4.61
March	1.78	2.02	2.34	3.25
April	1.09	1.06	1.63	1.26
May	1.01	1.42	1.33	1.91
June	0.93	0.91	1.21	1.45
July	0.63	0.64	0.83	0.95
August	0.52	0.37	0.94	0.77
September	0.74	0.51	0.69	0.56
October	0.84	1.21	1.09	1.09
November	1.21	0.79	1.49	1.63
December	1.35	1.67	1.82	2.14

4-10. Consider the small community described in Prob. 4-9. Assume that the build-out population for the community is 15,000, and that the residential wastewater flowrate will be 75 gal/cap·d. The commercial and light industrial flow in 1996 (250,000 gal/d) is 70 percent of what it will be at build-out. Due to high infiltration and inflow (I/I) rates, a sewer repair program will be implemented. The I/I contribution will be either 125, 100, 75, or 50 gal/cap·d (to be selected by instructor), depending on the degree of repair achieved. Estimate the ADWF, the AWWF, and the average annual flow that will be received at the community treatment plant at build-out. Justify the use of the AWWF as the nominal design capacity for the treatment plant.

4-11. Consider the treatment plant for the small community described in Prob. 4-9. Use the plotting position method described in Sec. 4-4 [see Eq. (4-15)] to plot the monthly influent values versus the corresponding probabilities for each year. Determine the nature of the distribution by plotting the values on arithmetic- and log-probability paper and checking for linearity. If the average annual flow to the treatment facility at build-out is estimated to be 2.0 Mgal/d, what will the peak monthly flow be? (*Hint:* the average annual flow occurs at the 50 percent line on the probability graph. Use the slope of the wettest year to pass a line through the 50 percent line at 2.0 Mgal/d, and read the flow value for the highest month from the graph.)

4-12. The data provided in the table below represent influent flow values to a wastewater treatment plant serving a small community. Use the data to conduct the statistical analyses described in Prob. 4-9, steps *a, b, c,* and *d*. If the data in the table below are not distributed either arithmetic-normally or log-normally, suggest a method for determining the required parameters.

Community population and average monthly influent flowrate data for the period from 1993 through 1996

Year:	1993	1994	1995	1996
Population:	**9600**	**9680**	**9760**	**9910**
Month	**Influent flowrate, Mgal/d**			
January	1.10	0.97	1.17	0.95
February	1.18	0.91	0.99	0.98
March	1.08	0.90	1.10	0.96
April	0.98	0.87	1.01	0.93
May	0.93	0.85	0.96	0.91
June	0.90	0.86	0.94	0.92
July	0.89	0.87	0.91	0.90
August	0.88	0.84	0.90	0.91
September	0.86	0.86	0.88	0.90
October	0.87	0.85	0.91	0.93
November	0.89	0.89	0.92	0.96
December	0.92	0.95	0.96	1.00

4-13. Use the information in the following table (choose one row only to be selected by instructor) to estimate the lifetime risk of developing cancer due to the consumption of water containing the indicated constituent. Assume a 70-year lifetime.

Constituent	Constituent concentration, mg/L	Body weight of consumer, lb	Daily volume ingested, L/d
Arsenic	0.15	135	2.0
Chloroform	0.32	205	2.5
Benzine	0.25	220	1.5
Heptachlor	0.20	200	1.8

4-14. For the four chemicals listed in Prob. 4-13 estimate the concentrations that result in a risk of 1/100,000. Use the information given in Prob. 4-13 for body weight, daily volume of water ingested, and lifetime.

4-15. Compare the typical health risk assessment to an ecological risk assessment. How are the two assessments different?

4-16. An existing constructed wetlands system needs to be upgraded to meet new permit requirements. The new effluent requirements are 10 mg/L of BOD and TSS. Determine the mean design effluent BOD and TSS concentrations required to meet 90 percent reliability for the monthly standard using the COR method. Average monthly effluent BOD and TSS data for the existing facility for a 1-year period are given below.

Month	BOD, mg/L	TSS, mg/L
January	12	20
February	8	12
March	7	9
April	9	10
May	8	5
June	15	12
July	14	4
August	10	5
September	18	6
October	16	5
November	6	7
December	3	5

REFERENCES

Baker, L. (1990) Personal communication, Jackson, CA.

Bounds, T. (1997) Personal communication, Orenco Systems, Inc., Sutherland, OR.

Cockerham L. G., and B. S. Shane (1994) *Basic Environmental Toxicology.* CRC Press, Boca Raton, FL.

Crites, R., C. Lekven, S. Wert, and G. Tchobanoglous (1997) A Decentralized Wastewater System for a Small Residential Development in California, *The Small Flows Journal,* Vol. 3, Issue 1, Morgantown, WV.

Crump, K. S. (1984) An Improved Procedure for Low-Dose Carcinogenic Risk Assessment from Animal Data, *Journal of Environmental Pathology, Toxicology, and Oncology,* Vol. 4, No. 4/5, pp. 339–349.

Federal Register (1989) Amendment to the Secondary Treatment Regulations: Percent Removal Requirements During Dry Weather Periods for Treatment Works Served by Combined Sewers, 40 CFR Part 133.

Gerba, C. P., J. B. Rose, and C. N. Haas (1995) Water-Borne Disease—Who Is at Risk? In *Water Quality Technology Proceedings,* pp. 231–254. American Water Works Association, Denver.

Haas, C. N. (1983) Estimation of Risk due to Low Levels of Microorganisms: A Comparison of Alternative Methodologies, *American Journal of Epidemiology,* Vol. 118, pp. 573–582.

Kolluru R. V., S. M. Bartell, R. M. Pitblado, and R. S. Stricoff (1996) *Environmental Assessment and Management Handbook,* McGraw-Hill, New York.

NRC (1991) *Frontiers in Assessing Human Exposure,* National Research Council, National Academy Press, Washington, DC.

Neely, W. B. (1994) *Introduction to Chemical Exposure and Risk Assessment,* Lewis Publishers, Boca Raton, FL.

Niku, S., E. D. Schroeder, and F. J. Samaniego (1979) Performance of Activated Sludge Processes and Reliability-Based Design, *Journal Water Pollution Control Association,* Vol. 51, p. 2841.

Pepper, I. L., C. P. Gerba, and M. L. Brusseau (eds.) (1996) *Pollution Science,* Academic Press, San Diego.

Qasim, S. R. (1985) *Wastewater Treatment Plants: Planning, Design, and Operation,* Holt, Rinehart and Winston, New York.

Standard Methods (1995) *Standard Methods for the Examination of Water and Waste Water,* 19th ed., American Public Health Association, Washington, DC.

Tanaka, H., T. Asano, E. D. Schroeder, and G. Tchobanoglous (1998) Estimating the Reliability of Wastewater Reclamation and Reuse Using Enteric Virus Monitoring Data, *Water Environment Research,* Vol. 70, No. 1.

Tchobanoglous, G., and F. L. Burton (1991) *Wastewater Engineering. Treatment, Disposal, Reuse.* 3rd ed., McGraw-Hill, New York.

U.S. EPA (1982) *Evaluation and Documentation of Mechanical Reliability of Conventional Wastewater Treatment Plant Components,* EPA 600/2-82-044, U.S. Environmental Protection Agency, Washington, DC.

U.S. EPA (1986a) *Risk Assessment and Management: Framework for Decision Making,* EPA 600/9-85-002, U.S. Environmental Protection Agency, Washington, DC.

U.S. EPA (1986b) *Guidelines for Carcinogen Risk Assessment,* U.S. Environmental Protection Agency, Federal Register, Vol. 51, No. 185, pp. 33 992–34 003, September 24, 1986.

U.S. EPA (1989) *Risk Assessment, Guidance for Superfund,* Vol. I, *Human Health Evaluation Manual* (Part A), Office of Emergency and Remedial Response, EPA 540/1-89-002, U.S. Environmental Protection Agency, Washington, DC.

U.S. EPA (1990) *Risk Assessment, Management and Communication of Drinking Water Contamination.* EPA 625/4-89/024, U.S. Environmental Protection Agency, Washington, DC.

U.S. EPA (1991) Health Effects Assessment Summary Tables, Annual FY 1991, Publication No. 9200.6-303 (91-1), Office of Solid Waste and Emergency Response, U.S. Environmental Protection Agency, Washington, DC.

U.S. EPA (1992) *Framework for Ecological Risk Assessment.* EPA 1630/R-92/001, U.S. Environmental Protection Agency, Washington, DC.

U.S. EPA (1996) *Integrated Risk Information System (IRIS), Electronic On-Line Database of Summary Health Risk Assessment and Regulatory Information on Chemical Substances. Assessment,* EPA 1630/R-92/001, U.S. Environmental Protection Agency, Washington, DC.

WPCF (1991) Design of Municipal Wastewater Treatment Plants, Manual of Practice No. 8, Water Pollution Control Federation, Alexandria, VA.

WCPH (1996) Total Resource Recovery Project, Final Report, Prepared for City of San Diego, Water Utilities Department, Western Consortium for Public Health, Oakland, CA.

WCPH (1997) Total Resource Recovery Project Aqua III San Pasqual Health Effects Study Final Summary Report, Prepared for City of San Diego, Water Utilities Department, Western Consortium for Public Health, Oakland, CA.

CHAPTER 5

Wastewater Pretreatment Operations and Processes

Wastewater, as discussed in Chap. 2, contains suspended solids of varying size. Wastewater from individual homes discharged to septic tanks may contain, in addition to wastewater and kitchen waste solids, diapers, toothbrushes, tampon holders, prophylactics, and other large items. These same items are found in conventional wastewater collection systems, along with a variety of other materials including sand and grit. The primary purposes of pretreatment are: (1) to condition the wastewater with respect to subsequent treatment processes, (2) to remove materials that may interfere with downstream treatment processes and equipment, and (3) to minimize the accumulation of material in downstream processes. The facilities used for the pretreatment of wastewater are considered in this chapter.

5-1 ROLE OF PRETREATMENT IN WASTEWATER TREATMENT

In its most general form, pretreatment can be thought of as a processing sequence by which the distribution of the sizes of particles found in wastewater is altered. The principal pretreatment operations and processes are identified in Table 5-1. The particle sizes in wastewater that are affected by the various pretreatment processes are also identified in Table 5-1 and illustrated in Fig. 5-1. More recently, pretreatment, in which materials such as cigarette butts, pieces of plastic, and prophylactics are removed from the waste stream, has assumed a greater importance where the biosolids (sludge) produced during treatment are composted. To meet more stringent product specifications, these materials must be screened from the cured compost, when they are not removed during pretreatment.

TABLE 5-1
Typical operations and processes used for the pretreatment of wastewater and treated effluent, and the size of particles affected

Operation/process	Application/occurrence	Particle size affected*
Communition	Used to cut up or grind large particles remaining after coarse screening into smaller particles of a more uniform size	6 mm
Filtration (pretreatment for membrane processes and disinfection)	Removal of particles that affect the performance of downstream processes	0.015–0.5 mm
Flotation	Removal of particles with specific gravity less than water	0.005–5 mm
Flow equalization	Used to equalize the flow and the characteristics of the wastewater	
Gravity separation	Removal of settleable solids and floating material	>0.040 mm
Gravity separation, accelerated	Removal of grit	0.15–1.0 mm
Grit removal	Removal of grit, sand, and gravel, usually following comminution	0.15–1.0 mm
Imhoff tank	Used for the removal of suspended materials from household wastewater by sedimentation and flotation	<0.040 mm
Membrane filtration (pretreatment for reverse osmosis)	Used for the removal of colloidal and subcolloidal material	0.06–100 μm
Mixing	Used to mix chemicals and to homogenize waste materials	
Oil and grease removal	Removal of oil and grease from individual discharges	
Primary effluent filtration (PEF)	Used for the removal of suspended materials following primary sedimentation	0.005–4 mm
Screening, coarse	Used to remove large particles such as sticks, rags, and other large debris from untreated wastewater by interception	>15 mm
Screening, fine	Removal of small particles	2.5–5.0 mm
Screening, micro	Removal of small particles	0.15–1.5 mm
Septic tank	Used for the removal of suspended materials from household wastewater by sedimentation and flotation	>0.040 mm

*Not all particle sizes are removed equally. Incidental removal of smaller particles will also occur.

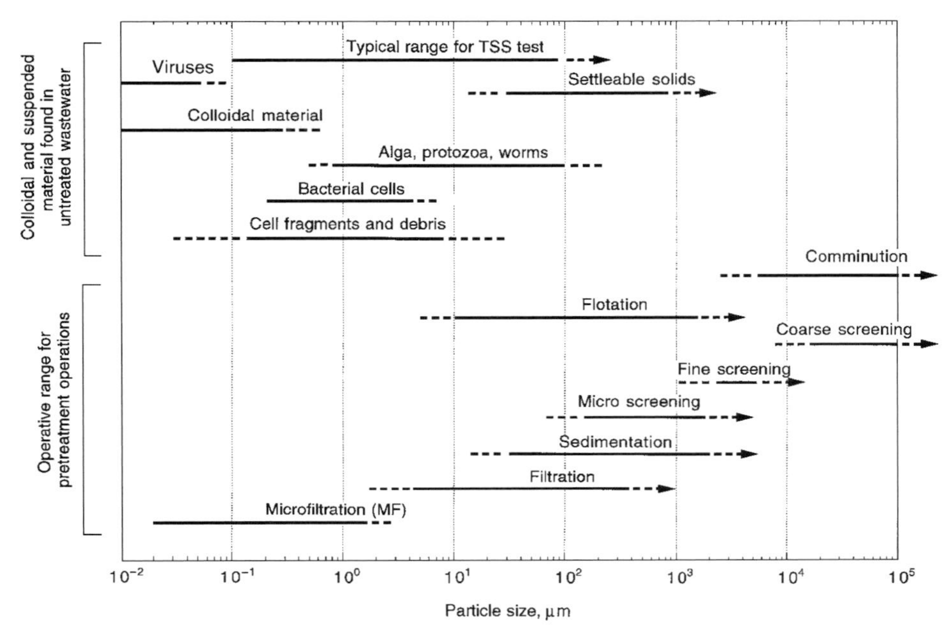

FIGURE 5-1
Typical particle sizes affected by pretreatment processes.

Pretreatment in Onsite Systems

The most serious operational problem encountered with individual onsite systems and in systems in which each individual septic tank is retained has been the carryover of solids and oil and grease due to poor design and the lack of proper septic-tank maintenance. The carryover of suspended material is most serious where a large central disposal field is to be used for the disposal of the septic-tank effluent without any further treatment. Recognizing that poor septic-tank maintenance will be the order of the day, some regulatory agencies have required the addition of a large septic or other solids separation unit before the collected septic-tank effluent can be disposed of in subsurface disposal fields. The use of screened filter vaults for septic tanks, as discussed in Sec. 5-10, has reduced the discharge of TSS and oil and grease significantly. The presence of oil and grease in the effluents from residential septic tanks, and especially from septic tanks servicing restaurants, has led to the failure of downstream treatment processes such as intermittent and recirculating sand filters (see Chap. 11).

Pretreatment in Centralized Treatment Systems

Pretreatment in centralized treatment systems involves coarse screening, comminution, grit removal, oil and grease removal, flow equalization, and TSS removal. Large debris found in wastewater has proven to be a problem with the operation of pumps and other mechanical equipment in wastewater treatment plants. Large floating materials including prophylactics tend to accumulate in sedimentation facilities forming an unsightly scum layer. Sand and grit have proven to be a problem because they tend to accumulate in downstream tankage. Oil and grease coat pumps and piping and reduce the effectiveness of biological treatment. Flow equalization, in which the flow rate to be treated is constant, is especially important for small treatment plants. Additional details on pretreatment operations may be found in WEF (1994).

5-2 COARSE SCREENING

Screening is typically the first unit operation encountered in a wastewater treatment plant. The types of screening devices now in use are summarized in Fig. 5-2. A coarse screen is a device used to intercept and retain the coarse solids found in wastewater. The screening element will typically consist of parallel bars or rods or wires of uniform size. A screen composed of parallel bars or rods is called a *bar rack* (or sometimes a *bar screen*). The term "screen" is also used to describe screening devices consisting of perforated plates, wedge wire elements, and wire cloth (see also Sec. 5-4). Additional details on the types of screening devices used are presented in Table 5-2. The materials removed by these devices are known as *screenings*.

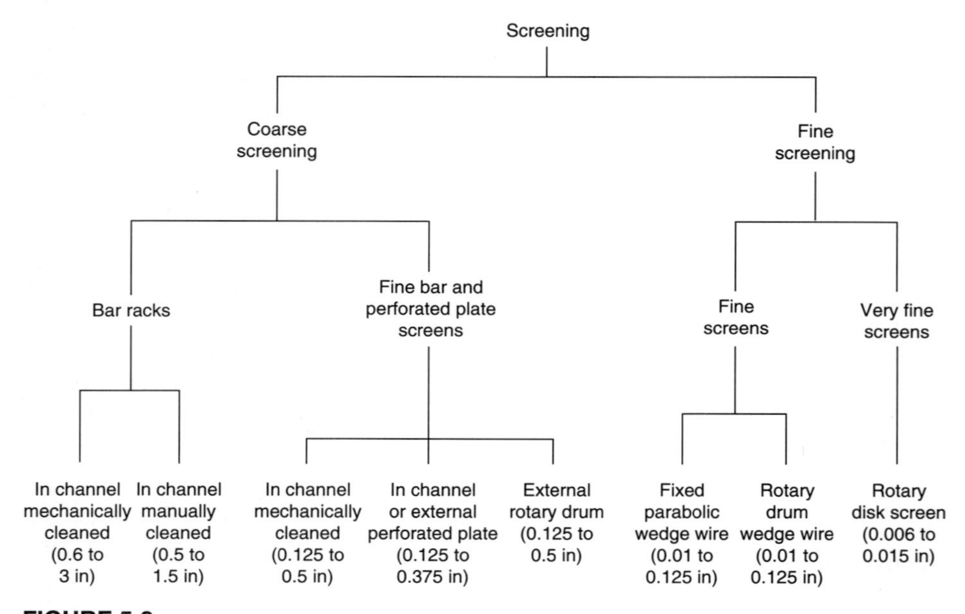

FIGURE 5-2
Definition sketch for the types of screens used for wastewater treatment.

TABLE 5-2

Description of screening devices used in wastewater treatment

Classification of screening device	Size classification	Size range of screen opening		Screen material	Application
		in	mm		
Bar screen (manually and mechanically cleaned)					
Manually cleaned	Coarse	1.0–2.0	25–50	Bars	Removal of coarse suspended solids and as a pretreatment step for fine screening
Mechanically cleaned	Coarse	0.6–3.0	15–75	Bars	As above
Fine bar or perforated coarse screen (mechanically cleaned)					
Fine bar	Fine coarse	0.125–0.50	3–12.5	Thin bars	Pretreatment
Perforated plate	Fine coarse	0.125–0.375	3–9.5	Perforated plate	Pretreatment
Rotary drum	Fine coarse	0.125–0.50	3–12.5	Stainless steel wedge wire	Pretreatment
Fine screen (mechanically cleaned)					
Fixed parabolic	Fine	0.01–0.125	0.25–3.2	Stainless steel wedge wire	Pretreatment
Rotary drum	Fine	0.01–0.125	0.25–3.2	Stainless steel wedge wire	Pretreatment
Rotary disk	Very fine (micro)	0.006–0.015	0.15–0.38	Stainless steel cloth	Primary treatment

Description

Coarse screens, as shown in Fig. 5-2 and Table 5-2, include manually and mechanically cleaned bar racks and mechanically cleaned fine bar and perforated-plate screens.

Bar racks. Typically, bar racks have clear openings (spaces between bars) of $\frac{1}{2}$ in (12.5 mm) or more. In wastewater treatment, bar racks are used to protect pumps, valves, pipelines, and other appurtenances from damage or clogging by rags and large objects. According to the method used to clean them, bar racks and screens are designated as manually cleaned or mechanically cleaned. Manually cleaned bar racks are typically used at small treatment plants. The screenings removed by racking are usually placed on a perforated plate to dewater. Mechanically cleaned bar racks use either endless chains, cables, or reciprocating cogwheel-driven mechanisms to drive the rakes that are used to remove the accumulated screenings. Typical examples of bar racks used for wastewater treatment are shown in Fig. 5-3. An example of an early bar screen is shown in Fig. 5-4*a*. It is interesting to note that modern bar screens (see Fig. 5-3) are essentially the same as the early bar screen shown in Fig. 5-4*a*.

(*a*)

(*b*)

FIGURE 5-3
Examples of bar racks used in wastewater treatment plants: (*a*) mechanically cleaned rack with a manually cleaned rack in the bypass channel and (*b*) mechanically cleaned.

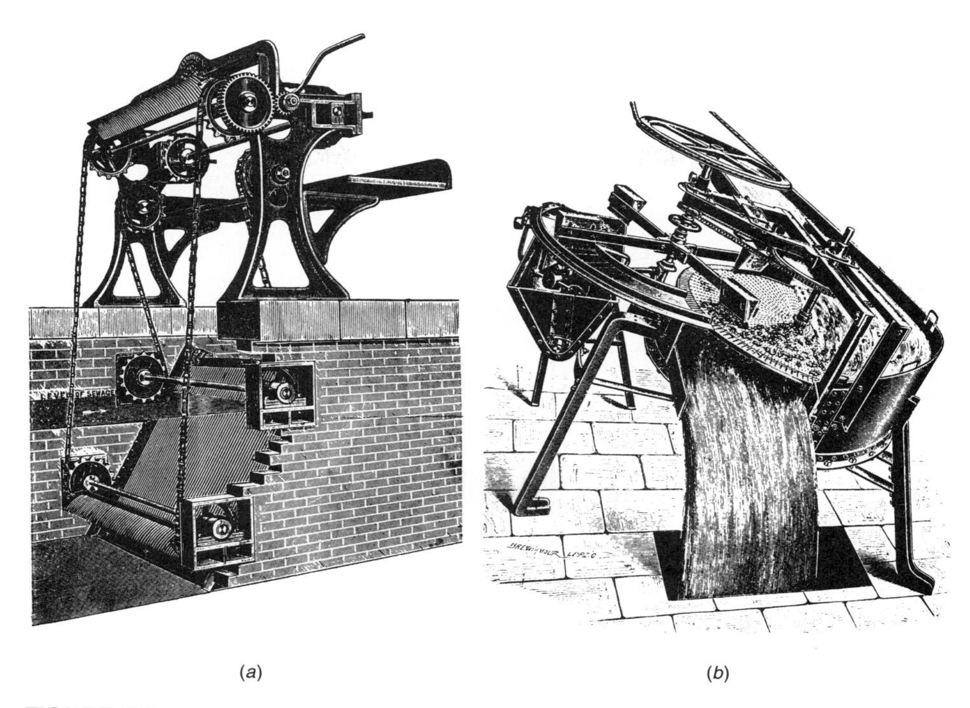

(*a*) (*b*)

FIGURE 5-4
Examples of screening devices from the early 1900s: (*a*) bar screen and (*b*) inclined fine screen.

Fine bar and perforated plate screens. During the past 10 years a number of screening devices have been developed that would have been classified previously as fine screens. The use of fine coarse screens is especially advantageous where the sludge produced during treatment is to be composted. Removal of small particulate solids from the influent wastewater results in the production of a higher-quality compost. The openings in fine coarse screens can vary from about 0.125 to 0.5 in (3.2 to 12.5 mm). Both in-channel and external units have been developed. In-channel fine coarse screens are cleaned mechanically, usually with a reciprocating rake or a helical screw. The perforated plate and the rotary drum are two examples of external screening devices. The openings on perforated plate and rotary drum screens typically vary from about 0.125 to 0.375 in (3.2 to 9.5 mm). Typical examples of fine coarse screens are presented in Fig. 5-5.

Coarse Screenings: Characteristics, Quantities, and Disposal

Coarse screenings, collected on bar racks of about $\frac{1}{2}$-in (12.5-mm) or greater spacing, consist of debris such as rocks, branches, pieces of lumber, leaves, paper, tree roots, plastics, and rags. Organic matter can collect as well. The accumulation of

(a)

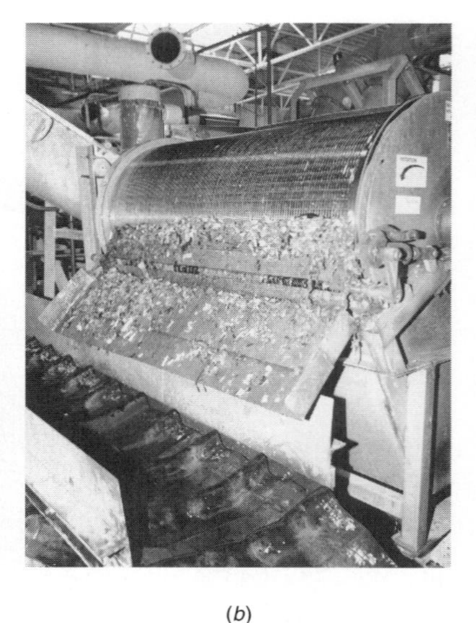

(b)

FIGURE 5-5
Examples of fine coarse screens used in wastewater treatment plants: (*a*) self-cleaning climber screen and (*b*) rotary drum screen.

oil and grease can also be a serious problem, especially in cold climates. The quantity and characteristics of screenings collected for disposal vary, depending on the type of bar rack, the size of the bar rack opening, the type of sewer system, and the geographic location. Typical data on the quantities of coarse screenings to be expected at centralized treatment plants served by conventional gravity sewers are reported in Table 5-3. Data on the moisture content and specific weight of screenings are also included in Table 5-3.

Fine screenings consist of materials that are retained on screens with openings less than about 0.5 in (12.5 mm). The materials retained on fine screens include small rags, paper, plastic materials of various types, razor blades, grit, undecomposed food waste, feces, etc. Compared to coarse screenings, the moisture content

TABLE 5-3
Typical information on the characteristics and quantities of coarse screenings removed from wastewater with bar racks

Size of opening between bars, in	Moisture content, %	Specific weight, lb/ft^3	Volume of screenings, ft^3/Mgal	
			Range	Typical
0.5	60–90	40–68	5–10	7
1.0	50–80	40–68	2–5	3
1.5	50–80	40–68	1–2	1.5
2.0	50–80	40–68	0.5–1.5	0.75

TABLE 5-4
Typical information on the characteristics and quantities of screenings removed from wastewater with fine bar and perforated plate screens

Operation	Size of opening, in	Moisture content, %	Specific weight, lb/ft^3	Volume of screenings, ft^3/Mgal	
				Range	Typical
Fine bar screens	0.5	80–90	40–60	6–15	10
Perforated plate*	0.25	80–90	40–60	4–8	6
Rotary drum*	0.25	80–90	40–60	4–8	6

*Following coarse screening.

is slightly higher and the specific weight of the fine screenings is slightly lower (see Table 5-4). Because putrescible matter, including pathogenic fecal material, is contained within screenings, they must be handled and disposed of properly. Fine screenings contain substantial grease and scum, which requires similar care, especially if odors are to be avoided.

Design Considerations

Typical design information on manually and mechanically cleaned bar racks is presented in Table 5-5. The analysis associated with the use of coarse screening devices involves the determination of the headloss through them. Hydraulic losses through bar racks are a function of approach velocity and the velocity through the bars. The headloss through bar racks can be estimated by using the following equation (Metcalf and Eddy, 1930).

$$h_L = \frac{1}{0.7}\left(\frac{V^2 - v^2}{2g}\right) \tag{5-1}$$

where h_L = headloss, ft (m)
0.7 = an empirical discharge coefficient to account for turbulence and eddy losses

TABLE 5-5
Typical design information for manually and mechanically cleaned bar racks*

Item	Unit	Manually cleaned	Mechanically cleaned
Bar size:			
Width	in	0.2–0.6	0.2–0.6
Depth	in	1.0–1.5	1.0–1.5
Clear spacing between bars	in	1.0–2.0	0.6–3.0
Slope from vertical	deg	30–45	0–30
Approach velocity	ft/s	1.0–2.0	2.0–3.25
Allowable headloss	in	6	6

*Adapted from Tchobanoglous and Burton (1991).

V = velocity of flow through the openings of the bar rack, ft/s (m/s)
v = approach velocity in upstream channel, ft/s (m/s)
g = acceleration due to gravity, ft/s^2 (m/s^2)

The headloss calculated by using Eq. (5-1) applies only when the bars are clean. Headloss increases with the degree of clogging. The buildup of headloss can be estimated by assuming that a portion of the open space in the upper portion of the bars in the flow path is clogged. The use of Eq. (5-1) is illustrated in Example 5-1.

EXAMPLE 5-1. HEADLOSS BUILDUP IN COARSE SCREENS. Determine the buildup of headloss through a bar screen when 50 percent of the flow area is blocked off by the accumulation of coarse solids. Assume the following conditions apply:

Approach velocity = 2 ft/s
Velocity through clean bar rack = 3 ft/s
Open area for flow through clean bar rack = 2 ft^2

Solution

1. Compute the clean water headloss through the bar rack using Eq. (5-1):

$$h_L = \frac{1}{0.7}\left(\frac{V^2 - v^2}{2g}\right)$$

$$= \frac{1}{0.7}\left[\frac{(3\ \text{ft/s})^2 - (2\ \text{ft/s})^2}{2 \times 32.2\ \text{ft/s}^2}\right] = 0.11\ \text{ft}$$

2. Estimate the headloss through the clogged bar rack (increasing the velocity by the ratio 100%/50%) using the orifice equation as given below.

$$h_L = \frac{1}{C}\left(\frac{V^2 - v^2}{2g}\right)$$

The velocity through the clogged bar rack is

$$V_c = 3\ \text{ft/s} \times 100/50 = 6\ \text{ft/s}$$

Assuming the flow coefficient for the clogged bar rack is approximately 0.6, the estimated headloss is

$$h_L = \frac{1}{0.6}\left[\frac{(6\ \text{ft/s})^2 - (2\ \text{ft/s})^2}{2 \times 32.2\ \text{ft/s}^2}\right] = 0.83\ \text{ft}$$

Comment. Where mechanically cleaned bar racks are used, the cleaning mechanism typically is actuated by the buildup of headloss. In some cases, the rack is cleaned at predetermined time intervals, as well as at the buildup of headloss.

5-3 FINE SCREENING

In the 1920s and earlier, fine screens were a common feature of wastewater treatment plants. A typical example of an early inclined disk screen, in which the screening element consisted of bronze or copper plate with milled slots, is shown in Fig. 5-4*b*.

The use of fine screens was abandoned because of the difficulty in cleaning the oils and grease that would accumulate on the screen. Since the early 1980s, when better screening materials and better screening devices became available, there has been a resurgence of interest in the field of wastewater treatment in the use of screens of all types. The applications range from the removal of coarse and fine solids from untreated wastewater to the removal of the residual suspended solids from biological treatment processes.

Description

Fine screens, as shown in Fig. 5-2, typically vary in size from 0.010 to 0.125 in (0.25 to 3.2 mm). Very fine screens vary in size from 0.006 to 0.015 in (0.15 to 0.38 mm). With the development of better screening materials and equipment, the use of fine screens for grit removal and as a replacement for (and a means of upgrading the performance of) primary sedimentation tanks is increasing. Although there are a variety of fine screens, three commonly used screens are considered in the following discussion: inclined-wedge wire screens, rotary drum screens, and rotary disk screens. Additional information on fine screens is presented in Table 5-2.

Inclined-wedge wire screens. One of the most common types of screens used for the pretreatment of wastewater is the inclined self-cleaning-type wedge wire screen (see Fig. 5-6). Wedge wire screens are available in sizes ranging from 0.01 to 0.125 in (0.25 to 3.2 mm).

FIGURE 5-6
Typical inclined-wedge wire screen.

FIGURE 5-7
View of rotary disk screen with 250-μm openings used as a replacement for primary sedimentation.

Rotary drum screens. The openings on rotary drum fine screens can vary from about 0.01 to 0.125 in (0.25 to 3.2 mm). With the exception of the size of the screen, rotary drum fine screens are similar to the fine coarse screen shown in Fig. 5-5*b*.

Rotary disk screens. In recent times very fine screens, typically with openings of about 0.01 in (0.25 mm), have been used as replacement for primary sedimentation facilities. A rotary disk screen has been used in the City of San Diego aquaculture facility for more than 10 years (see Fig. 5-7).

Fine Screenings: Characteristics, Quantities, and Disposal

The material removed with fine screens with small openings and with very fine screens has similar characteristics to primary sludge removed in primary sedimentation. Typical data on the removal of BOD and TSS with fine screens are reported in Table 5-6.

Design Considerations

Where fine screens are used as alternatives to primary sedimentation basins, the following (secondary) facilities must be sized appropriately to handle the solids and BOD_5 not removed by the screens, as compared to the use of primary sedimentation facilities. The clear-water headloss through screens may be obtained from

TABLE 5-6
Typical data on the removal of BOD and TSS with fine screens used to replace primary sedimentation*

Process	Size of openings, in	BOD removal, %	TSS removal, %
Fixed parabolic	0.0625	5–20	5–30
Rotary drum	0.01	25–50	25–45
Rotary disk screen	0.01	35–55	35–55

*The actual removal achieved will depend on the nature of the wastewater collection system and the wastewater travel time.

manufacturers' rating tables, or may be calculated by means of the common orifice formula:

$$h_L = \frac{1}{C(2g)}\left(\frac{Q}{A}\right)^2 \tag{5-2}$$

where h_L = headloss, ft (m)
C = coefficient of discharge for the screen
g = acceleration due to gravity, ft/s^2 (m/s^2)
Q = discharge through screen, ft^3/s (m^3/s)
A = effective open area of submerged screen, ft^2 (m^2)

Values of C and A depend on screen design factors, such as the size and milling of slots, the wire diameter and weave, and particularly the percent of open area, and must be determined experimentally. A typical value of C for a clean screen is 0.60. The headloss through a clean screen is relatively insignificant. The important determination is the headloss during operation, which depends on the size and quantity of solids in the wastewater, the size of the screen openings, and the method and frequency of cleaning. The approach to estimating the differential headloss through fine screens is the same as illustrated in Example 5-1 for coarse screens.

5-4 COMMINUTION

Comminutors are devices used to grind or cut up (comminute) the coarse solids found in wastewater without removing them from the flow. The cut-up solids, returned to the wastewater, are removed in the downstream treatment operations and processes. In onsite systems comminution is used where pressure sewers with grinder pumps are used (see Chap. 6). At wastewater treatment plants, comminutors are often used to eliminate the messy and offensive task of screenings handling and disposal. In cold climates, the use of comminutors precludes the need to prevent collected screenings from freezing. Grinders are also used for inline pipeline installations to shred solids, particularly ahead of wastewater and sludge pumps.

The suitability of using comminution devices at wastewater treatment plants is a debatable question. Comminutors have a reputation for unreliability and high

FIGURE 5-8
Rags accumulating on air diffusers in downstream process.

maintenance. Some designers feel that any material that can be removed by screening should not be allowed to remain in the waste stream, regardless of the form. Other designers maintain that once cut up, the coarse solids are handled more easily in the downstream processes. A disadvantage of comminutors is that comminuted solids often present downstream problems. For example, rags, which tend with agitation to recombine after comminution into ropelike strands, are particularly troublesome. Recombined rags can have a number of negative impacts, such as accumulating on air diffusers (see Fig. 5-8), and clogging pump impellers, sludge pipelines, and heat exchangers. Unfortunately much of the cut-up material will appear in the sludge, and will ultimately have to be removed if the sludge is composted.

Description

Different types of comminutors are available from a number of manufacturers. One type of comminutor (see Fig. 5-9*a*) consists of a vertical revolving-drum screen with $\frac{1}{4}$-in (6-mm) slots in small machines and $\frac{3}{8}$-in (10-mm) slots in large machines. Coarse material is cut by the cutting teeth and the shear bars on the revolving drum as the solids are carried past a stationary comb. The small sheared particles pass through the drum slots and out of a bottom opening through an inverted siphon and into the downstream channel.

Other types of comminuting devices consist of (1) a stationary semicircular screen grid mounted in a rectangular channel with rotating or oscillating circular cutting disks, (2) a unit containing two large-diameter vertical rotating shafts equipped with cutting blades (see Fig. 5-9*b*), and (3) a unit containing a conical-shaped screen grid, the axis of which is located parallel to the channel flow. This unit is also equipped with cutting blades. In all of these types, the screen grid intercepts the larger solids, while smaller solids pass through the space between the grid and

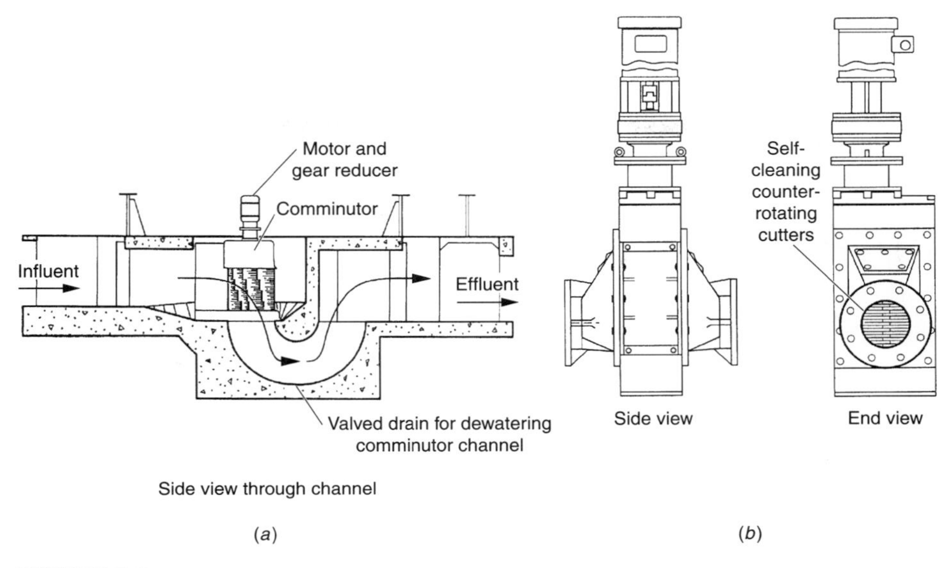

FIGURE 5-9
Typical comminutors used in: (*a*) open channel (adapted from Chicago Pump) (see also Fig. 5-10) and (*b*) inline connection (adapted from Disposable Waste Systems, Inc.).

cutting blades. The method of cutting or shredding the solids is the principal difference in each type.

Design Considerations

Comminutors are often constructed with a bypass arrangement so that a manual bar rack is used in case flow rates exceed the capacity of the comminutor, or in case there is a power or mechanical failure (see Fig. 5-10). Stop gates and provisions for draining should also be included to facilitate maintenance. Headloss through a comminutor usually ranges from several inches to 1 ft (0.3 m), and can approach 3 ft (0.9 m) in large units at maximum flowrates. Detailed design is not required because comminutors are provided as a complete self-contained unit. Recommended channel dimensions, capacity ranges, upstream and downstream submergence, and power requirements are obtained from the manufacturer. Because manufacturers' headloss characteristics are usually based on clean water, the ratings should be decreased by approximately 70 to 80 percent to account for partial clogging of the screen.

Where comminuting devices precede grit removal facilities, the cutting teeth are subject to high wear and require frequent sharpening or replacement. To prolong the life of the equipment, rock traps in the channel upstream of the comminutor should be provided to collect material that could jam the cutting blade and to reduce the wear on the cutting surfaces and on portions of the mechanism where there is a small clearance between moving and stationary parts.

(a)

(b)

FIGURE 5-10
Comminutor with bar rack bypass channel: (*a*) for very small treatment plant and (*b*) for small treatment plant.

5-5 FLOW EQUALIZATION

The variations that are observed in the influent-wastewater flowrate and strength at almost all wastewater treatment facilities were discussed previously in Chap. 4. Flow equalization is used to overcome the operational problems caused by flowrate variations, to improve the performance of the downstream processes, and to reduce the size and cost of downstream treatment facilities. In effect, flow equalization dampens flowrate variations so that a constant or nearly constant outlet flowrate is achieved.

Application of Flow Equalization

Flow equalization can be applied in a number of different situations, depending on the characteristics of the collection system and treatment objectives. Flow equalization is especially useful in small plants that experience high peak-to-average flow and organic loading ratios. If the peak-to-average flowrate ratio is 2 or less, the use of flow equalization may not be economically feasible. Two types of flow equalization basins are used: online and offline (see Fig. 5-11). In most wastewater applications, when the flow is equalized the strength of the wastewater will typically be variable. Equalization of the waste strength can be accomplished, but at a greater expense, as

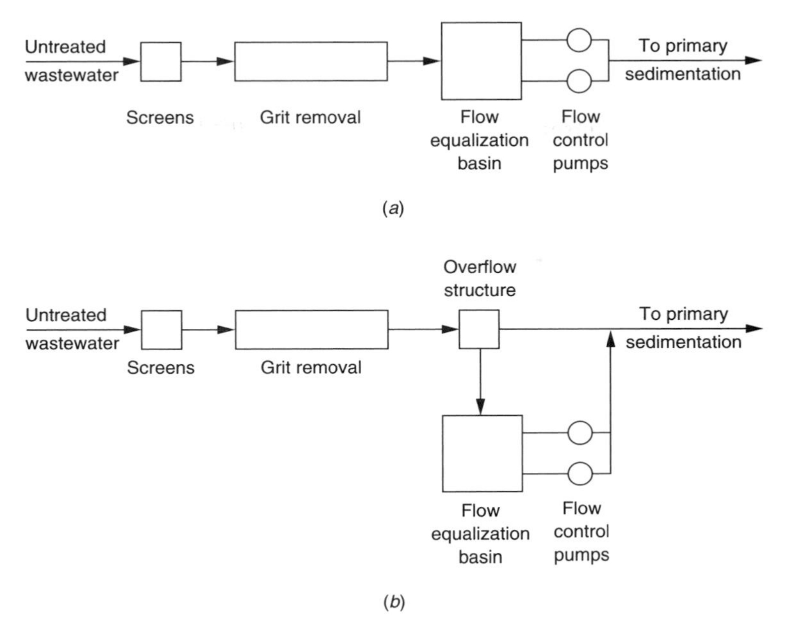

FIGURE 5-11
Flow diagrams for flow equalization: (*a*) online and (*b*) offline.

a larger equalization basin volume will typically be required. Equalization may also be used to dampen pH and toxicity variations in the influent wastewater.

Mass Balance Analysis

The sizing of a flow equalization facility is based on a mass balance analysis. In the mass balance method, the influent wastewater volume during a given time interval is compared to the average hourly flow volume based on a 24-h period. If the influent flow volume is less than the average hourly flow volume, wastewater is withdrawn for the equalization basin. If the influent flow volume is greater than the average hourly flow, the excess wastewater is stored in the equalization basin. The cumulative difference is used to estimate the required storage volume. The mass balance method for sizing flow equalization basins is illustrated in Example 5-2.

EXAMPLE 5-2. DETERMINATION OF VOLUME FOR FLOW EQUALIZATION BASIN. Calculate the required minimum volume for an offline flow equalization basin, using the average hourly flow values calculated in Example 4-1.

Solution

1. Create a spreadsheet to calculate the volume of the flow equalization basin. Enter the 24 one-hour time intervals used in Example 4-1 (e.g., 12 to 1 A.M.) in the first column.

Time interval (1)	Average hourly flow, Mgal/d (2)	Average hourly flow, gal/min (3)	Hourly inflow volume, gal (4)	Hourly volume to or from storage,* gal (5)	Cumulative volume to or from storage at end of interval gal† (6)
12–1 A.M.	0.220	152.78	9,167	−13,681	−13,681
1–2	0.160	111.11	6,667	−16,181	−29,861
2–3	0.135	93.75	5,625	−17,222	−47,083
3–4	0.135	93.75	5,625	−17,222	−64,306
4–5	0.165	114.58	6,875	−15,972	−80,278
5–6	0.280	194.44	11,667	−11,181	−91,458
6–7	0.525	364.58	21,875	−972	**−92,431**
7–8	0.745	517.36	31,042	8,194	−84,236
8–9	0.815	565.97	33,958	11,111	−73,125
9–10	0.800	555.56	33,333	10,486	−62,639
10–11	0.750	520.83	31,250	8,403	−54,236
11–12	0.680	472.22	28,333	5,486	−48,750
12–1 P.M.	0.605	420.14	25,208	2,361	−46,389
1–2	0.535	371.53	22,292	−556	−46,944
2–3	0.480	333.33	20,000	−2,847	−49,792
3–4	0.460	319.44	19,167	−3,681	−53,472
4–5	0.475	329.86	19,792	−3,056	−56,528
5–6	0.555	385.42	23,125	278	−56,250
6–7	0.825	572.92	34,375	11,528	−44,722
7–8	1.075	746.53	44,792	21,944	−22,778
8–9	1.060	736.11	44,167	21,319	−1,458
9–10	0.830	576.39	34,583	11,736	**10,278**
10–11	0.525	364.58	21,875	−972	9,306
11–12	0.325	225.69	13,542	−9,306	0
Total	13.160	9138.89	548,333		
Average	0.548	380.79	22,847		

*Negative values (−) correspond to withdrawals from storage. Positive values correspond to diversions to storage.

†Note: Last digit may not agree due to rounding.

2. Enter the average hourly flowrates, associated with each time period, in column 2 (as calculated previously, the flow unit is Mgal/d).
3. Convert the flowrate values in column 2 from Mgal/d to gal/min, and enter the results in column 3. For example, for the first time interval:

$$\left(\frac{0.220 \text{ Mgal}}{\text{d}}\right)\left(\frac{10^6 \text{ gal}}{\text{Mgal}}\right)\left(\frac{\text{d}}{24 \times 60 \text{ min}}\right) = 152.78 \text{ gal/min}$$

4. For each time period, convert the influent wastewater flowrate in column 3 to an hourly volume, and enter the value in column 4. Also calculate the average of the 24 hourly influent volumes.
   a. The inflow volume for the first time interval is

$$(152.78 \text{ gal/min})(60 \text{ min}) = 9167 \text{ gal}$$

   b. The average hourly volume = 548,333/24 = 22,847 gal.

5. For each time period, determine the volume withdrawn from or sent to storage by subtracting the average hourly volume from the inflow volume for the time period being considered. For the first time period:

$$\begin{array}{ccccc} \text{Inflow volume} \\ \text{during time period} & - & \text{average hourly} \\ \text{volume} & = & \text{flow to (+) or from (−)} \\ \text{storage} \end{array}$$

$$9167 - 22,847 = -13,680$$

   Positive values indicate flow to storage, while negative values indicate flow being removed from storage. Enter the hourly flow volumes in column 5.
6. Calculate the cumulative sum of the hourly volumes to or from storage (column 5), and enter the values in column 6.
7. Determine the required storage volume by identifying the maximum and minimum of cumulative values in column 6 (these values are highlighted in the table above). The difference between these two values is the minimum required storage volume to equalize the flow.

$$\text{Maximum value (9–10 P.M.)} = 10,278 \text{ gal}$$

$$\text{Minimum value (6–7 A.M.)} = -92,431 \text{ gal}$$

$$\text{Required storage volume} = 10,278 - (-92,431) = 102,700 \text{ gal}$$

Comment. A graphical approach known as the *Rippl diagram* method can also be used to estimate the required equalization volume. In the Rippl method, as illustrated in the diagram below, a cumulative inflow curve is plotted. Lines, with a slope equal to the average demand for the day, are drawn tangent to the maximum and minimum points along the cumulative inflow curve. The vertical distance between the two tangent lines corresponds to the required storage volume.

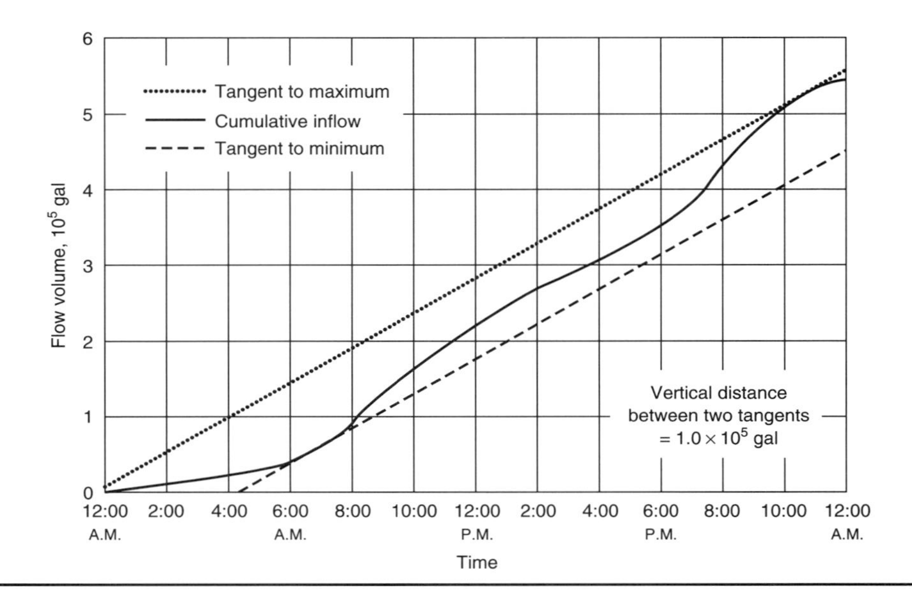

5-6 MIXING

Mixing is of fundamental importance in the design and operation of wastewater treatment plants. Most mixing operations at wastewater treatment facilities fall into three categories: (1) rapid mixing of chemicals, (2) mixing of the fluid in reactors and holding tanks, and (3) flocculation. For example, in the activated-sludge process, the contents of the aeration tank must be mixed and air or pure oxygen must be supplied to provide the microorganisms with oxygen. Diffused air is often used to fulfill both the mixing and oxygen requirements. Alternatively, mechanical turbine-aerator mixers may be used. Chemicals are also mixed with sludge to improve its dewatering characteristics. In aerobic digestion, mixing is used to accelerate the biological conversion process and to distribute the heat generated from biological conversion reactions to the contents of the digester uniformly.

Description

Mixing can be carried out in a number of different ways, including (1) hydraulic jumps in open channels, (2) Venturi flumes, (3) pipelines, (4) pumping, (5) static mixers, and (6) mechanical mixers. In the first four methods, mixing is accomplished as a result of turbulence that exists in the flow regime. In the last two, turbulence is induced through the input of energy. Typical devices used for mixing in wastewater treatment plants are shown in Fig. 5-12.

Propeller and turbine mixers. Mixers with small impellers operating at high speeds are best suited for dispersing small amounts of chemicals or gases in wastewater (see Fig. 5-12*a*). Mixers with slow-moving impellers are best for blending two fluid streams, or for flocculation.

High-speed induction mixer. The high-speed induction mixer, a relatively recent development, is an efficient mixing device for a variety of chemicals. The Water Champ for chlorine mixing, described in Chap. 12, Sec. 12-6, and shown in Fig. 12-42, consists of a motor-driven open propeller that creates a vacuum in the chamber directly above the propeller. The vacuum created by the impeller induces the chemical to be mixed directly from the storage container without the need for dilution water. The high operating speed of the impeller (3450 rev/min) provides a thorough mixing of the chemical that is being added to the water.

Static mixers. Inline static mixers contain internal vanes that bring about sudden changes in the velocity patterns as well as momentum reversals (see Fig. 5-12*b*). Inline static mixers are used most commonly for mixing of chemicals with wastewater. Over and under baffled channels are another form of static mixer, but are not recommended because mixing times less than 1 second are difficult to obtain.

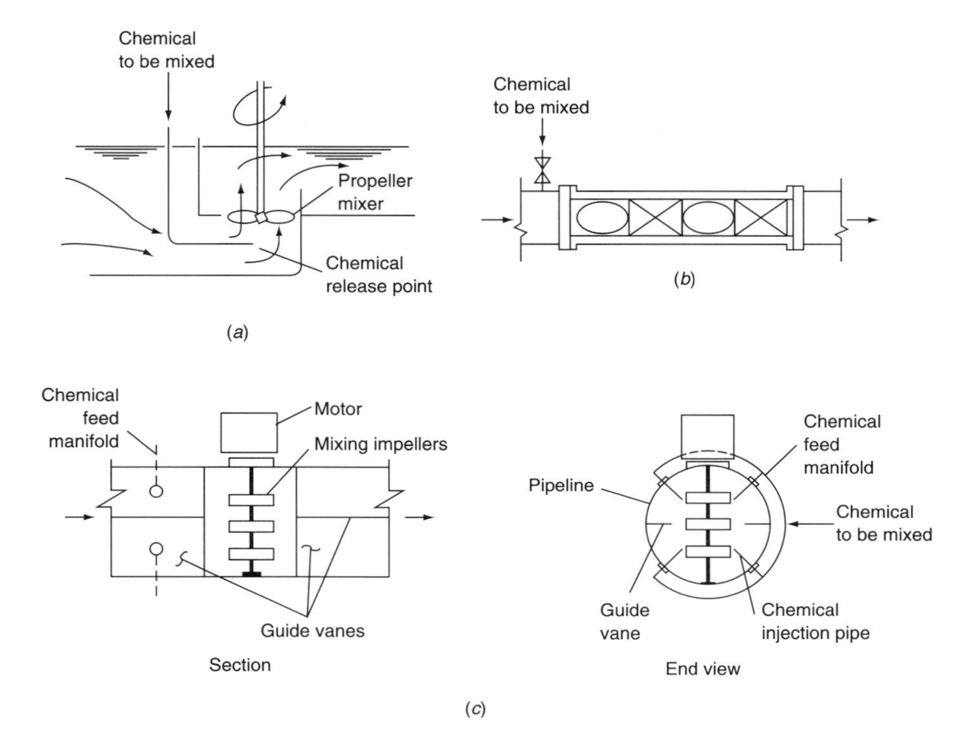

FIGURE 5-12
Typical devices used to mix chemicals with wastewater with mixing times equal to or less than 1 s: (*a*) propeller mixer in open channel, (*b*) inline static mixer for pipeline, and (*c*) inline mechanical mixer for pipelines (adapted from Tchobanoglous and Burton, 1991).

Paddle mixers. Paddle mixers generally rotate slowly, as they have a large surface to the liquid. Paddles are used as flocculation devices when coagulants, such as aluminum or ferric sulfate, and coagulant aids, such as polyelectrolytes and lime, are added to wastewater or sludges. Flocculation is promoted by gentle mixing brought about by the slow-moving paddles, which, as shown in Fig. 5-13, rotate the liquid and promote mixing. Increased particle contact will promote floc growth; however, if the mixing is too vigorous, the shear forces that are set up will break up the floc into smaller particles. Agitation should be controlled carefully so that the floc particles will be of suitable size and will settle readily. The production of a good floc usually requires a detention time of 10 to 20 minutes.

Pneumatic mixing. In pneumatic mixing, a gas (usually air or oxygen) is injected into the bottom of mixing or activated-sludge tanks and the turbulence caused by the rising gas bubbles serves to mix the fluid contents of the tank.

FIGURE 5-13
Typical paddle mixers used for flocculation.

Dissipation of Energy in Mixing

Based on the reasoning that more input power creates greater turbulence, and greater turbulence leads to better mixing, the power input per unit volume of liquid can be used as is a rough measure of mixing effectiveness. Camp and Stein (1943) studied the establishment and effect of velocity gradients in coagulation tanks of various types and developed the following equations that can be used for the design and operation of mixing systems.

$$G = \sqrt{\frac{P}{\mu V}} \tag{5-3}$$

where G = mean velocity gradient, 1/s
P = power requirement, ft·lb/s (W)
μ = dynamic viscosity, lb·s/ft^2 (N·s/m^2)
V = flocculator volume, ft^3 (m^3)

In Eq. (5-3), G is a measure of the mean velocity gradient in the fluid. As shown, the value of G depends on the power input, the viscosity of the fluid, and the volume of the basin. Multiplying both sides of Eq. (5-3) by the theoretical detention time $t_d = V/Q$ yields

$$Gt_d = \frac{V}{Q}\sqrt{\frac{P}{\mu V}} = \frac{1}{Q}\sqrt{\frac{PV}{\mu}} \tag{5-4}$$

where t_d = detention time, s, and Q = flowrate, ft^3/s (m^3/s). Typical values for G for various mixing operations are reported in Table 5-7. The power required for various types of mixers is considered in the following discussion. The use of Eq. (5-3) is illustrated in Example 5-3.

TABLE 5-7
Typical velocity gradient *G* and detention time values for wastewater treatment processes

	Range of values*	
Process	**Detention time**	***G* value, s^{-1}**
Mixing		
Typical rapid mixing operations in wastewater treatment	10–30 s	500–1500
Rapid mixing for effective initial contact and dispersion of chemicals	$\leq$1 s	1500–6000
Rapid mixing of chemicals in contact filtration processes	<1 s	2500–7500
Flocculation		
Typical flocculation processes used in wastewater treatment	30–60 min	50–100
Flocculation in direct filtration processes	2–10 min	25–150
Flocculation in contact filtration processes	2–5 min	25–200

*Adapted from Tchobanoglous and Burton (1991).

EXAMPLE 5-3. POWER REQUIREMENT TO DEVELOP VELOCITY GRADIENTS. Determine the theoretical power requirement to achieve a G value of 100/s in a tank with a volume of 10^5 ft^3 (2832 m^3). Assume the water temperature is 60°F (15.6°C). What is the corresponding value when the water temperature is 40°F (4.4°C)?

Solution

1. Determine the theoretical power requirement at 60°F using Eq. (5-3) rearranged as follows: μ at 60°F $= 2.359 \times 10^{-5}$ lb·s/ft^2 (see App. C-1).

$$P = G^2 \mu V$$

$$P = (100/\text{s})^2 \times 2.359 \times 10^{-5}\ \text{lb·s/ft}^2 \times 10^5\ \text{ft}^3$$

$$= 23{,}590\ \text{ft·lb}_f/\text{s}\ (32.0\ \text{kW})$$

2. Determine the theoretical power requirement at 40°F: μ at 40°F $= 3.229 \times 10^{-5}$ lb·s/ft^2 (see App. C-1).

$$P = (100/\text{s})^2 \times 3.229 \times 10^{-5}\ \text{lb·s/ft}^2 \times 10^5\ \text{ft}^3$$

$$= 32{,}290\ \text{ft·lb}_f/\text{s}\ (48.8\ \text{kW})$$

Design Considerations

The principal design consideration in mixing is the required power input. The power requirements for mixing using propeller and turbine mixers, paddle mixers, static mixers, and pneumatic mixing are delineated in the following discussion.

TABLE 5-8
Values of *k* for estimating power requirements for mixing with various mixers*

Impeller	Laminar range, Eq. (5-5)	Turbulent range, Eq. (5-6)
Propeller, square pitch, 3 blades	41.0	0.32
Propeller, pitch of two, 3 blades	43.5	1.00
Turbine, 6 flat blades	71.0	6.30
Turbine, 6 curved blades	70.0	4.80
Fan turbine, 6 blades	70.0	1.65
Turbine, 6 arrowhead blades	71.0	4.00
Flat paddle, 6 blades	36.5	1.70
Shrouded turbine, 2 curved blades	97.5	1.08
Shrouded turbine with stator (no baffles)	172.5	1.12

*From Rushton (1952).

Power requirements for propeller and turbine mixers. Effective mixing usually occurs in a turbulent flow regime in which inertial forces predominate. Using an analysis of inertial and viscous forces, Rushton (1952) has developed the following relationships for power requirements for laminar and turbulent conditions.

Laminar: $$P = k\mu n^2 D^3 \tag{5-5}$$

Turbulent: $$P = k\rho n^3 D^5 \tag{5-6}$$

where P = power requirement, ft·lb/s (W)
k = constant (see Table 5-8)
μ = dynamic viscosity of fluid, lb·s/ft^2 (N·s/m^2)
ρ = mass density of fluid, slug/ft^3 (kg/m^3)
D = diameter of impeller, ft (m)
n = rotational speed, rev/s

Values of k, as developed by Rushton (1952), are presented in Table 5-8. For the turbulent range, it is assumed that vortex conditions have been eliminated by four baffles at the tank wall, each 10 percent of the tank diameter, as shown in Fig. 5-14.

Equation (5-5) applies if the Reynolds number is less than 10, and Eq. (5-6) applies if the Reynolds number is greater than 10,000. For intermediate values of the Reynolds number, the article by Rushton (1952) should be consulted. The Reynolds number is given by

$$N_R = \frac{D^2 n \rho}{\mu} \tag{5-7}$$

where D = diameter of impeller, ft (m)
n = rotational speed, rev/s
ρ = mass density of liquid, slug/ft^3 (kg/m^3)
μ = dynamic viscosity, lb·s/ft^2 (N·s/m^2)

Where propeller or turbine mixers are used it is imperative that vortexing or mass swirling of the liquid be eliminated. Vortexing, in which the liquid to be mixed

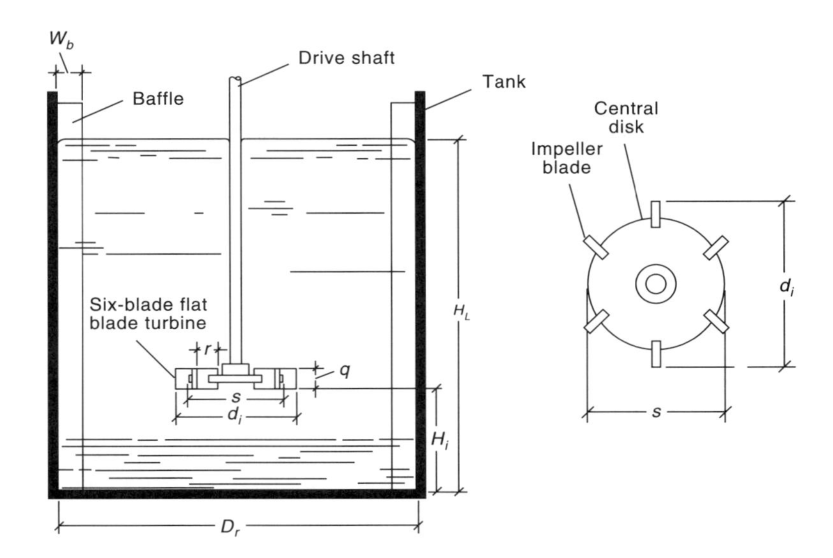

Notes: 1. The agitator is a six-blade flat turbine impeller
2. Impeller diameter, d_i = 1/3 tank diameter
3. Impeller height from bottom, H_i = 1.0 impeller diameter
4. Impeller blade width, q = 1/5 impeller diameter
5. Impeller blade length, r = 1/4 impeller diameter
6. Length of impeller blade mounted on the central disk = $r/2$ = 1/8 impeller diameter
7. Liquid height, H_L = 1.0 tank diameter
8. Number of baffles = 4 mounted vertically at tank wall and extending from the tank bottom to above the liquid surface
9. Baffle width, W_b = 1/10 tank diameter
10. Central disk diameter, s = 1/4 tank diameter

FIGURE 5-14
Definition sketch for vortex condition controlled by baffles (adapted from Holland and Chapman, 1966).

rotates with the impeller, causes a reduction in the difference between the fluid velocity and the impeller velocity and thus decreases the effectiveness of mixing. If the mixing vessel is fairly small, vortexing can be prevented by mounting the impellers off-center or at an angle with the vertical, or by having them enter the side of the basin at an angle. In circular and rectangular tanks the usual method used to limit vortexing is to install four or more vertical baffles extending approximately one-tenth the diameter out from the wall. These baffles effectively break up the mass rotary motion and promote vertical mixing.

Power requirements for paddle mixers. Power in a mechanical paddle system can be related to the drag force on the paddles as follows.

$$F_D = \frac{C_D A \rho v_p^2}{2} \tag{5-8}$$

$$P = F_D v_p = \frac{C_D A \rho v_p^3}{2} \tag{5-9}$$

where F_D = drag force, lb (N)
C_D = coefficient of drag of paddle moving perpendicular to fluid
A = cross-sectional area of paddles, ft^2 (m^2)
ρ = mass fluid density, slug/ft^3 (kg/m^3)
v_p = relative velocity of paddles with respect to the fluid, ft/s (m/s), usually assumed to be 0.6 to 0.75 times the paddle-tip speed
P = power requirement, ft·lb/s (W)

It has been found that with a paddle-tip speed of approximately 2 to 3 ft/s (0.6 to 0.9 m/s), sufficient turbulence is achieved without breaking up the floc. The application of Eq. (5-9) is illustrated in Example 5-4.

EXAMPLE 5-4. POWER REQUIREMENTS AND PADDLE AREA FOR A WASTEWATER FLOCCULATOR. Determine the theoretical power requirement and the paddle area required to achieve a G value of 50/s in a tank with a volume of 10^5 ft^3 (2831 m^3). Assume that the water temperature is 60°F (15.6°C), the coefficient of drag C_D for rectangular paddles is 1.8, the paddle-tip velocity v_p is 2 ft/s (0.6 m/s), and the relative velocity of the paddles is $0.75v_p$.

Solution

1. Determine the theoretical power requirement using Eq. (5-3). μ at 60°F = 2.359×10^{-5} lb·s/ft^2 (see App. C-1).

$$P = G^2 \mu V$$

$$P = (50/\text{s})^2 \times 2.359 \times 10^{-5} \frac{\text{lb·s}}{\text{ft}^2}(10^5 \text{ ft}^3)$$

$$= 5898 \text{ ft·lb}_f/\text{s} \ (8.0 \text{ kW})$$

2. Determine the required paddle area using Eq. (5-9). ρ at 60°F = 1.938 slug/ft^3 (see App. C-1).

$$A = \frac{2P}{C_D \rho v^3}$$

$$A = \frac{2 \times 5898 \text{ ft·lb/s}}{1.8\ (1.938 \text{ slug/ft}^3)(0.75 \times 2.0 \text{ ft/s})^3}$$

$$= 1002 \text{ ft}^2 \ (98.9 \text{ m}^2)$$

Power requirements for static mixers. The power consumed by static mixing devices can be computed using the following equation.

$$P = \gamma Q h \tag{5-10}$$

where P = power dissipated, ft·lb/s (kW)
γ = specific weight of water lb/ft^3 (kN/m^3)
Q = flowrate, ft^3/s (m^3/s)
h = headloss dissipated as liquid passes through device, ft (m)

Power requirements for pneumatic mixing. When air is injected in mixing or flocculation tanks or channels, the power dissipated by the rising air bubbles can be estimated with the following equation:

$$P = p_a V_a \ln\left(\frac{p_c}{p_a}\right) \tag{5-11}$$

where P = power dissipated, ft·lb/s (kW)
p_a = atmospheric pressure, lb/ft^2 (kN/m^2)
V_a = volume of air at atmospheric pressure, ft^3/s (m^3/s)
p_c = air pressure at the point of discharge, lb/ft^2 (kN/m^2)

Equation (5-11) is derived from a consideration of the work done when the volume of air released under compressed conditions expands isothermally. If the flow of air at atmospheric pressure is expressed in terms of ft^3/min (m^3/min) and the pressure is expressed in terms of feet (meters) of water, Eq. (5-11) can be written as follows:

$$P = KQ_a \ln\left(\frac{h + 34}{34}\right) \quad \text{U.S. customary units} \tag{5-12a}$$

$$P = KQ_a \ln\left(\frac{h + 10.33}{10.33}\right) \quad \text{SI units} \tag{5-12b}$$

where K = constant = 81.5 (1.689 in SI units)
Q_a = air flowrate at atmospheric pressure, ft^3/min (m^3/min)
h = air pressure at the point of discharge expressed as the height of a column of water, ft (m)

The velocity gradient G achieved in pneumatic mixing is obtained by substituting P from Eq. (5-12) into Eq. (5-3). The use of Eq. (5-12) is illustrated in Example 5-5.

EXAMPLE 5-5. AIR FLOW REQUIREMENT FOR PNEUMATIC MIXING. Determine the theoretical air flow requirement to mix pneumatically the contents of the reactor to achieve a G value of 50/s in a tank with a volume of 10^5 ft^3. Assume that the water temperature is 60°F (15.6°C). The length, width, and depth of the tank are 100, 50, and 20 ft, respectively.

Solution

1. The theoretical power requirement obtained previously in Example 5-4 is

$$P = 5898 \text{ ft·lb}_f\text{/s}$$

2. Determine the required air flow using Eq. (5-12*a*):

$$Q_a = \frac{P}{K \times \ln\left(\frac{h+34}{34}\right)} = \frac{5898}{81.5 \times \ln\left(\frac{20+34}{34}\right)} = 156 \text{ ft}^3\text{/min}$$

5-7 INTRODUCTION TO GRAVITY SEPARATION

The removal of suspended and colloidal materials from wastewater by gravity separation is one of the most widely used unit operations in wastewater treatment. A summary of gravitational phenomena is presented in Table 5-9. Sedimentation is the

TABLE 5-9
Types of gravitational phenomena involved in wastewater treatment*

Type of settling phenomenon	Description	Application/occurrence
Discrete particle settling (type 1)	Refers to the settling of particles in a suspension of low solids concentration by gravity in a constant acceleration field. Particles settle as individual entities, and there is no significant interaction with neighboring particles.	Removal of grit and sand particles from wastewater.
Flocculant settling (type 2)	Refers to a rather dilute suspension of particles that coalesce, or flocculate, during the settling operation. By coalescing, the particles increase in mass and settle at a faster rate.	Removal of a portion of the suspended solids in untreated wastewater in primary settling facilities, and in upper portions of secondary settling facilities. Also removes chemical floc in settling tanks.
Hindered settling, also called zone settling (type 3)	Refers to suspensions of intermediate concentration, in which interparticle forces are sufficient to hinder the settling of neighboring particles. The particles tend to remain in fixed positions with respect to each other, and the mass of particles settles as a unit. A solids-liquid interface develops at the top of the settling mass.	Occurs in secondary settling facilities used in conjunction with biological treatment facilities.
Compression settling (type 4)	Refers to settling in which the particles are of such concentration that a structure is formed, and further settling can occur only by compression of the structure. Compression takes place from the weight of the particles, which are constantly being added to the structure by sedimentation from the supernatant liquid.	Usually occurs in the lower layers of a deep sludge mass, such as in the bottom of deep secondary settling facilities and in sludge-thickening facilities.
Accelerated gravity settling	Removal of particles in suspension by gravity settling in an acceleration field.	Removal of grit and sand particles from wastewater.
Flotation separation	Removal of particles in suspension that are lighter than water by flotation.	Removal of greases and oils, light material that floats; thickening of sludges.

*Adapted from Tchobanoglous and Burton (1991).

term applied to the separation of suspended particles that are heavier than water, by gravitational settling. The terms *sedimentation* and *settling* are used interchangeably. A sedimentation basin may also be referred to as a sedimentation tank, settling basin, or settling tank. Accelerated gravity settling involves the removal of particles in suspension by gravity settling in an accelerated flow field. Flotation is another form of gravity separation in which particles lighter than water will float to the surface, to be removed by skimming. The fundamentals of gravity separation are introduced in this section. The design of facilities for the removal of grit and TSS is considered in Secs. 5-8 and 5-9, respectively.

Description

Sedimentation is used for the removal of grit, TSS in primary settling basins, biological floc in the activated-sludge settling basin, and chemical floc when the chemical coagulation process is used. It is also used for solids concentration in sludge thickeners. In most cases, the primary purpose is to produce a clarified effluent, but it is also necessary to produce sludge with a solids concentration that can be handled and treated easily.

On the basis of the concentration and the tendency of particles to interact, four types of gravitational settling can occur: (1) discrete particle, (2) flocculant, (3) hindered (also called *zone*), and (4) compression. Other gravitational separation processes include accelerated gravity settling and flotation. Because of the fundamental importance of the separation processes in the treatment of wastewater, the analysis of each type of separation process is discussed separately. In addition, tube settlers, used to enhance the performance of sedimentation facilities, are also described.

Particle Settling Theory

The settling of discrete, nonflocculating particles can be analyzed by means of the classic laws of sedimentation formulated by Newton and Stokes. Newton's law for the terminal velocity of a particle is obtained by equating the gravitational force of the suspended particle to the frictional resistance, or drag. The gravitational force, considering the effect of buoyancy, is given by

$$F_G = (\rho_s - \rho_w) g V_p \tag{5-13}$$

where F_G = gravitational force
ρ_s = density of particle
ρ_w = density of water
g = acceleration due to gravity
V_p = volume of particle

The frictional drag force depends on the particle velocity, fluid density, fluid viscosity, particle diameter, and the drag coefficient C_d (dimensionless), and is given by Eq. (5-14):

$$F_d = \frac{C_d A_p \rho_w v_p^2}{2} \tag{5-14}$$

where C_d = drag coefficient
A_p = cross-sectional or projected area of particles at right angles to v_p
v_p = particle velocity

Equating the gravitational force to the frictional drag force for spherical particles yields Newton's law:

$$v_{p(t)} = \sqrt{\frac{4g}{3C_d}\left(\frac{\rho_s - \rho_w}{\rho_w}\right)d_p} \approx \sqrt{\frac{4g}{3C_d}(\mathrm{sg}_p - 1)d_p} \tag{5-15}$$

where $v_{p(t)}$ = terminal velocity of particle
d_p = diameter of particle
sg_p = specific gravity of the particle

The coefficient of drag C_d takes on different values depending on whether the flow regime surrounding the particle is laminar or turbulent. The drag coefficient for various particles is shown in Fig. 5-15 as a function of the Reynolds number. As shown in Fig. 5-15, there are three more or less distinct regions, depending on the Reynolds number: laminar ($N_R < 1$), transitional ($N_R = 1$ to 2000), and turbulent ($N_R > 2000$). Although particle shape affects the value of the drag coefficient, for particles that are approximately spherical, the curve in Fig. 5-15 is approximated by the following equation (upper limit of $N_R = 10^4$):

$$C_d = \frac{24}{N_R} + \frac{3}{\sqrt{N_R}} + 0.34 \tag{5-16}$$

The Reynolds number for settling particles is defined as

$$N_R = \frac{\phi_p v_p d_p \rho_w}{\mu} = \frac{\phi_p v_p d_p}{\nu} \tag{5-17}$$

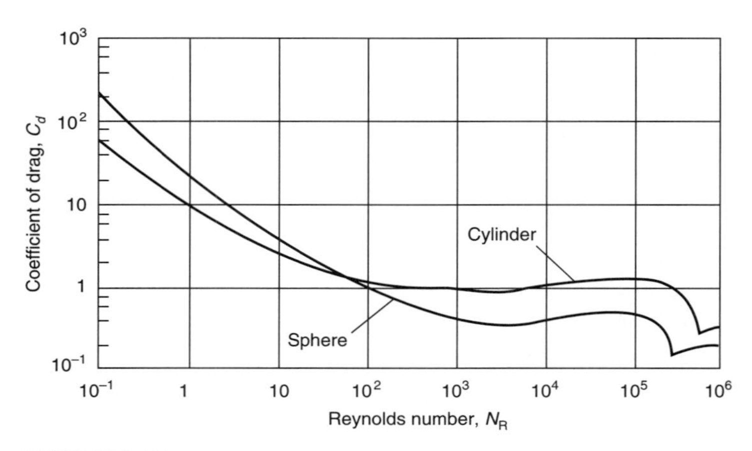

FIGURE 5-15
Coefficient of drag as a function of Reynolds number.

where ϕ_p = shape factor for particle, dimensionless
μ = dynamic viscosity
ρ_w = density of water
ν = kinematic viscosity, defined as ρ_w/μ

Other terms are as defined above.

The shape factor, a dimensionless number, is defined as the ratio of the surface area of an equivalent sphere to the surface area of the particle, for particles of the same volume:

$$\phi_p = \frac{\text{surface area of equivalent sphere}}{\text{surface area of particle}} \tag{5-18}$$

It should be noted that shape factor ϕ_p is also known as the sphericity factor, which is designated as ψ. Typical values for the shape factor for spherical particles, for rounded sand, and for crushed coal and angular sand, are 1.0, 0.82, and 0.73, respectively. Additional material on the sphericity of particles used for the filtration of treated wastewater may be found in Sec. 12-3, Chap. 12.

Settling in the laminar region. For Reynolds numbers less than about 1.0, viscosity is the predominant force governing the settling process, and the first term in Eq. (5-16) predominates. Assuming spherical particles, substitution of the first term of the drag coefficient equation [Eq. (5-16)] into Eq. (5-15) yields Stokes' law:

$$v_p = \frac{g(\rho_s - \rho_w)d_p^2}{18\mu} = \frac{g(\text{sg}_p - 1)d_p^2}{18\nu} \tag{5-19}$$

where ν = kinematic viscosity. Other terms are as defined above.

For laminar-flow conditions, Stokes found the drag force to be

$$F_d = 3\pi\mu v_p d_p \tag{5-20}$$

Stokes' law [Eq. (5-19)] can also be derived by equating the drag force found by Stokes to the effective weight of the particle [Eq. (5-13)].

Settling in the transition region. In the transition region, the complete form of the drag equation [Eq. (5-16)] must be used to determine the settling velocity, as illustrated in Example 5-6. Because of the nature of the drag equation, finding the settling velocity is an iterative process. As an aid in visualizing settling in the transition region, Fig. 5-16 has been prepared; the figure covers the laminar and the transition region for particle sizes of interest in environmental engineering.

Settling in the turbulent region. In the turbulent region, inertial forces are predominant, and the effect of the first two terms in the drag coefficient equation [Eq. (5-16)] is reduced. For settling in the turbulent region, a value of 0.4 is used for the coefficient of drag. If a value of 0.4 is substituted into Eq. (5-15) for C_d, the resulting equation is

$$v_p = \sqrt{3.33g\left(\frac{\rho_s - \rho_w}{\rho_w}\right)d_p} \approx \sqrt{3.33g(\text{sg}_p - 1)d_p} \tag{5-21}$$

The use of Eqs. (5-15) through (5-19) is illustrated in Example 5-6.

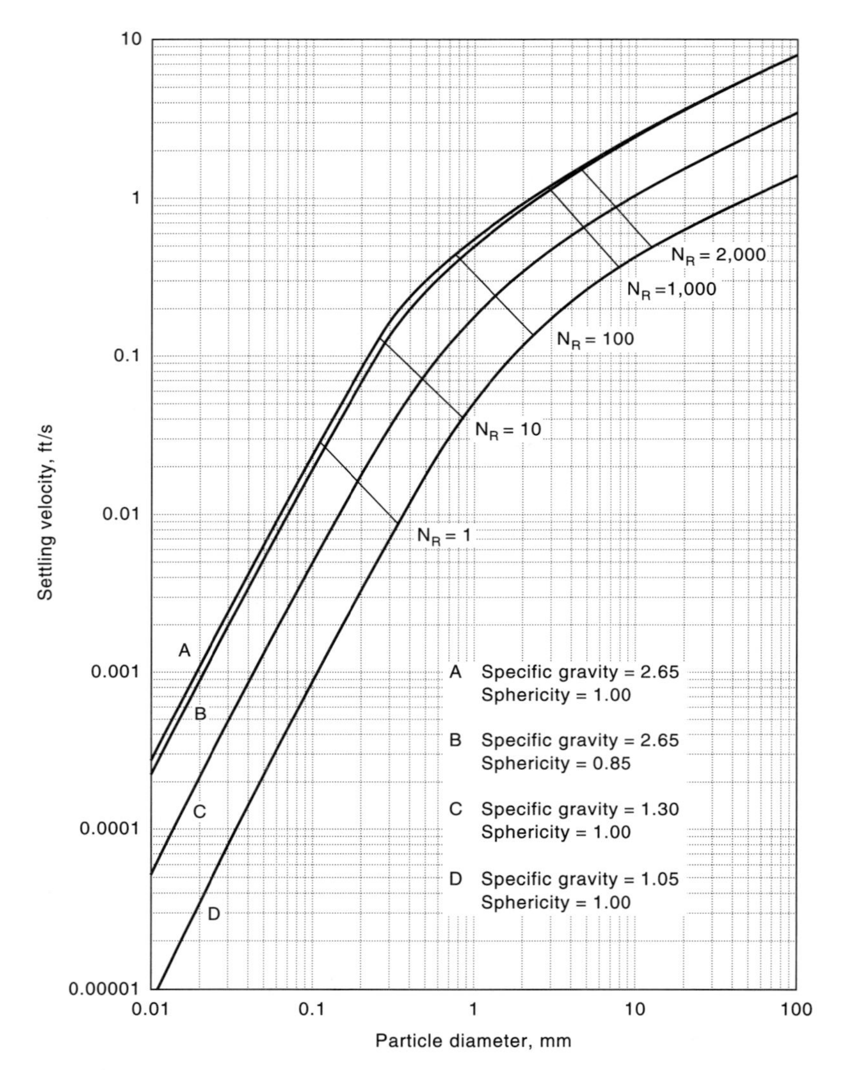

FIGURE 5-16
Settling velocities for various particle sizes under varying conditions: (*a*) settling velocity in ft/s versus particle size in mm.

EXAMPLE 5-6. DETERMINATION OF PARTICLE TERMINAL SETTLING VELOCITY. Determine the terminal settling velocity for a sand particle with an average diameter of 0.5 mm (0.00164 ft), a shape factor of 0.85, and a specific gravity of 2.65, settling in water at 68°F (20°C). Assume a kinematic viscosity value of 1.091×10^{-5} ft^2/s.

Solution

1. Determine the terminal settling velocity for the particle using Stokes' law [Eq. (5-19)]:

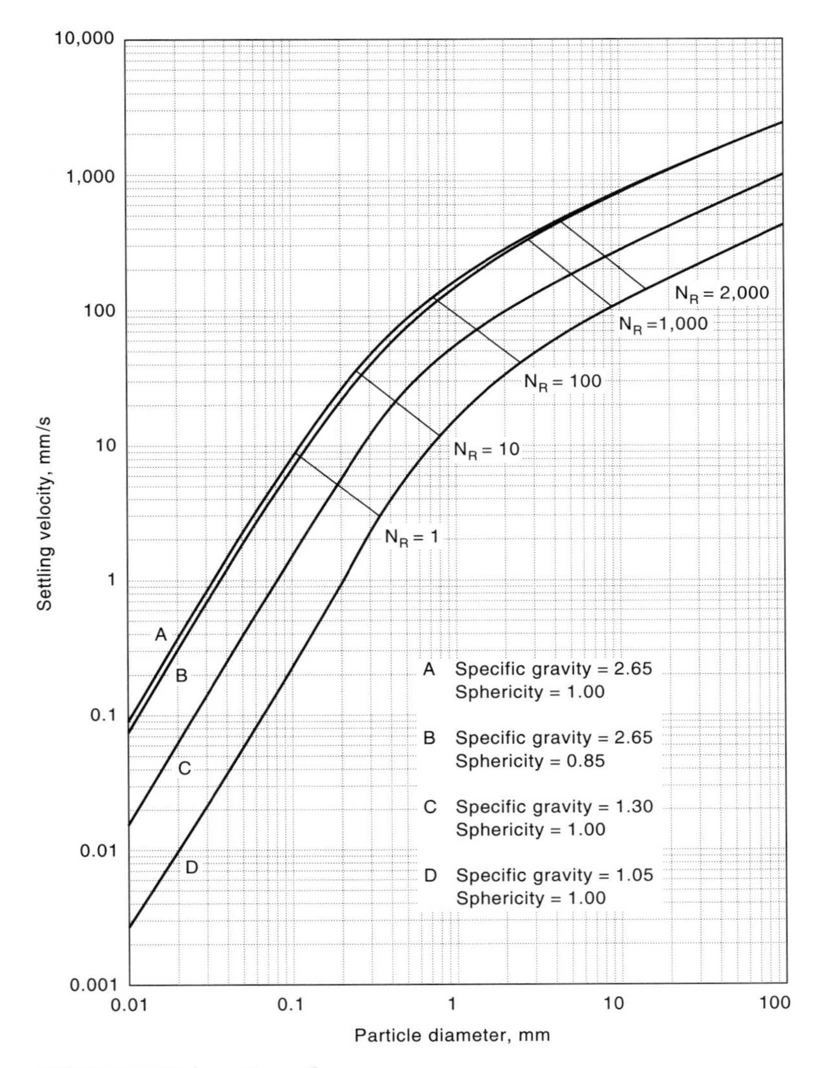

FIGURE 5-16 (*continued*)
Settling velocities for various particle sizes under varying conditions: (*b*) settling velocity in mm/s versus particle size in mm.

$$v_s = \frac{g(\text{sg} - 1)d_p^2}{18\nu}$$

$$= \left(\frac{32.2 \text{ ft}}{\text{s}^2}\right)\left(\frac{2.65 - 1}{18}\right)\left(\frac{0.00164 \text{ ft}}{1}\right)^2\left(\frac{\text{s}}{1.091 \times 10^{-5} \text{ ft}^2}\right) = 0.728 \text{ ft/s}$$

2. Check the Reynolds number [Eq. (5-17)]:

$$N_{\text{R}} = \frac{\phi v_s d_p}{\nu} = \left(\frac{0.85}{1}\right)\left(\frac{0.728 \text{ ft}}{\text{s}}\right)\left(\frac{0.00164 \text{ ft}}{1}\right)\left(\frac{\text{s}}{1.091 \times 10^{-5} \text{ ft}^2}\right) = 93.0$$

The use of Stokes' law is not appropriate for Reynolds numbers greater than 1.0. Therefore, Newton's law [Eq. (5-15)] must be used to determine the settling velocity in the transition region (see Fig. 5-16). The drag coefficient term in Newton's equation is dependent on the Reynolds number, which is a function of the settling velocity. Because the settling velocity is not known, an initial settling velocity must be assumed. The assumed velocity is used to compute the Reynolds number, which is used to determine the drag coefficient, which is used in the Newton equation to calculate the settling velocity. A solution is achieved when the initial assumed settling velocity is approximately equal to the settling velocity resulting from Newton's equation. The solution process is iterative, as illustrated below.

3. For the first assumed settling velocity, use the Stokes' law settling velocity calculated above. Using the resulting Reynolds number, also determined previously, compute the drag coefficient:

$$C_d = \frac{24}{N_R} + \frac{3}{\sqrt{N_R}} + 0.34 = \frac{24}{93.0} + \frac{3}{\sqrt{93.0}} + 0.34 = 0.909$$

4. Use the drag coefficient in Newton's equation to determine the particle settling velocity:

$$v_s = \sqrt{\frac{4g(\text{sg} - 1)d}{3C_d}} = \sqrt{\left(\frac{4}{3}\right)\left(\frac{32.2\text{ ft}}{\text{s}^2}\right)\left(\frac{2.65 - 1}{0.909}\right)\left(\frac{0.00164\text{ ft}}{1}\right)} = 0.357\text{ ft/s}$$

Because the initial assumed settling velocity (0.728 ft/s) does not equal the Newton's equation settling velocity (0.357 ft/s), a second iteration is necessary.

5. For the second iteration, assume a settling velocity value of 0.30 ft/s, and calculate the Reynolds number. Use the Reynolds number to determine the drag coefficient, and use the drag coefficient in Newton's equation to find the settling velocity.

$$N_R = \left(\frac{0.85}{1}\right)\left(\frac{0.30\text{ ft}}{\text{s}}\right)\left(\frac{0.00164\text{ ft}}{1}\right)\left(\frac{\text{s}}{1.091 \times 10^{-5}\text{ ft}^2}\right) = 38.3$$

$$C_d = \frac{24}{38.3} + \frac{3}{\sqrt{38.3}} + 0.34 = 1.45$$

$$v_s = \sqrt{\left(\frac{4}{3}\right)\left(\frac{32.2\text{ ft}}{\text{s}^2}\right)\left(\frac{2.65 - 1}{1.45}\right)\left(\frac{0.00164\text{ ft}}{1}\right)} = 0.283\text{ ft/s}$$

Although the assumed settling velocity (0.30 ft/s) and the calculated settling velocity (0.283 ft/s) still do not agree, they are closer. With successive iterations, the actual settling velocity is shown to be about 0.276 ft/s.

6. Confirm the accuracy of the particle settling velocity value of 0.276 ft/s:

$$N_R = \left(\frac{0.85}{1}\right)\left(\frac{0.276\text{ ft}}{\text{s}}\right)\left(\frac{0.00164\text{ ft}}{1}\right)\left(\frac{\text{s}}{1.091 \times 10^{-5}\text{ ft}^2}\right) = 35.3$$

$$C_d = \frac{24}{35.3} + \frac{3}{\sqrt{35.3}} + 0.34 = 1.52$$

$$v_s = \sqrt{\left(\frac{4}{3}\right)\left(\frac{32.2\text{ ft}}{\text{s}^2}\right)\left(\frac{2.65 - 1}{1.52}\right)\left(\frac{0.00164\text{ ft}}{1}\right)} = 0.276\text{ ft/s}$$

Because the settling velocity used to compute the Reynolds number agrees with the output settling velocity value from Newton's equation, the solution has been confirmed.

Discrete Particle Settling (Type 1)

In the design of sedimentation basins, the usual procedure is to select a particle with a terminal velocity v_c and to design the basin so that all particles that have a terminal velocity equal to or greater than v_c will be removed. The rate at which clarified water is produced is equal to

$$Q = Av_c \tag{5-22}$$

where A is the surface of the sedimentation basin. Rearranging Eq. (5-22) yields

$$v_c = \frac{Q}{A} = \text{overflow rate, gal/ft}^2\cdot\text{d (m}^3/\text{m}^2\cdot\text{d)}$$

Thus, the critical velocity is equivalent to the overflow rate or surface loading rate, a common basis of design for Type 1 settling; the flow capacity is independent of the depth.

For continuous-flow sedimentation, the length of the basin and the time a unit volume of water is in the basin (detention time) should be such that all particles with the design velocity v_c will settle to the bottom of the tank. The design velocity, detention time, and basin depth are related as follows:

$$v_c = \frac{\text{depth}}{\text{detention time}} \tag{5-23}$$

In practice, design factors must be adjusted to allow for the effects of inlet and outlet turbulence, short circuiting, sludge storage, and velocity gradients due to the operation of sludge-removal equipment. These factors are discussed in Sec. 5-9. In the above discussion ideal settling conditions have been assumed.

Type 1 idealized settling in three different types of settling basins is illustrated in Fig. 5-17. Full-scale settling basins used in practice are shown in Fig. 5-18. Particles that have a velocity of fall less than v_c will not all be removed during the time provided for settling. Assuming that the particles of various sizes are uniformly distributed over the entire depth of the basin at the inlet, it can be seen from an analysis of the particle trajectory in Fig. 5-19 that particles with a settling velocity less than v_c will be removed in the ratio

$$X_r = \frac{v_p}{v_c} \tag{5-24}$$

where X_r is the fraction of the particles with settling velocity v_p that are removed.

In most suspensions encountered in wastewater treatment, a large gradation of particle sizes will be found. To determine the efficiency of removal for a given settling time, it is necessary to consider the entire range of settling velocities present in the system. The settling velocities of the particles can be obtained by (1) use of sieve analysis and hydrometer tests combined with Eq. (5-17) or (2) use of settling column tests. The particle settling data are used to construct a velocity settling curve as shown in Fig. 5-19.

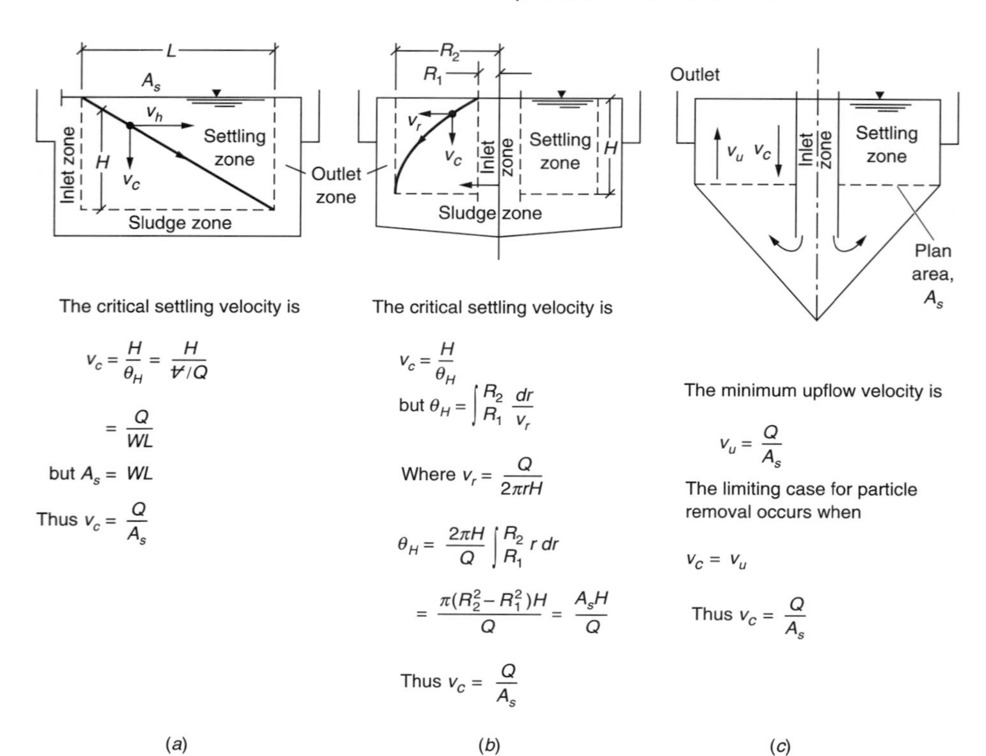

FIGURE 5-17
Definition sketch for the idealized settling of discrete particles in three different types of settling basins: (*a*) rectangular, (*b*) circular, and (*c*) upflow (adapted from Barnes et al., 1980).

FIGURE 5-18
Typical settling basins for small treatment plants: (*a*) rectangular and (*b*) circular (sludge drying beds in the background).

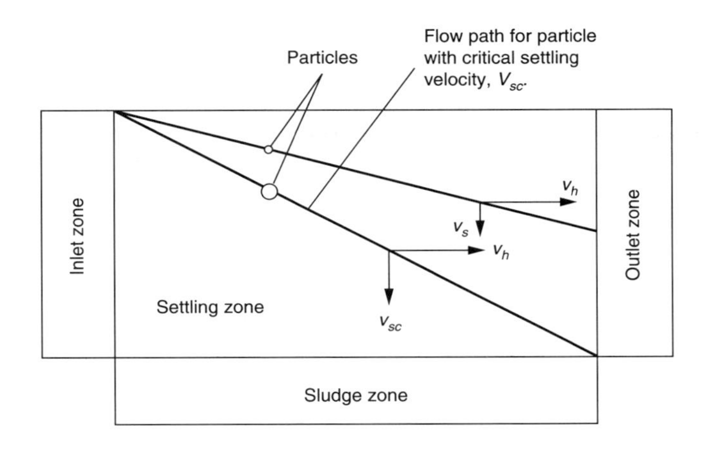

FIGURE 5-19
Definition sketch for the analysis of ideal discrete particle settling.

For a given clarification rate Q, where

$$Q = v_c A \tag{5-25}$$

only those particles with a velocity greater than v_c will be removed completely. The remaining particles will be removed in the ratio v_p/v_c. The total fraction of particles removed for a continuous distribution is given by Eq. (5-26):

$$\text{Fraction removed} = (1 - X_c) + \int_0^{X_c} \frac{v_p}{v_c} dx \tag{5-26}$$

where $1 - X_c$ = fraction of particles with velocity v_p greater than v^c and

$\int_0^{X_c} \frac{v_p}{v_c} dx$ = fraction of particles removed with v_p less than v_c

For discrete particles within a given settling velocity range, the following expression may be used

$$\text{Total fraction removed} = \frac{\sum_{i=1}^{n} \frac{v_{ni}}{v_c}(n_i)}{\sum_{i=1}^{n} n_i} \tag{5-27}$$

where v_n = average velocity of the particles in the ith velocity range and n_i = the number of particles in the ith velocity range. The use of Eq. (5-27) is illustrated in Example 5-7.

EXAMPLE 5-7. CALCULATION OF REMOVAL EFFICIENCY FOR A PRIMARY SEDIMENTATION BASIN (TYPE 1 SETTLING). Determine the removal efficiency for a sedimentation basin with a critical overflow velocity of 6.5 ft/h (1167 gal/ft^2·d) in treating a wastewater containing particles whose settling velocities are distributed as given in the table below. Plot the particle histogram for the influent and effluent wastewater.

Settling velocity, ft/h	Number of particles per liter $\times 10^{-5}$
0.0–1.5	20
1.5–3.0	40
3.0–4.5	80
4.5–6.0	120
6.0–7.5	100
7.5–9.0	70
9.0–10.5	20
10.5–12.0	10
Total	460

Solution

1. Create a table for calculating the percentage removal for each particle size. Enter the particle settling velocity ranges in column 1.

Settling velocity range, ft/h (1)	Average settling velocity, ft/h (2)	Number of particles in influent, $\times 10^{-5}$ (3)	Fraction of particles (4)	Number of particles removed, $\times 10^{-5}$ (5)	Particles remaining in effluent, $\times 10^{-5}$ (6)
0.0–1.5	0.75	20	0.115	2.3	17.7
1.5–3.0	2.25	40	0.346	13.8	26.2
3.0–4.5	3.75	80	0.577	46.2	33.8
4.5–6.0	5.25	120	0.808	96.9	23.1
6.0–7.5	6.75	100	1.000	100.0	0.0
7.5–9.0	8.25	70	1.000	70.0	0.0
9.0–10.5	9.75	20	1.000	20.0	0.0
10.5–12.0	11.25	10	1.000	10.0	0.0
Total		460		359.2	100.8

2. Calculate the average particle settling velocity for each velocity range by taking the average of the range limits, and enter the values in column 2. For the first velocity range, the average settling velocity is (0.0 + 1.5)/2 = 0.75 ft/h.
3. Enter the number of influent particles for each velocity range in column 3.
4. Calculate the removal fraction for each velocity range by dividing the average settling velocity by the critical overflow velocity (6.5 ft/h), and enter the result in column 4. For the first velocity range

$$\text{Fraction removed} = \frac{v_{n1}}{v_c} = \frac{0.75}{6.5} = 0.115$$

   Where the result is greater than 1.0, enter a value of 1.0, because all of the particles are removed.
5. Determine the number of particles removed by multiplying the number of influent particles by the percent removal (column 3 × column 4). Enter the values in column 5.
6. Calculate the particles remaining by subtracting the particles removed from the number of influent particles (column 3 − column 5). Enter the result in column 6.

7. Compute the removal efficiency by calculating the sum of particles removed and dividing the sum by the total number of particles in the influent.

$$\text{Total fraction removed} = \frac{\sum_{i=1}^{n} \frac{v_{ni}}{v_c}(n_i)}{\sum_{i=1}^{n} n_i} = \frac{359.2 \times 10^5}{460 \times 10^5} = 78.1\%$$

8. Plot the particle histogram for the influent and effluent wastewater.

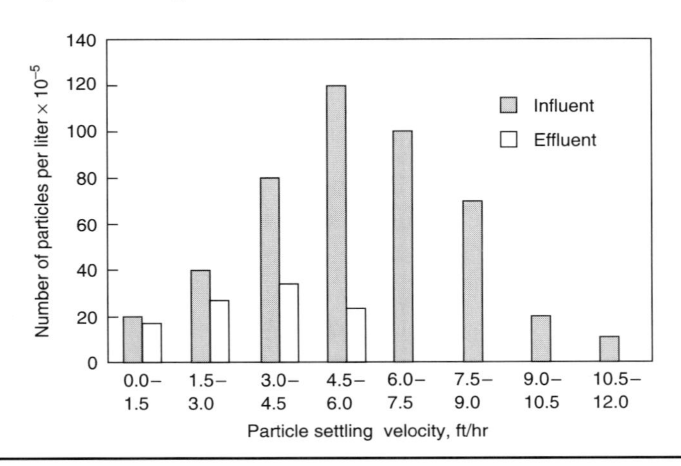

Flocculant Particle Settling (Type 2)

Particles in relatively dilute solutions will not act as discrete particles but will coalesce during sedimentation. As coalescence or flocculation occurs, the mass of the particle increases and it settles faster. The extent to which flocculation occurs is dependent on the opportunity for contact, which varies with the overflow rate, the depth of the basin, the velocity gradients in the system, the concentration of particles, and the range of particle sizes. The effects of these variables can be determined only by sedimentation tests.

To determine the settling characteristics of a suspension of flocculant particles, a settling column may be used. Such a column can be of any diameter but should be equal in height to the depth of the proposed tank. Satisfactory results can be obtained with a 6-in (150-mm) -diameter plastic tube about 10 ft (3 m) high. Sampling ports should be inserted at 2-ft (0.6-m) intervals. The solution containing the suspended matter should be introduced into the column in such a way that a uniform distribution of particle sizes occurs from top to bottom.

Care should be taken to ensure that a uniform temperature is maintained throughout the test to eliminate convection currents. Settling should take place under quiescent conditions. At various time intervals, samples are withdrawn from the ports and analyzed for suspended solids. The percent removal is computed for each sample analyzed and is plotted as a number against time and depth, as elevations are plotted on a survey grid. Curves of equal percent removal are drawn as shown in

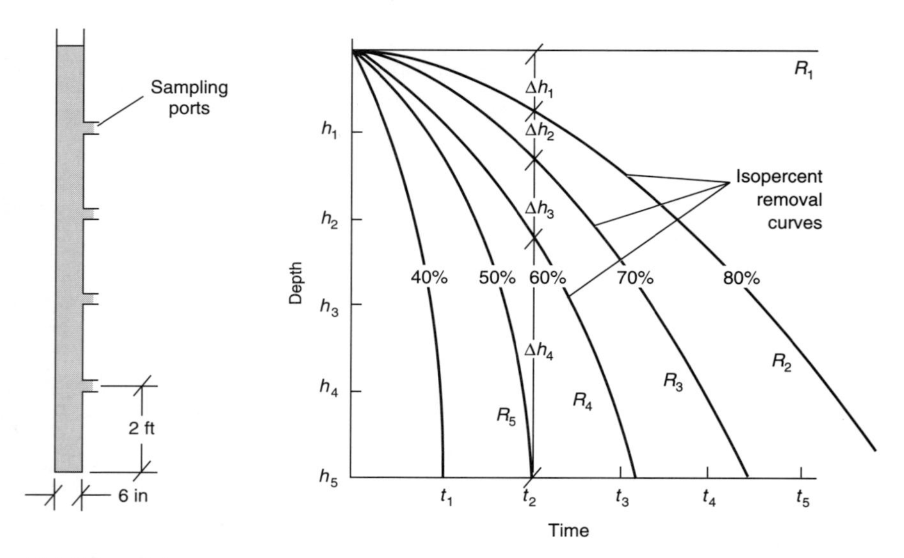

FIGURE 5-20
Definition sketch for the analysis of flocculant Type 2 settling.

Fig. 5-20. From the curves shown in Fig. 5-20, the overflow rate for various settling is determined by noting the value where the curve intersects the x axis. The settling velocity is

$$v_c = \frac{H}{t_c} \tag{5-28}$$

where H = the height of the settling column, ft (m), and t_c = the time required for a given degree of removal to be achieved, min. The fraction of particles removed is given by

$$R, \% = \sum_{h=1}^{n} \left(\frac{\Delta h_n}{H}\right)\left(\frac{R_n + R_{n+1}}{2}\right) \tag{5-29}$$

where
R = TSS removal, %
n = number of equal percent removal curve
Δh_1 = distance between curves of equal percent removal, ft
H = total height of settling column, ft
R_n = equal percent removal curve number n
R_{n+1} = equal percent removal curve number $n + 1$

Determination of the amount of material removed by using the curve given in Fig. 5-20 is illustrated in Example 5-8.

EXAMPLE 5-8. REMOVAL OF FLOCCULANT SUSPENDED SOLIDS (TYPE 2 SETTLING). Determine the removal efficiency for a basin 10 ft deep with an overflow rate equal to 1490 gal/ft^2·d (8.3 ft/h) using the following settling data, obtained at an existing treatment plant where chemicals are added to the influent to enhance the TSS removal.

Time, min	Percent total suspended solids removed at indicated depth (ft) 1.0	2.0	4.0	6.0	8.0
20	53				
30	64	54			
40	72	62	53	50	
50	80	68	59	55	54
60	86	76	65	60	59
70		82	71	65	64
80		88	76	71	69
90			82	76	73
100			88	82	78
110				87	83
120					88

Solution

1. Use the data provided in the table above to generate percent removal curves for 50, 60, 70, 80, and 90 percent TSS removal (the axes for the curves should be depth versus time). The percent removal curves are depicted in the figure given below.

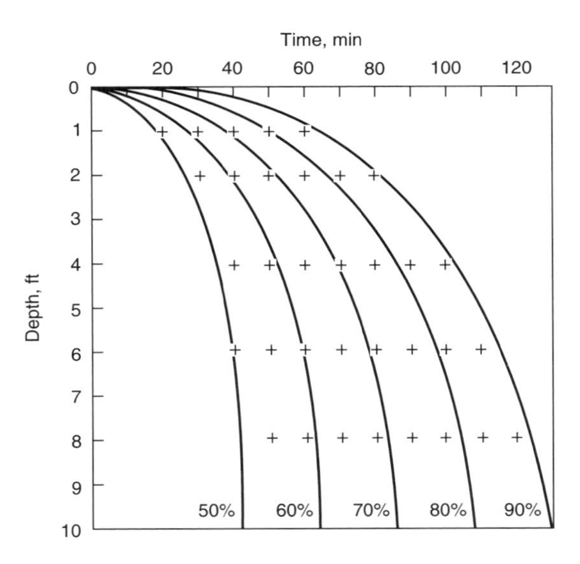

2. Determine the critical settling time by dividing the basin depth by the overflow rate, and estimate the percentage removal at the 10-ft depth at the critical settling time.
   a. Critical settling time:

$$\text{Settling time} = \frac{\text{basin depth}}{\text{overflow rate}} = \left(\frac{10\text{ ft}}{8.3\text{ ft/h}}\right)\left(\frac{60\text{ min}}{\text{h}}\right) = 72.3\text{ min}$$

   b. Percentage removal at the 10-ft depth:
   Interpolating between the 60 and 70 percent curves, a value of 64 percent removal is obtained for the calculated settling time of 72.3 min.

3. Determine the total percent removal using Eq. (5-29):

$$R, \% = \sum_{i=1}^{n} \left(\frac{\Delta h_i}{H}\right)\left(\frac{R_i + R_{i+1}}{2}\right)$$

Draw a vertical line from 64 percent at the 10-ft depth to the zero axis and determine the appropriate Δh values for the various percentage removal curves:

For 100 to 90%, Δh_1 = 1.34 ft (1.34 − 0.00)

For 90 to 80%, Δh_2 = 1.19 ft (2.53 − 1.34)

For 80 to 70%, Δh_3 = 2.16 ft (4.69 − 2.53)

For 70 to 64%, Δh_4 = 5.31 ft (10.0 − 4.69)

Using the above information for each Δh, the efficiency is computed as follows:

$\frac{\Delta h_i}{H} \times \frac{R_i + R_{i+1}}{2}$	=	% removal
$\frac{1.34}{10.0} \times \frac{90 + 100}{2}$	=	12.73
$\frac{1.19}{10.0} \times \frac{80 + 90}{2}$	=	10.12
$\frac{2.16}{10.0} \times \frac{70 + 80}{2}$	=	16.20
$\frac{5.31}{10.0} \times \frac{64 + 70}{2}$	=	35.58
1.00		74.63

The TSS removal for quiescent settling is 74.6 percent.

Comment. To account for the less-than-optimum conditions encountered in the field, the design settling velocity or overflow rate obtained from column studies often is multiplied by a factor of 0.65 to 0.85, and the detention times are multiplied by a factor of 1.25 to 1.5.

Plate and Tube Settlers

Plate and tube settlers are shallow settling devices consisting of stacked offset trays or bundles of small plastic tubes of various geometries (see Fig. 5-21*a*). They are used to enhance the settling characteristics of sedimentation basins. Plate and tube settlers have been used in primary, secondary, and tertiary sedimentation applications. Normal practice is to insert the plate or tube settlers in sedimentation basins (either rectangular or circular) of sufficient depth. The flow within the basin passes upward through the plate or tube modules and exits from the basin above the modules (see Fig. 5-21*b*). The solids that settle out within the plates or tubes move by means of gravity countercurrently downward and out of the tube modules to the basin bottom.

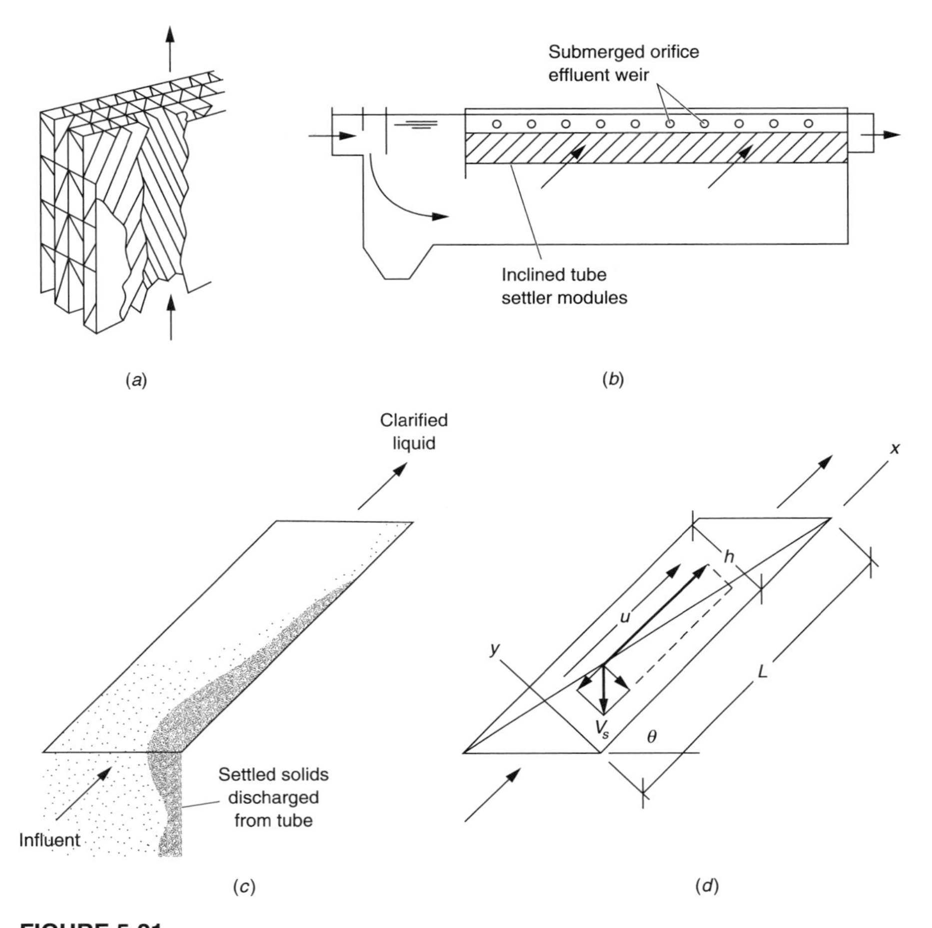

FIGURE 5-21
Plate and tube settlers: (*a*) module of inclined tubes, (*b*) tubes installed in a rectangular sedimentation tank, (*c*) operation, and (*d*) inclined tube details (adapted from Tchobanoglous and Schroeder, 1985).

To be self-cleaning, plate or tube settlers are usually set at an angle between 45 and 60° above the horizontal. When the angle is increased above 60°, the efficiency decreases. If the plates and tubes are inclined at angles less than 45°, sludge will tend to accumulate within the plates or tubes. To control biological growths and the production of odors (the principal problems encountered with their use), the accumulated solids must be flushed out periodically (usually with a high-pressure hose). The need for flushing poses a problem with the use of plate and tube settlers where the characteristics of the solids to be removed vary from day to day.

Referring to the definition sketch, the analysis of the plate and tube settlers is as follows. For the inclined-coordinate system presented in Fig. 5-21*c*, the velocity components for the particles within the tube settler are as follows:

$$V_{sx} = U - V_s \sin\theta \tag{5-30}$$

$$V_{sy} = -V_s \cos\theta \tag{5-31}$$

where V_{sx} = velocity component in x direction
U = fluid velocity in x direction
V_s = normal settling velocity of particle
θ = inclination angle for tube with horizontal axis
V_{sy} = settling velocity in y direction

The vertical component of the velocity, V_{sy}, is the critical velocity component, and the analysis for the removal is the same as the analysis presented previously for discrete particles.

Hindered (Zone) Settling (Type 3)

In systems that contain high concentrations of suspended solids, both hindered or zone settling (Type 3) and compression settling (Type 4) usually occur in addition to discrete (free) and flocculant settling. The settling phenomenon that occurs when a concentrated suspension, initially of uniform concentration throughout, is placed in a graduated cylinder, is illustrated in Fig. 5-22. Because of the high concentration of particles, the liquid tends to move up through the interstices of the contacting particles. As a result, the contacting particles tend to settle as a zone or "blanket,"

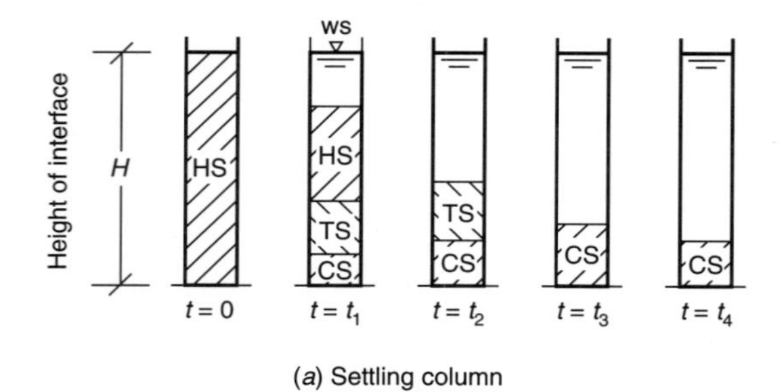

(a) Settling column

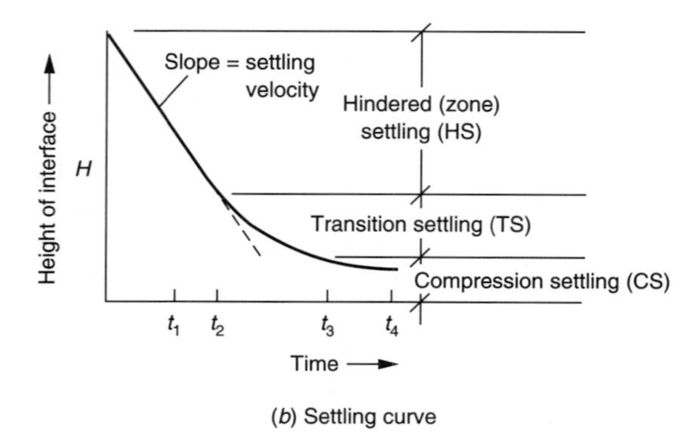

(b) Settling curve

FIGURE 5-22
Definition sketch for hindered (zone) Type 3 settling: (a) settling column; (b) settling curve (adapted from Reynolds and Richards, 1996).

maintaining the same relative position with respect to each other. The phenomenon is known as *hindered settling.* As the particles in this region settle, a relatively clear layer of water is produced above the particles in the settling region. The scattered, relatively light particles remaining in this region usually settle as discrete or flocculant particles, as discussed previously in this chapter. In most cases, an identifiable interface develops between the upper region and the hindered-settling region in Fig. 5-22. The rate of settling in the hindered-settling region is a function of the concentration of solids and their characteristics.

As settling continues, a compressed layer of particles begins to form on the bottom of the cylinder in the compression-settling region. The particles in this region apparently form a structure in which there is close physical contact between the particles. As the compression layer forms, regions containing successively lower concentrations of solids than those in the compression region extend upward in the cylinder. Thus, in actuality the hindered-settling region contains a gradation in solids concentration different from that found at the interface of the settling region to that found in the compression-settling region.

Because of the variability encountered, settling tests are usually required to determine the settling characteristics of suspensions where hindered and compression settling are important considerations. On the basis of data derived from column settling tests, two different design approaches can be used to obtain the required area for the settling/thickening facilities. In the first approach, the data derived from one or more batch settling tests are used. In the second approach, known as the *solids flux method,* data from a series of settling tests conducted at different solids concentrations are used. Both methods are described in the following discussion. It should be noted that both methods have been used where existing plants are to be expanded or modified. These methods are, however, seldom used in the design of small treatment plants.

Area requirement based on single-batch test results. For purposes of design, the final overflow rate selected should be based on a consideration of the following factors: (1) the area needed for clarification, (2) the area needed for thickening, and (3) the rate of sludge withdrawal. Column settling tests, as previously described, can be used to determine the area needed for the free-settling region directly. However, because the area required for thickening is usually greater than the area required for the settling, the rate of free settling rarely is the controlling factor. In the case of the activated-sludge process, where stray, light fluffy floc particles may be present, it is conceivable that the free flocculant settling velocity of these particles could control the design.

The area requirement for thickening is determined according to a method developed by Talmadge and Fitch (1955). A column of height H_o is filled with a suspension of solids of uniform concentration C_o. The position of the interface as time elapses and the suspension settles is given in Fig. 5-23. The rate at which the interface subsides is then equal to the slope of the curve at that point in time. According to the procedure, the area required for thickening is given by

$$A = \frac{Qt_u}{H_o} \tag{5-32}$$

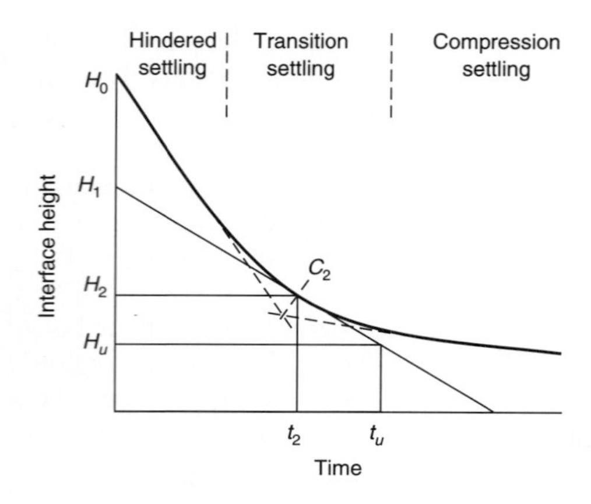

FIGURE 5-23
Graphical analysis of hindered (zone) Type 3 interface settling curves (after Talmadge and Fitch, 1955).

where A = area required for sludge thickening, ft^2 (m^2)
Q = flowrate into tank, ft^3/s (m^3/s)
H_o = initial height of interface in column, ft (m)
t_u = time to reach desired underflow concentration, s

The critical concentration controlling the sludge-handling capability of the tank occurs at a height H_2 where the concentration is C_2. This point is determined by extending the tangents to the hindered-settling and compression regions of the subsidence curve to the point of intersection and bisecting the angle thus formed, as shown in Fig. 5-23. The time t_u can be determined as follows:

1. Construct a horizontal line at the depth H_u that corresponds to the depth at which the solids are at the desired underflow concentration C_u. The value of H_u is determined by using the following expression:

$$H_u = \frac{C_o H_o}{C_u} \tag{5-33}$$

2. Construct a tangent to the settling curve at the point indicated by C_2.
3. Construct a vertical line from the point of intersection of the two lines drawn in steps 1 and 2 to the time axis to determine the value of t_u.

With this value of t_u, the area required for the thickening is computed using Eq. (5-32). The area required for clarification is then determined. The larger of the two areas is the controlling value. Application of this procedure is illustrated in Example 5-9.

EXAMPLE 5-9. CALCULATIONS FOR SIZING AN ACTIVATED-SLUDGE SETTLING TANK. The settling curve shown in the following diagram was obtained for an activated sludge with an initial solids concentration C_o of 3000 mg/L. The initial height of the interface in the settling column was at 2.5 ft. Determine the area required to yield a thickened sludge concentration C_u of 12,000 mg/L with a total flow of 0.1 Mgal/d. In addition, determine the solids loading in $lb/ft^3 \cdot d$ and the overflow rate in $gal/ft^2 \cdot d$.

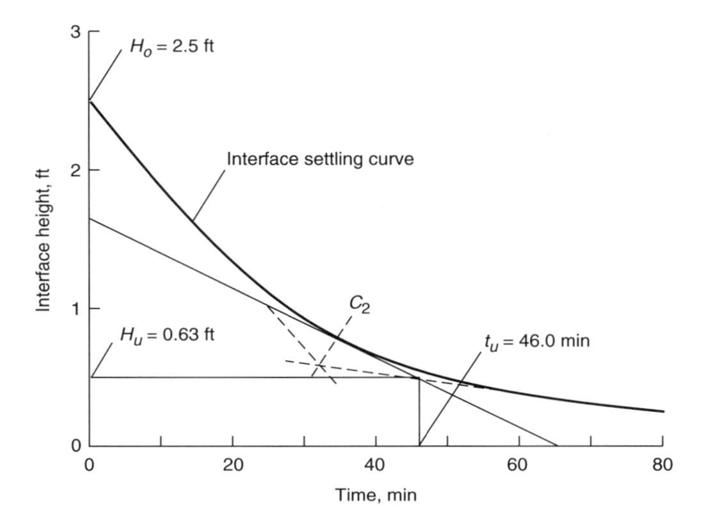

Solution

1. Determine the area required for thickening using Eq. (5-33). First determine the value of H_u:

$$H_u = \frac{C_o H_o}{C_u} = \frac{3000 \text{ mg/L} \times 2.5 \text{ ft}}{12{,}000 \text{ mg/L}} = 0.63 \text{ ft}$$

On the settling curve, a horizontal line is constructed at $H_u = 0.63$ ft. A tangent is constructed to the settling curve at C_2, the midpoint of the region between hindered and compression settling. Bisecting the angle formed where the two tangents meet determines point C_2. The intersection of the tangent at C_2 and the line $H_u = 0.63$ ft determines t_u. Thus $t_u = 46.0$ min, and the required area is

$$A = \frac{Qt_u}{H_o} = \frac{(0.1 \text{ Mgal/d})[1.55 \text{ ft}^3/\text{s}\cdot(\text{Mgal/d})](46.0 \text{ min})(60 \text{ s/min})}{2.5 \text{ ft}} = 171 \text{ ft}^2$$

2. Determine the area required for clarification.
   a. Determine the interface subsidence velocity v. The subsidence velocity is determined by computing the slope of the tangent drawn from the initial portion of the interface settling curve. The computed velocity represents the unhindered settling rate of the sludge.

$$v = \left(\frac{2.5 \text{ ft} - 1.0 \text{ ft}}{29.5 \text{ min}}\right)\left(\frac{60 \text{ min}}{\text{h}}\right) = 3.1 \text{ ft/h}$$

   b. Determine the clarification rate. Because the clarification rate is proportional to the liquid volume above the critical sludge zone, it may be computed as follows:

$$Q_c = 0.1 \text{ Mgal/d} \times [1.55 \text{ ft}^3/\text{s}\cdot(\text{Mgal/d})]\left(\frac{2.5 \text{ ft} - 0.63 \text{ ft}}{2.5 \text{ ft}}\right) = 0.116 \text{ ft}^3/\text{s}$$

   c. Determine the area required for clarification. The required area is obtained by dividing the clarification rate by the settling velocity:

$$A = \frac{Q_c}{v} = \frac{(0.116 \text{ ft}^3/\text{s}) \times 60 \text{ s/min} \times 60 \text{ min/h}}{3.1 \text{ ft/h}} = 135 \text{ ft}^2$$

3. The controlling area is the thickening area (171 ft^2) because it exceeds the area required for clarification (135 ft^2).

4. Determine the solids loading:

$$\text{Solids, lb/d} = 0.1 \times 8.34 \times 3000 \text{ mg/L}$$
$$= 2502 \text{ lb/d}$$
$$\text{Solids loading} = \frac{2502 \text{ lb/d}}{171 \text{ ft}^2}$$
$$= 14.6 \text{ lb/ft}^2\cdot\text{d}$$

5. Determine the hydraulic loading rate:

$$\text{Hydraulic loading rate} = \frac{100{,}000 \text{ gal/d}}{171 \text{ ft}^2}$$
$$= 585 \text{ gal/ft}^2\cdot\text{d } (23.8 \text{ m}^3/\text{m}^2\cdot\text{d})$$

Comment. An alternative approach for sizing the secondary clarifiers using the initial settling velocity of the sludge is given in Sec. 7-6, Chap. 7.

Area requirements based on solids flux analysis. An alternative method of determining the area required for hindered settling is based on an analysis of the solids (mass) flux (Coe and Clevenger, 1916). In the solids flux method of analysis it is assumed that a settling basin is operating at steady state. Within the tank, the downward flux of solids is brought about by gravity (hindered) settling and by bulk transport due to the underflow that is being pumped out and recycled. Although this method is well developed from laboratory measurements, full-scale clarifiers seldom operate as modeled by the solids flux method. For a more complete discussion and analysis of the solids flux method of analysis the following references are recommended: Dick and Ewing (1967), Dick and Young (1972), Keinath (1989), Wahlberg and Keinath (1988), and Yoshika et al. (1957).

Compression Settling (Type 4)

The volume required for the sludge in the compression region can also be determined by settling tests. The rate of consolidation in this region has been found to be proportional to the difference in the depth at time t and the depth to which the sludge will settle after a long period of time. The long-term consolidation can be modeled as a first-order decay function:

$$H_t - H_\infty = (H_2 - H_\infty)e^{-i(t-t_2)} \tag{5-34}$$

where H_t = sludge height at time t
H_∞ = sludge depth after long period, say 24 h
H_2 = sludge height at time t_2
i = constant for a given suspension

It has been observed that stirring serves to compact sludge in the compression region by breaking up the floc and permitting water to escape. Rakes are often used on sedimentation equipment to manipulate the sludge and thus produce better compaction.

Gravity Separation in an Accelerated Flow Field

Sedimentation, as described previously, occurs under the force of gravity in a constant acceleration field. The removal of settleable particles can also be accomplished by taking advantage of a changing acceleration field. A number of devices that take advantage of both gravitational and centrifugal forces and induced velocities have been developed for the removal of grit from wastewater. The principles involved are illustrated in Fig. 5-24. In appearance, the separator looks like a squat tin can with holes in the top and bottom. Wastewater, from which grit is to be separated, is introduced tangentially near the top and exits through the opening in the top of the unit. The liquid is removed at the top. Grit is removed through the opening in the bottom of the unit.

Because the top of the separator is enclosed, the rotating flow creates a free vortex within the separator. The most important characteristic of a free vortex is that the product of the tangential velocity times the radius is a constant:

$$Vr = \text{constant} \tag{5-35}$$

where V = tangential velocity, ft/s (m/s), and r = radius, ft (m).

The significance of Eq. (5-35) can be illustrated by the following example. Assume the tangential velocity in a separator with a 5-ft (1.5-m) radius is 3 ft/s (0.9 m/s). The product of the velocity times the radius at the outer edge of the separator is equal to 15 ft²/s (1.35 m²/s). If the discharge port has a radius of 1 ft (0.9 m), then the tangential velocity at the entrance to the discharge port is 15 ft/s (4.5 m/s). The centrifugal force experienced by a particle within this flow field is equal to the square of the velocity divided by the radius. Because the centrifugal force is also

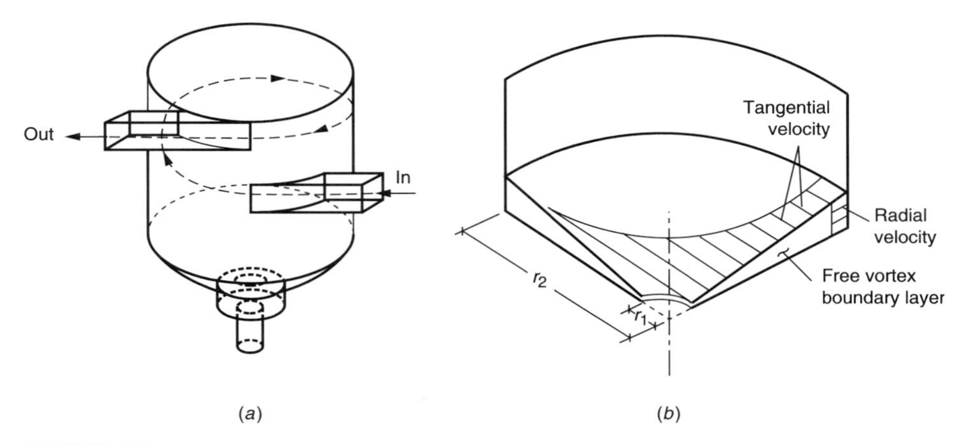

FIGURE 5-24
Definition sketch for accelerated gravity separation: (*a*) flow pattern within separator leading to the formation of a free vortex and (*b*) characteristics of a free vortex.

proportional to the inverse of the radius, a fivefold decrease in the radius results in a 125-fold increase in the centrifugal force.

Because of the high centrifugal forces near the discharge port, some of the particles, depending on their size, density, and drag, are retained within the body of the free vortex near the center of the separator, while other particles are swept out of the unit. Grit and sand particles will be retained while organic particles are discharged from the unit. Organic particles having the same settling velocity as sand will typically be from 4 to 8 times as large. The corresponding drag forces for these organic particles will be from 16 to 64 times as great. As a result, the organic particles tend to move with the fluid and are transported out of the separator. The particles held in the free vortex ultimately settle to the bottom of the unit under the force of gravity. Organic particles that sometimes settle usually consist of oil and grease attached to grit or sand particles.

Flotation

Flotation is a unit operation used to separate solid or liquid particles from a liquid phase. Separation is brought about by introducing fine gas (usually air) bubbles into the liquid phase. The bubbles attach to the particulate matter, and the buoyant force of the combined particle and gas bubbles is great enough to cause the particle to rise to the surface. Particles that have a higher density than the liquid can thus be made to rise. The rising of particles with lower density than the liquid can also be facilitated (e.g., oil suspension in water). Flotation, with bottom skimming for heavy particles, has been used as replacement for sedimentation where most of the suspended material is nearly the same density as water. The most common application of flotation is for the thickening of waste sludges. Flotation has also been used for the removal of algae from pond effluents, with varying degrees of success.

Description. In conventional municipal wastewater treatment, air is used as the flotation agent. Air bubbles are added or caused to form by one of the following three methods:

1. Injection of air while the liquid is under pressure, followed by release of the pressure (dissolved-air flotation)
2. Aeration at atmospheric pressure (induced air flotation)
3. Saturation with air at atmospheric pressure, followed by application of a vacuum to the liquid (vacuum flotation)

In all three of the above methods, the degree of removal can be enhanced through the use of various chemical additives. In septic and Imhoff tanks, unassisted natural flotation is used to remove oils and grease. Some gas-assisted flotation also occurs.

Dissolved-air flotation. Dissolved-air flotation (DAF) is the most common type of flotation system used in wastewater treatment. In dissolved-air flotation systems, air is dissolved in the wastewater under a pressure of several atmospheres, followed by release of the pressure. In small pressure systems, the entire flow may be pressurized by means of a pump to 40 to 50 lb/in^2 gage (275 to 350 kPa) with compressed air added at the pump suction (see Fig. 5-25*a*). The entire flow is held

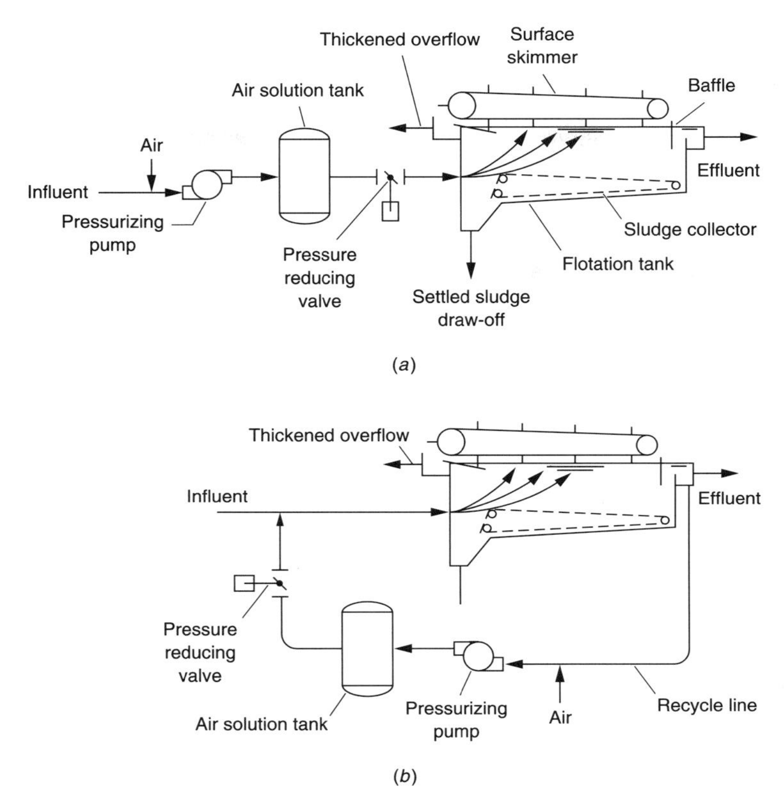

FIGURE 5-25
Definition sketch for dissolved air flotation: (*a*) direct flow and (*b*) recycled flow.

in a retention tank under pressure for several minutes to allow time for the air to dissolve. It is then admitted through a pressure-reducing valve to the flotation tank, where the air comes out of solution in minute bubbles throughout the entire volume of liquid.

In the larger units, a portion of the DAF effluent (15 to 100 percent) is recycled, pressurized, and semisaturated with air (Fig. 5-25*b*). The recycled flow is mixed with the unpressurized main stream just before admission to the flotation tank, with the result that the air comes out of solution in contact with particulate matter at the entrance to the tank. Pressure types of units have been used mainly for the treatment of industrial wastes and for the concentration of sludges.

Chemicals used to aid the flotation process function to create a surface or a structure that can easily absorb or entrap air bubbles. Inorganic chemicals, such as the aluminum and ferric salts and activated silica, can be used to bind the particulate matter together and, in so doing, create a structure that can easily entrap air bubbles. Various organic polymers can be used to change the nature of either the air-liquid interface or the solid-liquid interface, or both. These compounds usually collect on the interface to bring about the desired changes. Dissolved air flotation has been used to remove algae from lagoon effluent (Chap. 8) and to thicken sludge (Chap. 14).

5-8 GRIT REMOVAL

Grit is composed of sand, gravel, cinders, or other heavy solid materials that have subsiding velocities or specific gravities substantially greater than those of the organic putrescible solids in wastewater. Grit is removed from wastewater: (1) to protect mechanical equipment from abrasion and accompanying abnormal wear, (2) to reduce formation of heavy deposits in downstream processes, and (3) to reduce the frequency of digester cleaning caused by excessive accumulations of grit. The types of grit removal facilities are considered in the following discussion, which includes design considerations and the characteristics, quantities, and disposal of grit.

Description

Typically, grit chambers are located after the bar racks and before the primary sedimentation tanks. In some treatment plants, grit chambers precede the screening facilities. Generally, the installation of fine coarse screening facilities ahead of the grit chambers makes the operation and maintenance of the grit removal facilities easier. Three types of grit chambers are used: horizontal-flow, either of a channel-type, rectangular, or square configuration; aerated; and vortex-type.

Design Considerations

Design of grit chambers is commonly based on the removal of grit particles having a specific gravity of 2.65 and a wastewater temperature of 15.5°C (60°F). Design considerations for the various types of grit removal facilities are given in the following discussion. Typical design data for horizontal-flow grit chambers are presented in Table 5-10.

TABLE 5-10
Typical design information for horizontal-flow grit chambers*

		Value	
Item	**Unit**	**Range**	**Typical**
Detention time	s	45–90	60
Horizontal velocity	ft/s	0.8–1.3	1.0
Settling velocity for removal of:			
50-mesh material (0.30 m)	ft/min†	9.2–10.2	9.6
100-mesh material (0.15 mm)	ft/min†	2.0–3.0	2.5
Headloss in a control section as percent of depth in channel	%	30–40	36‡
Added length allowance for inlet and outlet turbulence	%	25–50	30

*Adapted from Tchobanoglous and Burton (1991).
† If the specific gravity of the grit is significantly less than 2.65, lower velocities should be used.
‡ For Parshall flume control.

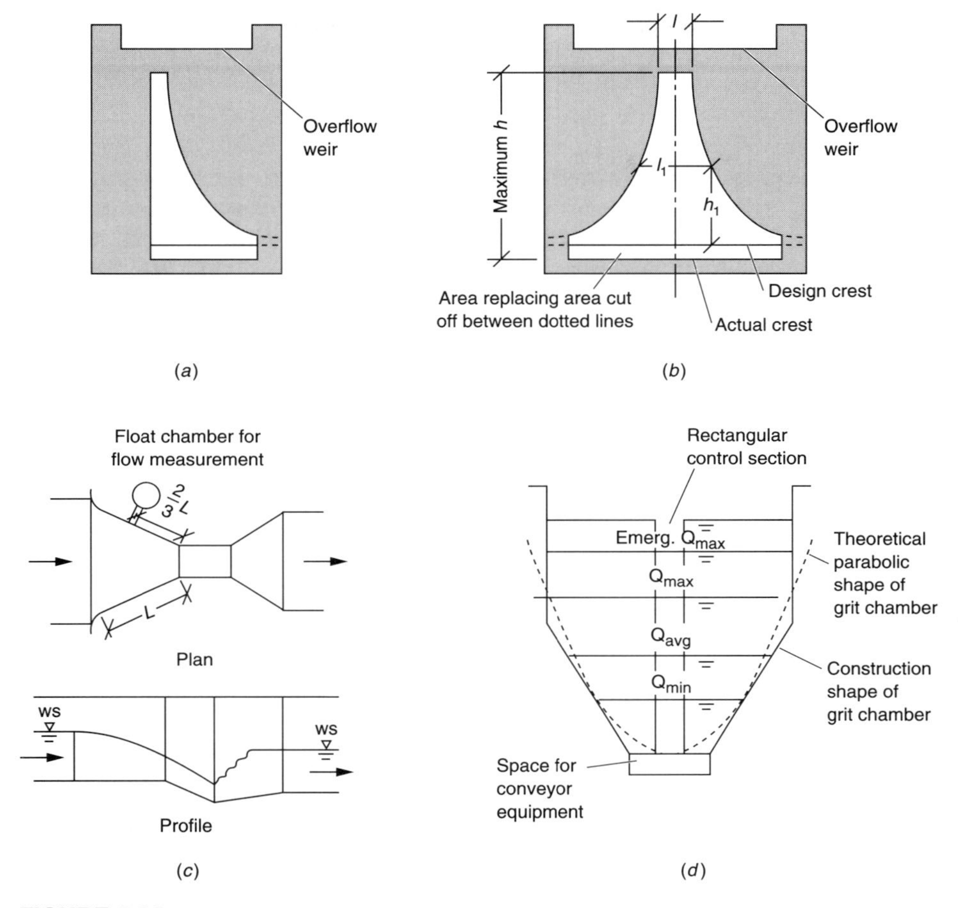

FIGURE 5-26
Typical control sections used with channel-type horizontal-flow grit chambers to maintain a constant flow through velocity with varying influent flowrates: (*a*) Sutro weir, (*b*) proportional weir, (*c*) Parshall flume, and (*d*) parabolic-shaped chamber with rectangular control section.

Channel-type horizontal-flow grit chambers. The oldest type of grit chamber used is the horizontal-flow, velocity-controlled channel-type. Grit is removed by maintaining a velocity as close to 1.0 ft/s (0.3 m/s) as practical and by providing sufficient time for grit particles to settle to the bottom of the channel. Under ideal conditions, the design velocity should allow the heavier grit to settle out while most of the organic particles will be transported through and out of the chamber. The velocity of flow is controlled by the dimensions of the unit and the use of special weir sections at the effluent end (see Fig. 5-26). Grit removal from horizontal-flow grit chambers is accomplished usually by a conveyor with scrapers, buckets, or plows. Screw conveyors or bucket elevators are used to elevate the removed grit for washing or disposal. In small plants, grit chambers are sometimes cleaned manually.

Rectangular horizontal-flow grit chambers. In the horizontal-flow type, the flow passes through the chamber in a horizontal direction and the straight line velocity of flow is controlled by the dimensions of the unit, special influent distribution gates, and the use of special weir sections at the effluent end.

Square horizontal-flow grit chambers. Square horizontal-flow grit chambers (see Fig. 5-27) have been in use since the 1930s. Influent to the units is distributed over the cross section of the tank by a series of vanes or gates, and the distributed wastewater flows across the tank and overflows a weir in a free discharge. In square grit chambers, the solids are raked by a rotating mechanism to a sump at the side of the tank. Settled grit may be moved up an incline by a reciprocating rake mechanism or a screw conveyor (see Fig. 5-28), or grit may be pumped from the tank through a cyclone degritter to separate the remaining organic material and concentrate grit. The concentrated grit then may be washed again. Organic solids, separated from the grit, are returned to the flow.

Square horizontal-flow grit chambers are designed on the basis of overflow rates that are dependent on particle size and the temperature of the wastewater. Nominally they are designed to remove 95 percent of the 100-mesh particles at peak flow. Where

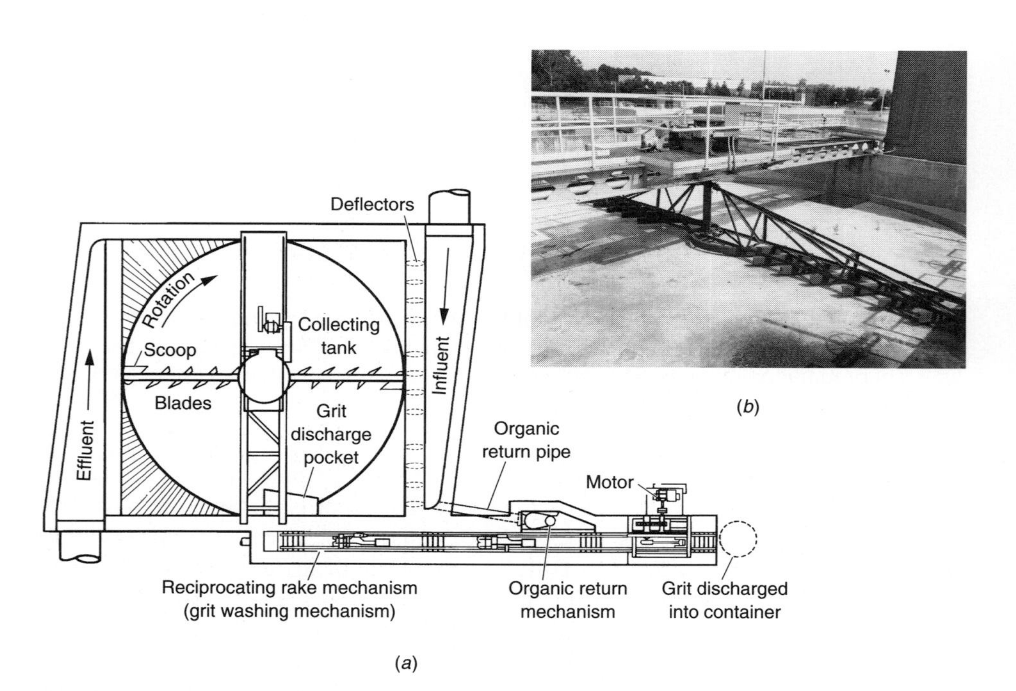

FIGURE 5-27
Typical square horizontal-flow grit chamber: (*a*) definition sketch (adapted from Dorr Oliver) and (*b*) view of typical unit.

(*a*)

(*b*)

FIGURE 5-28
Grit dewatering with a screw conveyor: (*a*) overall view of grit dewatering device and (*b*) close-up view of continuous screw.

square grit chambers are used, at least two units should be used. Additional design information is presented in Table 5-10.

Aerated grit chambers. Grit is removed in aerated grit chambers by causing the wastewater to flow in a spiral pattern as shown in Fig. 5-29. Because of their mass, grit particles will be accelerated (i.e., the spiral flow field is accelerated by the air) and will diverge from the streamlines, ultimately settling to the bottom. Two major factors contributing to the popularity of the aerated grit chambers, as compared to horizontal-flow grit chambers, are (1) there is minimal wear on the equipment and (2) separate grit washing facilities are not needed. In areas where industrial wastewater is discharged to the collection system, the potential release of VOCs from aerated grit chambers must be considered.

Aerated grit chambers are nominally designed to remove particles 70 mesh (0.21 mm) or larger, with 2- to 5-minute detention periods at the peak hourly rate of flow. The velocity of roll or agitation governs the size of particles of a given specific gravity that will be removed. If the velocity is too great, grit will be carried out of the chamber; if it is too small, organic material will be removed with the grit. With proper adjustment, almost 100 percent removal of the desired size will be obtained, and the grit will be well washed. The cross section of the tank is designed to create a spiral circulation pattern. The grit hopper, about 3 ft (0.9 m) deep with steeply sloping sides, is located along one side of the tank under the air diffusers (see Fig. 5-29). The

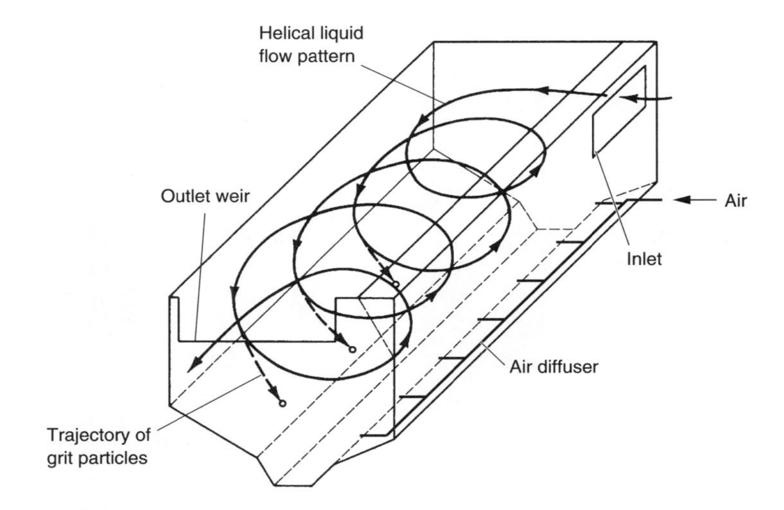

FIGURE 5-29
Aerated grit chamber with spiral flow pattern.

TABLE 5-11
Typical design information for aerated grit chambers*

Item	Unit	Value	
		Range	Typical
Detention time at peak flowrate	min	2–5	3
Dimensions:			
Depth	ft	7–16	10
Length	ft	25–65	40
Width	ft	8–23	12
Width-depth ratio	Ratio	1:1 to 5:1	1.5:1
Length-width ratio	Ratio	3:1 to 5:1	4:1
Air supply per foot of length	$ft^3/ft \cdot min$	3–8	5
Grit quantities	ft^3/Mgal	0.5–27	2

*Adapted from Tchobanoglous and Burton (1991), WEF (1992), and WPCF (1985).

diffusers are located about 1.5 to 2 ft (0.45 to 0.6 m) above the normal plane of the bottom. To improve grit removal effectiveness, influent and effluent baffles can be used for hydraulic control. Basic design data for aerated grit chambers are presented in Table 5-11. The design of aerated grit chambers is illustrated in Example 5-10.

EXAMPLE 5-10. DESIGN OF AN AERATED GRIT CHAMBER. Design an aerated grit chamber for the treatment of municipal wastewater. The average flowrate is 0.85 Mgal/d. Assume that the aerated grit chamber will be designed for the peak hour flowrate; use a peaking factor of 4.0.

Solution

1. Establish the peak hour flowrate for design. The peak design flowrate is

$$\text{Peak flowrate} = 0.85 \text{ Mgal/d} \times 4.0 = 3.40 \text{ Mgal/d}$$

2. Determine the volume of the grit chamber. Assume that the detention time at the peak hour flowrate is 3 min.

$$\text{Aeration chamber volume} = \text{peak flow} \times \text{detention time}$$

$$V = \left(\frac{3.4 \times 10^6 \text{ gal}}{\text{d}}\right)(3 \text{ min})\left(\frac{\text{d}}{24 \times 60 \text{ min}}\right)\left(\frac{\text{ft}^3}{7.48 \text{ gal}}\right)$$

$$= 947 \text{ ft}^3$$

3. Determine the dimensions of each grit chamber. Use a width to depth ratio of 1.2:1, and assume that the depth is 6 ft.
   a. Width $= 1.2(6 \text{ ft}) = 7.2 \text{ ft}$
   b.

$$\text{Length} = \frac{\text{volume}}{\text{width} \times \text{depth}} = \frac{947 \text{ ft}^3}{7.2 \text{ ft} \times 6 \text{ ft}} = 21.9 \text{ ft} \qquad \text{Use } 22.0 \text{ ft}$$

   c. Check the length to width ratio:

$$\frac{\text{L}}{\text{W}} = \frac{22 \text{ ft}}{7.2 \text{ ft}} = \frac{3.1}{1} \qquad \text{OK}$$

4. Determine the air supply requirement for both chambers. Assume that 5 $\text{ft}^3/\text{ft}\cdot\text{min}$ of length will be adequate.

$$\text{Air required (length basis)} = \left(\frac{5 \text{ ft}^3}{\text{ft}\cdot\text{min}}\right)(22.0 \text{ ft}) = 110 \text{ ft}^3/\text{min}$$

5. Estimate the average quantity of grit that must be handled. Assume a value of 7 ft^3/Mgal.

$$\text{Grit volume} = \left(\frac{7 \text{ ft}^3}{\text{Mgal}}\right)\left(\frac{0.85 \text{ Mgal}}{\text{d}}\right) = 6.0 \text{ ft}^3/\text{d}$$

Comment. In designing aerated grit chambers, the aeration system should be sized so that the air flow can be varied to control the size of grit removed and the cleanliness of the grit. The length of the grit chamber is sometimes increased by 10 to 15 percent to account for less-than-ideal inlet and outlet conditions.

Vortex-type grit chambers. The vortex-type grit chamber consists of a cylindrical tank in which the flow enters tangentially, creating a vortex flow pattern. Two types of devices are shown in Fig. 5-30. In one type (Fig. 5-30*a*), wastewater enters and exits tangentially. The adjustable axial propeller is used to produce a toroidal flow path for grit particles. The grit settles by gravity into the bottom hopper. Solids are removed from the hopper by a grit pump or an air lift pump. Grit removed from the unit can be processed further to remove any remaining organic material.

In the second type (Fig. 5-30*b*), a vortex is generated by the flow entering tangentially at the top of the unit. Effluent exits the center of the top of the unit from

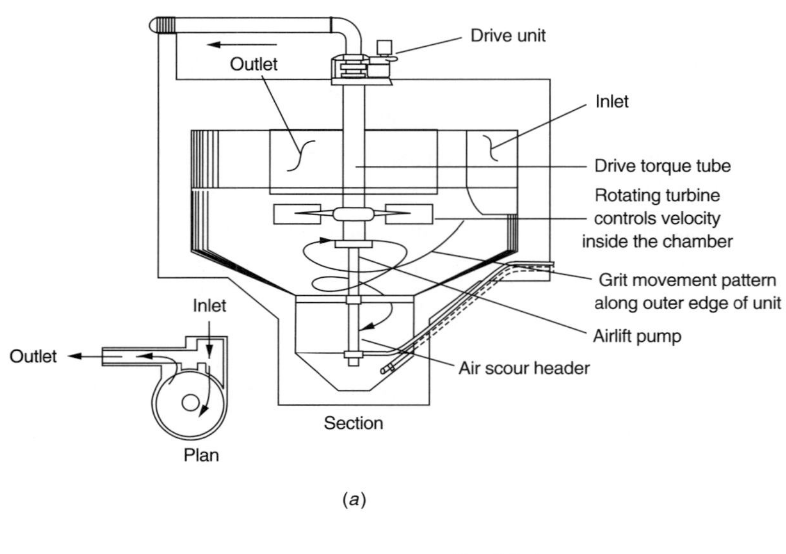

(*a*)

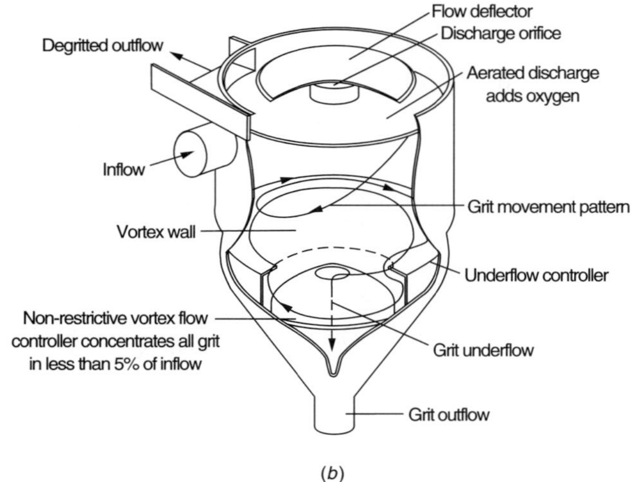

(*b*)

FIGURE 5-30
Typical vortex-type grit removal facilities: (*a*) Pista (adapted from Smith and Loveless) and (*b*) Teacup (adapted from Eutek, Inc).

a rotating cylinder, or "eye" of the fluid. Centrifugal and gravitational forces within this cylinder limit the release of particles with densities greater than water. Grit settles by gravity to the bottom of the unit, while organics, including those separated from grit particles by centrifugal forces, exit principally with the effluent. Typical design data are presented in Table 5-12. If more than two units are installed, special arrangements for flow splitting are required.

TABLE 5-12
Typical design information for vortex-type grit chambers*

Item	Unit	Value	
		Range	Typical
Detention time at average flowrate	s	20–30	30
Diameter			
Upper chamber	ft	4.0–24.0	
Lower chamber	ft	3.0–6.0	
Height	ft	9.0–16.0	
Removal rates			
50 mesh material (0.30 mm)	%	92–98	95+
70 mesh material (0.21 mm)	%	80–90	85+
100 mesh material (0.15 mm)	%	60–70	65+

*Adapted from Tchobanoglous and Burton (1991).

Grit: Characteristics, Quantities, and Disposal

In addition to the materials cited previously, grit includes eggshells, bone chips, seeds, coffee grounds, and large organic particles such as food wastes. The characteristics and quantities of grit removed from municipal wastewaters are considered below.

Characteristics of grit. Generally, what is removed as grit is predominantly inert and relatively dry. However, grit composition can be highly variable, with moisture content ranging from 13 to 65 percent, and volatile content from 1 to 56 percent. The specific gravity of clean grit particles reaches 2.7 for inerts, but can be as low as 1.3 when substantial organic material is agglomerated with inerts. A bulk density of 100 lb/ft^3 (1600 kg/m^3) is used commonly for grit. Particles identified as the cause of most downstream problems are typically 0.2 mm and larger. The actual size distribution of retained grit exhibits variation due to differences in collection system characteristics, as well as variations in grit removal efficiency. Generally, most grit particles are retained on a No. 100 mesh (0.15 mm) sieve, reaching nearly 100 percent retention in most instances.

Quantities of grit. The quantities of grit will vary greatly from one location to another, depending on the type of sewer system, the characteristics of the drainage area, the condition of the sewers, the frequency of street sanding to counteract icing conditions, the types of industrial wastes, the number of household garbage grinders served, and the sand content of the soil in the area. The quantity of grit can vary from 0.5 to more than 30 $ft^3/Mgal$ (0.0037 to 0.22 $m^3/10^3$ m^3). Because of the serious problems associated with grit, special studies should be undertaken to define the amounts of grit to be expected.

Grit separation and washing. Unwashed grit may contain 50 percent or more of organic material. Unless promptly disposed of, this material may attract insects and rodents. In both warm and cold climates, if unwashed grit is not disposed of

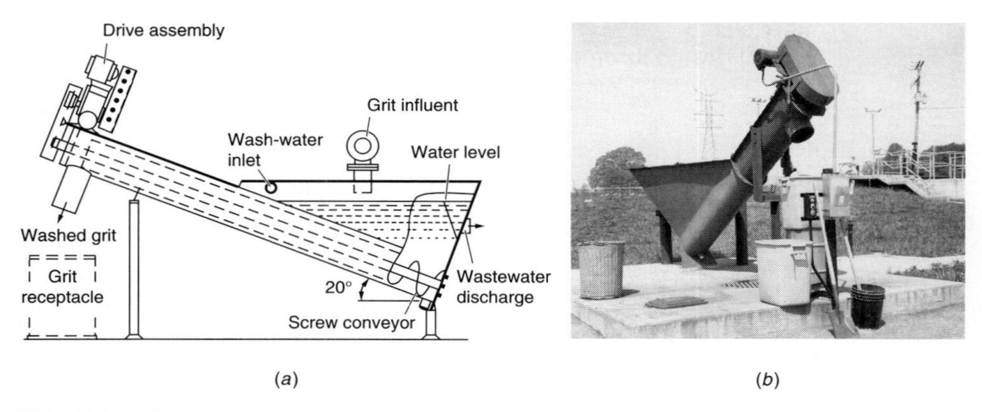

FIGURE 5-31
Typical washer used to remove organic material from grit: (*a*) definition sketch (adapted from Walker Process) and (*b*) view of typical unit.

promptly, it will undergo rapid decomposition, release malodorous compounds, and will attract insects. Insects, especially flies, are attracted by the release of volatile organic acids. Several grit separators and washers have been developed for the removal of organic material from grit. A typical example of a grit separation and washing unit is shown in Fig. 5-31.

Disposal of grit. Landfilling is the method used most commonly for the disposal of grit from small plants. As with screenings, some states require grit to be lime-stabilized before disposal in a landfill.

5-9 SEDIMENTATION

The objective of treatment by sedimentation is to remove readily settleable solids and floating material and thus reduce the suspended solids content. Primary sedimentation is used as a preliminary treatment step in the further processing of the wastewater. Efficiently designed and operated primary sedimentation tanks should remove from 50 to 70 percent of the suspended solids and from 25 to 40 percent of the BOD_5. The purpose of this section is to describe the various types of sedimentation facilities, to consider their performance, and to review important design considerations. Sedimentation tanks used for secondary treatment are considered in Chap. 7.

Description

In larger treatment plants (0.75 Mgal/d and larger), except for those with Imhoff tanks, either circular or rectangular mechanically cleaned sedimentation tanks of standardized design are used for the removal of TSS. It should be noted that primary sedimentation is often omitted in the design of small plants. The selection of the type of sedimentation unit for a given application is governed by the size of the installation, by rules and regulations of local control authorities, by local site conditions, and by the experience and judgment of the engineer. Two or more tanks should be provided so that the process may remain in operation while one tank is out of service for maintenance and repair work.

Rectangular tanks. The flow pattern in rectangular tanks is horizontal (as opposed to radial in circular tanks). Rectangular sedimentation tanks may use either chain-and-flight sludge collectors or traveling bridge–type collectors for the removal of the settled sludge (see Fig. 5-32). With flight collectors, the settled solids are scraped to sludge hoppers in small tanks and to transverse troughs in larger tanks. The transverse troughs are equipped with collecting mechanisms (cross collectors), usually either chain-and-flight or screw-type collectors, which convey solids to one or more sludge hoppers. One or more scraper blades are suspended from the bridge. It is also desirable to locate sludge-pumping facilities close to the hoppers where sludge is collected at the influent end of the tanks. One sludge-pumping station can conveniently serve two or more tanks.

Because flow distribution in rectangular tanks is critical, one of the following inlet designs is used: (1) full-width inlet channels with inlet weirs, (2) inlet channels with submerged ports or orifices, or (3) inlet channels with wide gates and slotted baffles. Inlet baffles are effective in reducing the high initial velocities and distribute flow over the widest possible cross-sectional area. Where full-width baffles are used, they should extend from 6 in (150 mm) below the surface to 12 in (300 mm) below the entrance opening.

(*a*)

(*b*)

FIGURE 5-32
Typical rectangular sedimentation tanks: (*a*) with chain and flight sludge collectors and (*b*) traveling bridge-type collector.

Scum is usually collected at the effluent end of rectangular tanks with the flights returning at the liquid surface. Several methods can be used to remove scum, including: (1) manually up an inclined apron, (2) with a horizontal, slotted pipe that can be rotated by a lever or a screw, (3) by a transverse rotating helical wiper, (4) with chain and flight collectors, and (5) with scum rakes where bridge-type sedimentation tank equipment is used. In installations where appreciable amounts of scum are collected, the scum hoppers are usually equipped with mixers to provide a homogeneous mixture prior to pumping. Scum is usually routed to digesters and disposed of with the sludge produced at the plant; however, separate scum disposal is used by many plants.

Circular tanks. The flow pattern in circular tanks is radial (as opposed to horizontal in rectangular tanks). To achieve a radial flow pattern, the wastewater to be settled can be introduced in the center or around the periphery of the tank, as shown in Fig. 5-33. Both flow configurations have proved to be satisfactory generally, although the center-feed type is used more commonly. In the center-feed design (Fig. 5-33*a*), the wastewater is transported to the center of the tank in a pipe suspended from the bridge, or encased in concrete beneath the tank floor. At the center of the tank, the wastewater enters a circular well designed to distribute the flow equally in all directions. The center well has a diameter typically between 15 and 20 percent of the total tank diameter and ranges from 3 to 8 ft (1 to 2.5 m) in depth. The sludge removal mechanism revolves slowly and may have two or four arms equipped with scrapers. The arms also support blades for scum removal. A typical center-feed circular clarifier equipped with a scraper mechanism for sludge removal is shown in Fig. 5-34.

In the peripheral feed design (Fig. 5-33*b*), a suspended circular baffle a short distance from the tank wall forms an annular space into which the wastewater is discharged in a tangential direction. The wastewater flows spirally around the tank

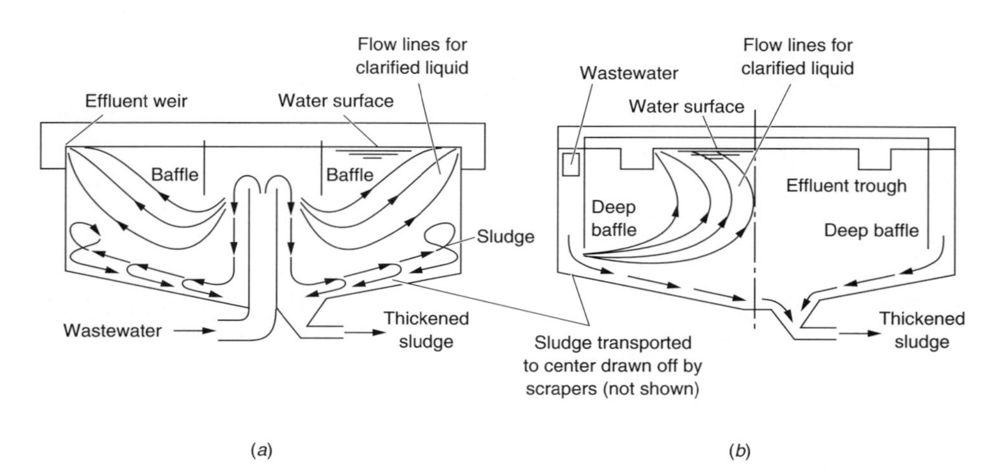

FIGURE 5-33
Typical circular sedimentation tanks: (*a*) center feed and (*b*) peripheral feed.

(a)

(b)

FIGURE 5-34
Typical center-feed circular sedimentation tank: (*a*) with scraper mechanism used in primary sedimentation and (*b*) with suction draw-off used with activated sludge.

and underneath the baffle, and the clarified liquid is skimmed off over weirs on both sides of a centrally located weir trough. Grease and scum are confined to the surface of the annular space.

Circular tanks 12 to 30 ft (3.6 to 9 m) in diameter have the sludge removal equipment supported on beams spanning the tank. Tanks 35 ft (10.5 m) in diameter and larger have a central pier that supports the mechanism and is reached by a walkway or bridge. The bottom of the tank is sloped at about 1 in/ft (1 in 12) to form an inverted cone, and the sludge is scraped to a relatively small hopper located near the center of the tank.

Performance of Sedimentation Tanks

The efficiency of sedimentation basins with respect to the removal of BOD and TSS is reduced by: (1) eddy currents formed by the inertia of the incoming fluid, (2) wind-induced circulation cells formed in uncovered tanks, (3) thermal convection currents, (4) cold or warm water causing the formation of density currents that move along the bottom of the basin and warm water rising and flowing across the top of the tank, and (5) thermal stratification in hot arid climates (Fair and Geyer, 1954). These subjects are considered in the following discussion.

BOD and TSS removal. Typical performance data for the removal of BOD and TSS in primary sedimentation tanks, as a function of the detention time and constituent concentration, are presented in Fig. 5-35, which is derived from observations of the performance of actual sedimentation tanks. The curvilinear relationships in the figure can be modeled as rectangular hyperbolas using the following relationship:

$$R = \frac{t}{a + bt} \tag{5-36}$$

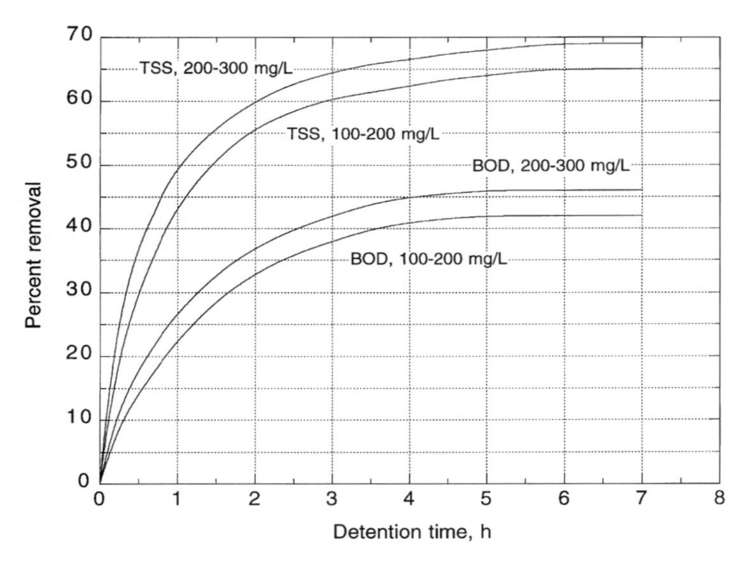

FIGURE 5-35
Typical BOD and TSS removal in primary sedimentation tanks (adapted from Greeley, 1938).

where R = expected removal efficiency, %
t = nominal detention time, h
a, b = empirical constants

Typical values for the empirical constants in Eq. (5-36) at 20°C are as follows:

Item	*a*, h	*b*
BOD	0.018	0.020
TSS	0.0075	0.014

Typical data on the specific gravity and solids concentration achieved in primary sedimentation are presented in Table 5-13.

TABLE 5-13
Typical values of specific gravity and solids concentration of sludge from primary sedimentation tanks

Type of sludge	Specific gravity	Solids concentration, %*	
		Range	Typical
Primary only:			
Medium strength wastewater	1.03	4–12	6
From combined sewer system	1.05	4–12	6.5
Primary and waste activated sludge	1.03	2–6	3
Primary and trickling filter humus sludge	1.03	4–10	5

*Percent dry solids.

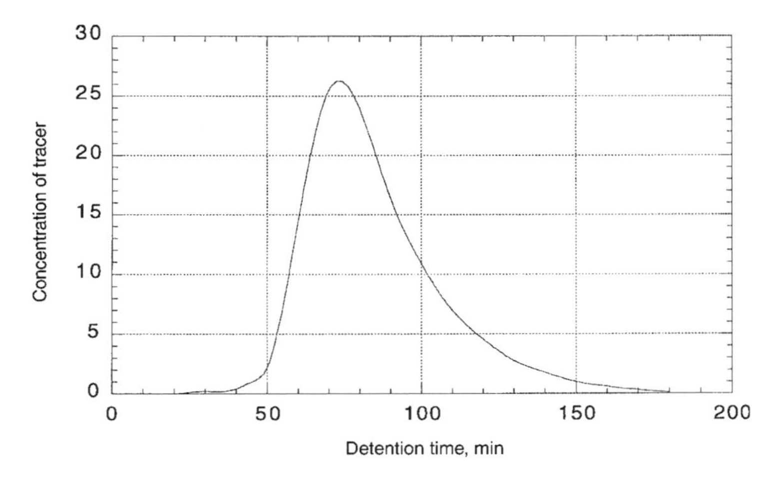

FIGURE 5-36
Typical time-concentration tracer response curve for sedimentation tanks.

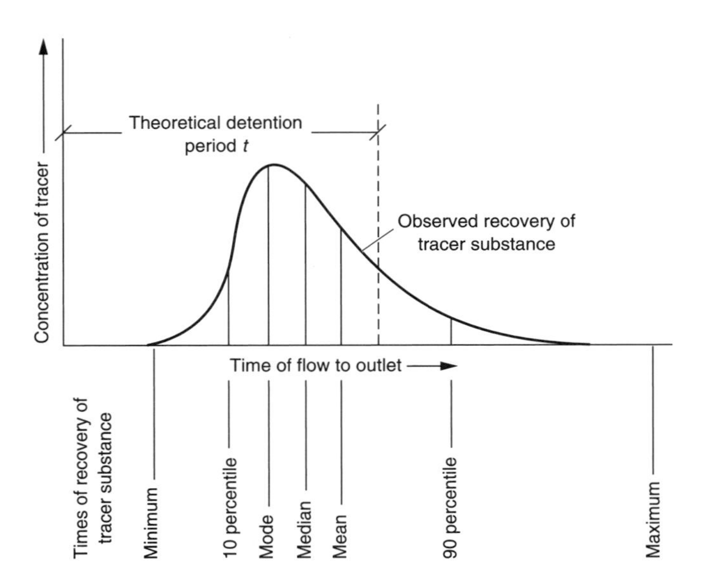

FIGURE 5-37
Identification of parameters used to analyze time-concentration tracer response curves (from Fair and Geyer, 1954).

Short circuiting and hydraulic stability. In an ideal sedimentation basin, a given block of entering water should remain in the basin for the full detention time. Unfortunately, in practice, sedimentation basins are seldom ideal and considerable short circuiting will be observed for one or more of the reasons cited above. A typical time-concentration tracer curve for a sedimentation basin is shown in Fig. 5-36. The stability of the basin can be assessed by repeating the tracer tests several times. If in the repeated tests the time-concentration curves are similar, then the basin is stable. If the time-concentration curves [also known as *residence time distribution* (RTD) curves] are not repeatable, the basin is unstable and the performance of the basin will be erratic (Fair and Geyer, 1954). The characteristic parameters of the time-concentration curve are depicted in Fig. 5-37 and defined in Table 5-14.

In 1932, Morrill (1932) suggested, on the basis of his studies of sedimentation basins, that the ratio of the 90 percentile to the 10 percentile value from the cumulative tracer curve could be used as a measure of the dispersion index, and that 1 over

TABLE 5-14

Various terms used to describe the performance of sedimentation tanks and other plug-flow reactors*

Term	Definition
T	Theoretical mean residence time (v_v/Q).
t_i	Time at which tracer first appears.
t_p	Time at which the peak concentration of the tracer is observed (mode).
t_g	Mean time to reach centroid of the RTD curve.
t_{10}, t_{50}, t_{90}	Time at which 10, 50, and 90 percent of the tracer has passed through the reactor.
t_{90}/t_{10}	Morrill dispersion index, MDI.
1/MDI	Volumetric efficiency as defined by Morrill (1932).
t_i/T	Index of short circuiting. In an ideal plug-flow reactor, the ratio is 1, and approaches 0 with increased mixing.
t_p/T	Index of modal retention time. Ratio will approach 1 in a plug-flow reactor, and 0 in a complete-mix reactor. For values of the ratio greater than or less than 1.0 the flow distribution in the reactor is not uniform.
t_g/T	Index of average retention time. A value of 1 would indicate that full use is being made of the volume. A value of the ratio greater than or less than 1.0 indicates the flow distribution is not uniform.
t_{50}/T	Index of mean retention time. The ratio t_{50}/θ is a measure of the skew of the RTD curve. In an effective plug-flow reactor, the RTD curve is very similar to a normal or Gaussian distribution. A value of t_{50}/θ less than 1.0 corresponds to an RTD curve that is skewed to the left. Similarly, for values greater than 1.0 the RTD curve is skewed to the right.

*Adapted from Morrill (1932), Fair and Geyer (1954), and U.S. EPA (1987).

the dispersion index is a measure of the volumetric efficiency. The dispersion index as proposed by Morrill is given by

$$\text{Morrill dispersion index, MDI} = \frac{P_{90}}{P_{10}} \tag{5-37}$$

where P_{90} = 90 percentile value from log-probability plot and P_{10} = 10 percentile value from log-probability plot. The percentile values are obtained from a log-probability plot of the cumulative tracer curve values.

The volumetric efficiency is given by

$$\text{Volumetric efficiency, \%} = \frac{1}{\text{MDI}} \times 100 \tag{5-38}$$

The determination of the Morrill dispersion index and the volumetric efficiency is illustrated in Example 5-11.

EXAMPLE 5-11. DETERMINATION OF MORRILL DISPERSION INDEX AND VOLUMETRIC EFFICIENCY FOR A RECTANGULAR SEDIMENTATION BASIN. The following data were obtained from a dye test of a rectangular primary sedimentation tank. Using these data, determine the Morrill dispersion index and the corresponding basin efficiency.

Time, min	Concentration, %	Cumulative, %
0	0.00	0.00
10	0.00	0.00
20	0.10	0.10
30	2.30	2.40
40	14.50	16.90
50	25.50	42.40
60	24.00	66.40
70	16.20	82.60
80	9.50	92.10
90	4.20	96.30
100	2.00	98.30
110	1.00	99.30
120	0.30	99.60
130	0.20	99.80
140	0.10	99.90
150	0.10	100.00

Solution

1. Plot the time (log scale) versus the cumulative concentration percentage (probability scale) on log-probability paper. The required plot is given below.
2. Determine the Morrill dispersion index. From the plot prepared in step 1:

$$\text{MDI} = \frac{P_{90}}{P_{10}} = \frac{80}{36} = 2.22$$

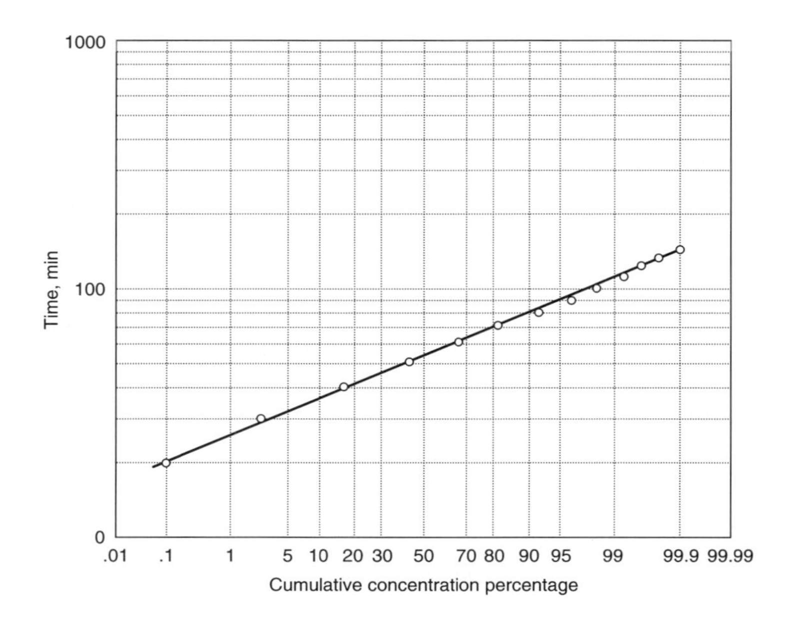

3. Determine the volumetric efficiency:

$$\text{Volumetric efficiency, \%} = \frac{1}{\text{MDI}} \times 100 = \frac{1}{2.22} \times 100 = 45\%$$

Temperature effects. Temperature effects can be significant in sedimentation basins. It has been shown that a 1°C temperature differential between the incoming wastewater and the wastewater in the sedimentation tank will cause a density current (see Fig. 5-38*b*) to form. The impact of the temperature effects on performance will depend on the material being removed and its characteristics.

Wind effects. Wind blowing across the top of open sedimentation basins can cause circulation cells to form (see Fig. 5-38*d*). When circulation cells form, the effective volumetric capacity of the basin is reduced. As with temperature effects, the impact of the reduced volume on performance will depend on the material being removed and its characteristics.

Overall design. Although a variety of environmental and design factors can affect the performance of sedimentation basins, it has been observed that the performance will be governed by the overall design, discussed below, which still remains an art form.

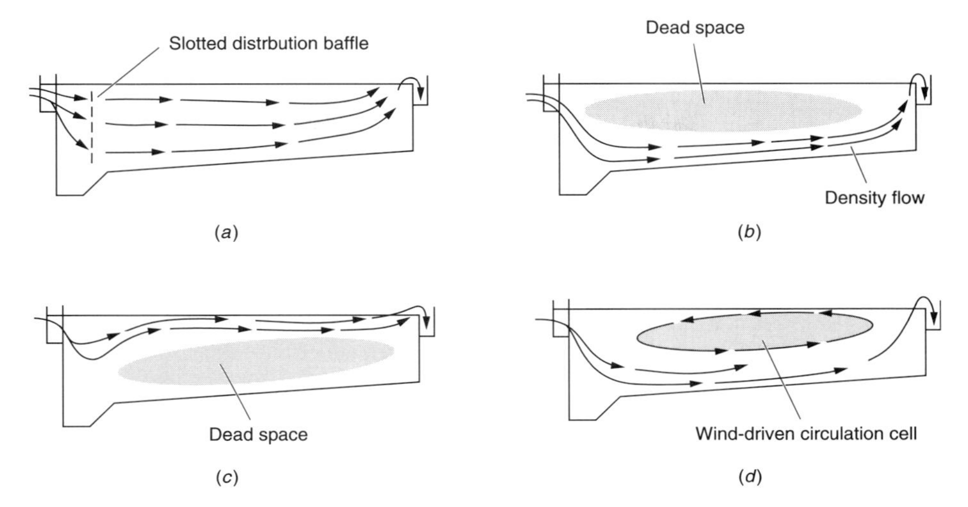

FIGURE 5-38
Typical flow patterns observed in rectangular sedimentation tanks: (*a*) ideal flow, (*b*) effect of density flow or thermal stratification (water in tank is warmer than influent), (*c*) effect of thermal stratification (water in tank is colder than influent), and (*d*) formation of wind-driven circulation cell.

Sludge: Characteristics, Quantities, and Disposal

Detailed information on the characteristics, quantities, and disposal of sludge removed during primary sedimentation is presented in Chap. 14, which deals with the management of septage and sludge.

Basis of Design

If all solids in wastewater were discrete particles of uniform size, uniform density, reasonably uniform specific gravity, and fairly uniform shape, the removal efficiency of these solids would be dependent on the surface area of the tank and time of detention. The depth of the tank would have little influence, provided that horizontal velocities were maintained below the scouring velocity. However, the solids in most wastewaters are not of such regular character but are heterogeneous in nature, and the conditions under which they are present range from total dispersion to complete flocculation. Design parameters for sedimentation are considered below. Typical design data for rectangular and circular sedimentation tanks are presented in Tables 5-15 and 5-16. Additional details on the analysis and design of sedimentation tanks may be found in WPCF (1985). A design procedure is illustrated in Example 5-12.

Detention time. Normally, primary sedimentation tanks are designed to provide $1\frac{1}{2}$ to $2\frac{1}{2}$ hours of detention, based on the average rate of wastewater flow. Tanks that provide shorter detention periods ($\frac{1}{2}$ to 1 h), with less removal of suspended

TABLE 5-15
Typical design information for primary sedimentation tanks followed by secondary treatment

Item	Unit	Value: Range	Value: Typical
Primary sedimentation tanks followed by secondary treatment			
Detention time	h	1.5–2.5	2.0
Overflow rate			
Average flow	gal/ft^2·d	740–1,230	1,000
Peak hourly flow	gal/ft^2·d	2,000–3,000	2,200
Weir loading	gal/ft·d	10,000–40,000	15,000
Primary settling with waste activated sludge return			
Detention time	h	1.5–2.5	2.0
Overflow rate			
Average flow	gal/ft^2·d	600–800	700
Peak hourly flow	gal/ft^2·d	1,200–1,700	1,500
Weir loading	gal/ft·d	10,000–40,000	20,000

TABLE 5-16
Typical design information for rectangular and circular sedimentation tanks used for primary and secondary treatment of wastewater

Item	Unit	Primary: Range	Primary: Typical	Secondary: Range	Secondary: Typical
Rectangular:					
Depth	ft	10–16	14	10–22	18
Length	ft	50–300	80–130	50–300	80–130
Width*	ft	10–80	16–32	10–80	16–32
Flight speed	ft/min	2–4	3	2–4	3
Circular:					
Depth	ft	10–16	14	10–22	18
Diameter	ft	10–200	40–150	10–200	40–150
Bottom slope	in/ft	$\frac{3}{4}$–2	1.0	$\frac{3}{4}$–2	1.0
Flight travel speed	rev/min	0.02–0.05	0.03	0.02–0.05	0.03

*If widths of rectangular mechanically cleaned tanks are greater than 20 ft, multiple bays with individual cleaning equipment may be used, thus permitting tank widths up to 80 ft or more.

solids, are sometimes used for preliminary treatment ahead of biological treatment units. It should be noted that primary sedimentation is omitted in many biological treatment processes used for small systems. In cold climates, increases in water viscosity at lower temperatures retard particle settling in clarifiers and reduce performance at wastewater temperatures below 20°C (68°F). A curve showing the increase in detention time necessary to equal the detention time at 20°C is presented in Fig. 5-39 (ASCE, 1989). For wastewater having a temperature of 10°C, for example,

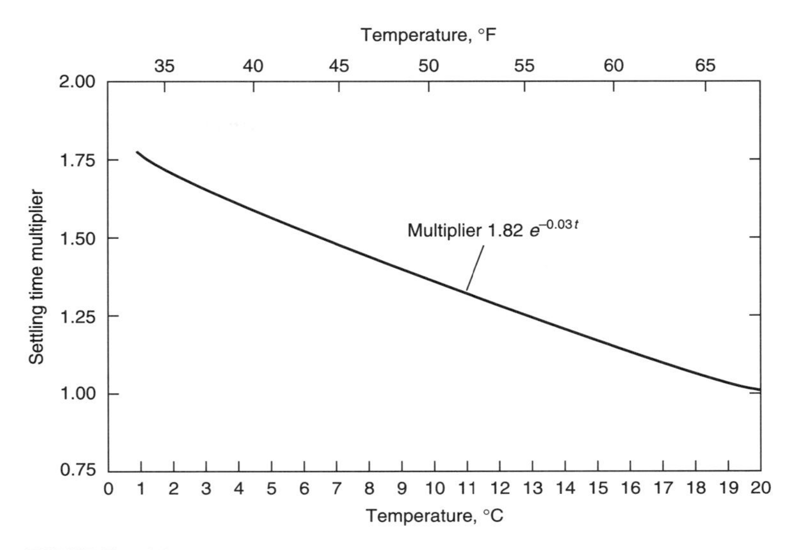

FIGURE 5-39
Curve of the increase in detention time required at cooler temperatures to achieve the same performance as achieved at 20°C (ASCE, 1989).

the detention period is 1.38 times that required at 20°C to achieve the same efficiency. Thus, in cold climates, safety factors should be considered in clarifier design to ensure adequate performance.

Surface loading rates. Sedimentation tanks are normally designed on the basis of a surface loading rate (commonly termed *surface overflow rate,* SOR) expressed as gallons per square foot of surface area per day, gal/ft^2·d (cubic meters per square meter of surface area per day, m^3/m^2·d). The selection of a suitable loading rate depends on the type of suspension to be separated. Typical values for various suspensions are reported in Table 5-15. Designs for municipal plants must also meet the approval of state regulatory agencies, many of which have adopted standards for surface loading rates that must be followed. When the area of the tank has been established, the detention period in the tank is governed by water depth. Overflow rates in current use result in nominal detention periods of 2.0 to 2.5 h, based on average design flow.

The effect of the surface loading rate and detention time on suspended solids removal varies widely depending on the character of the wastewater, proportion of settleable solids, concentration of solids, and other factors. It should be emphasized that overflow rates must be set low enough to ensure satisfactory performance at peak rates of flow, which may vary from over 3 times the average flow in small plants (see discussion of peak flowrates in Chap. 4).

Weir rates. In general, weir loading rates have little effect on the efficiency of primary sedimentation tanks unless effluent upflow velocities are extreme, due to limited weir lengths. Effects related to the placement of the weirs and the design of the tanks have generally proven to be more important. Typical weir loading rates are given in Table 5-15.

Scour velocity. To avoid the resuspension (scouring) of settled particles, horizontal velocities through the tank should be kept sufficiently low. Using the results from studies by Shields (1936), Camp (1946) developed the following equation for the critical velocity.

$$v_H = \left(\frac{8k(s-1)gd}{f}\right)^{1/2} \tag{5-39}$$

where v_H = horizontal velocity that will just produce scour
k = constant which depends on type of material being scoured
s = specific gravity of particles
g = acceleration due to gravity
d = diameter of particles
f = Darcy-Weisbach friction factor

Typical values of k are 0.04 for unigranular sand and 0.06 for more sticky, interlocking matter. The term f (the Darcy-Weisbach friction factor) depends on the characteristics of the surface over which flow is taking place and the Reynolds number. Typical values of f are 0.02 to 0.03. Either U.S. customary or SI units may be used in Eq. (5-39), so long as they are consistent, because k and f are dimensionless.

EXAMPLE 5-12. DESIGN OF A PRIMARY SEDIMENTATION BASIN. The average flowrate at a small municipal wastewater treatment plant is 0.5 Mgal/d. The highest observed peak daily flowrate is 1.25 Mgal/d. Design a rectangular (aspect ratio 4:1) primary clarifier for this wastewater treatment facility. Calculate the scour velocity, to determine if settled material will become resuspended. Estimate the BOD and TSS removal at average and peak flow. Use a surface overflow rate (SOR) of 900 gal/ft²·d at average flow (see Table 5-15) and a side water depth of 12 ft.

Solution

1. Calculate the required surface area. For average flow conditions, the required area is

$$A = \frac{Q}{\text{SOR}} = \frac{0.5 \times 10^6 \text{ gal/d}}{900 \text{ gal/ft}^2\cdot\text{d}} = 556 \text{ ft}^2$$

2. Determine the tank surface dimensions. For an aspect ratio of 4 to 1,

$$4L^2 = 556 \text{ ft}^2$$

L = 11.8 ft, so the idealized surface dimensions are 11.8 ft by 47.2 ft. However, for the sake of convenience, the surface dimensions will be rounded to 12 ft by 48 ft.

3. Determine the detention time and overflow rate at average flow. For the assumed side water depth of 12 ft,

$$\text{Tank volume} = 12 \text{ ft} \times 48 \text{ ft} \times 12 \text{ ft} = 6912 \text{ ft}^3$$

$$\text{Overflow rate} = \frac{Q}{A} = \frac{0.5 \times 10^6 \text{ gal/d}}{12 \text{ ft} \times 48 \text{ ft}} = 868 \text{ gal/ft}^2\cdot\text{d}$$

$$\text{Detention time} = \frac{\text{Vol}}{Q} = \left(\frac{6912\ \text{ft}^3}{0.5 \times 10^6\ \text{gal/d}}\right)\left(\frac{7.48\ \text{gal}}{\text{ft}^3}\right)\left(\frac{24\ \text{h}}{\text{d}}\right) = 2.48\ \text{h}$$

4. Determine the detention time and overflow rate at peak flow:

$$\text{Overflow rate} = \frac{Q}{A} = \frac{1.25 \times 10^6\ \text{gal/d}}{12\ \text{ft} \times 48\ \text{ft}} = 2170\ \text{gal/ft}^2\cdot\text{d}$$

$$\text{Detention time} = \frac{\text{Vol}}{Q} = \left(\frac{6912\ \text{ft}^3}{1.25 \times 10^6\ \text{gal/d}}\right)\left(\frac{7.48\ \text{gal}}{\text{ft}^3}\right)\left(\frac{24\ \text{h}}{\text{d}}\right) = 0.99\ \text{h}$$

5. Calculate the scour velocity (Eq. 5-39), using the following values:

Cohesion constant	$k = 0.05$
Specific gravity	$s = 1.25$
Acceleration due to gravity	$g = 32.2\ \text{ft/s}^2$
Diameter of particles	$d = 100\ \mu\text{m} = 3.28 \times 10^{-4}\ \text{ft}$
Darcy-Weisbach friction factor	$f = 0.025$

$$v_H = \left(\frac{8k(s-1)gd}{f}\right) = \left(\frac{(8)(0.05)(0.25)(32.2)(3.28 \times 10^{-4})}{0.025}\right)^{1/2} = 0.21\ \text{ft/s}$$

6. Compare the scour velocity calculated in the previous step to the peak flow horizontal velocity (the peak flow divided by the cross-sectional area through which the flow passes). The peak flow horizontal velocity through the settling tank is

$$v = \frac{Q}{A_x} = \left(\frac{1.25 \times 10^6\ \text{gal/d}}{12\ \text{ft} \times 12\ \text{ft}}\right)\left(\frac{\text{ft}^3}{7.48\ \text{gal}}\right)\left(\frac{\text{d}}{(24\ \text{h/d})(3600\ \text{s/h})}\right) = 0.013\ \text{ft/s}$$

The horizontal velocity value, even at peak flow, is substantially less than the scour velocity. Therefore, settled matter should not be resuspended.

7. Use Eq. (5-36) and the accompanying coefficients to estimate the removal rates for BOD and TSS at average and peak flow.

a. At average flow:

$$\text{BOD removal} = \frac{t}{a + bt} = \frac{2.48}{0.018 + (0.020)(2.48)} = 37\%$$

$$\text{TSS removal} = \frac{t}{a + bt} = \frac{2.48}{0.0075 + (0.014)(2.48)} = 59\%$$

b. At peak flow:

$$\text{BOD removal} = \frac{t}{a + bt} = \frac{0.99}{0.018 + (0.020)(0.99)} = 26\%$$

$$\text{TSS removal} = \frac{t}{a + bt} = \frac{0.99}{0.0075 + (0.014)(0.99)} = 46\%$$

5-10 SEPTIC TANKS

A septic tank is used to receive the wastewater discharged from individual residences and other nonsewered facilities. Septic tanks, as shown schematically in Fig. 5-40

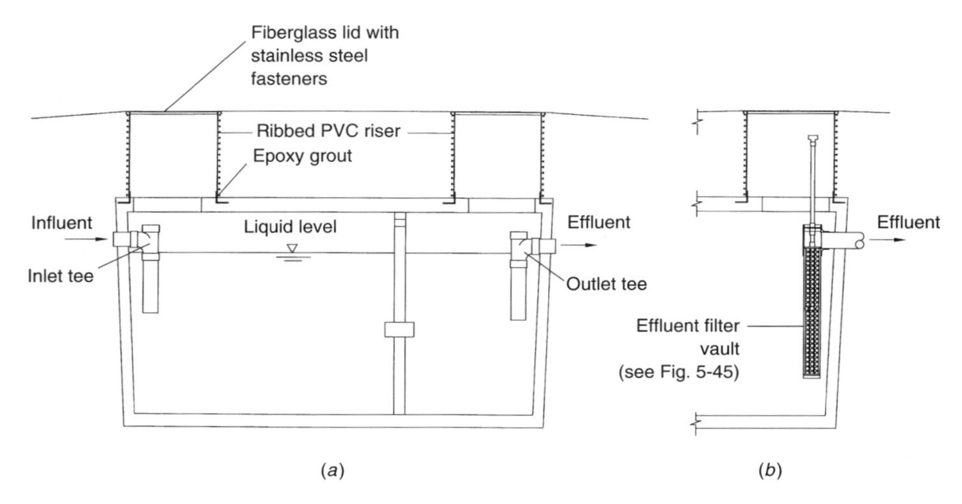

FIGURE 5-40
Schematic of septic tanks: (*a*) conventional two-compartment tank with outlet tee and (*b*) cutaway of single compartment tank with effluent filter vault (adapted from Orenco Systems, Inc.).

(*a*) (*b*)

FIGURE 5-41
Views of septic tanks: (*a*) monolithic concrete construction with glass panels for viewing flow and scum and sludge accumulation and (*b*) fiberglass construction.

and photographically in Fig. 5-41, are prefabricated tanks that serve as a combined settling and skimming tank, as an unheated unmixed anaerobic digester, and as a sludge storage tank. A septic tank followed by a soil absorption system constitutes what is known as a conventional onsite wastewater management system. Because of the fundamental importance of the septic tank in the onsite management of wastewater, a detailed description along with design and operational and maintenance considerations is presented in this section.

Description

Used principally for the treatment of wastes from individual residences, septic tanks of various sizes are used for establishments such as schools, summer camps, parks, trailer parks, and motels. The types of materials used for the construction, the functional operation, operational problems, and appurtenances for septic tanks are considered in the following discussion. Before we discuss these topics, it will be useful to review briefly the historical development of the septic tank.

Historical development of the septic tank. The modern septic tank can be traced back to about 1860 with the early work of Mouras in France (Dunbar, 1908). In fact, it is surprising to compare the modern septic tank shown in Fig. 5-40 to the modified Mouras tank known as the Fosse Mouras tank of the 1870s, shown in Fig. 5-42. The name *septic tank* is attributed to Donald Cameron, who so named the tank for the septic actions and conditions within the tank. Cameron applied for and was granted British patent No. 21,142 in 1895. A United States patent was issued in 1899. Much legal controversy followed, and a variety of tank configurations developed. An early type of septic tank system, developed by employees of the U.S. Public Health Service, is illustrated in Fig. 5-43. In describing their apparatus they wrote (Lumsden et al., 1915):

> This apparatus consists of the following parts:
>
> 1 A water-tight tank, barrel, or other container, to receive and liquefy the excreta.
> 2 A covered water-tight can, pot, barrel, or other vessel, to receive the effluent or outflow.
> 3 A connecting pipe about 2.5 inches in diameter, about 12 inches long, and provided with an open T at one end, both openings of the T being covered with wire screens.
> 4 A tight box, preferably zinc lined, which fits tightly on the top of the liquefying barrel. It is provided with an opening on top for the seat which has an automatically closing lid.
> 5 An antisplashing device, consisting of a small board placed horizontally under the seat about an inch below the level of the transverse connecting pipe. It is held in place by

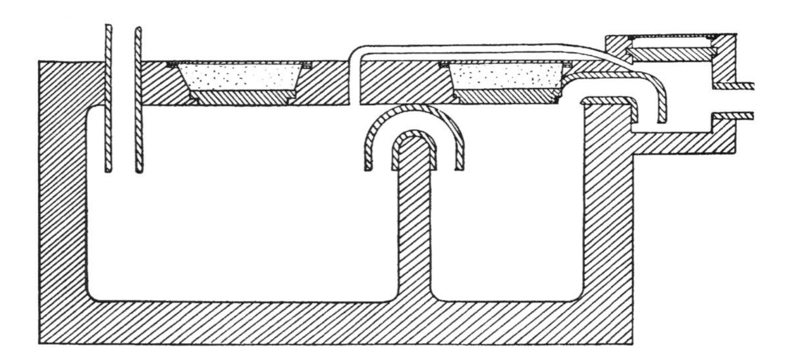

FIGURE 5-42
Modification of original septic tank credited to Fosse Mouras, circa 1860. (Dunbar, 1908).

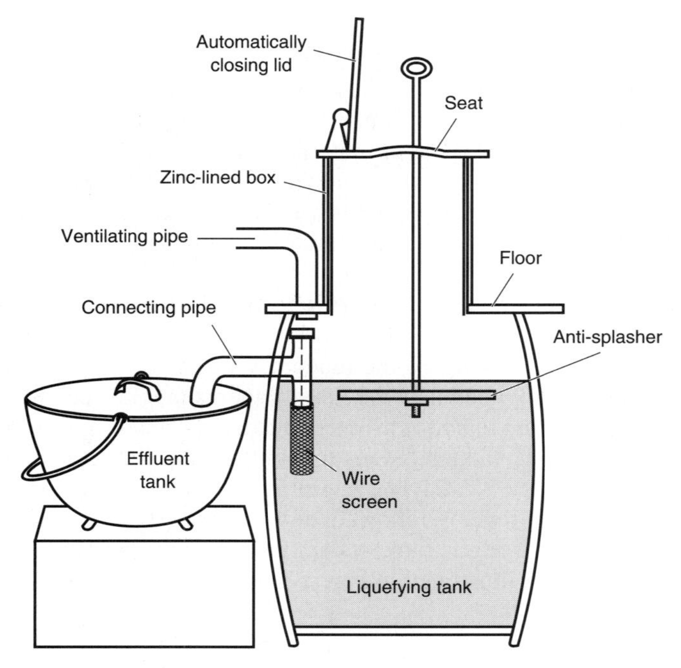

FIGURE 5-43
Early "L.R.S. privy with an ordinary vinegar barrel used as a liquefying tank and an iron pot for effluent tank" (from Lumsden et al., 1915).

a rod, which passes through a hole in the side of the seat and by which the board is raised and lowered. A layer of chips floated in the tank may be used instead of this antisplashing device.

6 A ventilating pipe, such as a stovepipe or wooden flue, connecting the space under the seat with the open air.

The liquefying tank is filled with water up to the point where it begins to trickle into the effluent tank. A pound or two of old manure should be added to the water to start fermentation . . . .

It should be noted that the use of a watertight tank and an effluent screen are two very important features of a modern septic tank.

Materials of construction. Typically, septic tanks are made of concrete or fiberglass, although other materials such as steel, redwood, and polyethylene have been used. The use of steel and redwood tanks is no longer accepted by most regulatory agencies. Polyethylene tanks have been used; their structural integrity is inferior to concrete and fiberglass tanks. Long-term creep, resulting in deformation, has been a problem with polyethylene tanks. Fiberglass tanks, being more expensive, are used in areas inaccessible to concrete tank delivery trucks. Regardless of the material of construction, a septic tank must be *watertight* and *structurally sound* if it is to function properly, especially where subsequent treatment units such as intermittent and recirculating packed bed filters or pressure sewers are to be used. The difference in cost between a low-cost septic tank versus a watertight and structurally sound septic tank is minimal. Should the low-cost tank leak, the cost of repairing it will far exceed the cost of a new tank.

Functional operation. Settleable solids in the incoming wastewater settle and form a sludge layer at the bottom of the tank. Oils and greases and other light materials float to the surface, where a scum layer is formed as floating materials accumulate. Settled and skimmed wastewater flows from the clear space between the scum and sludge layers to the disposal field or to a treatment unit, if one is used. The organic material retained in the bottom of the tank undergoes facultative and anaerobic decomposition and is converted to more stable compounds and gases such as carbon dioxide (CO_2), methane (CH_4), and hydrogen sulfide (H_2S). The sludge that accumulates in the septic tank is composed primarily of lint from the washing of clothes and the lignous material contained in toilet paper. While these materials will eventually be decomposed biologically, the rate is extremely slow, which accounts for the accumulation. It is interesting to note that the early septic tanks were known as liquefaction tanks, because it was observed that, in the absence of extraneous materials, essentially all of the solids discharged to a septic tank were liquefied. The use of lint traps and biodegradable toilet paper will limit the accumulation of sludge in a septic tank.

Although hydrogen sulfide is produced in septic tanks, odors are not usually a problem because the hydrogen sulfide combines with the metals in the accumulated solids to form insoluble metallic sulfides. Even though the volume of the solid material being deposited is being reduced continually by anaerobic decomposition, there is always a net accumulation of sludge in the tank. Material from the bottom of the tank that is buoyed up by the adhesion of decomposition gases will often stick to the bottom of the scum layer, increasing its thickness. Because the long-term accumulation of scum and sludge can reduce the effective settling capacity of the tank, the contents of the tanks must be pumped periodically. The various layers and zones that form in a septic tank are identified in Fig. 5-44.

Operational problems. Historically, the most serious operational problem with septic tanks has been the carryover of solids and oils and grease. The carryover of solids in septic tank effluent has led to a premature reduction in the hydraulic acceptance rate of the leachfield, causing the development of damp or wet spots in the vicinity of the leachfields, and ultimately the surfacing of the effluent. Extraneous groundwater entering nonwatertight septic tanks has resulted in: (1) the hydraulic overloading of the leachfields, also ultimately leading to the surfacing of the effluent; (2) the disruption of the anaerobic digestion process going on within the tank; and (3) the severe hydraulic overloading of downstream treatment processes, such as intermittent and recirculating packed bed filters. Thus, to repeat a previous comment, septic tanks must be watertight and structurally sound if they are to protect the environment and function properly.

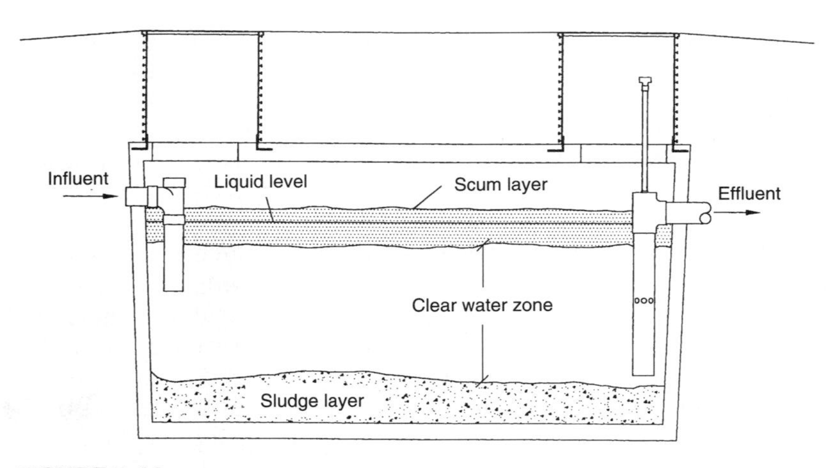

FIGURE 5-44
Definition sketch for the sludge, clear water, and scum zones that form in a septic tank.

Septic tank appurtenances. Two compartments have been used to limit the discharge of solids in the effluent from the septic tank. Based on measurements made in both single and double compartments, the benefit of a two-compartment tank appears to depend more on the design of the tank than the use of two compartments (Seabloom, 1982; Winneberger, 1984). A more effective way to eliminate the discharge of untreated solids involves the use of an effluent filter in conjunction with a single-compartment tank (see Fig. 5-45*a*). Operationally, effluent flows into the filter through the inlet holes located in the center of the vault chamber. Before passing into the center of the vault, the effluent must pass through a screen which is located on the inside of the vault. Because of the large surface area of the filter screen, clogging is not excessively rapid. If necessary, the screen can be removed and cleaned (see Fig. 5-45*b*). It should be noted that the effluent filter vault functions, in effect, as a second chamber. An advantage of the effluent filter vault is that it can be installed in both existing and new septic tanks to limit the discharge of gross untreated solids. An effluent filter used in conjunction with an effluent filter pump is shown in Fig. 5-46.

Design Considerations

Important design and operational considerations for septic tanks include: (1) tank configuration, (2) the structural integrity of the tank, (3) the watertightness of the tank, (4) the size of the tank, (5) septic tank appurtenances, (6) the use of large septic tanks, (7) routine inspection, and (8) septic tank pumping.

Tank configuration. Most concrete septic tanks in use currently are rectangular with an interior baffle to divide the tank and access ports to permit inspection

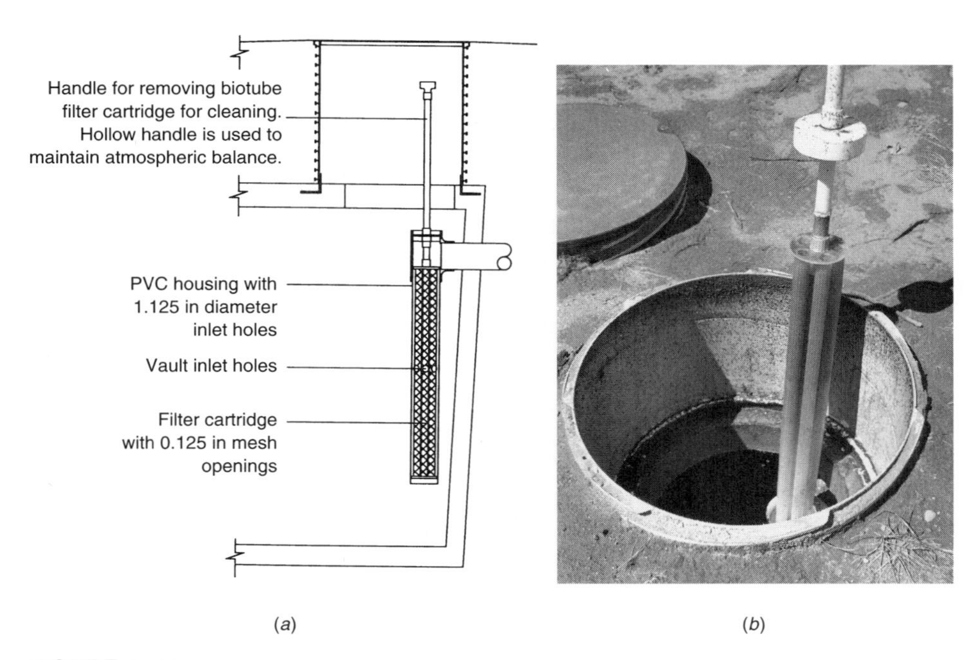

FIGURE 5-45
Typical effluent filter vault used for the removal of solids: (*a*) definition sketch (adapted from Orenco Systems, Inc.) and (*b*) filter cartridge removed for cleaning by hosing off the accumulated solids.

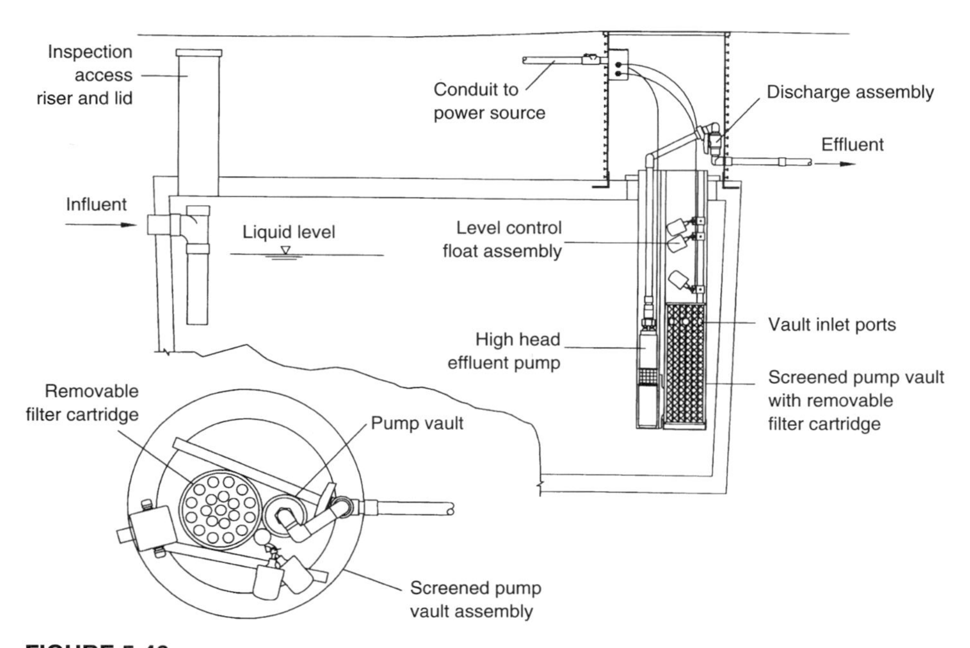

FIGURE 5-46
Effluent filter vault used in conjunction with the multistage high-head pump (adapted from Orenco Systems, Inc.).

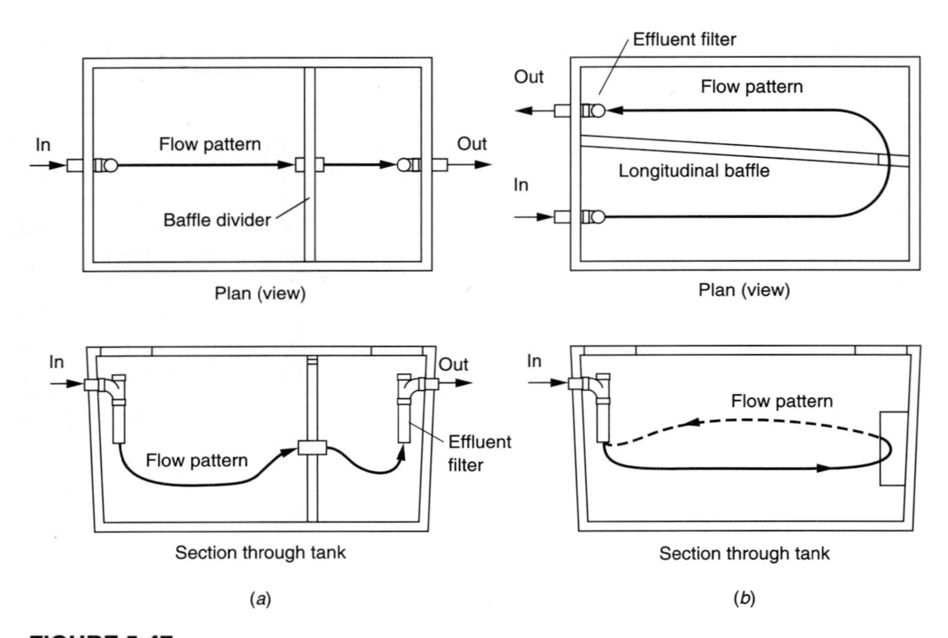

FIGURE 5-47
Flow patterns in septic tanks: (*a*) with interior baffle placed across tank, and (*b*) interior baffle placed longitudinally.

and cleaning (see Fig. 5-47*a*). The larger chamber formed by the interior baffle typically contains about two thirds of the tank volume. Although a divider is used, the rationale for its use is historical more than scientific. Both Seabloom et al. (1982) and Winneberger (1984) have found, on the basis of field measurements, that the performance of a single-compartment tank is equal to or exceeds the performance of a two-compartment tank of the same liquid volume. In fact, the divider in the tank actually limits the available surface area for scum and sludge accumulation. With respect to the dividing baffle, a far more rational placement is longitudinally as shown in Fig. 5-47*b*. Not only does the longitudinal placement of the baffle improve the removal of scum and sludge, but also the structural integrity of the tank is enhanced.

Structural integrity of the tank. The long-term performance of a septic tank will depend on its structural integrity. For concrete septic tanks, structural integrity is dependent on the method of construction, the placement of the reinforcing steel, and the composition of the concrete mix (Bounds, 1996). For maximum structural integrity, the walls and bottom of the tank should be poured monolithically. Where the walls and bottom are poured monolithically, the top should be cast in place with the reinforcing steel from the walls extending into the top slab. In some cases, a water seal is placed between the wall and the top. Set-on-top lids should be avoided, as they will separate where uneven settling occurs. Additional details and specifications for both concrete and fiberglass septic tanks have been reported by Bounds (1996).

Testing for watertightness. Watertight tanks, as noted previously, are a necessity for the protection of the environment and for the operation of subsequent processing and/or disposal facilities. Each tank should be tested for watertightness and structural integrity by completely filling the tank with water before and after installation. Hydrostatic testing is conducted at the factory by filling the tank with water and letting it stand for 24 hours. If no water loss is observed after 24 hours, the tank is acceptable. Because some water absorption may occur with concrete tanks, the tank should be refilled and allowed to stand for an additional 24 hours. If the water loss after the second 24-hour period is greater than 1 gal the tank should be rejected. To avoid the separation of the top of a concrete tank from the body of the tank during hydrostatic testing, the water level in the riser must be controlled. It is important that the above procedure be repeated once the tank is installed (Bounds, 1996).

Sizing septic tanks. Over the years a number of empirical relationships have been developed to estimate the required size of tank. The minimum size tank recommended in various codes is 750 gal. On the basis of the authors' experience with tanks of various sizes, the following recommendations are made for septic tank sizing to achieve effective performance with respect to the removal of BOD, TSS, oil and grease, and to minimize the frequency of pumping out the contents of the tank (see subsequent discussion).

One or two bedrooms	1000 gal
Three bedrooms	1500 gal
Four bedrooms	2000 gal
More than four bedrooms	An engineering assessment is required

An additional reason for using larger septic tanks is that it is often difficult to increase the size of the existing septic tank, which may be desirable if the ownership of the home changes or the home is expanded.

Use of large septic tanks. Although septic tanks are used primarily for individual residences and other community facilities, large septic tanks have been used to serve clusters of homes and commercial establishments as well as small communities. Typically, large septic tanks are designed as plug-flow reactors. As a rule of thumb, the volumetric capacity of large septic tanks should be equal to about 5 times the average flow. Taking into account the accumulation of scum and sludge based on average flow (50 gal/capita·d) and an appropriate peaking factor (PF), the following equations can be used to estimate the required volumetric capacity of large septic tanks.

Pump-out interval, yr	Volume, gal
3	$2.8Q_{ave} \times PF$
4	$3.2Q_{ave} \times PF$
5	$3.65Q_{ave} \times PF$
6	$4.0Q_{ave} \times PF$

In the above equations, the numerical value times the average flow is equal to the minimum septic tank volume required for the indicated pumpout frequency, based on the work of Bounds (1996). The peaking factor used in the above equations can be thought of as a factor of safety. A typical value for the peaking factor for the sizing of large septic tanks is 1.5. Using a peaking factor of 1.5, the tank volume required varies from 3.3 to 6.8 times the average flow corresponding to pump-out frequencies of 2 to 5 years, respectively. These values, based on scum and sludge accumulation, compare favorably to the rule of thumb value of 5 times the average flow. It should be noted that the above septic sizing criterion does not take into account extreme peak flows that occur at schools, churches, and some recreational facilities.

The minimum size tank should be 1500 gal. In large installations, parallel tanks are often used to provide redundancy, and to allow for maintenance. A typical example of a large septic tank serving a small community is shown in Fig. 5-48. Site constraints often govern the size of the septic tank that can be used. In locations where it may not be possible to locate a single tank, multiple tanks are used. If a single small tank must be used, it is imperative that the owner be alerted to the fact that the contents of the tank may have to be pumped more frequently than if a properly sized tank had been used.

Septic Tank Maintenance

Because septic tanks are buried and are thus out of sight, some homeowners forget that septic systems require periodic maintenance. Often, residents of populated areas served by gravity collection systems relocate to areas where septic systems are used, and assume they can flush any material and any flow volume into the system, just as they could with the gravity sewers they used before moving. However, septic systems can be affected by certain constituents, and their capacity to handle flow is finite. If abused, septic systems can fail, creating nuisance conditions and possible health risks. However, by following a few simple operation guidelines, septic systems can provide years of troublefree service.

Routine inspection. The routine inspection of septic tanks on an annual or semiannual basis is conducted to: (1) check for watertightness, (2) check for the entry of extraneous water into the tank, (3) check for breaks in the tightline where a tightline is used to connect the septic tank to the leachfield, and (4) monitor the accumulation of scum and sludge. Scum and sludge can be measured as shown in Fig. 5-49. To measure the thickness of the scum layer, the L-shaped tool is pushed through the scum layer, rotated, and lifted up to the bottom of the scum layer. The thickness of the layer is read from the scale on the tool. To measure the depth of the sludge layer, the light-extinction probe is lowered through the clear zone until the sludge is encountered, at which point the light goes off.

Septic tank pumping. As noted above, the long-term accumulation of scum and sludge will reduce the effective settling capacity of the tank. The settling capacity of the tank is restored by pumping out the contents of the tank. The frequency with which the contents of a tank should be pumped has been studied

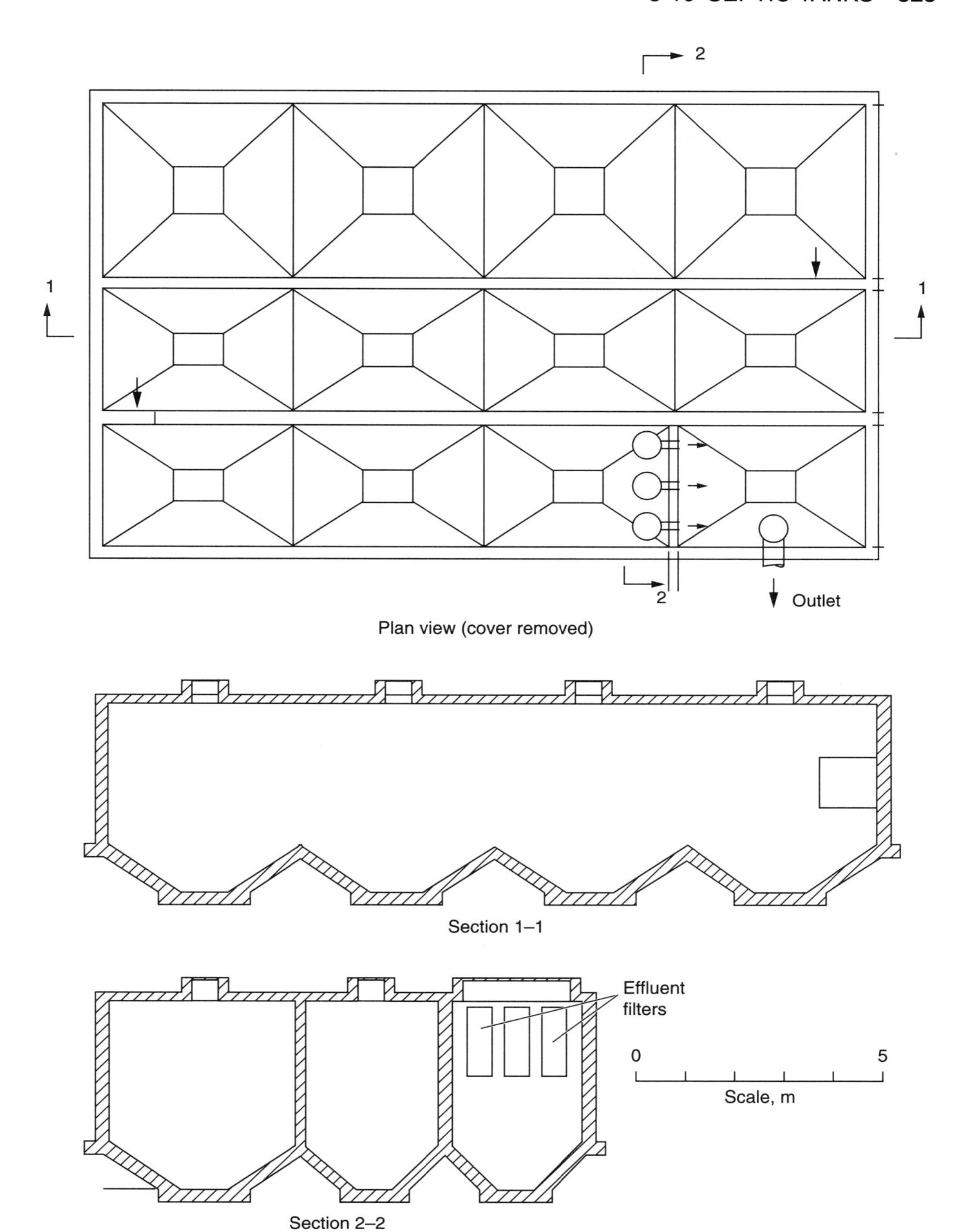

FIGURE 5-48
Typical septic tank used for a small community (courtesy Dialynas, 1997).

FIGURE 5-49
Probe used to determine thickness of scum layer and depth of sludge layer: (*a*) L-shaped scum layer probe resting on scum layer and (*b*) light-extinction probe for determining depth of sludge layer.

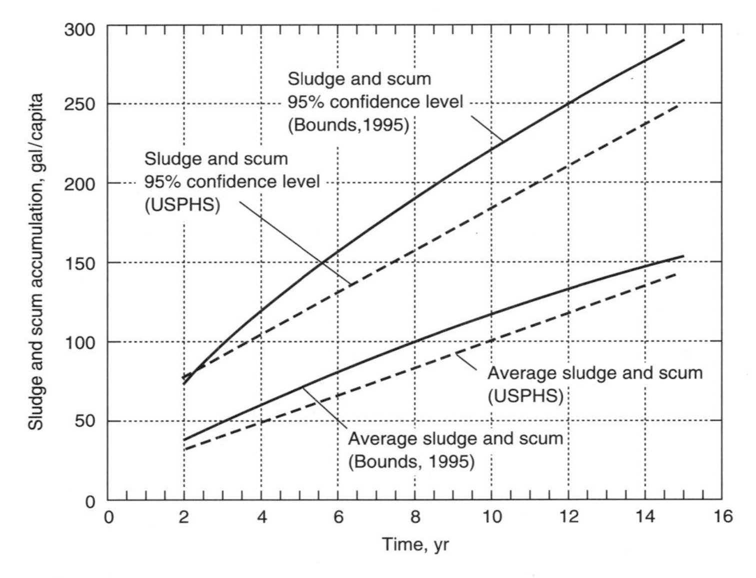

FIGURE 5-50
Analysis of septic tank septage pumping frequency (adapted from Bounds, 1995).

extensively by Bounds (1995). The results of his analysis are presented graphically in Fig. 5-50. Detailed information on the characteristics, quantities, and disposal of septage removed from septic tanks is presented in Chap. 14, which deals with the management of septage and biosolids.

5-11 OIL AND GREASE REMOVAL

Wastewater from restaurants, laundromats, and service stations typically contains significant amounts of oils and grease and detergents. When oils and greasy wastes enter a conventional septic tank, there is the possibility that they can be discharged along with the septic tank effluent to the soil absorption system and treatment system. Oils and greases, along with suspended solids, tend to accumulate on the surfaces of the soil absorption system, ultimately leading to a reduction in the infiltration capacity. Oils and greases are especially troublesome because of their persistence. Typical concentration values for oil and grease in wastewater from restaurants will vary from about 1000 to more than 2000 mg/L. To avoid problems with downstream treatment units in decentralized wastewater treatment and disposal systems, the effluent oil and grease concentration should be less than about 30 mg/L.

In recent times, the problems associated with the removal of oils and greases have been made more complex by the increase in the number of different types of oils and greases available for cooking (e.g., olive oil, canola, lard, etc.). The problem is complicated further because many of the oils are soluble at relatively low temperatures, which makes their removal more difficult. Typically, skimming or interceptor tanks are used to trap oils by flotation and grease by cooling and flotation. The contents of the tank serve as a heat exchanger cooling the incoming liquid, which helps to solidify the greases. For flotation to be effective, the interceptor tank must detain the fluid for an adequate period of time (typically greater than 30 minutes).

Oil and Grease Removal from Individual Discharges

Although a number of commercial oil and grease traps are available, they have not proven to be effective because of the limited detention time provided in such units. Also, most commercial units are rated on average flow and not the instantaneous peak flows observed in the field from restaurants and laundries. The use of conventional septic tanks as interceptor tanks has proven to be very effective. Depending on the tank configuration, some replumbing may be necessary when septic tanks are used as grease traps. Typically the inlet is situated below the water surface and the outlet is placed closer to the bottom of the tank (see Fig. 5-51). The larger volume provided by the septic tank has been beneficial in achieving the maximum possible separation of oils and greasy wastes. For restaurants, use of a series of three interceptor tanks (e.g., septic or similar-type tanks) has proven to be effective for the separation of oil and grease.

Volumes for grease interceptor tanks typically vary from 1.0 to 3.0 times the average daily flowrate. Although interceptor tanks with a volume corresponding to the

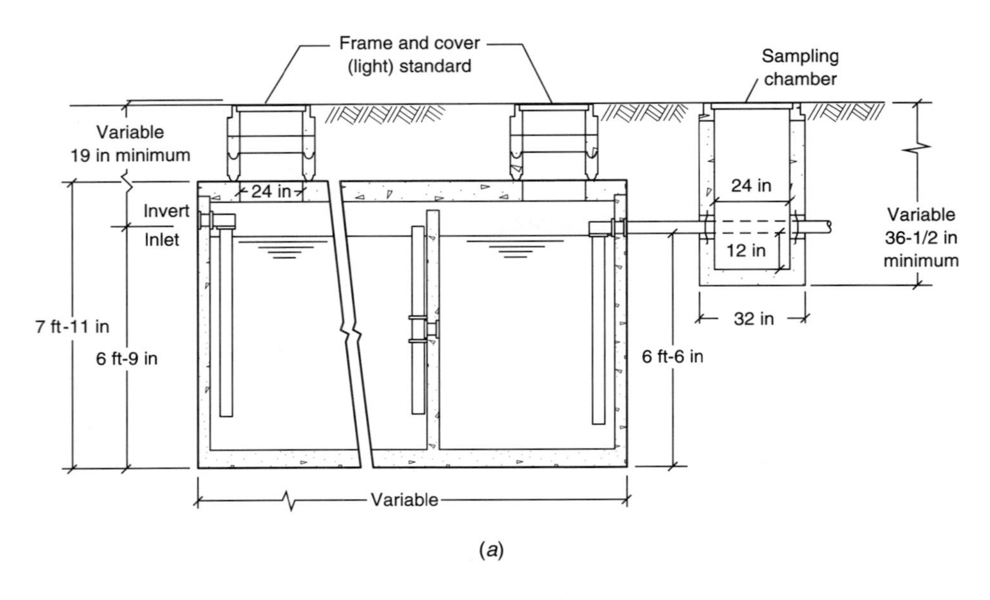

(*a*)

(*b*)

FIGURE 5-51
Typical commercial oil and grease traps: (*a*) schematic with external sampling chamber (adapted from Jensen Precast) and (*b*) view of 5000-gal tanks at fabrication facility.

average daily flow have been used, larger tank volumes are highly recommended. The minimum recommended tank size is 1500 gal. Depending on the specific activities at a given facility, the accumulated sludge and scum may have to be removed as often as every 3 to 6 months.

Oil and Grease Removal at Wastewater Treatment Plants

Where extensive amounts of oil and grease are encountered in centralized wastewater treatment plants, combined aerated grit and scum removal systems, such as shown in Fig. 5-52, are used. In the combined facilities shown in Fig. 5-52, wastewater enters directly into a spiral flow circulation pattern controlled by the air supply. The spiral circulation helps to scour and wash the grit and to direct it to the grit

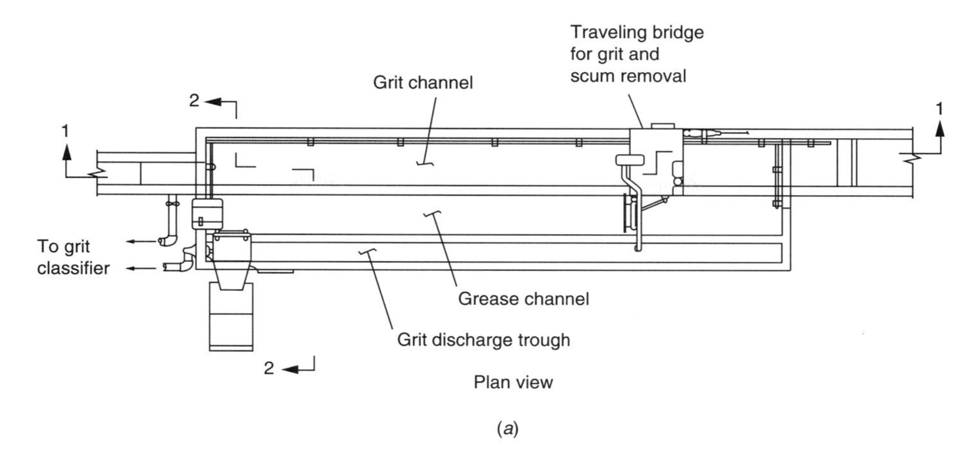

(*a*)

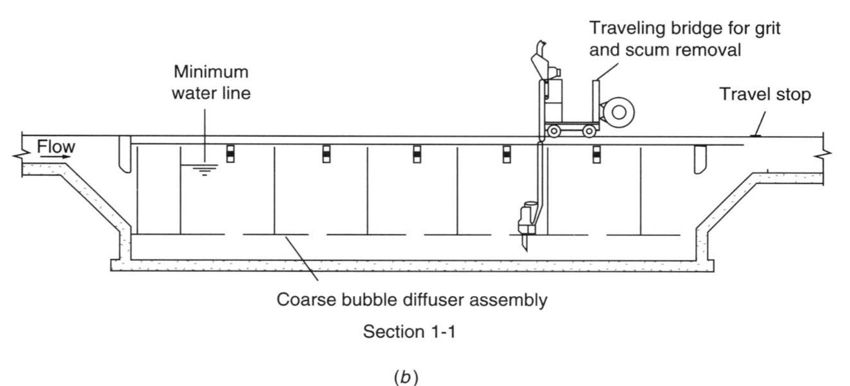

(*b*)

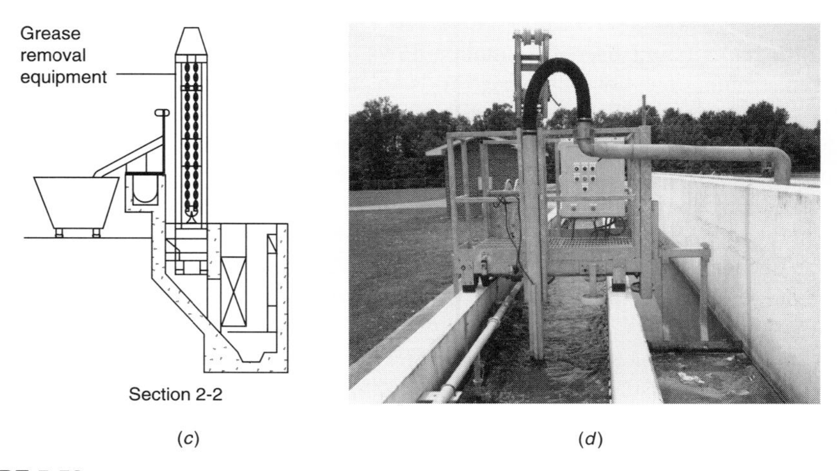

(*c*) (*d*)

FIGURE 5-52
Removal of oil and grease in an aerated grit tank: (*a*) plan view, (*b*) longitudinal section view, (*c*) cross section (courtesy Schreiber Corp.), and (*d*) view of typical unit.

hopper in the bottom of the tank. The degritted wastewater then flows through a baffle-controlled wall to a grease separation zone. Grease from the surface of the separation zone and grit from the hopper are removed by means of a traveling bridge equipped with a grit pump and surface skimmer. The size of the grit removed is controlled by adjusting the air supply. Maximum grit removal is typically assured with a minimum tank length of 50 ft (15.2 m).

5-12 IMHOFF TANKS

An Imhoff tank (see Fig. 5-53) consists of a two-story tank in which sedimentation is accomplished in the upper compartment and digestion of the settled solids is accomplished in the lower compartment. Imhoff tanks are used to receive wastes from residences and other facilities that are served with conventional gravity sewers or pressure sewers with grinder pumps. Imhoff tanks were used widely before the use of separate heated digestion tanks became common. Imhoff tanks are still used occasionally because they are simple to operate and do not require highly skilled supervision. There is no mechanical equipment to maintain, and operation consists of removing scum daily and discharging it into the nearest gas vent, reversing the flow of wastewater twice a month to even up the solids in the two ends of the digestion compartment, depending on the design, and drawing sludge periodically to the sludge-drying beds.

Description

Conventional unheated Imhoff tanks are either rectangular or circular, with the circular tanks being used for smaller flows. The removal of settleable solids and the anaerobic digestion of these solids in an Imhoff tank is similar to a septic tank. As shown in Fig. 5-53, solids pass through an opening in the bottom of the settling chamber into the unheated lower compartment for digestion. Scum accumulates in the sedimentation compartment and in the vents adjoining the sedimentation compartments. Gas produced in the digestion process in the lower compartment escapes through the vents. Because of the overhanging lip in the bottom of the sedimentation chamber, gases and gas-buoyed sludge particles rising from the sludge layer in the bottom of the tank are not released to the sedimentation compartment.

Over the years, several manufacturers have developed mechanized forms of the Imhoff tank usually consisting of a circular sedimentation tank mounted on top of a circular sludge-digestion tank, with several gas vents rising to the surface around the periphery of the unit. Digested sludge is also scraped mechanically to a central draw-off pipe. The mechanized tank can be equipped for scum collection at the surface of the tank and for scum stirring beneath the roof of the digestion compartment. Although effective, the mechanical simplicity of the unheated tank is lost.

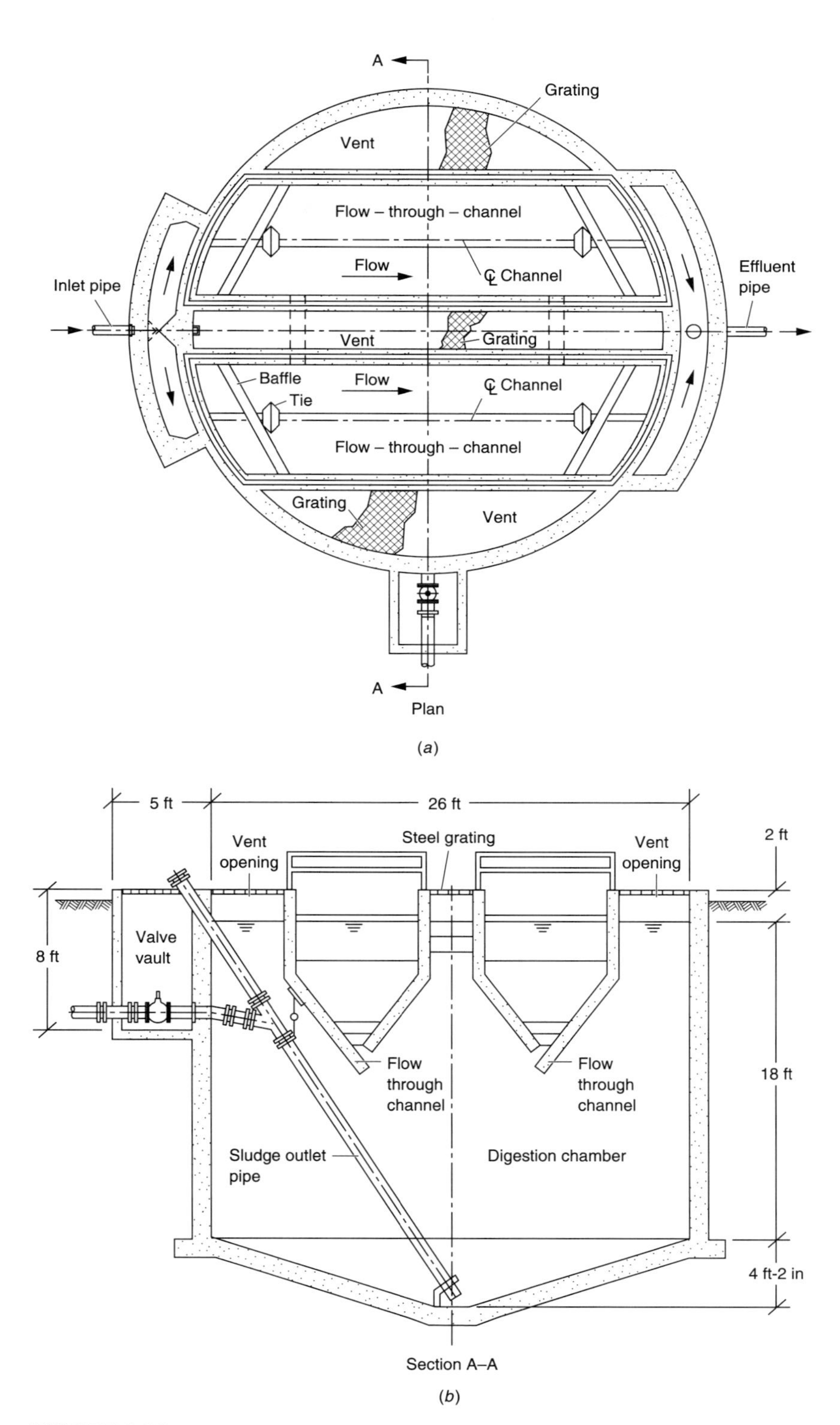

FIGURE 5-53
Typical circular Imhoff tank: (*a*) plan view and (*b*) section view.

TABLE 5-17
Typical design criteria for unheated Imhoff tanks

Design parameter	Unit	Value	
		Range	Typical
Settling compartment			
Overflow rate peak hour	gal/ft²·d	600–1000	800
Detention time	h	2–4	3
Length to width ratio		2:1:1–5:1	3:1
Slope of settling compartment	ratio	1.25:1–1.75:1	1.5:1
Slot opening	in	6–12	10
Slot overhang	in	6–12	10
Scum baffle			
Below surface	in	10–16	12
Above surface	in	12	12
Freeboard	in	18–24	24
Gas vent			
Area (percent of total area)	%	15–30	20
Width of gas vent opening*	in	18–30	24
Sludge digestion section			
Storage capacity (unheated)	month	4–8	6
Volume†	ft³/capita	2–3.5	2.5
Sludge withdrawal pipe	in	8–12	10
Depth below slot to top of sludge	ft	1–3	2
Total water depth (surface to tank bottom)	ft	24–32	30

*Minimum width of opening must be 18 in (450 mm) to allow a person to enter for cleaning.
†Based on a 6-month digestion period.

Design Considerations

Typical design criteria for Imhoff tanks are presented in Table 5-17. As reported in Table 5-17, the settling compartments of Imhoff tanks are customarily designed to have a surface overflow rating of 600 gal/ft^2 · d at the average rate of flow, and a detention period of about 3 h. The bottom of the settling compartment of the conventional unheated tank is usually sloped 1.4 vertical to 1.0 horizontal. The slot that permits solids to drop through to the digestion compartment has a minimum opening of 6 in (150 mm). The capacity of the unheated digestion compartment should provide for 6 months of sludge storage during the cold portion of the year. In most designs, sludge-drying beds are located adjacent to or nearby the Imhoff tank to minimize the problems associated with sludge handling.

5-13 OTHER SEPARATION PROCESSES

Other processes used for the separation of the particles found in wastewater include primary effluent filtration and effluent filtration.

Primary Effluent Filtration

With the development of new types of gravity filters, the filtration of primary effluent has now become practical. Primary effluent filtration is used following primary sedimentation. The filtered effluent can either be discharged to the ocean, treated further to remove additional BOD, or applied to rapid infiltration basins. The types of filters (see Chap. 12) that have proven successful include: (1) the pulsed-bed filter, (2) the continuous backwash filter, and (3) the Fuzzy filter. Additional details on primary effluent filtration can be found in Matsumoto et al. (1982) and England et al. (1994).

Effluent Filtration

Filtration is now used extensively for achieving supplemental removals of suspended solids (including particulate BOD) from wastewater effluents of biological and chemical treatment processes. Filtration is also used to remove chemically precipitated phosphorus. The use of rapid granular medium filters for effluent polishing following secondary treatment is gaining in popularity, especially in reuse applications. Effluent filtration is considered in detail in Chap. 12.

5-14 ODOR CONTROL

Odors can be generated from many sources, and their control is critical to successful wastewater management. For small and decentralized wastewater systems, odors are generally the result of anaerobic decomposition of organic matter or the reduction of sulfates to hydrogen sulfide gas. The principal sources of odors in small and DWM systems are (1) septic tank effluent containing hydrogen sulfide; (2) pumping stations, manholes, and cleanouts in collection systems; (3) headworks of treatment plants, including flow equalization, preaeration, screening, and grit removal; (4) septage receiving and handling; (5) organically overloaded biological treatment processes; (6) sludge thickening, conditioning, and dewatering; and (7) biosolids composting and land application. The sources of odor, general approaches to control odor in the gaseous form, and design considerations for bulk media biofilters are considered in this section.

Sources of Odors

The principal sources of odors and the relative potential for release of odor are presented in Table 5-18. Minimization of odors from these sources is the concern of the design engineer and the best management practices of the operator. Considerations for odor minimization are presented in this section and in the design sections for the collection and treatment components.

TABLE 5-18
Sources of odor in wastewater management systems*

Source	Odor potential
Onsite treatment	
Septic tank	Moderate
Imhoff tank	High
Collection system	
Air release valves	High
Cleanouts	High
Manholes	High
Pumping stations	High
Treatment plant	
Headworks	High
Screening	High
Flow equalization	High
Preaeration	High
Grit removal	High
Septage handling	High
Sidestream returns†	High
Primary clarifiers	High
Trickling filters	Moderate
Aeration	Low
Lagoons	Moderate
Secondary clarifiers	Low/moderate
Sludge handling	
Thickening/holding	High
Aerobic digestion	Moderate
Sludge storage basins	Moderate/high
Dewatering	High
Composting	High

*Adapted from WEF (1995).
†Sidestreams could include digester decant, dewatering flows, or backwash water.

Onsite components. Odor release from septic tanks, interceptor tanks, and holding tanks is generally minimized by sealing and burial of the tanks and by venting through plumbing with water traps. For Imhoff tanks or other pretreatment units that are onsite, a combination of venting and design for solids handling that does not expose the solids to the atmosphere is used.

Collection system components. The potential for odor release from collection systems is high. Odor minimization considerations are presented in Chap. 6 for four different types of collection of raw wastewater and septic tank effluent. Control of sulfide in wastewater collection systems is described more fully in ASCE (1989) and in U.S. EPA (1985).

Treatment plant components. The headworks and preliminary treatment operations have the highest potential for release of odor, especially for treatment plants that have long collection systems where anaerobic conditions can be created. Odor release can be minimized in treatment plants by not creating turbulence or headloss until the appropriate odor control method is used to manage the odor.

Sludge and septage handling. Until sludge and septage are stabilized organically, they will be a source of odor. The highest potential for odor release occurs when unstabilized sludge is turned, spread, or stored. Details on odor control for sludge, septage, and biosolids are presented in Chap. 14.

Odor Control Methods

The general classification of odor control methods is presented in Table 5-19, along with typical applications in wastewater management. Odor control methods are either designed to treat the odor-producing compounds in the wastewater stream, or to treat the foul air. The majority of the methods in Table 5-19 are meant to treat the foul air. To control the release of odorous gases from treatment facilities, it has become more common to cover the facilities as shown in Fig. 5-54.

As described in Chap. 2, the threshold odor number or dilutions to threshold (D/T) is the measure of odor concentration. The principal methods used to control odorous gases are chemical scrubbers, activated carbon, and bulk medium biofilters. Details on chemical scrubbers and activated carbon are provided in WEF (1995, 1997). Bulk medium filters are discussed below and their design is illustrated in Example 5-13.

Bulk medium filters. Soil, peat, compost, and similar bulk medium biofilters have been used to remove 90 to 99 percent of biodegradable odors (Bohn and Bohn, 1988). Soil and compost seem to be the most common media used. The requirements for the biofilter medium are (1) sufficient porosity and near-uniform particle size, (2) particles with large surface areas and significant pH-buffering capacities, and (3) the ability to support a large population of microflora (WEF, 1995).

Operation of biofilters. As odorous gases are passed upward through the bulk medium biofilter, two processes occur simultaneously: absorption/adsorption and bioconversion. Odorous gases are adsorbed on the surfaces of the biofilter medium particles and absorbed into the moist surface layer. Microorganisms, principally bacteria, actinomycetes, and fungi, attached to the filtering medium, oxidize the absorbed/adsorbed gases and renew the treatment capacity of the medium (Williams and Miller, 1992a).

Design criteria for biofilters. The design of bulk medium biofilters is typically based on a consideration of the gas residence time in the bed, the unit air loading rate, and the constituent elimination capacity which is defined as follows:

TABLE 5-19
Control methods for odorous gases in wastewater management

Classification	Methods	Applications and comments
Physical	Adsorption on activated carbon	Malodorous constituents are adsorbed onto activated carbon, which must be regenerated periodically.
	Containment*	Covers and hoods are used to contain and direct odorous gases to treatment facility.
	Dilution	Odors can be reduced by mixing with odor-free air. Discharge through tall stacks may be used to achieve atmospheric dispersion.
	Thermal oxidation	Combustion of off-gases at temperatures from 1200 to 1500 °F will eliminate odors.
Chemical	Caustic scrubbing	Odorous gases are passed through scrubbing tower to reduce odors.
	Recirculating liquid packed-bed scrubbers	Odorous gases are passed through liquid packed-bed scrubbers containing chemical oxidants such as sodium hypochlorite, chlorine solutions, hydrogen peroxide, and potassium permanganate.
Biological	Biological conversion	Biological processes in the wastewater can reduce odors by converting malodorous constituents through oxidation.
	Biological tower biofilters	Scrubbing towers, filled with bulk media and associated attached growth, are used to reduce odors.
	Compost filters*†	Gases can be passed through beds of compost to remove odors.
	Sand and soil filters*†	Gases can be passed through beds of sand or soil to remove odors.
	Trickling filters and activated-sludge aeration tanks	Malodorous gases can be pumped through trickling filters or aeration tank diffusers to reduce odors.

*Methods used most commonly for small treatment facilities.
†Methods used for odor control with alternative collection systems.

FIGURE 5-54
Covered primary sedimentation tanks to limit the release of odorous gases.

$$EC = \frac{Q(C_o - C_e)}{V} \tag{5-40}$$

where EC = constituent elimination rate, $g/m^3 \cdot s$
Q = volumetric flowrate, m^3/s
C_o = influent constituent concentration, g/m^3
C_e = effluent constituent concentration, g/m^3
V = volume of empty bed, m^3

Constituent elimination rates are determined experimentally and are usually reported as a function of the constituent loading rate (e.g., mg $H_2S/m^3 \cdot h$ for hydrogen sulfide). An essentially linear 1-to-1 constituent elimination rate has been reported by Yang and Allen (1994) for H_2S loading rates up to a maximum value of about 130 g $S/m^3 \cdot h$, beyond which the elimination rate becomes essentially constant at rate of 130 g $S/m^3 \cdot h$ with increased loading. It should be noted that H_2S is eliminated easily as it passes through a biofilter.

Typical design criteria for bulk medium (e.g., compost, sand) biofilters are presented in Table 5-20. Typical bulk medium odor control facilities are shown schematically in Fig. 5-55. During operation, biofilters tend to dry out unless moisture or humidity is added. Optimal physical characteristics of a filter material include a pH of between 6 and 8, airfilled pore space between 40 and 60 percent, and organic matter content of 35 to 55 percent (Williams and Miller, 1992a).

Some states regulate the design of compost biofilters. Massachusetts Department of Environmental Protection has a draft policy that limits application rates, specifies biofilter emission rates and odor sampling, and specifies setbacks from property lines. Under the policy, the odor emission limit at the surface of the biofilter is 50 dilutions to threshold. The maximum loading rate specified is 3 $ft^3/ft^2 \cdot min$ (Finn and Spencer, 1997).

TABLE 5-20
Design considerations for bulk medium biofilters*

Item	Units	Value
Oxygen concentration	parts oxygen/parts oxidizable gas	100
Moisture		
Compost filter	%	40–50
Soil filter	%	10–25
Temperature, optimum	°C	37
pH of medium	unitless	6–8
Gas residence time	s	30–60
Depth of medium	ft	3–5
Loading rate†	$ft^3/ft^2 \cdot min$	1.5–3
Back pressure, maximum	in of water	8

*Adapted from WEF (1995, 1997), Williams and Miller (1992b), and Finn and Spencer (1997).
† Depends on concentration of odorous gases.

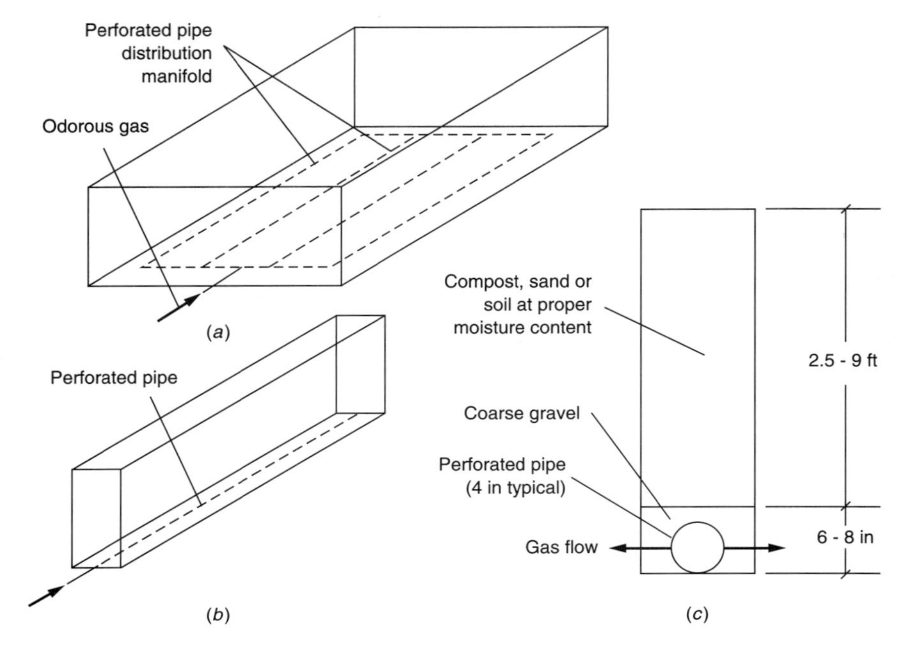

FIGURE 5-55
Typical configurations for bulk medium odor control facilities: (*a*) filter bed and (*b*) filter trench.

EXAMPLE 5-13. DESIGN OF ODOR CONTROL BIOFILTER. Determine the size of compost filter needed to scrub the air from an enclosed volume of 100 ft^3 using the design criteria given in Table 5-20. Assume 12 air changes per hour are needed. Assume a bed porosity of 40 percent. Will the volume selected be adequate if the air contains 10 ppm of H_2S in addition to other odorous constituents?

Solution

1. Estimate the air flow to be scrubbed:

$$\begin{aligned}\text{Flow} &= \text{volume/time} \\ &= 100\ \text{ft}^3 \times 12\ \text{changes per hour} = 1200\ \text{ft}^3/\text{h} \\ &= 1200\ \text{ft}^3/60\ \text{min/h} = 20\ \text{ft}^3/\text{min}\end{aligned}$$

2. Select a loading rate from Table 5-20; use 2.5 $ft^3/ft^2 \cdot min$.

3. Select a filter bed depth from Table 5-20; use 4 ft.

4. Calculate the area and volume needed for the filter bed:

$$\begin{aligned}\text{Area} &= \text{gas flow/ loading rate} \\ &= (20\ \text{ft}^3/\text{min})/(2.5\ \text{ft}^3/\text{ft}^2\cdot\text{min}) \\ &= 8\ \text{ft}^2 \\ \text{Volume} &= 8\ \text{ft}^2 \times 4\ \text{ft} = 32\ \text{ft}^3\end{aligned}$$

5. Check the air detention time:

$$\text{Detention time} = \text{volume/flowrate}$$
$$\text{Volume of air} = 8\ \text{ft}^2 \times 4\ \text{ft} \times 0.40 = 12.8\ \text{ft}^3$$
$$\text{Detention time} = 12.8\ \text{ft}^3/(20\ \text{ft}^3/\text{min})$$
$$= 0.64\ \text{min} = 38\ \text{s (OK since it's} > 30\ \text{s)}$$

6. Determine whether the volume of the biofilter determined in step 5 is adequate to treat the H_2S.
   a. Determine the concentration of H_2S in g/m^3 as outlined in Example 2-1:

$$\text{g/m}^3 = \left(\frac{10\ \text{L}^3}{10^6\ \text{L}^3}\right)\left(\frac{34.08\ \text{g/mole H}_2\text{S}}{22.4 \times 10^{-3}\ \text{m}^3/\text{mole of H}_2\text{S}}\right)$$
$$= 0.0152$$

   b. Determine the mass loading rate of S in g S/h

$$M_S = \left(\frac{1200\ \text{ft}^3}{\text{h}}\right)\left(\frac{28.3\ \text{L}}{\text{ft}^3}\right)\left(\frac{\text{m}^3}{10^3\ \text{L}}\right)\left(\frac{0.0152\ \text{g}}{\text{m}^3}\right)\left(\frac{32}{34.08}\right)$$
$$= 0.48\ \text{g S/h}$$

   c. Determine the required volume, assuming an elimination rate of 65 g S/m^3 · h, which incorporates a factor of safety of 2:

$$V = \frac{(0.48\ \text{g S/h})}{(65\ \text{gS/m}^3\text{h})} = 0.0074\ \text{m}^3\ (0.26\ \text{ft}^3)$$

   Because the volume of the bed (32 ft^3) is significantly greater, H_2S will not be an issue.

Comment. From the results of the computation carried out in step 6, it is clear why compost and soil filters are so effective in the elimination of H_2S.

PROBLEMS AND DISCUSSION TOPICS

5-1. An air flocculation system is to be designed. If a G value of 60 s^{-1} is to be used, estimate the air flowrate that will be necessary for a 6200 ft^3 flocculation chamber. Assume the depth of the flocculation basin is to be 12 ft.

5-2. Determine the settling velocity in feet per second of a sand particle in a flow regime with a Reynolds number equal to 275. Assume that the specific gravity of the sand is equal to 2.65 and the diameter is 0.04 in.

5-3. The following data have been obtained experimentally for a distribution of discrete particles in a wastewater.

Percent of initial conc. settling at rate $<v$	v, ft/s $\times 10^{-3}$
0	0
5	3.94
20	7.87
40	11.02
60	15.75
80	25.59
90	35.43
95	47.72

a. Estimate the removal in the settling zone of an ideal basin after 1 hour of settling of the suspension, if the depth of the basin is 10 ft.

b. What would be the concentration of suspended matter at a depth of 5 ft?

5-4. Determine the removal efficiency for a sedimentation basin with a critical velocity v_o of 6.5 ft/h in treating a wastewater containing particles whose settling velocities are distributed as given in the table below. Plot the particle histogram for the influent and effluent wastewater.

Velocity, ft/h	Number of particles
0.0–1.5	20
1.5–3.0	40
3.0–4.5	80
4.5–6.0	120
6.0–7.5	100
7.5–9.0	70
9.0–10.5	20
10.5–12.0	10

5-5. Demonstrate for an "ideal" grit basin receiving a waste containing discrete particles of uniform density that obey Newton's law, that the diameter of particles that are 100 percent removed is a function of $(Q/A)^2$.

5-6. The rate of flow through an ideal clarifier is 2.0 Mgal/d, the detention time is 1 h, and the depth is 10 ft. If a full-length movable horizontal tray is set 3 ft below the surface of the water, determine the percent removal of particles having a settling velocity of 3 ft/h. Could the removal efficiency of the clarifier be improved by moving the tray? If so, where should the tray be located and what would be the maximum removal efficiency? What effect would moving the tray have if the particle settling velocity were equal to 1 ft/h?

5-7. For a flocculant suspension, determine the removal efficiency for a basin 10 ft deep with an overflow rate v_o equal to 10 ft/h using the laboratory settling data presented in the following table.

Time, min	Percent total suspended solids removed at indicated depth (in ft)				
	1.5	3.0	4.5	6.0	7.5
20	61				
30	71	63	55		
40	81	72	63	61	57
50	90	81	73	67	63
60		90	80	74	68
70			86	80	75
80				86	81

5-8. Using a minimum of two references from the literature, contrast dissolved air flotation with sedimentation, on the basis of the following parameters. List all references.

a. Detention time

b. Surface loading rate

c. Power input

d. Efficiency

e. Most favorable application for each type

5-9. Demonstrate that the proportional weir (see Fig. 5-26*b*) can be used to control the velocity through a grit chamber, given that $l_1 \sqrt{h_1}$ for a proportional weir is constant.

5-10. If the average flow-through velocity in a horizontal grit chamber is 0.5, 0.75, 1.0, or 1.25 ft/s (to be selected by instructor) and the length of the grit chamber is 40, 50, 60 ft (to be selected by instructor), determine the size of grit particle that can be removed assuming the specific gravity of the grit is equal to 2.65, the depth of flow is 2.0, 2.5, 3.0, or 3.5 ft (to be selected by instructor), and the temperature of the wastewater is 18°C.

5-11. Given the following performance data, obtained from field measurements, for a sedimentation basin used to treat a wastewater having an initial suspended solids concentration of 250 mg/L, determine: (*a*) the constants in the following TSS removal efficiency versus time relationship:

$$R = \frac{t}{a + bt}$$

where R = percent TSS removal, t = time in hours, and a and b are constants, and (*b*) the residual TSS that would be expected for a detention period of 210 minutes. Use the constants from part *a*.

Settling time, h	0	0.5	1.0	1.5	2.0	2.5	3.0
TSS removal, %	0	37.5	48	52	59	60	62

5-12 A hydraulic study of the flow-through characteristics of three sedimentation basins was made using lithium chloride (LiCl) as a tracer by injecting a slug of the tracer at the inlet and measuring the concentration at the outlet. The results of this study are presented in the following table.

Time, min	Concentration of LiCl at outlet, mg/L		
	Basin 1	Basin 2	Basin 3
0	0	0	0
5	Trace	Trace	0
10	5	2	0
15	30	6	0
20	46	14	Trace
25	57	25	3
30	61	36	8
40	54	57	30
50	39	53	55
60	24	36	59
70	12	19	36
80	6	8	10
90	2	2	Trace
100	Trace	Trace	0

a. Plot a curve of the ratio of concentration of salt at the outlet to the inlet concentration (C/C_o, as the ordinate) against the ratio of the actual time over the theoretical detention time (t/t_o) for sedimentation basin 1, 2, or 3 (to be selected by instructor). Assume $C_o = 70$ mg/L and $t_o = 40$, 50, and 56 min for the three basins, respectively.

b. Calculate the t/t_o ratios for the mean, median, mode, and minimum times.

c. From part *b*, what can you say about the sedimentation basin with respect to short circuiting and dead spaces?

d. If a sedimentation basin has marked short circuiting and/or dead spaces, does this necessarily mean that it will be less efficient in removing particles than one without short circuiting?

e. Determine the Morrill dispersion index.

f. Determine the volumetric efficiency of the basin as defined by Morrill.

5-13. Using the data from Prob. 5-12, prepare a cumulative plot of the tracer leaving basin 1, 2, or 3 (to be selected by instructor). Using a cascade of complete-mix reactors, determine the number of reactors in series that are needed to model the cumulative dye tracer curve.

5-14. Design a circular radial-type sedimentation tank for a community with an expected population of 15,000 at the end of the period of design. Assume the minimum flow is 30 percent of the daily average flow. Mechanical sludge collection and skimming are to be utilized.

a. Select the tank dimensions (depth, diameter, and weir and scum baffle setting) to provide an overflow rate of 800 gal/ft^2·d at peak flow. Assuming standard tank dimensions to fit mechanisms which are made in diameters of whole feet, and in depths of half-feet from 7.5 to 12 ft SWD (side water depth), calculate the actual surface loading of a practical tank.

b. Determine the overflow rates at average and minimum flows.

c. Calculate detention for maximum and minimum design flows.

d. Determine the weir loading for each design flow.

e. Estimate the amount of dry solids and total volume of 96 percent sludge removed on an average day, given that the removal of TSS can be described with the following empirical expression.

$$R = \frac{100}{1.2 + 6.7 \times 10^{-4}\ \mathrm{SOR}}$$

where R = expected TSS removal, %

SOR = surface overflow rate, gal/ft^2·d

5-15. Prepare a table and compare the data from a minimum of six references with regard to the following primary sedimentation tank design parameters: (1) detention time (with and without preaeration), (2) expected BOD removal, (3) expected suspended solids removal, (4) mean horizontal velocity, (5) surface loading, gal/ft^2·d, (6) effluent weir overflow rate per unit length, (7) Froude number, (8) size of organic particle removed, (9) length-to-width ratio (rectangular tanks), (10) average depth. List all references.

5-16. Why is it important to have watertight septic tanks?

5-17. What impact will the use of kitchen food-waste grinders have on the accumulation of solids in a septic tank? Can you offer a practical example from another field to substantiate your argument?

5-18. What size of septic tank would you recommend for a cluster system serving six homes? Two of the homes have two bedrooms, two have three bedrooms, and two have five bedrooms. Would you recommend any limitations on the use of the cluster septic tank?

5-19. What type of experiment would you design to demonstrate the short-term and long-term benefits of septic tank additives?

5-20. What effect does aspect ratio, i.e., length-to-width ratio, have on the performance of a large community-type septic tank?

5-21. Can you suggest any changes, in addition to some of the changes proposed in the text with respect to the design of septic tanks, to improve their design and operation?

5-22. Estimate the solids accumulation in a septic tank for a small community of 1000 people using the information given in the chapter and appropriate assumptions. List all of the assumptions used in your analysis.

5-23. Determine the solidification temperature for three common cooking oils.

5-24. List the principal reasons why Imhoff tanks fell out of favor as wastewater pretreatment units. Were the reasons based mostly on science or public perception?

REFERENCES

ASCE (1989) Sulfide in Wastewater Collection and Treatment Systems, Manual of Practice No. 69, American Society of Civil Engineers, New York.

Barnes, D., P. J. Bliss, B. W. Gould, and H. R. Vallentine (1981) *Water and Wastewater Engineering Systems,* Pitman Publishing, Marshfield, MA.

Bohn, H. L., and R. K. Bohn (1988) Soil Beds Weed Out Air Pollutants, *Chemical Engineering,* Vol. 95, No. 6, pp. 73–76.

Bounds, T. R. (1995) Septic Tank Septage Pumping Intervals, in R. W. Seabloom (ed.) Proceedings 8th Northwest On-Site Wastewater Treatment Short Course and Equipment Exhibition, University of Washington, Seattle.

Camp, T. R. (1942) Grit Chamber Design, *Sewage Works Journal*, Vol. 14, p. 368.

Camp, T. R. (1946) Sedimentation and the Design of Settling Tanks, *Transactions American Society of Civil Engineers,* Vol. 111.

Camp, T. R., and P. C. Stein (1943) Velocity Gradients and Internal Work in Fluid Motion, *Journal Boston Society of Civil Engineers,* Vol. 30, p. 209.

Coe, H. S. and G. H. Clevenger (1916) Determining Thickener Unit Areas, *Transactions of American Institute of Mining Engineers,* Vol. 55, No. 3.

Dialynas, G. (1997) Personal communication, Iraklio, Crete.

Dick, R. I., and B. B. Ewing (1967) Evaluation of Activated Sludge Thickening Theories, *Journal of Sanitation Engineering Division.,* ASCE, Vol. 93, No. SA-4.

Dick, R. I., and K. W. Young (1972) Analysis of Thickening Performance of Final Settling Tanks, Proceedings of the 27th Industrial Waste Conference, Purdue University, Eng. Ext. Series 141, Purdue, IN.

Dunbar, Professor Dr. (1908) *Principles of Sewage Treatment,* Charles Griffen, London.

England, S. K., J. L. Darby, and G. Tchobanoglous (1994) Continuous-Backwash Upflow Filtration for Primary Effluent, *Water Environment Research,* Vol. 66, No. 2, pp. 145–152.

Fair, G. M., and J. C. Geyer (1954) *Water Supply and Waste-Water Disposal,* Wiley, New York.

Finn, L., and R. Spencer (1997) Managing Biofilters for Consistent Odor and VOC Treatment, *BioCycle*, Vol. 38, No. 1.

Greeley, S. A. (1938) Sedimentation and Digestion in the United States, in L. Pearse (ed.), *Modern Sewer Disposal: Anniversary Book of the Federation of Sewage Works Associations,* Lancaster Press, New York.

Holland, F. A., and Chapman, F. S. (1966) *Liquid Mixing and Processing in Stirred Tanks,* Reinhold, London.

Keinath, T. M. (1989) Operational Dynamics and Control of Secondary Clarifiers, *Journal of Water Pollution Control Federation,* Vol. 57, No. 7, p. 770.

Lumsden, L. L., C. W. Stiles, and A. W. Freeman (1915) Safe Disposal of Human Excreta at Unsewered Homes, Public Health Bulletin No. 68, U.S. Public Health Service, Government Printing Office, Washington, DC.

Matsumoto, M. R., T. M. Galeziiewski, G. Tchobanoglous, and D. S. Ross (1982) Filtration of Primary Effluent, *Journal of Water Pollution Control Federation,* Vol. 54, No. 12, pp. 1581–1591.

Metcalf, L. and H. P. Eddy (1930) *Sewerage and Sewage Disposal: A Textbook*, 2nd ed., McGraw-Hill, New York.

Morrill, A. B. (1932) Sedimentation Basin Research and Design, *Journal of American Water Works Association,* Vol. 24, p. 1442.

Reynolds, T. D., and P. A. Richards (1996) *Unit Operations and Processes in Environmental Engineering,* PWS Publishing Company, Boston.

Rushton, J. H. (1952) Mixing of Liquids in Chemical Processing, *Industrial Engineering Chemistry,* Vol. 44, No. 12.

Seabloom, R. W., D. A. Carlson, and J. Engeset (1982) Septic Tank Performance Compartmentation, Efficiency and Stressing, Proceedings Fourth Northwest Onsite Wastewater Short Course, University of Washington, Seattle.

Shields, A. (1936) Application of Similitude Mechanics and Turbulence Research to Bed-Load Movement, *Mitt. der Preuss. Versuchsanstalt für Wasserbau und Schiffbau,* No. 26, Berlin.

Talmadge, W. P., and E. B. Fitch (1955) Determining Thickener Unit Areas, *Industrial Engineering Chemistry,* Vol. 47, No. 1.

Tchobanoglous, G., and E. D. Schroeder (1985) *Water Quality: Characteristics, Modeling, Modification,* Addison-Wesley, Reading, MA.

Tchobanoglous, G., and F. L. Burton (1991) *Wastewater Engineering. Treatment, Disposal, Reuse,* 3rd ed., McGraw-Hill, New York.

U.S. EPA (1985) *Odor Control and Corrosion Control in Sanitary Sewerage Systems and Treatment Plants,* Design Manual, EPA-625/1-85-018, U.S. Environmental Protection Agency, Washington, DC.

U.S. EPA (1987) Design and Operation Considerations—Preliminary Treatment, EPA-430/09-87-007, U.S. Environmental Protection Agency, Washington, DC.

Wahlberg, E. J., and T. M. Keinath (1988) Development of Settling Flux Curves Using SVI, *Journal of Water Pollution Control Federation,* Vol. 60, p. 2095.

WEF (1992) *Design of Municipal Wastewater Treatment Plants,* Vol. I: Chaps. 1–12, WEF Manual of Practice No. 8, ASCE Manual and Report on Engineering Practice No. 76, Water Environment Federation, Alexandria, VA.

WEF (1994) *Preliminary Treatment for Wastewater Facilities,* WEF Manual of Practice OM-2, Water Environment Federation, Alexandria, VA.

WEF (1995) *Odor Control in Wastewater Treatment Plants*, WEF Manual of Practice No. 22, ASCE Manuals and Reports on Engineering Practice No. 82, Water Environment Federation, Alexandria, VA.

WEF (1997) Septage Handling, Manual of Practice No. 24, Water Environment Federation, Alexandria, VA.

WPCF (1985) *Clarifier Design*, WPCF Manual of Practice FD-10, Water Pollution Control Federation, Alexandria, VA.

Williams, T. O., and F. C. Miller (1992a) Odor Control Using Biofilters, *BioCycle*, Vol. 33, No. 10.

Williams, T. O., and F. C. Miller (1992b) Biofilters and Facilities Operations, *BioCycle*, Vol. 33, No. 11.

Winneberger, J. H. T. (1984) *Septic Tank Systems: A Consultant's Toolkit,* Butterworth Publishers, Boston, MA.

Yang, Y., and E. R. Allen (1994) Biofiltration Control of Hydrogen Sulfide I: Design and Operational Parameters, *Journal Air and Waste Management Association,* Vol. 44, pp. 863–868.

Yoshika, N., et al. (1957) Continuous Thickening of Homogeneous Flocculated Slurries, *Kagaku Kogaku,* Vol. 26, 1957 (also in *Chemical Engineering,* Vol. 21, Tokyo, 1957).

CHAPTER 6

Alternative Wastewater Collection Systems

The types of systems used for the collection of wastewater range from conventional gravity sewers to pressure sewers and vacuum sewers. In the 1960s and 1970s, when many rural communities examined collection systems for unsewered areas, the cost of conventional gravity sewers was found to be large (up to 4 times more) compared to the cost of treatment and disposal (U.S. EPA, 1991). To avoid these high costs, alternative sewer systems have been developed. The purpose of this chapter is to introduce the alternative types of collection systems that have been developed, to discuss their application, and to illustrate the design procedures for each. Following a general introduction to wastewater collection and a consideration of general design issues applicable to each, the alternative types of collection systems are considered separately.

6-1 WASTEWATER COLLECTION SYSTEMS

Wastewater collection systems typically convey wastewater from the area of generation to the point where the wastewater will receive treatment. A variety of collection system materials and design methodologies are now available. Selection of the appropriate sewer system will depend on the unique properties and characteristics of the community to be served.

Types of Wastewater Collection Systems

The various types of alternative sewers are described with respect to their basic components in this section. Definition sketches of the types of alternative collection systems are presented in Fig. 6-1. Conventional gravity sewers will also be described in this section, because they may be considered the standard to which "alternative" systems are compared.

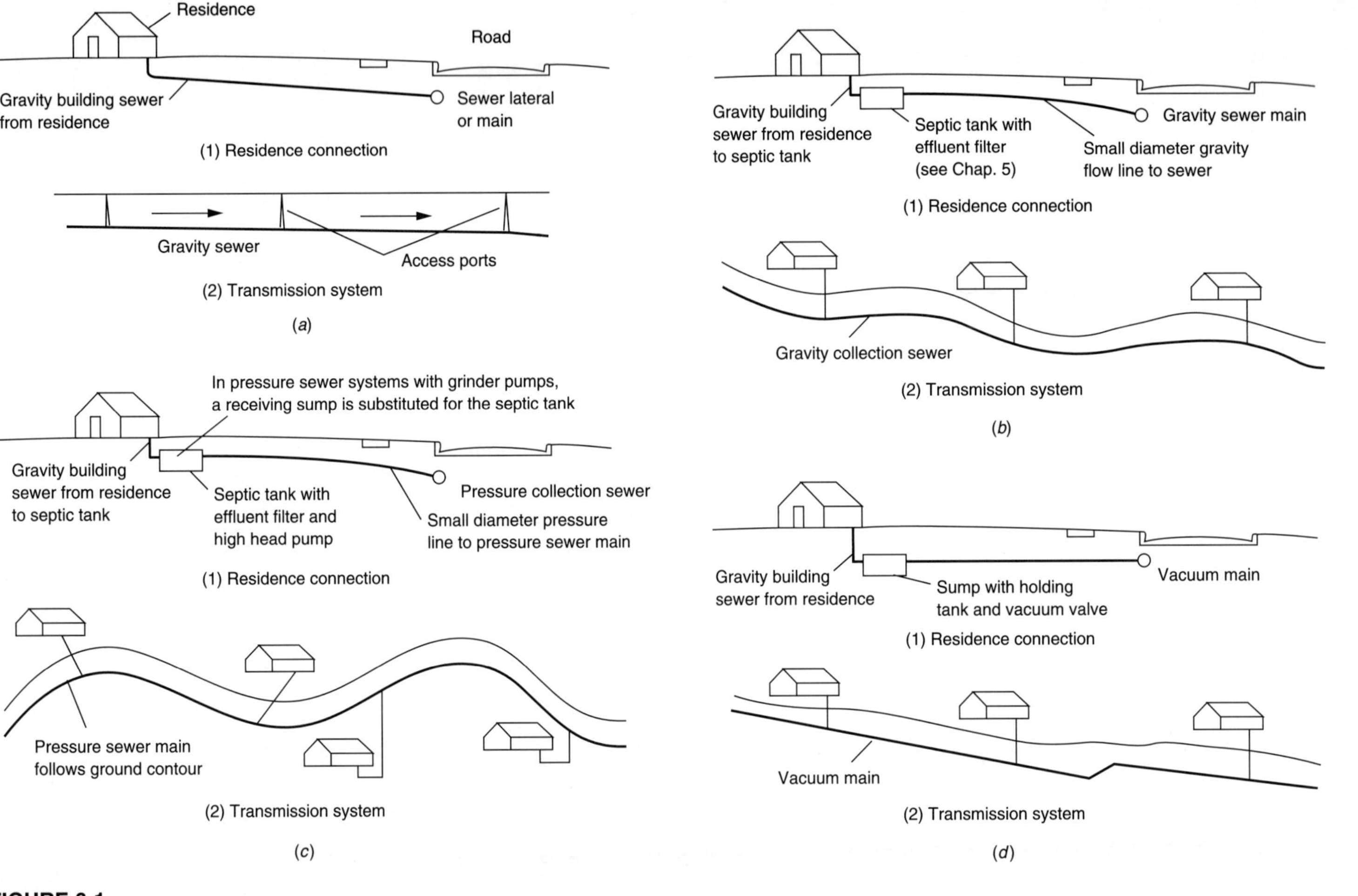

FIGURE 6-1
Definition sketch for sewer systems: (*a*) conventional gravity, (*b*) septic tank effluent gravity (STEG), (*c*) septic tank effluent pump (STEP) and pressure sewer with grinder pumps, and (*d*) vacuum sewer.

Conventional gravity sewers. The use of gravity sewers (see Fig. 6-1*a*), based on the empirical observation that water flows downhill, dates back to Minoan times (circa 3500 B.C.) (Angelakis and Spyridakis, 1996). Where gravity sewers are used, building sewers are connected directly without any pretreatment. The minimum diameter of conventional gravity sewers is usually 6 to 8 in (150 to 200 mm) to allow cleaning for accumulations of grease and solids. Because of the presence of solids, a constant minimum slope is required to keep velocities at or above 2 ft/s (0.6 m/s) to avoid the deposition of solids. Gravity sewers are connected to access ports (historically known as *manholes,* since the entry to the access port was sized so that a man could enter for the purpose of maintenance) at each change of grade or alignment. Access port spacings vary from 300 to 500 ft (90 to 150 m), depending on available sewer cleaning equipment and maintenance methods. One of the major problems with conventional gravity sewers is the infiltration (flow into) of extraneous flow during periods of high ground water, and the exfiltration (flow out) during dry weather periods. A comparison of conventional gravity sewers to alternative sewers is presented in Table 6-1. Details of gravity sewer design are available in standard references (Tchobanoglous, 1981; ASCE, 1992).

Septic tank effluent gravity sewers. In the septic tank effluent gravity (STEG) sewer, a small-diameter [1 to 2 in (25 to 50 mm)] plastic pipe is used to convey the effluent from a septic tank, equipped with an effluent filter, to a small-diameter collection system (see Fig 6-1*b*). Because there are no solids to settle in the

TABLE 6-1
Comparison of conventional gravity sewers to pressure sewers with septic tanks

Issue	Conventional gravity	Pressure sewers
Infiltration and inflow	Usually encountered	Avoided
Minimum velocities	Required to avoid solids deposition	Not required
Minimum diameter	6–8 in (150–200 mm)	2 in (50 mm)
Downhill slopes	Must be maintained at all times	Not required, follow the topography
Cleaning access to main lines	Access ports regularly spaced	Cleanouts and pigging ports
Trench depth	Minimum depth to 20–30 ft (6–9 m) depending on the slope of the sewer	Maintain minimum depth as with water transmission lines
Pump stations	Needed for low areas where downhill slopes cannot be maintained	Built in to each service or cluster of services
Conflicts with other buried utilities	May require redesign to avoid conflicts	Easily avoided
Ease of construction	Deep and wide trenches go in relatively slowly with traffic disruption	Narrow, shallow trenches go in relatively quickly with minimal traffic disruption

collection system, the collection system can be laid at a variable grade, just below the ground surface [e.g., 3 ft (0.9 m)]. As a result, STEG systems are also known as *small-diameter variable-grade gravity sewers*. Because the collection main is watertight, there is no infiltration in the system. Small-diameter gravity sewers have been used since 1961 in Australia, where they are referred to as effluent drains. The first STEG system in the United States was constructed in 1977 in Westboro, Wisconsin (WPCF, 1986). To take advantage of topography, many systems are constructed with a combination of STEG and septic tank effluent pump sewers as discussed in the following.

Septic tank effluent pump sewers. In the modern septic tank effluent pump (STEP) system, a high-head turbine pump is used to pump screened septic tank effluent into a pressurized collection system (see Fig. 6-1*c*). The size of discharge line leading from the septic tank is typically $1–1\frac{1}{2}$ in (25–38 mm). The minimum pipe size used for the pressurized collection main is typically 2-in (50-mm) -diameter plastic pipe. As with the STEG system, infiltration is not an issue because the collection main is watertight. STEP sewers usually are placed just below the frost penetration depth. Because the lines are under pressure they can follow the terrain, as a water transmission line does. Because of the shallow burial depth, construction problems resulting from high groundwater and rocky soil can be avoided. It is interesting to note that the idea for a pressure sewer (a sewer within a sewer) was first proposed in the late 1960s by Fair (1968) as a solution for the problem of combined sewer overflows.

Pressure sewers with grinder pumps. In pressure sewer systems with grinder pumps, a septic tank is not used. In its place, the discharge pump, located in a small pump basin, is equipped with chopper blades that cut up the solids in the wastewater so that they can be transported under pressure in a small diameter pipeline (see Fig. 6-1*c*). As a consequence, higher solids and oil and grease concentrations are encountered. As with the STEP system, infiltration is not an issue because the collection main is watertight. The depth of burial for pressure sewers with grinder pumps is similar to depths of STEP sewers.

Vacuum sewers. In vacuum sewers, a central vacuum source is used to maintain a 15- to 20-in (380- to 500-mm) vacuum of mercury on small-diameter collection mains to transport the wastewater from individual homes to a central location (see Fig. 6-1*d*). As with the other alternative collection systems, infiltration is not an issue because the collection vacuum main is watertight. Vacuum sewers were developed in the nineteenth century, with the U.S. patent dating back to 1888 (WPCF, 1986). Currently, two vacuum sewer companies are operating in the United States.

Use of Alternative Wastewater Collection Systems

The alternative collection systems have a number of common features, such as use of lightweight plastic pipe buried at shallow depths. All have suffered from some

TABLE 6-2
Relative characteristics of alternative sewer systems*

Sewer type or combination	Ideal topography	Construction cost in rocky, high-groundwater sites	Sulfide potential	Minimum slope or velocity required
Conventional gravity	Downhill	High	Moderate	Yes
STEP	Uphill, undulating	Low	High	No
STEG	Downhill	Moderate	High	No
Grinder pump (GP)	Uphill	Low	Mod.–high	Yes
Vacuum	Flat	Low	Low	Yes
STEG-STEP	Undulating	Low–mod.	High	No
Conventional-GP	Undulating	Mod.–high	Moderate	Yes
Conventional-vacuum	Undulating	Mod.–high	Low–mod.	Yes

*Adapted from WPCF (1986).

misuse and misapplication in early installations, as have most developing technologies. A comparison of the relative characteristics of different alternative sewer types and combinations is presented in Table 6-2.

Topography. As indicated in Table 6-2, the highest cost for conventional gravity sewers is where undulating terrain, high groundwater, or rocky conditions exist. The use of STEP, variable-grade STEG, or a combination of the two, in undulating terrain can be cost-effective, compared to conventional gravity. Vacuum sewers are suited to flat terrain, such as around a lake, or in marinas and harbors.

Population density and growth. Where existing population density is low but appreciable growth is anticipated, there should be consideration for initial conditions of the system operation compared to ultimate flowrates. STEP and STEG systems, because of their relative freedom from minimum velocity requirements, can handle a wide divergence between initial and ultimate design populations. In addition, STEP systems can be programmed to alternate doses into the collection system when the system approaches capacity. Grinder pump systems require minimum scouring velocities to be reached daily. Therefore, a low ratio of initial-to-final design population will require special facilities for flushing the mains.

Satellite wastewater management. The treatment of wastewater in satellite facilities is undertaken for a variety of reasons including: (1) economics, (2) a conscious decision to reuse water locally, or (3) a decision to avoid expanding a centralized treatment system. As communities continue to develop and expand, the distances from the new developments to existing wastewater treatment facilities become so great that connecting to the existing wastewater treatment facilities is no longer economically feasible. If onsite disposal is not possible, alternative collection systems can be used for the collection and transport of wastewater from individual residences and commercial and institutional developments to a nearby site for treatment and reuse. To allow for the local reuse of treated wastewater, a STEP/STEG system was used at Stonehurst, a small residential development in Contra Costa

County, California, even though the developer could have connected to a nearby sewer (Crites et al., 1997 and Chap. 12). In some communities, outlying developments have been provided with city water service but not with wastewater collection service. When municipal wastewater collection service is not provided, the wastewater must be collected, treated, and reused locally. The City of Austin, Texas, has decided to take this approach to peripheral development (CES, 1996). Decentralized wastewater management is an ideal application for pressure sewers, especially in new developments.

Sustainable development. In the proposed development shown in Fig. 6-2, each home, served with city water, would be provided with a septic tank with an effluent pump and a textile filter for wastewater treatment (see Sec. 11-5, Chap. 11). The treated effluent from each home is discharged by gravity through a small-diameter line [1 in (25 mm)] to the wooded green areas between the houses where it would be given additional treatment in a subsurface wetland (see Chap. 9). The subsurface wetland replaces the pressure discharge mainline used in pressure sewer wastewater collection systems. The flow from each wooded green area is discharged to a central subsurface wetland. Effluent from the central wetland is collected and disinfected with ultraviolet light (UV) before being discharged to a storage

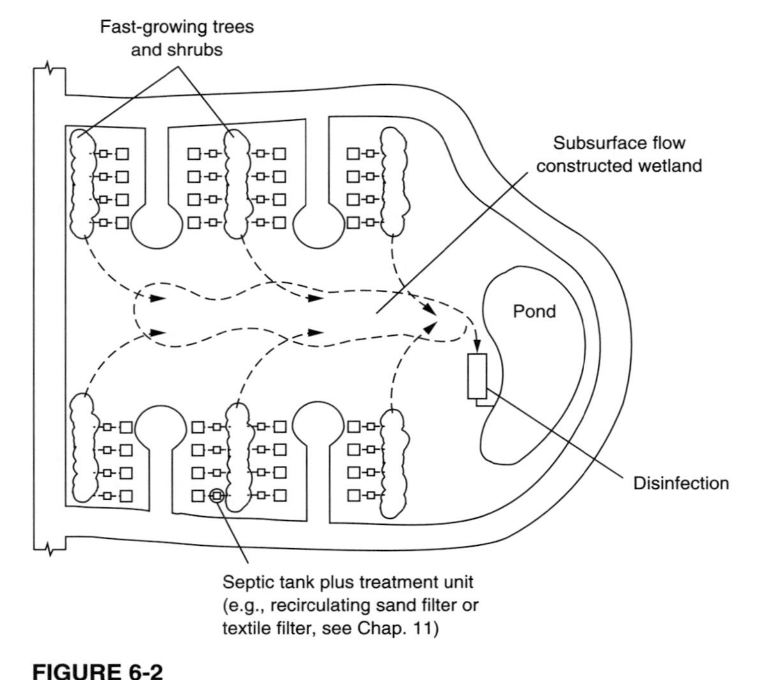

FIGURE 6-2
Application of a modified pressure sewer for the development of a new residential development (adapted from Van der Ryn Architects).

reservoir. In effect, the flow from each individual residence corresponds to the headwaters of a small stream which contribute to the formation of a larger stream, a copy of what happens in nature.

6-2 DESIGN CONSIDERATIONS FOR ALTERNATIVE WASTEWATER COLLECTION SYSTEMS

The design of a wastewater collection system requires systematic consideration of the community to be served. Besides evaluating site characteristics (e.g., topography, depth of soil, depth to water table, depth of freezing zone), the design engineer must quantify the amount and timing of wastewater that will be generated. Daily and seasonal flowrate fluctuations must be considered, as well as the potential for population growth. Once these data are collected, the engineer can use hydraulic principles to plan a collection system that will meet the needs of the community being considered. The subjects of design flowrates and collection system hydraulics are considered in this section.

Design Flowrates

Collection systems are designed on the basis of peak flowrates, which can be determined by considering the number of equivalent residences that are to be connected to the system.

Equivalent dwelling unit. As used in the design of alternative collection systems, an equivalent dwelling unit (EDU) is defined as a residence with a given number of residents. Thus, if an EDU were defined as a residence with 3.5 persons on average, a residence with 7 persons would correspond to two EDUs. An EDU may also represent the average household flowrate (gal/house·d) in a community.

Design peak flowrate. The design peak flowrate (DPF) is defined as that flowrate that would be expected in the collection system, assuming a given number of EDUs were discharging at the same time. DPF values are obtained from measurements made on existing alternative collection systems. The DPF per EDU is determined by dividing the measured peak flowrate by the total number of EDUs connected to the system. Typical values for systems with more than 50 EDUs range from 0.35 to 0.5 gal/min·EDU (1.3 to 1.9 L/min·EDU). The total DPF is given by

$$Q_{DP} = 0.5N \tag{6-1}$$

where Q_{DP} = design peak flowrate, gal/min
0.5 = measured discharge per EDU, gal/min·EDU
N = number of EDUs

For small numbers of EDUs, the use of Eq. (6-1) will underestimate the peak flowrate slightly. In some formulas that have been used, a constant flowrate term (10 to 20 gal/min) is added to Eq. (6-1) to account for the flowrate from one or more EDUs

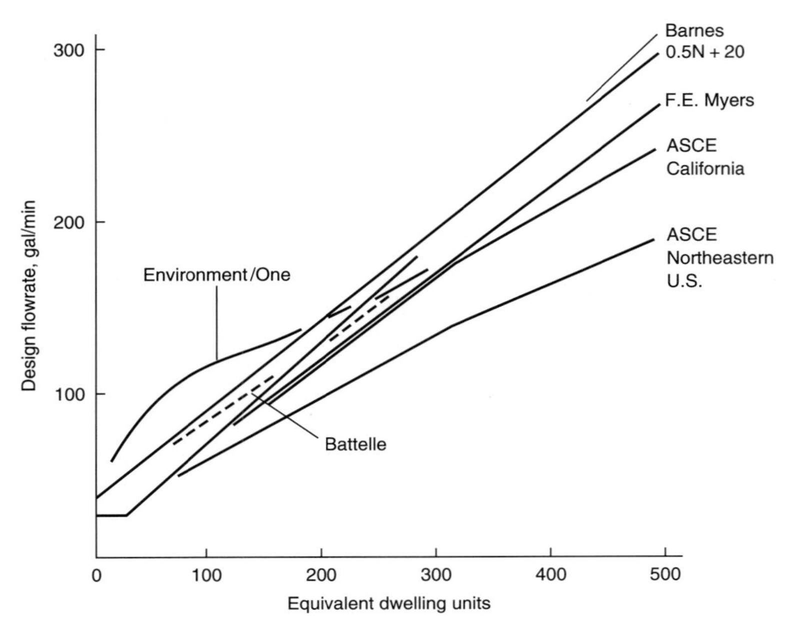

FIGURE 6-3
Graphical relationships between design peak flow and number of EDUs (from U.S. EPA, 1991).

connected to the collection system at its extremities. The 10 to 20 gal/min (38 to 76 L/min) values are the minimum flows for one STEP pump in operation. Where flow control orifices are used, the minimum flowrate drops to 8 to 10 gal/min (30 to 38 L/min). However, because the minimum pipeline size used is typically 2 in (50 mm), sufficient capacity is usually available in these cases. If other pipe sizes are used, Eq. (6-1) can be modified to reflect the discharge from individual EDUs. Other graphical relationships that have been used or proposed for calculating the peak flowrate based on the number of EDUs are illustrated in Fig. 6-3.

An alternative method for calculating the design peak flowrates for collection systems with onsite pumps is based on the number of onsite pumps most likely to be in simultaneous operation. Field data collected from operating grinder pump pressure sewer collection systems are presented in Table 6-3. When representative data are available, this method may also be applied to septic tank effluent pump (STEP) systems.

Hydraulics of Wastewater Collection Systems

Once the design peak flowrate has been determined, hydraulic principles are used to determine the proper pipe diameter for conveying the flowrate efficiently.

TABLE 6-3
Number of grinder pumps in operation*

Number of grinder pumps connected upstream	Maximum number of grinder pumps operating simultaneously
1	1
2–3	2
4–9	3
10–18	4
19–30	5
31–50	6
51–80	7
81–113	8
114–146	9
147–179	10
180–212	11
213–245	12
246–278	13
279–311	14
312–344	15

*Data from *Environment/One* (1992).

Velocity and headloss in pipes. In selecting a pipe size to convey a given design flowrate, the two governing parameters are velocity and headloss. Although various equations for calculating these parameters exist, the equation most often used in practice is the Hazen-Williams formula:

$$V = 1.318CR^{0.63}S^{0.54} \tag{6-2}$$

where V = velocity of flow, ft/s
C = Hazen-Williams coefficient
R = hydraulic radius (flow area divided by wetted perimeter), ft
S = slope of energy grade line, ft/ft

It should be noted that for round pipes flowing full:

$$R = D/4$$

$$S = h_f/L$$

where D = inside diameter of pipe, ft
h_f = headloss due to friction, ft
L = length of pipeline, ft

For the materials most commonly used in alternative collection system construction (e.g., PVC pipe), a Hazen-Williams coefficient of 150 may be used. For nonstandard pipeline materials, use the manufacturer's recommended value or refer to suggested values available in hydraulics textbooks and in the literature.

The flowrate, expressed in ft^3/s, obtained by multiplying the velocity in the pipe [Eq. (6-2)] by the cross-sectional area of the pipe, is given by

$$Q = \left(\frac{\pi D^2}{4}\right)(1.318C)\left(\frac{D}{4}\right)^{0.63}\left(\frac{h_f}{L}\right)^{0.54} \tag{6-3}$$

$= \text{flowrate, ft}^3\text{/s}$

Rearranging and solving for h_f yields:

$$h_f = 4.72(L)\left(\frac{Q}{C}\right)^{1.85}(D)^{-4.87} \tag{6-4}$$

When the flowrate is expressed in gal/min and the pipe diameter is expressed in inches, the following form of the Hazen-Williams formula is used:

$$h_f = 10.5(L)\left(\frac{Q}{C}\right)^{1.85}(D)^{-4.87} \tag{6-5}$$

where h_f = headloss through the distribution pipe, ft
L = length of pipe, ft
Q = flowrate, gal/min
C = Hazen-Williams coefficient
D = inside diameter of pipe, in

When the headloss form of the Hazen-Williams formula [Eq. (6-4) or (6-5)] is used, different pipe sizes are substituted in the formula to determine the resulting headloss. Alternatively, the equation can be used to calculate the slope of the energy grade line ($S = h_f/L$) by dividing both sides by the pipe length. The Hazen-Williams formula is based on actual flow area, so the inside diameter of the pipe, rather than the pipe nominal size, must be used. Nominal pipe sizes that are used typically in septic tank effluent collection systems are presented in Table 6-4 along with the actual inside diameters.

TABLE 6-4
Nominal pipe sizes with outside and inside diameters for PVC pipe

Nominal pipe size, in	Outside diameter, in	Inside diameter, in		
		Schedule 40	Schedule 80	Class 200
$\frac{1}{2}$	0.840	0.622	0.546	
$\frac{3}{4}$	1.050	0.824	0.742	0.930
1	1.315	1.049	0.957	1.189
$1\frac{1}{2}$	1.900	1.610	1.500	1.720
2	2.375	2.067	1.939	2.149
$2\frac{1}{2}$	2.875	2.469	2.323	2.601
3	3.500	3.068	2.900	3.166
4	4.500	4.026	3.826	4.072
6	6.625	6.065	5.761	5.993
8	8.625	7.981	7.625	7.805
10	10.750	10.020	9.564	9.728
12	12.750	11.938	11.376	11.538

For gravity flow in pipes and channels, Manning's equation has also been used:

$$V = \left(\frac{1.486}{n}\right)(R)^{2/3}(S)^{1/2} \tag{6-6}$$

Manning's equation is similar to the Hazen-Williams formula [Eq. (6-2)], but the roughness coefficient ("Manning's *n*") appears in the denominator instead of the numerator. Therefore, Manning's *n* values will increase for pipe surfaces that are rougher, whereas the value of Hazen-Williams coefficient *C* decreases with increasing roughness. Where Hazen-Williams coefficient *C* values of 140 to 160 are used with plastic pipe, appropriate corresponding Manning's *n* values are 0.013 to 0.009.

Both the Hazen-Williams and Manning equations are based on empirical data, and may be inaccurate when used for inappropriate applications. For more rigorous analysis of pipeline hydraulics, the Darcy-Weisbach equation may be used. Additional information on the Darcy-Weisbach equation may be found in Sanks et al. (1998) and Tchobanoglous (1981) and in textbooks dealing with fluid mechanics.

Pipe selection. The slope of the energy grade line is obtained by dividing the headloss by the pipe length. The slope of the energy grade line can be used as a measure of the suitability of the pipe diameter selected. In general, the slope of the energy grade line should fall between 0.005 and 0.015 ft/ft (0.5 and 1.5 percent). If the computed value is too low, the pipe is oversized and material costs will be excessive. If the value is too high, higher pumping costs due to excessive friction loss will occur.

Slopes and corresponding velocity values for various flowrates and pipe sizes are presented in Table 6-5. The values given in Table 6-5 can be used to make an initial pipe diameter selection. For example, it can be seen that for a flowrate of 75 gal/min (284 L/min), a 3-in (75-mm) pipe will carry the flow at 3.06 ft/s (0.93 m/s) with a slope of the energy grade line equal to 1.06 percent. These values fall within acceptable ranges. The velocity and slope values for the 2-in (50-mm) and 4-in (100-mm) pipe do not fall within recommended ranges; therefore, these pipe sizes should not be considered for conveying the 75 gal/min (284 L/min) flowrate. It should be remembered that intermediate pipe sizes, not considered in the table, may also be available.

The velocity should also be calculated, either by using Eq. (6-2) or by dividing the design flowrate by the pipe interior cross-sectional area. To avoid excessive friction loss, the velocity should be less than 5.0 ft/s (1.5 m/s). Unlike conventional sewers, there is no required minimum velocity. Some states have indicated that 1.0 to 1.5 ft/s (0.3 to 0.46 m/s) should be maintained during peak daily flowrate periods (U.S. EPA, 1991).

6-3 SEPTIC TANK EFFLUENT GRAVITY (STEG) SEWERS

Septic tank effluent gravity systems are also known as *small-diameter variable-grade gravity* sewers or *effluent drain* systems. These systems are described separately from STEP systems; however, in practice STEG systems are often combined STEP/STEG.

TABLE 6-5

Slope of the energy grade line and velocity at specified flows for various pipe sizes*

		2 in[†]		3 in		4 in		6 in		8 in	
EDUs	Flow, gal/min	Slope, %	Velocity, ft/s	Slope, %	Velocity, ft/s	Slope, %	Velocity, ft/s	Slope, %	Velocity, ft/s	Slope, %	Velocity, ft/s
20	10	0.17	0.88								
30	15	0.36	1.33								
50	25	0.92	2.21	0.14	1.02						
70	35	1.71	3.10	0.26	1.43	0.08	0.86				
100	50	3.32	4.42	0.50	2.04	0.15	1.23				
150	75	7.02	6.63	1.06	3.06	0.31	1.85	0.05	0.85		
200	100			1.81	4.08	0.53	2.46	0.08	1.14		
250	125			2.74	5.09	0.80	3.08	0.12	1.42		
300	150			3.83	6.11	1.13	3.70	0.17	1.71	0.05	1.01
350	175					1.50	4.31	0.23	1.99	0.06	1.17
400	200					1.92	4.93	0.29	2.27	0.08	1.34
500	250					2.90	6.16	0.44	2.84	0.12	1.68
600	300							0.62	3.41	0.17	2.01
700	350							0.82	3.98	0.23	2.35
800	400							1.05	4.55	0.29	2.68
900	450							1.31	5.12	0.36	3.02
1000	500							1.59	5.69	0.44	3.35
1200	600									0.62	4.02
1400	700									0.82	4.69
1600	800									1.05	5.36
1800	900									1.30	6.03
2000	1000									1.58	6.71

*Slope of energy grade line calculated using Hazen-Williams $C = 150$.

[†]Inside diameters for Class 200 PVC pipe (see Table 6-4) have been used.

Application of Septic Tank Effluent Gravity (STEG) Sewers

STEG systems were first developed in the 1960s in Australia where effluent from existing septic tanks was collected in 4-in (100-mm) pipes. The lack of solids in the septic tank effluent allowed the use of relatively flat gradients with a minimum velocity of 1.5 ft/s (0.45 m/s) (SAHC, 1986).

In the United States, a small-diameter STEG system was placed into operation in 1975 at Mt. Andrew, Alabama (Simmons et al., 1982). The system consisted of 2-in (50-mm) and 3-in (75-mm) PVC gravity lines with variable grades. The system was designed without access ports (manholes) and without a minimum velocity requirement. At Westboro, Wisconsin, a demonstration system was designed with uniform gradients and the Australian guideline of 4-in (100-mm) pipe size. A list of selected STEG systems is presented in Table 6-6.

Any site that can be sewered with conventional gravity sewers can usually be sewered with STEG. Advantages of STEG include smaller pipe size, shallower depth of burial, reduced overall gradient, ability to reverse (inflective) gradient for selected portions of the system, reduced infiltration/inflow, and no access ports. STEG systems are well suited to previously developed or undeveloped areas with low to moderate relief. Site constraints such as shallow soil, rolling (undulating) terrain, and shallow groundwater can be overcome with STEG systems.

Onsite System Components

The onsite components of a STEG system usually include a building sewer or house lateral, a septic tank or interceptor tank, and a service lateral. Access ports (manholes) are not provided for major junctions or high points. A schematic of a STEG system is provided in Fig. 6-4.

Building sewer. Raw wastewater is conveyed from the house or building plumbing to the septic tank in a building sewer or house lateral. The common size of

TABLE 6-6
Selected septic tank effluent gravity (STEG) systems*

Location	Date online	Number of connections or services	Minimum pipe size, in	Length of pipe per connection or service, ft
Mt. Andrew, Alabama	July 1975	31	2	81
Westboro, Wisconsin	Sept. 1975	87	4	217
Avery, Idaho	Sept. 1981	55	4	122
Miranda, California	Nov.1982	100	3	96
West Point, California	Nov. 1985	155	2	116
Zanesville, Ohio	Oct.1986	711	2	86
Muskingham County, Ohio	Nov.1986	767	2	117
Lake Sherwood, California	1990	30	2	110

*From U.S. EPA (1991).

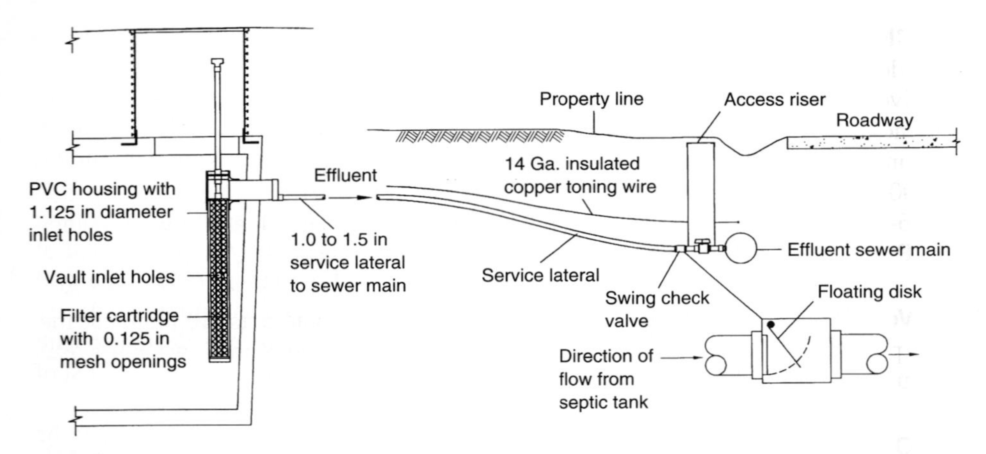

FIGURE 6-4
Schematic of septic tank effluent gravity (STEG) collection system components.

the building sewer is 4 in (100 mm). The lateral is laid at a constant slope, usually 2 percent. The lateral can be a significant source of groundwater infiltration if it is not watertight.

Septic tanks. Interceptor or septic tanks are used to remove solids and floatable materials that can clog the downstream piping. Details on septic tanks are provided in Chap. 5 and in the discussion of STEP systems. In STEG systems, an effluent filter is used for suspended solids control.

Service laterals. Service laterals convey the effluent from the septic tank to the collection main. The laterals are usually on private property and are PVC with diameters of 2 in (50 mm). The service laterals are not necessarily laid on a uniform grade or with a straight alignment. There is typically a check valve at the end of the lateral to isolate it from the mainline. In STEG systems care needs to be taken to prevent air binding of the service lateral. A uniform slope with no high points will generally suffice.

Collection System Components

The principal components of the offsite collection system are collector mains, cleanouts, vents and air release valves, and odor control measures.

Mainlines. Collector mains are fed by each connection and convey the STEG effluent to the treatment plant (see Fig. 6-4). Mains can range from 2 in (50 mm) to 8 in (200 mm) or larger, depending on the cumulative flowrate. Plastic (PVC) pipe is used with solvent-welded or rubber gasket joints.

Cleanouts or pigging ports. Because conventional access ports (manholes) provide opportunity for infiltration/inflow and sediment to enter the system, less expensive cleanouts are used to provide access for cleaning. Hydraulic flushing is a common method of STEG system cleaning. Cleanout access ports are located at minor junctions on the mainline, at changes in pipe size, at high points, and at intervals of 500 to 1000 ft (150 to 300 m). A typical mainline cleanout detail is shown in Fig. 6-5. An alternative to cleanouts is to install "pigging" ports, as described under STEP systems.

Vents and air release valves. Where inflective gradients are planned, the high points in the mains must be vented. Air release or combination air release/vacuum valves, as shown in Fig. 6-6, are often used together with a cleanout.

Odor control. Control of odors is important in STEG systems, as hydrogen sulfide generation can be expected when sulfate levels in the water supply are 30 to 50 mg/L or more, and if long sewer lines are required to serve remote areas. Control devices can include treatment of gases through soil filters, compost filters, carbon adsorption reactors in air release facilities, and scrubbing facilities. Other measures to maintain aerobic conditions in STEG systems include aeration, chlorination, and the addition of hydrogen peroxide.

Design Considerations

Design considerations for STEG sewers are presented in Table 6-7. Details are provided in U.S. EPA (1991). The layout and design of a STEG main is illustrated in Example 6-1.

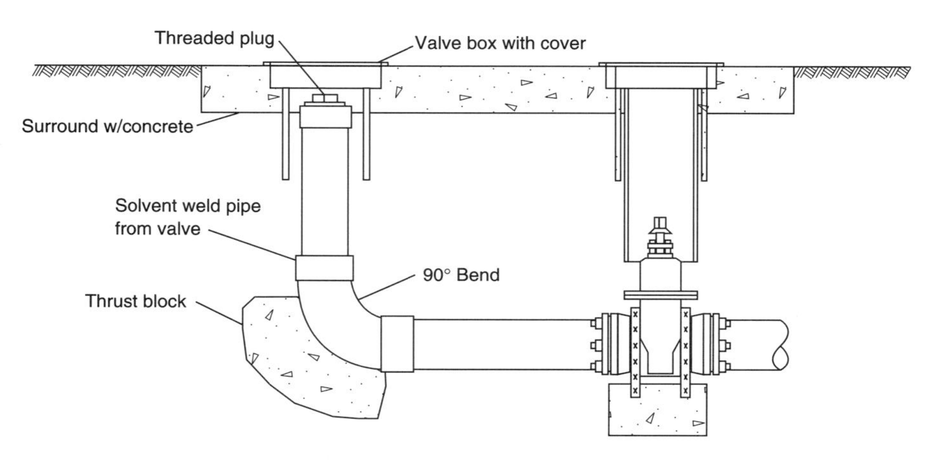

FIGURE 6-5
Detail of typical mainline cleanout access for STEG system.

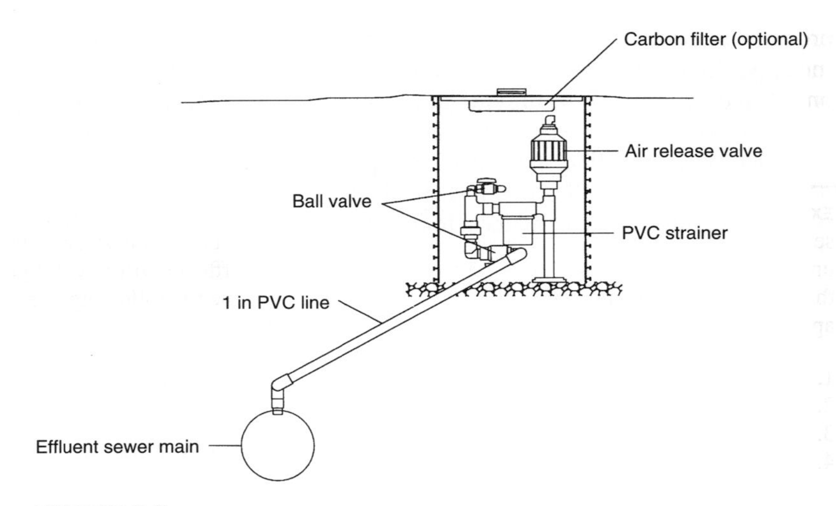

FIGURE 6-6
Typical automatic air release valve detail for STEG system.

TABLE 6-7
Typical design data for STEG sewer wastewater collection systems

Item	Unit	Range	Typical
Service lateral pipeline diameter	in	2.0–4.0	3.0
Collector main pipeline diameter	in	4.0–8.0	6.0
Trench depth*	in	24–36	30
Cleanout intervals†	ft	400–1000	500
Service connection discharge flow rate	gal/min	0.1–1.0	0.4

*Use frost depth in cold climate areas (when insulated or heat-traced piping not used).
† Pigging stations can be farther apart, depending on pipe size variation.

Construction Considerations

Construction of STEG sewers is similar to construction of conventional gravity sewers except that strict horizontal and vertical control of the main alignment is not required. As a result, trenchers can be used to produce a narrow, relatively inexpensive, shallow trench. In addition, obstacles discovered during construction can usually be avoided by changing either the horizontal or vertical (variable-grade) alignment.

Select or imported backfill is needed only for bedding and surrounding the pipe if the native trench spoil contains cobble or does not fill around the pipe snugly. Granular materials such as medium or coarse sand or pea gravel are used typically.

Operation and Maintenance

Operation and maintenance of STEG systems is similar to STEP systems except for the absence of the pumps. It is recommended that the mainlines be cleaned using polypropylene "pigs" as needed. Based on past experience, construction debris may need to be removed by flushing or pigging before the system is put into operation.

EXAMPLE 6-1. LAYOUT AND DESIGN OF STEG SEWER COLLECTION MAIN. Design a STEG sewer collection main to serve a small development of 55 EDUs (equivalent dwelling units), as shown in the plan view given below. Septic tank effluent will flow by gravity through the small-diameter gravity collection system. Assume the following conditions apply:

1. Design peak flowrate = (0.5 gal/EDU·min)(number of EDUs)
2. Pipeline material = Class 200 PVC pipe (see Tables 6-4 and 6-5)
3. Hazen-Williams coefficient = 150
4. Minimum nominal pipe size = 2 in

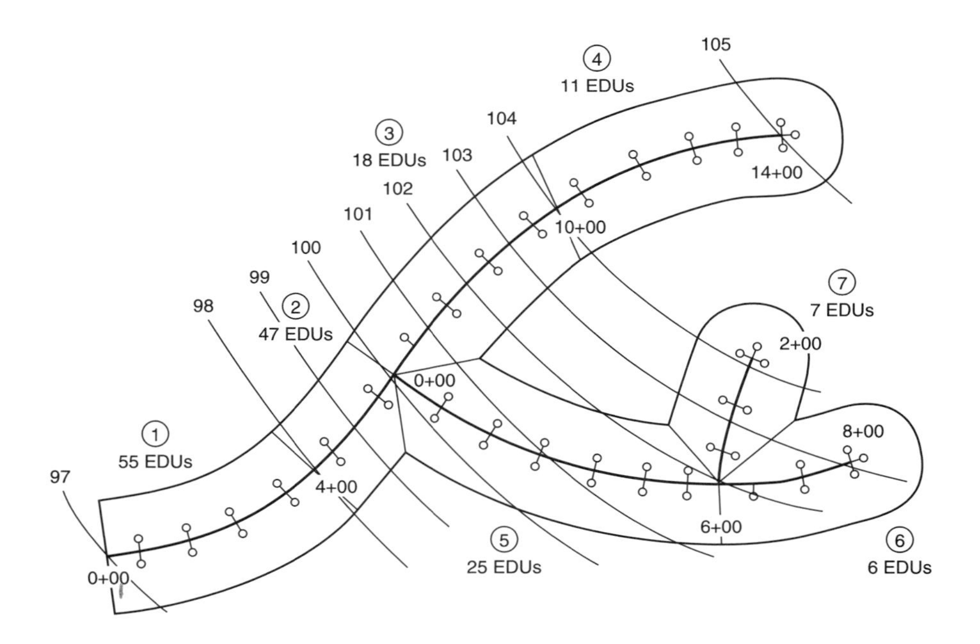

Solution

1. Prepare a profile for the sewer pipeline collection system, showing pipe stations and elevations. Using the profile diagram (as given below), divide the system into convenient pipeline sections. Separate pipes should be considered wherever branches or breaks in slope occur. Although STEG systems may be variable grade (strictly uniform gradients are not required), the calculation of flowrate is simplified by considering sections of uniform gradient separately, when possible. For this solution, the system has been divided into seven pipeline sections.

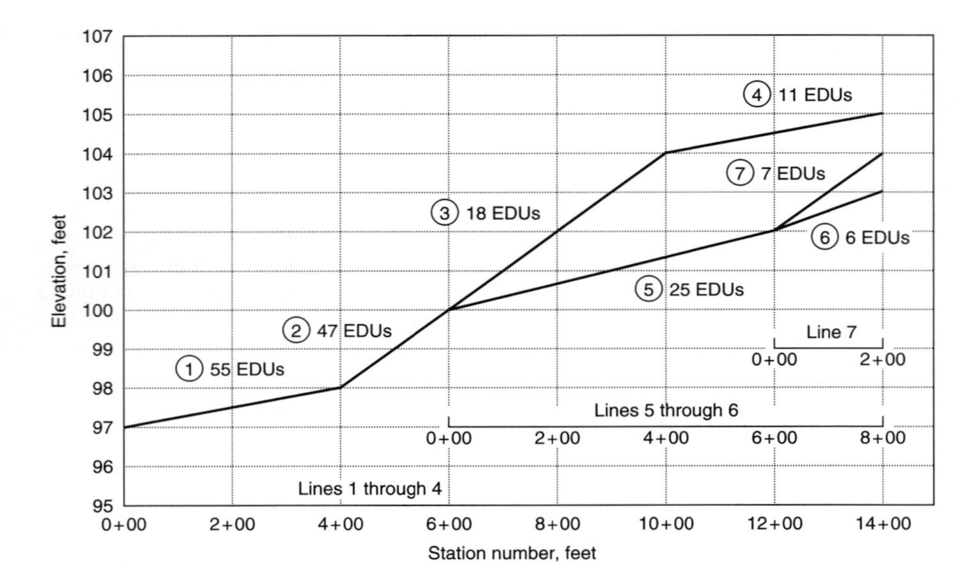

2. Set up a spreadsheet for the computation of sewer system pipe sizes. The pipe section numbers for the seven pipe sections, starting at the downstream end, are entered in column 1.

Hydraulic computations for a STEG sewer collection system

Pipe number (1)	Cumulative EDUs (2)	Design flow, gal/min (3)	Upstream station (4)	Downstream station (5)	Length, ft (6)	Upstream elevation, ft (7)	Downstream elevation, ft (8)
1	55	27.5	4+00	0+00	400	98	97
2	47	23.5	6+00	4+00	200	100	98
3	18	9.0	10+00	6+00	400	104	100
4	11	5.5	14+00	10+00	400	105	104
5	25	12.5	12+00	6+00	600	102	100
6	6	3.0	8+00	6+00	200	103	102
7	7	3.5	2+00	0+00	200	104	102

Pipe number (1)	Elevation drop, ft (9)	Slope, ft/ft (10)	Nominal diameter, in (11)	Inside diameter, in (12)	Velocity, ft/s (13)	Cross-sectional area, in^2 (14)	Pipe capacity, gal/min (15)	Ratio of design flow to pipe capacity (16)
1	1.0	0.0025	3.0	3.166	1.40	7.87	34.4	0.80
2	2.0	0.0100	2.0	2.149	2.32	3.63	26.3	0.89
3	4.0	0.0100	2.0	2.149	2.32	3.63	26.3	0.34
4	1.0	0.0025	2.0	2.149	1.10	3.63	12.4	0.44
5	2.0	0.0033	2.0	2.149	1.28	3.63	14.5	0.86
6	1.0	0.0050	2.0	2.149	1.60	3.63	18.1	0.17
7	2.0	0.0100	2.0	2.149	2.32	3.63	26.3	0.13

3. For each pipe, determine the cumulative number of EDUs served. EDUs that drain directly to the pipeline being considered, plus all upstream EDUs whose discharge

will also be conveyed by the pipeline, are included. The cumulative number of EDUs served is entered in column 2.

4. Using Eq. (6-1), calculate the design flowrate for each pipeline section. For example, the pipe in section 1 will carry flow from all 55 EDUs. Therefore, the design peak flowrate for section 1 is:

$$\text{Design peak flowrate} = (0.5\ \text{gal/EDU}\cdot\text{min})(55\ \text{EDUs}) = 27.5\ \text{gal/min}$$

   The design flowrate for each section is entered in column 3.
5. Enter the station number for the upstream end of each pipe section in column 4.
6. Enter the station number for the downstream end of each pipe section in column 5.
7. Calculate the length of each section pipeline by finding the distance between stations for each section (column 4 − column 5). These values are entered in column 6.
8. Enter the elevation for the upstream end of each pipe section in column 7.
9. Enter the elevation for the downstream end of each pipe section in column 8.
10. Calculate the drop in elevation (column 7 − column 8) and enter the result in column 9.
11. Calculate the slope by dividing the elevation drop by the length of the pipe in the section (column 9/column 6), and enter the result in column 10.
12. Enter an assumed nominal pipe diameter value for each pipe section (this value can be modified successively until a satisfactory solution is achieved). The assumed values are entered in column 11.
13. For the nominal pipe sizes selected, enter the inside diameter values (see Table 6-4) in column 12.
14. Use the Hazen-Williams formula [Eq. (6-2)] to calculate the velocity of flow in each pipe, assuming the pipe is flowing full. For the pipeline in section 1:

$$C = 150$$

$$d = 3.166\ \text{in}$$

$$S = 0.0025$$

$$R = \frac{d}{4} = \left(\frac{3.166\ \text{in}}{4}\right)\left(\frac{1\ \text{ft}}{12\ \text{in}}\right) = 0.06596\ \text{ft}$$

   Then

$$V = 1.318CR^{0.63}S^{0.54} = (1.318)(150)(0.06596)^{0.63}(0.0025)^{0.54} = 1.40\ \text{ft/s}$$

   The velocity values are entered in column 13. Note: Check to see that the calculated velocity falls within acceptable ranges for sewer design (ideally, the velocity should be less than 5 ft/s).
15. Calculate the pipe cross-sectional areas, and enter the values in column 14.
16. Multiply the pipe cross-sectional area (column 14) by the pipe flow velocity (column 13) to determine the capacity of the section pipeline when flowing full. Enter the values in column 15. For the pipe in section 1:

$$\text{Area} = 7.872\ \text{in}^2\left(\frac{1\ \text{ft}^2}{144\ \text{in}^2}\right) = 0.05467\ \text{ft}^2$$

$$\text{Velocity} = 1.40\ \text{ft/s}$$

$$\text{Capacity flow} = (0.05467\ \text{ft}^2)\left(\frac{1.40\ \text{ft}}{\text{s}}\right)\left(\frac{60\ \text{s}}{\text{min}}\right)\left(\frac{7.48\ \text{gal}}{\text{ft}^3}\right) = 34.4\ \text{gal/min}$$

17. Calculate the ratio of design flowrate (column 3) to the full capacity flowrate (column 15) by dividing column 3 by column 15 for each pipeline section. Enter the values in column 16. If the ratio is greater than 1, then surcharged conditions are likely to exist within the pipeline during moments of peak flow. If the ratio is much larger than 1, then a larger pipe size should be considered for that section. If the ratio is less than 1, then the pipe flows partially full.

6-4 SEPTIC TANK EFFLUENT PUMP (STEP) PRESSURE SEWERS

In effluent pressure sewer systems, wastewater is collected from individual houses or buildings in septic tanks or interceptor tanks, and then pumped into a pressure sewer. Where a septic tank is used to remove solids and grease before wastewater is pumped, the system is a STEP system. Pressure sewers with grinder pumps are considered in Sec. 6-5.

Application of STEP Pressure Sewers

STEP systems, as reported in Table 6-8, are distributed widely throughout the United States. STEP systems have been installed under a wide variety of site characteristics including shallow soils, high groundwater, rocky soils, and rolling terrain. Site considerations, design criteria, hydraulics, construction considerations, and operation and maintenance are described in the following paragraphs.

Onsite System Components

The principal onsite design components of a STEP system are the building sewer, septic tank, effluent screen vault, pump basins (usually for commercial facilities), effluent screens, pumps, service lateral, and valves. Design criteria and discussion of each component are provided in the following. The onsite components of a STEP system are illustrated in Fig. 6-7. Connection of the pressure line from the septic tank to the mainline is shown in Fig. 6-4.

Building sewers. Building sewers, also known as house laterals, connect the building plumbing to the septic tank. Typical building sewers are 4 in (100 mm) in diameter and sloped at a 2 percent grade toward the septic tank.

Septic tanks. Septic tanks (also called *interceptor tanks* in STEP systems) have been constructed from reinforced concrete, fiberglass, and polyethylene. Concrete septic tanks used in STEP systems need to be constructed specifically to be watertight. Fiberglass tanks are made of fiberglass-reinforced polyester (FRP) and should have an average thickness of 0.25 in (6 mm). Polyethylene septic tanks have a history of poor structural integrity and should be studied critically before they are allowed. A more complete discussion of septic tanks can be found in Chap. 5.

TABLE 6-8
Typical STEP system locations*

Location	Number of units	Date of first installation	Reason for use
Diamond Lake, Washington	525	1987	
Duncan Lake, Michigan	125	1989	
Elkton, Oregon	135	1989	Economics
McGrath, Alaska	60	1990	
Manila, California	350	1979	High groundwater
Martinez, California	47	1992	Undulating terrain
Missoula, Montana	1100	1990	Gravel soils, basements (8 ft depth at house)
Montesano, Washington	1125	1989	High groundwater, high I/I
Penn Valley, California	203	1989	High groundwater, rocky soil
Port St. Lucie, Florida	191	1973	
Priest Lake, Idaho	650	1970	High groundwater
Robbins, California	100	1997	High groundwater
Villa Verona, California	60	1989	Undulating terrain

*Orenco files; Nolte and Associates files.

Vaults and pump basins. Septic tank effluent is usually screened through 0.125-in (3-mm) openings prior to being pumped. The screen is typically housed in a vault within the septic tank. The pump vault provides storage for a working volume between the liquid levels for "pump on" and "pump off." A typical single-family pump will only need to cycle on 5 or 6 times per day. Pump vaults are typically made from polyvinyl chloride (PVC) pipe or FRP. An example of an external pump basin is shown in Fig. 6-8.

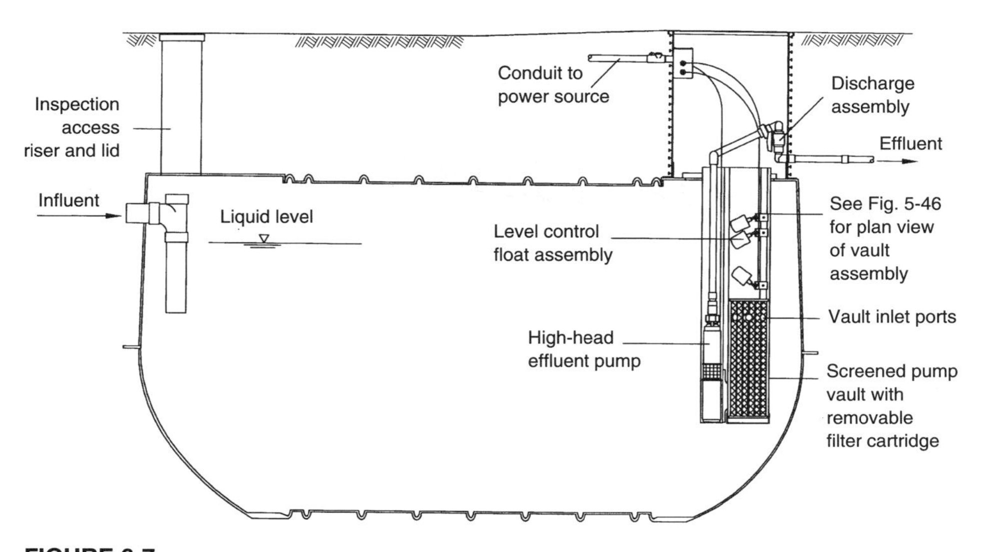

FIGURE 6-7
Schematic of septic tank effluent pump (STEP) collection system onsite components (adapted from Orenco Systems, Inc.) Connection to mainline sewer is as shown in Fig. 6-4.

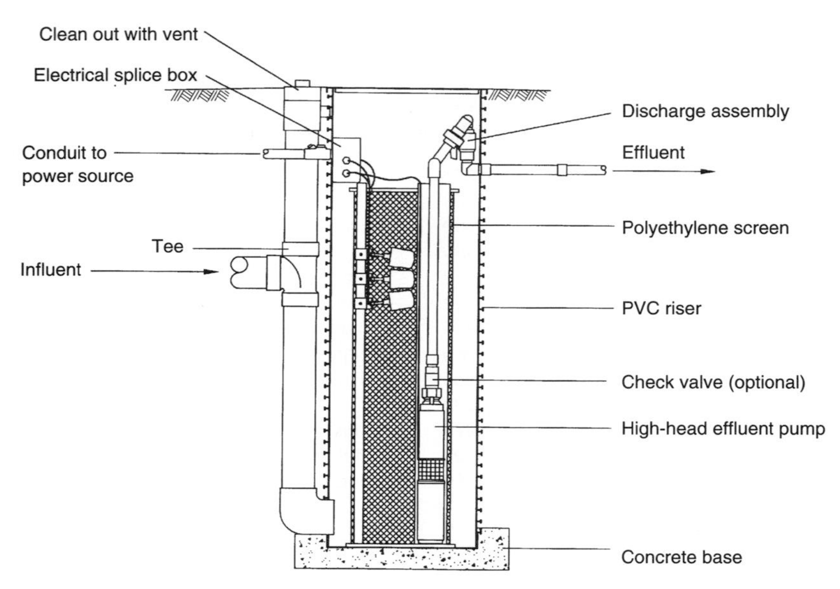

FIGURE 6-8
External pump basin for STEP collection system (adapted from Orenco Systems, Inc.).

Effluent screens. The effluent screen was introduced in the mid-1980s to screen out solids from septic tank effluent before pumping, thereby making it feasible to use turbine pumps. The effluent screen is shown in Fig. 5-46. Holes in the housing for the screen are located in the clear zone between the sludge layer and the scum layer; the holes allow the clearest liquid in the tank to enter the housing and flow through the $\frac{1}{8}$-in (3-mm) mesh screen. Effluent screens have a 12-ft^2 (1.1-m^2) surface area and the velocity is low enough so the screen does not clog. Effluent screens are typically cleaned when septic tanks are pumped.

Pumps. Submersible lightweight multiple-stage turbine pumps, as shown previously in Fig. 1-8, are the preferred choice for STEP system pumps. The pumps, constructed of stainless steel and plastic, are equipped with 0.5- to 1.5-horsepower motors. Because the pumps are lightweight (30 lb), they can be removed easily from the pump vault, should any maintenance be required. A typical head versus discharge curve for a turbine pump is shown in Fig. 6-9. To limit the discharge to 5 to 10 gal/min (0.3 to 0.6 L/s), a 0.25-in (6-mm) flow controller is installed in the pump discharge line. In addition, a 0.125-in (3-mm) hole is drilled in the pump discharge piping to allow for the expulsion of trapped air and to allow for recirculation of effluent back to the sump when the mainline pressure is too high to accept the full pump discharge. By using a flow controller in the pump discharge line, the discharge from individual onsite systems with widely varying static heads can be limited within a narrow range (see Fig. 6-9). As shown in Fig. 6-9, for two onsite systems with the same dynamic head and static heads differing by a factor of 2, the discharge is quite

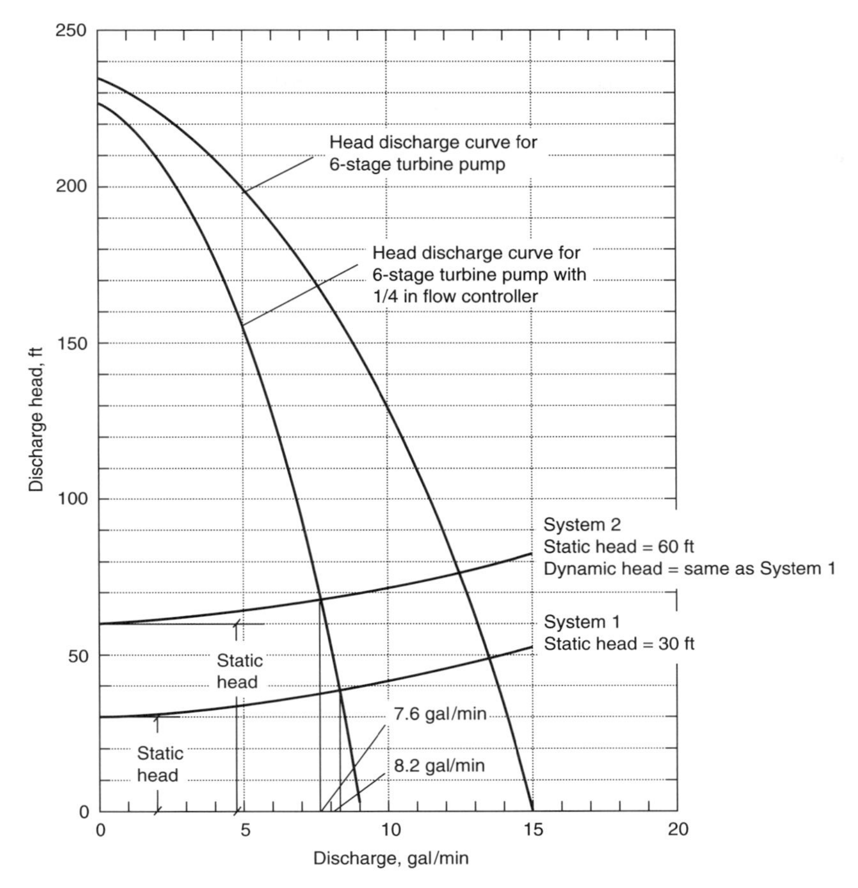

FIGURE 6-9
Typical head versus discharge curve for the multistage turbine pump used in a STEP collection system (pump curves adapted from Orenco Systems, Inc.).

similar (7.6 versus 8.2 gal/min). Because the individual onsite pumps are connected in parallel to the mainline, the discharge from any individual onsite pump when other pumps are operating can be determined by adding the discharge at constant head (Sanks et al., 1998 and Tchobanoglous, 1981). The advantage of using a turbine pump with a steep head versus discharge curve is that no controls are necessary for pump sequencing.

Electrical service. The pump control panel is commonly mounted on the house or building. A separate electrical service is not normally required.

Service lateral. The service lateral connects the effluent pump with the main line. Service laterals are 1 to 2 in (25 to 50 mm) in diameter and constructed of Class 200 or Schedule 40 PVC. A view of a typical individual STEP system septic tank with a discharge line to a pressure sewer is shown in Fig. 6-10.

FIGURE 6-10
Typical view of fiberglass septic tank with 1.25-in pressure discharge line to main pressure sewer.

Valves. The last components of the onsite system for a STEP installation are the check valves. The check valves are located at the pump outlet and at the edge of the property to isolate the onsite facilities from the main lines.

Collection System Components

The offsite components consist of mainlines, cleanouts and pigging ports, air release valves, and odor control facilities.

Mainlines. Mainlines are usually PVC and range in diameter (depending on flowrates and pressures) from 2 to 24 in (50 to 600 mm). For most smaller STEP systems, the maximum mainline diameters range from 2 to 3 in (50 to 75 mm). The mainline pipe design criterion is usually to keep the velocity at about 5 to 6 ft/s (1.5 or 1.8 m/s). Short runs of piping at greater than 5 ft/s (1.5 m/s) are acceptable if adequate head is available. A detail of the connection to the pressure main was presented in Fig. 6-4.

Pigging ports. Instead of access ports, STEP systems have periodic pigging ports. The ports are placed at pipeline terminations and where pipeline diameters change. The use of "pigs" to clean small-diameter pipelines is discussed in "Operations and Maintenance," below. A typical STEP collection system cleaning station is shown in Fig. 6-11.

Valves. Valves used in STEP systems include air release valves, pressure-sustaining valves, and isolation valves. To reduce the excessive friction headloss caused by trapped air and other gases (two-phase flow), air (gas) release valves (ARV), such as shown previously in Fig. 6-6, are used at high points and other loca-

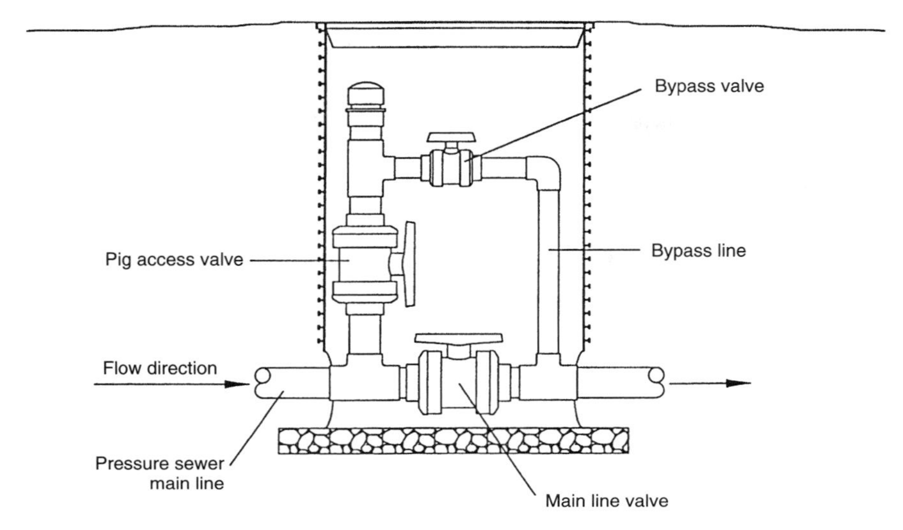

FIGURE 6-11
Typical cleaning station for launching cleaning device (pig) for STEP collection system.

tions on the mainline to allow the trapped gases to escape from the mainline. Odor control features are recommended for the valve boxes of air release valves. Pressure-sustaining devices are used to maintain full flow where the pressure is insufficient to maintain full flow (see Fig. 6-12). Two types of devices can be used to sustain the pressure: (1) standpipes, where the terrain is suitable, and (2) roll-seal valves. Air release valves are typically located at a point about 5 ft below the elevation of the downstream static hydraulic grade line (see Fig. 6-12*b*). If a line is oversized to allow for gravity flow, an air release valve is used at the highpoint and at a point 5 ft below the static hydraulic grade line (see Fig. 6-12*c*). Isolation valves are used on mainlines to isolate or close off a portion of the mainline for maintenance, cleaning, or repairs.

Odor control features. Some of the earlier STEP system designs failed to account for hydrogen sulfide generation and the release of odors. To overcome the potential for odor release at air release valves, activated carbon cartridges are often installed in valve boxes. At the end of a STEP system special features for odor control such as aeration, scrubbing, or soil or compost filtration can be used (see Chap. 5). During design an analysis of the expected water quality should be conducted to determine if sulfate concentrations will exceed 50 mg/L (see Chap. 2).

Design Considerations

Considerations in the planning of STEP systems include topography, density of service area buildings, and use of existing septic tanks. STEP systems have been

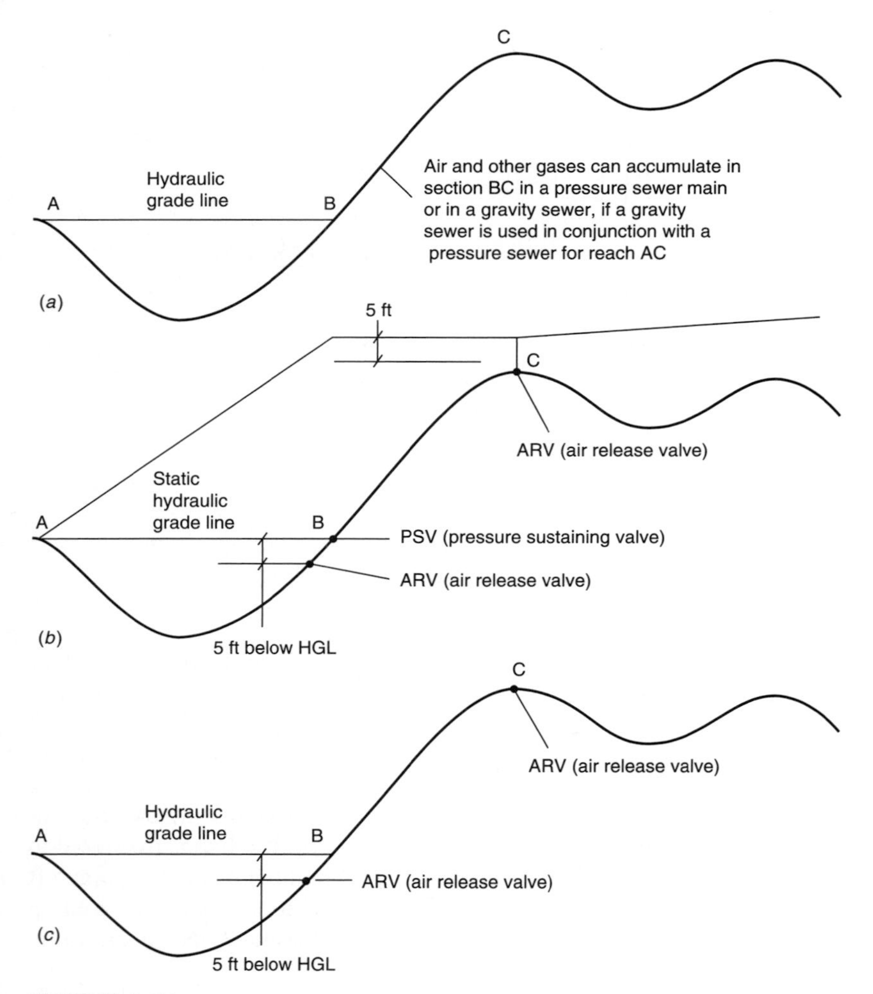

FIGURE 6-12
Definition sketch for the application of air release and pressure-reducing valves in STEP collection systems: (*a*) accumulation of air and other gases, (*b*) use of air release valves and pressure-sustaining valves in pressure sewers, and (*c*) use of air release valves in gravity section of pressure sewer.

planned and designed for existing unsewered communities where the housing density is already established. As a result, both existing and future land use and density must be considered in planning and design. Typical design data for STEP sewers are presented in Table 6-9. The layout and design of a STEP system is illustrated in Example 6-2.

Topography. STEP systems are particularly well suited to rolling or undulating terrain. Uphill collection is the ideal case for STEP systems. By taking advantage of effluent pumping and the resultant shallow depths of collection piping, STEP sys-

TABLE 6-9
Typical design data for STEP pressure sewer wastewater collection systems

Item	Unit	Range	Typical
Service lateral pipeline diameter	in	1.25–2.0	1.5
Collector main pipeline diameter	in	4.0–8.0	6.0
Trench depth*	in	24–36	30
Cleanout intervals	ft	400–1000	500
Pump discharge flow rate	gal/min	6–9	7

*Use frost depth in cold climate areas (when insulated or heat-traced piping not used).

tems can be cost-effective when compared to conventional gravity sewers in these topographic conditions. Very flat sites can also benefit from STEP by avoiding very deep sewers and the need for large pumping stations.

Cluster systems. The density of development will affect the approach to and economics of STEP systems. Sparse and low-density development is relatively expensive to sewer. If several buildings or houses can be clustered together, using individual septic tanks but one pump tank for every 4 to 10 homes, the economics of sewering will usually improve (Parker, 1997). A typical cluster pump station is as shown previously in Fig. 6-8.

Use of existing septic tanks. For some newer unsewered communities it may be possible to use existing septic tanks for pretreatment in a STEP system. The tests for suitability are age, materials of construction, and watertight condition. Concrete tanks of suitable size are usually the best candidates for retention. A test to determine if water is leaking into or out of the septic tank must be conducted, especially if high groundwater table conditions exist or are expected. If existing septic tanks are used in a STEP system, they should be followed by a separate pump basin with an effluent vault or screen for ease of construction and start-up, and to minimize solids and grease carryover. Experience with older unsewered communities, especially where high groundwater conditions exist, indicates that existing septic tanks should be abandoned and replaced with new watertight tanks.

Construction Considerations

The onsite portion of the STEP system is the most complex in terms of satisfying the needs of the new system, the onsite owner, and the existing facilities. Locations of the tanks, service lines, laterals, and electrical service are decided in the field and must be documented on paper and preferably on videotape. If the existing septic tank is to be used in the STEP system, the tank must be inspected and tested for watertightness. A separate pump tank is often used, as shown in Fig. 6-8, so that the existing tank and drainfield remain operational until the time of connection.

STEP systems can be installed at relatively shallow depths, which results in lower construction costs than for gravity sewers. Trenching may be done by a chain-type trencher or a wheel trencher. Pipe zone backfill should be a granular material

TABLE 6-10
Typical construction problems with STEP systems

1. Discharge plumbing not oriented as instructed to allow easy removal of screened vaults
2. Discharge elbows from effluent pumps broken
3. Risers completely separate from concrete tanks
4. Risers located in a drainage path
5. Risers installed upside down without bolt catchers for lids
6. Grade rings installed without bolt catchers for lids
7. Fiberglass lids have been driven over and broken
8. Tanks not tested adequately for watertightness by filling to top of the tanks and up into the installed risers
9. Control-alarm panels not located within direct view of the tanks
10. Terminal strips in control panels damaged
11. Float tethers altered from factory setting
12. Exterior discharge assembly not buried with sufficient cover to provide protection from freezing or physical damage

TABLE 6-11
Analysis of STEP system service calls for Glide and Elkton, Oregon*

	Glide		Elkton	
Maintenance requirement	**Total calls 1980–1996**	**Average per year**	**Total calls 1990–1996**	**Average per year**
Mechanical				
Effluent pump (EP)	272	16.0		
Turbine pump (TP)	14	1.3	1	0.1
Building sewer (BS)	92	5.4	2	0.3
Check valve (CV)	27	1.6		
Level control (LC)	632	37.2		
Control panel (CP)	131	7.7		
Service line (SL)	76	4.5		
Screen STEP (SC_{PV})	14	1.3		
Screen effluent filter (SC_{EF})	3	0.3	19	2.7
Filter valve (VA)	13	0.8	1	
Hose and valve assembly (HV)	53	3.1		
Physical				
Back pressure (BP)	296	17.4		
Air bound (AB)	291	17.1		
Sludge and scum (S&S)	6	0.4	91†	13
Clogging (CL)	63	3.7		
Power (PWR)	175	10.3	5	0.7
Infiltration and inflow (I/I)	96	5.6	16	2.3
Exfiltration (EX)	4	0.2		
No malfunction found on inspection (NMI)	160	9.4		
Other (siphoning, odors, etc.) (OT)	132	7.8	10	1.4
Total	2550	150	145	21
Hours per service call		0.9		

*Adapted from Bounds (1997).
† In 1996, a 6-year audit was performed on each tank (127 tanks). Typically, 10 percent of the tanks are monitored each year.

such as pea gravel, crushed rock, or sand. In some instances a slurry of sand can be used, especially when a narrow trench is used and prompt restoration of the pavement section is critical.

Potential problems with connection of existing onsite systems into new STEP facilities are listed in Table 6-10. The owners and design engineers should ensure that qualified inspectors be present during installation to avoid these problems.

Operation and Maintenance

Considerations for operation and maintenance (O & M) of STEP systems include frequency of septic tank pumping, need for pipeline cleaning, frequency of maintenance calls, and odor considerations.

Frequency of septic tank pumping. The need for and frequency of septic tank pumping is introduced in Chap. 5. For STEP systems, pumping is required to remove solids and grease to prevent clogging of the small-diameter mainlines. The frequency of pumping should be determined in each case, on the basis of actual accumulation rates. The tanks should be inspected after the first year of operation and subsequently every 2 to 3 years, depending on water use. Septage removal from most 1000-gal tanks should be scheduled when the sludge depth approaches 21 in (0.53 m) or the scum layer thickness approaches 10 in (0.25 m) (Bounds, 1992).

Need for pipeline cleaning. The mainlines do not require regular cleaning. The greatest need for "pigging" appears to occur after construction, to remove debris and gravel that has entered the mainline from construction of the line and onsite connections.

Frequency of maintenance calls. As STEP systems evolve, more reliable equipment is used in current designs. As a result, the frequency of maintenance calls decreases. The mean time between service calls (MTBSC) for two of the older systems (Port St. Lucie, Florida, and Manila, California), are 3.6 and 3.5 years, respectively (WPCF, 1986). At Glide, Oregon, the MTBSC is 6.2 years. A breakdown of service calls at Glide, Oregon, is presented in Table 6-11.

Odor considerations. Odors in STEP systems have occurred mainly at pump stations or air release valves (Rezek and Cooper, 1985). At Kalispell, Montana, the odors from the pump stations are vented to a drainfield for soil scrubbing. Odors from air relief valve boxes can be absorbed effectively onto activated carbon.

EXAMPLE 6-2. LAYOUT AND DESIGN OF STEP PRESSURE SEWER COLLECTION MAIN. Design a STEP pressure sewer collection system for the small community of 111 EDUs (equivalent dwelling units) depicted in the plan view shown below. Because highest elevation occurs at the discharge outlet to the wastewater treatment plant, the line will remain pressurized at all times. Assume the following conditions apply:

1. Design peak flowrate = (0.5 gal/EDU·min)(number of EDUs)
2. Pipeline material = Class 200 PVC pipe (see Tables 6-4 and 6-5)
3. Hazen-Williams coefficient = 150
4. Minimum nominal pipe size = 2 in

Solution

1. Prepare a profile diagram (as given on the next page) for the collection system, showing pipe stations and elevations. Using the profile and plan diagrams, divide the system into convenient pipeline sections. Separate pipes should be considered wherever branches occur, or where cumulative wastewater inputs from individual service connections become appreciable. Because pressure sewers can operate at variable grades, it is not necessary to consider separate pipe sections where breaks in slope occur. For the system presented in this example, the collection system has been divided into five sections.
2. Set up a spreadsheet for the computation of sewer system pipe sizes. The pipe section numbers for the five pipe sections, starting at the downstream end, are entered in column 1.

Hydraulic computations for a STEP pressure collection system

Pipe number (1)	Cumulative EDUs (2)	Design flow, gal/min (3)	Upstream station (4)	Downstream station (5)	Length, ft (6)	Nominal diameter, in (7)
1	111	55.5	15+00	0+00	1500	3.0
2	67	33.5	30+00	15+00	1500	2.5
3	21	10.5	40+00	30+00	1000	2.0
4	20	10.0	10+00	0+00	1000	2.0
5	19	9.5	10+00	0+00	1000	2.0

Pipe number (1)	Inside diameter, in (8)	Slope of EGL, % (9)	Cross-sectional area, in^2 (10)	Velocity, ft/s (11)	Headloss, ft (12)
1	3.166	0.61	7.87	2.26	9.1
2	2.601	0.62	5.31	2.02	9.4
3	2.149	0.18	3.63	0.93	1.8
4	2.149	0.17	3.63	0.88	1.7
5	2.149	0.15	3.63	0.84	1.5

3. For each pipe, determine the cumulative number of EDUs served. EDUs that pump directly to the pipeline being considered, plus all upstream EDUs whose flowrate will also be conveyed by the pipeline, are included. The cumulative number of EDUs served by each pipe section is entered in column 2.
4. Using Eq. (6-1), calculate the design flowrate for each pipeline section. For example, the pipe in section 1 will carry the flow from all 111 EDUs. Therefore, the design peak flowrate for section 1 is

$$\text{Design peak flowrate} = (0.5 \text{ gal/EDU·min})(111 \text{ EDUs}) = 55.5 \text{ gal/min}$$

The design flowrate for each section is entered in column 3.

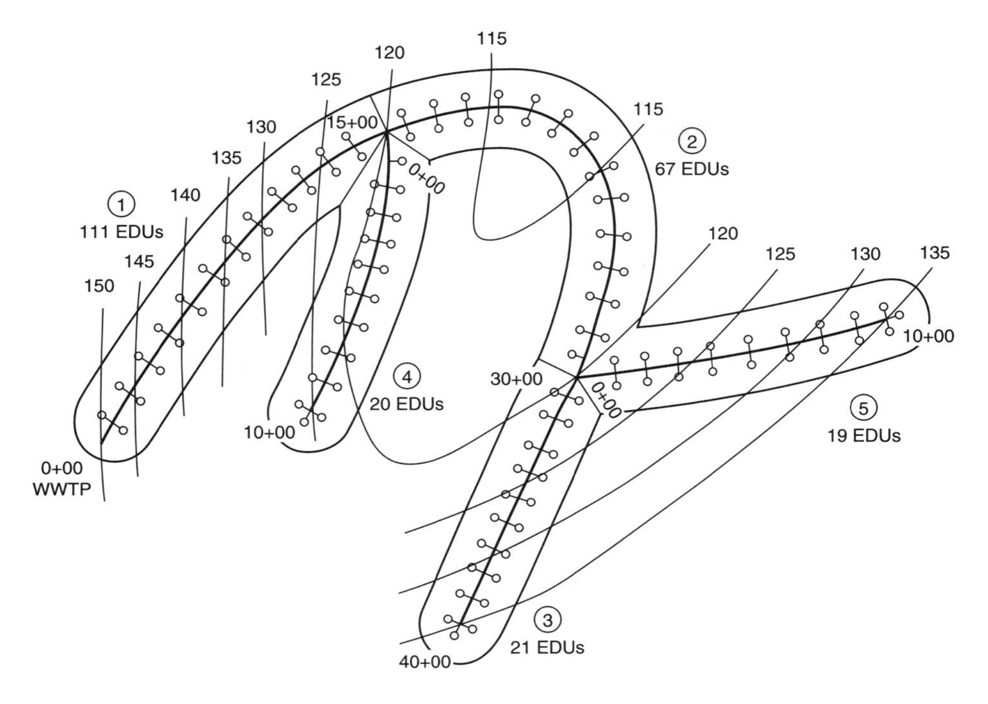

115
120
125
130
135
140
145
150
15+00
0+00
1
111 EDUs
2
67 EDUs
3
21 EDUs
4
20 EDUs
5
19 EDUs
10+00
30+00
40+00
0+00
WWTP

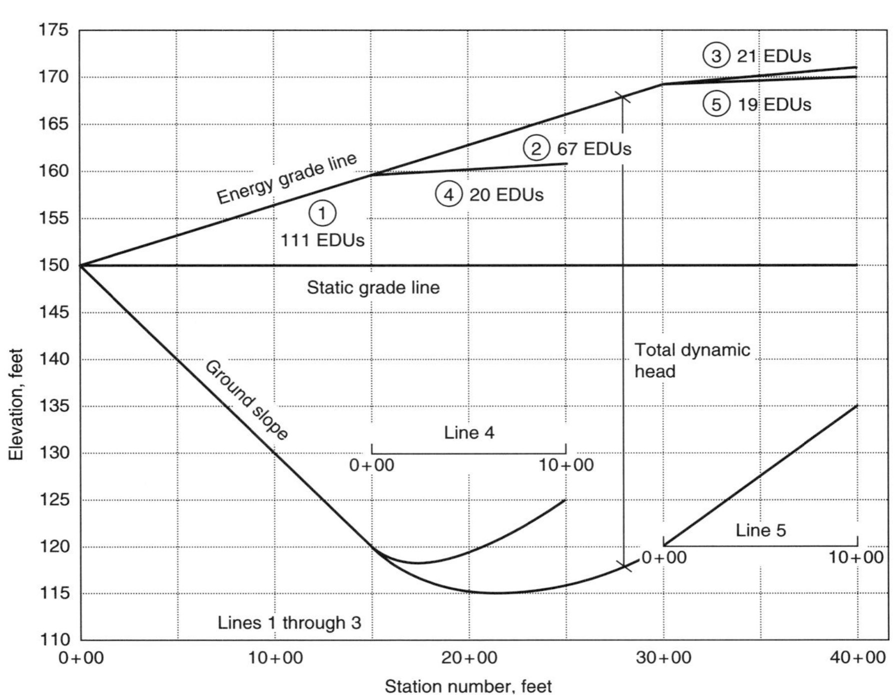

Elevation, feet
Station number, feet
Energy grade line
Static grade line
Ground slope
Total dynamic head
Line 4
Line 5
Lines 1 through 3
1 111 EDUs
2 67 EDUs
3 21 EDUs
4 20 EDUs
5 19 EDUs
0+00
10+00
20+00
30+00
40+00

5. Enter the station number for the upstream end of each pipe section in column 4.
6. Enter the station number for the downstream end of each pipe section in column 5.
7. Calculate the length of pipeline in each section (column 4 − column 5). The computed lengths are entered in column 6.
8. Enter an assumed nominal pipe diameter value for each pipe section (this value can be modified successively until a satisfactory solution is achieved). The assumed values are entered in column 7.
9. For the nominal pipe sizes selected, enter the inside diameter values (see Table 6-4) in column 8.
10. Use the Hazen-Williams formula [Eq. (6-5)] to calculate the slope of the energy grade line for each pipe. These values should fall within the range of 0.5 to 1.5 percent. If the value is too low, the pipe is oversized and materials costs will be excessive. If the value is too high, pumping costs due to excessive friction loss will occur. For the pipeline in section 1:

$$d = 3.166 \text{ in}$$

$$Q = 55.5 \text{ gal/min}$$

$$C = 150$$

$$\text{Slope} = \left(\frac{10.5}{3.166^{4.87}}\right)\left(\frac{55.5}{150}\right)^{1.85} = 0.0061 = 0.61\%$$

    The slope values are entered in column 9.
11. Using the pipe inside diameters (column 8), calculate the pipe cross-sectional areas, and enter the values in column 10.
12. Divide the design flowrate (column 3) by the pipe cross-sectional area (column 10) with appropriate units, to determine the velocity of flow for each pipe section. Enter the values in column 11. Check to see that the calculated velocity falls within acceptable ranges for sewer design (ideally, the velocity should be less than 5 ft/s).
13. Calculate the headloss due to friction for each pipe section by multiplying the energy grade line slope value (column 9) by the pipe length (column 6). If the slope value is expressed as a percentage, remember to divide the slope value by 100 before multiplying. The headloss values are entered in column 12.
14. Using the profile created in step 1, plot the energy grade lines for each pipe section. Begin at the downstream end, and add the headloss values to the static grade line. The height of the energy grade line represents the pump head necessary for flow to enter the system when peak flow conditions exist.

6-5 GRINDER PUMP PRESSURE SEWERS

Grinder pump systems are an alternative to STEP systems and traditional gravity-flow sewer systems. Like conventional sewers, grinder pump collection systems convey waste solids suspended in water. However, the grinder pump minces collected solids into small particles, converting the waste stream into a slurry that can be conveyed through small-diameter pipelines. The grinder pump also imparts pressure to the flow, making conveyance across uneven terrain or uphill possible.

Application of Grinder Pump Pressure Sewers

Conventional sewers flow by gravity through pipes installed at uniform downslope grade lines. When constant grades are followed for long distances, sewer pipe depths can become excessive. The costs and risks associated with pipeline installation in deep trenches (e.g., excessive soil excavation and transport, dewatering, shoring and sheeting, traffic disruption, restoration of pavement, and landscaping) can be reduced through the use of grinder pump collection systems. Because the grinder pump system is pressurized, there is no need to follow a constant grade line. Low-pressure (typically less than 60 lb/in^2 gage) grinder pump pipelines can be installed in shallow trenches that conform to varying surface contours, minimizing the impact and expense of sewer installation.

Grinder pump systems have been used since the 1970s in small rural communities, lakeside resorts, trailer parks, and other residential complexes. Their application is in areas with high groundwater, shallow soils, and rolling topography, and where population densities may not be high enough to support the installation costs of conventional gravity sewers. Evaluation of the suitability of grinder pump systems for a given location, as with all other wastewater collection systems, must be done on an individual basis.

Unlike conventional gravity-flow sewers, grinder pump collection systems are designed to be watertight. Flow in gravity sewers is augmented by infiltration of groundwater into leaking pipe joints and inflow of surface runoff into access ports. Because the grinder pump system is sealed, accommodation for infiltration and inflow is not necessary. Design pipe sizes for grinder pump systems are typically 1.5 to 8 in (37 to 200 mm), compared to 4 to 20 in (100 to 500 mm) for gravity systems with similar capacities. The use of smaller pipes and shallow trenches associated with grinder pump systems has made them a competitive alternative for many locations.

Onsite System Components

The principal onsite components in grinder pump systems are the grinder pumps and the grinder pump station. A diagram of a typical grinder pump receiving sump is shown in Fig. 6-13.

The grinder pump. Pressurization of wastewater is accomplished by passage through the grinder pump, typically a screw-type progressing cavity pump (see Fig. 6-13). Before entering the pump chamber, raw wastewater passes through stainless steel cutter bars and teeth that shred influent particles (e.g., toothpicks, cigarette filters, rags) into a fine slurry. Grinder pump hydraulic performance curves are typically very steep, indicating that they can perform at nearly constant output over a wide pressure range (see Fig. 6-14). Grinder pumps for residential connections are typically 0.5 to 2.0 hp, with discharge flowrates of 9 to 14 gal/min (34 to 53 L/min).

Grinder pumps are designed with safety features to enhance their performance and reduce operation and maintenance costs. High starting torque, necessary to

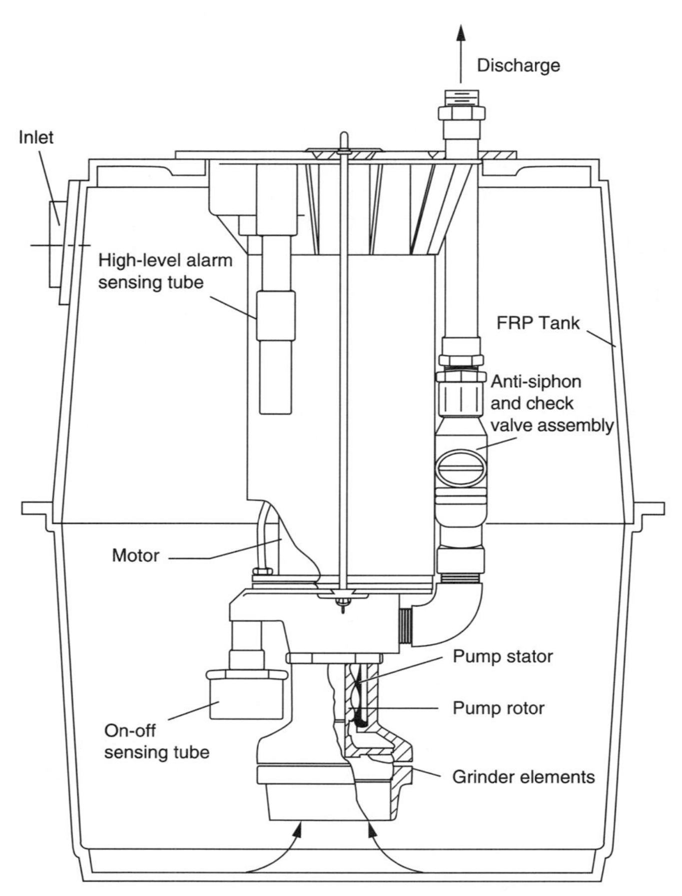

FIGURE 6-13
Typical grinder pump receiving sump and pump vault (from Environment/One Corp.).

overcome the initial resistance that occurs when the grinder pump has previously stopped while grinding hard material, is an inherent quality of progressing cavity pumps. Thermal overload protectors that automatically shut the pump off until it cools down are typically installed with grinder pumps. When power is restored after a power failure and several pumps turn on simultaneously, the combined pressure will not allow all of the pumps to discharge at the same time. Pumps that overheat will shut off automatically, and will be able to discharge after cooling off for a few minutes, when the line has more available capacity.

The grinder pump station. Grinder pumps are submersible and are usually installed in a wastewater storage tank whose volume is dependent on the type of service connection. Residential installations typically use individual grinder pumps in single-pump stations with approximately 30-gal (113-L) capacity. However, duplex and multiple-pump installations within stations with greater capacity have also been

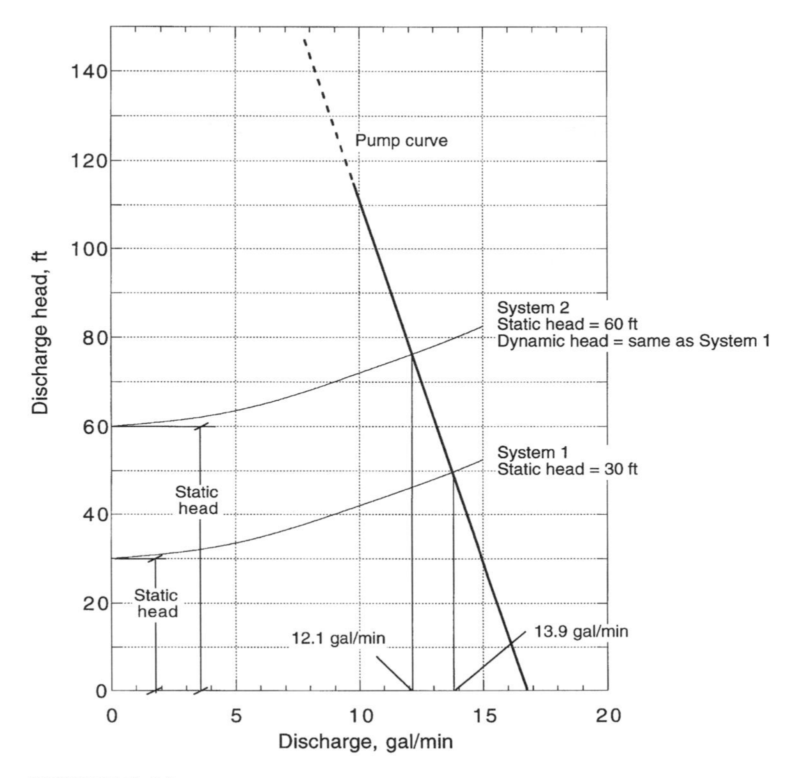

FIGURE 6-14
Typical head versus discharge curve for a grinder pump (pump curve adapted from Environment/One Corp.).

used. The pump station can be located within the basement or buried outside the home; a riser accessible from the surface is required for underground installations. Pump operation is controlled by pressure switches that monitor water level within the wastewater containment vessel. Additional level monitors control alarms for high water levels. The pump station is isolated from the pressure collection system by a hinged-flap check valve, which prevents backflow from the combined system from entering the individual pump station. When pump operation is initiated, the effluent pressure generated by the high-head grinder pump forces the check valve open, discharging the contents of the tank to the collection pipeline.

Collection System Components

As discussed earlier, grinder pump collection systems are composed of small-diameter pressurized pipelines, installed in shallow trenches. Although the pipeline from the household to the pump station may be 4 in (100 mm), the line exiting the pump basin is typically 1.5 to 2 in (37 to 50 mm). Because the grinder pump cuts suspended solids into minuscule fragments, pipeline clogging is uncommon.

To prevent settling of suspended matter, design pipeline velocities of 2 to 5 ft/s (0.6 to 1.5 m/s) are used. Valved cleanout ports are also installed every 1000 ft (300 m) on straight runs, and at terminal ends of pressure mains, to facilitate maintenance operations. Nevertheless, the pressurized collection system pipelines tend to be self-cleaning, because fluid velocities increase in areas where clogs reduce the flow cross-sectional area, increasing scour forces.

The use of plastic materials, such as PVC (polyvinylchloride) or HDPE (high-density polyethylene), is appropriate for the range of pressures associated with grinder pumps. These pipes are lightweight and easy to install, and are typically connected by either rubber gaskets or solvent welding. The hydraulic characteristics of plastic pipe are excellent, because of its smooth interior surface. Although plastic pipe material can degrade when exposed to sunlight for extended periods, buried pipeline will remain in good condition indefinitely. In areas where freezing conditions occur, the collection system pipeline should be buried sufficiently deep to prevent freezing.

Design Considerations

Typical design data for grinder pump pressure sewer systems are presented in Table 6-12. Details may be found in U.S. EPA (1991) and equipment manufacturers' design guides. The design and layout of a sewer collection main is illustrated in Example 6-3.

Construction Considerations

Construction considerations for grinder pump pressure sewers are similar to those for STEP systems. Grinder pump vaults should be located in an area of stable soil, accessible for construction, but not subject to vehicular traffic. In some designs the vault is located near the house so that the wiring can pass directly from the vault to the control panel without need for an electrical junction box in the pump vault (U.S. EPA, 1991).

TABLE 6-12
Typical design data for grinder pump pressure sewer wastewater collection sytems

Item	Unit	Range	Typical
Grinder pump power rating	kW	0.75–3.7	1.5
	hp	1–5	2
Grinder pump discharge pressure (gage)	lb/in^2	29–40	35
Grinder pump discharge	gal/min	5–25	12
Size of line from pump to pressure main	in	1–2	1.25
Size of pressure main	in	2–12	*

*Varies with location in system.

TABLE 6-13
Distribution of causes for call-out maintenance on selected grinder pump pressure sewer collection systems*

Category	Percent of occurrences
Electrically related	25–40
Pump related	20–25
Miscellaneous	20–30
Pump vault related	5–15
Piping related	5–15

*Adapted from U.S. EPA (1991).

Operation and Maintenance

Operational considerations for pressure sewers using grinder pumps are similar to those for STEP systems, except for maintenance of the grinder pumps. The causes for call-outs for maintenance of grinder pump pressure sewers were evaluated and are summarized in Table 6-13. The electrically related service calls were often caused by grease accumulation on the mercury float switches (U.S. EPA, 1991).

EXAMPLE 6-3. LAYOUT AND DESIGN OF GRINDER PUMP PRESSURE SEWER COLLECTION MAIN. Design a grinder pump pressure sewer collection system for the small community of 111 EDUs described in Example 6-2. Assumptions 1 to 3 used in Example 6-2 will apply to this problem, but the fourth assumption (minimum nominal pipe size = 2 in) will not be made. Unlike STEP systems, which convey clarified effluent that has undergone septic tank treatment, grinder pumps discharge raw wastewater with all the original solids. Grinder pumps reduce the size of solid particles and impart pressure to the flow, but they do not remove solids. Therefore, because of the increased load of grease and sludge, design pipeline velocities should be high enough to keep solid particles in suspension and provide scouring action to prevent clogging of the pipe. When flows are small, the use of pipe sizes under 2 in (50 mm) is required to accomplish the goal of maintaining minimum velocities. It is suggested that the design velocity be within the range of 3 to 5 ft/s (0.9 to 1.5 m/s).

Solution

1. Prepare a profile diagram (as given below) for the collection system, showing pipe stations and elevations. Using the profile and plan diagrams, divide the system into convenient pipeline sections. Separate pipes should be considered wherever branches occur, or where cumulative wastewater inputs from individual service connections become appreciable. Because pressure sewers can operate at variable grades, it is not necessary to consider separate pipe sections where breaks in slope occur. For the system presented in this example, the collection system has been divided into five sections.

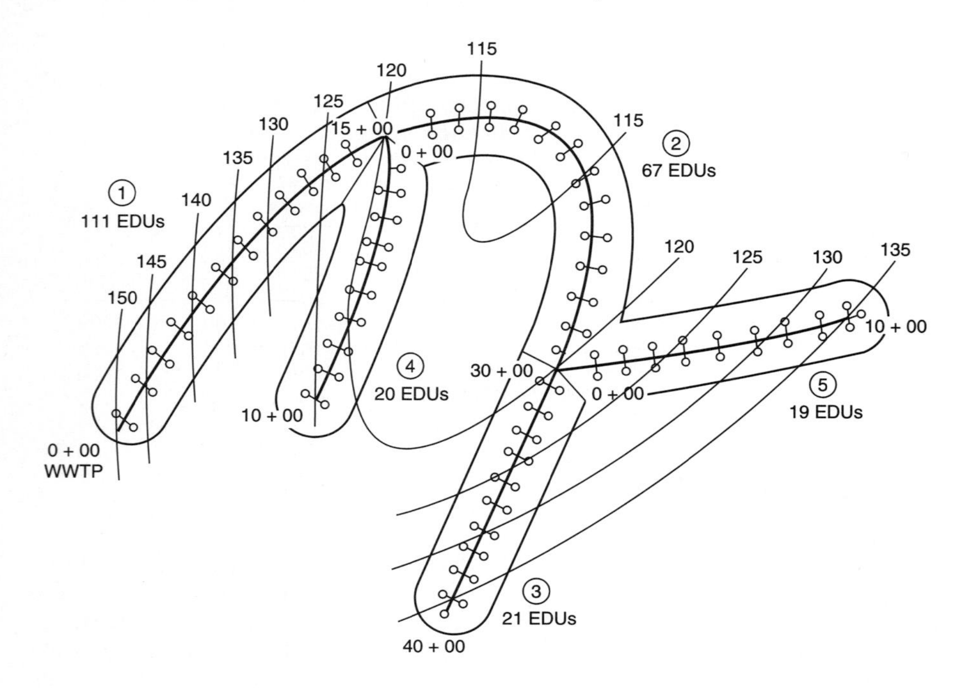

115
120
125
130
135
140
145
150
15 + 00
0 + 00
115
2
67 EDUs
1
111 EDUs
120
125
130
135
10 + 00
4
20 EDUs
30 + 00
0 + 00
5
19 EDUs
10 + 00
0 + 00
WWTP
3
21 EDUs
40 + 00

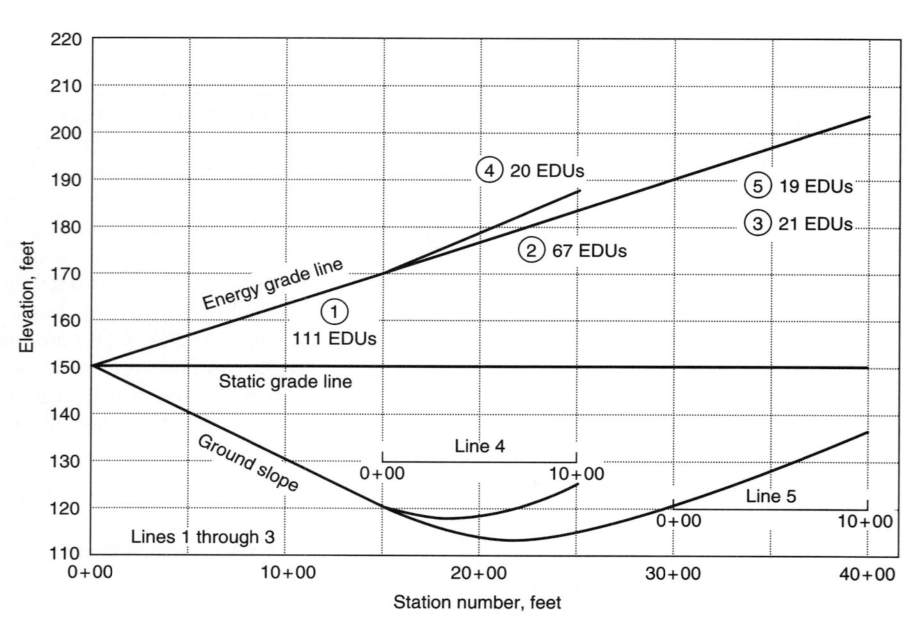

Elevation, feet
220
210
200
190
180
170
160
150
140
130
120
110
4 20 EDUs
5 19 EDUs
3 21 EDUs
2 67 EDUs
Energy grade line
1
111 EDUs
Static grade line
Ground slope
Line 4
0+00
10+00
Line 5
0+00
10+00
Lines 1 through 3
0+00
10+00
20+00
30+00
40+00
Station number, feet

2. Set up a spreadsheet for the computation of sewer system pipe sizes. The pipe section numbers for the five pipe sections, starting at the downstream end, are entered in column 1.

Hydraulic computations for a grinder pump pressure collection system

Pipe number (1)	Cumulative EDUs (2)	Maximum number operating (3)	Design flow, gal/min (4)	Upstream station (5)	Downstream station (6)	Length, ft (7)
1	111	8	88.0	15+00	0+00	1500
2	67	7	77.0	30+00	15+00	1500
3	21	5	55.0	40+00	30+00	1000
4	20	5	55.0	10+00	0+00	1000
5	19	5	55.0	10+00	0+00	1000

Pipe number (1)	Nominal diameter, in (8)	Inside diameter, in (9)	Slope of EGL, % (10)	Cross-sectional area, in² (11)	Velocity, ft/s (12)	Headloss, ft (13)
1	3.0	3.166	1.43	7.87	3.59	21.4
2	3.0	3.166	1.12	7.87	3.14	16.8
3	2.5	2.601	1.56	5.31	3.32	15.6
4	2.5	2.601	1.56	5.31	3.32	15.6
5	2.5	2.601	1.56	5.31	3.32	15.6

3. For each pipe, determine the cumulative number of EDUs served. EDUs that pump directly to the pipeline being considered, plus all upstream EDUs whose discharge will also be conveyed by the pipeline, are included. The cumulative number of EDUs served by each pipe section is entered in column 2.
4. Use the information provided in Table 6-3 to determine the maximum number of grinder pumps likely to be operating simultaneously. These values are based on the number of cumulative EDUs served (column 2). Enter the values in column 3.
5. Determine the peak design flowrate by multiplying the maximum number of grinder pumps operating simultaneously (column 3) by 11 gal/min. For the pipe number 1, the peak design flowrate is

   $$\text{Design peak flowrate} = (8 \text{ grinder pumps operating})(11 \text{ gal/min}) = 88.0 \text{ gal/min}$$

   These values are entered in column 4.
6. Enter the station number for the upstream end of each pipe section in column 5.
7. Enter the station number for the downstream end of each pipe section in column 6.
8. Calculate the length of pipeline in each section (column 5 − column 6). The computed lengths are entered in column 7.
9. Enter an assumed nominal pipe diameter value for each pipe section (this value can be modified successively until a satisfactory solution is achieved). To achieve adequate pipeline velocities where flowrates are small, the use of pipe sizes under 2 in is acceptable. The assumed values are entered in column 8.
10. For the nominal pipe sizes selected, enter the inside diameter values (see Table 6-4) in column 9.
11. Use the Hazen-Williams formula [Eq. (6-5)] to calculate the slope of the energy grade line for each pipe. Ideally, these values should fall within the range of 0.5 to

1.5 percent. However, because of the need to maintain adequate velocities in the collection system, slightly higher values may be necessary. For the pipeline in section 1:

$$d = 3.166 \text{ in}$$

$$Q = 88.0 \text{ gal/min}$$

$$C = 150$$

$$\text{Slope} = \left(\frac{10.5}{3.166^{4.87}}\right)\left(\frac{88.0}{150}\right)^{1.85} = 0.0143 = 1.43\%$$

The slope values are entered in column 10. When compared to the solution to Example 6-2, it can be seen that the energy grade line slope values calculated in this solution are higher as a result of the increased flowrates.

12. Using the pipe inside diameters (column 9), calculate the pipe cross-sectional areas, and enter the values in column 11.
13. Divide the design flowrate (column 4) by the pipe cross-sectional area (column 11) to determine the velocity of flow for each pipe section. Enter the values in column 12. Check to see that the calculated velocity falls within acceptable ranges for grinder pump sewer design (ideally, the velocity should be 3 to 5 ft/s).
14. Calculate the headloss due to friction for each pipe section by multiplying the energy grade line slope value (column 10) by the pipe length (column 7). If the slope value is expressed as a percentage, remember to divide the slope value by 100 before multiplying. The headloss values are entered in column 13.
15. Using the profile created in step 1, plot the energy grade lines for each pipe section. Begin at the downstream end, and add the headloss values to the static grade line. The height of the energy grade line represents the pump head necessary for flow to enter the system when peak flow conditions exist. Notice that the required pump head values are greater for the grinder pump pressure sewer collection system than for the STEP pressure sewer collection system.

6-6 VACUUM SEWERS

Vacuum sewers operate on the basis of differential air pressure created by a central vacuum pump station. Wastewater from each household or connection is discharged to a sump that is isolated from the main vacuum line by a vacuum/gravity valve that is normally closed. When a predetermined volume of wastewater collects in the sump, an air pressure actuator allows the vacuum/gravity valve to open. The differential pressure between the air pressure in the sump and the vacuum in the sewer propels the wastewater into the sewer leading to the vacuum station. The valve remains open long enough for a quantity of air to enter the collection system behind the wastewater. The spring-loaded valve is controlled pneumatically and requires no electricity to operate. The typical components of a vacuum valve pit are shown in Fig. 6-15.

The flow of wastewater within the vacuum line approximates the form of a spiral rotating hollow cylinder as it travels along the vacuum line. Eventually, the flow pattern disintegrates because of pipe friction, and the wastewater flows to low points

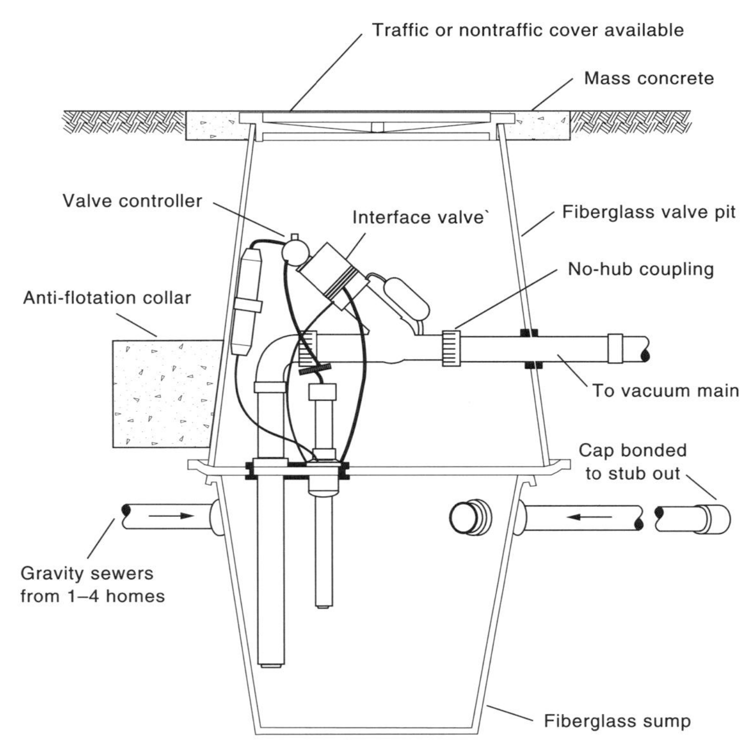

FIGURE 6-15
Typical components of vacuum collection system sump and valve pit (from AIRVAC).

in the vacuum line. The wastewater that accumulates at the low points is pushed over the sawtooth in the line (see Fig. 6-1*e*) by the next liquid cylinder that enters the system. The two-phase flow that occurs in a vacuum system is, at present, not understood completely (U.S. EPA, 1991). The vacuum in the collection system is maintained by a vacuum pump, located at a collection station such as shown in Fig. 6-16. Once the effluent reaches the vacuum collection station receiving tank, it is pumped under pressure to the treatment facility. Vacuum sewers are normally designed in conjunction with the manufacturer. There are currently two manufacturers of vacuum sewers operating in the United States: AIRVAC and Iseki.

Application of Vacuum Sewers

Site conditions that are considered conducive to the use of vacuum sewers include:

- Flat terrain
- Slightly rolling terrain with small elevation changes
- High water table
- Rocky or unstable soils

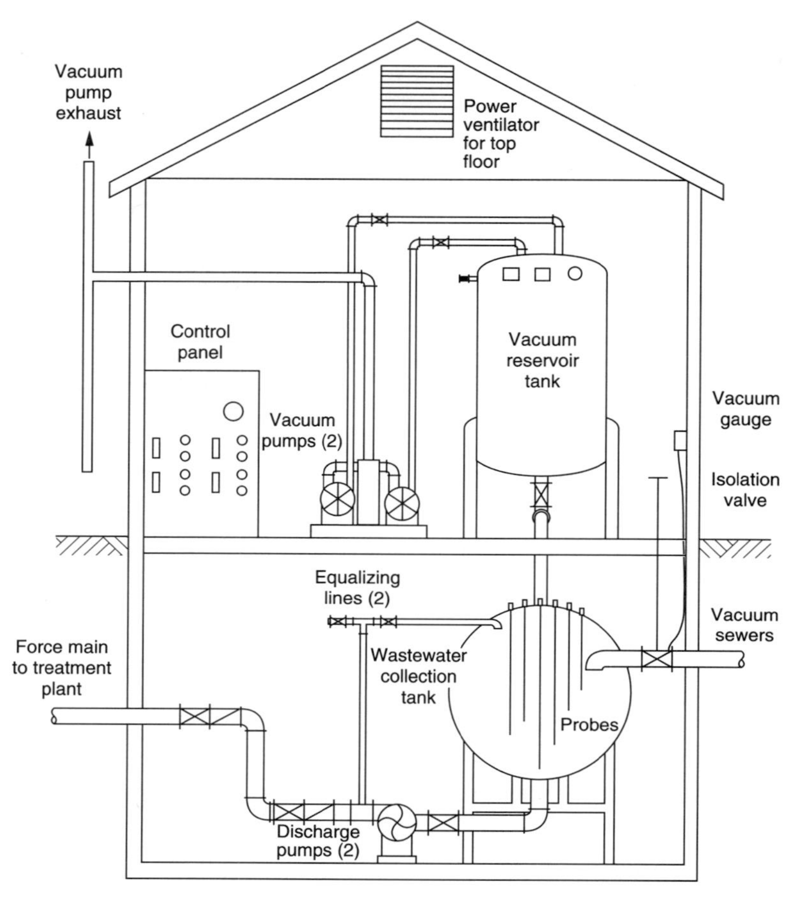

FIGURE 6-16
Typical vacuum station for vacuum collection system (from AIRVAC).

On the basis of past experience, a minimum of 70 to 100 connections are needed for a vacuum sewer system to be cost-effective (U.S. EPA, 1991). Below the minimum for a custom vacuum station, a less expensive package vacuum station can be considered for as few as 25 connections. The typical number of connections per custom vacuum station is 200 to 300.

Vacuum sewers have a maximum lift of about 15 to 20 ft (4.6 to 6.1 m). For level or slight upgrade transport a sawtooth configuration of collection piping is used, as shown in Fig. 6-17. The minimum slope on the vacuum mainline is 0.2 percent between lifts.

Onsite System Components

The elements of a vacuum sewer system include the service, the collection mains, and the vacuum station. The service includes the vacuum valve, an auxiliary vent, the valve sump, and (for larger flows) buffer tanks.

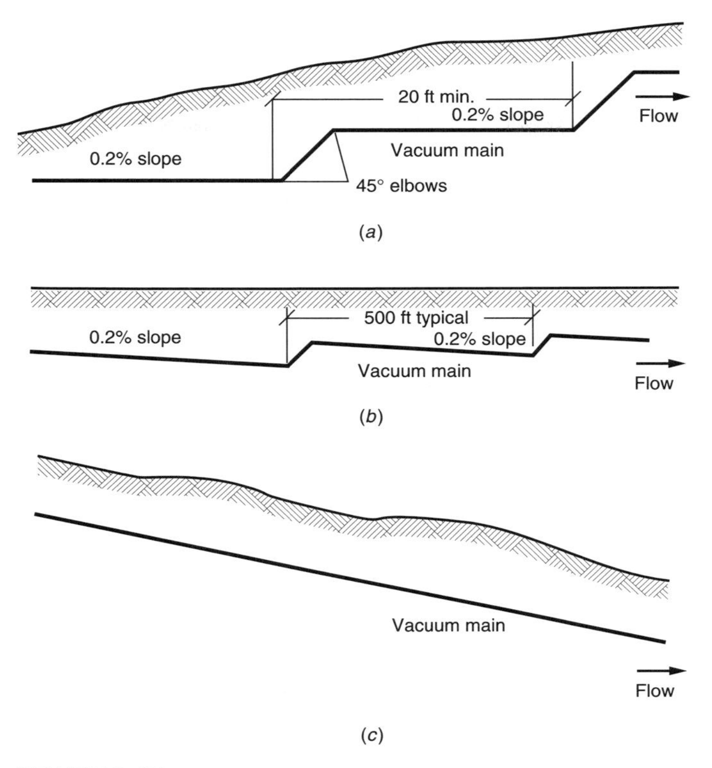

FIGURE 6-17
Typical configurations for vacuum collection system: (*a*) upgrade transport, (*b*) level grade transport, and (*c*) downgrade transport (adapted from AIRVAC).

Collection System Components

Offsite system components include the collection mains and the vacuum station. The collection mains are typically PVC and range from 3 to 8 in (75 to 200 mm) in diameter. The vacuum station is usually designed by the manufacturer and includes a standby generator.

Design flows are peak flowrates that are expected to occur once or twice a day. The equation for peak flowrate is (U.S. EPA, 1991)

$$Q_{DP} = 0.5N + 20 \tag{6-7}$$

where Q_{DP} = design peak flowrate, gal/min
0.5 = measured discharge per EDU, gal/min·EDU
N = number of equivalent dwelling units (EDUs)
20 = typical safety factor (larger values provide a greater safety factor), gal/min

Vacuum sewers are designed to operate on two-phase flows (air and liquid), with air being added for twice the time period as for the liquid. Tangential liquid velocities in the typical vacuum sewer are 15 to 18 ft/s (4.6 to 5.5 m/s), which are well above the minimum of 3 ft/s (0.9 m/s) required for self-cleaning. Friction loss charts for sidewall/diameter ratio (SDR) 21 PVC pipe and 2:1 air/liquid ratio have been developed by AIRVAC and are contained in their design manual (AIRVAC, 1989).

Design Considerations

Although the Hazen-Williams formula [Eq. (6-5)] is used typically to calculate friction headlosses, it must be remembered that flow in vacuum sewer systems is complex. The flow is unsteady (it occurs in surges, with intermediate quiescent phases), two-phase (initially, a foamy combination of air and wastewater enters the system), and may consist of both pressure-driven and gravity-driven flow. Therefore, the use of the Hazen-Williams equation can provide only an estimate of friction losses within the system. Use of empirical data, based on observation and experience with operating vacuum systems, is recommended for designing vacuum sewer wastewater collection systems.

Vacuum sewer valves operate pneumatically, according to pressure differences. For the valve to function properly, a clear passage through the collection system (above the water surface) should exist, so that vacuum pressures will be transmitted to the most distant points of the system. The use of excessively high lifts should be avoided, because they can create conditions where water fills the pipe, eliminating the required airway. For this reason, many short lifts are preferred to few high lifts. Ideally, as a conveyance pipeline descends between lifts (typically, at a slope of 0.2 percent or greater), the crown (interior top surface) of the pipe at the downstream end immediately before a lift should not descend lower than the invert (interior bottom surface) of the upstream end of the pipe after the previous lift. For example, when a 4-in (100-mm) pipe descends at a slope of 0.2 percent, the pipe drops 4 in after a distance of 167 ft (51 m). If a 1-in (25-mm) passageway is to be maintained above the surface of standing water collected in the low parts, then the 4-in (100-mm) pipe should descend only 3 in (75 mm) between lifts. For a slope of 0.2 percent, this will occur every 125 ft (38 m). Therefore, for ideal operation, 3-in (75-mm) lifts should occur every 125 ft. However, this may be impractical, and 12-in lifts located every 500 ft have been used successfully. Typical design data used in the design of vacuum sewer systems are presented in Table 6-14. The layout and design of a vacuum sewer collection main is illustrated in Example 6-4.

Construction Considerations

Construction of a vacuum sewer system is similar to a STEP/STEG system in that pipelines are buried at shallow depths. Vacuum service lines are buried at a minimum of 30 in (750 mm). Line changes can be made with vacuum sewers using 45° bends (no 90° bends). Adequate backfill compaction in the area of the abrupt change is critical.

Historical construction problems with vacuum sewers have been analyzed in a 1989 survey and categorized as short-term and long-term. Short-term problems were

TABLE 6-14
Typical design data for vacuum sewer wastewater collection systems

Item	Unit	Range	Typical
Height of water level on vacuum discharge valve	in	3–40	30
Air/liquid ratio		1–10	2
Vacuum maintained in collection system	in Hg	12–29	16
Vacuum maintained in collection system	ft H_2O	14–33	18
Trench depth	ft	3–5	4
Houses per collector sump		1–4	2
Service lateral pipeline diameter	in	2–4	3
Collection main pipeline diameter	in	4–10	6
Vacuum pump operation time	h/d	3–5	4

solvent welding in cold weather and onsite pit alignment. Solvent welding should be avoided under 39°F (4°C) or solvent-welded joints should be replaced by gasketed pipe and fittings (U.S. EPA, 1991). Pit alignment problems can be eliminated in detailed design by selecting the proper configuration of valve pit stubout and vacuum service line alignment.

The long-term problems identified were valve pit settlement and excessive use of fittings. Valve pit settlement can be solved by proper compaction. New systems are installed using gasketed pipe and solvent-welded fittings. The use of fittings can be minimized by proper planning and by installing the valve pits before the vacuum sewer lines are installed.

Operation and Maintenance

Early vacuum systems were plagued by consistent operational problems. The mean time between service calls (MTBSC) was typically less than 4 years (U.S. EPA, 1991). By contrast the systems surveyed in 1989 average an MTBSC of 10 years. Operating experience for four vacuum sewer systems is presented in Table 6-15.

TABLE 6-15
Operation and maintenance experience with vacuum sewers*

		Location			
Item	Unit	Westmoreland, Tennessee	Wheeling, West Virginia	Lake Chataqua, Celeron, New York	White House, Tennessee
Customers	Number	540	250	2500	360
Operation and maintenance labor	h/EDU·yr	4.4	2.4	0.6	1.5
Power	kWh/EDU·yr	460	160	190	180
Service calls	no./yr	48	24	40	24
MTBSC†	yr	10.2	8.3	22.5	10.8

*From U.S. EPA (1991).
†MTBSC = mean time between service calls.

EXAMPLE 6-4. LAYOUT AND DESIGN OF VACUUM SEWER COLLECTION MAIN. Design a vacuum sewer collection main to serve a small lakeside resort community of 60 EDUs, as shown in the plan view given below. The residences are located around the perimeter of the lake. Wastewater will be drawn from individual collection sumps, through the vacuum system, to the vacuum collection station shown in the diagram. Assume the following conditions apply:

1. Design peak flowrate = (0.5 gal/EDU·min)(number of EDUs) + 20 gal/min
2. Pipeline material = Class 200 PVC pipe (see Tables 6-4 and 6-5)
3. Hazen-Williams coefficient = 150
4. Pipeline friction headloss must be less than 5.0 ft
5. Pipeline profile change static headloss must be less than 8.0 ft

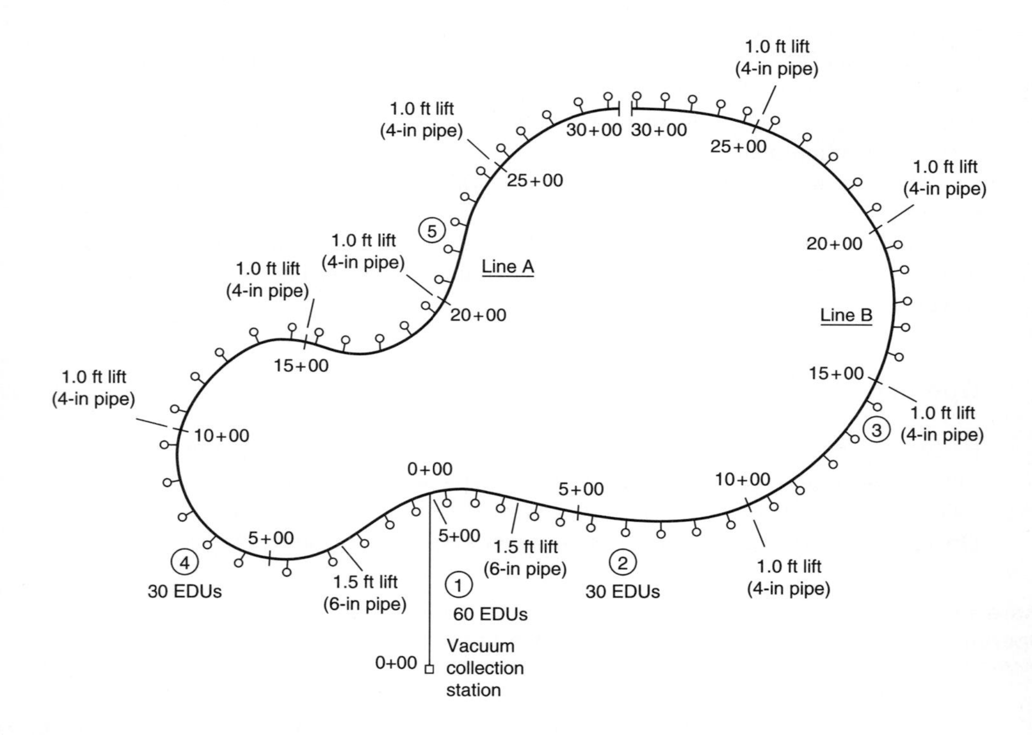

Solution

1. Use the community plan view to divide the vacuum sewer collection system pipeline into sections. It is recommended that flow distribution be divided as evenly as possible. Therefore, the lake perimeter pipeline has been split equally into two branches, each 3000 ft long. A spur pipeline runs from the branch junction to the vacuum collection station. The length of the spur pipeline is 500 ft, so the maximum flow distance is 3500 ft.
2. Calculate the design peak flowrate using Eq. (6-7). A total of 60 EDUs will be served by the vacuum sewer collection system, so the total design peak flowrate is

Design peak flowrate = (0.5 gal/EDU · min)(60 EDUs) + 20 gal/min = 50 gal/min

The flows associated with various distances from the vacuum collection station are given in the table below.

Station number	Cumulative number of EDUs served	Design peak flowrate, gal/min
30 + 00	0	20.0
25 + 00	5	22.5
20 + 00	10	25.0
15 + 00	15	27.5
10 + 00	20	30.0
5 + 00	25	32.5
0 + 00	30	35.0
0 + 00*†	60	50.0

*Combined flowrate from lines A and B.
† Station 0 + 00 corresponds to station 5 + 00 on line 1.

3. Determine the pipe sizes to be used. It is recommended that 3-in pipe be used only for the discharge from four EDUs or fewer, and that 4-in pipe be used for flowrates up to 38.0 gal/min. For this solution, 4-in pipe will be used for both branches. Because the use of 4-in pipe is not recommended for lengths over 2000 ft, only the first 2000 ft (from the point furthest from the vacuum collection station) will consist of 4-in pipe. For the remaining 1000 ft of each branch, 6-in pipe will be used. The use of 6-in pipe is recommended for flowrates up to 106 gal/min. Therefore, the spur pipeline from the branch junction to the vacuum collection station will also be 6 in.
4. Confirm the pipeline choices made by verifying that friction headlosses in the system will be less than 5 ft. Using Eq. (6-5) for the first 2000 ft of branch pipeline (diameter = 4 in),

$$h_f = (10.5)(2000\text{ ft})\left(\frac{30.0\text{ gal/min}}{150}\right)^{1.85}(4.072\text{ in})^{-4.87} = 1.15\text{ ft}$$

For the remaining 1000 ft of branch pipeline (diameter = 6 in),

$$h_f = (10.5)(1000\text{ ft})\left(\frac{35.0\text{ gal/min}}{150}\right)^{1.85}(5.993\text{ in})^{-4.87} = 0.12\text{ ft}$$

For the 500-ft spur pipeline (diameter = 6 in),

$$h_f = (10.5)(500\text{ ft})\left(\frac{50.0\text{ gal/min}}{150}\right)^{1.85}(5.993\text{ in})^{-4.87} = 0.11\text{ ft}$$

The combined friction headloss for the entire flow path is 1.38 ft, well below the friction headloss limit of 5.0 ft. Notice that the friction headlosses for the branch pipeline sections were calculated by using the maximum flowrates occurring in the pipe (the flowrate at the end of each pipe section) over the total section distances. Because the flowrate is actually incremental, the true friction headlosses are lower than the values calculated.

5. Determine the pipeline slope. A minimum pipeline slope of 0.2 percent is recommended. Where the groundslope is greater than 0.2 percent, the pipeline slope may follow the natural groundslope. However, as the terrain in this example is essentially flat, the pipeline will be installed at a slope of 0.2 percent (a drop of 12 in for every 500 ft).
6. Determine the distance between profile changes (lifts). As established in step 1, the vacuum system consists of two 3000-ft branches. When the branch length is added to the 500-ft spur length, the maximum flow path length is 3500 ft. Using the slope determined in the previous step, a total drop of 7 ft will occur over this length. To prevent pipe depths from becoming excessive, lifts will be used to raise the pipe level periodically.

   Each lift consists of connections made with two 45° elbows to raise the pipe level back to its original depth (see Fig. 6-17). Using profile changes, the pipeline assumes a sawtooth configuration. Wastewater moves by gravity flow down the longer pipe sections (sloped downward at 0.2 percent), and is pushed upward through the short lift sections by the flushing action caused by surges of wastewater entering the system.

   It is recommended that 1.0-ft lifts, occurring every 500 ft, be used for 4-in pipe, and that 1.5-ft lifts, occurring every 750 ft, be used for 6-in pipe. Lifts for the 4-in pipe will be installed at station numbers 30+00, 25+00, 20+00, and 15+00. Lifts for the 6-in pipe will be installed in both branches at station number 7+50. No profile changes will occur in the spur pipeline section. By installing these lifts, the overall pipeline drop will not exceed 1.5 ft, minimizing trench excavation and cover.
7. Confirm that the pipeline lifts do not reduce available head to unacceptable levels by verifying that the cumulative static headlosses (due to lifts) are less than 8.0 ft. Static headlosses can be calculated by subtracting the pipe diameter from the lift height. For the flow path through either branch, the static headlosses are:

$$\text{4-in pipeline static headloss} = (4 \text{ lifts})(12 - 4 \text{ in}) = 32 \text{ in}$$

$$\text{6-in pipeline static headloss} = (1 \text{ lift})(18 - 6 \text{ in}) = 12 \text{ in}$$

   The cumulative static headloss for each flow path is 44 in, which is below the limit of 8.0 ft.

PROBLEMS AND DISCUSSION TOPICS

6-1. List four geologic conditions that would favor the use of pressure sewers versus conventional gravity sewers.

6-2. What are the difficulties in using air release valves in pressure sewer systems?

6-3. What conditions would cause the release of dissolved gases in pressure sewers?

6-4. What problems are caused by the accumulation of gases in a pressure sewer?

6-5. Verify the hydraulic computations given in Table 6-5 for the 2-in, 3-in, or 4-in (to be determined by instructor) diameter pipelines.

6-6. Describe why a variable-grade gravity sewer can have zero or negative slopes with septic tank effluent.

6-7. Design a STEG system for the area shown below (alternative to be selected by instructor). Assume that Class 200 PVC pipe will be used, that the minimum pipe size is 2.0 in, and that the pipe Hazen-Williams coefficient is 150.

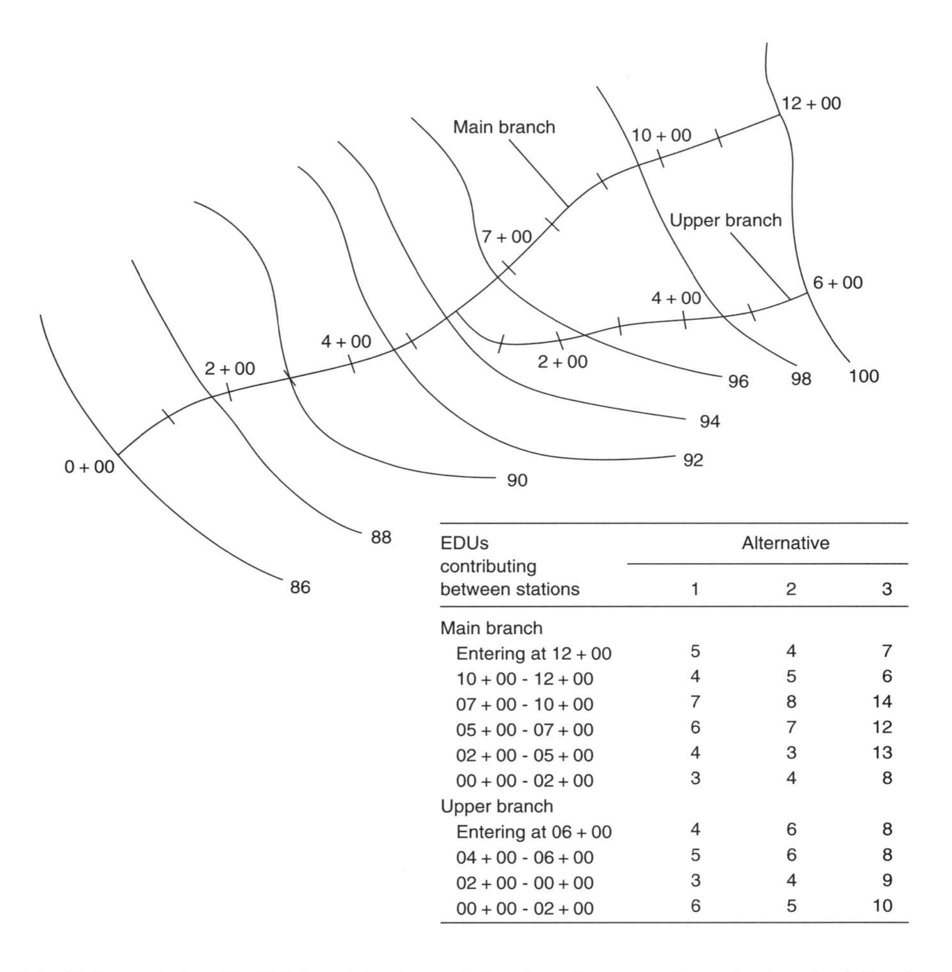

EDUs contributing between stations	Alternative 1	2	3
Main branch			
Entering at 12 + 00	5	4	7
10 + 00 - 12 + 00	4	5	6
07 + 00 - 10 + 00	7	8	14
05 + 00 - 07 + 00	6	7	12
02 + 00 - 05 + 00	4	3	13
00 + 00 - 02 + 00	3	4	8
Upper branch			
Entering at 06 + 00	4	6	8
04 + 00 - 06 + 00	5	6	8
02 + 00 - 00 + 00	3	4	9
00 + 00 - 02 + 00	6	5	10

6-8. If the static head is 100 ft and the dynamic head can be approximated using the following expression, estimate the discharge from an individual STEP system using the pump curve given in Fig. 6-9.

$$\text{Dynamic head} = 20\frac{Q^2}{2g}$$

$$Q = \text{discharge, gal/min}$$

6-9. Why are low-capacity high-head turbine pumps more suitable for use in STEP systems, as compared to conventional centrifugal pumps?

6-10. Design a STEP system for one of the subdivisions shown below (to be selected by instructor). Assume that Class 200 PVC pipe will be used (see Table 6-4), and that the minimum pipe size will be 1.5 in.

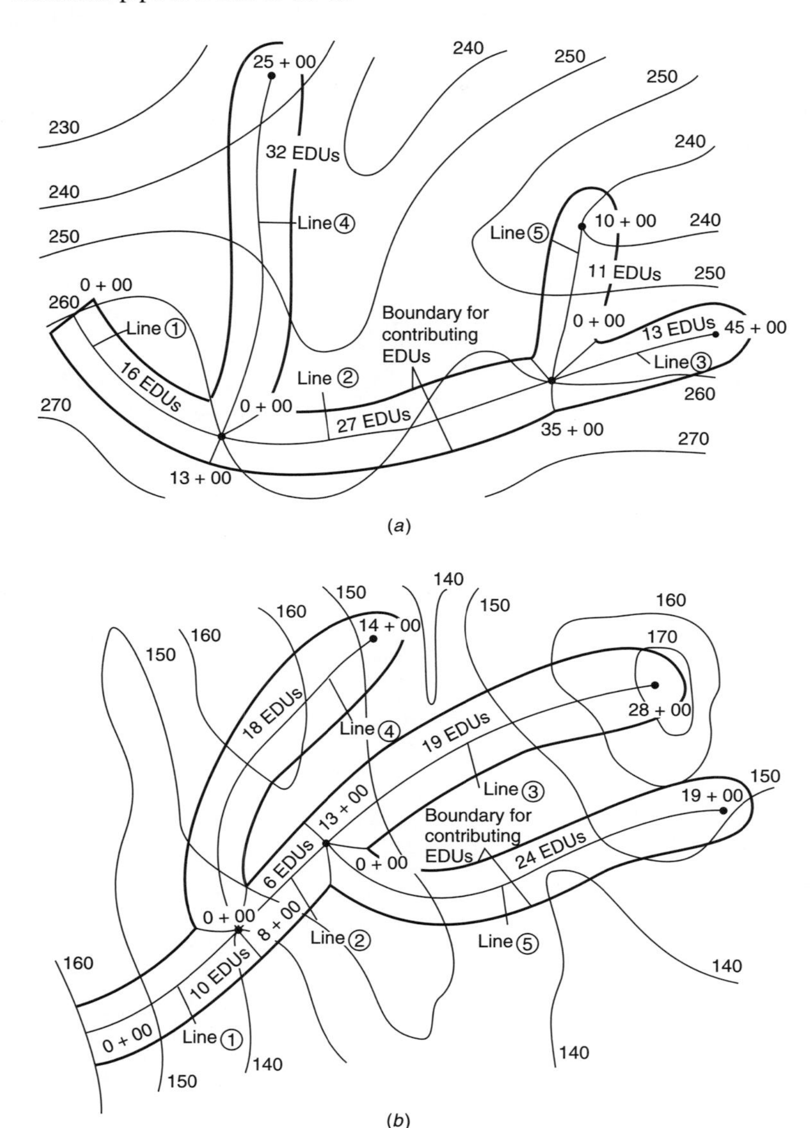

(*a*)

(*b*)

6-11. If the static head is 80 ft and the dynamic head can be approximated using the following expression, estimate the discharge from an individual grinder pump system using the pump curve given in Fig. 6-14.

$$\text{Dynamic head} = 16\frac{Q^2}{2g}$$

$$Q = \text{discharge, gal/min}$$

6-12. What role can microtunnelling play in STEP system construction?

6-13. Design a grinder pump collection system for one of the subdivisions used in Prob. 6-10 (to be selected by instructor). Assume that Class 200 PVC pipe will be used (see Table 6-4).

6-14. Why would a vacuum sewer system be favored over conventional gravity sewers to service small lots surrounding a lake?

REFERENCES

AIRVAC (1989) *Design Manual, Vacuum Sewerage Systems,* Rochester, IN.

Angelakis, A. N. and S. Spyridakis (1996). The Status of Water Resources in Minoan Times: A Preliminary Study. In A. N. Angelakis and A. Issar, Editors. *Diachronic Climatic Impacts on Water Resources in Mediterranean Region,* Springer-Verlag, Heidelberg, Germany.

ASCE (1992) Gravity Sanitary Sewer Design and Construction, Joint Task Force of ASCE and WPCF, ASCE Manuals and Reports on Engineering Practice No. 60, WPCF MOP No. FD-5, American Society of Civil Engineers, New York.

Bounds, T. (1992) Study Provides Formulas for Determining Septic Tank Pumping Intervals, *Small Flows,* Vol. 7, No. 4, National Small Flows Clearinghouse, West Virginia University, Morgantown, WV.

Bounds, T. (1997) Personal Communication, Orenco Systems, Inc. Sutherlin, OR.

CES (1996) Alternative Wastewater Management Project, City of Austin, TX, Community Environmental Services, Inc., Austin, TX.

Crites, R., C. Lekven, S. Wert, and G. Tchobanoglous (1997) Decentralized Wastewater System for a Small Residential Development in California, *Small Flows Journal,* Vol. 3, Issue 1, Morgantown, WV.

Environment/One (1992) Design Handbook: Low Pressure Sewer Systems Using *Environment/One Grinder Pumps,* Schenectady, NY.

Fair, G. M. (1968) Converted Sewer System, United States Patent 3,366,339, filed Nov. 26, 1965, and assigned by the inventor to the public.

Orenco Systems Inc. (1996) *Alternative Sewer Designs,* Sutherlin, OR.

Parker, M. (1997) Personal communication. Roseburg, OR.

Rezek, J. W., and I. A. Cooper (1985) *Investigations of Existing Pressure Sewer Systems,* EPA 600/2-85/051, Environmental Protection Agency, Cincinnati, OH.

SAHC (1986) *Public Health Inspection Guide No. 6: Common Effluent Drainage Schemes,* South Australian Health Commission, Adelaide, South Australia.

Sanks, R. L., G. M. Jones, B. E. Bosserman, and G. Tchobanoglous (eds.) (1998) *Pumping Station Design,* 2nd ed., Butterworths, Stoneham, MA.

Simmons, J. D., J. O. Newman, and C.W. Rose (1982) Small Diameter, Variable-Grade Gravity Sewers for Septic Tank Effluent, On-Site Sewage Treatment, Proceedings of the Third National Symposium on Individual and Small Community Sewer Treatment, American Society of Agricultural Engineers, ASAE Publication 1-82, pp. 130–138.

Tchobanoglous, G. (1981) *Wastewater Engineering: Collection and Pumping of Wastewater,* McGraw-Hill, New York.

U.S. EPA (1991) *Alternative Wastewater Collection Systems.* EPA 625/1-91/024, Center for Environmental Research Information, U.S. Environmental Protection Agency, Cincinnati, OH.

WPCF (1986) Alternative Sewer Systems, MOP No. FD-12, Water Pollution Control Federation, Alexandria, VA.

CHAPTER 7

Biological Treatment and Nutrient Removal

The principal objectives of biological treatment are to stabilize the organic matter and to coagulate and remove the nonsettleable colloidal solids found in domestic wastewater and septic tank effluent. Depending on local circumstances, additional objectives may include the removal of nutrients such as nitrogen and phosphorus, as well as trace organic compounds. At the most fundamental level, biological treatment involves: (1) the conversion of the dissolved and colloidal carbonaceous organic matter into various gases and cell tissue, (2) the formation of biological flocs composed of the cells and inorganic colloidal matter present in settled influent wastewater, and (3) the subsequent removal of the biological flocs by gravity settling. It should be noted, however, that if the cell tissue produced is not removed by settling, the cell tissue in the wastewater will still exert a BOD and the treatment will be incomplete.

Over the years it has been found that, with proper analysis and environmental control, almost all wastewaters can be treated biologically. Because of the importance of biological treatment with respect to the protection of the environment, it is the purpose of this chapter to introduce the concepts and practice of biological wastewater treatment. Subjects to be considered include an introduction to: (1) biological treatment methods; (2) bacterial metabolism, energetics, and growth, (3) biological treatment kinetics; (4) modeling of biological treatment kinetics; (5) biological nutrient removal; (6) aerobic suspended-growth process; (7) attached-growth process; (8) anaerobic suspended- and attached-growth and hybrid processes; and (9) pre-engineered (package) wastewater treatment plants.

7-1 INTRODUCTION TO BIOLOGICAL TREATMENT METHODS

As an introduction to biological wastewater treatment it will first be helpful to introduce some common terminology that is used to describe the various biological treatment methods, and then to consider the major classification and application of the treatment processes.

TABLE 7-1
Some useful definitions dealing with biological wastewater treatment

Term	Definition
Aerobic (oxic) processes	Biological treatment processes that occur in the presence of oxygen.
Anaerobic processes	Biological treatment processes that occur in the absence of oxygen.
Anoxic process	The process by which nitrate nitrogen is converted biologically to nitrogen gas in the absence of oxygen. This process is also known as anoxic denitrification.
Facultative processes	Biological treatment processes in which the organisms can function in the presence or absence of molecular oxygen.
Hybrid (combined) processes	Various combinations of aerobic, anoxic, and anaerobic processes grouped together to achieve a specific treatment objective.
Attached-growth processes	Biological treatment processes in which the microorganisms responsible for the conversion of the organic matter or other constituents in the wastewater to gases and cell tissue are attached to some inert medium, such as rocks, slag, or specially designed ceramic or plastic materials. Attached-growth treatment processes are also known as fixed-film processes.
Biological nutrient removal	The term applied to the removal of nitrogen and phosphorus in biological treatment processes.
Carbonaceous BOD removal	Biological conversion of the carbonaceous organic matter in wastewater to cell tissue and various gaseous end products. In the conversion, it is assumed that the nitrogen present in the various compounds is converted to ammonia.
Denitrification	The biological process by which nitrate is converted to nitrogen and other gaseous end products.
Hybrid processes	Term used to describe combined processes (e.g., combined suspended- and attached-growth processes).
Lagoon processes	A generic term applied to treatment processes that take place in ponds or lagoons with various aspect ratios and depths.
Nitrification	The two-stage biological process by which ammonia is converted first to nitrite and then to nitrate.
Stabilization	The biological process by which the organic matter in the sludges produced from the primary settling and biological treatment of wastewater is stabilized, usually by conversion to gases and cell tissue. Depending on whether this stabilization is carried out under aerobic or anaerobic conditions, the process is known as aerobic or anaerobic digestion.
Substrate	The term used to denote the organic matter or nutrients that are converted during biological treatment or that may be limiting in biological treatment. For example, the carbonaceous organic matter in wastewater is referred to as the substrate that is converted during biological treatment.
Suspended-growth processes	Biological treatment processes in which the microorganisms responsible for the conversion of the organic matter or other constituents in the wastewater to gases and cell tissue are maintained in suspension within the liquid.

Some Useful Definitions

Common terms used in the field of biological wastewater treatment and their definitions are presented in Table 7-1. The first five entries in Table 7-1 refer to the metabolic function of the processes. As reported, the principal processes used for the biological treatment of wastewater can be classified with respect to their metabolic function as aerobic processes, anaerobic processes, anoxic processes, facultative processes, and combined processes. The individual processes are accomplished in suspended-growth systems, attached-growth systems, or combinations thereof. For example, an aerobic treatment process may be a suspended or attached growth process.

Terminology used commonly to describe the types of treatment processes is presented in the second group of entries in Table 7-1. It should be noted that all of the biological processes used for the treatment of wastewater are derived from processes occurring in nature. The aerobic and anaerobic cycles discussed previously in Chap. 3 are typical examples. By controlling the environment of the microorganisms, the decomposition of wastes is increased. Regardless of the type of waste, the biological treatment process consists of controlling the environment required for optimum growth of the microorganisms involved.

Application of Biological Treatment Processes

The principal applications of the processes identified in Table 7-1 are, as summarized in Table 7-2, for: (1) the removal of the carbonaceous organic matter in wastewater, usually measured as biochemical oxygen demand (BOD), chemical oxygen demand (COD), or total organic carbon (TOC); (2) nitrification; (3) denitrification; (4) phosphorus removal; and (5) waste stabilization. The emphasis in this chapter will be on the removal of carbonaceous material and biological nutrient removal. Sludge stabilization is discussed in Chap. 14.

7-2 INTRODUCTION TO MICROBIAL METABOLISM, ENERGETICS, AND GROWTH

To appreciate the role and function of microorganisms in bringing about the biological treatment of wastewater, it will be helpful to consider first the nature of the metabolism, energetics, and growth of bacteria, the microorganisms of primary importance in biological treatment. Treatment kinetics are considered in the following section.

Bacterial Metabolism

The process by which microorganisms grow and obtain energy is complex and intricate; there are many pathways and cycles. Bacterial metabolism, which encompasses all of the chemical reactions that go on within a cell, is illustrated in a simplified form in Fig. 7-1. The two principal reactions that make up the metabolic process are known

TABLE 7-2
Major biological treatment processes used for wastewater treatment

Type	Common name	Use*
Aerobic processes:		
Suspended-growth	Activated-sludge processes	Carbonaceous BOD removal, nitrification
	Aerated lagoons	Carbonaceous BOD removal, nitrification
	Aerobic digestion	Stabilization, carbonaceous BOD removal
Attached-growth	Trickling filters	Carbonaceous BOD removal, nitrification
	Rotating biological contactors	Carbonaceous BOD removal, nitrification
	Packed-bed reactors	Carbonaceous BOD removal, nitrification
Hybrid (combined) suspended- and attached-growth processes	Trickling filter/activated sludge	Carbonaceous BOD removal (nitrification)
	Constructed wetland	Carbonaceous BOD removal (nitrification)
Anoxic processes:		
Suspended-growth	Suspended-growth denitrification	Denitrification
Attached-growth	Fixed-film denitrification	Denitrification
Anaerobic processes:		
Suspended-growth	Anaerobic contact processes	Carbonaceous BOD removal
	Anaerobic digestion	Stabilization, carbonaceous BOD removal
Attached-growth	Anaerobic fixed bed	Carbonaceous BOD removal, waste stabilization (denitrification)
Hybrid	Upflow anaerobic sludge blanket	Carbonaceous BOD removal, especially high-strength wastes
	Upflow sludge blanket/ fixed-bed reactor	Carbonaceous BOD removal
Combined aerobic, anoxic, and anaerobic processes:		
Suspended-growth	Single- or multistage processes, various proprietary processes	Carbonaceous BOD removal, nitrification, denitrification, and phosphorus removal
Combined suspended- and attached-growth	Single- or multistage processes	Carbonaceous BOD removal, nitrification, denitrification, and phosphorus removal
Lagoon processes:		
Aerobic lagoons	Aerobic lagoons	Carbonaceous BOD removal
Maturation (tertiary) lagoons	Maturation (tertiary) lagoons	Carbonaceous BOD removal (nitrification)
Facultative lagoons	Facultative lagoons	Carbonaceous BOD removal
Anaerobic lagoons	Anaerobic lagoons	Carbonaceous BOD removal (waste stabilization)

*Major uses are presented first; other uses are identified in parentheses.

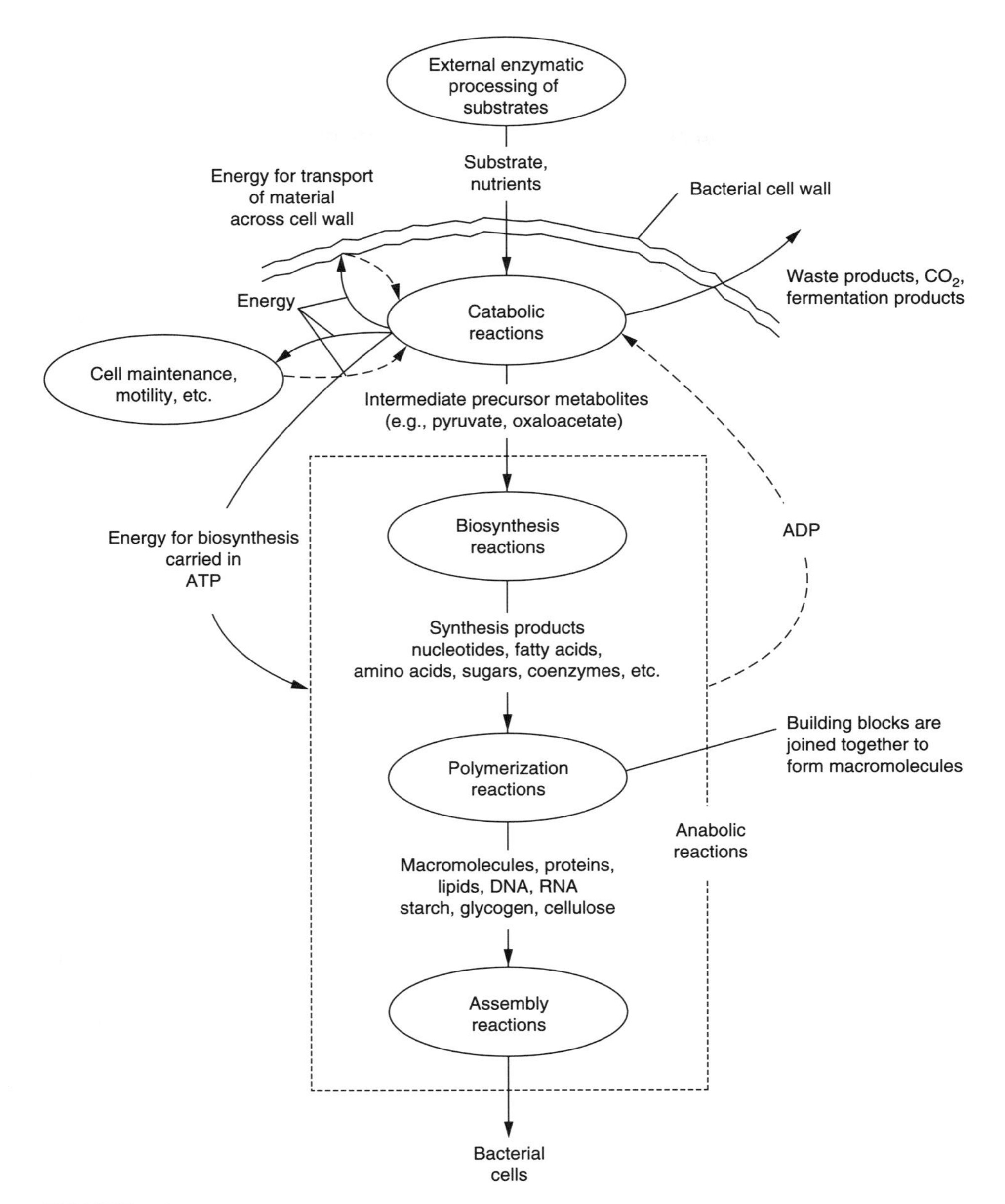

FIGURE 7-1
Schematic representation of bacterial metabolism composed principally of catabolic and anabolic reactions.

as: (1) catabolic reactions and (2) anabolic reactions. *Catabolic reactions* result in the breakdown of complex organic molecules into simpler substances, along with the release of energy. *Anabolic reactions* bring about the formation of more complex molecules, and usually require energy. The energy for the anabolic reactions is derived from catabolic reactions.

Importance of enzymes. Vital to the catabolic and anabolic reactions depicted in Fig. 7-1 are the actions of enzymes, which are organic catalysts produced by the living cell. Enzymes are proteins, or proteins combined either with an inorganic molecule or with a low-molecular-weight organic molecule. As catalysts, enzymes have the capacity to increase the speed of chemical reactions greatly without being altered themselves. There are two general types of enzymes, extracellular and intracellular. When the substrate or nutrient required by the cell is unable to enter the cell wall, the extracellular enzyme converts the substrate or nutrient to a form that can then be transported into the cell. Intracellular enzymes are involved in the catabolic and anabolic reactions within the cell (see Fig. 7-1).

Enzymes are known for their high degree of efficiency in converting substrate to end products. One enzyme molecule can change many molecules of substrate per minute to end products. Enzymes are also known for their high degree of substrate specificity. This high degree of specificity means that the cell must produce a different enzyme for every substrate it uses. An enzyme reaction can be represented by the following general equation:

$$\underset{\text{Enzyme}}{(E)} + \underset{\text{Substrate}}{(S)} \rightarrow \underset{\substack{\text{Enzyme-}\\\text{substrate}\\\text{complex}}}{(E)(S)} \rightarrow \underset{\text{Product}}{(P)} + \underset{\text{Enzyme}}{(E)} \qquad (7\text{-}1)$$

As illustrated, the enzyme functions as a catalyst by forming a complex with the substrate, which is then converted to a product and the original enzyme. At this point, the product may be acted upon by another enzyme. In fact, a sequence of complexes and products may be formed before the final end product is produced. In a living cell, the transformation of the original substrate to the final end product is accomplished by such an enzyme system. The activity of enzymes is substantially affected by pH and temperature, as well as the substrate concentration. Each enzyme has a particular optimum pH and temperature. The optimum pH and temperature of the key enzymes in the cell are reflected in the overall temperature and pH preferences of the cell.

Need for energy. Along with enzymes, energy is required to carry out the biochemical reactions in the cell. Energy is released in the cell by oxidizing organic or inorganic matter (catabolic reactions) or by a photosynthetic reaction. The energy released is captured and stored in the cell by certain organic compounds. The most common storing compound is adenosine triphosphate (ATP). The energy captured by this compound is used for cell synthesis, maintenance, and motility (see Fig. 7-1). When the ATP molecule has expended its captured energy to the anabolic reactions involved in cell synthesis and for cell maintenance, it changes to a lower energy state called adenosine diphosphate (ADP). This ADP molecule can again capture the energy released in the breakdown of organic or inorganic matter, resuming its

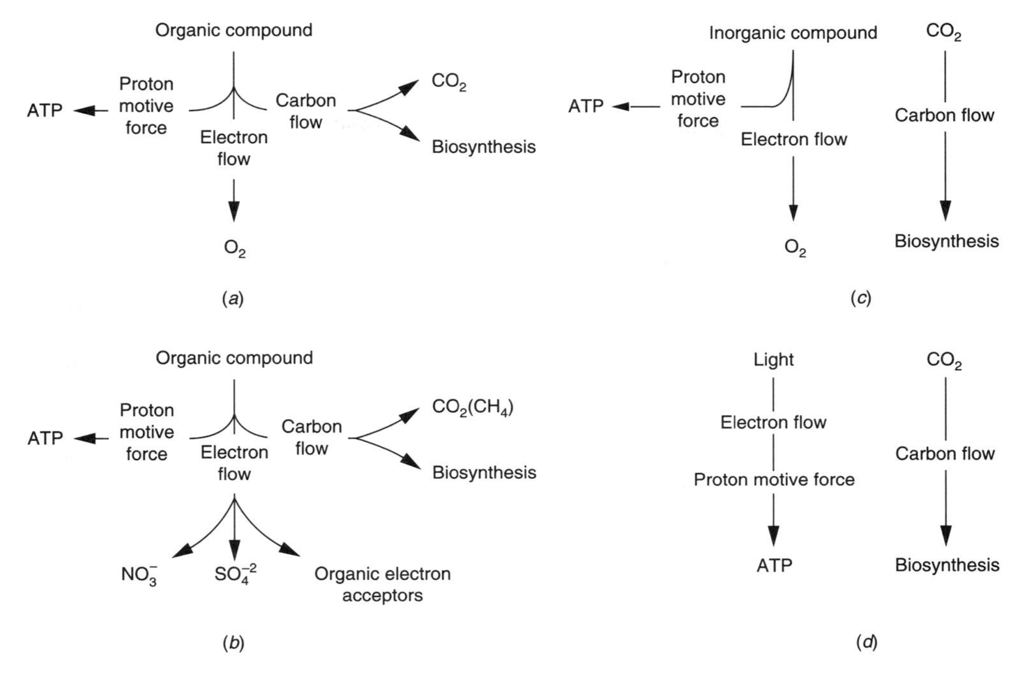

FIGURE 7-2
Flow of energy and carbon in (*a*) aerobic respiration, (*b*) anaerobic respiration, (*c*) chemolithotrophic metabolism, and (*d*) phototrophic metabolism (adapted from Madigan et al., 1996).

energized state as the ATP molecule. The diagram shown in Fig. 7-1 applies to aerobic, anaerobic, or facultative organisms.

For heterotrophic bacteria, only a portion of the organic waste is converted into end products. The energy obtained from this biochemical reaction is used in the synthesis of the remaining organic matter into new cells. As the organic matter in the wastewater becomes limiting, there will be a decrease in cellular mass, because of the utilization of cellular material without replacement. If this situation continues, eventually all that will remain of the cell is a relatively stable organic residue. The net decrease in cellular mass is termed *endogenous respiration*. The overall process was depicted previously in Fig. 2-16 in Chap. 2. The general flow of energy and carbon for aerobic and anaerobic respiration and chemolithotrophic and phototrophic metabolism is depicted in Fig. 7-2. As shown, the carbon source for the synthesis of new cells for chemolithotrophic and phototrophic organisms is carbon dioxide. The energy for cell synthesis is either from the energy given off from inorganic oxidation-reduction reactions or from light.

Nutrient requirements. Nutrients, rather than organic or inorganic substrate in the wastewater, may at times be the limiting material for cell synthesis and growth. Bacteria require nutrients, principally nitrogen and phosphorus, for growth. These nutrients may not always be present in sufficient quantities, as in the case of high-carbohydrate industrial wastes (e.g., sugar beets, sugar cane). Nutrient addition to the waste may be necessary for the proper growth of the bacteria and the subsequent degradation of the waste material.

Bioenergetics

Bioenergetics involves the application of thermodynamic principles to biological reactions and processes. The material presented in this section is intended only to serve as an introduction to this subject, which is beyond the scope of this book. Chemical reactions are accompanied by changes in energy. The amount of energy gained or lost during the reaction defines the total amount of energy involved. In thermodynamic terms free energy G^o, known as *Gibbs free energy,* is the energy available to do work. The change in energy due to reaction is termed ΔG^o. The superscript is used to denote the fact that the free energy values were obtained at standard conditions (i.e., pH $= 7.0$ and $T = 25°C$). If ΔG^o for a reaction is negative, energy is released and the reaction will proceed spontaneously as written. Such reactions are termed *exergonic*. If ΔG^o is positive, the reaction does not proceed as written, but the reverse reaction will occur. Such reactions are termed *endergonic*. Reactions with positive ΔG^o values can be completed if energy is added.

In living organisms, the utilization of energy is brought about by oxidation-reduction reactions that are linked together in that one compound loses electrons (electron donor) while another gains electrons (electron acceptor). Using hydrogen gas as the electron donor, the reaction is

$$H_2 \rightarrow 2H^+ + 2e^- \tag{7-2}$$

Equation (7-2) is known as a *half-reaction*. Because oxidation-reduction reactions are linked, a second reaction in which a substance is reduced must be found. If oxygen is used as the electron acceptor, the following reaction can be written.

$$0.5O_2 + 2e^- \rightarrow O^{-2} \tag{7-3}$$

When these two reactions are coupled, the resulting complete oxidation-reduction reaction is

$$\begin{array}{l} H_2 \rightarrow 2H^+ + 2e^- \\ 0.5O_2 + 2e^- \rightarrow O^{-2} \\ \hline H_2 + 0.5O_2 \rightarrow H_2O \end{array} \tag{7-4}$$

Consider the following oxidation-reduction reactions for the complete oxidation of carbohydrates.

$$\begin{array}{ll} 0.25CH_2O + 0.25H_2O \rightarrow 0.25CO_2 + H^+ + e^- & \Delta G^o = 41.84 \\ 0.25O_2 + H^+ + e^- \rightarrow 0.5H_2O & \Delta G^o = -78.14 \\ \hline 0.25CH_2O + 0.25O_2 \rightarrow 0.25CO_2 + 0.25H_2O & \Delta G^o = -36.30 \\ CH_2O + O_2 \rightarrow CO_2 + H_2O & \end{array} \tag{7-5}$$

TABLE 7-3
Typical electron acceptors in bacterial reactions commonly encountered in the management of wastewaters

Electron environment	Acceptor	Process
Aerobic	Oxygen, O_2	Aerobic metabolism
Anaerobic	Nitrate, NO_3^-	Denitrification
Anaerobic	Sulfate, SO_4^{-2}	Sulfate reduction
Anaerobic	Carbon dioxide, CO_2	Methanogenesis

Because the net free energy value is negative ($\Delta G = -36.30$), the above oxidation-reduction reaction will proceed as written. As noted previously in Chap. 2 (see Table 2-14), organisms that use chemicals as sources of energy and electrons are known as chemotrophs. Chemotrophs that use organic compounds as energy sources and electron donors are called *chemoorganotrophs*. Chemotrophs able to use inorganic reduced compounds such as ammonia (NH_3), hydrogen gas (H_2), and hydrogen sulfide (H_2S) as energy sources are known as *chemolithotrophs. Phototrophs* are able to use light as an energy source. For chemoorganotrophs, the electron acceptor under aerobic conditions is oxygen. Under anaerobic conditions, nitrate (NO_3^-), sulfate (SO_4^{-2}), and carbon dioxide (CO_2), as well as some organic compounds, can serve as electronic acceptors (see Table 7-3). It should also be noted that there are a number of other electron acceptors under anaerobic conditions (Madigan et al., 1997). Oxygen is the electron acceptor for chemolithotrophs.

McCarty (1971, 1975) has developed half-reactions and free energy values for a variety of electron donors and electron acceptors, a brief list of which is presented in Table 7-4. The various half-reactions can be used to describe the processes depicted in Fig. 7-2 by combining them according to the following relationship (McCarty, 1971, 1975).

$$R = f_s R_s + f_e R_a - R_d \tag{7-6}$$

where
R = overall balanced reaction
R_s = half-reaction for synthesis of cell tissue
R_a = half-reaction for electron acceptor
R_d = half-reaction for electron donor
f_s = fraction of electron donor used for cell synthesis
f_e = fraction of electron donor used for energy
$f_s + f_e = 1$

The minus sign in Eq. (7-6) means that the electron donor equation given in Table 7-4 must be reversed and then added to the other two equations. The value of f_s will vary with the type of electron donor as well as the type of reaction (heterotrophic versus autotrophic) and electron acceptor. A typical value of f_s where oxygen is the electron acceptor ranges from 0.6 for protein to 0.72 for carbohydrate. In the first equation given in Table 7-4, the term $C_5H_7O_2N$, first proposed by Hoover and Porges (1952), is used to represent bacterial cell tissue. Application of Eq. (7-6) is illustrated in Example 7-1. Additional details on bioenergetics may be found in McCarty (1971, 1975).

TABLE 7-4
Half-reactions for biological systems*

Reaction number	Half-reaction		$\Delta G^\circ(W)$,[†] kJ per electron equivalent
	Reactions for bacterial cell synthesis (R_s)		
	Ammonia as nitrogen source:		
1.	$\frac{1}{5}CO_2 + \frac{1}{20}HCO_3^- + \frac{1}{20}NH_4^+ + H^+ + e^-$	$= \frac{1}{20}C_5H_7O_2N + \frac{9}{20}H_2O$	
	Nitrate as nitrogen source:		
2.	$\frac{1}{28}NO_3^- + \frac{5}{28}CO_2 + \frac{29}{28}H^+ + e^-$	$= \frac{1}{28}C_5H_7O_2N + \frac{11}{28}H_2O$	
	Reactions for electron acceptors (R_a)		
	Oxygen:		
3.	$\frac{1}{4}O_2 + H^+ + e^-$	$= \frac{1}{2}H_2O$	−78.14
	Nitrate:		
4.	$\frac{1}{5}NO_3^- + \frac{6}{5}H^+ + e^-$	$= \frac{1}{10}N_2 + \frac{3}{5}H_2O$	−71.67
	Sulfate:		
5.	$\frac{1}{8}SO_4^{-2} + \frac{19}{16}H^+ + e^-$	$= \frac{1}{16}H_2S + \frac{1}{16}HS^- + \frac{1}{2}H_2O$	21.27
	Carbon dioxide (methane fermentation):		
6.	$\frac{1}{8}CO_2 + H^+ + e^-$	$= \frac{1}{8}CH_4 + \frac{1}{4}H_2O$	24.11
	Reactions for electron donors (R_d)		
	Organic donors (heterotrophic reactions)		
	Domestic wastewater:		
7.	$\frac{9}{50}CO_2 + \frac{1}{50}NH_4^+ + \frac{1}{50}HCO_3^- + H^+ + e^-$	$= \frac{1}{50}C_{10}H_{19}O_3N + \frac{9}{25}H_2O$	31.80
	Protein (amino acids, proteins, nitrogenous organics)		
8.	$\frac{8}{33}CO_2 + \frac{2}{33}NH_4^+ + \frac{31}{33}H^+ + e^-$	$= \frac{1}{66}C_{16}H_{24}O_5N_4 + \frac{27}{66}H_2O$	32.22
	Carbohydrate (cellulose, starch, sugars):		
9.	$\frac{1}{4}CO_2 + H^+ + e^-$	$= \frac{1}{4}CH_2O + \frac{1}{4}H_2O$	41.84
	Grease (fats and oils):		
10.	$\frac{4}{23}CO_2 + H^+ + e^-$	$= \frac{1}{46}C_8H_{16}O + \frac{15}{46}H_2O$	27.61
	Methanol:		
11.	$\frac{1}{6}CO_2 + H^+ + e^-$	$= \frac{1}{6}CH_3OH + \frac{1}{6}H_2O$	37.51

(*continued*)

TABLE 7-4
(*Continued*)

Reaction number	Half-reaction		$\Delta G_o(W)$,† kJ per electron equivalent
Inorganic donors (autotrophic reactions):			
12.	$FE^{+3} + e^-$	$= FE^{+2}$	−74.40
13.	$\frac{1}{2}NO_3^- + H^+ + e^-$	$= \frac{1}{2}NO_2^- + \frac{1}{2}H_2O$	−40.15
14.	$\frac{1}{8}NO_3^- + \frac{5}{4}H^+ + e^-$	$= \frac{1}{8}NH_4^+ + \frac{3}{8}H_2O$	−34.50
15.	$\frac{1}{6}NO_2^- + \frac{4}{3}H^+ + e^-$	$= \frac{1}{6}NH_4^+ + \frac{1}{3}H_2O$	−32.62
16.	$\frac{1}{6}SO_4^{-2} + \frac{4}{3}H^+ + e^-$	$= \frac{1}{6}S + \frac{2}{3}H_2O$	19.48
17.	$\frac{1}{8}SO_4^{-2} + \frac{19}{16}H^+ + e^-$	$= \frac{1}{16}H_2S + \frac{1}{16}HS^- + \frac{1}{2}H_2O$	21.28
18.	$\frac{1}{4}SO_4^{-2} + \frac{5}{4}H^+ + e^-$	$= \frac{1}{8}S_2O_3^{-2} + \frac{5}{8}H_2O$	21.30
19.	$H^+ + e^-$	$= \frac{1}{2}H_2$	40.46
20.	$\frac{1}{2}SO_4^{-2} + H^+ + e^-$	$= SO_3^{-2} + H_2O$	44.33

*Adapted from McCarty (1975) and Sawyer et al. (1994).
† Reactants and products at unit activity except $[H^+] = 10^{-7}$.

EXAMPLE 7-1. WRITE A BALANCED REACTION FOR THE OXIDATION OF GLUCOSE. Write a balanced reaction for the oxidation of glucose (carbohydrate) assuming that ammonia will serve as the nitrogen source for cell tissue. Assume also that oxygen is the electron acceptor and that f_s for the reaction is 0.7. What quantity of cells is produced per pound of glucose converted?

Solution

1. Write a balanced reaction using Eq. (7-6) and the half-reactions given in Table 7-4.

$$R = 0.7 \text{ (reaction 1)} + 0.3 \text{ (reaction 3)} - \text{reaction 9}$$

$$(0.7)R_s,\ 0.14CO_2 + 0.035HCO_3^- + 0.035NH_4^+ + 0.7H^+ + 0.7e^- = 0.035C_5H_7O_2N + 0.315H_2O$$

$$(0.3)R_a,\ 0.075O_2 + 0.3H^+ + 0.3e^- = 0.15H_2O$$

$$-R_d,\ 0.25CH_2O + 0.25H_2O = 0.25CO_2 + H^+ + e^-$$

$$R,\ \underset{\substack{0.25(30)\\7.5}}{0.25CH_2O} + 0.035HCO_3^- + 0.035NH_4^+ + 0.075O_2 = 0.11CO_2 + \underset{\substack{0.035(113)\\3.96}}{0.035C_5H_7O_2N} + 0.215H_2O$$

2. Determine the pounds of cells produced per pound of glucose oxidized:

$$\text{Cells produced} = 3.96 \text{ lb}/7.5 \text{ lb} = 0.53 \text{ lb/lb glucose converted}$$

Comment. Use of half-reactions to write balanced reactions makes it possible to study a variety of biological conversion processes.

Bacterial Growth

Bacteria can reproduce by binary fission, by a sexual mode, or by budding. Generally, they reproduce by binary fission; the original cell becomes two new organisms. The time required for each fission, termed the *generation time,* can vary from days to less than 20 min. For example, if the generation time is 30 min, one bacterium would yield 16,777,216 bacteria after a period of 12 h. This computed value is a hypothetical figure, for bacteria would not continue to divide indefinitely because of various environmental limitations, such as substrate availability, nutrient concentration, or even system size. The growth of bacteria in terms of numbers and bacterial mass in pure culture is considered below. Growth in mixed cultures is also considered.

Growth in terms of bacterial numbers. The general growth pattern of bacteria in a batch culture is shown in Fig. 7-3. Initially, a small number of organisms are inoculated into a fixed volume of culture medium, and the number of viable organisms is recorded as a function of time. The growth pattern based on the number of cells has four more or less distinct phases.

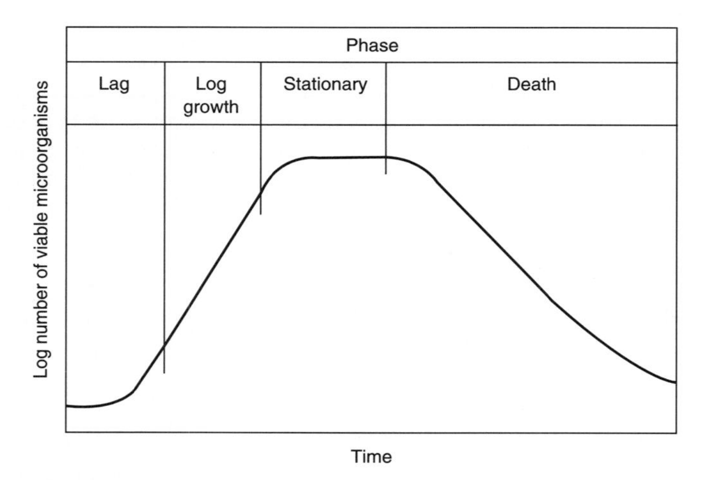

FIGURE 7-3
Typical bacterial growth curve in terms of log of number of viable organisms.

1. *The lag phase.* Upon addition of an inoculum to a culture medium, the lag phase represents the time required for the organisms to acclimate to their new environment and begin to divide.
2. *The log-growth phase.* During this period the cells divide at a rate determined by their generation time and their ability to process food (constant percentage growth rate).
3. *The stationary phase.* Here the population remains stationary. Reasons advanced for this phenomenon are: (*a*) that the cells have exhausted the substrate or nutrients necessary for growth and (*b*) that the growth of new cells is offset by the death of old cells.
4. *The log-death phase.* During this phase the bacteria death rate exceeds the production of new cells. The death rate is usually a function of the viable population and environmental characteristics. In some cases, the log-death phase is the inverse of the log-growth phase.

Growth in terms of bacterial mass. The corresponding growth pattern, in terms of the mass of microorganisms, as illustrated in Fig. 7-4, can be described as follows:

1. *The lag phase.* Again, bacteria require time to acclimate to their nutritional environment. The lag phase in terms of bacterial mass is not as long as the corresponding lag phase in terms of numbers because mass begins to increase before cell division takes place.
2. *The log-growth phase.* There is always an excess amount of food surrounding the microorganisms, and the rate of metabolism and growth is only a function of the ability of the microorganism to process the substrate.
3. *Declining growth phase.* The rate of increase of bacterial mass decreases because of limitations in the food supply.

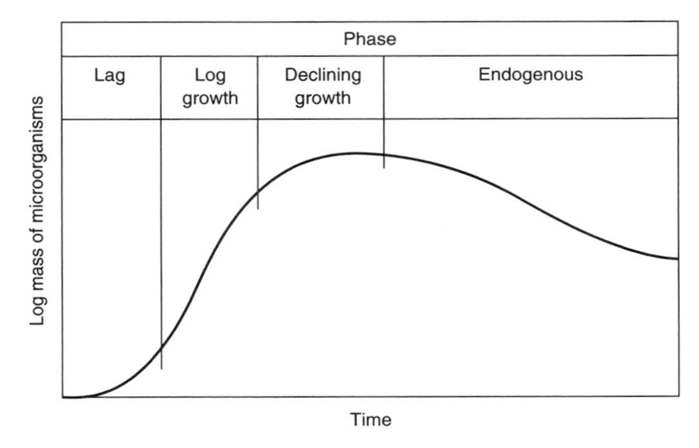

FIGURE 7-4
Typical bacterial growth curve in terms of log of mass of organisms.

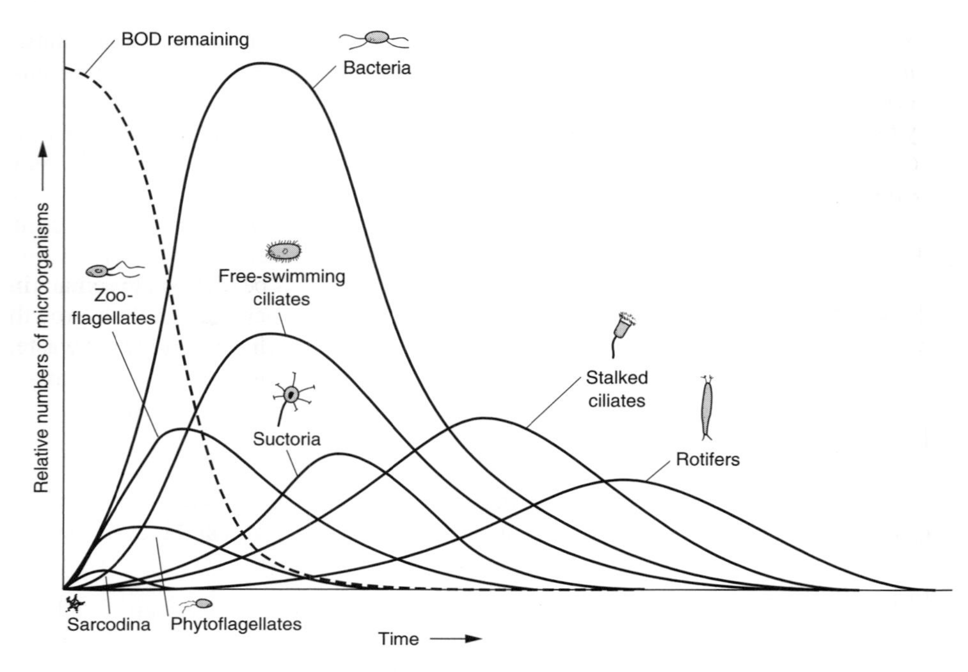

FIGURE 7-5
Relative growth of microorganisms stabilizing organic waste in a liquid environment.

4. *Endogenous phase.* The microorganisms are forced to metabolize their own protoplasm without replacement, because the concentration of available food is at a minimum. During this phase, a phenomenon known as *lysis* can occur in which the nutrients remaining in the dead cells diffuse out to furnish the remaining cells with food (known as *cryptic growth*).

Growth in mixed cultures. It is important to note that in the preceding discussion only a single population of microorganisms was considered. Most biological treatment processes are composed of complex, interrelated, mixed biological populations, with each particular microorganism in the system having its own growth curve. The position and shape of a particular growth curve in the system, on a time scale, will depend on the food and nutrients available and on environmental factors, such as temperature and pH and whether the system is aerobic or anaerobic. While bacteria are of primary importance, many other microorganisms take part in the stabilization of the organic waste (see Fig. 7-5).

7-3 INTRODUCTION TO BIOLOGICAL TREATMENT KINETICS

The general classes of microorganisms of importance in wastewater treatment have been discussed previously in Chap. 2. Bacterial metabolism, energetics, and growth are considered in Sec. 7-2. Although the characteristics of the environment needed for their growth have been described, nothing has been said about how to control the environment of the microorganisms. Environmental conditions can be controlled by pH regulation, temperature regulation, nutrient or trace-element addition, oxygen addition or exclusion, and proper mixing. Control of the environmental conditions will ensure that the microorganisms have a proper medium in which to grow. For example, flow equalization, as discussed in Chap. 5, can be utilized to equalize the flow rate as well as the organic loading rate.

To ensure that the microorganisms will grow, they must be allowed to remain in the system long enough to reproduce. The required period depends on their growth rate, which is related directly to the rate at which they metabolize or utilize the waste. Assuming that the environmental conditions are controlled properly, effective waste stabilization can be ensured by controlling the growth rate of the microorganisms. Biological growth kinetics and energetics are considered below.

Cell Growth

In both batch and continuous culture reactors in which the contents of the reactor are mixed completely, the rate of growth of bacterial cells can be defined by the following relationship:

$$r_g = \mu X \tag{7-7}$$

where r_g = rate of bacterial growth, mass/unit volume·time
μ = specific growth rate, time^{-1}
X = concentration of microorganism, mass/unit volume

Because $dX/dt = r_g$ for batch culture (see Sec. 3-4, Chap. 3), the following relationship is also valid for a batch reactor:

$$\frac{dX}{dt} = \mu X \tag{7-8}$$

Substrate-Limited Growth

In a batch culture, if one of the essential requirements (substrate and nutrients) for growth were present in only limited amounts, it would be depleted first and growth would cease. In a continuous culture, it has been found that the effect of a limiting substrate or nutrient can often be defined adequately using the following expression proposed by Monod (1942, 1949):

$$\mu = \mu_m \frac{S}{K_s + S} \tag{7-9}$$

where μ = specific growth rate, time^{-1}
μ_m = maximum specific growth rate, time^{-1}

S = concentration of growth-limiting substrate in solution, mass/unit volume

K_s = half-velocity constant, substrate concentration at one-half the maximum growth rate, mass/unit volume

The effect of substrate concentration on the specific growth rate is shown in Fig. 7-6.

If the value of μ from Eq. (7-9) is substituted in Eq. (7-7), the resulting expression for the rate of growth is

$$r_g = \frac{\mu_m XS}{K_s + S} \qquad (7\text{-}10)$$

Cell Growth and Substrate Utilization

In both batch and continuous-growth culture systems, a portion of the substrate is converted to new cells and a portion is oxidized to inorganic and organic end products. Because the quantity of new cells produced has been observed to be reproducible for a given substrate, the following relationship has been developed between the rate of substrate utilization and the rate of growth.

$$r_g = -Yr_{su} \qquad (7\text{-}11)$$

where r_g = rate of bacterial growth, mass/unit volume·time

Y = maximum yield coefficient, mg/mg (defined as the ratio of the mass of cells formed to the mass of substrate consumed, measured during any finite period of logarithmic growth)

r_{su} = substrate utilization rate, mass/unit volume·time

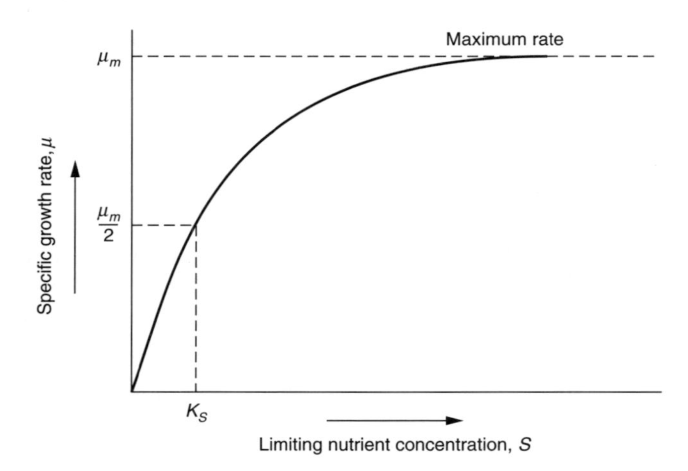

FIGURE 7-6
Effects of a limiting nutrient on the specific growth rate.

On the basis of laboratory studies, it has been concluded that yield depends on: (1) the oxidation state of the carbon source and nutrient elements, (2) the degree of polymerization of the substrate, (3) pathways of metabolism, (4) the growth rate, and (5) various physical parameters of cultivation.

If the value of r_g from Eq. (7-10) is substituted in Eq. (7-11), the rate of substrate utilization can be defined as follows:

$$r_{su} = -\frac{\mu_m XS}{Y(K_s + S)} \tag{7-12}$$

In Eq. (7-12), the term μ_m/Y is often replaced by the term k, defined as the maximum rate of substrate utilization per unit mass of microorganisms:

$$k = \frac{\mu_m}{Y} \tag{7-13}$$

If the term k is substituted for the term (μ_m/Y) in Eq. (7-12), the resulting expression is:

$$r_{su} = -\frac{kXS}{K_s + S} \tag{7-14}$$

Effects of Endogenous Metabolism

In bacterial systems used for wastewater treatment, the distribution of cell ages is such that not all the cells in the system are in the log-growth phase. Consequently, the expression for the rate of growth must be corrected to account for the energy required for cell maintenance. Other factors, such as death and predation, must also be considered. These factors are usually lumped together, and it is assumed that the decrease in cell mass caused by them is proportional to the concentration of organisms present. This decrease is often identified in the literature as the *endogenous decay*. In equation form, endogenous respiration can be represented as follows:

$$\underset{\substack{\text{(cells)}\\113\\1}}{C_5H_7O_2N} + \underset{\substack{160\\1.42}}{5O_2} \xrightarrow{\text{bacteria}} 5CO_2 + 2H_2O + NH_3 + \text{energy} \tag{7-15}$$

Although the endogenous respiration reaction shown above results in relatively simple end products and energy, stable organic end products are also formed. From Eq. (7-15), it can be seen that, if all of the cells were oxidized completely, the COD (also the UBOD) of the cells is equal to 1.42 times the concentration of cells.

The endogenous decay term can be formulated as follows:

$$r_d \text{ (endogenous decay)} = -k_d X \tag{7-16}$$

where k_d = endogenous decay coefficient, time^{-1}, and X = concentration of cells, mass/unit volume. When Eq. (7-16) is combined with Eqs. (7-10) and (7-11), the following expressions are obtained for the net rate of growth:

$$r'_g = \frac{\mu_m XS}{K_s + S} - k_d X \tag{7-17}$$

$$= -Yr_{su} - k_d X \tag{7-18}$$

where r'_g = net rate of bacterial growth, mass/unit volume·time.

The corresponding expression for the net specific growth rate is given by the following equation, which is the same as the expression proposed by Van Uden (1967):

$$\mu' = \mu_m \frac{S}{K_s + S} - k_d \tag{7-19}$$

where μ' = net specific growth rate, time^{-1}.

The effects of endogenous respiration on the net bacterial yield are accounted for by defining an observed yield as follows (Ribbons, 1970, Van Uden, 1967):

$$Y_{obs} = \frac{r'_g}{r_{su}} \tag{7-20}$$

Effects of Temperature

The temperature dependence of the biological reaction-rate constants is very important in assessing the overall efficiency of a biological treatment process. Temperature not only influences the metabolic activities of the microbial population, but also has a profound effect on such factors as gas-transfer rates and the settling characteristics of the biological solids. The effect of temperature on the reaction rate of a biological process is usually expressed in the following form:

$$r_T = r_{20}\theta^{(T-20)} \tag{7-21}$$

where r_T = reaction rate at T°C
r_{20} = reaction rate at 20°C
θ = temperature activity coefficient
T = temperature, °C

Values of θ for biological processes vary from about 1.02 to 1.09, with 1.04 being typical. Theta values for biological processes should not be confused with values given previously in Chap. 3 for the BOD determination.

Other Rate Expressions

In reviewing the kinetic expressions used to describe the growth of microorganisms and the removal of substrate, it is very important to remember that the expressions presented are empirical and are used for the purpose of illustration, and that they are not the only expressions available. Other expressions which have been used to describe the rate of substrate utilization include the following:

$$r_{su} = -k \tag{7-22}$$

$$r_{su} = -kS \tag{7-23}$$

$$r_{su} = -kXS \tag{7-24}$$

$$r_{su} = -kX\frac{S}{S_o} \tag{7-25}$$

$$r_{su} = -\frac{kS}{K_s + S_o} \tag{7-26}$$

What is fundamental in the use of any rate expression is its application in a mass balance analysis. In this connection, it does not matter if the rate expression selected has no relationship to those used commonly as described in the literature, so long as it describes the observed phenomenon. It is equally important to remember that specific rate expressions should not be generalized to cover a broad range of situations on the basis of limited data or experience.

7-4 INTRODUCTION TO THE MODELING OF BIOLOGICAL TREATMENT KINETICS

The primary purpose of this section is to illustrate the application of the growth kinetics considered in the previous section to the biological treatment of wastewater. In Sec. 7-2, cell growth and substrate utilization were defined in terms of a single well-defined substrate (food source). Historically, in wastewater treatment, the constituents that are normally measured include BOD_5, COD, and TSS. It should be noted that the use of BOD_5 and TSS as regulatory parameters has become institutionalized.

Unfortunately, because all three of these constituents are aggregate constituents, as described in Chap. 2, the individual components that make up these constituents are unknown. Thus, in the modeling of biological treatment kinetics, which follows, it is assumed that the influent substrate to be treated is soluble and does not contain suspended solids. The use of a soluble substrate is reasonable for the purpose of mathematical development. To develop a more fundamental understanding, the individual components composing the substrate must be defined. Efforts that have been made to model biological treatment on a more fundamental level are discussed at the end of this section. It should be noted that the modeling of biological treatment processes is extremely useful as a diagnostic tool in understanding and evaluating process operation and performance.

Mass Balances for Suspended-Growth Complete-Mix Processes without Recycle

Operationally, biological waste treatment can be accomplished by using a process flow diagram such as that shown in Fig. 7-7*a*. The treatment process depicted in

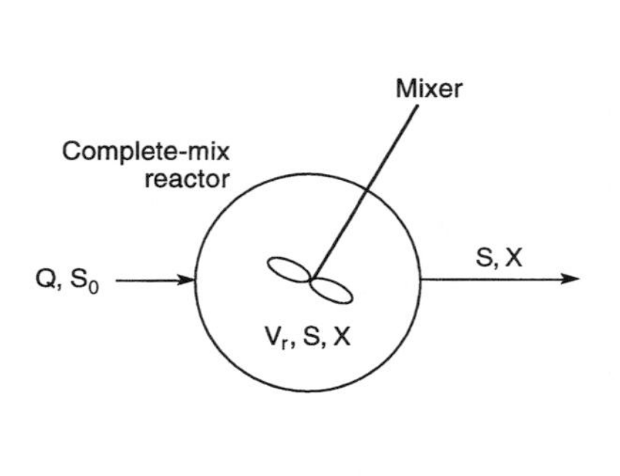

(a)

(b)

FIGURE 7-7
Complete-mix reactor without recycle: (*a*) definition sketch and (*b*) typical aerated lagoon.

Fig. 7-7*b* is known as the *aerated lagoon process*. Soluble organic waste is introduced into a reactor where an aerobic bacterial culture is maintained in suspension. The reactor contents are referred to as the mixed liquor.

Microorganism mass balance without recycle. A mass balance for the mass of microorganisms in a complete-mix reactor without recycle can be written as follows:

1. General word statement:

$$\begin{matrix}\text{Rate of accumulation} \\ \text{of microorganism} \\ \text{within the system} \\ \text{boundary}\end{matrix} = \begin{matrix}\text{rate of flow of} \\ \text{microorganism} \\ \text{into the system} \\ \text{boundary}\end{matrix} - \begin{matrix}\text{rate of flow of} \\ \text{microorganism} \\ \text{out of the system} \\ \text{boundary}\end{matrix} + \begin{matrix}\text{net growth of} \\ \text{microorganism} \\ \text{within the system} \\ \text{boundary}\end{matrix} \tag{7-27}$$

2. Simplified word statement:

$$\text{Accumulation} = \text{inflow} - \text{outflow} + \text{net growth} \tag{7-28}$$

3. Symbolic representation:

$$\frac{dX}{dt}V_r = QX_o - QX + V_r(r'_g) \tag{7-29}$$

where dX/dt = rate of change of microorganism concentration in the reactor measured in terms of mass (volatile suspended solids), mass VSS/unit volume·time

V_r = reactor volume

Q = flowrate, volume/time

X_o = concentration of microorganisms in influent, mass VSS/unit volume

X = concentration of microorganisms in reactor, mass VSS/unit volume

r'_g = net rate of microorganism growth, mass VSS/unit volume·time

In Eq. (7-29) it is assumed that the volatile fraction of the total suspended solids can be used as an approximation of the active biological mass (i.e., the volatile fraction is proportional to the activity of the microbial mass in question). Although a number of other measures, such as nitrogen, protein, DNA, and ATP content, have been used, the volatile suspended solids test is used principally because of its simplicity.

If the value of r'_g from Eq. (7-17) is substituted into Eq. (7-29), the result is:

$$\frac{dX}{dt}V_r = QX_o - QX + V_r\left(\frac{\mu_m XS}{K_s + S} - k_d X\right) \tag{7-30}$$

where S = substrate concentration in effluent from reactor, mg/L. If it is assumed that the concentration of microorganisms in the influent can be neglected and that steady-state conditions prevail ($dX/dt = 0$), Eq. (7-30) can be simplified to yield:

$$\frac{Q}{V_r} = \frac{1}{\theta} = \mu' = \frac{\mu_m S}{K_s + S} - k_d \tag{7-31}$$

where θ = hydraulic detention time, V/Q. In Eq. (7-31), the term $1/\theta$ corresponds to the net specific growth rate μ' [see Eq. (7-19)]. The term $1/\theta$ also corresponds to $1/\theta_c$, where θ_c, known as the *mean cell residence time,* is defined as the mass of organisms in the reactor divided by the mass of organisms removed from the system each day. In effect, θ_c corresponds to the average time the microorganisms remain in the system. For the reactor shown in Fig. 7-7, θ_c is given by the following expression:

$$\theta_c = \frac{V_r X}{QX} = \frac{V_r}{Q} \tag{7-32}$$

Substrate mass balance without recycle. Performing a substrate balance corresponding to the microorganism mass balance given in Eq. (7-30) results in the following expression [see also Eq. (7-14)]:

$$\frac{dS}{dt}V_r = QS_o - QS - V_r\left(\frac{kXS}{K_s + S}\right) \tag{7-33}$$

At steady state ($dS/dt = 0$), the resulting equation is:

$$(S_o - S) - \theta\left(\frac{kXS}{K_s + S}\right) = 0 \tag{7-34}$$

where $\theta = V_r/Q$.

The effluent microorganism and substrate concentrations may be obtained as follows. If Eq. (7-31) is solved for the term $S/(K_s + S)$ and the resulting expression is substituted into Eq. (7-34) and simplified by using Eq. (7-13), then the effluent steady-state microorganism concentration is given by:

$$X = \frac{\mu_m(S_o - S)}{k(1 + k_d\theta)} = \frac{Y(S_o - S)}{1 + k_d\theta} \tag{7-35}$$

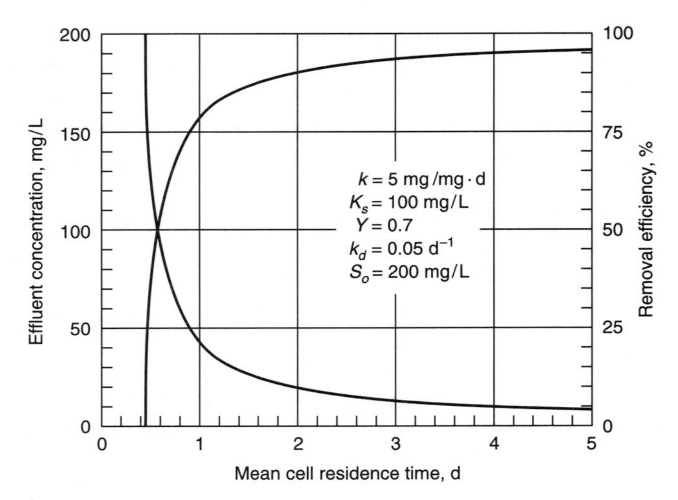

FIGURE 7-8
Effluent waste concentration and removal efficiency versus mean cell residence time for a complete-mix reactor without recycle ($\theta = \theta_c$).

Rearranging Eq. (7-31) and applying Eq. (7-13) gives the effluent substrate concentration as:

$$S = \frac{K_s(1 + \theta k_d)}{\theta(Yk - k_d) - 1} \tag{7-36}$$

Thus, if the kinetic coefficients are known, Eqs. (7-35) and (7-36) can be used to predict effluent microorganism and substrate concentrations (see Fig. 7-8). It is important to note that the effluent concentrations predicted from the above equations are based on a soluble waste and do not take into account any influent organic matter that may be nonbiodegradable or any volatile and inert suspended solids that may be present. Actual effluent substrate and suspended solids concentrations from the treatment process are dependent on the performance of the sedimentation tanks.

The observed yield, Y_{obs}, given by the following expression, is derived by substituting the value of X given by Eq. (7-35) for r'_g in Eq. (7-20), and by dividing by the term $(S_o - S)$, which corresponds to the value of r_{su} expressed as a concentration value.

$$Y_{obs} = \frac{Y}{1 + k_d\theta} \tag{7-37}$$

Mass Balances for Suspended-Growth Complete-Mix Processes with Recycle

In the complete-mix system, shown schematically in Fig. 7-9 and pictorially in Fig. 7-10, the contents of the reactor are mixed completely, and it is assumed that

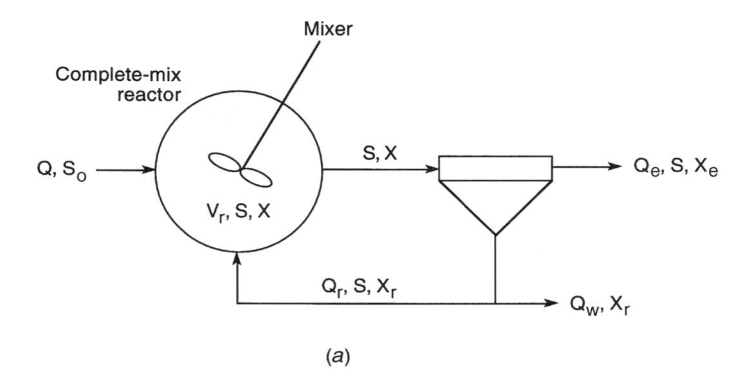

(a)

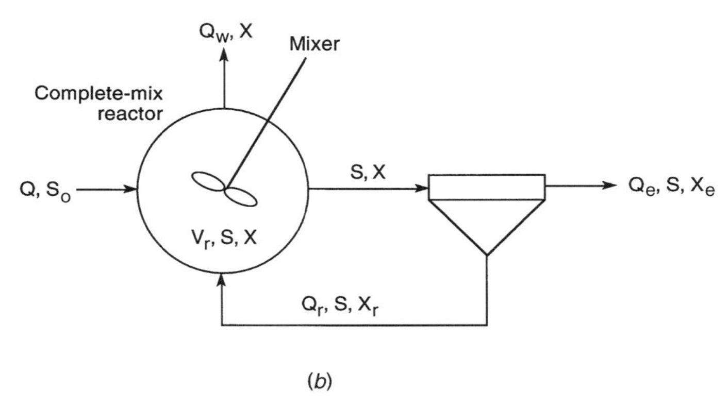

(b)

FIGURE 7-9
Definition sketch for a complete-mix reactor with recycle: (*a*) with wasting from the recycle line and (*b*) with wasting from the reactor.

(a)

(b)

FIGURE 7-10
Views of complete-mix reactors with surface aeration.

there are no microorganisms in the wastewater influent. The flow diagram depicted in Fig. 7-9 is known as the *complete-mix activated-sludge process*. As shown in Fig. 7-9, an integral part of the activated-sludge process is a solids separation unit (sedimentation tank) in which the cells from the reactor are separated (settled) and then returned to the reactor. Because of the presence of a sedimentation tank, two additional assumptions must be made in the development of the kinetic model for this system:

1. Waste stabilization by the microorganisms occurs only in the reactor unit. This assumption leads to a conservative model (in some systems a limited amount of waste stabilization may occur in the settling unit).
2. The volume used in calculating the mean cell residence time (discussed below) for the system includes only the volume of the reactor unit.

In effect, it is assumed that the sedimentation tank serves as a reservoir from which solids are returned to maintain a given solids level in the aeration tank. If the system is such that these assumptions do not hold true, then the model must be modified accordingly.

Microorganism mass balance with recycle. Referring to Fig. 7-9*a*, a mass balance for the microorganisms in the entire system can be written as

$$\frac{dX}{dt}V_r = QX_o - (Q_wX_r + Q_eX_e) + V_r(r_g') \tag{7-38}$$

$$\text{Accumulation} = \text{inflow} - \text{outflow} + \text{net growth} \tag{7-39}$$

where dX/dt = rate of change of microorganism concentration in the reactor measured in terms of mass (volatile suspended solids), mass VSS/unit volume·time
V_r = reactor volume
Q = flowrate, volume/time
X_o = concentration of microorganisms in influent, mass VSS/unit volume
Q_w = waste flowrate, volume/time
X_r = concentration of microorganisms in return line from the sedimentation tank, mass VSS/unit volume
Q_e = effluent flowrate, volume/time
X_e = concentration of microorganisms in effluent, mass VSS/unit volume
r_g' = net rate of microorganism growth, mass VSS/unit volume·time

Substituting Eq. (7-18) for the rate of growth, and assuming that the cell concentration in the influent is zero and steady-state conditions prevail ($dX/dt = 0$), yields:

$$\frac{Q_wX_r + Q_eX_e}{V_rX} = -Y\left(\frac{r_{su}}{X}\right) - k_d \tag{7-40}$$

The left hand side of Eq. (7-40) represents the inverse of the mean cell residence time as defined previously, thus:

$$\theta_c = \frac{V_rX}{Q_wX_r + Q_eX_e} \tag{7-41}$$

If wasting is from the reactor (see Fig. 7-9*b*) and the solids lost in the effluent are neglected, then θ_c can be defined as

$$\theta_c = \frac{V_r}{Q_w} \tag{7-42}$$

Substituting Eq. (7-41) for θ_c in Eq. (7-40) yields:

$$\frac{1}{\theta_c} = -Y\left(\frac{r_{\text{su}}}{X}\right) - k_d \tag{7-43}$$

The term r_{su} in the above equation is determined from the following expression:

$$r_{\text{su}} = -\frac{Q}{V_r}(S_o - S) = -\frac{(S_o - S)}{\theta} \tag{7-44}$$

where $(S_o - S)$ = mass concentration of substrate utilized, mg/L
S_o = substrate concentration in influent, mg/L
S = substrate concentration in effluent, mg/L
θ = hydraulic detention time, d

In Eq. (7-43), the term $(-r_{\text{su}}/X)$ is known as the *specific substrate utilization rate, U*. Using the definition of r_{su} given in Eq. (7-44), the specific utilization rate is calculated as follows:

$$U = -\frac{r_{\text{su}}}{X} = \frac{S_o - S}{\theta X} = \frac{Q}{V_r}\frac{S_o - S}{X} \tag{7-45}$$

From Eqs. (7-13) and (7-14), the specific substrate utilization rate U is also equal to:

$$U = \frac{\mu_m XS}{YX(K_s + S)} = \frac{kS}{(K_s + S)} \tag{7-46}$$

If the term U is substituted for the term $(-r_{\text{su}}/X)$ in Eq. (7-43), the resulting equation is:

$$\frac{1}{\theta_c} = YU - k_d \tag{7-47}$$

The mass concentration of microorganisms in the reactor, X, can be obtained by substituting the last expression for r_{su} in Eq. (7-44) into Eq. 7-43 and solving for X:

$$X = \frac{\theta_c}{\theta}\frac{Y(S_o - S)}{(1 + k_d\theta_c)} \tag{7-48}$$

The difference between Eqs. (7-35) and (7-48) is the term θ_c/θ, which accounts for the fact that the mass of organisms in the reactor is independent of the hydraulic residence time.

A term closely related to the specific utilization rate U, and commonly used in practice as a design and control parameter, is known as the *food-to-microorganism ratio* (F/M). The food-to-microorganism ratio is defined as follows:

$$\frac{F}{M} = \frac{S_o}{\theta X} \tag{7-49}$$

The terms U and F/M are related by the process efficiency as follows:

$$U = \frac{(F/M)E}{100} \tag{7-50}$$

where E is the process efficiency, defined as:

$$E = \frac{S_o - S}{S_o} \times 100 \tag{7-51}$$

where E = process efficiency, percent
S_o = influent substrate concentration
S = effluent substrate concentration

Substrate mass balance with recycle. Performing a substrate balance, the effluent substrate concentration is:

$$S = \frac{K_s(1 + \theta_c k_d)}{\theta_c(Yk - k_d) - 1} \tag{7-52}$$

It should be noted that Eq. (7-52) is the same as Eq. (7-36), developed for a complete-mix reactor without recycle, with the exception that θ is replaced by θ_c. The corresponding equation for the observed yield in a system with recycle is the same as Eq. (7-37), with θ_c substituted for θ:

$$Y_{\text{obs}} = \frac{Y}{1 + k_d\theta_c} \tag{7-53}$$

Application of the equations developed for a complete-mix reactor with recycle are illustrated in Example 7-2.

EXAMPLE 7-2. DESIGN OF CONVENTIONAL ACTIVATED-SLUDGE PROCESS USED TO TREAT A SOLUBLE WASTE. Wastewater from a bottle-washing plant contains a *soluble* organic waste having a COD of 300 mg/L. From extensive laboratory studies, the BOD_5 was found to be equal to 0.60 times the COD. The average flowrate of 1.0 Mgal/d is to be treated with a complete-mix activated-sludge process. The effluent BOD_5 and TSS are to be equal to or less than 30 mg/L 95 percent of the time. Assume that the temperature is 20°C and that the following conditions are applicable:

1. Influent volatile suspended solids to reactor are negligible.
2. Return sludge concentration (assumed) = 8000 mg/L of total suspended solids = 6400 mg/L volatile suspended solids.
3. Mixed-liquor suspended solids (MLSS) = 2500 mg/L (assumed).
4. Mixed-liquor volatile suspended solids (MLVSS) = 2000 mg/L = 0.80 total MLSS.
5. Mean cell residence time θ_c = 8 days.
6. Hydraulic regime of reactor = complete mix.
7. Kinetic coefficients Y = 0.46 lb cells/lb substrate (COD) consumed, $k_d = 0.06\text{d}^{-1}$.
8. It is estimated that 80 percent of the effluent solids are biodegradable. Assume the biological solids can be converted from a COD (same as UBOD) demand to a BOD_5 demand using the factor 0.6.

9. Waste contains adequate nitrogen and phosphorus and other trace nutrients for biological growth.
10. The coefficient of reliability (COR) for the process for both BOD and TSS, based on an analysis from similar plants in the vicinity, is 0.70 and 0.65, respectively. A detailed discussion of the coefficient of reliability may be found in Sec. 4-8, Chap. 4.

Using the above information determine the reactor volume, the sludge wasting rate, the recirculation ratio, and the hydraulic retention time for the reactor. Also determine the specific substrate utilization rate and the food-to-microorganism ratio.

Solution

1. Estimate design BOD_5 and TSS on the basis of COR values.

$$\begin{aligned}\text{Mean design BOD}_5 &= \text{COR} \times \text{BOD}_5 \text{ required}\\ &= 0.70 \times 30 \text{ mg/L} = 21 \text{ mg/L}\end{aligned}$$

$$\text{Mean design TSS} = 0.65 \times 30 = 19.5 \text{ mg/L}$$

2. Estimate the soluble BOD_5 in the effluent by calculating the particulate COD, converting the particulate COD to particulate BOD_5, and subtracting the particulate BOD_5 from the total BOD_5.
   a. Convert the effluent TSS to particulate COD [the volatile fraction is 0.80, and the conversion from cell mass to COD is 1.42, as shown in Eq. (7-15)]:

$$\text{Particulate COD} = 19.5(0.8)1.42 = 22.15 \text{ mg/L}$$

   b. Convert the particulate COD to particulate BOD_5:

$$\text{Particulate BOD}_5 = (0.6)22.15 = 13.29 \text{ mg/L}$$

   c. For the effluent, soluble BOD_5 = total BOD_5 − particulate BOD_5:

$$\text{Soluble BOD}_5 = 21.0 - 13.29 = 7.71 \text{ mg/L soluble BOD}_5$$

$$\text{Soluble COD} = 7.71/0.6 = 12.85 \text{ mg/L}$$

   Note that the total effluent COD = 22.15 + 12.85 = 35.00 mg/L, which equals total effluent BOD/0.6.

3. Calculate the biological treatment efficiency, based on soluble and total COD removal:

   The efficiency based on soluble COD removal would be

$$E_s = \frac{300 - 12.85}{300}(100) = 95.7\%$$

   The efficiency based on total COD removal would be

$$E_{\text{average}} = \frac{300 - 35}{300}(100) = 88.3\%$$

4. Compute the required volume using Eq. (7-48) by substituting V/Q for θ, the hydraulic detention time, and rearranging the equation as follows:

$$V = \frac{YQ\theta_c(S_o - S)}{X(1 + k_d\theta_c)}$$

$$V, \text{Mgal} = \frac{0.46(1\ \text{Mgal/d})(8\ \text{d})(300\ \text{mg/L} - 12.85\ \text{mg/L})}{(2000\ \text{mg/L})\left[1 + \left(0.06\text{d}^{-1}\right)(8\ \text{d})\right]}$$

$$= 0.36\ \text{Mgal}$$

5. Compute the sludge-production rate on a mass basis.
   a. From Eq. (7-53), the observed yield is

$$Y_{\text{obs}} = \frac{Y}{1 + k_d\theta_c} = \frac{0.46}{1 + 0.06(8)} = 0.31$$

   b. The biomass production rate is:

$$\text{lb VSS/d} = (Y_{\text{obs}}\ \text{lb/lb})[(S_o - S)\ \text{mg/L}](Q\ \text{Mgal/d})[8.34\ \text{lb/Mgal}\cdot(\text{mg/L})]$$

$$= 0.31(300 - 12.85)(1)(8.34) = 742\ \text{lbVSS/d}$$

6. Compute the biomass wasting rate if wasting is accomplished from the reactor, as shown in Fig. 7-9*a*, or from the recycle line, as shown in Fig. 7-9*b*. Take into account the solids lost in the plant effluent. Also assume that $Q_e = Q$.
   a. Determine the wasting rate from the reactor using Eq. (7-41) (because wasting is from the reactor, $X_r = X$):

$$\theta_c = \frac{V_r X}{Q_w X_r + Q_e X_e}$$

   can be solved for Q_w and converted to

$$Q_w = \frac{V_r X - Q_e X_e \theta_c}{X_r \theta_c}$$

$$= \frac{(0.36\ \text{Mgal})(2000\ \text{mg/L}) - (1.0\ \text{Mgal/d})(0.8 \times 19.5\ \text{mg/L})(8\ \text{d})}{(2000\ \text{mg/L})(8\ \text{d})}$$

$$= 0.0372\ \text{Mgal/d}$$

   b. Determine the wasting rate from the recycle line:

$$Q_w = \frac{V_r X - Q_e X_e \theta_c}{X_r \theta_c}$$

$$= \frac{(0.36\ \text{Mgal})(2000\ \text{mg/L}) - (1.0\ \text{Mgal/d})(0.8 \times 19.5\ \text{mg/L})(8\ \text{d})}{(6400\ \text{mg/L})(8\ \text{d})}$$

$$= 0.0116\ \text{Mgal/d}$$

   c. Determine the amount of sludge wasted in pounds per day:

$$\text{Wasted sludge, lb/d} = (X_r\ \text{mg/L}) \times (Q_w\ \text{Mgal/d}) \times 8.34$$

   For sludge wasting from reactor:

$$\text{Wasted sludge} = (2500\ \text{mg/L}) \times (0.0372\ \text{Mgal/d}) \times 8.34 = 776\ \text{lb/d}$$

   For sludge wasting from return line:

$$\text{Wasted sludge} = (8000\ \text{mg/L}) \times (0.0116\ \text{Mgal/d}) \times 8.34 = 774\ \text{lb/d}$$

   Note that in either case the weight of sludge wasted is the same, and that either wasting method will achieve a θ_c of 8 days for the system. The organic fraction of the sludge is $0.8(775) = 620$ lb/d, which, when combined with the VSS in the effluent (130 lb/d), approximately equals the biomass production rate (742 lb/d).

7. Compute the recycle ratio [see Eqs. (7-91) and (7-92)]:

$$\text{Aerator VSS conc} = 2000 \text{ mg/L}$$

$$\text{Return VSS conc} = 6400 \text{ mg/L}$$

From a mass balance on the reactor,

$$Q_r X_r = (Q + Q_r)X$$

The recycle ratio is obtained by rearranging the above equation.

$$\frac{Q_r}{Q} = \frac{X}{X_r - X} = \frac{2000}{6400 - 2000} = 0.45$$

8. Compute the hydraulic retention time (HRT) for the reactor:

$$\text{HRT} = \frac{V}{Q} = \frac{0.36 \text{ Mgal}}{1 \text{ Mgal/d}} = 0.36 \text{ d} = 8.6 \text{ h}$$

9. Check the specific substrate utilization rate and the food-to-microorganism ratio.
   a. From Eq. (7-45), the specific substrate utilization rate is:

$$U = \frac{S_o - S}{\theta X} = \frac{(300 - 12.85) \text{ mg/L}}{(0.36 \text{ d})(2000 \text{ mg/L})} = 0.40 \frac{\text{mg/L BOD}_5 \text{ utilized}}{(\text{mg/L MLVSS}\cdot\text{d})}$$

   b. From Eq. (7-49), the food-to-microorganism ratio is

$$F/M = \frac{S_o}{\theta X} = \frac{300 \text{ mg/L}}{(0.36 \text{ d})(2000 \text{ mg/L})} = 0.42 \frac{\text{mg BOD}_5\text{/L applied}}{(\text{mg MLVSS/L}\cdot\text{d})}$$

Comment. In this example a soluble waste was treated and the kinetic approach to design could be applied. As will be discussed in Sec. 7-6, in dealing with domestic wastewater the kinetic approach is of limited value because the measured parameters are lumped parameters (e.g., the BOD comprises soluble and particulate fractions).

Mass Balances for Suspended-Growth Plug-Flow Processes with Recycle

The plug-flow system with cellular recycle, shown schematically in Fig. 7-11*a* and pictorially in Fig. 7-11*b*, can be used to model certain forms of the activated-sludge process. The distinguishing feature of this recycle system is that the hydraulic regime of the reactor is of a plug-flow nature. In a true plug-flow model, all the particles entering the reactor stay in the reactor an equal amount of time. Some particles may make more passes through the reactor because of recycling, but while they are in the tank, they all pass through in the same amount of time.

A kinetic model of the plug-flow system is mathematically difficult, but Lawrence and McCarty (1970) made two simplifying assumptions that lead to a useful kinetic model of the plug-flow reactor:

1. The concentration of microorganisms in the influent to the reactor is approximately the same as that in the effluent from the reactor. This assumption applies only if $\theta_c/\theta > 5$. The resulting average concentration of microorganisms in the reactor is symbolized as $\overline{X}$.

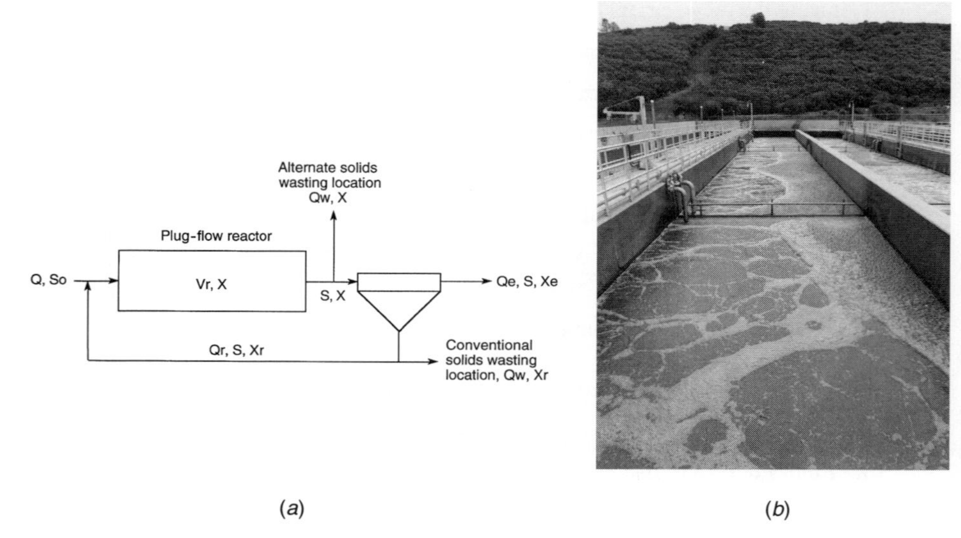

FIGURE 7-11
Plug-flow reactor with recycle: (*a*) definition sketch and (*b*) view of plug-flow reactor.

2. The rate of substrate utilization as the waste passes through the reactor is given by the following expression [compare to Eq. (7-14)]:

$$r_{su} = \frac{kS\overline{X}}{K_s + S} \tag{7-54}$$

Integrating Eq. (7-54) over the retention time of the waste in the tank and simplifying gives the following expression:

$$\frac{1}{\theta_c} = \frac{Yk(S_o - S)}{(S_o - S) + (1 + \alpha)K_s \ln (S_i/S)} - k_d \tag{7-55}$$

where S_o = influent concentration
S = effluent concentration
S_i = influent concentration to reactor after dilution with recycle flow
$= \dfrac{S_o + \alpha S}{1 + \alpha}$
α = recycle ratio

Other terms are as defined previously. Equation (7-55) is related to Eq. (7-52), which applies to complete-mix systems, with or without recycle. The main difference in the two equations is that in Eq. (7-55), θ_c is also a function of the influent waste concentration S_o.

Stability of Suspended-Growth Biological Treatment Processes

To assess process stability, it is helpful to consider the relationship between the effluent substrate concentration S and process efficiency E as illustrated in Fig. 7-12,

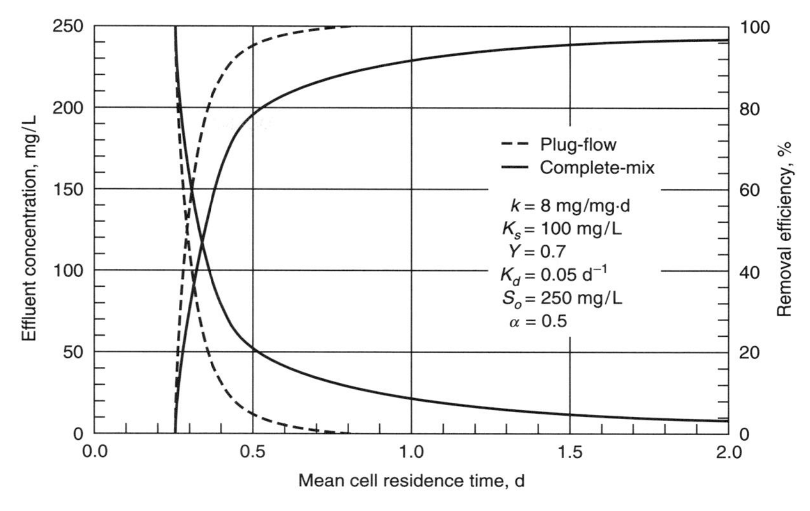

FIGURE 7-12
Effluent waste concentration and removal efficiency for complete-mix and plug-flow reactors with recycle versus mean cell residence time.

which has been prepared with a specific set of kinetic coefficients. For a specified waste, a given biological community, and a particular set of environmental conditions, the kinetic coefficients Y, k, K_s, and k_d are usually fixed. It is important to note that domestic wastewater may have significant variability in its composition and may not always be treated as a single waste type in evaluating the kinetic coefficients. Referring to Fig. 7-12, it is clear that the effluent concentration S and the treatment efficiency E are related directly to θ_c. It is also apparent from Fig. 7-12 that there is a certain minimum value of θ_c below which waste stabilization does not occur. This critical value of θ_c is called the *minimum mean cell residence time, θ_c^M*. Physically, θ_c^M is the residence time at which the cells are washed out or wasted from the system faster than they can reproduce. The minimum mean cell residence time can be calculated from the following equation, which is derived from Eqs. (7-13) and (7-31). It should be noted that, when washout occurs, the influent concentration S_o is equal to the effluent waste concentration S.

$$\frac{1}{\theta_c^M} = Y \frac{kS_o}{K_s + S_o} - k_d \tag{7-56}$$

In many situations encountered in waste treatment, S_o is much greater than K_s, so that Eq. (7-56) can be rewritten as:

$$\frac{1}{\theta_c^M} \approx Yk - k_d \tag{7-57}$$

Equations (7-56) and (7-57) can be used to determine the minimum mean cell residence time θ_c^M. Typical kinetic coefficients that can be used to solve for θ_c^M for the activated sludge process are given in Table 7-5. Obviously, biological treatment

TABLE 7-5
Typical kinetic coefficients for the activated-sludge process for domestic wastewater, neglecting influent particulate matter*

Coefficient	Basis‡	Value† Range	Value† Typical
k	d^{-1}	2–10	4
K_s	mg COD/L	15–70	40
	mg BOD_5/L	25–100	60
Y	mg VSS/mg COD	0.3–0.6	0.4
	mg VSS/mg BOD_5	0.4–0.8	0.6
k_d	d^{-1}	0.02–0.1	0.055

*Derived in part from Grady and Lim (1980), Lawrence and McCarty (1970), Orhan and Artan (1994).
†Values reported are for 20°C.
‡VSS = volatile suspended solids.

systems should not be designed with θ_c values equal to θ_c^M. To ensure adequate waste treatment, biological treatment systems are typically designed and operated with θ_c^d values from 2 to 20 times the θ_c^M value. In effect, the ratio of θ_c^d to θ_c^M can be considered to be a process safety factor (SF), as given below. The application of Eq. (7-58) is illustrated in Example 7-3.

$$\mathrm{SF} = \frac{\theta_c^d}{\theta_c^M} \tag{7-58}$$

EXAMPLE 7-3. FACTOR OF SAFETY FOR BIOLOGICAL TREATMENT PROCESS DESIGN. Determine the factor of safety for the activated-sludge process design in Example 7-2. Use assumed values of $k = 5.0\ d^{-1}$ and $K_s = 40$ mg COD/L (see Table 7-5).

Solution

1. Determine the minimum mean cell residence time using Eq. (7-56) and the data from Example 7-2:

$$\frac{1}{\theta_c^M} = Y\frac{kS_o}{K_s + S_o} - k_d$$

$$= \frac{(0.46)(5)(300)}{40 + 300} - 0.06 = 1.97$$

$$= \frac{1}{1.97} = 0.51\ \mathrm{d}$$

2. Determine the factor of safety using Eq. (7-58):

$$\mathrm{SF} = \frac{\theta_c^d}{\theta_c^M} = \frac{8}{0.51} = 15.7$$

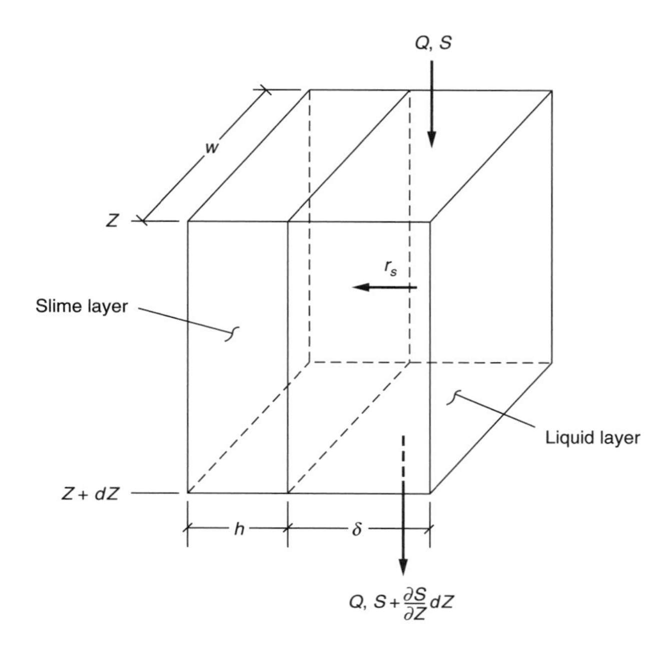

FIGURE 7-13
Definition sketch for the analysis of the attached-growth process (adapted from Atkinson et al., 1974b).

Mass Balance for Attached-Growth Process without Recycle

In predicting the performance of attached-growth processes (e.g., trickling filters), the organic and hydraulic loadings and the degree of treatment required are among the important factors that must be considered. In the following discussion, the theoretical mass balance approach proposed by Atkinson (1974a) and Atkinson et al. (1974b), is used to illustrate the modeling of attached growth. Atkinson and his coworkers and others proposed the following model to describe the rate of flux of organic material into the slime layer of an attached-growth process, assuming that diffusion into the slime layer controls the rate of reaction and that there is no concentration gradient across the liquid film (see Fig. 7-13).

$$r_s = -\frac{Ehk_o\overline{S}}{K_m + \overline{S}} \tag{7-59}$$

where r_s = rate of flux of organic material into the slime layer, ft/d
E = effectiveness factor ($0 \leq E \leq 1$)
h = thickness of slime layer, ft
k_o = maximum reaction rate, d^{-1}
$\overline{S}$ = average substrate (e.g., BOD) concentration in the bulk liquid in the volume element, mg/L
K_m = half-velocity constant, mg/L

Because the effectiveness factor E is approximately proportional to the BOD concentration in the liquid, Eq. (7-59) can be rewritten as follows:

$$r_s = \frac{fhk_o\overline{S}^2}{K_m + \overline{S}} \tag{7-60}$$

where f = proportionality factor. This model can be applied to the analysis of attached-growth processes by performing a mass balance analysis for the organic material contained in the liquid volume (see Fig. 7-13).

1. General word statement:

Rate of accumulation of substrate within the volume element	=	rate of flow of substrate into the volume element	−	rate of flow of substrate out of the volume element	+	rate of substrate flux into the slime layer from the volume element

(7-61)

2. Simplified word statement:

$$\text{Accumulation} = \text{inflow} - \text{outflow} + \text{generation} \tag{7-62}$$

3. Symbolic representation:

$$\frac{\partial \overline{S}}{\partial t} dV = QS - Q\left(S + \frac{\partial S}{\partial D} dD\right) + dDw\left(-\frac{fhk_o\overline{S}^2}{K_m + \overline{S}}\right) \tag{7-63}$$

where Q = volumetric flowrate, ft^3/d
w = width of section under consideration, ft
D = filter depth, ft

Assuming that steady-state conditions prevail ($\partial \overline{S}/\partial t = 0$), Eq. (7-63) can be simplified to yield:

$$Q\frac{dS}{dD} = -fk_o hw \frac{\overline{S}^2}{K_m + \overline{S}} \tag{7-64}$$

If it is now assumed that the value of the saturation coefficient K_m is small relative to the value of BOD, then Eq. (7-64) can be written as:

$$\frac{dS}{dD} = -\frac{fk_o hw\overline{S}}{Q} \tag{7-65}$$

Equation (7-65) can now be integrated between the limits of S_e and S_i and 0 and D to yield:

$$\frac{S_e}{S_i} = \exp\left[-(fk_o h)\frac{wD}{Q}\right] \tag{7-66}$$

where S_e = effluent concentration, mg/L, and S_i = influent concentration resulting after the untreated incoming wastewater is mixed with recycled effluent, mg/L. The use of Eq. (7-66) involves the determination of the coefficients f, h, and k_o for a given set of operating conditions. The effect of the wastewater temperature on the performance of the filter can be accounted for by adjusting the k value using Eq. (7-21). A commonly used value for the temperature coefficient θ is 1.035.

Values of Kinetic Coefficients

To apply the equations presented in this section, it is necessary to have available numerical values for the kinetic coefficients. Typical values for the kinetic coefficients are given in Table 7-5, where values for the half-saturation constant and the yield coefficient are given in terms of both COD and BOD. Historically, most kinetic coefficients have been based on BOD measurements. Although BOD values are used most commonly, a more rational approach is to develop and use COD-based coefficients. Using COD-based coefficient values, it is possible to avoid the problems encountered where there is a poor correlation between the BOD and COD values in the untreated wastewater. It should be recognized, however, that even the use of COD is less than satisfactory, because the components of the COD may not all be biodegradable to the same extent. The development of kinetics from bench-scale studies is illustrated in the following example.

EXAMPLE 7-4. DETERMINATION OF BIOLOGICAL GROWTH KINETIC COEFFICIENTS. Determine the values of the coefficients k, K_s, μ_m, Y, and k_d using the following data derived from a bench-scale activated-sludge study using a continuous-flow complete-mix reactor without recycle (see Fig. 7-7*a*).

Unit no.	S_o, mg/L COD	S, mg/L COD	$\theta = \theta_c$, d	X, mg VSS/L
1	400	13	3.2	123
2	400	24	2.0	127
3	400	33	1.6	127
4	400	47	1.3	124
5	400	66	1.1	119

Solution

1. Determine the coefficients K_s and k.
   a. Set up a computation table to determine the coefficients K_s and k using the following linear expression developed by equating Eq. (7-14) to Eq. (7-44):

$$\frac{X\theta}{S_o - S} = \frac{K_s}{k}\frac{1}{S} + \frac{1}{k}$$

Unit no.	$S_o - S$, mg/L	$X\theta$, mg VSS/L·d	$\frac{X\theta}{S_o - S}$, d	$\frac{1}{S}$, 1/(mg/L)
1	387	393.6	1.017	0.077
2	376	254.0	0.676	0.042
3	367	203.2	0.554	0.030
4	353	161.2	0.457	0.021
5	334	130.9	0.392	0.015

b. Plot the term $(X\theta/(S_o - S)$ versus $(1/S)$, as shown on the following page.

From the linear relationship given above, the y intercept equals $(1/k)$.

$$\frac{1}{k} = 0.25 \text{ d} \qquad k = 4.0 \text{ d}^{-1}$$

From the linear relationship given above, the slope of the curve in the data plot is equal to K_s/k:

$$\frac{K_s}{k} = \frac{1.00 - 0.25}{0.075 - 0.00} = 10.0 \text{ mg/L·d}$$

$$K_s = 10.0 \text{ mg/L·d} \times 4.0 \text{ d}^{-1}$$
$$= 40.0 \text{ mg/L}$$

2. Determine the coefficients Y and k_d.
   a. Set up a computation to determine the coefficients using Eqs. (7-43) and (7-44).

$$\frac{1}{\theta_c} = -Y\frac{r_{su}}{X} - k_d$$
$$= Y\frac{S_o - S}{X\theta} - k_d$$

Unit no.	$\frac{1}{\theta_c}$, d^{-1}	$\frac{S_o - S}{X\theta}$, d^{-1}
1	0.313	0.983
2	0.500	1.480
3	0.625	1.806
4	0.769	2.190
5	0.909	2.552

b. Plot the term $(1/\theta_c)$ versus $(S_o - S/X\theta)$, as shown on the following page.

From the linear relationship given above, the y intercept equals $(-k_d)$:

$$-k_d = -0.060 \text{ d}^{-1}$$
$$k_d = 0.060 \text{ d}^{-1}$$

From the linear relationship given above, the slope of the curve equals Y:

$$Y = \frac{0.85 - (-0.06)}{2.4 - 0.0} = 0.38$$

3. Determine the value of the coefficient μ_m using Eq. (7-13):

$$\mu_m = kY$$
$$= 4.0 \text{ d}^{-1} \times 0.38$$
$$= 1.52 \text{ d}^{-1}$$

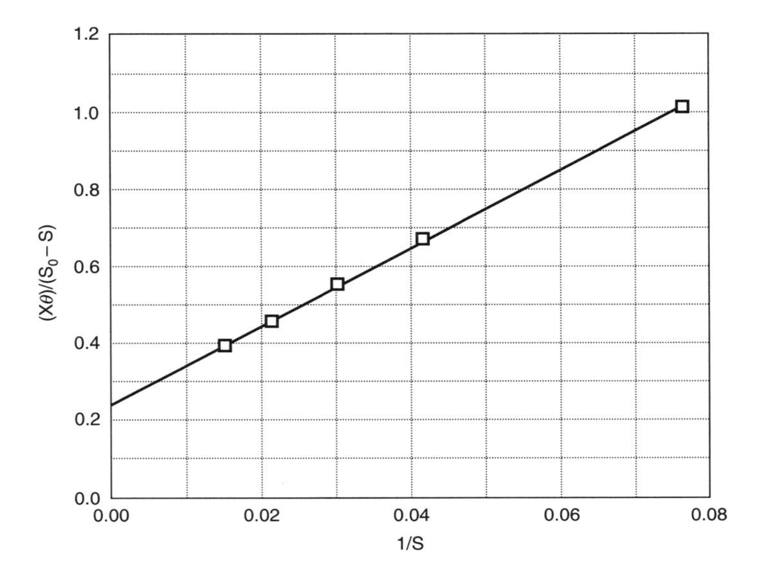

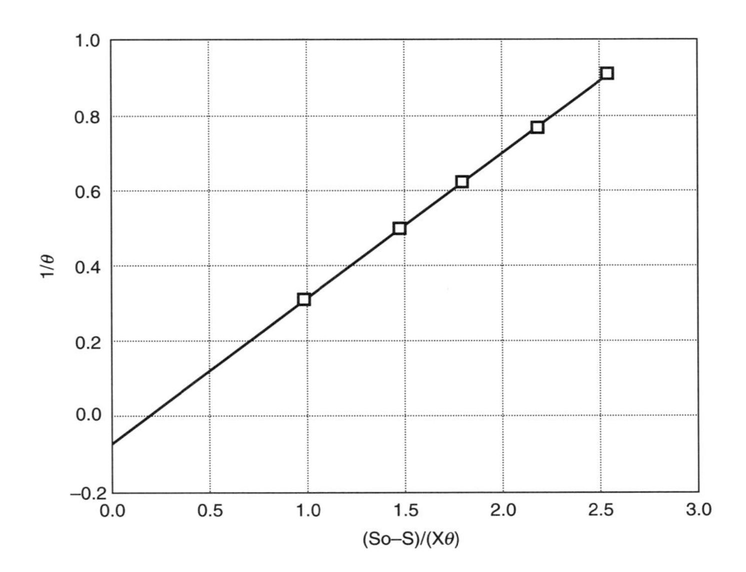

Comment. In this example, the kinetic coefficients were derived from data obtained in bench-scale complete-mix reactors without recycle. Similar data can be obtained for continuous-flow complete-mix reactors with recycle. An advantage of using reactors with recycle is that the mean cell residence time can be varied independently of the hydraulic detention time. A disadvantage is that small bench-scale reactors operated with solids recycle are difficult to control.

Advanced Modeling of the Activated-Sludge Process

Over the years, a number of models, such as those presented in this section, have been developed. Unfortunately, there has been little standardization with respect to terminology and parameters used. In 1983, in an effort to unify the many models, the International Association on Water Pollution Research and Control (IAWPRC), now known as the International Association on Water Quality (IAWQ), assembled a task group on *Mathematical Modeling for Design and Operation of Biological Wastewater Treatment*. The working group, composed of members from various countries, developed a model that can be used to predict the performance of the single-sludge activated-sludge process for carbonaceous oxidation, nitrification, and denitrification (see discussion in following section). Eight fundamental processes were utilized in the development of the Activated Sludge Model No. 1 (also known as the Task Group Model) (Henze et al., 1987).

Aerobic growth of heterotrophs
Anoxic growth of heterotrophs
Aerobic growth of autotrophs
Decay of heterotrophs
Decay of autotrophs
Ammonification of soluble organic nitrogen
Hydrolysis of entrapped organics
Hydrolysis of entrapped organic nitrogen

An important feature of the model is the use of COD instead of BOD to represent the organic matter. Further, the COD is divided into various subcategories such as soluble and particulate fractions which, in turn, can be inert, readily, and slowly biodegradable (see Chap. 2). The model was developed for use on an IBM-compatible PC. A second version of the model has been developed in which the number of factors considered has been increased. Unfortunately, both models require data on a large number of variables which are seldom available. To overcome this limitation, both models contain default values. To date, the greatest use that has been made of these models is to study various operational designs and schemes using the default values.

7-5 BIOLOGICAL NUTRIENT REMOVAL

Because both nitrogen and phosphorus can impact receiving water quality, the discharge of one or both of these constituents must often be controlled. Nitrogen may be present in wastewaters in various forms (e.g., organic, ammonia, nitrites, or nitrates). Most of the available nitrogen in both septic tank effluent and in municipal wastewater is in the form of organic or ammonia nitrogen. Typical total concentrations in septic tank effluent vary from 50 to 125 mg/L. The corresponding concentrations in municipal wastewater are about 25 to 35 mg/L. In wastewater treatment, about 20 percent of the total nitrogen settles out during primary sedimentation. During biological treatment, a major portion of the organic nitrogen is converted to ammonia nitrogen, a portion of which is incorporated into biological cells that are extracted from the treated wastewater stream before discharge, removing another 20 percent of the incoming nitrogen. The remaining 60 percent is normally discharged to the receiving waters.

Phosphorus is present in municipal wastewaters in organic form, as inorganic orthophosphate, or as complex phosphates. The complex phosphates represent about one-half of the phosphates in municipal wastewater and result from the use of these materials in synthetic detergents. Complex phosphates are hydrolyzed during biological treatment to the orthophosphate form (PO_4^{-3}). Of the total average phosphorus concentration of about 5 to 9 mg/L present in municipal wastewater, about 10 percent is removed as particulate material during primary sedimentation and another 10 to 20 percent is incorporated into bacterial cells during biological treatment. The remaining 70 percent of the incoming phosphorus normally is discharged with secondary treatment plant effluents.

Biological Nitrogen Removal

The two principal mechanisms for the biological removal of nitrogen, as illustrated in Fig. 7-14, are by assimilation and by nitrification-denitrification. Because nitrogen

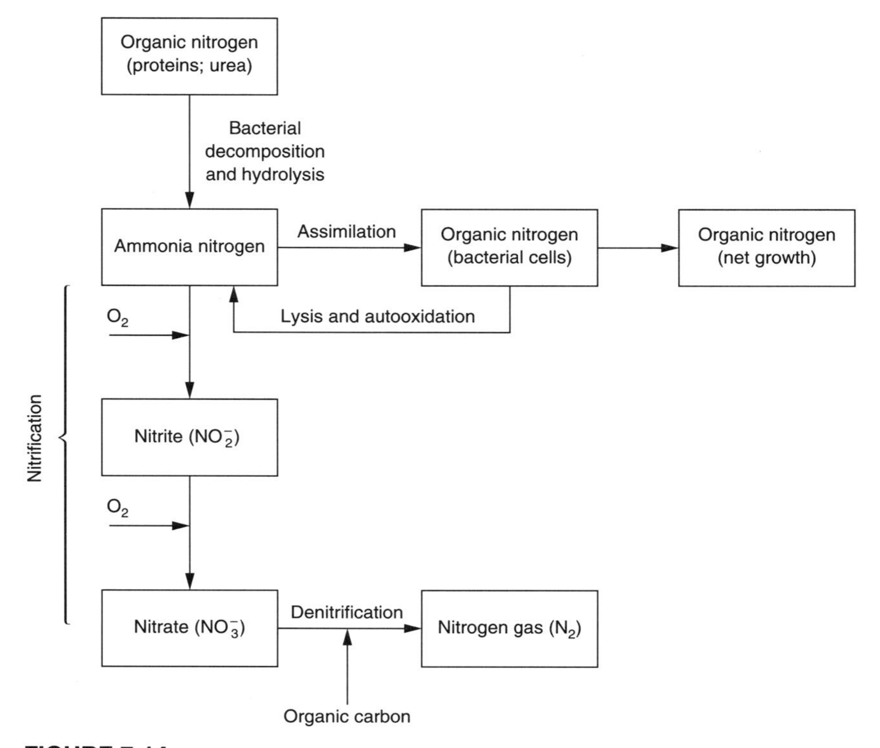

FIGURE 7-14
Definition sketch for the transformation of various forms of nitrogen in biological treatment processes (adapted from Sedlak, 1991).

is a nutrient, microorganisms in the treatment processes will assimilate ammonia-nitrogen and incorporate it into cell mass. Nitrogen can be removed from the waste-water by removing cells from the system. However, in most wastewaters there is more nitrogen than can be assimilated into cell tissue. In nitrification-denitrification, the removal of nitrogen is accomplished in two conversion steps. In the first step, ammonia is oxidized biologically to nitrate. In the second step, nitrate is reduced to nitrogen gas, which is vented from the system. Biological nitrogen removal by nitrification-denitrification removal can be accomplished in three ways:

1. Separate-stage carbon oxidation, nitrification, and denitrification
2. Combined carbon oxidation and nitrification and separate-stage denitrification
3. Combined carbon oxidation, nitrification, and denitrification

The basic principles of biological nitrification and denitrification are considered in the following discussion. Additional details and information on biological nitrification may be found in Sedlak (1991) and WEF (1992b).

Biological Nitrification

In biological nitrification, ammonia is oxidized in a two-step process: first to nitrite and then to nitrate. Nitrification stoichiometry, nitrification process variables, and nitrification applications are considered briefly in the following discussion.

Nitrification stoichiometry. The biological conversion of ammonia to nitrate can be described as follows:

Conversion of ammonia to nitrite (as typified by *Nitrosomonas*)

$$NH_4^+ + 1.5O_2 \rightarrow NO_2^- + 2H^+ + H_2O \tag{7-67}$$

Conversion of nitrite to nitrate by *Nitrobacter*

$$NO_2^- + 0.5O_2 \rightarrow NO_3^- \tag{7-68}$$

Overall conversion of ammonia to nitrate

$$NH_4^+ + 2O_2 \rightarrow NO_3^- + 2H^+ + H_2O \tag{7-69}$$

Along with obtaining energy, a portion of the ammonium ion is assimilated into cell tissue. The biomass synthesis reaction can be represented as follows:

$$4CO_2 + HCO_3^- + NH_4^+ + H_2O \rightarrow C_5H_7O_2N + 5O_2 \tag{7-70}$$

As noted previously in Sec. 7-2, the chemical formula $C_5H_7O_2N$ is used to represent the synthesized bacterial cells.

The half-reactions provided in Table 7-4 can be used to create an equation for the overall nitrification reaction. As demonstrated in Example 7-1, half-reactions for cell synthesis, oxidation of ammonia to nitrate, and reduction of oxygen to water can be combined to create Eq. (7-71) ($f_s = 0.10$). Due to rounding of the coefficients the equation does not balance exactly; however, the error introduced by rounding is negligible.

$$NH_4^+ + 1.731O_2 + 1.962HCO_3^- \rightarrow$$
$$0.038C_5H_7NO_2 + 0.962NO_3^- + 1.077H_2O + 1.769H_2CO_3 \quad (7\text{-}71)$$

From the above equation it will be noted that for each milligram of ammonia nitrogen converted, 3.96 mg of O_2 are utilized, 0.31 mg of new cells is formed, 7.01 mg of alkalinity are removed, and 0.16 mg of inorganic carbon is utilized. The oxygen required to oxidize 1.0 mg of ammonia nitrogen to nitrate (3.96 mg) is less than the theoretical value of 4.57 mg computed by Eq. (7-69) because the ammonia for cell synthesis is not considered in Eq. (7-69). Similarly, the alkalinity required for nitrification in Eq. (7-71) (7.01 mg/L) is less than the value of 7.14 mg calculated in Example 2-9 in Chap. 2, due to the conversion of some of the ammonia to cellular nitrogen. It should be recognized that the coefficient values in Eq. (7-71) are dependent on the value of f_s that is used.

Nitrification process variables. Nitrifying bacteria are sensitive organisms and extremely susceptible to a wide variety of inhibitors. From both laboratory studies and the operation of full-scale plants it has been found that the following factors affect the nitrification process: (1) concentration of ammonia and nitrite, (2) BOD_5/TKN ratio, (3) dissolved oxygen concentration, (4) temperature, and (5) pH. A variety of organic and inorganic agents can inhibit the growth and action of these organisms. High concentrations of ammonia and nitrous acid can be inhibitory. The concentration of nitrifying organism present will depend on the BOD_5/TKN ratio. Dissolved oxygen concentrations above 1 mg/L are essential for nitrification to occur. If DO levels drop below this value, oxygen becomes the limiting nutrient and nitrification slows or ceases. The effect of temperature on the growth of nitrifying bacteria is extremely significant. The effect of pH also is significant. A narrow optimal range between pH 7.5 to 8.6 exists, but systems acclimated to lower-pH conditions have successfully nitrified. Relationships that have been developed to quantify the aforementioned variables are summarized in Table 7-6. A generalized equation that can be used to estimate the growth rate μ_N, taking into account the relationships summarized in Table 7-6, is as follows:

$$\mu_N = \mu_{N\max}\left(\frac{N}{K_N + N}\right)\left(\frac{\text{DO}}{K_{\text{O2}} + \text{DO}}\right)(e^{0.098(T-15)})[1 - 0.833(7.2 - \text{pH})] \quad (7\text{-}78)$$

Typical kinetic coefficients for the nitrification process are given in Table 7-7. Application of Eq. (7-78) and the relationships and coefficients given in Tables 7-6 and 7-7 are illustrated in Example 7-5. The key concept in their application is to determine the minimum mean cell residence time subject to the most critical environmental constraints and the use of an appropriate safety factor. This approach is essentially the same as that used in the design of the suspended-growth activated-sludge process in a complete-mix reactor.

Nitrification processes. The principal nitrification processes, as reported in Table 7-2, may be classified as suspended-growth (see Fig. 7-15*a*) and attached-growth processes (see Fig. 7-15*b*). In the suspended-growth process, nitrification can be achieved either in a separate suspended-growth reactor following a conventional

TABLE 7-6

Effects of the major operational and environmental variables on the suspended-growth nitrification process*

Factor	Description of effect
Ammonia nitrite concentration	It has been observed that the concentration of ammonia and nitrite will affect the maximum growth rate of *Nitrosomonas* and *Nitrobacter*. Because the growth rate of *Nitrobacter* is considerably greater than that of *Nitrosomonas*, the rate of nitrification is usually modeled by using the conversion of ammonia to nitrite as the rate-limiting step: $$\mu_N = \mu_{N\max}\frac{N}{K_N + N} \quad (7\text{-}72)$$ Use $\mu_{N\max} = 0.45\ \text{d}^{-1}$ at 15°C
BOD_5/TKN	The fraction of nitrifying organisms present in the mixed liquor of a single-state carbon oxidation-nitrification process has been found to be related to the BOD_5/TKN ratio. The fraction of nitrifying organisms can be estimated by the following relationship: $$f_N = \frac{0.16(\text{NH}_3\ \text{removed})}{0.6(\text{BOD}_5\ \text{removed}) + 0.16(\text{NH}_3\ \text{removed})} \quad (7\text{-}73)$$
Dissolved oxygen concentration	The DO level has been found to affect the maximum specific growth rate μ_m of the nitrifying organisms. The effect has been modeled with the following relationship: $$\mu_N = \mu_{N\max}\frac{\text{DO}}{K_{O2} + \text{DO}} \quad (7\text{-}74)$$ A value of 1.3 mg/L can be used for K_{O2}.
Temperature	Temperature has a significant effect on nitrification rate constants. The overall nitrification rate decreases with decreasing temperature and is accounted for with the following two relationships: $$\mu_N = \mu_{N\max}e^{0.098(T-15)} \quad (7\text{-}75)$$ $$K_N = 10^{0.051T-1.158} \quad (7\text{-}76)$$ where T = °C
pH	It has been observed that the maximum rate of nitrification occurs between pH values of about 7.2 and 9.0. For combined carbon oxidation-nitrification systems, the effect of pH can be accounted for using the following relationship: $$\mu_N = \mu_{N\max}[1 - 0.833(7.2 - \text{pH})] \quad (7\text{-}77)$$

*Developed in part from U.S. EPA (1975) and Sedlak (1991).

activated-sludge treatment process, or in the same reactor used in the treatment of the carbonaceous organic matter. Nitrification can also be achieved in the same attached-growth reactor used for carbonaceous organic matter removal, or a separate reactor. Conventional and tower trickling filters, rotating biological contactors, and various submerged packed-bed reactors can be used for nitrification. In small systems, the single-reactor approach is used most commonly.

TABLE 7-7

Typical kinetic coefficients for the suspended-growth nitrification process*†

Overall coefficient	Basis	Value	
		Range	Typical†
μ_m	d^{-1}	0.4–2.0	0.9
K_N	NH_4^+-N, mg/L	0.2–3.0	0.5
Y	mg VSS/mg NH_4^+	0.1–0.3	0.16
k_d	d^{-1}	0.03–0.06	0.04

*Derived in part from U.S. EPA (1975), Sedlak (1991), and Tchobanoglous and Burton (1991).
†Values reported are for 20°C.

EXAMPLE 7-5. DESIGN OF SINGLE-STAGE SUSPENDED-GROWTH CARBON OXIDATION-NITRIFICATION PROCESS. Determine the concentration of ammonia in the effluent and the required hydraulic detention time for a single-stage activated-sludge process to achieve essentially complete nitrification when treating domestic wastewater. Assume the following conditions apply:

1. Influent flowrate = 0.1 Mgal/d (380 m^3/d)
2. BOD_5 after primary settling = 150 mg/L
3. TKN-N after primary settling = 30 mg/L
4. Minimum sustained temperature = 12°C
5. Dissolved oxygen to be maintained in the reactor = 2.0 mg/L
6. Buffer capacity of the wastewater is adequate to maintain the pH at or above a value of 7.2.
7. Use the kinetic coefficients given in Table 7-7.
8. Use a factor of safety of 2.0.

Solution

1. Determine the maximum growth rate for the nitrifying organisms under the stated operating conditions.
   a. The growth rate is determined from Eq. (7-78):

$$\mu_N = \mu_{N\max}\left(\frac{N}{K_N + N}\right)\left(\frac{\text{DO}}{K_{O2} + \text{DO}}\right)(e^{0.098(T-15)})[1 - 0.833(7.2 - \text{pH})]$$

   b. Substitute the known values and determine μ_N:

$$T = 12°\text{C}$$

$$\mu_{N\max} = 0.45\ \text{d}^{-1}\ \text{at}\ 15°\text{C}$$

$$N = 30\ \text{mg/L}$$

$$K_N = 10^{0.051\text{T}-1.158} = 10^{0.051(12)-1.158} = 0.28$$

$$\text{DO} = 2.0\ \text{mg/L}$$

$$K_{O2} = 1.3\ \text{mg/L}$$

$$\text{pH} = 7.2$$

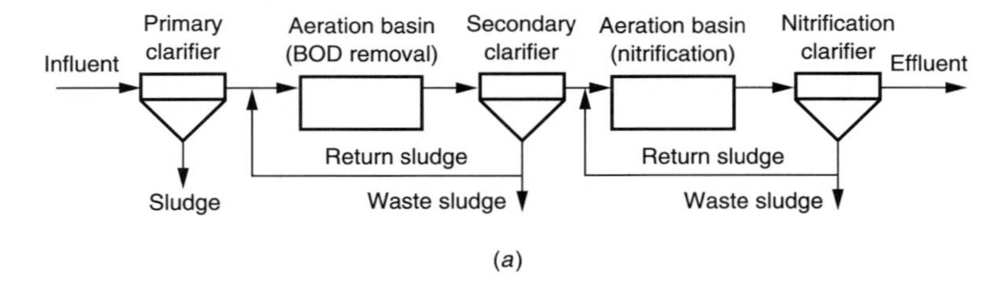

(*a*)

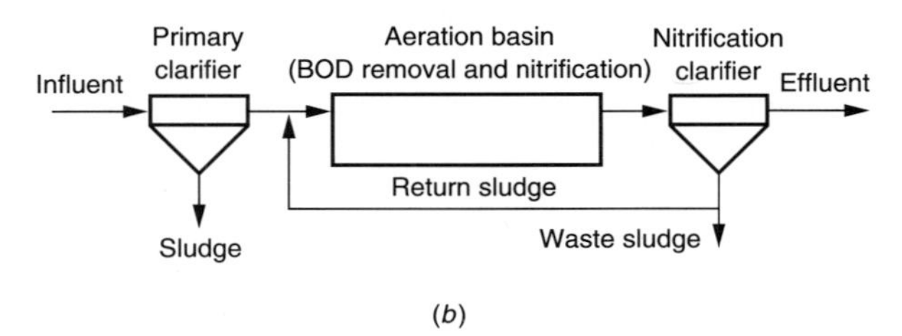

(*b*)

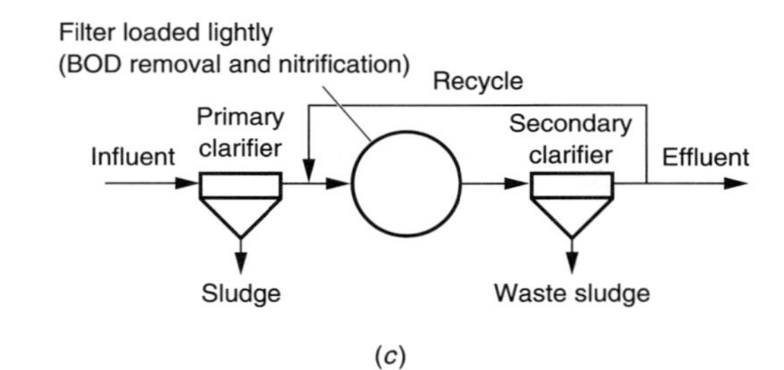

(*c*)

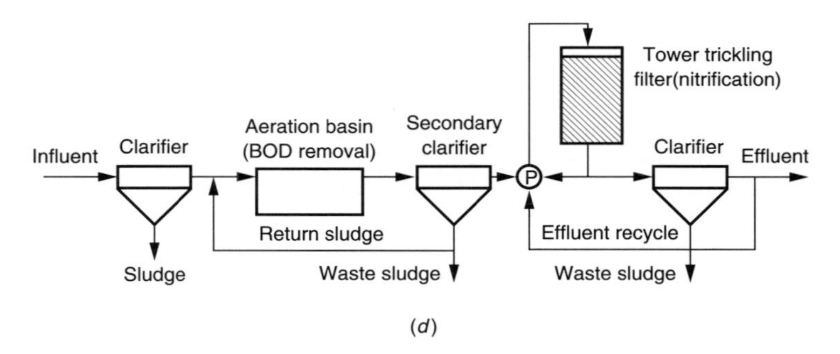

(*d*)

FIGURE 7-15

Processes used for nitrification: (*a*) suspended-growth separate-stage nitrification, (*b*) suspended-growth single-stage nitrification, (*c*) lightly loaded attached-growth reactor, (*d*) attached-growth reactor following activated-sludge process.

$$\mu_N = 0.45\left(\frac{30}{0.28 + 30}\right)\left(\frac{2.0}{1.3 + 2.0}\right)(e^{0.098(12-15)})[1 - 0.833(7.2 - 7.2)]$$

$$= 0.20 \text{ d}^{-1}$$

2. Determine the maximum rate of substrate utilization k using Eq. (7-13), substituting μ_N for μ_m and Y_N for Y.

$$k = \frac{\mu_N}{Y_N}$$

$$\mu_N = 0.20 \text{ d}^{-1} \qquad \text{(from step 1}b\text{ above)}$$

$$Y = 0.16 \qquad \text{(from Table 7-7)}$$

$$k = \frac{0.20 \text{ d}^{-1}}{0.16} = 1.25 \text{ d}^{-1}$$

3. Determine the minimum and design mean cell residence times.
   a. The minimum θ_c^M is determined by using Eq. (7-57):

$$\frac{1}{\theta_c^M} \approx Yk - k_d$$

$$Y = 0.16$$

$$k = 1.25 \text{ d}^{-1} \qquad \text{(from step 2)}$$

$$k_d = 0.04 \qquad \text{(from Table 7-7)}$$

$$\frac{1}{\theta_c^M} \approx 0.16(1.25 \text{ d}^{-1}) - 0.04 \approx 0.16 \text{ d}^{-1}$$

$$\theta_c^M = \frac{1}{0.16} = 6.25 \text{ d}$$

   b. Design θ_c (using a safety factor of 2.0):

$$\theta_c = \text{SF}(\theta_c^M) = 2.0(6.25 \text{ d}) = 12.5 \text{ d}$$

4. Determine the design substrate utilization factor U using Eq. (7-47).

$$\frac{1}{\theta_c} = YU - k_d$$

$$U = \left(\frac{1}{\theta_c} + k_d\right)\frac{1}{Y}$$

$$= \left(\frac{1}{12.5} + 0.04\right)\frac{1}{0.16} = 0.75 \text{ d}^{-1}$$

5. Determine the concentration of ammonia in the effluent using Eq. (7-46), substituting N for S.

$$U = \frac{kN}{K_N + N} = 0.75 \text{ d}^{-1}$$

$$k = 1.25 \text{ d}^{-1}$$

$$K_N = 0.28 \text{ mg/L}$$

Rearranging the above expression gives:

$$N = \frac{UK_N}{k - U} = \frac{0.75 \times 0.28}{1.25 - 0.75} = 0.42 \text{ mg/L}$$

6. Determine the substrate removal rate for the activated-sludge process using Eq. (7-47):

$$\frac{1}{\theta_c} = YU - k_d$$

$$\theta_c = 12.5 \qquad \text{(from step 3)}$$

$$Y = 0.6 \text{ lb VSS/lb BOD}_5 \qquad \text{(from Table 7-5)}$$

$$k_d = 0.055 \text{ d}^{-1} \qquad \text{(from Table 7-5)}$$

Rearranging the above expression gives:

$$U = \left(\frac{1}{\theta_c} + k_d\right)\frac{1}{Y}$$

$$= \left(\frac{1}{12.5} + 0.055\right)\frac{1}{0.6}$$

$$= 0.23 \text{ lb BOD}_5 \text{ removed/lb MLVSS}\cdot\text{d}$$

If it is assumed that the process efficiency is 90 percent, the corresponding value of the food-to-microorganism ratio is equal to 0.26 lb BOD_5 applied per lb MLVSS·d.

7. Determine the required hydraulic detention time for BOD oxidation and nitrification using Eq. (7-45):

$$U = \frac{S_o - S}{\theta X}$$

a. BOD_5 oxidation:

$$\theta_{\text{BOD}} = \frac{S_o - S}{UX}$$

$$S_o = 150 \text{ mg/L} \qquad \text{(from problem specification)}$$

$$S = 15 \text{ mg/L} \qquad \text{(assumed value)}$$

$$U = 0.23 \text{ d}^{-1} \qquad \text{(from step 6)}$$

$$X = \text{MLVSS, mg/L} \qquad (\text{assume } X = 2000 \text{ mg/L})$$

$$\theta = \frac{(150 \text{ mg/L} - 15 \text{ mg/L})}{0.23 \text{ d}^{-1} (2000 \text{ mg/L})} = 0.29 \text{ d} = 7.0 \text{ h}$$

b. Ammonia oxidation (nitrification):

$$\theta_N = \frac{N_o - N}{UX}$$

$$N_o = 30 \text{ mg/L} \qquad \text{(from problem specification)}$$

$$N = 0.42 \text{ mg/L} \qquad \text{(from step 5)}$$

$$U = 0.75 \text{ d}^{-1} \qquad \text{(from step 4)}$$

$$f_N = \frac{0.16(30 - 0.42)}{0.6(150 - 15) + 0.16(30 - 0.42)} = 0.055 \qquad \text{[see Eq. (7-73) in Table 7-6]}$$

$$X = 2000 \text{ mg/L} \times 0.055$$
$$= 110 \text{ mg/L}$$

$$\theta = \frac{(30 \text{ mg/L} - 0.42 \text{ mg/L})}{0.75 \text{ d}^{-1}(110 \text{ mg/L})} = 0.36 \text{ d} = 8.6 \text{ h}$$

Conclusion: Ammonia oxidation process controls the required hydraulic detention time.

Comment. In addition to the above computations, the alkalinity requirements should be checked. If the natural alkalinity of the wastewater is insufficient, it may be necessary to install a pH control system.

Biological Denitrification

In the past, the biological conversion of nitrate to nitrogen gas was often identified as anaerobic denitrification. However, the principal biochemical pathways are not anaerobic but rather a modification of aerobic pathways in which nitrate serves as the electron acceptor; therefore, the use of the term *anoxic* in place of *anaerobic* is considered appropriate. Denitrifying bacteria obtain energy for growth from the conversion of nitrate to nitrogen gas, but require a source of carbon for cell synthesis. Because nitrified effluents are usually low in carbonaceous matter, an external source of carbon is often required. In most biological denitrification systems, the incoming wastewater or cell tissue is used to provide the needed carbon. In the treatment of agricultural wastewaters that are deficient in organic carbon, methanol and other organic compounds have been used as a carbon source. Industrial wastes that are deficient in nutrients but contain organic carbon have also been used. Calculation of residence time for denitrification is illustrated in Example 7-6.

Denitrification stoichiometry. With methanol as the carbon source, the stoichiometry of separate-stage denitrification can be described as follows. The amount of methanol required can be determined by considering the overall reactions for the removal of nitrate, nitrite, and oxygen. The reactions, which can be derived from the half-reactions given in Table 7-4, are given in Eqs. (7-79) through (7-81). Coefficients for Eqs. (7-79), (7-80), and (7-81) were determined for f_s values of 0.1, 0.3, and 0.3, respectively. If different f_s values are used, the equation coefficients will vary (see Example 7-1) (McCarty, 1975).

For nitrate removal:

$$NO_3^- + 1.183CH_3OH + 0.273H_2CO_3 \rightarrow$$
$$0.091C_5H_7O_2N + 0.454N_2 + 1.820H_2O + HCO_3^- \qquad (7\text{-}79)$$

For nitrite removal:

$$NO_2^- + 0.681CH_3OH + 0.555H_2CO_3 \rightarrow$$
$$0.047C_5H_7O_2N + 0.476N_2 + 1.251H_2O + HCO_3^- \qquad (7\text{-}80)$$

For oxygen removal:

$$O_2^- + 0.952CH_3OH + 0.061NO_3^- \rightarrow 0.061C_5H_7NO_2 + 1.075H_2O + 0.585H_2CO_3 + 0.061HCO_3^- \quad (7\text{-}81)$$

Referring to Eq. (7-79), note that for each milligram of nitrate nitrogen converted, 2.70 mg of CH_3OH are utilized, 0.74 mg of new cells are formed, and 3.57 mg of alkalinity, expressed as $CaCO_3$, are formed. Considering the above reactions, the total methanol requirement is:

$$CH_3OH_{req} = 2.70\ (NO_3\text{-}N) + 1.56\ (NO_2\text{-}N) + 0.95\ DO \quad (7\text{-}82)$$

Since the COD equivalent of the methanol is approximately 1.5 mg COD/mg CH_3OH, Eq. (7-82) can be rewritten in terms of COD as follows:

$$COD_{req} = 4.05\ (NO_3\text{-}N) + 2.34\ (NO_2\text{-}N) + 1.43\ DO \quad (7\text{-}83)$$

It should also be noted that the oxygen equivalent of each milligram of nitrate nitrogen converted is 2.86, based on the use of this compound as an electron acceptor (see reactions 3 and 4 in Table 7-4). Thus, for each milligram per liter of NO_3-N converted, 2.86 mg/L of O_2 less are required. In practice, about 50 percent of the oxygen used for nitrification can be recovered through denitrification.

Denitrification process variables. Denitrification, as illustrated in Eqs. (7-79) through (7-81), occurs under anoxic conditions. Process variables affecting the denitrification process include: (1) concentration of nitrate, (2) concentration of carbon, (3) dissolved oxygen concentration, (4) temperature, and (5) pH. The approach used to model these variables is summarized in Table 7-8. It has been observed that the rate of denitrification depends primarily on the nature of concentration of the carbon source. Where the carbon concentration is not limiting, it has been observed that the specific rate of denitrification is zero order down to very low nitrate concentrations. Additional details and information on biological denitrification may be found in Sedlak (1991) and WEF (1992b).

Denitrification processes. As with nitrification, the principal denitrification processes may also be classified as suspended-growth and attached-growth (see Fig. 7-16). Suspended-growth denitrification is usually carried out in a plug-flow type of activated-sludge system (i.e., after any process that converts ammonia and organic nitrogen to nitrate). A variety of attached-growth processes have also been used for denitrification (see discussion in Sec. 7-8).

EXAMPLE 7-6. RESIDENCE TIME REQUIRED FOR SEPARATE-STAGE DENITRIFICATION. Determine the residence time required for separate-stage denitrification for the following conditions:

1. Influent nitrate to basin = 29.58(30 − 0.42) mg/L (see Example 7-5)
2. Effluent nitrate from basin = 2 mg/L
3. MLVSS = 2000 mg/L
4. Temperature = 12°C
5. Dissolved oxygen = 0.15 mg/L
6. $R_{DN_{20}} = 0.10\ d^{-1}$ (see Table 7-8)

TABLE 7-8
Effects of the major operational and environmental variables on the denitrification process*

Factor	Description of effect
Nitrate concentration	It has been observed that the concentration of nitrate will affect the maximum growth of the organisms responsible for denitrification. The effect of the nitrate concentration has been modeled by the following expression: $$\mu_{DN} = \mu_{DNmax}\frac{NO_3}{K_{NO_3} + NO_3} \quad (7\text{-}84)$$
Carbon concentration	The effect of the carbon concentration has also been modeled by a Monod-type expression: $$\mu_{DN} = \mu_{DNmax}\frac{C}{K_C + C} \quad (7\text{-}85)$$ where C = methanol concentration of carbon source, mg/L, and K_C = half-saturation constant for carbon source, mg/L.
Temperature	The effect of temperature is significant. It can be estimated by using the following expression: $$P = 0.25T^2 \quad (7\text{-}86)$$ where P = percent of denitrification growth rate at 20°C and T = temperature, °C.
pH	From available evidence, it appears that the optimum pH range is between about 6.5 and 7.5, and the optimum condition is around 7.0.
Specific denitrification rate	$$R_{DNT} = R_{DN20°C} \times 1.09^{(T-20)}(1 - DO) \quad (7\text{-}87)$$ where R_{DNT} = overall denitrification rate $R_{DN20°C}$ = specific denitrification rate, lb NO_3-N/lb MLVSS·d = 0.10 lb NO_3-N/lb MLVSS·d T = wastewater temperature, °C DO = dissolved oxygen in the wastewater, mg/L

*Developed in part from Eckenfelder and Grau (1992), Sedlak (1991), and U.S. EPA (1975).

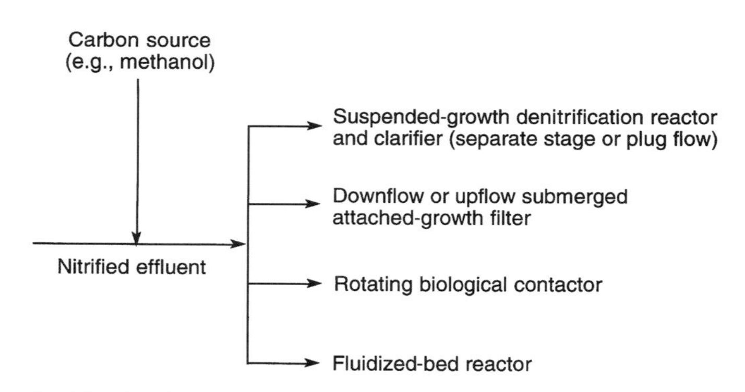

FIGURE 7-16
Alternative separate-stage denitrification processes using an external carbon source.

Solution

1. Calculate the denitrification rate for 12°C, using Eq. (7-87) in Table 7-8.

$$\begin{aligned} R_{DNT} &= R_{DN20} \times 1.09^{(T-20)}(1 - DO) \\ &= 0.1 \times 1.09^{(12-20)}(1 - 0.15) \\ &= 0.043 \text{ d}^{-1} \end{aligned}$$

2. Calculate the residence time using Eq. (7-45), substituting R_{DN20} for U:

$$U = \frac{N_o - N}{\theta X}$$

$$\begin{aligned} \theta &= \frac{N_o - N}{UX} \\ &= \frac{29.58 - 2}{0.043 \times 2000} \\ &= 0.32 \text{ d} \\ &= 7.7 \text{ h} \end{aligned}$$

Comment. The detention time computed by the above approach is also used to determine the time required in the cyclic activated-sludge processes used for denitrification, as described in the following section.

Combined Carbon Oxidation-Nitrification-Denitrification Processes

Because of the high cost of most organic carbon sources, a number of processes have been developed in which the carbon oxidation, nitrification, and denitrification processes are combined into a single process without any intermediate steps. Specific advantages of the single-stage process include: (1) reduction in the volume of air needed to achieve nitrification and BOD_5 removal, (2) potential elimination of the need for supplemental organic carbon sources (e.g., methanol) required for complete denitrification, (3) elimination of intermediate clarifiers required in staged nitrification-denitrification systems, (4) improved settling, and (5) improved process stability.

In the single-stage processes, either the endogenous decay of the organisms or the carbon in the wastewater is used to achieve denitrification. The approach that has been used to achieve denitrification using the carbon in the wastewater involves the creation of a series of alternating aerobic and anoxic stages (i.e., cyclic operation) without intermediate settling, as shown in Fig. 7-17. Cyclic operation is employed in all of the processes used for small treatment systems in which nitrogen is removed. Single-stage treatment processes suitable for small communities that can be operated in a cyclic mode include: (1) the oxidation ditch, (2) the Schreiber process, (3) the Biolac process, and (4) the sequencing batch reactor. The exact cyclic operation is based on past experience and local experimentation.

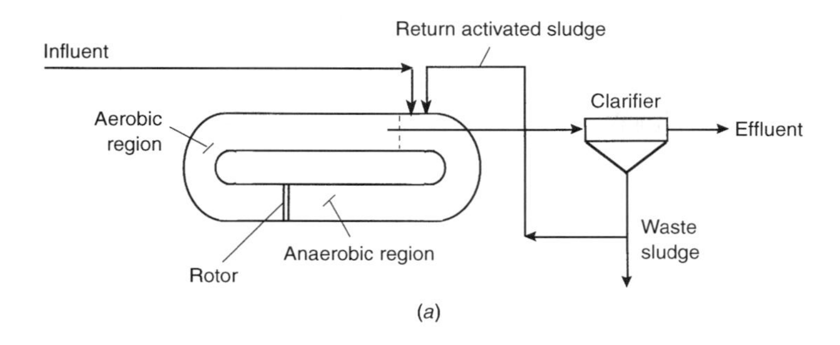

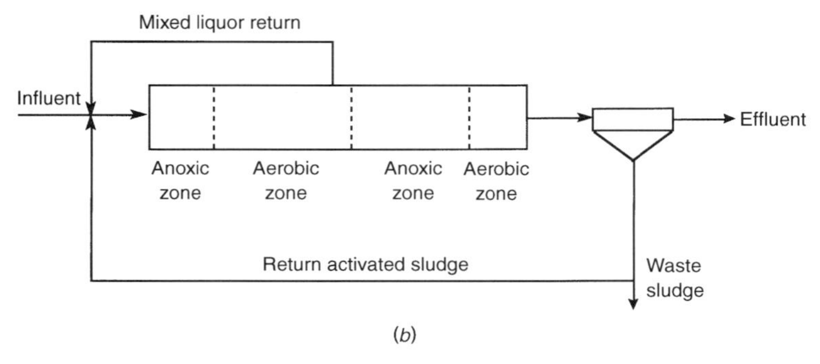

FIGURE 7-17
Combined single-stage nitrification-denitrification systems: (*a*) oxidation ditch and (*b*) four-stage plug-flow Bardenpho.

Biological Phosphorus Removal

Microbes utilize phosphorus for cell synthesis and energy transport. As a result, 10 to 30 percent of the influent phosphorus is removed during secondary biological treatment. Under certain operating conditions, more phosphorus than is needed may be taken up by the microorganisms. Phosphorus removal is accomplished by removing cells containing excess phosphorus. The basis for phosphorus removal in biological systems is based on the following observations (Sedlak, 1991):

1. A number of bacteria are capable of storing excess amounts of phosphorus as polyphosphates in their cells.
2. In the presence of simple fermentation products produced under anaerobic conditions (e.g., volatile fatty acids), these bacteria will assimilate them into storage products within the cells with the concomitant release of phosphorus.
3. Under aerobic conditions, energy is produced by the oxidation of storage products and polyphosphate storage within the cell increases.

In practice, biological phosphorus removal is accomplished by sequencing and producing the appropriate environmental conditions in the reactors. Under anaerobic conditions, a number of organisms respond to volatile fatty acids (VFAs) that are present in the influent wastewater by releasing stored phosphorus (see Fig. 7-18)

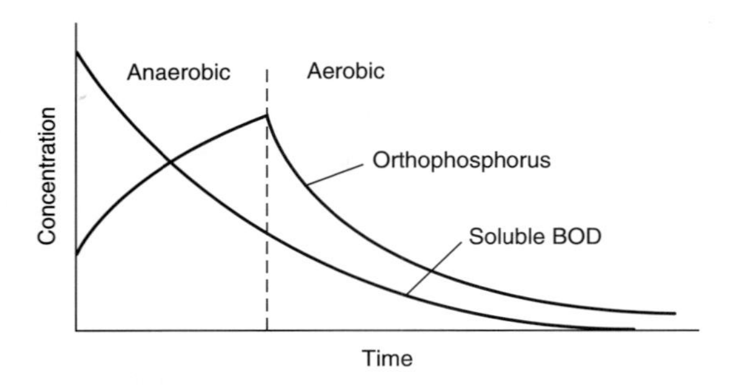

FIGURE 7-18
Fate of soluble BOD and phosphorus in biological nutrient removal reactor (adapted from Sedlak, 1991).

(Jenkins and Hermanowicz, 1997). When an anaerobic zone is followed by an aerobic (oxic) zone, the microorganisms exhibit phosphorus uptake above normal levels. Phosphorus not only is utilized for cell maintenance, synthesis, and energy transport, but also is stored for subsequent use by the microorganisms. The sludge containing the excess phosphorus is either wasted (Fig. 7-19*a*) or removed and treated in a side

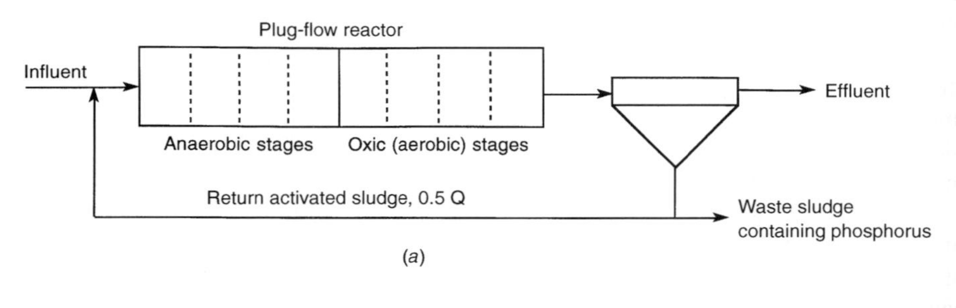

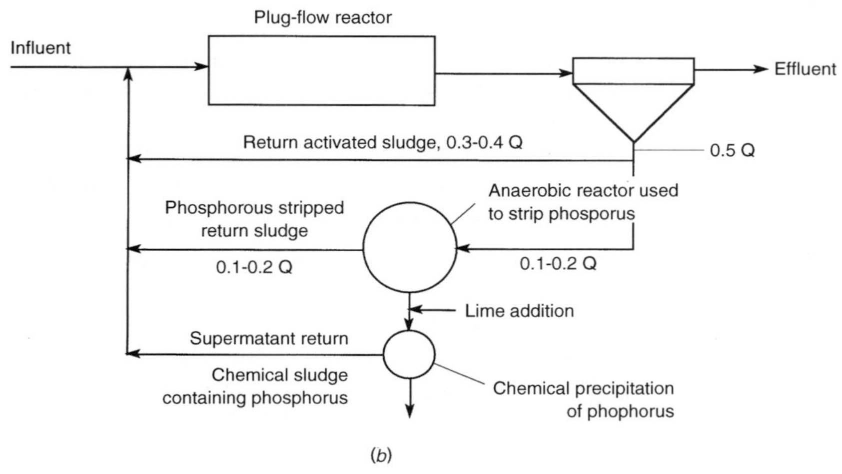

FIGURE 7-19
Alternative biological phosphorus removal systems: (*a*) A/O process, in which phosphorus is removed in the waste sludge and (*b*) PhoStrip process, in which phosphorus is removed in a separate phosphorus stripper.

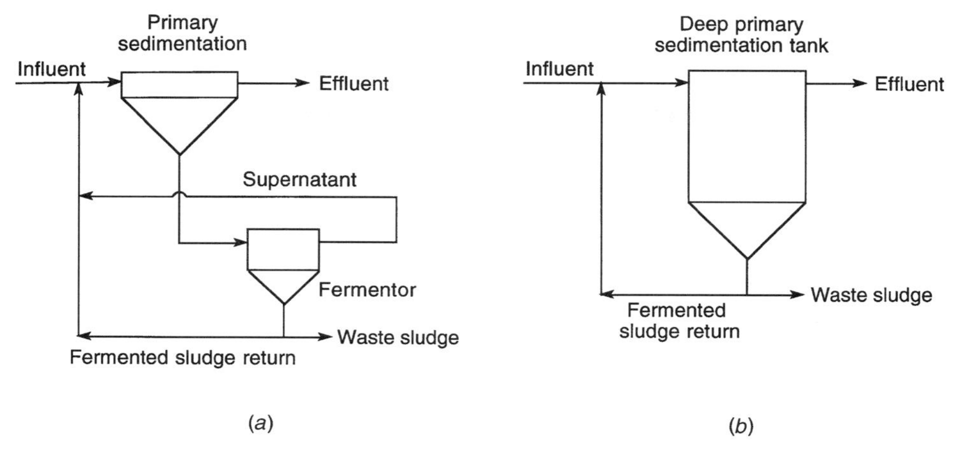

FIGURE 7-20
Use of an external fermenter to provide the needed volatile fatty acids for effective phosphorus removal: (*a*) with separate fermenter and (*b*) with deep primary sedimentation tank (adapted from Sedlak, 1991).

stream to release the excess phosphorus (Fig. 7-19*b*). Release of phosphorus occurs under anoxic conditions. Thus, biological phosphorus removal requires both anaerobic and aerobic reactors or zones within a reactor. Currently, a number of proprietary processes take advantage of one of these mechanisms.

Biological phosphorus removal can be accomplished in conjunction with treatment plants that nitrify and/or denitrify, with and without primary sedimentation. In plants not designed specifically to remove phosphorus, the removal of phosphorus can be accomplished through the use of an external fermenter (Daigger and Bowen, 1996). An external fermenter is used for the production of VFAs (see Fig. 7-20). In plants with primary sedimentation tanks, VFAs can be produced from the settled sludge by operating the sedimentation tank in a fermentation mode. The fermented primary sludge is recycled to the sedimentation tank inlet. Alternatively, primary sludge can be fermented in a gravity thickener with the overflow discharged to the biological reactor. In treatment plants without primary sedimentation facilities an inline fermentor can be added. Clearly a number of alternatives are possible.

Combined Biological Nitrogen and Phosphorus Removal

Where both nitrogen and phosphorus are to be removed, combination processes are used most commonly (see Fig. 7-21). The four processes shown in Fig. 7-21 are proprietary processes. The A^2/O process, a modification of the A/O process, provides an anoxic zone for denitrification. Nitrified MLSS are recycled to the head end of the anoxic stages. In the Bardenpho process, a sequence of anaerobic, anoxic, and aerobic steps is used to achieve both nitrogen and phosphorus removal. Nitrogen is removed by nitrification-denitrification, while phosphorus is removed by wasting sludge from the system. The UCT process, named for the University of Cape

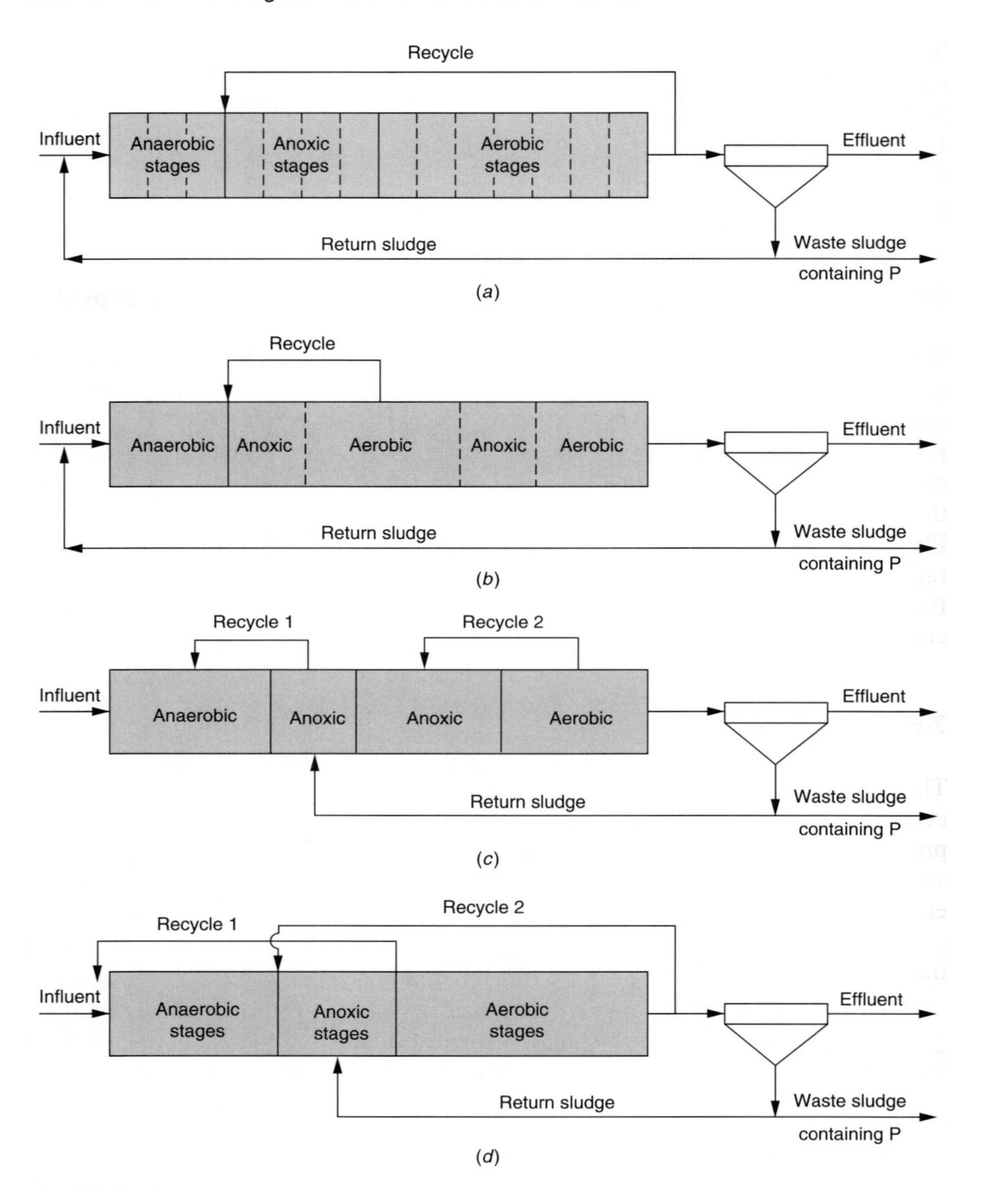

FIGURE 7-21
Combined biological nitrogen and phosphorus removal processes: (*a*) A^2/O process, (*b*) five-stage Bardenpho process, (*c*) UCT process, and (*d*) VIP process. *Note:* Nitrogen is released to the atmosphere in the anoxic stages.

Town, features the return activated sludge recycled to the anoxic stage, instead of the anaerobic stage, as in the A^2/O process. The VIP process is named for the Virginia Initiative Plant in Norfolk, Virginia. The VIP process is similar to the A^2/O and UCT processes except for the location of the recycle flows. It should be noted that a number of other processes are available for the combined removal of nitrogen and phosphorus (Sedlak, 1991).

Control of Combined Biological Nitrogen and Phosphorus Removal

With the recent development and demonstrated reliability of online ammonia, nitrate, and phosphorus monitors, the required process phases for biological nutrient removal can be established by monitoring changes in the concentrations of these constituents. For example, nitrate monitoring can be used to assess the rate of nitrification and concentration in the aerobic phase to ensure complete nitrification and can indicate the rate and concentration in the anoxic phase to assess the degree of denitrification. Phosphorus monitoring can be used to assess the phosphorus concentrations and the rate of release in the anaerobic phase and degree of uptake in the aerobic phase. Process performance can thereby be controlled, and the need for the addition of other carbonaceous sources such as acetic acid or methanol can also be determined.

7-6 AEROBIC SUSPENDED-GROWTH PROCESSES

The purpose of this section is to introduce the reader to the activated-sludge process. A suspended-growth process, activated-sludge is the most commonly used process for the biological treatment of wastewater. The material presented deals with a description of the basic activated-sludge process, classification of the different activated-sludge processes, practical process design considerations, operational problems, and the design of activated-sludge processes suitable for small communities.

Process Description

Developed in England in 1914 by Ardern and Lockett (1914), the activated-sludge process was so named because it involved the production of an activated mass of microorganisms capable of stabilizing a waste aerobically. Many versions of the original process are in use today, but fundamentally they are all similar. The two most common variants, the plug-flow and complete-mix processes, are shown in Fig. 7-22.

In the activated-sludge process, screened or screened and settled wastewater is mixed with varying amounts (20 to 100 percent) of concentrated underflow from the secondary clarifier. The mixture enters an aeration tank (see Figs. 7-9 and 7-11), where the organisms and wastewater are mixed together with a large quantity of air. Under these conditions, the organisms oxidize a portion of the waste organic matter to carbon dioxide and water to obtain energy and synthesize the other portion into new microbial cells utilizing the energy obtained from the oxidation (see Example 7-1).

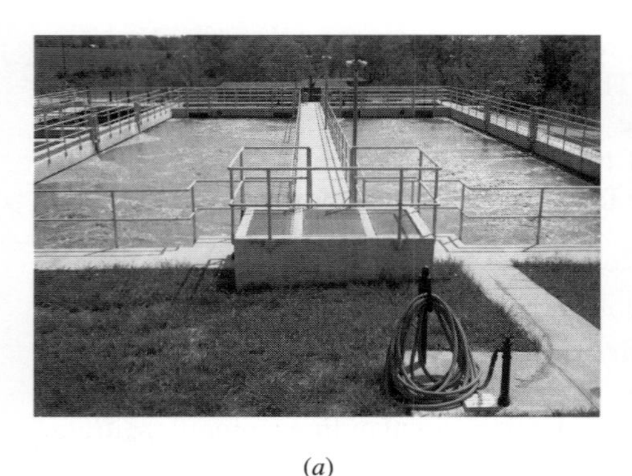

(*a*)

(*b*)

FIGURE 7-22
Views of most common types of activated-sludge systems: (*a*) plug-flow and (*b*) complete-mix.

The mixture then enters a settling tank where the flocculant microorganisms settle and are removed from the effluent stream. The settled microorganisms, or *activated sludge,* are then recycled to the head end of the aeration tank to be mixed again with wastewater. New activated sludge is being produced continuously in this process, and the excess sludge produced each day (waste activated sludge) must be disposed of together with the sludge from the primary treatment facilities. The effluent from a properly designed and operated activated-sludge plant is of high quality, usually having BOD_5 and TSS concentrations equal to or less than 10 mg/L.

Classification of Activated-Sludge Processes

Over the years, a number of variations of the basic activated-sludge process have been developed. The principal activated-sludge processes can be classified as shown in Fig. 7-23. Descriptions of the various processes are presented in Table 7-9. A comparison of the three general categories of activated-sludge processes is presented in Table 7-10. The principal types of activated-sludge processes used for small communities (i.e., less than 0.01 to 5 Mgal/d) are variations of the extended aeration process. Within the extended aeration category are a number of processes that can operate in a cyclic mode that can be used for the biological removal of nutrients. A number of extended aeration processes are considered in this section.

Practical Process Design Considerations

In Sec. 7-4, the kinetics of the activated-sludge process were considered on the assumption that the influent wastewater did not contain any suspended solids. However, because domestic wastewater does contain suspended solids, the kinetic relationships developed in Sec. 7-4 do not reflect adequately what occurs in real systems.

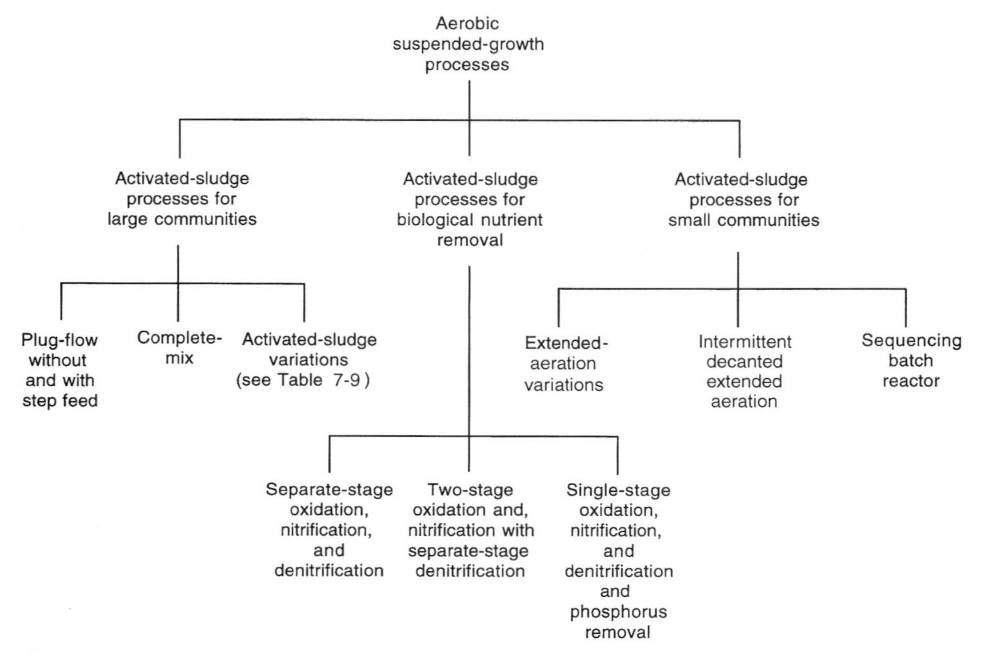

FIGURE 7-23
Definition sketch for the classification of activated-sludge processes.

The material presented in this section deals with the practical approach used to design the activated-sludge process. Important design considerations include: (1) process loading criteria, (2) sludge production, (3) oxygen requirements and supply, (4) reactor sizing and configuration, (5) energy requirements for mixing, and (6) secondary sedimentation. Application of the process design relationships to be discussed is illustrated in Example 7-7 presented at the end of this section.

Process loading criteria. Process loading criteria commonly used for the sizing of the activated-sludge process include the food-to-microorganism ratio (F/M), the mean cell residence time (MCRT), and volumetric loading rate. Each of these parameters is considered below. Typical process loading values for various activated-sludge processes are given in Table 7-11.

Food-to-microorganism ratio. The F/M ratio, expressed as pounds COD or BOD applied per pound of mixed liquor suspended solids (MLSS) per day, represents the mass of substrate applied to the aeration tank each day, versus the mass of suspended solids (microorganisms) in the aeration tank.

Mean cell residence time. The MCRT, expressed in days, is a measure of the average amount of time the biological solids remain in the aeration tank [see Eq. (7-41)]. The total concentration of biological solids maintained in the aeration tank normally varies between 800 and 6000 mg/L. Typically, 40 to 85 percent of the total suspended solids are assumed to be volatile.

TABLE 7-9
Description of activated-sludge processes and process modifications

Process	Description
	Activated-sludge processes for large communities
Conventional plug-flow	Settled wastewater and recycled activated sludge enter the head end of the aeration tank and are mixed by diffused air or mechanical aeration. Air application is generally uniform throughout tank length. During aeration period, adsorption, flocculation, and oxidation of organic matter occurs. Activated-sludge solids are separated in a secondary settling tank.
Plug-flow step feed	Step feed is a modification of the conventional plug-flow process in which the settled wastewater is introduced at several points in the aeration tank to equalize the *F*/*M* ratio, thus lowering peak oxygen demand. Generally three or more parallel channels are used. Flexibility of operation is one of the important features of this process.
Tapered aeration	Tapered aeration is a modification of the conventional plug-flow process. Varying aeration rates are applied over the tank length depending on the oxygen demand. Greater amounts of air are supplied to the head end of the aeration tank and the amounts diminish as the mixed liquor approaches the effluent end. Tapered aeration is usually achieved by using different spacing of the air diffusers over the tank length.
Modified aeration	Modified aeration is similar to the convention plug-flow process except that shorter aeration times and higher *F*/*M* ratios are used. BOD removal efficiency is lower than other activated-sludge processes.
Kraus process	Kraus process is a variation of the step aeration process used to treat wastewater with low nitrogen levels. Digester supernatant is added as a food source to a portion of the return sludge in a separate aeration tank designed to nitrify. The resulting mixed liquor is then added to the main plug-flow aeration system.
Complete-mix	Process is an application of the flow regime of a complete-mix reactor. Settled wastewater and recycled activated sludge are introduced typically at several points in the aeration tank. The organic load on the aeration tank and the oxygen demand are uniform throughout the tank length.
High-rate aeration	High-rate aeration is a process modification in which high MLSS concentrations are combined with high volumetric loadings. This combination allows high *F*/*M* ratios and long mean cell residence times with relatively short hydraulic detention times.
High-purity oxygen	High-purity oxygen is used instead of air in the activated-sludge process. Oxygen is diffused into covered aeration tanks and is recirculated.
	Activated-sludge processes for biological nutrient removal
Single-stage nitrification	In single-stage nitrification, both BOD and ammonia reduction occur in a single biological stage. Reactor configurations can be either a series of complete-mix reactors or plug-flow.
Separate-stage nitrification	In separate-stage nitrification, a separate reactor is used for nitrification, operating on a feed waste from a preceding biological treatment unit. The advantage of this system is that operation can be optimized to conform to the nitrification needs. An external source of carbon is normally required.
Single-stage nitrification/ denitrification	Reactor configurations can be either a series of complete-mix reactors or plug-flow.
Nitrogen and phosphorus removal	Phosphorus removal is accomplished in conjunction with processes for nitrification and denitrification by creating conditions suitable for the luxury uptake of phosphorus.

(*continued*)

TABLE 7-9
(Continued)

Process	Description
	Activated-sludge processes for small communities
Contact stabilization	Contact stabilization uses two separate tanks or compartments for the treatment of the wastewater and stabilization of the activated sludge. The stabilized activated sludge is mixed with the influent (either raw or settled) wastewater in a contact tank. The mixed liquor settled in a secondary settling tank and return sludge are aerated separately in a reaeration basin to stabilize the organic matter. Aeration volume requirements are typically 50 percent less than conventional plug flow.
Extended aeration	Extended aeration process is similar to the conventional plug-flow process except that it operates in the endogenous respiration phase of the growth curve, which requires a low organic loading and long aeration time. Process is used extensively for prefabricated package plants for small communities.
Oxidation ditch	The oxidation ditch consists of a ring- or oval-shaped channel and is equipped with mechanical aeration devices. Screened wastewater enters the ditch, is aerated, and circulates at about 0.8 to 1.2 ft/s (0.25 to 0.35 m/s). Oxidation ditches typically operate in an extended aeration mode with long detention and solids retention times. Secondary sedimentation tanks are used for most applications.
Intermittent decanted extended aeration	A single reactor in which all of the steps of the activated-sludge process occur. Flow into the reactor is continuous as compared to the sequencing batch reactor. Because the mixed liquor remains in the reactor during all of the treatment steps, separate secondary sedimentation facilities are not required. Sludge wasting occurs during the aeration portion of the cycle.
Sequencing batch reactor	The sequencing batch reactor is a fill and draw type reactor system involving one or two complete-mix reactors in which all steps of the activated-sludge process occur. Because the mixed liquor remains in the reactor during all of the treatment steps, separate secondary sedimentation facilities are not required. Typically, two reactors are used for sequencing.

TABLE 7-10
Comparison of aerobic suspended-growth treatment processes*

Factor	High-rate non-nitrifying	Conventional non-nitrifying	Low-rate nitrifying
Biomass concentration attainable	Low	Intermediate	High
Typical MCRT	0.75–2 d	3–8 d	High (>15–20 d)
Growth of filamentous microorganisms	Problematic	Controllable with anoxic and anaerobic selectors	Generally not a problem
Removal efficiency	Fair to good to excellent	Good to excellent	Excellent
Resistance to organic shock loadings	Fair	Good	Excellent

*Adapted in part from Eckenfelder and Grau (1992) and Orhan and Artan (1994).

Volumetric organic loading rate. Volumetric organic loading rates, expressed in terms of lb COD or BOD/10^3 $ft^3 \cdot d$, are based on experience with the activated-sludge process when used to treat domestic wastewater. Typical volumetric loading rates are given in Table 7-11.

Reactor sizing and configuration. The configuration of the reactor depends on the type of activated-sludge process selected. As noted previously, plug-flow and complete-mix reactors are the two most common. As more activated-sludge plants are designed to accomplish biological nutrient removal, plug flow is the preferred reactor configuration. The size of the reactor can be determined by using any of the three loading parameters discussed previously:

1. Volume based on F/M ratio:

$$V = \frac{(Q)(S_o)}{X(F/M)} \tag{7-88}$$

TABLE 7-11
Typical values for the design parameters for selected activated-sludge processes

Process modification	θ_c, d	F/M lb BOD_5/lb MLVSS·d	Volumetric loading rate, lb BOD_5/10^3 $ft^3 \cdot d$	MLSS, mg/L	V/Q, h	Q_r/Q
Conventional plug flow	3–15	0.2–0.6	20–40	1000–3000	4–8	0.25–0.75
Complete mix	0.75–15	0.2–1.0	50–120	800–6500	3.5	0.25–1.0
Step feed	3–15	0.2–0.5	40–60	1500–3500	3–5	0.25–0.75
Single-stage nitrification	8–20	0.10–0.20 (0.02–0.15)*	5–20	1500–3500	6–15	0.50–1.50
Separate-stage nitrification	15–100	0.05–0.20 (0.04–0.15)*	3–9	1500–3500	3–6	0.50–2.00
Contact stabilization	5–15	0.2–0.6	60–75	(1000–3000)† (4000–9000)‡	(0.5–1.0)† (3–6)‡	0.5–1.50
Extended aeration	20–40	0.04–0.10	5–15	2000–8000	18–36	0.5–1.50
Oxidation ditch	15–30	0.04–0.10	5–15	2000–8000	8–36	0.5–1.50
Intermittent decanted extended aeration	12–25	0.04–0.08	5–15	2000–8000	20–40	N/A
Sequencing batch reactor	10–30	0.04–0.10	5–15	2000–8000	12-50	N/A

*TKN/MLVSS.
†Contact unit.
‡Solids stabilization unit.
§MLSS varies depending on the portion of the operating cycle.

2. Volume based on mean cell residence time:

$$V = \frac{(\theta_c)(Q)(S_o)(Y)}{X} \tag{7-89}$$

3. Volume based on volumetric loading:

$$V = \frac{(Q)(S_o)(8.34)}{(L_{org})(10^6 \text{ gal/1.0 Mgal})(\text{ft}^3/7.48 \text{ gal})} \tag{7-90}$$

where V = reactor volume, Mgal
Q = wastewater flowrate, Mgal/d
S_o = influent substrate concentration, mg/L
θ_c = mean cell residence time, d
X = average mixed-liquor total suspended solids, mg/L
F/M = food-to-microorganism ratio
8.34 = conversion factor, lb/[Mgal·(mg/L)]
L_{org} = volumetric organic loading rate, lb COD or $BOD_5/10^3$ ft^3·d

Sludge recycling. The amount of sludge that must be recycled to maintain the MLSS can be determined by performing a mass balance around the reactor as follows:

$$Q(X_o) + Q_r(X_r) = (Q + Q_r)(X) \tag{7-91}$$

where Q = influent flowrate, Mgal/d
Q_r = recycle flowrate, Mgal/d
X_o = influent TSS concentration, mg/L
X_r = TSS concentration in recycle line, mg/L
X = MLSS in reactor, mg/L

If it is assumed that the influent TSS concentration is small relative to the other TSS concentrations, then:

$$\alpha \approx \frac{Q_r}{Q} \approx \frac{(X,\ \text{mg/L})}{(X_r,\ \text{mg/L} - X,\ \text{mg/L})} \tag{7-92}$$

where α is the recycle ratio. The recycle ratio must be established to assess the solids loading on the secondary sedimentation tank.

Sludge production. The quantity of sludge produced that must be wasted on a daily basis is a function of the characteristics of the wastewater, the mean cell residence time θ_c, and the endogenous decay coefficient. The observed yield, as given by Eq. (7-53), is of limited utility in that the particulate biodegradable constituents found in wastewater, as well as the inorganic constituents that are in the primary effluent, are not considered. From the operation of actual plants, the following values can be used to estimate the sludge production in activated-sludge plants treating domestic wastewater.

Activated-sludge process	MCRT, d	Yield, lb cell/lb COD applied	
		With primary sedimentation	Without primary sedimentation
High rate (non-nitrifying)	0.75–2	0.5–0.8	0.6–0.9
Conventional (non-nitrifying)	3–8	0.4–0.6	0.5–0.8
Low rate (nitrifying)	>15	0.3–0.5	0.5–0.7

Adapted in part from Eckenfelder and Grau (1992), Orhan and Artan (1994), and Tchobanoglous and Burton (1991).

Oxygen requirements and supply. The theoretical oxygen requirements for the activated-sludge process, taking into account carbonaceous oxidation, nitrification, and denitrification, can be estimated by the following relationship:

$$\text{lb O}_2 \approx (S_o - S) - P_x(1.42) + 4.6(\text{NO}_3)_f - 2.86(\text{NO}_3)_u \tag{7-93}$$

where S_o = influent substrate concentration, mg COD/L

S = effluent substrate concentration, mg COD/L

P_x = cells produced that are wasted, mg/L

1.42 = conversion factor for cells to COD

$(\text{NO}_3)_f$ = amount of nitrate formed, mg/L

2.86 = conversion factor for the oxygen equivalent of nitrate

$(\text{NO}_3)_u$ = amount of nitrate utilized, mg/L

As a practical matter, taking into account the particulate biodegradable constituents found in wastewater, the typical oxygen requirements for the activated-sludge process are as follows.

Activated-sludge process	MCRT, d	Oxygen requirements
High rate (non-nitrifying)	0.75–2	0.6–0.8 lb O_2/lb COD applied
Conventional (non-nitrifying)	3–8	0.7–0.9 lb O_2/lb COD applied
Low rate (nitrifying)	>15	0.8–1.1 lb O_2/lb COD applied plus 4.6–4.7 lb O_2/lb NO_3-N formed

Adapted in part from Eckenfelder and Grau (1992), Orhan and Artan (1994), and Tchobanoglous and Burton (1991).

The actual amount of oxygen required must be obtained by converting the computed value to a standard oxygen requirement which reflects the effects of salinity–surface tension (beta factor); temperature; elevation; diffuser depth, density, placement, and airflow (for diffused aeration systems); the desired oxygen operating level; and the effects of mixing intensity and basin geometry. The interrelationship of these factors is given by the following general expression.

$$\text{AOTR} = \text{SOTR}\left(\frac{\beta C_{\bar{s}TH} - C_L}{C_{s20}}\right)(1.024^{T-20})(\alpha)(F) \tag{7-94}$$

where AOTR = actual oxygen transfer rate under field conditions, lb O_2/h

SOTR = standard oxygen transfer rate in tap water at 20°C and zero dissolved oxygen, lb O_2/h

β = salinity–surface tension correction factor, typically 0.95 to 0.98

$= \dfrac{C_s\ \text{(wastewater)}}{C_s\ \text{(clean water)}}$

$C_{\bar{s}TH}$ = average dissolved oxygen saturation concentration in clean water in aeration tank at temperature T and altitude H, mg/L

$= (C_{sTH})\dfrac{1}{2}\left(\dfrac{P_d}{P_{atmH}} + \dfrac{O_t}{21}\right)$

C_{sTH} = oxygen saturation concentration in clean water at temperature T and altitude H (see App. E), mg/L

P_d = pressure at the depth of air release, lb/in^2

P_{atmH} = atmospheric pressure of altitute H (see App. D), lb/in^2

O_t = percent oxygen concentration leaving tank, usually 18 to 20

C_L = operating oxygen concentration, mg/L

C_{s20} = dissolved oxygen saturation concentration in clean water at 20°C and 1 atm, mg/L

T = operating temperature, °C

α = oxygen transfer correction factor for waste

$= \dfrac{K_La\text{(wastewater)}}{K_La\text{(clean water)}}$

F = fouling factor for fine and very fine diffusers, typically 0.65 to 0.9

Note that the AOTR and SOTR values given above can also be expressed as transfer efficiencies. The fouling factor F is used to account for both internal and external fouling. Internal fouling is caused by impurities in the compressed air, whereas external fouling is caused by the formation of biological slimes and inorganic precipitants. The oxygen necessary for the biological process can be supplied by using air or pure oxygen. Three methods of introducing oxygen to the contents of the aeration tank are used commonly: (1) mechanical aeration, (2) injection of diffused air, and (3) injection of high-purity oxygen.

Mechanical aeration. In mechanical aeration, rotating devices are used to mix the contents of the aeration basin and to introduce oxygen into the liquid by dispersing fine water droplets in the air so the oxygen can be adsorbed (see Fig. 7-24). Typical oxygen transfer rates and α values for mechanical aeration devices are reported in Table 7-12. When using Eq. (7-94) with mechanical aerators, no correction is made for average dissolved oxygen in the tank, and the value of the fouling factor is one.

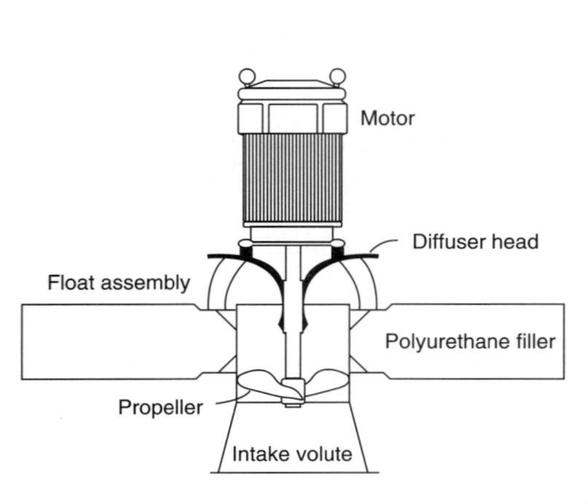

(*a*)

(*b*) (*c*)

FIGURE 7-24
Typical mechanical aeration devices: (*a*) high-speed floating aerator, (*b*) slow-speed turbine aerator on a fixed platform, and (*c*) slow-speed turbine aerator on floats.

TABLE 7-12
Typical ranges of oxygen transfer rates and α values for various types of mechanical aerators*

Aerator type	Transfer rate, lb O_2/hp·h SOTR[†]	Transfer rate, lb O_2/hp·h AOTR[‡]	α
High-speed surface aerator	2.0–3.0	1.2–1.8	0.6–0.9
Low-speed surface aerator	2.5–4.5	1.5–2.8	0.6–0.9
Rotor brush surface aerator	2.0–3.5	1.5–2.1	0.6–0.9
Turbine aerator	2.0–4.0	1.2–2.5	0.6–0.9
Static tube aerator	2.0–3.0	1.2–1.8	0.6–0.9

*Derived in part from Eckenfelder and Grau (1992) and Tchobanoglous and Burton (1991).
[†]Standard conditions: clean water at T = 20°C, P_{atm} = 14.7 lb_f/in^2, and initial dissolved oxygen = 0 mg/L
[‡]Field conditions: wastewater at T = 25°C, P_{atm} = 14.7 lb_f/in^2, β = 0.95, operating dissolved oxygen = 2 mg/L, α = 0.85, and F = 1.

Diffused air aeration. The injection of diffused air involves introducing air under pressure into the aeration tank through diffusion plates or other suitable devices (see Fig. 7-25). The air injected into the reactor serves to keep the contents of the reactor well mixed. In diffused air systems the actual oxygen transfer under field conditions is determined using Eq. (7-94). The required blower capacity can be determined from the following equation:

$$Q_{\text{air}} = \frac{W_{\text{oxygen}}}{(\text{AOTE})(\text{O}_2)(\gamma_{\text{air}})(1440 \text{ min/d})} \tag{7-95}$$

where Q_{air} = required air flow, ft^3/min
W_{oxygen} = oxygen requirements, lb/d
AOTE = actual oxygen transfer efficiency, expressed as a fraction
O_2 = fractional percent of oxygen in air by weight (0.2315)
γ_{air} = specific weight of air (0.075 lb/ft^3 at one atmosphere and 20°C)

Typical oxygen transfer efficiencies and α values for diffused air aeration devices are reported in Table 7-13. Additional details on fine pore aeration devices may be found in U.S. EPA (1989).

The blower power required can be determined from Eq. (7-96), where P_2 reflects the loss of head in the header, the aeration devices, and aeration system appurtenances and the operating pressure.

$$P_w = \frac{w_{\text{air}} R T_1}{550 ne}\left[\left(\frac{P_2}{P_1}\right)^{0.283} - 1\right] \tag{7-96}$$

where P_w = power requirement of each blower, hp
w_{air} = weight of air flow, lb/s
R = engineering gas constant for air, 53.5 ft·lb/(lb air)·°R (U.S. customary units)

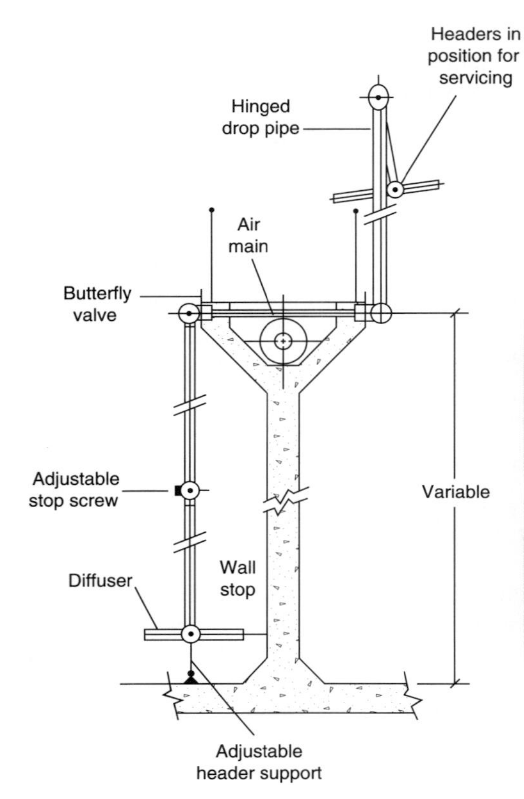

(*a*)

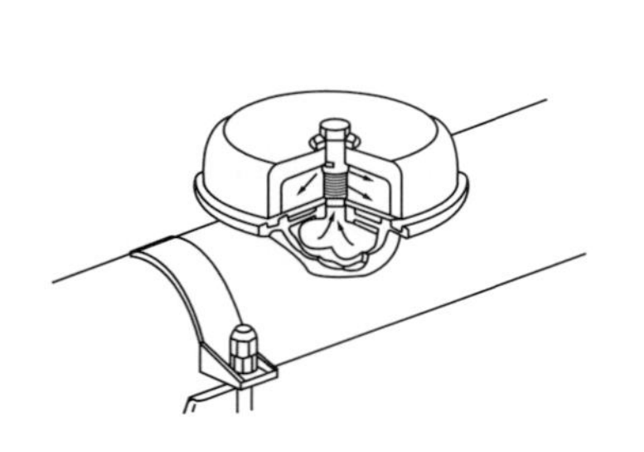

(*b*)

FIGURE 7-25
Typical diffused air aeration devices: (*a*) plastic-wrapped aeration tubes and (*b*) porous dome aerators.

TABLE 7-13
Typical ranges of oxygen transfer efficiencies and α values for various diffused air aeration systems used for various activated-sludge processes*

Activated-sludge process	MCRT, d	Diffuser type	SOTE†, %	AOTE‡, %	α
High rate	0.75–2	Coarse	5–15	4–6	0.4–0.8
(non-nitrifying)	0.75–2	Fine	20–30	8–12	0.3–0.6
Conventional	3–8	Coarse	5–15	4–6	0.4–0.8
(non-nitrifying)	3–8	Fine	20–30	8–12	0.3–0.6
	3–8	Very fine	25–35	10–14	0.3–0.6
Low rate	>15	Coarse	5–15	4–8	0.4–0.8
(nitrifying)	>15	Fine	20–30	8–14	0.3–0.6
	>15	Very fine	25–35	10–16	0.3–0.6

*Adapted in part from Tchobanoglous and Burton (1991).
†Standard conditions: clean water at 20°C, 14.7 lb_f/in^2, and initial dissolved oxygen = 0 mg/L. Reported values are for a diffuser depth of submergence of 15 ft.
‡Field conditions: wastewater at T = 25°C, P_{atm} = 14.7 lb_f/in^2, depth of diffuser submergence = 15 ft, β = 0.95, O_t = 19%, α = 0.5, and F = 0.85. Field values for low rate nitrifying processes using fine and very fine diffusers increased by about 15% to account for the effect of the increased biomass.

T_1 = absolute inlet temperature, °R
P_1 = absolute inlet pressure, lb_f/in^2 (atm)
P_2 = absolute outlet pressure, lb_f/in^2 (atm)
$n = (k-1)/k = 0.283$ for air
k = 1.395 for air
550 = ft·lb/s·hp
e = efficiency (usual range for compressors is 0.70 to 0.90)

High-purity oxygen. The use of high-purity oxygen requires the use of covered aeration tanks. High-purity oxygen generated at the site is injected into the aeration tank. In recent years the use of pure oxygen has declined. Pure oxygen is almost never used for small installations.

Energy requirement for mixing. Typical power requirements for maintaining a completely mixed flow regime with mechanical aerators vary from 0.75 to 1.50 hp/10^3 ft^3 (19 to 39 kW/10^3 m^3), depending on the design of the aerator and the geometry of the tank (i.e., square or rectangular). The corresponding G value can be computed by the methods presented in Chap. 5. In diffused air systems, the air requirement to ensure good mixing varies from 20 to 30 ft^3/10^3 ft^3·min (20 to 30 m^3/10^3 m^3·min) of tank volume, for a spiral roll aeration pattern. For a grid-type aeration system in which the diffusers are installed uniformly along the aeration basin bottom, mixing rates of 10 to 15 ft^3/10^3 ft^3·min (10 to 15 m^3/10^3 m^3·min) have been suggested (Tchobanoglous and Burton, 1991).

Secondary sedimentation. As noted in the introduction to this chapter, if the cell tissue produced during wastewater treatment is not removed by settling, the cell tissue in the wastewater will still exert a BOD and the treatment will be

FIGURE 7-26
Typical sampler used to obtain samples from a sedimentation tank to assess the settleability of activated sludge and the performance of secondary sedimentation tanks.

incomplete. Thus, the secondary sedimentation facilities are an integral part of the activated-sludge conversion process. Further, if the mixed liquor for the aerator cannot be settled and returned to the aeration tank, the process will not function properly. Operationally, the secondary sedimentation facilities must perform two functions: (1) clarification and (2) thickening. Both of these functions must be taken into account in the design of secondary sedimentation facilities. The area required for clarification will depend on the overflow rate, as discussed in Chap. 5. The area required for thickening will depend on the rate at which the mixed liquor will settle to the bottom of the sedimentation tank and thicken. The performance of existing sedimentation facilities can be assessed by collecting a series of samples in the secondary clarifier (see Fig. 7-26) (IAWQ, 1997).

Over the years, several measures have been developed to try to quantify the settling characteristics of activated sludge. Two commonly used measures are the sludge volume index (SVI) and the zone settling rate (Standard Methods, 1995). The SVI is determined by measuring the settled volume after 30 min and the corresponding suspended solids concentration. The numerical value is computed from the following expression:

$$\text{SVI} = \frac{\text{settled volume of sludge (mL/L)} \times 1000}{\text{suspended solids (mg/L)}} \tag{7-97}$$

Unfortunately, the test has no theoretical basis and is subject to significant errors. For example, if a sludge with a concentration of 10,000 mg/L did not settle at all after 30 minutes, the SVI value would be 100, which is considered a good-settling sludge (SVI values below 100). SVI values above 150 are typically associated with filamentous growth.

The zone settling rate test is used to measure the rate at which height of the interface settles. The zone settling velocity is usually determined by using a 4- to 8-in column equipped with an internal stirring mechanism, such as shown in Fig. 7-27. The value obtained for the zone settling velocity can be used for design by allowing a factor of safety to account for the nonideal conditions that exist in the field in actual clarifiers:

$$\mathrm{OR}_{\mathrm{design}} = \frac{(V_i)(179.5)}{\mathrm{SF}} \tag{7-98}$$

where OR = surface overflow rate, $\mathrm{gal/ft^2 \cdot d}$
V_i = settling velocity of interface, ft/h
179.5 = conversion factor from ft/h to $\mathrm{gal/ft^2 \cdot d}$ [(24 h/d) (7.48 $\mathrm{gal/ft^3}$)]
SF = safety factor = typically 1.75 to 2.5

If the interface settling velocity is unknown or a new plant is being designed, it can be estimated from the following equation (Wilson and Lee, 1982; Wilson, 1996):

$$V_i = V_{\max} \exp(-K \times 10^{-6} X) \tag{7-99}$$

where V_i = settling velocity of interface, ft/h
$V_{\max}$ = maximum settling velocity of interface, typically 23.0 ft/h
K = constant, typically 600 L/mg for activated-sludge mixed liquor with an SVI of 150
X = average mixed-liquor total suspended solids, mg/L

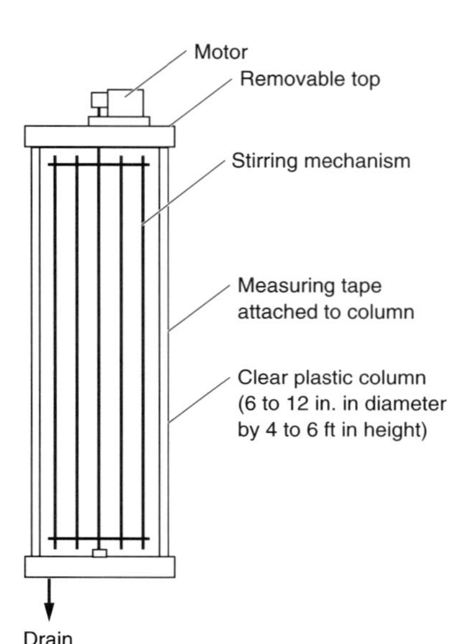

FIGURE 7-27
Typical settling column with internal stirring mechanism used to determine zone settling velocity.

TABLE 7-14
Typical design values for secondary clarifiers following various types of biological treatment*

Type of treatment	Overflow rate gal/ft²·d		Loading lb/ft²·h		Depth, ft
	Average	Peak	Average	Peak	
Air-activated sludge (excluding extended aeration)	400–800	1000–1200	0.8–1.2	2.0	12–20
Extended aeration	200–400	600–800	0.2–1.0	1.4	12–20
Extended aeration package plant	200–300	500–600	0.1–1.0	1.2	†
Trickling filtration	400–600	1000–1200	0.6–1.0	1.6	10–18
Rotating biological contactors	400–800	1000–1200	0.6–1.0	1.6	10–18
Secondary effluent	400–800	1000–1200	0.8–1.2	2.0	10–18
Nitrified effluent	400–600	800–1000	0.6–1.0	1.6	10–18

*Adapted in part from Tchobanoglous and Burton (1991).
†Varies depending on the design configuration.

The approach used most commonly for the design of secondary sedimentation facilities is to base the design on a consideration of the overflow rate and the solids loading rate. Typical values for these parameters are presented in Table 7-14. It should be noted that there are two schools of thought on the use of these sizing parameters. Parker et al. (1996b) have shown that if the solids in the sedimentation tank are managed properly, the overflow rate has little or no effect on the effluent quality of a wide range of overflow rates. Many designers consider the overflow rate of fundamental importance. Additional details on the design and performance of secondary sedimentation facilities may be found in IAWQ (1997). An excellent historical perspective may be found in Anderson (1945).

Operational Problems

Two of the most serious problems with the activated-sludge process are: (1) a phenomenon known as bulking, in which the sludge from the aeration tank will not settle, and (2) the development of a biological surface foam. Where extreme bulking exists, a portion of the suspended solids from the aerator will be discharged in the effluent. Bulking can be caused by: (1) the growth of organisms that can grow in a filamentous form (primarily *Sphaerotilus*) that will not settle (see Fig. 7-28) or (2) the growth of microorganisms that incorporate large volumes of water into their cell structure, making their density near that of water, thus causing them not to settle. In addition to the discharge of biological solids in the effluent, the large volume of sludge that must be handled is another problem. Foaming or frothing (see Fig. 7-29) is also commonly encountered with the activated-sludge process. Foaming is caused, most often, by the excessive growth of the organism *Nocardia*.

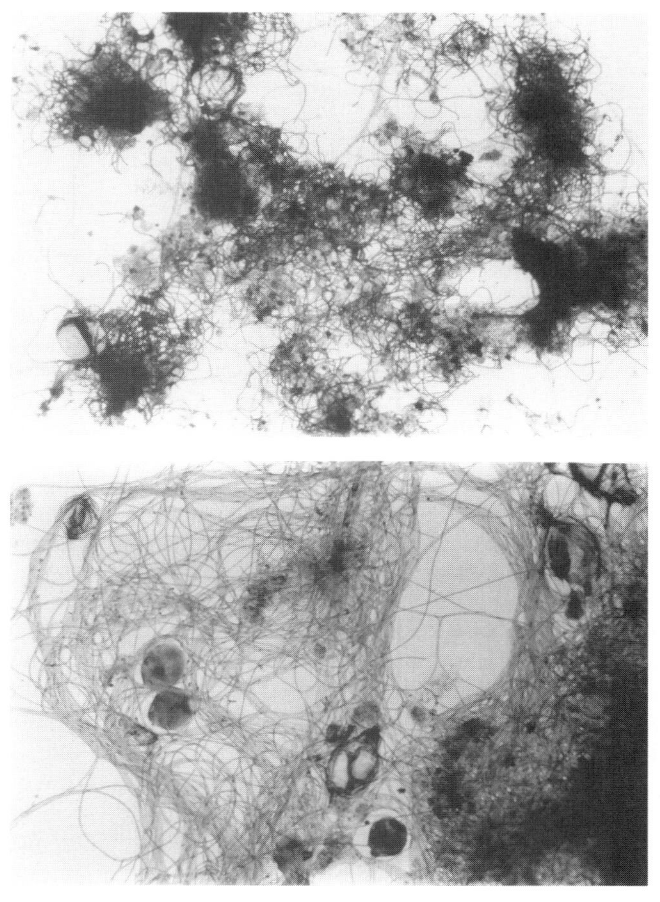

FIGURE 7-28
Typical examples of filamentous organisms that can develop in the activated-sludge process and affect the settleability of the MLSS.

FIGURE 7-29
Foaming due to the growth of *Nocardia* on surface of aeration basin.

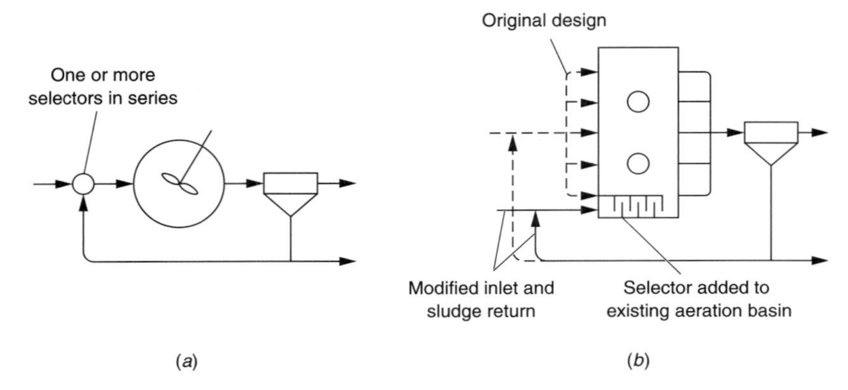

FIGURE 7-30
Biological selectors used for the control of filamentous organisms: (*a*) one or more selectors added to complete-mix activated-sludge reactor and (*b*) selector constructed into existing activated-sludge reactor (adapted from Chudoba, 1985, and Jenkins et al., 1986).

Control of filamentous organisms. Control of filamentous organisms has been accomplished in a number of ways including: (1) the addition of chlorine or hydrogen peroxide to the return waste activated sludge; (2) the alteration of the dissolved-oxygen concentration in the aeration tank; (3) the alteration of points of waste addition to the aeration tank to alter the F/M ratio; (4) the addition of major nutrients (i.e., nitrogen and phosphorus); (5) the addition of trace metals, nutrients, and growth factors; and, more recently, (6) the addition of inorganic talc and (7) the use of a selector (see Fig. 7-30). It should be noted that the use of a selector will not solve problems that are caused by dissolved oxygen or nutrient deficiencies (Albertson, 1987; Jenkins et al., 1986; Jenkins and Hermanowicz, 1997).

A selector is a small tank (20–60 min contact time), or series of tanks (typically three), in which the incoming wastewater is mixed with the return sludge under aerobic, anoxic, or anaerobic conditions. The use of a selector for the control of filamentous organisms was first proposed by Chudoba et al. (1973). The high substrate concentration in the selector favors the growth of nonfilamentous microorganisms (see Fig. 7-31). If the wastewater were added directly to a complete-mix reactor, the low substrate concentration in the reactor would favor the growth of filamentous organisms (Chudoba, 1985). For a three-part aerobic selector the recommended loading rates are as follows (Jenkins et al., 1986):

1. First compartment = 12 kg COD/kg MLSS·d
2. Second compartment = 6 kg COD/kg MLSS·d
3. Third compartment = 3 kg COD/kg MLSS·d

For anoxic selectors, the F/M values can be reduced because the selector uses both a feed-starve cycle and the ability to denitrify for selection against filamentous organisms. The recommended loadings are:

1. First compartment = 6 kg COD/kg MLSS·d
2. Second compartment = 3 kg COD/kg MLSS·d
3. Third compartment = 1.5 kg COD/kg MLSS·d

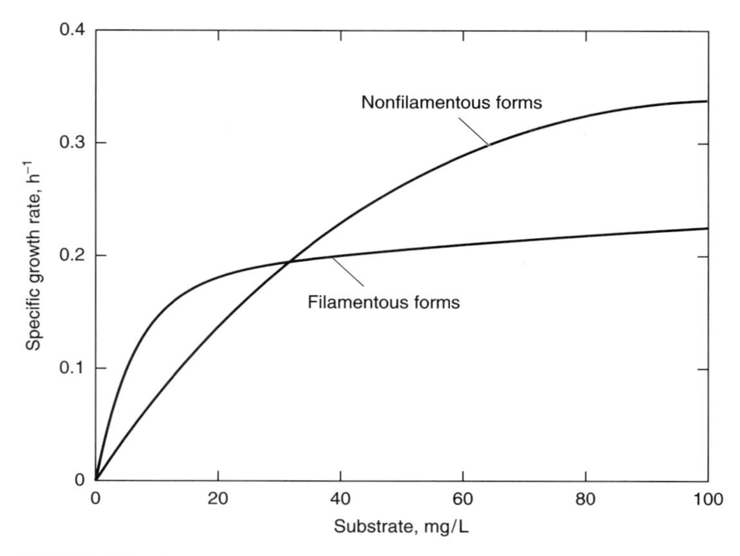

FIGURE 7-31
Typical growth curves for filamentous and nonfilamentous organisms as a function of the substrate concentration.

Control of *Nocardia*. *Nocardia*, a hydrophobic bacteria, which grows in a filamentous form, can lead to the development of a hydrophobic foam on the surface of biological treatment units (i.e., the aeration basin and secondary sedimentation tank). When the organisms grow in sufficient numbers, they tend to trap air bubbles which subsequently float to the surface and accumulate as scum. The problem with *Nocardia* is exacerbated by the fact that biological treatment facilities and secondary sedimentation tanks are baffled, which tends to retain these organisms and thus foster their continued growth. There are seldom any problems with *Nocardia* in unbaffled treatment facilities (see Fig. 7-32). The U.S. EPA requirement that all secondary sedimentation facilities must have surface skimmers typically leads to the accumulation of this organism. The problem is continual where *Nocardia* foam is recycled

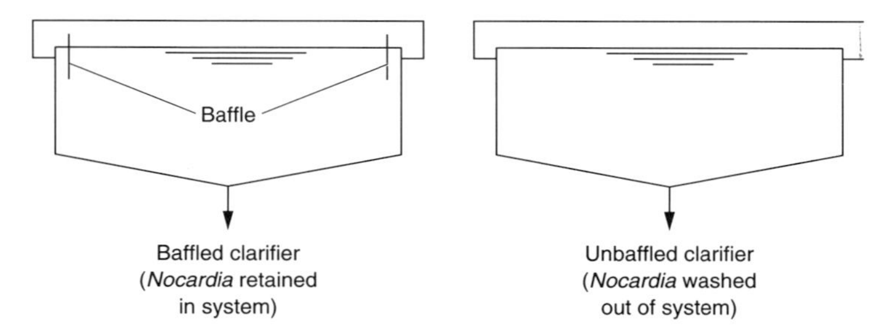

FIGURE 7-32
Control of *Nocardia* using unbaffled secondary sedimentation facilities (adapted from Jenkins, 1997).

internally (e.g., back to the headworks). Measures that can be used to control *Nocardia* include: (1) avoid trapping of the foam and recycling skimmed foam to the headworks or other intermediate points, (2) use a small amount of polymer to flocculate the organism, (3) use a dilute surface chlorine spray (about 0.5 mg/L), (4) use an anoxic selector, and (5) use an anaerobic selector. The last two remedies are readily available in plants designed to remove nutrients biologically (Jenkins, 1997).

Design of Low-Rate Cyclic Activated-Sludge Processes

Over a period of years of experimenting with various activated sludge process modifications, and on the basis of the results of extensive research, the design of activated sludge systems for small systems has evolved toward the use of processes that are lightly loaded. The principal reasons for the selection of such processes is that: (1) they produce a high-quality effluent, (2) they have proven to be stable and easy to operate, (3) they are energy-efficient, and (4) they can be used for biological nutrient removal. To achieve nutrient removal, all of the low-rate cyclic activated-sludge processes can be operated in a cyclic manner, allowing for the development of aerobic, anoxic, and anaerobic conditions within the reactor. Following a presentation of some background material, five of these commonly used processes, suitable for small communities, are introduced and reviewed briefly. The design of such processes is illustrated in Example 7-7.

Background. The low-load concept of using F/M (food-to-microorganism) ratios less than 0.10 for the sizing of biological reactors has become quite common especially when biological nutrient removal is a process goal. For domestic wastewater having an influent BOD_5 of 200 mg/L, the resulting hydraulic detention time in the aeration reactor will be 20 h or greater, depending on the actual F/M ratio selected. Typical F/M values for carbon removal and nitrification vary from 0.05 to 0.1. Corresponding values for carbon oxidation, nitrification, and denitrification will vary from 0.03 to 0.06. The MCRT for low-load systems will be 20 to 40 d or more resulting in a very stable waste activated sludge. It is not unusual to find new facilities that are only partially loaded with MCRT values up to 60 d. Most often the SVI will be less than 80, which is beneficial in ensuring TSS removal in the final clarifier. Typical volumetric BOD loadings in the reactor are less than 12 to 15 $lb/10^3\ ft^3$.

Most often, primary clarification is not considered in the process flow diagram. The principal reason is to eliminate the necessity of having to deal with the resulting primary sludge, along with the secondary sludge. Solids removed in primary sedimentation can be reduced significantly by aerobic and anaerobic digestion. The alternative is to deal with more total solids for disposal with oxic biological conversion. Typical waste production rates were given previously. For treatment facilities of less than about 2 Mgal/d (7570 m^3/d), primary sedimentation and anaerobic digestion are not usually justified. With low-strength influent BOD, the use of primary clarifiers has the potential of reducing the carbon source that is needed for biological nutrient removal. In addition to the BOD conversion in the aerobic phase, there is a need for carbonaceous material in the anoxic and anaerobic phases. To achieve biological nutrient removal, all of the low-rate processes can be operated in such a fashion so as to produce the two or three phases needed for biological nutrient removal.

The final clarifier is sized for the hydraulic loading. Typical clarifier designs use a surface settling rate of 300 to 400 gal/ft^2·d and limit the peak flow to 800 to 1000 gal/ft^2·d. Because of the long detention times (up to 6 hours or more at design flows), the method of settled solids removal from the sedimentation tank must be capable of returning the activated sludge back to the biological reactor quickly. The removal mechanism, such as helical scrapers, will accomplish collection of the settled solids and rapid transport at a uniform high concentration. Typical return activated sludge rates range from 50 to 150 percent of influent flow. The ability to maintain a low sludge blanket level (6 to 8 in) in the sedimentation tank is typical, due to the low SVI and the large surface area. The depth of the clarifier relates to accommodating the peak hydraulic flows for sludge storage and sufficient time for settling. Sufficient depth ensures low suspended solids in the effluent, a necessity when the goal is biological nutrient removal. Typical depths range from 12 to 20 ft (Carolan, 1997).

Oxidation ditch. The oxidation ditch (see Fig. 7-33) is a modification of the complete-mix extended aeration activated-sludge process carried out in a continuous channel reactor. The depth of the aeration channel typically varies from 4 to 12 ft. The shape of the channel is that of an elongated oval race track, although a number of different configurations have been developed. One or more rotor aerators, located across the channel, are used to provide the necessary oxygen to the liquid and to keep the contents of the channel mixed and moving. The oxidation ditch was developed in Netherlands in the 1950s at the Research Institute for Public Health Engineering under the leadership of Dr. Pasveer. The first oxidation ditch was put into operation

FIGURE 7-33
View of oxidation ditch activated-sludge process.

in Holland in 1954. The first oxidation ditch was put into operation in the United States in 1963 at Beaverton, Oregon. The senior author had an opportunity to work on the design of that system.

Since its introduction in 1963, the oxidation ditch has proved to be an extremely effective process and is now used extensively. The design of the oxidation ditch is usually based on a volumetric organic loading of approximately between 9 and 15 lb BOD/10^3 $ft^3 \cdot d$. Typically, MCRTs are greater than 20 to 30 d, and the MLSS is generally between 2000 to 8000 mg/L. Because of the long detention times and high MLSS and efficient aeration it has been possible to achieve complete nitrification, even during winter operation in most locations. By controlling the DO concentration and increasing the length of the ditch, it is possible to achieve carbon oxidation, nitrification, and denitrification in the oxidation ditch. Because the MLSS are moving continuously in the oxidation ditch, a small amount of untreated wastewater is introduced over the entire length of the oxidation ditch. Further, because of the large mass of organisms in the system, the oxidation ditch has been very effective in the treatment of organic shock loadings.

Countercurrent aeration system. The countercurrent aeration system (CCAS), a proprietary modification of the extended-aeration activated-sludge process developed by Schreiber, can provide the three process phases needed for biological nutrient removal (i.e., aerobic (oxic), anoxic, and anaerobic) in a single aeration reactor basin (see Fig. 7-34). The aeration system consists of a support bridge that rotates on top of a round tank. A drive unit moves the bridge on the outer end that travels on top of the tank wall. Fine-bubble membrane diffusers are supported from the bridge at 9 in above the bottom of the tank. The structural supports for the diffusers and the diffuser units travel with the bridge, thereby providing movement of the water that results in mixing of the reactor contents. The bridge travel is one revolution in 3 min or less, which imparts sufficient energy to keep the contents of the basin mixed without the necessity of using air through the diffusers. The process advantage is the separation of the air supply from the mixing energy demands of the reactor.

With the bridge rotating, the aerobic phase occurs when the air is on and dispersed through the reactor contents. Monitoring of the residual dissolved oxygen (DO) provides a means of controlling the airflow rate. A residual DO of 0.7 to 1.0 mg/L has proven sufficient for the biological process to reduce the BOD and provide for nitrification. Sufficient time for nitrification exists because of the low-load design conditions. With the bridge still rotating for mixing, the air supply can be stopped. The complete basin will then be deprived of an outside oxygen supply, resulting in anoxic conditions. As denitrification is completed, the reactor will not have any oxygen available, resulting in an anaerobic condition which enhances the phosphorus removal process.

A means of process control was developed over 10 years ago that correlates the process reactions to changes in turbidity values. Mixed liquor from the reactor is pumped, at about 1 gal/min, to a small cone clarifier where the solids and liquid fractions are separated. The liquid fraction is then passed through a scatter turbidity meter that monitors the change in turbidity values, which correlates with a change in suspended solids concentrations. The correlation is quite simple in that, with the

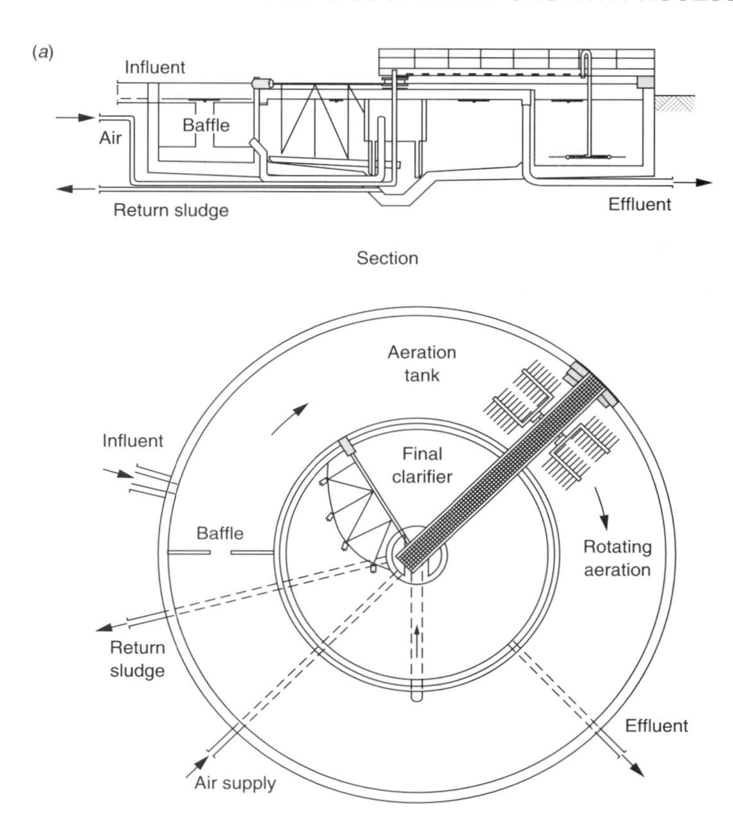

(b)

FIGURE 7-34
Countercurrent aeration system: (*a*) definition sketch (courtesy Schreiber Corp.) and (*b*) view of typical unit.

highest degree of treatment, the oxic phase, the suspended solids are the lowest and with a degraded degree of treatment, through the anoxic and anaerobic phases, the suspended solids will be the highest. From this relationship, the aeration on time can be controlled to create the three process phases of oxic, anoxic, and anaerobic for enhanced biological nutrient removal.

Biolac process. The Biolac process is a unique proprietary process that combines long solids retention times with submerged aeration in earthen basins. Fine-bubble membrane diffusers are attached to floating aeration chains which are moved across the basin by the air released from the diffusers (see Fig. 7-35). The diffusers are located about 1 ft (0.3 m) off the bottom of the basin and are individually controllable to permit flexibility in fine tuning the system to the oxygen demand of the wastewater. Aeration basins are typically 8 to 15 ft (2.4 to 4.6 m) deep.

The process can be designed for nitrification because the sludge age ranges from 40 to 70 d. The food-to-microorganism ratio ranges from 0.04 to 0.1 and the MLSS range is from 1500 to 5000 mg/L. A variation of the standard process, known as the *wave oxidation modification*, allows biological nitrification and denitrification to occur simultaneously by using timers to cycle the air flowrate to each aeration chain.

An integral clarifier can be used with an overflow rate of 400 $gal/ft^2 \cdot d$, or a conventional external clarifier can be used. Sludge removal is achieved by using airlift pumps. The standard aeration air flow design is 4 sft^3/min per 1000 ft^3 of basin volume or 15 to 18 hp/Mgal of basin volume. The aeration chain spacing is typically 12 to 20 ft (3.7 to 6.1 m), although the aeration can be tapered through a basin.

Intermittent decanted extended aeration. The intermittent decanted extended aeration (IDEA) process, developed in Australia over the past 30 years, is illustrated schematically in Fig. 7-36*a* (Boncardo, 1997, and Chong, 1997). Physically, the process is carried out in a single basin with continuous inflow of wastewater. An important difference between the IDEA process and conventional plants is that in conventional plants the biological treatment and sedimentation processes are carried out simultaneously in separate tanks, whereas in the IDEA process the processes are carried out sequentially in the same tank. Because both aeration and settling occur in the same basin, sludge does not have to be returned from the clarifier to maintain the sludge content in the aeration basin. In normal operation a 4-hour cycle is used. Each cycle is divided into three phases: (1) aeration, (2) settlement, and (3) decanting, each phase taking about 2 h, 1 h, and 1 h, respectively. Nitrification and BOD conversion occur during the aeration phase. Denitrification occurs during the settlement phase. Additional denitrification and the removal of clarified effluent occur during the decanting phase. Sludge wasting occurs during the aeration phase when the contents of the basin are mixed completely. Wasting sludge from the aeration basin simplifies the wasting process because the quantity wasted each day is a fraction of the basin volume [see Eq. (7-42)].

In systems serving very small communities (2000 persons or fewer), the decanted mixed liquor is discharged to a sludge storage lagoon for thickening and long-term storage. The supernatant is returned to the plant influent. Where storage lagoons are used, the solids are periodically allowed to dry out and are disposed of on land or in a landfill. In larger plants, the decanted mixed liquor is often discharged to a thickener. The thickened solids are dewatered by a belt filter press. The supernatant from the thickener and the filtrate from the belt press are returned to the plant influent. The dewatered solids are typically composted for use as a soil amendment.

(a)

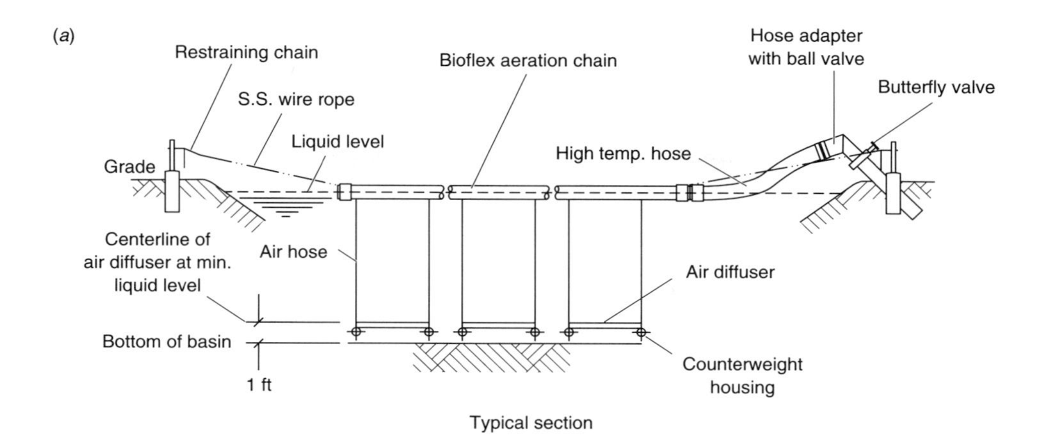

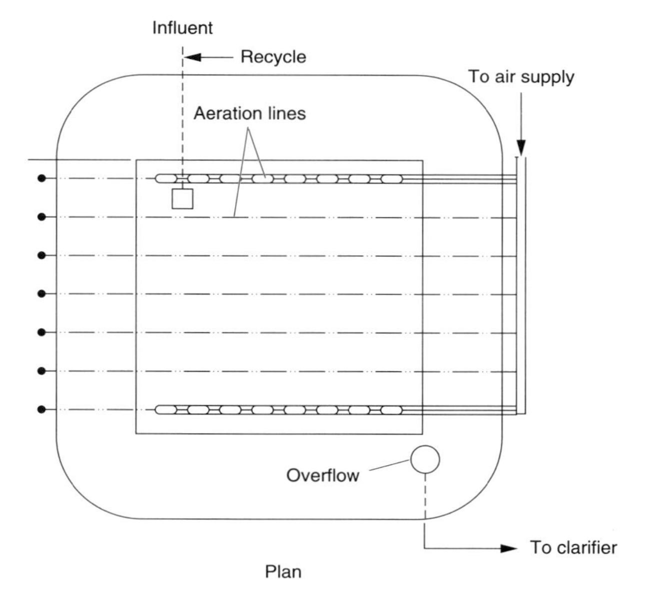

(b)

FIGURE 7-35
Biolac activated-sludge process: (*a*) definition sketch (courtesy Parkson Corp.) and (*b*) view of typical unit.

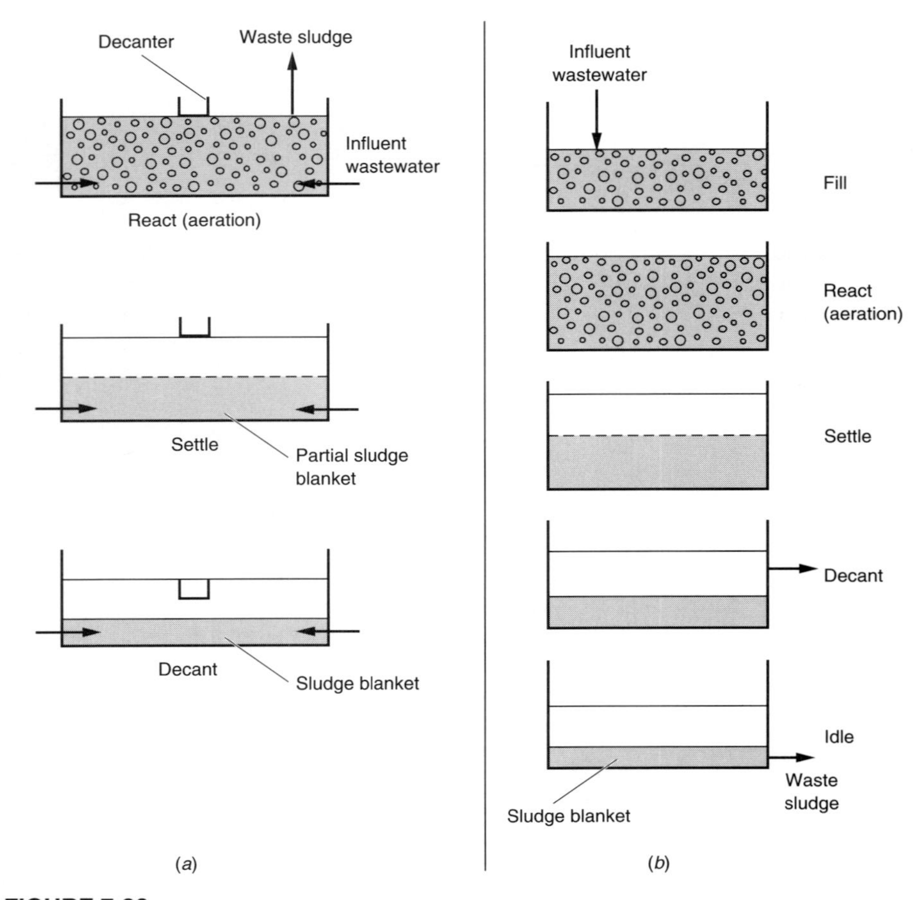

FIGURE 7-36
Operating cyclics for (*a*) intermittent decanted extended aeration (IDEA) (adapted from Chong, 1997) and (*b*) sequencing batch reactor (SBR) activated-sludge processes (adapted from U. S. EPA, 1986).

Typical design parameters for the process are as follows (Boncardo, 1997, and Chong, 1997):

Food-to-microorganism ratio	0.04 to 0.08
Solids retention time	12 to 25 d
Hydraulic retention time at ADWF	20 to 40 h
Weir loading rate (maximum rate)	100 to 140 × 10^3 gal/d·ft
Aeration capacity	2.4 lb O_2/lb BOD
Full aeration up to	3 to 4 × ADWF
Aeration time	12 h/d
Aerator type	Surface or diffused air
Operating cycles per day	6

The operating cycle is usually modified for prolonged wet-weather operation. The IDEA process has also been modified for the removal of phosphorus by creating conditions upstream of the biological reactor to promote the release and luxury uptake of phosphorus. The modification involves the addition of an anoxic and an anaerobic process in front of the aeration basin (Boncardo, 1997). The advantage of the IDEA process compared to the sequencing batch reactor process, a variant of the IDEA process described below, is that two tanks are not required and that the required controls are simplified.

Sequencing batch reactor. A sequencing batch reactor (SBR), a fill-and-draw activated-sludge treatment process, is a variant of the IDEA process described above. As currently implemented, SBR systems utilize five steps in common, which are carried out in sequence as follows: (1) fill, (2) react (aeration), (3) settle (sedimentation/clarification), (4) draw (decant), and (5) idle. Each of these steps is illustrated in Fig 7-36*b*. Depending on the treatment objectives, the fill operation can be fill only, fill and mix, or fill, mix, and aerate. A number of process modifications have been made in the times associated with each step to achieve specific treatment objectives. Sludge wasting is another important step in the SBR operation that greatly affects performance. Wasting is not included as one of the five basic process steps because there is no set time period within the cycle dedicated to wasting. The amount and frequency of sludge wasting is determined by performance requirements, as with a conventional continuous-flow system. In an SBR operation, sludge wasting usually occurs during the settle or idle phases. Some modifications of the SBR process also include continuous-flow modes of operation with multiple tanks. Additional details on SBRs may be found in U.S. EPA (1986).

EXAMPLE 7-7. DESIGN OF EXTENDED AERATION ACTIVATED-SLUDGE PROCESS. Design an extended aeration activated-sludge treatment process with diffused aeration to treat a wastewater with the following characteristics. Base the design on an F/M ratio of 0.04 d^{-1} to achieve an effluent with a nitrate concentration of 2.0 mg/L or less and an ammonia concentration less than 1.0 mg/L. Assume the plant will be operated at a θ_c value of 20 d. Also assume that the MLSS is 4000 and that the volatile fraction is equal to 0.65.

1. BOD_5 after primary sedimentation = 160 mg/L
2. COD after primary sedimentation = 250 mg/L
3. Organic nitrogen after primary sedimentation = 18 mg/L
4. Ammonia nitrogen after primary sedimentation = 22 mg/L
5. Phosphorus after primary sedimentation = 5 mg/L
6. TSS after primary sedimentation = 100 mg/L
7. Alkalinity = 350 mg/L as $CaCO_3$
8. Wastewater temperature during coldest month = 20°C
9. Average flowrate = 0.75 Mgal/d
10. Maximum day peaking factor = 2.5

Solution

1. Define expected effluent quality. On the basis of past experience with the extended aeration process, the following effluent characteristics are expected:

$$
\begin{aligned}
BOD_5 &= 10.0 \text{ mg/L} \\
TSS &= 10.0 \text{ mg/L} \\
NH_3\text{-N} &= <1.0. \text{ mg/L} \\
NO_3\text{-N} &= \leq 2.0. \text{ mg/L} \\
P &= 1.0 \text{ mg/L}
\end{aligned}
$$

2. Determine design loadings.

$$
\begin{aligned}
BOD_5 &= (160 \text{ mg/L})(0.75 \text{ Mgal/d})(8.34) = 1001 \text{ lb/d} \\
COD &= (250 \text{ mg/l})(0.75 \text{ Mgal/d})(8.34) = 1564 \text{ lb/d} \\
\text{Org-N} &= (18 \text{ mg/L})(0.75 \text{ Mgal/d})(8.34) = 113 \text{ lb/d} \\
NH_3\text{-N} &= (22 \text{ mg/L})(0.75 \text{ Mgal/d})(8.34) = 138 \text{ lb/d} \\
TSS &= (100 \text{ mg/L})(0.75 \text{ Mgal/d})(8.34) = 626 \text{ lb/d} \\
P &= (5 \text{ mg/L})(0.75 \text{ Mgal/d})(8.34) = 31.3 \text{ lb/d}
\end{aligned}
$$

3. Determine the size of the reactor required and check the actual detention time and the organic loading rate. A single sludge system is to be used with an F/M ratio of 0.04 d^{-1}.
   a. Determine the volume of the reactor using Eq. (7-88), assuming the MLSS will be equal to 4000 mg/L:

$$
V = \frac{(Q)(S_o)}{X(F/M)} = \frac{(0.75 \text{ Mgal/d})(160 \text{ mg/L})}{4000 \text{ mg/L}(0.04/\text{d})} = 0.75 \text{ Mgal}
$$

   b. Check the detention time:

$$
\theta = \frac{V}{Q} = \frac{0.75 \text{ Mgal/d } (24 \text{ h/d})}{0.75 \text{ Mgal}} = 24.0 \text{ h}
$$

   c. Check the volumetric organic loading rate per 10^3 ft^3:

$$
\begin{aligned}
L_{org} &= \frac{\text{lb BOD}_5/\text{d}}{(V \text{ ft}^3)(10^3 \text{ ft}^3/1000 \text{ ft}^3)} \\
&= \frac{1001}{(0.75 \text{ Mgal})(10^6 \text{ gal/Mgal})(1.0 \text{ ft}^3/7.48 \text{ gal})} \times \frac{1000 \text{ ft}^3}{10^3 \text{ ft}^3} \\
&= 10.0 \text{ lb BOD}_5/10^3 \text{ ft}^3
\end{aligned}
$$

4. Determine the oxygen requirement.
   a. Find the total oxygen required, assuming no return from denitrification and that all of the nitrogen is converted to nitrate (see Sec. 7-5):

$$
\begin{aligned}
O_2 \text{ req.} &= (1.0 \text{ lb } O_2/\text{lb COD})(\text{lb COD applied/d}) + 4.6(\text{lb } NO_3 \text{ produced/d}) \\
&= (1.0)(1564 \text{ lb/d}) + 4.6(113 + 138) = 2719 \text{ lb } O_2/\text{d} \\
&= 2719 \text{ lb } O_2/\text{d} = 113 \text{ lb } O_2/\text{h}
\end{aligned}
$$

   b. Determine the required standard oxygen transfer rate for a diffused air aeration system using Eq. (7-94) (see also Table 7-12):

$$
\text{AOTR} = \text{SOTR}\left(\frac{\beta C_{\bar{s}TH} - C_L}{C_{s20}}\right)(1.024^{T-20})(\alpha)(F)
$$

where

$$
\begin{aligned}
\text{AOTR} &= 113 \text{ lb } O_2/\text{h} \\
\beta &= 0.95 \\
T &= 20^\circ \text{ C} \\
P_d &= 21.2 \text{ lb/in}^2 \text{ (depth of diffusers} = 15 \text{ ft)}
\end{aligned}
$$

$$O_t = 19\%$$

$$C_{sTH} = 9.08 \text{ (see App. E)}$$

$$C_{\overline{s}TH} = C_{sTH}\,\frac{1}{2}\left(\frac{P_d}{14.7} + \frac{O_t}{21}\right)$$

$$= (9.08)\left(\frac{21.2}{14.7} + \frac{19}{21}\right) = 10.66$$

$$C_L = 2.0 \text{ mg/L}$$

$$\alpha = 0.6$$

$$F = 0.9 \text{ (assumed)}$$

$$\text{SOTR} = \text{AOTR}\left(\frac{C_{s20}}{\beta C_{\overline{s}TH} - C_L}\right)\left(\frac{1}{1.024^{T-20}}\right)\left(\frac{1}{\alpha}\right)\left(\frac{1}{F}\right)$$

$$= 113\left[\frac{9.08}{(0.95)(10.66) - 2.0}\right]\left(\frac{1}{1.024^{20-20}}\right)\left(\frac{1}{0.6}\right)\left(\frac{1}{0.9}\right)$$

$$= 234 \text{ lb } O_2\text{/h}$$

5. Determine area of secondary sedimentation tank and check the solids loading rate.
   a. Estimate the initial settling velocity using Eq. (7-99):

$$V_i = V_{\max} \exp(-K \times 10^{-6} X)$$

$$V_{\max} = 23.0 \text{ ft/h}$$

$$K = 500 \text{ L/mg}$$

$$X = 4000 \text{ mg/L}$$

$$V_i = 23 \exp(-500 \times 10^{-6} \times 4000) = 3.1 \text{ ft/h}$$

   b. Determine the overflow rate using Eq. (7-98) using a factor of safety of 2.0:

$$\text{OR}_{\text{design}} = \frac{(V_i)(179.5)}{\text{SF}} = \frac{(3.1)(179.5)}{2} = 278 \text{ gal/ft}^2\cdot\text{d} \qquad \text{Use } 280 \text{ gal/ft}^2\cdot\text{d}$$

   c. Determine the recycle ratio for Eq. (7-92), assuming the underflow concentration is 10,000 mg/L:

$$\alpha = \frac{(X, \text{mg/L})}{(X_u, \text{mg/L} - X, \text{mg/L})}$$

$$= \frac{(4000 \text{ mg/L})}{(10{,}000 \text{ mg/L} - 4000 \text{ mg/L})} = 0.67$$

   d. Determine the solids loading on the secondary sedimentation tank:

$$\text{SLR} = \frac{(1+\alpha)(Q)(X)(8.34)}{24A}$$

$$A = \frac{Q}{\text{OR}_{\text{design}}} = \frac{(0.75 \text{ Mgal/d})(10^6 \text{ gal/Mgal})}{280 \text{ gal/ft}^2\cdot\text{d}} = 2679 \text{ ft}^2$$

$$\text{SLR} = \frac{(1 + 0.67)(0.75)(4000)(8.34)}{24(2679)} = 0.65 \text{ lb/ft}^2\cdot\text{h}$$

6. Determine waste sludge production assuming a yield of 0.4 lb cell/lb COD applied for a process with primary sedimentation:

$$\text{Waste sludge} = (0.4 \text{ lb cell/lb COD})(\text{lb COD applied/d})$$
$$= (0.4)(1564) = 626 \text{ lb/d}$$

7. Determine the maximum growth rate for the nitrifying organisms under the stated operating conditions (see also Tables 7-6 and 7-7).
   a. The growth rate is determined using Eq. (7-78):

$$\mu_N = \mu_{N\max}\left(\frac{N}{K_N + N}\right)\left(\frac{\text{DO}}{K_{O2} + \text{DO}}\right)\left(e^{0.098(T-15)}\right)[1 - 0.833(7.2 - \text{pH})]$$

   b. Substitute the known values and determine μ_N:

$$T = 20°\text{C}$$
$$\mu_{N\max} = 0.45 \text{ d}^{-1} \text{ at } 15°\text{C}$$
$$N = 40 \text{ mg/L}$$
$$K_N = 10^{0.051T-1.158} = 10^{0.051(20)-1.158} = 0.73$$
$$\text{DO} = 2.0 \text{ mg/L}$$
$$K_{O2} = 1.3$$
$$\text{pH} = 7.2 \text{ (assumed)}$$
$$\mu_N = 0.45\left(\frac{40}{0.73 + 40}\right)\left(\frac{2.0}{1.3 + 2.0}\right)\left(e^{0.098(20-15)}\right)[1 - 0.8333(7.2 - 7.2)]$$
$$= 0.44 \text{ d}^{-1}$$

8. Determine the maximum rate of substrate utilization, k, using Eq. (7-13):

$$k = \frac{\mu_N}{Y_N}$$

$$\mu_N = 0.44 \text{ d}^{-1} \quad \text{(from step 7 above)}$$
$$Y = 0.16 \quad \text{(from Table 7-7)}$$
$$k = \frac{0.44 \text{ d}^{-1}}{0.16} = 2.75 \text{ d}^{-1}$$

9. Determine the design substrate utilization factor U using Eq. (7-47).

$$\frac{1}{\theta_c} = YU - k_d$$

$$k_d = 0.04 \quad \text{(from Table 7-7)}$$
$$U = \left(\frac{1}{\theta_c} + k_d\right)\frac{1}{Y}$$
$$= \left(\frac{1}{20} + 0.04\right)\frac{1}{0.16} = 0.56 \text{ d}^{-1}$$

10. Determine the concentration of ammonia in the effluent using Eq. (7-46):

$$U = \frac{kN}{K_N + N} = 0.56\ \text{d}^{-1}$$

$$k = 2.75\ \text{d}^{-1} \qquad \text{(from step 8 above)}$$

$$K_N = 0.73 \qquad \text{(from Table 7-7)}$$

$$0.56 = \frac{2.75N}{0.73 + N}$$

$$N = 0.187\ \text{mg/L} \qquad (0.187 < 1.0\ \text{mg/L; OK})$$

11. Determine the required hydraulic detention time for nitrification by solving a modified version of Eq. (7-45) for θ_N:

$$\theta_N = \frac{N_o - N}{UX}$$

$$N_o = 40\ \text{mg/L} \qquad \text{(from problem specification)}$$

$$N = 0.30\ \text{mg/L} \qquad \text{(from step 6)}$$

$$U = 0.56\ \text{d}^{-1}$$

$$f_N = \frac{0.16(40 - 0.187)}{0.6(160 - 10) + 0.16(40 - 0.187)} = 0.066 \quad \text{[see Eq. (7-73) in Table 7-6]}$$

$$X = 4000\ \text{mg/L} \times 0.65 \times 0.066$$

$$= 171.6\ \text{mg/L}$$

$$\theta = \frac{(40\ \text{mg/L} - 0.187\ \text{mg/L})}{0.56\ \text{d}^{-1}(171.6\ \text{mg/L})} = 0.41\ \text{d} = 9.9\ \text{h}$$

12. Calculate the denitrification rate for 20°C, using Eq. (7-87) in Table 7-8 (assume $R_{\text{DN20}} = 0.1\ \text{d}^{-1}$ and DO $= 0.15$ mg/L):

$$R_{\text{DN}T} = R_{\text{DN20}} \times 1.09^{(T-20)}(1 - \text{DO})$$

$$= 0.1 \times 1.09^{(20-20)}(1 - 0.15)$$

$$= 0.085\ \text{d}^{-1}$$

13. Calculate the required residence time for denitrification using Eq. (7-45), substituting the value of R_{DN20} for U:

$$U = \frac{N_o - N}{\theta X}$$

$$\theta = \frac{N_o - N}{UX}$$

$$= \frac{40 - 2}{0.085 \times 4000 \times 0.65}$$

$$= 0.17\ \text{d}$$

$$= 4.1\ \text{h}$$

14. The combined detention required for nitrification and denitrification is

$$D_{TN+DN} = 9.9\text{ h} + 4.1\text{ h} = 14.0\text{ h} \qquad (14.0\text{ h} < 24.0\text{ h; OK})$$

Comment. In step 3, no reduction was made in the required oxygen for the use of nitrate. If the utilization of nitrate is considered, the oxygen requirement could be reduced by about 50 percent of the amount required to nitrify the ammonia in the first place.

7-7 AEROBIC ATTACHED-GROWTH PROCESSES

In aerobic attached-growth processes, the microorganisms responsible for treatment are attached to a fixed medium, in contrast to the suspended-growth processes considered in the previous section. The first use of attached-growth processes similar to the ones used today can be traced back to the early 1890s (Dunbar, 1908). For many years, attached-growth processes were the most commonly used biological treatment process for the treatment of domestic wastewater (see Fig. 7-37). As more stringent treatment requirements have been applied, use of conventional attached-growth processes as typified by the trickling filter process as the sole means of treatment has declined. However, new hybrid processes have proven to be quite effective. Although the primary focus of the material presented in this section is on the trickling filter process, the other attached-growth processes are also discussed.

Process Classification and Characteristics

Aerobic attached-growth treatment processes currently in use may be classified, as shown in Fig. 7-38, as: (1) nonsubmerged attached-growth processes, (2) hybrid attached- and suspended-growth processes, and (3) submerged-growth processes.

FIGURE 7-37
View of early attached-growth process (Library of Congress).

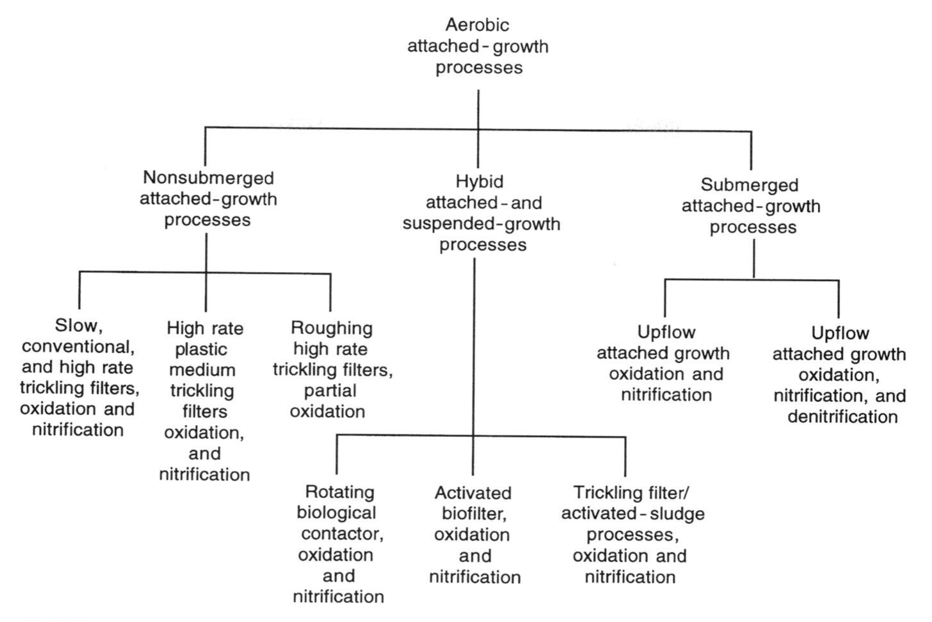

FIGURE 7-38
General classification of attached-growth processes.

Nonsubmerged attached-growth processes are most suitable for the treatment of soluble and other relatively dilute organic wastes. The hybrid processes can be used to treat wastes with both particulate and soluble constituents. Submerged attached-growth processes, a relatively recent development, are used for the treatment of domestic wastewaters including the oxidation of carbonaceous material, nitrification, and denitrification.

Process Operation and Microbiology

The following discussion of the process microbiology of attached-growth processes refers primarily to nonsubmerged trickling filters, although the microbiology of the hybrid processes is quite similar. The microbiology of the submerged attached-growth processes is not as well defined, as they are relatively new processes. To understand the role of the microorganisms in bringing about the treatment of wastewater, it is appropriate to consider the operation of a trickling filter.

Operationally, wastewater is applied to the surface of the filter intermittently by one or more rotary distributors, and percolates downward through the bed to underdrains, where it is collected and discharged through an outlet channel. A gelatinous biological film forms on the filter medium, and the fine suspended, colloidal, and dissolved organic solids are absorbed by this film where biochemical oxidation of the organic matter is accomplished by aerobic bacteria. The film eventually becomes quite thick with accumulated organic matter (typically in the form of cell tissue) and will slough off (or unload) from time to time and be discharged with the effluent. Therefore, effluent from trickling filters requires sedimentation to remove the solids that pass the filter. Continuous sloughing can be achieved with proper control of the hydraulic application rate (Albertson, 1989). Recirculation of trickling filter effluent or effluent from secondary settling facilities is a common feature of trickling filters. The rates of recirculation are generally adjusted as the wastewater flow changes to maintain approximately constant flow through the filters.

The biological conversion of organic matter in a trickling filter is brought about by a community of microorganisms including aerobic, anaerobic, and facultative bacteria; fungi; algae; and protozoans. Higher animals, such as worms, insect larvae, and snails, are also present. Variations in the individual population of the biological community occur throughout the filter depth with changes in organic loading, hydraulic loading, influent wastewater composition, pH, temperature, air availability, and other environmental factors.

Facultative bacteria are the predominating microorganisms in nonsubmerged trickling filters. Along with the aerobic and anaerobic bacteria, their role is to decompose the organic material in the wastewater. Within the slime layer, where adverse conditions prevail with respect to growth, filamentous forms will be found. In the lower reaches of the filter, nitrifying bacteria will be present. The fungi present are also responsible for waste stabilization, but their contribution is usually important only under low-pH conditions or with certain industrial wastes. At times, their growth can be so rapid that the filter clogs and ventilation becomes restricted.

Algae can grow only in the upper reaches of the filter where sunlight is available. Generally, algae do not take a direct part in waste degradation, but during the daylight hours they add oxygen to the percolating wastewater. From an operational standpoint, the algae are troublesome because they can cause clogging of the filter surface, which produces odors. The protozoa in the filter are predominantly of the ciliate group. As in the activated-sludge process, their function is not to stabilize the waste but to control the bacterial population. The higher animals, such as snails, worms, and insects, feed on the biological films in the filter and, as a result, help to keep the bacterial population in a state of high growth or rapid food utilization. The higher animal forms are not as common in high-rate tower trickling filters. Snails are especially troublesome in nitrifying filters where they have been known to consume most of the growth of nitrifying bacteria.

Nonsubmerged Attached-Growth Processes

Sometimes called *biological filters,* trickling filters were developed somewhat naturally from attempts to filter municipal and other wastewaters. Because microbial growth on the filter medium quickly clogged the filters, larger and larger media were used until a rock size of 2 to 4 in (50 to 100 mm) was reached, and clogging was minimal (Dunbar, 1908). To provide even greater porosity, new types of plastic media have been developed. Because of the high porosity and the reduced weight of the media, tower-type filters are now used commonly.

(*a*)

(*b*)

FIGURE 7-39
Views of conventional rock trickling filters typical of those used for small communities: (*a*) open sided and (*b*) closed sided.

TABLE 7-15
Typical design information for various types of nonsubmerged trickling filters*

Item	Low rate	Intermediate rate	High rate	High rate	Roughing
Filter medium	Rock/slag	Rock/slag	Rock/slag	Plastic	Plastic/redwood
Size, in	1–5/2–5	1–5/2–5	1–5/2–5	24×24×48	24×24×48[†‡]
Specific surface, ft^2/ft^3	12–30	12–30	12–30	24–60	24–60/12–15
Void space, %	40–55	40–55	40–55	92–97	92–97/70–80
Specific weight, lb/ft^3	50–90	50–90	50–90	2–6	2–6/9–12
Hydraulic loading rate					
$gal/ft^2 \cdot min$	0.02–0.06	0.06–0.16	0.16–0.64	0.2–1.20	0.8–3.2
Mgal/ac·d	1–4	4–10	10–40	15-90	50–200[§]
Organic loading rate					
Carbon removal, lb $BOD_5/10^3 ft^3 \cdot d$	5–25	15–30	30–80	50–200	100–500
Nitrification, lb $BOD_5/10^3 ft^3 \cdot d$	5–10	5–15	5–15	10–25	
Depth, ft	6–8	6–8	6–8	10–40	15–40
Recirculation ratio	0	0–1	1–2	1–2	1–4
Sloughing	Intermittent	Intermittent	Continuous	Continuous	Continuous
BOD_5 removal efficiency, %	80–90	50–80	65–90	65–90	40–70
Filter flies	Many	Some	Few	Few or none	Few or none

*Adapted in part from Sarner (1980) and Tchobanoglous and Burton (1991).
[†] Dimensions in inches of typical module of plastic filter medium.
[‡] 48 × 48 × 20 in for redwood, if available.
[§] Does not include recirculation.

FIGURE 7-40
View of top of tower trickling filter with plastic filter medium.

Conventional and tower trickling filters. A conventional trickling filter (see Fig. 7-39) consists of a bed of crushed rock, slag, or gravel, whose particles range from about 2 to 4 in (50 to 100 mm) in size. The bed is commonly 6 to 9 ft (2 to 3 m) deep, although shallower beds are sometimes used. Conventional filters are usually classified as low, standard, and high rate, depending on the hydraulic and organic loading rate and the rate of recirculation (see Table 7-15). Because no recirculation is used in the low-rate trickling filter, it is seldom used, but continues to find applications. The effluent BOD_5 and TSS concentrations from well operated and maintained standard-rate trickling filters are typically in the 20-mg/L range. Lightly loaded conventional trickling filters have proven to be effective for nitrification.

The tower trickling filter (see Fig. 7-40) is a modification of the conventional trickling filter process, in which specially designed high-porosity plastic modules are used as the fixed medium. The specific surface area of these plastic media per unit volume varies from about 24 to 60 ft^2/ft^3 (80–200 m^2/m^3), as compared to about 12 to 30 ft^2/ft^3 (40 to 100 m^2/m^3) for rock. Such filters are built from 15 to 40 ft (4.5 to 12 m) high. They are often used in conjunction with conventional activated-sludge facilities to reduce high seasonal loadings from canneries and similar activities. The effluent BOD_5 and TSS concentrations from a well-operated and well-maintained high-rate trickling filter is typically 30 mg/L or higher. Lightly loaded tower trickling filters have also proven to be effective for nitrification.

Important process design and operating parameters for conventional trickling filters are summarized in Table 7-16. Over the years, a number of investigators have proposed equations to describe the removals observed, including Atkinson et al., (1974b), Bruce and Merkens (1973), Eckenfelder (1963), Fairall (1956), Galler and Gotass (1966), Germain (1966), Logan et al. (1987a, 1987b), NRC (1946), Schultz (1960), and Velz (1948). Based on an extensive study of the operating records of trickling filter plants serving World War II military installations, the National

TABLE 7-16
Process design and operational considerations for conventional trickling filters

Item	Definition
Hydraulic loading rate	The hydraulic loading rate is an empirically-derived design and operating parameter that relates to ponding, surface shearing rate, and hydraulic detention time. Usually hydraulic loading rate is reported in units of volume of wastewater, including recycle, per unit cross-sectional area per day. Because most rock-medium trickling filters are between 1 and 2 m in depth, the volumetric loading rate used in some countries is easily translated into flow per unit area-time.
Organic loading rate	Waste-material loading on trickling filters is characterized by the organic loading rate in terms of kilograms of BOD_5 per cubic meter-day (kg $BOD_5/m^3 \cdot d$). There is no parameter for solids loading, and solids removal in trickling filters has not been characterized in any predictive manner.
Oxygen transfer	Air is usually supplied to trickling filters through natural drafts resulting from temperature differences between the ambient and the internal air. Deep plastic-medium filters often require the use of compressed air to supply a forced draft. Maximum oxygen transfer rates in natural-draft trickling filters are about 28 $g/m^2 \cdot d$, and this corresponds to uptake rates expected in the biofilm for applied UBOD concentrations of about 400 mg/L. Many industrial wastes are considerably stronger than 400 mg/L UBOD, and anoxic conditions can occur within the trickling filters, which can lead to the production of rather nasty odors. Odor problems have resulted in the covering of the filters, with venting of the odorous gas to soil filters or other odor control facilities.
Recycle	Effluent recycle in high-rate systems, including those with plastic filter media, may be from a point ahead of, or following, the sedimentation tank (Fig. 7-41). Presedimentation recycle provides an advantage in that sloughed cells are mixing with incoming wastewater, thus enhancing the reaction rate, but it also has the disadvantage of increasing the possibility of plugging the unit. Post sedimentation recycle increases the sedimentation tank loading and tends to dilute the wastewater without adding a reactant, but does not have the plugging potential of the presedimentation recycle format. Both configurations provide flow equalization.
Dosing rate	The dosing rate can be adjusted to obtain continual and uniform growth of biomass and sloughing of excess biomass as a function of the organic loading rate. The dosing rate in in/pass can be approximated by multiplying the organic loading rate expressed in $BOD/10^3 ft^3$ by a factor of 0.1 to 0.12.
Temperature	Temperature of wastewater is more important than the air temperature. The effect of temperature on the performance of filters is accounted for by adjusting the removal rate coefficient using Eq. (7-21) with a θ value of 1.035.
Operational problems	Operational problems can include excessive algae growth leading to filter clogging, development of odors due to the lack of oxygen transfer, breeding of flies, and infestations of snails and beetles.
Secondary sedimentation facilities	All the sludge from trickling filter settling tanks is removed to sludge processing facilities. The design of these tanks is similar to the design of primary settling tanks, except that the surface loading rate is based on the plant flow plus the recycle flow minus the underflow (often neglected). Suggested overflow rates and solids loading rates for settling tanks following trickling filters range from 400 to 600 $gal/ft^2 \cdot d$ and 0.6 to 1.0 $lb/ft^2 \cdot h$, respectively.

TABLE 7-17

Formulas used to compute the performance of conventional rock and plastic tower filters

Equation		Definition of terms
		NRC formulas
Single-stage or first-stage of a two-stage rock filter:		E_1 = efficiency of BOD removal for process at 20°C, including recirculation and sedimentation, percent
$E_1 = \dfrac{100}{1 + 0.0561\sqrt{\dfrac{W_1}{VF}}}$	(7-100)	W_1 = BOD loading to filter, lb/day
		V = volume of filter media, 10^3 ft^3
		F = recirculation factor
		R = recirculation ratio Q_r/Q
Recirculation factor:		Q_r = recirculation flow
		Q = wastewater flow
$F = \dfrac{1 + R}{(1 + R/10)^2}$	(7-101)	E_2 = efficiency of BOD removal for second-stage filter at 20°C, including recirculation and settling, %
Second-stage filter:		E_{1f} = fraction of BOD removed in first-stage filter
		W_2 = BOD loading applied to second-stage filter, lb/day
$E_2 = \dfrac{100}{1 + \dfrac{0.0561}{1 - E_1}\sqrt{\dfrac{W_2}{VF}}}$	(7-102)	
		First-order formulation
For plastic media tower filters:		S_e = Total BOD_5 of settled effluent from filter, mg/L
		S_i = Total BOD_5 of wastewater applied to the filter, mg/L
$\dfrac{S_e}{S_o} = \exp\left[-k_{20}D(Q_v)^{-n}\right]$	(7-103)	k_{20} = treatability constant corresponding to a filter of depth D at 20 °C, $(gal/min)^{0.5}$ ft for n = 0.5
		D = depth of filter, ft
		Q_v = volumetric flowrate applied per unit volume of filter, gal/min·ft^2 = Q/A
		Q = flowrate applied to filter without recirculation, gal/min
		A = cross-sectional area of filter, ft^2
		n = experimental constant, usually 0.5
		k_2 = treatability constant corresponding to a filter of depth D_2
		k_1 = treatability constant corresponding to a filter of depth D_1
		D_1 = depth of filter one, ft
		D_2 = depth of filter two, ft
$k_2 = k_1\left(\dfrac{D_1}{D_2}\right)^x$	(7-104)	x = 0.5 for vertical-flow plastic- and rock-medium filters
		= 0.3 for cross-flow plastic-medium filters

Research Council (NRC) developed empirical relationships to predict the performance of trickling filters (NRC, 1946). The formulas given in Table 7-17 are primarily applicable to single-stage and multistage rock systems, with varying recirculation rates (see Fig. 7-41). Because of the more predictable properties of a plastic medium, a number of more or less empirical first-order relationships have been developed to predict the performance of trickling filters packed with plastic media. Velz (1948) was one of the first researchers to propose a first-order relationship to define the performance of trickling filters. Others have been proposed by Eckenfelder (1963). One commonly used formulation, proposed by Germain (1966) and Schultz (1960)

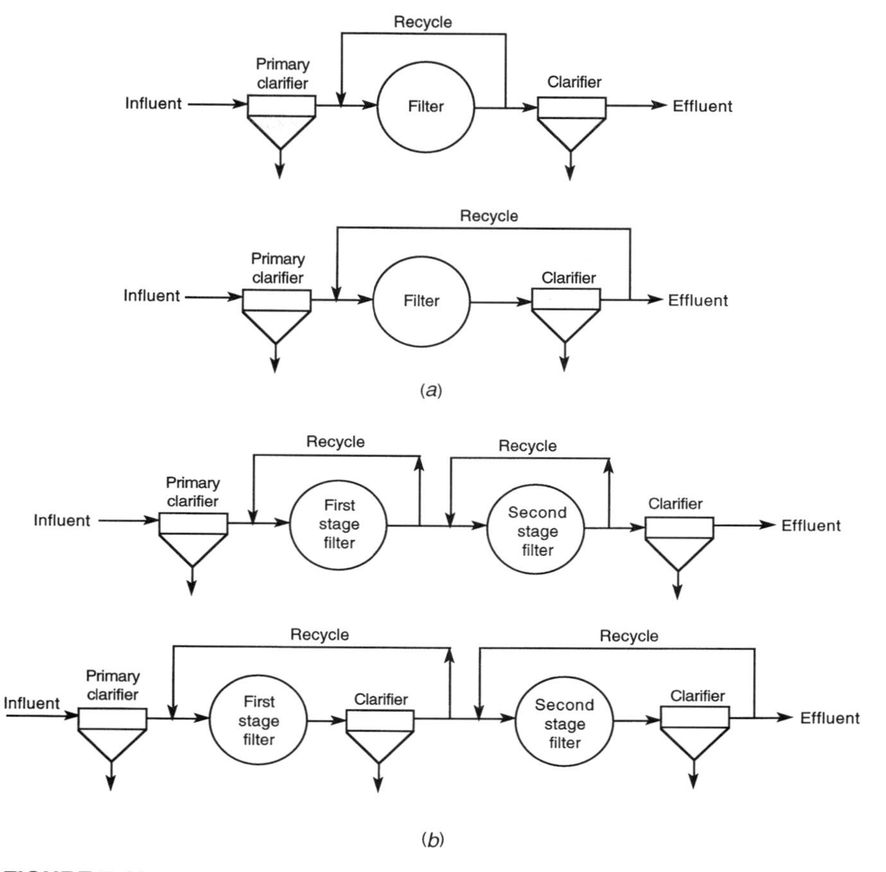

FIGURE 7-41
Standard-rate and high-rate trickling filter flow diagrams with various recirculation patterns: (*a*) single-stage filters and (*b*) two-stage filters.

to describe the observed performance of plastic packed trickling filters, is given in Table 7-17. It should be noted that none of the formulations given in Table 7-17 deal with nitrification. The application of these equations is illustrated in Example 7-8.

EXAMPLE 7-8. TRICKLING FILTER SIZING USING NRC AND FIRST-ORDER EQUATIONS. A two-stage trickling filter is to be used to treat a municipal waste having a BOD_5 of 150 mg/L after primary sedimentation. The desired effluent quality is 20 mg/L of BOD_5. If both of the filter depths are to be 6 ft and the recirculation ratio is 2:1, find the required filter diameters. Assume $Q = 0.75$ Mgal/d, wastewater temperature $= 20°C$, and $E_1 = E_2$. Prepare a second design for the same wastewater using a 30-ft-deep tower trickling filter. Assume the treatability constant, determined by using a 20-ft-deep test filter, is 0.085 $(gal/min)^{0.5}$ ft at 20°C.

Solution—Part 1: Two-stage conventional trickling filter process

1. Compute E_1 and E_2:
   If the intermediate BOD_5 concentration (effluent from filter 1) is designated as C, then the efficiencies for the two filters may be defined as follows:

$$E_1 = \frac{150 - C}{150} \qquad \text{and} \qquad E_2 = \frac{C - 20}{C}$$

   Setting $E_1 = E_2$, it can be determined that $C = \sqrt{3000} = 54.78$ mg BOD_5/L. Therefore,

$$E_1 = \frac{150 - 54.77}{150} = 0.635 = 63.5\% \qquad \text{and} \qquad E_2 = \frac{54.77 - 20}{54.77} = 0.635 = 63.5\%$$

   The individual efficiencies can also be determined by noting that $E_1 + E_2(1 - E_1) =$ overall efficiency.

2. Compute the recirculation factor using Eq. (7-101) (Table 7-17):

$$F = \frac{1 + R}{(1 + R/10)^2} = \frac{1 + 2}{(1 + 2/10)^2} = 2.08$$

3. Compute the volume of the first-stage filter using Eq. (7-100) (Table 7-17):

$$E_1 = \frac{100}{1 + 0.0561\sqrt{\dfrac{W_1}{VF}}}$$

   Rearranging and solving for V yields:

$$V = \frac{W_1(0.0561)^2}{F(100/E_1 - 1)^2}$$

   a. Compute the BOD_5 loading for the first-stage filter:

$$W_1 = (150 \text{ mg/L})(0.75 \text{ Mgal/d})(8.34) = 938 \text{ lb BOD}_5\text{/d}$$

   b. Compute the volume of the first filter:

$$V, 10^3 \text{ ft}^3 = \frac{938(0.0561)^2}{2.08(100/63.5 - 1)^2} = 4.30 \times 10^3 \text{ ft}^3$$

4. Compute the diameter of the first filter:

$$A = \frac{V}{d} = \frac{4300 \text{ ft}^3}{6 \text{ ft}} = 717 \text{ ft}^2 = \frac{\pi d^2}{4}$$

$$d = 30.2 \text{ ft; use 30 ft}$$

5. Compute the volume of the second-stage filter using Eq. (7-102) (Table 7-17):

$$E_2 = \frac{100}{1 + \dfrac{0.0561}{1 - E_1}\sqrt{\dfrac{W_2}{VF}}}$$

Rearranging and solving for V yields:

$$V,\ 10^3\ \text{ft}^3 = \frac{W_2(0.0561)^2}{F(1 - E_1)^2(100/E_2 - 1)^2}$$

a. Compute the BOD_5 loading for the second-stage filter:

$$W_1 = (54.77\ \text{mg/L})(0.75\ \text{Mgal/d})(8.34) = 342\ \text{lb BOD}_5\text{/d}$$

b. Compute volume of second filter:

$$V,\ 10^3\ \text{ft}^3 = \frac{342(0.0561)^2}{2.08(1 - 0.635)^2(100/63.5 - 1)^2} = 11.76 \times 10^3\ \text{ft}^3$$

6. Compute the diameter of the second filter:

$$A = \frac{V}{d} = \frac{11{,}760\ \text{ft}^3}{6\ \text{ft}} = 1960\ \text{ft}^2 = \frac{\pi d^2}{4}$$

$$d = 50\ \text{ft}$$

7. Compute the volumetric organic loading rate (L_{org}) for each filter:
   a. First-stage filter

$$L_{org} = \frac{(938\ \text{lb BOD}_5\text{/d})}{4.30 \times 10^3\ \text{ft}^3} = 218\ \text{lb BOD}_5/10^3\ \text{ft}^3\cdot\text{d}$$

   b. Second-stage filter

$$L_{org} = \frac{(342\ \text{lb BOD}_5\text{/d})}{11.76 \times 10^3\ \text{ft}^3} = 29.1\ \text{lb BOD}_5/10^3\ \text{ft}^3\cdot\text{d}$$

8. Compute the wastewater hydraulic loading rate (L_W) based on forward flow for each filter:
   a. First-stage filter

$$L_W = \frac{(0.75 \times 10^6\ \text{gal/d})}{(717\ \text{ft}^2)(1440\ \text{min/d})} = 0.73\ \text{gal/ft}^2\cdot\text{min}$$

   b. Second-stage filter

$$L_W = \frac{(0.75 \times 10^6\ \text{gal/d})}{(1960\ \text{ft}^2)(1440\ \text{min/d})} = 0.27\ \text{gal/ft}^2\cdot\text{min}$$

Comment—Part 1. To reduce construction costs, the two trickling filters are often made the same size. Where two filters of equal diameter are used the efficiencies will be unequal. In many cases, the hydraulic loading rate will be limited by local or state standards.

Solution—Part 2: Tower trickling filter process

1. Determine the surface area required for a 30-ft-deep filter using Eq. (7-103):

$$\frac{S_e}{S_o} = \exp[-k_{20}D(Q_v)^{-n}]$$

   a. Determine the treatability constant k_{20} value for a filter depth of 30 ft using Eq. (7-104):

$$k_{30} = k_{20}\left(\frac{D_{20}}{D_{30}}\right)^{0.5} = 0.085\left(\frac{20}{30}\right)^{0.5} = 0.069$$

b. Substituting Q/A for Q_v in Eq. (7-103) and rearranging yields:

$$A = Q\left(\frac{-\ln S_e/S_i}{k_{20}D}\right)^{1/n}$$

c. Substitute known values in the above expression and solve for the area A:

$$S_e = 20 \text{ mg/L}$$
$$S_i = 150 \text{ mg/L}$$
$$n = 0.5$$
$$k_{20} = 0.069 \text{ (gal/min)}^{0.5} \text{ ft for n} = 0.5$$
$$D = 20 \text{ ft}$$
$$Q = (0.75 \times 10^6 \text{ gal/d})/(1440 \text{ min/d}) = 521 \text{ gal/min}$$

$$A = 521\left(\frac{-\ln 20/150}{0.069(30)}\right)^{1/0.5} = 494 \text{ ft}^2$$

2. Compute the volumetric organic loading rate (L_{org}) for the filter:

$$L_{org} = \frac{938 \text{ lb BOD}_5}{494 \text{ ft}^2 \times 30 \text{ ft}} = 63.3 \text{ lb BOD}_5/10^3 \text{ ft}^3$$

3. Compute the wastewater hydraulic loading rate (L_W) for the filter, based on forward flow:

$$L_W = \frac{(0.75 \times 10^6 \text{ gal/d})}{(494 \text{ ft}^2)(1440 \text{ min/d})} = 1.05 \text{ gal/ft}^2\cdot\text{min}$$

Roughing filters. Roughing filters are specially designed trickling filters operated at high hydraulic loading rates requiring the use of high recycle rates. Because higher hydraulic loadings cause nearly continuous sloughing of the slime layer, the sloughed biological solids in the recycle stream, if present, may contribute to organic removal within the filter as in a suspended-growth process. If this mechanism is significant, process efficiency may be greater than predicted by an attached-growth model. Roughing filters are used primarily to reduce the organic loading on downstream processes and in seasonal nitrification applications where the purpose is to reduce the organic load so that a downstream biological process will nitrify dependably during the summer months. Roughing filters are normally designed with loading factors developed from pilot plant studies and data derived from full-scale installations, although the analysis presented for the tower trickling filter can be used (Tchobanoglous and Burton, 1991).

Operational problems with trickling filters. Operational problems in attached-growth systems include the potential for odors, nuisances such as filter flies and snails, and predators on the attached growth biofilm. Odors can generally be managed by not overloading the filters, and can be minimized by proper attention to air flow in the design (Schroeder and Tchobanoglous, 1976). Snails are an operational problem that can affect performance, besides being a nuisance. Snail control is not unlike the problem of filaments and *Nocardia* in activated sludge in that many plants suffer the problem and the solution varies with the facility. Snails can be discouraged from growing on the filter by raising the pH of the water to 9 or slightly above, by submerging the filter, by periodic drying followed by flushing, and by adding selective biocides in conjunction with the other methods (Parker et al., 1996a).

Hybrid Attached- and Suspended-Growth Processes

The development and use of hybrid attached- and suspended-growth aerobic systems has increased in recent years. The most common hybrid combined systems include: (1) the rotating biological contactor, (2) activated biofilter, (3) trickling filter followed by a solids contactor, (4) roughing filter followed by an activated-sludge process, (5) biofilter followed by an activated-sludge process, and (6) trickling filter followed by an activated-sludge process. Because of the complexity of the reactions occurring in the various hybrid processes, it is common to use the organic loading rates, expressed in terms of lb BOD_5 or COD/10^3 ft^3·d, for the design of these processes. Representative design loadings for these processes are reported in Table 7-18.

Rotating biological contactor. In the rotating biological contactor (RBC) process, a number of circular plastic disks are mounted on a central shaft (Fig. 7-42). These disks are submerged (from 40 to 80 percent) and rotated in a tank containing the wastewater to be treated. The microorganisms responsible for treatment become attached to the disks and rotate into and out of the wastewater. The oxygen necessary for the conversion of the organic matter adsorbed from the liquid is obtained by adsorption from the air as the slime layer on the disk is rotated out of the liquid. In some designs, air is added to the bottom of the tank to provide oxygen and to

TABLE 7-18
Typical design information for hybrid attached- and suspended-growth processes*

		Aeration basin		
Process combination	**Trickling filter loading**	$\theta_{c,d}$	***F/M*, lb BOD_5 applied/lb MLVSS·d**	**MLSS, mg/L**
Activated biofilter	Low†	N/A	N/A	1500–4000
Trickling filter/solids contact	Low	0.5–2.0	N/A	1000–3000
Roughing filter/ activated sludge	High‡	2–5	0.5–1.2	1500–3000
Biofilter/activated sludge	High	2–5	0.5–1.2	1500–4000
Trickling filter/activated sludge	High	4–8	0.2–0.5	1500–4000

*Adapted from U.S. EPA (1988).
†Typically less than 40 lb $BOD_5/10^3$ ft^3·d.
‡Typically greater than 100 lb $BOD_5/10^3$ ft^3·d.
N/A = Not applicable.

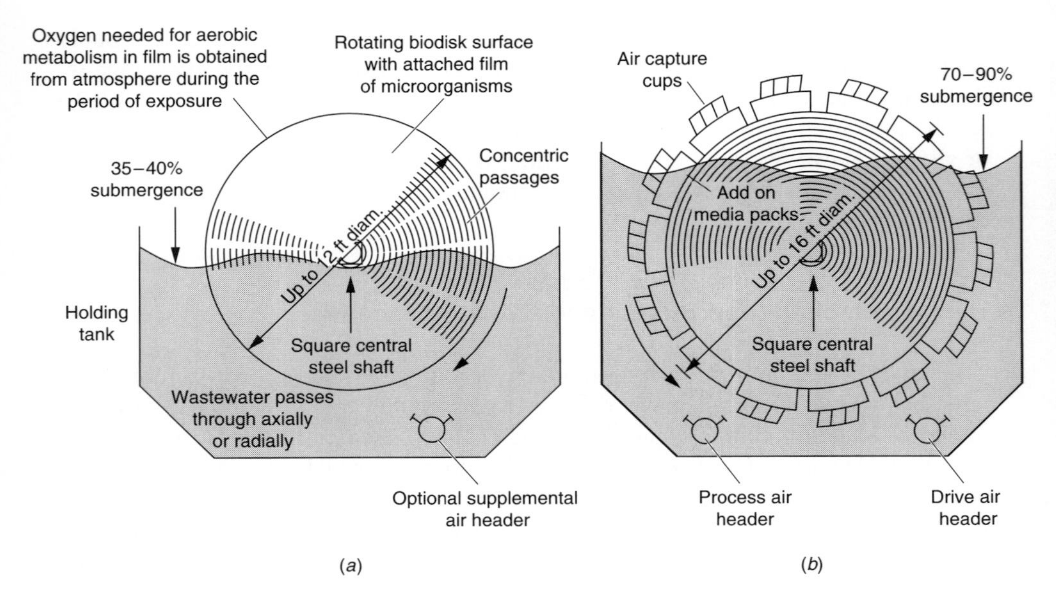

FIGURE 7-42
Definition sketch for rotating biological contactor: (*a*) conventional type with mechanical drive and optional air input and (*b*) submerged type with supplemental aeration and air cups for rotation.

rotate the disks when the disks are equipped with air capture cups. Conceptually and operationally, the bio-disk process is similar to the trickling filter process with a high rate of recirculation. Over the years, a number of small package plants have been developed using RBC disks. The most commonly used RBC is the submerged type with supplemental aeration as shown in Fig. 7-42*b*. RBCs have also been developed for the biological treatment of odors.

Activated biofilter. The activated biofilter (ABF) process resembles a high-rate trickling filter, except the secondary sludge is recycled to the trickling filter (see Fig. 7-43*a*). A separate suspended-growth process is not generally used, although one modification incorporates short-term aeration prior to secondary sedimentation. The return sludge is controlled to maintain a high concentration of suspended growth in the filter. The biofilter uses redwood media instead of other types of media. The advantages of this process are: (1) significantly higher levels of BOD removal can be achieved by producing a combination of attached and suspended growth and (2) BOD loadings 4 to 5 times higher than those used in conventional filters can be applied.

Trickling filter/solids contact process. The trickling filter/solids contact (TF/SC) process consists of a trickling filter, an aerobic contact tank, and a final clarifier (see Fig. 7-43*b*). Modifications to this system include a return sludge aeration tank and flocculating center well clarifiers. The trickling filters are sized to

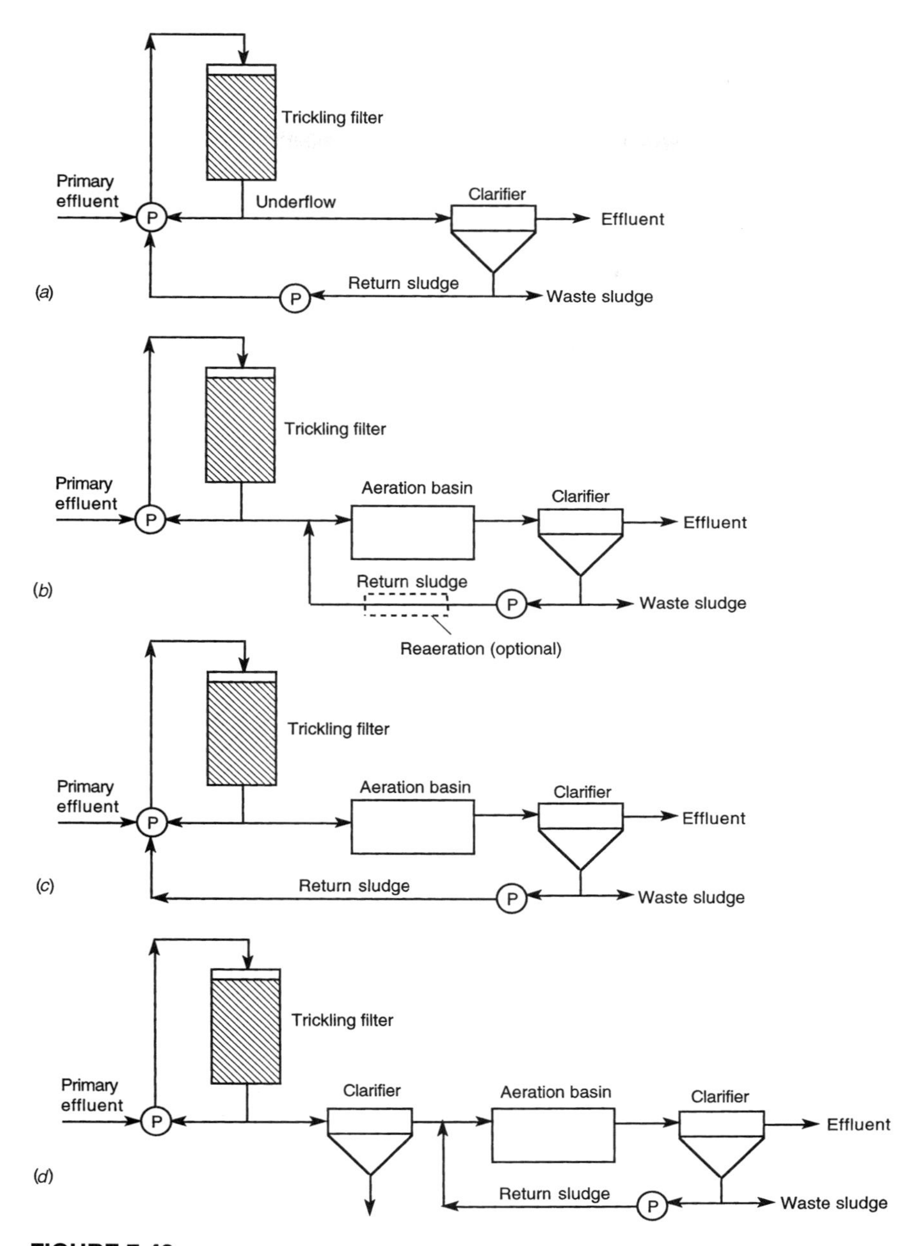

FIGURE 7-43
Typical combined attached- and suspended-growth aerobic treatment system flow diagrams: (*a*) activated biofilter, (*b*) trickling filter solids contact and roughing filter/activated sludge, (*c*) biofilter/activated sludge, and (*d*) series trickling filter/activated sludge.

remove the major portion of the BOD, typically 60 to 85 percent. The biological solids formed on the trickling filter are sloughed off and concentrated through sludge recirculation in the contact tank. In the contact tank, the suspended growth is aerated for less than 1 hour, causing the flocculation of the suspended solids and further removal of soluble BOD. When short solids contact times are used, a sludge reaeration tank is usually required. Because of the high level of dispersed solids in contact tank effluent, flocculating center well clarifiers have been found to be effective in maximizing solids capture.

Roughing filter/activated-sludge process. The roughing filter/activated-sludge (RF/AS) process configuration is similar to the TF/SC system (see Fig. 7-43*b*). The RF/AS system, however, operates at higher total organic loadings. The trickling filter removes a portion of the BOD and provides process stabilization, particularly when shock loads occur. The aeration basins are required to treat the balance of the organic loading not removed by the trickling filters.

Biofilter/activated-sludge process. The biofilter/activated-sludge (BF/AS) process is similar to the ABF process, except an aeration tank is used following the trickling filter (see Fig. 7-43*c*). Return activated sludge is recycled over the trickling filter. The average organic loading and aeration tank hydraulic retention times typically are similar to the RF/AS system. Suggested design procedures are provided in Arora and Humphries (1987), where they introduce the concept of system F/M by considering the biofilter and aeration basin as one integral treatment system. The system F/M values typically used for the design of the aeration basin for normal carbonaceous BOD removal are between 1.0 and 1.5, which is 3 to 4 times higher than the corresponding value for a conventional activated-sludge aeration basin not preceded by a biofilter. As a result, the aeration basin size is reduced to about one-fourth of that for a conventional system.

Series trickling filter/activated-sludge process. The trickling filter process followed by an activated-sludge process (see Fig. 7-43*d*) is often used to upgrade an existing activated-sludge system by adding an upstream trickling filter. An alternative arrangement involves the addition of an activated-sludge process downstream from an existing trickling filter. This system is also used to reduce the strength of wastewater where industrial and domestic wastewater are treated in common treatment facilities, and in applications where nitrification is required.

Submerged Attached-Growth Processes

A relatively recent development, submerged attached-growth reactors have been developed for the oxidation of carbonaceous material and nitrification and for oxidation, nitrification, and denitrification.

Submerged upflow attached growth. Submerged upflow attached-growth reactors (see Fig. 7-44) have been developed to achieve: (1) carbonaceous oxidation

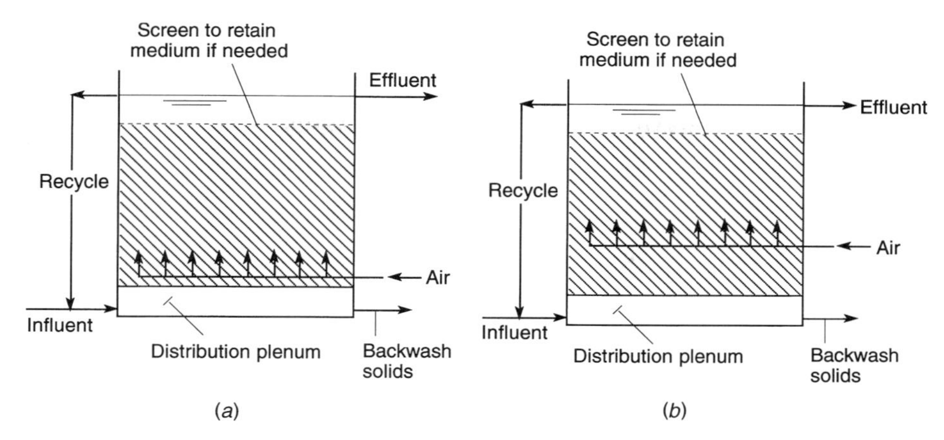

FIGURE 7-44
Definition sketch for submerged attached-growth process: (*a*) for carbon oxidation and nitrification and (*b*) for carbon oxidation, nitrification, and denitrification.

and nitrification (see Fig. 7-44*a*) and (2) carbonaceous oxidation, nitrification, and denitrification (see Fig. 7-44*b*). In the second type of upflow attached-growth process, the first portion of the reactor is anoxic (see Fig. 7-44*b*) to facilitate the denitrification of the nitrified effluent. One advantage of the submerged upflow reactors is that high organic loadings can be treated effectively.

Submerged upflow fluidized-bed attached growth. In the upflow fluidized-bed attached-growth reactor, the bacterial biomass is grown on a suitable medium (e.g., sand, expanded ceramic beads). The liquid to be treated, along with the oxygen required for aerobic conversion, is introduced into the reactor from a plenum in the bottom of the reactor. The rate at which the liquid is applied is sufficient to fluidize the bed. The shearing force of the liquid as it passes through the fluidized bed is sufficient to limit the growth of the biomass on the support medium. Depending on the degree of fluidization, upflow fluidized-bed attached-growth reactors are sometimes referred to as expanded beds. Expanded-bed reactors are essentially the same as fluidized beds, with the exception of the degree of bed expansion.

7-8 ANAEROBIC SUSPENDED- AND ATTACHED-GROWTH AND HYBRID PROCESSES

Anaerobic biological treatment involves the decomposition of organic and inorganic matter in the absence of molecular oxygen. The major applications have been, and remain today, the stabilization of concentrated sludges produced from the treatment of wastewater and the treatment of concentrated organic industrial wastes. More recently, the treatment of low-strength domestic wastewater has become feasible with hybrid anaerobic processes. Although conventional anaerobic digestion is typically

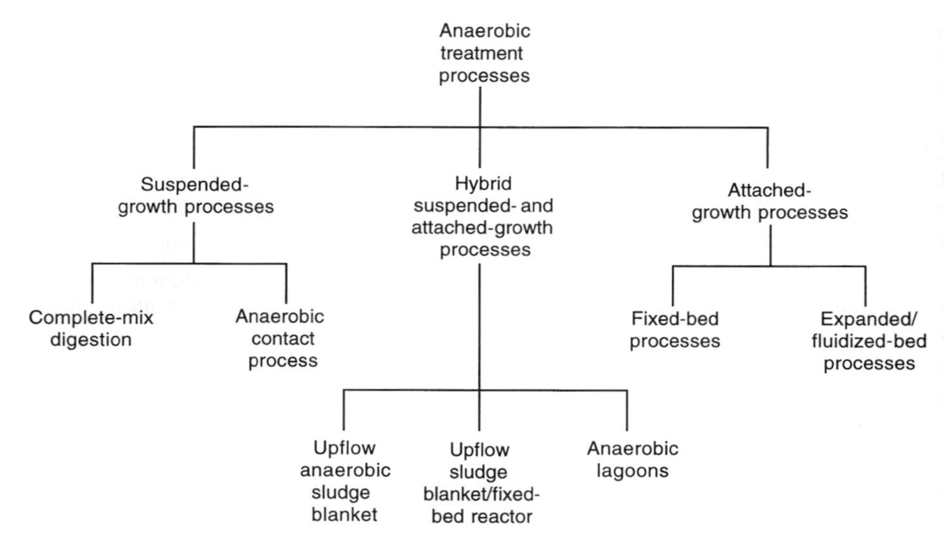

FIGURE 7-45
Classification of anaerobic treatment processes (adapted from Sutton, 1990).

used in wastewater treatment plants with a capacity greater than 5 Mgal/d, it is important to consider anaerobic processes because of their importance in the treatment of low-strength wastes (e.g., domestic wastewater) and their application to small flows. The following discussion of anaerobic treatment will deal with: (1) process classification and characteristics, (2) process microbiology, (3) anaerobic treatment process variables, (4) suspended-growth processes, (5) attached-growth processes, (6) hybrid processes, and (7) process design considerations.

Process Classification and Characteristics

Anaerobic treatment processes that are currently in use may be classified as shown in Fig. 7-45. The principal characteristics of these processes are compared in Table 7-19. The suspended-growth processes are typically used to treat waste containing particulate biodegradable materials, such as the sludge from primary and secondary treatment. The attached growth processes are most suitable for the treatment of soluble organic wastes, such as from food processing facilities. The hybrid processes can be used to treat wastes with both particulate and soluble constituents, although they function best with soluble waste. Following a discussion of the process microbiology, each of these classifications will be considered in greater detail.

Process Microbiology

The biological conversion of the organic matter under anaerobic conditions typically occurs in three steps (see Fig. 7-46). In the first step, one group of organisms

TABLE 7-19
Comparison of anaerobic suspended-growth, hybrid, and attached-growth treatment processes*

Factor	Suspended-growth	Hybrid systems	Attached-growth
Biomass concentration attainable	Low	High	High
Attainable MCRTs	Low	High	High
Suitable for wastewaters with particulates	Yes	Partial removal of particulates	Poor removal of particulates
Suitable for very concentrated wastewaters	Yes	No	No
Suitable for dilute wastewaters	No	Yes	Yes
Removal efficiency	Limited	High	High
Resistance to toxics and dynamic operating conditions	Limited due to short MCRTs	Longer MCRTs impart improved stability	Longer MCRTs impart improved stability
Maintenance of internal hydraulic integrity	Relatively simple with mechanical mixing	Generally satisfactory with effluent recycle and evolved biogas mixing	Excess biomass accumulation can negatively impact reactor hydraulics
Power requirements	Generally lowest	Higher if effluent recycle practiced	Can be high if support medium is fluidized

*Adapted from Speece (1983, 1996).

is responsible for hydrolyzing organic polymers and lipids to basic structural building blocks such as monosaccharides, amino acids, and related compounds that are suitable as a source of energy and cell carbon. In the second step, another group of anaerobic bacteria ferments the breakdown products to simple organic acids, the most common of which is acetic acid. This second group of nonmethanogenic microorganisms consists of facultative and obligate anaerobic bacteria. Collectively, these microorganisms are often identified in the literature as *acidogens* or *acid formers.* Among the nonmethanogenic bacteria that have been isolated from anaerobic digesters are *Clostridium* spp., *Peptococcus anaerobus, Bifidobacterium* spp., *Desulphovibrio* spp., *Corynebacterium* spp., *Lactobacillus, Actinomyces, Staphylococcus,* and *Escherichia coli.* Other physiological groups present include those producing proteolytic, lipolytic, ureolytic, or cellulytic enzymes (Hawkes, 1963; Higgins and Burns, 1975).

A third group of microorganisms converts the hydrogen and acetic acid formed by the acid formers to methane gas and carbon dioxide. The bacteria responsible for this conversion are strict anaerobes and are called methanogenic. Collectively, they are identified in the literature as *methanogens* or *methane formers.* Many of the

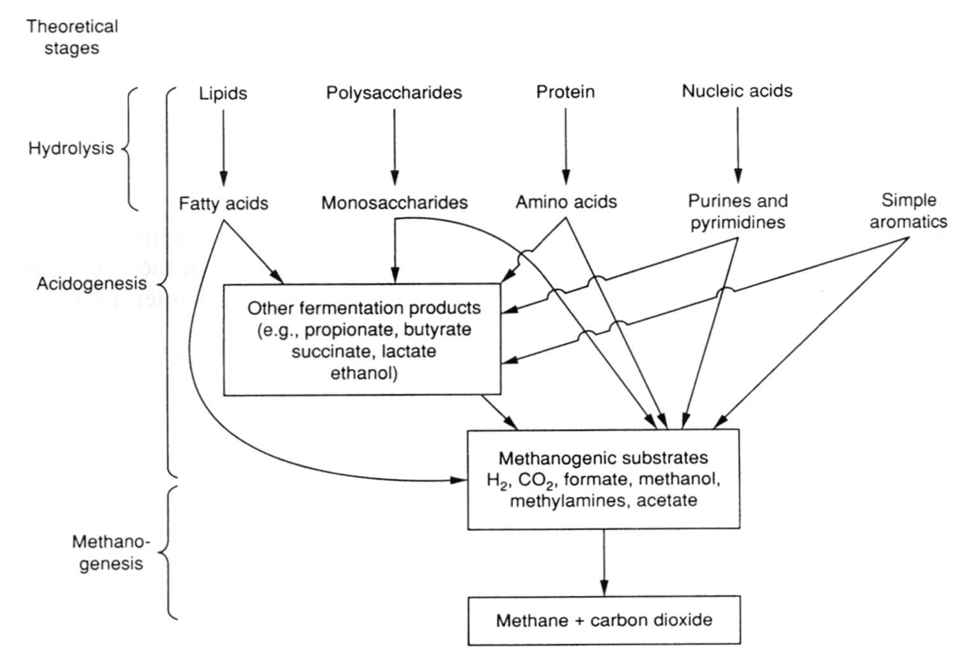

FIGURE 7-46
Pathways leading to the production of methane and carbon dioxide from the anaerobic digestion of the organic substrates (adapted from Holland et al., 1987).

methanogenic organisms identified in anaerobic digesters are similar to those found in the stomachs of ruminant animals and in organic sediments taken from lakes and rivers. The principal genera of microorganisms that have been identified include the bacterial rods (*Methanobacterium, Methanobacillus*) and spheres (*Methanococcus, Methanosarcina*) (Hawkes, 1963; Higgins and Burns, 1975). The most important bacteria of the methanogenic group are the ones that utilize hydrogen and acetic acid. They have very slow growth rates; as a result, their metabolism is usually considered rate limiting in the anaerobic treatment of an organic waste. Waste stabilization in anaerobic treatment is accomplished when methane and carbon dioxide are produced. Methane gas is highly insoluble, and its departure from solution represents actual waste stabilization.

It is important to note that methane bacteria can use only a limited number of substrates for the formation of methane. Currently, it is known that methanogens use the following substrates: CO_2, H_2, formate, acetate, methanol, methylamines, and carbon monoxide. Typical energy-yielding conversion reactions involving these compounds are as follows:

$$4H_2 + CO_2 \rightarrow CH_4 + 2H_2O \tag{7-105}$$

$$4HCOOH \rightarrow CH_4 + 3CO_2 + 2H_2O \tag{7-106}$$

$$CH_3COOH \rightarrow CH_4 + CO_2 \tag{7-107}$$

$$4CH_3OH \rightarrow 3CH_4 + CO_2 + 2H_2O \quad (7\text{-}108)$$

$$4(CH_3)_3N + 6H_2O \rightarrow 9CH_4 + 3CO_2 + 4NH_3 \quad (7\text{-}109)$$

In anaerobic treatment, the methanogens and the acidogens form a syntrophic (mutually beneficial) relationship. The methanogens are able to utilize the hydrogen produced by the acidogens because of their efficient hydrogenase enzymes. Because the methanogens are able to maintain an extremely low partial pressure of H_2, the equilibrium of the fermentation reactions is shifted toward the formation of more oxidized end products (e.g., formate and acetate). The utilization of the hydrogen, produced by the acidogens and other anaerobes, by the methanogens is termed *interspecies hydrogen transfer.* In effect, the methanogenic bacteria remove compounds that would inhibit the growth of acidogens. The amount of methane produced per pound of UBOD converted is 5.62 ft^3, as illustrated in Example 7-9.

EXAMPLE 7-9. CONVERSION OF ORGANIC MATERIAL TO METHANE GAS. Determine the amount of methane produced per pound of UBOD stabilized to methane and carbon dioxide. Assume that the starting compound is glucose ($C_6H_{12}O_6$).

Solution

1. Write a balanced equation for the conversion of glucose to CO_2 and CH_4 under anaerobic conditions, neglecting the growth of cells:

$$\underset{180}{C_6H_{12}O_6} \rightarrow \underset{132}{3CO_2} + \underset{48}{3CH_4}$$

Note that although the glucose has been converted, the methane has an oxygen requirement for complete conversion to carbon dioxide and water.

2. Write a balanced equation for the oxidation of methane to CO_2 and H_2O, and determine the pounds of methane formed per pound of UBOD:

$$\underset{48}{3CH_4} + \underset{192}{6O_2} \rightarrow 3CO_2 + 6H_2O$$

From the above equation and the equation in step 1, the UBOD per pound of glucose is (192/180) lb, and 1.0 lb of glucose yields (48/180) lb of methane, so that the ratio of the amount of methane produced per pound of UBOD converted is

$$\frac{\text{lb } CH_4}{\text{lb UBOD}} = \frac{48/180}{192/180} = 0.25$$

Therefore, for each pound of UBOD converted, 0.25 lb of methane is formed.

3. Determine the volume equivalent of the 0.25 lb of methane produced from the stabilization of 1.0 lb of UBOD:

$$V_{CH4} = (0.25 \text{ lb})\left(\frac{454 \text{ g}}{\text{lb}}\right)\left(\frac{1 \text{ mol}}{16 \text{ g}}\right)\left(\frac{22.4 \text{ L}}{\text{mol}}\right)\left(\frac{\text{ft}^3}{28.32 \text{ L}}\right)$$

$$= 5.61 \text{ ft}^3 \text{ of } CH_4 \text{ at standard conditions (32°F and 1 atm)}$$

Therefore, 5.61 ft^3 of methane is produced per pound of UBOD converted.

Anaerobic Treatment Process Variables

To maintain an anaerobic treatment system that will stabilize an organic waste efficiently, the nonmethanogenic and methanogenic bacteria must be in a state of dynamic equilibrium. To establish and maintain such a state, the reactor contents should be void of dissolved oxygen and free from inhibitory concentrations of such constituents as heavy metals and sulfides. Also, the pH of the aqueous environment should range from 6.6 to 7.6. Sufficient alkalinity should be present to ensure that the pH will not drop below 6.2, because the methane bacteria cannot function below this point. When digestion is proceeding satisfactorily, the alkalinity will normally range from 1000 to 5000 mg/L, and the volatile fatty acids will be less than 250 mg/L. A sufficient amount of nutrients, such as nitrogen and phosphorus, must also be available to ensure the proper growth of the biological community. Depending on the nature of the waste to be digested, growth factors may also be required. Temperature is another important environmental parameter. The optimum temperature ranges are the mesophilic, 30 to 38°C (86 to 100°F), and the thermophilic, 49 to 57°C (120 to 135°F). Most anaerobic treatment processes are operated in the mesophilic temperature range. While thermophilic operation of anaerobic processes is feasible technically, it has proven to be difficult to implement operationally. Additional details on the digester operation may be found in Malina and Pohland (1992), McCarty (1964, 1966), and Speece (1983, 1996).

Suspended-Growth Anaerobic Processes

The principal suspended-growth processes are the anaerobic digestion process and the anaerobic contact process. Of the processes identified in Fig. 7-45, anaerobic digestion is used most commonly in the field of wastewater treatment.

Anaerobic digestion process. In the anaerobic digestion process, the organic material in mixtures of primary settled and biological sludges is converted biologically, under anaerobic conditions, to a variety of end products including methane (CH_4) and carbon dioxide (CO_2). The process is carried out in an airtight reactor. Sludge, introduced continuously or intermittently, is retained in the reactor for varying periods of time. The stabilized sludge, withdrawn continuously or intermittently from the reactor, is reduced in organic content and is nonputrescible, and its pathogen content is greatly reduced.

The two types of anaerobic digesters that are commonly used are identified as *standard-rate* and *high-rate* digesters. In the standard-rate digestion process (see Fig. 7-47*a*), the contents of the digester are usually unheated and unmixed. Detention times for the standard-rate process vary from 30 to 60 days. In a high-rate digestion process (see Fig. 7-47*b*), the contents of the digester are heated and mixed completely. The required detention time for high-rate digestion is typically 20 days or less. A combination of these two basic processes is known as the *two-stage process* (see Fig. 7-47*c*). The primary function of the second stage is to separate the digested

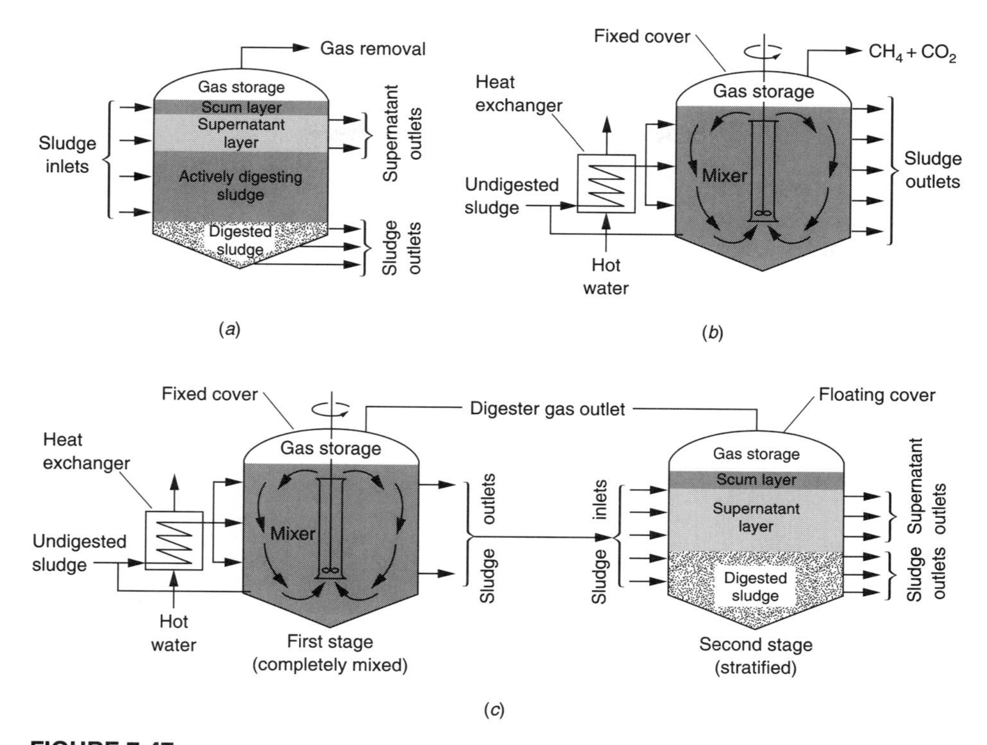

FIGURE 7-47
Typical anaerobic digesters: (*a*) conventional standard-rate single-stage process, (*b*) high-rate, complete-mix, single-stage process, and (*c*) two-stage process (adapted from Tchobanoglous and Burton, 1991).

solids from the supernatant liquor; however, additional digestion and gas production may occur. It should be noted that the mean cell residence time of the microorganisms in the complete-mix digester is equivalent to the hydraulic detention time of the liquid in the digester. As the operation temperature is increased, the minimum mean cell residence time is reduced significantly. Thus heating of the reactor contents lowers not only the mean cell residence time necessary to achieve adequate treatment, but also the hydraulic detention time, and a smaller reactor volume can be used (Tchobanoglous and Burton, 1991).

Anaerobic contact process. Many high-strength industrial wastes (typically greater than 1500 mg/L COD) can be stabilized effectively with the anaerobic contact process. As shown in Fig. 7-48*a*, untreated wastes are mixed with recycled sludge solids and then digested in a sealed complete-mix reactor. After digestion, the mixture is separated in a clarifier or vacuum flotation unit, and the supernatant is discharged as effluent, usually for further treatment. Settled anaerobic sludge is then recycled to seed the incoming wastewater. Because of the low synthesis rate of anaerobic microorganisms, the excess sludge that must be disposed of is minimal. The

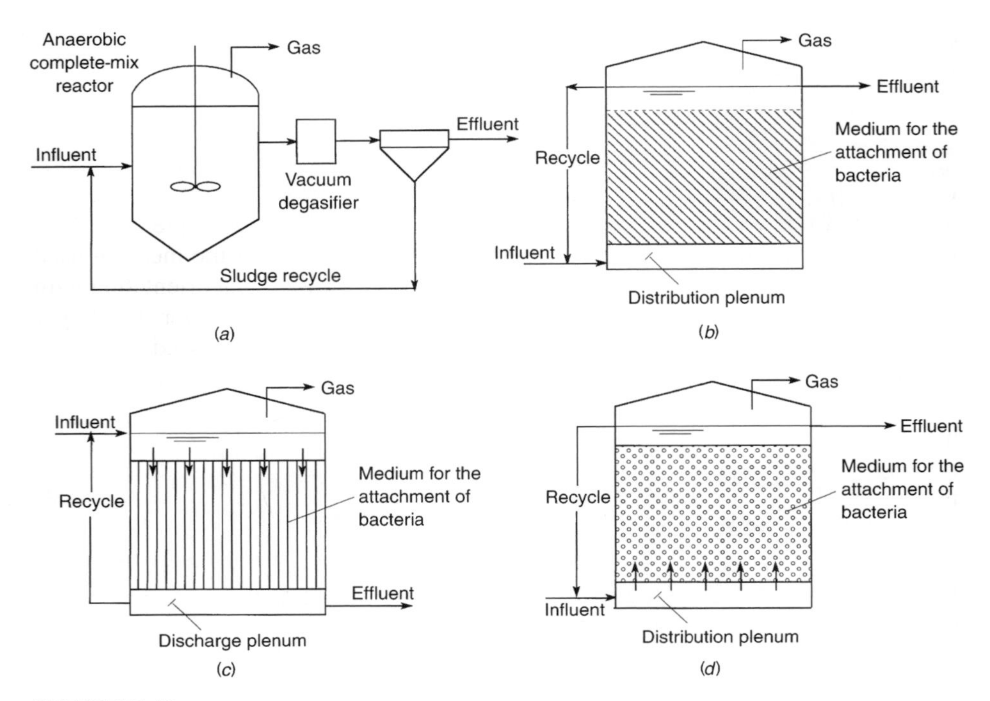

FIGURE 7-48
Typical reactor configurations used for anaerobic wastewater treatment: (*a*) anaerobic contact process, (*b*) upflow attached-growth process, (*c*) downflow attached-growth process, and (*d*) fluidized-bed attached-growth process (adapted from Speece, 1983).

anaerobic contact process has been used successfully for a variety of high-strength soluble wastes.

Attached-Growth Anaerobic Processes

One of the major limitations of the complete-mix anaerobic digestion process is the fact that from a cost-effectiveness standpoint only relatively low MCRTs can be used. To overcome this limitation a number of anaerobic attached-growth processes have been developed. Three such processes are considered in the following discussion: (1) the upflow process, (2) the downflow process, and (3) the fluidized-bed process.

Upflow attached-growth process. The most common anaerobic attached-growth treatment process is the upflow anaerobic filter process used for the treatment of carbonaceous organic wastes (see Fig. 7-48*b*). The anaerobic filter is a column filled with various types of solid media used for attachment of bacteria. The waste flows upward through the column, contacting the medium, on which anaerobic bacteria grow and are retained. Because the bacteria are retained on the medium and not washed off in the effluent, mean cell residence times on the order of 100 days can be obtained. Large values of θ_c can be achieved with short hydraulic retention times, so the anaerobic filter can be used for the treatment of low-strength wastes at ambient temperature.

Downflow attached-growth process. The downflow process is similar to the upflow process with the exception that the wastewater to be treated is introduced at the top of the reactor and the treated effluent is withdrawn from a plenum at the bottom of the reactor (see Fig. 7-48*c*). Another difference is that the medium used for the attachment of the bacterial biomass is arranged in a vertical orientation with larger clear spacing between the media as compared to the upflow filter. The larger spacing has resulted in less clogging. A variety of media have been used.

Fluidized-bed attached-growth process. In the fluidized-bed anaerobic reactor, the bacterial biomass is grown on a suitable medium (e.g., sand, expanded ceramic beads). The liquid to be treated is introduced into the reactor from a plenum in the bottom of the reactor (see Fig. 7-48*d*). The rate at which the liquid is applied is sufficient to fluidize the bed. The shearing force of the liquid as it passes through the fluidized bed is sufficient to limit the growth of the biomass on the support medium. Depending on the degree of fluidization, upflow reactors are sometimes referred to as *expanded beds*. Expanded-bed reactors are essentially the same as fluidized beds, with the exception that the degree of bed expansion is less.

Hybrid Anaerobic Processes

As the potential of anaerobic treatment has become understood more clearly, a number of hybrid anaerobic processes have been developed for both high- and low-strength wastes. The most common is the upflow anaerobic sludge blanket (UASB) process. A variation of this process is known as the upflow sludge blanket/fixed-bed reactor (USBFB). Still another variation is the so-called covered anaerobic lagoon process. Although there are many other processes, only these three will be discussed briefly in the following discussion. Additional details on these and other processes may be found in Speece (1996).

Upflow anaerobic sludge blanket reactor. The UASB was developed in the Netherlands (Lettinga et al., 1980). The process is widely used in Europe and in South America. The liquid to be treated is introduced in the bottom of the reactor (see Fig. 7-49*a*), where it flows upward through a sludge blanket composed of dense biologically-formed granules or particles. The sludge granules vary in size from $\frac{1}{16}$ to as large as $\frac{1}{4}$ in. In some cases the sludge blanket is flocculant. The gases produced under anaerobic conditions (principally methane and carbon dioxide) serve to mix the contents of the reactor as they rise to the surface. The rising gas also helps to form and maintain the granules. Material buoyed up by the gas strikes the degassing baffles and settles back down onto the bed from the quiescent settling zone above the sludge blanket. The gas is trapped in a gas collection dome located in the top of the reactor.

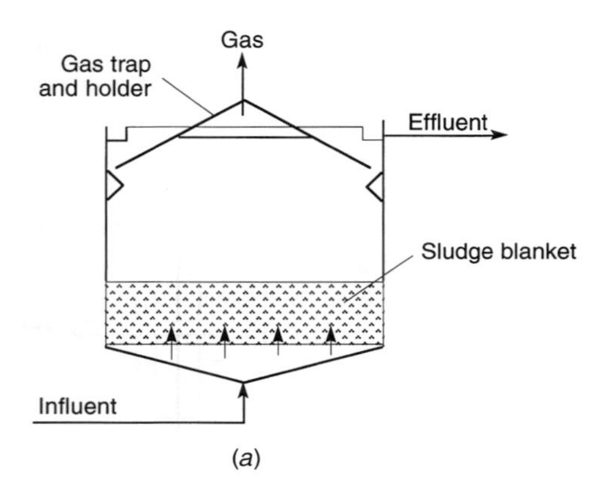

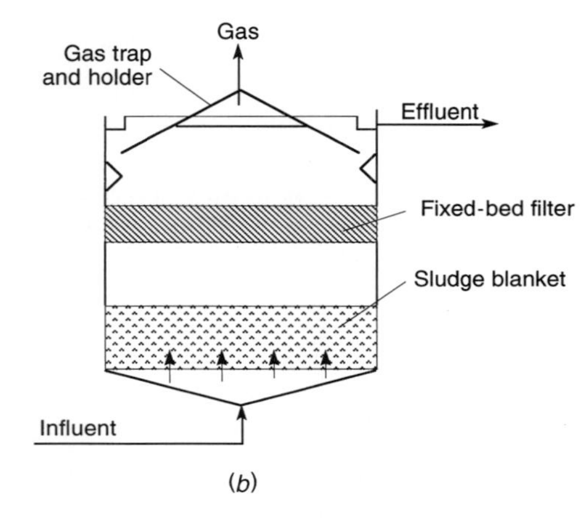

FIGURE 7-49
Hybrid anaerobic treatment processes: (*a*) upflow anaerobic sludge blanket and (*b*) upflow sludge blanket/fixed-bed reactor (adapted from Speece, 1983, and Malina and Pohland, 1992).

Upflow sludge blanket/fixed-bed reactor. The USBFB is a variant of the UASB. The major difference is that a fixed bed is incorporated above the sludge blanket (see Fig. 7-49*b*). The fixed bed is used to trap any solids that would otherwise pass out of the reactor. Because a bacterial biomass will develop on the fixed media, additional treatment occurs as the liquid passes through the fixed bed. A variant of USBFB is shown in Fig. 7-50. Untreated wastewater is introduced into the bottom of the insulated two-compartment septic tank. Treatment occurs as the wastewater flows up through the digested sludge. Additional treatment is achieved as the wastewater flows past the biomass on the suspended screens in the second compartment. Typical BOD removals achieved with a detention time of 18 h varied from 55 to 75 percent, regardless of the season (Living Technologies, Inc., 1997). The gas produced was vented through a compost filter.

Covered anaerobic lagoon. Still another process variation is the covered anaerobic lagoon. In this system, a hole is excavated and lined with a geomembrane. A geomembrane placed on styrofoam floats is used to cover the lagoon and to recover the gas. The recovered gas is utilized for the production of energy. A typical system

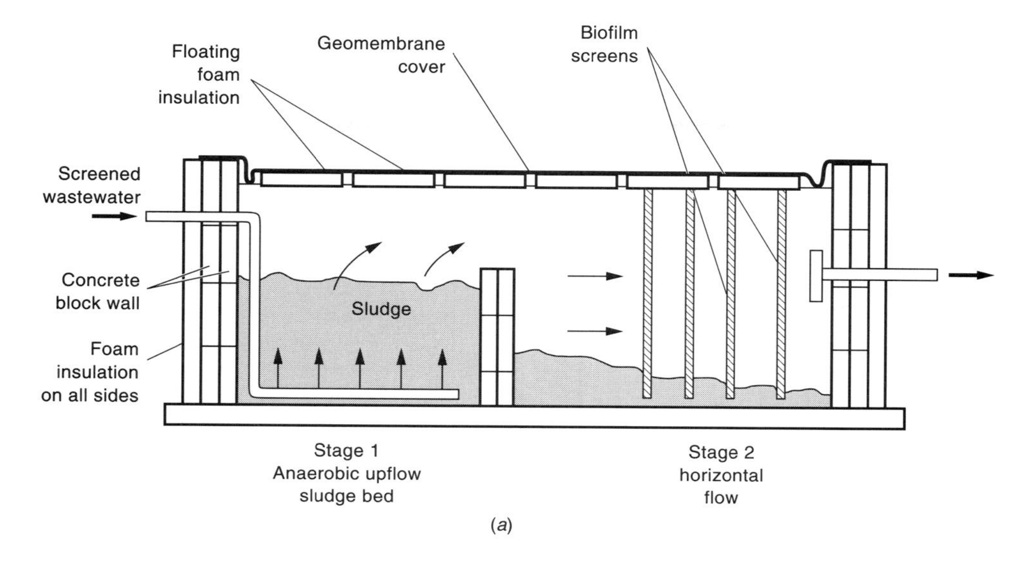

(*a*)

(*b*)

FIGURE 7-50
Anaerobic septic tank with suspended film for treatment of domestic wastewater: (*a*) definition sketch (adapted from Living Technologies, Inc., 1997) and (*b*) view of covered anaerobic septic tank at Frederick, MD.

used for the treatment of wastes from a piggery is shown in Fig. 7-51. The collected gas is used for the production of electricity. Digester gas is typically about 65 percent methane with a heating value of about 600 Btu/ft^3 (22,400 kJ/m^3). To avoid hydrogen sulfide corrosion problems, when digester gas is used in internal combustion engines, it is important to preheat the engine for about a half hour with natural gas before switching to digester gas. Similarly, if the engine is to be shut down, it must be operated for a half hour on natural gas to clean out the cylinder heads before shutdown.

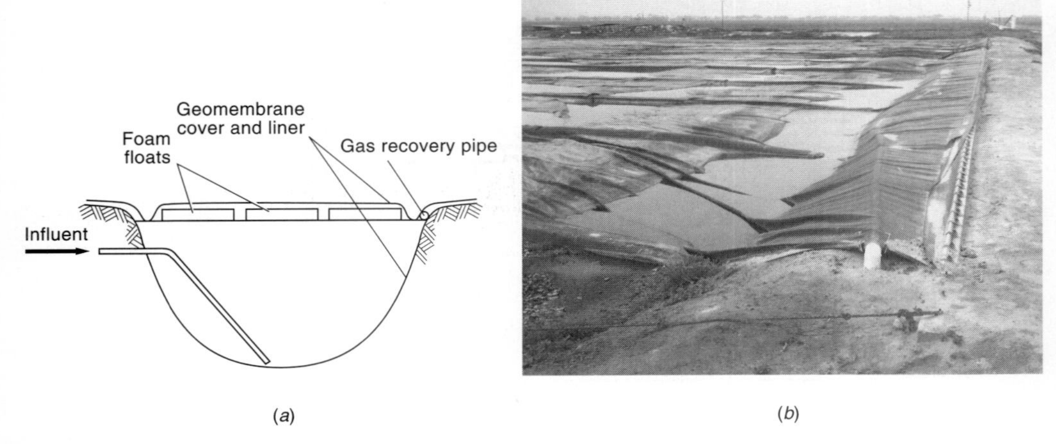

FIGURE 7-51
Digestion and gas recovery system for piggery wastes: (*a*) definition sketch and (*b*) view of typical unit.

Process Design Considerations

The design of anaerobic treatment processes depends on the type of system (i.e., complete-mix or attached-growth). For conventional complete-mix processes, the design is based on the analysis of a complete-mix reactor without recycle, as presented previously in Sec. 7-3. Typical kinetic coefficients for wastewater sludges are reported in Table 7-20. Care should be exercised in using the coefficients given in Table 7-20 because of the variability of individual sludges. Suggested θ_c values for the design of complete-mix anaerobic digesters are given in Table 7-21.

Alternatively, anaerobic treatment processes (both suspended-growth and attached-growth) can be sized on the basis of the volumetric organic and volatile

TABLE 7-20
Typical kinetic coefficients for the anaerobic sludge digestion of domestic wastewater sludges*

		Value†	
Coefficient	**Basis‡**	**Range**	**Typical**
k	d^{-1}	0.5–2	1.0
K_s	mg COD /L	500–2500	1500
Y	mg VSS/mg COD	0.05–0.15	0.1
k_d	d^{-1}	0.02–0.05	0.03

*Derived in part from Lawrence and McCarty (1969), Malina and Pohland (1992), Speece (1996), and WPCF (1987).
†Values reported are for 20 °C.
‡VSS = volatile suspended solids.

TABLE 7-21
Suggested mean cell residence times for use in the design of complete-mix anaerobic digesters*

Operating temperature, °C	θ_c^M	θ_c suggested for design, d
18	11	28
24	8	20
30	6	14
35	4	10
40	4	10

*From McCarty (1964, 1968).

solids loading rates (L_{org} and L_{VSS}) by using

$$V = \frac{CQF}{L_{org} \text{ or } L_{VSS}} \tag{7-110}$$

where V = volume of reactor, ft^3 (m^3)
C = concentration of organic or volatile material in wastewater, mg COD/L or mg VSS/L (g/m^3)
Q = wastewater flowrate, Mgal/d (m^3/d)
F = conversion factor, 8.34 lb/[Mgal·(mg/L)] (10^{-3} kg/g)
L_{org} = volumetric organic loading rate, lb COD/ft^3·d (kg COD/m^3·d)
L_{VSS} = volumetric volatile suspended solids loading rate, lb VSS/ft^3·d (kg VSS/m^3·d)

The selection of appropriate organic and volatile solids loading factors is most commonly based on the results of pilot plant studies. Typical values for these parameters are reported in Table 7-22. It should be noted that neither the L_{org} nor the L_{VSS} is

TABLE 7-22
Typical process and performance data for anaerobic processes*

Process	Input, COD, mg/L	Hydraulic detention time, h	Organic loading rate, lb COD/ft^3·d	Volatile solids loading rate, lb VSS/ft^3·d	COD removal, %
Anaerobic digestion†	1500–5000	360–480	0.1–0.3	0.06–0.2	45–65
Anaerobic contact	1500–5000	2–10	0.1–1.0		75–90
Upflow anaerobic sludge blanket (UASB)	5000–15,000	18–30	0.05–1.0		65–85
Upflow filter bed	10,000–20,000	18–30	0.1–1.25		75–90
Fluidized-bed	5000–10,000	8–16	0.3–1.8		80–95

*Adapted in part from Malina and Pohland (1992), Speece (1996), and WPCF (1987).
† For domestic wastewater sludge
Note: lb COD/ft^3·d × 16.0185 = kg COD/m^3·d

a fundamental parameter, but a parameter used for sizing the reactor. The sizing of a digester is illustrated in Example 7-10.

EXAMPLE 7-10. SIZING OF A CONVENTIONAL COMPLETE-MIX ANAEROBIC DIGESTER. Estimate the size of a complete-mix digester required to treat the sludge from a primary treatment plant designed to treat 1.0 Mgal/d (3800 m^3/d) of domestic wastewater. Check the organic loading rate and estimate the amounts of methane and total digester gas produced at standard conditions.

1. Operating temperature = 36°C.
2. The organic material removed per Mgal of wastewater = 1200 lb COD.
3. The TSS removed per Mgal of wastewater = 1200 lb. (*Note:* Solids are about 70 percent biodegradable, and the conversion factor from cell mass to COD is 1.42. Therefore, the overall conversion factor from TSS to COD is 0.70 × 1.42 = 1.0).
4. Moisture content of sludge (by weight) = 96%.
5. Specific gravity of sludge = 1.025.
6. Assume the following kinetic coefficients apply:

$$k = 1.4 \text{ d}^{-1}$$

$$K_s = 2000 \text{ mg COD/L}$$

$$Y = 0.1 \text{ mg VSS/mg COD}$$

$$k_d = 0.03 \text{ d}^{-1}$$

7. Use a factor of safety of 2.0 in estimating the required digester volume.
8. The sludge contains adequate nitrogen and phosphorus for biological growth.

Solution

1. Compute the daily sludge volume and COD concentration.
   a. Determine the sludge weight in pounds per day:

$$\text{Sludge solids weight} = 1200 \text{ lb/d}$$

$$\text{Moisture content} = 96\%, \text{ therefore } 4\% \text{ of sludge is solids}$$

$$\frac{\text{Sludge solids}}{\text{Sludge water}} = \frac{0.04}{0.96} = \frac{1200 \text{ lb/d}}{28{,}800 \text{ lb/d}}$$

$$\text{Total sludge weight} = 1200 + 28{,}800 = 30{,}000 \text{ lb/d}$$

   b. Determine the sludge volume, using the specific weight of water (62.4 lb/ft^3) and the sludge specific gravity provided above:

$$\text{Sludge volume} = \left(\frac{30{,}000 \text{ lb}}{\text{d}}\right)\left(\frac{\text{ft}^3}{1.025 \times 62.4 \text{ lb}}\right) = 469 \text{ ft}^3\text{/d}$$

   c. Determine the sludge solids concentration:

$$C_s = \frac{(1200 \text{ lb/d})}{(8.34)(0.003508 \text{ Mgal/d})} = 41{,}016 \text{ mg/L}$$

2. Compute the digester volume.
   a. Using Eq. (7-57) and the given kinetic coefficients, estimate the minimum hydraulic residence time and apply a safety factor of 2.0 to determine the design value.

$$\frac{1}{\theta_c^M} \approx Yk - k_d = (0.10)(1.4) - 0.03 = 0.11$$

$$\theta_c^M = \frac{1}{0.11} = 9.1 \text{ d}$$

$$\theta_d = (9.1 \times 2.0) = 18.2 \text{ d}$$

b. The required digester volume is:

$$V = Q\theta_d = (469.0 \text{ ft}^3/\text{d})(18.2 \text{ d})$$
$$= 8536 \text{ ft}^3 = 63{,}850 \text{ gal}$$

3. Determine the volumetric organic loading rate:

$$\text{lb COD/ft}^3\cdot\text{d} = \frac{1200 \text{ lb COD}}{8536 \text{ ft}^3} = 0.14 \qquad \text{(OK; see Table 7-22)}$$

4. Compute the sludge-production rate on a mass basis.
   a. Using Eq. (7-53), calculate the observed yield:

$$Y_{\text{obs}} = \frac{Y}{1 + k_d\theta_c} = \frac{0.10}{1 + 0.03(18.2)} = 0.065$$

b. Estimate the digester effluent COD concentration using Eq. (7-52) and the given data:

$$S = \frac{K_s(1 + \theta k_d)}{\theta(Yk - k_d) - 1}$$
$$= \frac{2000(1 + 18.2 \times 0.03)}{18.2(0.10 \times 1.4 - 0.03) - 1} = 3086 \text{ mg/L}$$

$$\text{lb COD/d} = (3086)(0.003508)(8.34) = 90.3 \text{ lb COD/d}$$

c. The biomass production rate is:

$$\text{lb VSS/d} = (Y_{\text{obs}} \text{ lb/lb})\left[(S_o - S) \text{ mg/L}\right](Q \text{ Mgal/d})[8.34 \text{ lb/Mgal}\cdot(\text{mg/L})]$$
$$= 0.065(41{,}016 - 3086)(0.003508)(8.34) = 72.1 \text{ lb VSS/d}$$

5. Compute the percent stabilization:

$$\%\text{ stabilization} = \frac{(1200 - 90.3) - 1.42(72.1)}{1200} \times 100 = 83.9\%$$

6. Compute the volume of methane produced per day at standard conditions.
   a. Determine the lb COD converted to gas per day:

$$\text{lb COD to gas} = \text{COD entering digester} - \text{COD leaving digester} - \text{COD converted to cells}$$

$= 1200 - 90.3 - 1.42(72.1) = 1007.3$ lb, where 1.42 is used to convert cell mass to COD (see Eq. (7-15)].

b. Convert the lb COD gas produced per day to volume of methane produced per day, using the conversion factor derived in Example 7-9:

$$V_{CH4} = 5.61(1007.3) = 5651 \text{ ft}^3/\text{d (at 32°F and 1.0 atm)}$$

7. Estimate the total gas production. Because digester gas is about two-thirds methane, the total volume of gas produced is:

$$V_T = \frac{5651 \text{ ft}^3/\text{d}}{0.67} = 8434 \text{ ft}^3/\text{d}$$

Comment. It should be noted that a similar digester volume would have been estimated using the kinetic coefficients given in Table 7-20. As noted in the text, great care should be exercised when using kinetic coefficients because of the great variability in the values reported in the literature. Because of the relatively small amount of gas produced as determined in step 6, the recovery of digester gas is generally not feasible in small treatment plants. Because of the greater complexity and operational attention required, the trend has been to use aerobic digesters in small treatment plants (typically less than 5 Mgal/d).

7-9 PRE-ENGINEERED (PACKAGE) WASTEWATER TREATMENT PLANTS

Commercially available prefabricated treatment plants known as *package plants* are often used for the treatment of wastewater for individual properties and small communities. Although package plants are available in capacities up to 1.0 Mgal/d (3800 m^3/d), they are used most commonly for wastewater flows in the range from 0.001 to 0.2 Mgal/d (3.8 to 760 m^3/d). Small package plants have also been developed for individual home use. Properly sized, operated, and maintained, these plants can usually provide satisfactory treatment for small wastewater flows.

When package plants employing biological treatment first came into use, it was believed that if they were operated to achieve complete oxidation, no excess biological sludge had to be wasted. As a result of this erroneous assumption, sludge would build up and periodically discharge from the system. This discharge phenomenon, termed *burping,* still occurs in small package plants operated without proper maintenance. In the following discussion the principal operational issues encountered with package plants, the types of package most commonly used, and suggested design requirements for package plants are reviewed.

Types of Package Plants

The most common type of biological treatment package plant for flows in the range of from 0.001 to 0.2 Mgal/d (3.8 to 760 m^3/d) is the extended aeration activated-sludge process. In addition, numerous hybrid systems involving both aerobic and anaerobic processes have been developed (see Fig. 7-52).

Design and Operational Issues with Package Plants

The major design and operational issues that affect the performance of package plants employing biological treatment (usually some type of activated-sludge process) include (U. S. EPA, 1977):

(*a*)

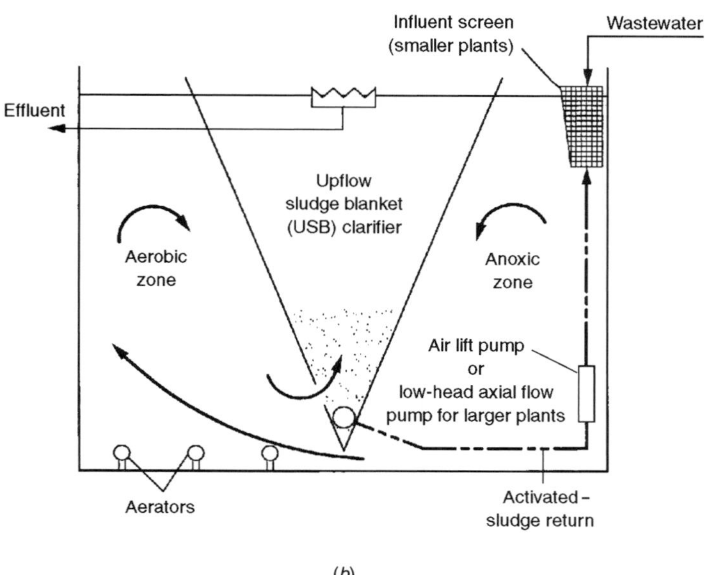

(*b*)

FIGURE 7-52
Typical package wastewater treatment plants: (*a*) Extended aeration plant with structure constructed of concrete. Prefabricated package plants constructed of steel are also available. (*b*) Schematic of small modular package plant available in sizes suitable for individual residences to small communities (courtesy Ecofluid Systems, Inc.).

1. Hydraulic shock loads—the large variations in flow from small communities, accentuated by the use of oversized pumps where wastewater is pumped.
2. Very large fluctuations in both flow and BOD loading.
3. Very small flows that make the design of self-cleansing conduits and channels difficult.

4. Adequate or positive sludge return, requiring provisions for a recirculation rate of up to 3:1, for extended aeration systems to meet all normal conditions.
5. Adequate provision for scum and grease removal from final clarifier.
6. Denitrification in final clarifier, with resultant solids carryover.
7. Inadequate removal and improper provision for handling and disposing of waste sludge.
8. Adequate control of MLSS in the aeration tank.
9. Adequate antifoaming measures.
10. Large and rapid temperature change.
11. Adequate control of air supply rate.
12. Adequate design under organic and solids loadings, which can cause poor treatment performance and odor problems.

Although the above factors are related more specifically to package plants employing biological treatment, many of the factors (e.g., 1, 2, and 3) also apply to package plants employing physical/chemical treatment. Measures that can be taken to address the above issues are discussed below.

Improving the Performance of Package Plants

The performance of most package plants can be improved by sizing the treatment facilities conservatively (especially the secondary settling facilities), and by specifying positive means for handling and pumping the side stream flows. Because of the uncertainties of field operation, it is recommended that the overflow rate at peak hourly flow be limited to 600 to 800 gal/ft^2·d (24 to 33 m^3/m^2·d). Positive and effective means should be provided for returning waste sludge to the aeration chamber. Although air lift pumps have been used for returning waste sludge, they are undesirable in this application because the rate of return cannot be adjusted easily or reliably. The secondary settling tank should also be equipped with scum collection facilities and an effective system for the removal of the accumulated scum. Determination of the critical criteria for a prefabricated package plant is considered in Example 7-11.

EXAMPLE 7-11. DEVELOPING SPECIFICATIONS FOR A PREFABRICATED PACKAGE PLANT. A prefabricated package plant is to be used to treat the wastewater from a small subdivision consisting of 650 individual family residences. The average occupancy has been estimated to be 3.1 persons per residence. Use a flow of 60 gal/person·d and a peaking factor of 2.5 for flow, BOD, and TSS. Select the type of package plant and size the principal components of the plant. The effluent from the package plant must meet EPA secondary standards. Assume the following conditions apply:

1. BOD per capita with ground-up kitchen waste = 0.220 lb/capita·d (see Table 4-11)
2. TSS per capita with ground-up kitchen waste = 0.250 lb/capita·d (see Table 4-11)
3. TKN-N per capita with ground-up kitchen waste = 0.032 lb/capita·d (see Table 4-11)
4. Wastewater flowrate = 60 gal/capita·d
5. Peak daily flow ratio = 2.5

6. Peak hourly flow ratio = 4.0 to 1.0
7. Peak daily ratio for N = 1.5
8. F/M ratio = 0.05
9. MLSS operating level = 3000 to 6000 mg/L
10. BOD_5 = 0.65 COD
11. 4.6 mg O_2/L is required to nitrify ammonia (see Sec. 7-5)

Solution

1. Estimate the average and peak wastewater flowrates.
   a. The total number of persons is

$$650 \times 3.1 \text{ persons/home} = 2015 \text{ persons}$$

   b. The corresponding average flowrate based on 60 gal/person·d is:

$$2015 \text{ persons} \times 60 \text{ gal/capita·d} = 120{,}900 \text{ gal/d}$$

   c. The corresponding peak daily flowrate, based on a factor of 2.5 (see Table 4-20), is

$$120{,}900 \text{ gal/d} \times 2.5 = 302{,}250 \text{ gal/d}$$

2. Estimate the daily average and peak BOD, TSS, and N mass loading rates to be treated.
   a. For a value of 0.22 lb/capita·d, the average BOD mass loading rate is:

$$\text{BOD, lb/d} = 2015 \times 0.22 \text{ lb/capita·d} = 443.3 \text{ lb/d}$$

   For a peaking factor of 2.5, the peak BOD mass loading rate is:

$$443.3 \text{ lb/d} \times 2.5 = 1108 \text{ lb/d}$$

   For an average flowrate of 120,900 gal/d, the corresponding BOD concentration is

$$\text{BOD, mg/L} = (443.3 \text{ lb/d})/(0.1209 \text{ Mgal/d} \times 8.34) = 440 \text{ mg/L}$$

   b. For a value of 0.25 lb/capita·d, the average TSS daily mass loading rate is:

$$\text{TSS, lb/d} = 2015 \times 0.25 \text{ lb/capita·d} = 503.8 \text{ lb/d}$$

   For a peaking factor of 2.5, the peak TSS mass loading rate is:

$$503.8 \text{ lb/d} \times 2.5 = 1260 \text{ lb/d}$$

   For an average flowrate of 120,900 gal/d, the corresponding TSS concentration is

$$\text{TSS, mg/L} = (503.8 \text{ lb/d})/(0.1209 \text{ Mgal/d} \times 8.34) = 500 \text{ mg/L}$$

   c. For a value of 0.032 lb/capita·d, the average N daily mass loading rate is

$$\text{N, lb/d} = 2015 \times 0.032 \text{ lb/capita·d} = 64.5 \text{ lb/d}$$

   For a peaking factor of 1.5, the peak TSS mass loading rate is:

$$64.5 \text{ lb/d} \times 1.5 = 96.8 \text{ lb/d}$$

   For an average flowrate of 120,900 gal/d, the corresponding TSS concentration is:

$$\text{N, mg/L} = (64.5 \text{ lb/d})/(0.1209 \text{ Mgal/d} \times 8.34) = 64.0 \text{ mg/L}$$

3. Select the type of treatment process.
   a. An extended aeration activated-sludge process package plant is recommended. The principal reasons for selecting an extended aeration activated-sludge process are: (1) excellent effluent quality, (2) the ability to remove nitrogen and phosphorus, (3) relatively low sludge yield, (4) relative simplicity, and (5) relative ease of operation.
   b. Assume that an inline perforated plate screen (described in Chap. 5) will be used for removing coarse solids, and that primary sedimentation facilities will not be used.

4. Size the principal treatment process components.
   a. Using a detention time of one day at average flow, the required aeration tank volume is equal to the average daily flow. Thus, the aeration tank volume is equal to:

$$\text{Volume} = (120{,}900 \text{ gal/d}) \times (1.0 \text{ d}) = 120{,}900 \text{ gal}$$

   b. The aeration system must be capable of providing the required amount of oxygen to meet the sustained peak demand. Thus, on the basis of the peak organic loading rate and assuming an oxygen transfer efficiency of 6 percent, determine the required capacity of the aeration system.

   The oxygen supplied must be sufficient for the oxidation of the carbonaceous organic material and for the conversion of ammonia to nitrate. Neglect any reduction in the amount of oxygen required due to denitrification.

$$O_2 \text{ required} = \frac{1108 \text{ lb/d}}{0.65} + (96.8)(4.6) = 2150 \text{ lb/d}$$

   Assume the specific weight of air at standard temperature and pressure is 0.0750 lb/ft^3 and contains 23.2 percent oxygen by weight. The air requirement is:

$$\text{Air required} = \frac{(2150 \text{ lb/d})}{(0.0750 \text{ lb/ft}^3)(0.232)(0.06)(1440 \text{ min/d})} = 1430 \text{ ft}^3\text{/min}$$

   c. For a peak hour factor of 4 and an overflow rate of 800 $\text{gal/ft}^2\cdot\text{d}$, the surface area required for the secondary settling tank is:

$$\text{Surface area} = \frac{(120{,}900 \text{ gal/d} \times 4)}{800 \text{ gal/ft}^2\cdot\text{d}} = 605 \text{ ft}^2$$

Comment. Because the hourly peaking factors are greater for small flows, the overflow rate at average flow is 200 $\text{gal/ft}^2\cdot\text{d}$. To avoid problems with rising sludge due to any residual denitrification, the secondary settling tank must be equipped with positive means for the continuous removal of the accumulated sludge.

PROBLEMS AND DISCUSSION TOPICS

7-1. Using the balanced reaction developed in Example 7-1, estimate the oxygen and nitrogen requirements per pound of glucose converted.

7-2. Using the half-reactions developed in Example 7-1, prepare a plot of cell yield versus f_s (use f_s values of 0.4, 0.6, 0.7, and 0.8). How significant is the value of f_s with respect to cell yield?

7-3. If the generation time for bacteria is 20 min, estimate the number of bacteria that would be present after 6 h.

7-4. If 65,536 bacteria are present and the generation time is equal to 30 min, estimate the time required to obtain this number by binary fission.

7-5. Prepare a plot similar to Fig. 7-8 using the following coefficients and data:

$$k = 5 \text{ mg/mg·d}$$
$$K_s = 90 \text{ mg/L}$$
$$Y = 0.6$$
$$k_d = 0.045 \text{ d}^{-1}$$
$$S_o = 600 \text{ mg/L}$$

7-6. If the kinetic coefficients for the activated-sludge process are $Y = 0.75$ mg VSS/mg BOD_5 and $k_d = 0.05 \text{ d}^{-1}$, determine the volume of a complete-mix aerated lagoon without recycle if $\theta_c = 8$ d and the MLSS is 2000 mg/L, of which 70 percent is volatile.

7-7. Determine the reactor volume, the sludge wasting rate, the recirculation ratio, and the hydraulic residence time for a complete-mix activated-sludge process used to treat a soluble sugar waste. Assume the conditions given in Example 7-2 apply with the following exceptions:

1. Return sludge concentration = 10,000 mg/L
2. Mixed liquor suspended solids = 3500 mg/L
3. Mean cell residence time = 15 d
4. The coefficient of reliability (COR) values for BOD and TSS are 0.65 and 0.6, respectively

7-8. If the wastewater in Example 7-2 was totally deficient in nitrogen and phosphorus, estimate the amount of each constituent that must be added each day for the process to function properly.

7-9. Derive Eq. (7-55) for a plug-flow reactor.

7-10. Determine the factor of safety for an activated-sludge process used to treat a waste with a soluble COD of 250 mg/L. Assume the following conditions and coefficients are applicable:

$$\theta_c^d = 5 \text{ d}$$
$$Y = 0.4 \text{ (COD basis)}$$
$$k = 6 \text{ mg/mg·d}$$
$$K_s = 50 \text{ mg COD/L}$$
$$k_d = 0.055 \text{ d}^{-1}$$
$$S_o = 600 \text{ mg/L}$$

7-11. Using the following data, determine the kinetic coefficients k, K_s, μ_m, Y, and k_d.

Unit no.	S_o, mg/L COD	S, mg/L COD	$\theta = \theta_c$, d	X, mg VSS/L
1	488	42	1.5	206
2	575	28	2.0	247
3	728	15	3.5	299
4	660	21	2.5	281
5	512	17	3.0	212

7-12. Determine the nitrogenous oxygen demand for cell tissue ($C_5H_7NO_2$) for the complete oxidation of organic nitrogen to nitrate.

7-13. Derive Eq. (7-71), assuming $f_s = 0.10$.

7-14. Determine the effluent ammonia concentration, the amount of alkalinity that must be added, and the required hydraulic residence time for a single-state activated-sludge process designed to nitrify the wastewater from a small residential development served with a pressure sewer. Assume the following conditions apply:

1. Influent flowrate = 0.050 Mgal/d
2. Influent BOD_5 = 200 mg/L
3. Influent TKN = 60 mg/L
4. Minimum DO in reactor = 1.5 mg/L
5. Influent alkalinity = 350 mg/L as $CaCO_3$
6. Kinetic coefficients given in Table 7-7
7. Factor of safety = 2.5

7-15. Solve Prob. 7-14, assuming the following conditions are different from the given conditions.

3. Influent TKN = 80 mg/L
5. Influent alkalinity = 400 mg/L as $CaCo_3$
7. Factor of safety = 2.0

7-16. Derive Eq. (7-79), assuming $f_s = 0.1$.

7-17. Derive Eq. (7-80), assuming $f_s = 0.3$.

7-18. Derive Eq. (7-81), assuming $f_s = 0.3$.

7-19. A nitrified effluent contains 15 mg/L of nitrate-nitrogen, 1.6 mg/L of nitrite-nitrogen, and 2 mg/L of dissolved oxygen. Determine the methanol requirement for denitrification. Will the effluent BOD affect the methanol requirement?

7-20. Verify the correctness of Eq. (7-82).

7-21. Determine the residence time required for a separate-stage denitrification process, assuming the following conditions apply:

1. Influent nitrate = 50 mg/L
2. Effluent nitrate = 5 mg/L
3. Temperature = 15°C
4. Dissolved oxygen = 0.1 mg/L
5. $R_{DN20} = 0.10\ d^{-1}$

7-22. Solve Prob. 7-21 assuming the temperature is 25°C.

7-23. The amount of oxygen required by a small activated-sludge treatment plant is 1500 lb/d. Assuming the following conditions apply, estimate the required blower horsepower.

1. AOTE = 15%
2. Temperature = 15°C
3. P_1 = 14.7 lb/in^2
4. P_2 = 19.9 lb/in^2

7-24. Solve Prob. 7-23 for the following temperatures: 0, 10, 20, 30, and 40°C, and prepare a plot of blower horsepower versus temperature.

7-25. Effluent from a community septic tank is treated in an activated-sludge plant equipped with two aeration tanks. If the aeration tanks are 45 by 45 ft by 15 ft deep and the MLSS are 2000 mg/L, determine the detention time, volumetric organic loading rate, and the F/M ratio for a flowrate of 0.75 Mgal/d and an influent BOD of 220 mg/L.

7-26. A complete-mix package plant with integral settling tank is to be used for an individual home. If the kinetic coefficients are Y = 0.8 lb VSS/lb BOD_5 and $k_d = 0.05\ d^{-1}$, estimate the volume of sludge that would accumulate in the system every 3 months. Assume that the total flow into the system is 200 gal/d with a BOD of 200 mg/L. The hydraulic retention time in the unit is 36 h.

7-27. A complete-mix extended aeration activated-sludge process is to be used for the treatment of septic tank effluent from a small community composed of 1500 residents. If the following conditions apply, estimate how often the sludge must be removed from the system if the MLSS in the aeration tank is allowed to vary between 6000 and 10,000 mg/L.

1. Flowrate = 60 gal/capita·d
2. Influent BOD_5 (without primary treatment) = 200 mg/L
3. Y = 0.7 lb VSS/BOD_5
4. $k_d = 0.06\ d^{-1}$
5. Design F/M ratio at 6000 mg/L MLSS = 0.06

7-28. Compare and contrast the intermittent decanted extended aeration and the oxidation ditch processes for the complete treatment of wastewater (with and without phosphorus removal) from small communities (less than 2500 residents).

7-29. An existing complete-mix activated-sludge treatment plant is designed as an extended aeration process with an F/M ratio of 0.05 and a MLSS concentration of 4000 mg/L. If the following conditions apply, estimate the additional flow that could be treated if the F/M ratio was increased to 0.6.

1. Influent BOD_5 (without primary treatment) = 250 mg/L
2. Return sludge concentration = 10,000 mg/L/L
3. Ratio of MLVSS/MLSS = 0.80
4. Y = 0.7 lb VSS/BOD_5
5. k_d = 0.06 d^{-1}

7-30. Using the NRC formulas and the data given in Part 1 of Example 7-8, estimate the removal efficiencies and the corresponding organic and hydraulic loading rates if both the first- and second-stage filters are to be the same diameter and depth.

7-31. Using the data given in Part 2 of Example 7-8, estimate the surface area required and the corresponding organic and hydraulic loading rates if the depth of the filter must be limited to 18 ft.

7-32. Estimate the amount of methane gas produced per pound of UBOD stabilized, assuming the starting compound is cell tissue ($C_5H_7NO_2$). Assume the following equation applies:

$$2(C_5H_7NO_2) + 6H_2O \rightarrow 5CO_2 + 5CH_4 + 6NH_3$$

7-33. Estimate the total volume of gas produced from 1000 lb of dry sludge at 55°C and 1 atmosphere pressure. Assume the volatile fraction of the sludge is 75 percent, that 60 percent of the volatile solids is converted to gas, and that the digester gas contains 55 percent methane.

7-34. Determine the detention time and volume of a complete-mix digester, for a sludge flowrate of 50,000 gal/d assuming the following coefficients apply:

1. Y = 0.09 mg VSS/mg COD
2. k_d = 0.035 d^{-1}
3. k = 1.5 d^{-1}

7-35. If the following coefficients apply, determine whether an 18-d detention time will provide a factor of safety of 2.0.

1. Y = 0.12 mg VSS/mg COD
2. k_d = 0.03 d^{-1}
3. k = 1.35 d^{-1}

7-36. A waste with the following characteristics is to be treated in a complete-mix anaerobic digester. Estimate the size of the digester, the sludge production rate, and the degree of stabilization.

1. Flowrate = 20,000 gal/d
2. Influent COD = 4000 mg/L
3. Y = 0.1 mg VSS/mg COD
4. k_d = 0.03 d^{-1}
5. k = 1.4 d^{-1}
6. K_s = 2000 mg COD/L

7-37. Prepare a set of specifications for a complete-mix extended aeration activated-sludge process for the treatment of septic tank effluent from 10 homes. The specifications should include average and peak flowrate and influent and effluent BOD (or COD), TSS, and N.

7-38. Assuming the following conditions apply, estimate how often sludge should be removed form the complete-mix activated-sludge process specified in Prob. 7-37. List clearly all of the assumptions used in your analysis.

1. $Y = 0.85$ mg VSS/mg BOD_5
2. $k_d = 0.055\ d^{-1}$

REFERENCES

Albertson, O. E. (1987) The Control of Bulking Sludges: From the Early Innovators to Current Practice, *Journal of Water Pollution Control Federation,* Vol. 59, p. 172.

Albertson, O. E. (1989) Optimizing Rotary Distributor Speed for Trickling Filters, Water Pollution Control Federation *Operators Forum,* Vol. 2, No. 1.

Albertson, O. E., and G. Davis (1984) Analysis of Process Factors Controlling Performance of Plastic Bio-Media, presented at the 57th Annual Meeting of the Water Pollution Control Federation, New Orleans, October.

Anderson, N. E. (1945) Design of Final Settling Tanks for Activated Sludge, *Sewage Works Journal,* Vol. 17, p. 50.

Ardern, E., and W. T. Lockett (1914) Experiments on the Oxidation of Sewage without the Aid of Filters, *Journal Society of Chemical Industries,* Vol. 33, pp. 523, 1122.

Arora, M. L., and M. B. Humphries (1987) Evaluation of Activated Biofiltration and Activated Biofiltration/Activated Sludge Technologies, *Journal of Water Pollution Control Federation,* Vol. 59, No. 4.

Atkinson, B. (1974a) *Biochemical Reactors,* Pion Limited, London.

Atkinson, B., I. J. Davies, and S. Y. How (1974b) The Overall Rate of Substrate Uptake by Microbial Films, Parts I and II, *Transactions of Institution of Chemical Engineers,* London.

Boncardo, G. S. (1997) Selection and Design of Sewage Collection, Treatment, and Reuse Systems, paper presented at the International Workshop on Sewage Treatment and Reuse for Small Communities, Iraklio, Crete.

Bruce, A. M., and J. C. Merkens (1973) Further Studies of Partial Treatment of Sewage by High-Rate Biological Filtration, *Journal of Institute of Water Pollution Control,* Vol. 72, No. 5, London.

Carolan, A. (1997) Personal communication. Trussville, AL.

Chong, R. (1997) Intermittent Extended Aeration Process, internal publication, Department of Public Works and Services, New South Wales, Sydney, Australia.

Chudoba, J. (1985) Control of Activated Sludge Filamentous Bulking: VI Formulation of Basic Principles, *Water Research,* Vol.19, p.1017.

Chudoba, J., P. Grau, and V. Ottova (1973) Control of Activated Sludge Filamentous Bulking: II Selection of Microorganisms by Means of a Selector, *Water Research,* Vol. 7, p. 1389.

Daigger, G. T., and P. T. Bowen (1996) Economic Considerations on the Use of Fermentors in Biological Nutrient Removal Systems, Proceedings of the Water Environment

Federation Conference Seminar on the Use of Fermentation to Enhance Biological Nutrient Removal, Alexandria, VA.

Dunbar, Professor Dr. (1908) *Principles of Sewage Treatment,* Charles Griffen, London.

Eckenfelder, W. W., Jr. (1963) Trickling Filtration Design and Performance, *Transactions American Society of Civil Engineers,* Vol. 128.

Eckenfelder, W. W. Jr. (1987) Biological Phosphorus Removal: State of the Art Review, *Pollution Engineering,* p. 88.

Eckenfelder, W. W. Jr., and P. Grau (eds.) (1992) *Activated Sludge Process Design and Control Theory and Practice,* Vol. 1, Water Quality Management Library, Technomic Publishing, Lancaster, PA.

Fairall, J. M. (1956) Correlation of Trickling Filter Data, *Sewage and Industrial Wastes,* Vol. 28, No. 9.

Galler, W. S., and H. B. Gotass (1966) Optimization Analysis for Biological Filter Design, *Journal Sanitation Engineering Division,* ASCE, Vol. 92, No. SA1.

Germain, J. E. (1966) Economic Treatment of Domestic Waste by Plastic-Medium Trickling Filter, *Journal of Water Pollution Control Federation,* Vol. 38, pp. 192–203.

Grady, C. P. L., Jr., and H. C. Lim (1980) *Biological Wastewater Treatment: Theory and Application,* Marcel Dekker, New York.

Hawkes, H. A. (1963) *The Ecology of Waste Water Treatment,* MacMillan, New York.

Henze, M., C. P. L. Grady, Jr., W. Gujer, G. v. R. Marias, and T. Matsuo (1987) *Activated Sludge Model No. 1,* IAWPRC Science and Technology Report No. 1, International Association for Water Pollution Research and Control, London.

Higgins, I. J., and R. G. Burns (1975) *The Chemistry and Microbiology of Pollution,* Academic, London.

Holland, K. T., J. S. Knapp, and J. G. Shoesmith (1987) *Anaerobic Bacteria,* Chapman and Hall, New York.

Hoover, S. R., and N. Porges (1952) Assimilation of Dairy Wastes by Activated Sludge, II: The Equation of Synthesis and Oxygen Utilization, *Sewage and Industrial Wastes,* Vol. 24.

IAWQ (1997) Theory, Modelling, Design, and Operation of Secondary Clarifiers, IAWQ Scientific and Technical Report, International Association on Water Quality, London.

Jenkins, D. (1997) Personal communication, Berkeley, CA.

Jenkins, D., M. G. Richard, and G. T. Daigger (1986) Manual on the Causes and Control of Activated Sludge Bulking and Foaming, prepared for Water Research Commission, Republic of South Africa and U.S. EPA, Cincinnati, OH, Water Research Commission, Pretoria, South Africa.

Jenkins, D., and S. W. Hermanowicz (eds.)(1997) *Proceedings Second International Conference on Microorganisms in Activated Sludge and Biofilm Processes,* International Association on Water Quality, Berkeley, CA.

Lawrence, A. W., and P. L. McCarty (1969) Kinetics of Methane Fermentation in Anaerobic Treatment, *Journal of Water Pollution Control Federation,* Vol. 41, No. 2, Part 2.

Lawrence, A. W., and P. L. McCarty (1970) A Unified Basis for Biological Treatment Design and Operation, *Journal of Sanitation Engineering Division,* ASCE, Vol. 96, No. SA3.

Lettinga, G., A. F. M. van Velsen, S. W. Hobma, W. J. de Zeeuw, and A. Klapwijk (1980) Use of the Upflow Sludge Blanket (USB) Reactor Concept for Biological Wastewater Treatment. *Biotechnology and Bioengineering,* Vol. 22, pp. 699–734.

Levine, A. D., G. Tchobanoglous, and T. Asano (1985) Characterization of the Size Distribution of Contaminants in Wastewater: Treatment and Reuse Implications, *Journal of Water Pollution Control Federation,* Vol. 57, No. 7, pp. 205–216.

Living Technologies, Incorporated (1997) Advanced Ecologically Engineered System, South Burlington, VT, Progress Report, Massachusetts Foundation for Excellence, Boston, MA.

Logan, B. E., S. W. Hermanowicz, and D. S. Parker (1987a) Engineering Implications of a New Trickling Filter Model, *Journal of Water Pollution Control Federation,* Vol. 59, No. 12, pp. 1017–1028.

Logan, B. E., S. W. Hermanowicz, and D. S. Parker (1987b) A Fundamental Model for Trickling Filter Process Design, *Journal of Water Pollution Control Federation,* Vol. 59, No. 12, pp. 1029–1042.

Madigan, M. T., J. M. Martinko, and J. Parker (1997) *Brock Biology of Microorganisms,* 8th ed., Prentice Hall, Upper Saddle River, NJ.

Malina, J. F., Jr., and F. G. Pohland (eds.) (1992) *Design of Anaerobic Processes for the Treatment of Industrial and Municipal Wastes,* Vol. 7, Water Quality Management Library, Technomic Publishing, Lancaster, PA.

McCarty, P. L. (1964) Anaerobic Waste Treatment Fundamentals, *Public Works,* Vol. 95, nos. 8–12.

McCarty, P. L. (1966) Kinetics of Waste Assimilation in Anaerobic Treatment, *Developments in Industrial Microbiology,* Vol. 7, American Institute of Biological Sciences, Washington, DC.

McCarty, P. L. (1968) Anaerobic Treatment of Soluble Wastes, in E. F. Gloyna and W. W. Eckenfelder, Jr. (eds.), *Advances in Water Quality Improvement,* University of Texas Press, Austin, TX.

McCarty, P. L. (1971) Energetics and Bacterial Growth, in S. D. Faust and J. V. Hunter (eds.), *Organic Compounds in Aquatic Environments,* Marcel Dekker, New York.

McCarty, P. L. (1975) Stoichiometry of Biological Reactions, *Progress in Water Technology,* Vol. 7, pp. 157–172.

Monod, J. (1942) Recherches sur la croissance des cultures bacteriennes, Herman et Cie., Paris.

Monod, J. (1949) The Growth of Bacterial Cultures, *Annual Review of Microbiology,* Vol. III.

NRC (National Research Council) (1946) Trickling Filters in Sewage Treatment at Military Installations, *Sewage Works Journal,* Vol. 18, No. 5.

Orhan, D., and N. Artan (1994) *Modeling of Activated Sludge Systems,* Technomic Publishing Co., Inc., Lancaster, PA.

Parker, D., T. Jacobs, E. Bower, D. W. Stowe, and G. Farmer (1996a) Maximizing Trickling Filter Nitrification Rates Through Biofilm Control: Research Review and Full Scale Application, in *Proceedings of the Third International IAWQ Special Conference on Biofilm Systems,* London.

Parker, D., et al. (1996b) Flocculator-Clarifiers Bring Performance Benefits to Large Treatment Plants, *Water Science and Technology,* Vol. 33, No. 12.

Ribbons, D. W. (1970) Quantitative Relationships between Growth Media Constituents and Cellular Yields and Composition, in J. W. Norris and D. W. Ribbons (eds.), *Methods in Microbiology,* Vol. 3A, Academic, London.

Sarner, E. (1980) *Plastic Packed Trickling Filters,* Ann Arbor Science Publishers, Inc., Ann Arbor, MI.

Sawyer, C. N., P. L. McCarty, and G. F. Parkin (1994) *Chemistry for Environmental Engineering,* 4th ed., McGraw-Hill, New York.

Schroeder, E. D., and G. Tchobanoglous (1976) Mass Transfer Limitations on Trickling Filter Design, *Journal Water Pollution Control Federation,* Vol. 48, No. 4.

Schultz, K. L. (1960) Load and Efficiency of Trickling Filters, *Journal of Water Pollution Control Federation,* Vol. 33, No. 3, pp. 245–260.

Sedlak, R. I. (ed.) (1991) *Phosphorus and Nitrogen Removal from Municipal Wastewater,* 2nd ed., The Soap and Detergent Association, Lewis Publishers, New York.

Speece, R. E. (1983) *Anaerobic Biotechnology for Industrial Wastewater Treatment, Environmental Science and Technology,* Vol. 17, No. 9, p. 416.

Speece, R. E. (1996) *Anaerobic Biotechnology for Industrial Wastewaters,* Archae Press, Nashville, TN.

Standard Methods (1995) *Standard Methods for the Examination of Water and Waste Water* (1995), 19th ed., American Public Health Association, Washington, DC.

Sutton, P. M. (1990) Anaerobic Treatment of High Strength Wastes: System Configurations and Selection, presented at Anaerobic Treatment of High Strength Wastes, University of Wisconsin, Milwaukee.

Tchobanoglous, G., and F. L. Burton (1991) *Wastewater Engineering Treatment, Disposal, Reuse,* 3rd ed., McGraw-Hill, New York.

Tchobanoglous, G., and E. D. Schroeder (1985) *Water Quality: Characteristics, Modeling, Modification,* Addison-Wesley, Reading, MA.

U.S. EPA (1975) Process Design Manual for Nitrogen Control, Office of Technology Transfer, U.S. Environmental Protection Agency, Washington, DC.

U.S. EPA (1977) Process Design Manual Wastewater Treatment Facilities for Sewered Small Communities, U.S. Environmental Protection Agency, EPA-625/1-77-009, Environmental Research Information Center, Cincinnati, OH.

U.S. EPA (1986) Sequencing Batch Reactors, EPA/625/8-86/011, U.S. Environmental Protection Agency, Cincinnati, OH.

U.S. EPA (1987) Phosphorus Removal, Design Manual, EPA/625/1-87/001, U.S. Environmental Protection Agency, Cincinnati, OH.

U.S. EPA (1989) Design Manual Fine Pore Aeration Systems, EPA/625/1-89/023, U.S. Environmental Protection Agency, Cincinnati, OH.

Van Uden, N. (1967) Transport-Limited Growth in the Chemostat and Its Competitive Inhibition; A Theoretical Treatment, *Archiv für Mikrobiologie,* Vol. 58.

Velz, C. J. (1948) A Basic Law for the Performance of Biological Beds, *Sewage Works Journal,* Vol. 20, No. 4.

WEF (1992a) *Design of Municipal Wastewater Treatment Plants,* Vol. I: Chaps. 1–12, WEF Manual of Practice No. 8, ASCE Manual and Report on Engineering Practice No. 76, Water Environment Federation, Alexandria, VA.

WEF (1992b) *Design of Municipal Wastewater Treatment Plants,* Vol. II: Chaps. 13–20, WEF Manual of Practice No. 8, ASCE Manual and Report on Engineering Practice No. 76, Water Environment Federation, Alexandria, VA.

WPCF (1983) Nutrient Control, Manual of Practice No. FD-7, Water Pollution Control Federation, Washington, DC.

WPCF (1987) Activated Sludge, Manual of Practice No. OM-9, Water Pollution Control Federation, Alexandria, VA.

WPCF (1987) Anaerobic Sludge Digestion, Manual of Practice No. 16, 2nd ed., Water Pollution Control Federation, Alexandria, VA.

Wilson, T. E., and J. S. Lee (1982) Comparison of Final Clarifier Design Techniques, *Journal of Water Pollution Control Federation,* Vol. 54, p. 1376.

Wilson, T. E., and T. M. Keinath (1990) Secondary Clarifier Design Standards, an issue paper of the ASCE Clarifier Research Technical Committee, ASCE Environmental Division Conference.

Wilson, T. E. (1996) A New Approach to Interpreting Settling Data, in *Proceedings WEFTEC 96, 69th Annual Conference and Exposition,* Vol. 1, Part 1, Wastewater Treatment Research, Water Environment Federation, Alexandria, VA.

CHAPTER 8

Lagoon Treatment Systems

Lagoon treatment systems are earthen basins or reservoirs that are engineered and constructed to treat wastewater. Work on lagoons in the 1940s led to the development of wastewater treatment lagoons as low-cost alternatives (McGauhey, 1968; Marais, 1970). In practice, the terms *lagoon* and *pond* are used interchangeably.

8-1 TYPES OF LAGOON SYSTEMS

Wastewater treatment lagoons range in depth from shallow to deep, and are often categorized by their aerobic (dissolved oxygen concentration) status and the source of that oxygen for bacterial assimilation of wastewater organics. The four major types of lagoon systems are classified with respect to the presence and source of oxygen in Table 8-1.

Another method of classifying lagoons, which can include all types of lagoons, is based on the frequency and duration of their effluent discharges:

- Total containment lagoons
- Controlled discharge lagoons
- Hydrograph controlled release lagoons
- Continuous discharge lagoons

The total containment or evaporation lagoon should be considered only where evaporation rate exceeds the precipitation rate on an annual basis. The controlled discharge concept is to select a time of year when stream flow conditions are satisfactory for accepting a lagoon effluent discharge. The hydrograph controlled release (HCR) lagoon, a variation of the controlled discharge concept, is designed with a discharge that is correlated to the stream flow. As with the controlled discharge lagoons, the HCR lagoon discharges only when the stream flow is above a minimum acceptable

TABLE 8-1
Classification of lagoon systems based on the presence and source of oxygen

Type of lagoon	Presence of oxygen
Aerobic	Photosynthesis provides oxygen for aerobic conditions throughout the water column.
Facultative	Surface zone is aerobic. Subsurface zone may be anoxic or anaerobic.
Partial-mix aerated	Surface aeration produces aerobic zone that ranges from half depth to total depth depending on oxygen input and lagoon depth.
Anaerobic	Entire depth is anaerobic.

value. Most HCR and controlled release lagoons are facultative lagoons. All types of lagoons can be operated as continuous discharge lagoons.

Applicability

Lagoon technology is used primarily for small rural communities; however, aerated lagoons and facultative lagoons are often used in medium-sized communities, especially in the West. Over 7000 lagoon systems are in use in the United States for treatment of municipal and industrial wastewater (Crites, 1992). These lagoon systems are used alone, or in combination with other wastewater treatment systems.

The advantages of lagoons include:

- Low capital costs.
- Minimum operations and operational skills needed.
- Sludge withdrawal and disposal needed only at 10- to 20-yr intervals.
- Compatibility with land and aquatic treatment processes.

The disadvantages of lagoons include:

- Large land areas may be required.
- High concentrations of algae may be generated, which can be problematic for surface discharge.
- Nonaerated lagoons often cannot meet stringent effluent limits.
- Lagoons can impact groundwater negatively if liners are not used, or if liners are damaged.
- Improperly designed and operated lagoons can become odorous.

Aerobic Lagoons

Aerobic lagoons are quite shallow to allow light to penetrate the full lagoon depth. As a result, aerobic lagoons have active photosynthetic activity through the entire water column during daylight hours. Typical lagoon depths range from 1 to 2 ft (0.3

to 0.6 m). Lagoons designed to maximize the photosynthetic activity of algae are also called *high-rate lagoons.* The use of the term *high rate* refers to the photosynthetic oxygen production rate of the algae present, not to the metabolic rate, which does not change. The photosynthetic oxygen allows bacteria to degrade organics aerobically. During daylight hours, dissolved oxygen and pH values rise and peak, followed by a drop in the hours of darkness. Detention times are relatively short, with 5 d being typical. Aerobic lagoons are used in combination with other lagoons (see Sec. 8-7) and are limited to warm, sunny climates (Reed et al., 1995).

Facultative Lagoons

Facultative lagoons are the most common and versatile of the different lagoon types. An example of a facultative lagoon is shown in Fig. 8-1. They are 5 to 8 ft (1.5 to 2.5 m) deep and are also referred to as *oxidation ponds* or *stabilization lagoons.* Treatment is accomplished by bacterial action in an upper aerobic layer and a lower layer that can be anoxic or anaerobic, depending on wind-induced mixing. Settleable solids are deposited on the lagoon bottom. The two layers are shown schematically in Fig. 8-2. Oxygen is provided by natural surface aeration and photosynthesis. Facultative lagoons can be used as controlled discharge lagoons, total containment lagoons, and storage lagoons for land treatment systems.

Partial-Mix Aerated Lagoons

Partial-mix aerated lagoons are deeper and more heavily loaded organically than facultative lagoons. Oxygen is supplied typically through floating mechanical aerators or submerged diffused aeration. Aerated lagoons are 6 to 20 ft (2 to 6 m) deep with

FIGURE 8-1
Example of facultative lagoon used for treatment of domestic wastewater.

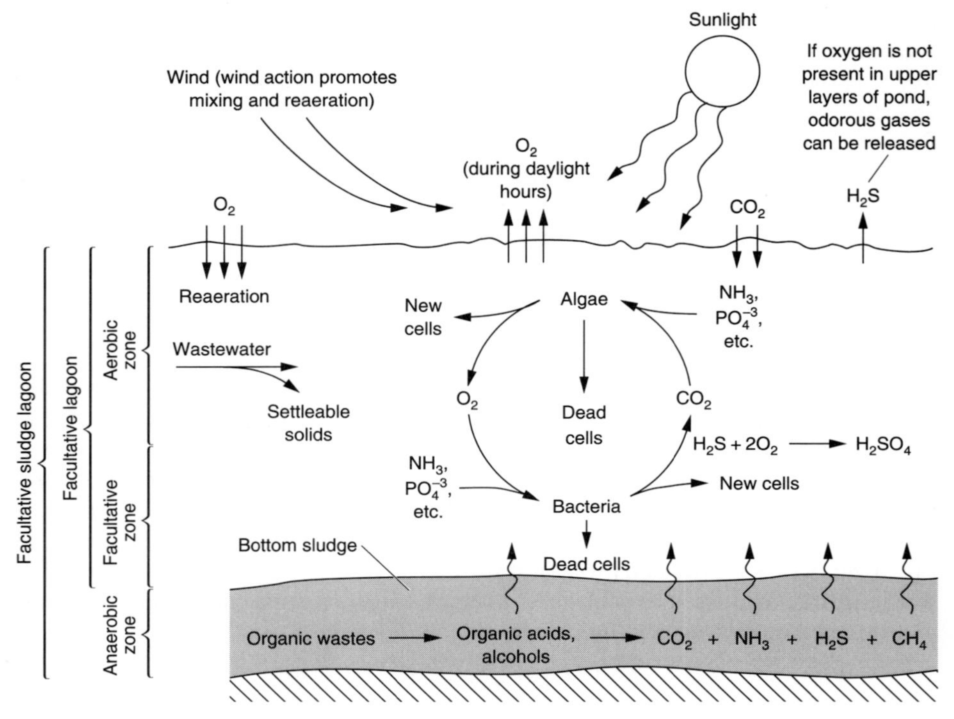

FIGURE 8-2
Definition sketch for the interactions occurring in a facultative lagoon (from Tchobanoglous and Schroeder, 1985).

detention times that range from 3 to 20 d. The principal advantage of aerated lagoons is that they require less land area than other lagoon systems. Partial-mix aerated lagoons are popular in the United States because they have the advantages of facultative lagoons (minimal sludge generation and handling), but have a lesser land requirement.

Anaerobic Lagoons

Anaerobic lagoons are used for high-strength, typically industrial wastewater in remote rural areas. They have no aerobic zones, are 15 to 30 ft (5 to 10 m) in depth, and have detention times of 20 to 50 d. Because of the potential for odor production, anaerobic lagoons need to be covered or sited away from populated areas.

Comparison of Lagoon Types

Because of their widespread use, facultative lagoons, partial-mix aerated lagoons, anaerobic lagoons, and combinations of the lagoon systems will be discussed in the remaining sections of this chapter. The following discussion will include design criteria, process performance, and physical design factors.

8-2 FATE PROCESSES FOR WASTEWATER CONSTITUENTS

The mechanisms for removal of BOD, TSS, nitrogen, phosphorus, and pathogens are presented in this section.

BOD Removal

Lagoons are low-mass biological reactors. For all but anaerobic lagoons, the soluble BOD is reduced by bacterial oxidation. Particulate BOD is removed by sedimentation. In facultative and anaerobic lagoons, anaerobic biological conversion occurs. BOD removal in lagoons depends on the detention time and lagoon water temperature.

Total Suspended Solids Removal

The influent suspended solids are removed by sedimentation in lagoon systems. Algal solids that develop during treatment become the majority of the effluent suspended solids. Effluent suspended solids can range as high as 140 mg/L for aerobic lagoons to 60 mg/L for aerated lagoons. If slow-rate land treatment reuse or land application follow the lagoons, algal suspended solids are of little concern. However, because most algal solids are difficult to remove from water and effluent standards often cannot be met, additional processes may be needed to remove the solids.

Alternative processes that can be used to upgrade lagoon effluents for TSS removal include:

- Intermittent sand filters
- Microstrainers
- Rock filters
- Dissolved air flotation (DAF)
- Floating aquatic plants
- Constructed wetlands

A discussion of the first four processes is presented at the end of this chapter, and the latter two processes are described in Chap. 9. Intermittent sand filters are described in more detail in Chap. 11.

Nitrogen Removal

Nitrogen removal in lagoons appears to be the result of a combination of mechanisms including volatilization of ammonia (which is pH-dependent), algal uptake,

TABLE 8-2
Nitrogen removal in facultative lagoons*

Location	Theoretical detention time, d	Median water temp., °C	Median pH	Median alkalinity, mg/L	Influent nitrogen, mg/L	Median removal, %
Corrine, Utah first 3 cells	42	10	9.4	555	14.0	46
Eudora, Kansas 3 cells	231	14.7	8.4	284	50.8	82
Kilmichael, Mississippi 3 cells	214	18.4	8.2	116	35.9	80
Peterborough, New Hampshire 3 cells	107	11	7.1	85	17.8	43

*From Reed et al. (1995).
Note: Samples taken after 3rd cell for % removal.

nitrification/denitrification, sludge deposition, and adsorption onto bottom soils. Nitrogen removal in four facultative lagoons is presented in Table 8-2.

Phosphorus Removal

Without the addition of chemicals for precipitation, the removal of phosphorus in lagoons is minimal. Chemical addition using alum or ferric chloride has been used effectively to reduce phosphorus to below 1 mg/L (Reed et al., 1995).

Application of chemicals can be on either a batch or continuous-feed basis. For controlled release lagoons, the batch process is appropriate. The State of Minnesota has 11 facultative lagoon systems that use the addition of liquid alum directly into secondary cells via motorboat to meet a spring and fall discharge limitation of 1 mg/L (Surampalli et al., 1993).

For continuous-flow applications, a mixing chamber is often used between the last two lagoons or between the last lagoon and a clarifier. In Michigan, both aerated lagoons and facultative lagoons have been used with continuous-flow applications. Influent phosphorus concentrations for 21 treatment facilities ranged from 0.5 to 15 mg/L with an average of 4.1 mg/L, and the effluent target is 1 mg/L (Surampalli et al., 1993).

Pathogen Removal

Significant removal of bacteria, parasites, and viruses occurs in multiple cell lagoons with long detention times. Removal of pathogens in lagoons is due to natural die-off, predation, sedimentation, and adsorption. Helminths and parasitic cysts and eggs settle to the bottom in the quiescent zone of lagoons. Facultative lagoons with three

TABLE 8-3
Fecal coliform removal in lagoon systems*

Location	Number of cells	Detention time, d	Fecal coliform, MPN/100 mL	
			Influent	Effluent
Facultative lagoons				
Peterborough, New Hampshire	3	107	4.3×10^6	3.6×10^5
Eudora, Kansas	3	231	2.4×10^6	2.0×10^2
Kilmichael, Mississippi	3	214	12.8×10^6	2.3×10^4
Corrine, Utah	7	42	1.0×10^6	7.0×10^0
Partial-mix aerated lagoons				
Windber, Pennsylvania	3	48	1×10^6	3.0×10^2
Edgerton, Wisconsin	3	73	1×10^6	3.0×10^1
Pawnee, Illinois	3	144	1×10^6	3.3×10^1
Gulfport, Mississippi	2	18	1×10^6	1.0×10^5

*Adapted from Reed et al. (1995).

cells and about 20 days detention time, and aerated lagoons with a separate settling cell prior to discharge, will provide more than adequate helminth and protozoon removal. Data on removal of fecal coliforms are presented in Table 8-3.

Comparison of Lagoon Design Features and Performance

A comparison of design features and typical performance is presented in Table 8-4. Actual performance can vary significantly from location to location and from season to season.

TABLE 8-4
Design features and typical performance of lagoon systems*

Feature	Type of lagoon		
	Aerobic	Facultative	Partial-mix aerated
Treatment goals	Secondary	Secondary, preliminary to land treatment	Secondary, polishing, preliminary to land or aquatic treatment
Climate needs	Warm	None	None
Detention time, d	10–40	25–180	5–20
Depth, ft	3–5	5–8	10–20
Organic loading rate, lb/ac·d	35–110	20–60	40–360
Effluent characteristics, mg/L			
BOD, mg/L	20–40	30–40	20–40
TSS, mg/L	80–140	40–100	30–60

*Adapted from Reed et al. (1995).

8-3 PRELIMINARY TREATMENT OF WASTEWATER

The level of treatment required prior to the lagoon ranges from none to primary treatment. The advantage of preliminary treatment is to minimize the floatable material that can cause a nuisance in the first lagoon. For institutional wastewaters it is common to use screens or macerating devices to reduce the size of rags and other floatable material. For residential wastewater it is common to use only enough screening to protect headworks and pumping stations from large objects and rags. Flow monitoring, sampling, and pumping stations are typically the only preliminary devices prior to wastewater discharge into lagoons (Reed et al., 1995).

8-4 FACULTATIVE LAGOONS

Facultative lagoons are generally designed based on BOD loading rate. The concept is to design the lagoons for a long enough detention time and a low enough organic loading rate to achieve aerobic conditions in the surface layer of the lagoon. Settled solids are digested in the lower anaerobic zone of the facultative lagoon. During the spring and fall, the thermal turnover of the lagoon contents will often result in the settled solids being resuspended. Resuspension of settled solids can also occur if the water temperature in the organic sediments increases to above 22°C, because of anaerobic gas production. Although suspended solids are present in lagoon effluents, they consist of suspended algae and do not include suspended organic matter associated with the influent wastewater (Reed et al., 1995).

Lagoon Dynamics

A number of variables affect the performance of facultative lagoons, including algal growth and decay, wind mixing, temperature, and thermal turnover. To appreciate the complex interactions that occur in facultative lagoons, it is useful to consider the growth and decay of several types of algae.

Facultative and aerobic lagoons typically generate large concentrations of algae. The dynamics of growth and decay, and the dominance of four classes of freshwater algal types, affect the levels of suspended solids in most lagoon effluents. The four classes of algae are:

1. *Green.* Common green algae are those of the *Chlorella* group found in lagoons.
2. *Motile green. Euglena* is a member of this particular group of algae. The presence of motile green algae usually means the lagoon is overloaded organically.
3. *Yellow green or golden brown.* Of this group of algae, the most important are the diatoms.
4. *Blue-green.* These unicellular algae have the ability, when dying or dead, to form dense mats on the water surface that do not settle readily. They can also fix nitrogen from the atmosphere. Blue green algae include *Anabaena, Anacystis,* and *Oscillatoria.*

The dominant form of algae in a lagoon often changes with the seasons. When blue-green algae predominate, problems with settleability of the algal solids often occur. The variation of algal species in a facultative lagoon in Davis, California, is shown in Table 8-5. Note that the peak concentration of the blue-green algae *Anacystis* in late April and early May corresponds with an effluent suspended solids concentration peak of 178 mg/L.

TABLE 8-5
Variation of algal species in facultative lagoon effluent Davis, California, 1975*

	Jan 8	Jan 29	Feb 7	Mar 4	Mar 31	Apr 18	Apr 21	Apr 30
Algal genera								
Aphanizomenon	12,400[†]	12,700	12,000	11,800	11,000	8400	3300	1200
	130[‡]	130	130	160	160	160	160	160
Anacystis	0	0	0	0	0	200	2100	12,700
						5	5	5
Ankistrodesmus	200[¶]	300[¶]	400[¶]	1200[¶]	400[¶]	100[¶]	0	0
Chlamydomonas	0	0	0	0	0	0	0	0
Chlorella	0	0	0	0	0	0	0	0
Golenkinia	Trace	400	700	200	200	800	500	Trace
		5	5	7	7	10	10	7
Scenedesmus	400	500	700	700	1100	2100	3000	1600
	3 × 7	3 × 7	5 × 10	5 × 10	5 × 10	7 × 15	7 × 15	7 × 15
Pond quality								
COD, mg/L	128	108	111	134	146	210	243	302
SS, mg/L	54	61	51	50	55	130	131	166

	May 20	Jun 12	Jul 23	Sep 4	Sep 29	Oct 15	Oct 22
Algal genera							
Aphanizomenon	600	300	0	200	5200	25,600	15,600
	50	25		10	20	25	25
Anacystis	11,500	9400	7900	8300	900[§]	100[§]	0
	25	60	60	70			
Ankistrodesmus	0	0	0	100	1200	600	100
				10	20	25	15
Chlamydomonas	0	0	100	400	1100	800	600
			3	3	3	3	3
Chlorella	0	0	Trace	200	47,000	1400	0
				5	5	5	
Golenkinia	Trace	0	0	0	0	0	0
	5						
Scenedesmus	0	0	0	200	200	300	300
				3 × 7	3 × 7	3 × 7	3 × 7
Pond quality							
COD, mg/L	327	315	285	240	193	194	185
SS, mg/L	178	161	150	142	88	81	66

*From Stowell (1976).
[†] First row of data for a genus indicates colonies or cells per mL.
[‡] Second row of data for a genus indicates average size, μm.
[§] Very irregular shape.
[¶] Fusiform shape.

Temperature changes in facultative (nonaerated) lagoons may result in vertical stratification of the lagoon profile during certain seasons of the year, if wind action is not sufficient to keep the lagoon mixed. The stratification is caused by an increase in water density with depth because of the decrease in temperature. During summer, the upper zone of the lagoon profile warms and the density of the upper water decreases, resulting in two stratified layers. When the fall temperatures drop, stratification is decreased and the lagoon can be mixed by wind action. As a result, the lower zones, with little oxygen, are mixed with the upper zones in what is known as the *fall turnover*. Odor events can occur during the turnover, until aerobic surface conditions are reestablished.

Design Procedures

Several methods or models are available for the design of facultative lagoons. In each method a number of design factors are included: detention time, temperature, and reaction rate. The methods are:

- Areal loading rate method
- Plug flow with axial dispersion
- Multiple correlation method
- Complete-mix model
- Plug-flow model

Areal loading rate method. Areal BOD loading rates are based on average air temperatures for the coldest month of the year. The recommended BOD loading rates for various ranges of temperatures are given in Table 8-6. Individual states have maximum BOD loading rates for lagoons, and the designer should check with the appropriate state agency for current standards.

The first lagoon in a series of lagoons is usually limited to a BOD loading of 35 lb/ac·d (40 kg/ha·d). For warm climates, in which the minimum monthly air temperature is higher than 59°F (15°C), loadings on the primary lagoon cell as high as 80 lb/ac·d (90 kg/ha·d) may be allowed.

The areal loading rate method is used by many of the western states, and requires the least amount of design data. It is based on operational experiences in different geographical areas of the United States. This empirical database takes into account overturns, rising sludge, and seasonal algal concentration variations that can affect

TABLE 8-6
Area loading rate values for different temperatures*

	BOD loading rate	
Average winter air temperature, °C	kg/ha·d	lb/ac·d
>15	45–90	40–80
0–15	22–45	20–40
<0	11–22	10–20

*From U.S. EPA (1983).

treatment. The areal loading method is recommended over the other methods unless comparable systems are available upon which to base rate constants for the other models.

Plug flow with axial dispersion. The plug flow with axial dispersion model is introduced in Chap. 3, and the design equation used is presented as Eq. (3-63). Recognizing that the flow regime through lagoon systems is somewhere between plug flow and complete mix, Thirimurthi proposed the arbitrary flow model developed by Wehner and Wilhelm (Thirimurthi, 1974). Because dispersion is related to wind mixing, the level of dispersion will vary with changing wind conditions.

The dimensionless term k_T is shown in Fig. 3-14 plotted against the percent BOD remaining for dispersion numbers ranging from zero for an ideal plug flow unit to infinity for a completely mixed unit. Dispersion numbers that have been measured in wastewater lagoons range from 0.1 to 2.0, with most values less than 1.0 (Reed et al., 1995). A typical dispersion factor for planning purposes is 0.5. The selection of a value for D will significantly affect the detention time required to produce a given effluent quality.

The value of k_T to use depends on the temperature. At 20°C the value for k_T is 0.15 d^{-1}. Use Eq. (3-22) to calculate the value k for other temperatures. Use a θ value of 1.04.

Multiple correlation method. The multiple correlation method developed by Gloyna is applicable only for 80 to 90 percent removal efficiency for BOD and requires lagoon temperature, an algal toxicity factor, and a sulfide oxygen demand factor (Gloyna, 1976). This method is complex and is not described further.

Complete-mix model. The complete-mix model was developed by Marais and Shaw (Marais and Shaw, 1961). It is based on first-order kinetics and has a proposed upper limit for the BOD_5 concentration of 55 mg/L to avoid anaerobic conditions and odors. The lagoon depth of the first cell is adjusted for influent values above 55 mg/L.

Plug-flow model. The plug-flow model is used to relate BOD removal, detention time, and apparent first-order removal rates:

$$C_e/C_o = \exp[-k_T t] \tag{8-1}$$

where C_e = effluent BOD_5 concentration, mg/L
C_o = influent BOD_5 concentration, mg/L
k_T = apparent first-order removal rate, d^{-1}
t = hydraulic detention time, d

A typical value of k_T is 0.1 for 20°C.

Removal rate constants also vary with the water temperature as presented in Eq. (3-22). The value for θ is 1.04. Design of a lagoon based on the plug-flow model is illustrated in Example 8-1.

EXAMPLE 8-1. DESIGN OF A FACULTATIVE LAGOON. Design a facultative lagoon for an average flow of 946 m^3/d (0.25 Mgal/d) to achieve an effluent BOD of 30 mg/L with an influent BOD of 220 mg/L. The coldest month has an average air temperature of 5°C and an average water temperature of 7°C. Compare the detention times for both the area loading rate method and the plug-flow with axial dispersion method using the Wehner and Wilhelm equation. Use a dispersion factor of 0.5.

Solution

1. For the area loading rate method, select an organic loading rate of 35 kg/ha·d from Table 8-6 for the temperature range of 0 to 15°C.
2. Calculate the lagoon area by dividing the BOD load by the value of 35 kg/ha·d:

$$\text{BOD loading} = \left(\frac{946\ \text{m}^3}{\text{d}}\right)\left(\frac{220\ \text{mg}}{\text{L}}\right)\left(\frac{1000\ \text{L}}{\text{m}^3}\right)\left(\frac{\text{kg}}{10^6\ \text{mg}}\right) = 208.1\ \text{kg/d}$$

$$\text{Area} = \frac{\text{BOD load}}{\text{loading rate}} = \frac{208.1\ \text{kg/d}}{35\ \text{kg/ha·d}} = 5.95\ \text{ha}$$

3. Use a lagoon depth of 2 m to calculate the detention time:

$$t = \frac{\text{volume}}{\text{flowrate}} = \left(\frac{5.95\ \text{ha} \times 2\ \text{m}}{946\ \text{m}^3/\text{d}}\right)\left(\frac{10{,}000\ \text{m}^2}{\text{ha}}\right) = 126\ \text{d}$$

4. For the plug-flow with axial dispersion method, use the dispersion factor of 0.5 and enter Fig. 3-14 at 13.6 percent. The corresponding value from Fig. 3-14 for kt is 3.3.
5. Calculate the value of k for a temperature of 7°C.

$$k = 0.15(1.04^{(7-20)}) = 0.090\ \text{d}^{-1}$$

6. Calculate the detention time by dividing the value of kt by the value of k.

$$t = \frac{kt}{k} = \frac{3.3}{0.090\ \text{d}^{-1}} = 36.7\ \text{d}$$

Nitrogen Removal

Two equations, Eq. (8-2) and Eq. (8-3), have been developed to relate nitrogen removal to pH, detention time, and temperature for facultative lagoons. The primary mechanism for nitrogen removal in facultative lagoons is ammonia volatilization.

$$N_e = N_o \exp\{-k_T[t + 60.6(\text{pH} - 6.6)]\} \tag{8-2}$$

where N_e = effluent total nitrogen, mg/L
N_o = influent total nitrogen, mg/L
k_T = temperature-dependent removal coefficient, $\text{d}^{-1} = k_{20}\theta^{(T-20)}$
k_{20} = removal coefficient at 20°C, $\text{d}^{-1} = 0.0064$
θ = 1.039
T = water temperature, °C
t = time, d
pH = 7.3 exp[0.0005(Alk)]
Alk = expected influent alkalinity, mg/L

$$N_e = N_o \left[\frac{1}{1 + t(0.000576T - 0.00028)\ \exp[(1.08 - 0.042T)(\text{pH} - 6.6)]} \right] \quad (8\text{-}3)$$

where all terms are as defined above.

The first-order nitrogen removal relationships can be used to predict the detention time necessary for nitrogen removal. They can also be used to predict the nitrogen removal that can be expected in lagoons that may be used as preapplication treatment for land treatment. The first equation [Eq. (8-2)] is based on plug flow, and the second equation [Eq. (8-3)] is based on a complete-mix reactor.

Equation (8-2) (Reed, 1985) was developed for facultative lagoons, assuming plug flow. The pH value is calculated from the alkalinity concentration, using the relationship provided above. The water temperature is used to adjust the k_T value. The k_T value is an apparent k value that includes dispersion. The relationship was validated using lagoon data from Canada and Gulfport, Mississippi, to cover a range of pH values from 6.4 to 9.5, temperatures from 1 to 28°C, and detention times from 5 to 231 d (Reed, 1985).

Equation (8-3) (Pano and Middlebrooks, 1982) was also developed for facultative lagoons, assuming complete-mix reactors. Both relationships can be used to predict the removal of total nitrogen, but Pano and Middlebrooks (1982) also have an equation for estimating ammonia removal. The calculation of nitrogen removal is illustrated in Example 8-2.

EXAMPLE 8-2. NITROGEN REMOVAL IN FACULTATIVE LAGOONS. A facultative lagoon has a theoretical detention time of 100 days. The water temperature averages 22°C and the pH is 7.6. Estimate the effluent total nitrogen concentration if the influent total nitrogen is 30 mg/L. Compare the results from the two equations.

Solution

1. Use Eq. (3-22) to calculate the value of k_{22} for Eq. (8-2):

$$k_T = k_{20}(\theta)^{T-20}$$

$$k_{22} = 0.0064\,(1.039)^2 = 0.0069$$

2. Use Eq. (8-2) to calculate the value of N_e:

$$\begin{aligned} N_e &= N_o \exp\{-k_T[t + 60.6\,(\text{pH} - 6.6)]\} \\ &= 30 \exp\{-0.0069[100 + 60.6\,(7.6 - 6.6)]\} \\ &= 30 \exp(-1.108) = 9.9 \text{ mg/L} \end{aligned}$$

3. Calculate the value of N_e using Eq. (8-3):

$$N_e = N_o \left[\frac{1}{1 + t(0.000576T - 0.00028) \exp[(1.08 - 0.042T)(\text{pH} - 6.6)]} \right]$$

$$= 30 \left[\frac{1}{1 + 100\,(0.000576 \times 22 - 0.00028) \exp[(1.08 - 0.042 \times 22)(7.6 - 6.6)]} \right]$$

$$= 30 \left[\frac{1}{1 + 1.2392 \exp(0.156)} \right] = 12.3 \text{ mg/L}$$

Comment. Clearly the two models predict somewhat different results. A conservative approach would be to use the higher concentration of 12.3 mg/L predicted by Eq. (8-3).

8-5 PARTIAL-MIX AERATED LAGOONS

Aeration of lagoon systems is usually provided to meet the oxygen requirements for BOD removal and, in some cases, nitrification. Aerated lagoons described in this chapter are partially mixed, not the complete mix that would be associated with the activated-sludge process. A combination of lagoons with varying degrees of mixing, known as dual-power multicellular (DPMC) aerated lagoons, is described at the end of this section. Examples of partial-mix aerated lagoons are shown in Fig. 8-3.

Design Procedure

Even though partial-mix lagoons are not mixed completely, it is common practice to use the complete-mix model and first-order reaction-rate kinetics to approximate their performance. The design equation for a series of n equal-sized lagoons is

$$t = (n/k)[(C_o/C_n)^{1/n} - 1] \tag{8-4}$$

where C_n = effluent BOD concentration from cell n, mg/L
C_o = influent BOD concentration to first cell, mg/L
k = BOD removal-rate constant, d^{-1}
t = hydraulic detention time, d
n = number of cells in the series

The apparent reaction-rate constant is assumed to be the same for each cell, and the lagoons should be of equal size. A k value of 0.276 d^{-1} at 20°C is recommended

(*a*)

(*b*)

FIGURE 8-3
Examples of partial-mix aerated lagoons used for: (*a*) domestic wastewater and (*b*) food processing wastewater.

in the Ten States Recommended Standards (Ten States, 1990). The influence of temperature on the reaction rate is shown in Eq. (3-22). Use a θ value of 1.036.

The water temperature of the lagoon can be estimated by

$$T_w = \frac{AfT_a + QT_i}{Af + Q} \tag{8-5}$$

where T_w = lagoon water temperature, °C (°F)
A = surface area of lagoon, m^2 (ft^2)
f = proportionality factor = 0.5
T_a = ambient air temperature, °C (°F)
Q = wastewater flowrate, m^3/d (gal/d)
T_i = influent water temperature, °C (°F)

To calculate the lagoon temperature, estimate the lagoon area from the detention time by using Eq. (8-4). If the temperature is significantly below 20°C, a few iterations will be needed to correct the k factor, determine the appropriate lagoon area, and calculate the lagoon water temperature. The procedure is illustrated in Example 8-3.

The number of cells used in series affects the detention time needed for treatment. For example, if the detention time needed with one cell is 20.2 d, the needed detention times, if multiple cells are used, are as follows:

$n = 2 \quad t = 11.0$ d
$n = 3 \quad t = 9.4$ d
$n = 4 \quad t = 8.7$ d
$n = 5 \quad t = 8.2$ d
$n = 6 \quad t = 8.0$ d

As can be seen in the above array, the benefit of multiple cells decreases quickly after about 4 cells.

EXAMPLE 8-3. DESIGN OF A PARTIAL-MIX AERATED LAGOON. Determine the lagoon temperature for a 4-cell, partial-mix aerated lagoon with a flowrate of 2270 m^3/d (0.6 Mgal/d), an influent water temperature of 15°C (59°F), and an ambient air temperature for the coldest month of 5°C (41°F). Assume the BOD reduction required is from 175 to 20 mg/L, and that the BOD removal rate constant $k = 0.28\ d^{-1}$.

Solution

1. Solve Eq. (8-4) for a 4-cell lagoon system:

$$t = n/k[(C_o/C_n)^{1/n} - 1] = 4/0.28[(175/20)^{1/4} - 1] = 10.3 \text{ d}$$

2. Calculate the surface area of the lagoon using a depth of 3 m (10 ft):

$$\text{Area} = \frac{\text{volume}}{\text{depth}} = \frac{tQ}{d} = \frac{(10.3 \text{ d})(2270 \text{ m}^3/\text{d})}{3 \text{ m}} = 7794 \text{ m}^2$$

3. Calculate the lagoon temperature using Eq. (8-5):

$$T_w = \frac{AfT_a + QT_i}{Af + Q} = \frac{(7794)(0.5)(5) + (2270)(15)}{(7794)(0.5) + 2270} = 8.7\ °\text{C}$$

4. Correct the k factor for temperature. Use 8°C for T.

$$k_T = k_{20}(1.036)^{T-20} = 0.276(1.036)^{-12} = 0.18\ \mathrm{d}^{-1}$$

5. Recalculate the lagoon area for the new k value:

$$t = (10.3\ \mathrm{d})\left(\frac{0.28}{0.18}\right) = 16.0\ \mathrm{d}$$

$$\text{Area} = \left(\frac{16\ \mathrm{d} \times 2270\ \mathrm{m^3/d}}{3}\right) = 12{,}100\ \mathrm{m}^2 = 1.2\ \mathrm{ha}$$

6. Recalculate the lagoon temperature:

$$T_w = \frac{AfT_a + QT_i}{Af + Q} = \frac{(12{,}100)(0.5)(5) + (2270)(15)}{(12{,}100)(0.5) + 2270} = 7.7°\mathrm{C}\ (46°\mathrm{F})$$

Comment. Because the value of 7.7 is close to the assumed 8 for the reaction-rate coefficient, the lagoon temperature is 7.7°C.

Aeration Requirements

Aeration of partial-mix lagoons is accomplished by either floating surface mechanical aerators or submerged diffused aeration systems. The key to the successful use of partial-mix aerated lagoons is to apply the aeration to meet the oxygen demand of the wastewater and to apply that aeration in the upper portion of the lagoon, allowing the lower portion to remain quiescent to enhance sedimentation and anaerobic decomposition. The requirements for oxygen can be calculated using the following form of Eq. (7-94):

$$\mathrm{SOTR} = \frac{\mathrm{AOTR}}{[(\beta C_{sTH} - C_L)/C_{s20}]1.025^{T-20}(\alpha)} \tag{8-6}$$

where SOTR = standard oxygen transfer rate to tap water at 20°C and zero dissolved oxygen, lb/h (kg/h)

AOTR = actual oxygen transfer rate required, under field conditions, to treat the wastewater, lb/h (kg/h)

β = salinity-surface tension correction factor, typically 0.95 to 0.98.

C_{sTH} = dissolved oxygen saturation concentration in clean water at temperature T and altitude H (see App. E), mg/L

C_L = operating dissolved oxygen concentration (typically 2 mg/L)

C_{s20} = dissolved oxygen saturation concentration in clean water at 20°C and 1 atm, mg/L

T = wastewater temperature, °C

α = oxygen transfer correction factor for waste, typically 0.6 to 0.9 (see also Table 7-12)

Surface aeration equipment can transfer from 1.5 to 3 lb/(hp-h) (0.9 to 1.9 kg/kW-h) of oxygen into wastewater. The manufacturers of diffused aeration equipment often recommend an oxygen transfer value of 4.4 lb/(hp-h) (2.7 kg/kW-h) (Reed et al., 1995). These values can be used for preliminary selection of aeration equipment; however, more detailed analyses should be made of power requirements during design. The calculation of aeration requirements is illustrated in Example 8-4.

EXAMPLE 8-4. AERATION REQUIREMENTS FOR A PARTIAL-MIX AERATED LAGOON. Design a partial-mix aerated lagoon to treat 1514 m^3/d (0.4 Mgal/d). Select a 5-cell system to reduce the BOD from 200 to 30 mg/L under the following conditions:

$$k_{20} = 0.28\ \text{d}^{-1}$$

$$T_a = 5°\text{C (lowest average winter air temperature)}$$

$$T = 25°\text{C (summer air temperature)}$$

$$T_i = 20°\text{C}$$

$$\text{Elevation} = 100\ \text{m}$$

$$d = 3\ \text{m}$$

Solution

1. Determine the detention time and volume per cell.
   a. Assume a lagoon temperature of 10°C and calculate the detention time using Eq. (8-4):

$$k = 0.28(1.036)^{(10-20)} = 0.196\ \text{d}^{-1}$$

$$t = (5/0.196)[(200/30)^{1/5} - 1] = 11.8\ \text{d}$$

   b. Determine the volume per cell based on 5 cells:

$$\text{Volume per cell} = \left(\frac{11.8\ \text{d}}{5\ \text{cells}}\right)\left(\frac{1514\ \text{m}^3}{\text{d}}\right) = 3573\ \text{m}^3$$

2. Calculate the lagoon dimensions. Use a rectangular lagoon with a length-to-width ratio of 2:1 and Eq. (8-7). Use a sideslope of 3:1.

$$V(6/d) = (2W)(W) + (2W - 2(3)(3)(W - 2(3)(3)) + 4[2W - 3(3)][W - 3(3)]$$

$$2V = 2W^2 + (2W - 18)(W - 18) + 4(2W - 9)(W - 9)$$

$$2(3573) = 12W^2 - 162W + 648$$

$$W = 30\ \text{m}$$

$$L = 60\ \text{m}$$

$$A = 1800\ \text{m}^2$$

3. Check the lagoon water temperature using the lagoon area and Eq. (8-5):

$$T_w = \frac{AfT_a + QT_i}{Af + Q} = 9.1°\text{C}$$

   The computed temperature is close enough to the assumed 10°C so that a second iteration is not needed.

4. Determine the oxygen requirements for the lagoon system based on the organic loading in each cell and using Eq. (8-6). The maximum oxygen requirements will occur in the summer months because the lagoon temperature will be highest then.

a. Calculate the summer lagoon temperature:

$$T_w = \frac{AfT_a + QT_i}{Af + Q} = 23.6°\text{C}$$

At 23.6°C the tap water oxygen saturation value is 8.46 mg/L (value interpolated from the values in App. E).

b. Calculate the organic loading on the first cell:

$$C_oQ = 200 \text{ mg/L } (1514 \text{ m}^3/\text{d})(0.001 \text{ kg/g}) = 303 \text{ kg/d}$$

c. Calculate the oxygen demand. The oxygen demand is assumed to be 1.5 times the organic loading.

$$N_a = 1.5(303) = 455 \text{ kg/d}$$

d. Calculate the equivalent oxygen transfer using Eq. (8-6):

Calculate the oxygen saturation value. Assume $\alpha = 0.8$ and $\beta = 0.95$.

$$\text{SOTR} = \frac{455 \text{ kg/d}}{[(0.95 \times 8.46 - 2.0)/.908](1.025)^{(23.6-20)}(0.8)} = 783 \text{ kg/d}$$

The loadings on the remaining cells can be calculated similarly using both Eqs. (8-4) and (8-6).

5. Calculate the aeration power needed for the first cell. Use 1.9 kg O_2/kW-h as the oxygen transfer.

$$\text{Power} = 783/1.9(24 \text{ h/d}) = 17.2 \text{ kW} \qquad \text{(use 20 hp, if available)}$$

Comment. A square lagoon will generally provide the best *L*/*W* ratio (for earthwork and aeration costs) for partial-mix aerated lagoons.

Dual-Power Multicellular Aerated Lagoons

The concept of dual-power multicellular (DPMC) aerated lagoons is to combine a series of partially mixed lagoons. The first cell has a depth of 10 ft (3 m) and is aerated with a surface aerator at a rate of 30 hp/Mgal (6 W/m^3). The subsequent three cells are aerated at a rate of 5 hp/Mgal (1 W/m^3). Detention time in the first cell is 1.5 to 2 days, and the overall detention time of all four cells is 4.5 to 5 d (Rich, 1980).

Complete mixing in aerated lagoons requires between 75 to 150 hp/Mgal, typically 100 hp/Mgal (20 W/m^3) of aeration. The recommended 30 hp/Mgal (6 W/m^3) is more than the 10 hp/Mgal (2 W/m^3) of most partial-mix aerated lagoons; however, it is less than the mixing required to keep all solids in suspension. The combination of two levels of aeration meets the oxygen requirement for biological conversion while minimizing the algae production by the turbulence of the mixing. A number of DPMC systems are in operation in South Carolina (Rich, 1996).

8-6 ANAEROBIC LAGOONS

Anaerobic lagoons are used for pretreatment of high-strength wastes in rural areas. As introduced in Chap. 7, anaerobic treatment is carried out by a wide variety of bacteria categorized into two groups, acid formers and methane formers. The two major groups of bacteria must operate cooperatively to ensure that the organic carbon is converted to methane.

Advantages of anaerobic treatment in comparison to aerobic treatment are (1) high degree of waste stabilization is possible, (2) low production of waste biological sludge, (3) low nutrient requirements, (4) no oxygen requirements, and (5) methane production. Disadvantages of anaerobic treatment are (1) incomplete BOD removal, (2) relatively high temperature requirements, and (3) potential for odor production (Hammer and Jacobson, 1970).

Systems are designed on the basis of surface loading rate, volumetric loading rate, and hydraulic detention time. In climates where the temperature exceeds 22°C, a 50 percent BOD reduction can be obtained using the following design criteria (WHO, 1987):

- Depth between 8 and 16 ft (2.5 and 5 m)
- Hydraulic detention time of 5 d
- Volumetric loading up to 0.0187 $lb/ft^3 \cdot d$ (0.3 $kg/m^3 \cdot d$)

8-7 COMBINATIONS OF LAGOONS

Combinations of lagoons can be attractive from the standpoint of both wastewater treatment and energy efficiency. Lagoon types can be combined, as illustrated by the integrated lagoon systems approach, and they can be combined for different purposes, such as controlled discharge and storage prior to land treatment.

Advanced Integrated Wastewater Lagoon Systems

The advanced integrated wastewater lagoon system [or advanced integrated wastewater pond system (AIWPS)] concept combines multiple lagoons with recycle [see Fig. 8-4 (Oswald, 1991)]. The system consists of a deep primary, facultative lagoon followed by a shallow aerobic lagoon (see Fig. 8-5). The primary lagoon has fermentation pits for anaerobic treatment of the settled solids. The fermentation pits must be unaerated and unmixed (see Fig. 8-5*b*), and will then serve as upflow anaerobic digesters. An example of an advanced integrated lagoon system is provided to illustrate the concept.

The City of St. Helena, California, located in the Napa Valley north of San Francisco, installed an integrated lagoon system in 1966 to treat 0.5 Mgal/d (1890 m^3/d) of municipal wastewater (see Fig. 8-6). Three shallow floating aerators have been added to the primary lagoon to supplement the recycled, algae-laden lagoon 2 water that serves to provide an aerobic cap on the primary lagoon. A third lagoon is

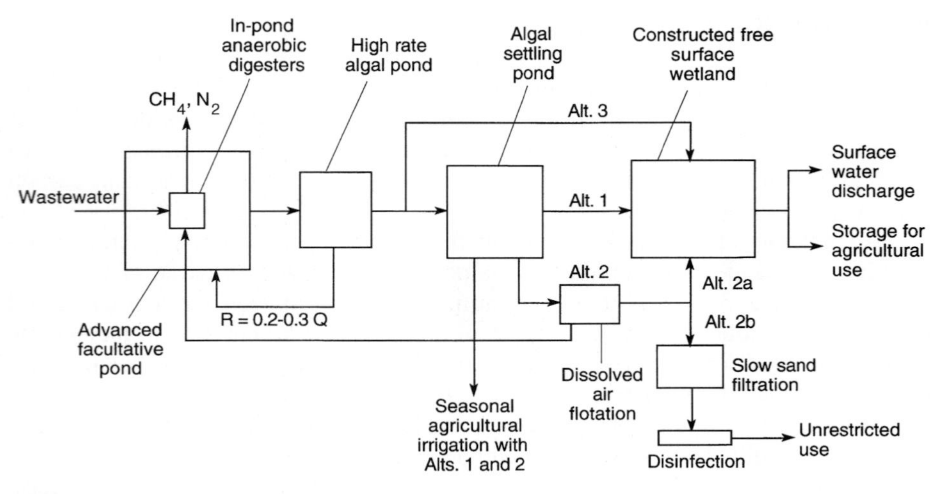

FIGURE 8-4
Advanced integrated wastewater pond system definition sketch.

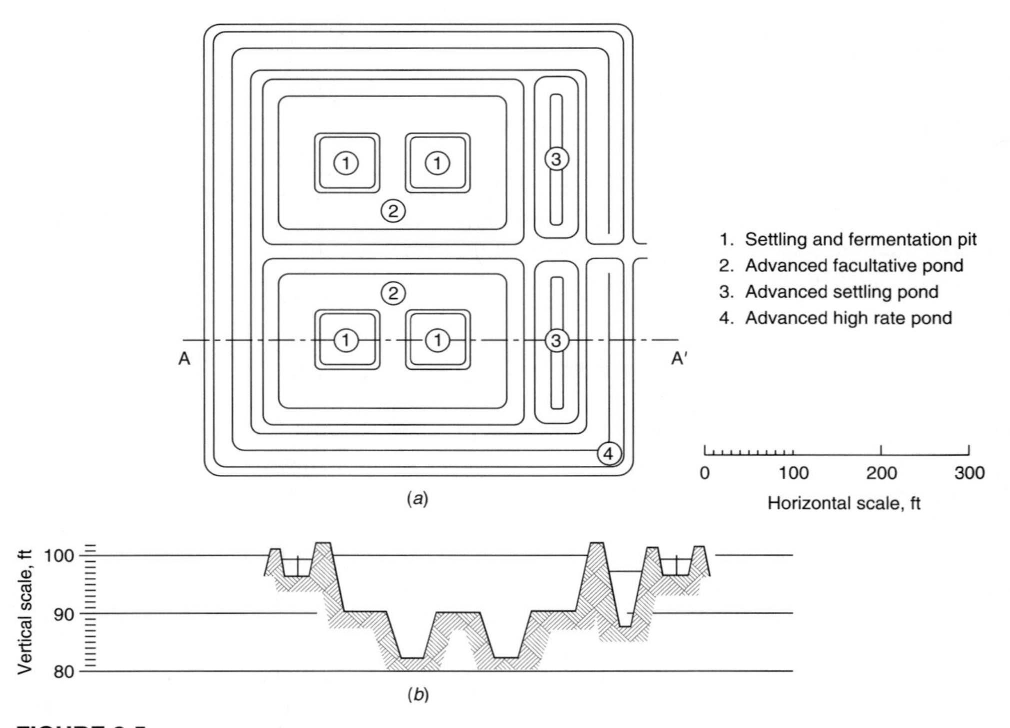

FIGURE 8-5
Typical example of advanced integrated wastewater pond system: (*a*) plan view and (*b*) section through pond at A-A′ (from Oswald, 1991).

FIGURE 8-6
Aerial view of St. Helena AIWPS showing primary pond in the background and final polishing ponds in the foreground (courtesy B. Green).

used in series for settling, and the treated effluent is chlorinated prior to irrigation. The fourth and fifth lagoons are maturation or storage lagoons. Design factors and performance of the St. Helena system are summarized in Table 8-7. The BOD loading rate based on design flowrate and a typical influent BOD of 300 mg/L is 345 lb/ac·d (390 kg/ha·d).

TABLE 8-7
Design factors and performance for St. Helena, California, lagoon system

Design factor	Unit	Value
Design flowrate	m^3/d	1892
Average flowrate (1994)	m^3/d	1514
Primary lagoon		
Aeration	hp	5
BOD loading	kg/ha·d	390
Depth	m	3
Area	ha	1.2
Detention time	d	19
High-rate aerobic lagoon		
Depth	m	1
Area	ha	2.1
Detention time	d	10
Setting lagoon		
Depth	m	3
Area	ha	1
Detention time	d	15
BOD_5 influent (1994)	mg/L	250–300
BOD_5 effluent (1994)	mg/L	15–40
TSS influent (1994)	mg/L	200–250
TSS effluent (1994)	mg/L	20–40

After 27 years of operation there has been little accumulation of sludge in the primary lagoon. No sludge has been removed from the lagoon and the depth of sludge in 1993 was about 1 ft (0.3 m) on average, with maximum sludge depths near the wastewater inlets of about 4 ft (1.2 m).

Controlled Discharge Lagoons

Controlled discharge lagoons are a variation of complete-retention lagoons used in the north central United States. Lagoons with seasonal discharge generally operate with the following design criteria:

1. Minimum detention time of 6 months above a minimum depth of 2 ft (0.6 m).
2. Overall organic loading of 20 to 25 lb/ac·d (22 to 28 kg/ha·d).
3. Water depth of 6 ft (2 m) or less in the first cell and 8 ft (2.5 m) or less in the next two to three cells.
4. At least three cells in series for reliability.

The discharge must be timed to coincide with acceptable stream flow and quality. The following steps are typically taken for controlled discharge:

1. Isolate the cell to be discharged, typically the final one in the series, by valving off the inlet line from the preceding cell.
2. Analyze the cell contents for constituents of concern in the discharge permit.
3. Avoid any conflicting duties that affect lagoon levels during the proposed discharge period.
4. Monitor conditions in the receiving water and request approval from regulatory agency for discharge.
5. Begin discharge when approval is received. Isolate the subsequent cells and draw them down until only the primary cell remains. Interrupt the discharge for a week while untreated wastewater is fed into one of the drawn-down cells. The primary cell is now isolated and can be drawn down 24 in (0.6 m), and then the internal series flow pattern without discharge is resumed.
6. During the discharge periods, effluent samples should be taken frequently to confirm that the discharge limits are being met.

The hydrograph controlled release (HCR) lagoon is a variation of the controlled discharge concept. The HCR method was developed in the southern United States to correlate peak stream flows with lagoon discharge.

8-8 DESIGN OF PHYSICAL FACILITIES

The physical design of a wastewater lagoon is as important as the conceptual or process design. Elements of the physical design include lagoon configurations, recirculation, lagoon lining and sealing, lagoon embankments, wind and temperature effects, and lagoon hydraulics.

Lagoon Configurations

Lagoon configurations can include either series or parallel operations. The advantage of series operation is improved treatment because of reduced short circuiting. The advantage of parallel configuration is that the loading can be distributed more uniformly over a larger area. Combinations of parallel and series operation can be accomplished along with recirculation.

Most partial-aerated lagoons are square, whereas facultative lagoons may be square or rectangular with a length-to-width ratio of 3:1 to 4:1. Lagoons are usually constructed with earthen, lined basins with side slopes of 2.5 to 3.5 to 1. The volume of the cell can be calculated by the following equation:

$$V = [LW + (L - 2sd)(W - 2sd) + 4(L - sd)(W - sd)]d/6 \tag{8-7}$$

where V = volume of cell or lagoon, ft^3 (m^3)
L = length of cell or lagoon at water surface, ft (m)
W = width of cell or lagoon at water surface, ft (m)
s = slope factor (s:1 is horizontal distance to 1 vertical)
d = depth of cell or lagoon, ft (m)

Lagoon Recirculation

The benefits of lagoon recirculation include mixing, dilution, and aeration. The ability to recirculate is a positive design feature even if the need for recirculation is infrequent. Both recirculation within a lagoon (intralagoon recirculation) and recirculation from one lagoon to another (interlagoon recirculation) have been used with success. These types of recirculation are illustrated in Fig. 8-7. Mixing and dilution are the primary reasons to use recirculation. Low-head, high-volume pumps can be used to accomplish recirculation.

Lagoon Lining and Sealing

Lagoons are usually lined to prevent seepage of partially treated wastewater to the groundwater. Methods of lining or sealing are categorized into three groups:

1. Natural and chemical treatment liners
2. Compacted earth or soil-cement liners
3. Geomembrane liners

Natural sealing of lagoons has been found to result from three mechanisms:

1. Physical clogging of soil pores by settled solids
2. Chemical clogging of soil pores
3. Biological clogging due to microbial growth

Although these mechanisms will tend to seal off wastewater lagoons in a year or so, most regulatory agencies will require more positive means of lagoon sealing if

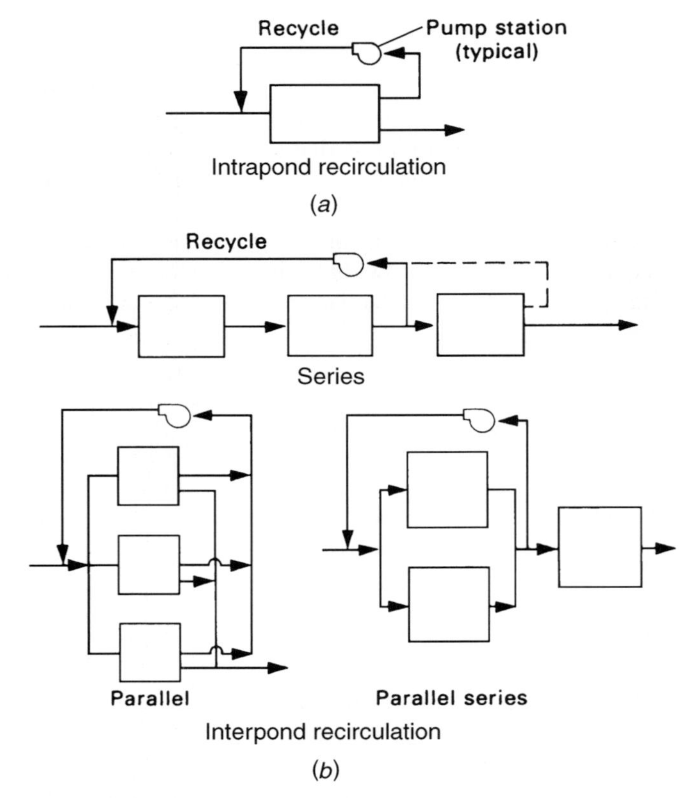

FIGURE 8-7
Definition sketch for lagoon recirculation: (*a*) intralagoon (within a lagoon) and (*b*) interlagoon (between lagoons).

groundwater protection is of concern. Lining materials such as bentonite, asphalt, soil cement, and geomembrane liners can reduce seepage to rates that are acceptable to regulatory agencies (U.S. EPA, 1983).

Lagoon Embankments

Lagoon embankments, dikes, or levees are earthen berms that must be constructed for stability and protected against wave damage, erosion due to weather, and burrowing by rodents. Depending on the type of soil used in the embankments, a design side slope must be determined with recommendations usually coming from soils engineers.

Most lagoons are constructed by excavating the lagoon bottom to produce fill to construct the levee defining the upper portion of the lagoon. Ideally the cut into the lagoon bottom will provide the fill needed to construct the levees. It is not uncommon for each cubic yard of cut to produce only 0.7 to 0.8 cubic yards of fill because

FIGURE 8-8
Typical example of pond embankment protection using 6- to 12-in riprap.

of shrinkage (levee compaction) and rejection of cut material not suitable for use as levee fill. The actual design of the levees and their slopes, keyed foundations, compaction, liners, and erosion protection should be done by a qualified soils engineer.

Protection of embankments should extend from 1 ft or more (0.3 m) below the minimum water level to at least 1 ft (0.3 m) above the maximum water level (see Fig. 8-8). Asphalt, geomembrane, polyethylene, concrete, riprap, or low-growing grasses can be used to provide protection against wave action. Riprap can be used to protect the liner from damaging ultraviolet radiation, so the combination of a liner with riprap can protect against erosion and rodent burrowing. Rodents cause the most problems with unlined lagoons and riprap on unlined lagoons can make both weed and rodent control more difficult (Reed et al., 1995). Muskrats and nutria will try to burrow into earthen levees. If infestations occur, varying the lagoon water depth can discourage muskrats from burrowing in the embankments (U.S. EPA, 1977).

Wind and Temperature Effects

Wind generates a circulatory flow pattern in lagoons. The lagoon inlet and outlet axis should be aligned perpendicular to the prevailing wind direction to minimize short circuiting due to wind.

Lagoons that are stratified because of temperature differences between the influent and the lagoon contents operate differently in summer and winter. In summer, the influent is generally colder than the lagoon so it sinks to the lagoon bottom and flows toward the outlet. In the winter, the reverse is true and the inflow rises to the surface and flows toward the outlet. The use of baffles, multiple inlets and outlets, and recirculation tends to minimize the adverse effects of stratification.

Lagoon Hydraulic Design

Important elements of lagoon hydraulics include inlet and outlet structures, baffling, wind effects, and lagoon stratification.The typical hydraulic design of existing lagoons involves the use of an inlet pipe located in the center of the lagoon. It has been shown in hydraulic and performance studies, however, that a center-feed inlet device is not the most efficient method of introducing wastewater into a lagoon (Mangelson and Watters, 1972).

The influent wastewater should be distributed in multiple outlets, by a diffuser. The inlets and outlets should be placed so that the velocity profile of flow through the lagoon is uniform between successive inlets and outlets (Reed et al., 1995). An inlet device that produces uniform distribution is a pipe laid on the lagoon bottom with multiple ports or nozzles all pointing in one direction and at a slight angle above the horizontal (U.S. EPA, 1983). Port headloss is designed for about 1 ft (0.3 m) at average flow, resulting in a velocity of 8 ft/s (2.4 m/s).

Outlet devices should also be multiple to avoid short circuiting. In deeper lagoons (more than 5 ft), the outlet device should be designed for multiple-depth withdrawal, and all withdrawals should be at least 1 ft (0.3 m) below the water surface. Transfer pipes should be numerous and large enough to limit peak headloss to about 3 to 4 in (75 to 100 mm) with the pipes flowing full. Headlosses in supply and return channels should be no more than 10 percent of the headlosses in the transfer pipes. When such a ratio is maintained, uniform distribution is assured (U.S. EPA, 1983).

If a single inlet device is used, the use of baffles can greatly enhance the distribution. Commercial float-supported plastic baffling is available for lagoons. If baffling is used, the cross-sectional area of flow should remain as close as possible to constant so that the velocity remains constant. Baffling has the additional advantage of enhancing mixing as the flow goes around the baffle.

8-9 UPGRADING LAGOON EFFLUENT

Lagoon treatment systems are not very efficient in producing effluents with low suspended solids concentrations. A number of technologies have been used to upgrade lagoon effluents for suspended solids removal including intermittent sand filters (see also Chap. 11), wetland and aquatic treatment systems (Chap. 9), and land treatment systems (Chap. 10). Technologies that are not discussed in other chapters, such as rock filters, microstrainers, high-rate sand filters, and dissolved air flotation are discussed in this section.

Intermittent Sand Filters

Intermittent sand filters (ISFs), as described in Chap. 11, are biological and physical treatment units. Intermittent sand filters can upgrade lagoon effluent by filtering out the suspended solids. The algae collect on the surface of the sand filter as the wastewater is applied and treated. The accumulation of solids occurs in a 2- to 3-in (50- to 80-mm) layer that must be removed periodically.

The depth of sand in the filter is at least 18 in (450 mm) plus a sufficient depth for at least 1 year of cleaning cycles (Reed et al., 1995). A single cleaning event may remove 1 to 2 in (25 to 50 mm) of sand. A typical bed depth is 36 in (900 mm).

The sand for single-stage ISFs should have an effective size between 0.20 and 0.30 mm with a uniformity coefficient of less than 5.0. Less than 1 percent of the sand particles should be smaller than 0.1 mm. Clean, pit-run concrete sand is suitable for use in intermittent sand filters.

For lagoon effluent treatment, typical hydraulic loading rates for an ISF range from 1.2 to 1.8 ft/d (0.37 to 0.56 m/d). For high concentrations of algae, greater than 50 mg/L, the hydraulic loading should be reduced to 0.6 to 1.2 ft/d (0.19 to 0.37 m/d) to increase the run time between cleaning events. The lower end of the range should be used for filters in cold-weather locations to avoid the possible need for bed cleaning during the winter months.

The total filter area required for an ISF is determined by dividing the average flow rate by the design hydraulic loading rate. One spare filter should be added to ensure continuous operation because it may take several days for a cleaning event. A minimum of three units is preferred. In small systems that use manual cleaning, the individual bed should not be larger than 1000 ft^2 (90 m^2). In larger systems with mechanical cleaning equipment, the individual beds can be as large as 55,000 ft^2 (5000 m^2). Further details may be found in Russell et al. (1980) and U.S. EPA (1983). A typical ISF is shown in Fig. 8-9.

Rock Filters

Rock filters remove suspended solids by sedimentation as lagoon effluent flows horizontally through the void spaces in the rocks (see Fig. 8-10). The accumulated algae are then degraded biologically. Design parameters for rock filters at Veneta, Oregon; West Monroe, Louisiana; Eudora, Kansas; and California, Missouri are presented in Table 8-8 (Middlebrooks, 1993).

The advantages of the rock filter are its simplicity of operation and its relatively low construction cost. Odor problems can occur, especially in wastewater containing significant concentrations of sulfate (greater than 50 mg/L). In addition, the design life for the filters and cleaning procedures have not yet been established (Middlebrooks, 1988). As shown in Table 8-8, there are several systems that have been in operation for over 20 yr.

Microstrainers

Although early experience with algae removal by microstrainers was largely unsuccessful, a 1-μm polyester fabric has been developed with reported success (Middlebrooks, 1993). The first full-scale microstrainer application to lagoon effluent was

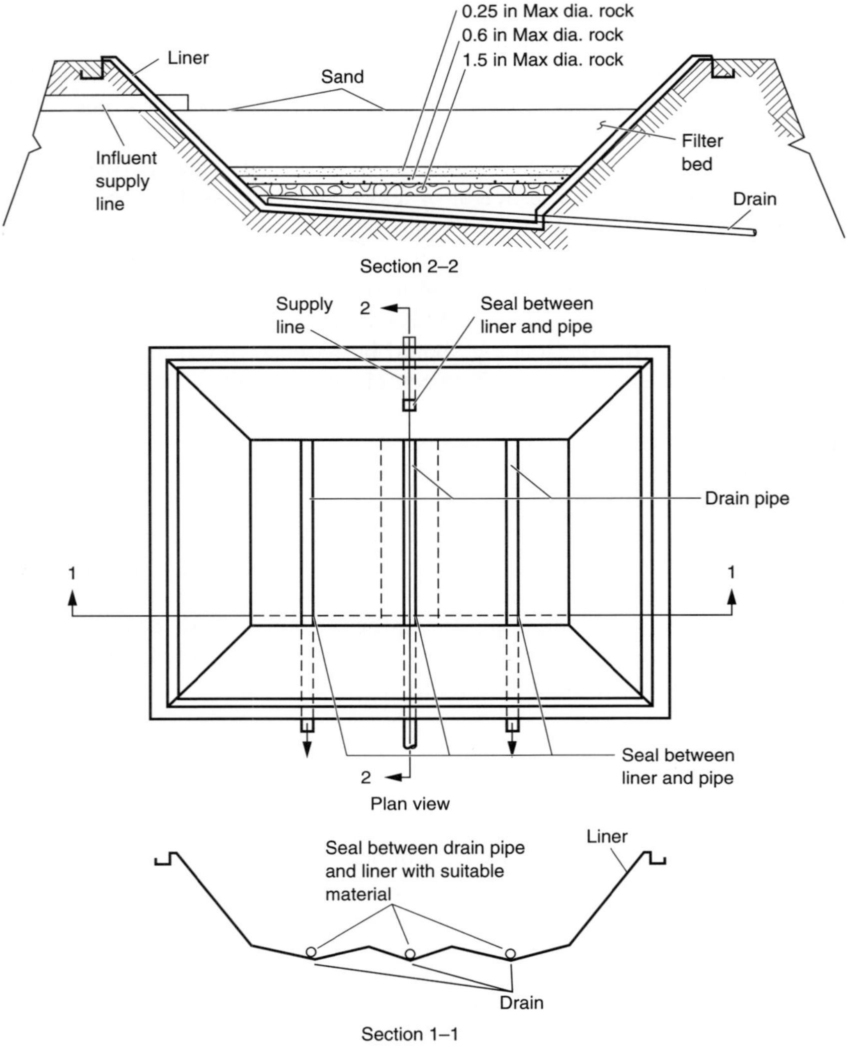

0.25 in Max dia. rock
0.6 in Max dia. rock
1.5 in Max dia. rock
Liner
Sand
Influent supply line
Filter bed
Drain

Section 2–2

Supply line
Seal between liner and pipe
Drain pipe
Seal between liner and pipe

Plan view

Seal between drain pipe and liner with suitable material
Liner
Drain

Section 1–1

FIGURE 8-9
Definition sketch for intermittent sand filter used for treatment of lagoon effluent (adapted from U.S. EPA, 1983).

at Camden, South Carolina, in 1981 (Harrelson and Cravens, 1982). Typical design criteria for microstrainers include surface loading rates of 295 to 400 ft/d (90 to 120 m/d) and headlosses up to 24 in (600 mm). Other process variables include drum speed and backwash rate and pressure. The service life of the screen is reported to be about 1.5 yr, which is considerably less than the manufacturer's prediction of 5 yr. Difficulties with screen binding and short run times were experienced with the Camden system. Careful study and evaluation of current manufacturer's experience should be conducted prior to selecting microstrainers for lagoon upgrading.

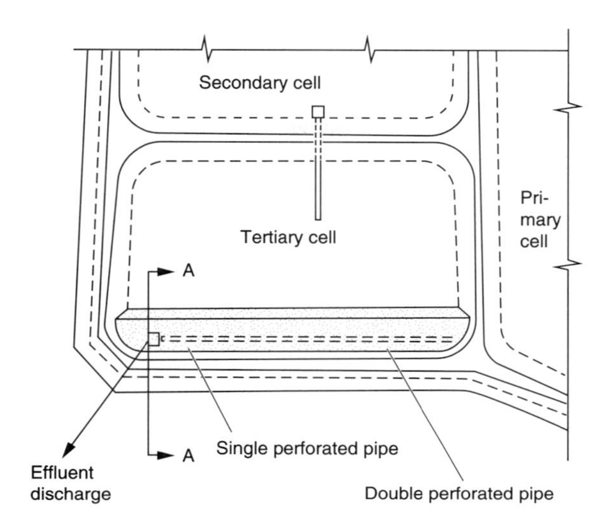

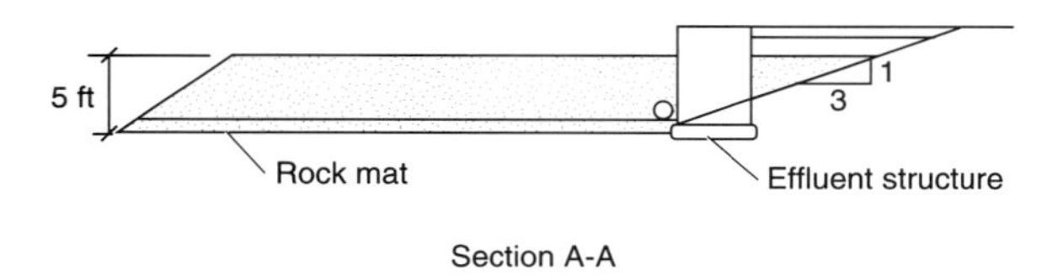

FIGURE 8-10
Definition sketch for rock filter used for the removal of TSS from lagoon effuent (adapted from U.S. EPA, 1983).

TABLE 8-8
Rock filter design parameters

Parameter	Veneta, Oregon	West Monroe, Louisiana	Eudora, Kansas	California, Missouri	State of Illinois*
Year installed	1975	1981	1970	1974	
Depth, ft	6.6	5.9	4.9	5.5	1 above water surface
Rock size, in	3–8	2–5	1	2.4–5	3.2–6†
Hydraulic loading rate, ft/d	1.0	1.2	1.3–4.0‡	1.3	2.6

*Guidelines require postaeration ability. Chlorination of postaeration cell encouraged.
† Rock to be free of fines and soft weathering stone; no flat rock.
‡ 1.3 ft/d in winter and spring, up to 4.0 ft/d in the summer.

High-Rate Sand Filters

The performance of rapid sand filters in treating algae is mixed, depending on the level of pretreatment, the hydraulic loading rate, the size and nature of the algae, and the level of coagulants used. Removals typically range from less than 20 percent to more than 70 percent, depending on sand size and the variables mentioned previously, including the time of year. Dual-media, deep-bed, continuously backwashed, and pulsed-bed filters (see Chap. 12) have all been used to treat algal-laden lagoon waters (U.S. EPA, 1983). Without coagulation, algae are too small, have too low an affinity for sand, and are not removed effectively by direct filtration. Hydraulic loading rates typically less than 2 gal/ft^2·min (120 mm/d) are needed for effective removal.

Dissolved Air Flotation

The flotation process involves the formation of the fine gas bubbles that become physically attached to the algal solids, causing them to float to the surface where they are physically removed. Chemical coagulation results in the formation of a floc-bubble matrix that allows more efficient separation to take place in the flotation tank. In dissolved air flotation (DAF), a portion of the influent (or recycled effluent) is pumped to a pressure tank where the liquid is mixed with high-pressure air to supersaturate the liquid. The pressurized stream is then mixed with the influent, the pressure is released, and fine bubbles are formed.

The design factors for DAF treatment include surface loading rate, design pressure, air/solids ratio, percent recycle, and coagulant dosage. The performance of DAF units treating lagoon effluent is presented in Table 8-9. Surface loading rates range typically from 2 to 2.7 gal/ft^2·min (120 to 160 mm/d). Design pressures range from 25 to 80 lb/in^2 (175 to 560 kPa). The air/solids ratio is defined as the weight of air bubbles added to the process divided by the weight of suspended solids entering the tank. Typical air/solids ratios range from 0.05 to 0.10. The recycle percentage can range from 25 to 100, as shown in Table 8-9.

TABLE 8-9
Algae removal by dissolved air flotation[a]

Location	Coagulant	Dose, mg/L	Surface loading rate, gal/ft^2·d	Detention time, min	Effluent TSS, mg/L	Removal, %
Stockon, California	Alum Acid	225 to pH 6.4	2.6	17[b]	20	81
Lubbock, Texas	Lime	150	[c]	12[d]	0–50	>79
El Dorado, Alaska	Alum	200	3.9[e]	8[e]	36	92
Logan, Utah	Alum	300	1.3–2.4[f]	[c]	4	96
Sunnyvale, California	Alum Acid	175 to pH 6.0–6.3	1.95[g]	11[g]	30	80

[a]From U.S. EPA (1983).
[b]Including 33% pressurized (35–60 lb/in^2) recycle.
[c]No data available.
[d]Including 30% pressurized (50 lb/in^2) recycle.
[e]Including 100% pressurized recycle.
[f]Including 25% pressurized (45 lb/in^2) recycle.
[g]Including 27% pressurized (55–70 lb/in^2) recycle.

8-10 AQUACULTURE

Aquatic animals that have been attempted in wastewater treatment lagoons include fish, daphnia, and brine shrimp. Although aquatic animals have been cultured successfully (Dinges, 1982) the improvement in water quality has been generally minimal (Reed et al., 1995).

Fish

Fish usually require a pH range between 6.5 and 9, dissolved oxygen of 2 mg/L or more, and low concentrations (less than 1 mg/L) of un-ionized ammonia (NH_3). Because of these requirements, fish culture is most successful in the latter or final cells of multicell treatment lagoon systems.

Fish culture was studied at two full-scale wastewater treatment systems, and the results are presented in Table 8-10. The Quail Creek experiment was conducted in 1973 (Coleman et al., 1974), and the Benton Services Center work was conducted between 1975 and 1980 (Henderson, 1979). Although these experiments were successful at growing fish, the water quality benefits were less impressive than the net production of fish. Currently, federal and state regulators will not allow direct human consumption of such fish even though no contamination has been reported in microbiological studies (Coleman et al., 1974; Henderson, 1979). The market for the fish would be pet food supplement, bait fish, or fertilizer.

At Arcata, California, coho and chinook salmon have been cultured successfully in a mixture of lagoon effluent and seawater (Allen and Dennis, 1974). The objective was to raise fish, not improve water quality.

TABLE 8-10
Fish culture in wastewater lagoons*

Item	Quail Creek, Oklahoma	Benton Services Center, Arkansas
Number of lagoon cells	6	6
Aerated cells	1–2	None
Fish stocked	Channel catfish, tilapia, fat-head minnows, golden shiners	Silver carp, bighead carp, grass carp, buffalo fish, channel catfish
Cells stocked	3–6	3–6
Total detention time, d	140	72
Detention time in stocked cells, d	90	48
Stocking rate, lb/ac	26	380
Net production, lb/ac·month	39	390
Effluent BOD, mg/L	6	9.4
Effluent TSS, mg/L	12	17

*Adapted from Coleman et al. (1974).

Smith (1991) proposed a system of alternating lagoon cells that culture zooplankton with lagoon cells that culture filter-feeding fish. Under this approach, both aquaculture and water quality benefits can be sustained. To accomplish water quality improvement, the waste products from the animals must be treated as part of the overall system design.

Daphnia

Daphnia and brine shrimp are filter-feeding crustaceans that can remove algae and other suspended solids from wastewater. Daphnia are water fleas that can clarify lagoon effluent at population densities of 500/L or more. Short-term tests at Giddings, Texas (2 months) and Las Virgenes, California (9 months) demonstrated sporadic improvement, depending on temperature and wastewater pH. If the wastewater pH exceeds 8, the daphnids are susceptible to ammonia toxicity. Because daphnia tend to "bloom" and destroy all the algae and then die out, they are not considered to be dependable enough to be part of a wastewater system design. The main limitation to the use of brine shrimp is the requirement for a saline water.

PROBLEMS AND DISCUSSION TOPICS

8-1. Estimate the amount of oxygen that can be transferred across the surface of a lagoon if the dissolved oxygen is maintained at 2 mg/L. Assume a completely mixed lagoon and a water temperature of 20°C.

8-2. If a large lagoon system develops a 50 percent coverage with cattails, will the total amount of oxygen that can be transferred to the liquid be increased, decreased, or remain the same? Justify your answer.

8-3. It has been observed that, in side-by-side lagoons, one of the lagoons may have significant wave action, whereas the other may be quiescent. How would you explain this phenomenon?

8-4. Design a lagoon system composed of three equal-sized lagoons in series to reduce the coliform count of septic tank effluent from 10^6/100 mL to 10/100 mL. Assume a coefficient of natural decay is 0.0025 d^{-1}.

8-5. Assuming a dispersion coefficient for a facultative lagoon is 0.25, determine the size of lagoon required to reduce the BOD from 120 to 20 mg/L. Assume the first-order BOD decay coefficient is 0.15 What size ideal plug-flow lagoon would be required to achieve the same results?

8-6. Design an oxidation lagoon to treat 1 Mgal/d of wastewater from a small community. Assume that average influent BOD is 200 mg/L and the required effluent BOD is to be equal to or less than 20 mg/L.

8-7. The combined flow from a STEP system averages 0.5 Mgal/d and has a BOD of 150 mg/L and is discharged to a facultative lagoon with an area of 5 ac and a depth of 8 ft. Determine the hydraulic detention time and the lagoon organic loading rate.

8-8. Determine the average lagoon temperature for a 10-ac lagoon in January. Use an ambient air temperature of 5°C, and an influent water temperature of 8°C.

8-9. The rainfall in an area is 30 in/yr and the evaporation rate is 72 in/yr. Determine the lagoon area required for a complete retention system. Assume a flow of 0.25 Mgal/d and a seepage rate of 1 in/mo.

8-10. Compare the volume of two lagoons. Lagoon 1 has a side slope of 2:1 and lagoon 2 has a side slope of 3:1. Both lagoons have a surface area of 2 ac and a depth of 10 ft.

8-11. Determine the temperature of a 30-ac pond if the air temperature is 40°F. Use a flowrate of 0.5 Mgal/d and an influent water temperature of 65°F.

8-12. Given the following information, and assuming a seepage of 1.2 in/mo, determine the average detention time for a lagoon with a surface area of 4 ac and a depth of 6 ft.

Month	Rainfall, in	Evaporation, in
January	4.0	1.6
February	4.5	1.0
March	2.6	2.5
April	1.2	3.5
May	0.8	5.0
June	0.2	7.0
July	0.0	9.0
August	0.0	10.0
September	0.2	8.0
October	0.4	4.5
November	1.6	2.0
December	2.0	1.6

REFERENCES

Allen, G. H., and L. Dennis (1974) Report on Pilot Aquaculture Systems Using Domestic Wastewaters for Rearing Pacific Salmon Smolts, in *Proceedings Wastewater Use in the Production of Food and Fiber,* EPA/660/2-74-041. U.S. Environmental Protection Agency, Washington, DC.

Coleman, M. S., J. P. Henderson, H. G. Chichester, and R. L. Carpenter (1974) Aquaculture as a Means to Achieve Effluent Standards, in *Proceedings Wastewater Use in the Production of Food and Fiber,* EPA/660/2-74-041. U.S. Environmental Protection Agency, Washington, DC.

Crites, R. W. (1992) Chap. 13, Natural Systems, in *Design of Municipal Wastewater Treatment Plants,* Vol. II, WEF Manual of Practice No. 8, Water Environment Federation, Alexandria, VA, pp. 831–867.

Dinges, R. (1982) *Natural Systems for Water Pollution Control,* Van Nostrand Reinhold, New York.

Gloyna, E. F. (1976) Facultative Waste Stabilization Pond Design. Ponds as a Waste Treatment Alternative, Water Resources Symposium No. 9, University of Texas, Austin.

Hammer, M. J., and C. D. Jacobson (1970) Anaerobic Lagoon Treatment of Packinghouse Wastewater, *Proceedings 2nd International Symposium for Waste Treatment Lagoons,* Federal Water Quality Administration (FWQA), Kansas City, MO.

Harrelson, M. E., and J. B. Cravens (1982) Use of Microscreens to Polish Lagoon Effluent, *Journal of Water Pollution Control Federation,* Vol. 54, No. 1, pp. 36–42.

Henderson, S. (1979) Utilization of Silver and Bighead Carp for Water Quality Improvements, *Proceedings Aquaculture Systems for Wastewater Treatment,* EPA/430/9-80-006. U.S. Environmental Protection Agency, Washington, DC.

McGauhey, P. H. (1968) *Engineering Management of Water Quality,* McGraw-Hill, New York.

Mangelson, K. A., and G. Z. Watters (1972) Treatment Efficiency of Waste Stabilization Ponds, *Journal of Sanitary Engineering Division,* ASCE, Vol. 98, No. SA2.

Marais, G. V. R. (1970) Dynamic Behavior of Oxidation Ponds, *Proceedings 2nd International Symposium for Waste Treatment Lagoons,* FWQA, Kansas City, MO.

Marais, G. V. R., and V. A. Shaw (1961) A Rational Theory for the Design of Sewage Stabilization Ponds in Central and South Africa, *Transactions of South African Institution of Civil Engineers,* Vol. 3, p. 205.

Middlebrooks, E. J. (1988) Review of Rock Filters for the Upgrade of Lagoon Effluents, *Journal of Water Pollution Control Federation,* Vol. 60, No. 9, pp. 1657–1662.

Middlebrooks, E. J. (1993) Upgrading Pond Effluents: An Overview. Waste Stabilization Ponds and the Reuse of Pond Effluents, *Proceedings of the 2nd International Association of Water Quality International Specialist Conference,* Oakland, CA.

Oswald, W. J. (1991) Introduction to Advanced Integrated Wastewater Ponding Systems, *Water Science Technology,* Vol. 24, No. 5, pp. 1–7.

Pano, A., and E. J. Middlebrooks (1982) Ammonia Nitrogen Removal in Facultative Wastewater Stabilization Ponds, *Journal of Water Pollution Control Federation,* Vol. 54, No. 4, pp. 344–351.

Reed, S. C. (1985) Nitrogen Removal in Wastewater Stabilization Lagoons, *Journal of Water Pollution Control Federation,* Vol. 57, No. 1, pp. 39–45.

Reed, S. C., R. W. Crites, and E. J. Middlebrooks (1995) *Natural Systems for Waste Management and Treatment,* 2nd ed., McGraw-Hill, New York.

Rich, L. G. (1980) *Low-Maintenance Mechanically Simple Wastewater Treatment Systems,* McGraw-Hill, New York.

Rich, L. G. (1996) Low-Tech Systems for High Levels of BOD and Ammonia Removal, *Public Works,* Vol. 127, No. 4, pp. 41–42.

Russell, J. S., E. J. Middlebrooks, and J. H. Reynolds (1980) *Wastewater Stabilization Lagoon—Intermittent Sand Filter Systems,* EPA 600/2-80-032, U.S. Environmental Protection Agency, Cincinnati, OH.

Smith, D. W. (1991) Sewage Treatment with Complementary Filter Feeders: A New Method to Control Excessive Suspended Solids and Nutrients, presented at the 64th Annual Conference of Water Pollution Control Federation, Toronto.

Stowell, R. (1976) *A Study of the Screening of Algae from Stabilization Ponds.* Masters Thesis, Department of Civil Engineering, University of California, Davis.

Surampalli, R. Y., S. K. Banerji, C. J. Pycha, and E. R. Lopez (1993) Phosphorus Removal in Ponds, *Proceedings of the 2nd International Association of Water Quality International Specialist Conference,* Oakland, CA.

Tchobanoglous, G., and E. D. Schroeder (1985) *Water Quality: Characteristics, Modeling, Modification,* Addison-Wesley, Reading, MA.

Ten States (1990) Ten States Recommended Standards for Wastewater Facilities, Report of the Wastewater Committee of the Great Lakes—Upper Mississippi River Board of State Public Health and Environment Managers, Health Education Services, Albany, NY.

Thirimurthi, D. (1974) Design Criteria for Waste Stabilization Ponds, *Journal of Water Pollution Control Federation,* Vol. 46, No. 9, pp. 2094–2106.

U.S. EPA (1983) *Municipal Wastewater Stabilization Ponds—Design Manual,* EPA-625/1-83-015, U.S. Environmental Protection Agency, Office of Water Program Operations, CERI, Cincinnati, OH.

U.S. EPA (1977) *Operations Manual for Stabilization Lagoons.* EPA-430/9-77-012, NTIS No. PB-279443, Office of Water Program Operations, U.S. Environmental Protection Agency, Washington, DC.

WHO (1987) *Wastewater Stabilization Ponds, Principles of Planning and Practice,* WHO Technical Publication 10, World Health Organization, Regional Office for the Eastern Mediterranean, Alexandria.

CHAPTER 9

Wetlands and Aquatic Treatment Systems

Wetlands and aquatic treatment systems are those that use aquatic plants and animals for the treatment of municipal and industrial wastewater. Aquatic treatment covers a broad range of system types including a variety of constructed wetlands systems, floating aquatic plant systems, and combinations of both floating aquatic and wetland systems. The material to be presented is organized into sections dealing with: (1) types of and application of wetland and aquatic systems, (2) treatment kinetics in constructed wetlands and aquatic systems, (3) free-water-surface constructed wetlands, (4) subsurface-flow constructed wetlands, (5) floating aquatic plant systems using water hyacinth, (6) floating aquatic plant systems using duckweed, (7) combination systems, (8) design procedures for constructed wetlands and aquatic systems, (9) management of constructed wetlands and aquatic systems, and (10) emerging technologies.

9-1 TYPES OF AND APPLICATION OF WETLANDS AND AQUATIC SYSTEMS

The principal types of wetlands and aquatic systems and their applications are introduced in this section. Aquatic systems that have been researched and demonstrated, but not applied in full-scale systems in the United States, are considered in Sec. 9-10, which deals with emerging technologies.

Types of Systems

The principal types of wetlands and aquatic treatment systems considered in this chapter include:

- Free-water-surface (FWS) constructed wetlands
- Subsurface-flow (SF) constructed wetlands
- Floating aquatic plant systems
- Combination systems

These systems, introduced in the following paragraphs, are considered separately in greater detail in Secs. 9-3 through 9-6.

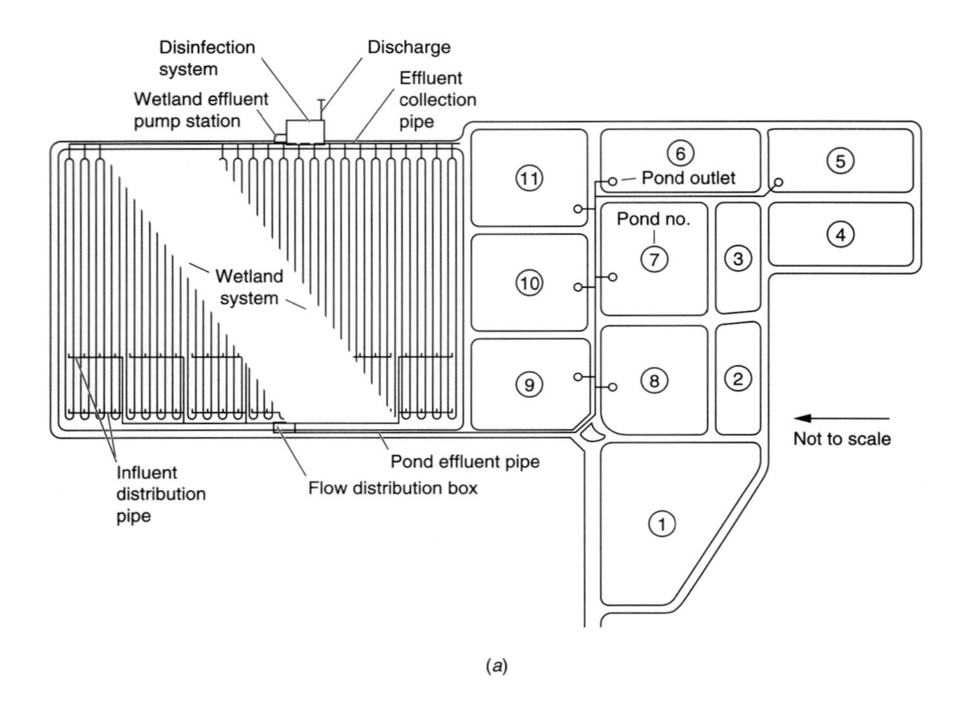

(*a*)

(*b*)

FIGURE 9-1
Free water surface (FWS) constructed wetland at Gustine, California: (*a*) definition sketch and (*b*) view of plug-flow channels.

Free-water-surface constructed wetlands. In a free-water-surface constructed wetland (marsh or swamp), the emergent vegetation is flooded to a depth that ranges from 4 to 18 in (100 to 450 mm). Typical vegetation for FWS systems includes cattails, reeds, sedges, and rushes. A FWS system consists typically of channels or basins with a natural or constructed impermeable barrier to prevent seepage. Some FWS systems are designed for complete retention of the applied wastewater through seepage and evapotranspiration. Wastewater is treated as it flows through the vegetation by attached bacteria and by physical and chemical processes. A typical free-water-surface constructed wetland at Gustine, California is shown in Fig. 9-1.

Subsurface-flow constructed wetlands. In a subsurface-flow constructed wetland (see Fig. 9-2) the wastewater is treated as it flows laterally through the porous medium. Emergent vegetation is planted in the medium, which ranges from coarse gravel to sand. The depth of the bed ranges from 1.5 to 3.3 ft (0.45 to 1 m) and the slope of the bed is typically 0 to 0.5 percent.

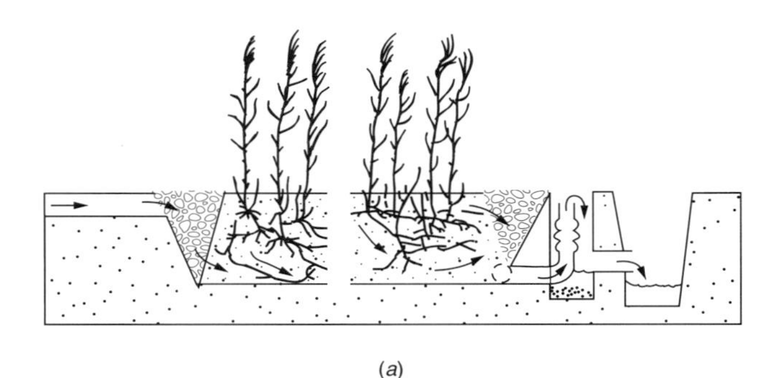

(*a*)

(*b*)

FIGURE 9-2
Typical SF constructed wetland: (*a*) definition sketch and (*b*) view of SF wetland at Mesquite, Nevada, before planting.

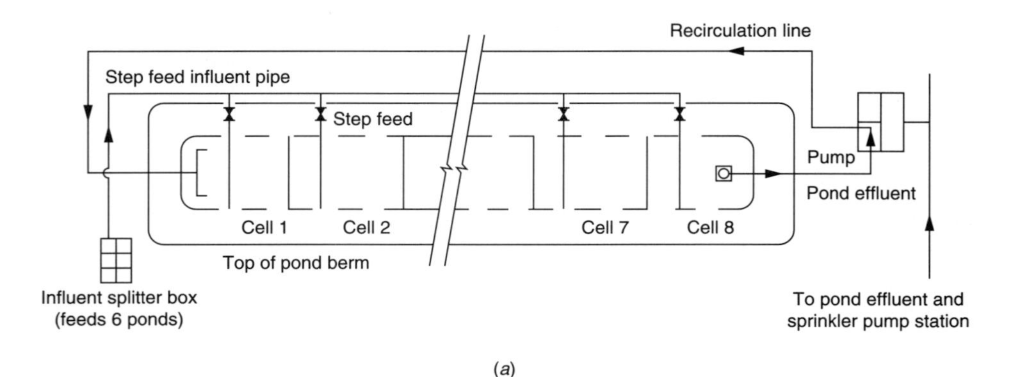

(*a*)

(*b*)

FIGURE 9-3
Floating plant aquatic treatment system at San Diego, California: (*a*) definition sketch for step-feed plug-flow channel with effluent recirculation to cell 1 (adapted from WCPH, 1996), and (*b*) view of water hyacinth system.

Floating aquatic plant systems. The two principal types of floating aquatic plant systems are the water hyacinth and duckweed systems (see Fig. 9-3). The water hyacinth (or similar plant) system involves floating or suspended plants with relatively long roots in ponds 2 to 4 ft (0.6 to 1.2 m) deep. The root structure serves as a medium for the attached growth of bacteria. The duckweed, on the other hand, has very short roots [usually less than 0.4 in (10 mm) long] and therefore functions as a surface shading system.

Combination systems. Aquatic and wetland systems can be used in combination, usually in series, to achieve specific water quality objectives. For example, a duckweed or hyacinth system could be used prior to a constructed wetland to minimize algae concentrations. A combined aerated aquatic treatment system with a constructed wetland has been studied for the treatment of septage at Harwich, Massachusetts (Nolte and Associates, 1989).

TABLE 9-1
Representative applications of constructed wetlands and aquatic treatment systems

Objective	Constituent removed/objective
Acid mine drainage	Metals and acidity
Advanced treatment	Nitrogen and phosphorus
Advanced treatment	Heavy metals and refractory organics
Combined secondary and advanced treatment	Organic matter (e.g., BOD_5), total suspended solids (TSS), pathogens, nitrogen, and phosphorus
Habitat development	Enhanced environmental resources
Irrigation return water	Nitrogen and phosphorus
Landfill leachate	Organic matter
Reclamation and water reuse	Organic matter, total suspended solids (TSS), and pathogens to restrictive standards (e.g., turbidity $\leq$ 2 NTU, SS $\leq$ 5 mg/L, and total coliform $\leq$ 2.2 organisms/100 mL)
Secondary treatment	Organic matter (e.g., BOD_5), total suspended solids (TSS), and pathogens
Septage treatment	Organic matter (e.g., BOD_5), total suspended solids (TSS), pathogens, nitrogen, and phosphorus
Stormwater treatment	Organic matter (e.g., BOD_5), total suspended solids (TSS), pathogens, nitrogen, phosphorus, and heavy metals and refractory organics

Application of Constructed Wetlands and Aquatic Systems

Constructed wetlands and aquatic systems have been used in a number of applications for the treatment of wastewaters with diverse characteristics. The principal types of applications are reported in Table 9-1. Three different applications, as discussed below, include tertiary treatment, stormwater treatment, and habitat development. However, as shown in Table 9-1, constructed wetlands have been used in a variety of other applications.

Sacramento County, California. The Sacramento Regional County Sanitation District operates the FWS constructed wetland shown in Fig. 9-4. The objectives of the project are to demonstrate advanced treatment for removal of metals, ammonia, and toxicity from the secondary effluent. The project results in the treatment of 1 Mgal/d (3785 m^3/d) on 15 ac (6.0 ha).

Stormwater wetlands. Wetlands for stormwater treatment and flow attenuation are becoming increasingly popular. An example of a stormwater wetlands is presented in Fig. 9-5.

Arcata, California. Wetlands at Arcata serve treatment, habitat enhancement, and educational benefits. The constructed wetland is used to treat 2.3 Mgal/d

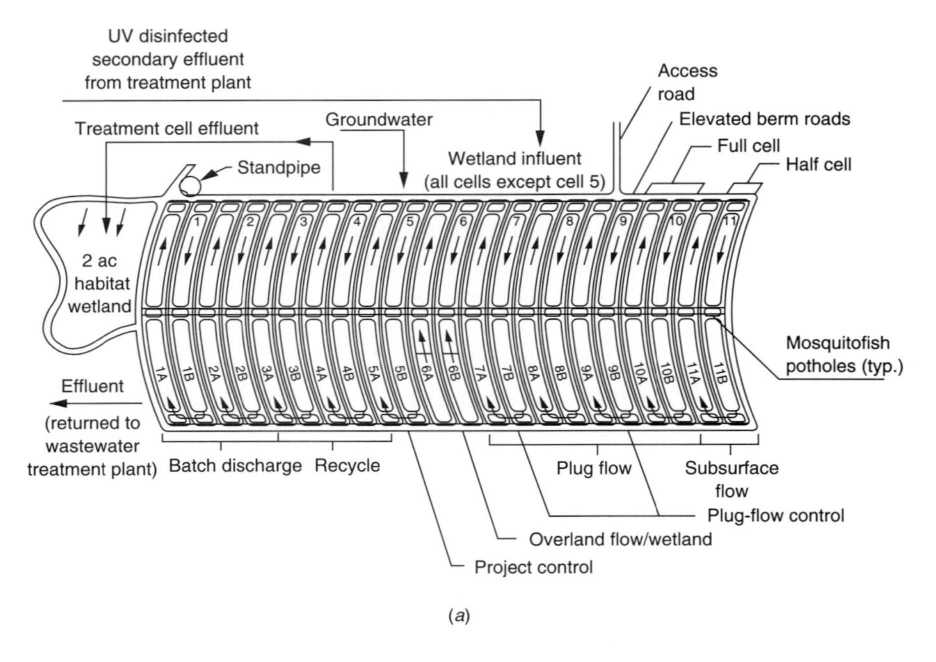

(a)

(b)

FIGURE 9-4
Sacramento Regional CSD Demonstration Wetland: (a) definition sketch and (b) view of the wetland.

(8.7 m^3/d) of effluent from a facultative lagoon. The effluent from the wetlands is then discharged into 31 ac (12.5 ha) of enhancement wetlands (marshes). The enhancement wetlands at Arcata feature a combination of 50 percent rooted vegetation and 50 percent open water (Gearheart and Finney, 1996). Views of the FWS wetlands at Arcata are shown in Fig. 9-6.

(*a*)

(*b*)

FIGURE 9-5
Typical views of constructed wetlands used for stormwater treatment: (*a*) small Australian system and (*b*) highway runoff system near Davis, California.

FIGURE 9-6
View of Arcata, California, FWS constructed wetlands.

9-2 TREATMENT KINETICS AND EFFLUENT VARIABILITY IN CONSTRUCTED WETLANDS AND AQUATIC SYSTEMS

Constituent removal mechanisms, constituent transformations, the types of reaction rates and their determination, the impact of plant decay, and the nature and variability of the effluent from constructed wetlands and aquatic systems are the topics considered in this section. The purpose is to provide a perspective for the analysis of the treatment performance of these systems, which are considered separately in Secs. 9-3 through 9-6. It will be apparent, after reading this section, that much additional information must be gathered and analyzed before the design of these systems can be considered scientific and routine.

Constituent Removal Mechanisms and Transformations

The principal removal and/or transformation mechanisms in various wetland systems are summarized in Table 9-2. Constituents considered are organic matter (e.g., BOD), suspended solids, nitrogen, phosphorus, metals, trace organics, and

TABLE 9-2
Summary of principal removal and transformation mechanisms in constructed wetlands for the constituents of concern in wastewater

Constituent	Free water system	Subsurface flow	Floating aquatics
Biodegradable organics	Bioconversion by aerobic, facultative, and anaerobic bacteria on plant and debris surfaces of soluble BOD, adsorption, filtration, and sedimentation of particulate BOD	Bioconversion by facultative and anaerobic bacteria on plant and debris surfaces	Bioconversion by aerobic, facultative, and anaerobic bacteria on plant and debris surfaces
Suspended solids	Sedimentation, filtration	Filtration, sedimentation	Sedimentation, filtration
Nitrogen	Nitrification/denitrification, plant uptake, volatilization	Nitrification/denitrification, plant uptake, volatilization	Nitrification/denitrification, plant uptake, volatilization
Phosphorus	Sedimentation, plant uptake	Filtration, sedimentation, plant uptake	Sedimentation, plant uptake
Heavy metals	Adsorption of plant and debris surfaces, sedimentation	Adsorption of plant roots and debris surfaces, sedimentation	Adsorption of plant roots, sedimentation
Trace organics	Volatilization, adsorption, biodegradation	Adsorption, biodegradation	Volatilization, adsorption, biodegradation
Pathogens	Natural decay, predation, UV irradiation, sedimentation, excretion of antibiotics from roots of plants	Natural decay, predation, sedimentation, excretion of antibiotics from roots of plants	Natural decay, predation, sedimentation

pathogens. An understanding of the removal mechanisms is of great importance in the development of models that can be used to predict process performance. As shown in Table 9-2, it is difficult to separate constituent removal and transformation processes, as both occur simultaneously in these systems. Definition of the removal mechanisms for individual constituents is complicated further because the constituent may be present in several forms, which will vary with the degree of treatment (e.g., soluble, colloidal, and particulate BOD and organic and ammonia nitrogen).

Constituent transformations that occur in wetlands and aquatic systems are related to the carbon and nutrient cycles, considered previously in Chap. 2. In all wetlands and aquatic systems, both aerobic and anaerobic conditions occur to varying degrees at the same time. For example, the aerobic zone in FWS systems will usually be limited to the open water zones and a very limited upper portion of the water column. If the organic loading that is applied with the wastewater is large, the aerobic zone may extend for only a short distance into the water column. The development of an oxygen sag is quite common in FWS systems. Because both the aerobic and anaerobic conditions exist in FWS systems, both the aerobic and anaerobic carbon cycles are operative. Further, because the relative dominance of aerobic to anaerobic conditions will vary throughout the year, especially in northern climates, it is difficult to predict which cycle is dominant with respect to the treatment of organic material.

Volume- versus Area-Based Reaction Rates for the Removal of Constituents in Constructed Wetlands

In reviewing the constructed wetlands literature dealing with the removal of BOD, as well as other constituents, care must be taken to determine whether the rate constant is based on volume or on the surface area of the control volume. For example, with reference to Fig. 9-7, a volume-based removal-rate coefficient as proposed by Reed et al. (1995) will be given as follows. It should also be noted that in the following analysis it is assumed that the BOD is contributed from a single soluble constituent.

$$r_{BOD} = -k\text{BOD} \qquad (9\text{-}1)$$

where r_{BOD} = rate of BOD loss per unit time per unit volume, $ML^{-3}T^{-1}$
k = rate coefficient for BOD removal, T^{-1}
BOD = carbonaceous BOD concentration, ML^{-3}

An area-based removal model has been proposed by Kadlec and Knight (1996):

$$r_{BOD} = -k_A(A/V)(\text{BOD}) = -(k_A/H)(\text{BOD}) \qquad (9\text{-}2)$$

where r_{BOD} = rate of BOD loss per unit time per unit volume, $ML^{-3}T^{-1}$
k_A = rate coefficient for BOD removal, LT^{-1}
A = surface area, L^2
V = volume, L^3
BOD = carbonaceous BOD concentration, ML^{-3}
H = depth, L

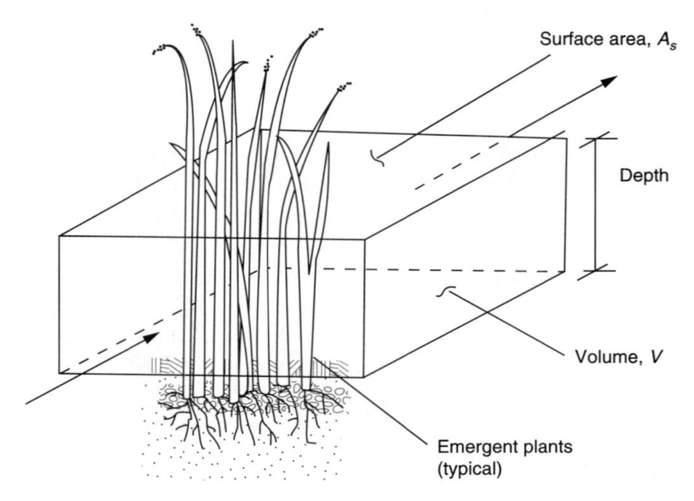

FIGURE 9-7
Definition sketch for modeling the removal of BOD and TSS in an FWS constructed wetland.

If the depth of the water in the wetlands does not change, then the two kinetic rate coefficients can be related directly. The difficulty in using either of the rate coefficients occurs when values developed for one water depth are applied to another depth. Further, neither coefficient is very reflective of all of the transformations occurring within the wetland. Clearly, the removal-rate constant for BOD must be related to the plant surface area below the water surface, and to the plant detrital material present in the wetland. More focused research needs to be done to determine how to model what is actually occurring with aggregate or lumped parameters in these systems. Because of the limited understanding of the actual removal mechanisms, the removal-rate coefficients now used for the design of constructed wetlands are *apparent* coefficients, and do not necessarily have any theoretical basis.

Actual (Nonideal) versus Ideal Flow in Constructed Wetlands

Ideal plug flow is typically assumed in the analysis and design of constructed wetlands. Unfortunately, it has been observed that plug-flow conditions seldom exist in the field. What normally occurs is that preferential-flow channels develop within the wetland as illustrated in Fig. 9-8. The nonideal flow conditions that occur in practice can be modeled (1) by using Eq. (3-13), developed for first-order kinetics and a plug-flow reactor with axial dispersion, and (2) by simulating the actual flow by using a number of complete-mix reactors in series. From dye measurements, it has been found that a cascade of four to six complete-mix reactors in series can be used to model the actual performance of constructed wetlands designed as plug-flow reactors. The impact of nonideal flow on the performance of an assumed plug-flow reactor is considered in Example 9-1.

FIGURE 9-8
Typical example of preferential-flow channels that develop in FWS constructed wetlands, leading to axial dispersion.

EXAMPLE 9-1. ESTIMATE THE APPARENT REMOVAL-RATE CONSTANT FOR A FWS WETLAND. Estimate the apparent removal-rate constant for a FWS constructed wetland which takes into account axial dispersion. A plug-flow reactor has been designed with the following dimensions: width 200 ft, length 400 ft, and depth 1.25 ft. The flowrate is equal to 20,000 ft^3/d. Assume the first-order removal-rate constant for soluble BOD in the wetland, based on experiments conducted at a depth of 1.25 ft, is equal to 1.2 d^{-1}. If the influent soluble BOD value is equal to 300 mg/L, estimate the theoretical and actual BOD to be expected in the effluent, and the apparent removal-rate constants. Neglect the BOD contributed by the system constituents. The void ratio n (porosity) is 0.75.

Solution

1. Determine the theoretical detention time:

$$t = \frac{V}{Q} = \frac{ndA}{Q} = \frac{(0.75)(1.25\text{ ft})(200\text{ ft})(400\text{ ft})}{20,000\text{ ft}^3/\text{d}} = 3.75\text{ d}$$

2. Determine the theoretical effluent BOD from the plug-flow reactor assuming ideal plug flow:

$$\begin{aligned}\text{BOD}_{\text{eff}} &= \text{BOD}_{\text{inf}}e^{-kt} \\ &= 300e^{-1.2\times 3.75} = 3.3\text{ mg/L}\end{aligned}$$

3. Estimate the actual effluent BOD concentration. Assume the actual hydraulic performance of the constructed wetland can be modeled as a cascade of four equal-volume complete-mix reactors. Using Eq. (3-52), estimate the effluent BOD concentration from the plug-flow reactor.

$$\frac{C_4}{C_o} = \frac{1}{(1 + kV/4Q)^4}$$

where C_4 = effluent BOD concentration from the 4th reactor in series, mg/L
C_o = influent BOD concentration = 300 mg/L
k_o = overall BOD removal rate constant = 1.2 d^{-1}
V = total volume of wetland = 75,000 ft^3 (200 ft × 400 ft × 1.25 ft × 0.75)
4 = number of complete-mix reactors in series
Q = flow rate = 20,000 ft^3/d

Substituting and solving for C_4, which corresponds to the expected effluent BOD from the plug-flow reactor, yields

$$C_4 = \frac{300}{(1 + 1.2 \times 75{,}000)/(4 \times 20{,}000)^4} = 14.7 \text{ mg/L}$$

4. Determine the apparent BOD removal-rate constant, assuming a plug-flow model is used to estimate the effluent BOD from the wetland:

$$\frac{\text{BOD}_{\text{eff}}}{\text{BOD}_{\text{inf}}} = e^{-k_{\text{apparent}} \times t}$$

The detention time t is equal to 3.75 d [75,000 ft^3/(20,000 ft^3/d)]. Substituting the appropriate influent and effluent BOD values, the value of the apparent BOD removal-rate constant is

$$\frac{14.7}{300} = e^{-k_{\text{apparent}} \times 3.75}$$

$$\ln \frac{14.7}{300} = -3.016 = -k_{\text{apparent}} \times 3.75$$

$$k_{\text{apparent}} = 3.016/3.75 = 0.804 \text{ d}^{-1}$$

Comment. The above computations illustrate the importance of taking into account axial dispersion in constructed wetlands. Further, because of the limited data that are available in the literature, and the varying conditions that exist in constructed wetlands, the removal-rate constants for BOD and TSS currently used for the design of FWS constructed wetlands are apparent removal-rate constants.

Analysis of Constituent Removal-Rate Constants

The removal of wastewater constituents can be modeled mathematically as described in Chap. 2. Some of the problems encountered in developing models for constructed wetlands and aquatic systems are highlighted in the following discussion, especially with respect to the removal of BOD and TSS. Modeling BOD and TSS removal is complicated further because both are lumped constituents comprising multiple-size particles (see discussion in Chap. 2). For example, the kinetics involved in the removal of soluble and particulate BOD are quite different. Similarly, the mechanisms involved in the removal of settleable and colloidal suspended solids are quite different.

Modeling the removal of BOD. One of the difficulties encountered in modeling the removal of BOD in constructed wetlands and aquatic systems is that the influent BOD may be soluble, colloidal, and/or particulate. In addition, the removal can occur via aerobic/anoxic/anaerobic biological mechanisms and by flocculation/sedimentation. As a consequence, the value of the BOD removal-rate constant will depend on the distribution of the BOD between the three fractions. As reported in Table 2-10 in Chap. 2, the value of the BOD removal-rate constant can vary by a factor of 4 between particle sizes varying from 0.01 to 100 μm. In addition, aerobic and anoxic/anaerobic zones exist simultaneously in the wastewater column. Thus, the BOD removal-rate constants used in the design wetlands, as reported in the literature, are overall removal-rate constants, and should be modified to reflect the nature of the BOD in specific applications.

Another issue that occurs in modeling the removal of BOD in constructed wetlands, resulting from the presence of colloidal and/or particulate BOD composed of particles of varying size, is that the BOD removal-rate constant will vary as the wastewater passes through the wetland as illustrated in Fig. 3-12 in Chap. 3. As shown in Fig. 3-12, as the large particles are removed, by mechanisms such as flocculation/sedimentation, entrapment, and straining by chance contact, the removal-rate coefficient for the remaining smaller particles is reduced, even though the particles themselves may be easier to degrade. To account for the fact that the treatment response decreases as the most responsive constituents are removed, a retarded-rate expression should be used (see discussion in Chap. 3). The typical form of a retarded rate expression is

$$k = \frac{k_{\text{overall}}}{(1 + rt)^n} \tag{9-3}$$

where k = removal-rate constant at time t, 1/d
k_{overall} = initial overall removal-rate constant at time $t = 0$, 1/d
r = coefficient of retardation, 1/d
t = time, $t = L/v$
n = exponent related to the constituent being removed, unitless
L = length, ft
v = velocity, ft/d

When the r or n values are equal to zero, the value of k/k_o is equal to 1, and the overall removal-rate coefficient is constant. For example, the overall BOD removal-rate coefficient would be constant if all of the BOD were soluble or colloidal or particulate of a specified size. For this case, the value of the exponent n is equal to 0. For typical wastewater that contains soluble, colloidal, and particulate BOD, the value of the exponent n is approximately 1.0. For typical wastewater, the coefficient of retardation, which varies with plant density, is approximately equal to 0.2d^{-1}. Here again, sufficient data are not available in the literature that can be used to apply the retarded-removal-rate coefficient with confidence. The importance of the coefficient of retardation will depend on the distribution of the BOD components between the soluble, colloidal, and suspended fractions.

Modeling the removal of TSS. From the above discussion, it is clear that a retarded removal-rate coefficient should be used for modeling the removal of TSS.

The TSS modeling problem is further complicated because of the flocculation of particles that can occur anywhere within the wetland, which increases locally the overall removal-rate constant. In most wetlands, the removal-rate coefficient for TSS is continually changing as the wastewater flows through the wetland. The estimation of TSS removal is illustrated in Example 9-2.

EXAMPLE 9-2. ESTIMATE THE TSS REMOVAL IN A FWS WETLAND. Estimate the removal of TSS in the plug-flow reactor considered in Example 9-1 (width 200 ft, length 400 ft, and depth 1.25 ft). The flowrate is equal to 20,000 ft^3/d. Assume the overall first-order removal-rate constant for TSS in the wetland is equal to 1.25/d. If the influent TSS value is equal to 160 mg/L, estimate the actual TSS to be expected in the effluent, assuming the coefficient of retardation is equal to 0.2 d^{-1}. Compare the TSS effluent value with retardation to the corresponding value without retardation. Assume no flocculation occurs in the wetland, and that the void ratio is 0.75.

Solution

1. Estimate the actual effluent TSS concentration. Assume the actual hydraulic performance of the constructed wetland can be modeled as a cascade of four equal-volume complete-mix reactors. Combining Eqs. (3-52) and (9-3), estimate the effluent TSS concentration from the plug-flow reactor.

$$\frac{TSS_4}{TSS_o} = \frac{1}{\left(1 + \dfrac{k_o}{1 + rt} \times \dfrac{V}{4Q}\right)^4} = \frac{1}{\left(1 + \dfrac{0.25 k_o t}{1 + rt}\right)^4}$$

where TSS_4 = effluent TSS concentration from the fourth reactor in series, mg/L
TSS_o = influent TSS concentration = 160 mg/L
k_o = unretarded overall TSS removal-rate constant = 1.25 d^{-1}.
r = coefficient of retardation, 0.2 d^{-1}
t = detention time, d = $V/Q = ndA/Q$
V = total volume of wetland = 75,000 ft^3
4 = number of complete-mix reactors in series
Q = flowrate = 20,000 ft^3/d

Substituting and solving for TSS, which corresponds to the expected effluent TSS from the plug-flow reactor, yields:

$$\frac{TSS_4}{TSS_o} = \frac{1}{\left[1 + \dfrac{0.25 \times 1.25 \times 3.75}{1 + (0.2 \times 3.75)}\right]^4} = 20.6 \text{ mg/L}$$

2. Estimate the effluent TSS concentration without retardation:

$$TSS_4 = \frac{160}{[1 + (0.25 \times 1.25 \times 3.75)]^4} = 7.2 \text{ mg/L}$$

Comment. The importance of taking into account both axial dispersion and retardation in constructed wetlands is clearly illustrated in this example. The first-order removal-rate constant was assumed for purposes of illustrating the concept.

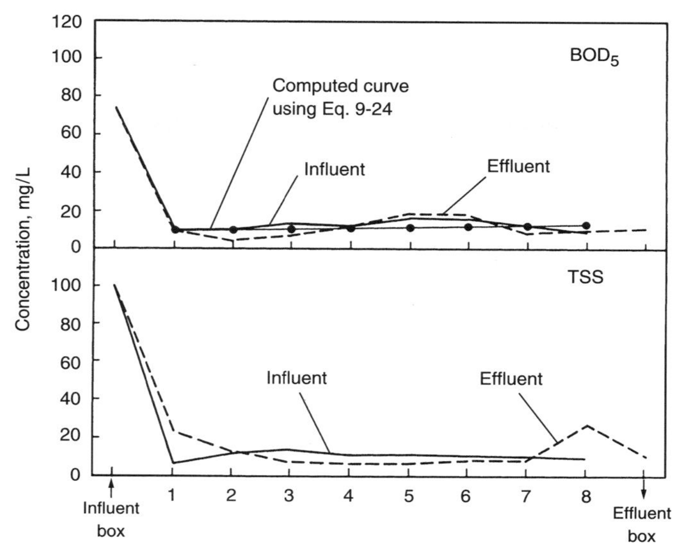

FIGURE 9-9
Removal of BOD and TSS with distance, in water hyacinth treatment system at Aqua III in San Diego, CA (detention time = 6.4 d).

Impact of detention time on observed removal-rate constants for BOD and TSS. Another observation that has been made in a number of wetlands and aquatic systems is that both BOD and TSS are removed extremely rapidly near the influent end of the constructed wetland or aquatic system. Data for the removal of BOD with distance in one of the plug-flow water hyacinth ponds at the San Diego aquaculture project are shown in Fig. 9-9 (WCPH, 1996). The key finding at San Diego was that, under aerobic conditions, secondary treatment was attained within the first 50 ft, and that the remaining 350 ft of the pond provided minimal treatment, if any. This finding led to the development of a pond operating system which incorporated step feed with effluent recycle, as discussed in Sec. 9-5.

If, for practical purposes, the BOD removal during the first 50 ft is modeled as a first-order function (Eq. 3-4), then the apparent removal-rate constant for the first segment of the plug-flow reactor (detention time = 0.8 d) would be

$$\ln\left[\frac{C}{C_o}\right] = k_{\text{apparent}} \times t$$

$$\ln\left[\frac{29}{140}\right] = k_{\text{apparent}} \times 0.8 \text{ d}$$

$$k_{\text{apparent}} = 1.97 \text{ d}^{-1}$$

If, on the other hand, the effluent value from the entire plug-flow pond had been used in the analysis, the corresponding value of the apparent removal-rate constant would be

$$\ln\left[\frac{12}{140}\right] = k_{\text{apparent}} \times 6.4 \text{ d}$$

$$k_{\text{apparent}} = 0.38 \text{ d}^{-1}$$

The difference between these values is significant. This type of problem pervades most of the information reported in the literature on constructed wetlands, where only input and output values are used in determining the apparent removal-rate constant, especially where varying factors of safety have been incorporated into the design of the system.

Effect of temperature. It is also instructive to consider the effect of temperature in the above situations. Bacterial activity responsible for BOD removal is temperature-dependent, with θ values for constructed wetlands ranging from 1.02 to 1.06. It has been observed that bacterial populations in natural systems can acclimate to colder temperatures and maintain their mass in spite of slower activity rates (Vela, 1974). With lower temperatures, the removal of influent BOD occurs farther down the wetland than when water temperatures are higher. In constructed wetlands with excessive detention times, the effect of temperature on BOD and TSS removal will not be observed, as illustrated below.

If it is assumed that the wastewater temperature is 10°C and the temperature coefficient is 1.06, then, from Eq. (3-22) the value of the two removal-rate coefficients, determined above, would be

$$\frac{k_2}{k_1} = \theta^{(T_2 - T_1)}$$

$$\frac{k_2}{1.97/\text{d}} = 1.06^{(10-20)} \qquad k_2 = 1.10 \text{ d}^{-1}$$

$$\frac{k_2}{0.38/\text{d}} = 1.06^{(10-20)} \qquad k_2 = 0.21 \text{ d}^{-1}$$

In the first case, a value of 29 mg/L will be reached after 1.43 d. A value of 12 mg/L, corresponding to the effluent value observed during the summer, would have been reached during the cold period in 2.23 d.

$$\ln\left[\frac{29}{140}\right] = -1.10 \times t$$

$$t = 1.43 \text{ d}$$

$$\ln\left[\frac{12}{140}\right] = -1.10 \times t$$

$$t = 2.23 \text{ d}$$

The implications of this simple analysis are also significant. For example, if the effluent from the plug-flow pond (detention time = 6.4 d) were sampled during the period when the wastewater temperature was 10° C, it would be concluded that temperature has no effect on the process. As with the BOD analysis, temperature effects are often misinterpreted when only input and output values are used in the analysis. This situation is especially problematic when effluent values from overdesigned systems are used. In addition, the input and output values used for the determination of the temperature effects are confounded statistically, because they also include the effects of natural variation. This situation is considered subsequently in the discussion dealing with the selection of design values for BOD that take variability into account.

Modeling the removal of other wastewater constituents. For practical purposes, the modeling of the removal of nitrogen in constructed wetlands is accomplished by assuming that the organic nitrogen in the influent will convert into the form of ammonia nitrogen. Use of the temperature correction factor to adjust the removal-rate coefficient for nitrogen is also appropriate, because both nitrification and denitrification are highly temperature-sensitive. A similar approach is often used for other constituents that may be lumped (or have distributed parameters).

Impact of Vegetation Decay in Wetlands and Aquatic Systems

An important characteristic of both natural and constructed wetlands, especially the free-water-surface type, is related to the growth and the short- and long-term impact of the decay of plant vegetation in these systems. When plant mass dies back and is submerged in water, water-soluble organic substances are transferred to the liquid by leaching. The leached material consists primarily of amino acids, sugars, and nonvolatile and volatile aliphatic acids. In general, these materials are readily metabolized within the wetland (Westermann, 1993). Further (long-term) degradation of the plant material in the system will depend on the ratios of the major polymers (lignin, cellulose, and hemicellulose) in the thatch layer, the structure of the lignocellulose, and the physicochemical character of the wetland (see Westermann, 1993).

The importance of this discussion is that, in constructed wetlands, as well as in aquatic treatment systems, it has been observed that the effluent from such systems will contain varying concentrations of organic matter, without the application of wastewater. Typical concentration values for the organic material in the effluent, expressed in terms of BOD, are in the range from 2 to 10 mg/L, with typical values from 3 to 5 mg/L. This effluent value has often led to the formulation of BOD removal models on the false premise that the effluent BOD is residual influent BOD (see discussion of BOD in Sec. 2-1, Chap. 2).

Composition of Effluent BOD

The BOD in the effluent from constructed wetlands and aquatic systems is composed of the BOD resulting from plant decay, as discussed above, and the residual BOD remaining from the original influent BOD. As noted in Sec. 2-1 in Chap. 2, the residual BOD derived from the influent BOD will typically comprise cell tissue and cell fragments, especially in systems with long detention times. The total BOD in the effluent is given by

$$BOD_{ECW} = BOD_{PD} + BOD_{RIW} \tag{9-4}$$

where BOD_{ECW} = effluent BOD from constructed wetland, mg/L
BOD_{PD} = BOD resulting from plant decay, mg/L
BOD_{RIW} = residual BOD from influent wastewater, mg/L

Because both BOD_{PD} and BOD_{RIW} have been observed to vary throughout the year, this variability must be considered in the design of constructed wetlands and aquatic systems. At present, limited data are available on the variability of the plant decay contribution (BOD_{PD}) with season. What data are available are contradictory. In some systems, the BOD contribution from plant decay increases during the summer, whereas in other systems it increases during the winter. For this reason, it is recommended that a typical value be used for estimating BOD_{PD} until more information becomes available.

Design of Constructed Wetlands Taking into Account Variability

As noted in Chap. 3 (Sec. 3-7), because of the variations observed in effluent quality, a treatment process should be designed to produce an average effluent concentration below the permit requirements. In Eq. (9-4), the effluent BOD (BOD_{RIW}) value reflects the effects of temperature, axial dispersion, and natural process variability. Because most of the effluent BOD data that have been collected to date are confounded statistically, the variability due to axial dispersion versus temperature versus natural causes is unknown. The variability of BOD_{RIW} including the combined effects of axial dispersion, temperature, and natural variability can be assessed by analyzing the long-term average monthly performance data from operating systems using the coefficient of reliability as outlined in Chap. 3. Typical values for the coefficient of variation for the different types of wetlands subject to different temperature ranges

TABLE 9-3
Typical coefficients of variation for constructed wetlands subject to different wastewater temperature variations*

		Coefficient of variation V_x for removal of	
Type of wetland	Temperature range, °C	BOD	TSS
Free water surface	5–20	0.40–0.65	0.30–0.50
	10–25	0.25–0.40	0.20–0.40
Subsurface flow	5–20	0.25–0.30	0.25–0.50
	10–25	0.25–0.40	0.20–0.40
Water hyacinths	10–20	0.20–0.25	0.20–0.25
	15–25	0.15–0.20	0.15–0.25

*The data presented in this table should be used with caution, as there is great variability in the performance of these systems.

are given in Table 9-3. The determination of the BOD design value for a constructed wetland is illustrated in Example 9-3.

EXAMPLE 9-3. DETERMINE THE EFFLUENT BOD DESIGN CONCENTRATION VALUE TAKING INTO ACCOUNT THE VARIABILITY IN THE PERFORMANCE OF A CONSTRUCTED WETLAND. Using the following average monthly effluent BOD data from the Ouray, Colorado, FWS constructed wetland wastewater treatment facility, determine the coefficient of reliability and the appropriate BOD design value, if the effluent from a similar treatment facility is to be equal to or less than 30 mg/L 90 percent of the time.

	BOD, mg/L	
Month	**1994**	**1995**
January	10	12
February	10	8
March	11	7
April	14	9
May	19	8
June	19	15
July	24	14
August	24	10
September	15	18
October	12	16
November	1	6
December	11	3

Solution

1. Use the coefficient of reliability approach introduced in Sec. 3-7 in Chap. 3 to determine the appropriate design value.
2. Determine the statistics for the given data using a standard statistical package.

Parameter	**Value**
Minimum	1
Maximum	24
Sum	296
Points	24
Mean	12.3
Median	11.5
RMS	13.6
Standard deviation	5.8
Variance	33.9
Standard error	1.2
Skewness	0.3
Kurtosis	−0.2

3. Determine the coefficient of reliability using Eq. (4-23):

$$\text{COR} = (V_x{}^2 + 1)^{1/2} \times \exp\left\{-Z_{1-\alpha}\left[\ln\left(V_x{}^2 + 1\right)\right]^{1/2}\right\}$$

a. Determine the value of V_x using the results of the statistical analysis:

$$V_x = \frac{\sigma_x}{m_x} = \frac{5.8}{12.3} = 0.472$$

b. The value of $Z_{1-\alpha}$ for a cumulative probability of 90 percent from Table 4-24 is 1.282.

c. Determine the coefficient of reliability:

$$\text{COR} = (0.472^2 + 1)^{1/2} \times \exp\left\{-1.282\left[\ln\left(0.472^2 + 1\right)\right]^{1/2}\right\} = 0.622$$

4. Determine the appropriate design value for BOD. Using Eq. (4-22) and the COR value determined in step 3, the design value is

$$m_x = (\text{COR})X_s = 0.622 \times 30 \text{ mg/L} = 18.7 \text{ mg/L}$$

Comment. As the variability in the effluent quality increases, the COR value becomes smaller, and a more conservative design value must be used to achieve the proposed level of treatment.

9-3 FREE-WATER-SURFACE CONSTRUCTED WETLANDS

The use of constructed wetlands with water levels above the ground surface has ranged from achieving secondary treatment, to polishing of secondary effluent, to providing wildlife habitat and reuse of the water. The material presented in this section deals with a description of the process, constituent removal and transformation mechanisms, process performance, and process design considerations. General design considerations and the management for these systems are discussed on Secs. 9-8 and 9-9, respectively.

Process Description

A free-water-surface (FWS) system consists typically of channels or basins with a natural or constructed impermeable barrier to prevent seepage. Plants in free-water-surface constructed wetlands serve a number of purposes. Stems, submerged leaves, and litter serve as support media for the growth of attached bacteria. Leaves above the water surface shade the water and reduce the potential for algal growth. Oxygen is transported from the leaves down into the root zone, which supports the plant growth. A limited amount of oxygen may leak out of the submerged stems to support attached bacterial growth. Pretreatment for FWS wetlands usually consists of settling (septic tanks or Imhoff tanks), screening with a rotary disk filter, or stabilization lagoons. Because the major sources of oxygen are surface reaeration in open water from the atmosphere and attached-growth algae, the BOD loading generally needs to be kept below 100 lb/ac·d.

TABLE 9-4
Typical characteristics of emergent plants used in constructed wetlands

Common name	Scientific name	Temperature, °C		pH range for effectiveness	Maximum salinity tolerance, ppt
		Desirable	Seed germination		
Bulrush	*Scirpus* spp.	16–27		4–9	20
Cattail	*Typha latifolia*	10–30	12–24	4–10	30
Common arrowhead	*Sagittaria latifolia*				
Common reed	*Phragmites australis*	12–23	10–30	2–8	45
Rush	*Juncus* spp.	16–26		5–7.5	20
Sedge	*Carex* spp.	14–32		5–7.5	
Yellow flag	*Iris pseudacorus*				

Source: Stephenson et al. (1980)
Note: ppt = parts per thousand

Site selection. Site features for potential FWS sites are similar to those for wastewater treatment ponds. Slopes of 0 to 3 percent are most favorable. Soils should be slowly permeable. Compacted clay or synthetic liners may be required to limit percolation. Groundwater levels can be relatively high without causing any concern because percolation is restricted or eliminated.

Vegetation types. Emergent plants most frequently used in FWS include: cattails, bulrush, reeds, arrowhead, and sedges. Characteristics of these plants are summarized in Table 9-4 (Stephenson et al., 1980). More details on emergent plants are available in the reference works by Mitsch and Gosselink (1993) and Hammer (1992). In addition to the plants listed in Table 9-4, arrow arum (*Peltandra* spp.) and pickerelweed (*Pontederia* spp.) have been used in constructed wetlands. Other locally grown emergent vegetation can also be considered.

Constituent Removal and Transformation Mechanisms

High removals of BOD and TSS can be expected from FWS wetlands, along with significant removals of nitrogen, metals, trace organics, and pathogens. The degree of removal usually is dependent on detention time and temperature. The operative removal mechanisms for FWS constructed wetlands are described below.

BOD removal. Soluble and particulate BOD are removed by different mechanisms in FWS constructed wetlands. Soluble BOD is removed by biological activity and adsorption on the plant and detritus surfaces and in the water column. The low velocities and emergent plants facilitate flocculation/sedimentation and entrapment of the particulate BOD. Organic solids, removed by sedimentation and filtration, as discussed below, will exert an oxygen demand, as does the decaying vegetation. As a result, the influent BOD is removed rapidly with length down the wetland cell. The observed BOD in the wetland will also reflect the detrital and benthic demand, which leads to a "background" concentration.

Total suspended solids removal. The principal removal mechanisms for TSS are flocculation and sedimentation in the bulk liquid, and filtration (mechanical straining, chance contact, impaction, and interception) in the interstices of the detritus. Most of the settleable solids are removed within 50 to 100 ft of the inlet. Optimal removal of TSS requires a full stand of vegetation to facilitate sedimentation and filtration and to avoid regrowth of algae. Algal solids may take 6 to 10 d of detention time for removal.

Nitrogen removal. Nitrogen removal in constructed wetlands is accomplished by nitrification and denitrification. Plant uptake accounts for only about 10 percent of the nitrogen removal. Nitrification and denitrification are microbial reactions that depend on temperature and detention time. Nitrifying organisms require oxygen and an adequate surface area to grow on and, therefore, are not present in significant numbers in either heavily loaded systems (BOD loading $>$ 100 lb/ac·d) or in newly constructed systems with incomplete plant cover. On the basis of field experience with FWS systems, one to two growing seasons may be needed to develop sufficient vegetation to support microbial nitrification. Denitrification requires adequate organic matter (plant litter or straw) to convert nitrate to nitrogen gas. The reducing conditions in mature FWS constructed wetlands resulting from flooding are conducive to denitrification. If nitrified wastewater is applied to a FWS wetland, the nitrate will be denitrified within a few days of detention.

Phosphorus removal. The principal removal mechanisms for phosphorus in FWS systems are adsorption, chemical precipitation, and plant uptake. Plant uptake of inorganic phosphorus is rapid; however, as plants die, they release phosphorus, so that long-term removal is low. Phosphorus removal depends on soil interaction and detention time. In systems with zero discharge or very long detention times, phosphorus will be retained in the soil or root zone. In flow-through wetlands with detention times between 5 and 10 d, phosphorus removal will seldom exceed 1 to 3 mg/L. Depending on environmental conditions within the wetland, phosphorus, as well as some other constituents, can be released during certain times of the year, usually in response to changed conditions within the system such as a change in the oxidation-reduction potential (ORP).

Metals removal. Heavy metal removal is expected to be very similar to that of phosphorus removal, although limited data are available on actual removal mechanisms. The removal mechanisms include adsorption, sedimentation, chemical precipitation, and plant uptake. As with phosphorus, metals can be released during certain times of the year, usually in response to change in the oxidation-reduction potential within the system.

Trace organics removal. Although limited data are available on removal of trace organics, the FWS process is similar to overland flow (see Chap. 10) where removals of 88 to 99 percent have been reported (Reed et al., 1995). Removal mechanisms include volatilization, adsorption, and biodegradation.

Pathogen removal. Pathogenic bacteria and viruses are removed in constructed wetlands by adsorption, sedimentation, predation, and die-off from exposure to sunlight (UV) and unfavorable temperatures.

Process Performance

The performance expectations for FWS constructed wetlands are presented in Tables 9-5 through 9-8. Performance depends, of course, on design criteria, wastewater characteristics, and operations.

BOD and TSS removal. Operating data from a number of FWS constructed wetlands for removal of BOD and TSS are presented in Table 9-5. Removals are

TABLE 9-5
Typical BOD and TSS removals observed in FWS constructed wetlands

	BOD, mg/L		TSS, mg/L		
Location	**Influent**	**Effluent**	**Influent**	**Effluent**	**Reference**
Arcata, California	26	12	30	14	Gearheart et al., 1989
Benton, Kentucky	25.6	9.7	57.4	10.7	U.S. EPA, 1993
Cannon Beach, Oregon	26.8	5.4	45.2	8.0	U.S. EPA, 1993
Ft. Deposit, Alabama	32.8	6.9	91.2	12.6	U.S. EPA, 1993
Gustine, California	75	19	102	31	Crites, 1996
Iselin, Pennsylvania	140	17	380	53	Watson et al., 1979
Listowel, Ontario	56.3	9.6	111	8	Herskowitz et al., 1987
Ouray, Colorado	63	11	86	14	Andrews, 1996
West Jackson Co., Mississippi	25.9	7.4	40.4	14.1	U.S. EPA, 1993
Sacramento Co., California	23.9	6.5	8.9	12.2	Nolte and Associates, 1997

TABLE 9-6

Typical ammonia and nitrogen removals observed in FWS constructed wetlands

Location	Type of effluent	Ammonia, mg/L		Total nitrogen, mg/L	
		Influent	Effluent	Influent	Effluent
Arcata, California*	Oxidation pond	12.8	10		11.6
Iselin, Pennsylvania†	Oxidation pond	30	13		
Jackson Bottoms, Oregon	Secondary	9.9	3.1		
Listowel, Ontario‡	Primary	8.6	6.1	19.1	8.9
Pembroke, Kentucky	Secondary	13.8	3.35		
Sacramento Co., California§	Secondary	14.1	7.2	16.8	9.1

*Full-scale operation from August 1986 to May 1988 (Gearheart et al., 1989).
†Full-scale operation from March 1983 to September 1985 (Watson et al., 1987).
‡System 4, pilot operation from 1980 to 1984 (Herskowitz et al., 1987).
§Demonstration wetland, May 1995 to November 1995.

TABLE 9-7

Removal of metals in constructed wetlands at Sacramento Regional County Sanitation District*

Metal	Average total recoverable concentration, μg/L	
	Influent	Effluent
Antimony (Sb)	0.43	0.19
Arsenic (As)	1.75†	2.38
Cadmium (Cd)	0.10	0.07
Chromium (Cr)	1.05	1.32
Copper (Cu)	8.90	4.14
Lead (Pb)	0.85	0.25
Mercury (Hg)	11.39 ng/L‡	4.57 ng/L‡
Nickel (Ni)	6.85	8.34
Silver (Ag)	0.34	0.05
Zinc (Zn)	37.0	7.4

*Results for the period July 1996 to December 1996 (Nolte and Associates, 1997). The system began receiving treated wastewater in May 1994.
†Arsenic in the influent dropped from 2.63 μg/L in 1995 to 1.75 μg/L in 1996.
‡ng/L = nanograms per liter.

TABLE 9-8
Removal of fecal coliform in FWS constructed wetland systems

Location	Unit	Influent	Effluent*	Detention time, d
Iselin, Pennsylvania; cattails and grasses†				
Winter season (Nov.–April)	No./100 mL	1.7×10^6	4.3×10^3	6
Summer season (May–Oct.)	No./100 mL	1.0×10^6	723	6
Arcata, California; bulrush wetland‡				
Winter season (Nov.–April)	No./100 mL	4.3×10^3	900	1.9
Summer season (May–Oct.)	No./100 mL	1.8×10^3	80	1.9
Listowel, Ontario; cattails‡				
Winter season (Nov.–April)	No./100 mL	5.56×10^5	1.4×10^3	7–14
Summer season (May–Oct.)	No./100 mL	1.98×10^5	400	7–14

*Undisinfected.
†Sand bed, subsurface flow.
‡Free water surface.

typically 60 to 80 percent for BOD and 50 to 90 percent for TSS (depending on the nature and concentration of the influent TSS).

Ammonia removal. As shown in Table 9-6, the degree of nitrification in FWS systems is relatively incomplete, ranging from 25 percent in Arcata to 56 percent at Iselin. The data should not be construed as indicating that nitrification cannot be complete in FWS constructed wetlands, because nitrification has not been required generally in effluent permit limitations. Most of the systems that have been monitored were designed for BOD and TSS removal with detention times between 5 and 10 d. At Sacramento County, seasonal nitrification has averaged 75 percent for a detention time of 10 d (Crites et al., 1997).

Nitrogen removal. Nitrogen removal is limited by the ability of the FWS system to nitrify. When nitrogen is present in the nitrate form, nitrogen removal is generally rapid and complete. The removal of nitrate depends on the concentration of nitrate, the detention time, and the available organic matter. Because the water column is nearly anoxic in many wetlands treating municipal wastewater, the reduction of nitrate will occur within a few days. At Sacramento County, the nitrate does not accumulate as the ammonia is nitrified. Nitrate-nitrogen concentrations average 0.44 mg/L and range from 0.08 mg/L in the summer to 0.94 mg/L in the spring (Nolte and Associates, 1997).

Phosphorus removal. Phosphorus removal in wetlands depends on the loading rate and the detention time. Because plants take up phosphorus over the growing season and then release some of it during senescence, reported removal data must be questioned as to when the system was sampled and how long the system had been in operation. At Sacramento County, the typical spring/summer uptake of phosphorus is about 0.5 mg/L, resulting in an annual removal of 14 percent (Nolte and Associates, 1997).

TABLE 9-9
Typical design criteria and expected effluent quality for FWS constructed wetlands

Item	Unit	Value
Design parameter		
Detention time	d	2–5 (BOD) 7–14 (N)
BOD loading rate	lb/ac·d	<100
Water depth	ft	0.2–1.5
Minimum size	ac/Mgal·d	5–10
Aspect ratio		2:1 to 4:1
Mosquito control		Required
Harvesting interval	yr	3–5
Expected effluent quality*		
BOD_5	mg/L	<20
TSS	mg/L	<20
TN	mg/L	<10
TP	mg/L	<5

*Expected effluent quality based on a BOD loading equal to or less than 100 lb/ac·d and typical settled municipal wastewater.

Metals removal. Metals removal depends on detention time, influent metal concentrations, and metal speciation. Removal data for heavy metals in the Sacramento County Demonstration Wetlands are presented in Table 9-7.

Pathogen removal. Fecal coliform removals of 99.9 percent (3 log reduction) have been reported at Iselin, Pennsylvania, and at Listowel, Ontario (Watson et al., 1989). Virus removal at Arcata, California, ranged from 90 to 99 percent (Gersberg et al., 1989). Removals of pathogen indicator organisms are presented in Table 9-8.

Process Design Considerations

The principal process design criteria for FWS constructed wetlands are detention time, organic loading rate, required surface area, and water depth. Hydraulic loading rate is a common basis of comparison, either in in/d or ac/Mgal·d, but both rates are calculated from the area and the flow. Other design considerations include aspect (length-to-width) ratio, hydraulic considerations, thermal considerations, and vegetation harvesting. Typical process design criteria are presented in Table 9-9.

Detention time for BOD. The required detention time, taking into account axial dispersion and temperature effects, can be determined theoretically by

$$t = \frac{V}{Q} = \left[\frac{1}{(C_n/C_o)^{1/n}} - 1\right] \times \frac{n}{k_o} \tag{9-5}$$

where t = detention time for BOD removal, d
V = total volume of wetland, ft^3 (gal)
Q = flowrate, ft^3/d (gal/d)
C_n = effluent BOD concentration from the nth reactor in series, mg/L
C_o = influent BOD concentration, mg/L
n = number of complete-mix reactors in series
k_o = overall BOD removal-rate constant, corrected for temperature, 1/d

The value of C_n is the residual BOD value from the influent BOD. The actual total BOD concentration in the effluent consists of the residual BOD value from the influent BOD plus the BOD from plant decay. Typically, four reactors in series are used most commonly to account for axial dispersion in plug-flow reactors. The value of k_o is usually based on controlled pilot-scale experiments in which axial dispersion is not an issue.

Because insufficient data are available to determine the overall removal rate constant k_o or COR values, it is recommended that the apparent k factor in Eq. (9-6) be used for design. Derived from field observations, the empirical temperature-corrected apparent BOD removal-rate constant k_{apparent} is 0.678 d^{-1}. It should be noted that Eqs. (9-5) and (9-6) will yield approximately the same answer, if k = 1.01 d^{-1} for Eq. (9-5), because apparent removal-rate constants are derived from systems with varying amounts of axial dispersion.

$$t = -\frac{\ln C/C_o}{k_{\text{apparent}}} \tag{9-6}$$

If adequate statistical data are available for similar systems in similar climatic conditions, the design detention time can be computed by using the coefficient of reliability concept as outlined in Chap. 3. The BOD design value for the wetland is

$$\text{BOD}_{\text{DES}} = \text{BOD}_{\text{RIW}} \times (\text{COR}) \tag{9-7}$$

where BOD_{RIW} = residual average monthly BOD from influent wastewater, mg/L, and COR = coefficient of reliability (see Sec. 3-7 in Chap. 3 and Example 9-2).

Organic loading rate. As a general rule, the organic loading rate (OLR) should not exceed 100 lb BOD/ac·d (110 kg BOD/ha·d), if aerobic conditions near the water surface are to be maintained and odors are to be minimized. The organic loading rate can be checked by the following expression:

$$L_{\text{org}} = \frac{(C)(d_w)(\eta)(F_1)}{t \times F_2} \tag{9-8}$$

where L_{org} = organic loading rate, lb BOD/ac·d (kg BOD/ha·d)
C = BOD concentration in influent, mg/L (g/m^3)
d_w = depth of flow, ft (m)
η = plant based void ratio, 0.65 to 0.75 typically
F_1 = conversion factor, 8.34 lb/[Mgal·(mg/L)] (0.001 kg/g)
t = detention time, d
F_2 = conversion factor, 3.07 ac·ft/Mgal (10^{-4} ha/m^2)

Required surface area. Once the detention time is calculated, the net area of the wetland can be determined from

$$A = (Q_{\text{ave}})(t)(3.07)/(d_w)(\eta) \quad (9\text{-}9)$$

where Q_{ave} = average daily flow through the wetland, Mgal/d, and A = area, ac. Other terms are as described previously.

The average flow through the wetland can be estimated by the following equation:

$$Q_{\text{ave}} = \frac{Q_{\text{in}} + Q_{\text{out}}}{2} \quad (9\text{-}10)$$

The average flow must be used to account for the influence of evapotranspiration, seepage losses, and precipitation. Evapotranspiration values for wetland plants are typically equal to the potential evapotranspiration from an open water surface (Reed et al., 1995). The calculation of required area is illustrated in Example 9-4. The two design methods are compared in Example 9-5.

Water balance. In the arid west, where evapotranspiration exceeds precipitation on an annual basis, it may be necessary to conduct a water balance to determine the effect of evapotranspiration on detention time and effluent water quality. For reuse wetlands, the net loss of water is by evapotranspiration and percolation. The water balance approach is detailed in Chap. 10, Sec. 10-3, and consists of monthly tabulations of inflow, precipitation, evapotranspiration, seepage or percolation, and outflow or storage. The flowrate values are converted to inches per month (in/mo) or millimeters per month (mm/mo) to allow the outflow values to be calculated. If the evapotranspiration rate is a large portion of the inflow (greater than 25 percent), the effects on water quality, particularly trace metals such as selenium, should be evaluated.

Aspect ratio. The surface dimensions can be determined by the following expression:

$$w = \left(\frac{A}{R_A}\right)^{1/2} \quad (9\text{-}11)$$

where w = width of FWS wetland, ft (m)
A = area of FWS wetland, ft^2 (m^2)
R_A = aspect ratio, length/width

To minimize short circuiting of wastewater from the inlet to the outlet, relatively large aspect ratios (length-to-width) of rectangular basins have been proposed. If large aspect ratios (greater than 10:1) are used, a relatively large hydraulic gradient is needed to prevent backup and overflow problems in the wetland cells. Aspect ratios of 2:1 to 4:1 have been used (Reed et al., 1995).

EXAMPLE 9-4. DETERMINE AREA REQUIRED FOR BOD REMOVAL IN A FWS CONSTRUCTED WETLAND. Determine the area required for a FWS constructed wetland used to treat primary treated wastewater with an influent BOD of 100 mg/L. To be assured of odorless operation, the maximum organic loading rate is to be equal to or less than 100 lb BOD/ac·d. Assume the overall first-order BOD removal rate constant is $1.0\ d^{-1}$ at 20°C. The average wastewater temperature during the coldest month is about 10°C. The average water depth is to be 1.25 ft. Because of evaporation, the effluent flowrate is equal to $0.8 \times$ the influent flow rate. Assume that the plant porosity is 0.70 and that the average BOD in the effluent due to plant decay is 5 mg/L. The combined effluent BOD (from plant decay and residual from influent) must be 25 mg/L or less. The observed temperature coefficient and COR values for a similar FWS system in the same temperature regime are 1.02 and 0.50 (at 99 percent), respectively. Compare the detention time to that calculated by using Eq. (9-6).

Solution

1. Determine the maximum allowable residual BOD from the influent wastewater that can be present in the effluent from the wetland using Eq. (9-4):

$$\text{BOD}_{\text{RIW}} = \text{BOD}_{\text{ECW}} - \text{BOD}_{\text{PD}}$$
$$= 25\ \text{mg/L} - 5\ \text{mg/L} = 20\ \text{mg/L}$$

2. Determine the required detention time using Eq. (9-5), for a cascade of four reactors in series, to account for axial dispersion in an ideal plug-flow reactor:

$$t = \frac{V}{Q} = \left[\frac{1}{(C_n/C_o)^{1/n}} - 1\right] \times \frac{n}{k_o}$$

a. Determine the temperature-corrected overall BOD removal-rate constant using Eq. (3-15):

$$k_{o(10)} = k_{o(20)} \times 1.02^{(10-20)}$$
$$= 1.0/\text{d} \times 1.02^{(10-20)} = 0.82\ \text{d}^{-1}$$

b. Determine the required detention time:

$$t = \left[\frac{1}{(20/100)^{0.25}} - 1\right] \times \frac{4}{0.82} = 2.4\ \text{d}$$

3. Check the required detention time using the COR approach.

a. Determine the required effluent concentration based on the COR value:

$$\text{BOD}_{\text{DES}} = \text{BOD}_{\text{RIW}} \times (\text{COR}) = 20 \times 0.50 = 10\ \text{mg/L}$$

b. Determine the detention time based on first-order BOD removal kinetics [Eq. (9-6)]:

$$t = -\frac{\ln(C/C_o)}{k_o}$$
$$= -\frac{\ln(10/100)}{1.0\ \text{d}} = 2.3\ \text{d}$$

4. Determine the detention time using the apparent k value of 0.678 at 20°C:

a. Calculate k using Eq. (3-15):

$$k = 0.678 \times 1.02^{(10-20)}$$
$$= 0.556\ \text{d}^{-1}$$

b. Determine the detention time based on k_{apparent} [Eq. (9-6)]:

$$t = -\ln(C/C_o)/k$$
$$= 2.5 \text{ d}$$

Use 2.5 d detention time as the most conservative of the three approaches.

5. Check the organic loading rate using Eq. (9-8):

$$L_{\text{org}} = \frac{(C)(d_w)(\eta)(F_1)}{t \times F_2}$$

$$= \frac{(100)(1.25)(0.7)(8.34)}{2.5 \times 3.07} = 95 \text{ lb BOD/ac·d}$$

6. Determine the area required using Eq. (9-9):

$$A = (Q_{\text{ave}})(t)(3.07)/(d_w)(\eta)$$

a. Determine the value of Q_{ave}:

$$Q_{\text{ave}} = \frac{1.0 + 0.8(1)}{2} = 0.9 \text{ Mgal/d}$$

b. Determine the area:

$$A = (0.9)(2.5)(3.07)/(1.25)(0.7) = 7.9 \text{ ac}$$

Comment. The value of k_o in Eq. (9-5) is assumed, whereas the k_{apparent} value of 0.678 has been derived from a number of operating systems. The use of Eq. (9-6) is recommended for design.

If the coefficient of reliability (COR) is extremely low, a larger area will be required. However, as will be shown subsequently, the area required for the removal of nitrogen will far exceed the area required for BOD and TSS removal.

EXAMPLE 9-5. COMPARISON OF DESIGN METHODS. Compare the design approach used for a FWS constructed wetland at Ouray, Colorado, to the design approach using the coefficient of retardation COR as outlined in Sec. 9-2 and in Chap. 3. The summer temperature of the water in the constructed wetland is 15.9°C. The average influent BOD is 73 mg/L. The required effluent BOD is equal to or less than 30 mg/L. The overall apparent BOD removal-rate constant used in the design was 0.678 d^{-1} at 20°C.

Solution—Existing design from Ouray, Colorado (Andrews, 1996)

1. Determine the maximum allowable residual BOD from the influent wastewater that can be present in the effluent from the wetland using Eq. (9-4):

$$\text{BOD}_{\text{RIW}} = \text{BOD}_{\text{ECW}} - \text{BOD}_{\text{PD}}$$
$$= 30 \text{ mg/L} - 4 \text{ mg/L} = 26 \text{ mg/L}$$

2. The required detention time was determined from Eq. (9-6):

$$t = -\frac{\ln C/C_o}{k_{\text{apparent}}}$$

where C = effluent BOD concentration = 26 mg/L

C_o = influent BOD concentration = 73 mg/L

k_{apparent} = overall BOD removal-rate constant, corrected for temperature, d^{-1}

a. The apparent overall BOD removal rate constant is temperature-corrected by using Eq. (3-15):

$$k_{15.9} = k_{20} \times 1.06^{(15.9-20)}$$

$$= (0.678/\text{d}) \times 1.06^{-4.1} = 0.53\ \text{d}^{-1}$$

b. The required detention time was then computed as follows:

$$t = -\frac{\ln(26/73)}{0.53} = 1.95\ \text{d}$$

Solution—Design based on COR approach using actual performance data (Andrews, 1996)

3. Estimate the overall BOD removal-rate constant using Eq. (9-5) for a cascade of four reactors in series to account for axial dispersion:

$$k_o = \left[\frac{1}{(C_n/C_o)^{1/n}} - 1\right] \times \frac{n}{t}$$

a. Determine the temperature-corrected overall BOD removal-rate constant:

$$k_o = \left[\frac{1}{(26/73)^{0.25}} - 1\right] \times \frac{4}{1.95} = 0.6\ \text{d}^{-1}$$

b. Estimate the 20°C overall BOD removal-rate constant.

$$k_{o(20)} = \frac{k_{o(10)}}{1.06^{(15.9-20)}} = \frac{0.60}{0.787} = 0.76\ \text{d}^{-1}$$

4. Determine the required detention time using the COR approach.

a. Determine the required effluent BOD design concentration. Using the actual performance data from the Ouray system, the COR value, based on meeting a specified effluent limit 90 percent of the time, was found to be equal to 0.622 (see Example 9-3).

$$\text{BOD}_{\text{DES}} = \text{BOD}_{\text{RIW}} \times (\text{COR}) = 26 \times 0.622 = 16.2\ \text{mg/L}$$

b. Determine the detention time based on first-order BOD removal kinetics:

$$t = -\frac{\ln C/C_o}{k_o}$$

$$= -\frac{\ln(16.2/73)}{0.76/\text{d}} = 1.98\ \text{d}$$

Comment. A factor of safety of 15 to 25 percent is typically applied to the calculated detention time when Eq. (9-5) is used. In this case, the 1.95-d detention time determined in step 2 would be increased from 2.2 to 2.4 d. With the COR approach, the factor of safety is built in at the 90 percent level of reliability. For example, if the effluent reliability limit was 99 percent, the corresponding detention time would be equal to 2.4 d.

Loading rates for TSS removal. The removal of TSS is the result of physical interactions within the wetland. The removal of TSS has been related to the hydraulic loading rate, as given by the following empirical equation.

$$C_e = C_o[0.1139 + 8.4 \times 10^{-4}(L_W)] \quad (9\text{-}12)$$

where C_e = effluent TSS, mg/L
C_o = influent TSS, mg/L
L_W = wastewater hydraulic loading rate, in/d

Detention time for nitrogen removal. Longer detention times are necessary, typically, for nitrification and nitrogen removal than for BOD removal. In addition, the BOD loading rate must be relatively low so that the bacteria responsible for nitrification can obtain adequate oxygen to function.

The following first-order plug-flow equation can be used to predict ammonia nitrogen removal (Reed et al., 1995):

$$\frac{N_e}{N_o} = e^{-kt} \quad (9\text{-}13)$$

where N_e = effluent ammonia-nitrogen concentration, mg/L
N_o = influent ammonia-nitrogen concentration, mg/L
k = 0.2187 d^{-1} (at 20°C)
t = detention time, d

As noted previously in Sec. 9-2, in applying Eq. (9-13) all of the organic and ammonia nitrogen present is assumed to be in the form of ammonia. As more information is developed, it may be possible to model the various fractions of nitrogen (e.g., soluble, colloidal, and particulate).

If it is necessary to approximate the ammonia removal for conditions of temperature below 10°C, the recommended procedure would be to reduce the k value to 0.0389. For temperatures of 1°C and higher, use a θ value of 1.048 in Eq. (3-22) (Reed et al., 1995).

Nitrate removal can also be predicted from Eq. (9-13) by using a k factor of 1.0 and a θ value of 1.15. Nitrate removal is relatively rapid, provided that ample carbon is available in the system.

Total nitrogen removal can be estimated by combining the steps of ammonia transformation (nitrification) and nitrate removal (denitrification). To check the calculated total nitrogen removal, use Eq. (9-14). To obtain 50 percent removal of nitrogen, Eq. (9-14) would predict a hydraulic loading rate of 2 in/d or less (assuming the total nitrogen concentration in the influent is less than 20 mg/L) (WPCF, 1989). The empirical relationship represented by Eq. (9-14) is the result of a regression analysis with an r^2 value of 0.79.

$$N_t = 0.193N_o + 3.94(L_W) - 1.75 \quad (9\text{-}14)$$

where N_t = effluent total nitrogen concentration, mg/L
N_o = influent total nitrogen concentration, mg/L
L_W = wastewater hydraulic loading rate, in/d

The hydraulic loading rate can be used to calculate the net wetland area from

$$A = \frac{QF}{L_W} \quad (9\text{-}15)$$

where A = net wetland area, ac (ha)
Q = average flow, Mgal/d (m^3/d)
F = conversion factor, 36.8 ac·in/Mgal (0.1 ha·mm/m^3)
L_W = wastewater hydraulic loading rate, in/d (mm/d)

Hydraulic loading rates have ranged from 0.3 to 2.5 in/d (7.5 to 62.5 mm/d) for operating FWS constructed wetlands. Detention times for significant nitrogen removal (10 mg/L removal or more) should be in the range of 8 to 14 days or more. Nitrogen removal and nitrification will be reduced when water temperatures fall below 10°C and cannot be expected when water temperatures fall below 4°C.

Loading rates for phosphorus removal. A first-order (area-based) rate constant of 10 m/yr (27.4 mm/d) has been proposed for estimating phosphorus removal in constructed wetlands (Kadlec and Knight, 1996).

$$C_e/C_o = \exp(-k_A/L_{\overline{W}}) \tag{9-16}$$

where C_e = effluent phosphorus, mg/L
C_o = influent phosphorus, mg/L
k_A = 1.07 in/d (27.4 mm/d)
$L_{\overline{W}}$ = average annual wastewater hydraulic loading rate, in/d (mm/d)

To calculate the land requirement, use

$$A = -Q_{\text{ave}} \ln(C_e/C_o)F/k_A \tag{9-17}$$

where A = surface area of wetland, ft^2 (m^2)
Q_{ave} = average wastewater flowrate through the wetland, ft^3/d (m^3/d)
F = conversion factor, 12 in/ft (1000 mm/m)

Water depth. Water depth can range from 4 to 18 in (100 to 450 mm). For warm water conditions the 4 to 8 in (100 to 200 mm) range is typical. Storage of wet-weather flows can be accommodated for short periods (30 d or so) at depths of 20 to 40 in (500 to 1000 mm).

Hydraulic considerations. In FWS wetlands, the headloss from vegetation must also be considered. The hydraulic gradient can be estimated by using the following modified form of Manning's equation:

$$v = \frac{1}{n}(d_w^{2/3})(s^{1/2}) \tag{9-18}$$

where v = liquid flow velocity, ft/s (m/s)
n = Manning's coefficient, s/$ft^{1/3}$ (s/$m^{1/3}$)
d_w = depth of water in wetland, ft (m)
s = hydraulic gradient or slope of the water surface, ft/ft (m/m)

The resistance factor a depends on the density of the vegetation and litter. The factor is related to Manning's n as follows:

$$n = \frac{a}{d_w^{1/2}} \tag{9-19}$$

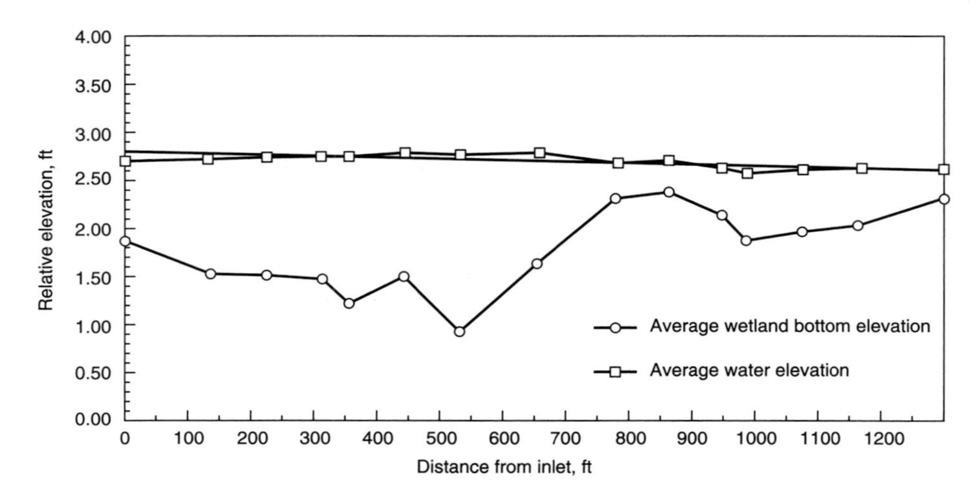

FIGURE 9-10
Hydraulic profile along centerline of FWS constructed wetland at Sacramento, California.

where a = resistance factor, $\text{s}\cdot\text{ft}^{1/6}$ ($\text{s}\cdot\text{m}^{1/6}$)
= 0.487 for sparse vegetation and $d_w > 1.3$ ft
= 1.949 for moderately dense vegetation in a wastewater wetland for $d_w = 1.0$ ft
= 7.795 for very dense vegetation and litter layer, $d_w < 1.0$ ft

The application of Eqs. (9-18) and (9-19) is illustrated in Example 9-6. A typical profile along the centerline of one of the plug-flow channels in the Sacramento Regional constructed wetland is shown in Fig. 9-10.

EXAMPLE 9-6. HYDRAULICS OF FWS CONSTRUCTED WETLANDS. A FWS wetland has an aspect ratio of 3:1 and a wastewater hydraulic loading rate of 3 in/d. For a flow of 0.5 Mgal/d and a depth of 1 ft, calculate the area, dimensions, and flow velocity. Calculate Manning's n and the hydraulic gradient. Assume the value of the resistance factor a is 1.949.

Solution

1. Determine the field area using Eq. (9-15):

$$A = \frac{QF}{L_W} = \frac{0.5(38.6\ \text{ac}\cdot\text{in/Mgal})}{3\ \text{in/d}} = 6.1\ \text{ac}$$

2. Calculate the width of the cell using Eq. (9-11):

$$w = (A/R_A)^{1/2} = \left[\frac{(6.1)(43,560\ \text{ft}^2/\text{ac})}{3}\right]^{1/2}$$

$$W = 298\ \text{ft}$$

$$L = 3W = 894\ \text{ft}$$

3. Calculate the velocity:

$$v = \frac{Q}{d_w w} = \frac{(0.5 \text{ Mgal/d})(3.07 \text{ ac·ft/Mgal})(43,560 \text{ ft}^2\text{/ac})}{(1 \text{ ft})(298 \text{ ft})}$$

$$= 224 \text{ ft/d}$$

$$= 0.0026 \text{ ft/s}$$

4. Calculate Manning's n:

$$n = \frac{a}{d_w^{1/2}} = \frac{1.949 \text{ s·ft}^{1/6}}{1^{1/2}} = 1.949 \text{ s·ft}^{1/3}$$

5. Calculate the hydraulic gradient:

$$v = \frac{1}{n}(d_w^{2/3})(s^{1/2})$$

$$s^{1/2} = \frac{v}{(1/n)(d_w^{2/3})}$$

$$= \frac{0.0026}{(1/1.949)(1^{2/3})}$$

$$s = 0.0000257 \text{ ft/ft}$$

6. Determine the headloss for a length of 894 ft:

$$h_L = sL = 0.0000257(894 \text{ ft}) = 0.023 \text{ ft}$$

Comment. No losses of flow through seepage or evaporation were assumed in the calculation of the velocity and the gradient. Flow losses will reduce the velocity and the resulting gradient. The headloss of 0.023 ft is acceptably small.

Thermal considerations. In very cold climates wetlands will be impacted by temperatures below 1°C (33.8°F). Constructed wetlands can operate successfully in cold climates as evidenced by the numerous systems in Canada (Pries, 1996). Where water temperatures between 1 and 4°C (34 and 40°F) are expected for more than a month, data from existing plants should be analyzed using the methods outlined in Chap. 3. For cold-climate design, the thermal models in Reed et al. (1995) should be consulted.

Vegetation harvesting. Periodic harvesting of the emergent vegetation may be required to maintain hydraulic capacity, promote active growth, or avoid mosquito production. Harvesting for nutrient removal is not practical, and is not recommended. Harvesting activity will affect performance, so the harvested cell should be taken out of service before and for several weeks after harvesting. Harvested vegetation can be burned, chopped and composted, or chopped and used as mulch.

Vegetation planting and establishment. An important aspect of design is the preparation of a strong specification for vegetation planting and establishment. Planting can be done by seeding or transplanting most species. Contractors are generally not expert in planting wetland vegetation and initial plantings may fail.

Experience has shown that 0.5 to 2 years may be required before vegetation in a constructed wetland becomes established fully.

Physical Features of FWS Wetlands

The principal physical features of FWS constructed wetlands include inlet and outlet structures, recirculation, and liners.

Inlet and outlet structures. Uniform distribution of wastewater across the head end of the FWS wetland is critical to a successful system (see Fig. 9-11). Gated pipe, weirs, or drilled holes in distribution pipelines can be used to spread the wastewater across the inlet end of the wetland. Features of outlet structures include adjustable weirs with stop logs and submerged outlet header pipes with control valves. The ability to vary the water depth and to drain the basin should be provided. Basins should be sloped at 0.4 to 0.5 percent grade to facilitate draining.

Recirculation. The ability to recirculate partially or fully treated effluent back to the head end of the basin is an important consideration. Recirculation can reduce organic and solids concentrations, provide more dissolved oxygen to the inlet point,

(*a*)

(*b*)

FIGURE 9-11
Typical (*a*) inlet and (*b*) outlet devices at the Sacramento Regional CSD free-water-surface (FWS) constructed wetland.

and improve overall performance. Recirculation is most effective when combined with step feed (see Sec. 9-5).

Liners. A constructed wetland may need a liner to seal the bottom and sides and thereby prevent or minimize seepage. Depending on the site selected, soil type, groundwater depth and quality, level of pretreatment, and regulatory considerations, a natural or synthetic liner may be required. Bentonite clay is a typical earthen liner while geomembrane liners are also available (Kays, 1986).

9-4 SUBSURFACE-FLOW CONSTRUCTED WETLANDS

A constructed wetland with the flow beneath the surface of a gravel or sand medium is known as a subsurface-flow (SF) system. The process description, constituent removal and transformation mechanisms, performance expectations, and process design considerations are presented and discussed in this section. General design considerations and the management for these systems are discussed in Secs. 9-8 and 9-9, respectively.

Process Description

Subsurface-flow systems have also been termed *rock-reed filters, microbial rock plant filters, vegetated submerged beds, marsh beds, tule beds,* and *hydrobotanical systems.* In Germany a similar type of system that uses native soil and reeds is known as the *root zone method.* Subsurface-flow systems have the advantages of smaller land area requirements and avoidance of odor and mosquito problems, as compared to free-water-surface (FWS) systems. Disadvantages of SF systems are the increased cost due to the gravel media and the potential for clogging of the media. Pretreatment for SF wetlands typically consists of primary treatment.

Site selection. The SF wetland takes less space than a comparable FWS system and generally has a sloped bottom of 0 to 0.5 to percent. If soils are permeable (greater than 0.2 in/h) it may be necessary to install a liner below the bed medium.

Vegetation types. The vegetation in SF systems is similar to FWS wetlands and tends to be bulrush, reeds, and in some cases cattails. The purpose of the vegetation is to provide oxygen into the root zone and add to the surface area for biological growth in the root zone. The actual transport of oxygen to the root zone and then into the water column is limited (Brix, 1993). The roots also release organics as they decay, which supports denitrification. The aboveground portion of the vegetation provides little benefit except for nutrient uptake and plant growth. Plant harvesting is not necessary (Gersberg et al., 1985).

Bed medium. The subsurface flow wetland medium is usually gravel, although in early systems sand was also used. The gravel size has varied from 0.12 in

TABLE 9-10
Typical medium characteristics for SF wetlands

Medium type	Effective size d_{10}, mm	Effluent porosity η	Hydraulic conductivity, ft/d
Medium sand	1	0.30	1640
Coarse sand	2	0.32	3280
Gravelly sand	8	0.35	16,400
Medium gravel	32	0.40	32,800
Coarse gravel	128	0.45	328,000

Note: d_{10} is the diameter of a particle in a weight distribution of particles that is smaller than all but 10% of the particles.

to 1.25 in (3 to 32 mm), with inlet zone gravel size as large as 2 in (50 mm). The inlet zone should have the largest-diameter medium to minimize the clogging potential. At Sydney, Australia, the medium in the inlet zone is 1.2 in to 1.6 in (30 to 40 mm) in diameter while in the remainder of the bed the medium is 0.2 in to 0.4 in (5 to 10 mm). Characteristics of SF media are presented in Table 9-10.

Constituent Removal and Transformation Mechanisms

As with FWS wetlands, SF wetlands can be expected to produce a high-quality effluent in terms of BOD, TSS, and pathogens. The principal removal mechanisms are biological conversion, physical filtration and sedimentation, and chemical precipitation and adsorption as described in the FWS wetlands section. Lesser removal of nitrogen, phosphorus, metals, and trace organics than for BOD and TSS should be expected, with removals dependent on detention time, media characteristics, loading rates, and management practices.

BOD removal. Removal of BOD is accomplished biologically and physically. Removal of BOD occurs primarily under anaerobic conditions; however, a portion of the BOD is converted by facultative organisms. The rate of removal is related to detention time and temperature as described under "Process Design Considerations."

Suspended solids removal. The mechanisms for TSS removal are similar to those for TSS removal from FWS systems. The lack of a free water surface in SF wetlands avoids the wind currents and resuspension of solids, resulting in the potential for a lower effluent TSS concentration. The majority of the solids settle out or are trapped within the first 10 to 20 percent of the bed flow distance. Observations at a number of operating SF systems indicated clogging of the inlet zone resulting in surface flow down a portion of the flow path. The clogging appears to be the result of the high solids and organic loading occurring at the entry zone of the bed. The most severe clogging has occurred with long narrow beds receiving algae-laden effluent from facultative ponds. The algae are trapped in the medium near the inlet and the decomposing algae add to the organic load.

Nitrogen removal. Nitrogen removal is accomplished by nitrification/denitrification. Although SF wetlands have the ability to denitrify the available nitrate-nitrogen, the limitation on nitrogen removal is the nitrification step. The subsurface flow regime is nearly anaerobic, except for the top few inches and aerobic microsites near the plant roots. Nitrification requires a supply of oxygen, either from the plant roots, surface reaeration, effluent recirculation, or batch loading to induce oxygen flow down into the media between applications. Supplemental aeration using subsurface tubing can be used to provide oxygen at a point in the flow path where the BOD has been reduced below 30 mg/L, so that the oxygen provided is of use to the nitrifying bacteria.

Phosphorus removal. The mechanisms for phosphorus removal are essentially the same as for FWS wetlands. Special media are required to effect substantial removal of phosphorus by adsorption. As in the FWS systems, phosphorus can be released during certain times of the year, usually in response to changes in the environmental conditions within the system.

Metals removal. The removal mechanisms for metals include adsorption, sedimentation, chemical precipitation, and plant uptake. As in the FWS systems, metals can be released during certain times of the year, usually in response to change in the oxidation-reduction potential (ORP) within the system.

Trace organics removal. Removal mechanisms for trace organics are similar to those for FWS wetlands except that volatilization and photodecomposition are limited.

Pathogen removal. Removal of bacteria and viruses occurs by adsorption, filtration, sedimentation, and predation.

Process Performance

The performance expectations for SF constructed wetlands are considered in the following discussion. As with the FWS system, process performance depends on design criteria, wastewater characteristics, and operations.

BOD removal. Performance data for BOD removal are presented in Table 9-11. Removal of BOD appears to be faster and somewhat more reliable with SF wetlands than for FWS wetlands, partly because the decaying plants are not in the water column, thereby producing slightly less organic matter in the final effluent.

TSS removal. SF wetlands are efficient in removal of suspended solids, with effluent TSS levels below 10 mg/L, typically.

Nitrogen removal. Although the SF system at Santee was able to remove 86 percent of the nitrogen from primary effluent, other SF systems have reported

TABLE 9-11
Total BOD removal observed in SF wetlands

Location	Pretreatment	Concentration, mg/L		Removal, %	Nominal detention time, d
		Influent	Effluent		
Benton, Kentucky*	Oxidation pond	23	8	65	5
Mesquite, Nevada†	Oxidation pond	78	25	68	3.3
Santee, California‡	Primary	118	1.7	88	6
Sydney, Australia§	Secondary	33	4.6	86	7

*Full-scale operation from March 1988 to November 1988 at 80 mm/d (Watson et al., 1989).
†Full-scale operation, January 1994 to January 1995.
‡Pilot-scale operation, 1984, operated at 50 mm/d (Gersberg et al., 1985).
§Pilot-scale operation at Richmond, New South Wales, near Sydney, operated at 40 mm/d from December 1985 to February 1986 (Bavor et al., 1987).

removals of from 20 to 70 percent. When detention times exceed 6 to 7 d, an effluent total nitrogen concentration of about 10 mg/L can be expected, assuming a 20 to 25 mg/L influent nitrogen concentration. If the applied wastewater has been nitrified (using extended aeration, overland flow, or recirculating sand filters), the removal of nitrate through denitrification can be accomplished with detention times of 2 to 4 d.

Phosphorus removal. Phosphorus removal in SF wetlands is largely ineffective because of limited contact between adsorption sites and the applied wastewater. Depending on the loading rate, detention time, and media characteristics, removals may range from 10 to 40 percent for input phosphorus in the range from 7 to 10 mg/L. Crop uptake is generally less than 10 percent (about 0.5 lb/ac·d)(0.55 kg/ha·d).

Metals removal. There are limited data available on metals removal using municipal wastewater in SF systems. In acid mine drainage systems, removal of iron and manganese is significant. Total iron has been shown to be reduced from 14.3 to 0.8 mg/L and total manganese from 4.8 to 1.1 mg/L (Brodie et al., 1989). At Santee, California, removal of copper, zinc, and cadmium was 99 percent, 97 percent, and 99 percent, respectively, during a 5.5-d detention time (Gersberg et al., 1984).

Pathogen removal. A removal of 99 percent (2 log) of total coliform was found when primary effluent was applied at 2 in/d (detention time 6 d) at Santee, California (Gersberg et al., 1989).

Process Design Considerations

Important process design criteria include detention time, required surface area, BOD and solids loading rates, and medium depth. Representative process design criteria are presented in Table 9-12. The design procedure is illustrated in Example 9-7.

TABLE 9-12
Typical design criteria and expected effluent quality for SF constructed wetlands

Item	Unit	Value
Design parameter		
Detention time	d	3–4 (BOD) 6–10 (N)
BOD loading rate	lb/ac·d	<100
TSS entry loading rate	$lb/ft^2 \cdot d$	0.008
Water depth	ft	1–2
Medium depth	ft	1.5–2.5
Mosquito control		Not needed
Harvest schedule		Not needed
Expected effluent quality*		
BOD_5	mg/L	<20
TSS	mg/L	<20
TN	mg/L	<10
TP	mg/L	<5

*Expected effluent quality based on a BOD loading equal to or less than 100 lb/ac·d and typical municipal wastewater.

Detention time for BOD removal. The detention time is determined by using Eq. (9-6). The value of the apparent removal rate constant at 20°C is about 1.1 d^{-1}. The overall BOD loading on SF wetlands should not exceed about 100 lb/ac·d (112 kg/ha·d). These rates will not be exceeded in practice with primary effluent applied at up to 2 in/d (50 mm/d).

Like FWS wetlands, SF wetlands experience some regeneration of BOD due to decay, primarily from the roots because the decaying vegetation stays on the surface on the bed and remains out of the water column. Depending on the time of year, there will be some accretion from the surface vegetation. The root decay will generate 2 to 3 mg/L of BOD_{PD}.

Required surface area. Once the detention time is calculated the net area of the wetland can be determined from

$$A_s = \frac{(Q_{ave})(t)(3.07)}{(\eta)(d_w)} \tag{9-20}$$

where Q_{ave} = average daily flow through the wetland, Mgal/d, and A_s = surface area, ac. Other terms are as described previously. The average flow through the wetland can be estimated from Eq. (9-10).

Aspect ratio. The surface dimensions of the SF wetland can be determined by using Eq. (9-11) as given previously in the discussion of FWS systems. Aspect ratios should be determined in conjunction with Darcy's law (Eq. 9-23).

Suspended solids entry zone loadings. If an aspect ratio greater than about 4:1 is used, the influent solids loading may be of concern. To avoid clogging

TABLE 9-13
Comparison of the behavior of sand, gravel, and rock filters operated at various suspended solids loading rates*

Material	Typical particle size, mm	Nominal TSS loading rate, $g/m^2 \cdot d$	Performance
Sand	0.17	5	Clogging in > 5 years
		10	Clogging in 50 days
		30	Clogging in < 10 days
	0.40	10	Clogging in > 0.5 years
		30	Clogging in 35 days
		70	Clogging in 10 days
	0.68	20	Clogging in > 0.5 years
		40	Clogging in 50 days
		80	Clogging in 20 days
Gravel	5–10 (inlet)	40	Infiltration for 3+ years
	5–10 (w/g)	200	Clogging in 3 months
	40 (inlet)	18	Infiltration for 3+ years
	40 (inlet-primary)	80–160	Infiltration for 1+ year
Rock	9–25	13–464†	Clogging in 11 months
	10–50	113–629†	Infiltration for 17+ months, but poor TSS removal
	63–127	102‡	Infiltration for 14+ months, but poor TSS removal

*From Bavor and Schulz (1993).
†Represents loadings with 50 mg/L algal solids.
‡Represents loadings with 69 mg/L algal solids.
Notes: The loading rates were those estimated to apply per square meter of surface available for infiltration. The data for sand and rock filters are adapted from Middlebrooks et al. (1982). Gravel filters were at Eudora, Kansas, and California, Missouri. Surface areas were estimated from the volumetric loading rates and estimates of the open surface in the illustrated designs. Gravel size at the water/gravel interface is noted as w/g.

of the inlet zone with suspended solids, the entry-zone solids loading values must be checked. Although suspended solids loading limits have not been developed in this country, experience in Australia has led to the recommendation that entry-zone TSS limits not exceed $0.008\ lb/ft^2 \cdot d$ (Bavor et al., 1989), where the area used in the loading rate is the cross-sectional area of the entry zone. Soil clogging experience as a function of medium size is compared in Table 9-13.

The entry zone organic loading rate can be computed as follows:

$$L_{TSS} = \frac{\text{constituent mass loading, lb/d}}{\text{entry zone cross-sectional area, } w\, d_m,\ \text{ft}^2} \tag{9-21}$$

where w = width of SF wetland, ft, and d_m = depth of medium, ft.

Depth of medium. The depth of the medium may range from 18 to 30 in (450 to 750 mm). Typical rooting depths range from 6 to 12 in. To obtain rooting depths of 12 in or more, the water depth must be systematically lowered over a number of growing seasons to force the roots to penetrate deeper. The depth of the medium does not have to be much deeper than the rooting depth. The water level is kept 3 to 6 in (75 to 150 mm) below the top of the medium.

Detention time for nitrogen removal. Some SF systems have been designed for ammonia removal, with the previously mentioned problem of inadequate oxygen availability. On the basis of pilot studies in Australia, the following relationship between detention time and ammonia removal was developed (Bavor et al., 1987):

$$A = \frac{Q(\ln N_o - \ln N_e)}{k(d_w)(\eta)(F)} \tag{9-22}$$

where A = surface area of SF wetland for NH_4-N removal, ac (ha)
Q = average flow through the wetland, ft^3/d (m^3/d)
N_o = influent NH_4-N concentration, mg/L
N_e = effluent NH_4-N concentration, mg/L
k = $0.107\ d^{-1}$ (for 20°C)
d_w = depth of liquid in bed, ft (m)
η = effective porosity of bed medium expressed as a decimal
F = conversion factor, 43,560 ft^2/ac (10,000 m^2/ha)

The temperature dependence of k can be calculated using Eq. (3-22) and a θ value of 1.15.

Hydraulic considerations. Headloss through SF wetlands can be estimated from Darcy's law:

$$A = d_w w = \frac{Q}{kS} \tag{9-23}$$

where A = cross-sectional area of inlet zone, perpendicular to the flow path, ft (m)
d_w = depth of liquid in bed, ft (m)
w = bed width, ft (m)
Q = flow into system, ft^3/d (m^3/d)
k = hydraulic conductivity from Table 9-10 (or measured in field, preferably), ft/d (m/d)
S = slope, expressed as a decimal (headloss)

In using Eq. (9-23), the measured value of k should be used when available and multiplied by a safety factor of 10 percent to account for root and tuber growth. In the absence of measured data, use the values in Table 9-10 multiplied by 10 percent. For sloped beds, use the bottom slope, which can vary from 0 to 1 percent or more. When a flat bed is used and the gradient is controlled with an overflow weir, use 0.001 for S.

Vegetation establishment. For very small systems [less than 2 ac (0.8 ha)] vegetation can often be transplanted from nearby sources or obtained commercially. Rhizome cuttings should be 4 in (100 mm) long and have a growing shoot at the end of the cutting. The root end of the cutting should be placed about 2 in (50 mm) below the media surface. The bed should then be flooded with water to the surface or sprinkled frequently. If flooding is used, the water level must be maintained carefully during this period so that the exposed plant shoots are not submerged.

Planting densities for the most commonly used species are 3-ft (1-m) centers for cattails, 1.5-ft (0.5-m) centers for reeds and bulrush (Reed et al., 1995). For beds larger than 2 ac (0.8 ha) it may be more economical to hydroseed the vegetation. In any case, the vegetation should be allowed to become established with 3 to 6 months of growth before wastewater applications begin.

Physical Features of SF Wetlands

Important physical features of SF wetlands include inlet and outlet structures, recirculation, and bed liners. To provide for operational flexibility, each system should have multiple cells (minimum of 2).

Inlet and outlet structures. The inlet system must be designed so that the influent flow is distributed uniformly over the length of the entry zone. Typical devices for influent distribution are gated pipe, slotted pipe, or troughs with V-notch weirs. The first 10 ft (3 m) of the entry zone is usually filled with large rock (2 to 4 in or 50 to 100 mm) to minimize clogging. If a step feed operation is desired, a second influent distributor can be placed parallel to the entry zone distributor at a distance (50 ft or 15 m or more) down the flow path.

Outlet devices should consist of perforated pipes submerged to the bottom of the bed with valves or adjustable-level outlet pipes to control the water depth. An example outlet device is shown schematically in Fig. 9-12.

Recirculation. The ability to recirculate treated effluent to dilute influent concentrations, improve treatment, and avoid overloading can be built into SF

FIGURE 9-12
Typical inlet device for subsurface-flow (SF) constructed wetland at Hardin, Kentucky.

systems by using recirculation pumps and piping. If the SF effluent must be pumped to its final reuse/discharge point, recirculation pumping is very inexpensive and is recommended.

Bed liners. If the soil is permeable, a bed liner will usually be required to prevent loss of water to the groundwater. The liner may consist of native clay, bentonite, asphalt, or geomembrane liners (Kays, 1986). A smooth-surfaced 30-mil plastic membrane liner is used typically (Reed et al., 1995).

EXAMPLE 9-7. DESIGN OF SF CONSTRUCTED WETLAND. A community of 2000 generates a flow of 0.16 Mgal/d of septic tank effluent. Septic tank effluent characteristics are 130 mg/L of BOD and 20 mg/L ammonia nitrogen. Design a SF wetland to produce an effluent BOD of 10 mg/L. Determine the detention time needed to reduce the ammonia concentration to 6 mg/L. Use a 9°C temperature for the septic tank effluent during the coldest month. Use k values of 1.1 d^{-1} and 0.107 d^{-1} for BOD removal and nitrification, respectively.

Solution

1. Calculate the k_T value using Eq. (3-22) for a temperature 9°C:

$$\frac{k_T}{k_{20}} = 1.06^{(T-20)}$$

$$k_T = 1.1(1.06^{(9-20)}) = 0.58\ \text{d}^{-1}$$

2. Calculate the detention time for BOD removal using Eq. (9-6):

$$t = -\frac{\ln(C/C_o)}{k_{\text{apparent}}}$$

$$= \frac{\ln(10/130)}{0.58} = 4.42\ \text{d}$$

3. Check the organic loading rate using Eq. (9-8):

$$L_{\text{org}} = \frac{(C)(d_w)(\eta)(F_1)}{t \times F_2}$$

$$= \frac{(130)(1.25)(0.4)(8.34)}{4.42 \times 3.07} = 40\ \text{lb BOD/ac}\cdot\text{d}$$

4. Select the depth of the medium, d_w. From Table 9-12 select 1.5 ft as the medium depth.

5. Determine the field area for the SF bed using Eq. (9-20) with an effective porosity of 0.40 (see Table 9-10), a medium depth of 1.5 ft, and a liquid depth of 1.25 ft:

$$A_s = \frac{(Q_{\text{ave}})(t)(3.07)}{(\eta)(d_w)}$$

$$= \frac{(0.16)(4.42)(3.07)}{(0.40)(1.25)} = 4.34\ \text{ac}$$

6. Calculate the k value for nitrification using Eq. (3-22).

$$k_T = 0.107(1.06^{(9-20)})$$

$$= 0.056\ \text{d}^{-1}$$

7. Calculate the surface area for nitrification using Eq. (9-22):

$$A = \frac{Q(\ln N_o - \ln N_e)}{k(d_w)(\eta)(F)}$$

$$Q = (0.16 \text{ Mgal/d})(133{,}690 \text{ ft}^3\text{/Mgal})$$
$$= 21{,}390 \text{ ft}^3\text{/d}$$

$$A = \frac{21{,}390[\ln(20) - \ln(6)]}{0.056(1.25)(0.4)(43{,}560)}$$
$$= 21.0 \text{ ac}$$

8. Calculate the detention time corresponding to the 21.0 ac.

$$t = \frac{Ad_w\eta}{Q}$$
$$= (21.0 \text{ ac})(1.25 \text{ ft})(0.40)/(0.16 \text{ Mgal/d})(3.07 \text{ ac}\cdot\text{ft/Mgal})$$
$$= 21.4 \text{ d}$$

Note: This detention time is excessively long. Supplemental aeration, an RSF, and other supplemental treatment will be necessary to make this SF system cost-effective under these conditions.

9. Calculate the cross-sectional area from Darcy's law [Eq. (9-23)] using $k = 32{,}800$ from Table 9-10 multiplied by 10 percent. Use $S = 0.01$.

$$A = \frac{Q}{kS}$$
$$= \frac{(0.16 \text{ Mgal/d})(133{,}690 \text{ ft}^3\text{/Mgal})}{(32{,}800)(0.10)(0.01)}$$
$$= 652 \text{ ft}^2$$

10. Calculate the dimensions of the system.
    a. Calculate the width:

$$w = A/d$$
$$= 652 \text{ ft}^2/1.25 \text{ ft}$$
$$= 521.6 \text{ ft}$$

    b. Calculate the length:

$$L = \text{bed area/width}$$
$$= (21.0 \text{ ac})(43{,}560 \text{ ft}^2\text{/ac})/521.6 \text{ ft}$$
$$= 1754 \text{ ft}$$

    c. Calculate the aspect ratio:

$$R = L/w$$
$$= 1754/521.6$$
$$= 3.36$$

Comment. If COR data are available, the method used for the FWS can be employed. The background BOD of 3 mg/L, however, is close to the design value of 10 mg/L.

9-5 FLOATING AQUATIC PLANT SYSTEMS—WATER HYACINTHS

The two most commonly used floating aquatic plants are water hyacinths and duckweed. The use of water hyacinths is considered here. Duckweed systems are considered in the following section. The material presented in this section deals with a description of the process, constituent removal and transformation mechanisms, performance expectations, and process design considerations. General design considerations and the management for these systems are discussed in Secs. 9-8 and 9-9, respectively.

Process Description

The two principal types of water hyacinth wastewater treatment systems can be described as: (1) aerobic nonaerated and (2) aerobic aerated. *Nonaerated* aerobic systems are typically shallow arbitrary-flow ponds or plug-flow ponds (channels) covered with water hyacinths, operated without and with effluent recycle and step feed (see Figs. 9-13*a* through *f*). *Aerated* aerobic systems are similar to nonaerated

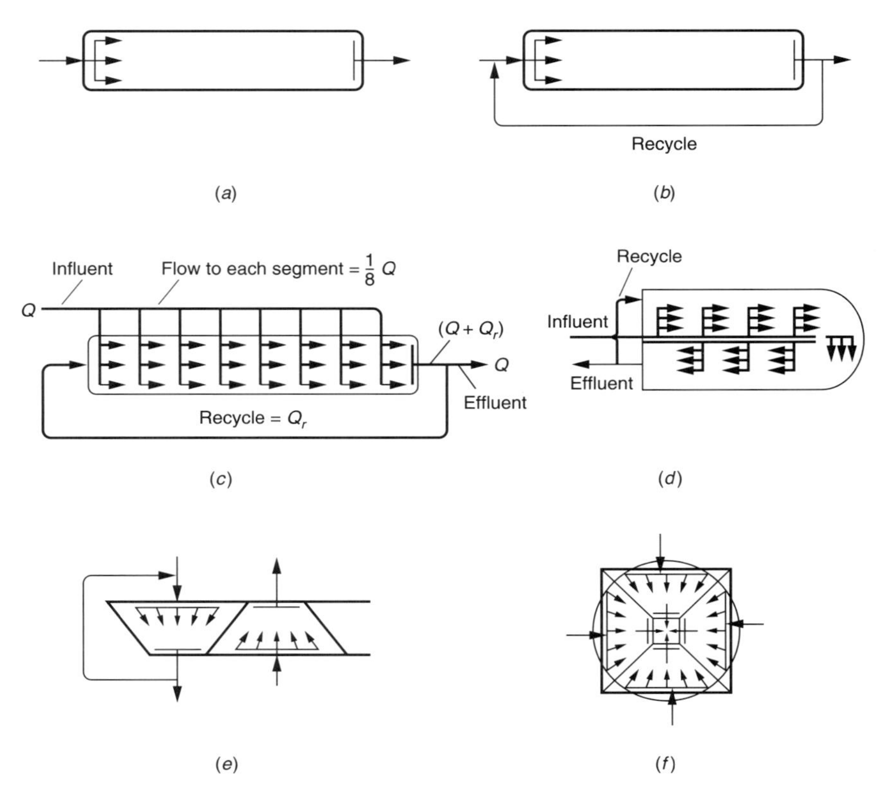

FIGURE 9-13
Alternative configurations for water hyacinth treatment basins.

plug-flow systems, with the exception that supplemental air is provided, and the operating water depths are usually greater (see Figs. 9-13*b* through *e*). An important advantage of an *aerated* system is that higher organic loading rates are possible, and reduced land area is required. *Facultative/anaerobic* ponds, of various flow configurations, employing water hyacinths have also been used. Because such systems are operated at very high organic loading rates, odors and increased mosquito populations are common. As a result of odor and mosquito problems, facultative/anaerobic water hyacinth systems are seldom, if ever, used in the United States.

Site selection. Level to slightly sloping, uniform topography is preferred for the construction of water hyacinth treatment systems. Although ponds and channels may be constructed on steeper sloping or uneven sites, the amount of earthwork required will affect the cost of the system.

Climate. Because of their sensitivity to cold temperatures, the use of water hyacinths is restricted to the southern portions of California, Arizona, Texas, Mississippi, Louisiana, Alabama, and Georgia, and all of Florida. Water temperatures as low as 10°C (50°F) can be tolerated if the air temperature does not drop below 5 to 10°C (41 to 50°F). Combined systems of several aquatic plants (e.g., duckweed, pennywort, and water hyacinth) may be suitable for locations with greater climatic variations (DeBusk and Reddy, 1987).

Water hyacinths. The water hyacinth (*Eichhornia crassipes*) is a perennial, freshwater aquatic macrophyte. The plant grows rapidly, especially in wastewater. Individual plants range from 20 to 47 in (500 to 1175 mm) from the top of the lavender flowers to the root tips. It has been estimated that starting with 10 individual plants, water hyacinths can spread and cover a 1-ac pond within 8 months (Reed et al., 1995). Water hyacinths cannot tolerate cold weather. Under freezing conditions leaves and flowers die, but the plant can regenerate unless the rhizome tip freezes. Water hyacinths die at about −6°C and cannot persist where winter temperatures average 1°C or less.

Constituent Removal and Transformation Mechanisms

High removals of BOD and TSS can be expected from properly designed water hyacinth treatment systems, with lesser efficiencies demonstrated for nutrients, metals, and pathogens. The operative removal mechanisms are described below.

Biochemical oxygen demand removal. A portion of the BOD in the influent wastewater is removed along with the TSS by sedimentation from the water column, as the wastewater flows through the treatment reactor. Another portion of the BOD associated with the suspended solids that will not settle by gravity is removed by filtration along with the TSS as wastewater flows through the roots of the water hyacinths. Soluble BOD is removed by adsorption as the wastewater flows past the water hyacinth roots. As with the removal of TSS, transport of the wastewater to the root zone is a critical design consideration with respect to the removal of soluble BOD in water hyacinth treatment systems. Soluble BOD is also removed by bacterial conversion in the water column. In time, a portion of the BOD associated with the organic fraction of the TSS accumulated in the root zone and the adsorbed soluble BOD will be converted by the organisms attached to the roots, using oxygen transported to the roots by the plant. As noted above, the roots of the water hyacinth plant will senesce and drop to the bottom of the pond or channel, carrying with them the accumulated suspended solids and bacteria. The material that accumulates on the bottom of the reactor undergoes long-term anaerobic decomposition and consolidation. There is also some release of organic material in the form of intermediate- and short-chain organic acids resulting from the first-stage anaerobic decomposition of the solids accumulated on the pond bottom. The removal of the soluble BOD represented by these acids is as described above.

Total suspended solids removal. A portion of the TSS in the influent wastewater is removed by sedimentation from the water column, as the wastewater flows through the treatment pond or channels. Another portion of the suspended solids that will not settle by gravity is removed by filtration as wastewater flows through the roots of the water hyacinths. Because filtration is such an important removal mechanism, transport of the wastewater to the root zone is a critical design consideration in water hyacinth treatment systems. In time, a portion of the organic fraction of the TSS accumulated in the root zone will be converted by the organisms attached to the roots. With the further passage of time, the TSS on the roots will continue to accumulate. Ultimately, the roots of the water hyacinth plant will senesce and drop to the bottom of the pond or channel, carrying with them the accumulated suspended solids. Additional filtration occurs as the roots drop to the bottom of the pond. The material that accumulates on the bottom of the reactor undergoes long-term anaerobic decomposition and consolidation.

Nitrogen removal. Biological nitrification-denitrification is the principal mechanism involved in the removal of nitrogen. A portion of the organic nitrogen is removed by sedimentation. Nitrogen is also taken up by plant growth and can be removed by plant harvesting, but not effectively. Some nitrogen is also lost by volatilization, where aeration is provided. The principal location where nitrification-denitrification occurs is in the root zone. Thus, it is very important for the wastewater, containing various forms of nitrogen, to flow past the water hyacinth roots where the bacteria responsible for the conversion of nitrogen are located.

Phosphorus removal. Adsorption to wastewater solids and plant material, adsorption to organic matter in sludge layer, and plant uptake are the principal means by which phosphorus is removed from wastewater. Limited amounts of phosphorus are removed where routine harvesting of water hyacinth plants is practiced. Adsorption to the organic matter in the sludge layer is the ultimate fate of the phosphorus which remains in the system. Where there are effluent limitations on phosphorus, phosphorus should be removed in a preapplication or posttreatment step, because phosphorus removal in water hyacinth treatment systems is limited, and often erratic.

Heavy metals removal. The removal of heavy metals occurs primarily through adsorption to wastewater solids and plant material. Limited plant uptake has been observed. Relatively small amounts of heavy metals are removed where water hyacinth plants are harvested. Adsorption to the organic matter in the sludge layer is the ultimate fate of heavy metals which remain in the system. As with the FWS and SF wetlands, metals can be released from the sediments, usually in response to changes in the oxidation-reduction potential (ORP) within the system.

Trace organics removal. Adsorption to wastewater solids and plant material, limited plant uptake, and biological conversion in the root zone are the principal removal mechanisms for the priority organic pollutants. As with the heavy metals, priority organic pollutants are removed where water hyacinth plants are harvested. Adsorption to the organic matter in the sludge layer is the ultimate fate of priority organic pollutants which remain in the system.

Pathogens removal. The removal of pathogens occurs by sedimentation and filtration, as described above, and natural decay within the water column. Of the operative removal mechanisms, natural decay appears to be the most effective. In systems with short hydraulic detention times, increased organism counts have been observed.

Process Performance

Typical performance data for a floating aquatic plant system using water hyacinths are presented in Tables 9-14 through 9-17. Long-term and monthly performance

TABLE 9-14
Overall constituent removal performance summary for water hyacinth wastewater treatment cells at San Diego (Aqua III), October 1994 through September 1995*

		Value		
Constituent	Unit	Influent	Effluent	Reduction, %
BOD	mg/L	148	12.6†	91
TSS	mg/L	131	9.7	93
TOC	mg/L	72	14	81
Turbidity	NTU	88	13.5	84
TS	mg/L	1322	1183	11
NH_4-N	mg/L	21	9.5	55
NO_3-N	mg/L	0.05	1.4	
TKN	mg/L	31	13.9	46
TN	mg/L	31	15.3	51
PO_4	mg/L	5.1	3.4	33
SO_4	mg/L	283	309	−9

*From WCPH (1996).
†CBOD value measured in effluent.

TABLE 9-15

Monthly BOD loading and performance summary for Aqua III water hyacinth ponds, October 1994 through September 1995

Month	Flow, Mgal/d	BOD loading, lb/ac·d	BOD, mg/L		TSS, mg/L		Turbidity, NTU	
			Influent	Effluent	Influent	Effluent	Influent	Effluent
October 1994	0.86	164	159	8.4	140	6.5	68.2	6.1
November	1.00	203	171	10.2	140	4.2	72.7	7.5
December	0.93	187	164	17.9	141	5.9	80.6	17.7
January 1995	0.86	134	125	10.3	116	10.3	83.0	13.3
February	1.05	183	140	13.2	131	10.9	93.2	14.6
March	1.02	148	119	9.3	113	8.1	82.5	10.4
April	1.16	214	152	8.6	134	6.9	95.8	12.2
May	1.22	217	147	12.0	137	8.2	95.6	13.4
June	1.16	227	142	13.7	148	9.1	95.2	13.3
July	1.14	235	144	13.8	135	14.0	93.3	15.2
August	1.16	256	149	16.3	119	14.0	91.6	16.9
September	1.19	289	158	17.0	120	18.2	95.1	20.8
Average	1.06	205	148	12.6	131	9.7	87.2	13.5

TABLE 9-16

Overall metals removal performance summary for water hyacinth wastewater treatment cells at San Diego (Aqua III), October 1994 through September 1995

Constituent	Unit	Influent	Effluent	Percent reduced
Arsenic	μg/L	2.5	1.5	40
Cadmium	μg/L	1.2	0.1	92
Chromium	μg/L	2.0	1.3	35
Copper	μg/L	42.5	9.3	78
Lead	μg/L	8.0	0.6	93
Mercury	μg/L	0.1	0.1	0
Nickel	μg/L	4.4	3.7	16
Selenium	μg/L	2.1	2.2	0
Zinc	μg/L	24.0	2.4	90

data for BOD and TSS are considered in the following discussion for the San Diego system.

Long-term performance data. Long-term performance data for the principal constituents of concern in wastewater, for the period from October 1994 through September 1995, are summarized in Table 9-14. The average pond organic loading rate during the period was 205 lb BOD/ac·d (230 kg/ha·d). As shown in Table 9-14, the effluent quality from the ponds is impressive with respect to the removal of BOD, TSS, TOC, and turbidity. As expected, significant removals of nitrogen and phosphorus were not achieved. Although not reported here, the performance data for Aqua II, the previous water hyacinth system located in Mission Valley, were essentially the

TABLE 9-17
Water hyacinth productivity data for Aqua III in San Diego, August 1994 through September 1995*

Month	Temperature, °F		Solar radiation, watts/m²	Amount harvested†		Productivity‡	
	Air	Water		Yd³	Wet weight, lb	Wet lb/ac·d	Dry ton/ac·yr
August 1994	77	74	450	3661	922,572	4261	51
September	72	68	430	2745	691,740	3302	39
October	63	64	360	1590	400,680	1851	22
November	51	55	340	960	241,920	1155	14
December	51	53	300	594	149,688	691	8
January 1995	54	55	340	0	0	0	0
February	61	58	380	200	50,400	258	3
March	58	62	430	950	239,400	1106	13
April	59	63	520	1150	289,800	1383	16
May	62	65	550	2134	537,768	2484	29
June	67	68	530	5638	1,420,776	6781	80
July	74	72	490	4060	1,023,120	4726	56
August	76	74	470	3324	837,648	3869	46
September	74	71	450	2399	604,548	2885	34

*From WCPH (1996).
†Total wet weight based on an average specific weight of 252 lb/yd³. Specific weight is based on average of weights measured in a 1-yd³ box.
‡Productivity calculated as total weight harvested divided by 6.98 ac of pond surface area and total days in a month. The solids content of water hyacinths is 6.5%

same, with the exception that the concentration of the influent BOD was a bit higher (WCPH, 1996).

Monthly performance data for BOD, TSS and turbidity. Monthly performance data for BOD and TSS, for the period from October 1994 through September 1995, are summarized in Table 9-15, and illustrated graphically in Fig. 9-14. Monthly pond organic loading rates are also given in Table 9-14. The most striking thing about the data shown in Fig. 9-14 is the relatively small monthly variation in the effluent values. Properly designed and operated water hyacinth treatment systems have proven to be extremely stable (robust). When the performance and stability of the water hyacinth system at Aqua III, as shown in Fig. 9-14, are compared to comparably sized activated-sludge systems, the water hyacinth treatment system exhibits significantly more consistent effluent values as evidenced by the slope of the probability curve and the performance or stability coefficient defined as the ratio of the P_{80} to P_{10} values. The markedly steeper slopes observed with activated-sludge systems are indicative of a less stable system. Reasons for the greater stability of the water hyacinth system are, most probably, related to physiological and structural diversity provided to the treatment volume by the plant roots.

Metals removal. Typical data on the removal of metals in the water hyacinth ponds at Aqua III in San Diego are reported in Table 9-16. The limited removals achieved with nickel and arsenic are consistent with the data reported in Table 9-7 for Sacramento County.

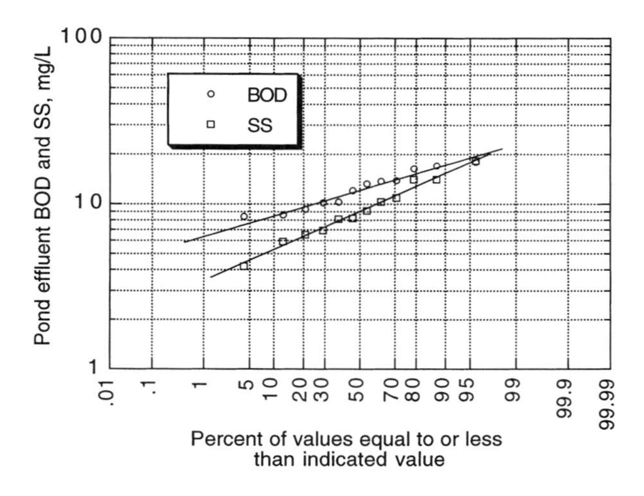

FIGURE 9-14
Performance and stability of water hyacinth system in San Diego.

Water hyacinth growth and harvesting. Water hyacinth plant growth is described in two ways: (1) as the percentage of pond surface covered over a given time period and (2) as the plant density in units of wet plant mass per unit of surface area. Under normal conditions, loosely packed water hyacinths can cover the water surface at relatively low plant densities, about 2 lb/ft^2 (10 kg/m^2) wet weight. Plant densities as high as 16 lb/ft^2 (80 kg/m^2) wet weight can be reached. As in other biological processes, the growth rate of water hyacinths is dependent on temperature. Both air and water temperatures are important in assessing plant vitality. Typical data on the quantity of water hyacinths harvested at Aqua III in San Diego are reported in Table 9-17 (WCPH, 1996).

Process Design Considerations

Objectives of different water hyacinth systems are presented in Table 9-18. Design criteria for water hyacinth systems that are designed to meet the objectives in

TABLE 9-18
Objectives of different types of water hyacinth systems

Type of influent wastewater	Treatment objective	Typical BOD loading rates, lb/ac·d
Primary effluent	Secondary treatment	50–100*
Primary effluent	Advanced secondary treatment	250–450†
Facultative pond effluent	Secondary treatment	45–90
Secondary effluent	Nutrient removal	10–50

*Organic loading rates at 200 lb/ac·d and more have been used; however, there is increased risk of odors and mosquito nuisance.
†Aeration should be provided for BOD loadings above 100 lb/ac·d.

TABLE 9-19
Typical design criteria and expected effluent quality from nonaerated and aerated water hyacinth wastewater treatment systems*

		Typical design criteria	
Item	**Unit**	**Secondary aerobic (nonaerated)**	**Secondary aerobic (aerated)**
Design parameter			
Influent wastewater		Fine-screened or settled	Fine-screened or settled
Influent BOD_5	mg/L	130–180	130–180
Organic loading rate	lb/ac·d	60–80	250–450
Water depth	ft	1.5–2.5	4–4.5
Detention time	d	10–30	4–8
Hydraulic loading rate	Mgal/ac·d	0.03–0.07	0.16–0.30
Application mode		Step feed	Step feed
Aeration	ft^3/min·Mgal	None	400-425
Type of aerator		None	Fine bubble
Channel cross section		Trapezoidal	Trapezoidal
Channel top width†	ft	20–30	20–30
Channel side slopes		1:1	1:1
Channel lining	Type	Geomembrane	Geomembrane
Liner thickness	mil	40–80	40–80
Pond geometry		Horseshoe shape	Horseshoe shape
Recirculation ratio	Qr/Q	0–2	1–2
Water temperature	°C	>10	>10
Harvest schedule		Annually to seasonally	Monthly to weekly
Expected effluent quality‡			
BOD_5	mg/L	<25	<15
TSS	mg/L	<25	<15
TN	mg/L	<20	<15
TP	mg/L	<7	<5

*Adapted in part from WCPH (1996) and WPCF (1989).
†Top width will depend on the method and equipment used to harvest the water hyacinth.
‡Based on typical domestic wastewater and a loading between 80 and 450 lb/ac·d.

Table 9-18 are presented in Table 9-19. Design considerations, including BOD loading rate, water depth, detention time, mosquito control, and vegetation harvesting are discussed below. The design of a water hyacinth system is illustrated in Example 9-8.

BOD loading rate. The range of loading rates for BOD_5 for water hyacinth systems is from 60 to as high as 450 lb/ac·d, depending on the system configuration and whether supplemental aeration is used. At Walt Disney World, Florida, a water hyacinth system without aeration was loaded with primary effluent up to 400 lb/ac·d. Mosquito and odor problems became significant above 200 lb/ac·d. Average loadings on a water hyacinth system should not exceed 100 lb/ac·d unless aeration is used.

Modeling BOD removal kinetics. On the basis of the results of studies conducted at San Diego, it was demonstrated that, for a plug-flow reactor with step feed

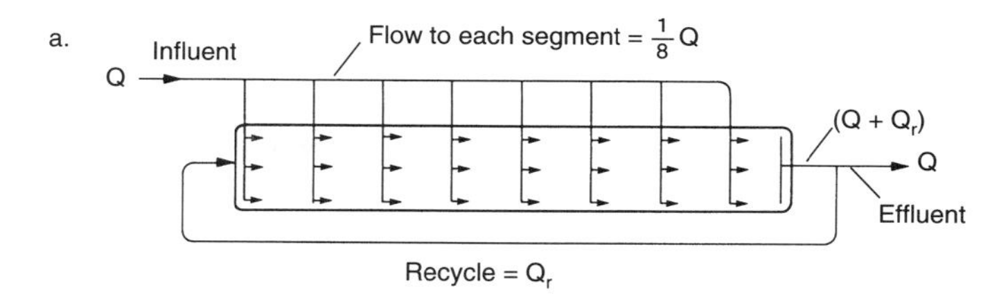

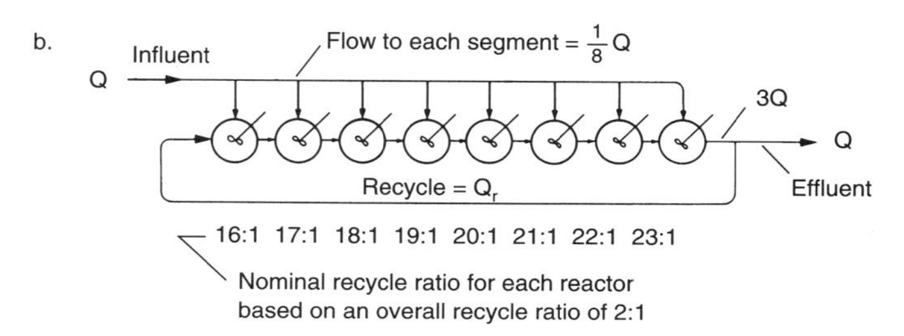

FIGURE 9-15
Schematic for modeling the water hyacinth ponds at San Diego.

and recycle, BOD removal can be modeled by first-order kinetics, assuming each segment of the reactor, corresponding to a feed point, can be modeled as a complete-mix reactor as shown in Fig. 9-15 (Tchobanoglous et al., 1989). The steady-state materials balance for the first complete-mix reactor in the series of eight reactors as shown in Fig. 9-15 is given by

$$\text{Accumulation} = \text{inflow} - \text{outflow} + \text{generation}$$

$$0 = Q_r(C_8) + 0.125Q(C_o) - (Q_r + 0.125Q)(C_1) + (-k_T)(C_1)V_1 \tag{9-24}$$

where Q_r = recycle flow, Mgal/d
C_8 = concentration of BOD_5 in effluent from reactor 8 in series, mg/L
$0.125Q$ = inflow to each individual cell (Q/8), Mgal/d
C_o = concentration of BOD_5 in influent, mg/L
C_1 = concentration of BOD_5 in effluent from reactor 1 in series, mg/L
k_T = overall first-order removal-rate constant at temperature T, 1/d
V_1 = volume of first reactor in series, Mgal

The estimated value of k_T to be used in the above expression for BOD_5 removal for an aerated water hyacinth system is on the order of 1.95 d^{-1} at 20°C (Tchobanoglous et al., 1989). Other values that have been reported for nonaerated systems in the literature range from 0.7 to about 1.25. The above expression can be used iteratively to check the design of a pond system on the basis of surface loading rate.

Water depth. The critical concern with respect to water depth is to control the vertical mixing in the pond so that the wastewater to be treated will come into contact with the plant roots where the bacteria that accomplish the treatment are located

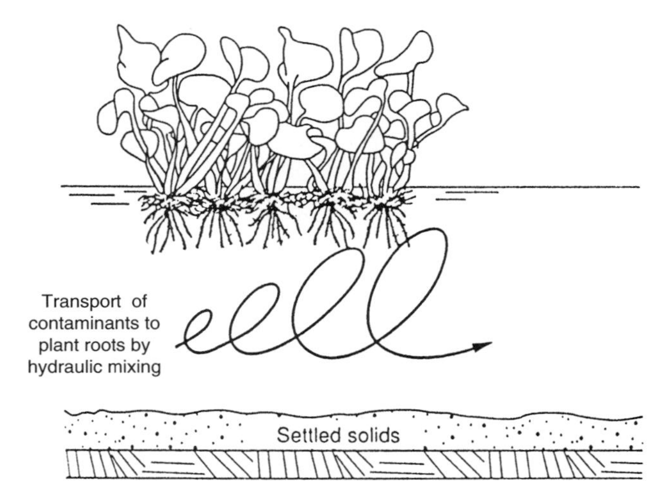

FIGURE 9-16
Water hyacinth roots as sites for bacterial growth.

(see Fig. 9-16). The poor performance observed in some of the early water hyacinth systems was a result of the fact that the operating depth was too deep to promote vertical mixing. Density currents, which allowed the incoming wastewater to flow along the bottom to the outlet with little or no treatment, were also a problem in the early deep systems. Typical operating depths for nonaerated and aerated water hyacinth systems range from 1.5 to 2.5 ft (0.45 to 0.75 m) and 4 to 4.5 ft (1.2 to 1.4 m), respectively. To accommodate variable operating conditions, water hyacinth systems should be designed with an outlet structure that allows the operating depth to be varied.

In aerated water hyacinth systems, greater liquid depths can be used, because the aeration devices also serve as air lift pumps which cause a circulation flow to develop in the pond as shown in Fig. 9-16. The circulation pattern allows the wastewater to come in contact with the plant roots. In addition to the creation of circulation patterns, the added oxygen in aerated systems has made it possible to use organic loading rates 4 times as high as those used for the design of nonaerated systems, successfully. Where aeration devices are used, it is extremely important to use devices that produce fine bubbles, as large bubbles exert a relatively large buoyant force which tends to lift the roots of the water hyacinth plants out of the water column. When fine bubbles are used, the bubbles are intercepted by the roots of the plants. In turn, oxygen is extracted from the air bubble until the buoyant force exceeds the force of adhesion. Where fine-bubble diffusers are used, the measured oxygen transfer efficiency is greater than would be predicted from the operating depth.

Detention time. The detention time needed for BOD removal can be estimated by Eq. (9-6). For systems (duckweed or water hyacinth) in which algae removal is important, a detention time of about 20 days is usually necessary to break the growth cycle of the algae. Aerated water hyacinth systems can perform with detention times of 4 to 10 d, depending on the organic loading and effluent objectives.

Mosquito control. One of the most effective methods for the control of mosquitoes, developed at Aqua II and Aqua III in San Diego, is the use of sprinklers to prevent mosquito ovaposition (WCPH, 1996). The use of sprinklers for the control of mosquitoes is considered in Sec. 9-9. If water sprinkling is to be used for mosquito control, then provisions must be made in the design to account for the water required for spraying.

Vegetation management. The need for water hyacinth harvesting depends on water quality objectives, the growth rates of the plants, mosquito control strategy, and the effects of predators such as weevils. If nutrient removal by plant uptake is a system objective, frequent harvesting is necessary. Design considerations related to water hyacinth harvesting include the type of equipment to be used for harvesting, the required area for storing and processing the harvested plants, and area required for composting, if this method of processing is adopted.

Physical Features of Floating Aquatic Plant Systems Using Water Hyacinths

Physical features of aquatic plant systems include basin configurations, inlet and outlet structures, and aeration. Details of levee and basin construction are available in Stephenson et al. (1980).

Pond configuration. Typical pond configurations used for water hyacinth systems were shown previously in Fig. 9-13. Most of the early water hyacinth systems involved rectangular basins operated in series, similar to stabilization ponds (Figs. 9-13*a*, *b*). In later designs, recycle and step feed (Figs. 9-13*c*, *d*) are employed to (1) reduce the concentration of the organic constituent at the plant root zone, (2) improve the transport of wastewater to the root zone, and (3) reduce the formation of odors. The use of a wraparound design (Fig. 9-13*e*) shortens the required length of the step feed and recycle lines and reduces recycle pumping costs. The wraparound design was also used at Sacramento, California (see Fig. 9-4).

Inlet and outlet structures. Inlet structures range from concrete or wooden weir boxes to manifold pipes with multiple outlets. The objective is to provide a low-maintenance system that will distribute wastewater and solids evenly into the basin without clogging. A subsurface discharge is preferred for inlets, while interbasin and extrabasin transfer structures can be either surface or subsurface. Outlet devices should be located as far from the inlet as possible to avoid short circuiting. If variable operating depths are planned, the outlet should be capable of removing effluent from various depths including periodic draining of the basin.

Supplemental aeration. Water hyacinth systems can benefit from supplemental aeration. If a high-rate system, such as the San Diego water hyacinth system, is selected, the use of aeration with fine-bubble plate diffusers is appropriate (DeBusk et al., 1989).

EXAMPLE 9-8. AQUATIC TREATMENT SYSTEM USING WATER HYACINTHS. A city of 4000 with a STEP-type wastewater collection system is currently connected to an oxidation treatment pond that is overloaded and not meeting its discharge standards of 30 mg/L BOD and TSS. The city has a 4-ac parcel of land. If the per capita flow is 80 gal/d and the BOD is 120 mg/L, determine if the 4-ac parcel is large enough for an effective aerated-type water hyacinth treatment system to upgrade the pond effluent to meet the discharge standards. If the 4-ac parcel is sufficient, lay out a typical water hyacinth system. To accommodate the available harvesting equipment, the maximum width at the top of the water hyacinth ponds should be 26 ft including 1.0 ft of freeboard. Because the ponds will be lined, use a 1:1 side slope. If the removal-rate coefficient is 1.95 d^{-1}, estimate the effluent quality.

Solution

1. Calculate the wastewater flowrate and daily BOD mass loading.
   a. Wastewater flowrate:

$$\begin{aligned} Q &= 4000 \text{ people} \times 80 \text{ gal/capita·d} = 320{,}000 \text{ gal/d} \\ &= 0.32 \text{ Mgal/d} \end{aligned}$$

   b. BOD mass loading rate:

$$\text{BOD, lb/d} = 0.32 \text{ Mgal/d} \times 120 \text{ mg/L} \times 8.34 = 320 \text{ lb/d}$$

2. Select design parameters for the aerated water hyacinth system using the information given in Table 9-19:

$$\begin{aligned} &\text{Organic loading rate} = 275 \text{ lb/ac·d} \\ &\text{Depth} = 4 \text{ ft} \\ &\text{Maximum width of water surface} = 24 \text{ ft} \\ &\text{Application mode} = \text{step feed with recycle of 1 to 1} \end{aligned}$$

3. Determine the required surface area:

$$A = \frac{320 \text{ lb/d}}{275 \text{ lb/ac·d}} = 1.16 \text{ ac}$$

4. Determine the physical characteristics and the volume of the water hyacinth ponds.
   a. Determine the total number of ponds required for a pond length of 300 ft:

$$\text{No. of ponds} = \frac{(1.16 \text{ ac})(43{,}560 \text{ ft}^2\text{/ac})}{(300 \text{ ft})(24 \text{ ft})} = 7$$

   b. Determine the pond bottom width:

$$\text{Bottom width} = 24 \text{ ft} - (2 \times 4) = 16 \text{ ft}$$

   c. Determine the pond volume, neglecting end corrections:

$$\begin{aligned} V_{\text{Total}} &= 7 \times \left(\frac{24 + 16}{2}\right)(4 \text{ ft})(300 \text{ ft}) = 168{,}000 \text{ ft}^3 \\ &= 1.26 \text{ Mgal} \end{aligned}$$

5. Determine the detention time:

$$t = \frac{1.26 \text{ Mgal}}{0.32 \text{ Mgal/d}} = 3.94 \text{ d}$$

6. Determine the area required for the pond system, assuming an additional area equal to that for the ponds will be required for access roads and processing facilities.

$$\text{Total area required} = \frac{(7)(24\text{ ft} + 2 \times 1.0\text{ ft})(300\text{ ft})(2)}{43{,}560\text{ ft}^2/\text{ac}} = 2.5\text{ ac}$$

The available 4-ac area is insufficient for a water hyacinth treatment system.

7. Estimate the effluent from the water hyacinth ponds using Eq. (9-24):

$$0 = Q_r(C_8) + 0.125Q(C_o) - (Q_r + 0.125Q)(C_1) + (-k_T)(C_1)V_1$$

a. Assume the following conditions apply:

Q_r = recycle flow = 0.32/7 = 0.0457 Mgal/d

C_8 = assumed concentration of BOD_5 in effluent from reactor 8 in series = 20 mg/L

$0.125Q$ = inflow to each individual cell (Q/8) = 0.00594 Mgal/d

C_o = concentration of BOD_5 in influent = 120 mg/L

C_1 = concentration of BOD_5 in effluent from Reactor 1 in series = unknown, mg/L

k_T = first-order reaction-rate constant = 1.95 d^{-1}

V_1 = volume of first reactor in series = $(168{,}000\text{ ft}^3)/(7 \times 8)$ = 3000 ft^3 = 22,440 gal

b. Determine the value of C_1:

$$C_1 = \frac{(0.047\text{ Mgal/d})(20\text{ mg/L}) + (0.00594\text{ Mgal/d})(120\text{ mg/L})}{(0.047 + 0.00594) + (1.95 \times 0.02244)}$$

$$= 17.0\text{ mg/L}$$

c. Solving iteratively for C_8 yields a value of about 17 mg/L. Therefore, the effluent from the proposed water hyacinth system will meet the treatment objectives.

9-6 FLOATING AQUATIC PLANT SYSTEMS—DUCKWEED

Floating aquatic plant systems using duckweed have been used in wastewater treatment for a variety of purposes including secondary treatment, advanced secondary treatment, and nutrient removal. The most widely used option for duckweed systems, achieving secondary treatment using enhanced sedimentation, will be the focus of this section.

Process Description

Duckweed systems have been designed primarily to upgrade facultative pond effluents. The process performance and design considerations for duckweed systems are presented in this section.

TABLE 9-20
Nutrient composition of water hyacinths and duckweed grown in wastewater

	Dry weight, %	
Constituent	**Water hyacinth**	**Duckweed**
Crude protein	18.1	38.7
Nitrogen (N)	2.9	5.9
Phosphorus (P)	0.6	0.6

Source: WPCF (1989).

Duckweed

Duckweed (*Lemna* spp.) are small freshwater plants with leaves (fronds) that are 0.04 to 0.12 in (1 to 3 mm) in width and roots that are less than 0.4 in (10 mm) long. Duckweed grow faster than water hyacinth (reportedly 30 percent faster) (WPCF, 1989) and are higher in protein (see Table 9-20). Duckweed can form a surface mat on a pond by doubling the area covered in 4 days. The plants do not transfer oxygen into the water, leaving the duckweed pond effluent anoxic. Duckweed are very sensitive to wind drifting and therefore require baffles to keep the plants in place. Duckweed is more cold-tolerant than water hyacinth. Water temperatures of

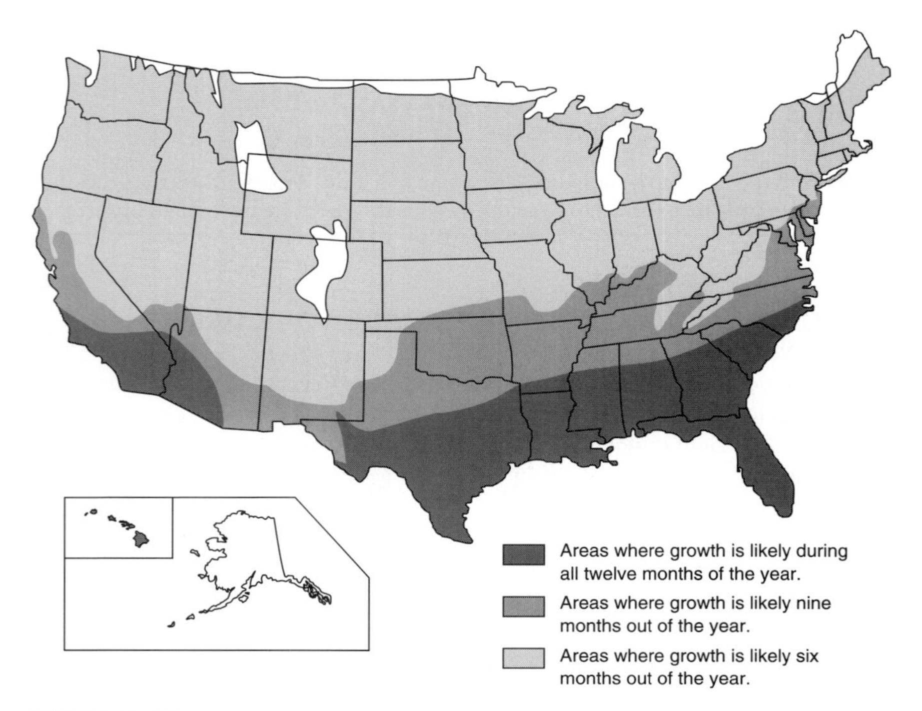

FIGURE 9-17
U.S. zones for the growth of duckweed (adapted from Leslie, 1983).

7°C or higher are needed to sustain the growth. As shown in Fig. 9-17, duckweed can be grown for at least 9 months of the year in temperate climes and 6 months of the year in nearly all U.S. climes.

Constituent Removal and Transformation Mechanisms

Removals of BOD and TSS are generally quite good, with lesser efficiencies demonstrated for nutrients, metals, and pathogens. The operative removal mechanisms are described below.

BOD and TSS removal. BOD removal in duckweed systems occurs as a result of biological activity that is similar to the reactions that occur in facultative ponds. The duckweed cover the surface of the ponds and limit the growth of algae, thereby reducing the oxygen in the water column for aerobic bacterial activity. In addition, the duckweed limits the wind-aided reaeration from the atmosphere, thereby further limiting the BOD removal. As a consequence, the BOD loading should be limited to 25 lb/ ac·d (27.5 kg/ha·d) or less.

Nitrogen removal. Nitrogen is removed in aquatic treatment systems by microbial nitrification-denitrification, and to a lesser extent, by plant uptake and harvest. In duckweed systems, denitrification will occur readily; however, nitrification requires an input of oxygen.

Phosphorus removal. Plant uptake and harvest is the pathway for phosphorus removal from duckweed systems.

Metals and trace organics removal. Metal removal mechanisms include plant uptake, chemical precipitation, and adsorption.

Pathogen removal. The removal of entering bacteria and viruses in aquatic plant systems is similar to the mechanisms that are operative in ponds—natural die-off, sedimentation, predation, adsorption, and exposure to ultraviolet light.

Process Performance

Performance expectations for floating aquatic plant systems using duckweed are reviewed in the following discussion.

BOD removal. High levels of BOD and TSS removal are generally expected from duckweed systems. To achieve secondary treatment, most duckweed systems are coupled either with facultative or aerated ponds. In 1995 there were 35 operational wastewater treatment facilities designed specifically as duckweed systems. Most are designed to achieve secondary treatment. The removal of BOD and TSS in duckweed systems is shown in Table 9-21.

TABLE 9-21
Typical effluent BOD and TSS values observed in duckweed systems

Location	Design flow, Mgal/d	Detention time, d	Effluent BOD, mg/L	Effluent TSS, mg/L	Permit*
Arkadelphia, Arkansas	3.0	10	12	15	30/90
Ellaville, Georgia	0.2	20	13	10	20/30
Four Corners, Louisiana	0.16	24	3	3	10/15
Kyle, Texas	0.89	12	18	13	30/30
Mamou, Louisiana	0.8	30	5	8	10/15
Nokesville, Virginia	0.05	12	6	5	12/12
White House, Tennessee	0.8	27	3	4	10/30

*Average BOD/TSS for most stringent season.

Suspended solids removal. The duckweed plants play a major role in the removal of suspended solids. The surface mat blocks the sunlight and the mats enhance sedimentation by creating quiescent conditions. The rate at which the suspended solids settle depends on the nature of the solids. Algal cells take a relatively long period of time (6 to 10 d) to die and begin to settle.

Nitrogen removal. Nitrogen can be removed either by plant uptake and harvesting or by nitrification-denitrification. To remove nitrogen by plant harvest, optimum growth must be achieved and frequent harvest must be accomplished. The density of the plants at the water surface depends on the temperature, availability of nutrients, and frequency of harvest. The typical density on a wastewater pond may range from 0.25 to 0.75 lb/ft^2 (1.2 to 3.6 kg/m^2) (Reed et al., 1995). The optimum growth rate is about 0.1 $lb/ft^2 \cdot d$ (0.49 $kg/m^2 \cdot d$).

Annual harvest amounts range from 5.9 to 17.3 ton/ac (13 to 38 mt/ha) with 10 ton/ac (22 mt/ha) being typical. Assuming that the nitrogen content is 5.9 percent of the dry matter, 98 lb/ac·mo of nitrogen can be removed. Assuming that a 12-acre system, 5 ft deep, is used, the harvest of 98 lb/acre·mo of nitrogen in the duckweed would amount to 4.7 mg/L removal of nitrogen from the duckweed pond. Because nitrogen removal via plant harvest is not practical, the Lemna Corporation has developed a submerged media nitrification reactor with supplemental aeration.

Phosphorus removal. Phosphorus removal can be achieved by plant harvest, but only to the same limited extent as for nitrogen. Generally, less than 1 mg/L of phosphorus can be removed by plant uptake and harvest. If wastewater phosphorus concentrations are low and removal requirements are minimal, then harvesting, as practiced in the Devils Lake, North Dakota, system, may be suitable. If significant phosphorus removal is required, however, the use of chemical precipitation with alum, ferric chloride, or other chemicals in a separate treatment step may be more cost-effective.

Metals removal. Duckweed has been shown to accumulate 27 μg zinc, 10 μg lead, and 5.5 μg nickel per mg of duckweed when exposed to 10 mg/L of the three metals. Because the metals concentrations in municipal wastewater are very low, the metals concentrations in duckweed are similarly low.

TABLE 9-22
Typical design criteria and expected effluent quality for duckweed systems

Item	Unit	Value
	Design parameter	
Influent wastewater		Facultative pond effluent
BOD loading rate	lb/ac·d	20–25
Detention time	d	20–30
Water depth	ft	5–8
Hydraulic loading rate	gal/ac·d	<55,000
Harvest schedule		Monthly for secondary treatment, weekly for nutrient removal
	Expected effluent quality*	
Secondary treatment		
BOD_5	mg/L	<30
TSS	mg/L	<30
TN	mg/L	<15
TP	mg/L	<6
Nutrient removal		
BOD_5	mg/L	<10
TSS	mg/L	<10
TN	mg/L	<5
TP	mg/L	<2

*Expected effluent quality based on loadings equal to or less than given values.

Process Design Considerations

Process design criteria and expected water quality for duckweed systems are presented in Table 9-22. Design considerations include detention time, BOD loading rates, and water depth.

Detention time. For duckweed systems in which algae removal is important, a detention time of about 20 days is usually necessary to break the growth cycle of the algae.

BOD loading rate. The range of loading rates for BOD_5 is from 20 to 25 lb/ac·d for duckweed systems. The Lemna Corporation, which offers proprietary floating plastic barriers (see Fig. 9-18) and harvesting equipment (see Fig. 9-19), suggests that wastewater entering the duckweed portion of the facility be partially treated to a BOD level of 60 mg/L or less by facultative ponds, aerated ponds, or mechanical treatment plants. To achieve a 20-mg/L BOD in the effluent, Lemna suggests a target influent of 40 mg/L, 20-d detention time, and a pond sizing of 12 ac/Mgal·d (12.8 $m^2/m^3 \cdot d$), based on a minimum pond depth of 5 ft (1.5 m). To achieve a final BOD of 10 mg/L, it is suggested that a target influent BOD of 30 mg/L, a 28-d hydraulic detention time, and a pond sizing of 17.5 ac/Mgal·d) (18.4 $m^2/m^3 \cdot d$) be used.

FIGURE 9-18
Floating plastic barriers for control of duckweed.

FIGURE 9-19
Floating harvester used to harvest duckweed.

Water depth. With duckweed the depth can be 5 to 8 ft (1.5 to 2.4 m) or deeper because there is no root-bacteria contact to be achieved.

Physical Features of Floating Aquatic Plant Systems Using Duckweed

Physical features of aquatic plant systems using duckweed include basin configurations and inlet and outlet structures. Details of levee and basin construction are available in Stephenson et al. (1980).

Basin configurations. Duckweed ponds can be designed like water hyacinth basins or as large ponds with floating baffles. Floating baffles are used to reduce the effects of the wind. Long narrow basins can also be used, and emergent plants can be added to avoid wind effects.

Inlet and outlet structures. Inlet structures range from concrete or wooden weir boxes to manifold pipes with multiple outlets. The objective is to provide a low-maintenance system that will distribute wastewater and solids evenly into the basin without clogging. A subsurface discharge is preferred for inlets, while interbasin and extrabasin transfer structures can be either surface or subsurface. Outlet devices should be located as far from the inlet as possible to avoid short circuiting. If variable operating depths are planned, the outlet should be capable of removing effluent from various depths, including periodic draining of the basin.

9-7 COMBINATION SYSTEMS

There are a number of systems that combine natural treatment processes. The earlier meadow-marsh-pond system demonstrated on Long Island in the 1970s led to combinations of overland flow and constructed wetlands and to combinations of aquatic plant pond systems with constructed wetlands. The solar aquatic system at Harwich, Massachusetts (see Chap. 14), is an example of the latter combination for the treatment of septage.

Frederick, Maryland

The Advanced Ecologically Engineered System (AEES) at Frederick, Maryland, is one of several related projects in the United States intended to provide advanced treatment of municipal wastewater. The AEES technology is also called a "Living Machine" by Dr. John Todd, who developed the concept. The facility at Frederick was constructed in 1993 and treated 40,000 gal/d of screened and degritted wastewater. The schematic flow diagram and several views of the system are presented in Fig. 9-20. The detention time was 3.6 d through the system (Reed et al., 1996). Approximately 85 percent of the BOD removal in the system was accounted for in the anaerobic upflow filter. Performance data for the system are presented in Table 9-23.

Benton, Kentucky

A constructed wetland system at Benton, Kentucky, was designed for the removal of BOD, TSS, and ammonia from an existing treatment pond. When ammonia removal fell short of expectations, a retrofit design was proposed by Reed (Reed et al., 1995). The design involved the addition of a recirculating gravel filter capable of nitrifying the wastewater. The nitrified effluent was then reintroduced into the constructed wetland to accomplish the denitrification step. Design factors and process performance data are presented in Table 9-24.

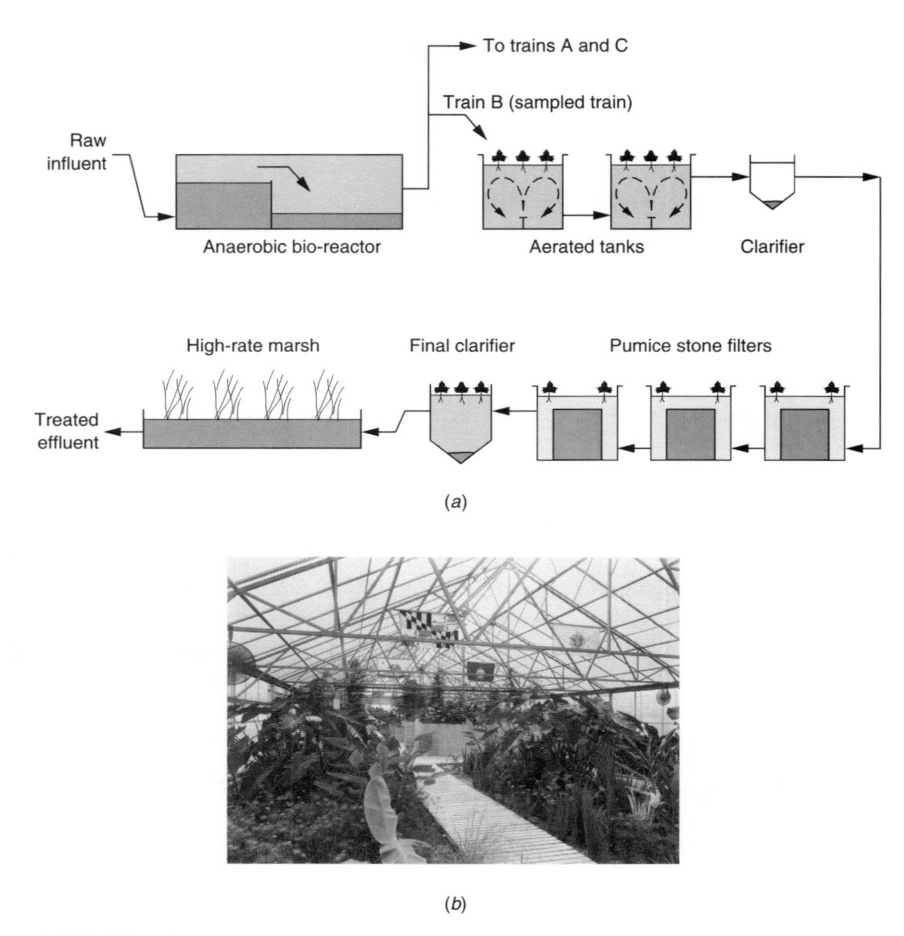

FIGURE 9-20
Frederick, Maryland, aquatic treatment system: (*a*) flow diagram and (*b*) view of treatment system inside greenhouse. High-rate marsh is shown in the foreground. Anaerobic bio-reactor is shown in Fig. 7-50 in Chap. 7.

TABLE 9-23
Performance data for the Frederick, MD Living Machine, March 1995 to March 1996*

Item	Unit	Process value	Observed mean value
BOD_5	mg/L	≤ 10	4.8
COD (total)	mg/L	≤ 50	27
TSS	mg/L	≤ 10	1.4
NH_4-N	mg/L	≤ 1	1.7
NO_3-N	mg/L	≤ 5	7.0
Nitrogen (total)	mg/L	≤ 10	10.6
Phosphorus (total)	mg/L	≤ 3	6.0

*From Living Technologies, Inc (1997).

TABLE 9-24
Design parameters and performance data for recirculating gravel filter at Benton, Kentucky*

Item	Unit	Value
Design parameter		
Flowrate	Mgal/d (m^3/d)	1.0 (3785)
Medium size	in (mm)	0.25 (7)
Medium depth	ft (m)	2 (0.6)
Hydraulic loading rate	gal/ft^2·d (m/d)	140 (5.7)
Ammonia loading	m^2/kg·d	1230
Recirculation ratio		3–1
Wastewater temperature	°C	12–19
Effluent quality†		
TKN influent	mg/L	10–20
TKN effluent	mg/L	1.4–4
Ammonia removal	%	77

*Adapted from Askew et al. (1994).
†Expected effluent quality based on loadings equal to or less than given values.

9-8 DESIGN PROCEDURES FOR CONSTRUCTED WETLANDS

Design considerations and overall design procedures are considered in this section. Design procedures for constructed wetlands and floating aquatic plant systems are presented in Table 9-25.

Process Design Procedure for FWS Wetlands

The procedure for process design of FWS constructed wetlands involves the following steps:

1. Determine the limiting effluent requirements for BOD, TSS, and nitrogen or phosphorus.
2. Determine the allowable effluent BOD by subtracting 5 mg/L for BOD related to plant decay.
3. Select an appropriate apparent BOD removal-rate constant and correct for the critical temperature.
4. Calculate the detention time to achieve the desired level of BOD removal.
5. Alternatively, if nearby performance data are available, determine the coefficient of reliability (COR) for the percent reliability required, and calculate the k value for the overall BOD removal required. Use Eq. (9-6) to determine the required detention time.
6. If BOD and TSS are the only parameters to be removed, the organic loading rate should be checked, and the larger of the two areas should be selected.
7. Determine the detention time required for nitrogen or ammonia removal.

TABLE 9-25
Principal steps in the design of constructed wetlands and aquatic plant systems

FWS wetlands	SF wetlands	Water hyacinths	Duckweed
Define treatment requirements	Define treatment requirements	Define treatment requirements	Define treatment requirements
Characterize wastewater	Characterize wastewater	Characterize wastewater	Characterize wastewater
Gather background information	Gather background information	Gather background information	Gather background information
Site evaluation	Site evaluation	Site evaluation	Site evaluation
Determine pretreatment level	Determine pretreatment level	Determine pretreatment level	Determine pretreatment level
Select vegetation	Select vegetation	Determine design parameters	Determine design parameters
Determine design parameters	Determine design parameters	Vector control measures	Detailed design of system
Vector control measures	Detailed design of system components	Detailed design of system	Determine monitoring requirements
Detailed design of system components		Determine monitoring requirements	
Determine monitoring requirements			

8. Select the largest detention time for design, based on the limiting design parameter.
9. Determine the required area. Increase the area by 15 to 25 percent for a factor of safety.
10. Select an aspect ratio consistent with the site constraints and determine the dimensions of the wetland.
11. Check the headloss to ensure adequate head between the influent and effluent ends.

Process Design Procedure for SF Wetlands

The procedure for process design of SF constructed wetlands involves the following steps:

1. Determine the limiting effluent requirements for BOD, TSS, and nitrogen. Reduce the target effluent values by the expected plant decay concentrations.
2. Determine the detention time using first-order kinetics with plug flow.
3. Calculate the required area for nitrogen or ammonia removal.
4. Determine the field area for the SF bed, based on the largest required detention time. Increase the area by 15 to 25 percent for a factor of safety.

5. Calculate the cross-sectional area needed to hydraulically accept the flow, using Darcy's law.
6. Once the cross-sectional area has been determined, calculate the width by dividing the area by the depth.
7. Calculate the bed length to achieve the needed surface area of the bed.
8. Check the bed dimensions for reasonableness. The length-to-width ratio can range from 0.2:1 up to 2:1. Adjust the length or width as necessary to ensure a reasonable length, in case of heavy precipitation.

9-9 MANAGEMENT OF CONSTRUCTED WETLANDS AND AQUATIC SYSTEMS

Operation and maintenance considerations in the management of constructed wetlands and floating aquatic plant systems are described in this section.

Management of Constructed Wetlands

Issues involved in the management of FWS wetlands include mosquito control, vegetation harvesting, wildlife considerations, and monitoring. For SF wetlands the wildlife and monitoring elements apply, and vegetation management is included.

Mosquito control. With FWS wetlands mosquito control is essential. The provisions cited for mosquito control in water hyacinth systems are also applicable to FWS constructed wetlands, including stocking with mosquitofish, maintenance of aerobic conditions, use of biological controls, and the encouragement of predators. At Arcata, California, the FWS wetland actually produces less mosquito larvae than the previous unused marshy area because of the encouragement of habitat for swallows and mosquitofish.

At Sacramento County, the following combination of management techniques has been successful in controlling mosquitoes (Williams et al., 1996):

1. Mosquitofish stocking.
2. Daily monitoring for mosquito larvae from April through October.
3. Applications of *Bacillus thuringensis israelensis* (Bti) when needed.
4. Vegetation management to maintain open water and pathways for mosquitofish to get to the mosquito larvae.

Vegetation harvesting. Harvesting of the emergent vegetation is practiced to maintain hydraulic capacity, promote active growth, and avoid mosquito growth. Harvesting for nutrient removal is not practical and is not recommended. Harvesting will affect performance, so the harvested cell should be taken out of service before and for several weeks after harvesting. Harvested vegetation can be burned, chopped and composted, chopped and used as mulch, or digested (Hayes et al., 1987). A vegetation control strategy for a typical FWS wetlands is presented in Table 9-26.

TABLE 9-26
Vegetation control strategy for FWS constructed wetlands*

Issue	Outcome
Operating goal	Process performance
Problem identification	Clogging of flow paths, odors from decomposition, short circuiting, low density, poor plant health
Causative factors	Aggressive growth, lack of vegetative management, excessive water depth, poor water flow patterns, seasonal variation, grazing
Management strategies	Reduce water depth, soil enhancement, supplemental planting, controlled burns, periodic harvesting
Lead time	Growing season
Evaluation of control	Vegetation surveys, vegetation maps, photographic records

*Adapted from Tchobanoglous (1993).

Wildlife considerations. Wildlife, including ducks, shorebirds, raptors, field birds, deer, jackrabbits, and muskrats will be attracted to wetlands. In Florida, alligators and snakes have been reported in constructed wetlands. Ducks need open water, which may not be compatible with the need for thick vegetation to achieve secondary treatment. Burrowing animals, such as nutria, can also create problems with berms (Crites and Lesley, 1997). If wildlife habitat enhancement is a project goal, islands raised above deeper water should be considered. These habitat islands can support upland vegetation and provide nesting trees for birds (Wilhelm et al., 1989).

Monitoring. Monitoring needs can include flow, surface water quality, and groundwater quality. Variable-height weirs can be used to monitor flow out of the wetland and to provide a convenient sampling point. Surface water sampling points should be located at catwalks or boardwalks to allow sampling without disturbing the flow. A summary of suggested monitoring parameters is presented in Table 9-27.

Vegetation management for SF wetlands. Harvesting is not necessary for SF wetland vegetation; however, development of the roots into the media is important. After initial establishment, the water level needs to be dropped so that the roots will extend, eventually to the bottom of the media.

Management of Floating Aquatic Plant Systems

Both water hyacinth and duckweed systems require management to avoid odors and to harvest the plants. For water hyacinth systems the management issues are mosquito control, vegetation harvesting, and sludge removal. For duckweed systems the issues are vegetation harvesting and sludge management. Mosquito problems with duckweed systems do not occur because the pond surface is effectively sealed off by the plants, and the female mosquitoes cannot reach the water to lay their eggs.

TABLE 9-27
Summary of suggested monitoring parameters for constructed wetlands*

Parameter	Project phase (Pre- or during construction or ongoing)	Location	Frequency of collection
Water quality†‡			
Dissolved oxygen	Ongoing	In, out, along profile	Weekly
Hourly dissolved oxygen	Ongoing	Selected locations	Quarterly
Temperature	Pre, ongoing	In, out, along profile	Daily/weekly
Conductivity	Pre, ongoing	In, out	Weekly
pH	Pre, ongoing	In, out	Weekly
BOD	Pre, ongoing	In, out, along profile	Weekly
SS	Pre, ongoing	In, out, along profile	Weekly
Nutrients	Pre, ongoing	In, out, along profile	Weekly
Chlorophyll A	Ongoing	Within wetland, along profile	Annually
Metals (Cd, Cr, Cu, Pb, Zn)	Pre, ongoing	In, out, along profile	Quarterly
Bacteria (total and fecal coliform)	Pre, ongoing	In, out	Monthly
EPA priority pollutants	Pre, ongoing	In, out, along profile	Annually
Other organics	Pre, ongoing	In, out, along profile	Annually
Biotoxicity	Pre, ongoing	In, out	Semiannually
Sediments			
Redox potential	Pre, ongoing	In, out, along transects	Quarterly
Salinity	Pre, ongoing	In, out, along transects	Quarterly
pH	Pre, ongoing	In, out, along transects	Quarterly
Organic matter	Pre, post	In, out, along transects	Quarterly
Vegetation			
Plant coverage	Ongoing	Within wetland, along transects	Quarterly
Identification of plant species	Ongoing	Within wetland, along transects	Annually
Plant health	Ongoing	Within wetland	Observe weekly

(*continued*)

TABLE 9-27
(Continued)

Parameter	Project phase (Pre- or during construction or ongoing)	Location	Frequency of collection
Biota			
Plankton (zooplankton tow)	Ongoing	Within wetland, along transects	Quarterly
Invertebrates	Ongoing	Within wetland, along transects	Annually
Fish	Ongoing	Within wetland, along transects	Annually
Birds	Pre, ongoing	Within wetland, along transects	Quarterly
Endangered species	Pre, during, ongoing	Within wetland, along transects	Quarterly
Mosquitoes	Pre, during, ongoing	Within wetland, selected locations	Weekly during critical months
Wetland development			
Flowrate	Ongoing	In, out	Continuous
Flowrate distribution	Ongoing	Within wetland	Annually
Water surface elevations	Ongoing	Within wetland	Semiannually
Marsh surface elevations	Ongoing	Within wetland	Quarterly

*Adapted from Tchobanoglous (1993).
†Water quality for pre- and during construction refers to the wastewater that is to be applied to wetland.
‡Permitting agencies may not require all parameters to be tested, or to be tested at the same frequency.

Mosquito control. In many areas of the United States, the propagation of mosquitoes in aquatic systems may be the critical factor in their acceptance or rejection. A typical vector control strategy is outlined in Table 9-28. The objective of mosquito control is to suppress mosquito populations below the threshold level required for disease transmission or nuisance tolerance levels. Strategies that have been used to control mosquito populations include (WCPH, 1996):

1. Stocking ponds with mosquitofish (*Gambusia* spp.).
2. More effective pretreatment to reduce the total organic loading on the aquatic system, to help maintain aerobic conditions.
3. Step feed of influent waste stream with recycle (see Fig. 9-13).
4. More frequent plant harvesting.
5. Water spraying in the evening hours (see Fig. 9-21).
6. Application of chemical control agents (larvicides).
7. Diffusion of oxygen (with aeration equipment).
8. Biological control agents (e.g., Bti).

Fish used for control of mosquitoes (typically *Gambusia* spp.) will die under the anaerobic conditions that exist in organically overloaded ponds. In addition to inhibited fish populations, mosquitoes may develop in dense water hyacinth systems

TABLE 9-28
Operational issue—vector control strategy for water hyacinth treatment systems

Issue	Outcome
Operating goal	Control of mosquitoes
Problem identification	Increased counts in resting box, emergence traps, dip samples
Causative factors	Execessive plant growth, lack of predators
Management strategies	Draw water surface down, use biological controls, use conventional controls (Bear oil 2000, Bti)
Lead time	2 to 3 weeks, depending on sampling frequency
Evaluation of control	Reduced larval count

FIGURE 9-21
Sprinkler system used to control the breeding and production of mosquitoes.

when plants have been allowed to grow tightly together. Pockets of water form as the plants bridge together that are accessible to the mosquitoes but not the fish.

One of the most effective methods for the control of mosquitoes, developed at Aqua II and Aqua III in San Diego, is the use of sprinklers to prevent mosquito oviposition. For example, *Culex* spp. mosquitoes oviposit on still bodies of water and require about 20 to 35 minutes to complete oviposition. The use of sprinklers both disrupts mosquito flight patterns and effectively reduces oviposition by disturbing or killing the female mosquitoes before or shortly after they have landed. The sprinklers are operated from about 8 P.M. to 6 A.M. The sprinkler coverage pattern is shown in Fig. 9-21 (WCPH, 1996).

Vegetation management in hyacinth systems. At Aqua III, water hyacinths are harvested by a truck-mounted hydraulic crane with an 85-ft boom and

FIGURE 9-22
Articulated clamshell for harvesting water hyacinths.

a 5-ft open type clamshell (Fig. 9-22). Harvested water hyacinths are composted on site. Because high moisture content tends to reduce the effectiveness of the compost process, the harvested water hyacinths are chopped by a tub grinder (see Fig. 9-23) and spread out in a thin layer to reduce the moisture content before composting (see Fig. 9-24). After about 5 d, when the moisture content has been reduced to about 60 percent, the water hyacinths are formed into windrows for the composting process.

No bulking agent is required and temperatures greater than 160°F can be maintained for 15 d, without the addition of supplementary moisture. The overall

FIGURE 9-23
Tub grinder used to chop water hyacinths for predrying before composting.

FIGURE 9-24
Water hyacinth spread out to dry, to reduce the initial moisture content before composting in windrows.

volume reduction, brought about by processing, drying, and composting, between the volume of the newly harvested wet water hyacinths and the final volume of compost remaining after the composting process, has averaged 99 percent (i.e., a reduction of 100 to 1) (WCPH, 1996). Such a high volume reduction is possible, because the water hyacinths contain little lignin, and both bacterial and fungal organisms are involved in the composting process. The final compost meets the US EPA 503 pathogen and metals requirements for Class A compost (WCPH, 1996).

Sludge management in hyacinth systems. The solids that accumulate in aquatic systems include plant detritus, inorganic solids, and biological sludge. These solids are usually removed infrequently (annually or less frequently as needed). Sludge accumulation in the San Diego water hyacinth system averaged 3.3 in over 13 months of operation (Tchobanoglous et al., 1989).

Duckweed management. The need for duckweed harvesting depends on water quality objectives and the growth rates of the plants. Monthly harvesting is typical during the growing season. If nutrient removal by plant uptake is a system objective, harvesting frequencies as high as once per week may be required. Management alternatives for harvested duckweed include composting, use as animal feed, and land application.

Sludge management in duckweed systems. The solids that accumulate in aquatic systems include plant detritus, inorganic solids, and biological sludge. These solids are usually removed infrequently (annually or less frequently as needed).

9-10 EMERGING TECHNOLOGIES

Emerging technologies are those that have shown promise in small research and demonstration projects. These include vertical-flow wetlands, batch-flow wetlands, submerged vegetation aquatic plant systems, and algal turf scrubbers.

Vertical-Flow Wetlands

In vertical-flow constructed wetlands the applied water flows through the gravel bed in a manner similar to flow in a planted rapid infiltration system. During the loading period, air is forced out of the bed; during the drying period, atmospheric air is drawn into the bed, which increases oxygenation of the bed. Diffusion of atmospheric oxygen into the bed is rapid because the diffusion of oxygen is approximately 10,000 times faster in air than in water. Vertical-flow wetlands have been used in Europe and in reed bed treatment and dewatering of biosolids (Brix, 1993).

Batch-Flow Wetlands

Batch-flow wetlands and aquatic plant systems have the same appeal as vertical-flow wetlands. Both are attempts to increase the oxygen levels in the root zone and detritus areas in the beds. Batch-flow wetlands (8 d of filling and 2 d emptying) are being tested at Sacramento Regional County Sanitation District (Crites et al., 1996).

Submerged Vegetation Aquatic Plant Systems

Submerged vegetation has been studied by various researchers (Kozak and Bishop, 1987; Eighmy et al., 1987). Previous efforts were limited by the need to aerate the vegetated ponds. Recent research at Contra Costa County shows promise for nitrifying secondary effluent (Bouey, 1996).

Algal Turf Scrubbers

The algal turf scrubber is an attempt to grow and harvest attached-growth algae (periphyton) for nutrient removal. The device has been operated in pilot studies for phosphorus removal in the Everglades nutrient removal project and at Patterson, California.

PROBLEMS AND DISCUSSION TOPICS

9-1. Constructed wetlands can be designed by using a first-order equation based on detention time or by using surface area loading rates. Compare and contrast these two methods for the design.

9-2. An FWS wetland is designed to treat 0.15 Mgal/d of facultative treatment pond effluent from a BOD of 80 mg/L to a BOD of 20 mg/L. Determine the detention time and net field area required for treatment. Use a water depth of 12 in. Use a water temperature of 20°C.

9-3. An FWS wetland is proposed to treat either lagoon effluent (minimum temperature of 5°C) or an Imhoff tank effluent (minimum temperature of 8° C). Compare needed detention times for 85 percent BOD removal.

9-4. A secondary effluent needs 85 percent ammonia removal. If the ammonia nitrogen concentration is 20 mg/L in the secondary effluent, calculate the hydraulic loading rate needed for a FWS wetland. If the wetland water depth is 4 in, what is the detention time?

9-5. For the two temperatures in Prob. 9-3, calculate the detention time needed for 80 percent nitrogen removal by water hyacinths.

9-6. Using the following data from a Gustine, California, FWS wastewater treatment system, determine the coefficient of reliability for 92 percent.

	BOD, mg/L	
Month	**1994**	**1995**
January	18	15
February	9	15
March	12	10
April	17	8
May	24	10
June	19	15
July	26	18
August	21	29
September	23	26
October	29	25
November	25	29
December	30	24

9-7. Using the following data from a San Diego water hyacinth wastewater treatment system, determine the coefficient of reliability for 99 percent.

Month	BOD, mg/L
January	8.4
February	10.2
March	17.9
April	10.3
May	13.2
June	9.3
July	8.6
August	12.0
September	13.7
October	13.8
November	16.3
December	17.0

9-8. Determine the detention time, lagoon volume, surface area, and depth for a duckweed system designed to upgrade a facultative treatment pond effluent (TSS = 90 mg/L) to achieve 30 mg/L TSS. The flow is 0.1 Mgal/d.

9-9. A water hyacinth system is being designed to treat septic tank effluent (BOD = 130 mg/L) to secondary treatment levels (BOD = 30 mg/L). Calculate the required detention time to achieve the required BOD removal if the k value is 0.6 at 20°C. What detention time is needed if the minimum average water temperature is 12°C?

9-10. An SF wetland is being designed for 90 percent BOD removal. What detention time is needed for a water temperature of 15°C? Use a gravel medium.

9-11. An SF wetland has a BOD loading of 80 lb/ac·d. If the medium is gravel with a depth of 24 in, how deep should the water level be during operation? What cross-sectional area should the entry zone have to avoid clogging if the applied TSS concentration is 90 mg/L?

9-12. What role does the diversity of vegetation species play in the observed treatment performance of aquatic treatment systems?

9-13. What techniques have been used successfully to manage and control mosquitoes in wetlands and aquatic treatment systems? Why are mosquitoes not a problem in duckweed systems?

REFERENCES

Andrews, T. (1996) Personal communication, Ouray, CO, FWS wetlands performance data.

Askew, G. L., M. W. Hines, and S. C. Reed (1994) Constructed Wetlands and Recirculating Gravel Filter Systems: Full-Scale Demonstration and Testing, *Proceedings of the 7th International Symposium on Individual and Small Community Sewage Systems,* ASAE, pp. 85–94, Atlanta, GA.

Bavor, H. J., D. J. Roser, and S. A. McKersie (1987) Nutrient Removal Using Shallow Lagoon-Solid Matrix Macrophyte Systems, in K. R. Reddy and W. H. Smith (eds.), *Aquatic Plants for Water Treatment and Resource Recovery,* pp. 228–236, Magnolia Publishing, Orlando, FL.

Bavor, H. J., D. J. Roser, P J. Fisher, and I. C. Smalls (1989) *Joint Study on Sewage Treatment Using Shallow Lagoon-Aquatic Plant Systems,* Vol. 2, Treatment of Secondary Effluent, Water Research Laboratory, Hawkesbury Agricultural College, Richmond, NSW, Australia.

Bavor, H. J., and T. J. Schulz (1993) Sustainable Suspended Solids and Nutrient Removal in Large-Scale, Solid Matrix, Constructed Wetland Systems, in G. A. Moshiri (ed.), *Constructed Wetlands for Water Quality Improvement,* pp. 219–225, Lewis Publishers, Boca Raton, FL.

Bouey, J. (1996) Personal communication, San Ramon, CA.

Brix, H. (1993) Wastewater Treatment in Constructed Wetlands: System Design, Removal Processes, and Treatment Performance, in G.A. Moshiri, (ed.), *Constructed Wetlands for Water Quality Improvement,* pp. 9–22, Lewis Publishers, Boca Raton, FL.

Brodie, G. A., D. A. Hammer, and D. A. Tomljanovich (1989) Treatment of Acid Drainage with a Constructed Wetland at the Tennessee Valley Authority 950 Coal Mine, in D. A. Hammer (ed.), *Constructed Wetlands for Wastewater Treatment,* pp. 201–209, Lewis Publishers, Chelsea, MI.

Crites, R.W. (1996) Constructed Wetlands for Wastewater Treatment and Reuse, presented at the Engineering Foundation Conference, Environmental Engineering in the Food Processing Industry, XXVI, Santa Fe, NM.

Crites, R. W., G. D. Dombeck, and C. R. Williams (1996) Two Birds with One Wetland: Constructed Wetlands for Effluent Ammonia Removal and Reuse Benefits, *Proceedings of WEFTEC 96,* Water Environment Federation, Dallas, TX.

Crites, R. W., and D. Lesley (1997) Reclamation and Reuse Using Constructed Wetlands, presented at the Pacific Northwest Pollution Control Association Conference, Seattle, WA.

Crites, R. W., G. D. Dombeck, R.C. Watson, and C. R. Williams (1997) Removal of Metals and Ammonia in Constructed Wetlands, *Water Environment Research,* Vol. 69, No. 2.

DeBusk, T. A., and K. R. Reddy (1987) Wastewater Treatment Using Floating Aquatic Macrophytes: Contaminant Removal Processes and Management Strategies, in K. R. Reddy and W. H. Smith (eds.), *Aquatic Plants for Water Treatment and Resource Recovery,* Magnolia Publishing, Orlando, FL.

DeBusk, T. A., K. R. Reddy, and K. S. Clough (1989) Effectiveness of Mechanical Aeration in Floating Aquatic Macrophyte-Based Wastewater Treatment Systems, *Journal of Environmental Quality,* Vol. 18, pp. 349–354.

Eighmy, T. T., L. S. Jahnke, and P. L. Bishop (1987) Productivity of Photosynthetic Characteristics of *Elodea Nuttallii* Grown in Aquatic Treatment Systems, in K. R. Reddy and W. H. Smith (eds.), *Aquatic Plants for Water Treatment and Resource Recovery,* pp. 453-462, Magnolia Publishing, Orlando, FL.

Gearheart, R. A., F. Klopp, and G. Allen (1989) Constructed Free Surface Wetlands to Treat and Receive Wastewater: Pilot Project to Full Scale, in D. A. Hammer (ed.), *Constructed Wetlands for Wastewater Treatment,* pp. 121–137, Lewis Publishers, Chelsea, MI.

Gearheart, R. A., and B. A. Finney (1996) Criteria for Design of Free Surface Constructed Wetlands Based Upon a Coupled Ecological and Water Quality Model, presented at the Fifth International Conference on Wetland Systems for Water Pollution Control, Vienna, Austria.

Gersberg, R. M., R. A. Gearheart, and M. Ives (1989) Pathogen Removal in Constructed Wetlands, in D. A. Hammer (ed.), *Constructed Wetlands for Wastewater Treatment,* pp. 431–445, Lewis Publishers, Chelsea, MI.

Gersberg, R. M., S. R. Lyons, B. V. Elkins, and C. R. Goldman (1984) The Removal of Heavy Metals by Artificial Wetlands, *Proceedings Water Reuse Symposium III,* Vol. 2, American Water Works Association Research Foundation, pp. 639–648.

Gersberg, R. M., B. V. Elkins, R. Lyons, and C. R. Goldman (1985) Role of Aquatic Plants in Wastewater Treatment by Artificial Wetlands, *Water Research,* Vol. 20, pp. 363–367.

Hammer, D. A. (1992) *Creating Freshwater Wetlands,* Lewis Publishers, Boca Raton, FL.

Hayes, T. D., H. R. Isaacson, K. R. Reddy, D. P. Chynoweth, and R. Biljetina (1987) Water Hyacinth Systems for Water Treatment, in K. R. Reddy and W. H. Smith (eds.), *Aquatic Plants for Water Treatment and Resource Recovery,* pp. 121–139, Magnolia Publishing, Orlando, FL.

Herskowitz, J., S. Black, and W. Lewandowski (1987) Listowel Artificial Marsh Treatment Project, in K. R. Reddy and W. H. Smith (eds.) *Aquatic Plants for Water Treatment and Resource Recovery,* pp. 248–254, Magnolia Publishing, Orlando, FL.

Kadlec, R., and R. Knight (1996) *Treatment Wetlands,* Lewis Publishers, Boca Raton, FL.

Kays, W. B. (1986) *Construction of Linings for Reservoirs, Tanks, and Pollution Control Facilities,* 2nd ed., Wiley-Interscience, New York.

Kozak, P. M., and P. L. Bishop (1987) The Effect of Mixing and Aeration on the Productivity of *Myriophyllum Heterophyllum* Mich x (Water Milfoil) During Aquatic Wastewater

Treatment, in K. R. Reddy and W. H. Smith (eds.), *Aquatic Plants for Water Treatment and Resource Recovery,* pp. 445–452, Magnolia Publishing, Orlando, FL.

Leslie, M. (1983) Water Hyacinth Wastewater Treatment Systems: Opportunities and Constraints in Cooler Climates, EPA 600/2-83-075, U.S. Environmental Protection Agency, Washington, DC.

Living Technologies, Inc. (1997) Advanced Ecologically Engineered System, South Burlington, VT, Progress Report, Massachusetts Foundation for Excellence, Boston, MA.

Middlebrooks, E. J., C. H. Middlebrooks, J. H. Reynolds, G. Z. Watters, S. C. Reed, and D. B. George (1982) *Wastewater Stabilization Lagoon Design Performance and Upgrading,* MacMillan, New York.

Mitsch, W. J., and J. G. Gosselink (1993) *Wetlands,* Van Nostrand Reinhold, New York.

Nolte and Associates (1989) Harwich Septage Treatment Pilot Study—Evaluation of Technology for Solar Aquatic Septage Treatment System, prepared for Ecological Engineering Associates, Marion, MA.

Nolte and Associates (1997) Sacramento Regional Wastewater Treatment Plant Demonstration Wetlands Project—1996 Annual Report, prepared for the Sacramento Regional County Sanitation District, Elk Grove, CA.

Pries, J. (1996) Constructed Treatment Wetland Systems in Canada, presented at the Constructed Wetlands in Cold Climates Symposium, Niagara-on-the-Lake, Ontario, Canada.

Reed, S. C., R. W. Crites, and E. J. Middlebrooks (1995) *Natural Systems for Waste Management and Treatment,* 2nd ed., McGraw-Hill, New York.

Reed, S. C., J. Salisbury, L. Fillmore, and R. K. Bastian (1996) An Evaluation of the Living Machine Wastewater Treatment Concept, *Proceedings of WEFTEC 96,* Water Environment Federation, Dallas, TX.

Stephenson, M., G. Turner, P. Pope, J. Colt, A. Knight, and G. Tchobanoglous (1980) The Use and Potential of Aquatic Species for Wastewater Treatment, App. A, The Environmental Requirements of Aquatic Plants, Publication No. 65, California State Water Resources Control Board, Sacramento, CA.

Tchobanoglous, G., F. Maitski, K. Thompson., and T. H. Chadwick (1989) Evolution and Performance of City of San Diego Pilot-Scale Aquatic Wastewater Treatment System Using Water Hyacinths, *Research Journal of Water Pollution Control Federation,* Vol. 61, No. 11/12, pp. 1625–1635.

Tchobanoglous, G. (1993) Constructed Wetlands and Aquatic Plant Systems: Research, Design, Operational, and Monitoring Issues, in G. A. Moshiri (ed.), *Constructed Wetlands for Water Quality Improvement,* pp. 23–34, Lewis Publishers, Boca Raton, FL.

U.S. EPA (1988) *Design Manual of Constructed Wetlands and Aquatic Plant Systems for Municipal Wastewater Treatment.* EPA 625/1-88/022, U.S. Environmental Protection Agency, Center for Environmental Research Information. Cincinnati, OH.

U.S. EPA (1993) *Constructed Wetlands for Wastewater Treatment and Wildlife Habitat.* EPA 832-R-93-005, Municipal Technology Branch, Washington, DC.

Vela, G. R. (1974) Effect of Temperature on Cannery Waste Oxidation, *Journal of Water Pollution Control Federation,* Vol. 46, No. 1, pp. 198–202.

WPCF (1989) *Natural Systems for Wastewater Treatment,* Manual of Practice, FD-16, Water Pollution Control Federation, Alexandria, VA.

Watson, J. T., F. D. Diodato, and M. Launch (1987) Design and Performance of the Artificial Wetlands Wastewater Treatment Plant at Iselin, Pennsylvania, in K. R. Reddy and W. H Smith (eds.), *Aquatic Plants for Water Treatment and Resource Recovery,* pp. 263–271, Magnolia Publishing, Orlando, FL.

Watson, J. T., S. C. Reed, R. H. Kadlec, R. L. Knight, and A. E. Whitehouse (1989) Performance Expectations and Loading Rates for Constructed Wetlands, in D. A. Hammer (ed.), *Constructed Wetlands for Wastewater Treatment,* pp. 319–351, Lewis Publishers, Chelsea, MI.

Westermann, P. (1993) Wetland and Swamp Ecology, in T. E. Ford (ed.), *Aquatic Microbiology: An Ecological Approach,* Blackwell Scientific Publications, Boston.

WCPH (1996) Total Resource Recovery Project, Final Report, City of San Diego, CA, Western Consortium for Public Health with EOA, Inc., Oakland, CA.

Wilhelm, M., S. R. Lawry, and D. D. Hardy (1989) Creation and Management of Wetlands Using Municipal Wastewater in Northern Arizona: A Status Report, in D. A. Hammer (ed.), *Constructed Wetlands for Wastewater Treatment,* pp. 179–185, Lewis Publishers, Chelsea, MI.

Williams, C. R., R. D. Jones, and S. A. Wright (1996) Mosquito Control in a Constructed Wetland, *Proceedings of WEFTEC 96,* Water Environment Federation, Dallas, TX.

CHAPTER 10

Land Treatment Systems

Land treatment is the controlled application of wastewater onto the land surface to achieve a designed degree of treatment through natural physical, chemical, and biological processes within the plant-soil-water matrix. Different levels of wastewater treatment can be achieved with either municipal or industrial wastewater, depending on the site characteristics, loading rates, wastewater characteristics, and design objectives.

10-1 LAND TREATMENT PROCESSES

The three land treatment processes are (1) slow rate (SR), (2) rapid infiltration (RI), and (3) overland flow (OF). Each of these processes is introduced in this section. Design details for each process are presented in separate sections following a discussion of removal mechanisms.

Slow Rate

The slow-rate process is the oldest and most widely used land treatment technology (Jewell and Seabrook, 1979). The process evolved from "sewage farming" in Europe in the 1840s to wastewater irrigation in the United States in the 1880s. The effectiveness of land treatment was established in the 1860s in England (Jewell and Seabrook, 1979). In a survey of 143 wastewater facilities in 1899, slow-rate land treatment systems were the most frequently used form of treatment (Rafter, 1899). Slow-rate land treatment was rediscovered at Penn State in the 1960s (Sopper and Kardos, 1973) and received considerable research interest from 1970 through 1980. Most of the references that are still current are from the 1970s. There are currently about 1200 SR systems in the United States (Reed and Crites, 1984). A selected list

TABLE 10-1
General information on selected slow-rate systems*

Location	Crop	Year started	Flow, Mgal/d	Hydraulic loading rate, ft/yr
Coleman, Texas	Coastal Bermuda grass	1930	0.4	5.6
Dickinson, North Dakota	Brome grass	1957	1.5	2.0
Fremont, Michigan	Alfalfa, oats, rye	1975	0.3	8.5
Hart, Michigan	Forest, grass	1974	0.7	7.9
Kennett Square, Pennsylvania	Forest	1973	0.05	6.9
Kerman, California	Alfalfa, sugar beets, barley, oats, cotton, almonds	1976	0.5	2.9
Lake of the Pines, California	Forest, pasture	1978	0.6	8.7
Ravenna, Michigan	Grass	1969	0.09	3.9
Santa Anna, Texas	Alfalfa, pasture	1966	0.15	3.1
Snoqualmie Pass, Washington	Forest	1983	1.1	8.8
Sweetwater, Texas	Pasture, oats, wheat	1958	1.0	3.9
Wayland, Michigan	Alfalfa, timothy, clover	1971	0.25	3.6
West Dover, Vermont	Forest	1975	1.6	5.6
Winters, Texas	Coastal Bermuda grass, Sudan grass	1924	0.5	13.1
Wolfeboro, New Hampshire	Forest	1976	0.3	4.3

*Adapted from Weston Inc. (1982), Crites and Reed (1986).

of small SR systems is presented in Table 10-1. A typical SR system is shown in Fig. 10-1.

Rapid Infiltration

Rapid infiltration is a land treatment process that resembles intermittent sand filtration (Chap. 11). After pretreatment, the wastewater is applied to level spreading basins containing permeable soil. Wastewater is treated by physical, chemical, and biological mechanisms as it percolates through the soil. The potential hydraulic pathways in RI systems are shown in Fig. 10-2. A majority of the 320 RI systems in the United States use the natural drainage of percolate into nearby surface waters via the pathway in Fig. 10-2*b*. In a few cases underdrains are used to transport the percolate to surface water (Fig. 10-2*c*). Recovery wells, although used at the large RI

FIGURE 10-1
Typical slow-rate system used for woodland irrigation (poplar trees in the winter).

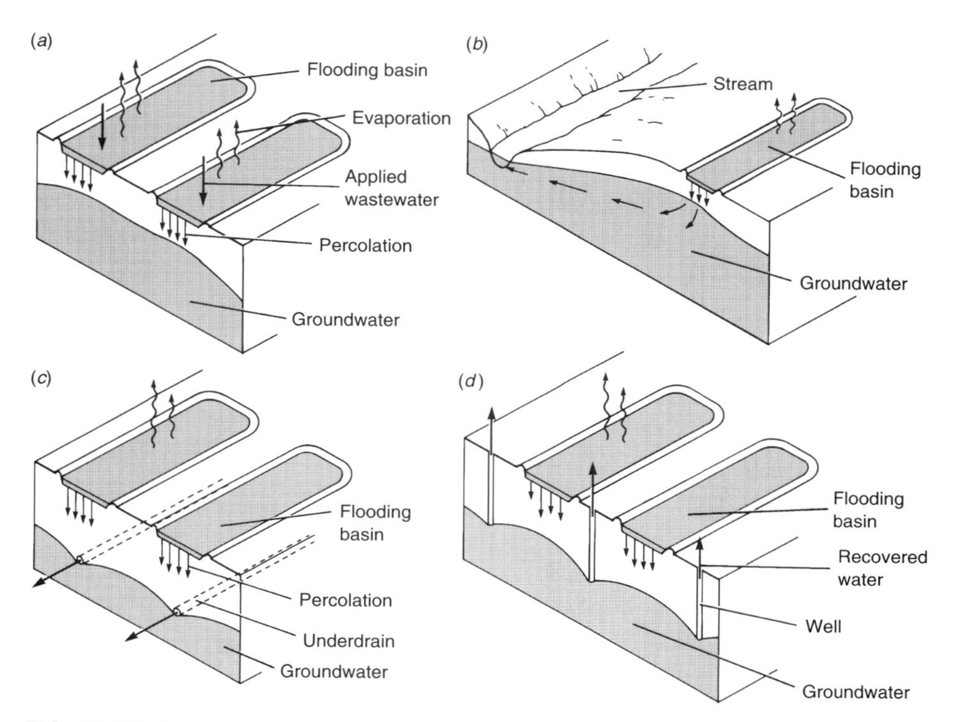

FIGURE 10-2
Definition sketch for potential hydraulic pathways for rapid infiltration: (*a*) hydraulic pathway to groundwater, (*b*) pathway to shallow groundwater and surface water, (*c*) recovery by underdrains, and (*d*) recovery by wells.

TABLE 10-2
General information on selected rapid infiltration systems

Location	Treatment	Year started	Flow, Mgal/d	Hydraulic loading rate, ft/yr
Calumet, Michigan	None	1887	1.6	116
Darlington, South Carolina	Aerated lagoon	1982	1.6	91
Ft. Devens, Massachusetts	Primary	1941	1.0	100
Hollister, California	Secondary	1946*	1.0	50
Lake George, New York	Secondary	1939	1.1	140
West Yellowstone, Montana	Secondary	1966	1.0	550

*Primary treatment until 1980.

system at Phoenix, Arizona, are seldom used for small RI systems. A list of selected RI systems is presented in Table 10-2.

The key to wastewater treatment by RI is the intermittent application of wastewater to the basin surface. Continuous flooding not only greatly reduces percolation rates with time, but does not allow aerobic treatment processes to occur in the soil. Multiple basins are used to allow rotation of the basin being flooded and provide the necessary resting/drying time. A schematic of a typical RI system is shown in Fig. 10-3.

Overland Flow

The overland-flow process was developed to take advantage of slowly permeable soils such as clays. The process was pioneered in the United States at Napoleon,

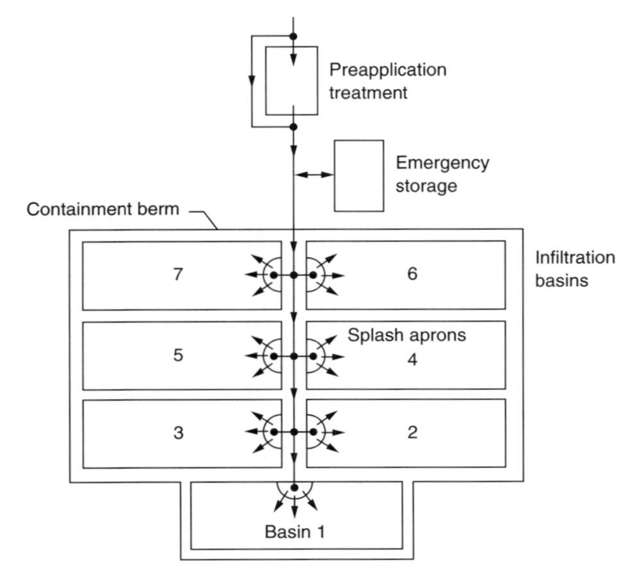

FIGURE 10-3
Typical schematic for rapid infiltration system.

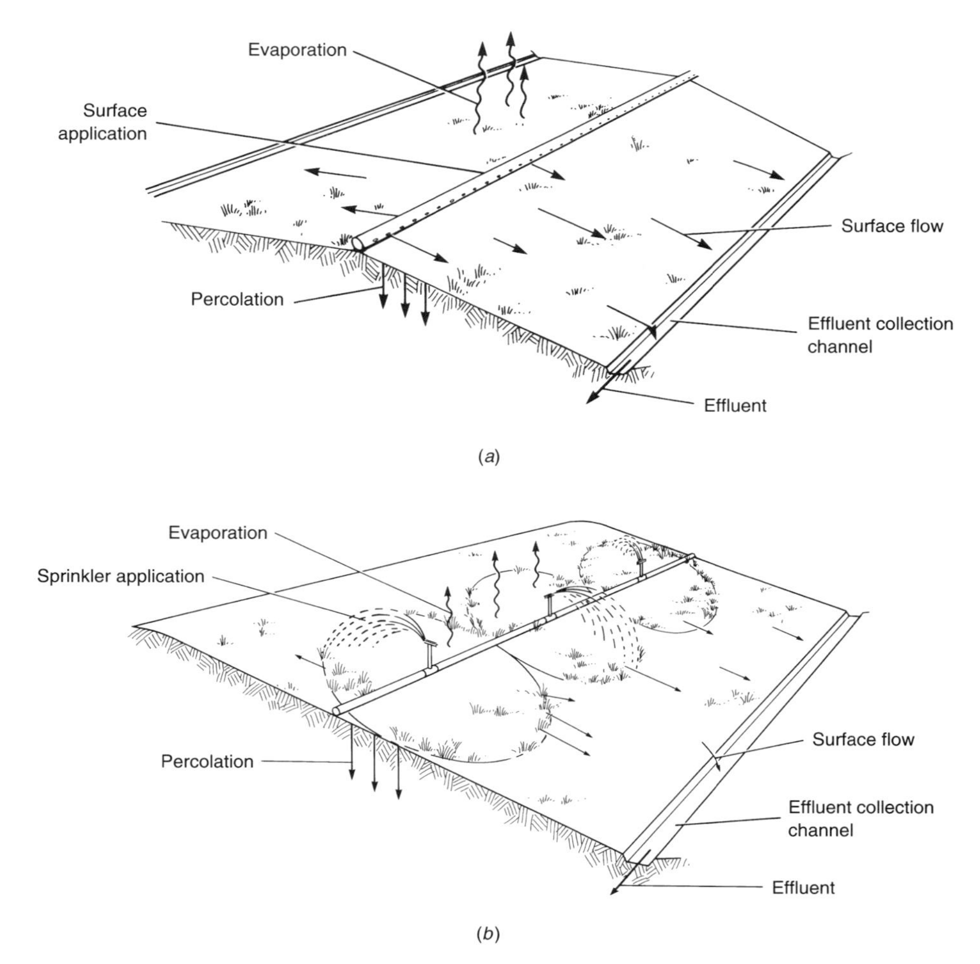

FIGURE 10-4
Definition sketch for overland flow: (*a*) surface application at the top of the slope and (*b*) sprinkler application partway down the slope.

Ohio, in 1954 and developed further by the Campbell Soup Company at Paris, Texas (Pound and Crites, 1973). Treatment occurs in OF systems as wastewater flows down vegetated, gently graded smooth slopes that range from 2 to 10 percent in grade. Typical slopes are 2 to 4 percent in grade and 120 to 150 ft (36 to 45 m) long. Treated effluent is collected at the bottom of the slope. The process is shown schematically in Fig. 10-4.

Research at Ada, Oklahoma, was conducted in the early 1970s by EPA on the OF process and its applicability to treatment of screened raw municipal wastewater and primary effluent (Thomas et al., 1974). As a result of this and other research (Martel et al., 1982; Smith and Schroeder, 1985), over 50 OF systems have been designed to treat municipal wastewater (Reed et al., 1995). A list of selected municipal overland-flow systems is presented in Table 10-3.

TABLE 10-3
General information on selected municipal overland-flow systems

Location	Effluent source	Flow, Mgal/d	Field, area, ac	Hydraulic loading rate, ft/yr
Carbondale, Illinois	Oxidation pond	0.010	0.15	75
Davis, California	Primary sedimentation	4	200	22
Easley, South Carolina	Oxidation pond	0.020	1.3	21
Falkner, Mississippi	Oxidation pond	0.035	2.6	15
Mesquite, Nevada	Aerated pond	0.8	10	90

Comparison of Processes

Typical design features of the three processes are presented in Table 10-4. The soil matrix is used for treatment in the SR and RI processes, after infiltration of the wastewater into the soil. The soil surface, vegetation, and soil biota are used for treatment in the OF process.

Preferred site characteristics for the three land treatment processes are presented in Table 10-5. Suitable characteristics for SR sites are broad in their range as compared to OF and RI; as a result, sites for SR are generally more plentiful. As shown in Table 10-5, the soil should have a rapid permeability, such as sand, and be relatively deep (10 ft or 3 m is preferred).

TABLE 10-4
Comparison of land treatment process design features

	Land treatment process		
Feature	**Slow rate**	**Rapid infiltration**	**Overland flow**
Minimum pretreatment	Primary sedimentation*	Primary sedimentation†	Grit removal and comminution†
Annual loading rate, ft/yr	2–18	18–360	10–70‡
Typical annual loading rate, ft/yr	5	100	30
Field area for 1 Mgal/d at typical rate, ac	224	11	37
Range of field area for 1 Mgal/d, ac	60–560	3–60	16–112
Application method	Sprinkler or surface	Usually surface	Sprinkler or surface
Disposition of applied wastewater	Evapotranspiration and percolation	Mostly percolation	Surface runoff, evapotranspiration, and some percolation
Need for vegetation	Required	Sometimes used to stabilize soil	Required

*With restricted public access; crops not used for direct human consumption.
†With restricted public access.
‡Higher rates possible with higher levels of pretreatment.

TABLE 10-5
Preferred site characteristics for treatment

	Land treatment process		
Characteristic	**Slow rate**	**Rapid infiltration**	**Overland flow**
Soil depth	>2 ft	>5 ft	>0.5 ft
Soil permeability	Slow to moderately rapid	Rapid	Slow
Permeability range	0.06 to 20 in/h	>2 in/h	<0.2 in/h
Depth to groundwater	2 to 3 ft (minimum)	10 ft (lesser depths are acceptable during operation or if underdrains are provided)	Not critical
Grade	Less than 20% on cultivated land; less than 40% on noncultivated land	Less than 10% to avoid excess earthwork	Finish slopes of 2 to 8%
Climatic restrictions	Storage often needed for cold and wet weather	None (possibly modify operation in cold weather)	Storage often needed for cold weather

10-2 REMOVAL MECHANISMS

Removal mechanisms are described in the following paragraphs for all three types of land treatment. In most cases the removal of constituents is accomplished by a combination of biological, physical, and chemical means.

BOD Removal

The soil is a biological filter containing large quantities of bacteria (Reed, 1972). Removal of BOD is accomplished by absorption of dissolved organics and bacterial oxidation. The upper soil column is teeming with microorganisms. Estimates of 10^7 bacteria, 10^6 actinomycetes, and 10^5 fungi per gram of soil are typical values (Miller, 1973). These microorganisms are responsible for BOD removal from the applied wastewater. For overland flow, the bacterial growth in both the upper soil and in the plant litter are responsible for the removal.

Biological growth of organisms is sensitive to temperature. As shown in Fig. 10-5, the organisms achieve optimum growth at relatively high temperatures, but continue to reproduce even at very low temperatures. The heavy line indicates microliters of oxygen consumed per hour per gram of wet soil, given as a ratio of consumption at the given temperature to consumption at 30°C. The finer line represents bacterial growth, based on bacteria obtained from the Paris, Texas, overland-flow system.

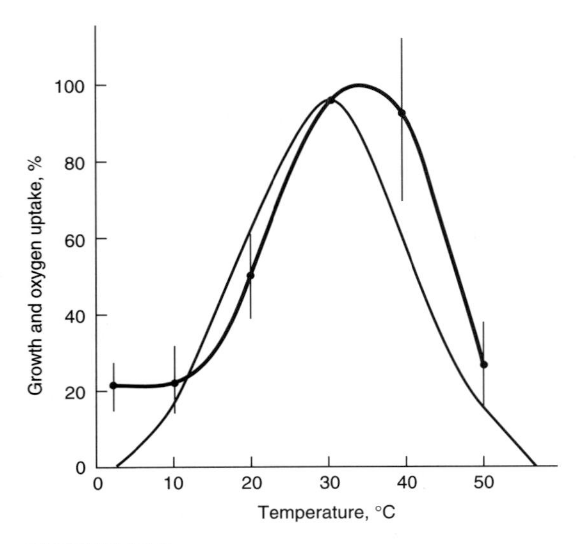

FIGURE 10-5
Effect of temperature on oxygen consumption and growth of bacteria from the water treatment field. The heavy line indicates μL O_2 consumed/g·h wet soil and is given as the ratio of consumption at the given temperature to consumption at 30°C; the bars show the standard deviation of measurements at that temperature while the point at 30°C represents the average of 65 determinations. The finer line represents growth; standard deviations were of the same magnitude as those shown for oxygen uptake (from Vela, 1974).

Total Suspended Solids Removal

Suspended solids are removed by filtration and entrapment in slow-rate and rapid-infiltration systems, and by sedimentation, filtration, and entrapment in overland-flow systems. Intermittent applications allow the removed solids to desiccate, degrade if they are organic solids, and become part of the soil matrix. Inorganic solids become part of the soil over time.

Nitrogen Removal

Nitrogen removal in SR systems occurs as a result of crop uptake, nitrification/denitrification, and, to a lesser extent, ammonia volatilization and incorporation (soil storage). Nitrification and denitrification are two processes that occur even in aerobic soils because the aerobic conditions promote nitrification and anoxic microsites in the soil allow denitrification (Reed and Crites, 1984).

Nitrification and denitrification are the mechanisms for removal of ammonia and nitrate from the wastewater in RI systems. Ammonia adsorption also plays an important role in retaining ammonia in the soil long enough for biological conversion. Nitrification and denitrification are affected by low temperatures and proceed slowly at temperatures of 36 to 41°F (3.6 to 5°C). In addition, denitrification requires an adequate carbon source and the absence of available oxygen.

The overland-flow process is well suited to achieving nitrogen removal by nitrification and denitrification. A double layer is formed on the soil surface and within the vegetation with the upper zone being aerobic and the lower layer being oxygen-deficient. Nitrification in OF systems occurs in the upper zone and, under proper conditions, denitrification occurs in the lower zone. The proper conditions include warm temperatures, a BOD:N ratio of 3:1 or more, and a loading period of 6 to 12 h/d with a resting period of 12 to 18 h/d.

Phosphorus Removal

Chemical immobilization of phosphorus and plant uptake are the principal mechanisms of phosphorus removal. For slow-rate systems, chemical precipitation and absorption account for the majority of the removal while plant uptake may account for 20 to 25 percent of the total phosphorus removal. For RI systems, chemical precipitation and adsorption are the mechanisms for removal. The same mechanisms function in overland flow; however, limited soil contact restricts the level of phosphorus removal.

Metals Removal

Removal of trace metals from wastewater is accomplished by soil adsorption, precipitation, ion exchange, and complexation in SR systems. Fine-textured and organic soils have a large capacity for metals removal and can be expected to achieve nearly complete removal. For RI systems, even in sandy soils, metals removal of 90 percent or more is typical. Metals removal in OF systems depends on the form of the metal (dissolved or particulate) and is generally low, comparable to phosphorus removal.

Trace Organics Removal

The mechanisms responsible for removal of trace organics include photodecomposition, volatilization, sorption, and degradation. The mechanisms are operative in all three land treatment processes. The importance of each mechanism will vary with local environmental conditions and wastewater application rates.

Pathogen Removal

Removal of microorganisms, including pathogenic bacteria, viruses, and helminths, is accomplished by soil filtration, adsorption, desiccation, radiation, predation, and exposure to other adverse environmental conditions. Because of their large size, helminths and protozoa are removed on the soil surface by filtration. Bacteria are removed from wastewater by filtration and adsorption, with typical removals of 99.9 percent or more (U.S. EPA, 1981). Virus removal is principally by adsorption.

10-3 SLOW-RATE SYSTEMS

Slow-rate systems can encompass a wide variety of different land treatment facilities, ranging from hillside spray irrigation to agricultural irrigation, and from forest irrigation to golf course irrigation. The design objectives can include wastewater treatment, water reuse, nutrient recycling, open space preservation, and crop production.

Design Objectives

Slow-rate systems can be classified as Type 1 (slow infiltration) or Type 2 (crop irrigation), depending on the design objective. When the principal objective is wastewater treatment, the system is classified as Type 1. For Type 1 systems the land area is based on the limiting design factor (LDF), which can either be the soil permeability or the loading rate of a wastewater constituent such as nitrogen. Type 1 systems are designed to use the most wastewater on the least amount of land. The term *slow infiltration* indicates that Type 1 systems are similar in concept to rapid infiltration, but have substantially lower hydraulic loading rates.

Type 2 systems are designed to apply sufficient water to meet the crop irrigation requirement. The area needed for a Type 2 system depends on the crop water use, not on the soil permeability or the wastewater treatment needs. Water reuse and crop production are the principal objectives. The area needed for Type 2 systems is generally larger than for a Type 1 system for the same wastewater flow. For example, for 1 Mgal/d (3785 m^3/d) of wastewater flow, a Type 1 system would typically require 60 to 150 ac (24 to 60 ha) compared to 200 to 500 ac (80 to 200 ha) for a Type 2 system.

Management alternatives. Unlike rapid infiltration and overland flow, slow-rate systems can be managed in several different ways. The other two land treatment systems require that the land be purchased and the system managed by the wastewater agency. For slow-rate systems the three major options are (1) purchase and management of the site by the wastewater agency, (2) purchase of the land and leaseback to a farmer, and (3) contracts between the wastewater agency and farmers for use of private land for the slow-rate process. The latter two options allow farmers to manage the slow-rate process and harvest the crop. A representative list of small SR systems that use each of the different management alternatives is presented in Table 10-6.

Site Selection

The selection process for an SR site involves identification, characterization, evaluation, and ranking of alternative sites. The process can be relatively simple in rural areas when areas of 40 ac or less are involved. When areas of 100 ac (40 ha) or more are needed, the process should be somewhat more rigorous with alternatives being evaluated on a numerical rating or economic comparison.

TABLE 10-6
Management alternatives used in selected SR systems*

Purchase and management by agency	Flow, Mgal/d	Agency purchase and lease to farmer	Flow, Mgal/d	Farmer contract	Flow, Mgal/d
Dinuba, California	1.5	Coleman, Texas	0.4	Camarillo, California	3.8
Fremont, Michigan	0.3	Kerman, California	0.5	Dickinson, North Dakota	1.5
Kennett Square, Pennsylvania	0.05	Lakeport, California	0.5	Mitchell, South Dakota	2.4
Lake of the Pines, California	0.6	Modesto, California	20.0	Quincy, California	0.75
Oakhurst, California	0.25	Perris, California	0.8	Petaluma, California	4.2
West Dover, Vermont	1.6	Winters, Texas	0.5	Sonoma Valley, California	2.7
Wolfeboro, New Hampshire	0.3	Santa Rosa, California	15.0	Sonora, California	1.2

*Adapted from Weston Inc. (1982), Crites and Reed (1986), and California SWRCB (1990).

Site identification. The process of site identification should start with a U.S. Geological Survey (USGS) topographic map on which potential sites can be located. Starting with parcels adjacent to the wastewater treatment facilities, each potential site should be identified and marked on the USGS map. County assessors' maps can then be obtained to determine parcel size, ownership, and zoning. A visit to each site will allow observation of current land use, crops or vegetation, and general topography. Once potential sites have been identified, site characteristics must be analyzed.

Site characteristics. The important characteristics of SR sites are topography (slope and relief), stormwater runoff, soils, geology, groundwater, and land use. Regional factors such as climate and water rights should also be characterized.

Topography. Topographic features of importance include the slope and relief of the site and any watercourses passing through or around the site. As shown in Table 10-5, the maximum slope (grade) should not exceed about 20 percent for an agricultural site (15 percent if the crop requires cultivation or tilling). Pasture grasses can be grown on slopes of 15 to 20 percent, depending on the erosion potential of the soil and runoff control constraints. Sprinkler application on wooded (forested) sites with slopes of 30 to 40 percent can be accomplished.

The difference in elevation between the high point and the low point on the site is the relief. Relief should be documented so that the effect on pumping and distributing wastewater across the site can be determined.

Stormwater runoff and flooding. The potential for stormwater draining onto the site should be evaluated. Any surface drainage features should be investigated to determine the flow capacity and the potential to result in flooding or backup of drainage water onto the site.

Soils. A detailed soil survey should be obtained from the local office of the Natural Resources Conservation Service (NRCS) [formerly Soil Conservation Service (SCS)] of the U.S. Department of Agriculture. The detailed survey will contain information on the soil types on the site as well as the physical, chemical, and hydraulic properties of each soil type. Physical characteristics of importance are soil depth, texture, structure, and particle size information. Important chemical characteristics include pH, electrical conductivity (EC), exchangeable sodium percentage (ESP), cation exchange capacity (CEC), and organic matter. Soil chemical levels and their interpretation are presented in Table 10-7. Hydraulic properties such as infiltration rate and permeability are two very important soil characteristics.

Soil surveys contain maps that show soil series boundaries and soil textures to a depth of 5 ft (1.5 m). In addition, descriptions of the physical characteristics, chemical properties, drainage, slope, erosion potential, suitability for locally grown crops, and interpretive and management information are provided.

The physical and hydraulic properties depend on the relative percentages of sand, silt, and clay. Sand particles range from 0.05 mm up to 2.0 mm in size; silt

TABLE 10-7
Soil chemical levels and their interpretation in evaluating sites for SR treatment*

Soil property	Description
pH of saturated soil paste	
<4.2	Too acid for most crops to do well
4.2–5.5	Suitable for acid-tolerant crops
5.5–8.4	Suitable for most crops
>8.4	Too alkaline for most crops, indicates a possible sodium problem
CEC, meq/100 g dry soil	
1–10	Sandy soils (limited adsorption)
12–20	Silt loam (moderate adsorption)
>20	Clay and organic soils (high adsorption)
Exchangeable cations, % of CEC (desirable range)	
Sodium	<5
Calcium	60–70
Potassium	5–10
ESP, % of CEC	
<5	Satisfactory
>10	Reduced permeability in fine-textured soils
>20	Reduced permeability in coarse-textured soils
EC_e, mmhos/cm at 25% of saturation extract	
<2	No salinity problems
2–4	Restricts growth of very salt-sensitive crops
4–8	Restricts growth of many crops
8–16	Restricts growth of all but salt-tolerant crops
>16	Only a few very salt-tolerant crops make satisfactory yields

*Adapted from U.S. EPA (1981).

particles range from 0.002 mm up to 0.05 mm; and clay particles are smaller than 0.002 mm. Fine-textured soils have large percentages of silt and clay. Coarse-textured soils have typically a large percentage of sand.

Hydraulic properties of soils depend on the texture and structure. Fine-textured soils (clayey) do not drain quickly and retain soil moisture for long periods of time. The slow drainage makes crop management difficult, but the retention of moisture supports plant growth and the biological treatment mechanisms in the soil. Coarse-textured (sandy) soils drain quickly and have low moisture retention. The rapid drainage makes crop management easier and allows higher wastewater applications, but the low moisture retention may require frequent applications to maintain a healthy crop. Medium-textured (loamy) soils are generally best suited for SR systems because loamy soils have a balance between drainage and wastewater renovation.

Soil structure is the aggregation of individual soil particles. A soil is well structured if the aggregates resist disintegration when the soil is wetted or tilled. The large pores in a well-structured soil conduct water and air, thereby making well-structured soils desirable for infiltration and percolation.

Soil depth is important for wastewater treatment as well as for crop root development. For SR systems the soil can be as shallow as 1 ft (0.3 m) for grass crops, but 5 ft (1.5 m) is preferred for complete wastewater treatment. Retention of wastewater constituents such as phosphorus, heavy metals, and viruses is a function of residence time of wastewater in the soil and the degree of contact between soil colloids and wastewater constituents.

Hydraulic properties are most important for Type 1 SR systems; however, infiltration rates must be considered for design of all SR systems. The infiltration rate is the rate at which water passes through the soil surface. The vertical permeability or percolation rate is the rate at which water moves through the soil profile. Infiltration rates are not constant for a given soil but will vary with the water content of the soil and with the temperature, presence of solids, and level of soil compaction. As the moisture content approaches saturation, the infiltration rate will approach a minimum value.

The NRCS has defined permeability classes for soil as presented in Table 10-8. The saturated permeability can be estimated for a given soil from the NRCS class or

TABLE 10-8
NRCS permeability classes for saturated soil*

Soil permeability, in/h	Permeability class
<0.06	Very slow
0.06–0.2	Slow
0.2–0.6	Moderately slow
0.6–2.0	Moderate
2.0–6.0	Moderately rapid
6.0–20.0	Rapid
>20	Very rapid

*From U.S. EPA (1981).

it can be measured in the field. If the NRCS class is used to assign a soil permeability, the lowest permeability should generally be used.

Geology. Geologic discontinuities that can provide short-circuit pathways for the percolate to reach groundwater are important to locate during planning. If an adequate soil depth (5 ft or 1.5 m or more) is present, discontinuities are less critical for SR sites than for RI sites. The geology of each site should be studied to determine the presence of impermeable layers that would restrict percolate flow and create shallow perched groundwater.

Groundwater. The depth to groundwater and its quality are important site considerations. Permanent groundwater should be distinguished from localized perched groundwater. If perched groundwater is present, the reasons for its presence and its flow direction should be determined. Shallow groundwater can interfere with both percolation of treated effluent and with growth of certain crops. A depth to groundwater of 3 to 4 ft (0.9 to 1.2 m) or more is preferred. Lesser depths will restrict drainage. Seasonal high groundwater may be acceptable if storage or stream discharge is practiced when the high-groundwater conditions occur.

Land use. Existing and proposed land uses and the zoning of the potential site and adjacent lands are important to SR site selection. Conflicting land use is a major reason for abandonment of several SR systems. An appropriate site should be remote from impending residential or commercial development.

Slow-rate land treatment systems can conform with several land use objectives including:

1. Protection of open space that is used for land treatment
2. Reclamation of land by using renovated water to establish vegetation on disturbed landscape
3. Formation of buffer areas around major public facilities, such as airports
4. Augmentation of parklands by irrigating such lands with renovated water
5. Management of flood plains by using such areas for slow-rate land treatment

Climate. Climatic factors such as evapotranspiration (ET), precipitation, temperature, and wind are important for slow-rate systems. The water balance, growing season, number of days of potential operation, type of crop, storage needs, and stormwater runoff are affected by climatic factors. The needed climatic data can be obtained from three publications of the National Oceanic and Atmospheric Administration (NOAA):

1. *Monthly Summary of Climatic Data*
2. *Climatic Summary of the United States*
3. *Local Climatological Data*

The local office of NOAA or the National Climatic Center in Asheville, North Carolina 28801, can be contacted for these publications.

Site Investigations

Field investigations are important in the site selection process. For slow-rate sites the investigations include site inspection, soil profile evaluations, and infiltration rate measurements.

Site inspection. Inspection of the site is crucial to evaluate the topography, crops or vegetation, soil, drainage features, and the land use of the surrounding land. The species of natural vegetation growing in an unirrigated area can be used as an indication of the soil characteristics and the depth to existing groundwater.

If the site has been farmed or irrigated, it is important to know past farming practices, including crops grown, irrigation rates, drying times required between applications, and use of fertilizers or soil amendments. The farmer or irrigator should be interviewed, if possible, or the local farm advisor should be contacted. Any springs, wells, or perennial wet spots should be identified.

Evaluation of soil profile. It is generally advisable to evaluate the actual soil profile and compare the results to the NRCS detailed soil survey. To evaluate the soil profile it is recommended that backhoe pits be dug in several locations on the site. Typical backhoe pits on a potential site are shown in Fig. 10-6.

Backhoe pits are preferred over soil borings because they (1) allow a wide, direct view of the soil profile conditions such as hardpan or clay pan layers, fractured rock, or rock that is close to the surface, (2) can reveal mottling or bluish/grayish color streaks that indicate that high groundwater conditions have occurred, and (3) allow direct and accurate samples to be taken at critical depths in the soil profile. For SR sites, a backhoe depth of 4 to 6 ft (1.2 to 1.8 m) is usually adequate.

FIGURE 10-6
Examples of backhoe pits used in detailed soil investigations to assess the soil profile and its suitability for land treatment.

Infiltration rate measurement. Field measurement of infiltration rates is more important for RI systems than for SR systems; however, there are instances in which the infiltration rate may be a factor in SR site selection. Techniques for measuring infiltration rates include the flooding basin, air entry permeameter, cylinder infiltrometer, and sprinkler infiltrometer. For SR sites the appropriate techniques are usually the cylinder infiltrometer or the sprinkler infiltrometer. The sprinkler infiltrometer test is described briefly in the EPA Design Manual (U.S. EPA, 1981), whereas the other techniques are described in Sec. 10.4.

Site selection criteria. If the number of potential sites is small, an economic analysis of the acceptable sites is sufficient to select the best site. If the number of acceptable sites is large or if the best site is not obvious, a ranking procedure can be used. The important factors in the ranking of sites—soil depth, depth to groundwater, soil permeability, site slope, and land use—should be evaluated from composite site maps. The ranking factors and suggested numerical values are presented in Table 10-9. The numerical totals for each site are then determined and the sites with the highest ranking numbers can then be compared economically.

TABLE 10-9
Slow-rate site selection criteria and rankings[a, b]

Characteristic	Agricultural sites	Forest sites
Soil depth,[c] ft		
1–2	E[d]	E[d]
2–5	3	3
5–10	8	8
>10	9	9
Minimum depth to groundwater, ft		
<4	0	0
4–10	4	4
>10	6	6
Soil permeability,[e] in/h		
<0.06	1	1
0.06–0.2	3	3
0.2–0.6	5	5
0.6–2.0	8	8
2.0–20.0	8	8
>20.0	1	1
Slope, percent		
0–5	8	8
5–10	6	8
10–15	4	6
15–20	1	5
20–30	0	4
30–35	E[d]	2
>35	E[d]	0
Existing or planned land use		
Industrial/commercial	0	0
High-density residential	0	0
Low-density residential	1	1
Forested	1	4
Agricultural/open space	4	3
Overall suitability rating[f]		
Low	<15	<15
Moderate	15–25	15–25
High	25–35	25–35

[a]Adapted from U.S. EPA (1981).
[b]The higher the maximum number in each characteristic, the more important the characterisic; the higher the ranking, the greater the suitability.
[c]Depth of the soil profile to bedrock.
[d]Excluded.
[e]Permeability of the most restrictive layer in the soil profile.
[f]Sum of values for each characteristic.

Treatment Performance

Slow-rate systems are very effective in removing BOD, TSS, nitrogen, phosphorus, metals, trace organics, and pathogens. In climates where evapotranspiration (ET) exceeds precipitation, the salinity of the applied wastewater will increase as water passes through the soil, depending on the hydraulic loading rate and the balance of ET and precipitation. The expected performance of SR systems is described in the following paragraphs.

BOD loading and removal. On the basis of successful operations where high-strength food processing wastewater is applied, BOD loading rates of up to 450 lb/ac·d (500 kg/ha·d) can be used. Most municipal SR systems are loaded at less than 10 lb/ac·d (11 kg/ha·d) of BOD, which is an order of magnitude lower than the capacity of the soil. As presented in Table 10-10, BOD removal in SR systems regularly exceeds 98 percent.

Total suspended solids loading and removal. Filtration through the soil is responsible for TSS removals of 99 percent or more. Percolate SS values of 1 mg/L or less are usually below 25 lb/ac·d. Industrial SR systems often operate with SS loadings of 100 lb/ac·d (110 kg/ha·d) using sprinkler application (Reed and Crites, 1984).

Nitrogen removal. Design for nitrogen is accomplished using a nitrogen balance that matches the expected removal plus a percolate nitrate nitrogen of less than 10 mg/L (drinking water standard) at the project boundary against the nitrogen loading. Nitrogen removal in SR systems is presented in Table 10-11.

The SR systems in Table 10-11 use forage crops to remove much of the applied nitrogen. Nitrogen removal in forested SR systems is presented in Table 10-12. Nitrogen uptake by crops is not constant; it depends on the crop yield and the nitrogen content of the harvested crop. Nitrogen uptake rates of forage, field, and forest crops are presented in the system design section. To achieve the nitrogen uptake

TABLE 10-10
BOD removal for SR systems*

Location	Flow Mgal/d	BOD loading, lb/ac·d	Applied BOD, mg/L	Percolate BOD, mg/L	Removal, %
Dickinson, North Dakota	10	3	42	<1	>98
Hanover, New Hampshire	Pilot				
Primary effluent		12	101	1.4	98.6
Secondary effluent		4	36	1.2	96.7
Roswell, New Mexico	4.0	3	43	<1	>98
San Angelo, Texas	5.5	11	119	1.0	99.1
Yarmouth, Massachusetts	Pilot	10	85	<2	>98

*From U.S. EPA (1981), WPCF (1983), and Giggey et al. (1989).

TABLE 10-11
Nitrogen removal for agricultural crop SR systems*

Location	Crop	Total nitrogen applied		Percolate total nitrogen, mg/L	Removal, %
		lb/ac·yr	mg/L		
Dickinson, North Dakota	Brome grass	134	11.8	3.9	67
Hanover, New Hampshire					
Primary effluent	Reed canary grass	543	28.0	9.5	66
Secondary effluent		525	26.9	7.3	73
Pleasanton, California	Pasture	560	27.6	2.5	91
Roswell, New Mexico	Corn	414	66.2	10.7	84
San Angelo, Texas	Sorghum, oats, grass	739	35.4	6.1	83
Yarmouth, Massachusetts	Reed canary grass	400	30.8	1.8	94

*From U.S. EPA (1981), WPCF (1983), and Giggey et al. (1989).

TABLE 10-12
Nitrogen removal for forested SR systems*

Forest type (location)	Nitrogen loading, lb/ac·yr	Nitrate nitrogen in percolate, mg/L
Douglas fir seedlings (Northwest)	312	5.3
Poplar (Northwest)	356	0.1
Poplar (Lake States)	92	2.8
Red Pine (Lake States)	117	5.4
Southern mixed hardwood	610	8.0
Eastern mixed hardwood	134	4.6
Mixed hardwood (Clayton County, Georgia)	272	<0.1
Mixed hardwood (Penn State, Pennsylvania)	276	3.1
Hardwood, balsam, hemlock, and spruce (West Dover, Vermont)	240	6.1
Mixed hardwood and pine (Wolfeboro, New Hampshire)	82	1.6

*From WPCF (1983), Nutter (1986), and Sopper (1986).

rates expected for the crop, an excess of nitrogen must be applied so that the crop can compete effectively with the soil microorganisms for available nitrogen.

Biological nitrogen removal in SR systems occurs by nitrification/denitrification. The loss to denitrification depends on the BOD:N ratio and the soil temperature, pH, and moisture. Intermittent application, which is characteristic of SR systems, also serves to enhance nitrification followed by denitrification. The typical loss to denitrification is 25 percent of the applied nitrogen. If the BOD:N ratio is high (20 or 40:1), denitrification can be responsible for 80 percent or more nitrogen loss. Food processing wastewater with BOD:N ratios of 5 to 10:1 can expect 50 percent of the nitrogen loading to be lost to denitrification.

Ammonia volatilization losses of 10 percent can be expected if the soil pH is above 7.8 and the CEC is low (low absorption of ammonium by the soil). A wastewater that has a pH above 7.8 and is applied by sprinklers may lose ammonia directly to the atmosphere.

Nitrogen can also be stored in the soil. In arid regions and for sites initially low in organic matter, the nitrogen applied can be stored at rates up to 200 lb/ac·yr (220 kg/ha·yr). For soils that are rich in organic matter (75 percent) the nitrogen storage will be insignificant and should be ignored in design. For soils with less than 2 percent organic matter, nitrogen storage can be a significant loss for the first 3 to 4 years of operation. Eventually, equilibrium is reached and net storage of nitrogen stops. Unless the storage of nitrogen is significant in the short term, it is most conservative to assume that net storage will be zero.

Phosphorus removal. Phosphorus removal at various SR sites is presented in Table 10-13. Phosphorus adsorption occurs rapidly onto soils, with chemical precipitation being a slower process. Because soils characteristically have very reactive surfaces containing iron, aluminum, and calcium, insoluble phosphates are formed with these species in the soil. Acidic conditions favor complexes with aluminum and iron. Alkaline soil conditions favor phosphate complexes with calcium.

TABLE 10-13
Phosphorus removal for SR systems*

Location	Total applied phosphate, mg/L as P	Percolate total phosphate, mg/L as P
Camarillo, California	11.8	0.2
Dickinson, North Dakota	6.9	0.05
Hanover, New Hampshire	7.1	0.03
Roswell, New Mexico	7.95	0.39
Tallahassee, Florida	10.5	0.1
Helen, Georgia	13.1	0.22
State College, Pennsylvania	5.6	0.08
Clayton County, Georgia	4.9	0.02
West Dover, Vermont	4.2	0.4
Wolfeboro, New Hampshire	3.3	0.02

*Adapted from WPCF (1983), Giggey et al. (1989), Nutter (1986), Sopper (1986), Reed and Crites (1986).

Laboratory-scale phosphorus adsorption tests can be used to estimate the short-term adsorption of candidate soils (U.S. EPA, 1981). Actual removals will be 2 to 5 times higher because of slower chemical precipitation processes that renew adsorption sites. Addition of soil amendments can be used to renew the phosphorus removal capacity of soils.

Phosphorus removal can be estimated conservatively by the following equation (Reed and Crites, 1984):

$$P_x = Pe^{-kt} \tag{10-1}$$

where P_x = total phosphorus at a distance x on the flow path, mg/L
P = total phosphorus in the applied wastewater, mg/L
k = 0.048 d^{-1}
t = detention time, d
= $x(0.40)/k_x S$ [see Darcy's law, Eq. (9-23)]
x = distance along flow path, ft (m)
k_x = hydraulic conductivity in soil in direction x, ft/d (m/d)
S = hydraulic gradient; $S = 1$ for vertical flow, H/L for horizontal flow

Metals removal. Some trace elements can be toxic to plants and/or consumers of plants. In most cases, maintenance of soil pH at or above 6.5 will retain trace elements as unavailable insoluble compounds. If the soil pH falls below 6.5, the metals tend to become more soluble and therefore able to be leached deeper into the soil or to groundwater. Maximum loadings of many trace elements have been suggested for soils with low trace element retention capacities as presented in Table 10-14.

Trace organics removal. At Muskegon, Michigan, the SR system receives stable organics from many industrial sources and effectively removes them from the wastewater. Of the 59 organic pollutants identified in the raw wastewater, the percolate contained only low levels of 10 organic compounds. Although trace organics are generally not of concern for small systems, SR is an effective process for trace organics removal.

Pathogen removal. The SR process is effective in removing bacteria and viruses from wastewater (U.S. EPA, 1981). Because of this effective removal and the practice of disinfection prior to irrigation, removal of microorganisms is not a limiting factor in the design of SR systems.

Preapplication Treatment

Preliminary treatment for an SR system can be provided for a variety of reasons, including public health protection, nuisance control, distribution system protection, or soil and crop considerations. For Type 1 systems, preliminary treatment, except for solids removal, is deemphasized because the SR process can usually achieve final water quality objectives with minimal pretreatment. Public health and nuisance control guidelines for Type 1 SR systems have been issued by EPA (U.S. EPA, 1981) and are given in Table 10-15.

TABLE 10-14

Suggested maximum applications of trace elements to soils without further investigation[a, b]

Element	Mass application to soil, lb/ac	Typical concentration, mg/L[c]
Aluminum	4070	10
Arsenic	82	0.2
Beryllium	82	0.2
Boron	605	1.4[d]
Cadmium	8	0.02
Chromium	82	0.2
Cobalt	41	0.1
Copper	164	0.4
Fluoride	820	1.8
Iron	4070	10
Lead	4070	10
Lithium	—	2.5[e]
Manganese	164	0.4
Molybdenum	8	0.02
Nickel	164	0.4
Selenium	16	0.04
Zinc	1640	4

[a]Adapted from U.S. EPA (1981).
[b]Values based on the tolerances of sensitive crops, mostly fruits and vegetables, grown on soils with low capacities for retaining elements in unavailable forms.
[c]Based on reaching maximum mass application in 20 years at an annual application rate of 8 ft/yr (1.8 m/yr).
[d]Boron exhibits toxicity to sensitive plants at values of 0.75 to 1.0 mg/L.
[e]Lithium toxicity limit is suggested at 2.5 mg/L concentration for all crops, except citrus, which uses a 0.075-mg/L limit. Soil retention is extremely limited.

TABLE 10-15

Pretreatment guidelines for slow-rate systems*

1. Primary treatment—acceptable for isolated locations with restricted public access.
2. Biological treatment by lagoons or in-plant processes, plus control for fecal coliform count to less than 1000 MPN/100 mL—acceptable for controlled agriculture irrigation except for human food crops to be eaten raw.
3. Biological treatment by lagoons or in-plant processes, with additional BOD or SS control as needed for aesthetics, plus disinfection to log mean of 200 MPN/100 mL (EPA fecal coliform criteria for bathing waters)—acceptable for application in public access areas such as parks and golf courses.

*From U.S. EPA (1981).

Type 2 systems are designed to emphasize reuse potential and require greater flexibility in the handling of wastewater. To achieve this flexibility, preliminary treatment levels are usually higher. In many cases, Type 2 systems are designed for regulatory compliance following preliminary treatment so that irrigation can be accomplished by other parties such as private farmers.

Distribution system constraints. Preliminary treatment is generally required to prevent problems of capacity reduction, plugging, and localized generation of odors in the distribution system. For this reason, a minimum primary treatment (or its equivalent) is recommended for all SR systems to remove settleable solids and oil and grease. For sprinkler systems, it is further recommended that the size of the largest particle in the applied wastewater be less than one-third the diameter of the sprinkler nozzle to avoid plugging.

Water quality considerations. The total dissolved solids (TDS) in the applied wastewater can affect plant growth, soil characteristics, and groundwater quality. Guidelines for interpretation of water quality for salinity and other specific constituents for SR systems are presented in Table 10-16. The term *restriction on*

TABLE 10-16
Guidelines for interpretation of water quality*

Problem and related constituent	Degree of restriction on use: None	Slight to moderate	Severe	Crops affected
Salinity, mg/L, as TDS	<450	450–2000	>2000	Crops in arid areas affected by high TDS, impacts vary
Permeability†				
SAR = 0–3 and TDS	>450	130–450	<130	All crops
= 3–6	>770	200–770	<200	
= 6–12	>1200	320–1200	<320	
= 12–20	>1860	800–1860	<800	
= 20–40	>3200	1860–3200	<1860	
Specific ion toxicity				
Sodium, mg/L	<70	>70		Tree crops and woody ornamentals
Chloride, mg/L	<140	140–350	>350	
Boron, mg/L	<0.7	0.7–3.0	>3.0	Fruits trees and some field crops
Residual chlorine, mg/L	<1.0	1.0–5.0	>5.0	Ornamental, only if overhead sprinklers are used

*Adapted from Ayers and Westcot (1985).
†Evaluate impacts on soil permeability using a combination of sodium adsorption ratio (SAR) and TDS. Worst impacts are on clay soils with high SAR values combined with low TDS.

use does not indicate that the effluent is unsuitable for use, rather it means there may be a limitation on the choice of crop or need for special management.

Sodium can adversely affect the permeability of soil by causing clay particles to disperse. The potential impact is measured by the sodium absorption ratio (SAR), which is a ratio of sodium concentration to the combination of calcium and magnesium. The SAR is defined as

$$\text{SAR} = \frac{[\text{Na}]}{\sqrt{\left(\frac{[\text{Ca}] + [\text{Mg}]}{2}\right)}} \tag{10-2}$$

where SAR = sodium absorption ratio, unitless
[Na] = sodium concentration, meq/L (mg/L divided by 23)
[Ca] = calcium concentration, meq/L (mg/L divided by 20)
[Mg] = magnesium concentration, meq/L (mg/L divided by 12.15)

In Type 2 SR systems the leaching requirement must be determined from the salinity of the applied water and the tolerance of the crop to soil salinity. Leaching requirements range from 10 to 40 percent, with typical values being 15 to 25 percent. Specific crop requirements for soil-water salinity must be used to determine the needed leaching requirement (Reed and Crites, 1984; Reed et al., 1995).

Design Procedure

A flowchart of the design procedure for slow-rate systems is presented in Fig. 10-7. The procedure is divided into a preliminary and a final design phase. Determinations made during the preliminary design phase include (1) crop selection, (2) preliminary treatment, (3) distribution system, (4) hydraulic loading rate, (5) field area, (6) storage needs, and (7) total land requirement.

When the preliminary design phase is completed, economic comparisons can be made with other wastewater management alternatives. The text will focus on preliminary or process design with references to detailed design procedures (U.S. EPA, 1981; Pair, 1983; Hart, 1975; USDA, 1983).

Crop Selection

The selection of the type of crop in a slow-rate system can affect the level of preliminary treatment, the selection of the type of distribution system, and the hydraulic loading rate. The designer should consider economics, growing season, soil and slope characteristics, and wastewater characteristics in selecting the type of crop. Forage crops or tree crops are usually selected for Type 1 systems while higher-value crops or landscape vegetation are often used in Type 2 systems.

Type 1 system crops. In Type 1 SR systems the crop must be compatible with high hydraulic loading rates, have a high nutrient uptake capacity, a high

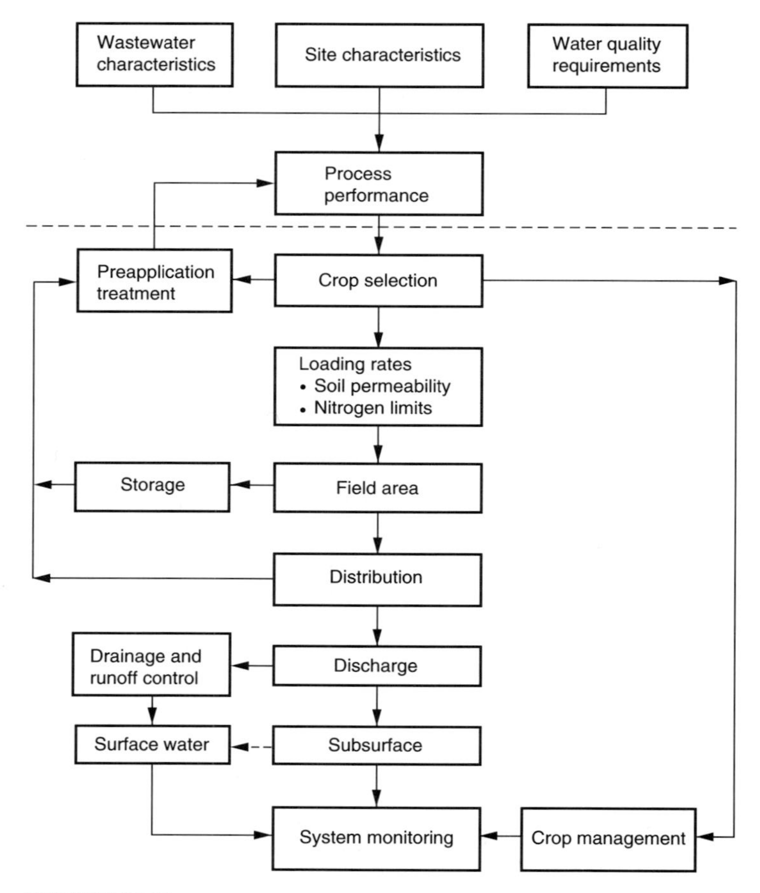

FIGURE 10-7
Flow diagram of design procedure for slow-rate systems.

consumptive use of water, and a high tolerance to moist soil conditions. Other characteristics of value are tolerance to wastewater constituents (such as TDS, chloride, boron) and limited requirements for crop management.

The nitrogen uptake rate is a major design variable for design of Type 1 systems. Typical nitrogen uptake rates for forage, field, and tree crops are presented in Table 10-17. The largest nitrogen removal can be achieved with perennial grasses and legumes. Legumes, such as alfalfa, can fix nitrogen from the air; however, they will preferentially take nitrate from the soil solution if it is provided. Use of legumes (clovers, alfalfa, vetch) in Type 1 systems should be limited to well-draining soils because legumes generally do not tolerate high soil moisture conditions.

The most common tree crops for Type 1 systems are mixed hardwoods and pines. Tree crops provide revenue potential as firewood, pulp, or biomass fuel. Tree species with high growth response such as eucalyptus and hybrid poplars will maximize nitrogen uptake.

TABLE 10-17
Typical nitrogen uptake values for selected crops*

Crop	Nitrogen uptake, lb/ac·yr	Crop	Nitrogen uptake, lb/ac·yr
Forage crops		Tree crops	
Alfalfa	200–600	Eastern forests	
Bromegrass	115–200	Mixed hardwoods	200
Coastal Bermuda grass	350–600	Red pine	100
Kentucky bluegrass	175–240	White spruce	200
Quackgrass	210–250	Pioneer succession	200
Orchard grass	220–310	Aspen sprouts	100
Reed Canary grass	300–400	Southern forests	
Ryegrass	160–250	Mixed hardwoods	250
Sweet clover†	155	Loblolly pine	200–250
Tall fescue	130–290	Lake states forest	
Field crop		Mixed hardwoods	100
Barley	110	Hybrid poplar	140
Corn	155–180	Western forest	
Cotton	65–100	Hybrid poplar	270
Grain sorghum	120	Douglas fir	200
Potatoes	200		
Soybeans	220		
Wheat	140		

*Adapted from U.S. EPA (1981).
†Legume crops can fix nitrogen from the air but will take up most of their nitrogen from applied wastewater.

Type 2 system crops. Crop irrigation or water reuse systems can use a broad variety of crops and landscape vegetation including trees, grass, and field and food crops. Field crops often include corn, cotton, sorghum, barley, oats, and wheat.

Hydraulic Loading Rates

Hydraulic loading rates for SR systems are expressed in units of inches per week (in/wk) [millimeters per week (mm/wk)] or feet per year (ft/yr) [meters per year (m/yr)]. The basis of determination varies from Type 1 to Type 2.

Hydraulic loading for Type 1 SR systems. The hydraulic loading rate for a Type 1 system is determined by using the water balance equation:

$$L_W = \mathrm{ET} - \mathrm{Pr} + P \tag{10-3}$$

where L_W = wastewater hydraulic loading rate, in/mo (mm/mo)
ET = evapotranspiration rate, in/mo (mm/mo)
Pr = precipitation rate, in/mo (mm/mo)
P = percolation rate, in/mo (mm/mo)

The water balance is generally used on a monthly basis. The design values for precipitation and ET are generally chosen for the wettest year in 10, to be conservative. For slow-rate systems the surface runoff (tailwater) is usually captured and reapplied. An exception is the forested Type 1 system where surface and subsurface seepage is allowed by the regulatory agency. Seepage (the surfacing of groundwater) may occur on or off the site without causing water quality problems.

The design percolation rate is based on the permeability of the limiting layer in the soil profile. For Type 1 systems the permeability is often measured in the field using either cylinder infiltrometers, sprinkler infiltrometers, or the basin flooding technique. The range of soil permeability is usually contained in the detailed soil survey from NRCS. Although the given range is often wide, i.e., 0.2 to 0.6 in/h (5 to 15 mm/h), the lower value is often used in preliminary planning. The design percolation rate is calculated from the soil permeability, taking into account the variability of the soil conditions and the overall cycle of wetting (application) and drying (resting) of the site.

$$P\text{ (daily)} = K(0.04 \text{ to } 0.10)(24 \text{ h/d}) \tag{10-4}$$

where P = design percolation rate, in/d (mm/d)
K = permeability of limiting soil layer, in/h (mm/h)
0.04 to 0.10 = adjustment factor to account for the resting period between applications and the variability of the soil conditions

On the basis of either NRCS permeability data or field test results, it is recommended that the daily design percolation rate should range from 4 to 10 percent of the total rate. Selection of the adjustment factor depends on the site and the degree of conservativeness desired. For most SR systems the wetting period is 5 to 15 percent of a given month. If the soil is wet for only 5 percent of the time, only that percent of the time (in a given month) should be used as percolation time. The 4 percent factor should be used when the soil type variation is large, when the wet/dry ratio is small (5 percent or less), and the soil permeability is less than 0.2 in/h (5 mm/h). The high percentages, up to 10 percent, can be used where soil permeabilities are higher, the soil permeability is more uniform, and the wet/dry ratio is higher than 7 percent.

Hydraulic loading for Type 2 SR systems. For crop irrigation systems, the hydraulic loading rate is based on the crop irrigation requirements. The loading rate can be calculated by

$$L_W = \left(\frac{\text{ET} - \text{Pr}}{1 + \text{LR}}\right)\left(\frac{100}{E}\right) \tag{10-5}$$

where L_W = wastewater hydraulic loading rate, in/yr (mm/yr)
ET = crop evapotranspiration rate, in/yr (mm/yr)
Pr = precipitation rate, in/yr (mm/yr)
LR = leaching requirement, fraction
E = irrigation efficiency, percent

The leaching requirement depends on the crop, the total dissolved solids (TDS) of the wastewater, and the amount of precipitation. The leaching requirement is typically 0.10 to 0.15 for low-TDS wastewater and a tolerant crop such as grass. For higher-TDS wastewater (750 mg/L or more) the leaching requirement can range from 0.20 to 0.30.

The irrigation efficiency is the fraction of the applied wastewater that corresponds to the crop ET. The higher the efficiency, the less the water that percolates through the root zone. Sprinkler systems usually have efficiencies of 70 to 80 percent, while surface irrigation systems usually have efficiencies of 65 to 75 percent.

Design Considerations

Design considerations for both types of SR systems are described in the following. Considerations for nitrogen loading, organic loading, land requirements, storage requirements, distribution systems, application cycles, surface runoff control, and underdrainage are presented.

Nitrogen loading rate. The limiting design factor (LDF) for many SR systems is the nitrogen loading rate. The total nitrogen loading (nitrate-N, ammonia-N, and organic-N) is important because the soil microorganisms will convert organic nitrogen to the plant-available inorganic forms. Limitations on the total nitrogen loading rate are based on meeting a maximum nitrate nitrogen concentration of 10 mg/L in the receiving groundwater at the boundary of the project (usually 20 to 100 ft or 6 to 30 m downgradient of the wetted field area). To make certain that the groundwater nitrate nitrogen concentration limit is met, the usual practice is to set the percolate nitrate nitrogen concentration at 10 mg/L prior to commingling of the percolate with the receiving groundwater.

The nitrogen loading rate must be balanced against crop uptake of nitrogen, denitrification, and the leakage of nitrogen with the percolate. The nitrogen balance is

$$L_n = U + fL_n + C_p P F \tag{10-6}$$

where L_n = nitrogen loading rate, lb/ac·yr (kg/ha·yr)

U = crop uptake of nitrogen, lb/ac·yr (kg/ha·yr)

f = fraction of applied nitrogen lost to nitrification/denitrification, volatilization, and soil storage (see Table 10-18)

C_p = concentration of nitrogen in percolate, mg/L (g/m^3)

P = percolate flow, in/yr (m/yr)

F = conversion factor, 0.226 lb·L/mg·ac·in (10 kg·m^2/g·ha)

TABLE 10-18
Denitrification loss factors for SR systems*†

Type of wastewater	*f* factor	
	Warm climates	Cold climates
High-strength wastewater (food processing)	0.8	0.5
Primary municipal wastewater	0.5	0.25
Secondary effluent	0.25	0.15–0.2
Tertiary effluent	0.15	0.10

*Adapted from Reed and Crites (1984).

†The loss factor is the fraction of applied nitrogen that is lost to denitrification, volatilization, and soil storage.

By combining the nitrogen balance and water balance equations, the hydraulic loading rate that will meet the nitrogen limits can be calculated:

$$L_{Wn} = \frac{C_p(\mathrm{Pr} - \mathrm{ET}) + 4.4U}{C_w(1 - f) - C_p} \qquad (10\text{-}7)$$

where L_{wn} = hydraulic loading rate controlled by nitrogen, in/yr (m/yr), and C_w = concentration of nitrogen in applied wastewater, mg/L. Other terms are as defined previously.

Crop uptake of nitrogen can be estimated from Table 10-17. The fraction of applied nitrogen that is lost to denitrification, volatilization, and soil storage depends on the wastewater characteristics and the temperature. The fraction will be highest for warm climates and high-strength wastewaters with BOD-to-nitrogen ratios of 5 or more (see Table 10-18).

Organic loading rate. Organic loading rates do not limit municipal SR systems but may be important for industrial SR systems. Loading rates for BOD often exceed 100 lb/ac·d (110 kg/ha·d) and occasionally exceed 300 lb/ac·d (330 kg/ha·d) for SR systems applying screened food processing and other high-strength wastewater. A list of industrial SR systems with organic loading rates in the above range is presented in Table 10-19. Odor problems have been avoided in these systems by providing adequate drying times between wastewater applications. Organic loading rates beyond 450 lb/ac·d (500 kg/ha·d) of BOD should generally be avoided unless special management practices are used (Reed et al., 1995).

Land requirements. The land requirements for a slow-rate system include the field area for application, plus land for roads, buffer zones, storage ponds, and preapplication treatment. The area can be calculated by

$$A = \frac{Q}{L_W F} \qquad (10\text{-}8)$$

where A = field area, ac (ha)
Q = annual flowrate, Mgal/yr (m^3/yr)

TABLE 10-19
BOD loading rates at existing industrial slow-rate systems*

Location	Industry	BOD loading rate, lb/ac·d
Almaden Winery, McFarland, California	Winery stillage	420
Anheuser Busch, Houston, Texas	Brewery	360
Bronco Wine, Ceres, California	Winery	128
Contadina, Hanford, California	Tomato processing	84–92
Frito-Lay, Bakersfield, California	Potato processing	84
Harter Packing, Yuba City, California	Peach and tomato	351
Hilmar Cheese, Hilmar, California	Cheese processing	420
Ore-Ida Foods, Plover, Wisconsin	Potato processing	190
Tri Valley Growers, Modesto, California	Tomato processing	200

*Adapted from Reed and Crites (1984).

F = conversion factor, 0.027 Mgal/ac·in (10.0 m^3/ha·mm)
L_W = wastewater design hydraulic loading rate, in/yr (mm/yr)

The design hydraulic loading rate can be based on soil permeability, crop irrigation requirements, or nitrogen loading rate. Modification to the land requirement based on storage will be discussed in the section on storage. The calculation of land requirements is illustrated in Example 10-1.

EXAMPLE 10-1. LAND AREA FOR A SLOW-RATE SYSTEM. Calculate the land requirements for a Type 1 slow-rate system for a community of 1000 persons. There is a moderately warm climate and a design wastewater flowrate of 65,000 gal/d. A partially mixed aerated lagoon produces an effluent with 50 mg/L BOD and 30 mg/L total nitrogen. A site has been located which has a limiting soil permeability K of 0.2 in/h. The selected mix of forage grasses will take up 300 lb/ac·yr of nitrogen. The water balance of evapotranspiration and precipitation shows a net evapotranspiration of 18 in/yr.

Solution

1. Calculate the annual design percolation rate using Eq. (10-4):

$$\begin{aligned} P\text{ (annual)} &= K(0.07)(24\text{ h/d})(365\text{ d/yr}) \\ &= 613K \\ &= 613(0.2) = 123\text{ in/yr} \end{aligned}$$

2. Calculate the wastewater loading rate L_W using Eq. (10-3):

$$\begin{aligned} L_W &= (\text{ET} - \text{Pr}) + P \\ &= (\text{net ET}) + P \\ &= 18\text{ in} + 123\text{ in} \\ &= 141\text{ in/yr} \end{aligned}$$

3. Calculate the field area based on soil permeability limits using Eq. (10-8):

$$Q = \frac{65{,}000\text{ gal}}{\text{d}} \times \frac{365\text{ d}}{\text{yr}} \times \frac{1\text{ Mgal}}{10^6\text{ gal}} = \frac{23.7\text{ Mgal}}{\text{yr}}$$

$$A = \frac{\text{Q}}{0.027 \times L} = \frac{23.7\text{ Mgal/yr}}{0.027 \times 141\text{ in/yr}} = 6.2\text{ ac}$$

4. Calculate the hydraulic loading rate controlled by nitrogen, using a percolate nitrate-nitrogen limit of 10 mg/L (Eq. 10-7). Use a denitrification percentage of 25 percent.

$$L_{Wn} = \frac{C_p(\text{Pr} - \text{ET}) + 4.4U}{C_w(1 - f) - C_p} = \frac{10(-18) + 4.4(300)}{30(1 - 0.25) - 10} = 91.2\text{ in/yr}$$

5. Calculate the field area based on nitrogen limits using Eq. (10-8):

$$A = \frac{\text{Q}}{0.027 \times L_{\text{wn}}} = \frac{23.7\text{ Mgal/yr}}{0.027 \times 91.2\text{ in/yr}} = 9.6\text{ ac}$$

6. Calculate the organic loading rate, assuming the 9.6 ac based on nitrogen limits will be the required field area:

$$\text{BOD in wastewater} = \frac{0.065 \text{ Mgal}}{\text{d}} \times \frac{50 \text{ mg}}{\text{L}} \times 8.34 = 27.1 \text{ lb/d}$$

$$\text{BOD loading} = \frac{27.1 \text{ lb/d}}{9.6 \text{ ac}} = 2.8 \text{ lb/ac·d}$$

Therefore, the BOD loading is not limiting because it is much less than 450 lb/ac· d.

7. Determine the field area required:

   Because the area for nitrogen limits (9.6 ac) is larger than the area required for soil permeability (6.2 ac), the required field area is 9.6 ac.

Comment. Nitrogen is the limiting design factor for this example.

Storage requirements. Wastewater is usually stored during periods when it is too wet or too cold to apply to the fields. Except for forested sites, where year-round application is possible, most systems also store wastewater during crop harvesting, planting, or cultivation. Storage for cold weather is generally required by most state regulatory agencies unless it can be shown that groundwater quality standards will not be violated by winter applications and that surface runoff will not occur as a result of wastewater application.

The conservative estimate of the storage period is to equate it to the nongrowing season for the crop selected. A more exact site-specific method is to use the water balance as shown in Example 10-2.

EXAMPLE 10-2. STORAGE REQUIREMENTS FOR A SLOW-RATE SYSTEM. Estimate the storage requirements for the SR system from Example 10-1 using the water balance approach. The monthly precipitation and evapotranspiration data are presented in the following table. The temperatures are too cold in January for wastewater application. The maximum percolation rate is 10.3 in/mo.

Month (1)	Crop ET* (2)	10-year rainfall† (3)
January	1.1	7.2
February	2.0	7.0
March	2.7	4.5
April	3.9	3.0
May	5.6	0.4
June	7.0	0.1
July	8.6	0.1
August	7.4	0.2
September	5.9	0.6
October	3.7	1.2
November	2.0	4.0
December	1.2	4.8
Total	51.1	33.1

*Forage crop evapotranspiration.

†Average distribution of rainfall for the wettest year in 10.

Solution

1. Determine the available wastewater each month:

$$\text{Available wastewater, in/mo} = \frac{\text{monthly flow}}{\text{area}}$$

$$= \left(\frac{65{,}000\ \text{gal}}{\text{d}}\right)\left(\frac{365\ \text{d}}{\text{yr}}\right)\left(\frac{\text{yr}}{12\ \text{mo}}\right)\left(\frac{1}{9.6\ \text{ac}}\right)\left(\frac{\text{ac}}{43{,}560\ \text{ft}^2}\right)\left(\frac{\text{ft}^3}{7.48\ \text{gal}}\right)\left(\frac{12\ \text{in}}{\text{ft}}\right)$$

$$= 7.6\ \text{in/mo}$$

2. Complete the water balance table as shown below.

Month (1)	Crop ET[a] (2)	10-year rainfall[b] (3)	Design percolation[c] (4)	Wastewater loading[d] (5)	Available wastewater[e] (6)	Change in storage[f] (7)	Cumulative storage (8)
January	1.1	7.2	6.1	0.0	7.6	+7.6	8.5
February	2.0	7.0	10.3	5.3	7.6	+2.3	10.8[g]
March	2.7	4.5	10.3	8.5	7.6	−0.9	9.9
April	3.9	3.0	8.1	9.0	7.6	−1.4	8.5
May	5.6	0.4	3.7	8.9	7.6	−1.3	7.2
June	7.0	0.1	2.0	8.9	7.6	−1.3	5.9
July	8.6	0.1	0.4	8.9	7.6	−1.3	4.6
August	7.4	0.2	1.7	8.9	7.6	−1.3	3.3
September	5.9	0.6	3.6	8.9	7.6	−1.3	2.0
October	3.7	1.2	6.4	8.9	7.6	−1.3	0.7
November	2.0	4.0	10.3	8.3	7.6	−0.7	0.0
December	1.2	4.8	10.3	6.7	7.6	+0.9	0.9
Total	51.1	33.1	73.2	91.2	91.2		

[a]Forage crop evapotranspiration.
[b]Average distribution of rainfall for the wettest year in 10.
[c]Maximum percolation rate is 10.3 in/mo.
[d]Loading rate is limited by percolation rate from November through March (January has zero loading due to cold weather); loading rate for April through October is limited by the annual nitrogen loading.
[e]Based on 65,000 gal/d and a field area of 9.6 ac.
[f]Available wastewater minus the wastewater loading.
[g]February is maximum month.

3. The design percolation rate (column 4 data) is 10.3 in/mo when that much rainfall or wastewater is applied. From April to October the wastewater loading is limited by the nitrogen loading and the design percolation rate is the difference between the wastewater loading (column 5) and the net evapotranspiration (ET − precipitation) (column 2 minus column 3).
4. The wastewater loading is limited by the nitrogen balance from April through October; by the precipitation and percolation rates for November, December, February, and March; and by the cold weather in January.
5. Determine the change in storage by subtracting the wastewater loading (column 5) from the available wastewater (column 6). Enter the amount in column 7.
6. The cumulative storage calculations (column 8) start with the first positive month for storage in the fall/winter (December). The maximum month for storage is February with a 10.8-in value. This depth is converted to millions of gallons as follows:

$$\text{Storage volume} = (10.8\text{ in})(9.6\text{ ac})\left(\frac{\text{ft}}{12\text{ in}}\right)\left(\frac{43{,}560\text{ ft}^2}{\text{ac}}\right)\left(\frac{7.48\text{ gal}}{\text{ft}^3}\right)\left(\frac{\text{Mgal}}{10^6\text{ gal}}\right)$$

$$= 2.82\text{ Mgal}$$

7. Convert the required storage volume into equivalent days of flow:

$$\text{Days of storage} = \frac{\text{volume, Mgal}}{\text{flow, Mgal/d}} = \frac{2.82\text{ Mgal}}{0.065\text{ Mgal/d}} = 43.4\text{ d}$$

Comment. The estimated storage volume from the above procedure can be adjusted during final design to account for the net gain or loss in volume from precipitation, evaporation, and seepage. In the wettest year in 10, the storage volume should be reduced to zero at one point in time during the year. To estimate the area needed for the storage pond, divide the required volume in ac-ft by a typical depth, such as 10 ft. The net precipitation falling on the surface area can then be added to the storage volume. Typical seepage rates allowed by state regulations range from 0.062 to 0.25 in/d. These state standards for pond seepage are becoming more stringent, and compaction or lining requirements are becoming more common. Therefore, a conservative approach would be to assume zero seepage.

Distribution techniques. The three principal techniques used for effluent distribution are sprinkler, surface, and drip application. Sprinkler distribution is often used in the newer SR systems (see Fig. 10-8), in most industrial wastewater (high solids content), and in all forested systems. Surface application includes border strip, ridge-and-furrow, and contour flooding. Drip irrigation should be attempted only with high-quality filtered effluent. A comparison of suitability factors for distribution systems is presented in Table 10-20.

Selection of the distribution technique depends on the soil, crop type, topography, and economics. Of the sprinkler systems, the portable hand move and solid set

FIGURE 10-8
View of sprinkler irrigation.

TABLE 10-20
Comparison of suitability factors for distribution systems

Distribution system	Suitable crops	Minimum infiltration rate, in/h	Maximum slope, %
	Sprinkler systems		
Portable hand move	Pasture, grain, alfalfa, orchards, vineyards, vegetable, and field crops	0.10	20
Wheel line (side roll)	All crops less than 3 ft high	0.10	10–15
Solid set	No restriction	0.05	No restriction
Center pivot or linear	All crops except trees	0.20	15
Traveling gun	Pasture, grain, alfalfa, field crops, vegetables	0.30	15
	Surface systems		
Graded borders, narrow (border strip), 15 ft wide	Pasture, grain, alfalfa, vineyards	0.30	7
Graded borders, wide, up to 100 ft	Pasture, grain, alfalfa, orchards	0.30	0.5–1.0
Straight furrows	Vegetables, row crops, orchards, vineyards	0.10	3
	Drip systems		
Drip tube or microjets	Orchards, landscape, vineyards, vegetables	0.02	No restriction

are most common for small systems because of the relatively high flowrates needed for the other systems. Continuous-move systems usually require 300 to 500 gal/min (1135 to 1890 L/min) to operate.

Application cycles. Sprinkler systems operate between once every 3 days and once every 10 days or more. Surface application systems operate once every 2 to 3 weeks. For all systems the total field area is divided into subsections or sets which are irrigated sequentially over the application cycle. For Type 1 systems the application schedule depends on the climate, the crop, and the soil permeability. For Type 2 (crop irrigation) systems the schedule depends on the crop, the climate, and the soil moisture depletion.

Surface runoff control. The surface runoff of applied wastewater from SR systems, known as *tailwater,* must be contained onsite. Collection of tailwater and its return to the distribution system or to the storage pond is an integral part of the design of surface application systems. Sprinkler systems on steep slopes or on slowly permeable soil may also use tailwater collection and recycle. A typical tailwater return system consists of a perimeter collection channel, a sump or pond, a pump, and a return forcemain to the storage or distribution system. Tailwater volumes range from 15 percent of applied flows for slowly permeable soils to 25 to 35 percent for moderately permeable soils (Hart, 1975).

Storm-induced runoff does not need to be retained onsite; however, stormwater runoff should be considered in site selection and site design. Erosion caused by stormwater runoff can be minimized by terracing steep slopes, contour plowing, no-till farming, and grass border strips. Provided that effluent application is stopped before the storm, the stormwater can be allowed to drain off the site.

Underdrainage. There are instances in which subsurface drainage is needed for SR systems to lower the water table and prevent waterlogging of the surface soils. The existence of a water table within 5 ft indicates the possibility of poor subsurface drainage and should cause the need for underdrains to be examined. For small SR sites [less than 10 ac (4 ha)] the need for underdrains may make the site uneconomical to develop.

Underdrains usually consist of 4 to 6 in (100 to 150 mm) of perforated plastic pipe buried 6 to 8 ft (1.8 to 2.4 m) deep. In sandy soils, drain spacings are 300 to 400 ft (91 to 122 m) apart in a parallel pattern. In clayey soils, the spacings are much closer, with 50 to 100 ft (15 to 30 m) apart being typical. Procedures for designing underdrains are described in Van Schilfgaarde (1974), USDA (1972), and USDI (1978).

Construction Considerations

In most instances the slow-rate site can be developed according to local agricultural practices (Crites, 1998). Local extension services, NRCS representatives, or agricultural engineering experts should be consulted. One of the key concerns is to pay attention to the soil infiltration rates. Earthworking operations should be conducted in such a way as to minimize soil compaction, and soil moisture should generally be substantially below optimum during these operations. High-flotation tires are recommended for all vehicles, particularly for soils with high percentages of fines. Deep ripping may be necessary to break up hardpan layers, which may be present below normal cultivation depths.

Operation and Maintenance

Proper operation of an SR system requires management of the applied wastewater, the crop, and the soil profile. Applied wastewater must be rotated around the site through the application cycle, allowing time for drying maintenance, cultivation, and crop harvest. The soil profile must also be managed to maintain infiltration rates, avoid soil compaction, and maintain soil chemical balance.

Compaction and surface sealing can reduce the soil infiltration or runoff. The causes can include (WPCF, 1989):

1. Compaction of the surface soil by harvesting or cultivating equipment.
2. Compaction from grazing animals when the soil is too wet. (Wait 2 to 3 days after irrigation to allow grazing by animals.)
3. A clay or silt crust can develop on the surface as the result of precipitation or wastewater application.
4. Surface clogging as a result of suspended solids application.

The compaction, solids accumulation, and crusting of surface soils may be broken up by cultivating, plowing, or disking when the soil surface is dry. At sites where clay pans (hard, slowly permeable soil layers) have formed, it may be necessary to plow to a depth of 2 to 6 ft (0.6 to 1.8 m) to mix the impermeable soil layers with more permeable surface soils.

A check of the soil chemical balance is required periodically to determine if the soil pH and percent exchangeable sodium are in the acceptable range (see Table 10-7). Soil pH can be adjusted by adding lime (to increase pH) or gypsum (to decrease pH). Exchangeable sodium can be reduced by adding sulfur or gypsum followed by leaching to remove the displaced sodium.

10-4 RAPID-INFILTRATION SYSTEMS

Rapid infiltration (RI) has the highest hydraulic loading rate of any land treatment system. The site selection criteria for RI are also more stringent. Design objectives, site selection procedures, treatment performance, design considerations, construction considerations, and operation and maintenance for RI systems are described in this section.

Design Objectives

The principal design objective for RI systems is wastewater treatment. Other design objectives can include recharge of streams by interception of shallow groundwater; recovery of water by walls or underdrains, with subsequent reuse; groundwater recharge; and temporary (seasonal) storage of renovated water in an aquifer. Most RI systems are designed to treat wastewater and avoid a direct discharge to a surface watercourse.

Site Selection

Site selection is very important to the success of an RI project because failure of RI systems is most often related to improper or insufficient site evaluation (Reed et al., 1985). The important factors in site evaluation and selection are the soil depth, soil permeability, depth to groundwater, and groundwater flow direction.

Soils investigation. Potential RI sites are located by using NRCS detailed soil surveys or other soil maps or surveys. As indicated in Tables 10-4 and 10-5, RI sites need deep, permeable soil without a high groundwater table. Once a potential site is located, it is necessary to conduct field investigations of the soil profile. Soil investigations can include backhoe pits, soil borings, and groundwater wells.

Backhoe pits are excavated normally to a depth of 8 to 10 ft (2.4 to 3 m). Pits should be located on each major soil type. The number of pits will vary with the site types. For example, a 20-ac (8-ha) site may need 6 to 10 backhoe pits to define the variability of the soil profile. Backhoe pits are excavated so that a soil scientist can walk into the pit and observe the soil profile. The various soil layers can be identified visually, and the presence of fractured near-surface rock, hardpan, mottling, layers or lenses of gravel or clay, or other anomalies can be identified and recorded. If the pit extends into groundwater, it can also be used for in-place testing of lateral soil permeability. Soil samples can be taken from each soil layer and analyzed for particle size, pH, and EC. Once observations are complete, level benches can be excavated at different depths in the soil profile (coinciding with different soil layers) to allow infiltration testing (U.S. EPA, 1984).

Soil borings are used to characterize the deeper soils (greater than 10 ft) and to determine depth to bedrock and groundwater. All borings should penetrate below the water table if it is within 30 to 50 ft (9 to 15 m) of the surface. Fewer borings are needed typically than backhoe pits, with 1 soil boring per 5 ac (2 ha) being typical.

Groundwater investigations. Investigation of groundwater conditions is important to RI system design. The depth to groundwater, thickness and permeability of the aquifers, and groundwater quality are important to determine. Because of the expense of drilling wells, the site and the RI process should be well established as the preferred wastewater management alternative prior to drilling. Existing onsite and nearby wells should be surveyed and sampled, and well logs should be analyzed prior to drilling onsite wells.

Once the RI site appears to be acceptable, groundwater wells should be drilled. The EPA recommendation is three wells for a complete RI site investigation (U.S. EPA, 1984). If the general groundwater flow direction has been identified, the wells should be located so that one is in the middle of the basin (RI) area, one is upgradient, and the third well is downgradient near the project boundary.

Infiltration tests. A critical element of RI site evaluation is to conduct field measurements of infiltration rate. The limiting rate of hydraulic flow in an RI system may be the basin surface, a subsurface layer, or the lateral flow away from the site. All three elements must be considered and measured, if possible. The surface and subsurface permeability can be measured using infiltration tests located at the elevation that will correspond to the basin surface and at critical depths in the subsurface.

The backhoe pits and soil borings can be used to estimate the presence of restrictions to vertical flow and to locate layers that need to be tested for infiltration rate (permeability or hydraulic conductivity).

There are a number of infiltration tests, but the preferred tests for RI systems are the flooded basin technique and the cylinder infiltrometer (Reed and Crites, 1984). The flooding basin test is preferred because it more closely approximates the conditions and size of a rapid infiltration basin. By making the basins 10 ft (3 m) in diameter, sidewall effects are minimized and measured rates are more reproducible. Basins are flooded to achieve soil saturation, then refilled to a depth of 8 to 12 in (200 to 300 mm), and the infiltration rate is measured by monitoring the water drop. At least three tests should be conducted per site and the steady-state minimum infiltration rate should be selected for design.

Treatment Performance

Rapid infiltration is an effective process for BOD, TSS, and pathogen removal. Removal of phosphorus and metals depends on travel distance and soil texture. Nitrogen removal can be significant when systems are managed for that objective.

BOD and TSS removal. Typical values of BOD loadings and BOD removals for RI systems are presented in Table 10-21. Suspended solids are typically 1 to 2 mg/L in the percolate from RI systems as a result of filtration through the soil profile.

BOD loadings on industrial RI systems range from 100 to 600 lb/ac·d (112 to 667 kg/ha·d). BOD loadings beyond 300 lb/ac·d (336 kg/ha·d) require careful management to avoid odor production. Suspended solids loadings of 100 to 200 lb/ac·d (112 to 224 kg/ha·d) or more require frequent disking or scarifying of the basin surface to avoid plugging of the soil. For example, a 150 lb/ac·d (168 kg/ha·d) loading of TSS at Hollister, California, required disking after each 3-week application/drying cycle.

Nitrogen removal. Nitrification/denitrification is the principal mechanism for removal of ammonia and nitrate from the wastewater in RI systems. Ammonia adsorption also plays an important role in retaining ammonia in the soil long enough for biological conversion. Nitrification and denitrification are affected by low temperatures and proceed slowly at temperatures of 36 to 41°F (3.6 to 5°C). In

TABLE 10-21
BOD removal for rapid infiltration systems

Location	Applied wastewater BOD, lb/ac·d*	Applied wastewater BOD, mg/L	Percolate concentration, mg/L	Removal, %
Boulder, Colorado	48†	131†	10†	92
Brookings, South Dakota	11	23	1.3	94
Calumet, Michigan	95†	228†	58†	75
Ft. Devens, Massachusetts	77	112	12	89
Hollister, California	156	220	8	96
Lake George, New York	47	38	1.2	97
Milton, Wisconsin	138	28	5.2	81
Phoenix, Arizona	40	15	0–1	93–100
Vineland, New Jersey	43	154	6.5	96

*Total lb/ac·yr applied divided by number of days in the operating season.
†COD basis.

addition, denitrification requires an adequate carbon source and the absence of available oxygen.

Experience with nitrification has been that rates of up to 60 lb/ac·d (67 kg/ha·d) can be achieved under favorable moisture and temperature conditions. Total nitrogen loadings should be checked to verify that they are not in excess of the 50 to 60 lb/ac·d (56 to 67 kg/ha·d) range. Ammonia will be retained in the upper soil profile when temperatures are too low [below 36°F (2.2°C)] for nitrification.

Nitrogen removal is a function of detention time, BOD:N ratio (adequate carbon source), and anoxic conditions. Detention time is related to hydraulic loading rate through the soil profile. For effective nitrogen removal (80 percent or more), the loading rate should not exceed 6 in/d. The BOD:N ratio needs to be 3:1 or more to ensure adequate carbon to drive the denitrification reaction. Secondary effluent will have a BOD:N ratio of about 1:1, while primary effluent usually has a BOD:N ratio of 3:1. To overcome the low BOD:N ratio in secondary effluent a longer application period (7 to 9 days) is necessary. Typical removals of total nitrogen and percolate concentration of nitrate nitrogen and total nitrogen are presented in Table 10-22.

Phosphorus removal. Phosphorus removal in RI is accomplished by adsorption and chemical precipitation. The adsorption occurs quickly and the slower-occurring chemical precipitation replenishes the adsorption capacity of the soil. Typical phosphorus removals for RI are presented in Table 10-23, including travel distances through the soil.

If phosphorus removal is critical, a phosphorus adsorption test using the specific site soil can be conducted (Reed and Crites, 1984). To conduct an adsorption test, about 10 g of soil is placed in containers containing known concentrations of phosphorus in solution. After periodic shaking, up to 5 days, the solution is decanted and analyzed for phosphorus. The difference in concentrations is attributed to adsorption onto the soil particles. The detailed procedure is presented by Enfield and Bledsoe (1975). Actual phosphorus retention at an RI site (long term) will be 2 to 5 times greater than the values obtained in the 5-day phosphorus adsorption test (U.S. EPA,

TABLE 10-22
Nitrogen removal for rapid infiltration systems*

Location	Applied total nitrogen		Percolate, mg/L		Total N removal, %
	lb/ac·d	mg/L	Nitrate-N	Total N	
Calumet, Michigan	20.7	24.4	3.4	7.1	71
Dan Region, Israel	28.9	13.0	6.5	7.2	45
Ft. Devens, Massachusetts	37.0	50.0	13.6	19.6	61
Hollister, California	14.9	40.2	0.9	2.8	93
Lake George, New York	12.5	12.0	7.0	7.5	38
Phoenix, Arizona	40.0	18.0	5.3	5.5	69
W. Yellowstone, Montana	115.6	28.4	4.4	14.1	50

*Adapted from Crites (1985a).

TABLE 10-23
Phosphorus removal for rapid-infiltration systems*

Location	Average concentration in applied wastewater, mg/L	Distance of travel, ft		Average concentration in renovated water, mg/L	Removal, %
		Vertical	Horizontal		
Boulder, Colorado[†]	6.2[†]	8–10	0	0.2–4.5	40–97
Brookings, South Dakota[‡]	3.0[‡]	2.6	0	0.45	85
Calumet, Michigan[†]	3.5[†]	10–30	0–400	0.1–0.4	89–97
	3.5[†]	[§]	5580[§]	0.03	99
Ft. Devens, Massachusetts[‡]	9.0[‡]	50	100	0.1	99
Hollister, California[‡]	10.5[‡]	22	0	7.4	29
Lake George, New York[‡]	2.1[‡]	10	0	<1	>52
	2.1[‡]	[§]	1970[§]	0.014	99
Phoenix, Arizona[†]	8–11[†]	30	0	2–5	40–80
	7.9[†]	20	100	0.51	94
Vineland, New Jersey[‡]	4.8[‡]	6.5–60	0	1.54	68
	4.8[‡]	13–52	850–1700	0.27	94

*Adapted from U.S. EPA (1981).
[†] Total phosphate measured.
[‡] Soluble phosphate measured.
[§] Seepage.

1981). If the travel distance to the critical point for phosphorus removal is known, the "worst case" phosphorus concentration can be calculated from Eq. (10-1).

Heavy metals removal. Heavy metal removal, by the same mechanisms as described for slow rate, will range from 50 to 90 percent for RI. Metals applied at very low concentrations (below drinking water standards) may not be affected by passage through sand (Crites, 1985b).

Trace organics. Trace organics are removed in RI systems by volatilization, sorption, and degradation. Removals depend on the constituent, the applied concentration, the loading rate, and the presence of easily degradable organics to serve as a primary substrate (Crites, 1985b). Removals have been studied at Phoenix, Arizona, Ft. Devens, Massachusetts, and Whittier Narrows, California, and have ranged from 10 to 96 percent.

Pathogens. Pathogens are filtered out by the soil and adsorbed onto clay particles and organic matter. Fecal coliform are removed by 2 to 4 orders of magnitude in many RI systems (U.S. EPA, 1981). At the RI site in Phoenix, Arizona, 99.99 percent virus removal was achieved after travel through 30 ft (10 m) of sand at a loading rate of 300 ft/yr (100 m/yr) (Crites, 1985b).

Preapplication Treatment

Once the overall treatment objective has been established, the appropriate level of preapplication treatment should be determined. For RI the minimum preapplication treatment is primary sedimentation or the equivalent. For small systems a short-detention-time pond is recommended. Long-detention-time facultative or aerobic ponds are not recommended because of their propensity to produce high concentrations of algae. The algae produced in stabilization ponds will reduce infiltration rates in RI systems significantly. If facultative or stabilization ponds are to be used with RI, it is recommended that an aquatic treatment or constructed wetland system be used between the pond and the RI basins to reduce TSS levels.

Design Procedure

The process design procedure is outlined in Table 10-24 for a typical RI system. If the hydraulic pathway is toward a surface water, the limiting design factor will be related to surface water quality requirements, which are usually BOD and TSS. Groundwater discharges are more often controlled by pathogen and nitrate requirements. If nitrogen removal is a process design consideration, the following six steps should also be followed after the annual hydraulic loading is calculated (step 6 in Table 10-24). If necessary, the steps 5 and 6 below may be repeated for different levels of preapplication treatment to achieve the required level of overall treatment.

1. Calculate the mass of ammonia that can be adsorbed by the soil cation exchange capacity.
2. Calculate the length of the application period that can be used without exceeding the mass of ammonia that can be adsorbed.
3. Compare the nitrogen loading rate to the 50 to 60 lb/ac·d (56 to 67 kg/ha·d) limit for nitrification.
4. Balance the ammonia adsorption with the available oxygen to establish the application and drying periods for nitrification.

TABLE 10-24
Process design procedure for rapid infiltration

Step	Description
1.	Characterize the soil and groundwater conditions with field measurements.
2.	Predict the hydraulic pathway of percolate.
3.	Select the infiltration rate from field data.
4.	Determine overall treatment requirements.
5.	Select the appropriate level of preapplication.
6.	Calculate the annual hydraulic loading rate.
7.	Calculate the field (basin) area.
8.	Check for groundwater mounding.
9.	Select the final hydraulic loading cycle.
10.	Determine the number of basins.

5. Balance the nitrate nitrogen produced by nitrification against the applied BOD to ensure an adequate BOD:N ratio. Revise application/drying cycle if necessary.
6. If necessary, consider reducing the soil infiltration rate to increase the detention time for higher nitrogen removal. Reduction of infiltration can be accomplished by incorporating silt or finer-textured topsoil, reducing the depth of flooding, or compacting the soil.

Design Considerations

Design considerations for RI systems include hydraulic loading rates, nitrogen loading rates, organic loading rates, land requirements, hydraulic loading cycle, infiltration system design, and groundwater mounding. The procedures involved in RI design are illustrated in Example 10-3.

Hydraulic loading rates. The hydraulic characteristics of the soil and aquifer system usually determine the design hydraulic loading rate of a site. In some instances the nitrogen or BOD loading may control the area needed; however, the limiting design factor (LDF) for RI systems is usually the infiltration/percolation rate.

The design hydraulic loading rate is the measured clean water infiltration rate multiplied by a design factor. The design factor depends on the procedure used for measuring the infiltration rate, on the variability of the infiltration test results, on the variation in soil characteristics over the site, and on the conservatism of the designer. Design factors account for the cyclical (intermittent loading and drying) nature of RI applications, the variability of site conditions and test measurements, and a long-term decrease in infiltration rates as a result of wastewater loadings. Typical design factors are presented in Table 10-25.

Nitrogen loading rates. Where nitrogen removal is important, the total nitrogen rate should be kept below 60 lb/ac·d (67 kg/ha·d). To determine the nitrogen loading rate from the hydraulic loading rate, use

$$L_n = \frac{L_W C F}{D} \tag{10-9}$$

TABLE 10-25
Typical design factors used to convert measured infiltration rates to RI hydraulic loading rates

	Design factors, percent of measured value	
Test procedure	**Conservative range, high variability of field data or site conditions**	**Less conservative, low variability of field data or site conditions**
Basin flooding test	5–7	8–10
Cylinder infiltrometer, air entry permeameter	1–2	2–4

where L_n = nitrogen loading rate, lb/ac·d (kg/ha·d)
L_W = wastewater hydraulic loading rate, in/yr (m/yr)
C = wastewater nitrogen concentration, mg/L (g/m^3)
F = conversion factor, 0.226 lb·L/mg·ac·in (10 kg·m^2/g·ha)
D = number of operating days per year

For a typical municipal wastewater with 40 mg/L of total nitrogen and an RI system operated 365 days/year, the 60-lb/ac·d (67-kg/ha·d) nitrogen loading rate corresponds to a hydraulic loading rate of 200 ft/yr (61 m/yr).

Organic loading rates. The limit on organic loading rates depends on the climate, the nature of the wastewater, and the remoteness of the site. From experience with food processing and winery wastewater, the BOD loading rate should generally be less than 600 lb/ac·d (667 kg/ha·d). For municipal systems a limit of about 300 lb/ac·d (336 kg/ha·d) is recommended.

Land requirements. The field area (basin bottoms) for an RI system can be calculated by

$$A = \frac{QF}{L_W} \tag{10-10}$$

where A = field area, ac (ha)
Q = wastewater flowrate, Mgal/yr (m^3/yr)
F = conversion factor, 3.07 ac·ft/Mgal (10^{-4} ha/m^3)
L_W = wastewater hydraulic loading rate, ft/yr (m/yr)

The limiting design factor (LDF) for an RI system must be determined by calculating the field area from Eq. (10-10) and comparing that value to the field area required on the basis of nitrogen or organic loading rates. The field area based on nitrogen or organic loading rates is calculated by

$$A = \frac{CQF}{L_W} \tag{10-11}$$

where A = field area, ac (ha)
C = concentration of nitrogen or BOD, mg/L (g/m^3)
Q = wastewater flowrate, Mgal/d (m^3/d)
F = conversion factor, 8.34 lb/[Mgal·(mg/L)] (0.001 kg/g)
L_W = limiting loading rate, lb/ac·d (kg/ha·d)

In addition to the field area, land requirements for an RI system include basin sideslopes, berms, access roads, and land for preapplication treatment.

EXAMPLE 10-3. PROCESS DESIGN FOR RAPID INFILTRATION. Determine the field area required for a rapid-infiltration system treating 0.5 Mgal/d of municipal wastewater. Preapplication in a 1-day detention time pond reduces the BOD to 150 mg/L and the total nitrogen to 40 mg/L. The site has a high variability of infiltration rates, and the minimum rate (basin flooding test) used for design is 3.0 in/h. Year-round application is possible and a nitrate-nitrogen limit in the downgradient groundwater is 10 mg/L. Discharge is to groundwater which is present at a depth of 50 ft under the site.

Solution

1. Calculate the hydraulic loading rate [Eq. (10-4)] using a design factor of 6 percent from Table 10-25:

$$L_W = 0.06\ (24\ \text{h/d})(365\ \text{d/yr})(3.0\ \text{in/h})(1\ \text{ft}/12\ \text{in})$$
$$= 131\ \text{ft/yr}$$

2. Compare the hydraulic loading to the acceptable nitrogen loading rates using Eq. (10-9):

$$L_n = \frac{L_W C(F \times 12\ \text{in/ft})}{D}$$
$$= \frac{(131)(40)(2.7)}{365}$$
$$= 38.8\ \text{lb/ac·d}$$

Because this nitrogen loading rate of 39 lb/ac·d is well below the 60-lb/ac·d limit, nitrogen loading is not limiting for the field area determination.

3. Calculate the BOD loading in lb/d:

$$\text{BOD load} = (8.34)(0.5\ \text{Mgal/d})(150\ \text{mg/L})$$
$$= 625\ \text{lb/d}$$

4. Calculate the field area needed on the basis of hydraulic loading rate:

$$A = 3.06(0.5\ \text{Mgal/d})(365\ \text{d/yr})/131\ \text{ft/yr}$$
$$= 4.3\ \text{ac}$$

5. Calculate the field area needed on the basis of a 300 lb/ac·d BOD loading rate:

$$A = \frac{626\ \text{lb/d}}{300\ \text{lb/ac·d}} = 2.1\ \text{ac}$$

6. Determine the field area required. Because the area needed for the hydraulic loading rate is greater than the area needed for the BOD loading rate, the field area required is 4.3 ac.

Hydraulic loading cycle. In RI systems the hydraulic loading cycle consists of an application (flooding) period followed by a drying (resting) period. This intermittent cycle is key to the successful performance of an RI system. Application periods range from 1 to 9 d while drying periods range from 5 to 20 d.

Hydraulic loading cycles can be selected to maximize infiltration rates, maximize nitrification, or maximize nitrogen removal. Hydraulic loading cycles for these three objectives are presented in Table 10-26. For mild climates, the shorter drying periods and longer application periods are used. For cold or wet climates, the longer drying periods are necessary.

Infiltration system design. Although sprinklers and subsurface perforated pipe may be used for distribution, the most common method is shallow level

TABLE 10-26
Hydraulic loading cycles for RI*

Loading objective	Wastewater applied	Season	Application, d	Drying/ resting, d
Maximize infiltration rates	Primary	Summer	1–2	6–7
		Winter	1–2	7–12
	Secondary	Summer	2–3	5–7
		Winter	1–3	6–10
Maximize nitrification	Primary	Summer	1–2	6–9
		Winter	1–2	7–13
	Secondary	Summer	2–3	5–6
		Winter	1–3	6–10
Maximize nitrogen removal	Primary	Summer	1–2	10–13
		Winter	1–2	13–20
	Secondary	Summer	7–9	9–12
		Winter	9–12	12–16

*Adapted from U.S. EPA (1981).

spreading basins (Fig. 10-9). Gravity distribution is used through pipes or ditches and the basins are divided by berms to allow periodic drying and scarification of the basin surface.

The number of basins depends on the site topography, hydraulic loading cycle, and wastewater flow. For small systems the basins are generally 0.5 to 2 ac in size. The minimum number of basins is typically 3 to 4 and may be as many as 10 to 15, depending on the loading cycle (Reed and Crites, 1984).

Groundwater mounding. During the application period, the applied wastewater percolates through the soil profile and reaches either a slowly permeable layer or the groundwater table. When the water cannot flow vertically it tends to form a

(*a*)

(*b*)

FIGURE 10-9
Views of infiltration basins: (*a*) basins at Del Rey, California, drying and prepared for infiltration and (*b*) basins near Colton, California, in use for infiltration and recovery.

temporary mound at the interface with the groundwater before it can move laterally with the groundwater. If the magnitude of the mound amounts to 1 to 2 ft of overall rise, the mounding is of little consequence. If the mound rises to within 2 ft of the basin surface, however, the treatment performance and rate of infiltration will diminish. Groundwater mounding equations (Reed and Crites, 1984) and nomographs (U.S. EPA, 1981) can be used to determine the impact of groundwater mounding.

Construction Considerations

Construction of rapid infiltration basins must be conducted carefully to avoid compacting the infiltration surface. Basin surfaces should be located in cut sections, with excavated material being placed and compacted in the berms. The berms do not need to be higher than 3 to 4 ft (1 to 1.3 m) in most cases. Erosion off the berm slopes should be avoided because erodible material is often fine-textured and can blind or seal the infiltration surface.

Operation and Maintenance

The operation and maintenance considerations for RI systems include maintenance of the infiltration rate, avoiding freezing and ice formation conditions in cold climates, and avoidance of solids clogging.

Cold climate operation. Storage is generally not provided for RI, even where cold winters would limit operation of SR or OF systems (Reed et al., 1995). Proper thermal protection is needed for pumps, piping, and valves (Reed and Crites, 1984). Wastewater can continue to be land-applied in RI basins throughout subfreezing weather provided the soil profile does not freeze with moisture in it. Approaches that can be used to avoid critical ice formation include:

1. Design of one basin with excess freeboard to accept continuous loading for up to 2 or 3 weeks during extreme conditions. This basin would then be rested for an extended period during warmer weather and the basin surface scarified.
2. Ridge and furrow surface application combined with a floating ice cover. The ice gives thermal protection to the soil and is supported on the ridges as wastewater infiltrates in the flowing furrows.
3. Retention of snow on the basins to insulate the soil, by snow fences.
4. Use wastewater (perhaps bypassed from the headworks) with minimal preapplication treatment to retain the available heat in the wastewater.

System management. It is essential that RI systems be operated with an application and a drying period. The drying period is critical to effective treatment, restoration of aerobic conditions, and maintenance of infiltration rates. The length of time required to dry each basin of visible water should be recorded and any increasing trend in needed drying time should be noted. An increase in the needed drying time can signal the need for basin surface maintenance. Such maintenance can include disking, scarifying (tilling or breaking up the surface), or scraping off surface solids.

10-5 OVERLAND-FLOW SYSTEMS

Overland flow is a fixed-film biological treatment system in which the grass and vegetative litter serve as the matrix for biological growth. Process design objectives, system performance design criteria and procedures, and land and storage requirements are described in this section.

Design Objectives

Overland flow (OF) can be used as a pretreatment step to a water reuse system or can be used to achieve secondary treatment, advanced secondary treatment, or nitrogen removal, depending on discharge requirements. Because OF produces a treated surface water, a discharge permit is required (unless the water is reused). In most cases the discharge permit will limit the discharge concentrations of BOD and total suspended solids (TSS), and that is the basis of the design approach in this chapter.

Site Selection

Overland flow is best suited to sites with slowly permeable soil and sloping terrain. Sites with moderately permeable topsoil and impermeable or slowly permeable subsoils can also be used. In addition, moderately permeable soils can be compacted to restrict deep percolation and ensure a sheet flow down the graded slope.

Overland flow may be used at sites with existing grades of 0 to 12 percent. Slopes can be constructed from level terrain (usually at least a 2 percent slope is constructed). Steep terrain can be terraced to a finished slope of 8 to 10 percent. At the wastewater application rates in current use, the site grade is not critical to performance while it is within the range of 2 to 8 percent. Site grades of less than 2 percent will need special attention to avoid low spots that will lead to ponding. Grades above 8 percent have an increased risk of short circuiting, channeling, and erosion.

Treatment Performance

Overland-flow systems are effective in removing BOD, TSS, nitrogen, and trace organics. They are less effective in removing phosphorus, heavy metals, and pathogens. Performance data and expectations are described in this section.

BOD loading and removal. In municipal systems the BOD loading rate typically ranges from 5 to 20 lb/ac·d. Biological oxidation accounts for the 90 to

TABLE 10-27
BOD removal for overland-flow systems

Location	Wastewater type	Application rate,* gal/ft·min	Slope length, ft	BOD, mg/L	
				Influent	Effluent
Ada, Oklahoma	Raw wastewater	0.10	120	150	8
	Primary effluent	0.13	120	70	8
	Secondary effluent	0.27	120	18	5
Easley, South Carolina	Raw wastewater	0.29	180	200	23
	Pond effluent	0.31	150	28	15
Hanover, New Hampshire	Primary effluent	0.17	100	72	9
	Secondary effluent	0.10	100	45	5
Melbourne, Australia	Primary effluent	0.32	820	507	12

*Application rate is average flow, gal/min, divided by the width of the slope, ft.

95 percent removal of BOD normally found in OF systems. On the basis of experience with food processing wastewater, the BOD loading rate can be increased to 100 lb/ac·d (110 kg/ha·d) for most wastewater without affecting BOD removal. BOD removals from four overland-flow systems are presented in Table 10-27 along with the application rate and slope length. A typical BOD concentration in the treated runoff water is about 10 mg/L.

Suspended solids removal. Overland flow is effective in removing biological and most suspended solids, with effluent TSS levels commonly being 10 to 15 mg/L. Algae are not removed effectively in most OF systems because many algal types are buoyant and resist removal by filtration or sedimentation. If effluent TSS limits are 30 mg/L or less, the use of facultative or stabilization ponds that generate high algae concentrations is not recommended prior to overland flow.

If OF is otherwise best suited to a site with an existing pond system, design and operational procedures are available to improve algae removal. The application rate should not exceed 0.12 gal/ft·min (0.10 m^3/m·h) for such systems, and a nondischarge mode of operation can be used during algae blooms. In the nondischarge mode, short application periods (15 to 30 min) are followed by 1- to 2-h rest periods. The OF systems at Heavener, Oklahoma, and Sumrall, Michigan, operate in this manner during algae blooms (WPCF, 1989).

Nitrogen removal. The removal of nitrogen by OF systems depends on nitrification/denitrification and crop uptake of nitrogen. The removal of nitrogen in several OF systems is presented in Table 10-28. As shown in Table 10-28 denitrification can account for 60 to 90 percent of the nitrogen removed with denitrification rates of 800 lb/ac·yr or more.

Up to 90 percent removal of ammonia was reported at 0.13 gal/ft·min (0.10 m^3/m·h) at the OF system at the city of Davis, California, where oxidation lagoon effluent was applied (Kruzic and Schroeder, 1990). At Sacramento County,

TABLE 10-28
Nitrogen removal for overland-flow systems*

Parameter	Ada, Oklahoma	Hanover, New Hampshire	Utica, Mississippi
Type of wastewater	Screened raw wastewater	Primary effluent	Pond effluent
Application rate, gal/ft·min	0.10	0.17	0.087
BOD:N ratio	6.3	2.3	1.1
Total nitrogen, lb/ac·yr			
Applied	1070	850	590
Removed	980	790	445–535
Crop uptake	100	190	220
Nitrification-denitrification	880	600	225–325
Removal, mass basis, %	92	94	75–90
Total nitrogen, mg/L			
Applied	23.6	36.6	20.5
Runoff	2.2	5.4	4.3–7.5
Removal, concentration basis, %	91	85	63–79

*Adapted from U.S. EPA (1981).

California, secondary effluent was nitrified at an application rate of 0.70 gal/ft·min (0.54 m^3/m·h). Ammonia concentrations were reduced from 14 to 0.5 mg/L (Nolte and Associates, 1997).

At Garland, Texas, nitrification studies were conducted with secondary effluent to determine whether a 2-mg/L summer limit for ammonia and a 5-mg/L winter limit could be attained. Application rates ranged from 0.43 to 0.74 gal/ft·min (0.33 to 0.57 m^3/m·h). Winter values for effluent ammonia ranged from 0.03 to 2.7 mg/L and met the effluent requirements. The recommended application rate for Garland was 0.56 gal/ft·min (0.43 m^3/m·h) for an operating period of 10 h/d and a slope length of 200 ft (61 m) with sprinkler application (Zirschky et al., 1989).

Phosphorus and heavy metal removal. Phosphorus removal in OF is limited to about 40 to 50 percent because of the lack of soil-wastewater contact. If needed, phosphorus removal can be enhanced by the addition of chemicals such as alum or ferric chloride.

Heavy metals are removed in OF by the same general mechanisms as in phosphorus absorption and chemical precipitation. Heavy metal removal will vary with the constituent metal from about 50 to about 80 percent (WPCF, 1989).

Trace organics. Trace organics are removed in OF systems by a combination of volatilization, absorption, photodecomposition, and biological degradation. If removal of trace organics is a major concern, Reed et al. (1995) and Jenkins et al. (1980) should be reviewed.

Pathogens. Overland flow is not very effective in removing microorganisms. Fecal coliforms will be reduced by about 90 percent when raw or primary effluent is applied; however, minimal removal occurs when secondary effluent (with much lower coliform levels than primary) is applied (U.S. EPA, 1981). Enteric virus removals up to 85 percent have been observed with overland flow.

Preapplication Treatment

The usual treatment prior to OF application is primary settling. For small systems Imhoff tanks or 1- to 2-day detention aerated ponds are recommended. Static or rotating fine screens have also been used successfully at Davis, California, and Hall's Summit, Louisiana (WPCF, 1989).

Design Criteria

The principal design criteria for the OF process are application rate and slope length. Other design criteria include hydraulic loading rate, application period, and organic loading rate. The relationship between hydraulic loading rate and application rate is

$$L_W = \frac{qPF}{Z} \tag{10-12}$$

where L_W = wastewater hydraulic loading rate, in/d (m/d)
q = application rate per unit width of slope, gal/ft·min (m^3/m·min)
P = application period, h/d
F = conversion factor, 96.3 min·ft^2·in/h·gal (60 min·h)
Z = slope length, ft (m)

Application rate. The application rate used for design of municipal OF systems depends on the limiting design factor (usually BOD) the preapplication treatment, limitations, and the climate. A range of suggested application rates is presented in Table 10-29. The lower end of the range shown in Table 10-29 should be used for cold climates and the upper end for warm climates. The relationship between application rate and slope length is shown in Fig. 10-10.

As mentioned previously, facultative ponds are not recommended as preapplication treatment for OF. If OF is used in conjunction with facultative ponds, however, the application rate should not exceed 0.12 gal/ft·min (0.10 m^3/m·h).

TABLE 10-29
Suggested application rates for OF systems, gal/ft·min

Preapplication treatment	Stringent requirements, BOD = 10–15 mg/L	Less stringent requirements, BOD = 30 mg/L
Screening, septic tank, tank, short-term aerated cell (1-d detention)	0.1–0.15	0.25–0.33
Sand filter, trickling filter, secondary	0.2–0.25	0.30–0.45

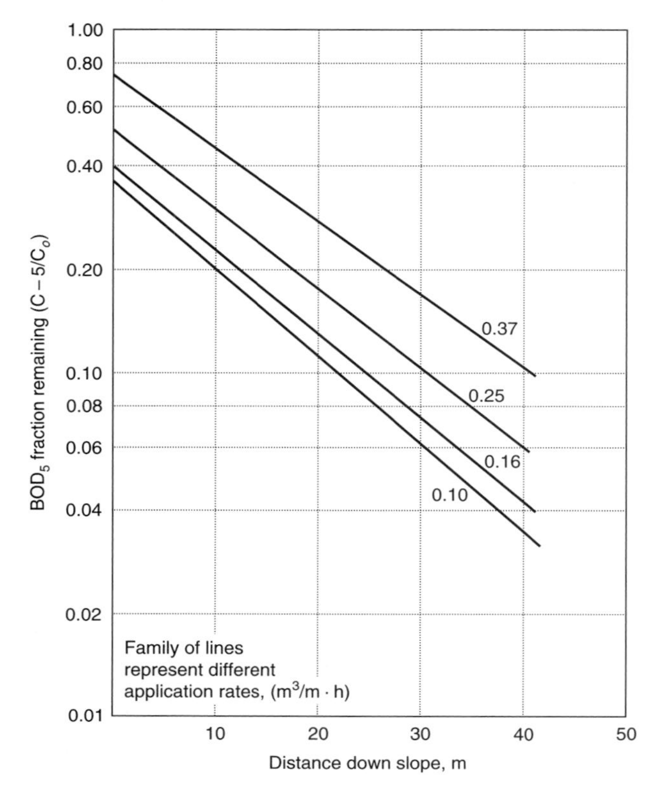

FIGURE 10-10
Overland-flow application rates and slope lengths (Smith and Schroeder, 1985).

Slope length. Slope lengths in OF practice have ranged typically from 100 to 200 ft (30 to 60 m). The longer the slope has been, the greater has been the removal of BOD, TSS, and nitrogen. The recommended slope length depends on the method of application. For gated pipe or spray heads that apply wastewater at the top of the slope, a slope length of 120 to 150 ft (36 to 45 m) is recommended. For sprinkler application, the slope length should be between 150 ft and 200 ft (45 to 61 m). The minimum slope length for sprinkler application (usually positioned one-third the distance down the slope) should be the wetted diameter of the sprinkler plus about 65 to 70 ft (20 to 21 m).

Hydraulic loading rate. The hydraulic loading rate, expressed in inches per day (in/d) or inches per week (in/wk), was the principal design parameter in the EPA Design Manual (U.S. EPA, 1981). Selecting the application rate, however, and calculating the resultant hydraulic loading rate has a more rational basis. Using the application rate approach allows the designer to consider varying the application rate and application period to accomplish a reduction or increase in hydraulic loading.

Application period. A range of application periods has been used successfully, with 6 to 12 h/d being most common. A typical application period is 8 h/d. With an 8-h/d application period, the total field area is divided into three sections. The application period can then be increased to 12 h/d for grass harvest or system maintenance. The application period can be increased to 24 h/d for a short time (3 to 5 days) without adverse impacts on performance.

Design Procedure

The procedure for design of OF systems is to establish the limiting design parameter; select the application rate, application period, and slope length; calculate the hydraulic loading rate; and calculate the field area required. The storage volume, if any, must also be determined, and the field area increased to account for stored volume.

Municipal wastewater—secondary treatment. A relationship between BOD removal and application rates has been developed (Reed et al., 1995; U.S. EPA, 1981; WPCF, 1989). If a system is to be designed for BOD removal only, the use of the "rational" model (Fig. 10-10) will predict a higher application than listed in Table 10-29. For small systems it is recommended that the application rate be selected from Table 10-29.

Industrial wastewater—secondary treatment. For industrial wastewater with BOD concentrations of 400 to 2000 mg/L or more, the organic loading rate is often limiting. The procedure for process design is as follows:

1. Calculate the BOD load from the concentration and flow:

$$\text{BOD load} = QCF \tag{10-13}$$

where BOD load = daily BOD load, lb/d (kg/d)
Q = wastewater flowrate, Mgal/d (m^3/d)
C = BOD concentration, mg/L (g/m^3)
F = conversion factor, 8.34 lb/[Mgal·(mg/L)] (0.001 kg/g)

2. Calculate the land area from

$$A = \text{BOD load}/100 \tag{10-14}$$

where A = field area, ac, and 100 = limiting loading of BOD, lb/ac·d.

When the BOD of the applied wastewater exceeds approximately 800 mg/L, the oxygen transfer from the atmosphere through the fixed film becomes limiting. The BOD removal rate will decline unless effluent recycling is practiced. In some industrial applications, effluent recycle to dilute the BOD in the raw wastewater has been used. For example, the BOD from a food processing wastewater was reduced from 1800 mg/L down to 500 mg/L with an effluent recycle ratio of 3:1 (Reed et al, 1995).

Design Considerations

Design considerations for OF include land area requirements, storage requirements, vegetation selection, distribution systems, and runoff collection.

Land requirements. The field area required for OF depends on the flow, the needed storage, and the loading rate. For an OF system that operates without storage, the field area is calculated by

$$A = \frac{QF}{L_W} \tag{10-15}$$

where A = field area, ac (ha)
Q = design wastewater flowrate, Mgal/d (m^3/d)
F = conversion factor, 36.8 ac·in/Mgal (0.1 ha·mm/m^3)
L_W = wastewater hydraulic loading rate, in/d (mm/d)

If wastewater must be stored because of cold weather, the field area is determined by

$$A = \frac{(365Q + V_S)(F)}{DL_W} \tag{10-16}$$

where A = field area, ac (ha)
Q = design wastewater flowrate, Mgal/d (m^3/d)
V_S = net gain or loss in storage volume from precipitation, evaporation, and seepage, M/gal (m^3)
D = number of operating days per year
L_W = wastewater hydraulic loading rate, in/d (mm/d)
F = conversion factor = 36.8 ac·in/Mgal (0.1 ha·mm/m^3)

Storage requirements. Storage of wastewater may be required for cold weather, wet weather, or crop harvesting. Cold weather storage is the most common with operations ceasing when temperatures fall below 32°F (0°C). Design storage days should be estimated from climatic records for the site.

Storage for wet weather is generally not necessary. Rainfall effects on BOD removal are minimal and storage is not required during normal rainfall events. If storage is not necessary for cold or wet weather, it is usually advisable to provide a week of storage to accommodate emergencies such as equipment problems, crop harvest, or maintenance. The storage pond should be located offline so that it contains pretreated effluent for a minimum time and is drained as soon as application is possible.

EXAMPLE 10-4. PROCESS DESIGN FOR OVERLAND FLOW. Calculate the field area for an overland-flow system to treat 0.5 Mgal/d of septic tank effluent. Preapplication is by a community septic tank and the discharge BOD limit is 30 mg/L. The climate is moderately warm, with 25 d of winter storage required.

Solution

1. Select an application rate appropriate for the degree of preapplication treatment, climate, and BOD removal requirement. Select 0.30 gal/ft·min from Table 10-29.
2. Select a slope length and application period. Select 150 ft for slope length and 8 h/d for the application period.
3. Calculate the wastewater hydraulic loading rate using Eq. (10-12):

$$L_W = \frac{qPF}{Z}$$

$$= \frac{(0.30)(8)(96.3)}{150}$$

$$= 1.5 \text{ in/d}$$

4. Calculate the number of operating days per year.

$$D = 365 - 25 = 340 \text{ d/yr}$$

5. Use a net gain from precipitation on the storage pond of 2 Mgal. Calculate the field area using Eq. (10-16):

$$A = \frac{[(365)(Q) + V_S]\, 36.8}{DL_W}$$

$$= \frac{[365(0.5) + 2]\, 36.8}{(340)(1.5)}$$

$$= 13.3 \text{ ac}$$

Vegetation selection. Perennial water-tolerant grass is recommended for OF vegetation. The function of the grass is to provide a support medium for microorganisms, prevent channeling and erosion, ensure thin sheet flow, and allow for filtration and sedimentation of solids in the vegetative litter. Nutrient uptake is another role—less critical in most cases. The grass crop is harvested usually three or more times per season and either sold for hay or green chop.

A mixture of grasses is recommended and the mixture should include warm-season and cool-season species. Warm-season grasses that have been used successfully include common and coastal Bermuda grass, Dallis grass, and Bahia grass. Cool season grasses include Reed Canary grass, tall fescue, redtop, Kentucky bluegrass, and orchard grass. Some grasses such as Reed Canary need a nurse crop for a year or two, such as rye grass, before they become well established. Local agricultural advisers should be consulted for the best mix of grasses for the particular site.

Distribution system. Surface application by gated pipe is an economically attractive alternative to impact sprinklers for small systems. For municipal wastewater, the surface application technique offers lower energy demand and avoids spray drift and the attendant setback distances of sprinklers. For industrial wastewater, the standard solid set sprinklers are recommended.

Runoff collection. Treated runoff is collected in grass-lined open drainage channels at the toe of the slope. These channels are typically V-type channels with sideslopes of 4:1 or more. Runoff channels should be sloped at 0.5 to 1 percent to avoid ponding.

Construction Considerations

The slope or terrace of an OF system must be graded to a uniform smoothness with no low spots or reversals in grade. Finish slopes of 1 to 8 percent are usually acceptable, although 2 percent is considered minimum in some states. Cross slopes should not exceed 0.5 percent, especially when finish slopes are 1 to 2 percent.

Where extensive cut-and-fill operations are necessary, the slope should be watered and allowed to settle after rough grading. Any depressions should then be filled, and the slope should be final-graded, disked, and landscaped.

Operation and Maintenance

Operation and maintenance considerations include fine tuning of the application cycle, vegetation harvesting, and maintenance of the slope and runoff collection channels.

PROBLEMS AND DISCUSSION TOPICS

10-1. A slow-rate land treatment system is to be loaded at a rate of 2.2 in/wk. If the average wastewater flowrate is 0.5 Mgal/d, estimate the field area needed for application. If the application season is reduced from 52 to 44 wk/yr, how much more field area is required?

10-2. A food processing wastewater contains 2000 mg/L of BOD and 60 mg/L of total nitrogen and has an average flow of 0.75 Mgal/d. Calculate the field area for the maximum recommended BOD loading and the field area for nitrogen loading if the crop is field corn and the denitrification factor is 0.8.

10-3. Using the ET and rainfall data in the table given in Example 10-2, calculate the storage required for a design percolation rate of 5.0 in/mo. All other conditions in Example 10-2 apply.

10-4. An irrigated pasture has an irrigation interval of 10 days. Cattle will graze the pasture after 3 d of drying. Lay out the irrigation sets and indicate the pattern by which the cattle will be rotated.

10-5. A community has the choice of applying facultative pond effluent or septic tank effluent to an overland-flow site. Which one is the more appropriate pretreatment for overland flow? Why?

10-6. Solve Example 10-4 for a design flow of 1.5 Mgal/d and a required effluent BOD of 15 mg/L.

10-7. An operator of an overland-flow system stops applying to the slopes when runoff occurs. Explain the impacts on treatment performance and the hydraulic balance.

10-8. A rapid-infiltration site has the following measured infiltration rates using the basin infiltration test; 2.5, 3.7, 2.9, and 3.1 in/h. Are the measured rates in inches per hour suitable for a rapid-infiltration site? Calculate the average infiltration rate and the hydraulic loading rate.

10-9. A community has a septic tank effluent gravity collection system and the rapid-infiltration site as characterized in Prob. 10-8. Assume that the depth to groundwater is 30 ft and that the average flowrate is 0.3 Mgal/d. Calculate the field area required for rapid infiltration. Calculate the nitrogen and BOD loading rates if the concentration of nitrogen is 40 mg/L and the applied BOD is 100 mg/L.

10-10. How much nitrogen removal would be expected for the loading rates in Prob 10-9. List your assumptions and qualify your answer.

10-11. Lay out and design a rapid-infiltration system for the conditions in Prob. 10-9. Select a loading cycle to maximize nitrification.

10-12. Prepare a one-page abstract of the article by J. C. Lance, F. D. Whisler, and R. C. Rice (1976), Maximizing Denitrification during Soil Filtration of Sewage Water, *Journal of Environmental Quality,* Vol. 5, p. 102.

REFERENCES

Ayers, R. S., and D. W. Westcot (1985) *Water Quality for Agriculture, FAO Irrigation and Drainage,* Paper 29, Revision 1, Food and Agriculture Organization of the United Nations, Rome.

California SWRCB (1990) California Municipal Wastewater Reclamation in 1987, California State Water Resources Control Board, Office of Water Recycling, Sacramento, CA.

Crites, R. W. (1985a) Micropollutant Removal in Rapid Infiltration, in T. Asano (ed.), *Artificial Recharge of Groundwater,* Butterworth Publishers, Stoneham, MA, pp. 579–608.

Crites, R. W. (1985b) Nitrogen Removal in Rapid Infiltration Systems, *Journal of Environmental Engineering Division* ASCE, Vol. 111, No. 6, pp. 865–873.

Crites, R. W., and S. C. Reed (1986) *Technology and Costs of Wastewater Application to Forest Systems,* The Forest Alternative for Treatment and Utilization of Municipal and Industrial Wastes, University of Washington Press, Seattle.

Crites, R. W. (1998) Slow Rate Land Treatment, in *Natural Systems for Wastewater Treatment, Manual of Practice,* Water Environment Federation, Alexandria, VA.

Enfield, C. G., and B. E. Bledsoe (1975) *Kinetic Model from Orthophosphate Reactions in Mineral Soils,* EPA-660/2-75-022, U.S. Environmental Protection Agency, Office of Research and Development, Ada, OK.

Giggey, M. D., R. W. Crites, and K. A. Brantner (1989) Spray Irrigation of Treated Septage on Reed Canarygrass, *Journal of Water Pollution Control Federation,* Vol. 61, pp. 333–342.

Hart, W. E. (1975) *Irrigation System Design,* Colorado State University, Department of Agricultural Engineering, Fort Collins, CO.

Jenkins, T. F., D. C. Leggett, and C. J. Martel (1980) Removal of Volatile Trace Organics from Wastewater, *Journal Environmental Science Health,* Vol. A15, p. 211.

Jewell, W. J., and B. L. Seabrook (1979) *History of Land Application as a Treatment Alternative,* U.S. Environmental Protection Agency, EPA 430/9-79-012, Washington, DC.

Kruzic, A. P., and E. D. Schroeder (1990) Nitrogen Removal in the Overland Flow Wastewater Treatment Process—Removal Mechanisms, *Research Journal of Water Pollution Control Federation,* Vol. 62, No. 7, pp. 867–876.

Martel, C. J., T. F. Jenkins, C. J. Diener, and P. L. Butler (1982) *Development of a Rational Design Procedure for Overland Flow Systems,* CRREL Report 82-2, U.S. Army Corps of Engineers, Hanover, NH.

Miller, R. H. (1973) The Soil as a Biological Filter, in W. E. Sopper and L. T. Kardos, *Recycling Treated Municipal Wastewater and Sludge through Forest and Cropland,* The Pennsylvania State University Press, University Park, PA.

Nolte and Associates (1997) Demonstration Wetlands Project, 1996 Annual Report, Sacramento Regional Wastewater Treatment Plant, Sacramento County, CA.

Nutter, W. L. (1986) Forest Land Treatment of Wastewater in Clayton County, Georgia: A Case Study, in D. W. Cole, C. L. Henry, and W. L. Nutter, *The Forest Alternative for Treatment and Utilization of Municipal and Industrial Wastes,* University of Washington Press, Seattle.

Pair, C. H. (ed.) (1983) *Irrigation,* 5th ed. Irrigation Association, Silver Spring, MD.

Pound, C. E., and R. W. Crites (1973) *Wastewater Treatment and Reuse by Land Application,* EPA 660/2-73-0066, U.S. Environmental Protection Agency, Washington, DC.

Rafter, G. W. (1899) Sewage Irrigation, Part II, *Water Supply and Irrigation Papers,* U.S. Geological Survey, No.22, U.S. Government Printing Office, Alexandria, VA.

Reed, S. C., and R. W. Crites (1986) Forest Land Treatment with Municipal Wastewater in New England, in D.W. Cole, C.L. Henry, and W.L. Nutter, *The Forest Alternative for Treatment and Utilization of Municipal and Industrial Wastes,* University of Washington Press, Seattle.

Reed, S. C., and R. W. Crites (1984) *Handbook of Land Treatment Systems for Industrial and Municipal Wastes,* Noyes Publications, Park Ridge, NJ.

Reed, S. C., R. W. Crites, and E. J. Middlebrooks (1995) *Natural Systems for Waste Management and Treatment,* 2nd ed., McGraw-Hill, New York.

Reed, S. C., R. W. Crites, and A. T. Wallace (1985) Problems with Rapid Infiltration—A Post Mortem Analysis, *Journal of Water Pollution Control Federation,* Vol. 57, No. 8, pp. 854–858.

Reed, S. C. (1972) *Wastewater Management by Disposal on the Land,* USA Cold Regions Research and Engineering Laboratory (CRREL) Special Report 171, Hanover, NH.

Smith, R. G., and E. D. Schroeder (1985) Field Studies of the Overland Flow Process for the Treatment of Raw and Primary Treated Municipal Wastewater, *Journal of Water Pollution Control Federation,* Vol. 57, No. 7, pp. 785–794.

Sopper, W. E. (1986) Penn State's Living Filter: Twenty-three Years of Operation, in D. W. Cole, C. L. Henry, and W. L. Nutter, *The Forest Alternative for Treatment and Utilization of Municipal and Industrial Wastes,* University of Washington Press, Seattle.

Sopper, W. E., and L. T. Kardos (1973) *Recycling Treated Municipal Wastewater and Sludge through Forest and Cropland,* The Pennsylvania State University Press, University Park, PA.

Thomas, R. E., K. Jackson, and L. Penrod (1974) *Feasibility of Overland Flow for Treatment of Raw Domestic Wastewater,* U.S. Environmental Protection Agency, EPA 660/2-74-087.

USDA (1983) Sprinkler Irrigation, Chap. 11, Sec. 15, in *Irrigation SCS National Engineering Handbook,* U.S. Department of Agriculture, Soil Conservation Service, U.S. Government Printing Office, Washington, DC.

USDA (1972) *Drainage of Agricultural Land. A Practical Handbook for the Planning, Design, Construction, and Maintenance of Agricultural Drainage Systems,* U.S. Department of Agriculture, Soil Conservation Service, U.S. Government Printing Office, Washington, DC.

USDI (1978) *Drainage Manual,* U.S. Department of the Interior, Bureau of Reclamation, U.S. Government Printing Office, Washington, DC.

U.S. EPA (1981) *Process Design Manual for Land Treatment of Municipal Wastewater,* EPA 625/1-81-013, U.S. Environmental Protection Agency, Cincinnati, OH.

U.S. EPA (1984) *Process Design Manual for Land Treatment of Municipal Wastewater, Supplement on Rapid Infiltration and Overland Flow,* EPA 625/1-81-0139, Center for Environmental Research Information (CERI), U.S. Environmental Protection Agency, Cincinnati, OH.

Van Schilfgaarde, U. (ed.) (1974) *Drainage for Agriculture.* American Society of Agronomy, Madison, WI.

Vela, G. R. (1974) Effect of Temperature on Cannery Waste Oxidation, *Journal of Water Pollution Control Federation.* Vol. 46, No. 1, pp. 198–202.

Weston Inc. (1982) *Operation and Maintenance Considerations for Land Treatment Systems,* EPA-600/2-82-039, Municipal Environmental Research Laboratory (MERL), U.S. Environmental Protection Agency, Cincinnati, OH.

WPCF (1983) *Nutrient Control, Manual of Practice,* No. FD-7, Water Pollution Control Federation, Alexandria, VA.

WPCF (1989) *Natural Systems, Manual of Practice,* No. FD-16, Water Pollution Control Federation, Alexandria, VA.

Zirschky, J., D. Crawford, L. Norton, S. Richards, and D. Deemer (1989) Meeting Ammonia Limits Using Overland Flow, *Journal of Water Pollution Control Federation,* Vol. 61, No. 12, pp. 1225–1232.

CHAPTER 11

Intermittent and Recirculating Packed-Bed Filters

Packed-bed filters are biological and physical treatment units with a long history of use in wastewater management. Packed-bed filters used for the treatment of wastewater include: (1) conventional and high-rate trickling filters, discussed previously in Chap. 7; (2) high-rate granular and porous medium filters used for effluent filtration; and (3) low-rate granular and porous medium filters. High-rate filters used for effluent filtration from centralized treatment facilities are considered in Chap. 12, which deals with effluent repurification and reuse. The focus of this chapter is on low-rate filters used for the treatment of wastewater from individual homes and other small decentralized facilities. The most commonly used low-rate packed-bed filters are: (1) intermittent (single-pass) sand filters and (2) recirculating (multipass) granular medium filters. Both types of filters are described in this chapter, along with other types of packed-bed filters. Before considering the individual types of packed-bed filters, it will be useful to consider the functional elements that make up these units and a description of the operative removal mechanisms.

11-1 FUNCTIONAL FEATURES OF PACKED-BED FILTERS

The most commonly used packed-bed treatment units comprise the following basic elements: (1) a container with a liner for holding the medium, (2) an underdrain system for removing the treated liquid, (3) the filtering medium, (4) a distribution and dosing system for applying the liquid to be treated onto the filtering medium, and (5) supporting appurtenances. Each of these features is considered briefly in the following discussion. A cutaway view of a packed-bed filter, without the container, is shown in Fig. 11-1.

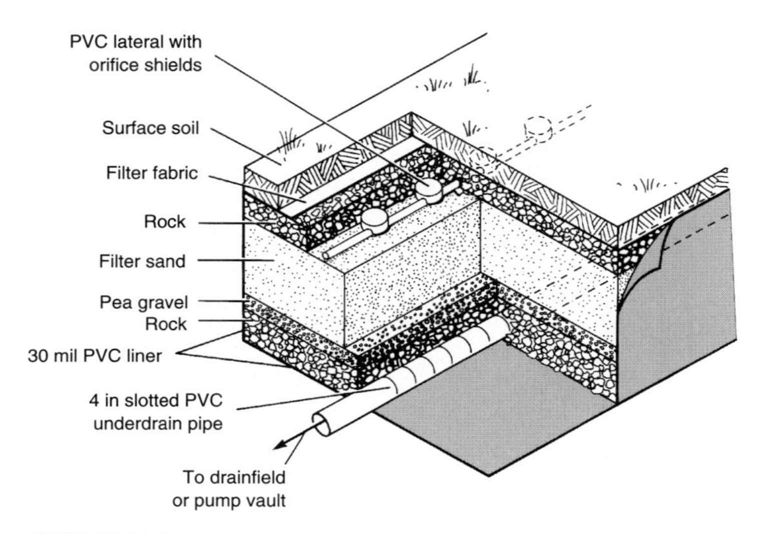

FIGURE 11-1
Cutaway view of the basic elements of a packed-bed filter including geomembrane liner (shown without container), underdrain pipe, support gravel, filter sand, lateral distribution pipe with orifice shield in gravel layer, and cover layer.

Container for Medium

Lined earthen pits, lined wooden structures, and concrete structures (see Fig. 11-2) have been used to contain the filtering material. The choice depends on the type of packed-bed filter (PBF) process to be used, the site conditions with respect to the placement of the unit, and economics. The most common liner material is 30-mil (0.76-mm) PVC or polypropylene. Bottomless filters have been used where the underlying soils have rapid permeability and adequate separation from the groundwater. Bottomless filters are significantly less expensive than contained filters because the liner can be omitted.

Underdrain System

The underdrain system is used to collect the treated liquid and transport it to a pump basin or to the watertight line leading to the disposal field. The most commonly used underdrain systems include the use of slotted or perforated pipe, with the slotted pipe being favored. The slots, cut halfway through the pipe, are 0.25 in (6 mm) wide to prevent biomat bridging. Where slotted pipe is used, the slots are placed facing upward, as shown in Fig. 11-3. The slots are placed upward so that they are not blocked by settlement into the liner. In intermittent PBFs and recirculating PBFs, the underdrain is surrounded with a granular medium of sufficient size that will not clog the slot openings.

(*a*) (*b*)

FIGURE 11-2
Examples of containers used for packed-bed filters: (*a*) plywood used to support PVC liner for buried intermittent sand filter and (*b*) concrete structure for recirculating sand filter.

FIGURE 11-3
Underdrain system for large packed-bed filters.

Filtering Medium

Over the years, a number of different filtering media have been investigated and used, including activated carbon, anthracite, bark, garnet, glass, gravel, ilmenite, lath (bundled), mineral tailings, peat, plastic, plastic foam, shale, and expanded shale. To date, the most commonly used medium for intermittent packed-bed filters (IPBF) is sand. Crushed recycled glass has also proven to be an effective filter material for IPBFs. Typical filter mediums for recirculating packed-bed filters

(RPBF) are coarse sand and fine gravel. Gravel and expanded shale are also used commonly for RPBFs. In the plastic foam filter, developed by Jowett and McMaster (1995), open-cell foam is used as the filtering medium.

Distribution and Dosing System

A distribution system is required to apply the liquid to be filtered uniformly over the filtering medium. Both gravity- and pressure-dosed systems have been used. Methods of application that have been used include spray nozzles, tilting buckets, special molded plastic sheets, and pressure-dosed manifold distribution systems. The most common method for dosing IPBFs is with a distribution manifold with evenly spaced $\frac{1}{8}$-in (3-mm) orifices. The size of the manifold distribution pipe is determined such that the difference in discharge between orifices is not greater than 10 percent (see Fig. 11-4). The dosing manifold, with orifices facing upward, is usually placed on a gravel layer. The orifices are usually covered with small orifice shields (see Fig. 11-5). The orifice shields prevent the gravel medium around the distribution system from blocking the orifices and also prevent discharge from breaking through the surface. In very cold climates, the dosing system must be designed to allow for drainback of any wastewater remaining in the distribution system after the completion of the dosing cycle, or the orientation of every fourth orifice must be downward.

Single-pass dosing. Dosing of the liquid to be treated onto the packed bed can be intermittent single-pass or multipass. In single-pass dosing, the liquid to be treated is applied to the packed bed only once. To optimize the treatment effective-

FIGURE 11-4
Typical distribution system for packed-bed filters. The distribution system is being tested hydraulically to check the uniformity of the discharge between orifices.

FIGURE 11-5
View of orifice shield used to protect orifice in distribution lateral from clogging.

ness, as discussed subsequently, the total liquid volume to be treated is applied equally anywhere from 12 to 72 doses per day. Because the liquid is applied periodically, single-pass systems are known commonly as *intermittent* systems (e.g., intermittent sand filter). Increasing dosing frequency from 4 to 24 times/d was found to improve performance, particularly with larger-diameter sands (Furman et al., 1955). In more detailed research conducted at the University of California at Davis, it was found that dosing frequencies of 12/d were significantly better for ISF performance than 4 times/d (Nor, 1991; Darby et al., 1996). The impact of the number of doses can be understood more clearly after reading the material dealing with removal mechanisms presented in the following section.

Multipass dosing. In multipass dosing, a portion of the liquid which has passed through the filter is diverted to reuse or disposal. The remaining liquid is returned to a recirculation tank where it is mixed with effluent from the septic tank and reapplied to the filter medium. Because a portion of the liquid is passed through the filter a number of times, multipass systems are commonly known as *recirculation* systems (e.g., recirculating gravel filter). The typical recirculation ratios for multipass systems vary from about 3:1 to 5:1, based on forward flow from the septic tank. The principal effect of recirculation is to reduce the total organic loading applied to the filter with each dose, and to increase the dissolved oxygen in the filter. Typical dosing frequencies vary from 48 to 120 times per day (e.g., less than 2 to 3 minutes on every 10 to 25 minutes).

Filter Appurtenances

The types of filter appurtenances required will depend on whether the packed bed is to function as a single- or multipass unit. The principal appurtenances for single-pass units include the necessary pump basins, pumps, and electrical control systems. Control systems for IPBFs are considered in Sec. 11-3. In addition to the appurtenances required for single-pass filters, multipass units require facilities for holding the flow for recirculation and for splitting the flow that is to be sent to the disposal field, and that which is to be mixed with the incoming flow for reapplication to the filter. Flow splitting can be accomplished in a number of different ways, as detailed in Sec. 11-4.

11-2 REMOVAL MECHANISMS IN PACKED-BED FILTERS

The constituent removal mechanisms operative in packed-bed filters, the failure mode, and the modeling of constituent removal in filters are considered in the following discussion.

Removal Mechanisms

The removal mechanisms operative in single- and multipass systems for the removal of BOD, TSS, turbidity, nitrogen, total and fecal coliform bacteria, and viruses are considered in the following discussion. Although the mechanisms are for the most part similar for both filters, there are differences brought about by the added liquid that is applied in multipass filters.

Single-pass filters. In single-pass filters, the process variables that affect the removal performance for BOD, TSS, oil and grease (definition includes fats; see Chap. 2), turbidity, nitrogen, bacteria, and viruses include the size of the filter medium, the hydraulic application rate per dose, and the solids and organic application rate per dose. Shortly after a filter is put into service, a thin bacterial film or slime layer begins to develop in the upper layers over the grains of the filter medium. The bacterial film is of fundamental importance in the operation of the filter, because it retains, by means of *absorption,* the soluble and small colloidal matter and microorganisms found in settled wastewater. The retained material is decomposed and oxidized during the rest period between doses.

Soluble organic matter is taken up almost instantaneously, while the absorbed colloidal material is solubilized enzymatically. The solubilized material is then transferred across the cell membrane and converted to end products. With each subsequent dose, some of the end products are transported farther into the bed, eventually being removed from the bottom of the filter. Larger particles are retained within the filter by means of filtration, with mechanical straining and chance contact being the principal removal mechanisms. As with the soluble and colloidal matter, the coarse organic solids are also processed between doses, and in the early morning hours when the solids and organic loadings to the filter are reduced. As the filter matures, the film layer will develop throughout the filter. Because the larger solids are removed in the upper portion of the filter, the distribution of solids within the filter is nonlinear, with the largest accumulation occurring in the upper 4 to 8 in (100 to 200 mm).

The importance of filter medium size and the hydraulic application rate can be understood by referring to Fig. 11-6. As shown, when the volume of liquid applied is sufficient to fill the pore space, some organic material, colloidal particles, and

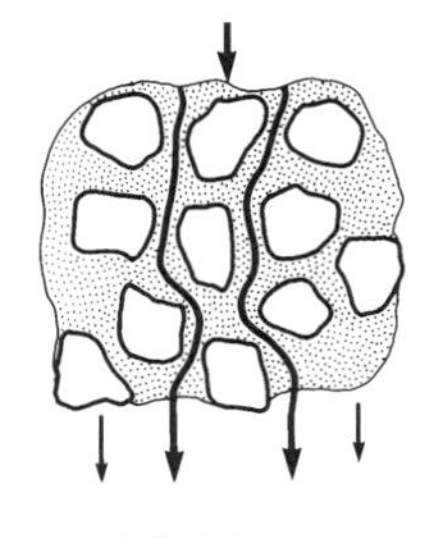

(*a*) 1 dose/d

Applied liquid dose fills the interstices of filter medium allowing some influent particles to pass through the filter untreated.

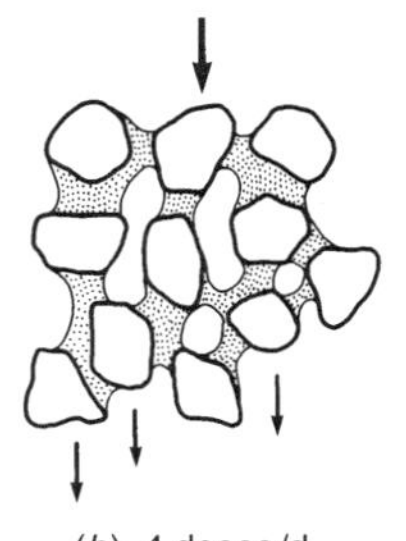

(*b*) 4 doses/d

Applied liquid dose partially fills the interstices of the filter medium.

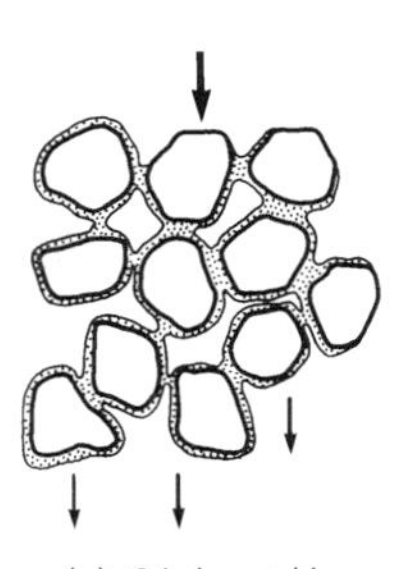

(*c*) 24 doses/d

Applied liquid dose flows around filter medium in a thin film. Most interstices are open.

FIGURE 11-6
Effect of hydraulic application rate on flow through filter: (*a*) 1 dose/d, (*b*) 4 doses/d, and (*c*) 24 doses/d.

microorganisms can pass through the filter untreated (Fig. 11-6*a*). As the volume applied per dose is reduced, partially unsaturated flow occurs (Fig. 11-6*b*). If the volume applied per dose is reduced further, the liquid will flow around the filter medium in a thin film (Fig. 11-6*c*). When film flow occurs, the soluble and colloidal material in the wastewater is absorbed, and oxygen from the air in the open interstitial space is transferred across the thin liquid film to the aerobic bacteria responsible for oxidation of the organic carbonaceous materials. It should be noted that the concentration of oxygen in air is about 250 mg/L at 20°C. Because absorption and the high oxygen concentration in air play such an important part in the operation of both IPBFs and RPBFs, it is imperative that the applied liquid flow over the filter medium in a thin film, if the single-pass filter is to function most effectively.

When the flow is in a thin film, the oxidation of carbonaceous materials will occur in the upper portions of the filter bed. Simultaneously, ammonia will be converted to nitrate (nitrification). In turn, the nitrate will be converted to nitrogen gas (denitrification) in anoxic microsites within the filter. Biological denitrification has been shown to occur under anoxic conditions and, for certain bacteria, under aerobic conditions (Robertson and Kuenen, 1990). The organisms responsible for denitrification utilize adsorbed carbon in particulate matter as the energy source. To sustain the performance of the filter, the microorganisms in the filter must be maintained in the endogenous growth rate. If too much organic material is applied, the bacterial growth rate will increase and the accumulation of material can occur, ultimately leading to failure, as described later.

Flow in a thin film is especially important if viruses are to be removed. In a recent study, it has been demonstrated that the formation of a uniform dense bacterial film, produced by increasing the number of doses per day, will have a significant effect on virus removal (Emerick et al., 1997a). Although sintered glass was used as the filtering medium, a similar effect was observed with sand, as shown in Fig. 11-7 (Emerick

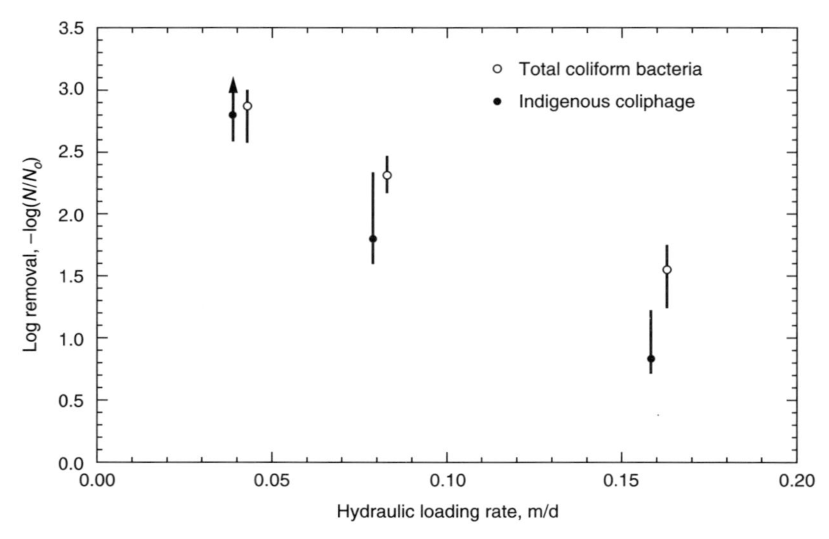

Note: Dosing frequency = 24 doses/d
Error bars indicate 95% confidence intervals.
Arrow indicates non detectable amounts of coliphage in at least one sample, resulting in unknown upper bound.

FIGURE 11-7
Impact of dosing rate on the removal of male-specific coliphage in intermittent sand filters (Emerick et al., 1997).

et al., 1997b). In the study reported on in Fig. 11-7, indigenous coliphage found in wastewater were used to study the removal of viruses. Because the concentration of indigenous coliphage was on the order of 10^3/100 mL, it is not known what total removals may have been possible. Although not studied, it appears that the development of a dense uniform bacterial film may be the result of a more uniform supply of nutrients to the upper region of the filter through the application of multiple doses. It is interesting to note that Buswell, in his book *The Chemistry of Water and Sewage Treatment,* published in 1928, recognized and discussed at length the importance of intermittent dosing on the performance of filters (Buswell, 1928).

Multipass filters. In multipass filters, the process variables that affect the removal performance for BOD, TSS, oil and grease, and turbidity are the same as described above for single-pass filters. The recirculated volume is of great significance because it serves to dilute the effluent from the septic tank, such that the organic matter applied in each dose, and absorbed in the bacterial film, can be processed more easily by the bacteria between doses. Because the depth of the water layer as it flows through the filter is greater than that in an intermittent sand filter, it is possible for small colloidal-sized particles, including bacteria and viruses, to pass through the filter without being absorbed. The organic matter in the influent is distributed to a greater depth within the filter because of the added liquid volume. The added liquid also serves to flush out of the filter partially decomposed organic materials, bacterial waste products, and other debris retained in the filter from the previous dose. The material flushed from the filter tends to accumulate in the bottom of the recirculation tank. In highly loaded multipass filters, a solids trap should be included to remove the material flushed from the filter before the effluent is discharged.

Failure of Single- and Multipass Filters

Both single- and multipass filters can fail when the hydraulic, solids, and/or organic application rates exceed certain limits, specific to each type of filter. The typical failure mode is manifested by surface ponding between liquid applications. In some cases, the applied liquid overflows the enclosure. Failure usually occurs as the result of the accumulation of untreated organic and inorganic solid materials and oil and grease from the influent, and residual cell tissue that has not undergone endogenous respiration to such an extent that the pore space becomes clogged. Other causes of failure include clogged underdrains and the presence of too many fines in the filter medium.

When the amount of dissolved and particulate organic material applied exceeds the amount of food needed to maintain the microorganisms in the slime layer in a low growth rate, an increased growth rate will occur. For the purpose of illustration, assume only soluble organic constituents are applied to the filter. Further, assume the microorganisms on the surface of the filter medium are capable of absorbing an excess amount of organic matter. During the time until the next dose, the organisms will process the stored organic matter, converting some of it to cell tissue. If there is not sufficient time for the organisms to reduce their mass through endogenous respiration between liquid applications (doses), the mass of organisms and unprocessed organic matter in the slime layer will gradually increase with time. As the film thickness increases, the rate of oxygen transfer and, in turn, biological activity is decreased, ultimately leading to the failure of the filter.

In a similar manner, if particulate organic matter and oil and grease are applied and trapped in the slime layer, the organisms in the slime layer must be able to solubilize and process the applied constituents before the next dose. If they are unable to do so, a gradual accumulation of solids will occur within the filter, ultimately leading to failure as evidenced by ponding. A solids balance for a filter is presented in Example 11-1 to illustrate that, if the filter is loaded lightly, it will take a very long time to clog.

EXAMPLE 11-1. SOLIDS BALANCE FOR AN INTERMITTENT SAND FILTER. Prepare a solids balance for an intermittent sand filter to assess how long it would take to fill half of the void space in the upper 6 in (150 mm) of the filter bed, assuming that 60 percent of the material retained in the filter is accumulated in the upper portion of the filter. Assume the following conditions also apply. The values for the influent BOD and TSS are based on the use of an effluent filter as described in Chap. 5.

1. Filter depth = 0.6 m (2 ft)
2. Porosity for sand = 40%
3. Average hydraulic loading rate = 40 $L/m^2 \cdot d$ (1.0 $gal/ft^2 \cdot d$)
4. Influent BOD = 130 mg/L

5. Effluent BOD = 5 mg/L
6. Influent TSS = 30 mg/L
7. Effluent TSS = 5 mg/L
8. Influent VSS = 80%
9. Influent oil and grease = 15 mg/L
10. Effluent oil and grease = 3 mg/L
11. Assume 60% of oil and grease is biodegraded
12. Portion of influent TSS associated with influent BOD = 80%
13. Net biological yield = 0.2 mg cells/mg BOD converted (organisms in filter are assumed to be in the maintenance phase)
14. Net biological yield = 0.3 mg cells/mg oil and grease converted
15. Specific gravity of the accumulated solids = 1.125

Solution

1. Determine the mass of TSS applied to the filter, for a unit surface area of 1.0 m^2:

$$\begin{aligned} M_{\text{TSS}} &= (30 \text{ mg/L})(40 \text{ L/m}^2\cdot\text{d})(1.0 \text{ m}^2)(365 \text{ d/yr})(1/1000 \text{ mg/g}) \\ &= 438 \text{ g/yr} \end{aligned}$$

2. Determine the mass of influent TSS remaining after degradation:

$$M_{\text{in}} = (438 \text{ g/yr})(1 - 0.8) = 87.6 \text{ g}$$

3. Determine the mass of biological solids produced from the BOD:

$$\begin{aligned} M_{\text{bio}} &= (130 \text{ mg/L} - 5 \text{ mg/L})(0.2)(40 \text{ L/m}^2\cdot\text{d})(1.0 \text{ m}^2)(365 \text{ d/yr})(1 \text{ g}/1000 \text{ mg}) \\ &= 365.0 \text{ g/yr} \end{aligned}$$

4. Determine the mass of TSS in the effluent:

$$\begin{aligned} M_{\text{ef}} &= (5 \text{ mg/L})(40 \text{ L/m}^2\cdot\text{d})(1.0 \text{ m}^2)(365 \text{ d/yr})(1 \text{ g}/1000 \text{ mg}) \\ &= 73.0 \text{ g/yr} \end{aligned}$$

5. Determine the mass of oil and grease applied to the filter:

$$\begin{aligned} M_{\text{O\&G}} &= (15 \text{ mg/L})(40 \text{ L/m}^2\cdot\text{d})(1.0 \text{ m}^2)(365 \text{ d/yr})(1 \text{ g}/1000 \text{ mg}) \\ &= 219.0 \text{ g/yr} \end{aligned}$$

6. Determine the mass of biological solids produced from the oil and grease (O&G):

$$\begin{aligned} M_{\text{bio(O\&G)}} &= (219 \text{ g/yr})(0.6)(0.3) \\ &= 39.4 \text{ g/yr} \end{aligned}$$

7. Determine the mass of oil and grease remaining after degradation:

$$\begin{aligned} M_{\text{deg}} &= (219.0 \text{ g/yr})(1 - 0.6) \\ &= 87.6 \text{ g/yr} \end{aligned}$$

8. Determine the mass of oil and grease in the effluent:

$$\begin{aligned} M_{\text{ef(O\&G)}} &= (3 \text{ mg/L})(40 \text{ L/m}^2\cdot\text{d})(1.0 \text{ m}^2)(365 \text{ d/yr})(1 \text{ g}/1000 \text{ mg}) \\ &= 43.8 \text{ g/yr} \end{aligned}$$

9. Determine the mass of solids accumulated in the filter:

$$\begin{aligned} \text{Accumulation} &= \text{inflow (remaining after degradation)} - \text{outflow} + \text{generation} \\ &= 87.6 \text{ g/yr (TSS)} + 87.6 \text{ g/L (O\&G)} - 73.0 \text{ g/yr (TSS)} \\ &\quad - 43.8 \text{ g/yr (O\&G)} + 365 \text{ g/yr (biological solids from BOD)} \\ &\quad + 39.4 \text{ g/yr (biological solids from O\&G)} \\ &= 462.8 \text{ g/yr} \end{aligned}$$

10. Determine time to fill half the filter void space in the upper 6 in (150 mm) of the filter.
   a. Volume of solids accumulated per year:
   $$V_{yr} = [(462.8 \text{ g/yr})/1.125] \times (1.0 \text{ L}/1000 \text{ g}) = 0.41 \text{ L/yr}$$
   b. Void volume of upper 150 mm of filter:
   $$V_{void} = (1.0 \text{ m}^2 \times 0.15 \text{ m})(0.40)(1000 \text{ L/m}^3) = 60 \text{ L}$$
   c. Time to fill half of the void space in the upper 6 in (150 mm) of filter:
   $$t_{fill} = (60 \text{ L} \times 0.5)/(0.41 \text{ L/yr} \times 0.6) = 122 \text{ yr}$$

Comment. In the above computations it is assumed that the oil and grease are soluble and are not reflected in the TSS and in the conventional BOD test. The computations presented in this example make it clear that, if a filter fails, it has been overloaded so much that the biological yield increases significantly, and that TSS and other absorbed constituents in the film are not degraded completely.

Modeling Filter Performance

The performance of single- and multipass filters has been modeled by first-order kinetics as follows:

$$\frac{C}{C_o} = e^{-k_{ab}t} \tag{11-1}$$

where C = effluent concentration, mg/L
C_o = influent concentration, mg/L
k_{ab} = first-order absorption removal-rate coefficient, 1/s
t = nominal travel time through filter, s

Because the first-order removal rate coefficient is an *absorption coefficient,* as opposed to a *kinetic coefficient,* its value will vary with each filter, depending on the hydraulic loading rate and the degree to which the slime layer has developed. Typical travel times through intermittent sand filters are on the order of 20 to 30 s, but these values will also vary with the thickness of the film. Assuming a C/C_o value of 0.05 and a typical travel time of 30 s for a sand filter with a depth of 2 ft, the corresponding value of k_{ab} is 0.1 s^{-1}. Compared to a typical BOD kinetic removal-rate constant of 0.23 d^{-1} (base e), the rate of removal by absorption is about 37,000 times as fast.

11-3 SINGLE-PASS (INTERMITTENT) PACKED-BED FILTERS

Intermittent sand filters have been used for both treatment of individual home wastewater and community wastewater, in either a centralized or decentralized mode, for well over 100 years. In this section, the early development and history of use of IPBFs, a description of a modern IPBF, typical applications for IPBFs, performance assessment, design criteria, construction considerations, and operation and maintenance are considered. Multipass filters are considered in the following section.

Early Development and History of Use

The first attempt to treat wastewater with sand and gravel filters can be traced back to Ealing and Chorley, England, in the late 1860s (Frankland, 1870). It should be noted that in the 1860s, the germ theory had not yet been developed and the role of microorganisms in bringing about the conversion of organic matter was unknown. Following the early experiments at Ealing and Chorley, Sir Edward Frankland, a member of the Royal Commission, conducted experiments in 6-ft (1.8-m) glass cylinders filled with various types of granular media including: (1) a coarse porous gravel from the much cited Beddington (sewage) irrigation meadows near Croydon, (2) sand from the red sandstone near Hambrook, (3) soil from Barking (sewage) irrigation farm, (4) a light yellowish-brown loamy marl from Dursley, and (5) peaty soil from Leland (Dunbar, 1908). The columns were dosed, at rates varying from 3.5 to 13 gal/yd^3, every morning and evening with wastewater from the city of London. The results of the experiments were very satisfactory and the Royal Commission came to the conclusion that "with the best of the filters (No. 1) 44,000 gallons of sewage could be thoroughly purified on one acre, if drains were laid at a depth of 6 feet." It should be noted that in the filtration experiments conducted by Frankland and later at Lawrence, Massachusetts, as described below, the formation of a surface mat was a common operational feature of the filters. The degree of mat formation and surface clogging depended on the degree of pretreatment the wastewater had received. Interestingly, the loading rate established in these early experiments is essentially the same as the rate used currently for the design of these filters where a high level of treatment is required. It is also interesting to note that these early experiments form the first basis for the modern practice of the biological treatment of wastewater and of land treatment.

Frankland's experiments might have been forgotten were it not for the Massachusetts State Board of Health, which became interested in the subject, and in the 1870s constructed the Experiment Station at Lawrence, Massachusetts, with the express purpose of conducting experiments into the treatment of wastewater using sand and gravel filters along the lines suggested by Frankland. By the late 1870s the concept of ISF was put into practice in Massachusetts. The first community to construct an ISF system was Lenox, Massachusetts, in 1876 (Mancl and Peeples, 1991). Between 1891 and 1937 the Massachusetts State Board of Health monitored the performance of over 26 community ISF systems. Data on a selected number of these 26 systems are presented in Table 11-1. Despite the excellent performance of these sand filters, increasing population and the resultant increased land requirements led most of the communities to change to trickling filters or activated sludge. In 1920, the U.S. Public Health Service published a report describing the treatment of wastewater from single homes and small communities (Frank and Rhynus, 1920). Intermittent sand filters were featured prominently (see Fig. 11-8), along with extensive data on their performance.

The use of sand filters waned until the 1940s, when pilot-scale studies were conducted at the University of Florida. These experimental ISF systems were more shallow and the sand size was coarser than was used in the Massachusetts systems. The sand sizes ranged from 0.26 to 0.46 mm, and the bed depths were 18 to 30 in (0.46 to

TABLE 11-1
Historical use of intermittent sand filters*

Location (Massachusetts)	Year started†	Loading rate, gal/ft²·d	Filter depth, in	Effective sand size, mm	Pretreatment
Andover	1902	0.8	48–60	0.15–0.2	Septic tank
Concord	1899	1.9		0.10–0.24	None
Farmington	1890	1.3	70	0.06–0.12	None
Gardner	1891	2.8	60	0.12–0.18	Settling tank
Marlborough	1891	1.7	55–70	0.14–0.15	Settling tank
Pittsfield	1901	2.3	48	0.15–0.18	Solids separation
Spencer	1897	1.4	48	0.18–0.34	None
Stockbridge	1899	0.5	36–55	0.17–0.27	None

*Adapted in part from Mancl and Peeples (1991).
†All systems reported in operation in the 1937 survey.

0.76 m) (Grantham et al., 1949). In the 1970s and 1980s the use of ISF for individual homes increased, with the work done at the University of Wisconsin and the state of Oregon leading the way. Additional details on the design and application of ISFs may be found in Anderson et al. (1986), Boyle et al. (1981), Grantham et al. (1949), Sauer et al. (1976), and U.S. EPA (1980).

Typical Modern Intermittent Sand Filter

A typical example of a modern ISF system is shown in Fig. 11-9. Perhaps the most important difference between the filter shown in Fig. 11-9 and the filters used by Frankland and at the Lawrence, Massachusetts, experimental station is that the modern filter is designed to operate without the formation of a surface-clogging layer of material. The application of the wastewater to be treated in multiple small doses is also different. Other features found in the modern embodiment of the old concept include a dosing manifold, with small orifices facing upward, that can be flushed; an optional air line that can be used to introduce air into the bottom of the filter; and a programmable electrical control panel.

Types of Applications

The principal applications for intermittent sand filters include treating septic tank effluent, polishing and nitrifying secondary effluents, and treating facultative pond effluent. The use of ISF systems to treat wastewater prior to irrigation without chlorine disinfection has also been reported (Hathaway and Mitchell, 1984; Scherer and Mitchell, 1982; and Mote et al., 1991).

Individual systems. Intermittent sand filters are used primarily for individual single family residences, but have been used for clusters of homes. They appear to

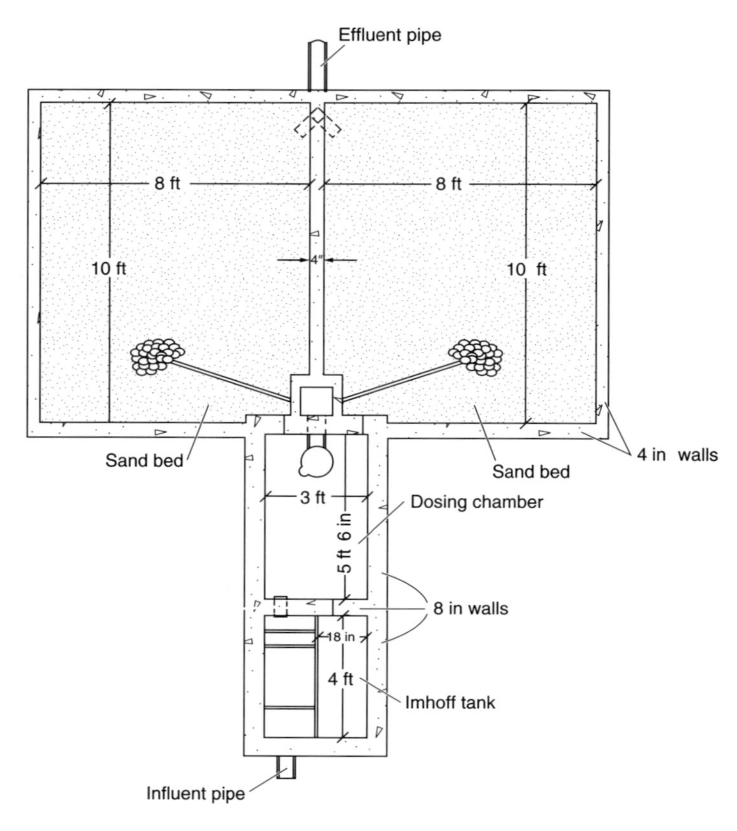

(*a*) Plan view

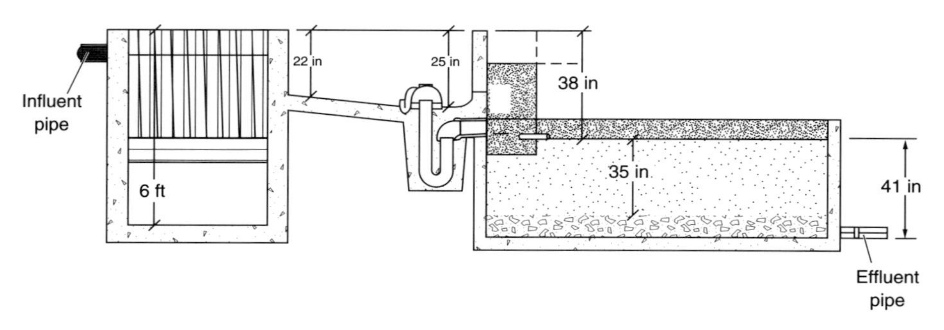

(*b*) Cross section

FIGURE 11-8

Plan and section views of a typical intermittent sand filter recommended by the U.S. Public Health Service for five persons, circa 1915–1920 (Frank and Rhynus, 1920): (*a*) plan view and (*b*) cross section.

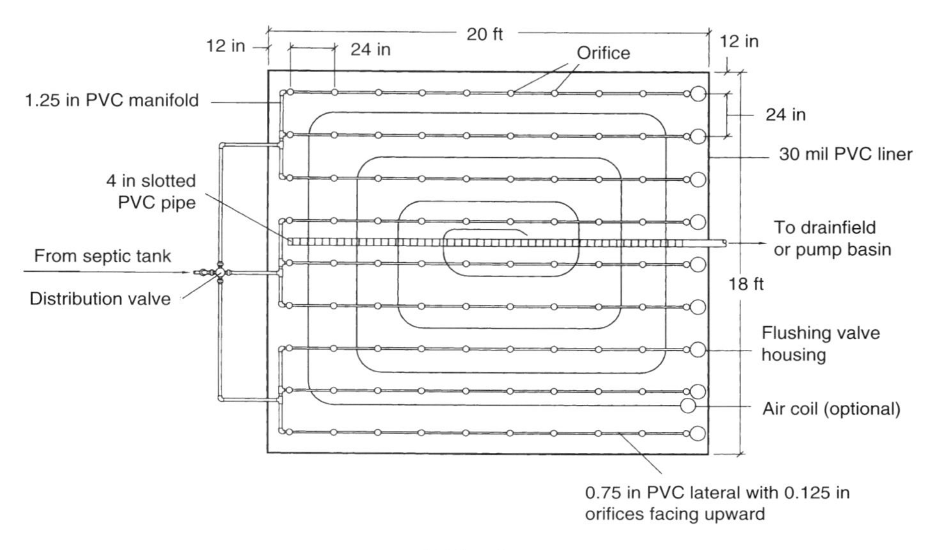

(*a*) Plan view

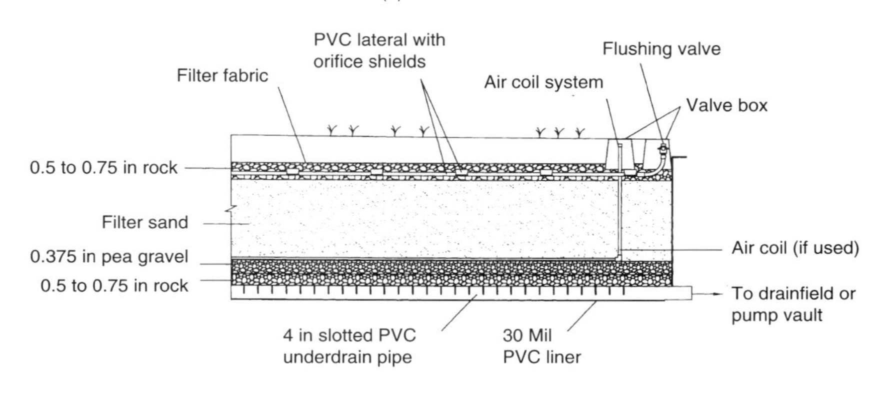

(*b*) Typical cross section

FIGURE 11-9
Schematic of modern intermittent sand filter: (*a*) plan view and (*b*) typical cross section (courtesy Orenco Systems, Inc.).

be economical for flows up to 3000 gal/d (Ball, 1996). The use of the effluent from IPBFs for recycling applications has proven to be quite successful because of the high quality of the effluent that is produced. Drip irrigation is the most common method of applying the treated effluent in reuse applications. It should be noted, however, that drip irrigation is not yet allowed in many parts of the country, although this method of application has been used successfully for a number of years in the southeastern part of the United States.

TABLE 11-2
Historical performance of Massachusetts ISF systems*

Location	Period	Loading rate, gal/ft²·d	BOD_5 influent, mg/L	BOD_5 effluent, mg/L
Brockton	1897–1903	0.9	314	6.2
Framingham	1893–1902		259	6.3
Gardner	1893–1902	2.8	122	11.4
Leicester	1897–1903	1.9	321	13.1
Marlborough	1894–1903	1.7	139	9.5
Natick	1897–1903	1.2	82	7.7
Spencer	1898–1903	1.4	116	6.9

*Adapted in part from Mancl and Peeples (1991).

Lagoon upgrading applications. The use of ISF systems to treat lagoon effluents was reported by Marshall and Middlebrooks (1974) and by Rich et al., (1995). Additional details on the use of sand filters for the treatment of lagoon effluent are discussed in Chap. 8.

Treatment Performance

Intermittent sand filters, with frequent small doses, produce a high-quality effluent that is low in BOD, TSS, turbidity, and ammonia. Properly designed ISFs (e.g., lightly loaded, small doses, and numerous orifices) have proven very effective in the removal of total and fecal coliform and viruses (Emerick et al., 1997). Historical data on the long-term removal of BOD and ammonia nitrogen by ISFs are presented in Tables 11-2 and 11-3, respectively. The performance of ISF systems studied from 1949 to 1997 is summarized in Table 11-4. ISF systems are also effective in removing bacteria, with 99.9 percent removal measured. Fecal coliform removal for ISF systems is summarized in Table 11-5. The removal of virus (F^+ specific coliphage) as a function of the dosing rate is illustrated in Fig. 11-7, presented previously.

TABLE 11-3
Historical performance of Massachusetts ISF systems for ammonia nitrogen removal*

Location	Period	Loading rate, gal/ft²·d	Ammonia, mg/L Influent	Ammonia, mg/L Effluent	Effluent nitrate-N, mg/L	Percent nitrogen removal
Brockton	1897–1903	0.9	40.7	1.5	12.9	77
Framingham	1893–1902		27.3	2.7	11.6	74
Gardner	1893–1902	2.8	21.2	7.5	5.5	58
Leicester	1897–1903	1.9	31.8	9.2	9.0	55
Marlborough	1894–1903	1.7	30.9	7.9	8.1	62
Natick	1897–1903	1.2	12.4	2.3	4.9	59
Spencer	1898–1903	1.4	16.0	2.1	6.9	65

*Adapted in part from Mancl and Peeples (1991).

TABLE 11-4
Summary of ISF performance with respect to the removal of BOD and nitrogen

	BOD_5			Total nitrogen		
Study and location	**Influent mg/L**	**Effluent mg/L**	**% removal**	**Influent**	**Effluent**	**% removal**
Grantham et al. (1949), Florida*	148	14	90	37	32	14
Furman et al. (1955), Florida	57	4.8	92	30	16	47
Sauer (1976), Wisconsin	123	9	93	nd†	nd	nd
Ronayne et al. (1984), Oregon	217	3.2	98	58	30	48
Effert et al. (1988), Ohio‡	127	4	97	42	38	10
Stinson Beach, California (1987–1990)	203	11	94	57	41	28
Nor (1991), Davis, California	82	0.5	99	14	7.2	47
Town of Paradise, California (1992)	148	6	96	38	19	50

*Relatively high hydraulic loadings (1.7 to 4 gal/ft^2·d) and shallow beds (18 to 30 in).
†nd = no data.
‡High hydraulic loading rates (2 to 10 gal/ft^2·d).

Design Considerations

Factors of importance in the design of ISF systems include type and size of filter medium, depth of filter bed, hydraulic loading rate, organic loading rate, and dosing frequency and duration. These design considerations are examined in the following material. Typical design criteria for ISF systems are summarized in Table 11-6. Corresponding design criteria for multipass filters are given in the following section. A range of values is given in Table 11-6 because the specific design values used will depend on treatment objectives. For example, if the effluent is to be disposed of in leachfields which are in a clayey soil, then only BOD and TSS removal are of importance to prevent the formation of biological mats, and a larger-size filter medium and higher hydraulic and organic loading rates can be used. If, on the other hand, the removal of viruses is important, then lower hydraulic and organic loading rates must be used (see Fig. 11-7).

Types and sizes of filter media. Sand, as noted previously, is the most commonly used filter medium. Most sands are mined from river deposits. The gradation of sand as well as other granular filter media is determined by using a series of

TABLE 11-5
Fecal coliform removal in ISF systems

		Fecal coliform, No./100 mL		
Study and location	**Type of system and pretreatment**	**Influent**	**Effluent**	**% removal**
Sauer (1976), Wisconson	Home/septic tank	590,000	650	99.9
Ronayne et al. (1984), Oregon	Home/septic tank	260,000	407	99.8
Effert et al. (1985), Ohio	Home/septic tank	219,000	1600	99.3
Effert et al. (1985), Ohio	Home/aerobic unit	5890	123	97.9

TABLE 11-6
Design criteria for intermittent sand filters

	U.S. customary units			SI units		
Design factor	**Unit**	**Range**	**Typical**	**Unit**	**Range**	**Typical**
Filter Medium						
Material			Durable washed sand			
Effective size	mm	0.25–0.75	0.35	mm	0.25–0.75	0.35
Depth	in	18–36	24	mm	450–900	600
Uniformity coefficient		<4	3.5		<4	3.5
Underdrain						
Type			Slotted or perforated drain pipe			
Size	in	3–4	4	mm	75–100	100
Slope	%	0–0.1	0	%	0–0.1	0
Pressure distribution						
Pipe size	in	1–2	1.5	mm	25–50	38
Orifice size	in	$\frac{1}{8}$–$\frac{1}{4}$	$\frac{1}{8}$	mm	3–6	3
Head on orifice	ft	3–6	5	m	1–2	1.6
Lateral spacing	ft	1.0–4	2	m	0.5–1.2	0.6
Orifice spacing	ft	1.0–4	2	m	0.5–1.2	0.6
Design parameters						
Hydraulic loading*	gal/ft²·d	1–1.5	1.25	L/m²·d	40–60	50
Organic loading	lb/ft²·d	0.0005–0.002	<0.001	kg BOD/m²·d	0.0025–0.01	<0.005
Dosing						
Frequency	times/d	12–48	18 min	times/d	12–48	18 min
Volume/orifice	gal/orif·dose	0.15–0.30	0.25	L/orif·dose	0.6–1.1	0.9
Dosing tank volume	flow/d	0.5–1.5	1.0	flow/d	0.5–1.5	1.0

*Based on peak flow.

TABLE 11-7
Designation and size of opening of U.S. sieve sizes

Sieve size or number	Size of opening		Sieve size or number	Size of opening	
	in	**mm**		**in**	**mm**
$\frac{3}{8}$ in	0.375*	9.51*	25	0.0280*	0.710*
$\frac{1}{4}$ in	0.250*	6.35*	30	0.0234	0.595
4	0.187	4.76	35	0.0197*	0.500*
6	0.132	3.36	40	0.0165	0.420
8	0.0937	2.38	45	0.0138*	0.350*
10	0.0787*	2.00*	50	0.0117	0.297
12	0.0661	1.68	60	0.0098*	0.250*
14	0.0555*	1.41*	70	0.0083	0.210
16	0.0469	1.19	80	0.0070*	0.177*
18	0.0394*	1.00*	100	0.0059	0.149
20	0.0331	0.841			

*Size does not follow the ratio $(2)^{0.5}$.

graded sieves (i.e., screens) of diminishing size in the ratio of $2^{0.5}$ (see Table 11-7). The quantity of sand in the bottom pan and retained on each sieve is weighed and the successive amounts held between adjacent sieves are added to obtain the cumulative weight. The cumulative weights are then converted to percentages of weight equal to or less than the size of the overlying sieve. The cumulative percentages are then plotted versus the size of the sieve opening on arithmetic-log or probability-log paper to obtain the size distribution. The analysis of sieve data is illustrated in Example 11-2. Additional information on the characterization of various filter materials with respect to size, surface area, and chemical properties may be found in Dallavalle (1948).

The effective size d_{10} and uniformity coefficient (UC) are the principal characteristics of the filtering material that affect the design and operation of low-rate intermittent and recirculating and high-rate granular and porous medium filters. For solid filtering mediums such as sand, the effective size d_{10} is defined as the 10 percent size by weight, determined by a wet-test sieve analysis (ASTM C117-95). The wet-test sieve analysis is the recommended method because the amount of fine material cannot be determined adequately with the dry-test method. It is interesting to note that the 10 percent size by weight corresponds quite closely to the median value by count. The uniformity coefficient is defined as the 60 percent size divided by the 10 percent size:

$$\text{UC} = \frac{d_{60}}{d_{10}} \tag{11-2}$$

Grain size affects the time of passage of a liquid through the medium. The uniformity coefficient is used to determine whether the individual particles are of similar size or whether there is a wide range of particle sizes. Determination of the effective size and the uniformity coefficient is illustrated in Example 11-2.

EXAMPLE 11-2. SAND SIZE ANALYSIS. Given the following data, obtained from a wet-test sieve analysis of washed river sand, determine the effective size and the uniformity coefficient.

Sieve size or number	Size of opening, mm	Percent retained
$\frac{3}{8}$ in	9.53	0
4	4.75	6.2
8	2.38	16.7
16	1.18	24.8
30	0.60	24.1
50	0.30	18.2
100	0.15	7.5
Pan		2.5

Solution

1. Compute the cumulative percent passing by weight and plot the resulting values versus sieve size opening.

a. Compute the cumulative percent passing:

Sieve size or number	Size of opening, mm	Percent retained	Cumulative percent passing
$\frac{3}{8}$ in	9.53	0	100
4	4.75	6.2	93.8
8	2.38	16.7	77.1
16	1.18	24.8	52.3
30	0.60	24.1	28.2
50	0.30	18.2	10.0
100	0.15	7.5	2.5
Pan		2.5	

b. Plot the cumulative percent passing versus the corresponding sieve size. Two different methods of plotting the data are presented below.

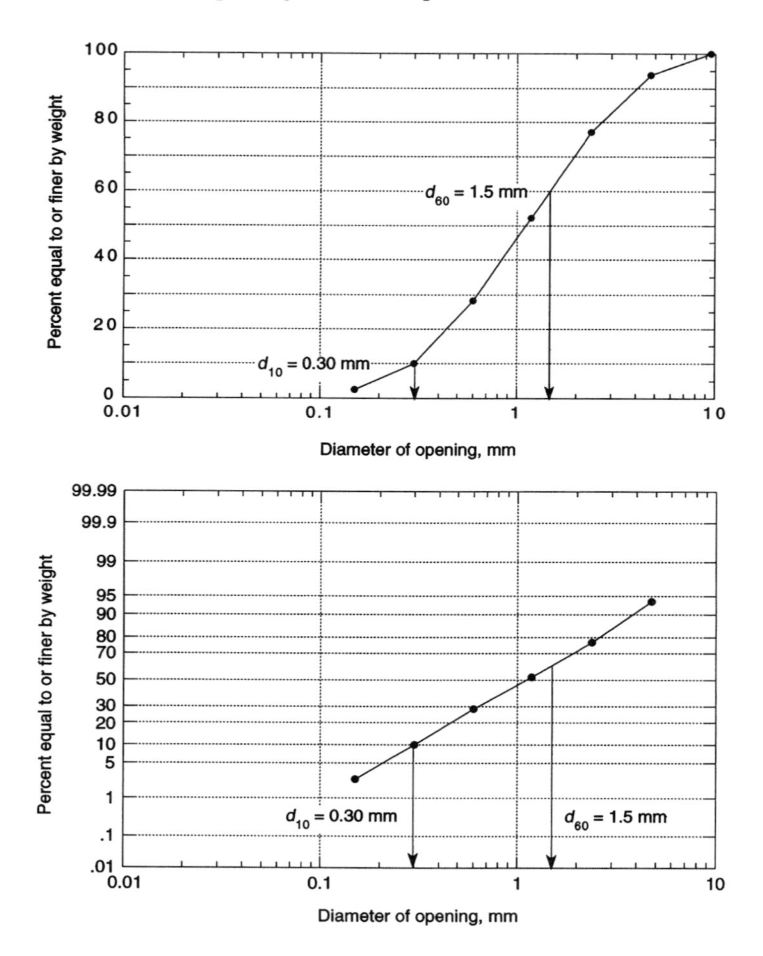

2. Determine the effective size and the uniformity coefficient.
   a. The d_{10} and d_{60} sizes obtained from the plot are:

$$d_{10} = 0.30 \text{ mm}$$

$$d_{60} = 1.50 \text{ mm}$$

   b. The UC value is:

$$\text{UC} = \frac{d_{60}}{d_{10}} = \frac{1.5}{0.3} = 5.0$$

The importance of the effective size can be appreciated by considering Hazen's formula for the velocity of flow of water through a porous medium under saturated flow conditions.

$$v_h = C(d_{10})^2 \frac{h}{L}\left(\frac{T + 10°}{60}\right) \tag{11-3}$$

where v_h = superficial (approach) filtration velocity, m/d
C = coefficient of compactness (varies from 700 to 1000 for new sand, and from 500 to 700 for sand that has been in use for a number of years)
d_{10} = effective size of medium, mm
h = headloss, m (ft)
L = depth of filter bed or layer, m (ft)
T = temperature, °F

Two very important issues in the selection of a sand for use in ISFs are: (1) the use of a durable sand and (2) sand that is free of dust, fine organic particles, fine silt, and clay particles. Dirty sand has been the cause of poor operation and clogging of ISF systems on numerous occasions. In time, the fine material will migrate to the bottom of the filter where it can form a restrictive layer. The impact of fine material is illustrated in Example 11-3. Ideally, the sand should be washed thoroughly and kiln-dried; however, it is sometimes difficult, if not impossible, to find kiln-dried sand. Recommended sand sizes are given in Table 11-8.

TABLE 11-8
Typical sand gradation for ISF*

Sieve size or number	Size of opening, mm	Cumulative percent passing	
		Range	Typical
$\frac{3}{8}$ in	9.5	100	100
4	4.75	40–100	99
10	2.0	62–100	84
16	1.18	45–82	54
30	0.6	25–55	25
50	0.3	5–20	6
60	0.25	0–10	4
100	0.15	0–4	Trace

*From Orenco Systems, Inc.

FIGURE 11-10
Jar test used to assess cleanliness of filtering material. If the layer of solids that forms on the top of the filter medium after shaking and settling for 30 minutes exceeds $\frac{1}{16}$ in, the filter material should be rejected.

The cleanliness of the sand or gravel can be assessed qualitatively in the field with a simple jar test as illustrated in Fig. 11-10. A quart jar is filled half full of the material to be tested and water is added to fill the jar. The material is then shaken vigorously and then allowed to settle. If, after settling for 30 minutes, a perceptible layer of material [e.g., greater than $\frac{1}{16}$ in (1.5 mm)] has accumulated on top of the filter material, the sand is not clean enough to be used as filtering material.

EXAMPLE 11-3. EFFECT OF SAND SIZE ON FILTRATION RATE. An unwashed filter sand with an effective size of 0.3 mm is to be used in an intermittent sand filter. If the size of silt present in the sand is 0.03 mm, estimate the difference in acceptance rates, under saturated flow conditions, if the silt were to migrate and form a layer on the bottom of the filter.

Solution

1. Determine the velocity of flow through a layer of each material using Eq. (11-3), assuming saturated flow and a unit value for the term h/L. Use a C value of 800 and assume the temperature of the wastewater is 50°F.

$$v_h = C(d_{10})^2 \frac{h}{L}\left(\frac{T + 10^\circ}{60}\right)$$

a. The velocity for the 0.3-mm sand is:

$$v_h = 800(0.3)^2(1)\left(\frac{50 + 10^\circ}{60}\right) = 72 \text{ m/d}$$

b. The velocity for the 0.03-mm silt is:

$$v_h = 800(0.03)^2(1)\left(\frac{50 + 10°}{60}\right) = 0.72 \text{ m/d}$$

Comment. The deleterious effect of fine silt that may be present in filter sand can be appreciated from the above computations, even if the actual flow-through were unsaturated. Similarly, the presence of excessive amounts of fine sand can also have deleterious effects on the performance of intermittent sand filters.

Depth of filter bed. Sand depths have varied from 18 to 48 in (450 to 1200 mm) with 24 in (500 mm) being used most commonly. Historically, deeper sand beds were used because the top 1 to 2 in of sand was removed periodically; however, the added depth is usually not warranted. When shallow 18-in (450-mm) beds are used, it has been found that the removal of BOD and TSS remains high while the degree of nitrification that can be achieved is reduced significantly. Additional discussion on the role of sand depth on treatment efficiency of ISFs may be found in Peeples et al. (1991).

Hydraulic loading and application rate. Typical hydraulic loading rates (L_W) for ISFs, based on *peak flow,* are in the range from 1 to 2 gal/ft^2·d (40 to 80 mm/d). Although higher hydraulic loading rates have been used, loading rates of 4 gal/ft^2·d (160 mm/d) resulted in filter clogging for a fine sand (0.29 mm) that was dosed 24 times/d (Nor, 1991; Darby et al., 1996). Hydraulic loading rates between 2 and 6 gal/ft^2·d (80 and 240 mm/d) have been used with larger sand sizes. Experimental loadings of 10 gal/d·ft^2 (400 mm/d) are being studied in Oregon (Ball, 1996).

Peak flow is used for design because the long-term performance of the filter is dependent on restricting the amount of organic matter added per dose. If the average flow were used, the amount of organic matter added per dose would exceed the recommended value 50 percent of the time. The recommended peaking factor for ISFs is 2.5. As noted in Chap. 4, in many states, the required design flows are based on an allowance of 120 to 150 gal/d per bedroom, which in theory accounts for peak flow. Again, as noted in Chap. 4, bedrooms do not generate wastewater, people do. Thus, to make the design of ISFs more rational, and to be consistent with the design of other wastewater management facilities, it is recommended that a per capita design allowance, based on *peak flow,* be used for design. For example, if a per capita allowance of 50 gal/d is used, then the approximate relationship between the number of bedrooms and the number of persons is as shown in Table 11-9, which indicates that the use of an allowance is less conservative for one or two bedrooms. The corresponding allowances for the state of Washington are reported in Table 11-9.

Although hydraulic loading rates are appropriate for the design of ISFs, a more appropriate term when considering the performance of ISFs is the hydraulic application rate (HAR), expressed in millimeters of liquid per dose (Emerick et al., 1997). The HAR is defined as

$$\text{HAR, mm/dose} = \frac{\text{Hydraulic loading rate, } L_W\text{, mm/d}}{\text{Dosing frequency, DF, dose/d}} \tag{11-4}$$

TABLE 11-9
Comparison of design flow based on a per capita allowance of 50 gal/d times a peaking factor versus design flow based on a per bedroom allowance*

Number of bedrooms	Number of persons	Flowrate, gal/capita·d†	Peaking factor	Design flow based on peak per capita flow, gal/d	Design flow based on per bedroom allowance, gal/d
1	2	50	2.5	250	150
2	3	50	2.5	375	300
3	4	50	2.5	500	450
4	5	50	2.5	625	600

*Adapted from Table 4-20.
†Flowrate per capita will vary with local conditions.

The importance of the HAR can be appreciated by reviewing the information presented in Table 11-10. At a L_W of 40 mm/d (1 gal/ft^2·d), if the filter were dosed once per day, the volume of the dose would represent 217 percent of the field capacity of the medium. Clearly, most of the applied liquid will move down through and out of the filter. If, on the other hand, the filter were dosed 24 times/d the volume of the dose would represent about 9 percent of field capacity. At 9 percent of

TABLE 11-10
Analysis of volume per dose for various hydraulic loading rates and dosing frequencies for intermittent sand filters*

Hydraulic loading rate, mm/d (gal/ft²·d)	Dosing frequency, times/d	Hydraulic application rate		Field capacity filled,† %
		mm/dose	gal/ft²·dose	
40 (1)	1	40	1	217
	2	20	0.5	107
	4	10	0.25	53
	8	5	0.12	26
	12	3.3	0.083	18
	24	1.67	0.042	9.0
81 (2)	1	81	2	427
	2	40	1	217
	4	20	0.5	107
	8	10	0.25	53
	12	6.75	0.12	26
	24	3.38	0.083	18
163 (4)	1	163	4	855
	2	82	2	427
	4	41	1	217
	8	20	0.5	107
	12	14	0.33	71
	24	6.79	0.17	36

*For 1 ft² of surface area and depth of 1.25 ft.
†Five percent as volumetric water content (water volume/total volume) from Bouwer (1978).

the field capacity, the applied liquid will flow over the filter medium in a thin layer, maximizing the opportunity for absorption of constituents from the applied liquid and the transfer of oxygen from the air in the interstices. It should also be noted that the field capacity of a filtering medium will increase as the film thickness within the filter increases.

Organic loading rate. The organic loading rate is composed of the soluble and particulate organic matter applied to the filter. The organic loading rate is typically expressed, on an area basis, as lb BOD or COD/$ft^2 \cdot d$ (kg BOD or COD/$m^2 \cdot d$). In some references, the organic loading rate is expressed, on a volumetric basis, as lb BOD or COD/$ft^3 \cdot d$ (kg BOD or COD/$m^3 \cdot d$). The amount of organic material added to the filter with each dose must be such that the microorganisms in the biological film can process the mass of organic matter added without accruing additional mass between doses (see Example 11-4). Although the appropriate organic loading rates are not well defined, typical values are in the range from 0.0005 to 0.002 lb BOD/$ft^2 \cdot d$ (0.0025 to 0.01 kg BOD/$m^2 \cdot d$).

EXAMPLE 11-4. ORGANIC LOADING OF INTERMITTENT SAND FILTER. Estimate the amount of material that can be processed in an intermittent sand filter with a porosity of 0.40. Assume the bacterial film occupies 5 percent of the void space in the filter bed and that the specific gravity of the biological film is 1.125.

Solution

1. Determine the mass of the biological film in the sand filter, based on 1.0 ft^3 of filter bed.
   a. Volume of bacterial film:

$$\begin{aligned}\text{Volume of film} &= (1.0\ \text{ft}^3/\text{ft}^3)(7.48\ \text{gal/ft}^3)(3.785\ \text{L/gal})(0.4)(0.05)\\ &= 0.57\ \text{L/ft}^3\end{aligned}$$

   b. Mass of bacterial film:

$$\text{Mass} = (0.57\ \text{L/ft}^3)(1.125)(1.0\ \text{kg/L}) = 0.64\ \text{kg/ft}^3 = 640\ \text{g/ft}^3$$

2. Assuming the endogenous respiration coefficient is $0.05d^{-1}$ (see Chap. 7) and that active microorganisms make up 10 percent of the bacterial film, the amount of waste that can be processed per day due to endogenous respiration is:
   a. Mass of active bacteria in the bacterial film:

$$\text{Mass of bacteria} = (640\ \text{g/ft}^3)(0.1) = 64.0\ \text{g/ft}^3$$

   b. Mass of organic matter that can be processed to maintain the bacteria is:

$$\text{Mass of organic matter} = (64.0\ \text{g/ft}^3)(0.05/\text{d}) = 3.2\ \text{g/ft}^3\cdot\text{d}$$

3. Convert the amount of organic matter processed per day to an organic loading rate applied per square foot:

$$\begin{aligned}\text{Organic loading rate} &= (3.2\ \text{g/ft}^3\cdot\text{d})/(1.0\ \text{ft}^2 \times 454\ \text{g/lb}) = 0.007\ \text{lb/ft}^2\cdot\text{d}\\ &= 0.034\ \text{kg/m}^2\cdot\text{d}\end{aligned}$$

Comment. The above computation illustrates the point that the performance of intermittent sand filters is extremely sensitive to the active mass of microorganisms in the film within the filter.

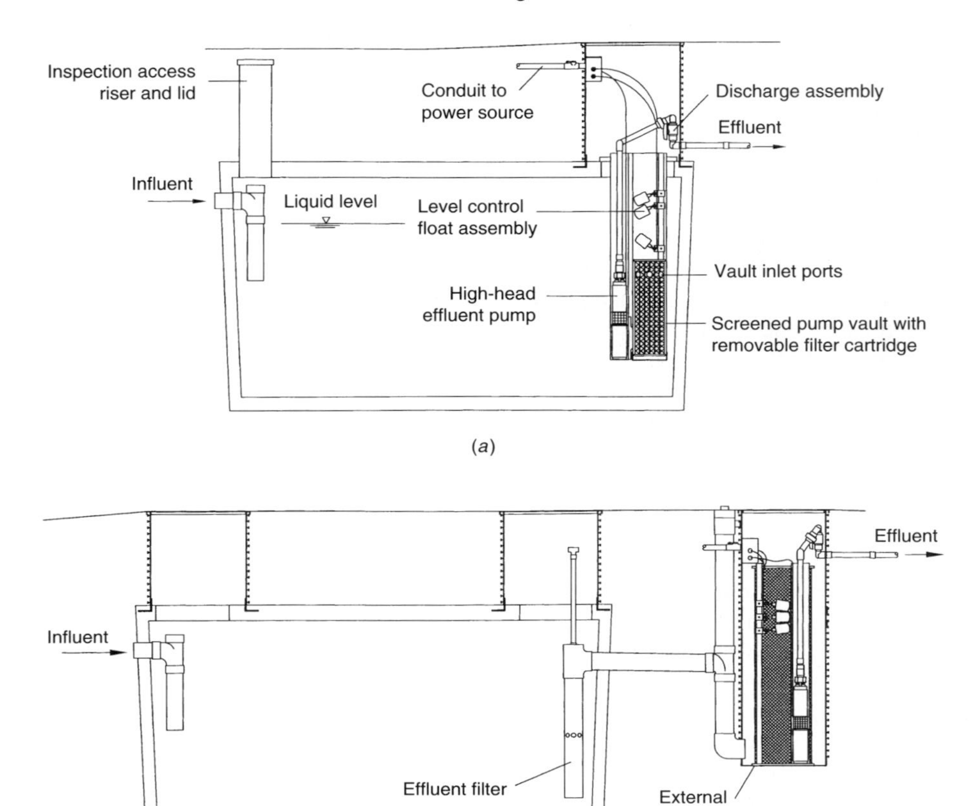

FIGURE 11-11
Location of filter dosing pump: (*a*) in septic tank filter vault and (*b*) in external pump basin (courtesy Orenco Systems, Inc.). Pumps used for effluent dosing are typically the same as those used in STEP systems (see Chap. 6).

Dosing frequency. As noted previously, increasing dosing frequency from 1 to 4 times/d to more than 12 times/d has improved the performance of sand filters, particularly those with larger-diameter sands. It is recommended that a minimum of 18 doses/d be used for normal septic tank effluent. Should the BOD concentration in the septic tank effluent be elevated (e.g., greater than 200 mg/L), it is recommended that the number of doses be increased to 24 times/d. When programmed dosing is used, the timer is typically set to dose the filter 18 times/d. To enhance treatment in ISFs, the amount of flow discharged per orifice per dose should be limited to about 0.25 gal/d (Bounds, 1996).

The filter dosing pump can be located either in the septic effluent filter vault or in a separate external pump basin (see Fig. 11-11). The dosing cycle is controlled from a control panel (see Fig. 11-12) using float switches and one or more programmable

FIGURE 11-12
Typical control panel used to house the electrical elements that control the operation of an intermittent sand filter. See Fig. 11-13 for typical schematic wiring diagram.

timers (see Fig. 11-13). By controlling the on and off times with the programmable timer, the filter is dosed more or less uniformly throughout the day. Further, short-term hydraulic surges (e.g., the discharge from a washing machine) are stored in the septic tank or external pump basin. By limiting the applied dose and maintaining a fixed off time, the performance of the filter is enhanced (Bounds, 1996). Although dosing controlled only by float switches is often used, it is not recommended, because the resulting wastewater doses and resting times are unequal, which can lead to premature clogging of the filter.

When the dosing frequency is set at 18 times/d, the corresponding total flow is equal to 1.5 times the average flow. Thus, at average flow the filter will be dosed 12 times/d. When an extreme event occurs (e.g., peak flow greater than 1.5 times the average flow) the normal cycle is completed and, depending on the adjustment of the upper alarm float, 40 to 60 gal will be dosed. The timer can also be set to dose a fixed volume on an hourly basis over the remaining 6 h until the next timed cycle is initiated. A programmable control unit is currently under development which will make it possible to adjust the volume per dose on a 24-h basis (Bounds, 1996). An effective alternative is to increase the size of the septic tank, or the external pump basin, so that they will function as equalization basins. The timer is then set to dose the filter 24 times per day, based on average flow. In many states the storage volume required in the external pump basin is 2 times the daily average flowrate. Extreme flow events are handled by pumping a fixed flow in response to the high-water alarm float.

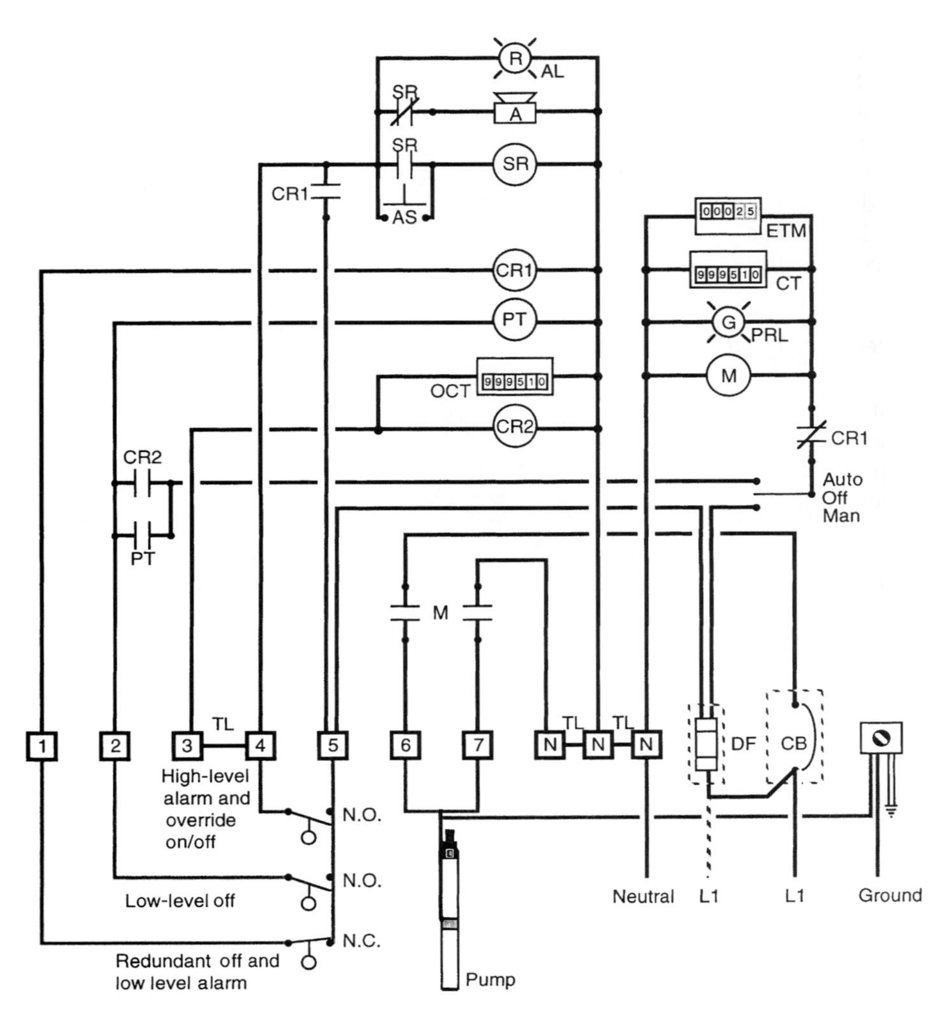

— = Factory wire
— = Field wire
- - - = Alternate field wire
A = Audio alarm, 115 Vac
AL = Alarm light
AS = Audio silence switch
CB = Circuit breaker
CR1 = Redundant off relay
CR2 = PT override relay
DF = Disconnect fuse

M = Motor contactor
OCT = High-level override event counter
PT = Programmable timer
SR = Silence control relay
TL = Terminal link

Options

ETM = Elapsed time meter
CT = Event counter
PRL = Pump run light

FIGURE 11-13
Typical schematic wiring diagram for an intermittent sand filter (courtesy Orenco Systems, Inc.).

Meters and counters included in a typical control panel include: an elapsed time meter for the septic tank pump, an elapsed time meter for the filter pump, a dose counter, and an override counter. The elapsed time meters are used to record the total time the effluent and filter pumps have been running. The dose counter is used to record the number of doses that have been applied to the filter. The override counter is used to record the number of times the pump had to be turned on to accommodate flows in excess of the normal daily flow. Additional details on pumps and controls may be found in Bounds (1996).

System Hydraulics for Intermittent Filters

The hydraulic analysis for an intermittent packed-bed filter involves determination of the system head capacity curve for the given system configuration so that a proper pump can be selected and the difference in discharge between orifices can be determined.

System head discharge curve. The elements that must be considered in the development of the system head discharge curve are summarized below, and identified graphically in Fig. 11-14.

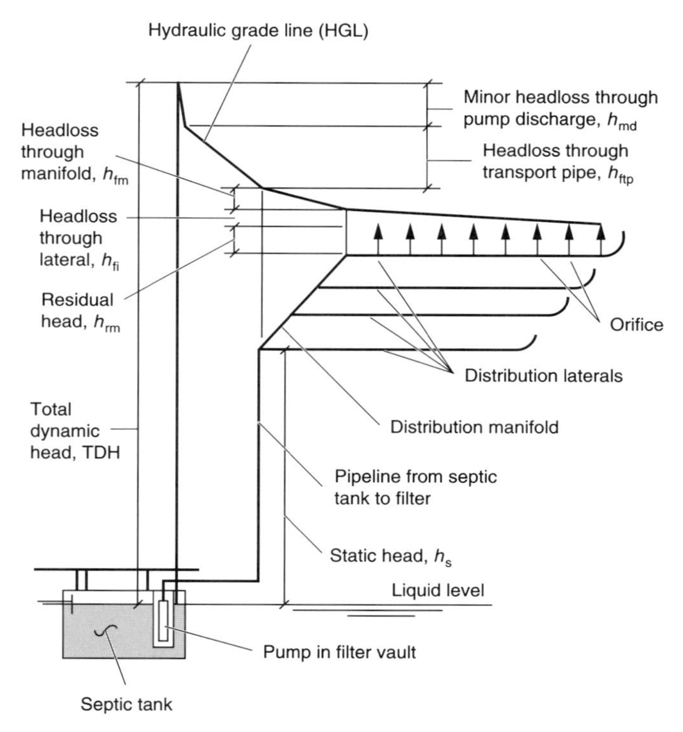

FIGURE 11-14
Definition sketch for the energy gradient in an intermittent sand filter system.

1. System static head, h_s
2. Minor headloss in the pump discharge assembly, h_{md}
3. Friction loss in the transport pipe from the septic tank filter vault (or external pump basin) to the filter, h_{ftp}
4. Headloss in the manifold pipe to which the transport pipe and filter pipe laterals are connected, h_{fm}
5. Headloss in the laterals, h_{fl}
6. Residual head at furthermost orifice in lateral, h_{rm}

The total dynamic head (TDH) for the system is

$$\text{TDH} = h_s + h_{md} + h_{ftp} + h_{fm} + h_{fl} + h_{rm} \tag{11-5}$$

The minor headlosses are computed as a function of the velocity head. The headloss due to friction can be computed by the Darcy-Weisbach equation or the Hazen-Williams equation. The Hazen-Williams equation, commonly used for plastic pipe, as given previously in Chap. 6 is:

$$h_{fp} = 10.5(L)\left(\frac{Q}{C}\right)^{1.85}(D^{-4.87}) \tag{6-5}$$

where h_{fp} = headloss through pipe, ft
L = length of pipe, ft
Q = pipe discharge, gal/min
C = Hazen-Williams discharge coefficient, 150 for plastic pipe
D = inside diameter of pipe, in (see Table 6-4, Chap. 6)

For a distribution lateral, it can be shown that the headloss between the first and last orifice in a series of evenly spaced orifices is approximately equal to one-third of the headloss that would occur if the total flow were to pass through the same length of distribution pipe without orifices (Fair and Geyer, 1954). Thus:

$$h_{fdp} = 1/3h_{fp} \tag{11-6}$$

where h_{fdp} = actual headloss through distribution lateral with orifices, ft.

Variations in orifice discharge. The difference in discharge between orifices in a distribution manifold can be assessed as follows. Assume the discharge from any orifice is to be held to a value mq_1, where m is a decimal value less than 1 and q_1 is the discharge from the first orifice. The discharge from orifice n can be computed by the following equation.

$$q_n = 2.45C(D^2)\sqrt{2gh_n} \tag{11-7}$$

where q_n = discharge from orifice n, gal/min
C = orifice discharge coefficient [usually = 0.63 for holes drilled in plastic pipe in the field; Ball (1996)]
D = diameter of orifice, in
g = acceleration due to gravity, 32.2 ft/s^2
h_n = head on orifice n, ft

The head on orifice n is equal to

$$h_n = \left[\frac{1}{(2.45CD^2)^2 2g}\right] q_n^2 = kq_n^2 = k(mq_1)^2 = m^2 kq_1^2 = m^2 h_1 \quad (11\text{-}8)$$

where k = constant
q_1 = discharge from first orifice, gal/min
q_n = discharge from the nth orifice, gal/min
h_1 = head on first orifice, ft
m = constant with decimal value less than 1.0

The head lost between orifices 1 and n, which corresponds to the headloss in the distribution pipe between orifices 1 and n, computed from Eq. (11-8), is

$$\Delta h_{(1-n)} = h_{\text{fdp}} = h_1 - h_n \quad (11\text{-}9)$$

The difference in discharge between orifices 1 and n, for a given distribution pipe and orifice size, can now be determined by using Eqs. (11-6) through (11-9). If the computed value of m is too low (e.g., less than 0.95 or some other acceptable value, typically 0.95 to 0.90), the size of the distribution pipe can be increased. Use of these equations is illustrated in Example 11-5. Additional details on pressure dosing may be found in Ball (1996). It should also be noted that it is important that the orifices be drilled with a drill press. Often, contractors will use a hand-held drill to drill the orifices in the field, which usually will result in orifices with large enough differences in diameters to cause poor distribution (Ball, 1996).

Construction Considerations

Because contractors often lack adequate training in the proper installation of ISFs, the engineer or designer should provide inspection to ensure that proper materials and construction techniques are used. Important construction considerations include sand cleanliness, the use of proper liner material, proper placement and assembly of the individual components, and proper construction and testing of the distribution manifold. The sand in an ISF is usually retained by a PVC liner (see Fig. 11-15). A liner thickness of 30 mil (0.76 mm) is recommended as being tough enough to withstand handling during construction, and is easily repaired if a puncture occurs. Heavier liners are difficult to work with and do not offer improved performance. PVC liners can be patched, much like an inner tube, with cement available from liner suppliers (Ball, 1991). The easiest approach to the construction of ISFs is to purchase a kit containing all the necessary components for a predetermined size of filter from a supplier.

Operation and Maintenance

Properly sized intermittent sand filter systems require some limited operation and maintenance. Routine monitoring involves the periodic checking of the various counters and elapsed-time meters. Noting the elapsed time the effluent pump has been on is extremely helpful in identifying the presence of leaky fixtures (e.g., the

(*a*)

(*b*)

FIGURE 11-15
Views of intermittent sand filter under construction: (*a*) buried and (*b*) aboveground.

discharge from a leaky toilet can amount to 2200 gal/d). Similarly, noting the elapsed time the filter pump has been on is extremely helpful in identifying the presence of extraneous flows (Jones, 1997). Other potential operating problems can be identified by checking the counters and meters. A detailed listing of probable causes for readings higher or lower than expected has been prepared by Bounds (1996). The use of telemetry for the monitoring and control of pump functions from a centralized location is discussed in Chap. 15.

The discharge head on the orifices should be checked annually, and the distribution manifold should be flushed every year. A typical manifold flushing arrangement is shown in Fig. 11-16. When the orifice shield is removed, the height to which water rises from the orifice in the lateral is used to check the head on the orifices. The manifold is flushed by opening the valve and allowing the solid material retained in the distribution manifold to be flushed onto the surface of the filter.

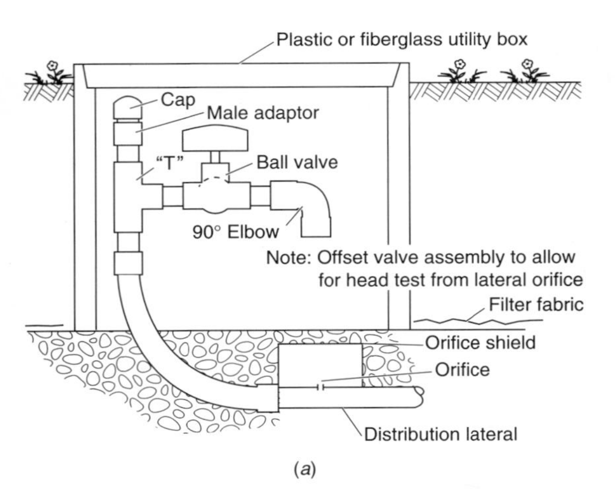

(*a*)

(*b*)

FIGURE 11-16
Typical distribution lateral flushing system: (*a*) schematic (courtesy P. Tanner) and (*b*) view of typical flushing station.

EXAMPLE 11-5. DESIGN OF AN INTERMITTENT SAND FILTER FOR A SINGLE DWELLING. Size and lay out an intermittent sand filter and distribution system for an individual three-bedroom residence in a nonsewered area. Determine organic loading rate, the flowrate, the duration of an individual dose, and the discharge per orifice. Check the difference in the discharge between the orifices. If the difference in the discharge between the orifices is greater than 5 percent $[(1 - m) \times 100]$ in either case, the distribution system should be resized. The septic tank is sized to provide a storage capacity of 2.5 times the average flow. Finally, estimate the total dynamic head. Assume the following conditions apply:

1. Average occupancy = 3.0 persons/d
2. Average wastewater flow = 50 gal/capita·d
3. Assumed peaking factor = 2.5 (see Table 4-20)
4. Size of septic tank = 1500 gal (4.5 m^3)
5. Effluent BOD from septic tank = 130 mg/L
6. Effluent TSS from septic tank = 30 mg/L
7. Hydraulic application rate = 1.25 gal/ft^2·d (51 L/m^2·d) based on peak design flow
8. Sand filter dose rate per day = 24 times/d
9. Distribution system orifice size = $\frac{1}{8}$ in (3.2 mm)
10. Orifice discharge head = 5 ft minimum (1.5 m)
11. Distance from septic tank to sand filter = 60 ft
12. Difference in elevation between septic tank and sand filter = 10 ft

Solution

1. Determine the size of the sand filter.
   a. Determine the average and peak flow rates:

$$Q_{\text{ave}} = (3.0 \text{ persons}) \times 50 \text{ gal/capita·d} = 150 \text{ gal/d}$$

$$Q_{\text{peak}} = 150 \text{ gal/d} \times 2.5 \text{ (see Table 4-20)} = 375 \text{ gal/d}$$

   b. Determine the required area of the sand filter:

$$\text{Area} = (375 \text{ gal/d})/(1.25 \text{ gal/ft}^2{\cdot}d) = 300 \text{ ft}^2$$

   c. Use a filter 17 ft × 18 ft. Check area: (17 × 18) = 306 ft^2, OK.

2. Determine the organic loading rate.

$$L_{org} = \frac{(1.25 \text{ gal/ft}^2{\cdot}\text{d})(3.785 \text{ L/gal})(130 \text{ mg/L})}{(1000 \text{ mg/g})(454 \text{ g/lb})}$$

$$= 0.00135 \text{ lb BOD/ft}^2{\cdot}\text{d}$$

3. Lay out sand filter and the effluent distribution system. Use a spacing of 1.5 ft between distribution pipes and orifices, and a wallspace of 1 ft.
   a. Determine the number of laterals for a nominal spacing of 1.5 ft:

$$\text{Number of laterals} = (17 - 2)/1.5 + 1 = 11 \text{ laterals}$$

   b. Determine the number of orifices for a nominal spacing of 1.5 ft:

$$\text{Number of orifices} = (18 - 1.5)/1.5 = 11 \text{ orifices/lateral}$$

   c. The layout of the sand filter and distribution system is shown below.

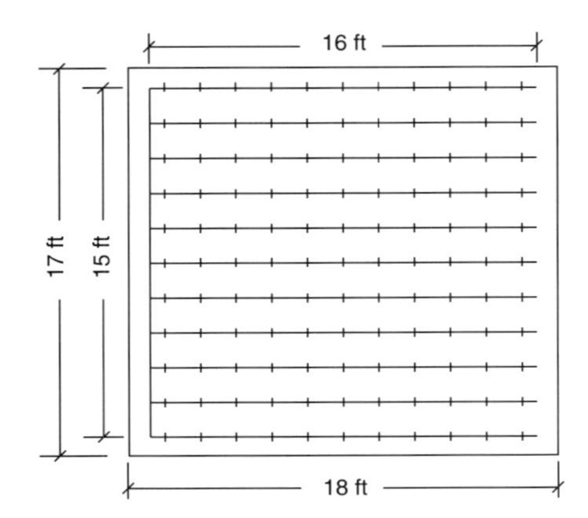

4. Determine the quantity of flow per dose and the flow in each lateral.
   a. Determine quantity of flow discharged per dose:

$$\text{Flow/dose} = (150 \text{ gal/d})/(24 \text{ dose/d}) = 6.25 \text{ gal/dose}$$

   b. Determine quantity of flow discharged per lateral per dose:

$$\text{Flow/lateral} = (6.25 \text{ gal/dose})/11 \text{ laterals} = 0.57 \text{ gal/lateral}\cdot\text{dose}$$

5. Determine the rate of discharge from an orifice, assuming 5 ft of residual head at the orifice. The rate of discharge from an orifice is computed by Eq. (11-7):

$$\begin{aligned} q_n &= 2.45C(D^2)\sqrt{2gh_n} \\ &= 2.45(0.63)(0.125)^2\sqrt{2(32.2)5} = 0.43 \text{ gal/orifice}\cdot\text{min} \end{aligned}$$

6. Determine the duration of the flow and the volume of flow discharged per orifice.
   a. The total flowrate into each lateral, based on 11 orifices/lateral, is:

$$\begin{aligned} \text{Flowrate/lateral} &= (0.43 \text{ gal/min}\cdot\text{orifice})(11 \text{ orifice/lateral}) \\ &= 4.73 \text{ gal/min}\cdot\text{lateral} \end{aligned}$$

$$Q_T = (4.73 \text{ gal/min}\cdot\text{lateral})(11 \text{ laterals}) = 52.0 \text{ gal/min}$$

   b. Determine the duration of flow:

$$\begin{aligned} \text{Duration} &= (6.25 \text{ gal/dose})/(52.0 \text{ gal/min}) \\ &= 0.12 \text{ min/dose} = 7.2 \text{ s/dose} \end{aligned}$$

   c. Determine discharge volume per orifice per dose:

$$\begin{aligned} \text{Volume/dose} &= (0.57 \text{ gal/lateral}\cdot\text{dose})/(11 \text{ orifices/lateral}) \\ &= 0.05 < 0.25 \text{ gal/orifice}\cdot\text{dose}, \quad \text{OK (see Table 11-6)} \end{aligned}$$

7. Determine the headloss in a lateral distribution pipe. As a first try, use a nominal 0.75-in (19-mm) diameter plastic pipe [actual ID = 0.930 in (24 mm); see Table 6-4]. It should be noted that in a float-controlled pumped system, the peak flow will affect only the length of time the dosing pump is on. If the operation of the pump is controlled with a programmable timer, the peak surges, as discussed previously, are stored in the septic tank and approximately equal doses are applied throughout the day.

a. The headloss in pipe without orifices is determined from Eq. (6-5):

$$h_{fp} = 10.5(L_{1-n})\left(\frac{Q}{C}\right)^{1.85} D^{-4.87}$$

$$= 10.5(15.5)\left(\frac{4.73}{150}\right)^{1.85} 0.930^{-4.87} = 0.39 \text{ ft}$$

b. The headloss in the lateral pipe with orifices is determined from Eq. (11-6*b*):

$$h_{fdp} = \tfrac{1}{3}h_{fp} = \Delta h_{(1-n)}$$

$$= \tfrac{1}{3}(0.39) = 0.13 \text{ ft}$$

8. Determine the difference in the discharge between the first and last orifice in each lateral.
   a. The head on the first orifice, from Eq. (11-9), is:

$$\Delta h_{(1-n)} = h_1 - h_n$$

$$h_1 = h_n + \Delta h_{(1-n)}$$

   b. Determine the value of m from Eq. (11-8):

$$h_n = m^2 h_1$$

$$m = \sqrt{\frac{5.0}{5.0 + 0.13}} = 0.987$$

   The difference in the discharge between the first and last orifice in each lateral is about 1.3 percent $[(1 - 0.987) \times 100]$, which is below the 5 percent value specified. If the orifices are drilled with a hand drill in the field, differences in the discharge between orifices as high as 5 percent can be expected. Improved accuracy is obtained with a drill press.

9. Estimate the total dynamic head. Assume a nominal 1.5-in (38-mm) diameter plastic pipe [actual ID = 1.720 in (44 mm)] will be used to supply the filter laterals.
   a. The headloss in the 1.5-in pipe from the septic tank to the sand filter, computed by Eq. (6-5), is:

$$h_{fp} = 10.5\,(L)\left(\frac{Q}{C}\right)^{1.85} D^{-4.87}$$

$$L = 60 \text{ ft}$$

$$Q_T = 52.0 \text{ gal/min}$$

$$h_{fp} = 10.5\,(60)\left(\frac{52.0}{150}\right)^{1.85} (1.720)^{-4.87} = 6.3 \text{ ft}$$

   b. Estimate the headloss in the 1.5-in distribution manifold used to connect the laterals:

$$\text{Length of distribution manifold} = 15 \text{ ft}$$

$$\text{Flow in the distribution manifold} = 52.0 \text{ gal/min}$$

   Headloss in distribution manifold without laterals:

$$h_{fp} = 10.5(15)\left(\frac{52.0}{150}\right)^{1.85} (1.720)^{-4.87} = 1.6 \text{ ft}$$

The headloss in distribution manifold with laterals:

$$h_{\text{fdp}} = \tfrac{1}{3}1.6 = 0.5 \text{ ft}$$

c. The total dynamic head is:

Friction loss in pipe from septic tank	=	6.3 ft
Hose and valve assembly (estimated)	=	5.0 ft
Fittings (estimated)	=	1.0 ft
Friction loss in manifold	=	0.5 ft
Friction loss in lateral	=	0.1 ft
Residual head on orifices	=	5.0 ft
Elevation difference (assumed, from low water level in pump tank to manifold centerline)	=	10.0 ft
TDH (estimated)		27.9 ft

Comment. In some cases, the presence of air in the piping system, leading from the septic tank to the filter, can cause high headloss, especially if there are high spots in the piping system. Use of small-diameter pipe and a high-head pump will force the air out, eliminating the problem. A pump should be selected to provide the required flowrate at a minimum head of 5 to 10 ft on the orifices. Higher pressures (up to 30 ft) are usually not a problem. Some designers include a ball valve to adjust the residual head [typically 5 to 10 ft (1.5 to 3 m)]. The use of ball valves is not recommended because they do not hold their settings, and it is easy for someone to unknowingly change the setting.

11-4 MULTIPASS (RECIRCULATING) PACKED-BED FILTERS

Multipass (recirculating) filters are similar to single-pass (intermittent) filters, with the exception that a portion of the treated effluent from the filter is returned to a recirculation tank where it is used to dilute the effluent from the septic tank before being applied to the filter. By diluting the strength of the septic tank effluent, higher application rates can be used.

Development and History of Use

Recirculating sand filters evolved from ISFs in the 1970s when Hines introduced their use in Illinois (Hines and Farveau, 1974). The RSF used by Hines consisted of a septic tank, recirculation storage tank, and an open sand filter. The contents of the recirculation tank were pumped over the filter two to eight times each hour. A valve (or flow splitter) in the recirculation tank allowed filtered effluent to either enter the tank or be routed to reuse/disposal, depending on the liquid level in the tank. In the mid-1970s RSF systems were used to treat septic tank effluent prior to disposal in roadside ditches. In West Virginia, the sand medium was replaced with bottom ash, a hard, durable by-product from coal-fired power plants (Swanson and Dix, 1988). In Oregon, sand was replaced with fine gravel (Ronayne et al., 1984). The significance of these alternative media types is that surface clogging is

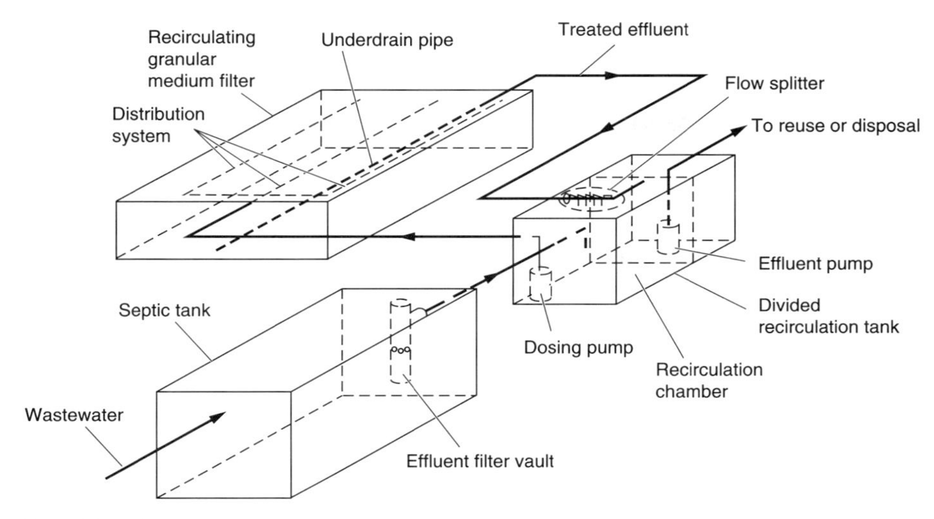

FIGURE 11-17
Schematic flow diagram for a recirculating granular medium filter.

greatly reduced or eliminated and higher loading rates are possible. A schematic flow diagram for a modern recirculating granular medium filter is presented in Fig. 11-17. As shown in Fig. 11-17, the recirculating granular medium filter requires the use of an additional tank as compared to the ISF. Additional details on the design and application of RSFs may be found in Piluk and Peters (1994).

Types of Applications

Recirculating sand and fine-gravel filters are used to treat septic tank effluent from individual homes, clusters of homes, institutions, and small communities with flows up to 1.0 Mgal/d (3785 m^3/d). They have also been used to nitrify pond effluent prior to discharge to constructed wetlands. RPBFs have also been used to treat septic tank effluent prior to UV disinfection and water reuse (see Chap. 12).

Individual systems. In Anne Arundel County, Maryland, RGFs have been used extensively for individual family residences (Piluk and Peters, 1994; Piluk, 1996–97). Typical examples are shown in Fig. 11-18. An innovation that has been developed in the application of these systems is the precasting of the concrete enclosure for the RGF and the addition of the piping and filter medium at the location where the tank is constructed (see Fig. 11-19). The completed unit is delivered to the construction site ready to be hooked up to the septic tank and the disposal system.

Stonehurst development. Stonehurst is a 47-lot subdivision located near the City of Martinez in Contra Costa County, California, approximately 25 miles east of San Francisco. The subdivision is located in a hilly, rural area without a wastewater collection system. An overall view of the rolling terrain is as shown previously in

(a)

(b)

FIGURE 11-18
Views of typical recirculating gravel filters: (*a*) typical example for small community and (*b*) small household unit in fabrication yard.

Fig. 1-1. The wastewater management system that was designed and constructed incorporates a number of innovative technologies (see Fig. 11-20) (Crites et al., 1997). The system includes the use of 1500-gal two-compartment watertight septic tanks with screened effluent filter vaults, high-head effluent pumps, a small-diameter variable-grade sewer, two pressure sewers, a recirculating granular medium filter

FIGURE 11-19
View of prefabricated recirculating gravel packed-bed filter delivered to home site, ready to install.

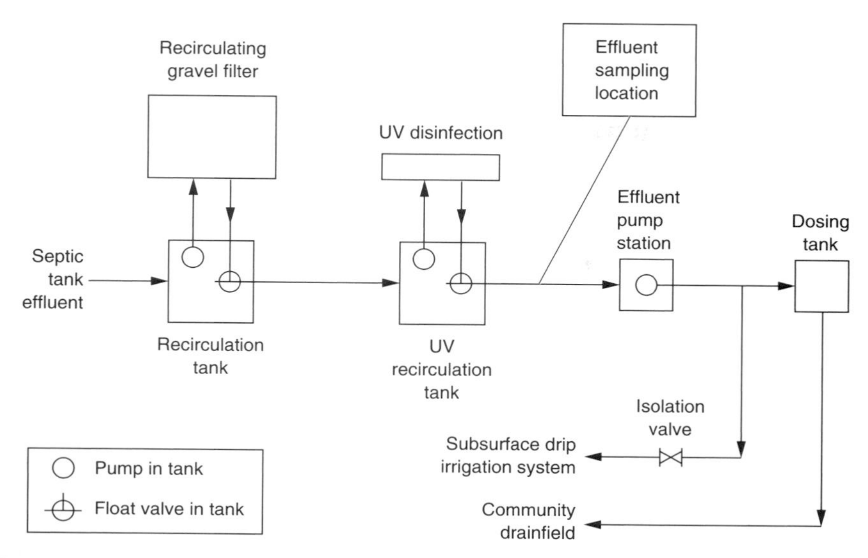

FIGURE 11-20
Schematic of Stonehurst treatment system incorporating recirculating fine gravel filters.

(see Fig. 11-21), a UV (ultraviolet light) disinfection unit, a subsurface drip irrigation system for reuse of treated effluent, and a community leachfield for wintertime disposal, when needed. The disinfected wastewater is required to have a 30-day average BOD and TSS of 15 mg/L, settleable solids concentration of 0.1 mL/L, and a total coliform concentration equal to or less than 23 MPN/100 mL.

FIGURE 11-21
View of recirculating gravel filter for Stonehurst development near Martinez, California. Building in background houses the UV disinfection system.

TABLE 11-11
Typical performance of recirculating filters for BOD and TSS removal*

	Filter medium		BOD, mg/L		TSS, mg/L	
Locations	d_{10}, mm	UC	Influent	Effluent	Influent	Effluent
Elkton, Oregon	3.5	1.8	141	6	32	6
Orcas Village, Washington	2.0	1.75	166	4	113	5
South Prairie, Washington	4.3	1.6	181	4	34	4

*Courtesy Orenco Systems, Inc.

Treatment Performance

The removal of BOD in recirculating filters depends on the loading rate and the size of the filter medium. Recirculating filters can produce an effluent with less than 10 mg/L of BOD and TSS when loading rates are less than 5 gal/ft^2·d (200 mm/d) and the medium size is 3 mm or less. Typical removals for BOD are presented in Table 11-11 for a range of filter medium sizes varying from sand to pea gravel. Performance data for the RGF used for the Stonehurst development described above are presented in Table 11-12.

Recirculating filters can produce a high-quality, partially nitrified effluent. Nitrogen removal with recirculating filters is typically 40 to 50 percent, as shown in

TABLE 11-12
Performance data for Stonehurst wastewater treatment system
For the 28-month period from June 1994 through September 1996. Effluent samples taken after recirculating gravel filter and UV disinfection as shown in Fig. 11-22*

				Observed values			
Constituent†	**Unit**	**Sample reporting period**‡	**Total number of samples to date**	**Range**	**Arithmetic mean**	**Geometric mean**	**Median**
BOD_5	mg/L	Monthly	56	0 to <5		<5	<5
COD	mg/L	Monthly	120	1.0–18.0		4.7	5.0
TSS	mg/L	Monthly	118	2.0–15.0		5.0	4.9
pH	Unitless	Monthly	120	6.96–8.65	§	§	7.61
Total coliform	MPN/100 mL	Monthly	118	<2–12.5	§	§	<2
NH_4	mg/L	Quarterly	9	0–15	¶	¶	0.0
NO_3	mg/L	Quarterly	9	3.55–37.0		12.5	12.2
TKN	mg/L	Quarterly	9	0–3	¶	¶	0.4
Oil and grease	mg/L	Quarterly	9	0–12	¶	¶	0.0
TDS	mg/L	Quarterly	9	340–770	630		656
EC	μmhos/cm	Quarterly	9	433–1200	894		1000

*Crites et al. (1997).
†TDS = total dissolved solids, EC = electrical conductivity.
‡Monthly values for BOD_5 are based on an average of two samples per month. Monthly values for TSS, COD, pH, and total coliform are based on an average of at least four samples per month.
§ Mean values cannot be reported for pH, which is a logarithmic function, and total coliform, which is based on a Poisson distribution.
¶Unable to define nature of distribution, because of the number of zero values reported.

TABLE 11-13
Nitrogen removal by recirculating filters

Location	Medium size, mm	Loading rate, gal/ft^2·d*	Total nitrogen, mg/L Influent	Total nitrogen, mg/L Effluent	Percent removal
Oregon (Ronayne, 1984)	1.2	1.5	58	32	45
Paradise, California (Nolte and Associates, 1992)	3.0	4.3	63	35	44
Paradise, California (Nolte and Associates, 1992)	3.0	2.5	57	26	54
Florida (Sandy et al., 1988)	2.4	3.8	55	9.6	82

*(gal/ft^2·d) × 40 = mm/d

Table 11-13. Nitrification and denitrification are responsible for the removal. Increased nitrogen removal can be achieved by using flooded underdrains to enhance denitrification, following the filter with an anaerobic filter, or adding a supplemental carbon source. If the ammonia level is high (over 60 mg/L), and alkalinity levels are low, nitrification can reduce the pH excessively. As ammonia is nitrified, 7.14 mg of alkalinity, expressed as calcium carbonate ($CaCO_3$), is destroyed for every 1 mg of ammonia oxidized to nitrate (see Example 2-9 in Chap. 2). If the community water supply is a soft, low-alkalinity water, there may be insufficient alkalinity to buffer the pH during nitrification. If the pH is reduced through the filter, the overall performance of the bacteria for removal of BOD and ammonia can be impaired. A source of alkalinity, such as sodium or calcium carbonate, can be added. Alternatively, the denitrification process, which restores about half of the alkalinity lost in nitrification, can be maximized. Alternatives to maximize nitrogen removal with recirculating sand filters are discussed in Sec. 11-6.

Design Considerations

Important factors in the design of RPBFs include type and size of filter medium, depth of filter bed, hydraulic loading rate, organic loading rate, dosing frequency and rate, distributed dosing, and modular design. These design considerations are examined in the following material. Typical design criteria for recirculating filters are presented in Table 11-14. The design of an RPBF is illustrated in Example 11-6.

Type and size of filter media. The types of filter media used for RPBFs range from coarse sand to bottom ash to fine gravel. The finest medium reported was a sand with an effective size of 0.3 mm (Belicek, 1986). Excessive clogging and short filter runs were observed with this sand size, and the investigators recommended an increased sand size of 0.6 mm. Screened bottom ash has been used in sizes ranging from 0.9 to 2.4 mm. The range of the medium sizes for gravel filters is from 1 to 6 mm. A typical value for a fine gravel size is 2.5 mm. When fine gravel larger than 3 mm is used, a sharp drop in the degree of nitrification has been observed (Ball and Denn, 1997).

TABLE 11-14
Design criteria for recirculating packed-bed filters

	U.S. customary units			SI units		
Design factor	**Unit**	**Range**	**Typical**	**Unit**	**Range**	**Typical**
Filter medium						
Material			Durable, washed granular medium			
Effective size	mm	1–5	2.5	mm	1–5	2.5
Depth	in	18–36	24	mm	450–900	600
Uniformity coefficient		<2.5	2.0		<2.5	2.0
Underdrains						
Type			Slotted or perforated drain pipe			
Size	in	3–4	4	mm	75–100	100
Slope	%	0–0.1	0	%	0–0.1	0
Pressure distribution						
Pipe size	in	1–2	1.5	mm	25–50	38
Orifice size	in	$\frac{1}{8}$–$\frac{1}{4}$	$\frac{1}{8}$	mm	3–6	3
Head on orifice	ft	3–6	5	m	1–2	1.6
Lateral spacing	ft	1.5–4	2	m	0.5–1.2	0.6
Orifice spacing	ft	1.5–4	2	m	0.5–1.2	0.6
Design parameters						
Hydraulic loading*	gal/ft^2·d	3–5	4	mm/d	120–200	160
BOD loading	lb/ft^2·d	0.002–0.008	<0.005	kg BOD/m^2·d	0.01–0.04	<0.025
Recirculation ratio		3:1–5:1	4:1		3:1–5:1	4:1
Dosing times						
Time on	min	<2–3	<2–2	min	<2–3	<2–2
Time off	min	15–25	20	min	15–25	20
Dosing						
Frequency	times/d	48–120		times/d	48–120	
Volume/orifice	gal/orif·dose	1–3	2	L/orif·dose	3.8–11.4	7.6
Dosing tank volume	flow/d	0.5–1.5	1	flow/d	0.5–1.5	1

*Based on peak flow.

Depth of filter bed. Bed depths for RPBFs have varied from 24 to 48 in (600 to 1200 mm). Historically, deeper bed depths were used; however, it has been found that bed depths beyond 24 in are usually not warranted. The depth of most RGFs is typically 24 in (600 mm).

Hydraulic loading rate. The hydraulic loading rate for recirculating filters, based on peak flow, ranges from 3 to 6 gal/ft^2·d (120 to 240 mm/d), depending on the size of the filtering medium. The typical hydraulic loading rate for recirculating filters, used to treat septic tank effluent from individual homes, is 5 gal/ft^2·d (200 mm/d), based on forward peak flow from the septic tank.

Organic loading rate. As with the intermittent filter, the organic loading rate comprises the soluble and particulate organic matter applied to the filter. Although the appropriate organic loading rates are not well defined for recirculating granular medium filters, typical values range from 0.002 to 0.008 lb BOD/ft^2·d (0.01 to 0.04 kg BOD/m^2·d). Higher organic loading rates have been used by adjusting the dosing frequency (Parker, 1996).

Recirculation ratio and dosing frequency. Recirculation ratios are typically 4 or 5 to 1, based on forward flow. The dosing interval can be set by a variable timer, and is typically less than 1 min to 5 min every 12 to 25 min. For higher-strength wastewaters, the dosing frequency has been increased to as high as 2 min every 6 min (Parker, 1996).

Flow splitting. Because most of the flow from a recirculating filter is returned to the recirculation basin to be mixed with the flow from the septic tank, some method must be available to divide the flow. Several methods have been used to accomplish the required flow split including, as shown in Fig. 11-22, the use of: (1) a splitter basin, (2) a recirculating orifice/ball float valve, and (3) the use of a flow-splitting baffle in the filter (Ball and Denn, 1997). Although the simple ball float valve was used in early designs, its continued use is not recommended because, if the recirculation dilution tank is full, a portion of the flow from the septic tank may pass through the filter only once before being discharged. Where a splitter basin is used (see

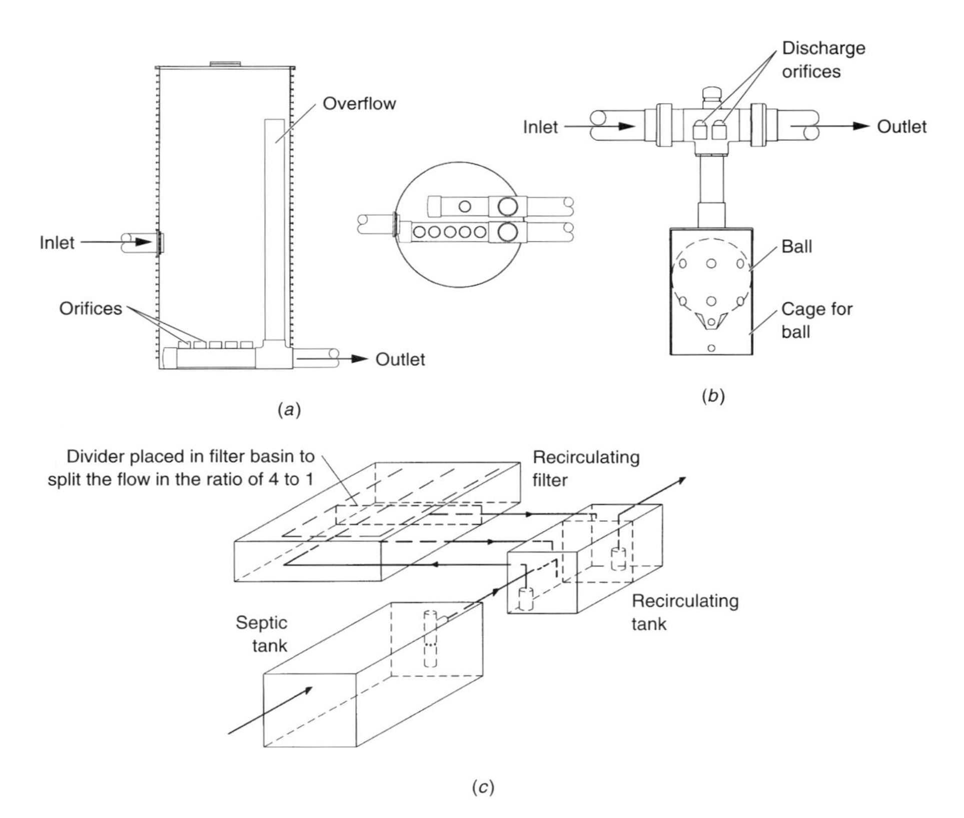

FIGURE 11-22
Methods available to split flow in recirculating packed-bed filters: (*a*) a splitter basin, (*b*) a recirculating orifice/ball float valve, and (*c*) a flow-splitting baffle in the filter (*a, b* from Ball and Denn, 1997).

Fig. 11-22*a*), a float in the recirculation tank is used to shut off an electrically actuated valve on the discharge line, so that all of the flow is returned to the recirculation tank.

The recirculating orifice/ball float valve (see Fig. 11-22*b*) combines the advantages of the simple ball float valve and the flow splitter, without the need for any electrical controls. When the flow is low, all of the flow is returned to the recirculating tank. If the tank is full, only a small portion of the flow is diverted to the discharge line through the last orifice (Ball and Denn, 1997). Another mechanical method of flow splitting involves the uses of a baffle in the filter (see Fig. 11-22*c*). The amount of flow recirculated to the recirculation tank depends on the placement of the divider baffle. Although the system is straightforward, odor problems can develop if the occupants of the home leave for a few days, and the recirculating tank is pumped dry.

Distributed dosing. When the design flows increase and more distribution laterals and orifices are required, a point will be reached where the flow from the required number of orifices will exceed the capacity of a single pump. Where the required flow from the orifices exceeds the pump capacity, the filter area can be divided into sections and a mechanical distributing valve can be used to dose each section of the filter sequentially (see Fig. 11-23). Typically, a single pump can pressurize up to six separate sections, depending on the dosing frequency (Ball and Denn, 1997). In some designs, to provide redundancy, two pumps may be used, with each pump pressurizing one or more of the sections. Where two pumps are used, the discharge piping from the two pumps is interconnected with appropriate valving. With interconnected pumps, should one pump fail, the remaining pump can be used to dose all of the sections until the faulty pump can be replaced.

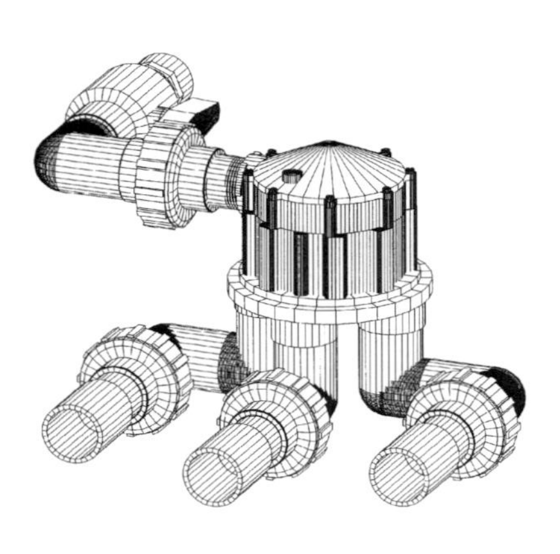

(*a*)

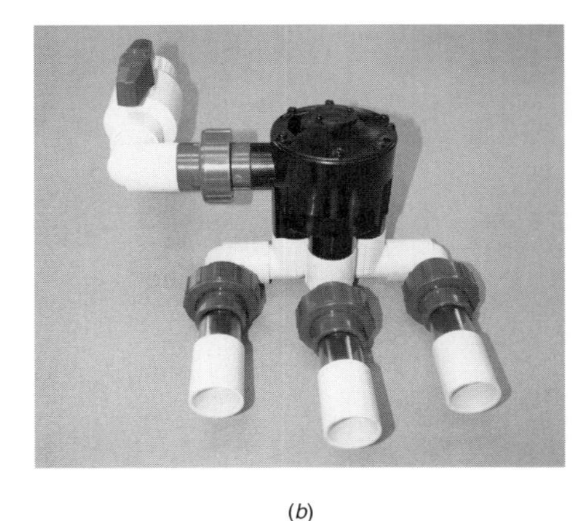

(*b*)

FIGURE 11-23
Mechanical distribution valve used to dose different sections of packed-bed filters: (*a*) CAD rendering (courtesy C. Jordan) and (*b*) photographic view.

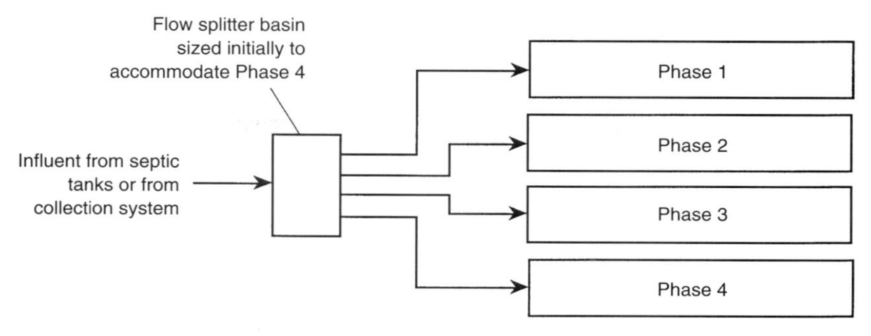

FIGURE 11-24
Diagram of modular filter design (adapted from Ball and Denn, 1997).

Modular design. Where the filter system is to be expanded to treat increased future flows, a modular design is used. In a typical arrangement, additional filter units would be added in stages. The necessary pumps would be added in an existing flow recirculation basin (see Fig. 11-24), designed to accommodate anticipated future flows.

Cold climate design considerations. Buried distribution systems are necessary in cold climates to avoid freezing. It is suggested that a 1.0-ft (0.3-m) layer of 25-mm stone be used to cover the distribution piping. It is also recommended that the septic tank and recirculating tanks be insulated and that a coarse medium be used (d_{10} = 2 to 2.5 mm, maximum size 5 mm, and UC < 2) (Loudon et al., 1984, 1989).

Construction Considerations

Construction considerations for recirculating filters are similar to those for ISF. Gradation and lack of fines in the medium as well as durability specifications for the filter material are important. A typical recirculating fine gravel filter is shown under construction in Fig. 11-25. As noted previously, prefabricated RPBFs have been developed. The use of prefabricated units reduces construction errors in the field assembly of these units.

Operation and Maintenance

Recirculating filters can be operated without surface clogging if the solids loading rate and medium size are in balance. Using the design criteria in Table 11-14 will result in operating systems that require minimal operation and maintenance. The counters and meters discussed in the previous section also apply to recirculating filters.

(a)

(b)

FIGURE 11-25
Views of large recirculating packed-bed filter under construction. (a) exposed underdrain system being prepared for sand placement and (b) sand being placed using a conveyor system fed from a sand stock pile.

EXAMPLE 11-6. DESIGN OF RECIRCULATING GRAVEL FILTER FOR A SINGLE DWELLING. Size and lay out a recirculating gravel filter and distribution system for an individual three-bedroom residence in a nonsewered area. Check the organic loading rate. Determine the total volume of flow per dose. Calculate the orifice flowrate and the total flowrate. Determine the on and off cycle times. The recirculation tank is sized to provide a storage capacity of 2.5 times the average flow. Assume the following conditions apply.

1. Average occupancy = 3.0 persons/d
2. Average wastewater flow = 50 gal/capita·d
3. Assumed peaking factor = 2.5 (see Table 4-20)
4. Size of septic tank = 1500 gal
5. Effluent BOD from septic tank = 130 mg/L
6. Effluent TSS from septic tank = 30 mg/L
7. Hydraulic application rate = 5.0 gal/ft^2·d (204 L/m^2·d) based on influent peak flow
8. Size of recirculation tank = 800 gal
9. Size of effluent pump basin = 400 gal
10. Filter dose rate per day = 48 times/d minimum
11. Distribution system orifice size = $\frac{1}{8}$ in (3 mm)
12. Orifice discharge head = 5 ft minimum (1.5 m)
13. Recirculation ratio = 4:1

Solution

1. Determine the size of the filter.
   a. Determine the average flow and peak flowrate:

$$Q_{\text{ave}} = (3.0) \text{ persons} \times 50 \text{ gal/capita·d} = 150 \text{ gal/d}$$

$$Q_{\text{peak}} = 150 \text{ gal/d} \times 2.5 \text{ (see Table 4-20)} = 375 \text{ gal/d}$$

   b. Determine the required area of the sand filter:

$$\text{Area} = (375 \text{ gal/d})/(5.0 \text{ gal/ft}^2\text{·d}) = 75 \text{ ft}^2$$

   c. Use a filter 8 ft × 10 ft. Check area (8 × 10) = 80 ft^2. OK.

2. Determine the organic loading rate:

$$L_{org} = \frac{(5.0\ \text{gal/ft}^2\cdot\text{d} \times 3.785\ \text{L/gal} \times 130\ \text{mg/L})}{(1000\ \text{mg/g} \times 454\ \text{g/lb})}$$

$$= 0.0054\ \text{lb BOD/ft}^2\cdot\text{d} \qquad \text{OK (see Table 11-13)}$$

3. Lay out gravel filter and the effluent distribution system. Use a spacing of 1.0 ft (0.3 m) between distribution pipes, orifices, and sidewalls.
   a. Determine the number of laterals based on a nominal spacing of 1.0 ft:

$$\text{Number of laterals} = (8 - 1.0)/1.0 + 1 = 8$$

   b. Determine the number of orifices:

$$\text{Number of orifices} = (10 - 1.0)/1 = 9$$

   c. The layout of the gravel filter and distribution system is shown below.

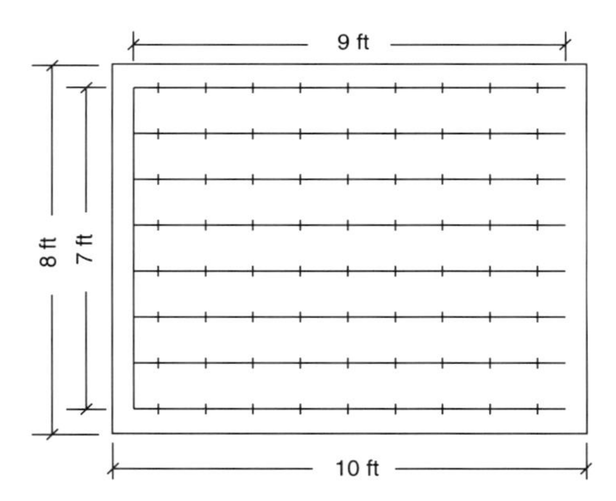

4. Determine the quantity of flow per dose and the flowrate in each lateral.
   a. Determine quantity of flow discharged per dose:

$$\text{Flow/dose} = (150\ \text{gal/d} \times 5)/(48\ \text{dose/d}) = 15.6\ \text{gal/dose}$$

   b. Determine the discharge per lateral.

$$\text{Flow/lateral} = (15.6\ \text{gal/dose})/8\ \text{laterals} = 1.95\ \text{gal/lateral}\cdot\text{dose}$$

5. Using Eq. (11-7), determine the rate of discharge from an orifice, assuming 5 ft of residual head at the orifice:

$$q_n = 2.45C(D^2)\sqrt{2gh_n}$$

$$= 2.45(0.63)(0.125)^2\sqrt{2(32.2)5} = 0.43\ \text{gal/min}\cdot\text{orifice}$$

6. Determine the duration of the flow and the volume of flow discharged at average flow.
   a. The total flowrate, based on 9 orifices/lateral, with 8 laterals, is

$$\text{Flowrate/lateral} = (0.43\ \text{gal/min}\cdot\text{orifice})(9\ \text{orifice/lateral})$$

$$= 3.9\ \text{gal/min}\cdot\text{lateral}$$

$$Q_T = (3.9\ \text{gal/min}\cdot\text{lateral})(8\ \text{laterals}) = 31\ \text{gal/min}$$

b. Determine the duration of flow:

$$\text{Duration} = (15.6 \text{ gal/dose})/(31 \text{ gal/min})$$
$$= 0.50 \text{ min/dose} = 30 \text{ s/dose}$$

7. Determine the on and off cycle times at average flow:

On time = 30 s

Off time = [(1440 min/d)/(48 dose/d)] − [(0.50 min/dose)] = 29.5 min

11-5 OTHER PACKED-BED FILTERS

Other types of packed filters that are being developed include an absorbent plastic medium filter and peat filters. Both of these two filters are single-pass filters.

Absorbent Plastic-Medium Filter

The Waterloo biofilter uses absorbent plastic foam medium for aerobic treatment of wastewater. The filter was developed in Canada at the University of Waterloo by Jowett (Jowett and McMaster, 1994, 1995). The filter medium, typically supplied in 3- or 4-in cubes, is characterized by high porosity and a large surface area (see Fig. 11-26). The depth of the filter is typically 4 ft (1.25 m). The high porosity and large surface area allow for high hydraulic loading rates while maintaining aerobic conditions in the medium.

FIGURE 11-26
View of foam cubes used in absorbent plastic-medium filter.

TABLE 11-15
Comparison of the performance of the Waterloo biofilter under conditions of natural convection and forced-air ventilation*

Constituent	Unit	Influent	Effluent	Reduction
Natural convection				
BOD_7	mg/L	234	14.4	93.5
NH_4-N	mg/L	8.1	10.2	
NO_3-N	mg/L	0.5	24.7	
Total coliform	CFU/100 mL	2.8×10^8	3.1×10^4	99.985
Forced ventilation				
BOD_7	mg/L	110	1.7	98.4
TSS	mg/L	115	5.0	93.9
NH_4-N	mg/L	2.3	0.04	
NO_3-N	mg/L	0.4	33.8	
Total coliform	CFU/100 mL	7.3×10^8	1.3×10^4	99.998

*From Jowett and McMaster (1994).

The wastewater can flow vertically without clogging the medium while allowing air to flow in a separate pathway through the medium. When water moves through an absorbent medium, it uses the interiors of the particles rather than the surrounding voids and is transferred from particle to particle across the points of contact. The large interparticle voids remain open for air circulation. Air flow around the particles occurs and oxygen is transferred directly to the bacterial film, without having to diffuse through water.

Although the Waterloo biofilter is still in the development stage, it represents a future trend in optimizing the size and loading rates for biological filters. In column studies, the filter was loaded at 20 gal/ft^2·d (800 L/m^2·d), based on forward flow, which is 4 times higher than a recirculating gravel filter and 20 times higher than typical rates for ISFs. A comparison of the performance of the Waterloo biofilter under conditions of natural convection and forced ventilation is presented in Table 11-15.

Field units for treatment of septic tank effluent were tested in Ontario. The units consisted of 155 ft^3 (4.4 m^3) polyethylene or concrete tanks, with a filter surface area of about 35 ft^2 (3.3 m^2). The septic tank effluent is pumped through perforated distribution pipes that are set above the filter medium, flows slowly down and through the medium, and is discharged out the bottom. A typical filter is shown in Fig. 11-27.

Textile Packed-Bed Filter

The textile filter (also known as the textile bioreactor) was developed as an alternative to sand and gravel filters. By utilizing nonwoven textile chips (i.e., small pieces of cut textile, sometimes called *squares* or *coupons*) instead of a granular medium, hydraulic loading rates can be increased significantly, thus reducing space requirements for the filter. The textile filter can be operated as a single-pass filter or a recirculating filter, depending on flow and design considerations. The applied

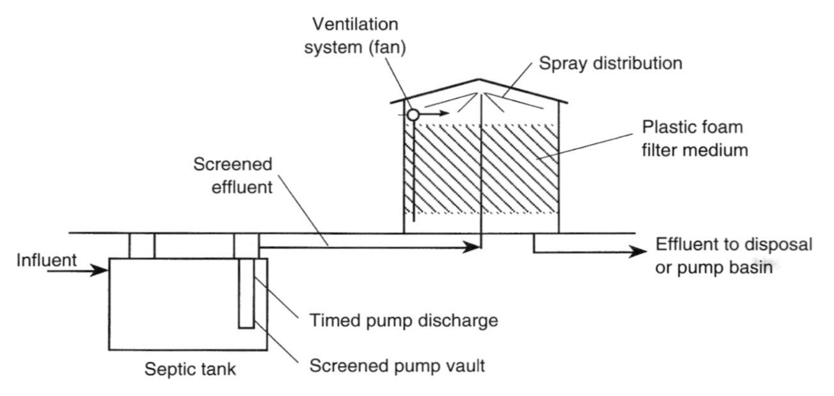

FIGURE 11-27
Flow diagram for absorbent plastic-medium filter (adapted from Jowett and McMaster, 1994).

water percolates both through and between the textile chips as shown in Fig. 11-28. Nonwoven textile chips offer a number of advantages for wastewater treatment. The complex fiber structure of the textile material offers an extremely large surface area for biomass attachment. Compacted or uncompacted, the total porosity of the filter bed is over 80 percent. The measured field capacity (i.e., the water-holding capacity) for compacted textile beds is about 40 percent and somewhat less for uncompacted beds. The corresponding hydraulic conductivities exceed 4 in/s (100 mm/s). Such high hydraulic conductivities reduce solids accumulation within the filter bed.

In terms of treatment, the water-holding capacity of the textile material appears to be a key factor. It has been shown that COD removal in the textile filter is related to retention time of the wastewater in the textile chips and filter height. Water retention in the textile filter is mostly due to capillary effects in the micropores found within the structure of the textile material and on the height over which capillary forces are exercised. Ultimately, the water-holding capacity is determined by the type of textile material used and the degree of compaction. When operated as a single-pass inter-

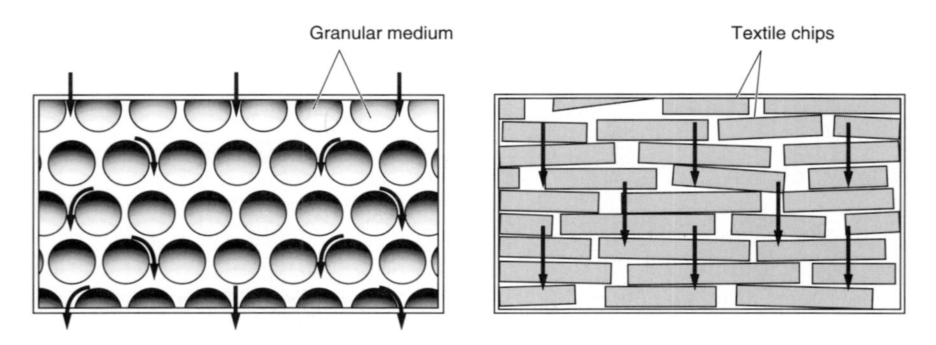

FIGURE 11-28
Definition sketch for flow through filter beds: (*a*) granular medium and (*b*) textile chip (adapted from Roy, 1997).

TABLE 11-16
Experimental data for standard nonlayered and layered textile filters used to treat typical residential wastewater*

Type of operation	Hydraulic loading rate, gal/ft²·d	Effluent quality, mg/L†			Fecal coliform, MPN/100 mL
		BOD	COD	TSS	
Standard nonlayered textile filter‡					
Single pass	5	6.4	54	5	7012
Single pass	10	12.4	110	10	74,000
Recirculation	15	3.9	20	2.0	200
Layered textile filter (with capillary breaks)§					
Single pass	30	9	85	22	9920
Recirculation	45	7	36	11	3480

*From Auger (1997), Chenier (1997), and Roy (1997).
†Average influent characteristics over 18 months: BOD = 162 mg/L, COD = 315 mg/L, and TSS = 92 mg/L.
‡Average over 18 months.
§Average over 6 months.

mittent filter, the nonlayered textile filter can treat up to 10 gal/ft^2·d (400 L/m^2·d) of septic tank effluent. If nonlayered, the reactor is operated as a recirculating filter; 15 gal/ft^2·d (600 L/m^2·d) can be applied. Effluent quality meets secondary standards, and some constituents meet or exceed advanced treatment standards (see Table 11-16). A typical installation is shown in Fig. 11-29 (Roy, 1997).

Recent developments that take full advantage of the hydraulic properties of the textile material show that hydraulic loading rates can be increased significantly. On the basis of the results of experimental studies, an entirely new filter concept was developed in which the filter bed is subdivided in hydraulically independent layers. Each layer affords complete capillary effects, such that a large volume of water is retained in the filter between pump cycles. When operated as a recirculating filter

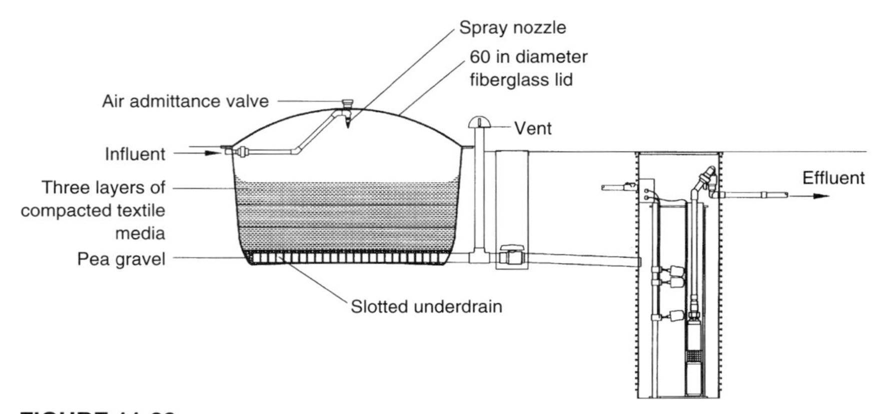

FIGURE 11-29
Schematic of textile packed-bed filter with external pump basin (courtesy Orenco Systems, Inc.).

TABLE 11-17
Typical design criteria for layered textile filter with capillary break between every 4 in of textile medium used to treat typical residential wastewater*

Type of operation	Hydraulic loading rate, gal/ft^2·d		Filter depth, in	Dose limit,¶ gal/ft^2
	Design†‡	Peak§		
Single pass	15	30	24	0.5
Recirculation	30	45	12	0.5

*Bounds (1996).
†Based on typical average daily flows (e.g., 50 gal/capita·d).
‡Average residential septic tank effluent from a septic tank with an effluent filter: BOD = 130 mg/L, TSS = 40 mg/L, and O and G = 20 mg/L.
§Peak residential septic tank effluent from a septic tank with an effluent filter: BOD = 200 mg/L, TSS = 60 mg/L, and O and G = 25 mg/L.
¶Hydraulic application rate, in both cases, should be limited to 0.8 in/dose (20 mm/dose).

at a hydraulic loading rate of 45 gal/ft^2·d (1800 L/m^2·d) and a 5 to 1 recirculation ratio, three 4-in (100-mm) layers have proven to be very effective (see Table 11-16) (Auger, 1997; Chenier, 1997; and Roy, 1997).

To reduce space requirements and construction costs, the final design of the layered filter is still being optimized with respect to layer thickness, number of layers, nature of the textile fibers, and compression rate. Current (1997) recommended design criteria for layered intermittent and recirculating textile filters used for treating typical residential septic tank effluent are presented in Table 11-17. The textile medium must be placed in compacted layers not less than 4 in (100 mm) deep.

Peat Filters

The use of peat filters as an alternative method of biological and physical treatment of septic tank effluent has been reported since 1984 (Brooks et al., 1984). These filters were used in Maine and Ontario (see Fig. 11-30). The peat is a permeable, absorbent medium that serves: (1) to filter the wastewater, (2) as a substrate for biological treatment, and (3) to reduce phosphorus concentration. Loading rates are usually the same as for intermittent sand filters, 1 gal/ft^2·d (4 cm/d). The results of monitoring several onsite peat filters are reported in Table 11-18. Commercial peat filters are available for individual home use. One manufacturer markets a 150 ft^2 (13.9 m^2) filter for the treatment of 500 gal/d (1892 L/d) of septic tank effluent. The manufacturer emphasizes the ability of the filter to reduce BOD, TSS, ammonia, and bacteria.

The importance of the source and characteristics of the peat have been demonstrated, based on the monitoring results obtained in Maine and in Ontario. The specifications of the State of Maine require a von Post decomposition of H-4, a pH of 3.5 to 4.5, and a moisture content of 50 to 60 percent. It was observed that peat from different sources provided various levels of treatment, leading to the additional requirement for a nitrate analysis of the peat leachate extract.

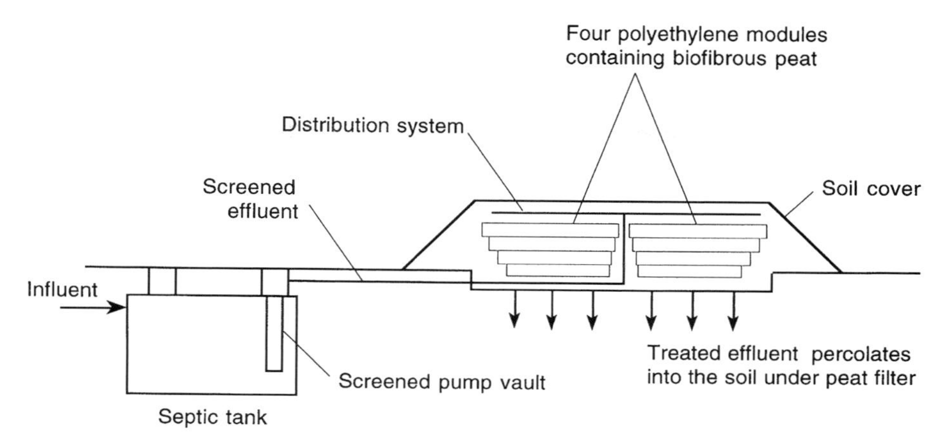

FIGURE 11-30
Definition sketch for peat filter (adapted from Bord na Mona Corp.).

TABLE 11-18
Summary performance data for peat filters used to treat individual household wastewater*

Constituent	Unit	Influent conc.	Effluent conc.	% reduction
BOD_5	mg/L	250–280	<10	96
TSS	mg/L	190	<10	95
NH_3 as N	mg/L	50	<5	90
Total N as N	mg/L	60	<10	80
Total P as P	mg/L	7.7	3.2	58
Total coliform	No./100 mL	3.0×10^6	6.0×10^3	99.9

*Adapted from McKee and Brooks (1994) and Henry (1995).

11-6 COMBINATION TREATMENT FOR NITROGEN REMOVAL

As discussed above, both intermittent and recirculating filters will nitrify septic tank effluent. The degree to which the nitrate will be denitrified depends on the availability of a carbon source and the alkalinity of the wastewater. In the literature dealing with nitrogen removal in intermittent and recirculating filters, much of the reported information is confounded statistically because nonstandardized sampling and analysis protocols are used and the alkalinity of the wastewater is not reported. To enhance the degree of denitrification or total nitrogen removal, a variety of flow diagrams have been developed (see Chap. 13).

Attempts have been made to recycle the effluent from IPBFs through the septic tank to enhance denitrification. Denitrification is enhanced; however, ammonium and organic nitrogen will remain in the septic tank effluent to limit the overall nitrogen removal. An alternative approach for the removal of nitrogen has been developed by Orenco Systems, Inc. (1996). The approach involves the removal of nitrogen in the septic tank by installing a small low-rate trickling filter above the septic tank (see Chap. 13). Additional details on the removal of nitrogen in ISFs and RSFs may be found in Sandy et al. (1987) and Whitmyer et al. (1991).

PROBLEMS AND DISCUSSION TOPICS

11-1. Demonstrate that the air at atmospheric pressure and 20°C contains about 279 mg/L of oxygen.

11-2. Verify that Eq. (11-6) is correct.

11-3. Determine the effective size and uniformity coefficient for the following sands (to be selected by the instructor).

Sieve size or number	Percent retained		
	A	B	C
$\frac{3}{8}$	0	0	0
10	1	0.1	28
16	15	3.9	22
30	30	26	20
50	29	48	18
60	19	19	8
100	2	2.9	3
Pan	4	0.1	1

11-4. Washed sand is available in the following size ranges. Determine the effective size and uniformity coefficient if equal parts of each size of sand were mixed. Prepare a size distribution plot for the final sand.

Size designation	Sand retained between indicated sieve sizes
6	6 × 10
8	8 × 14
12	12 × 25
20	20 × 40
30	30 × 60
50	50 × 100

11-5. Using the data from Prob. 11-4, determine what mixture of the size ranges should be used to produce a sand with: (1) an effective size of 0.30 and a uniformity coefficient of 2.50, (2) an effective size of 0.38 and a uniformity coefficient of 3.0, and (3) an effective size of 0.42 and a uniformity coefficient of 3.5 (to be selected by the instructor).

11-6. Determine the effective surface area in a cubic meter of sand composed of spherical particles 0.5 mm in diameter. Assume 1 percent of the area is taken up by particle-particle contact.

11-7. If the flow to a lateral is 5.0 gal/min and the lateral contains 26 orifices spaced 3 ft apart on center, determine the minimum theoretical size for the lateral so that the discharge from the first and last orifices will not vary by more than 5 percent, 10 percent, or 15 percent (to be selected by the instructor). What practical lateral size would you recommend and what will be the corresponding difference (percent) between the flow through the first and last orifices?

11-8. Lay out and size an intermittent sand filter and distribution system for a four-bedroom house. Assume that the following conditions apply:

Item	Unit	A	B	C
Average occupancy	persons/d	3.0	4.0	6.0
Size of septic tank	gal	1200	1500	2000
Loading rate	gal/ft^2·d	0.75	1.5	1.25
Dosing rate	times/d	12	16	18
Orifice size	in	$\frac{1}{8}$	$\frac{1}{8}$	$\frac{1}{8}$
Orifice discharge head	ft	8	10	5

11-9. Using data given in Example 11-5, estimate the pump discharge using the unrestricted pump curve given in Fig. 6-9. Assume the elevation difference between the septic tank and the filter is 60 ft and the length of pipe required from the septic tank to the filter is 90 ft.

11-10. Size and prepare a layout sketch of a recirculating sand filter for a community with a population of 1000 persons. Use a flowrate of 60 gal/capita·d. Assume the hydraulic application rate, based on forward flow, is 5.0 gal/ft^2·d (200 L/m^2·d).

11-11. Prepare a one-page abstract of the article by J. L. Darby, G. Tchobanoglous, M. A. Nor, and D. Maciolek (1996), Shallow Intermittent Sand Filtration: Performance Evaluation, *Small Flows Journal,* Vol. 2, Issue 1.

11-12. Prepare a one-page abstract of the article by R. W. Emerick, R. Test, G. Tchobanoglous, and J. L. Darby (1997), Shallow Intermittent Sand Filtration: Microorganism Removal, *Small Flows Journal,* Vol. 3, Issue 1.

11-13. An intermittent sand filter is dosed either once, 4 times, 12 times, or 24 times per day. Prepare a table and indicate what effect these dosing frequencies will have on BOD, TSS, NH_3, total N, total P, bacteria, and virus removal.

11-14. Prepare a one-page abstract of the article by M. Hines and R. E. Farveau (1974), Recirculating Sand Filters: An Alternative to Traditional Sewage Absorption Systems, in *Proceedings ASAE National Symposium of Home Sewage Disposal,* Chicago, ASAE Publication Proc-175, pp. 130–137, American Society of Agricultural Engineers, St. Joseph, MI.

REFERENCES

Anderson, D. L., R. L. Siegrist, and R. J. Otis (1986) Technology Assessment of Intermittent Sand Filters, Municipal Environmental Research Laboratory. U.S. Environmental Protection Agency, Cincinnati, OH.

Auger, R. (1997) Research Leader, Texel Inc., personal communication, Saint-Elzéar, Québec, Canada.

Ball, E. S. (1996) Pressure Dosing: Attention to Detail, in R. W. Seabloom (ed.), *Proceedings 8th Northwest On-Site Wastewater Treatment Short Course and Equipment Exhibition,* University of Washington, Seattle.

Ball, H. L. (1991) Sand Filters: State of the Art and Beyond, *Proceedings 6th National Symposium on Individual and Small Community Sewage Systems,* American Society of Agricultural Engineers, pp. 1011–1113, Chicago, IL.

Ball, J. L., and G. D. Denn (1997) Design of Recirculating Sand Filters Using a Standardized Methodology, in M. S. Bedinger, A. I. Johnson, and J. S. Fleming (eds.), *Site Characterization and Design of On-Site Septic Systems,* ASTM STP 898, American Society for Testing and Materials, Philadelphia.

Belicek, J., J. F. Zaal, and R. L. Kent (1986) A Recirculating Intermittent Sand Filter System for On-Site Wastewater Treatment, *Environment Canada,* Cat. No. En 44-14/1986.

Bounds, T. R. (1996) Pumps, Controls and Regulations, in R. W. Seabloom (ed.), *Proceedings 8th Northwest On-Site Wastewater Treatment Short Course and Equipment Exhibition,* University of Washington, Seattle.

Bouwer, H. (1978) *Groundwater Hydrology,* McGraw-Hill, New York.

Boyle, W. C., R. L. Siegrist, and C. C. Saw (1981) Treatment of Residential Graywater with Intermittent Sand Filtration, *Proceedings of Alternative Wastewater Treatment: Low-Cost Small Systems, Research and Development,* pp. 277–300.

Brooks, J. L., C. A. Rock, and R. A. Struchtemeyer (1984) The Use of Peat for On-Site Waste Water Treatment: 2. Field Studies, *Journal of Environmental Quality,* Vol. 13, No. 4, p. 524.

Buswell, A. M. (1928) *The Chemistry of Water and Sewage Treatment,* The Chemical Catalog Company, New York.

Chenier, R. (1997) Project Manager, Hydro-Québec, personal communication, Montréal.

Crites, R., C. Lekven, S. Wert, and G. Tchobanoglous (1997) A Decentralized Wastewater System for a Small Residential Development in California, *The Small Flows Journal,* Vol. 3, Issue 1, Morgantown, WV.

Dallavalle, J. M. (1948) *Micromeritics: The Technology of Fine Particles,* 2nd ed., Pitman, New York.

Darby, J. L., G. Tchobanoglous, M. A. Nor, and D. Maciolek (1996) Shallow Intermittent Sand Filtration: Performance Evaluation, *The Small Flows Journal,* Vol. 2, Issue 1. Morgantown, WV.

Dunbar, Professor Dr. (1908) *Principles of Sewage Treatment,* Charles Griffen & Company, London.

Effert, D., J. Morand, and M. Cashell (1985) Field Performance of Three Onsite Effluent Polishing Units, *Proceedings 4th National Symposium on Individual and Small Community Sewage Systems,* American Society of Agricultural Engineers, December 10–11, 1984, pp. 351–361, New Orleans.

Emerick, R. W., G. Tchobanoglous, and J. L. Darby (1997a) Use of Sintered glass as a Medium in Intermittently Loaded Wastewater Filters: Removal and Fate of Virus, Proceedings of the Water Environment Federation 70th Annual Conference and Exposition, Chicago, IL.

Emerick, R. W., R. Test, G. Tchobanoglous, and J. L. Darby (1997b) Shallow Intermittent Sand Filtration: Microorganism Removal, *Small Flows Journal,* Vol. 3, Issue 1, Morgantown, WV.

Fair, G. M., and J. C. Geyer (1954) *Water Supply and Waste-Water Disposal,* John Wiley & Sons, New York.

Frank, L. C., and C. P. Rhynus (1920) The Treatment of Sewage from Single Houses and Small Communities, *Public Health Bulletin No. 101,* U.S. Public Health Service, Washington, DC.

Frankland, Sir E. (1870) River Pollution Commission of Great Britain, First Report, London.

Furman, T. S., W. T. Calaway, and G. R. Grantham (1955) Intermittent Sand Filters—Multiple Loadings, *Sewage and Industrial Wastes.* Vol. 27, No. 3, pp. 261–276.

Grantham, G. R., D. L. Emerson, and A. K. Henry (1949) Intermittent Sand Filter Studies, *Sewage Works Journal,* Vol. 21, No. 6, pp. 1002–1016.

Hathaway, R. J., and D. T. Mitchell (1984) Sand Filtration of Septic Tank Effluent for All Seasons Disposal by Irrigation, *Proceedings 4th National Symposium on Individual and Small Community Sewage,* pp. 343–360.

Henry, H. (1995) Treatment of Septic Tank Effluent Using the Puraflo Peat Filter, Bord na Mona, Ireland.

Hines, M., and R. E. Farveau (1974) Recirculating Sand Filters: An Alternative to Traditional Sewage Absorption Systems, in *Proceedings of the National Home Sewage Disposal Symposium,* Chicago, American Society of Agricultural Engineers, Publication 175, pp. 130–137, St. Joseph, MI.

Jones, B. (1997) Personal communication, Stinson Beach, CA.

Jowett, E. C., and M. L. McMaster (1994) Absorbent Aerobic Biofiltration for Onsite Wastewater Treatment—Laboratory and Winter Field Results, *Proceedings 7th International Symposium on Individual and Small Community Sewage Systems,* pp. 424–435, American Society of Agricultural Engineers, Atlanta, GA.

Jowett, E. C., and M. L. McMaster (1995) Onsite Wastewater Treatment Using Unsaturated Absorbent Biofilters, *Journal of Environmental Quality,* Vol. 24, No. 1, pp. 86–95.

Loudon, T. L., D. B. Thompson, and L. E. Reese (1984) Cold Climate Performance of Recirculating Sand Filters, *Proceedings 4th National Symposium on Individual and Small Community Sewage Systems,* pp. 333–341, American Society of Agricultural Engineers, New Orleans.

Loudon, T. L., S. R. Wert, A. J. Gold, and T. A. McCarl (1989) Recirculating Sand Filters for Cold Regions, presented at the International Summer Meeting, ASAE, Canadian Society of Agricultural Engineering, Paper 89-2173, Québec, Canada.

Mancl, K. M., and J. A. Peeples (1991) One Hundred Years Later: Reviewing the Work of the Massachusetts State Board of Health on the Intermittent Sand Filtration of Wastewater from Small Communities, *Proceedings 6th National Symposium on Individual and Small Community Sewage Systems,* American Society of Agricultural Engineers, pp. 22–30, Chicago.

Marshall, G. R., and E. J. Middlebrooks (1974) Intermittent Sand Filtration to Upgrade Existing Wastewater Treatment Facilities, PRJEW 115-2, Utah Water Research Laboratory, Utah State University, Logan, UT.

McKee, J. A., and J. L. Brooks (1994) Peat Filters for Onsite Wastewater Treatment, *Proceedings 7th International Symposium on Individual and Small Community Sewage Systems,* pp. 526–535, American Society of Agricultural Engineers, Atlanta, GA.

Mote, C. R., J. W. Kleene, and J. S. Allison (1991) On-Site Domestic Wastewater Renovation Utilizing a Partially Saturated Sand Filter for Nitrogen Removal and a Lawn Sprinkler for Effluent Discharge, *Proceedings 6th National Symposium on Individual and Small Community Sewage Systems,* pp. 173–181, Chicago, IL.

Nolte and Associates (1992) Literature Review of Recirculating and Intermittent Sand Filter Operation and Performance, prepared for Town of Paradise, CA, and Central Valley Regional Water Quality Control Board, Sacramento, CA.

Nor, M. A. (1991) *Performance of Intermittent Sand Filters: Effects of Hydraulic Loading Rate, Dosing Frequency, Media Size, and Uniformity Coefficient,* thesis, Department of Civil Engineering, University of California at Davis.

Orenco Systems, Inc. (1996) Equipment Catalog.

Parker, M. (1996) Personal communication, Roseburg, OR.

Peeples, J. A., K. M. Mancl, and D. L. Widrig (1991) An Examination of the Role of Sand Depth on the Treatment Efficiency of Pilot Scale Intermittent Sand Filters, *Proceedings 6th National Symposium on Individual and Small Community Sewage Systems,* pp. 114–124, American Society of Agricultural Engineers, Chicago, IL.

Piluk, R. J. (1996–97) Personal communication.

Piluk, R. J., and E. C. Peters (1994) Small Recirculating Sand Filters for Individual Homes, *Proceedings 7th International Symposium on Individual and Small Community Sewage Systems,* American Society of Agricultural Engineers, Atlanta, GA.

Rich, L. G., G. E. Slagle, and D. V. Gore (1995) Low-Tech Treatment Produces High-Tech Effluent, *Water Environment & Technology,* Vol. 7, No. 3.

Robertson, L. A., and J. G. Kuenen (1990) Combined Heterotrophic Nitrification and Aerobic Denitrificaton on *Thiosphaera pantotropha* and Other Bacteria, *Antonie van Leeuwenhoek,* Vol. 67, pp. 139–162 (Netherlands).

Ronayne, M. A., R. C. Paeth, and S. A. Wilson (1984) Oregon Onsite Experimental Systems Program, EPA/600/2-84-157, Oregon Department of Environmental Quality. U.S. Environmental Protection Agency, Office of Research and Development, Washington, DC.

Roy, C. (1997) Vice President Research and Development, EAT Environment, Inc., personal communication, Québec, Canada.

Sandy, A. T., W. A. Sack, and S. P. Dix (1988) Enhanced Nitrogen Removal Using a Modified Recirculating Sand Filter (RSF^2), *Proceedings 5th National Symposium on Individual and Small Community Sewage Systems,* pp. 161–170, American Society of Agricultural Engineers, Chicago.

Sauer, D. K. (1976) Treatment Systems Required for Surface Discharge of Onsite Wastewater, *Proceedings 3rd National Conference on Individual Onsite Wastewater Systems,* Ann Arbor Science Publishers, Ann Arbor, MI.

Sauer, D. K., W. C. Boyle, and R. J. Otis (1976) Intermittent Sand Filtration of Household Wastewater, *Journal of Environmental Engineering Division,* ASCE, Vol. 102, No. EE4, pp. 789–803.

Scherer, B. P., and D. T. Mitchell (1982) Individual Household Surface Disposal of Treated Wastewater Without Chlorination, *Proceedings of 3rd National Symposium on Individual and Small Community Sewage Treatment,* ASAE Publication 1-82, American Society of Agricultural Engineers, St. Joseph, MI.

Swanson, S. W., and S. P. Dix (1988) Onsite Batch Recirculation Bottom Ash Filter Performance, *Proceedings 5th National Symposium on Individual and Small Community Sewage Systems,* pp. 132–141, American Society of Agricultural Engineers, Chicago, IL.

U.S. EPA (1980) Design Manual: Onsite Wastewater Treatment and Disposal Systems, EPA 625/1-80-012, US Environmental Protection Agency. Office of Water Program Operations, Cincinnati, OH.

Whitmyer, R. W., R. A. Apfel, R. J. Otis, and R. L. Meyer (1991) Overview of Individual Onsite Nitrogen Systems, *Proceedings 6th National Symposium on Individual and Small Community Sewage Systems,* pp. 143–164, American Society of Agricultural Engineers, Chicago.

CHAPTER 12

Effluent Repurification and Reuse

The uneven distribution of precipitation, continued population growth, contamination of both surface and groundwaters, and periodic droughts have forced water agencies to search for new and reliable water sources. The use of recycled or repurified wastewater for many nonpotable uses has proven to be the most reliable of sources. The purpose of this chapter is to introduce the reuse options and delineate the types of treatment required for repurification. The reuse options most suited to decentralized systems will be emphasized in this chapter.

Land treatment (Chap. 10), which involves water reuse, overlaps both agricultural irrigation and groundwater recharge, and therefore the water quality improvements that occur during soil/aquifer treatment are emphasized in Chap. 10, while the benefits of reuse of the water and nutrients are described in this chapter. Considerations for effluent reuse are discussed in this chapter, along with recent state regulations and representative examples of reuse.

Treatment processes used for reclamation or repurification of wastewater are described in this chapter. For small decentralized systems, the treatment processes can range from intermittent or recirculating granular medium filters (see Chap. 11) to membrane processes. Residual solids removal is often needed for higher levels of reuse, prior to disinfection with either chlorine or ultraviolet (UV) radiation.

12-1 EFFLUENT REUSE

A wide range of options for water reuse exists. For small and decentralized wastewater systems, agricultural and landscape irrigation are the most common forms of

FIGURE 12-1
Typical agricultural irrigation reuse system using treated municipal effluent.

water reuse. In this section, effluent reuse options are introduced and described, and guidelines for reuse are summarized.

Effluent Reuse Options

Types of effluent reuse include:

- Agricultural irrigation
- Landscape irrigation
- Industrial reuse
- Recreational impoundments
- Groundwater recharge
- Habitat wetlands
- Miscellaneous uses
- Augmentation of potable supplies

These reuse options are introduced in the following discussion, and the options that are most appropriate for decentralized systems are explored in greater detail in the latter part of the chapter.

Agricultural irrigation. Crop irrigation is one of the oldest and most common types of effluent reuse. Conceptually, it is identical to Type 2 slow-rate land treatment (see Chap. 10). In California, 63 percent of the total wastewater reuse is for agricultural irrigation (California SWRCB, 1990). Crops irrigated include trees, pasture grass, corn, alfalfa, and other feed, fodder, and fiber crops. Food crops have also been irrigated with tertiary disinfected effluent (Asano et al., 1992). A typical example of an agricultural irrigation reuse system is shown in Fig. 12-1.

Landscape irrigation. Landscape irrigation, also referred to as urban reuse, includes irrigation of:

- Parks
- Playgrounds
- Golf courses
- Freeway medians
- Landscaped areas around commercial, office, and industrial developments
- Landscaped areas around residences

Many landscape irrigation projects involve dual distribution systems—one distribution network for potable water and another for reclaimed water. The recycled water distribution system becomes the third water utility, after the wastewater and potable water systems, and is operated, maintained, and managed like the potable water system. The oldest municipal dual distribution in the United States is in St. Petersburg, Florida (U.S. EPA, 1992a). The system provides recycled water for a variety of uses, including a resource recovery powerplant and irrigation of school yards, a baseball stadium, residential lawns, commercial developments, and industrial parks.

Industrial reuse. Reuse of treated wastewater for industrial process or cooling water has been practiced at many locations throughout the United States (U.S. EPA, 1992a). The principal uses that industry has made of recycled water are cooling water, process water, boiler-feed water, and irrigation and maintenance of plant grounds. Cooling water, either for cooling towers or cooling ponds, creates the single largest demand for water in many industries and is the predominant industrial application (WPCF, 1989). Issues of concern in cooling water use include scaling, corrosion, biological growth, and fouling. Examples of industrial reuse are at Odessa, Texas; Fort Collins, Colorado; Lakeland, Florida; and Burbank, California (U.S. EPA, 1992a; Asano and Mujeriego, 1988).

Recreational impoundments. Recreational impoundments may serve a variety of functions from aesthetic, noncontact uses, to boating and fishing, to swimming. The required level of treatment will vary with the intended use of the water and the degree of public contact. The appearance of the recycled water is also of concern because the nutrients in the recycled water will stimulate the growth of algae and aquatic weeds. Removal of phosphorus and possibly nitrogen is usually needed to prevent algae growth in recreational reservoirs. Without nutrient control, there is a high potential for algae blooms, resulting in odors, an unsightly appearance, and eutrophic conditions.

Recycled water impoundments can be incorporated into urban landscape developments. Artificial lakes and golf course storage ponds and water traps can be supplied with recycled water. Examples of recreational impoundments include Las Colinas, Texas; Santee, California; Lubbock, Texas; and the Tillman Water Reclamation plant in Los Angeles (U.S. EPA, 1992a; WPCF, 1989).

(a)

(b)

FIGURE 12-2
Examples of habitat wetlands: (*a*) Arcata, California, and (*b*) Davis, California.

Groundwater recharge. Groundwater recharge helps provide a loss of identity between recycled water and groundwater. The loss of identity has an important, positive psychological impact where reuse is planned. Restrictions and reluctance to use recycled water can be overcome by groundwater recharge and subsequent recovery and use of the groundwater.

Purposes of groundwater recharge include:

- Establishment of seawater intrusion barriers
- Provision for further treatment for future reuse
- Provision for underground storage
- Augmentation of potable and nonpotable aquifers
- Control or prevention of ground subsidence

Groundwater recharge can be accomplished by either surface spreading or by injection. Surface spreading techniques are described in Sec. 10-4, Chap. 10. Injection is described later in this chapter.

Habitat wetlands. Natural or created habitat wetlands can make beneficial use of recycled water (see Fig. 12-2). Wetlands provide many valuable functions, including flood attenuation, wildlife and waterfowl habitat, productivity to support food chains, aquifer recharge, and water quality improvement. The distinction between a "constructed" wetland (see Chap. 9) and a "created" wetland is that the constructed wetland is intended as a treatment unit that can be modified or abandoned after its useful life has been completed. A created wetland, on the other hand, becomes a wetland area to be maintained and protected for its wildlife benefits in perpetuity.

Reclaimed water has been applied to wetlands for a variety of reasons, including:

- Creation, restoration, and enhancement of habitat
- Provision for additional treatment prior to discharge to receiving water
- Provision for a wet-weather disposal alternative for recycled water

Examples of habitat wetlands include Orlando, Florida; Showlow, Arizona; and Arcata, California. The Arcata wetlands consist of three 10-acre (4 ha) marshes and have attracted more than 200 species of birds, provided a fish hatchery for salmon, created a tourist attraction for the City of Arcata, and directly contributed to the development of the Arcata Marsh and Wildlife Sanctuary (U.S. EPA, 1992a).

Miscellaneous uses. A variety of miscellaneous uses have been made of reclaimed water:

- Flushing of toilets
- Supply for public or commercial laundries
- Fire fighting
- Construction water
- Flushing of sanitary sewers
- Snow making
- Washing aggregate and making concrete

Augmentation of potable supplies. Potable supplies can be augmented with reclaimed water; however, for small systems, the prospects are usually limited. Indirect potable reuse has been practiced in Fairfax County, Virginia, and in Clayton County, Georgia (Reed and Crites, 1984). Pipe-to-pipe direct reuse is only practiced at Windhoek, Namibia, and there only intermittently (U.S. EPA, 1992a). Research on direct potable reuse is being conducted at Denver, Colorado; Tampa, Florida; and San Diego, California (see Sec. 12-10) (Asano and Tchobanoglous, 1995).

Guidelines and Standards for Reuse

The state standards for water reuse were surveyed in September 1990 and are presented in U.S. EPA (1992a). The recommended EPA guidelines for water reuse are summarized in this section. As part of the 1990 survey of state regulations, all states were asked to provide an inventory of their existing reuse projects. The states responded with a list of 1900 reuse projects in 34 states.

EPA recommended guidelines. EPA has suggested guidelines for the following categories of reuse:

- Urban reuse
- Restricted access area irrigation
- Agricultural reuse—food crops
- Agricultural reuse—nonfood crops
- Recreational impoundments
- Landscape impoundments
- Construction uses
- Industrial reuse
- Groundwater recharge
- Indirect potable reuse

TABLE 12-1
Summary of EPA suggested guidelines for water reuse*

Level of treatment	Types of reuse	Reclaimed water quality	Reclaimed water monitoring	Setback distances
1. Disinfected tertiary[†]	Urban reuse[‡] Food crop irrigation Recreational impoundments	pH = 6–9 $BOD_5 \leq 10$ mg/L Turb. ≤ 2 NTU *E. coli* = none Res. $Cl_2 \geq 1$ mg/L	pH = weekly BOD = weekly Turb. = cont. *E. coli* = daily Res. Cl_2 = cont.	50 ft (15 m) to potable water supply wells[§]
2. Disinfected secondary	Restricted access area irrigation Food crop irrigation (commercially processed) Nonfood crop irrigation Landscape impoundments (restricted access) Construction Wetlands habitat	pH = 6–9 $BOD_5 = 30$ mg/L TSS = 30 mg/L *E. coli* = 200/100 mL Res. $Cl_2 \geq 1$ mg/L	pH = weekly BOD = weekly TSS = daily *E. coli* = daily Res. Cl_2 = cont.	100 ft (30 m) to areas accessible to the public (if spray irrigation) 300 ft (90 m) to potable water supply well

*From U.S. EPA (1992a).
[†] Filtration of secondary effluent.
[‡] Uses include landscape irrigation, vehicle washing, toilet flushing, use in fire protection, and commercial air conditioners.
[§] Setback increases to 500 ft (150 m) if impoundment bottom is not sealed.

For each category of reuse, there are suggested levels of treatment, minimum quality of reclaimed water, reclaimed water monitoring, and setback distances (U.S. EPA, 1992a). The guidelines are summarized in Table 12-1 for the two principal levels of treatment—disinfected tertiary (filtered water) and disinfected secondary.

Secondary effluent, undisinfected, is acceptable for injection into nonpotable aquifers and for some industrial uses. Primary effluent is acceptable for groundwater recharge of nonpotable aquifers using surface spreading.

Summary of state regulations. The state regulations or guidelines current to March 1992 are summarized in Table 12-2. There were 18 states with regulations and 18 states with guidelines on water reuse. Both the regulations and the guidelines vary considerably from state to state. States such as California, Arizona, Florida, and Texas have developed regulations that strongly encourage water reuse as a water resources conservation strategy. These states have developed comprehensive regulations specifying water quality requirements, treatment processes, or both for the full spectrum of reuse applications.

California has recently updated its reclamation criteria for reuse, and the regulations are summarized in Table 12-3. The acceptable uses for disinfected tertiary effluent are listed in Table 12-4.

TABLE 12-2
Summary of state reuse regulations and guidelines*

		Regulations or guidelines for each type			
State	**Regulations (R) or guidelines (G)**	**Unrestricted urban reuse**	**Restricted urban reuse**	**Nonfood crops**	**Food crops**
Alabama	G			×	
Arizona	R	×	×	×	×
Arkansas	G	×	×	×	×
California	R	×	×	×	×
Colorado	G	×	×	×	×
Delaware	G	×	×	×	
Florida	R	×	×	×	×
Georgia	G	×	×	×	
Hawaii	G	×	×	×	×
Idaho	R	×	×	×	×
Illinois	R†	×	×	×	
Indiana	R			×	×
Kansas	G	×	×	×	×
Maryland	G		×	×	
Michigan	R			×	×
Missouri	R		×	×	
Montana	G	×	×	×	×
Nebraska	G		×	×	

*From U.S. EPA (1992a).
† Draft or proposed.

(*continued*)

TABLE 12-2
(Continued)

State	Regulations (R) or guidelines (G)	Regulations or guidelines for each type: Unrestricted urban reuse	Restricted urban reuse	Nonfood crops	Food crops
Nevada	G*	×	×	×	×
New Jersey	R			×	
New Mexico	G	×	×	×	×
New York	G			×	
North Carolina	R		×		
North Dakota	G			×	
Oklahoma	G		×	×	
Oregon	R	×	×	×	×
South Carolina	G	×	×	×	
South Dakota	G	×	×	×	
Tennessee	R	×	×	×	
Texas	R	×	×	×	×
Utah	R	×	×	×	×
Vermont	R			×	
Washington	G	×	×	×	×
West Virginia	R			×	×
Wisconsin	R			×	
Wyoming	R	×	×	×	×

*From U.S. EPA (1992a).
† Draft or proposed.

TABLE 12-3
California wastewater reclamation criteria*

Category of reclaimed water	Criteria for: Total coliform, MPN/100 mL	Turbidity, NTU	Suitable uses
Disinfected tertiary†	2.2	2 average 5 maximum	All uses shown in Table 12-4
Disinfected secondary—2.2	2.2	N/A	All uses shown in Table 12-4 except irrigation of parks and playgrounds‡, food crops contacted by reclaimed water, nonrestricted impoundments
Disinfected secondary—23	23	N/A	Same restrictions as disinfected secondary—2.2, except no food crop irrigation, no restricted impoundment and no washing of yards
Undisinfected secondary§	N/A	N/A	Drip or surface irrigation of fodder, fiber, seed orchard and tree crops and sugar beets (commercially processed food crops)

*From California DHS (1994).
† Filtered through natural undisturbed soils or filter media, such as sand or diatomaceous earth.
‡ Urban areas such as parks, playgrounds, school yards, residential yards, and golf courses associated with residences.
§ Undisinfected wastewater means wastewater in which the organic matter has been stabilized, is non-putrescible, and contains dissolved oxygen.
N/A = not applicable.

TABLE 12-4

Representative uses and application methods for reclaimed water in California

General use	Conditions in which use is allowed[a]			
	Disinfected tertiary reclaimed water	**Disinfected secondary—2.2 reclaimed water**	**Disinfected secondary—23 reclaimed water**	**Undisinfected secondary reclaimed water**
All water uses other than potable use or food preparation[b]; and other than groundwater recharge (governed by other regulations)	Allowed[b]	Not allowed	Not allowed	Not allowed
Irrigation of:				
Parks, playgrounds, school yards, residential yards, and golf courses associated with residences	Spray, drip, or surface	Not allowed	Not allowed	Not allowed
Restricted-access golf courses, cemeteries, freeway landscapes	Spray, drip, or surface	Spray, drip, or surface	Spray, drip, or surface	Not allowed
Nonedible vegetation at other areas with limited public exposure[c]	Spray, drip, or surface	Spray, drip, or surface[c]	Spray, drip, or surface[c]	Not allowed
Sod farms	Spray, drip, or surface	Spray, drip, or surface	Spray, drip, or surface	Not allowed
Ornamental plants for commercial use	Spray, drip, or surface	Spray, drip, or surface	Spray, drip, or surface	Not allowed
All food crops	Spray, drip, or surface	Not allowed	Not allowed	Not allowed
Food crops that are above ground and not contacted by reclaimed water	Spray, drip, or surface	Drip or surface	Not allowed	Not allowed
Pasture for milking animals and other animals	Spray, drip, or surface	Spray, drip, or surface	Spray, drip, or surface	Not allowed
Fodder (e.g., alfalfa), fiber (e.g., cotton), and seed crops not eaten by humans	Spray, drip, or surface	Spray, drip, or surface	Spray, drip, or surface	Drip or surface
Orchards and vineyards bearing food crops	Spray, drip, or surface	Drip or surface	Drip or surface	Drip or surface
Orchards and vineyards not bearing food crops during irrigation	Spray, drip, or surface	Spray, drip, or surface	Spray, drip, or surface	Drip or surface

(continued)

TABLE 12-4
(Continued)

General use	Conditions in which use is allowed[a]			
	Disinfected tertiary reclaimed water	**Disinfected secondary—2.2 reclaimed water**	**Disinfected secondary—23 reclaimed water**	**Undisinfected secondary reclaimed water**
Christmas trees and other trees not bearing food crops	Spray, drip, or surface	Spray, drip, or surface	Spray, drip, or surface	Drip or surface
Food crop which must undergo commercial pathogen-destroying processing before consumption (e.g., sugar beets)	Spray, drip, or surface	Spray, drip, or surface	Spray, drip, or surface	Drip or surface
Impoundments:				
Supply for nonrestricted recreational impoundment	Allowed	Not allowed	Not allowed	Not allowed
Supply for restricted recreational impoundment	Allowed	Allowed	Not allowed	Not allowed
Supply for basins at fish hatcheries	Allowed	Allowed	Not allowed	Not allowed
Landscape impoundment without decorative fountain	Allowed	Allowed	Allowed	Not allowed
Other uses:				
Flushing toilets and urinals and priming drain	Allowed	Not allowed	Not allowed	Not allowed
Supply for commercial and public laundries for clothing and other linens	Allowed	Not allowed	Not allowed	Not allowed
Air conditioning and industrial cooling utilizing cooling towers[d]	Allowed	Not allowed	Not allowed	Not allowed
Industrial process with exposure of workers[e]	Allowed	Not allowed	Not allowed	Not allowed
Industrial cooling not utilizing cooling towers, spraying, or feature that creates aerosols or other mist	Allowed	Allowed	Allowed	Not allowed
Industrial process without exposure of workers	Allowed	Allowed	Allowed	Not allowed
Industrial boiler feed	Allowed	Allowed	Allowed	Not allowed

Fire fighting by dumping from aircraft	Allowed	Allowed	Allowed	Not allowed
Fire fighting other than by dumping from aircraft	Allowed	Not allowed	Not allowed	Not allowed
Water jetting for consolidation of backfill material around potable water pipelines during water shortages	Allowed	Not allowed	Not allowed	Not allowed
Water jetting for consolidation of backfill material around pipelines for reclaimed water, sewage, storm drainage, and gas and conduits for electricity	Allowed	Allowed	Allowed	Not allowed
Dampening soil for compaction at construction sites, landfills, and elsewhere	Allowed	Allowed	Allowed	Not allowed
Washing aggregate and making concrete	Allowed	Allowed	Allowed	Not allowed
Dampening roads and other surfaces for dust control	Allowed	Allowed	Allowed	Not allowed
Dampening brushes and street surfaces in street sweeping	Allowed	Allowed	Allowed	Not allowed
Flushing sanitary sewers[f]	Allowed	Allowed	Allowed	Not allowed
Washing yards, lots, and sidewalks	Allowed	Allowed	Not allowed	Not allowed
Supply for decorative fountain	Allowed	Not allowed	Not allowed	Not allowed

[a]See Section 60302, California Code of Regulations, Title 22, Reclamation Criteria.
[b]Disinfected tertiary effluent is suitable for all water uses that are not for potable use or food preparation, do not involve incorporation of reclaimed water into drink or food for humans, and do not conflict with provisions of the California Code of Regulations, federal regulations, statute, or other law.
[c]Disinfected secondary—2.2 reclaimed water and disinfected secondary—23 reclaimed water are suitable for irrigation of landscape vegetation and nonedible plants where: (*a*) the public would have access and exposure to irrigation water similar to that which would occur at a golf course or cemetery; and (*b*) children do not have direct access and exposure to irrigation water. There is no concern regarding access and exposure when disinfected tertiary reclaimed water is used.
[d]The industrial process generates mist or could involve facial contact with reclaimed water.
[e]The industrial process does not generate mist or involve facial contact with reclaimed water.
[f]The regulatory agency may approve, for flushing sanitary sewers, the use of disinfected wastewater, notwithstanding the fact that the median concentration of total coliform bacteria is higher than that of disinfected secondary—23 reclaimed water. For example, suitable for such use is wastewater that is always disinfected so that the median concentration of total coliform bacteria does not exceed 240/100 mL.

12-2 TREATMENT OPERATIONS AND PROCESSES FOR RECLAMATION AND REPURIFICATION

Treatment processes emphasized in this book are intended for (1) individual systems, (2) small community systems, and (3) self-contained systems. Individual reuse systems are presented in Chap. 13. Small community and self-contained systems for reuse are also presented in Chap. 13.

Treatment operations and processes described in this section include packed-bed filtration, membrane filtration (microfiltration, ultrafiltration, nanofiltration, and reverse osmosis), chemical precipitation for phosphorus removal, disinfection with chlorine, and disinfection with ultraviolet radiation. Other operations and processes are occasionally used for repurification of tertiary effluent (such as carbon adsorption); however, these operations and processes are normally associated with large, centralized wastewater treatment systems (Tchobanoglous and Burton, 1991).

Typical Reclamation Flow Diagrams

In California, for unrestricted urban reuse, where public exposure is likely in the reuse application, the use of a well oxidized, filtered, and disinfected effluent is

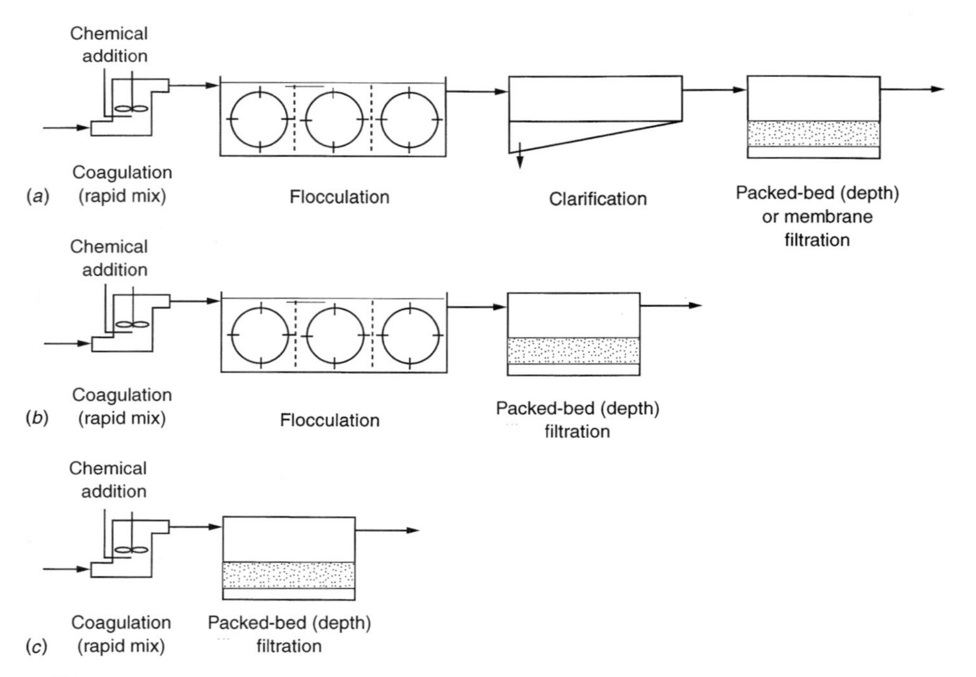

FIGURE 12-3
Reclamation flow diagrams to comply with California reclamation standards: (*a*) full treatment, (*b*) direct filtration, and (*c*) contact filtration.

TABLE 12-5
Typical design criteria for tertiary treatment prior to reuse*

Design parameter	Unit	Value: Range	Value: Typical
Coagulation—rapid mix			
Hydraulic detention time	s	0.5–5	<1
Flocculation			
Hydraulic detention time	min	10–30	20
Velocity gradient *G*	s^{-1}	20–100	40
Mixing energy × detention time (G·t)	Unitless	20,000–150,000	50,000
Sedimentation			
Peak overflow rate	gal/ft^2·d	800–1000	800
Filtration			
Rate with one filter out of service	gal/ft^2·in	4–6	5
Chlorination			
Rapid-mix detention time	s	<1	<1
Peak-flow modal contact time	min	30–120	90
Reactor design		Plug-flow	
UV disinfection			
Dosage at peak wk flow	mW·s/cm^2	100–160	140
Reactor design		Plug-flow	

*Adapted from Richard, Asano, and Tchobanoglous (1992).

required currently. There are three principal flow diagrams for tertiary filtration, as shown in Fig. 12-3—full treatment, direct filtration, and contact filtration. Typical design criteria for tertiary treatment prior to reuse are presented in Table 12-5.

Full treatment. The full or complete treatment flow diagram is essentially a water treatment flow diagram that involves coagulation, flocculation, clarification, filtration, and disinfection (see Fig. 12-3*a*). The treatment performance for this flow diagram in terms of TSS and pathogens is substantial, and virus-free water can be attained following disinfection using secondary effluent as the feed source (Asano et al., 1992). Full treatment has also proven to be effective for the removal of protozoan oocysts and cysts. A typical example of a full-treatment system for the production of reclaimed water is shown in Fig. 12-4.

Direct filtration. Direct filtration is the full treatment flow diagram with the clarifier removed (see Fig. 12-3*b*). The flocculation basin is used to develop a floc following the addition of metal salts and/or polymer. A turbidity value of 10 NTU in the conventional secondary effluent is typically used as the economic dividing line when deciding between the use of full treatment versus contact or direct filtration (Tchobanoglous and Burton, 1991). When lagoon effluent is being filtered,

FIGURE 12-4
Typical effluent treatment plant for reclamation and reuse applications. Lime flocculating clarifier is shown on left, lime storage tower in background, and dual-medium filters are shown on the right.

higher influent turbidities are expected, and the use of chemicals is usually required.

If the turbidity of the conventional secondary effluent is less than 7 to 9 NTU, direct or contact filtration can be used to meet the stringent turbidity requirement of 2 NTU imposed by the California Wastewater Reclamation Criteria (California DHS, 1994) without the need for chemicals. When the turbidity of the secondary effluent is above 7 to 9 NTU, a filtered effluent turbidity value of 2 NTU will usually require chemical addition and the filters may need to be operated at lower loading rates. Direct filtration is only partially effective for the removal of protozoan oocysts and cysts.

Contact filtration. In contact filtration, flocculation and clarification facilities are omitted and the system relies on inline coagulation prior to filtration (see Fig. 12-3*c*). In the *Pomona Virus Study* (San. Dist. LAC, 1977), it was demonstrated that, with adequate disinfection contact time, the equivalent of full treatment virus kill could be achieved. Contact filtration is also only partially effective for the removal of protozoan oocysts and cysts.

Typical Repurification Flow Diagrams

With the technology now available, repurification systems can be developed to produce any quality of finished water that is desired. The constraints include cost, energy use, by-product disposal, and social acceptance. Repurification for indirect

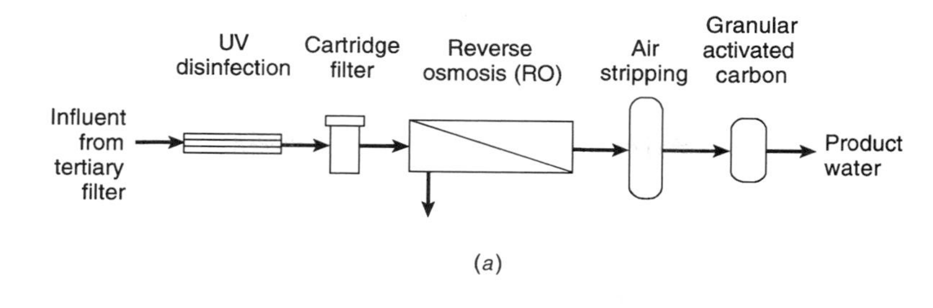

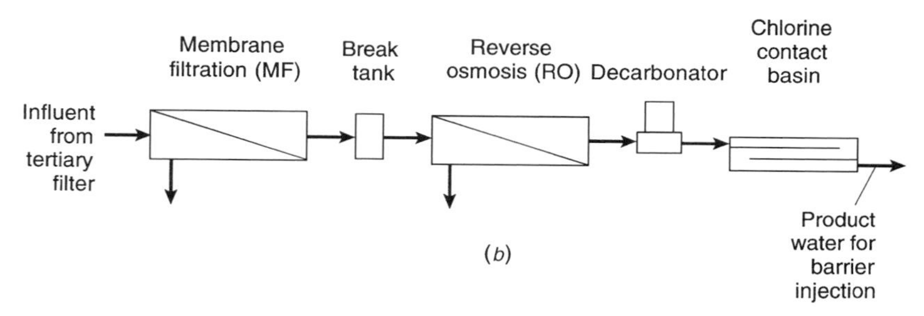

FIGURE 12-5
Typical flow diagrams for wastewater repurification and reuse: (*a*) for indirect potable reuse and (*b*) for use as barrier water for the control of salt water intrusion.

reuse as a potable water supply requires the highest level of treatment, the lowest concentrations of organics, and a "pathogen-free" water. Typical flow diagrams for repurification and reuse of wastewater are presented in Fig. 12-5. Most of the components of the flow diagrams illustrated in Fig. 12-5 are considered in the following sections.

Comparison of Performance for Various Repurification Flow Diagrams

Performance expectations for various technologies are of importance with respect to their use in repurification and reuse applications. Treatment levels achievable with various combinations of unit operations and processes used for wastewater treatment are reported in Table 12-6. Treatment technologies for individual and small systems are reported in the last four lines of Table 12-6. It is interesting to note that flow diagrams that contain septic tanks and intermittent sand filters will result in effluent qualities that are comparable to or superior to activated sludge with filtration. Where effluent disinfection is required, the use of chlorine and related compounds and conventional low-pressure UV radiation are used. The use of membrane processes, such as reverse osmosis, can be added to any flow diagram to reduce the mineral content of the effluent. Clearly, technology is now available for the production of high-quality water from wastewater, regardless of system size.

TABLE 12-6

Treatment levels achievable with various combinations of unit operations and processes used for wastewater repurification*

Treatment process	Typical effluent quality, mg/L except turbidity, NTU						
	TSS	BOD_5	COD	Total N	NH_3-N	PO_4-P	Turb
Activated sludge + filtration	4–6	< 5–10	30–70	15–35	15–25	4–10	0.3–5
Activated sludge + filtration + carbon adsorption	< 5	< 5	5–20	15–30	15–25	4–10	0.3–3
Activated sludge/nitrification, single stage	10–25	5–15	20–45	20–30	1–5	6–10	5–15
Activated sludge/nitrification-denitrification separate stages	10–25	5–15	20–35	5–10	1–2	6–10	5–15
Metal salt addition to activated sludge + nitrification-denitrification + filtration	< 5–10	< 5–10	20–30	3–5	1–2	< 1	0.3–2
Biological phosphorus removal†	10–20	5–15	20–35	15–25	5–10	< 2	5–10
Biological nitrogen and phosphorus removal† + filtration	< 10	< 5	20–30	< 5	< 2	< 2	0.3–2
Activated sludge + filtration + carbon adsorption + reverse osmosis	< 1	< 1	5–10	< 2	< 2	< 1	0.01–1
Activated sludge/nitrification-denitrification and phosphorus removal + filtration + carbon adsorption + reverse osmosis	< 1	<1	2–8	< 0.1–0.5	< 0.1–0.5	< 0.1–0.5	0.01–1
Septic tank with effluent filter vault	25–40	80–120	120–260	40–80	30–60	8–12	10–20
Septic tank with internal trickling filter	20–40	40–60	60–100	10–20	8–16	8–12	8–20
Septic tank with effluent filter vault + intermittent sand filtration	0–5	0–5	10–40	10–20	0–2	6–10	0.01–2
Septic tank + absorbent biofilter	5–15	5–15	30–80	10–20	8–16	6–10	1–2

*Adapted, in part, from Tchobanoglous and Burton (1991).

† Removal process occurs in the main flowstream as opposed to sidestream treatment.

12-3 REMOVAL OF RESIDUAL SOLIDS BY PACKED-BED DEPTH FILTRATION

Secondary effluent typically contains 5 to 30 mg/L of TSS. For most water reuse applications, disinfection will be required. For stringent effluent disinfection requirements, such as 2.2 MPN/100 mL of total coliforms, the turbidity and suspended solids must be reduced to avoid shielding of pathogens and indicator organisms by the solids. Where the treated effluent is to be repurified by reverse osmosis, the residual TSS must be removed to avoid membrane fouling. The process now used most commonly to remove the residual TSS from treated effluent is packed-bed (depth) filtration (see Fig. 12-3). In the future, it is anticipated that membrane filtration will replace depth filtration, as it is also effective for the removal of *Cryptosporidium* oocysts and *Giardia* cysts.

To understand the filtration process the material presented in this section deals with: (1) the definition of filtration and classification of filtration processes, (2) a general description of the filtration process, (3) a review of available filtration technologies, (4) a review of filter hydraulics, (5) modeling the filtration process, (6) filtration process variables, (7) filter design considerations, (8) hydraulic control of granular medium filters, (9) effluent filtration with chemical addition, and (10) need for pilot plant studies. Because the literature dealing with filtration is so voluminous, the information presented in this section is meant to serve as an introduction to the subject.

Definition of Filtration and Classification of Filtration Processes

Filtration is the process by which particulate material present in a liquid is removed from the liquid by means of a porous medium or membrane which retains the particulate matter but allows the liquid to pass. A general classification of the filtration processes commonly used in wastewater engineering is presented in Fig. 12-6.

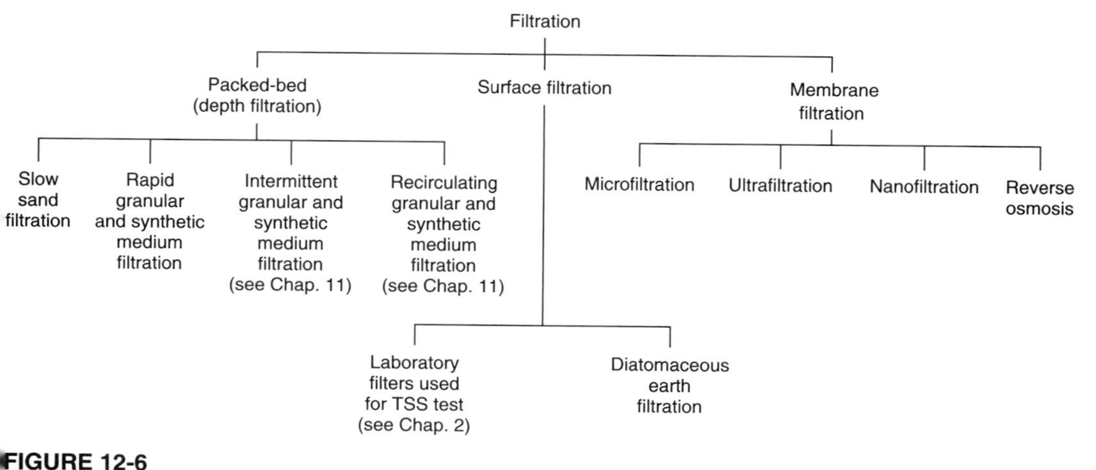

FIGURE 12-6
Classification of filtration processes used in wastewater management.

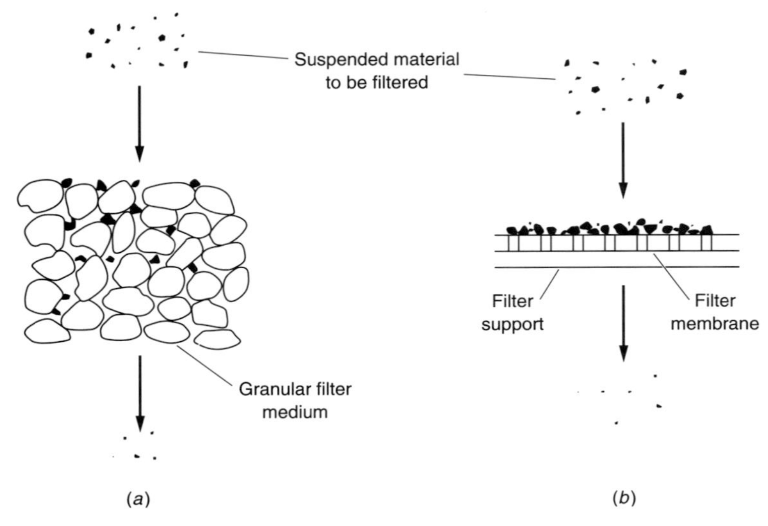

FIGURE 12-7
Definition sketch for filtration processes: (*a*) depth filtration and (*b*) surface filtration.

The two principal filtration processes (depth and surface) are illustrated graphically in Fig. 12-7. In depth filtration, as shown in Fig. 12-7*a*, the removal of suspended material occurs within and on the surface of the filter bed. In surface filtration (see Fig. 12-7*b*) the suspended material is removed by straining through a thin membrane or other straining surface.

Historically, the first filtration process developed for the treatment of wastewater was the slow sand filter (typical filtration rates 0.75 to 1.5 $gal/ft^2 \cdot d$), discussed previously in Chap. 11. The rapid packed-bed (depth) filter (typical filtration rates 2.0 to 10.0 $gal/ft^2 \cdot min$), the subject of this section, was developed to treat larger volumes of water in a facility with a smaller footprint. The membrane filters used to determine the concentration of TSS (see Fig. 2-5) and a kitchen colander are also examples of surface filters. Membrane technologies for the removal of suspended and dissolved solids are considered in Sec. 12-4.

Description of the Filtration Process

To introduce the subject of depth filtration through a porous medium, it is useful to describe: (1) the physical features of a conventional packed-bed filter, (2) the *filtration process* in which suspended material is removed from the liquid, (3) the particle removal mechanisms which bring about the removal of suspended material within the filter, and (4) the *backwash process* in which the material that has been retained within the filter is removed. Historically, the packed-bed filter in which the complete filtration cycle (filtration and backwashing) occurs sequentially is the most commonly used. However, during the past 20 years, a number of new types of fil-

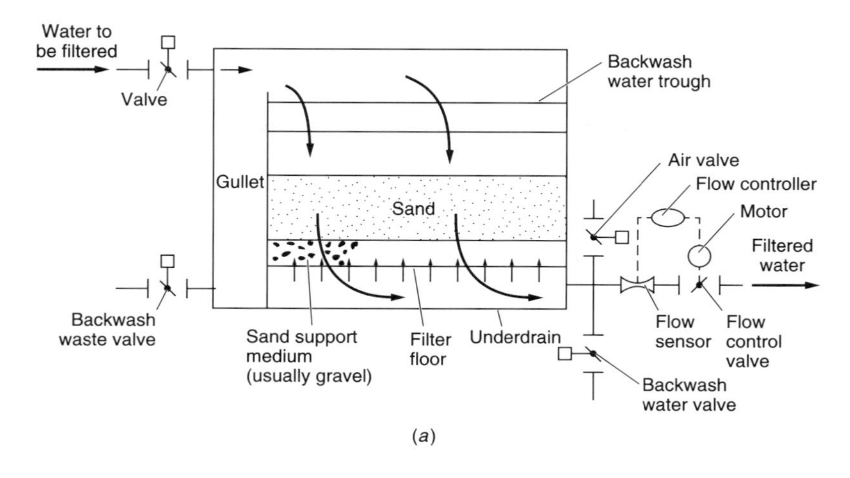

(*a*)

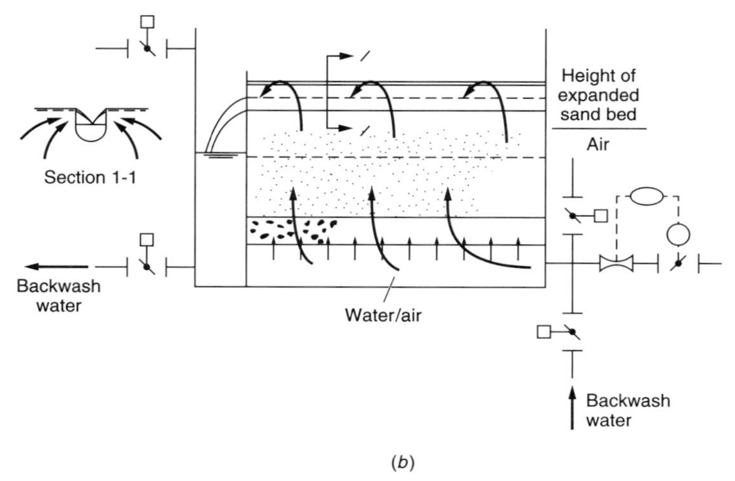

(*b*)

FIGURE 12-8
General features and operation of a conventional rapid packed-bed filter: (*a*) flow during filtration cycle and (*b*) flow during backwash cycle (adapted from Tchobanoglous and Schroeder, 1985).

ters have been developed in which these two processes (filtration and backwashing) occur either semicontinuously or continuously. These technologies are considered subsequently in the discussion of filter technologies.

Physical features of a conventional rapid granular-medium filter. The general features of a rapid granular-medium filter are illustrated in Fig. 12-8. As shown, the filtering medium (sand in this case) is supported on a gravel layer, which, in turn, rests on the filter underdrain system. Filtered water, collected in the underdrain, is discharged to a storage reservoir or to the distribution system. The underdrain system is also used to reverse the flow to backwash the filter. The water to be filtered enters the filter from an inlet channel. The hydraulic control of the filter is described in a subsequent section.

The filtration process. During filtration, wastewater containing suspended matter is applied to the top of the filter bed (Fig. 12-8*a*). As the water passes through the granular medium, the suspended matter in the wastewater is removed by a variety of mechanisms, including mechanical and chance contact straining, impaction, interception, adsorption, flocculation, and sedimentation. With the passage of time, as material accumulates within the interstices of the granular medium, the headloss through the filter starts to build up beyond the initial value, as shown in Fig. 12-9. After some period of time, the operating headloss or effluent turbidity reaches some predetermined headloss or turbidity value, and the filter must be cleaned. Ideally, the time required for the headloss buildup to reach the preselected terminal value should correspond to the time when the suspended solids in the effluent reach the preselected terminal value for acceptable quality (Tchobanoglous and Schroeder, 1985).

Particle-removal mechanisms. The principal mechanisms that are believed to contribute to the removal of material within a granular- and synthetic-medium filter are identified and described in Table 12-7. The major removal mechanisms (the first six listed in Table 12-7) are illustrated pictorially in Fig. 12-10. Straining has been identified as the principal mechanism that is operative in the removal of suspended solids during the filtration of settled secondary effluent from biological treatment processes (Tchobanoglous and Eliassen, 1970). Other mechanisms are probably also operative even though their effects are small and, for the most part, masked by the straining action. These other mechanisms include interception, impaction, and adhesion. In fact, it is reasonable to assume that the removal of some of the smaller particles shown in Fig. 12-10 must be accomplished in two steps involving: (1) the transport of the particles to the surface where they will be removed and (2) the

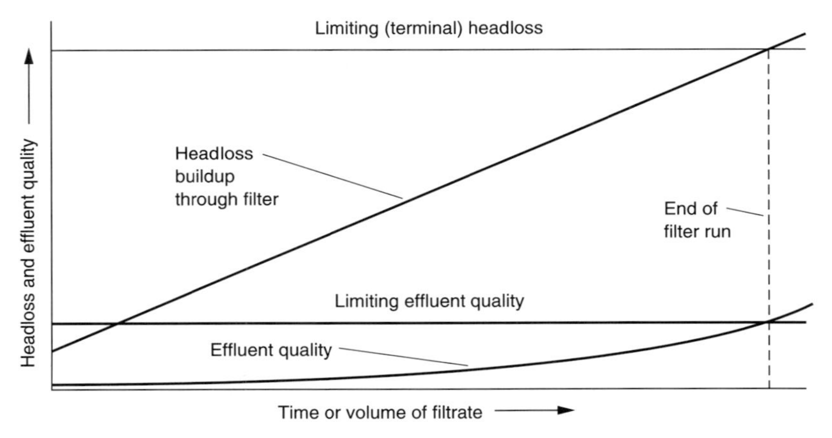

FIGURE 12-9
Definition sketch for length of filter run based on headloss and effluent turbidity.

TABLE 12-7
Principal mechanisms contributing to removal of material within a packed-bed filter*

Mechanism	Description
1. Straining	
a. Mechanical	Particles larger than the pore space of the filtering medium are strained out mechanically.
b. Chance contact	Particles smaller than the pore space are trapped within the filter by chance contact.
2. Sedimentation	Particles settle on the filtering medium within the filter.
3. Impaction	Heavy particles will not follow the flow streamlines.
4. Interception	Many particles that move along in the streamline are removed when they come in contact with the surface of the filtering medium.
5. Adhesion	Particles become attached to the surface of the filtering medium as they pass by. Because of the force of the flowing water, some material is sheared away before it becomes firmly attached and is pushed deeper into the filter bed. As the bed becomes clogged, the surface shear force increases to a point at which no additional material can be removed. Some material may break through the bottom of the filter, causing the sudden appearance of turbidity in the effluent.†
6. Flocculation	Large particles overtake small particles and join them, forming larger particles. These particles are then removed by one or more of the above removal mechanisms.
7. Chemical adsorption *a.* Bonding *b.* Chemical interaction 8. Physical adsorption *a.* Electrostatic forces *b.* Electrokinetic forces *c.* van der Waals forces	Once a particle has been brought in contact with the surface of the filtering medium or with other particles, either one of these mechanisms, or both, may be responsible for holding it there.
9. Biological growth	Biological growth within the filter will reduce the pore volume and may enhance the removal of particles with any of the above removal mechanisms (1 through 5).

*From Tchobanoglous and Burton (1991).
† Turbidity breakthrough does not occur with filters that operate continuously.

removal of particles by one or more of the operative removal mechanisms. O'Melia and Stumm (1967) have identified these two steps as transport and attachment.

Backwash process. When either the terminal headloss or the limiting effluent quality values are reached, the filter must be cleaned. Most granular-medium filters are cleaned by reversing the flow through the filter bed (see Fig. 12-8*b*). Filtered water is pumped through the filter bed at a rate sufficient to partially expand the bed, causing the particles of the filtering medium to abrade against each other. The suspended matter arrested within the filter is removed by the shear forces created by backwash water as it moves up through the bed. The suspended solids removed from the filter are removed with the washwater in the washwater troughs. Backwash hydraulics are considered in the section dealing with filter hydraulics.

Filtration Technologies

A number of filtration technologies are now available for the filtration of treated wastewater. The principal types of packed-bed filters that have been used for the filtration of wastewater are identified in Table 12-8 and illustrated in Fig. 12-11. As shown in Table 12-8, the filters can be classified in terms of their operation as semicontinuous or continuous. Within each of these two classifications there are a number of different types of filters depending on bed depth (e.g., shallow, conventional, and deep bed), the type of filtering medium used (mono-, dual-, and multimedium), whether the filtering medium is stratified or unstratified, the type of operation (downflow or upflow), and the method used for the management

TABLE 12-8
Comparison of principal types of packed-bed filters*

Type of filter (common name)	Type of filter operation	Filter bed details			Typical direction of fluid flow
		Type of filter bed	Filtering medium	Typical bed depth, in	
Conventional	Semicontinuous	Monomedium (stratified or unstratified)	Sand or anthracite	30	Downward
Conventional	Semicontinuous	Dual-medium (stratified)	Sand and anthracite	36	Downward
Conventional	Semicontinuous	Multimedium (stratified)	Sand, anthracite, and garnet	36	Downward
Deep bed	Semicontinuous	Monomedium (stratified or unstratified)	Sand or anthracite	72	Downward
Deep bed	Semicontinuous	Monomedium (stratified)	Sand or anthracite	72	Upward
Deep bed	Continuous	Monomedium (unstratified)	Sand	72	Upward
Pulsed bed	Semicontinuous	Monomedium (stratified)	Sand	11	Downward
Fuzzy Filter	Semicontinuous	Monomedium (unstratified)	Synthetic fiber	24†	Upward
Traveling bridge	Continuous	Monomedium (stratified)	Sand	11	Downward
Traveling bridge	Continuous	Dual-medium (stratified)	Sand and anthracite	16	Downward

*Adapted from Tchobanoglous and Burton (1991).
† Compressed depth.

of solids (i.e., surface or internal storage). For the mono- and dual-medium semicontinuous filters, a further classification is based on the driving force (e.g., gravity or pressure). Another important distinction that must be noted for the filters identified in Table 12-8 is whether they are proprietary or individually designed (to be discussed later). The following six types of porous medium filters commonly used for wastewater filtration are described below: (1) conventional downflow filters, (2) deep-bed downflow filters, (3) deep-bed upflow continuous-backwash filter, (4) pulsed-bed filter, (5) synthetic-medium filter, and (6) traveling-bridge filters. These filter technologies are described briefly in this section.

Conventional downflow filters. Single-, dual-, or multimedium filter materials are utilized in conventional downflow filters. Typically sand or anthracite is used

Backwash operation	Flowrate through filter	Solids storage location	Remarks	Type of design
Batch	Constant/ variable	Surface and upper bed	Rapid headloss buildup	Individual
Batch	Constant/ variable	Internal	Dual-medium design used to extend filter run length	Individual
Batch	Constant/ variable	Internal	Multimedium design used to extend filter run length	Individual
Batch	Constant/ variable	Internal		Individual
Batch	Constant	Internal		Proprietary
Continuous	Constant	Internal	Sand bed moves in countercurrent direction to fluid flow	Proprietary
Batch	Constant	Surface and upper bed	Air pulses used to break up surface mat and increase run length	Proprietary
Batch	Constant	Internal		Proprietary
Semicontinuous	Constant	Surface and upper bed	Individual filter cells backwashed sequentially	Proprietary
Semicontinuous	Constant	Surface and upper bed	Individual filter cells backwashed sequentially	Proprietary

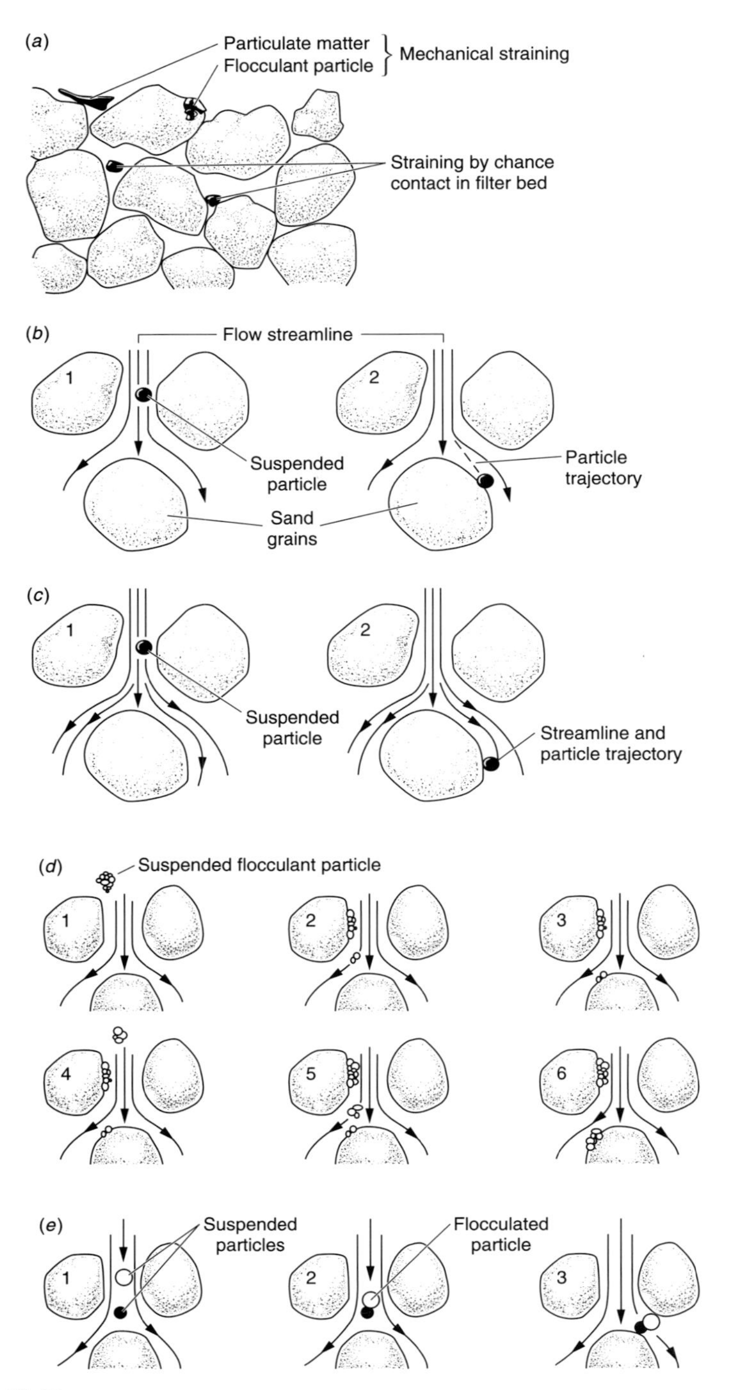

FIGURE 12-10
Definition sketch for removal of total suspended solids in a granular-medium filter: (*a*) by straining, (*b*) by sedimentation and inertial impaction, (*c*) by interception, (*d*) by adhesion, and (*e*) by flocculation (adapted from Tchobanoglous and Schroeder, 1985).

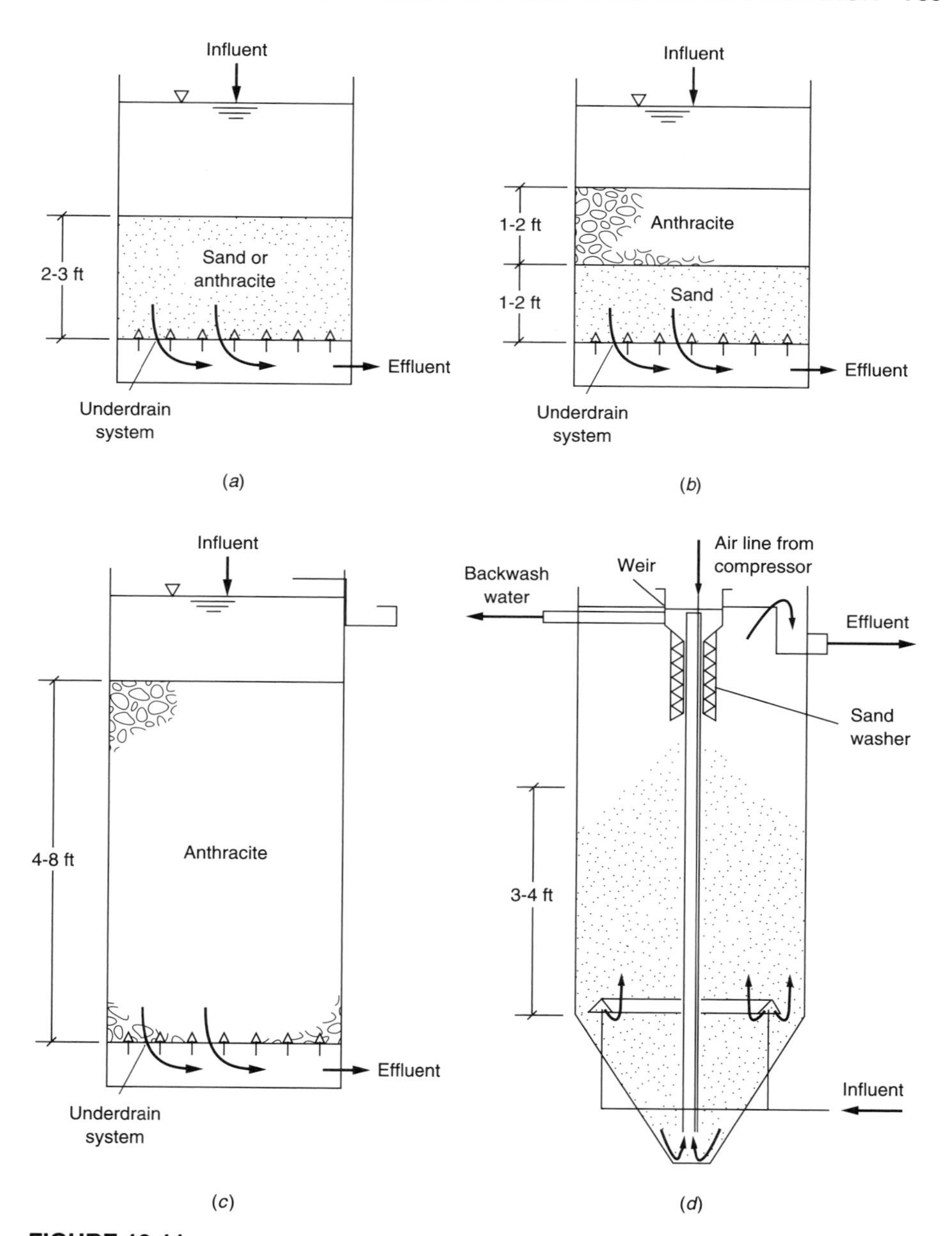

FIGURE 12-11
Definition sketches for the principal types of granular-medium filters: (*a*) conventional monomedium downflow filter, (*b*) conventional dual-medium downflow filter, (*c*) conventional monomedium deep-bed downflow filter, (*d*) continuous backwash deep-bed upflow filter.

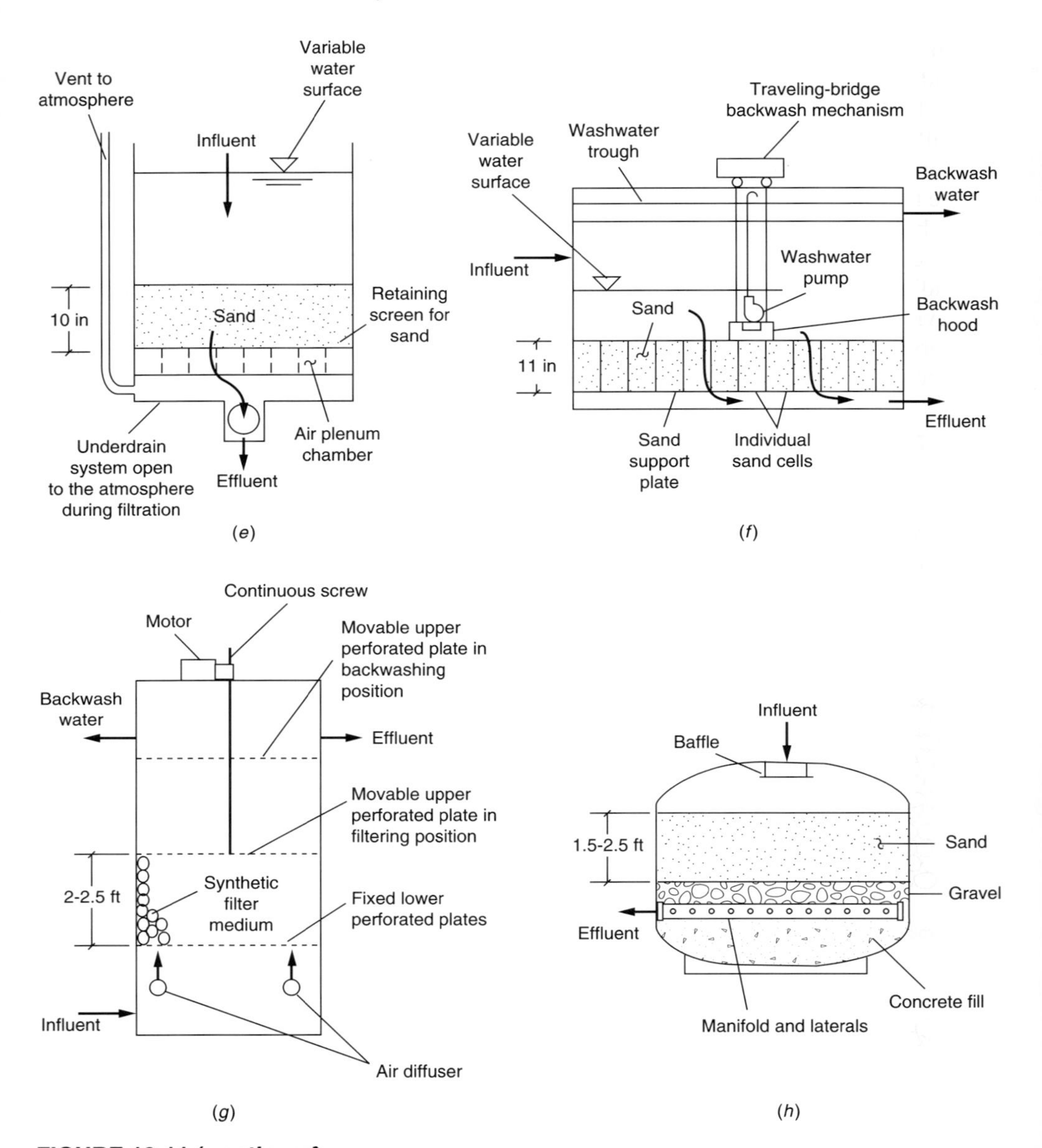

FIGURE 12-11 (*continued*)
Definition sketches for the principal types of granular-medium filters: (*e*) pulsed-bed filter, (*f*) traveling-bridge filter, (*g*) synthetic-medium filter, and (*h*) pressure filter (adapted in part from Tchobanoglous and Burton, 1991).

as the filtering material in single-medium filters (see Fig. 12-11*a*). Dual-medium filters usually consist of a layer of anthracite over a layer of sand (see Fig. 12-11*b*). Other combinations include: (1) activated carbon and sand, (2) resin beads and sand, and (3) resin beads and anthracite. Multimedium filters typically consist of a layer of anthracite over a layer of sand over a layer of garnet or ilmenite. Other combinations include (1) activated carbon, anthracite, and sand; (2) weighted, spherical resin beads, anthracite, and sand; and (3) activated carbon, sand, and garnet.

Dual- and multimedium and deep-bed monomedium filter beds were developed to allow the suspended solids in the liquid to be filtered to penetrate farther into the filter bed, and thus use more of the solids-storage capacity available within the filter (see Fig. 12-12). By comparison, in shallow monomedium beds, most of the removal has been observed to occur in the upper few millimeters of the bed. The penetration of the solids farther into the bed also permits longer filter runs because the buildup of headloss is reduced. The operation of conventional downflow filters is as described previously, and is illustrated in Fig. 12-8*a*. Typical design data for monomedium filters and dual- and multimedium filters are presented in Tables 12-9 and 12-10, respectively.

Deep-bed downflow filters. The deep-bed downflow filter is similar to the conventional downflow filter with the exception that the depth of the filter bed and the size of the filtering medium (usually anthracite) are greater than the corresponding values in a conventional filter. Because of the greater depth and larger medium size, more solids can be stored within the filter bed and the run length can be extended.

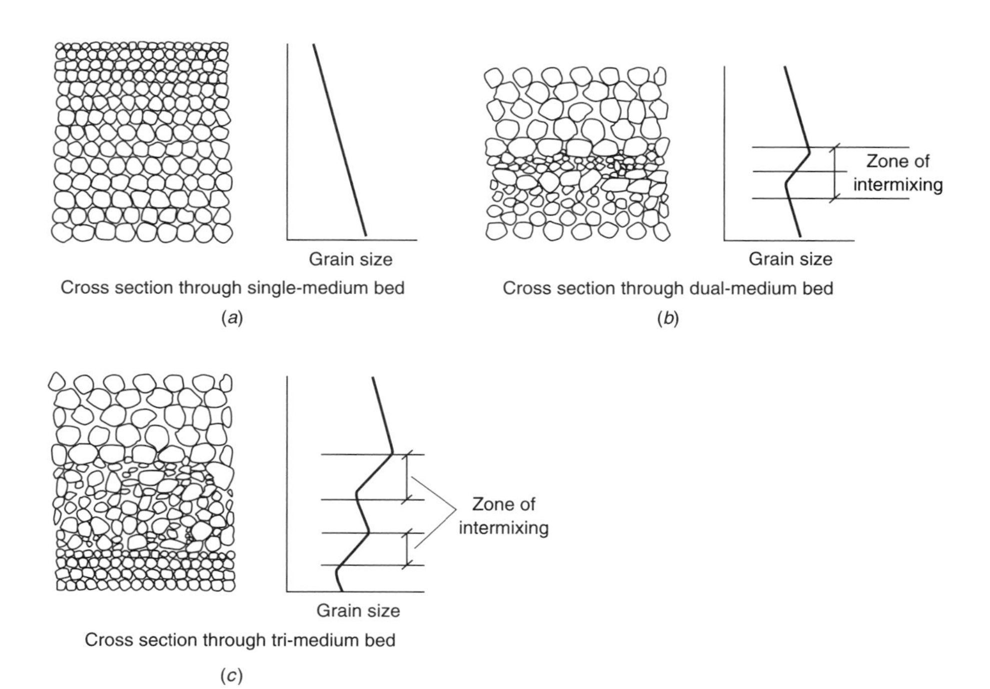

FIGURE 12-12
Schematic diagram of filter beds illustrating potential increase in storage capacity: (*a*) single-medium, (*b*) dual-medium, and (*c*) multimedium.

TABLE 12-9
Typical design data for packed-bed filters with mono-media*

Characteristic	Unit	Value	
		Range	Typical
Shallow-bed (stratified)			
Anthracite (ρ = 1.4–1.75, α = 0.56–0.60)			
Depth	in	12–20	16
Effective size	mm	0.8–1.5	1.2
Uniformity coefficient		1.3–1.8	1.6
Sand (ρ = 2.55–2.65, α = 0.40–0.46)			
Depth	in	10–12	11
Effective size	mm	0.35–0.6	0.45
Uniformity coefficient		1.2–1.6	1.5
Filtration rate	gal/ft^2·min	2–6	3
Conventional (stratified)			
Anthracite			
Depth	in	24–36	30
Effective size	mm	0.8–1.8	1.3
Uniformity coefficient		1.3–1.8	1.6
Sand			
Depth	in	20–36	24
Effective size	mm	0.4–0.8	0.5
Uniformity coefficient		1.2–1.6	1.5
Filtration rate	gal/ft^2·min	2–6	3
Deep-bed (unstratified)			
Anthracite			
Depth	in	36–84	60
Effective size	mm	1.5–3.0	2.0
Uniformity coefficient		1.3–1.8	1.6
Sand			
Depth	in	36–72	48
Effective size	mm	1.5–2.0	1.8
Uniformity coefficient		1.2–1.6	1.5
Filtration rate	gal/ft^2·min	2–10	5
Fuzzy Filter			
Synthetic medium			
Depth	in	24–36	30
Effective size	mm	25–30	28
Uniformity coefficient		1.1–1.2	1.1
Filtration rate	gal/ft^2·min	15–30	20

*Adapted from Tchobanoglous (1988).

Deep-bed upflow continuous backwash filters. In this filter, shown schematically in Fig. 12-13, the wastewater to be filtered is introduced into the bottom of the filter where it flows upward through a series of riser tubes and is distributed evenly into the sand bed through the open bottom of an inlet distribution hood. The water then flows upward through the downward-moving sand. The clean filtrate exits

TABLE 12-10
Typical design data for dual- and multimedium packed-bed filters*

Characteristic	Unit	Value	
		Range	Typical
Dual-medium			
Anthracite (ρ = 1.4–1.75, α = 0.56–0.60)			
Depth	in	12–30	24
Effective size	mm	0.8–2.0	1.3
Uniformity coefficient		1.3–1.8	1.6
Sand (ρ = 2.55–2.65, α = 0.40–0.46)			
Depth	in	6–12	12
Effective size	mm	0.4–0.7	0.55
Uniformity coefficient		1.2–1.6	1.5
Filtration rate	gal/ft^2·min	2–10	5
Multimedium			
Anthracite (top layer of quadmedial filter)			
Depth	in	8–20	16
Effective size	mm	1.0–2.0	1.5
Uniformity coefficient		1.5–1.8	1.6
Anthracite (second layer of quadmedial filter)			
Depth	in	4–16	8
Effective size	mm	1.0–1.6	1.1
Uniformity coefficient		1.5–1.8	1.6
Anthracite (top layer of trimedial filter)			
Depth	in	8–20	16
Effective size	mm	1.0–2.0	1.4
Uniformity coefficient		1.4–1.8	1.6
Sand			
Depth	in	8–16	10
Effective size	mm	0.4–0.6	0.45
Uniformity coefficient		1.3–1.8	1.6
Garnet (ρ = 3.8–4.3, α = 0.42–0.55) or ilmenite (ρ = 4.5, α = 0.40–0.5)			
Depth	in	2–6	4
Effective size	mm	0.2–0.5	0.3
Uniformity coefficient		1.5–1.8	1.6
Filtration rate	gal/ft^2·min	2–10	5

*Adapted from Tchobanoglous and Schroeder (1985) and Tchobanoglous (1988).

from the sand bed, overflows a weir, and is discharged from the filter. At the same time, sand particles, along with trapped solids, are drawn downward into the suction of an airlift pipe which is positioned in the center of the filter. A small volume of compressed air, introduced into the bottom of the airlift, draws sand, solids, and water upward through the pipe by creating a fluid with a density less than 1.

Impurities are scoured (abraded) from the sand particles during the turbulent upward flow. Upon reaching the top of the airlift, the dirty slurry spills over into the central reject compartment. With the filtrate weir set above the reject weir, a steady stream of clean filtrate flows upward, countercurrent to the movement of sand, through the washer section. The upflow liquid carries away the solids and reject

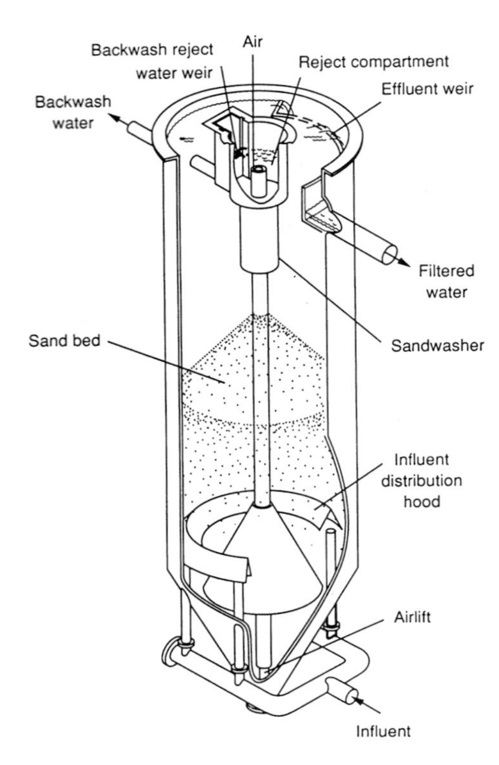

FIGURE 12-13
Schematic view of deep-bed upflow continuous backwash filter (courtesy Parkson Corporation).

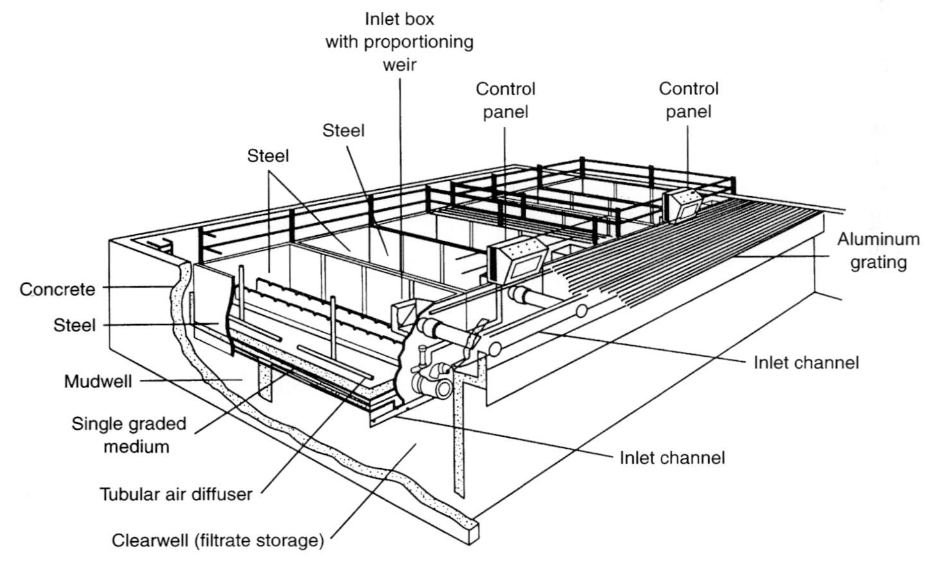

FIGURE 12-14
Schematic view of pulsed-bed filter (courtesy U.S. Filter).

water. Because the sand has a higher settling velocity than the removed solids, the sand is not carried out of the filter. The sand is cleaned further as it moves down through the washer. The cleaned sand is redistributed onto the top of the sand bed, allowing for a continuous uninterrupted flow of filtrate and reject water.

Pulsed-bed filters. The pulsed-bed filter (see Fig. 12-14) is a downflow gravity filter with an unstratified shallow layer of fine sand as the filtering medium. The shallow bed is used for solids storage, as opposed to other shallow-bed filters where solids are principally stored on the sand surface. An unusual feature of this filter is the use of an air pulse to disrupt the sand surface and thus allow penetration of suspended solids into the bed. The air pulse process involves forcing a volume of air, trapped in the underdrain system, up through the shallow filter bed to break up the surface mat of solids and renew the sand surface. When the solids mat is disturbed, some of the trapped material is suspended into the admixture over the sand, but most of the solids are entrapped within the filter bed. The intermittent air pulse causes a folding over of the sand surface, burying solids within the medium and regenerating the filter bed surface. The filter continues to operate with intermittent pulsing until a terminal headloss limit is reached. The filter then operates in a conventional backwash cycle to remove solids from the sand. It should be noted that, during normal operation, the filter underdrain is not flooded as it is in a conventional filter.

Synthetic-medium filters. Synthetic-medium filters are new to wastewater reclamation. A synthetic-medium filter developed and marketed by the Schreiber Corporation features a highly porous synthetic filter medium developed in Japan. The porous synthetic filter medium has a diameter of approximately 1.25 in (31 mm). On the basis of displacement tests, the porosity of the uncompacted quasi-spherical filter medium itself is estimated to be about 88 to 90 percent, and the porosity of the filter bed made up of the filter medium is approximately 94 percent. Unusual features of the filter include the fact that the porosity of the filter bed can be modified by compressing the filter medium and the fact that size of the filter bed is increased mechanically to backwash the filter (see Fig. 12-15). The filter medium also represents a departure from conventional filter media in that the fluid to be filtered flows through the medium as opposed to flowing around the filtering medium, as in sand and anthracite filters. The filter has been designated the Fuzzy Filter. As a result of the high porosity, filtration rates of 10 to 30 gal/ft^2·min (400 to 1200 L/m^2·min) have been pilot tested.

In the filtering mode, secondary effluent is introduced in the bottom of the filter. The influent wastewater flows upward through the filter medium, retained by two porous plates, and is discharged from the top of the filter. To backwash the filter, the upper porous plate is raised mechanically. While flow to the filter continues, air is introduced alternately from the left and right sides of the filter below the lower porous plate, causing the filter medium to move in a rolling motion. The filter medium is cleaned by the shearing forces as the wastewater moves past the filter and by abrasion as the filter medium rubs against itself. Wastewater containing the solids removed from the filter is diverted for subsequent processing. To put the filter back into operation after the backwash cycle has been completed, the raised porous plate

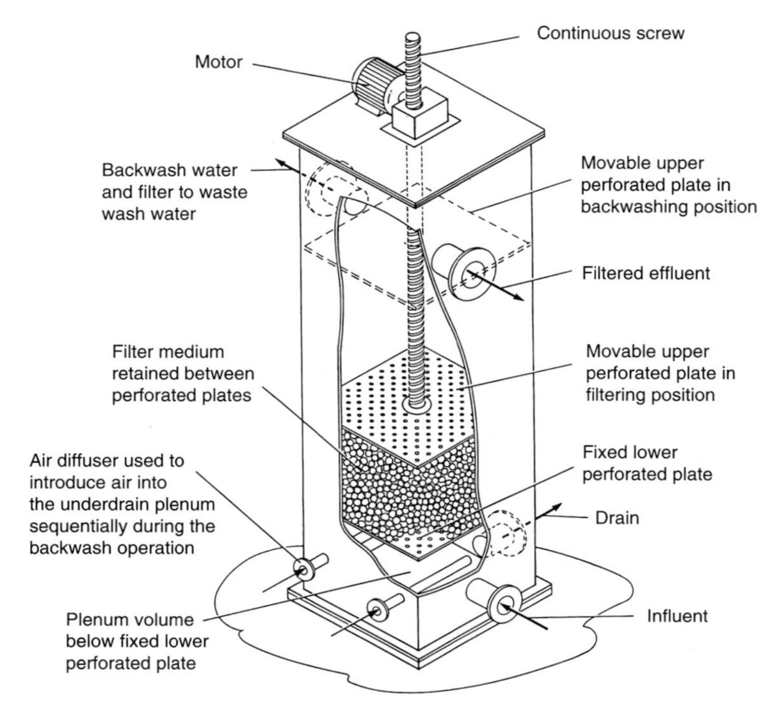

FIGURE 12-15
Schematic diagram of the Fuzzy Filter (courtesy Schreiber Corporation).

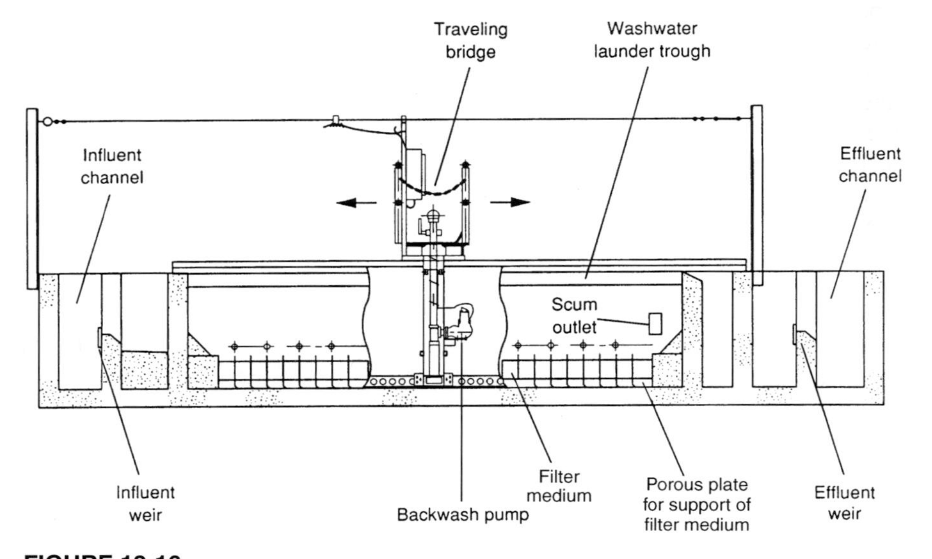

FIGURE 12-16
Traveling-bridge filter (courtesy IDI).

is returned to its original position. After a short flushing cycle, the filtered effluent valve is opened, and filtered effluent is discharged.

Traveling-bridge filters. The traveling-bridge filter (see Fig. 12-16) is a continuous-downflow, automatic backwash, low-head, packed-bed filter. The bed of the filter is divided horizontally into long independent filter cells. Each filter cell contains approximately 11 in (279 mm) of medium. Treated wastewater flows through the medium by gravity and exits to the clearwell plenum via a porous-plate polyethylene underdrain. Each cell is backwashed individually by an overhead traveling-bridge assembly, while all other cells remain in service. Water used for backwashing is pumped directly from the clearwell plenum up through the medium and deposited in a backwash trough. During the backwash cycle, wastewater is filtered continuously through the cells that are not being backwashed. The backwash mechanism includes a surface wash pump to assist in breaking up of the surface matting and "mudballing" in the medium.

Filter Hydraulics

During the past 60 years considerable effort has been devoted to the modeling of the filtration process. The models fall into two general categories: those models used to predict the clean-water headloss through a filter bed and the filter backwash expansion, and those models used to predict the performance of filters for the removal of suspended solids. Filter hydraulics are considered in the following discussion.

Clean-water headloss. Over the years a number of equations have been developed to describe the flow of clean water through a porous medium (Carman, 1937; Fair and Hatch, 1933; Hazen, 1925; Kozeny, 1927; Rose, 1945). The equations developed by these workers are summarized in Table 12-11. In most cases the equations for the flow of clean water through a porous medium are derived from a consideration of the Darcy-Weisbach equation for flow in a closed conduit, and dimensional analysis. The summation term in Eqs. (12-2), (12-6), and (12-8) is included to account for the stratification that occurs in filters. To account for stratification, the mean size of the material retained between successive sieve sizes is assumed to correspond to the mean size of the successive sieves (see Table 11-7 for sieve sizes), assuming that the particles retained between sieve sizes are substantially uniform (Fair and Hatch, 1933).

In applying the equations given in Table 12-11, some confusion exists over the definition of the shape factor. The definition of the shape factor will depend on whether particle surface, volume, or a linear dimension is of importance in the application. In Chap. 5, the shape factor for a particle ϕ_p is defined as the ratio of the surface area of an equivalent sphere to the surface area of the particle, for particles of the same volume. For spherical-shaped particles the specific surface area is:

$$\frac{A_p}{V_p} = \frac{\pi d^3}{\pi d^3/6} = \frac{6}{d} \tag{12-10}$$

TABLE 12-11

Formulas used to compute the clear-water headloss through a granular porous medium

Equation	
Carman-Kozeny (Carman, 1937)	
$h = \frac{f}{\phi}\frac{1-\alpha}{\alpha^3}\frac{L}{d}\frac{V_s^2}{g}$	(12-1)
$h = \frac{1}{\phi}\frac{1-\alpha}{\alpha^3}\frac{LV_s^2}{g}\sum f\frac{p}{d_g}$	(12-2)
$f = 150\frac{1-\alpha}{N_R} + 1.75$	(12-3)
$N_R = \frac{\phi d V_s \rho}{\mu}$	(12-4)
Fair-Hatch (Fair and Hatch, 1933)	
$h = k\nu S^2\frac{(1-\alpha)^2}{\alpha^3}\frac{L}{d^2}\frac{V_s}{g}$	(12-5)
$h = k\nu\frac{(1-\alpha)^2}{\alpha^3}\frac{LV_s}{g}\left(\frac{6}{\phi}\right)^2\sum\frac{p}{d_g^2}$	(12-6)
Rose (Rose, 1945)	
$h = \frac{1.067}{\phi}C_d\frac{1}{\alpha^4}\frac{L}{d}\frac{V_s^2}{g}$	(12-7)
$h = \frac{1.067}{\phi}\frac{LV_s^2}{\alpha^4 g}\sum C_d\frac{p}{d_g}$	(12-8)
$C_d = \frac{24}{N_R} + \frac{3}{\sqrt{N_R}} + 0.34$	(12-9)
Hazen (Hazen, 1925)	
$h = \frac{1}{C}\frac{60}{T+10}\frac{L}{(d_{10})^2}V_h$	(11-3)

Definition of terms

- C = coefficient of compactness (varies from 600 for very closely packed sands that are not quite clean to 1200 for very uniform clean sand)
- C_d = coefficient of drag
- d = grain size diameter, ft (m)
- d_g = geometric mean diameter between sieve sizes d_1 and d_2, $\sqrt{d_1 d_2}$, ft (m)
- d_{10} = effective grain size diameter, mm
- f = friction factor
- g = acceleration due to gravity, 32.2 ft/s^2 (9.81 m/s^2)
- h = headloss, ft (m)
- k = filtration constant, 5 based on sieve openings, 6 based on size of separation
- L = depth of filter bed or layer, ft (m)
- N_R = Reynolds number
- p = fraction of particles (based on mass) within adjacent sieve sizes
- S = shape factor (varies between 6.0 for spherical particles to 8.5 for crushed materials)
- T = temperature, °F
- V_h = superficial (approach) filtration velocity, m/d
- V_s = superficial (approach) filtration velocity, ft/s (m/s)
- α = porosity
- μ = viscosity, lb·s/ft^2 (N·s/m^2)
- ν = kinematic viscosity, ft^2/s (m^2/s)
- ρ = density, slug/ft^3 = lb·s^2/ft^4 (kg/m^3)
- ϕ = particle shape factor (1.0 for spheres, 0.82 for rounded sand, 0.75 for average sand, 0.73 for crushed coal and angular sand)

For irregularly shaped particles of the same volume the specific surface area is:

$$\frac{A_p}{V_p} = \frac{6}{\phi d} = \frac{S}{d} \tag{12-11}$$

where ϕ = shape factor and S = area-to-volume shape factor. In the literature, S has been identified as a shape factor (Fair et al., 1968). The area-to-volume shape factor is used in the Fair-Hatch equation [Eq. (12-5)]. Computation of the clean-water headloss through a filter is illustrated in Example 12-1.

EXAMPLE 12-1. DETERMINATION OF CLEAN-WATER HEADLOSS IN A PACKED-BED FILTER. Determine the clean-water headloss in a filter bed composed of 30 in of uniform sand with the size distribution given below for a filtration rate of 4.0 gal/ft^2·min. Assume that the operating temperature is 20°C. Use the Rose equation [Eq. (12-8)] given in Table 12-11 for computing the headloss. Assume the porosity of the sand in the various layers is 0.40, and use a ϕ value of 0.85 for the sand.

Sieve size or number	Percent of sand retained	Geometric mean size, ft*†
8–10	1	0.00716
10–12	3	0.00601
12–18	16	0.00425
18–20	16	0.00301
20–30	30	0.00232
30–40	22	0.00163
40–50	12	0.00116

*Using sieve size data from Table 11-7 in Chap 11.
† Geometric mean size $= \sqrt{d_1 d_2}$

Solution

1. Set up a computation table to determine the summation term in Eq. (12-8):

$$h = \frac{1.067}{\phi}\frac{LV_s^2}{\alpha^4 g}\sum C_d \frac{p}{d_g}$$

Sieve size or number (1)	Fraction of sand retained (2)	Geometric mean size, ft (3)	Reynolds number (4)	C_d (5)	$C_d(p/d)$ (6) = (5)(2)/(3)
8–10	0.01	0.00716	4.97	6.51	9.09
10–12	0.03	0.00601	4.17	7.56	37.74
12–18	0.16	0.00425	2.95	10.22	384.75
18–20	0.16	0.00301	2.09	13.90	738.87
20–30	0.30	0.00232	1.61	17.61	2,277.16
30–40	0.22	0.00163	1.13	24.40	3,293.25
40–50	0.12	0.00116	0.81	33.30	3,444.83
Summation					10,185.69

a. Determine the C_d value for each geometric mean as illustrated below:

$$C_d = \frac{24}{N_R} + \frac{3}{\sqrt{N_R}} + 0.34$$

$$N_R = \frac{\phi d V_s \rho}{\mu} = \frac{\phi d V_s}{\nu}$$

$$d = 0.00716 \text{ ft}$$

$$V_s = \frac{4 \text{ gal/ft}^2\cdot\text{min}}{7.48 \text{ gal/ft}^3} = 0.534 \text{ ft/min} = 0.0089 \text{ ft/s}$$

$$\nu = 1.091 \times 10^{-5} \text{ ft}^2\text{/s (see App. C)}$$

$$N_R = \frac{(0.85)(0.00716\ \text{ft})(0.0089\ \text{ft/s})}{(1.091 \times 10^{-5}\ \text{ft}^2/\text{s})}$$

$$N_R = 4.97$$

b. Determine C_d:

$$C_d = \frac{24}{4.97} + \frac{3}{\sqrt{4.97}} + 0.34 = 6.51$$

2. Determine the headloss through the stratified filter bed using Eq. (12-8) (see step 1):

$$L = 30\ \text{in} = 2.5\ \text{ft}$$

$$V_s = 0.534\ \text{ft/min} = 8.9 \times 10^{-3}\ \text{ft/s}$$

$$\phi = 0.85$$

$$\alpha = 0.40$$

$$g = 32.2\ \text{ft/s}^2$$

$$h = \frac{1.067}{(0.85)} \frac{(2.5\ \text{ft})(8.9 \times 10^{-3}\ \text{ft/s})^2}{(0.4)^4(32.2\ \text{ft/s}^2)}(10{,}185.69/\text{ft}) = 3.07\ \text{ft}$$

Backwash hydraulics. To understand what happens during the backwash operation, it will be helpful to refer to Fig. 12-17 in which the pressure drop across a packed bed, as the backwash velocity through it increases, is illustrated. Between points *A* and *B*, the bed is stable, and the pressure drop and Reynolds number N_R are related linearly. At point *B*, the pressure drop essentially balances the weight of the filter. Between points *B* and *C* the bed is unstable, and the particles adjust their position to present as little resistance to flow as possible. At point *C*, the loosest possible arrangement is obtained in which the particles are still in contact. Beyond point *C*, the particles begin to move freely but collide frequently so that the motion is similar to that of particles in hindered settling. Point *C* is referred to as the *point of fluidization.* By the time point *D* is reached, the particles are all in motion; and,

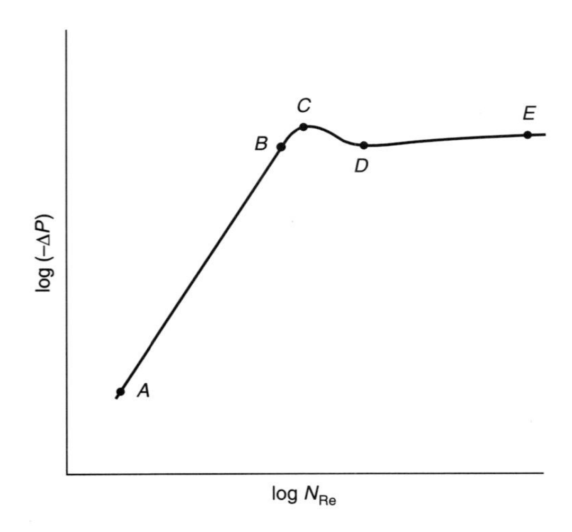

FIGURE 12-17
Schematic diagram illustrating the fluidization of a packed-bed filter (adapted from Foust et al., 1960).

beyond this point, increases in N_R result in very small increases in ΔP as the bed continues to expand and the particles move in more rapid and more independent motion. Ultimately, the particles will stream with the fluid, and the bed will cease to exist at point E.

To expand a filter bed composed of a uniform filter medium hydraulically, the headloss must equal the buoyant mass of the granular medium in the fluid. Mathematically this relationship can be expressed as:

$$h = l_e(1 - \alpha_e)\left(\frac{\rho_s - \rho_w}{\rho_w}\right) \tag{12-12}$$

where h = headloss required to expand the bed
l_e = the depth of the expanded bed
α_e = the expanded porosity
ρ_s = density of the medium
ρ_w = density of water

Because the individual particles are kept in suspension by the drag force exerted by the rising fluid, it can be shown from settling theory (see Sec. 5-7) that:

$$C_D A_p \rho_w \frac{v^2}{2} \phi(\alpha_e) = (\rho_s - \rho_w) g v_p \tag{12-13}$$

where v = face velocity of backwash water, m/s, and $\phi(\alpha_e)$ = correction factor to account for the fact that v is the velocity of the backwash water and not the particle-settling velocity v_p. Other terms are as defined previously.

From experimental studies (Fair, 1951; Richardson and Zaki; 1954) it has been found that the expanded-bed porosity can be approximated by using the following relationships, assuming the Reynolds number is approximately 1:

$$\phi(\alpha_e) = \left(\frac{v_s}{v}\right)^2 = \left(\frac{1}{\alpha_e}\right)^9 \tag{12-14}$$

Thus

$$\alpha_e = \left(\frac{v}{v_s}\right)^{0.22} \tag{12-15}$$

or

$$v = v_s \alpha_e^{4.5} \tag{12-16}$$

where v_s = settling velocity of particle. However, because the volume of the filtering medium per unit area remains constant, $(1 - \alpha)l$ must be equal to $(1 - \alpha_e)l_e$ so that:

$$\frac{l_e}{l} = \frac{1 - \alpha}{1 - \alpha_e} = \frac{1 - \alpha}{1 - (v/v_s)^{0.22}} \tag{12-17}$$

Where the filter medium is stratified, the smaller particles in the upper layers expand first. To expand the entire bed, the backwash velocity must be sufficient to lift the largest particle. To account for filter bed stratification, Eq. (12-17) is modified, assuming that particles retained between sieve sizes are substantially uniform (Fair and Hatch, 1933).

$$\frac{l_e}{l} = (1 - \alpha)\sum \frac{p}{(1 - \alpha_e)} \tag{12-18}$$

where p = fraction of filter medium retained between sieve sizes. Thus, the required backwash velocity and expanded depth can be estimated from Eqs. (12-17) and (12-18), respectively, as illustrated in Example 12-2. Additional details on filter bed expansion may be found in Amirtharajah (1978), Cleasby and Fan (1982), and Dharmarajah and Cleasby (1986); Leva (1959); and Richardson and Zaki (1954).

EXAMPLE 12-2. DETERMINATION OF REQUIRED BACKWASH VELOCITIES FOR FILTER CLEANING. A stratified sand bed with the size distribution given below is to be backwashed at a rate of 30 in/min. Determine the degree of expansion and whether the proposed backwash rate will expand all of the bed. Assume the following data are applicable:

Sieve size or number	Percent of sand retained	Geometric mean size, mm
8–10	1	2.18
10–12	3	1.83
12–18	16	1.30
18–20	16	0.92
20–30	30	0.71
30–40	22	0.50
40–50	12	0.35

1. Granular medium is sand
2. Porosity, α = 0.4
3. Specific gravity of sand = 2.65
4. Depth of filter bed = 36 in
5. Temperature = 20°C

Solution

1. Set up computation table to determine the summation term in Eq. (12-18):

$$\frac{l_e}{l} = (1 - \alpha)\sum \frac{p}{(1 - \alpha_e)}$$

Sieve size or number (1)	Percent of sand retained* (2)	Geometric mean size, mm (3)	v_s, ft/s (4)	v/v_s (5)	α_e (6)	$\frac{p}{1-\alpha_e}$ (7)
8–10	1	2.18	1.0	0.042	0.50	2.00
10–12	3	1.83	0.9	0.047	0.51	6.12
12–18	16	1.30	0.7	0.060	0.54	34.78
18–20	16	0.92	0.51	0.082	0.58	38.10
20–30	30	0.71	0.4	0.105	0.61	76.92
30–40	22	0.50	0.28	0.15	0.66	64.71
40–50	12	0.35	0.18	0.23	0.72	42.86
Summation						265.49

*For ease of computation, the percentage value is used instead of the decimal fractional value.

a. Determine the particle settling velocity using Fig. 5-16 in Chap. 5. Alternatively the particle settling velocity can be computed as illustrated in Example 5-6. The settling velocity values from Fig. 5-16 are entered in the computation table.

b. Determine the values of v/v_s and enter the computed values in the computation table. The backwash velocity is:

$$v = 30 \text{ in/min} = 0.042 \text{ ft/s}$$

c. Determine the values of α_e and enter the computed values in the computation table:

$$\alpha_e = \left(\frac{v}{v_s}\right)^{0.22} = \left(\frac{0.042}{1.0}\right)^{0.22} = 0.50$$

d. Determine the values for column 7 and enter the computed values in the computation table:

$$\frac{p}{1-\alpha_e} = \frac{1}{1-0.5} = 2.0$$

2. Determine the expanded bed depth using Eq. (12-18):

$$\frac{l_e}{l} = (1-\alpha)\sum \frac{p}{(1-\alpha_e)}\frac{1}{100}$$

$$= (2.5 \text{ ft})(1-0.4)(265.5)(1/100) = 3.98 \text{ ft}$$

3. Because the expanded porosity of the largest size fraction (0.50) is greater than the normal porosity of the filter material (0.40), the entire filter bed will be expanded.

Comment. The expanded depth needs to be known to establish the minimum height of the washwater troughs above the surface of the filter bed. In practice, the bottom of the backwash-water troughs is set from 2 to 6 in (50 to 150 mm) above the expanded filter bed. The width and depth of the troughs should be sufficient to handle the volume of backwash water used to clean the bed, with a minimum freeboard of 2.0 (50 mm) at the upper end of the trough.

Modeling the Filtration Process

The time-space removal of suspended solids in a packed-bed filter can be modeled by considering a mass balance on the volume element shown in Fig. 12-18. The basic mass balance for the volume element is:

$$\underbrace{\frac{\partial q}{\partial t}\Delta V}_{\substack{\text{Accum. of solids}\\ \text{in nonactive}\\ \text{pore space}}} + \underbrace{\overline{\alpha}(t)\frac{\partial C}{\partial t}\Delta V}_{\substack{\text{accum. of solids}\\ \text{in active}\\ \text{pore space}}} = \underbrace{QC|_x}_{\text{inflow}} - \underbrace{QC|_{x+\Delta x}}_{\text{outflow}} \qquad (12\text{-}19)$$

where $\partial q/\partial t$ = change in quantity of solids deposited within the filter with time, $\text{g/m}^3\cdot\text{s}$

ΔV = differential volume, m^3

$\overline{\alpha}(t)$ = average porosity (variable with time)

$\partial C/\partial t$ = change in average concentration of solids in pore space with time, $\text{g/m}^3\cdot\text{s}$

Q = filtration rate, m^3/s

C = concentration of suspended solids, g/m^3

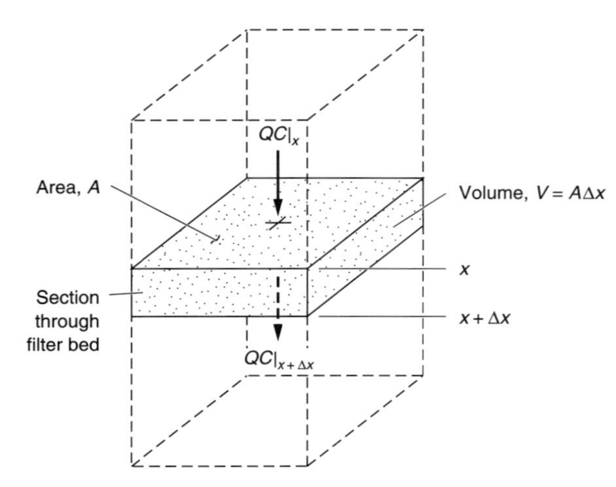

FIGURE 12-18
Definition sketch for the analysis of the filtration process (from Tchobanoglous and Schroeder, 1985).

Substituting Ax for AV and Av for Q where v is the filtration velocity (in cubic meters per square meter-second, or $m^3/m^2 \cdot s$), and taking the limit as Ax approaches zero results in:

$$-v\frac{\partial C}{\partial x} = \frac{\partial q}{\partial t} + \overline{\alpha}(t)\frac{\partial C}{\partial t} \tag{12-20}$$

The first term represents the difference between the mass of suspended solids entering and leaving the section, the second term represents the time rate of change in the mass of suspended solids present on the solid portion of the filter, and the third term represents the time rate of change in the suspended-solids concentration in the fluid portion of the filter volume. Because the quantity of fluid contained within the bed is usually small compared to the volume of liquid passing through the bed, the third term can be neglected and Eq. (12-20) can be simplified to yield:

$$-v\frac{\partial C}{\partial x} = \frac{\partial q}{\partial t} \tag{12-21}$$

Equation (12-21) is the one found most commonly in the literature dealing with filtration theory.

To solve Eq. (12-21), an additional independent equation is required. The most direct approach is to derive a relationship that can be used to describe the change in concentration of suspended matter with distance, such as:

$$\frac{\partial C}{\partial x} = \phi(V_1, V_2, V_3, \ldots) \tag{12-22}$$

in which V_1, V_2, and V_3 are the variables governing the removal of suspended matter from solution. The change in concentration of suspended solids with distance observed in filtering activated sludge effluent without any chemical additions is illustrated in Fig. 12-19. Because the removal curves in Fig. 12-19 remain constant with time, straining is the dominant removal mechanism (Tchobanoglous and Eliassen, 1970). Further, because the removal curves remain constant with time, Eq. (12-22) can be written as an ordinary differential equation.

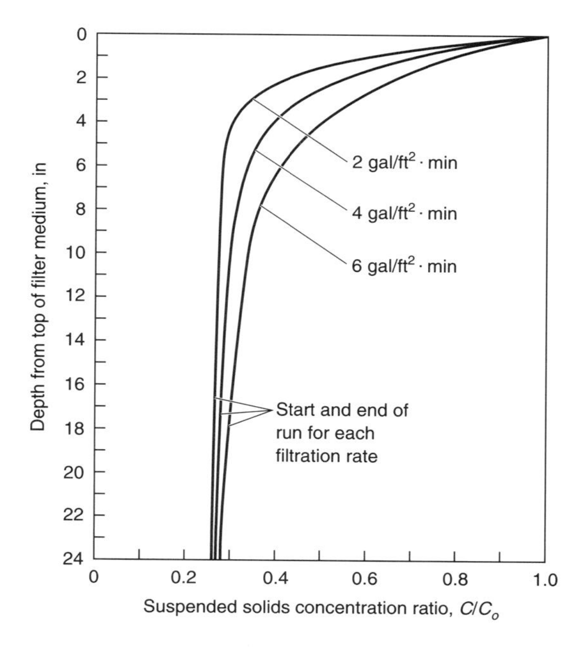

FIGURE 12-19
Normalized suspended solids removal curves as a function of depth and time observed during the filtration of settled effluent from an activated-sludge treatment process: (*a*) using a single-medium packed-bed filter.

From the shape of the normalized curves in Fig. 12-19, it can be concluded that the rate of change of concentration with distance must be proportional to some removal coefficient that is changing with the degree of treatment or removal achieved in the filter. For example, the entire particle-size distribution in the influent wastewater is passed through the first layer. The probability of removing particles from the waste stream is p_1. In the second layer, the probability of removing particles is p_2, where p_2 is less than p_1, assuming that some of the larger particles will be removed by the first layer. Continuing this argument, it can be reasoned that the rate of removal must always be changing as a function of the degree of treatment. The curves given in Fig 12-19*a* can be modeled by the following expression (Tchobanoglous and Eliassen, 1970):

$$\frac{dC}{dx} = \left[\frac{1}{(1 + Rx)^n}\right] r_o C \tag{12-23}$$

where C = concentration, mg/L
x = distance, in
R = retardation factor, in^{-1}
r_o = initial removal rate, in^{-1}
n = empirical constant related to particle size

In Eq. (12-23), the term within brackets is sometimes called a *retardation factor* (see discussion in Chap. 3). When the exponent n is equal to zero, the term within the brackets is equal to 1; under these conditions, Eq. (12-23) represents a logarithmic curve. When n equals 1, the value of the term within brackets drops off rapidly in the first 5 in (125 mm), and then more gradually as a function of distance. It appears that the exponent n may be related to the distribution of particle sizes in the

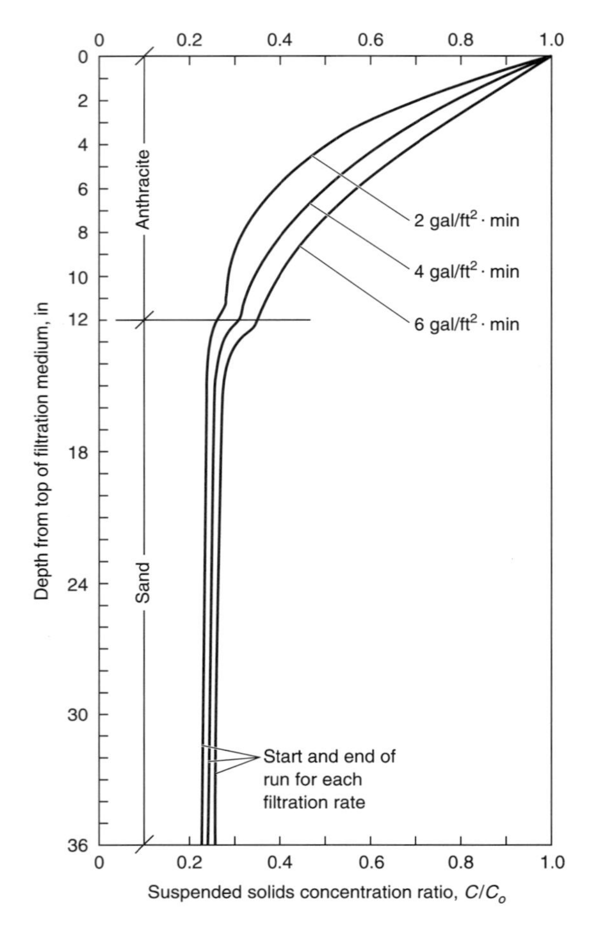

FIGURE 12-19 (*continued*) Normalized suspended solids removal curves as a function of depth and time observed during the filtration of settled effluent from an activated-sludge treatment process: (*b*) using a dual-medium packed-bed filter (adapted from Tchobanoglous and Eliassen, 1970).

influent. For example, in dealing with a uniform filter medium and filtering particles of one size, it would be expected that the value of the exponent n would be equal to zero and that the initial removal could be described as a first-order removal function. It should be noted that this equation was verified only for filtration rates up to 10 gal/ft^2·min (400 L/m^2·min) (Tchobanoglous, 1969). Additional details may be found in Tchobanoglous and Eliassen (1970).

Development of Headloss during Filtration

In the past, the most commonly used approach to determine the headloss in a clogged filter was to compute it with a modified form of the equations used to evaluate the clean-water headloss (see Table 12-11). In all cases, the difficulty encountered in using these equations is that the porosity must be estimated for various degrees of clogging. Unfortunately, the complexity of this approach renders most of these formulations useless or, at best, extremely difficult to use. An alternative approach is to relate the development of headloss to the amount of material removed by the filter.

The headloss would then be computed by the following expression:

$$H_t = H_o + \sum_{i=1}^{n} (h_i)_t \tag{12-24}$$

where H_t = total headloss at time t, m
H_o = total initial clean-water headloss, m
$(h_i)_t$ = headloss in the ith layer of the filter at time t, m

From an evaluation of the incremental headloss curves for uniform sand and anthracite when treated wastewater effluent is filtered, the buildup of headloss in an individual layer of the filter was found to be related to the amount of material contained within the layer (Tchobanoglous and Eliassen, 1970). The form of the resulting equation for headloss in the ith layer is:

$$(h_i)_t = a(q_i)_t^b \tag{12-25}$$

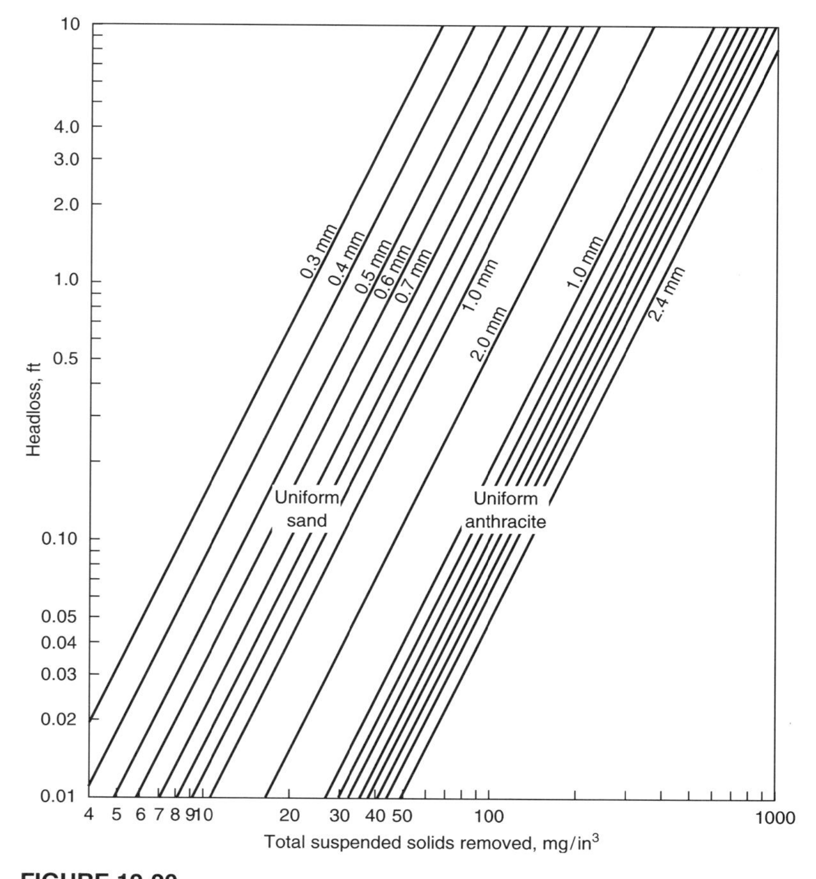

FIGURE 12-20
Headloss versus suspended solids removed for various sizes of uniform sand and anthracite (from Tchobanoglous, 1969).

where $(q_i)_t$ = cumulative amount of material deposited in the ith layer at time t, mg/cm^3, and a, b = constants. In Eq. (12-25), it is assumed that the buildup of headloss is a function only of the amount of material removed. This assumption is valid only for wastewater effluent filtration without chemical addition. Representative data for sand and anthracite for treated wastewater effluent are presented in Fig. 12-20. The determination of the buildup of headloss during the filtration process using the data presented in that figure is illustrated in Example 12-3.

The removal of suspended material by straining can be identified by noting: (1) the variation in the normalized concentration-removal curves through the filter as a function of time and (2) the shape of the headloss curve for the entire filter or an individual layer within the filter. If straining is the principal removal mechanism, the shape of the normalized removal curve will not vary significantly with time (see Fig. 12-19), and the headloss curves will be curvilinear.

EXAMPLE 12-3. ESTIMATION OF HEADLOSS BUILDUP DURING FILTRATION AND FILTER RUN LENGTH. Using the normalized total suspended solids removal ratio curves shown in Fig. 12-19, derived from filtration pilot plant studies conducted at an activated sludge wastewater treatment plant, estimate the headloss buildup as a function of the length of run. Compare the headloss development for the single-medium (sand) filter to that for the dual-medium (anthracite and sand) filter.

Biological treatment process

1. Mean cell residence time θ_c = 6 d
2. Average suspended solids concentration in effluent from secondary settling tank = 10 mg/L
3. Particle size distribution in effluent is similar to that shown in Fig. 12-20.

Pilot plant

1. Type of filter bed: single medium and dual media
2. Filter media: sand and anthracite and sand
3. Filter medium characteristics:

Sieve size or number	Size of opening, mm	Size of opening, in	Sand: Percent retained	Sand: Percent passing	Anthracite: Percent passing	Anthracite: Percent passing
4	4.76	0.187			0	100
6	3.36	0.132	0	100	10	90
8	2.38	0.0937	4	96	20	70
10	2.00	0.0787			20	50
12	1.68	0.0661	6	90	30	20
16	1.19	0.0469	20	70	20	0
20	0.841	0.0331	25	45		
30	0.595	0.0234	35	10		
40	0.420	0.0165	10	0		

4. Filter-bed depth
   a. Sand filter = 2 ft (0.6 m)
   b. Anthracite and sand filter = 1.0 ft (0.3 m) and 2.0 ft (0.6 m), respectively

5. Filtration rates = 2.0, 4.0, and 6.0 $gal/ft^2 \cdot min$ (81.5, 163.0, and 244.5 $L/m^2 \cdot min$)
6. Temperature = 20°C
7. General observation: average concentration-ratio curves plotted in Fig. 12-19*a* and *b* did not vary significantly with time.

Solution—Part 1: Headloss buildup for sand filter

1. To analyze the concentration ratio curves, rewrite Eq. (12-21) in a form suitable for numerical analysis:

$$-v\frac{\Delta C}{\Delta X} = \frac{\Delta q}{\Delta t}$$

where v = filter application rate, $L/in^2 \cdot min$
C = total suspended solids concentration at a given depth, mg/L
X = vertical position within filter, in
Δq = mass of solids removed per unit volume, mg/in^3
Δt = length of filter run time, min

Recognizing that $\Delta C = C_2 - C_1$, and that $-\Delta C = C_1 - C_2$, we find

$$-v\frac{\Delta C}{\Delta X} = -v\frac{C_2 - C_1}{X_2 - X_1} = v\frac{C_1 - C_2}{X_2 - X_1} = \frac{\Delta q}{\Delta t}$$

Rearranging gives

$$\Delta q = \frac{v(C_1 - C_2)\Delta t}{X_2 - X_1}$$

Once Δq values have been calculated for a given interval, the associated headloss values can be determined from the curves given in Fig. 12-20. The headloss values are then summed to calculate the total headloss.

2. Set up a computation table and determine the value of $-\Delta C(C_1 - C_2)$ for designated depths within the filter. The required computations are summarized in the following table (values shown are for the application rate 4 $gal/ft^2 \cdot min$). A 1-in interval is used, and computations are conducted starting at the filter surface, to the depth where the change in the concentration ratio becomes negligible (at about 12 in). As shown, values of C/C_o from the normalized removal ratio curve corresponding to the depths indicated in row 1 are entered in row 2. The value of the concentration at each depth, calculated by multiplying the concentration ratio by 10 mg/L (the influent concentration), is entered in row 3. The difference in concentration over the interval ($-\Delta C = C_1 - C_2$) is entered in row 4. For example, for the first interval the suspended solids concentration changes from 10.00 to 6.90, so $C_1 - C_2 = 10.00 - 6.90 = 3.10$ mg/L.

ΔC values for 1-in intervals in sand filter for a filtration rate of 4 $gal/ft^2 \cdot min$

Depth, in (1)	0	1	2	3	4	5	6	7	8	9	10	11	12
C/C_o (2)	1.000	0.690	0.530	0.440	0.380	0.345	0.320	0.308	0.298	0.290	0.288	0.282	0.280
C, mg/L (3)	10.00	6.90	5.30	4.40	3.80	3.45	3.20	3.08	2.98	2.90	2.88	2.82	2.80
$-\Delta C$, mg/L (4)	3.10	1.60	0.90	0.60	0.35	0.25	0.12	0.10	0.08	0.02	0.06	0.02	

3. Set up a computation table to determine the buildup of suspended solids and headloss within each layer of the filter. For the sand filter, use filter run times of 2, 4, and 6 h. As shown in the table below, the filter depth intervals are entered in column 1, and the particle sizes associated with the depth intervals are entered in column 2. The interval particle sizes can be estimated by assuming that the particles will be stratified, with the largest particles at the bottom. For sand, the smallest particle (0.42 mm) can be assumed to lie at the filter surface. At a depth of 1 in, one twenty-fourth of the total depth of 24 in has been passed. Therefore, the particle size associated with a depth of 1 in is the sieve size that 4.2 percent ($\frac{1}{24}$) of the sand will pass (see filter media characteristics above). By plotting the sand particle distribution data, the 4.2 percent passing particle size is estimated to be 0.47 mm (see figure given below). The particle size that best represents the interval overall is calculated by taking the geometric mean of the interval boundary sizes:

$$d = \sqrt{d_1 d_2} = \sqrt{0.42 \text{ mm} \times 0.47 \text{ mm}} = 0.44 \text{ mm}$$

The remaining particle sizes for sand are entered in column 2.

Calculation of Δh values for 1-in intervals in sand filter for a filtration rate of 4 gal/ft²·min

			Run time, h					
			2		4		6	
Depth, in (1)	d_p, mm (2)	$-\Delta C$, mg/L (3)	Δq, mg/in³ (4)	Δh, ft (5)	Δq, mg/in³ (6)	Δh, ft (7)	Δq, mg/in³ (8)	Δh, ft (9)
0								
	0.44	3.10	39.10	1.30	78.19	6.00	117.29	18.00
1								
	0.50	1.60	20.18	0.23	40.36	1.20	60.54	3.00
2								
	0.57	0.90	11.35	0.06	22.70	0.28	34.05	0.62
3								
	0.61	0.60	7.57	0.02	15.13	0.09	22.70	0.21
4								
	0.63	0.35	4.41		8.83	0.02	13.24	0.06
5								
	0.66	0.25	3.15		6.31	0.01	9.46	0.02
6								
	0.70	0.12	1.51		3.03		4.54	
7								
	0.71	0.10	1.26		2.52		3.78	
8								
Total				1.61		7.60		21.91

4. The necessary computations for a filtration rate of 4 gal/ft²·min (160 L/m²·min) are summarized in the table created in the previous step. Although the required computations for the other filtration rates are not shown, they are calculated using the same procedure. The values of $-\Delta C$ shown in column 3 are taken from column 5

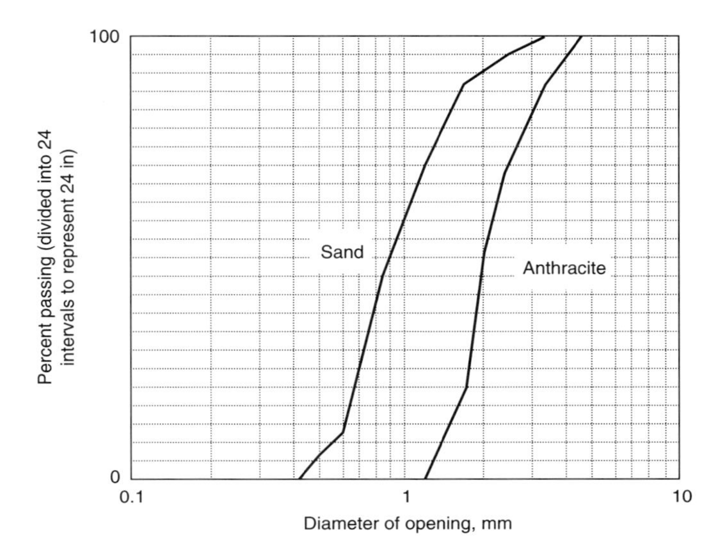

of the table prepared in step 2. The values of Δq shown in columns 4, 6, and 8 are computed using the difference equation given in step 1. To illustrate, for the sand layer between 0 and 1 in from the top of the column, the value of Δq after 2 hours is as follows:

$$\Delta q = \frac{v(C_1 - C_2)\Delta t}{\Delta X}$$

where

$$v = 4 \text{ gal/ft}^2\cdot\text{min} = 0.1051 \text{ L/in}^2\cdot\text{min}$$

$$C_1 - C_2 = 3.10 \text{ mg/L}$$

$$\Delta X = 1 \text{ in} - 0 \text{ in} = 1 \text{ in}$$

$$\Delta t = (2 \text{ h} \times 60 \text{ min/h} - 0) = 120 \text{ min}$$

$$\Delta q = \left(\frac{0.1051 \text{ L}}{\text{in}^2\cdot\text{min}}\right)\left(\frac{3.10 \text{ mg}}{\text{L}}\right)\left(\frac{120 \text{ min}}{1 \text{ in}}\right) = 39.1 \text{ mg/in}^3$$

The values for Δq are entered in columns 4, 6, and 8 for run times of 2, 4, and 6 h, respectively.

5. Using Fig. 12-20, find the values of incremental headloss Δh for each interval. The headloss values are dependent on the particle size and the Δq value, calculated in the previous step, for the interval. For example, the value of incremental headloss Δh for the sand layer between 0 and 1 in is obtained from Fig. 12-20 by finding the Δh value associated with the value of Δq and particle size for this layer. For the interval particle size of 0.44 mm and $\Delta q = 39.1$ mg/in^3, the Δh value from the figure can be read as 1.30 ft. The headloss values in the remaining sand layers are determined in a similar manner, and are entered in columns 5, 7, and 9 for run times of 2, 4, and 6 h, respectively. To simplify computations, negligible headloss values ($\Delta h < 0.01$ ft) have been ignored. Once all of the headloss values are entered, the entire column is summed to obtain the total headloss in the filter bed for the given run time. The total headloss for other filtration rates is determined in the same manner. Summary

data for the run times and flowrates are as follows:

Time, h	Headloss, ft		
	2.0 gal/ft²·min	4.0 gal/ft²·min	6.0 gal/ft²·min
2	0.84	1.61	2.44
4	3.55	7.60	10.66
6	7.02	21.91	27.52

6. Plot curves of headloss versus run time for the three flowrates. The curves, which are plotted for the data obtained in the previous steps, are shown in the following figure.

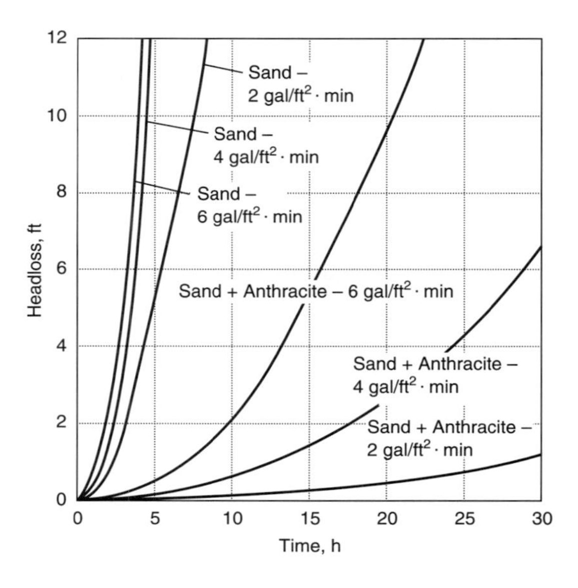

Solution—Part 2: Headloss buildup for anthracite and sand filter

7. To determine the headloss buildup for the anthracite and sand filter, follow the steps used in Part 1. Use the concentration ratio curves provided in Fig. 12-19*b*, and use filter run times of 10, 20, and 30 h. Plot curves of headloss versus run time for the three flowrates. To compare filter performance, determine the time required to reach a cumulative filter headloss of 10 ft, at an application rate of 6.0 gal/ft^2·min, for both the single- (sand) and dual- (sand and anthracite) medium filters. The results are provided in the following table and illustrated in the plot presented in Part 1, step 6 above.

Time, h	Headloss, ft		
	2.0 gal/ft²·min	4.0 gal/ft²·min	6.0 gal/ft²·min
10	0.12	0.54	2.09
20	0.54	2.62	9.61
30	1.22	6.60	21.03

Comment. The beneficial effects of using a dual medium versus a single-medium filter can be appreciated by comparing the plots prepared in steps 6 and 7. As shown in these plots, the run time to a headloss of 10 ft, at a filtration rate of 6 gal/ft^2·min (163 L/m^2·min), increased from 4 h for the sand filter to 20 h for the dual-medium filter. The extension in run time is of importance from a water production standpoint, as well as an economic standpoint. It should be noted that even though run times greater than 24 h are possible, the dual-medium filter should be backwashed once every day to avoid the buildup of grease and bacterial polymer balls, and thus maintain the long-term effectiveness of the filter.

Filtration Process Variables

The principal variables that must be considered in the design of filters are identified in Table 12-12. In the application of filtration to the removal of residual total suspended solids from treated effluents, it has been found that the nature of the particulate matter in the influent to be filtered, the size of the filter material or materials, the characteristics of the filter bed, and the filtration flowrate are perhaps the most important of the process variables (Table 12-12, items 1 through 4).

Influent characteristics. The most important influent characteristics are the total suspended solids concentration, particle size and distribution, and floc strength. Typically, the suspended solids concentration in the effluent from activated-sludge and trickling filter plants varies between 6 and 30 mg/L. Because this concentration

TABLE 12-12
Principal variables in the design of packed-bed filters*

Variable	Significance
1. Influent wastewater characteristics *a.* Total suspended solids concentration *b.* Floc or particle size and distribution *c.* Floc strength *d.* Floc or particle charge *e.* Fluid properties	Affect the removal characteristics of a given filter-bed configuration. To a limited extent the listed influent characteristics can be controlled by the designer.
2. Filter medium characteristics *a.* Effective size d_{10} *b.* Uniformity coefficient (UC) *c.* Type, grain shape, density, and composition	Affect particle removal efficiency and headloss buildup.
3. Filter-bed characteristics *a.* Porosity *b.* Stratification *c.* Bed depth	Porosity affects the amount of solids that can be stored within the filter. Bed depth affects initial headloss, length of run.
4. Filtration rate	Used in conjunction with variables 1, 2, and 3 to compute clear-water headloss.
5. Allowable headloss	Design variable.
6. Required effluent quality	Usually fixed regulatory requirement.

*Adapted in part from Tchobanoglous and Eliassen (1970) and Tchobanoglous and Schroeder (1985).

usually is the principal parameter of concern, turbidity is often used as a means of monitoring the filtration process. Within limits, it has been shown that the suspended solids concentrations found in treated wastewater can be correlated to turbidity measurements. A typical relationship for the effluent from a complete-mix, activated-sludge process is:

$$\text{TSS, mg/L} = (2.3 \text{ to } 2.7) \times (\text{turbidity, NTU}) \tag{12-26}$$

Typical data on the particle size distribution in the effluent from two activated-sludge plants were shown previously in Fig. 2-10 in Chap. 2, page 41. As illustrated, the particles are distributed into two distinct size ranges, with small particles varying in areal size (equivalent circular diameter) from 1 to 5 μm and large particles varying in size from 15 to 100 μm. The most significant observation relating to particle size is that the distribution of sizes is bimodal. This observation is important because the removal mechanisms for each size range will be different. For example, it seems reasonable to assume that the removal mechanism for particles 1.0 μm in size would be different from that for particles 80 μm in size or larger. A plot of the relative particle removal efficiency for the particles found in primary effluent is shown in Fig. 12-21. With respect to disinfection with either chlorine or UV radiation, the large particles are of concern as they can potentially shield microorganisms (see discussion of disinfection in Sec. 12-6).

Floc strength will vary with the type of process and the mode of operation. For example, the residual floc from the chemical precipitation of biologically processed wastewater may be considerably weaker than the residual biological floc before

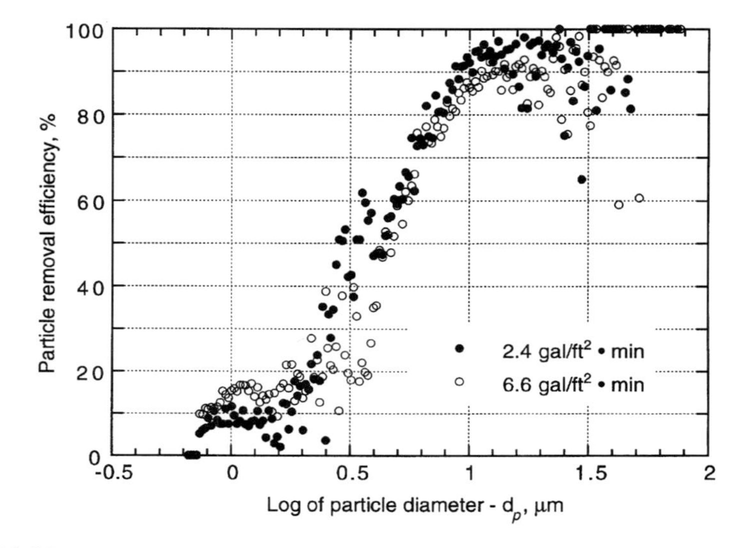

FIGURE 12-21
Relative particle size removal efficiency for the filtration of primary effluent using a deep-bed upflow continuous backwash filter at two different filtration rates (England et al., 1994).

precipitation. Further, the strength of the biological floc will vary with the mean cell residence time, increasing with longer mean cell residence time (see Chap. 7). The increased strength derives in part from the production of extracellular polymers as the mean cell residence time is lengthened. At extremely long mean cell residence times (20 days and longer), it has been observed that the floc strength will decrease, because more of the particles are small.

Filter medium and filter-bed characteristics. The effective size d_{10} and the uniformity coefficient (UC) are the principal filter medium characteristics that affect the filtration operation. Grain size affects both the clear-water headloss and the buildup of headloss during the filter run. If too small a filtering medium is selected, much of the driving force will be wasted in overcoming the frictional resistance of the filter bed (see Example 12-1). On the other hand, if the size of the medium is too large, many of the small particles in the influent will pass directly through the bed. Typical data on the characteristics of the filtering media used in both mono-, dual-, and multimedium filters are given in Tables 12-8 and 12-9. Filter-bed porosity, stratification, and bed depth will affect the filter hydraulics.

Filtration rate. The rate of filtration is important because it affects the real size of the filters that will be required. For a given filter application, the rate of filtration will depend primarily on the strength of the floc and the size of the filtering medium. For example, if the strength of the floc is weak, high filtration rates will tend to shear the floc particles and carry much of the material through the filter. It has been observed that filtration rates in the range of 2 to 10 gal/ft^2·min (80 to 400 L/m^2·min) will not affect the quality of the filter effluents because biological floc is strong. Filtration rates as high as 30 gal/ft^2·min (1200 L/m^2·min) have been used with the Fuzzy Filter.

Design Considerations for Effluent Filters

Design considerations for effluent filters include: (1) number and size of filter units, (2) selection of the type of filter, (3) performance characteristics of different types of filters, (4) selection and characterization of filtering materials, (5) filter backwashing systems, (6) filter appurtenances, (7) filter operating strategies, and (8) filter operating problems.

Number and size of filter units. One of the first decisions to be made in the design of a granular-medium filtration system is determining the number and size of filter units that will be required. The surface area required is based on the peak filtration and peak plant flowrates. The allowable peak filtration rate is usually established on the basis of regulatory requirements. Operating ranges for a given filter type are based on past experience, the results of pilot plant studies, and manufacturers' recommendations. The number of units generally should be kept to a minimum to reduce the cost of piping and construction, but it should be sufficient to ensure: (1) that the backwash flowrates do not become excessively large and (2) that when one filter unit is taken out of service for backwashing, the transient loading on the remaining units will not be so high that material contained in the filters will be dislodged. Transient loadings due to backwashing are not an issue with filters that backwash continuously. The sizes of the individual units should be consistent with the sizes of equipment available for use as underdrains, wash-water troughs, and surface washers. Typically, width-to-length ratios for individually designed gravity filters vary from 1:1 to 1:4. For proprietary and pressure filters, it is common practice to use standard sizes that are available from manufacturers.

Selection of the type of filter. The principal types of granular-medium filters that have been used for the filtration of wastewater have been discussed previously (see also Table 12-7). With proprietary filters, the manufacturer is responsible for providing the complete filter unit and its controls, based on basic design criteria and performance specifications. In individually designed filters, the designer is responsible for working with several suppliers in developing the design of the system components. Contractors and suppliers then furnish the materials and equipment in accordance with the engineer's design.

Performance characteristics of different types of filters. The critical question associated with the selection of any granular- or synthetic-medium packed-bed filter is whether it will perform as anticipated. Insight into the performance of packed-bed filters can be gained from a review of the data presented in Fig. 12-22, in which the results of testing seven different types of pilot-scale filters on the effluent from the same activated-sludge plant are shown. The principal conclusions to be reached from an analysis of the data presented in Fig. 12-21 are that: (1) given a high-quality filter influent (turbidity less than 5 to 8 NTU), all of the filters tested produced

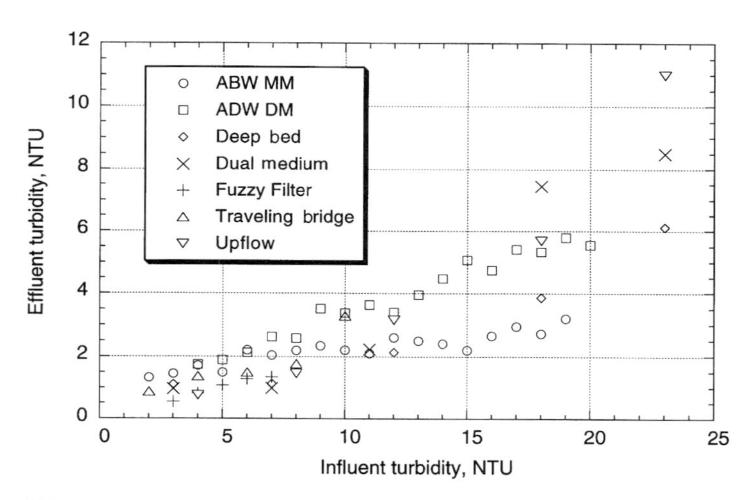

FIGURE 12-22
Effluent versus influent turbidity removal performance data for seven packed-bed filters used for wastewater treatment operated at 4 or 5 gal/ft^2·min, with the exception of the Fuzzy Filter which was operated at 20 gal/ft^2·min.

an effluent with an average turbidity of 2 NTU or less, and (2) when the influent turbidity was greater than about 5 to 8 NTU, chemical addition was required with all of the filters to achieve an effluent turbidity of 2 NTU or less. Using the relationship between turbidity and total suspended solids given in Eq. (12-26), an influent turbidity of 5 to 8 NTU corresponds to a suspended solids concentration of 12 to 18 mg/L. A comparison of three proprietary filters for a 5-Mgal/d water reclamation facility at Sacramento Regional County Sanitation District (CSD) is presented in Table 12-13.

For new wastewater treatment plants, extra care should be devoted to the design of the secondary settling facilities. With properly designed and operated settling facilities and the findings presented in Fig. 12-22, the decision on what type of filtration system should be used is often based on plant-related variables, such as the space available, duration of filtration period (seasonal versus year-round), the time available for construction, and costs. For existing plants that do not function well and must be retrofitted with effluent filtration, it may be appropriate to consider the type of a filter that can continue to function even when heavily overloaded. The pulsed-bed filter and both downflow and upflow deep-bed coarse-medium filters have been used in such applications.

Selection and characterization of filtering materials. Once the type of filter to be used has been selected, the next step is to specify the characteristics of the filtering medium, or media if more than one is used. Typically, the effective grain size

TABLE 12-13
Comparison of proprietary sand filters at Sacramento Regional CSD*

	Alternative filter		
Characteristic	**Pulsed bed**	**Traveling bridge**	**Continuous backwash**
Backwash clearwell	Filtered-effluent clearwell	Filtered-effluent plenum	None
Requirements for backwash equalization	Modest	Small	None
Characteristics of filter medium			
Type	Sand	Sand	Sand
Effective diameter, mm	0.45	0.60	1.4
Uniformity coefficient, UC	1.5	1.5	1.5
Depth, in	11	11–16	40
Filter height, ft	14	8	18
Ease of operation and maintenance	Modest	Modest	Simple
Percent backwash	5–10	10–15	5–10
Backwash effectiveness	Very good	Good	Good
Ability to tolerate upstream process upsets	Good	Modest	Very good
Potential for filter binding at high chemical doses	Modest	High	Low
Filter headloss, ft	8.5	4.0	3.0

*From Crites et al. (1996).

(d_{10}) and uniformity coefficient (UC), the specific gravity, solubility, hardness, and depth of the various materials used in the filter bed are specified. Sometimes it is advantageous to specify the 99 percent passing size and the 1 percent passing size to define the gradation curve for each filter medium more accurately (see Chap. 11 for the analysis of granular filter materials). In addition, during conceptual design, it is necessary: (1) to determine the type of underdrain system required to support the filtering materials, and (2) to determine the submergence requirements of the filter bed to minimize or prevent negative heads from occurring in the filter, especially in conventional downflow filters.

Filter backwashing systems. A filter bed can function properly only if the backwashing system cleans the material removed within the filter effectively. The methods commonly used for backwashing granular-medium filter beds include:

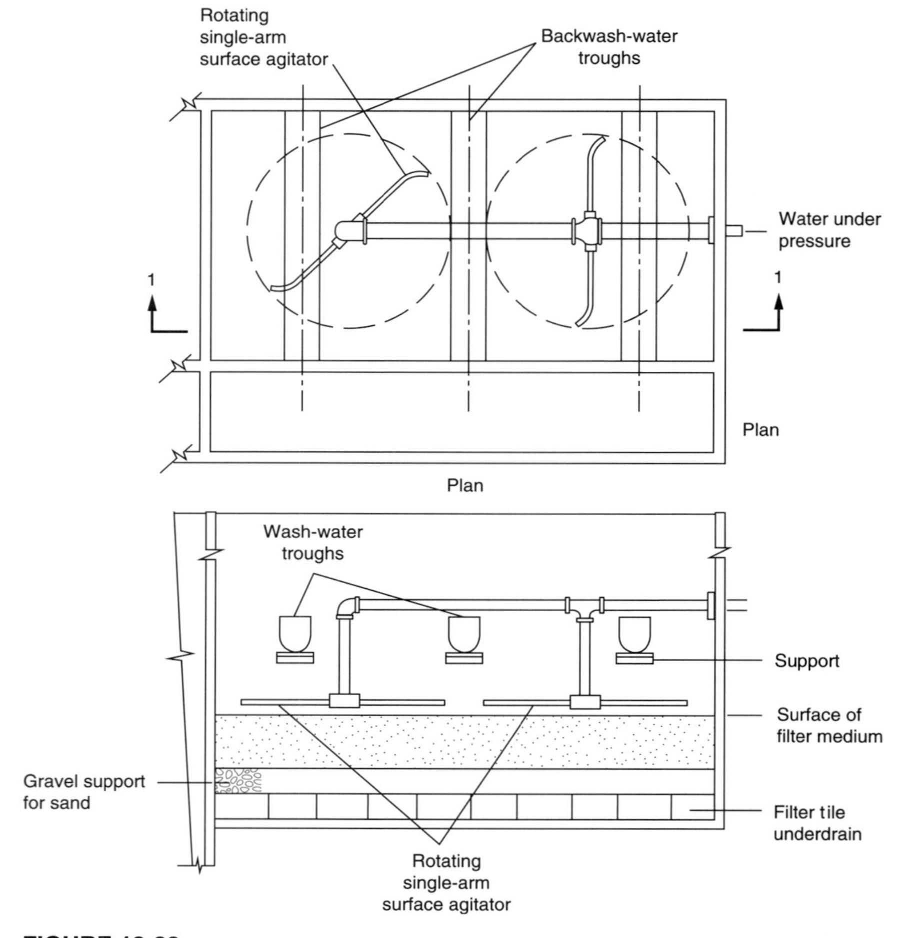

FIGURE 12-23
Surface washing facilities used to clean conventional granular medium filters (adapted from Tchobanoglous and Schroeder, 1985).

(1) water backwash with auxiliary surface water-wash agitation, (2) water backwash with auxiliary air scour, and (3) combined air-water backwashing. With the first two methods, fluidization of the granular medium is necessary to achieve effective cleaning of the filter bed at the end of the run. With the third method, fluidization is not necessary. Typical backwash flowrates required to fluidize various filter beds can be estimated as illustrated in Example 12-3.

Surface washers are often used to provide the shearing force required to clean the grains of the filtering medium used for wastewater filtration. Surface washers for filters can be fixed or mounted on rotary sweeps, as shown in Fig. 12-23. According to data on a number of systems, rotary sweep washers appear to be the most effective. Operationally, the surface washing cycle is started about 1 or 2 min before the water backwashing cycle is started. Both cycles are continued for about 2 min, at which time the surface wash is terminated. Water usage is as follows: for a single-sweep surface backwashing system, from 0.5 to 1.0 gal/ft^2·min (20 to 40 L/m^2·min); for a dual-sweep surface backwashing system, from 1.5 to 2.0 gal/ft^2·min (60 to 80 L/m^2·min) (Bishop and Behrman, 1976).

The use of air to scour the filter provides a more vigorous washing action than water alone. Operationally, air is usually applied for 3 to 4 min before the water backwashing cycle begins. In some systems, air is also injected during the first part of the water-washing cycle. Typical air flowrates range from 3 to 5 ft^3/ft^2·min (10 to 16 m^3/m^2·min).

The combined air-water backwash system is used in conjunction with the single-medium unstratified filter bed. Operationally, air and water are applied simultaneously for several minutes. The specific duration of the combined backwash varies with the design of the filter bed. Ideally, during the backwash operation, the filter bed should be agitated sufficiently so that the grains of the filter medium move in a circular pattern from the top to the bottom of the filter as the air and water rise up through the bed. Some typical data on the quantity of water and air required are reported in Table 12-14. To eliminate the possibility of air binding within conventional filters, a 2- to 3-min water backwash at subfluidization velocities is used to remove any air bubbles that may remain in the filter bed at the end of the combined air-water backwash.

TABLE 12-14

Typical amounts of air and water required for backwash process*

	Medium characteristics		Backwash rate	
Medium	**Effective size, mm**	**Uniformity coefficient**	**Water, gal/ft^2·min**	**Air, ft^3/ft^2·min†**
Sand	1.00	1.40	10	43
	1.49	1.40	15	65
	2.19	1.30	20	86
Anthracite	1.10	1.73	7	22
	1.34	1.49	10	43
	2.00	1.53	15	65

*Adapted from Tchobanoglous and Schroeder (1985).

†Air at 70°F (21°C) and 1.0 atm.

Filter appurtenances. The principal filter appurtenances are (1) the underdrain system used to support the filtering materials, collect the filtered effluent, and distribute the backwash water and air (where used) and (2) the wash-water troughs used to remove the spent backwash water from the filter.

The type of underdrain system to be used depends on the type of backwash system (see Fig. 12-24). In conventional water backwash filters without air scour, it is common practice to place the filtering medium on a support consisting of several layers of graded gravel. The design of a gravel support for a granular medium is delineated in the American Water Works Association (AWWA) Standard for Filtering Material B100-89. As the effluent bacterial disinfection requirements become more stringent, special attention must be devoted to the design of the underdrain

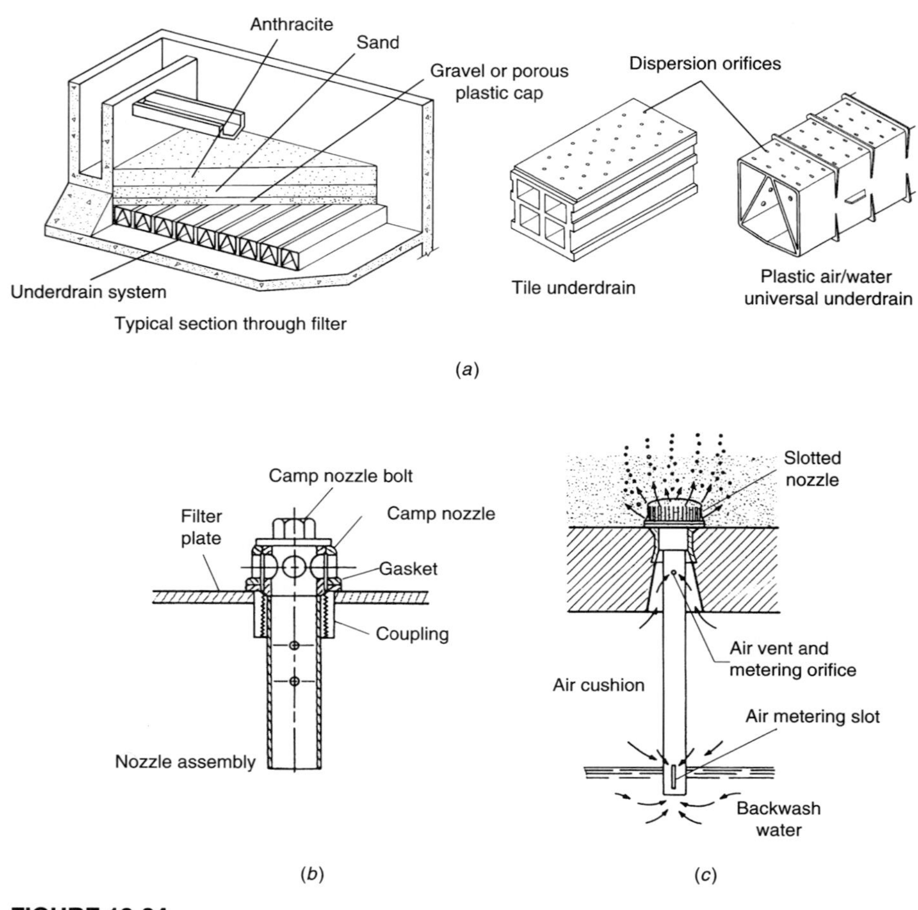

FIGURE 12-24
Typical underdrain systems used for wastewater filters: (*a*) underdrain system used with gravel or porous plastic cap support layer for filter media (courtesy Leopold) and (*b, c*) air-water nozzle used in filters without gravel support layer (courtesy Walker Process and Infilco-Degremont, Inc., respectively).

system and piping system leading to the disinfection facilities to avoid the growth of microorganisms on solid surfaces. The sloughing of clumps of microorganisms has been found to affect the disinfection process significantly, especially where reuse standards must be met (e.g., 2.2 MPN/100 mL).

Wash-water troughs are constructed of fiberglass, plastic, sheet metal, or of concrete, with adjustable weir plates. The particular design of the trough will depend to some extent on the other equipment used in the design and construction of the filter. Loss of filter material during backwashing is a common operating problem. To reduce this problem, baffles can be placed on the underside of the washwater troughs.

Filter operating strategies. Flow through the filters may be controlled from a water level upstream of the filters or from the water level in each filter. These water levels are used in conjunction with rate-of-flow controllers or a control valve to limit or regulate the flowrate through a filter. Filter hydraulic operating parameters requiring monitoring include filtered water flowrate, total headloss across each filter, surface wash- and backwash-water flowrates, and air flowrate if an air/water backwash system is employed. Water quality parameters that are usually monitored in filtered water include turbidity, suspended solids, particle size, and transmittance. Signals from effluent turbidity monitors and effluent flowrate are often used to pace the chemical feed system, where required.

Filter operating problems. The principal problems encountered in wastewater filtration and the control measures that have proved to be effective are reported in Table 12-15. Because these problems can affect both the performance and operation of a filter system, care should be taken in the design phase to provide the necessary facilities that will minimize their impact. When filtering secondary effluent containing residual biological floc, semicontinuous filters should be backwashed at least once every 24 h to avoid the formation of mudballs and the buildup of grease. In most cases, the frequency of backwashing will be more often.

Hydraulic Control of Granular-Medium Filters

Although a wide variety of hydraulic control schemes have been used in the design of filtration systems, the two most commonly used for conventional filters can be categorized as: (1) constant head–constant rate or variable head–constant rate and (2) either constant-head or variable-head declining rate.

Constant-rate filtration. In the past, constant-rate filtration was used most commonly. In this method of operation, the total operating head on the filter is fixed and the flow through the filter is controlled at a constant rate. The rate is controlled with a throttling valve on the discharge side of the filter (Fig. 12-25*a*). A variation that is often used is to allow the water level to build up over the surface of the filter bed while maintaining a constant rate of flow through the filter (Fig. 12-25*b*). Constant flow to the individual filters is maintained by means of a weir. With the pulsed-bed filter (Fig. 12-25*c*), constant flow is achieved by pumping or by the use of an influent flow-splitting weir.

TABLE 12-15

Summary of commonly encountered problems in the filtration of wastewater and control measures for those problems*

Problem	Description/control
Turbidity breakthrough[†]	Unacceptable levels of turbidity are recorded in the effluent from the filter, even though the terminal headloss has not been reached. To control the buildup of effluent turbidity levels, chemicals and polymers have been added to the filter. The point of chemical or polymer addition must be determined by testing.
Mudball formation	Mudballs are an agglomeration of biological floc, dirt, and the filtering medium or media. If the mudballs are not removed, they will grow into large masses that often sink into the filter bed and ultimately reduce the effectiveness of the filtering and backwashing operations. The formation of mudballs can be controlled by auxiliary washing processes such as air scour or water surface wash concurrent with, or followed by, water wash.
Buildup of emulsified grease	The buildup of emulsified grease within the filter bed increases the headloss and thus reduces the length of filter run. Both air scour and water surface wash systems help control the buildup of grease. In extreme cases, it may be necessary to steam-clean the bed or to install a special washing system.
Development of cracks and contraction of filter bed	If the filter bed is not cleaned properly, the grains of the filtering medium become coated. As the filter compresses, cracks develop, especially at the sidewalls of the filter. Ultimately, mudballs may develop. This problem can be controlled by adequately backwashing and scouring.
Loss of filter medium or media (mechanical)	In time, some of the filter material may be lost during backwashing and through the underdrain system (where the gravel support has been upset or the underdrain system has been installed improperly). The loss of the filter material can be minimized through the proper placement of washwater troughs and underdrain system. Special baffles have also proved effective.
Loss of filter medium or media (operational)	Depending on the characteristics of the biological floc, grains of the filter material can become attached to it, forming aggregates light enough to be floated away during the backwashing operations. The problem can be minimized by the addition of an auxiliary air and/or water scouring system.
Gravel mounding	Gravel mounding occurs when the various layers of the support gravel are disrupted by the application of excessive rates of flow during the backwashing operation. A gravel support with an additional 2- to 3-in (50- to 75-mm) layer of high-density material, such as ilmenite or garnet, can be used to overcome this problem.

*Adapted in part fromTchobanoglous and Burton (1991).

[†] Turbidity breakthrough does not occur with filters that operate continuously.

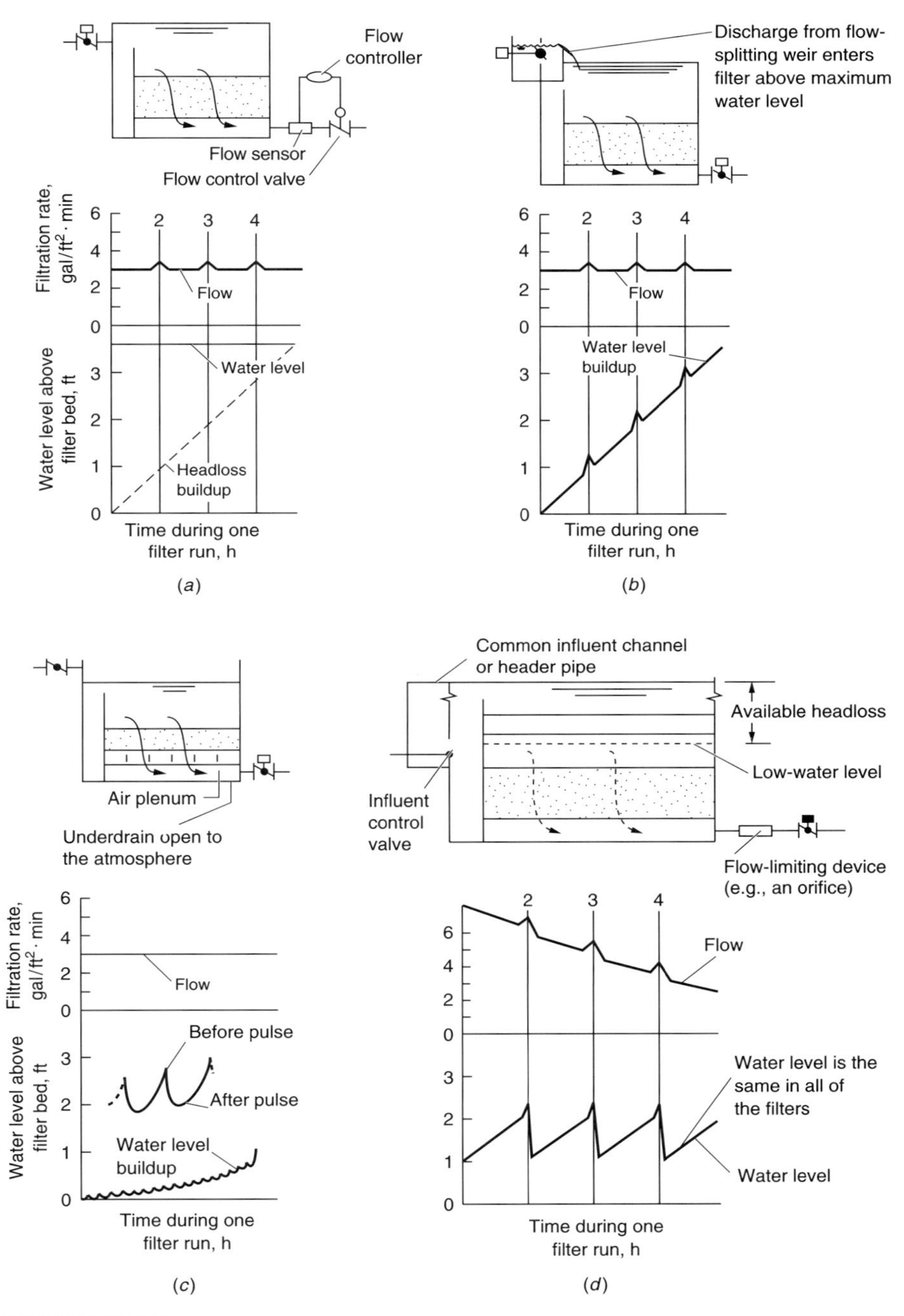

FIGURE 12-25
Definition sketch for filter operation: (*a*) fixed head, (*b*) variable head, (*c*) variable head with pulsed-bed filter and variable-rate filtration, and (*d*) variable flow, variable head. Curves for filters in *a, b,* and *d* are for the operation of one filter in a bank of four filters. The numbers represent the filter that is backwashing during the filter run. In practice, the time before backwashing will not be the same for all of the filters (adapted from Tchobanoglous and Schroeder, 1985).

Declining-rate filtration. In declining-rate filtration, as the name implies, the rate of flow though the filter is allowed to decline as the rate of headloss builds up with time. Declining-rate filtration systems are either influent-controlled or effluent-controlled. With influent control, a large influent channel or header is used to provide water to the individual filters, each of which is equipped with an isolation valve (Fig. 12-25*d*). During operation the water level is essentially the same in all of the filters because the large header acts as a reservoir. Thus, the flow to each filter is equal to the flow that can be processed by each filter at a given moment. When a filter is taken out of service for backwashing, the level on the remaining filters rises slightly to maintain the same rate of flow through the system.

When a bank of filters that has been out of service is brought back into service, the operating water level is below the wash-water trough and the filters operate as constant-rate filters, the flowrate being controlled by the setting of the influent control valves. Because extremely high filtration rates are possible when the filters are clean, a flow-restricting device (usually an orifice) is placed in the filter effluent discharge piping. When the water level in the filter is above the level of the wash-water troughs, the filters operate as variable-head declining-rate filters. In general, the clean-water headloss through the piping system, filtering media, and underdrains is about 0.80 to 1.20 m, so that the actual operating low-water level is above the wash-water troughs.

Effluent Filtration with Chemical Addition

Depending on the quality of the settled secondary effluent, chemical addition has been used to improve the performance of effluent filters. Chemical addition has also been used to achieve specific treatment objectives including the removal of specific constituents such as phosphorus, metal ions, and humic substances. Chemicals commonly used in effluent filtration include a variety of organic and inorganic polymers, alum, and ferric chloride. Use of organic polymers and the effects of the chemical characteristics of the wastewater on alum addition are considered in the following discussion.

Use of organic polymers. Organic polymers are typically classified as long-chain organic molecules with molecular weights varying from 10^4 to 10^6. With respect to charge, organic polymers can be cationic (positively charged), anionic (negatively charged), or nonionic (no charge). Polymers are added to settled effluent to bring about the formation of larger particles by bridging. Because the chemistry of the wastewater has a significant effect on the performance of a polymer, the selection of a given type of polymer for use as a filter aid generally requires experimental testing. Common test procedures for polymers involve adding an initial dosage (usually 1.0 mg/L) of a given polymer and observing the effects. Depending upon the effects observed, the dosage should be increased by 0.5 mg/L increments or decreased by 0.25 mg/L increments (with accompanying observation of effects) to obtain an operating range. After the operating range is established, additional testing can be done to establish the optimum dosage.

When lower-molecular-weight polymers, intended to serve as alum substitutes, are used, the dosage is considerably higher (10 mg/L or more) than with higher-molecular-weight polymers (0.25 to 1.25 mg/L). Whenever chemical addition is considered, special attention must be given to the design of initial mixing facilities, if maximum effectiveness is to be achieved. Typical mixing times of less than 1 second and G values greater than 2500 are recommended.

Effects of chemical characteristics of wastewater on alum addition. As with polymers, the chemical characteristics of the treated wastewater effluent can have a significant impact on the effectiveness of aluminum sulfate (alum) when it is used as an aid to filtration. For example, the effectiveness of alum is dependent on pH (see Fig. 12-26). Although Fig. 12-26 was developed for water treatment applications, it has been found to apply to most wastewater effluent filtration uses with minor variations. As shown in Fig. 12-26, the approximate regions in which the different phenomena, associated with particle removal in conventional sedimentation and filtration processes, are operative are plotted as a function of the alum dose and the pH of the treated effluent after alum has been added. For example, optimum particle removal by sweep floc occurs in the pH range of 7 to 8 with an alum dose of 20 to 60 mg/L. Generally, for many wastewater effluents that have high pH values

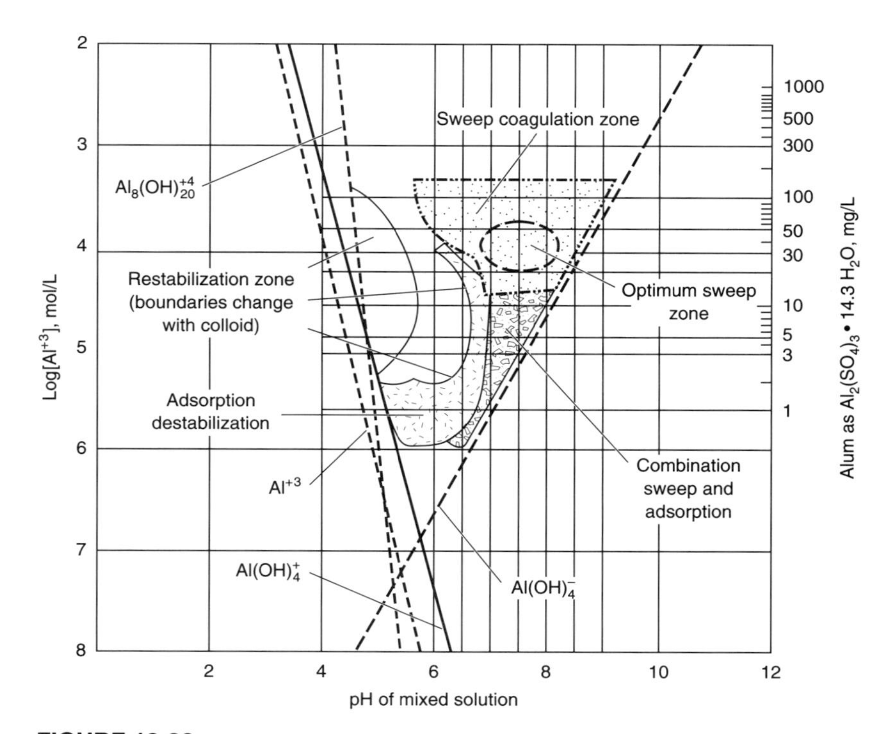

FIGURE 12-26
Typical operating ranges for alum coagulation (adapted from Amirtharajah, 1978).

(e.g., 7.3 to 8.5), low alum dosages in the range of 5 to 10 mg/L will not be effective. Because nitrification, alum addition, and chlorination can reduce the pH, an upward adjustment of pH may be necessary for some wastewaters. Although it is possible to operate at low alum dosages, proper pH control will be required.

Need for Pilot-Plant Studies

Although the information presented in this section is useful to the understanding of the nature of the filtration operation as it is applied to the filtration of treated wastewater, it must be stressed that there is no generalized approach to the design of full-scale filters. The principal reason is the inherent variability in the characteristics of the wastewater to be filtered (see Table 12-12). For example, changes in the degree of flocculation of the suspended solids in the secondary settling facilities will significantly affect the particle sizes and their distribution in the effluent which, in turn, will affect the performance of the filter. Further, because the characteristics of the effluent suspended solids will also vary with the organic loading on the process as well as with the time of day, filters must be designed to function under a wide range of operating conditions. The best way to ensure that the filter configuration selected for a given application will function properly is to conduct laboratory and pilot-plant studies (see Fig. 12-27). Because of the many variables that can be analyzed, care must be taken to change no more than one variable at a time to avoid

(*a*)

(*b*)

FIGURE 12-27

Typical pilot-plant filters: (*a*) Parkson DynaSand upflow continuous backwash filter and (*b*) Schreiber Fuzzy Filter. Pictures taken at University of California, Davis, campus wastewater treatment plant.

confounding the results in a statistical sense. Ideally, to assess seasonal variations in the characteristics of the effluent to be filtered, testing should be carried out at several intervals throughout a full year.

The performance of full-scale filters in reclamation applications can be predicted with reasonable accuracy by a simple laboratory filter test (see Fig. 2-5, Chap. 2). Using either a 5- or 8-μm polycarbonate membrane-type filter, a sample of the effluent to be filtered is passed through the test filters. (*Note:* some experimentation will be required to determine which pore size is most appropriate for a given treatment plant. A coarse prefilter is sometimes placed over the membrane filter to reduce the effects of auto filtration.) The turbidity of the filtrate is then measured. For example, if the turbidity of the filtrate is equal to or less than 2 NTU, the performance of the full-scale filters will be equal to or less than 2 NTU. The laboratory filter test can also be used to predict when chemicals must be added to maintain an effluent turbidity of 2 NTU or less. Typically, chemical addition is initiated when the laboratory turbidity test data reaches a value of about 1.8 to 1.9 NTU.

12-4 REMOVAL OF RESIDUAL SOLIDS BY MEMBRANE FILTRATION

Filtration, as defined in Sec. 12-3, involves the separation (removal) of particulate matter from a liquid. In membrane filtration, the range of particle sizes is extended to include dissolved constituents. The role of the membrane, as shown in Fig. 12-28, is to serve as a selective barrier that will allow the passage of certain constituents and will retain other constituents found in the liquid (Cheryan, 1986). The liquid that passes through the semipermeable membrane is known as *permeate* (also known as the *permeating stream* or *product stream*), and the liquid containing the retained constituents is known as the *retentate* (also known as the *concentrate, retained phase,* or *waste stream*). The rate at which the permeate flows through the membrane is known as the *rate of flux,* typically expressed as gal/ft^2·d. To introduce membrane technologies and their application, the following subjects are considered in this section: (1) membrane classification, (2) membrane configurations, (3) application of membrane technologies, and (4) the need for pilot-plant studies.

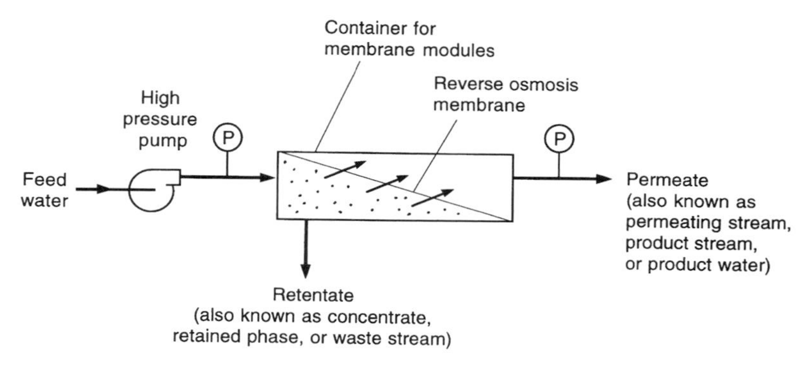

FIGURE 12-28
Definition sketch for the function of a membrane.

Membrane Process Classification

Membrane processes include dialysis, electrodialysis, microfiltration (MF), ultrafiltration (UF), nanofiltration (NF), and reverse osmosis (RO). Membrane processes can be classified in a number of different ways including: (1) the nature of the driving force, (2) the separation mechanism, (3) the nominal size of the separation achieved, and (4) the type of material from which the membrane is made. The general characteristics of membrane processes are reported in Table 12-16.

Driving force. Dialysis involves the transport of constituents through a semipermeable membrane on the basis of concentration differences. Electrodialysis involves the use of an electromotive force and ion selective membranes to accomplish the separation of charged ionic species. The distinguishing characteristic of the last four membrane processes considered in Table 12-16 (MF, UF, NF, and RO) is the application of hydraulic pressure to bring about the desired separation.

Removal mechanisms. Membranes used for the treatment of water and wastewater typically consist of a thin skin having a thickness of about 0.20 to 0.25 μm, supported by a more porous structure about 100 μm in thickness. The separation of particles in MF and UF is accomplished primarily by straining (sieving), as shown in Fig. 12-29*a*. In NF and RO, small particles are rejected by the water layer adsorbed on the surface of the membrane, which is known as a *dense membrane* (see Fig. 12-29*b*). Ionic species are transported across the membrane by diffusion through the pores of the macromolecule composing the membrane. Typically, NF can be used to reject constituents as small as 0.001 μm, whereas RO can reject particles as small as 0.0001 μm. Straining is also important in UF membranes, especially at the larger pore size openings.

Size of separation. The pore sizes in membranes are identified as macropores (greater than 50 nm), mesopores (2 to 50 nm), and micropores (less than 2 nm). Because the pore sizes in RO membranes are so small, the membranes are defined as dense. The classification of membrane processes on the basis of the size of separation is shown in Fig. 12-30 and in Table 12-16. In Fig. 12-30, it can be seen that there is considerable overlap in the sizes of particles removed, especially between NF and RO. Nanofiltration is used most commonly in water softening operations, in place of chemical precipitation.

Membrane materials. Membranes can be made from a number of different organic and inorganic materials. The membranes used for wastewater treatment are typically organic. The principal types of membranes used include polypropylene, cellulose acetate, aromatic polyamides, and thin-film composites. The choice of membrane and system configuration is based on minimizing membrane clogging and deterioration, typically determined in pilot-plant studies.

TABLE 12-16

General characteristics of membrane processes

Membrane process	Membrane driving force	Typical separation mechanism	Operating structure (pore size)	Typical operating range, μm	Permeate description	Retentate description
Dialysis	Concentration difference	Diffusion	Mesopores (2–50 nm)	—	Water + small molecules	Large molecules
Electrodialysis	Electromotive force	Ion exchange	Ion exchange	—	Water + ionic solutes	Nonionic solutes
Microfiltration	Pressure	Sieve	Macropores (>50 nm)	0.06–2	Water + dissolved solutes	Large suspended particles
Ultrafiltration	Pressure	Sieve	Mesopores (2–50 nm)	0.002–0.3	Water + small molecules	Large molecules
Nanofiltration	Pressure	Sieve + solution/diffusion + exclusion	Micropores (<2 nm)	0.001–0.01	Water	Solutes
Reverse osmosis	Pressure	Solution/diffusion + exclusion	Dense	0.0001–0.002	Water	Solutes

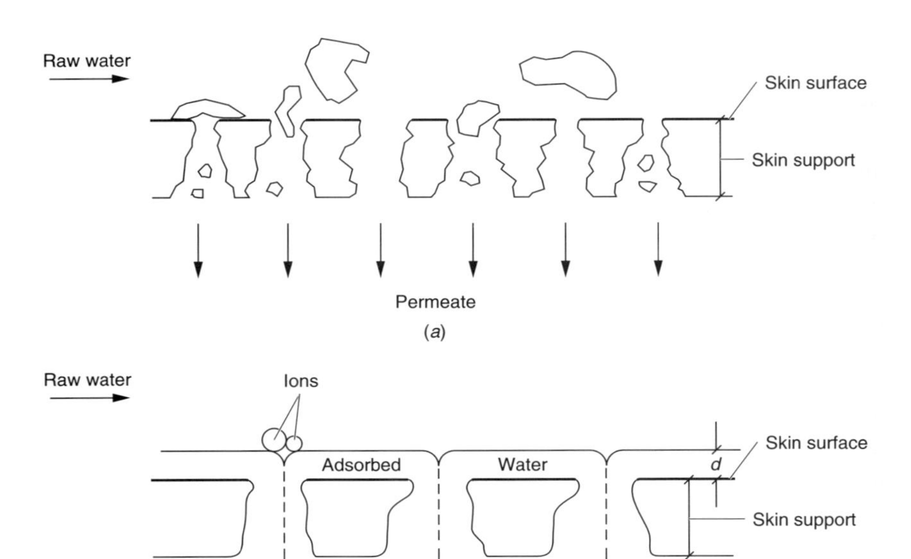

FIGURE 12-29
Definition sketch for the removal of wastewater constituents: (*a*) removal of large molecules and particles by sieve mechanism and (*b*) rejection of ions by adsorbed water layer (adapted from Tchobanoglous and Schroeder, 1985).

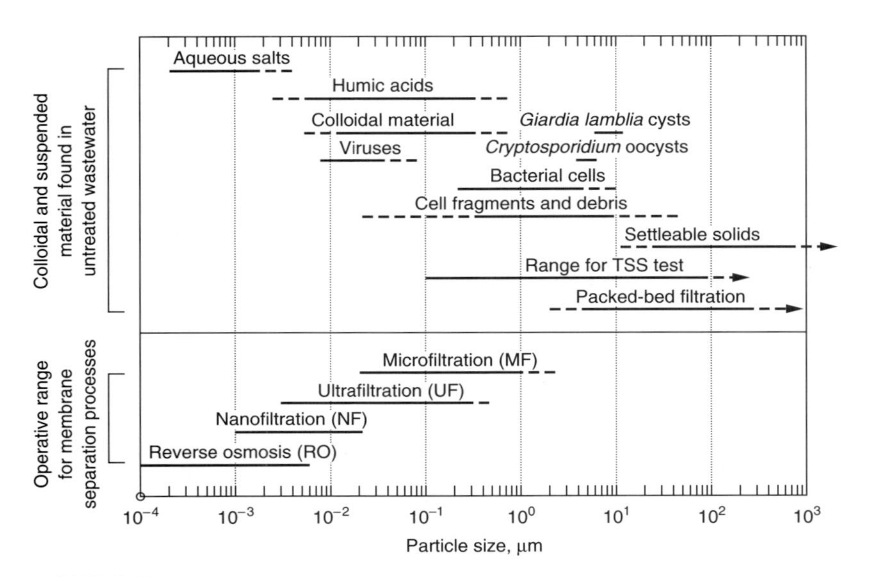

FIGURE 12-30
Constituents found in wastewater and operating ranges for membrane technologies and conventional filtration processes.

Membrane Configurations

In the membrane field, the term *module* is used to describe a complete unit comprising the membranes, the pressure support structure for the membranes, the feed inlet and outlet permeate and retentate ports, and the overall support structure. The principal types of membrane modules are: (1) plate and frame, (2) spiral wound, (3) hollow fiber, (4) tubular, and (5) rotating disk and cylinder (AWWA, 1996). A definition sketch for the first three of these membranes is shown in Fig. 12-31.

Plate and frame. Plate and frame membrane modules (see Fig. 12-31*a*) are comprised of a series of flat membrane sheets and support plates. The water to be treated passes between the membranes of two adjacent membrane assemblies. The plate supports the membranes and provides a channel for the permeate to flow out of the unit.

Spiral wound. In the spiral-wound membrane, a flexible permeate spacer is placed between two flat membrane sheets. The membranes are sealed on three sides. The open side is attached to a perforated pipe. A flexible feed spacer is added and the flat sheets are rolled into a tight circular configuration. The term *spiral* derives from the fact that the flow in the rolled bundle follows a spiral pattern (see Fig. 12-31*b*).

Hollow fiber. The hollow-fiber membrane module, as shown in Fig. 12-31*c* and *d,* consists of a bundle of hundreds to thousands of hollow fibers. The feed can be applied to the outside of the fiber (outside-in flow) or to the inside of the fiber (inside-out flow).

Tubular modules. In the tubular configuration, the membrane is cast on the inside of a support tube. A number of tubes are then placed in an appropriate container. The feed flows through the tube and the permeate is collected on the outside of the tubes.

Rotating disk and cylinder. Rotating disk and cylinder membrane modules have been developed to reduce the buildup of material on the membrane.

Membrane Operation

The operation of membrane processes is relatively simple. A pump is used to pressurize the feed solution and to circulate it through the module. A valve is used to maintain the pressure of retentate. The permeate is withdrawn, typically at atmospheric pressure. As constituents in the feed water accumulate on the membranes (a process often termed *membrane fouling*), the pressure builds up on the feed side, the membrane flux (i.e., flow through membrane) starts to decrease, and the percent rejection also starts to decrease (see Fig. 12-32). When the perfomance has deteriorated to a given level, the membrane modules are taken out of service and backwashed and/or cleaned chemically.

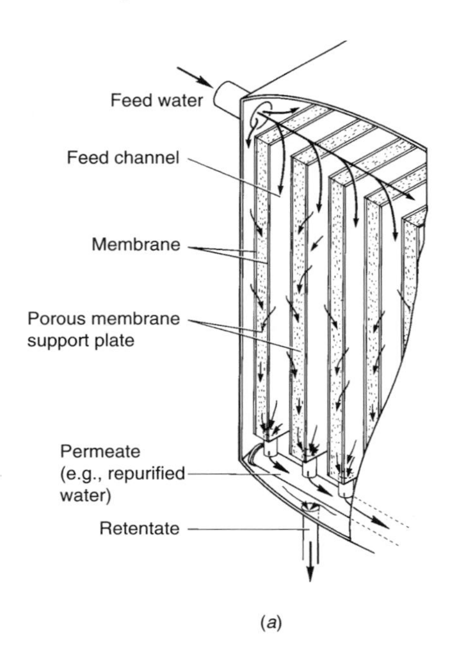

(*a*)

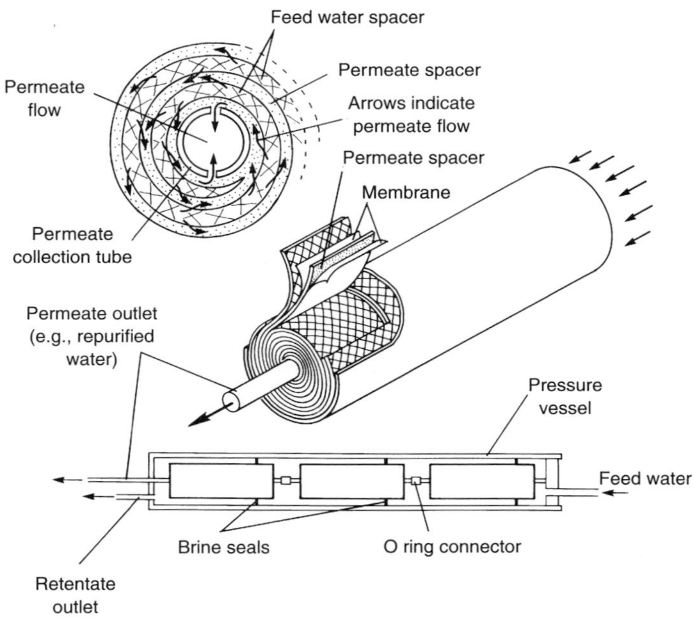

(*b*)

FIGURE 12-31
Typical membrane elements used in membrane technologies: (*a*) parallel plate, (*b*) spiral wound thin film composite.

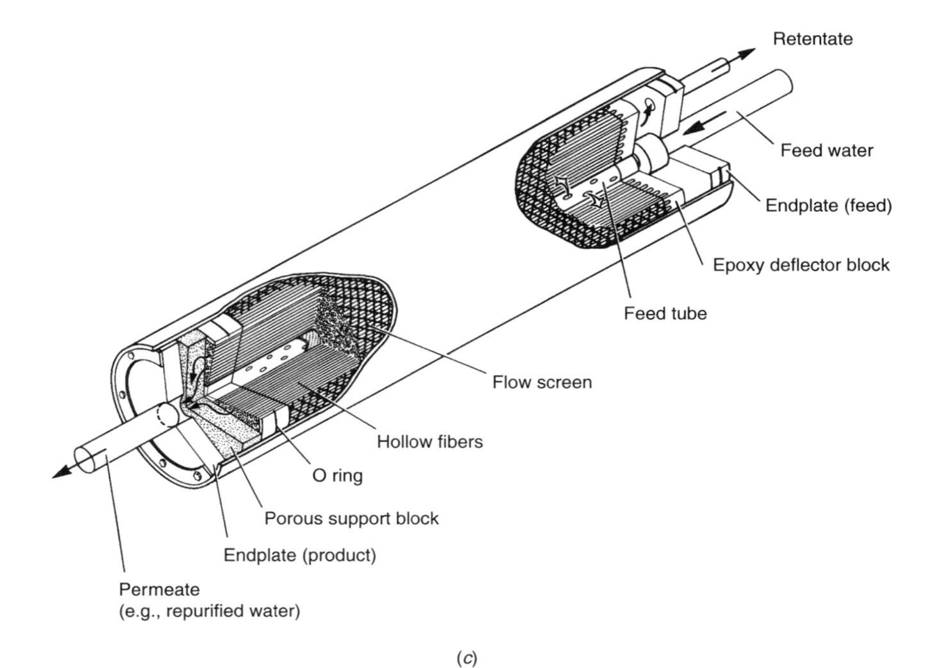

(*c*)

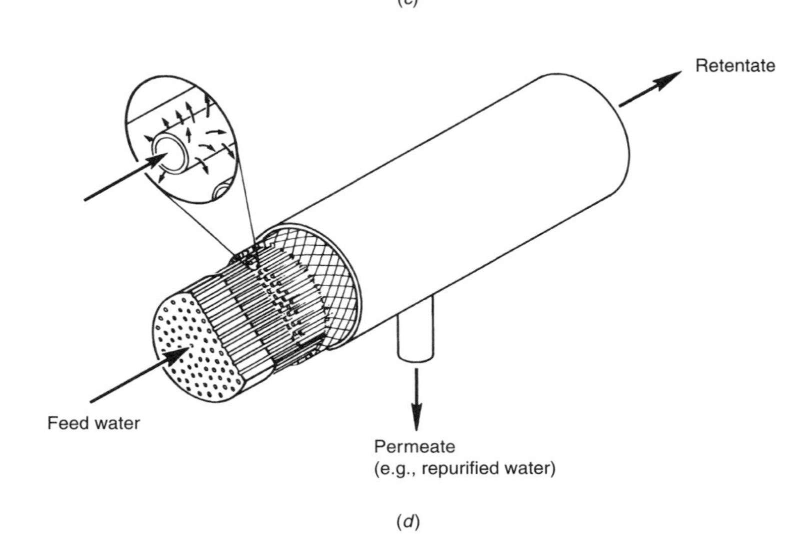

(*d*)

FIGURE 12-31 (*continued*)
Typical membrane elements used in membrane technologies: (*c*) hollow fiber with flow from the outside to the inside of the fiber, and (*d*) hollow fiber with flow from the inside to the outside of the fiber.

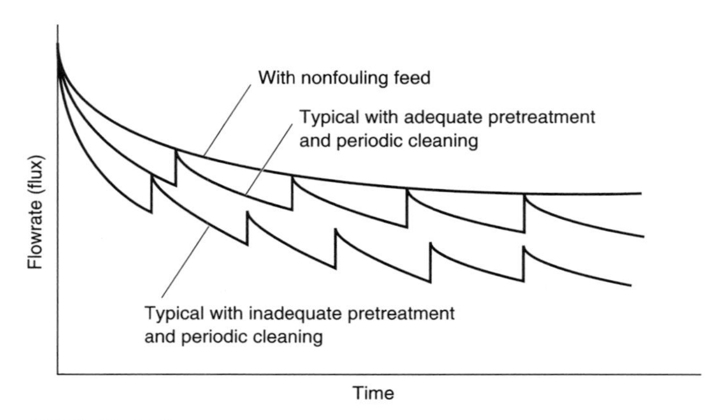

FIGURE 12-32
Definition sketch for the performance of a membrane filtration system.

Membrane fouling. The term *fouling* is used to describe the potential deposition and accumulation of constituents in the feed stream on the membrane. Membrane fouling is an important consideration in the design and operation of membrane systems, as it affects pretreatment needs, cleaning requirements, operating conditions, cost, and performance. Constituents in wastewater that can bring about membrane fouling are identified in Table 12-17. Fouling of the membrane, as reported in

TABLE 12-17
Constituents in wastewater that can affect the performance of membranes through the mechanism of fouling

Type of membrane fouling	Responsible constituents	Remarks
Fouling (cake formation sometimes identified as biofilm formation)	Metal oxides Organic and inorganic colloids Bacteria Microorganisms Concentration polarization	Damage to membranes can be limited by controlling these substances (for example, by the use of microfiltration before reverse osmosis).
Scaling (precipitation)	Calcium sulfate Calcium carbonate Calcium fluoride Barium sulfate Metal oxide formation Silica	Scaling can be reduced by limiting salt content, by adding acid to limit the formation of calcium carbonate, and by other chemical treatments (e.g., the addition of antiscalants).
Damage to membrane	Acids Bases pH extremes Free chlorine Bacteria Free oxygen	Damage to membranes can be limited by controlling these substances. Extent of damage depends on the nature of the membrane.

Table 12-17, can occur in three general forms: (1) a buildup of the constituents in the feed water on the membrane surface (commonly identified as cake formation), (2) formation of chemical precipitates due to the chemistry of the feed water, and (3) damage to the membrane due to the presence of chemical substances that react with the membrane, or biological agents that colonize the membrane.

Control of membrane fouling. Typically, three approaches are used to control membrane fouling: (1) pretreatment of the feed water, (2) membrane backflushing, and (3) chemical cleaning of the membranes. Pretreatment is used to reduce the TSS and bacterial content of the feed water. Often the feed water will be conditioned chemically to limit chemical precipitation within the units. The most commonly used method of eliminating the accumulated material from the membrane surface is backflushing with water and/or air. Chemical treatment is used to remove constituents that are not removed during conventional backwashing. Chemical precipitates can be removed by altering the chemistry of the feed water and by chemical treatment. Damage of the membrane by deleterious constituents typically cannot be reversed.

Assessing need for pretreatment for NF and RO. To assess the treatability of a given wastewater with NF and RO membranes, a variety of fouling indexes have been developed over the years. The three principal indices are the silt density index (SDI), the modified fouling index (MFI), and the miniplugging factor index (MPFI). Fouling indexes are determined from simple membrane tests. The sample must be passed through a 0.45-μm Millipore filter with a 47-mm internal diameter at 30 lb/in^2 gage to determine any of the indexes. The time to complete data collection for these tests varies from 15 min to 2 h, depending on the fouling nature of the water. The most widely used index is the SDI. The SDI is defined as follows.

$$\text{SDI} = \frac{100[1 - (T_i/T_f)]}{T_t} \tag{12-27}$$

where T_i = time to collect the initial sample of 500 mL, min
T_f = time to collect final sample of 500 mL, min
T_t = total time for running the test, min

The silt density index is a static measurement of resistance which is determined by samples taken at the beginning and end of the test. The SDI does not measure the rate of change of resistance during the test. Recommended maximum values of the SDI for NF and RO membranes are 4 and 3, respectively. The calculation of the SDI is demonstrated in Example 12-4.

EXAMPLE 12-4. SILT DENSITY INDEX FOR REVERSE OSMOSIS. Determine the silt density index for a water from the following test data:

Test run time = 30 min
Initial 500 mL = 2 min
Final 500 mL = 10 min

Solution

1. Calculate the SDI using Eq. (12-27).

$$SDI = \frac{100[1 - (T_i/T_f)]}{T_t}$$

$$= \frac{100[1 - (2/10)]}{30} = 2.67$$

2. Compare the SDI to the acceptable criteria.

Comment. Calculated SDI of 2.67 is less than 3, therefore, no further pretreatment is needed.

Application of Membranes

With evolving health concerns (see Chap. 2) and the development of new and lower-cost membranes, the application of membrane technologies in the field of environmental engineering has increased dramatically within the past five years. The increased use of membranes is expected to continue well into the future. In fact, it has been suggested that conventional filtration technology, such as described in Sec. 12-3, may be considered obsolete within 20 years. The principal applications of the various membrane technologies for the removal of the constituents found in wastewater are summarized in Table 12-18. Typical operating ranges in terms of operating pressures and flux rates, along with the types of membranes used, are reported in Table 12-19.

TABLE 12-18
Application of membranes for the removal of constituents found in wastewater

	Type of membrane				
Constituent	**MF**	**UF**	**NF**	**RO**	**Comments**
Biodegradable organics		✓	✓	✓	
Hardness			✓	✓	
Heavy metals			✓	✓	
Nitrate			✓	✓	
Priority organic pollutants		✓	✓	✓	
Synthetic organic compounds			✓	✓	
TDS			✓	✓	
TSS	✓	✓			Removed as pretreatment for NF and RO.
Bacteria	✓	✓	✓	✓	Used for membrane disinfection. Removed as pretreatment for NF and RO with MF and UF.
Protozoan oocysts and cysts	✓	✓	✓	✓	
Viruses			✓	✓	Used for membrane disinfection.

TABLE 12-19

Characteristics of membrane processes used in wastewater treatment applications

Membrane process	Typical operating range,* μm	Operating pressure		Rate of flux		Membrane details	
		lb/in^2	kPa	gal/ft^2·d	L/m^2·d	Type	Configuration
Microfiltration	0.06–2	4–30	28–210	10–40	405–1600	Polypropylene, acrylonitrile, nylon, and polytetrafluoroethylene	Spiral wound, hollow fiber, plate and frame
Ultrafiltration	0.005–0.1	25–150	170–1000	10–20	405–815	Cellulose acetate, aromatic polyamide and polytherurea thin film composites	Spiral wound, hollow fiber, plate and frame
Nanofiltration	0.001–0.01	75–250	515–1720	5–20	200–815	Cellulose acetate, aromatic polyamides	Spiral wound, hollow fiber
Reverse osmosis	0.0001–0.002	150–400	1000–2750	8–12	320–490	Cellulose acetate, aromatic polyamide, polyetherurea, and thin film composites	Spiral wound, hollow fiber

*Depends on type of membrane.

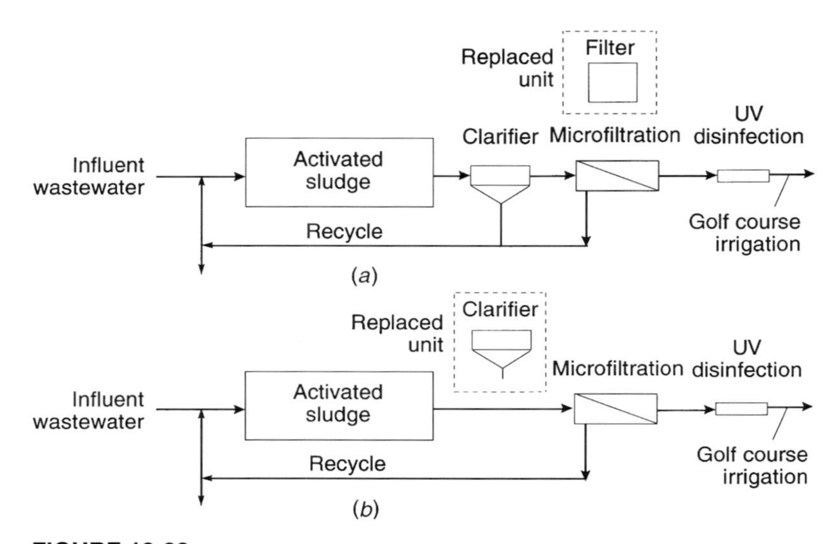

FIGURE 12-33
Typical flow diagrams involving the use of microfiltration: (*a*) as a replacement for granular medium filtration and (*b*) as a replacement for secondary clarification.

Microfiltration. Microfiltration (MF) can be used in a variety of ways in water reuse systems (see Fig. 12-33 and 12-34*a*). MF can be used to replace packed-bed filtration to remove turbidity, residual TSS, eukaryotic parasites, and most bacteria. MF can also be used in place of secondary clarifiers (see Fig 12-33*b*), in which case the MF unit can be located externally or internally. If the MF unit is located internally, the term *membrane bioreactor* is used to describe the process. In water repurification systems, MF can serve as pretreatment for advanced processes and reverse osmosis (see Fig. 12-34*a*). MF membranes are the most numerous on the market and the least expensive. As reported in Table 12-19, they are commonly made of polypropylene, acrylonitrile, nylon, and polytetrafluoroethylene.

Ultrafiltration. Ultrafiltration (UF) membranes are effective at excluding TSS, eukaryotic and prokaryotic microorganisms, some viruses, and dissolved compounds with high molecular weight, such as colloids, proteins, and carbohydrates. The membranes do not remove sugar or salt. UF is used typically in repurification applications as a pretreatment step for reverse osmosis (see Fig. 12.34*b*). Typical operating data are presented in Table 12-19.

Reverse osmosis. When two solutions having different solute concentrations are separated by a semipermeable membrane, a difference in chemical potential will exist across the membrane (Fig. 12-35). Water will tend to diffuse through the membrane from the lower-concentration (higher-potential) side to the higher-concentration (lower-potential) side. In a system having a finite volume, flow continues until the pressure difference balances the chemical potential difference. This balancing pressure difference is termed the *osmotic pressure* and is a function of the solute characteristics and concentration and temperature. If a pressure gradient opposite in direction and greater than the osmotic pressure is imposed across the

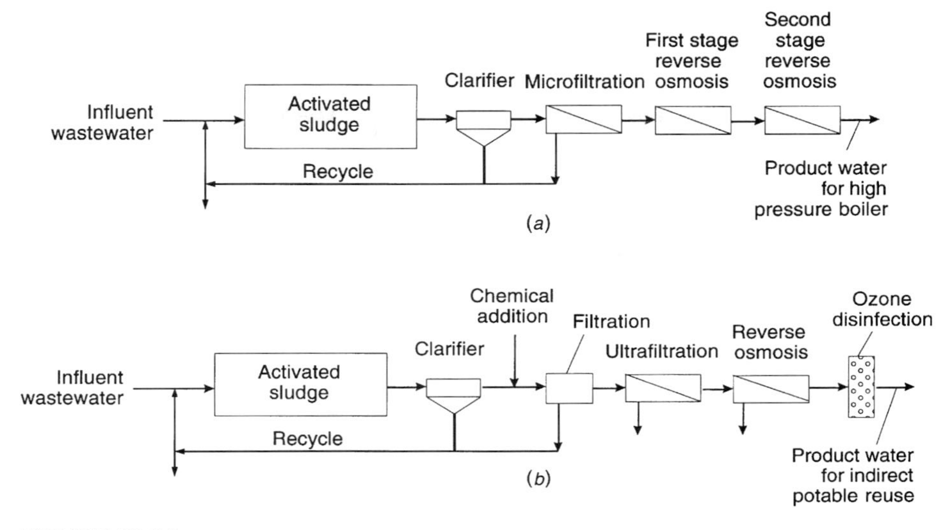

FIGURE 12-34
Typical flow diagrams involving the use of reverse osmosis: (*a*) for the production of water for high-pressure boilers and (*b*) repurification for indirect potable reuse.

membrane, flow from the more concentrated to the less concentrated region will occur and is termed *reverse osmosis* (see Fig. 12-35*c*).

Reverse osmosis (RO) is used for desalination and removal of dissolved solids from wastewater. The membranes exclude ions, but require relatively high pressures to produce the deionized water. Nanofiltration, also known as "loose" RO, can reject particles as small as 0.001 μm. Typical flow diagrams involving the use of reverse osmosis are shown in Figs. 12-5 and 12-34. The RO units at Aqua III at San Pasqual in San Diego County used for the repurification of pretreated water are shown in

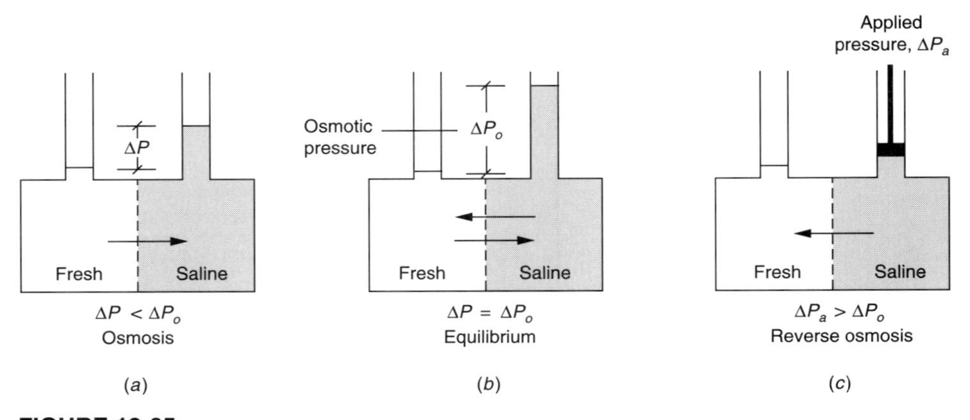

FIGURE 12-35
Definition sketch of osmotic flow: (*a*) osmotic flow, (*b*) osmotic equilibrium, and (*c*) reverse osmosis.

FIGURE 12-36
Reverse osmosis units at the San Pasqual treatment facility in San Diego County, California.

Fig. 12-36. Corresponding performance data are presented in Table 12-20. Disinfection of the RO feed water is usually practiced to minimize the bacterial growth on the membrane. Care must be taken with polyamide and thin-film composite membranes because they are sensitive to chemical oxidants. The need for cleaning depends on how long the flux rate is maintained.

Pilot Studies for Membrane Applications

Because every wastewater is unique with respect to its chemistry, it is difficult to predict beforehand how a given membrane process will perform. As a result, the selection of the best membrane for a given application is usually based on the results of pilot studies. Membrane fouling indexes can be used to assess the need for pretreat-

TABLE 12-20
Performance summary for Aqua III reverse osmosis system for the period from October 1994 through September 1995*†

Constituent	RO influent, mg/L	RO effluent, mg/L	Reduction, %
TOC	7.1	0.6	92
TS	1110	55	95
NH_3-N	9.8	0.7	93
NO_3-N	2.6	0.8	69
PO_4	1.49	0.1	93
SO_4	368	0.1	99
Cl	284	13	95

*From WCPH (1996).
†Typical flux rate during test period was 6.3 gal/ft^2·d.

ment. In some situations, manufacturers of membranes will provide a testing service to identify the most appropriate membrane for a specific water or wastewater.

The elements that make up a pilot plant include:

1. The pretreatment system
2. Tankage for flow equalization and cleaning
3. Pumps for pressurizing the membrane, recirculation, and backflushing with appropriate controls (e.g., variable-frequency drives)
4. The membrane test module
5. Adequate facilities for monitoring the performance of the test module
6. An appropriate system for backflushing the membranes.

The information collected should be sufficient to allow for the design of the full-scale system. Information typically required includes: (1) pretreatment requirements including chemical dosages, (2) general long-term operating characteristics, (3) rates of flux correlated to operating times, (4) cleaning frequency including protocol and chemical requirements, and (5) post-treatment requirements.

Typical water quality measurements may include:

Turbidity	Temperature
Particle counts	Heterotrophic plate count (see Chap. 2)
Total organic carbon	Other bacterial indicators
Nutrients	Specific constituents that can limit recovery such as silica, barium, calcium, and sulfate
Heavy metals	Biotoxicity
Organic priority pollutants	
Total dissolved solids	
pH	

The specific parameters selected for evaluation will depend on the final use to be made of the repurified water. Pilot test facilities at Aqua III, used to evaluate various membrane treatment options in connection with the production of repurified water for indirect potable reuse, are shown in Fig. 12-37.

12-5 REMOVAL OF PHOSPHORUS BY CHEMICAL METHODS

The principal chemical method for phosphorus removal is chemical precipitation. The chemical precipitation of phosphorus is accomplished by the addition of the salts of multivalent metal ions that form precipitates of sparingly soluble phosphates. The multivalent metal ions used most commonly are calcium [Ca^{+2}], aluminum [Al^{+3}], and iron [Fe^{+3}]. Because the chemistry of phosphate precipitation with calcium is quite different than with aluminum and iron, the two different types of precipitation are considered separately in the following discussion.

Phosphorus Precipitation with Calcium

When calcium, usually in the form of lime $Ca(OH)_2$, is added to water it reacts with the natural bicarbonate alkalinity to precipitate $CaCO_3$. As the pH value of the

(*a*)

(*b*)

(*c*)

FIGURE 12-37
Typical membrane filter pilot units: (*a*) microfiltration, (*b*) ultrafiltration, and (*c*) reverse osmosis. The test units are being evaluated in connection with water repurification for indirect potable reuse at the San Pasqual treatment facility in San Diego County, California.

wastewater increases beyond about 10, excess calcium ions will then react with the phosphate, as shown in Eq. (12-28), to precipitate hydroxylapatite $Ca_{10}(PO_4)_6(OH)_2$.

Phosphate precipitation with calcium:

$$10Ca^{+2} + 6PO_4^{-3} + 2OH^- \leftrightharpoons \underset{\text{Hydroxylapatite}}{Ca_{10}(PO_4)_6(OH)_2} \tag{12-28}$$

Because of the reaction of lime with the alkalinity of the wastewater, the quantity of lime required will, in general, be independent of the amount of phosphate present and will depend primarily on the alkalinity of the wastewater. The quantity of lime required to precipitate the phosphorus in wastewater is typically about 1.4 to 1.5 times the total alkalinity expressed as $CaCO_3$. Because a high pH value is required to precipitate phosphate, coprecipitation is usually not feasible. When lime is added to raw wastewater or to secondary effluent, pH adjustment is usually required before subsequent treatment or disposal. Recarbonation with carbon dioxide (CO_2) is used to lower the pH value.

Phosphate Precipitation with Aluminum and Iron

The basic reactions involved in the precipitation of phosphorus with aluminum and iron are as follows.

Phosphate precipitation with aluminum:

$$Al^{+3} + H_nPO_4^{3-n} \leftrightharpoons AlPO_4 + nH^+ \tag{12-29}$$

Phosphate precipitation with iron:

$$Fe^{+3} + H_nPO_4^{3-n} \leftrightharpoons FePO_4 + nH^+ \tag{12-30}$$

In the case of alum and iron, 1 mole will precipitate 1 mole of phosphate; however, these reactions are deceptively simple and must be considered in light of the many competing reactions and their associated equilibrium constants, and the effects of alkalinity, pH, trace elements, and ligands found in wastewater. Because of the many competing reactions, Eqs. (12-29) and (12-30) cannot be used to estimate the required chemical dosages directly. Therefore, dosages are generally established on the basis of bench-scale tests and occasionally by full-scale tests, especially if polymers are used (Tchobanoglous and Burton, 1991).

Application of Phosphorus Precipitation

A typical treatment flow diagram (see Fig. 12-3*a*) would include a tertiary treatment unit following aerobic secondary treatment, in which the metal salt is added. The resulting chemical precipitate will settle to the tank bottom where it can be removed as a sludge. Alternatively, metal salts can be added to secondary or primary settling tanks for removal with the organic solids. Phosphorus removal in packed-bed filtration is also possible if the alum or ferric chloride is mixed and coagulated prior to filtration (see Fig. 12-3*c*). This approach, used successfully at a number of locations, avoids the need for a clarifier, and the phosphorus can be collected with the filter backwash (Crites et al., 1988).

12-6 WASTEWATER DISINFECTION

Disinfection refers to the selective destruction of disease-causing organisms as opposed to sterilization, which is the destruction of all organisms. The disinfection of treated wastewater is of fundamental importance in the management of this resource. Disinfection technologies considered in this section include: (1) chlorination and dechlorination using liquid chlorine and sulfur dioxide (chlorination), (2) chlorination and dechlorination using liquid sodium hypochlorite and sodium bisulfite (hyperchlorination), (3) ultraviolet radiation (UV disinfection), and (4) ozone treatment (ozonation). Before discussing the individual technologies it will be helpful to consider: (1) the organisms of concern, (2) the alternative disinfection technologies available, (3) the mechanisms of disinfection, (4) the factors affecting the action of disinfectants, and (5) the modeling of the disinfection process. Following the discussion of the individual technologies, the advantages and disadvantages of each are reviewed.

Organisms of Concern

The principal infectious agents that occur in wastewater, as noted in Chap. 2, are classified into three broad groups: bacteria, eukaryotic parasites (principally protozoa and helminths), and viruses. Diseases caused by waterborne bacteria include typhoid, cholera, paratyphoid, and bacillary dysentery (see Table 2-20). Among the parasites, the protozoan organisms *Cryptosporidium parvum, Cyclospora,* and *Giardia lamblia* are of greatest concern. The most important helminthic parasites that may be found in wastewater are intestinal worms, including the stomach worm *Ascaris lumbricoides,* the tapeworms *Taenia saginata* and *Taenia solium,* the whipworm *Trichuris trichiura,* the hookworms *Ancylostoma duodenale* and *Necator americanus,* and the threadworm *Strongyloides stercoralis.* Diseases caused by waterborne viruses include poliomyelitis and infectious hepatitis.

Alternative Disinfection Technologies

Disinfection is most commonly accomplished by: (1) chemical agents, (2) physical agents, (3) mechanical means, and (4) radiation. The individual agents and means that have been used under each category are as follows.

Chemical agents

1. Chlorine and its compounds
2. Bromine
3. Iodine
4. Ozone
5. Phenol and phenolic compounds
6. Alcohols
7. Heavy metals and related compounds
8. Dyes
9. Soaps and synthetic detergents
10. Quaternary ammonium compounds
11. Hydrogen peroxide
12. Various alkalies and acids

Physical agents
1. Heat (e.g., boiling)
2. Light (ultraviolet radiation)

Mechanical means
1. Individual treatment processes

Radiation
1. Electromagnetic
2. Acoustic
3. Particle

With respect to the above listing, the disinfection technologies considered in this section include three which involve the use of chemical agents (chlorine, hypochlorite, and ozone) and one which involves the use of a physical agent (UV radiation). The characteristics of an ideal disinfectant and the actual characteristics of the disinfectants considered in this section are summarized in Table 12-21.

Mechanisms of Disinfectants

The principal mechanisms that have been put forth to explain the action of disinfectants are: (1) damage to the cell wall, (2) alteration of cell permeability, (3) alteration of the colloidal nature of the protoplasm, (4) inhibition of enzyme activity, and (5) damage to a cell's DNA and RNA. Damage or destruction of the cell wall will result in cell lysis and cell death. Alteration of the selective permeability of the cell membrane will allow vital nutrients, such as nitrogen and phosphorus, to escape. Heat, radiation, and highly acid or alkaline agents alter the colloidal nature of the protoplasm. Heat will coagulate the cell protein and acids or bases will denature proteins, producing a lethal effect. Oxidizing agents, such as chlorine, can alter the chemical arrangement of enzymes and inactivate the enzymes. If the cells' DNA and RNA are damaged (e.g., through the formation of double bonds by UV radiation) the organism will be unable to reproduce, and will ultimately die. The modes of action of the disinfection technologies considered in this section are summarized in Table 12-22. In addition to the modes of action summarized in Table 12-22, a variety of natural mechanisms are also at work in the environment including sunlight, temperature, salinity, exposure to heavy metals, predation by bacteriophage, predation by parasites, and old age.

Factors Affecting the Action of Disinfectants

In applying the disinfection agents or means that have been described, the following factors must be considered: (1) initial mixing, (2) contact time, (3) concentration and type of chemical agent, (4) intensity and nature of physical agent, (5) temperature, (6) number of organisms, (7) types of organisms, and (8) characteristics of the wastewater. Additional details on these factors are presented in Table 12-23. Of the factors listed in Table 12-23, the characteristics of the wastewater will generally have the greatest impact on the disinfection process, as they are less controllable.

TABLE 12-21

Comparison of ideal and actual characteristics of chlorine, sodium hypochlorite, UV radiation, and ozone as disinfectants

Characteristic	Properties/ response	Chlorine	Sodium hypochlorite*	UV radiation	Ozone
Availability	Should be available in large quantities and reasonably priced	Low cost	Moderately low cost	Moderately high cost	Moderately high cost
Deodorizing ability	Should deodorize while disinfecting	High	Moderate	N/A	High
Homogeneity	Solution must be uniform in composition	Homogeneous	Homogeneous	Homogeneous	N/A
Interaction with extraneous material	Should not be absorbed by organic matter other than bacterial cells	Oxidizes organic matter	Active oxidizer	Absorbance of UV radiation	Oxidizes organic matter
Noncorrosive and nonstaining	Should not disfigure metals or stain clothing	Highly corrosive	Corrosive	N/A	Highly corrosive
Nontoxic to higher forms of life	Should be toxic to microorganisms and nontoxic to humans and other animals	Highly toxic to higher life forms	Toxic	Toxic	Toxic
Penetration	Should have the capacity to penetrate through surfaces	High	High	Moderate	High
Safety concern	Should be safe to transport, store, handle, and use	High	Moderate	Low	Moderate
Solubility	Must be soluble in water or cell tissue	Slight	High	N/A	High
Stability	Loss of germicidal action on standing should be low	Stable	Slightly unstable	Must be generated	Unstable, must be generated
Toxicity to micro-organisms	Should be highly toxic at high dilutions	High	High	High as used	High as used
Toxicity at ambient temperatures	Should be effective in ambient temperature range	High	High	High	High

*An additional problem with sodium hypochlorite is the formation of precipitates in the feed line.

Note: N/A = not applicable

TABLE 12-22
Mechanisms of disinfection by chlorine, UV, and ozone

Chlorine	UV	Ozone
1. Oxidation. 2. Reactions with available chlorine. 3. Protein precipitation. 4. Modification of cell wall permeability. 5. Hydrolysis and mechanical disruption.	1. Photochemical damage to RNA and DNA (e.g., formation of double bonds) within the cells of an organism. 2. The nucleic acids in microorganisms are the most important absorbers of the energy of light in the wavelength range of 240–280 nm. 3. Because DNA and RNA carry genetic information for reproduction, damage of these substances can effectively inactivate the cell.	1. Direct oxidation/destruction of cell wall with leakage of cellular constituents outside of cell. 2. Reactions with radical by-products of ozone decomposition. 3. Damage to the constituents of the nucleic acids (purines and pyrimidines). 4. Breakage of carbon-nitrogen bonds leading to depolymerization.

TABLE 12-23
Factors affecting the action of disinfectants

Factor	Description
Initial mixing	Critical step in disinfection with chemicals. Disinfectant must be dispersed throughout the liquid to be disinfected. If initial mixing is prolonged, the disinfectant may react with compounds in wastewater, which may reduce its effectiveness.
Contact time	Time during which organisms in the fluid are exposed directly to chemical agent or intensity (UV radiation).
Concentration and type of chemical agent	Dose = concentration × time for chemical agents.
Intensity and nature of physical agent	Dose = intensity × time.
Temperature	Affects reactivity and the ionization constants for chemical agents.
Number of organisms	Important in historical development of disinfection kinetics. Free-swimming organisms less important than shielded organisms within particles or bacterial clumps.
Types of organisms	Different organisms have varying resistance to disinfecting agents.
Characteristics of the wastewater	Significant influence on effectiveness of disinfecting agents. Compounds in wastewater can react with chemical disinfectants or absorb energy (e.g., UV radiation).

The impact of these factors is considered in the discussion of the individual disinfection technologies.

Modeling the Disinfection Process

The effectiveness of disinfectants in wastewater treatment can be modeled with a relationship of the following general form:

$$N = \phi(N_o, W_f, D^n) \tag{12-31}$$

where N = number of organisms remaining after disinfection
N_o = number of organisms present before disinfection
W_f = water quality factor
D = dose of disinfectant
n = empirical coefficient

The characteristics of the wastewater are taken into account in the water quality factor. The characteristics that must be considered include: (1) the size, size distribution, and concentration of the particles that compose the TSS; (2) the nature of the organisms, the number of organisms, and their distribution within the liquid (e.g., free-swimming or particle-associated); and (3) the presence of chemical compounds that can react with the disinfectant and/or absorb radiant energy such as UV radiation. The impact of wastewater characteristics on chlorine, UV, and ozone disinfection are summarized in Table 12-24. The dose of disinfectant can be defined as follows for chemical agents [Eq. (12-32)] and physical agents [Eq. (12-33)], respectively:

$$\text{Dose}_{\text{chemical agent}} = C \times t \tag{12-32}$$

$$\text{Dose}_{\text{physical agent}} = I \times t \tag{12-33}$$

where C = concentration of chemical agent (e.g., chlorine)
t = contact time
I = intensity of physical agent (e.g., UV radiation)

The dose concept was introduced previously in Chap. 4 in connection with risk assessment. In applying Eqs. (12-32) and (12-33), initial mixing and the contact time become critical factors in the disinfection process (i.e., how well was the disinfectant mixed with the fluid to be disinfected, and what portion of the fluid remained in contact with the disinfectant for a given time period).

In 1908, Chick proposed the following simplified first-order model to describe the decrease of organisms with time for a given concentration of disinfectant, assuming the individual organisms are discrete entities within the fluid (Chick, 1908):

$$N = N_o e^{-kt} \tag{12-34}$$

where N = number of organisms remaining after disinfection
N_o = number of organisms present before disinfection
k = decay constant which reflects the impact of the factors affecting the action of the disinfectant
t = time

TABLE 12-24
Comparison of impact of wastewater characteristics on chorine, UV, and ozone disinfection

Wastewater characteristic	Chlorine disinfection	UV disinfection	Ozone disinfection
Ammonia	Combines with chlorine to form chloramines.	No or minor effect.	No or minor effect, can react at high pH.
BOD, COD, etc.	Organic compounds that make up the BOD and COD can exert a chlorine demand. The degree of interference depends on their functional groups and their chemical structure.	No or minor effect, unless humic materials constitute a large portion of the BOD.	Organic compounds that make up the BOD and COD can exert an ozone demand. The degree of interference depends on their functional groups and their chemical structure.
Hardness	No or minor effect.	Affects solubility of metals that may absorb UV radiation. Can lead to the precipitation of carbonates on quartz tubes.	No or minor effect.
Humic materials	Reduce effectiveness of chlorine.	Strong absorbers of UV radiation.	Affects the rate of ozone decomposition and the ozone demand.
Iron	No or minor effect.	Strong absorber of UV radiation.	No or minor effect.
Nitrite	Oxidized by chlorine.	No or minor effect.	Oxidized by ozone.
Nitrate	No or minor effect.	No or minor effect.	Can reduce effectiveness of ozone.
pH	Affects distribution between hypochlorous acid and hypochlorite ion.	Can affect solubility of metals and carbonates.	Effects the rate of ozone decomposition.
TSS	Shielding of embedded bacteria.	Absorption of UV radiation and shielding of embedded bacteria.	Increase ozone demand and shielding of embedded bacteria.

If the model proposed by Chick is combined with the concept proposed by Watson (1908), the Chick-Watson model as proposed by Haas and Kara (1984) is:

$$N = N_o e^{-kC^n t} \tag{12-35}$$

where N = number of organisms remaining after disinfection
N_o = number of organisms present before disinfection
k = decay constant which reflects the water quality characteristics cited above
C = concentration of chemical agent
n = empirical constant
t = time

Unfortunately, in treated wastewater which contains TSS, the relationships cited above do not apply, especially where stringent effluent microorganism standards must be met. The effect of effluent TSS on disinfection performance is illustrated in Fig. 12-38. It should be noted that Eqs. (12-34) and (12-35) apply to the straight line portion of the curve given in Fig. 12-38. Referring to Fig. 12-38, as the concentration of TSS increases, it becomes more difficult to achieve low organism concentrations at a given dose, due to shielding of the organisms by particles. Where chlorine or one of its compounds is used as the disinfectanct, the effects of particle shielding can be overcome by increasing the chemical concentration or the particles must be modified (e.g., reduced in size below some critical size by means of granular medium or membrane filtration, for example, or by reducing the number of embedded target pathogens). With current UV radiation technology, increasing the intensity will not improve the disinfection performance with respect to particle associated organisms (see discussion in UV section below). To improve the disinfection performance of UV radiation, the particles must be modified as described above.

Disinfection Using Liquid Chlorine and Dechlorination with Sulfur Dioxide

Chlorine, of all the chemical disinfectants, is perhaps the one most commonly used throughout the world. The reason for the ubiquitous use of chlorine is that it satisfies most of the requirements specified in Table 12-21. The most common chlorine compounds used in wastewater treatment plants are chlorine gas (Cl_2), calcium hypochlorite [$Ca(OCl)_2$], sodium hypochlorite (NaOCl), and chlorine dioxide (ClO_2). Calcium or sodium hypochlorite is often used at large facilities, primarily for reasons of safety as influenced by local conditions. The use of calcium hypochlorite is considered following the discussion of the use of chlorine.

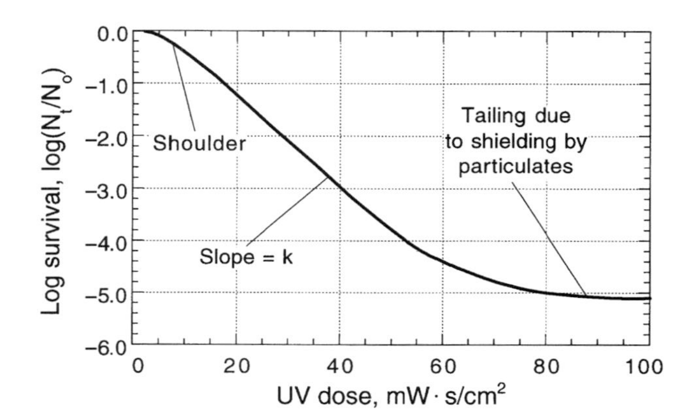

FIGURE 12-38
Effect of total suspended solids in secondary effluent on the performance of chlorine and UV disinfection.

Reactions of chlorine in water. When chlorine in the form of Cl_2 gas is added to water, two reactions take place: hydrolysis and ionization. Hydrolysis may be defined as:

$$Cl_2 + H_2O \rightarrow HOCl + H^+ + Cl^- \tag{12-36}$$

The stability constant for this reaction is:

$$k = \frac{[HOCl][H^+][Cl^-]}{[Cl_2]} = 4.5 \times 10^{-4} \text{ at } 25°C \tag{12-37}$$

Ionization may be defined as:

$$HOCl \rightarrow H^+ + OCl^- \tag{12-38}$$

The ionization constant for this reaction is:

$$k_i = \frac{[H^+][OCl^-]}{[HOCl]} = 2.9 \times 10^{-8} \text{at } 25°C \tag{12-39}$$

The quantity of HOCl and OCl^- present in a water is called the *free available chlorine.* The relative distribution of these two chemical species is very important because the germicidal efficiency of HOCl is significantly greater (up to 100 times) than that of OCl^-. The percentage distribution of HOCl at various pH and temperature values, as shown in Fig. 12-39, can be approximated by using Eq. (12-39) and the data given in Table 12-25:

$$\frac{[HOCl]}{[HOCl] + [OCl]} = \frac{1}{1 + [OCl]/[HOCl]} = \frac{1}{1 + k_i/[H]} \tag{12-40}$$

Reactions of chlorine with ammonia. The effluent from most treatment plants typically contains significant amounts of nitrogen, usually in the form of ammonia, or nitrate if the plant is designed to achieve nitrification. Because hypochlorous acid is a very active oxidizing agent, it will react readily with ammonia in the wastewater to form three types of chloramines in the successive reactions:

$$NH_3 + HOCl \rightarrow NH_2Cl \text{ (monochloramine)} + H_2O \tag{12-41}$$

$$NH_2Cl + HOCl \rightarrow NHCl_2 \text{ (dichloramine)} + H_2O \tag{12-42}$$

$$NHCl_2 + HOCl \rightarrow NCl_3 \text{ (trichloride)} + H_2O \tag{12-43}$$

These reactions are dependent on the pH, temperature, contact time, and the ratio of chlorine to ammonia. The two species that predominate, in most cases, are monochloramine (NH_2Cl) and dichloramine ($NHCl_2$). The chlorine in these compounds is called *combined available chlorine.* Although they are slow-reacting, chloramines also serve as disinfectants.

Breakpoint chlorination. The maintenance of a chlorine residual (combined or free) for the purpose of wastewater disinfection is complicated by the fact that free chlorine not only reacts with ammonia, as described above, but also is a strong oxidizing agent. The main reason for adding enough chlorine to obtain a free chlorine

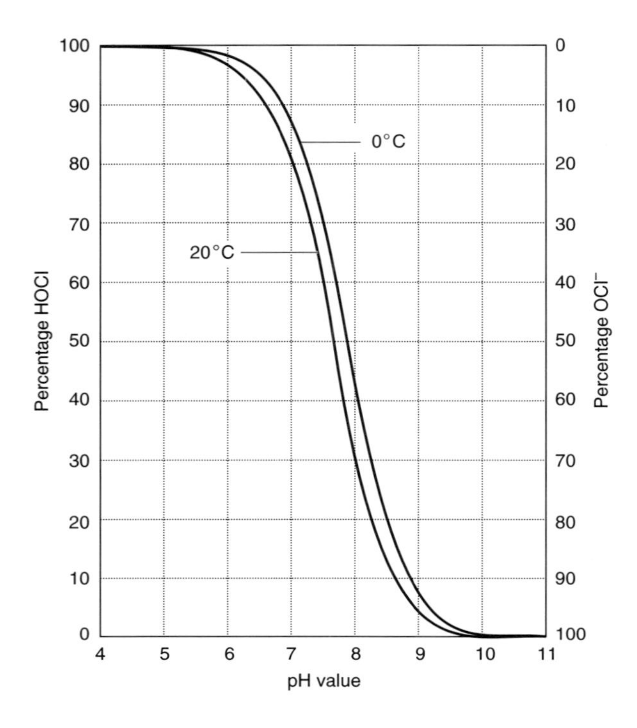

FIGURE 12-39
Distribution of hypochlorous acid and hypochlorite ion in water as a function of pH and temperature.

residual is to increase the reliability of the disinfection process. The stepwise phenomena that result when chlorine is added to wastewater containing ammonia and other readily oxidizable substances are illustrated in Fig. 12-40*a*. As chlorine is added, readily oxidizable substances, such as Fe^{+2}, Mn^{+2}, H_2S, and organic matter (Sung, 1974), react with the chlorine and reduce most of it to the chloride ion (point *A* in Fig. 12-40). After meeting this immediate demand, the chlorine continues to react with the ammonia to form chloramines between points *A* and *B*. For mole ratios of chlorine to ammonia less than 1, monochloramine and dichloramine will be formed.

TABLE 12-25
Values of the ionization constant of hypochlorous acid at different temperatures*

Temperature, °C	$K_i \times 10^8$, mol/L
0	1.5
5	1.7
10	2.0
15	2.3
20	2.6
25	2.9

*Computed by using equation from Morris (1966).

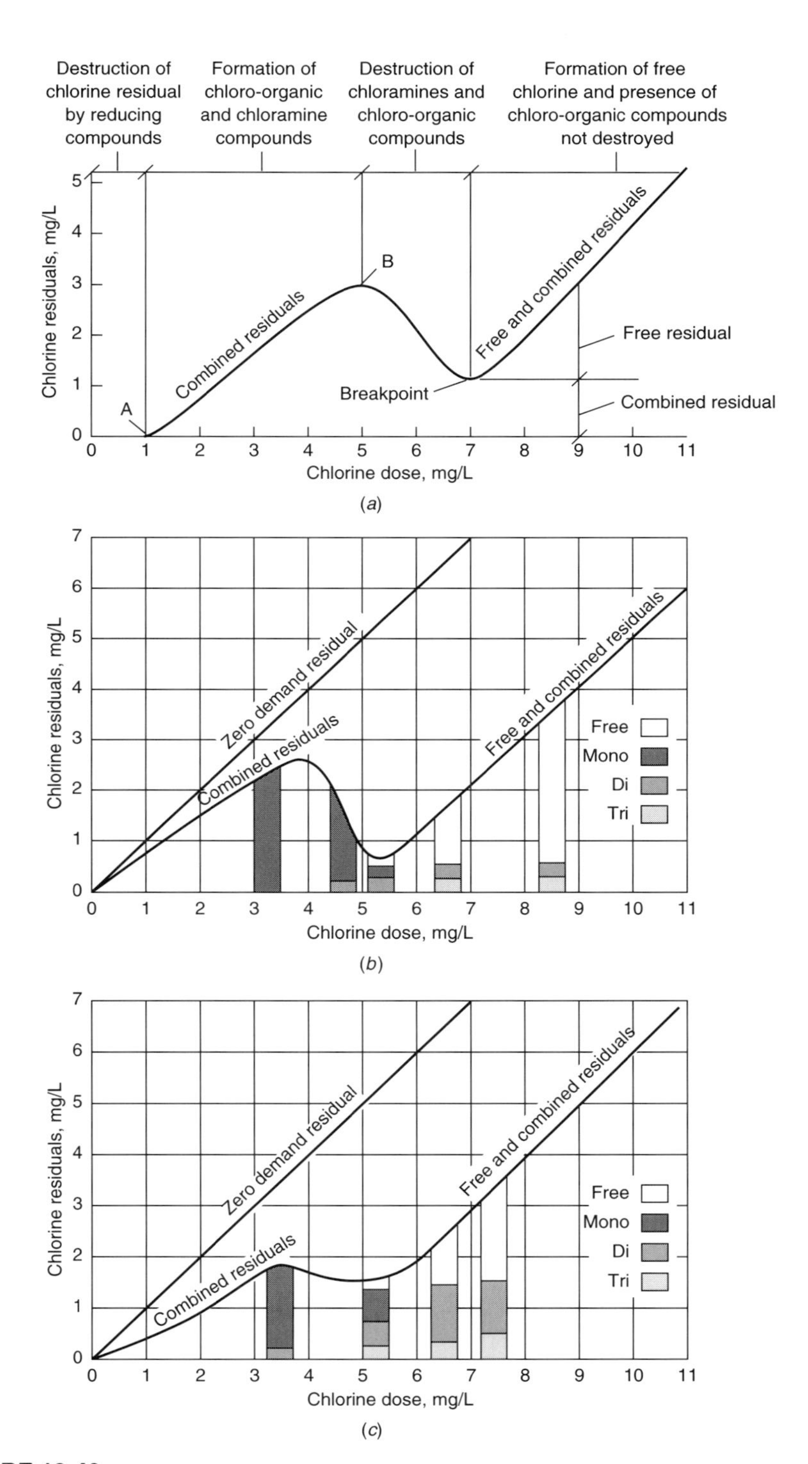

FIGURE 12-40

Curves obtained during breakpoint chlorination of wastewater: (*a*) generalized definition curve, (*b*) for wastewater containing ammonia nitrogen, and (*c*) for wastewater containing ammonia nitrogen and organic nitrogen (b and c adapted from White, 1992).

The distribution of these two chloramine forms is governed by their rates of formation, which are dependent on pH and temperature. Between point *B* and the breakpoint, some chloramines will be converted to nitrogen trichloride [see Eq. (12-43)]; the remaining chloramines will be oxidized to nitrogen (N_2) and nitrous oxide (N_2O), and the chlorine will be reduced to the chloride ion. With continued addition of chlorine, most of the chloramines will be oxidized at the breakpoint. Continued addition of chlorine past the breakpoint will result, as shown in Fig. 12-40*a*, in a directly proportional increase in the free available chlorine (i.e., unreacted hypochlorous acid and hypochlorite ion). Possible reactions to account for the appearance of N_2 and N_2O and the disappearance of chloramines during breakpoint chlorination are as follows (Saunier, 1976, and Saunier and Selleck, 1976):

$$NH_4^+ + HOCl \rightarrow NH_2Cl + H_2O + H^+ \tag{12-44}$$

$$NH_2Cl + HOCl \rightarrow NHCl_2 + H_2O \tag{12-45}$$

$$0.5NHCl_2 + 0.5H_2O \rightarrow 0.5NOH + H^+ + Cl^- \tag{12-46}$$

$$0.5NHCl_2 + 0.5NOH \rightarrow 0.5N_2 + 0.5HOCl + 0.5H^+ + 0.5Cl^- \tag{12-47}$$

The overall reaction, obtained by summing Eqs. (12-44) through (12-47), is:

$$NH_4^+ + 1.5HOCl \rightarrow 0.5N_2 + 1.5H_2O + 2.5H^+ + 1.5Cl^- \tag{12-48}$$

Occasionally, odor problems have developed during breakpoint chlorination operations because of the formation of nitrogen trichloride and related compounds [see Eq. (12-43)]. The presence of additional compounds that will react with chlorine, such as organic nitrogen, may greatly alter the shape of the breakpoint curve. Breakpoint chlorination curves for wastewater containing only ammonia nitrogen and ammonia and organic nitrogen are shown in Figs. 12-40*b* and 12-40*c*, respectively. Referring to Eqs. (12-44) through (12-48), it will be noted that alkalinity is consumed in the breakpoint reaction. For most wastewater effluents, the pH change is usually not significant. The theoretical weight ratio of chlorine to ammonia nitrogen at the breakpoint and the alkalinity required are determined in Example 12-5.

EXAMPLE 12-5. ANALYSIS OF CHLORINE BREAKPOINT STOICHIOMETRY. Determine the stoichiometric weight ratio of chlorine to ammonia nitrogen at the breakpoint and the amount of alkalinity required for each mg/L of ammonia nitrogen oxidized at the breakpoint.

Solution

1. Determine the molecular weight ratio of hypochlorous acid (HOCl), expressed as Cl_2, to ammonia (NH_3), expressed as N, using the overall reaction for the breakpoint phenomenon given by Eq. (12-48):

$$\underset{\substack{(17)\\(14)}}{NH_4^+} + \underset{\substack{1.5(52.45)\\1.5(2\times35.5)}}{1.5HOCl} \rightarrow 0.5N_2 + 1.5H_2O + 2.5H^+ + 1.5Cl^-$$

$$\text{Molecular ratio} = \frac{Cl_2}{NH_3\text{-N}} = \frac{1.5(2 \times 35.5)}{14} = 7.61$$

2. Determine the alkalinity required per 1.0 mg/L of ammonia nitrogen oxidized at the breakpoint. Assuming the pH of the solution is alkaline, the following expression can be written to describe the oxidation of ammonia:

$$NH_4^+ + Cl_2 \rightarrow 0.5N_2 + 4H^+ + 2Cl^-$$

Assuming lime will be used to neutralize the acidity, the required alkalinity ratio is:

$$2CaO + 2H_2O \rightarrow 2Ca^{+2} + 4OH^-$$

$$\text{Alkalinity required} = \frac{2(100 \text{ mg/millimole of } CaCO_3)}{14 \text{ mg/millimole of } NH_4^+ \text{ as N}} = 14.3$$

The stoichiometrically determined amount of alkalinity required per mg/L of ammonia nitrogen oxidized is 14.3 mg/L.

Comment. The ratio computed in step 2 will vary somewhat, depending on the actual reactions involved. In practice, the actual ratio typically has been found to vary from 8:1 to 10:1. Similarly, in step 3, the stoichiometric ratio of 14.3 mg/L of alkalinity, expressed as $CaCO_3$, for each 1.0 mg/L of ammonia nitrogen that is oxidized in the breakpoint chlorination process, will also depend on the actual reactions involved. In practice, it has been found that about 15 mg/L of alkalinity is actually required because of the hydrolysis of chlorine.

Germicidal effectiveness of various chlorine compounds. The germicidal efficiency of the three chlorine compounds [hypochlorous acid (HOCl), hypochlorite ion (OCl), or monochloramine (NH_2Cl)] are compared in Fig. 12-41. As shown in Fig. 12-41, for a given concentration and contact time, the germicidal efficiency of hypochlorous acid (HOCl) is significantly greater than that of either the hypochlorite ion (OCl) or monochloramine (NH_2Cl). It should be noted, however, that given an adequate contact time, monochloramine is nearly as effective as chlorine in achieving disinfection. The characteristics of the wastewater will play a significant role in the effectiveness of chlorine disinfection, as well as with other disinfection technologies (see Table 12-24). Sung (1970) was one of the first to identify the effects of the organic constituents in wastewater on the chlorination process. The relative germicidal effectiveness of chlorine, UV, and ozone for different organisms is presented in Table 12-26.

In the early 1970s, Collins conducted extensive experiments on the disinfection of various wastewaters. On the basis of the results of his research, Collins proposed the following model (Collins and Selleck, 1972).

$$\frac{N}{N_o} = \frac{1}{(1 + 0.23Ct)^3} \tag{12-49}$$

where N = number of organisms remaining after disinfection at time t
N_o = number of organisms present before disinfection
C = concentration of chemical agent
t = contact time

From the form of Eq. (12-49), it appears the plug-flow reactor kinetics are being modeled as a cascade of three reactors in series (see reactor analysis in Chap. 3). The term Ct will be recognized as the dose of the disinfectant [see Eq. (12-49]). Use

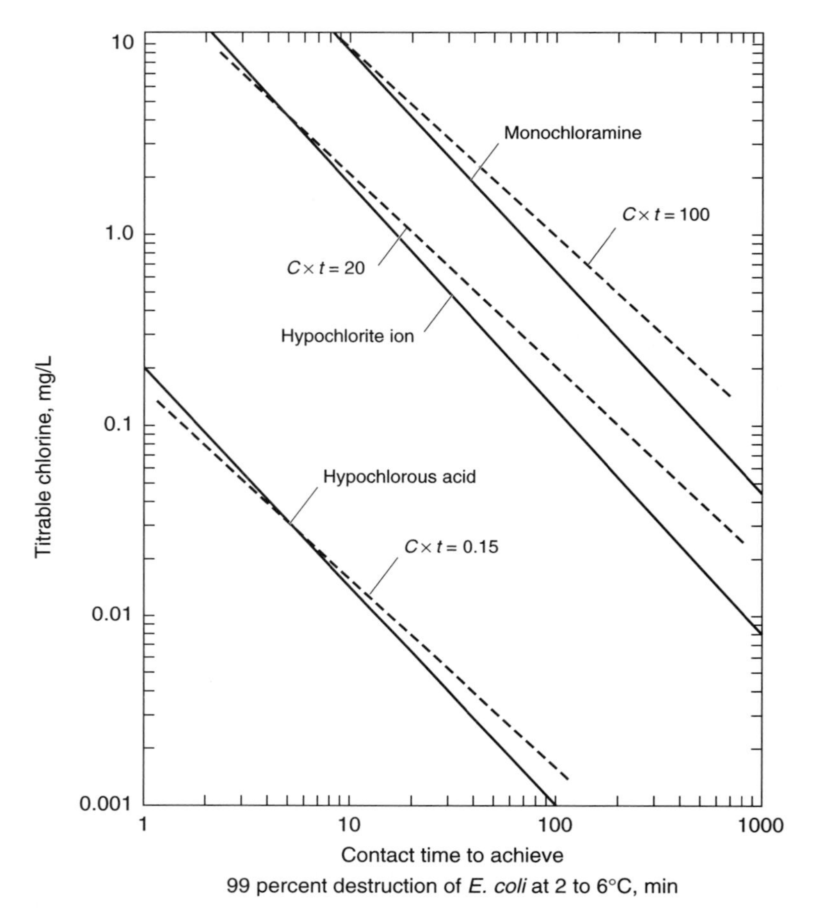

FIGURE 12-41
Comparison of germicidal efficiency of hypochlorous acid, hypochlorite ion, and monochloramine for 99 percent destruction of *E. coli* at 2 to 6°C with *Ct* values added for purposes of comparison (from Butterfield et al., 1943).

of the *Ct* value to evaluate the performance of disinfection facilities has become widespread. Various *Ct* (concentration × time) values have been superimposed on the original values in Fig. 12-41 for the purpose of comparison. A refinement of the original Collins model for the disinfection of secondary effluent as proposed by White (1992) is:

$$\frac{N}{N_o} = \left(\frac{Rt}{b}\right)^n \tag{12-50}$$

where R = chlorine residual remaining at the end of time t above
n = slope of experimental curve
b = value of x intercept when $N/N_o = 1$ or $\log N/N_o = 0$
Other terms as defined above for Eq. (12-49)

TABLE 12-26
Estimated relative effectiveness of chlorine, UV, and ozone disinfection for representative microorganisms of concern in wastewater*

	Dosage relative to total coliform dosage†		
Organism	**Chlorine**	**UV light**	**Ozone**
Bacteria			
Fecal coliform	0.9–1	0.9–1	0.9–1
Pseudomonas aeruginosa		1.5–2	
Salmonella typhi	1	0.9–1	
Staphylococcus aureus	2.5	1.5–2	
Total coliform	1	1	1
Viruses			
Adenovirus	0.5–1	0.6–0.8	
Coxsackie A2	6–7	0.8–1	
F specific bacteriophage	5–6	2–4	
Polio type 1	6–7	2–4	
MS-2 coliphage	5–7	6–8	2–4
Norwalk	5–6		
Protozoa			
Acanthamoeba castellanii		10–12	
Cryptosporidium parvum oocysts	8–10	10–20	6–8
Giardia lamblia cysts	6–8	7–14	4–6
Other			
Nematode eggs		10–12	

* Adapted in part from Darby et al. (1995).
† Relative doses based on discrete nonclumped single organisms in suspension at pH7-8 and a temperature of 20-25°C. If the organisms are clumped or particle-associated, the relative dosages have no meaning.

Typical values for the coefficients n and b for secondary effluent are -2.8 and 4.0, respectively. Because of the variability of the chemical composition of the secondary effluent, the constants must be determined for each wastewater.

The required chemical dosage for disinfection can be estimated by: (1) considering the chlorine demand of the wastewater, (2) assuming an allowance for decay during the chlorine contact time, and (3) determining the required chlorine residual from Eq. (12-50). Typical values for the chlorine demand of various wastewaters based on a contact time of 15 to 30 min are as follows (adapted in part from White, 1992).

Primary effluent	10 to 25 mg/L
Trickling filter effluent	5 to 15 mg/L
Activated-sludge effluent	4 to 10 mg/L
Filtered activated-sludge effluent	4 to 8 mg/L
Nitrified effluent	4 to 8 mg/L
Septic tank effluent	10 to 30 mg/L
Intermittent sand filter effluent	2 to 6 mg/L

Typical decay values for chlorine residual are on the order of 2 to 4 mg/L for contact time of about 1 hour. Estimation of the chlorine dosage is illustrated in Example 12-6.

EXAMPLE 12-6. ESTIMATE THE REQUIRED CHLORINE DOSAGE FOR A TYPICAL SECONDARY EFFLUENT. Estimate the chlorine dosage for a secondary effluent assuming the following conditions apply:

1. Effluent total coliform count $= 10^7/100$ mL
2. Required effluent total coliform count = 23/100 mL
3. Effluent chlorine demand = 10 mg/L
4. Demand due to decay during chlorine contact = 2.5 mg/L
5. Required chlorine contact time = 60 min

Solution

1. Estimate the required chlorine residual using Eq. (12-50):

$$\frac{N}{N_o} = \left(\frac{Rt}{b}\right)^n$$

Use the typical values given above for the coefficients:

$$\frac{23}{10^7} = \left(\frac{Rt}{4.0}\right)^{-2.8}$$

$$\left(\frac{23}{10^7}\right)^{-1/2.8} = \left[\left(\frac{Rt}{4.0}\right)^{-2.8}\right]^{-1/2.8}$$

$$(234.3)4 = R(60)$$

$$R = 15.6 \text{ mg/L}$$

2. The required chlorine dosage is:

$$\text{Chlorine dosage} = 10 \text{ mg/L} + 2.5 \text{ mg/L} + 15.6 \text{ mg/L} = 28.1 \text{ mg/L}$$

Dechlorination with sulfur compounds. Dechlorination involves the removal of the free and combined chlorine residual that remains after chlorination to reduce the toxic effects of chlorinated effluents. Where effluent toxicity requirements are applicable, or where dechlorination is used as a polishing step following the breakpoint chlorination process for the removal of ammonia nitrogen, sulfur dioxide (SO_2) is often used for dechlorination. Other chemicals that have been used include sodium bisulfite ($NaHSO_3$), sodium metabisulfite ($Na_2S_2O_5$), and sodium thiosulfate ($Na_2S_2O_3$). In fact, sodium bisulfite is beginning to replace sulfur dioxide as the chemical of choice for dechlorination. Granular activated carbon (GAC) has also been used for dechlorination (see Prob. 12-3).

Sulfur dioxide gas successively removes free chlorine, monochloramine, dichloramine, nitrogen trichloride, and poly-*n*-chlor compounds. When sulfur dioxide is added to wastewater, the following reactions occur:

Reaction with chlorine:

$$SO_2 + HOCl + H_2O \rightarrow Cl^- + SO_4^{-2} + 3H^+ \qquad (12\text{-}51)$$

Reaction with monochloramine:

$$SO_2 + NH_2Cl + 2H_2O \rightarrow Cl^- + SO_4^{-2} + NH_4^+ + 2H^+ \qquad (12\text{-}52)$$

Because the reactions of sulfur dioxide with chlorine and chloramines are nearly instantaneous, contact time is not usually a factor and contact chambers are not used, but rapid and positive mixing at the point of application is an absolute requirement.

Design considerations. Of the factors that affect the disinfection process (see Table 12-23), the designer has greatest control over initial mixing, contact time, and the chlorine feed rate. For small treatment facilities, chlorine will typically be obtained in 100- and 150-lb cylinders or ton containers. Chlorine can be added to wastewater as a liquid solution, or by direct injection. Where chlorine is added as a liquid solution, the gaseous chlorine from the storage containers must first be dissolved and then added to the liquid stream to be disinfected. Effective initial mixing of the chlorine solution with the wastewater can be accomplished by use of a turbulent flow regime or mechanical means such as: (1) hydraulic jumps in open channels, (2) venturi flumes, (3) pipelines, (4) pumps, (5) static mixers, or (6) vessels (chambers) with the aid of mechanical mixing devices (see Fig. 5-12, Chap. 5). Where the mixing is to be accomplished by turbulence, the G value for the mixing regime should vary from 500 to 1000 s^{-1}.

In the second approach, chlorine gas is injected directly into the liquid without the use of any dilution water. One such device used to inject chlorine gas, known as the Water Champ, is illustrated in Fig. 12-42. Operationally, the motor-driven open impeller creates a vacuum in the chamber surrounding the impeller. The vacuum is transmitted to the chlorinator. The high-speed impeller induces, by means of the vacuum, the chemical to be mixed from the storage container without the need for dilution water. The high operating speed of the impeller (3450 rev/min) results

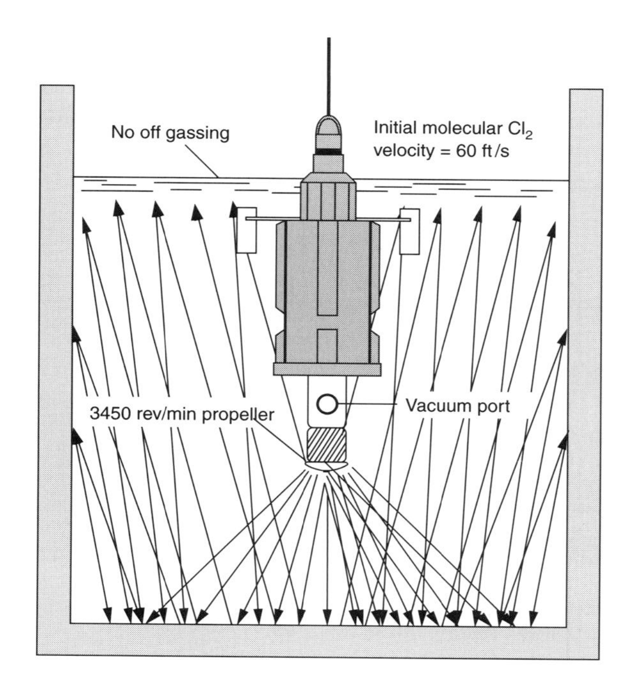

FIGURE 12-42
A submersible vacuum-type molecular chlorine vapor induction unit showing open channel cross section of diffusion and mixing. The unit is known as the Water Champ (courtesy Gardiner Equipment Company, Inc.).

in extremely effective mixing. The Water Champ inductor has revolutionized the injection of chlorine gas and other chemicals, as did the chlorine injection system when it was introduced with vacuum-operated chlorinators in 1920 (White, 1992).

Because of the importance of contact time, careful attention should be given to design of the contact chamber so that at least 80 to 90 percent of the wastewater is retained in the basin for the specified contact time. The best way to achieve the desired retention time is by using a plug-flow around-the-end type of contact chamber, or a series of interconnected basins or compartments. To avoid the development of dead zones with respect to flow and to minimize short circuiting, length-to-width ratios (L/W) of at least 10 to 1 and preferably 40 to 1 should be used. In addition, the height-to-width ratio should be less than 1. Short circuiting may also be minimized by reducing the velocity of the wastewater entering contact tanks. Baffles similar to those used on rectangular sedimentation tanks may be used for inlet velocity control. The placement of longitudinal baffles and turning vanes can also reduce short circuiting and improve the actual detention time. Typical chlorine contact basins for small treatment plants are shown in Fig. 12-43.

Under ideal conditions, the horizontal velocity at minimum flow in a chlorine contact basin should be sufficient to scour the bottom, or at least to limit the deposition of sludge solids that may have passed through the settling tank. In reality, with long detention times, solids deposition is a "fact of life." Horizontal velocities should be at least 6.5 to 15 ft/min (2 to 4.5 m/min). If the time of travel in the discharge or outfall sewer at the maximum design flow is equal to or exceeds the required contact time, it may be possible to eliminate the chlorine contact chambers. In some small plants, chlorine contact basins have been constructed of large-diameter sewer pipe. For most treatment plants, two or more contact basins should be used to facilitate maintenance (usually the removal of accumulated solids). Provisions should also be included for draining the basins and for scum removal.

The chlorine feed rate may be controlled in several ways, including: (1) manual control, (2) automatic flow proportioning (open-loop) control, (3) automatic residual (closed-loop) control, and (4) compound-loop control. In manual control (the simplest method), the chlorine feed rate is changed by the operator to suit current

FIGURE 12-43
Views of typical plug-flow chlorine contact basins for small flows.

conditions. The required dosage is usually determined by measuring the chlorine residual after 15 min of contact time, and adjusting the dosage to obtain a given residual. A second method is to pace the chlorine flowrate to the wastewater flowrate as measured by a primary meter such as a magnetic meter, Parshall flume, or flow tube. A third method is to control the chlorine dosage by measurement of the chlorine residual. Finally, compound-loop control system signals obtained from the wastewater flow meter and from the residual recorder provide more precise control of chlorine dosage and residual. A compound-loop system for both chlorination with chlorine and dechlorination with sulfur dioxide is shown in Fig. 12-44. In the control diagrams in Fig. 12-44, amperometric monitoring of the combined chlorine residual is the key control parameter.

Environmental impacts of chlorine disinfection. As noted above, chlorination is the most commonly used method for the destruction of pathogenic and other harmful organisms that may endanger human health. Unfortunately, certain organic constituents in wastewater can interfere with the chlorination process. Many of these organic compounds may react with chlorine to form toxic compounds that can have long-term adverse effects on the beneficial uses of the waters to which they are discharged. For example, trihalomethanes (THMs) are formed by the reactions between organic compounds such as humic and fluvic acids and chlorine. Even though the toxicity of the residual chlorine compounds is reduced significantly or eliminated by dechlorination, the long-term environmental impacts of the discharge of the compounds remaining after chlorination/dechlorination are unknown.

Disinfection with Sodium Hypochlorite and Dechlorination with Sodium Bisulfite

Concerns over the safety associated with the handling and storage of liquid and gaseous chlorine have led to the use of calcium hypochlorite [$Ca(OCl)_2$] and sodium hypochlorite ($NaOCl$). It should be noted, however, that some large plants that used either calcium and sodium hypochlorite have switched back to chlorine because of the problems associated with handling large volumes of these chemicals.

Preparation of sodium hypochlorite. Sodium hypochlorite is prepared by reacting chlorine with caustic soda according to the following reaction:

$$Cl_2 + 2NaOH \rightarrow NaOCl + NaCl + H_2O + \text{heat} \qquad (12\text{-}53)$$

As given by the above reaction 1.0 lb of chlorine residual reacts with 1.128 lb of caustic soda to produce 1.05 lb of sodium hypochlorite and 0.83 lb of sodium chloride.

Reactions of sodium hypochlorite in water. When sodium hypochlorite ($NaOCl$) is added to water, the same two reactions described above for chlorine take place; namely, hydrolysis and ionization.

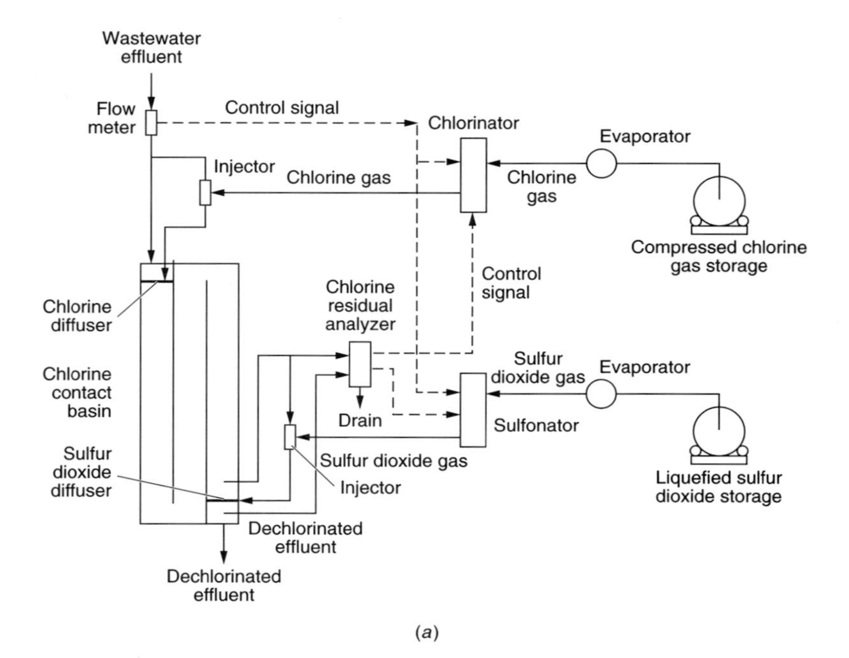

(*a*)

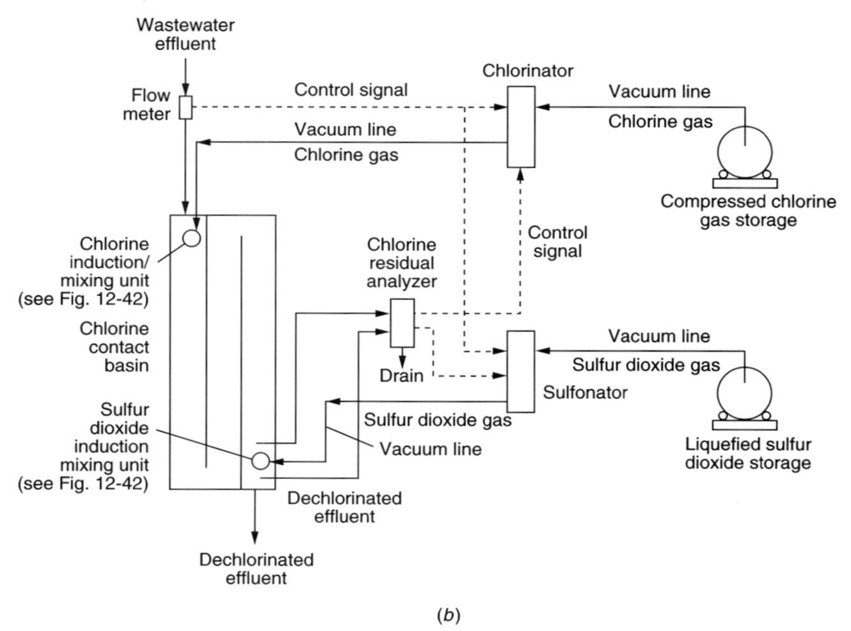

(*b*)

FIGURE 12-44
A compound-loop control system for chlorination with chlorine and dechlorination with sulfur dioxide: (*a*) injection of liquid chlorine and (*b*) injection of chlorine gas by induction.

Hydrolysis may be defined as:

$$NaOCl + H_2O \rightarrow HOCl + Na^+ + OH^- \tag{12-54}$$

Ionization may be defined as:

$$HOCl \rightarrow H^+ + OCl^- \tag{12-55}$$

The ionization reaction is the same as for chlorine [see Eq. (12-38)].

Germicidal effectiveness of sodium hypochlorite. The germicidal effectiveness of sodium hypochlorite is the same as that for chlorine, as the active agents are the same, although there is some evidence that a given wastewater will have a higher chlorine demand when liquid chlorine is used. Here again, the characteristics of the wastewater will play a significant role in the effectiveness of the disinfection process (see Table 12-24).

Dechlorination with sodium bisulfite. When sodium bisulfite ($NaHSO_3$) is added to wastewater, the following reaction occurs:

$$NaHSO_3 + Cl_2 + H_2O \rightarrow NaHSO_4 + 2HCl \tag{12-56}$$

For each 1.0 lb of chlorine residual removed, 1.46 lb of sodium bisulfite is required. Also, for each 1.0 lb of chlorine removed, 1.38 lb of alkalinity, expressed as $CaCO_3$, will be consumed.

Environmental impacts of sodium hypochlorite disinfection. Although some of the final by-products formed when sodium hypochlorite is used may be different from the by-products formed when chlorine is used, the same general concerns about the long-term environmental impacts of the discharge of the unknown compounds remain.

Ultraviolet (UV) Radiation Disinfection

The germicidal properties of the radiation emitted from ultraviolet (UV) light sources have been used in a wide variety of applications since the use of UV was pioneered in the early 1900s. First applied to high-quality water supplies, the use of ultraviolet light as a wastewater disinfectant has evolved during the last 10 years with the development of new lamps, ballasts, and ancillary equipment. With the proper dosage, ultraviolet radiation has proven to be an effective bactericide and viricide for wastewater, with minimal or no formation of toxic compounds.

Source of UV radiation. To produce UV energy, special UV lamps that contain mercury vapor are charged by striking an electric arc. The energy generated by the excitation of the mercury vapor contained in the lamp results in the emission of UV light. In general, UV disinfection systems are categorized according to the internal operating parameters of the UV lamp as *low-pressure, low- and high-intensity* and *medium-pressure, high-intensity* systems. Mercury-argon lamps are used, most

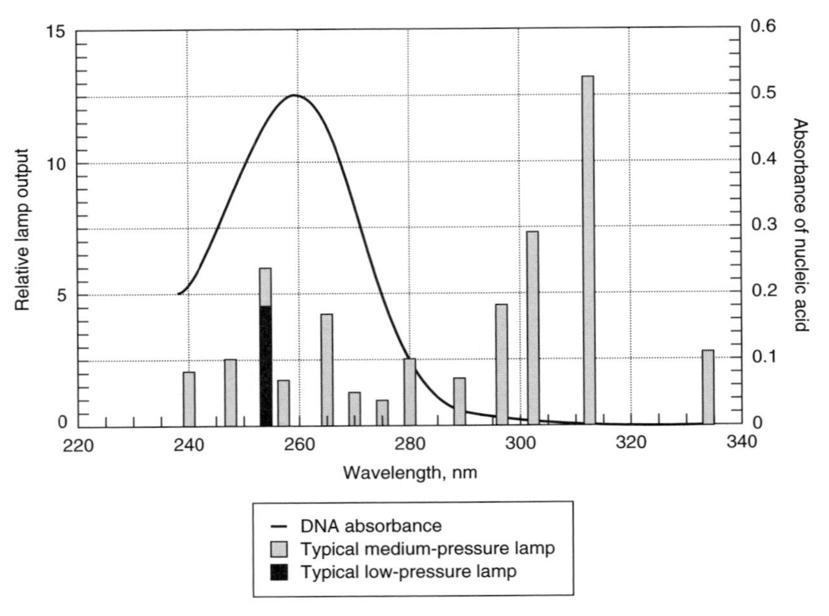

FIGURE 12-45
Ultraviolet lamp radiation spectra for both low-pressure, low- and high-intensity and medium-pressure, high-intensity and DNA adsorption.

commonly, to generate the UV C-region wavelengths. Low-pressure, low- and high-intensity UV lamps generate essentially monochromatic radiation at a wavelength of 254 nm, which is close to the 260-nm wavelength that is most effective for microbial inactivation (see Fig. 12-45). The low-pressure, high-intensity lamps generate about 2 to 4 times the UV intensity of the low-pressure, low-intensity lamps.

Medium-pressure, high-intensity UV lamps generate polychromatic radiation. While the medium-pressure, high-intensity lamp is less efficient at generating the 254 nm and other germicidal wavelengths (only 20 to 40 percent of the total energy of a high-intensity UV lamp is in the germicidal wavelength range), it does generate approximately 15 to 20 times higher germicidal UV intensity than its low-pressure, low-intensity counterpart.

Low-pressure UV systems. The principal components of a low-pressure, low-and high-intensity UV disinfection system used for wastewater are illustrated in Fig. 12-46. As shown, both horizontal (parallel) and vertical (perpendicular) flow UV disinfection systems are used. Examples of typical low-pressure UV disinfection systems for wastewater are shown in Fig. 12-47. The design flowrate is usually divided equally among a number of open channels. Each channel typically contains two or more banks of UV lamps in series, and each bank comprises a specified number of modules (or racks of UV lamps). Each module contains a specified number of UV lamps encased in quartz sleeves. The number of UV lamps per module is 2, 4, 8, 12, or 16. A spacing of 3 in (75 mm) between the centers of UV lamps is

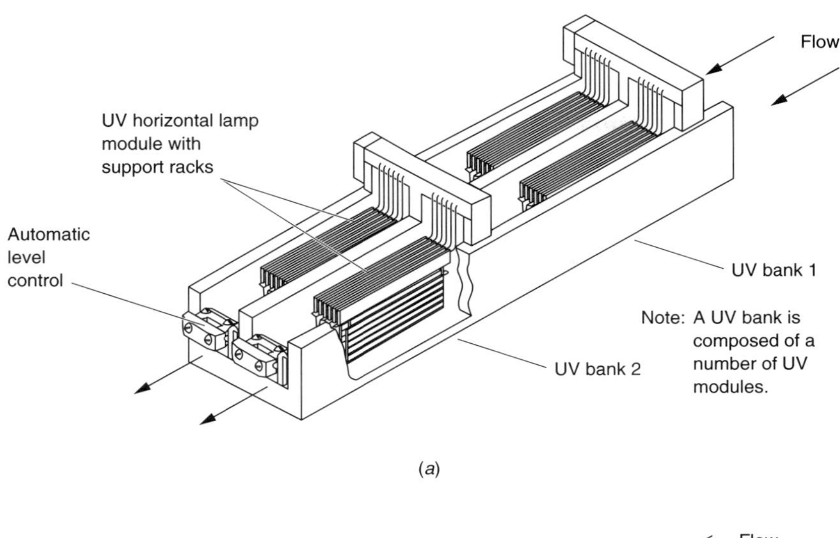

(*a*)

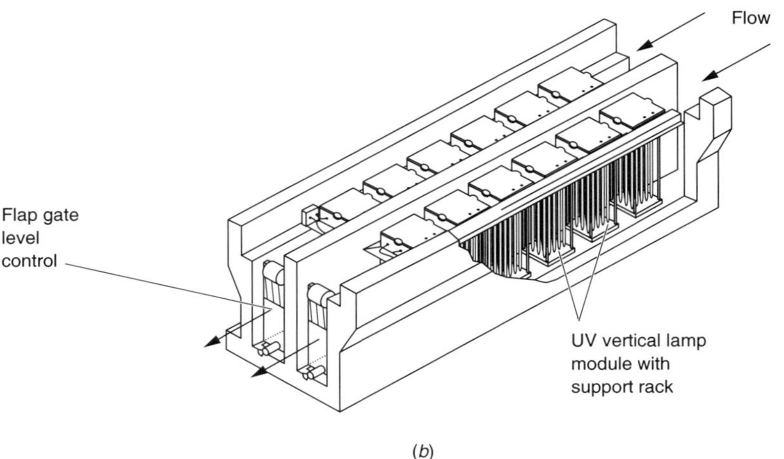

(*b*)

FIGURE 12-46
Isometric cut-away views of typical UV disinfection systems with cover grating removed: (*a*) horizontal lamp system parallel to flow (adapted from Trojan Technologies, Inc.) and (*b*) vertical lamp system perpendicular to flow (adapted from Infilco Degremont, Inc.).

currently the most frequently used lamp configuration by UV disinfection system manufacturers. A weighted flap gate (or similar device) is used to control the depth of flow through each disinfection channel. To overcome the effect of fouling, which reduces lamp output, the lamp modules used in low-pressure, low-intensity UV systems must be removed periodically from the flow channel and the lamps cleaned either manually or mechanically.

The low-pressure, low-intensity UV lamps are of a slim-line design and operate optimally at a lamp wall temperature of 40°C and an internal pressure of

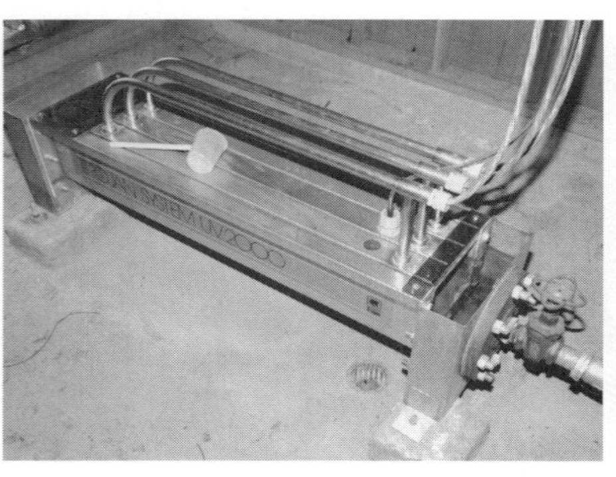

(*a*)

(*b*)

FIGURE 12-47
Views of low-pressure, low-intensity UV disinfection systems: (*a*) single bank unit, composed of 3 modules with 2 lamps per module, for 47 homes and (*b*) larger facility for pumped groundwater following surface spreading.

7×10^{-3} torr. Quartz sleeves are used to isolate the UV lamps from direct water contact and to control the lamp wall temperature. Because there is an excess of liquid mercury in the low-pressure, low-intensity UV lamp, the mercury vapor pressure is controlled by the coolest part of the lamp wall. If the lamp wall does not remain at its optimum temperature of 40°C, some of the mercury in the lamp condenses back to its liquid state, thereby decreasing the number of mercury atoms available to release photons of UV; hence UV output declines. In most cases, effluent has a cooling effect on the quartz sleeve, and additional power must be provided by the ballast to compensate for lamp wall heat loss. The quartz sleeve buffers the effluent temperature extremes to which the UV lamps are exposed, thereby maintaining a more or less uniform UV lamp output.

Approximately 88 percent of the lamp output is monochromatic at 253.7 nm, making it an efficient choice for disinfection processes. Although low-intensity lamps produce small amounts of UV at the 185-nm and 365-nm wavelengths, the lamp quartz sleeve is doped to prevent the emission of the 185-nm wavelength, which produces ozone. In time, the output of low-pressure, low-intensity UV lamps decreases because of a reduction in the electron pool within the UV lamp and the aging of the quartz sleeve (see Fig. 12-48). The useful life of a low-pressure, low-intensity UV lamp will vary from 10,000 to 13,000 h, depending on the number of on-off cycles per day. The useful life of the quartz sleeve is about 4 to 8 years.

Low-pressure, high-intensity UV lamps are similar to the low-pressure, low-intensity lamps, with the exception that a mercury-indium amalgam is used in place of mercury. Use of the mercury amalgam allows greater UV C output (from 2 to 4 times the output of conventional low-intensity lamps), greater stability over a broad temperature range, and greater lamp life (25 percent greater than other low-pressure lamps). The amalgam in the low-pressure, high-intensity UV lamps is used to maintain a constant level of mercury atoms. Low-pressure, high-intensity and medium-pressure, high-intensity UV disinfection systems are provided with mechanical wipers.

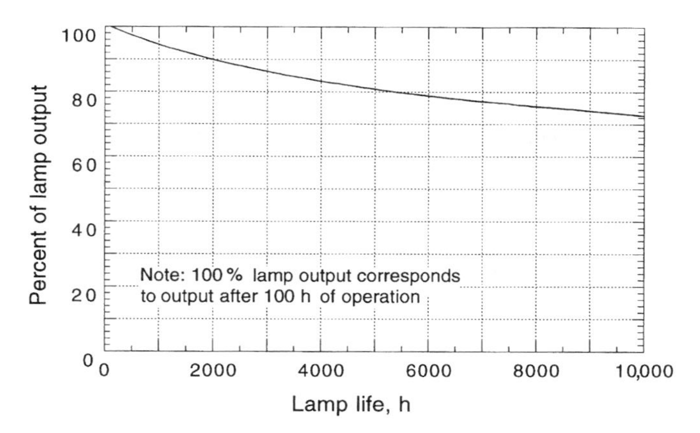

FIGURE 12-48
Typical reduction in output of low-pressure, low-intensity UV lamp because of aging.

Medium-pressure UV systems. Medium-pressure high-intensity UV lamps have been developed over the last decade, and they are tending to supplant low-pressure lamps, especially for higher wastewater flows. The lamps in a medium-pressure, high-intensity, UV system are arranged in modules, but are positioned in a flow reactor with a fixed geometry (see Fig. 12-49*a*). The high-intensity UV lamp (see Fig. 12-49*b*) operates at a higher mercury vapor pressure of 102 to 104 torr, and at a higher lamp wall temperature of between 600 to 900°C. Because of the higher operating temperature, the UV output of these lamps is unaffected by effluent temperature. This higher intensity is advantageous in treating poor-quality wastewaters with lower bulk fluid transmittance values and fewer particle-associated coliforms such as lagoon effluents and combined sewer overflows (CSOs). Because the high-intensity UV lamps operate at temperatures at which all the mercury is vaporized, the UV output can be modulated across a range of power settings without significantly changing the spectral distribution of the lamp. Further, because of the high operating temperature, mechanical wiping of the quartz sleeve is essential to avoid the formation of a fixed film on the surface of the sleeve. Although there are a number of manufacturers of high-intensity UV lamps, most of the manufacturers do not market complete UV disinfection systems. The particular UV lamp selected by UV system manufacturers is chosen on the basis of an integrated design approach in which the UV lamp, ballast, and reactor design are interdependent.

Germicidal effectiveness of UV radiation. Ultraviolet light is a physical rather than a chemical disinfecting agent. Radiation with a wavelength of around 254 nm, characteristic of low-pressure systems, penetrates the cell wall of the microorganism and is absorbed by cellular materials including DNA and RNA (see Fig. 12-50), which either prevents replication or causes death of the cell to occur. The required UV dose is determined based on collimated beam studies (see Fig. 12-51) or from pilot-scale studies (Darby et al., 1991) (see Fig. 12-52). In

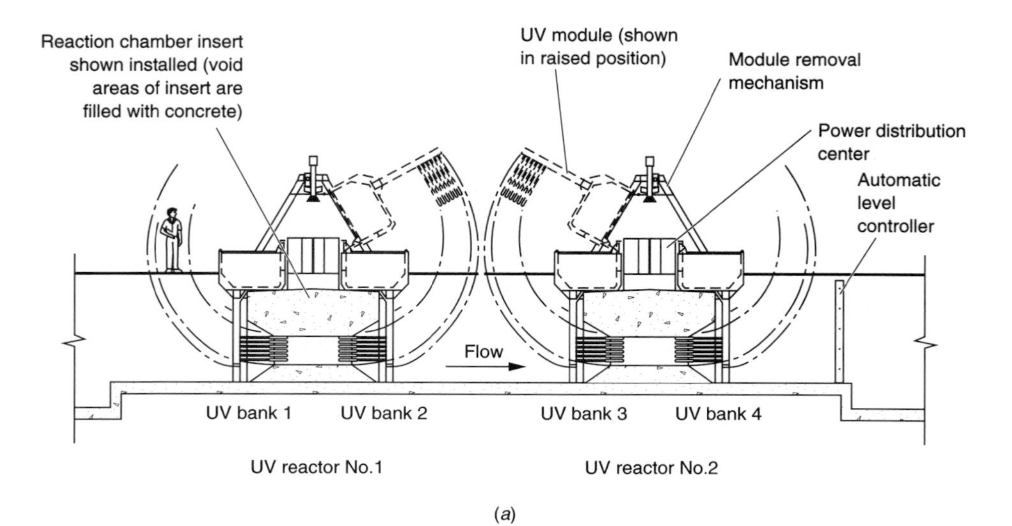

(*a*)

(*b*)

FIGURE 12-49
Views of medium-pressure, high-intensity UV disinfection system: (*a*) schematic view through UV reactor (courtesy Trojan Technologies, Inc.) and (*b*) pictorial view of UV system with lamps removed from the UV reactor.

collimated beam studies, the UV dose is controlled either by varying the UV intensity (e.g., by lengthening or shortening the tube length) or varying the exposure time. Because the geometry is fixed, the depth-averaged UV intensity within the Petri dish sample volume can be computed using Beer's law, and with the exposure time known, the UV dose can be determined. The UV dose is then correlated to the

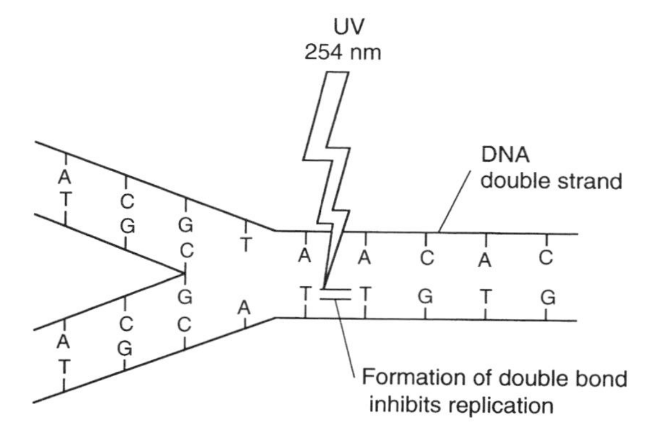

FIGURE 12-50
Formation of double bonds in microorganisms exposed to ultraviolet radiation (adapted from US EPA, 1986).

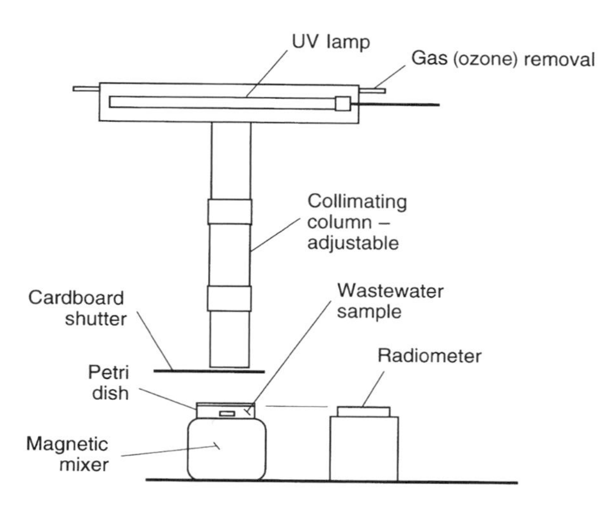

FIGURE 12-51
Collimated beam device used to develop dose-response curves for UV disinfection.

(a)

(b)

FIGURE 12-52
Typical pilot-scale test unit used to determine disinfection performance with UV radiation: (*a*) low-pressure, low-intensity and (*b*) medium-pressure, high-intensity.

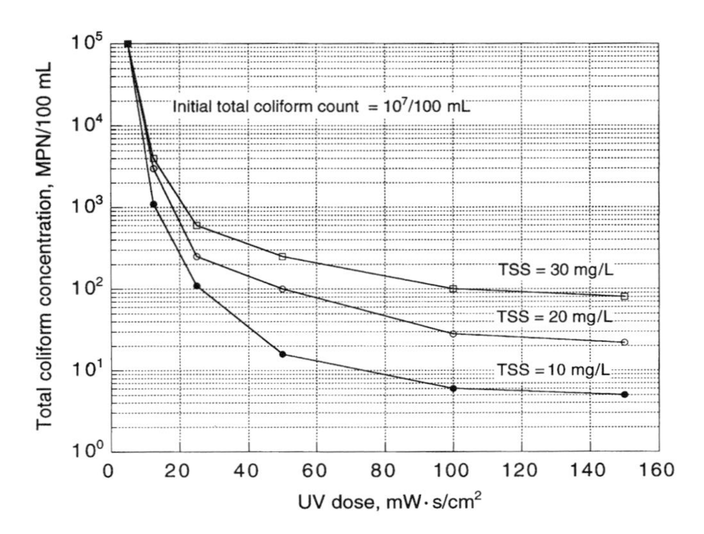

FIGURE 12-53
Typical dose-response curves for UV disinfection of wastewater effluents containing different TSS concentrations, developed from data obtained by using a collimated beam device.

microorganism inactivation results, typically measured by the MPN test (see Chap. 2). The results are reported in the form of a dose-response curve (see Fig. 12-53). To assess the inherent and process variability, as discussed in Chap. 4, the collimated beam test must be repeated a number of times to obtain statistical significance. Because the radiation output of the medium-pressure high-intensity UV lamps is polychromatic (see Fig. 12-45), the germicidal effectiveness of these lamps must be established on the basis of collimated beam or pilot-scale testing. The results can then be correlated to the performance of low-pressure monochromatic UV lamps for comparative purposes.

The UV dose required to achieve a 5 to 6 log reduction in the number of *dispersed* coliform organisms (total coliform/100 mL) is in the range from 10 to 40 mW·s/cm^2. The relative effectiveness of UV radiation for the disinfection of various nonclumped and/or non-particle-associated organisms is summarized in Table 12-26. Unfortunately, in wastewater, a number of coliform organisms are either clumped or particle-associated (Loge et al., 1997). As a result, the UV dose required for organisms that are in clumps or are particle associated will depend on the size and number of particles and the chemical characteristics of the wastewater. The impact of TSS on the effectiveness of UV disinfection is illustrated in Fig. 12-53. As shown, as the TSS concentration increases, the effectiveness of the UV dose is decreased. Further, it is important to note that increasing the dose in the tailing region (right hand side of graph) will result in limited improvement in the effectiveness of UV disinfection. The impact of intensity is considered below.

The effect of UV intensity is illustrated in Fig. 12-54. As shown, increasing the UV intensity tenfold has little effect on the particle associated coliforms. The reason that there is little or no effect, is that the adsorption of UV radiation by

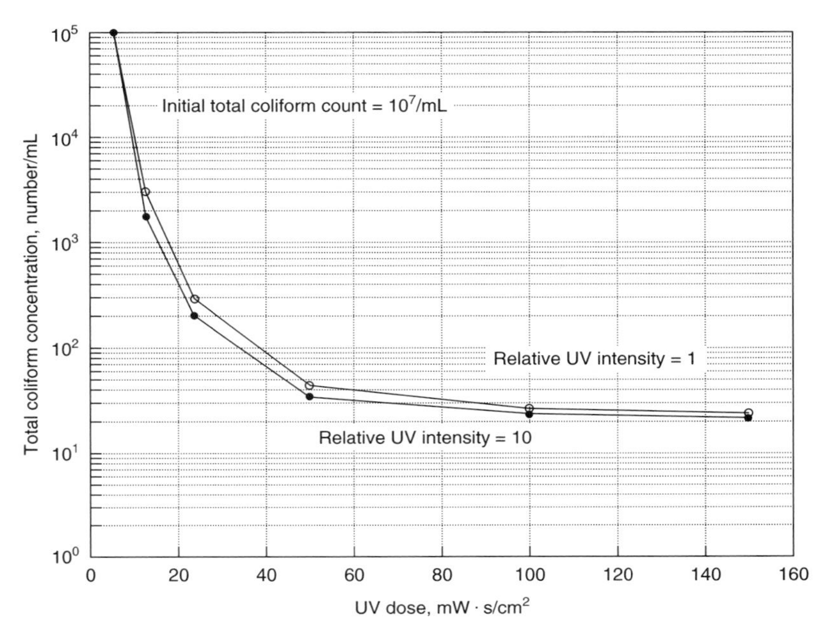

FIGURE 12-54
Impact of UV intensity on the effectiveness of UV disinfection of effluent from an air activated-sludge wastewater treatment plant.

wastewater particles is up to 10,000 times or more greater than that of the bulk liquid medium. Thus, particles, larger than some critical size, will effectively shield the embedded microorganisms (Loge et al., 1997). Because the effectiveness of UV disinfection is governed primarily by the number of particle associated coliforms, to improve the performance of a UV disinfection system either the number of particles with associated coliform must be reduced (e.g., by modifying the operation of the upstream processes) or the particles themselves must be removed (e.g., by improved clarifier design or filtration). Currently, to meet the stringent coliform requirements for body contact wastewater reuse applications (i.e., equal to or less than 2.2 MPN/100 mL), effluent filtration is required. In the future, it is anticipated that appropriate modifications will be made in the design and operation of biological wastewater treatment plants to produce an effluent with particles that can be disinfected more effectively.

Modeling UV disinfection. As noted previously in Table 12-24, water quality characteristics will affect the performance of disinfection systems. The relative significance of the wastewater characteristics will vary from plant to plant. Additionally, the variability of these characteristics and their relative influence within a single plant must also be considered. As a consequence, it is generally recognized that it is unlikely that a purely deterministic approach can be developed to predict the performance of a UV disinfection system using current measures of wastewater quality.

To deal with the many variables that can affect UV disinfection, an empirical regression model has been developed to predict the performance of UV disinfection systems in the tailing region (see Fig. 12-38) (Emerick and Darby, 1993; Darby et al., 1995; Loge et al., 1996a). The general form of the model is

$$N = W_f(I \times t)^n \tag{12-57}$$

where N = total coliform concentration after exposure to UV light, MPN/100 mL
W_f = water quality factor
I = average intensity of UV radiation in UV reactor, mW/cm^2
t = exposure time, s (assuming approximate plug-flow conditions)
n = empirical coefficient related to dose

The water quality factor is defined as follows

$$W_f = A(\mathrm{TSS})^a(\mathrm{UFT})^b(N_o)^c\beta^d \tag{12-58}$$

where TSS = total suspended solids, mg/L
UFT = unfiltered transmittance of liquid at 254 nm, %
N_o = influent total coliform concentration, MPN/100 mL
β = particle size distribution
A, a, b, c, d = empirical coefficients

The average UV intensity I is determined for a given lamp configuration (see Fig. 12-55) using point source summation (PSS) method developed by Jacob and Dranoff (1970), and later expanded by Qualls and Johnson (1983), Suidan and Severin (1986), and Scheible (1987). A computer program for computing the average UV intensity with the PSS model is available from the U.S. EPA (U.S. EPA, 1992b). It should be noted that, for any given wastewater treatment plant, not all of the variables included in Eq. (12-58) may be statistically significant, whereas for some treatment plants additional variables may be required. For

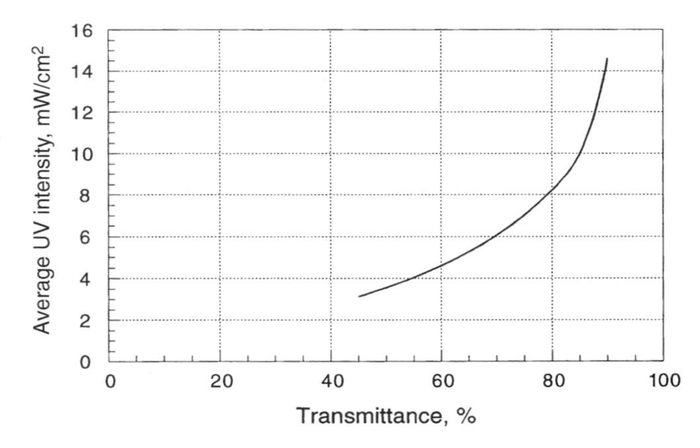

FIGURE 12-55
Typical curve of average UV intensity as a function of filtered wastewater transmittance for rectangular lamp arrays, calculated using the PSS method.

example, if the UFT value never varied, it would not be included in the final formulation. The multiple-correlation approach allows for the consideration of all of the variables that may affect the UV disinfection process. A graphical representation of Eq. (12-58) is presented in Fig. 12-56 for a wastewater in which the water quality factor was found to be equal to:

$$W_f = A(\text{TSS})^a(\text{UFT})^b \tag{12-59}$$

As illustrated in Fig. 12-56, the required UV dose ($I \times t$) will depend on the effluent discharge requirements. For example, the UV dosages required for a typical activated-sludge effluent with an unfiltered transmittance value of 65 percent and a TSS content of 20 mg/L to achieve various levels of total coliform disinfection are as follows.

Discharge requirement, MPN/100 mL	UV dosage range, mW·s/cm^2
1000	10–15
240	20–30
23	75–85

The lowest concentration of coliform organisms that can be achieved will depend on the number of coliforms associated with particles greater than some critical size. To achieve an effluent total coliform standard of 2.2 MPN/100 mL, effluent filtration will normally be needed to remove the residual TSS, which in turn will reduce the number of particle associated coliforms.

Design of UV reactor. The design of a UV disinfection system requires two general steps: (1) determination of the number of lamps required for disinfection and (2) determination of an optimal process configuration (e.g., the number of lamps per module, modules per bank, banks per channel, and the overall number of channels). Factors that affect the minimum number of UV lamps necessary for disinfection are: (1) the aging and fouling characteristics of the UV lamp/quartz sleeve assembly, (2) wastewater quality and its variability, and (3) the nature of the discharge permit itself and the level of confidence desired in meeting that permit. Wastewater discharge permits vary in both the stated coliform discharge limit (e.g., 23, 240, 1000 coliform per 100 mL) and in the method used to calculate the limit (e.g., 7- or 30-day running median value not to be exceeded more than once in a 30-day period).

Currently, both nonprobabilistic (deterministic) and probabilistic approaches are used to determine the number of UV lamps required. The nonprobabilistic approach is currently used by most UV manufacturers. This method is also used in the California Wastewater Reclamation Criteria (NWRI, 1993 and Tchobanoglous et al., 1994). Although not used as commonly, the probabilistic approaches are favored because they take into account the inherent variability of the disinfection process, the variability in the wastewater quality characteristics, and the probabilistic nature of most wastewater discharge permits. A detailed discussion of the various design

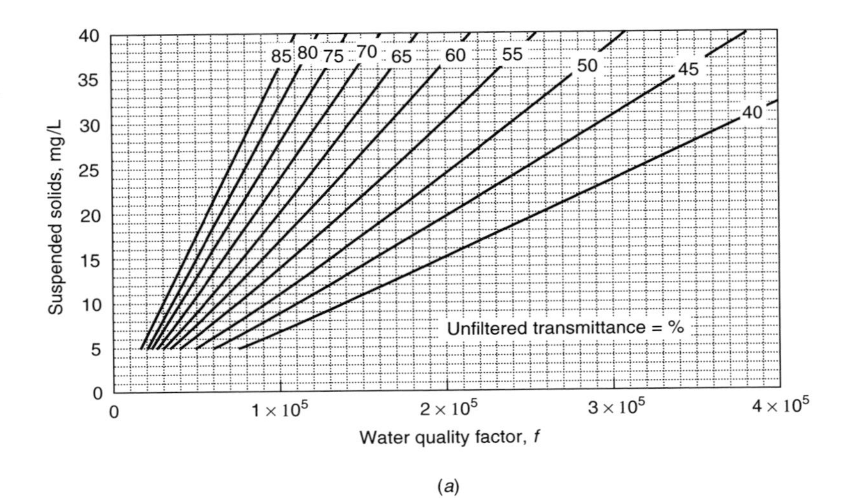

(*a*)

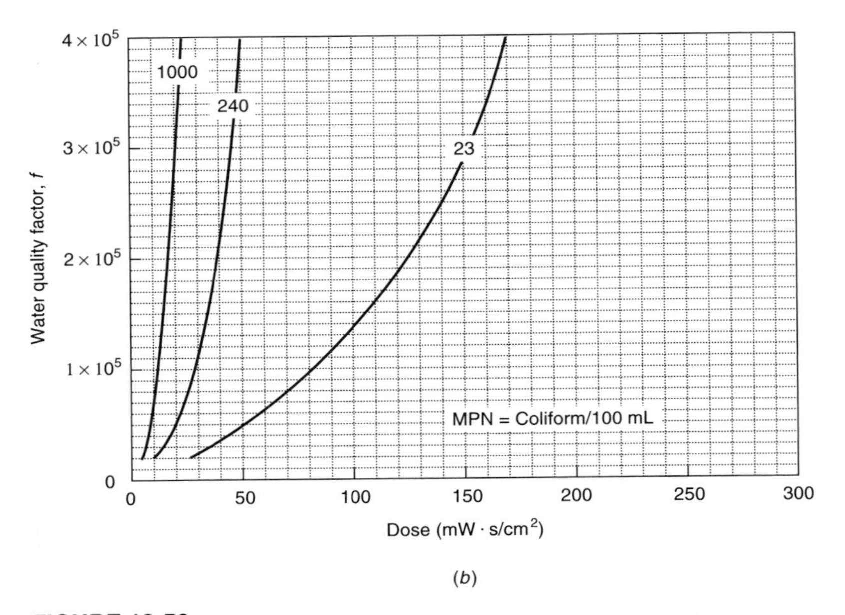

(*b*)

FIGURE 12-56
Prediction of effluent coliform concentration following UV disinfection: (*a*) determination of the water quality factor, based on TSS and UFT (lines of constant transmittance are indicated within the figure), and (*b*) estimation of UV dose based on water quality factor and desired effluent coliform concentration (adapted from Emerick and Darby, 1993).

TABLE 12-27

Steps involved in the deterministic approach to the design of UV disinfection systems*

Step	Requirement
1.	Plot the log of the effluent coliform concentration, obtained from a collimated beam study, as a function of UV dose (see Fig. 12-53).
2.	Determine the UV dose that corresponds to the lowest discharge limit stated in the WWTP permit. For example, as shown in Fig. 12-53 for an effluent with a TSS concentration of 20 mg/L, a dose of approximately 50 $mW \cdot s/cm^2$ is necessary to meet a permitted effluent coliform concentration of 100 MPN/100 mL.
3.	Adjust the required UV dose to account for lamp aging and fouling anticipated in the full-scale facility. Increase the UV dose by 30 percent to account for lamp aging and an additional 30 percent to account for lamp fouling. Additional safety or adjustment factors can be applied to further increase the UV dose to account for the uncertainties associated with variable wastewater characteristics. UV manufacturers with extensive experience in system design typically have a vast and varied array of data sources available to aid in determining appropriate adjustment factors for various design situations. This same data base is typically not available to municipalities and their consultants.
4.	Use the PSS method (or equivalent) to determine the average UV intensity in the UV disinfection system (see Fig. 12-55).
5.	Using the average intensity determined in step 4, determine the exposure time necessary to provide the required UV dose: $$\text{Exposure time, s} = \frac{\text{UV dose, mW}\cdot\text{s/cm}^2}{\text{average intensity, mW/cm}^2}$$
6.	Based on the design flowrate to be treated, determine the volume of flow that must be retained for the required exposure time: $$\text{Total volume, L} = (\text{exposure time, s}) \times (\text{design flowrate, L/s})$$
7.	Determine the number of lamps required for disinfection by dividing the total volume of flow to be disinfected by the volume of wastewater that can be treated per lamp. The volume of wastewater treated per lamp is constant for a given lamp spacing and is calculated by subtracting the cross-sectional area of the quartz sleeve from a square area with dimensions equal to the centerline lamp spacing [75 mm (3 in) is the current industry standard] and multiplying by the effective arc length of a UV lamp [typically 1470 mm (58 in)]: $$\text{Number of lamps required for disinfection} = \frac{\text{total volume, L}}{\text{volume per lamp, L/lamp}}$$
8.	Assuming various system configurations, select a configuration that meets disinfection criteria with the minimum number of lamps.
9.	Compute the headloss per channel using a value of 2.2 for the headloss coefficient K for each bank: $$\text{Headloss per channel} = K\,\frac{(\text{velocity, m/s})^2 \times (1000\text{ mm/m})}{2\,(9.81\text{ m/s}^2)}\,(\text{no. banks/channel})$$ If the headloss across a UV channel is greater than 50 to 60 mm, either add additional channels and/or increase the lamp array size to reduce the headloss to an acceptable value.

*Adapted from Loge et al. (1996b).

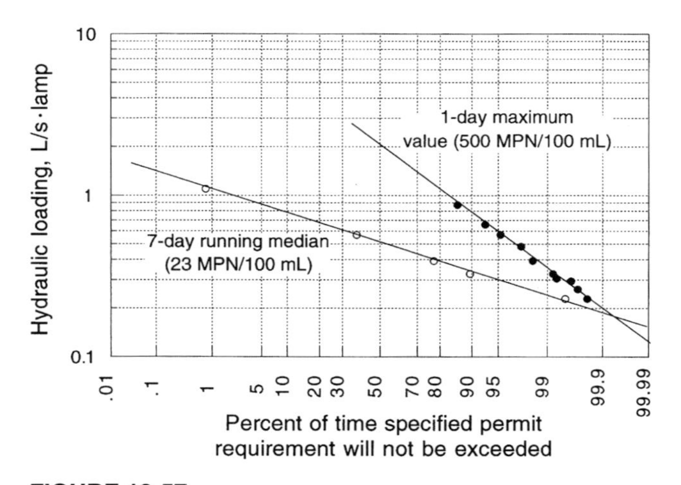

FIGURE 12-57
Percent of time a specified effluent permit requirement will not be exceeded, as a function of the UV lamp hydraulic loading rate for expanded water quality data sets.

procedures, discussed briefly below, may be found in Darby et al. (1995), Loge et al. (1996b), and Tchobanoglous et al. (1996).

In the most common nonprobabilistic approach, used by many UV manufacturers, the results obtained from a bench-scale collimated beam study are used to determine the minimum UV dose necessary to meet the most stringent wastewater treatment plant effluent discharge permit requirements. The probabilistic nature of the permit requirements is ignored. The variability in wastewater characteristics is accounted for indirectly through the application of various safety or adjustment factors. The actual steps used in this design approach are outlined in Table 12-27. Application of the design procedure outlined in Table 12-27 is illustrated in Example 12-7.

In one of several probabalistic approaches, described in detail elsewhere (Darby et al., 1995; Loge et al., 1996b, Tchobanoglous et al., 1996), the results of a pilot-scale field test are used to calibrate an empirical UV disinfection model. The model is then used in conjunction with a statistical technique (Monte Carlo analysis) to account for the uncertainty in the empirical model predictions, the variability in wastewater characteristics, and the probabilistic nature of permit requirements to determine the number of lamps required for disinfection (see Fig. 12-57). The steps involved in this approach are summarized in Table 12-28.

Environmental impacts of UV radiation disinfection. Because ultraviolet light is not a chemical agent, no toxic residuals are produced. However, certain chemical compounds may be altered by the ultraviolet radiation. On the basis of the evidence to date, it appears that the compounds formed are harmless or are broken down into more innocuous forms. At present, disinfection with ultraviolet light must be considered to have no adverse environmental impacts.

TABLE 12-28
Steps involved in the probabilistic approach to the design of UV disinfection systems*

Step	Requirement
1.	Calibrate the following empirical equation using the wastewater quality and disinfection performance data obtained from a pilot-scale UV disinfection study. $$N = A(TSS)^a(N_o)^b(\text{UFT})^c(I)^n(t)^n \quad (12\text{-}60)$$ where N = coliform density after exposure to UV light, MPN/100 mL TSS = total suspended solids, mg/L UFT = unfiltered transmittance at 253.7 nm, % N_o = influent coliform density, MPN/100 mL I = average intensity of UV light, mW/cm^2 t = exposure time, s (assuming approximate plug flow conditions) A, a, b, c, n = empirical coefficients (site specific) The advantage of a flexible empirical model is that statistical techniques can be used to determine the significance of variables thought to impact the performance of a UV disinfection system, instead of relying on human intuition in prejudging the variables of importance.
2.	A wastewater quality data set characterizing the significant model variables is collected from the WWTP to characterize the variability in wastewater quality. If a long-term data set is unavailable, it can be simulated by Monte Carlo techniques.
3.	Assume a range of flowrates including values above and below those used in the pilot system and determine the average detention time through the pilot-scale UV disinfection system for each flowrate.
4.	Convert the flowrates selected in step 3 into hydraulic loading rates by the following equation: $$\text{Hydraulic loading rate, L/s}\cdot\text{lamp} = \frac{\text{flowrate/channel, L/s}}{\text{number of lamps/channel}}$$
5.	A log-transformed version of the calibrated model is used to predict the effluent coliform concentration after exposure to UV light for each of the flowrates selected in step 3 for each of the days in the water quality data set. $$\log N = \log A + a\log(\text{TSS}) + b\log(N_o) + n\log(I) + n\log(t) + e \quad (12\text{-}61)$$ The error term e is a normally distributed random number with mean of zero and standard deviation equal to the adjusted root mean square error (RMSE′) of the calibrated model (see Darby et al., 1995, or Loge et al., 1996b, for the equation used to calculate the RMSE′). The error term accounts for the uncertainty associated with each empirical model prediction.
6.	The predicted coliform concentrations are analyzed according to the permit requirements at the WWTP (e.g., 7- or 30-day running median). The percent of time each permit requirement is not violated is then plotted as a function of UV hydraulic loading rate (step 4) on log-probability paper (see Fig. 12-57). The maximum allowable UV hydraulic loading rate is defined as the UV loading at which there is an acceptable likelihood the specified permit requirement will not be violated.
7.	The UV hydraulic loading can be adjusted to reflect reductions in lamp output associated with lamp aging and fouling anticipated in the full-scale UV reactor, as described previously for the nonprobabilistic approach (e.g., hydraulic loading rate × 0.7 × 0.7). The number of lamps required for disinfection, given a desired level of confidence in meeting the permit conditions is determined as follows: $$\text{Number of lamps} = \frac{\text{design flowrate, L/s}}{\text{allowable hydraulic loading rate per lamp, L/s}\cdot\text{lamp}}$$
8.	Assuming various system configurations, select a configuration that meets disinfection criteria with the minimum number of lamps.
9.	Compute the headloss per channel using a value of 2.2 for the headloss coefficient K for each bank: $$\text{Headloss per channel} = K\,\frac{(\text{velocity, m/s})^2 \times (1000\text{ mm/m})}{2\,(9.81\text{ m/s}^2)}\,(\text{no. banks/channel})$$ If the headloss across a UV channel is greater than 50 to 60 mm, either add additional channels and/or increase the lamp array size to reduce the headloss to an acceptable value.

*Adapted from Loge et al. (1996b).

EXAMPLE 12-7. DETERMINISTIC APPROACH TO UV DISINFECTION SYSTEM DESIGN. Using the following data, design a UV disinfection system to treat a wastewater nominal flow of 0.75 Mgal/d.

Wastewater characteristics:

1. Design flowrate (peak week) = 43.8 L/s (1.0 Mgal/d)
2. Average unfiltered transmittance = 63.8%
3. Average suspended solids = 6.1 mg/L
4. Average influent total coliform concentration = 1.2×10^6 MPN/100 mL
5. Average UV intensity (based on the PSS method of computation for a transmittance value of 63.8%) = 6.45 mW/cm^2

UV facilities:

1. Lamp type: low-pressure, low-intensity
2. Lamp configuration: lamps spaced on 75-mm centers (current industry standard)
3. Diameter of quartz sleeve encasing UV lamp = 23 mm (0.91 in)
4. Cross-sectional area of quartz sleeve = 4.15×10^{-4} m^2
5. Effective arc length of UV lamp = 1470 mm (58 in)

Regulatory requirements:

1. 23 MPN/100 mL based on a 7-d running median
2. 500 MPN/100 mL not to be exceeded any given day

Solution

1. Determine the minimum allowable UV dose:
   a. The log of the microorganism inactivation data obtained with the collimated beam is plotted for various UV doses (use test results given in Fig. 12-53 for a TSS value of 10 mg/L).
   b. The UV doses for permit requirements of 23 and 500 MPN/100 mL are 45 and 15 mW·s/cm^2, respectively. A UV dose of 45 mW·s/cm^2, corresponding to the most stringent discharge limit, is used in the following steps to determine the process configuration of the UV disinfection system.
   c. The UV dose is increased by 30 percent to account for lamp aging and an additional 30 percent to account for lamp fouling. The resulting UV dose is:

$$\text{UV dose} = \frac{45 \text{ mW·s/cm}^2}{0.7 \times 0.7} = 92 \text{ mW·s/cm}^2$$

2. Determine the number of UV lamps required.
   a. Using average UV dose, calculate the detention time required to meet the discharge requirement:

$$\text{Detention time} = \frac{92 \text{ mW·s/cm}^2}{6.45 \text{ mW/cm}^2} = 14.3 \text{ s}$$

   b. Determine the volume of flow that must be retained within the effective arc length of the UV lamp for the detention time determined in *a* above:

$$\text{Total volume, L} = (14.3 \text{ s}) \times (43.8 \text{ L/s}) = 626.3 \text{ L}$$

   c. Determine the number of lamps required for disinfection. The volume per lamp is obtained by subtracting the cross-sectional area of the lamp from a square area

75 mm× 75 mm and multiplying by the effective length of the UV lamp (typically 147 cm)

$$\text{Volume/lamp} = \left[(75\text{ mm})^2 - \frac{3.14}{4}(23\text{ mm})^2\right] \times 1470\text{ mm} \times \frac{1.0\text{ L}}{10^6\text{ mm}^3}$$

$$= 7.65\text{ L/lamp}$$

$$\text{Number of lamps} = \frac{\text{total volume, L}}{(\text{volume/lamp, L/lamp})}$$

$$\text{Number of lamps} = \frac{626.3\text{ L}}{(7.65\text{ L/lamp})} = 81$$

3. Determine UV disinfection system configuration.
   a. Select a configuration that meets disinfection criteria with the minimum number of lamps. Assume number of banks per channel = 3. Assume a lamp array $N_L \times N_M$, where N_L is the number of lamps per module and N_M is the number of modules per bank. For example, assume $N_L = 4$ and $N_M = 4$. Calculate the number of channels, N_C, given N_B, N_L, N_M, and the number of lamps required for disinfection:

$$\text{Number of channels} = \frac{\text{number of lamps required for disinfection}}{[(N_L \times N_M)\text{ lamps/bank}] \times N_B}$$

$$= \frac{81}{[(4 \times 4)\text{ lamps/bank}] \times 3\text{ banks}} = 1.69 \qquad \text{Use 2}$$

   b. Recompute the total number of lamps in the selected configuration:

$$\text{Total lamps} = N_L \times N_M \times N_B \times N_C$$

$$= [(4 \times 4)\text{ lamps/bank}] \times (3\text{ banks/channel}) \times 2\text{ channels}$$

$$= 96$$

   c. Calculate the number of excess lamps:

$$\text{Excess lamps} = \text{total lamps} - \text{lamps required for disinfection}$$

$$= 96 - 81 = 15$$

4. Check whether the headloss for the selected configuration is acceptable.
   a. Determine the channel cross-sectional area:

$$\text{Cross-sectional area of channel} = (4 \times 0.075\text{ m}) \times (4 \times 0.075\text{ m})$$

$$= 0.09\text{ m}^2$$

   b. Determine the net channel cross-sectional area by subtracting the cross-sectional area of the quartz sleeves (4.15×10^{-4} m^2/lamp):

$$\text{Net channel area} = 0.09\text{ m}^2 - [(4 \times 4)\text{ lamps/bank}] \times 4.15 \times 10^{-4}\text{ m}^2\text{/lamp}$$

$$= 0.083\text{ m}^2$$

   c. Determine the velocity in the channel:

$$\text{Velocity} = \frac{0.0438\text{ m/s}}{2\text{ channels} \times (0.083\text{ m}^2\text{/channel})} = 0.26\text{ m/s}$$

d. Determine the headloss per UV channel:

$$h_L = 2.2 \frac{(0.26 \text{ m/s})^2}{2 \times 9.81 \text{m/s}^2} \times 3 = 0.023 \text{ m} = 23 \text{ mm}$$

Because the headloss is less than 50 mm, the process configuration is acceptable. If the headloss were unacceptable, more channels would have to be added to reduce the headloss. Although the spacing could be altered, it is not commonly done.

Ozone Disinfection

Ozone (O_3) is an extremely strong oxidant that has been used for the disinfection of water and wastewater. Because ozone is chemically unstable, it decomposes to oxygen very rapidly after generation. As a result, ozone must be generated onsite, at or near the point of use.

Generation of ozone. The most efficient method of producing ozone today is by electrical discharge. Ozone is generated either from air or pure oxygen by applying a high voltage across the gap of narrowly spaced electrodes. The high-energy corona created by this arrangement dissociates one oxygen molecule, which reforms with two other oxygen molecules to create two ozone molecules. The gas stream generated by this process from air will contain about 0.5 to 3 percent ozone by weight, and from pure oxygen about twice that amount, or 1 to 6 percent ozone.

Reactions of ozone in water. When ozone is added to water, some of the chemical properties of ozone may be described by its decomposition reactions, which are thought to proceed as follows:

$$O_3 + H_2O \rightarrow HO_3^+ + OH^- \quad (12\text{-}62)$$

$$HO_3^+ + OH^- \rightarrow 2HO_2 \quad (12\text{-}63)$$

$$O_3 + HO_2 \rightarrow HO + 2O_2 \quad (12\text{-}64)$$

$$HO + HO_2 \rightarrow H_2O + O_2 \quad (12\text{-}65)$$

The free radicals formed, HO_2 and HO, have great oxidizing powers and are probably the active form in the disinfection process. These free radicals also possess the oxidizing power to react with other impurities in aqueous solutions.

Effectiveness of ozone. Ozone is an extremely reactive oxidant and it is generally believed that bacterial kill through ozonation occurs directly because of cell wall disintegration (cell lysis). Ozone is also a very effective viricide and is generally believed to be more effective than chlorine. Ozonation does not produce dissolved solids, and is not affected by the ammonium ion or the influent pH (unless highly elevated). For these reasons, ozonation is considered a viable alternative to either chlorination or hypochlorination, especially where dechlorination may be required. Because ozone decomposes rapidly, no chemical residual persists in the treated effluent that requires removal, as chlorine residuals do.

TABLE 12-29
Advantages and disadvantages of chlorine, UV, and ozone disinfection

Advantages	Disadvantages
Chlorine	
1. Well-established technology 2. Effective disinfectant 3. Chlorine residual can be maintained 4. Combined chlorine residual can also be provided by adding ammonia 5. Germicidal chlorine residual can be maintained in long transmission lines 6. Relatively inexpensive (cost is increasing with implementation of Uniform Fire Code regulations)	1. Residual toxicity of treated effluent must be reduced through dechlorination 2. Formation of trihalomethanes and other chlorinated hydrocarbons 3. Increased safety regulations, especially in light of the new Uniform Fire Code 4. Less effective in inactivating some viruses, spores, cysts at low dosages used for coliform organisms 5. TDS level of treated effluent is increased 6. Chloride content of the wastewater is increased 7. Release of volatile organic compounds from chlorine contact basins 8. Chemical scrubbing facilities may be required to meet Uniform Fire Code regulations 9. Acid generation; pH of the wastewater can be reduced if alkalinity is insufficient
UV	
1. Effective disinfectant 2. No residual toxicity 3. More effective than chlorine in inactivating most viruses, spores, cysts 4. Improved safety 5. Requires less space	1. No immediate measure of whether disinfection was successful 2. No residual effect 3. Less effective in inactivating some viruses, spores, cysts at low dosages used for coliform organisms 4. Relatively expensive (price is coming down as new and improved technology is brought to the market) 5. Large number of UV lamps required
Ozone	
1. Effective disinfectant 2. More effective than chlorine in inactivating most viruses, spores, cysts 3. Requires less space 4. Contributes dissolved oxygen	1. No immediate measure of whether disinfection was successful 2. No residual effect 3. Less effective in inactivating some viruses, spores, cysts at low dosages used for coliform organisms 4. Safety concerns 5. Corrosive 6. Relatively expensive 7. Highly operation and maintenance sensitive

Environmental impacts of ozone disinfection. Unlike the other chemical disinfecting agents previously discussed, the use of ozone produces mainly beneficial impacts on the environment. Although ozone residuals can be acutely toxic to aquatic life, ozone dissipates rapidly and ozone residuals will normally not be found by the time the effluent is discharged into the receiving water. Several investigators have reported that ozonation can produce some toxic mutagenic and/or carcinogenic compounds. These compounds are usually unstable, however, and are present only for a matter of minutes in the ozonated water. An additional benefit associated with the use of ozone for disinfection is that the dissolved oxygen concentration of the effluent will be elevated to near saturation levels, as ozone rapidly decomposes to oxygen after application.

Comparison of Alternative Disinfection Technologies

Comparison of the four disinfection technologies based on ideal and actual characteristics was presented previously in Table 12-21. A comparison of the germicidal effectiveness of the disinfection technologies was presented previously in Table 12-26. As shown, the dosages of chlorine required for the disinfection of viruses and protozoa are considerably higher than those required for total coliform. The same situation is also true for UV and ozone disinfection, but less so. The advantages and disadvantages of using chlorine, UV, and ozone for disinfection are summarized in Table 12-29.

12-7 AGRICULTURAL IRRIGATION

For small systems, agricultural or silvicultural irrigation will generally be an attractive alternative for water reuse, especially in rural areas. As described in Chap. 10 under Type 2 crop irrigation systems, the irrigation requirements are based on crop needs for water and nutrients. In this section, water quality considerations, treatment needs, long-term effects, farmer contracts, and representative case study examples are discussed.

Water Quality Considerations

Water quality considerations for reuse by agricultural irrigation include nutrients, salinity, sodium adsorption ratio, and trace elements. Restrictions on irrigation water quality were discussed in Chap. 10 and summarized in Table 10-15.

Nutrients. The primary nutrients are nitrogen, phosphorus, and potassium (N, P, and K). Typical levels of these nutrients in recycled water are presented in Table 12-30. The fertilizer mass loading of an effluent can be calculated by:

$$L_M = C L_W F \tag{12-66}$$

TABLE 12-30
Fertilizer content of typical secondary effluent

Nutrient	Typical concentration, mg/L	Fertilizer content, lb/ac·ft
Nitrogen	10–30	27–82
Phosphorus	6–20	16–54
Potassium	15–25	41–68

where L_M = mass loading, lb/ac·yr (kg/ha·yr)
C = nutrient concentration, mg/L $(g/m)^3$
L_w = hydraulic loading, ft/yr (m/yr)
F = conversion factor, 2.72 L·lb/ac·ft·mg (10 m^2·kg/ha·g)

Secondary nutrients include calcium, magnesium, and sulfur. Micronutrients include zinc, iron, copper, boron, and molybdenum. Secondary and micronutrients are often found in adequate concentrations in secondary effluent. More details on plant nutrient requirements may be found in the Western Fertilizer Handbook (California Fertilizer Association, 1995).

Salinity. Salts are added to the soil continually during irrigation. A typical irrigation water with 500 mg/L of total dissolved solids, applied at 3 ft/yr, results in the addition of 4080 lb/ac·yr of salts. Plants remove water, but not much salt. If evaporation exceeds precipitation on an annual basis, salts will accumulate in the soil, unless leaching occurs. Problems occur if the added salts accumulate to harmful levels in the root zone. These problems can include yield reduction due to the overall salinity or specific ion effects from constituents such as sodium or chloride.

Leaching of salts through the root zone must occur if irrigation of any crop is to be successful and sustainable. If rainfall does not equal evapotranspiration, the leaching must be accomplished by applying more water than the crop needs. The excess applied water or leaching fraction removes salt from the root zone.

Leaching requires drainage. Drainage, in turn, means that salts and water move through the root zone and into the underground strata and groundwater, or seep into surface streams.

Leaching requirements increase with an increase in the TDS content of the applied water and with an increase in the salt sensitivity of the crop. To determine the leaching requirement (LR), use:

$$LR = \frac{EC_W}{EC_{DW}} \tag{12-67}$$

where LR = leaching requirement as a decimal
EC_W = electrical conductivity of applied water, dS/m
EC_{DW} = electrical conductivity of drainage water dS/m

The EC in the recycled water can be measured in decisiemens per meter (dS/m), which is equivalent to millimhos per centimeter (mmhos/cm), or estimated from

TABLE 12-31
Values of EC_{DW} for crops with no yield reduction*

Crop	Electrical conductivity of drainage water (EC_{DW}), dS/m
Bermuda grass	13
Barley	12
Sugar beets	10
Cotton	10
Wheat	7
Tall fescue	7
Soybeans	5
Corn	5
Alfalfa	4
Orchard grass	3

*From Ayers (1977).

the TDS concentration (mg/L) by dividing the TDS by 640. The EC_{DW} values for various crops with no yield reduction for the crop are presented in Table 12-31. The use of Eq. (12-67) is illustrated in Example 12-8.

The leaching requirement is used in the establishment of the hydraulic loading rate to increase the net loading rate, as shown in the following equation from Chap. 10:

$$L_W = \left(\frac{\text{ET} - \text{Pr}}{1 + \text{LR}}\right)\left(\frac{100}{\text{E}}\right) \tag{10-5}$$

where L_W = wastewater hydraulic loading rate, ft/yr (mm/yr)
ET = crop evapotranspiration, ft/yr (mm/yr)
Pr = precipitation rate, ft/yr (mm/yr)
LR = leaching requirement, decimal
E = irrigation efficiency, % (65 to 80)

EXAMPLE 12-8. DETERMINE LEACHING REQUIREMENT FOR IRRIGATION. An effluent is to be used in an arid climate to irrigate alfalfa. The effluent TDS is 750 mg/L. Determine the leaching requirement to avoid yield reduction resulting from salinity effects.

Solution

1. Calculate the EC_W of the effluent:

$$EC_W = 750/640 = 1.17 \text{ dS/m}$$

2. Determine the EC_{DW} for alfalfa. From Table 12-31, the value is 4 dS/m.
3. Calculate the LR using Eq. (12-67):

$$\text{LR} = \frac{1.17}{4} = 0.29$$

Sodium adsorption ratio. Sodium can have an adverse impact on the permeability of clay soils when the ratio of sodium to calcium and magnesium is high. This adverse impact can occur as sodium ions tend to deflocculate the clay soil structure. As the clay particles deflocculate, they form a platelike structure, and the permeability is thereby reduced. The sodium adsorption ratio (SAR), as defined previously in Chap. 10, is:

$$\text{SAR} = \frac{[\text{Na}]}{\sqrt{\dfrac{[\text{Ca}] + [\text{Mg}]}{2}}} \tag{10-2}$$

where SAR = sodium adsorption ratio, unitless
[Na] = sodium concentration, meq/L
[Ca] = calcium concentration, meq/L
[Mg] = magnesium concentration, meq/L

An SAR of 10 or less should be acceptable for soils with a significant clay content (15 percent clay or more). Soils with little clay or nonswelling clays can tolerate an SAR up to 20, particularly if the TDS is 800 mg/L or more. An adjusted SAR has been used in the past (Ayers, 1977); however, current analysis is that the conventional SAR is sufficient. The calculation of SAR is illustrated in Example 12-9.

EXAMPLE 12-9. CALCULATE SAR FOR WASTEWATER USED FOR IRRIGATION. An effluent contains 155 mg/L of sodium, 26 mg/L of calcium, 16 mg/L of magnesium, and a TDS of 600 mg/L. Calculate the SAR and estimate the hazard for a clay soil.

Solution

1. Convert the concentrations of sodium, calcium, and magnesium to milliequivalents per liter:

$$\text{meq Na} = 155/23 = 6.74$$

$$\text{meq Ca} = 26/20 = 1.30$$

$$\text{meq Mg} = 16/12.15 = 1.32$$

2. Calculate the SAR using Eq. (10-2):

$$\text{SAR} = \frac{[\text{Na}]}{\sqrt{\dfrac{[\text{Ca}] + [\text{Mg}]}{2}}} = \frac{6.74}{\sqrt{\dfrac{1.30 + 1.32}{2}}} = 5.89$$

3. Compare the SAR and TDS to values in Table 10-15. For an SAR of 5.9 (within the range of 3 to 6) and a TDS of 600 (within the range of 200 to 770), the hazard is slight to moderate.

Comment. The common remedial measure for excess sodium is to treat the soil with gypsum (calcium sulfate), sulfur, or sulfuric acid. After treatment, water is applied to leach out the sodium. A measure of the effectiveness of treatment is to monitor the exchangeable sodium percentage (ESP) of the soil. The exchangeable sodium percentage is the ratio of the exchangeable sodium on the cation exchange capacity of the soil to the remaining cations. When the ESP is 15 or more, additional treatment is required. When the ESP is 5 or less, conditions are acceptable (Reed and Crites, 1984).

TABLE 12-32
Guidelines for trace element concentrations in treated wastewater effluent for unrestricted long-term application on any soil

Constituent	Maximum concentration, mg/L
Aluminum	5.0
Arsenic	0.1
Beryllium	0.1
Boron	0.75
Cadmium	0.01
Chromium	0.1
Cobalt	0.05
Copper	0.2
Fluoride	1.0
Iron	5.0
Lead	5.0
Lithium	2.5
Manganese	0.2
Molybdenum	0.01
Nickel	0.2
Selenium	0.02
Vanadium	0.1
Zinc	2.0

Trace elements. Trace elements can become toxic to plants or animals consuming the crop at relatively high concentrations. A listing is provided in Table 12-32 of the constituents of concern and the concentration below which no toxic effects have been observed.

Typical Pretreatment

The principal reasons for treatment prior to irrigation with wastewater are to protect public health and to avoid a public nuisance. Secondary treatment (oxidized wastewater) plus disinfection will generally protect public health, with proper crop selection and wastewater management. As shown in Tables 12-3 and 12-4, food crop irrigation generally requires disinfected tertiary effluent, unless the crops are processed commercially prior to use, in which case disinfected secondary effluent is usually acceptable. Secondary effluent is also stabilized adequately to be stored in reservoirs for long periods of time.

Long-Term Effects

In 1993, the U.S. EPA published new regulations on land application of sludge (see Chap. 14) in which the cumulative limits on long-term trace element loadings were

TABLE 12-33

Expected operating life of agricultural reuse sites based on trace element accumulation in the soil*

Element	Cumulative loading limit, lb/ac[†]	Median concentration in effluent, mg/L	Operating life, yr[‡]
Arsenic	36.6	<0.005	683
Cadmium	34.8	<0.005	650
Chromium	2677	0.02	12,500
Copper	1339	0.04	3125
Lead	268	0.008	3125
Mercury	15.2	0.0005	2833
Molybdenum	16.1	0.007	214
Nickel	375	0.004	8750
Selenium	89.2	<0.005	1667
Zinc	2499	0.04	5833

*Adapted from Page and Chang (1985) and U.S. EPA (1993).

† EPA cumulative loading limits for trace elements in land-applied municipal wastewater sludge.

‡ Accumulation of column 1 loadings based on irrigation of column 2 element concentration with an annual liquid loading of 4 ft (1.2 m).

revised. The cumulative limits are listed in Table 12-33, along with median values of each element found in secondary effluent. If the secondary effluent typified by the median values in Table 12-33 was applied at a hydraulic loading rate of 4 ft/yr (1.2 m/yr), the expected operating life of the system would range from 214 yr (for molybdenum) to 12,500 yr (for chromium).

With modern pretreatment programs in place for industrial wastewaters, it is unlikely that trace element concentrations in effluent will cause problems with agricultural irrigation of municipal recycled water. Research on long-term effects of wastewater irrigation have shown that it is both safe and sustainable (Johnson et al., 1974; Hinesly, 1978; and U.S. EPA, 1981).

Farmer Contracts

Contracts with farmers have been developed by sanitation agencies and cities over the years to allow for the reuse of recycled water for irrigation of private land. The contracts or agreements include a description of the proposed reuse, the responsibilities of the parties involved, and the terms and conditions for payments (if any) and termination.

Representative Reuse Examples

Two representative agricultural irrigation systems are described in this section. A pasture irrigation system is operated on private farmland in Petaluma, California. In Lodi, California, an agricultural crop irrigation system is operated on city-owned land.

Petaluma, California. In the 1970s the City of Petaluma, located in Sonoma County north of San Francisco, discharged secondary effluent into the Petaluma River. The California Regional Water Quality Control Board adopted discharge requirements that limited the discharge of secondary effluent to the months of December through April. For the remaining months of May through November, the effluent needed to be stored or reused. It was decided that a program would be embarked upon to enlist the local farmers in a water reuse project (Crites, 1982).

The process for selecting sites for effluent irrigation was as follows:

1. Prepare information packages for the farmers.
2. Make appointments to see the farmers.
3. Describe the project in face-to-face meetings and leave the package for review.
4. Follow up with the farmers to determine interest.
5. Obtain letters of interest from the farmers.
6. Evaluate individual sites for acceptability.
7. Prepare alternatives by combining adjacent sites.
8. Select a cost-effective combination of sites.

The information packages contained the following:

1. An introduction letter from the city.
2. An economic analysis from the farming perspective.
3. Description of irrigation techniques.
4. Water quality data on the recycled water.
5. Regional board regulatory requirements.
6. A sample customer contract.
7. A blank letter of intent.

The resultant system consisted of six individual sites totaling 550 ac (222 ha) of private farmland. Solid-set sprinkler and traveling water gun (see Fig. 12-58) irrigation systems are used on rolling pastureland. The design factors for the Petaluma reuse system are summarized in Table 12-34.

After 10 years of operation the Petaluma system has been expanded to 8 farmers and 775 ac (314 ha). The key original farmer, Dan Silacci, has become the overall

FIGURE 12-58
City of Petaluma irrigation of pasture land using a traveling water gun.

TABLE 12-34
Design factors for Petaluma, California, irrigation system*

Factor	Value	
	SI units	U.S. customary units
Flow: 1982	11,355 m^3/d	3 Mgal/d
1995	15,140 m^3/d	4 Mgal/d
Irrigated area: 1982	222 ha	550 ac
1995	312 ha	770 ac
Hydraulic loading rate	0.78 m/yr	2.56 ft/yr
Number of customers: 1982	6	6
1995	8	8
Crops	Pasture grass	Pasture grass
Type of irrigation	Sprinklers	Sprinklers
Year started	1977	1977

*From Crites (1982).

manager of all the irrigation, and has a long-term contract with the City of Petaluma. Petaluma also irrigates two golf courses with recycled water.

Lodi, California. Effluent from the City of Lodi's White Slough Water Pollution Control Facility has been used to grow crops for over 50 years. The city irrigates 900 ac (365 ha) of city-owned farmland with a combination of secondary effluent and cannery process water. The reuse system is shown in Fig. 12-59. Crops include alfalfa hay and field corn. These two crops also have biosolids applied during various phases of the growing season. In recent years, the city has also supplied recycled water to produce steam for a 49-megawatt power generator, and to replenish mosquitofish-rearing ponds. Additionally, the city has agreed to supply disinfected

FIGURE 12-59
City of Lodi wastewater irrigation system showing effluent flowing in irrigation ditch to irrigate alfalfa.

TABLE 12-35
Design factors for Lodi, California, irrigation system*

	Value	
Factor	**SI units**	**U.S. customary units**
Flow, 1996	8700 m^3/d	2.3 Mgal/d
Irrigated area, 1995	365 ha	900 ac
Hydraulic loading rate	0.88 m/yr	2.9 ft/yr
Number of customers	5	5
Crops	Alfalfa, forage crops, pasture, field crops	Alfalfa, forage crops, pasture, field crops
Type of irrigation	Surface and sprinkler	Surface and sprinkler
Year started	1946	1946

*Courtesy F. Forkas, 1997.

tertiary effluent to irrigate a proposed soccer complex. Design factors for the reuse system at Lodi are presented in Table 12-35.

12-8 LANDSCAPE IRRIGATION

Landscape irrigation includes urban area reuse of recycled water for golf courses, parks, playgrounds, freeway medians and interchanges, and residential lawns and shrubbery. This section includes discussions of water quality considerations, treatment needs, golf course irrigation, dual water systems, and representative reuse examples.

Water Quality Considerations

Water quality considerations for landscape irrigation are very similar to those for agricultural irrigation, as previously described. Because landscape irrigation usually involves noncultivated perennial vegetation, an additional concern is for specific ion toxicity, specifically for sodium, chloride, and boron. Guidelines for water quality criteria for these constituents are presented in Table 10-15 (Chap. 10). The analysis of trace element effects is illustrated in Example 12-10.

EXAMPLE 12-10. TRACE ELEMENTS IN LANDSCAPE IRRIGATION. Determine the life expectancy of a landscape irrigation system before the buildup of trace metals causes a problem. Use a limit for arsenic of 82 lb/ac from Table 10-14 and an arsenic concentration of 0.010 mg/L in the applied effluent. Use an annual hydraulic application of 3.5 ft/yr.

Solution

1. Calculate the annual loading of arsenic from the application using Eq. (12-66):

$$\begin{aligned} L_M &= CL_W F \\ &= (0.010 \text{ mg/L})(3.5 \text{ ft/yr})\ 2.72 \\ &= 0.0952 \text{ lb/ac}\cdot\text{yr} \end{aligned}$$

2. Calculate the years of application:

$$\text{Life} = \frac{82 \text{ lb/ac}}{0.0952 \text{ lb/ac}\cdot\text{yr}} = 861 \text{ yr}$$

Comment. The application of effluent to landscape vegetation can proceed for many years without harmful effects from applied metals.

Typical Pretreatment

The principal reasons for treatment prior to landscape irrigation with wastewater are to protect public health and to avoid a public nuisance. The degree of treatment depends on the method of irrigation (drip versus sprinkler) and on the potential for public contact. As shown in Tables 12-3 and 12-4, landscape irrigation of areas with public access generally requires disinfected tertiary effluent. Effluent from packed-bed filters can be used for subsurface drip irrigation.

Golf Course Irrigation

Golf course irrigation with recycled water is a long-established practice in the southwest, with documented history dating back to 1948 in California (Crites, 1984). In California there are over 60 golf courses irrigated with recycled water, including those at Petaluma and La Contenta.

The advantages of golf course irrigation are that reuse is spread out over a large part of the year, thereby reducing the winter storage needs; the courses use large quantities of water, ranging from 300 to over 500 acre-ft/yr (370,000 to 620,000 m^3/yr); and turf grasses are usually tolerant of relatively high concentrations of total dissolved solids.

Dual Water Systems

Dual water systems consist of nonpotable, recycled water distribution pipelines that are parallel to the potable domestic water supply pipelines. In new communities the installation of dual water systems is cost-effective, as in the community of Mission Viejo, California. There are guidelines for dual water or nonpotable water systems (AWWA, 1992). The guidelines include recommendations for pressure, minimum depth, minimum separation, and pipeline identification. These recommendations are summarized in Table 12-36.

TABLE 12-36
Guidelines for nonpotable water distribution lines*

Item	Recommendation
Pressure	10 lb/in^2 less than the potable supply line
Minimum depth	36 in below finished street grade
Minimum separation	10 ft horiziontal†
	1 ft vertical (below)
Pipe identification	
Color	Purple (Pantone 522)
Identification tape	"CAUTION: NONPOTABLE WATER—DO NOT DRINK"

*From AWWA (1992).
†Where conditions do not allow the full 10 ft of separation, the local health department can approve a lesser distance.

Representative Reuse Examples

Five representative examples of landscape irrigation are presented. The first, the Stonehurst project near Martinez, California, is a true decentralized system. The second is a commercial building water recycling project in Santa Monica, California. The third is a small community reuse system. The fourth and fifth are dual water systems.

Stonehurst development. A 47-house development was constructed in a rural area of Contra Costa County near the City of Martinez, California. The existing homeowners between the development and the nearest sewer system were opposed to the extension of the sewer. As a consequence, the developer obtained approval for 47 individual onsite systems. Instead of constructing individual leachfields, the developer installed a decentralized system that consists of individual septic tanks, a combination STEP/STEG collection system, a recirculating fine-gravel filter, a UV disinfection system, and a drip irrigation system. During the winter the effluent is discharged to a community leachfield. The design factors are presented in Table 12-37, and the flow diagram is presented in Fig. 11-20, Chap. 11. Performance data for the Stonehurst treatment system are also given in Table 11-12, Chap. 11 (Crites et al., 1997).

Water Gardens, Santa Monica, California. Recycling of water from office buildings has been practiced in Japan and in the United States. One example is the Water Gardens development, a four-story office building complex in Santa Monica. Wastewater collected from the offices is recycled and reused for landscape irrigation, water features, and, ultimately, urinal and toilet flushing. The treatment processes include pretreatment, biological oxidation, membrane filtration (ultrafiltration), activated carbon, ultraviolet radiation, and chlorination. The flow diagram is shown in Fig. 12-60.

At full building occupancy, approximately 26,000 gal/d (98 m^3/d) of wastewater from the sinks, showers, toilets, and urinals will pass through a 26,000-gal trash trap to remove large solids and then into a 60,000-gal equalization and emergency storage reservoir.

TABLE 12-37
Design factors for Stonehurst, California, reclamation and reuse system*

Factor	Unit	Value
Design flow	gal/d	14,100
Gravel filter		
Hydraulic loading rate	gal/ft²·d	3
Recirculation ratio		5
Size of medium, d_{10}	mm	3
Depth of medium	ft	2
Disinfection (see Fig. 12-47a)		
Type		UV
Number of modules		3
Lamps per module		2
Minimum UV dose	mW·s/cm²	95
Drip irrigation system		
Area irrigated	ac	1
Vegetation		Shrubs, groundcover
Emitter type		Nonclog, buried

*From Crites et al. (1997).

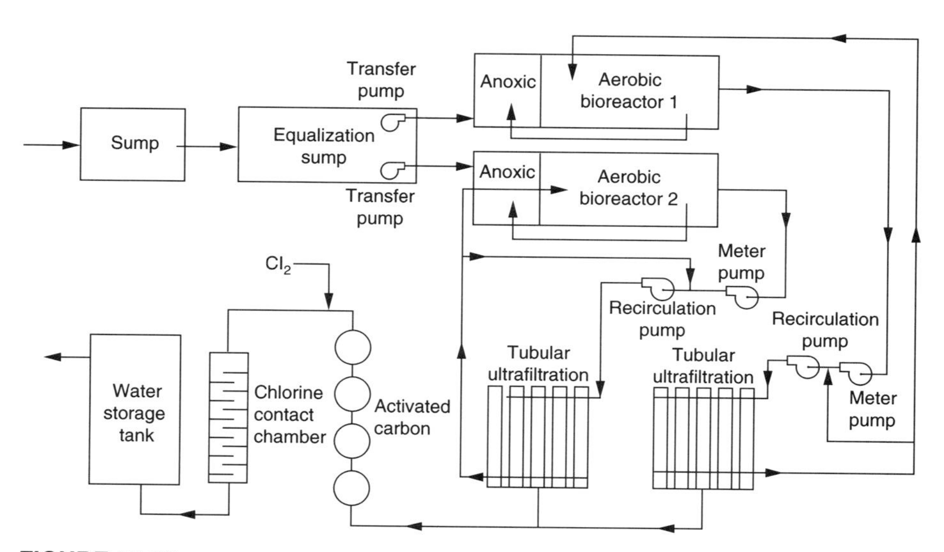

FIGURE 12-60
Flow diagram of water reclamation and reuse at the Water Gardens, Santa Monica, California (adapted from Jordan and Senthilnathan, 1996).

TABLE 12-38
Design factors for the Water Gardens, Santa Monica, California, reclamation and reuse system*

Item	Unit	Value
Design flow	gal/d	20,000
Initial year of operation		1992
Influent BOD_5	mg/L	600
Effluent BOD_5	mg/L	5
Influent TSS	mg/L	600
Effluent TSS	mg/L	5
Influent nitrogen	mg/L	150
Effluent nitrogen	mg/L	4
Effluent nitrogen limit	mg/L	10
Effluent turbidity	NTU	0.15
Effluent turbidity limit	NTU	2
Effluent total coliform	MPN/100 mL	<2
Effluent total coliform limit	MPN/100 mL	<2

*From Jordan and Senthilnathan (1996).

Biological treatment is accomplished by a 20,000-gal/d (76 m^3/d) activated-sludge system. Solids are separated from the treated liquid by means of membrane filters, allowing large concentrations (20,000 to 30,000 mg/L) of mixed-liquor suspended solids to be maintained in the reactor. By eliminating the clarifiers, the system requires much less area than conventional activated sludge.

Treated wastewater is pumped through tubular cross-flow membrane filters, composed of synthetic organic polymers, that have an average pore size of approximately 0.005 μm. The permeate from the ultrafiltration units is then passed through granular activated-carbon adsorption for color removal and a combination of UV and chlorine disinfection. The performance data for operations from 1992 through 1996 are presented in Table 12-38. During that period the system has never been out of compliance (Jordan and Senthilnathan, 1996).

The Water Gardens system was manufactured by Thetford Systems (now owned by Zenon Municipal Systems) and was called the Cycle-Let system. Thetford installed approximately 27 systems between 1974 and 1982, mostly for small commercial facilities on the east coast of the United States (Irwin, 1994).

La Contenta, California. La Contenta is a community in the Calaveras County foothills, consisting of some 700 single-family lots surrounding an 18-hole championship golf course. The community attracted additional development of 2700 homes and needed to collect and treat the resulting wastewater. The collection system is a conventional gravity sewer system, and the treatment plant is a biological secondary treatment system, followed by a conventional rapid sand filter and chlorine disinfection. Treated effluent is stored during the winter in an open reservoir and recycled for golf course irrigation during the summer (see Fig. 12-61). The design criteria for the treatment plant and reuse system are presented in Table 12-39.

FIGURE 12-61
Golf course irrigation at La Contenta, California.

TABLE 12-39
Design factors for La Contenta, California, irrigation system*

Factor	Value	
	SI units	U.S. common units
Treatment	Extended aeration, filtration, chlorination	Extended aeration, filtration, chlorination
Flow	1514 m^3/d	0.4 Mgal/d
Irrigated area	48 ha	120 ac
Hydraulic loading rate	1.1 m/yr	3.74 ft/yr
Number of customers	1	1
Crops	Turf grass	Turf grass
Type of irrigation	Sprinkler	Sprinkler
Year started	1992	1992

*Courtesy J. Scroggs, 1997.

Mission Viejo, California. Mission Viejo is a master-planned, recently incorporated city in Orange County, California. At buildout the population will exceed 90,000 and the developed area will be over 10,000 ac (4040 ha). Water reuse began in 1965 with the construction of the first 0.5-Mgal/d (1900-m^3/d) recycling plant. The recycled water was used for golf course irrigation.

The reasons for water reuse in Mission Viejo included both wastewater disposal and water supply augmentation (Roohk et al., 1989). Recycling wastewater meant avoiding the cost of an ocean outfall and reducing the importation of fresh water for landscape irrigation. The Mission Viejo recycling system has three major functions: (1) satisfy the landscape irrigation needs of the community, (2) provide for wastewater disposal, and (3) control non-point-source pollution of the downgradient groundwater basins. As part of the reuse system, a streamflow barrier was constructed that intercepted all nonstorm streamflow and transported it either to the

TABLE 12-40
Design factors for Mission Viejo, California, landscape irrigation system*

Factor	Value	
	SI units	U.S. common units
Flow, design	19,300 m^3/d	5.0 Mgal/d
Irrigated area	770 ha	1900 ac
Hydraulic loading rate	0.9 m/yr	2.95 ft/yr
Number of customers	85	85
Irrigated areas	Centralized landscape, parks, school yards, golf courses, freeway corridors, greenbelts	
Type of irrigation	Sprinkler	Sprinkler
Year started	1965	1965

*From Roohk et al. (1989).

recycling system or to the ocean, depending on the time of year. When demand for recycled water is high, the recycled streamflow is screened, disinfected, and blended with the tertiary effluent from the recycling plants. Seasonal storage facilities in Upper Oso Reservoir include a dam with spillway, combined inlet/outlet works, and a reservoir destratification/aeration system. The reservoir serves both for winter storage of recycled water, when demands are reduced, and for storage of stormwater runoff. The reuse system provides water to irrigate all of the centralized landscape, including parks, school yards, residential greenbelts, arterial slopes, freeway corridors, and golf courses. The ultimate irrigated area is projected to be 1900 ac (770 ha). The design factors for the Mission Viejo recycling system are presented in Table 12-40.

St. Petersburg, Florida. The City of St. Petersburg in southern Florida has been a leader in water reuse on the east coast since 1977. Water pollution control limitations led the city to its initial decision to reuse its recycled water. However, while pollution control was the initial impetus for reuse, the greatest benefit has been water conservation. St. Petersburg now operates one of the largest urban reuse systems in the world, providing a dual water system to more than 7000 residential homes and businesses (Johnson, 1990).

The dual water system was retrofitted into the community and now supplies recycled water for irrigation of individual homes, condominiums, parks, school grounds, and golf courses. In addition, recycled water is used for cooling tower makeup and supplemental fire protection. The dual distribution system consists of an extensive network of over 260 mi (420 km) of pipe. The design factors for the St. Petersburg reuse system are presented in Table 12-41.

Operational storage is provided in covered storage tanks located at the four treatment plants. Instead of seasonal storage, the city has 10 deep wells to dispose of excess recycled water. The presence of the injection wells also allows the city to reject recycled water that does not meet water quality standards or treatment performance.

TABLE 12-41
Design factors for St. Petersburg, Florida, dual water system*

Factor	Value	
	SI units	U.S. customary units
Flow	79,480 m^3/d	21 Mgal/d
Pipe network	420 km	260 mi
Type of use	Irrigation, cooling tower makeup, fire protection	
Number of customers	7000	7000
Irrigated areas	Private yards, parks, schoolyards, golf courses,	
Type of irrigation	Sprinkler	Sprinkler
Year started	1977	1977

*From Johnson (1990).

The quality requirements for reuse are a minimum of 4 mg/L of chlorine residual, and a maximum of 2.5 NTU of turbidity and 5 mg/L of suspended solids. In addition, a maximum chloride concentration of 600 mg/L is specified. On a yearly average, approximately 60 percent of the recycled water produced is injected into the deep wells (U.S. EPA, 1992a).

12-9 GROUNDWATER RECHARGE

Groundwater recharge with fresh water has been practiced for over 200 years worldwide (Asano, 1985). In Los Angeles County, California, recycled water has been used to recharge groundwater and provide indirect potable reuse since 1962.

TABLE 12-42
Groundwater recharge projects utilizing injection*

Project	Capacity, ac·ft/yr	Purpose	Status
Orange County Water District, Water Factory 21, California	16,800	Saltwater barrier	Operational since October 1976.
Palo Alto, California	2200	Saltwater barrier	Began operation in 1977. Suspended operation in 1982.
Nassau County, New York	2200	Saltwater barrier	Injection wells operated for 83 days in 1983.
El Paso, Texas	13,400	Groundwater replenishment	Operational since 1985.
West Coast Basin Barrier Project, California	5000	Saltwater barrier	Operation to begin July 1995.
Alamitos Barrier Project, California	3000	Saltwater barrier	Permit pending (September 1996).

*From Geselbracht (1996).

Recycled water from the Whittier Narrows plant has been used at an average rate of 26,500 ac·ft/yr (33 million m^3/yr), which represents 16 percent of the total inflow to the Montebello Forebay groundwater basin (Nellor et al., 1985). After a 5-yr study of the health aspects of groundwater recharge, it was concluded that the project did not demonstrate any measurable adverse impacts on the groundwater or on the health of the population ingesting the groundwater (Nellor et al., 1985).

For small and decentralized systems, groundwater recharge occurs on an individual home basis, with the systems at Stinson Beach and Paradise, California, as examples. As described in Chap. 10, most groundwater recharge occurs by surface spreading (see Fig. 10-2); however, direct injection has been used in some cases, usually where suitable and affordable land is not available. Direct injection involves the pumping of the recycled water into the recharge zone through a well. The receiving groundwater is usually a well-confined aquifer, often a coastal aquifer with a threat of seawater intrusion. A listing of groundwater recharge projects using direct injection is presented in Table 12-42.

12-10 INDIRECT POTABLE REUSE

Indirect potable reuse is primarily used for large systems; however, because some remote decentralized systems in water-short areas may decide to employ it, the technology and representative examples are provided for completeness. Indirect potable reuse has been practiced in a planned manner at Windhoek, Namibia, since 1969. Examples of indirect potable reuse in the United States include an emergency system at Chanute, Kansas; studies at Denver, Colorado (Lauer and Rogers, 1996) and Tampa, Florida; an existing system at Clayton County, Georgia; and a planned system at San Diego, California.

Windhoek, Namibia

Planned potable water reuse was implemented in Windhoek in a very arid area of southwest Africa in 1969. The City of Windhoek constructed a 1.2-Mgal/d (4600-m^3/d) water repurification plant to augment the city's water supply. The secondary treatment plant was followed by maturation (aerobic) ponds, recarbonation, flotation removal of algae, foam fractionation, chemical coagulation, breakpoint chlorination, rapid sand filtration, and activated carbon filtration. During the 1970s the plant contributed 13.5 percent of the total water consumption during 6 to 8 months of operation annually, while meeting all recommended World Health Organization standards. By the late 1970s, following treatment plant improvements, the plant operated for about 200 d/yr, accounting for as much as 50 percent of the total potable supply during periods of critical water shortage. The system remains in operation and public health studies are continuing with no reported adverse health effects (U.S. EPA, 1992a and WCPH, 1996).

Chanute, Kansas

As the result of a severe 5-year drought in Kansas from 1952 to 1957, the City of Chanute was forced to implement an emergency potable water reuse system. Because of the severe water shortage, the city recycled its secondary effluent into the Neosha River, a little more than 1 mile upstream of the city's water treatment plant intake. Over a 5-month period the recirculation resulted in seven cycles. After several cycles, however, the recycled water contained publicly unacceptable levels of taste, odor, and color, and the practice was discontinued (U.S. EPA, 1992a).

Clayton County, Georgia

An indirect potable water reuse system was placed into operation in Clayton County, Georgia, in 1979. The system consists of secondary treatment followed by slow-rate land treatment, irrigation of 2650 ac (1070 ha) of forest. The shallow groundwater and runoff from the spray irrigation system recharges Pates Creek, which flows into the water supply reservoir for Clayton County. In 1987, the system treated 14.1 Mgal/d (53,370 m^3/d). Operating the land treatment has significantly increased the flow in Pates Creek, without violating water quality standards (Reed and Bastian, 1991).

San Diego, California

A significant indirect potable water reuse project is being planned and designed for the City of San Diego (Gagliardo et al., 1996). A complete treatment system is proposed that includes a water hyacinth secondary treatment system (see Chap. 9), tertiary filters, UV disinfection, and reverse osmosis as shown previously in Fig. 12-5*a*. The repurified water is to be discharged into a drinking water supply reservoir. The design flow is 30 Mgal/d (113,600 m^3/d).

A health effects study was conducted to determine if an advanced water treatment system could reliably reduce contaminants of public health concern to levels such that the health risks posed by any proposed potable use of the treated water are no greater than those associated with the present water supply (WCPH, 1996). The health advisory committee concluded that the health risk associated with the use of the repurified water as a raw water supply is less than or equal to that of the existing water supply (WCPH, 1996).

12-11 FUTURE OF WATER REUSE

Water reuse is expected to increase in the future for both centralized and decentralized wastewater management systems. For centralized systems, water reuse frees up potable water supply for municipal and industrial uses. Conversions of landscape irrigation, fire fighting supply, and other nonpotable uses are expected to increase as demand for potable supply increases. For individual and small decentralized systems, landscape irrigation will continue to be the principal reuse option. In isolated commercial and industrial facilities, toilet flushing in buildings with dual plumbing systems, and use for water features and landscape irrigation will continue. To maximize the reuse of treated wastewater at or near the point of generation, satellite reclamation plants will continue to increase in number. Satellite plants, either connected to downstream collection systems (as in Los Angeles County for their upstream reclamation plants), or as stand-alone facilities as planned for Paradise, CA, will employ a variety of technologies, depending on the particular reuse opportunities.

Improvements in technology, such as for membrane systems, will continue, and, as a result, more water reuse will occur, because of the lower costs of reclamation. Indirect potable reuse is expected to increase in larger communities because the infrastructure for water distribution already exists. Where repurified water is to be reused in applications involving the potential for direct human contact, it is anticipated that treatment with one of the membrane technologies before disinfection with either chlorine, ozone, or UV radiation, or with other appropriate technologies, will be required to reduce the potential risk associated with pathogenic organisms, both known and unknown, as described in Sec. 2-7 in Chap.2.

PROBLEMS AND DISCUSSION TOPICS

12-1. A community generates 300 ac-ft/yr of recycled water. Competing uses for the water include a 120-ac golf course and a 100-ac wetland. The water demand for the golf course is 3 ft/yr and the water demand for the wetland is 2 ft/yr. Can both demands be met?

12-2. Using the equations developed by Fair and Hatch and Rose, determine the headloss through a 24-, 30-, or 36-in sand bed. Assume that the sand bed is composed of spherical unisized sand with a diameter of 0.45, 0.5, or 0.6 mm, the porosity of the sand is 0.40, and the filtration velocity is 2, 4, or 6 gal/ft^2·min. The temperature is 18°C. Specific values to be selected by instructor.

12-3. The data in the following table were obtained from a pilot-plant study on the filtration-settled secondary effluent from an activated-sludge treatment plant. Using these data, estimate the length of run that is possible with and without the addition of polymer if the maximum allowable headloss is 10 ft, the filtration rate is 4.0 gal/ft^2·min, and the influent suspended solids concentration is 10 mg/L. Uniform sand with a diameter of 0.45 mm and a depth of 2 ft was used in the pilot filters.

Normalized filter removal data for problem 12-3

	Concentration ratio C/C_o			Concentration ratio C/C_o	
Depth, in	With polymer addition	Without polymer addition	Depth, in	With polymer addition	Without polymer addition
0	1.00	1.00	14	0.10	0.33
2	0.46	0.70	16	0.10	0.32
4	0.29	0.57	18	0.10	0.31
6	0.20	0.49	20	0.10	0.31
8	0.15	0.44	22	0.10	0.31
10	0.13	0.39	24	0.10	0.31
12	0.11	0.36			

12-4. The normalized suspended solids removal-ratio curves, given in the following figure, were derived from a filtration pilot-plant study conducted at an activated-sludge plant. Using these curves, estimate: (1) the headloss buildup as a function of run length and (2) the run length to a terminal headloss of 10 ft as a function of the filtration rate for each sand size. The filtration rate was 4 gal/ft^2·min.

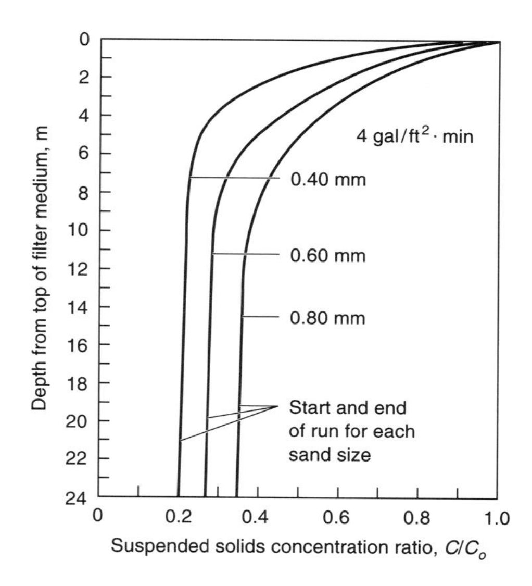

12-5. Compare the performance, energy needs, and reliability of conventional packed-bed filters, and microfilters. Cite at least two references.

12-6. A reverse osmosis membrane is operated at a pressure of 400 lb/in^2. Determine the cost of power, per acre-foot of water, if the combined efficiency of the pressurizing system (pump and motor) is 60 percent and the cost of power is $0.10/kWh.

12-7. Determine the amount of iron salt used as a coagulant in phosphorus removal that would be required to reduce the effluent phosphorus from 10 to 2 mg/L in the effluent from a community septic tank system.

12-8. Determine the amount of natural alkalinity needed for the treatment of settled effluent for phosphorus removal at an alum dosage of 45 mg/L if the flowrate is 0.1 Mgal/d. Estimate the amount of sludge produced, assuming a specific gravity of dry solids is 2.1 and that the aluminum hydroxide sludge contains 4 percent solids.

12-9. Determine the quantity of chlorine, in pounds per day, necessary to disinfect a daily average treated effluent flow of 1.0 Mgal/d and determine the size of the contact tank. Use a dosage of 15 mg/L, and size the chlorine contact chamber for a contact time of 30 min at a peak hourly flow, which is assumed to be 2.5 times the average flow.

12-10. Determine the amount of activated carbon that would be required per year to dechlorinate treated effluent containing a chlorine residual of 5 mg/L (as Cl_2) from a plant with an average flowrate of 1.0 Mgal/d. The chemical reaction is:

$$C + 2Cl_2 + 2H_2O \rightarrow 4HCl + CO_2$$

What dosage of sulfur dioxide would be required?

12-11. Design a low-pressure, low-intensity UV disinfection system, using three banks per channel, to treat a flow of 0.1, 0.25, 0.5, or 1.0 Mgal/d. Assume the average UV intensity, based on the PSS method of computation and the transmittance of the water, is 4, 8, or 10 mW/cm^2. Also assume the UV lamp spacing is 75 mm on centers; the diameter of the quartz sleeves encasing the UV lamp is 23 mm; the effective UV arc length is 1470 mm; the required UV dose is 80, 100, 120, or 140 $mW \cdot s/cm^2$; and the K value for estimating the headloss through the UV system is 2.2. Assume the required UV dose for permit requirements of 23 and 240 MPN/100 mL are 65 and 25 $mW \cdot s/cm^2$, respectively. Specific values to be selected by instructor.

12-12. Determine the power requirement for a UV installation that delivers 140 $mW \cdot s/cm^2$ for 6 months of the year and 100 $mW \cdot s/cm^2$ for the remainder of the year for a flow of 1 Mgal/d, assuming a loading of 27 L/W.

12-13. A crop is irrigated with recycled water whose salinity is expressed as EC_W = 1.0 dS/m. If the crop is irrigated to achieve a leaching fraction of 0.15, estimate the salinity of the deep percolate water and the appropriate leaching factor to maintain crop yield, assuming a significant loss of yield occurs when the TDS of the soil water exceeds 5000 mg/L.

12-14. Given the following waters with various chemical characteristics, determine the sodium adsorption ratio and comment on the suitability of the water for irrigation of clay soils. Water number to be selected by instructor.

	Water				
Constituent	**1**	**2**	**3**	**4**	**5**
Na	160	200	230	300	110
Mg	15	20	12	16	22
Ca	24	20	16	24	28
EC	1.0	1.1	0.6	1.8	0.8

12-15. Given the following analysis of effluent from a community recirculating gravel filter, determine whether an infiltration problem will develop in a clay soil if this water is used for irrigation.

$$\begin{aligned} Na &= 150 \text{ mg/L} \\ Mg &= 25 \text{ mg/L} \\ Ca &= 20 \text{ mg/L} \\ TDS &= 500 \text{ mg/L} \\ EC &= 0.75 \text{ dS/m} \end{aligned}$$

12-16. What are the important characteristics to be considered if septic tank effluent is to be applied in a reuse application involving drip irrigation?

12-17. What septic tank effluent characteristics would be of special importance if the effluent from an intermittent sand filter (ISF) is to be used for the culture of fish?

12-18. Should septic tank effluent discharging to soil absorption systems be considered disposal or reuse?

12-19. The City of San Diego has introduced the term "repurified" water to describe water that has undergone secondary and advanced treatment, including reverse osmosis and ion exchange. Do you believe that the use of the term *repurified* is appropriate? Defend your argument.

12-20. Prepare a one-page abstract of the article "Using Public Opinion Surveys to Measure Public Acceptance of a Recycled Water Program," by F. V. Filice (1996), *Water Reuse Conference Proceedings,* pp. 641–651, American Water Works Association, Water Environment Federation, San Diego.

REFERENCES

Amirtharajah, A. (1978) Optimum Backwashing of Sand Filters, *Journal of Environmental Engineering Division,* ASCE, Vol. 104, No. EE5, pp. 917–932.

Asano, T. (1985) *Artificial Recharge of Groundwater,* Butterworth Publishers, Boston.

Asano, T., and R. Mujeriego (1988) Evaluation of Industrial Cooling Systems Using Reclaimed Wastewater, *Water Science Technology,* Vol. 20, No. 10, pp. 163–174.

Asano, T., D. Richards, R. W. Crites, and G. Tchobanoglous (1992) Evolution of Tertiary Treatment Requirements in California, *Water Environment and Technology,* Vol. 4, No. 2, pp. 36–41.

Asano, T., and G. Tchobanoglous (1995) Drinking Repurified Wastewater, *Journal of Environmental Engineering,* ASCE, Vol. 121, No. 8.

AWWA (1992) *Guidelines for Distribution of Nonpotable Water,* American Water Works Association, California/Nevada Section.

AWWA (1996) *Membranes,* American Water Works Association, Denver, CO.

Ayers, R. S. (1977) Quality of Water for Irrigation, *Journal Irrigation Division ASCE,* Vol. 103, No. IR2, pp. 135–154.

Ayers, R. S., and D. W. Westcot (1985) *Water Quality for Agriculture,* FAO Irrigation and Drainage Paper No. 29, Revision 1, Food and Agriculture Organization of the United Nations, Rome.

Bishop, S. L., and B. W. Behrman (1976) Filtration of Wastewater Using Granular Media, paper presented at the 1976 Thomas R. Camp Lecture Series on Wastewater Treatment and Disposal, Boston Society of Civil Engineers, Boston.

Butterfield, C. T., E. Wattie, S. Megregian, and C. W. Chambers (1943) Influence of pH and Temperature on the Survival of Coliforms and Enteric Pathogens When Exposed to Free Chlorine, U.S. Public Health Service Report No. 58. Washington, D.C.

California DHS (1994) Wastewater Reclamation Criteria, Title 22, Division 4, California Administrative Code, California Department of Health Services, Sacramento, CA.

California Fertilizer Association (1995) *Western Fertilizer Handbook,* 8th ed., Interstate Publishers, Sacramento, CA.

California SWRCB (1990) *California Municipal Wastewater Reclamation in 1987,* California State Water Resources Control Board, Office of Water Recycling, Sacramento, CA.

Carman, P. C. (1937) Fluid Flow Through Granular Beds, *Transactions of Institute of Chemical Engineers,* London, Vol. 15, p. 150.

Cheryan, M. (1986) *Ultrafiltration Handbook,* Technomic Publishing Co., Inc., Lancaster, PA.

Chick, H. (1908) Investigation of the Laws of Disinfection, *Journal of Hygiene,* Vol. 8, p. 92.

Cleasby, J. L., and K. Fan (1982) Predicting Fluidization and Expansion of Filter Media, *Journal of Environmental Engineering Division,* ASCE, Vol. 107, No. EE3, pp. 455–472.

Collins, H. F., and R. E. Selleck (1972) Process Kinetics of Wastewater Chlorination, SERL Report No. 72-5, Sanitary Engineering Research Laboratory, University of California, Berkeley.

Crites, R. W. (1982) *Innovative and Alternative Treatment in Petaluma, California,* presented at the Hawaii Water Pollution Control Association Annual Conference, Honolulu.

Crites, R. W. (1984) *Wastewater Reuse by Golf Course Irrigation in California, Water Reuse Symposium III,* American Water Works Association, pp. 299–310, San Diego.

Crites, R. W., T. J. Mingee, and D. Richard (1988) *Advanced Wastewater Treatment in Del Norte County*, presented at the Annual Conference of the California Water Pollution Control Association, Sacramento.

Crites, R. W., D. Richard, and G. Tchobanoglous (1996) Wastewater Filtration Alternatives in California: Past History, Current Performance, and Future Trends, *1996 Water Reuse Conference Proceedings,* pp. 775–785, American Water Works Association, San Diego.

Crites, R. W., C. Lekven, S. Wert, and G. Tchobanoglous (1997) A Decentralized Wastewater System for a Small Residential Development in California, *Small Flows Journal,* Vol. 3, Issue 1, Morgantown, WV.

Darby, J., K. Snider, and G. Tchobanoglous (1991) UV Disinfection of Wastewater: A Pilot-Scale Study of Performance, *Proceedings of the 1991 Specialty Conference, Environmental Engineering,* ASCE, Reno, NV.

Darby, J. L., M. Heath, J. Jacangelo, F. Loge, P. Swain, and G. Tchobanoglous (1995) Comparative Efficiencies of UV Irradiation to Chlorination: Guidance for Achieving Optimal UV Performance, Project 91-WWD-1, Water Environment Research Foundation, Alexandria, VA.

Dharmarajah, A. H., and J. L. Cleasby (1986) Predicting the Expansion of Filter Media, *Journal of American Water Works Association,* Vol. 798, No. 12, pp. 66–76.

Emerick, R. W., and J. L. Darby (1993) Ultraviolet Light Disinfection of Secondary Effluents: Predicting Performance Based on Water Quality Parameters, *Proceedings of the Planning, Design, and Operation Effluent Disinfection Systems Specialty Conference,* Water Environment Federation, Whippany, NJ.

England, S. K., J. L. Darby, and G. Tchobanoglous (1994) Continuous-Backwash Upflow Filtration for Primary Effluent, *Water Environment Research,* Vol. 66, no. 66.

Fair, G. M. (1951) The Hydraulics of Rapid Sand Filters, *Journal of Institute of Water Engineers,* Vol. 5, p. 171.

Fair, G. M., J. C. Geyer, and D. A. Okun (1968) *Water and Wastewater Engineering,* Vol. 2, John Wiley & Sons, New York.

Fair, G. M., and L. P. Hatch (1933) Fundamental Factors Governing the Streamline Flow of Water Through Sand, *Journal American Water Works Association,* Vol. 25, No. 11, p. 1551.

Foust, A. S., L. A. Wenzel, C. W. Clump, L. Maus, and L. B. Andersen (1960) *Principles of Unit Operations*, John Wiley & Sons, New York.

Gagliardo, P. F., P. Findley, T. G. Richardson, and K. Weinberg (1996) Optimization of Reclamation and Repurification at San Diego North City, *1996 Water Reuse Conference Proceedings,* pp. 71–85, American Water Works Association, San Diego.

Geselbracht, J. (1996) Microfiltration/Reverse Osmosis Pilot Trials for Livermore, California, Advanced Water Reclamation, *1996 Water Reuse Conference Proceedings,* pp. 187–203, American Water Works Association, San Diego.

Haas, C. N., and S. B. Kara (1984) Kinetics of Microbial Inactivation by Chlorine: I. Review of Results in Demand-Free System, *Water Research* (G.B.), Vol. 18, p. 1443.

Hazen, A. (1905) *The Filtration of Public Water-Supplies,* 3rd ed., John Wiley & Sons, New York.

Hinesly, T. D., R. E. Thomas, and R. G. Stevens (1978) *Environmental Changes from Long-Term Land Application of Municipal Effluents,* U.S. Environmental Protection Agency, Washington, DC.

Irwin, J. (1994) Use of Membranes in Urban, Onsite Reclamation, A Case Study, *1994 Water Reuse Symposium Proceedings,* pp. 711–720, American Water Works Association, Water Environment Federation, Dallas, TX.

Jacques, R. S., and D. Williams (1996) Enhance the Feasibility of Reclamation Projects through Aquifer Storage and Recovery, *1996 Water Reuse Conference Proceedings,* pp. 15–28, American Water Works Association, San Diego.

Jacob, S. M., and Dranoff, J. S. (1970) Light Intensity Profiles in a Perfectly Mixed Photoreactor, *Journal of American Institute of Chemical Engineers,* Vol. 16, p. 359.

Johnson, J. D., and R. G. Qualls (1984) Ultraviolet Disinfection of a Secondary Effluent: Measurement of Dose and Effects of Filtration, EPA-600/2-84/160, PB85-114023, U.S. Environmental Protection Agency, Cincinnati, OH.

Johnson, R. D., R. L. Jones, T. D. Hinesly, and D. J. David (1974) *Selected Chemical Characteristics of Soils, Forages, and Drainage Water from the Sewage Farm Serving Melbourne, Australia,* prepared for the Department of the Army, Corps of Engineers, Washington, DC.

Johnson, W. D. (1990) Operating One of the World's Largest Urban Reclamation Systems—What We've Learned, *National Water Supply Improvement Association 1990 Biennial Conference Proceedings,* Vol. I, Walt Disney World Village, FL.

Jordan, E. J., and P. R. Senthilnathan (1996) Advanced Wastewater Treatment with Integrated Membrane BioSystems, presented at American Institute of Chemical Engineers 1996 Spring National Meeting, New Orleans.

Kozeny, J. (1927) Über Grundwasserbewegung, *Wasserkraft und Wasserwirtschaft*, Vol. 22, Nos. 4, 6, 7, 8, 10.

Lauer, W. C., and S. E. Rogers (1996) The Demonstration of Direct Potable Water Reuse: Denver's Pioneer Project, *1996 Water Reuse Conference Proceedings,* pp. 269–289, American Water Works Association, San Diego.

Leva, M. (1959) *Fluidization,* McGraw-Hill, New York.

Loge, F. J., R. W. Emerick, M. Heath, J. Jacangelo, G. Tchobanoglous, and J. L. Darby (1996a) Ultraviolet Disinfection of Wastewater Secondary Effluents: Prediction of Performance and Design, *Water Environment Research,* Vol. 68, No. 5, pp. 900–916.

Loge, F., J. L. Darby, and G. Tchobanoglous (1996b) UV Disinfection of Wastewater: A Probabilistic Approach to Design, *Journal of Environmental Engineering,* ASCE, Vol. 122, No. 12, pp. 1078–1084.

Loge, F., R. W. Emerick, C. R. Williams, W. Kido, G. Tchobanoglous, and J. Darby (1997) Impact of Particle Associated Coliform on UV Disinfection, *Proceedings of the Water Environment Federation 70th Annual Conference and Exposition,* Chicago, IL.

Morris, J. C. (1966) The Acid Ionization Constant of HOCL from 5°C to 35°C, *Journal of Physical Chemistry,* Vol. 70, p. 3798.

NWRI (1993) *UV Disinfection Guidelines for Wastewater Reclamation in California and UV Disinfection Research Needs Identification,* National Water Research Institute, prepared for the California Department of Health Services, Sacramento, CA.

Nellor, M. H., R. B. Baird, and J. R. Smyth (1985) Health Aspects of Groundwater, in T. Asano (ed.), *Artificial Recharge of Groundwater,* Butterworth Publishers, Boston.

O'Melia, C. R., and W. Stumm (1967) Theory of Water Filtration, *Journal of American Water Works Association,* Vol. 59, No. 11.

Page, A. L., and A. C. Chang (1985) Fate of Wastewater Constituents in Soil and Groundwater: Trace Elements, in *Irrigation with Reclaimed Municipal Wastewater: A Guidance Manual,* Lewis Press, Chelsea, MA.

Qualls, R. G., and J. D. Johnson (1983) Bioassay and Dose Measurement in UV Disinfection, *Applied and Environmental Microbiology,* Vol. 45, No. 3, pp. 872–877.

Reed, S. C., and R. Bastian (1991) Potable Water via Land Treatment and AWT, *Water Environment & Technology*, Vol. 3, No. 8, pp. 40–47.

Reed, S. C., and R. W. Crites (1984) *Handbook of Land Treatment Systems for Industrial and Municipal Wastes,* Noyes Publications, Park Ridge, NJ.

Richard, D., T. Asano, and G. Tchobanoglous (1992) *The Cost of Wastewater Reclamation in California,* Department of Civil and Environmental Engineering, University of California, Davis.

Richardson, J. F., and W. N. Zaki (1954) Sedimentation and Fluidisation, Part I, *Transactions of the Institute of Chemical Engineers,* Vol. 32, pp. 35–53

Roohk, D. L., T. L. Stephenson, and R. W. Crites (1989) Reclaimed Water Use in Mission Viejo, *California, Proceedings 16th Annual Water Resources Conference,* ASCE. Sacramento, CA.

Rose, H. E. (1945) An Investigation of the Laws of Flow of Fluids Through Beds of Granular Materials, *Proceedings Institute of Mechanical Engineers,* Vol. 153, p. 141.

San. Dist. LAC (1977) *Pomona Virus Study,* Sanitation Districts of Los Angeles County, prepared for the California State Water Resources Control Board. Sacramento, CA

Saunier, B. M. (1976) *Kinetics of Breakpoint Chlorination and Disinfection,* Ph.D. dissertation, University of California, Berkeley.

Saunier, B. M., and R. E. Selleck (1976) The Kinetics of Breakpoint Chlorination in Continuous Flow Systems, paper presented at the American Water Works Association Annual Conference, New Orleans.

Scheible, O. K. (1987) Developement of a Rationally Based Design Protocol for the Ultraviolet Light Disinfection Process, *Journal Water Pollution Control Federation,* Vol. 59, No. 1, pp. 25–31.

Suidan, M. T., and B. F. Severin (1986) Light Intensity Models for Annular UV Disinfection Reactors, *Journal American Institute of Chemical Engineers,* Vol. 32, No. 11, pp. 1902–1909.

Sung, R. D. (1974) Effects of Organic Constituents in Wastewater on the Chlorination Process, Ph.D. dissertation, Department of Civil Engineering, University of California, Davis, CA.

Tchobanoglous, G. (1969) *A Study of the Filtration of Treated Sewage Effluent,* Ph.D dissertation, Stanford University, Stanford, CA.

Tchobanoglous, G. (1988) Filtration of Secondary Effluent for Reuse Applications, Presented at the 61st Annual Conference of the Water Pollution Control Federation, Dallas, TX.

Tchobanoglous, G., and F. L. Burton (1991) *Wastewater Engineering, Treatment, Disposal, Reuse*, 3rd ed., McGraw-Hill, New York.

Tchobanoglous, G., and R. Eliassen (1970) Filtration of Treated Sewage Effluent, *Journal of Sanitation Engineering Division,* ASCE, Vol. 96, No. SA2.

Tchobanoglous, G., R. Hultquist, and F. Soroushian (1994) Ultraviolet Disinfection Guidelines for Wastewater Reclamation in California, *1994 Water Reuse Symposium Proceedings,* American Water Works Association, Water Environment Foundation, pp. 659–669, Dallas, TX.

Tchobanoglous, G., F. Loge, J. L. Darby, and M. Devries (1996) UV Design: Comparison of Probabilistic and Deterministic Design Approaches, *Water Science and Technology*, Vol. 33, pp. 251–260.

Tchobanoglous, G., and E. D. Schroeder (1985) *Water Quality: Characteristics, Modeling, Modification,* Addison-Wesley, Reading, MA.

U.S. EPA (1973) National Academy of Science—National Academy of Engineering, *Water Quality Criteria 1972: A Report of the Committee on Water Quality Criteria,* EPA-R3-73-033, U.S. Environmental Protection Agency, Washington, DC.

U.S. EPA (1980) *Long Term Effects of Slow-Rate Land Application of Municipal Wastewater: A Critical Summary,* U.S. Environmental Protection Agency, Washington, DC.

U.S. EPA (1981) *Process Design Manual: Land Treatment of Municipal Wastewater,* EPA 625/1-81-013, Center for Environmental Research Information, U.S. Environmental Protection Agency, Cincinnati, OH.

U.S. EPA (1986) *Design Manual: Municipal Wastewater Disinfection,* EPA/625/1-86/021, Office of Research and Development, Water Engineering Research Laboratory, Center for Environmental Research Information, U.S. Environmental Protection Agency, Cincinnati, OH.

U.S. EPA (1992a) *Manual—Guidelines for Water Reuse,* EPA/625/R-92/004, U.S. Environmental Protection Agency, Washington, DC.

U.S. EPA (1992b) *User's Manual for UVDIS, Version 3.1, UV Disinfection Process Design Manual,* EPA G0703, Risk Reduction Engineering Laboratory, U.S. Environmental Protection Agency, Cincinnati, OH.

U.S. EPA (1993) *Standards for the Use and Disposal of Sewage Sludge,* 40 CFR Part 503; U.S. Environmental Protection Agency, Washington DC.

Watson, H. E. (1908) A Note on the Variation of the Rate of Disinfection with Change in the Concentration of the Disinfectant, *Journal of Hygiene,* Vol. 8, p. 536.

WCPH (1996) Total Resource Recovery Project, Final Report, Western Consortium for Public Health, prepared for City of San Diego, Water Utilities Department, San Diego.

WEF (1996) *Wastewater Disinfection,* Manual of Practice FD-10, Water Environment Federation, Alexandria, VA.

White, G. C. (1992) *The Handbook of Chlorination and Alternative Disinfectants,* 3rd ed., Van Nostrand Reinhold, New York.

WPCF (1989) *Water Reuse Manual of Practice,* 2nd ed., Water Pollution Control Federation, Alexandria, VA.

CHAPTER 13

Effluent Disposal for Decentralized Systems

Effluent disposal options for decentralized systems range from soil absorption in conventional gravity leachfields to water reuse after high-tech membrane treatment. Individual onsite systems are the most prevalent wastewater management systems in the country. In this chapter, the various types of onsite wastewater systems, wastewater disposal options, site evaluation and assessment procedures, cumulative areal nitrogen loadings, nutrient removal alternatives, disposal of variously treated effluents in soils, design criteria for onsite disposal alternatives, design criteria for onsite reuse alternatives, correction of failed systems, and role of onsite management systems are described.

13-1 TYPES OF ONSITE SYSTEMS

While there are many types of onsite systems, most involve some variation of subsurface disposal of septic tank effluent. The four major categories of onsite systems are

- Conventional onsite systems
- Modified conventional onsite systems
- Alternative onsite systems
- Onsite systems with additional treatment

The most common onsite system is the conventional onsite system that consists of a septic tank and a soil absorption system (see Fig. 13-1). The septic tank, described in Chap. 5 along with the other forms of pretreatment, is the wastewater pretreatment unit used prior to onsite treatment and disposal. Modified conventional onsite systems include shallow trenches and pressure-dosed systems. Alternative

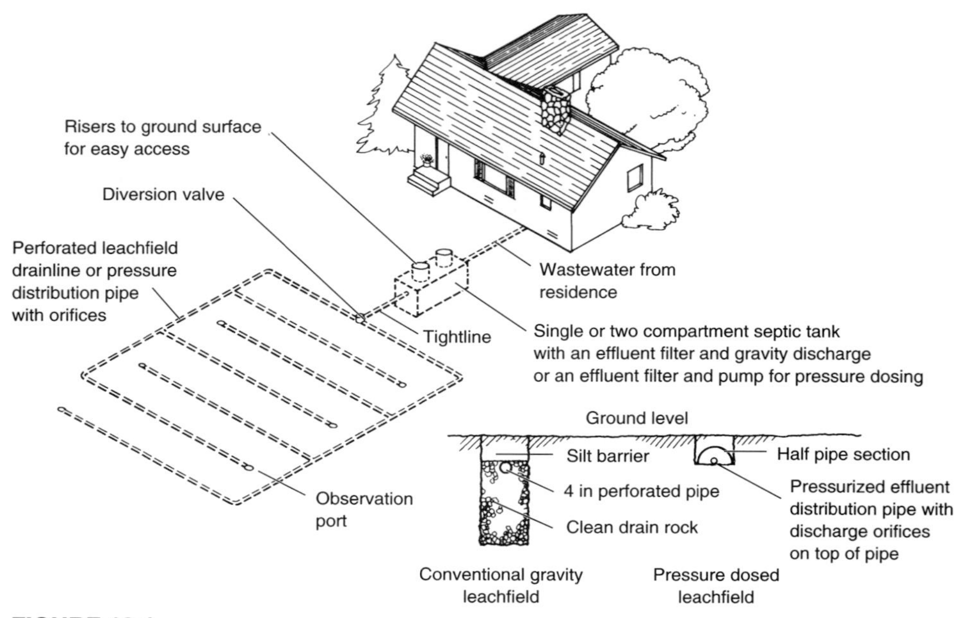

FIGURE 13-1
Conventional onsite system consisting of septic tank with gravity-fed soil absorption system.

TABLE 13-1
Types of onsite wastewater disposal/reuse systems

Disposal/reuse system	Remarks
Conventional systems	
Gravity leachfields—conventional trench	Most common system
Gravity absorption beds	
Modified conventional systems	
Gravity leachfields	
Deep trench	To get below restrictive layers
Shallow trench	Enhanced soil treatment
Pressure-dosed	
Conventional trench	To reach uphill fields
Shallow trench	Uphill and shallow sites
Drip application	Following additional treatment of septic tank effluent; to optimize use of available land area
Alternative systems	
Sand-filled trenches	Added treatment
At-grade systems	Less expensive than mounds
Fill systems	Import soil
Mound systems	
Evapotranspiration systems	Zero discharge
Evaporation ponds	See Chap. 8
Constructed wetlands	Requires a discharge or subsequent infiltration
Reuse systems	
Drip irrigation	Usually follows added treatment
Spray irrigation	Requires disinfection
Graywater reuse	
Other systems	
Holding tanks	Seasonal use alternative
Surface water discharge	Allowed in some states following added treatment

onsite disposal systems include mounds, evapotranspiration systems, and constructed wetlands. Additional treatment of septic tank effluent is sometimes needed, and intermittent and recirculating packed-bed filters, described in Chap. 11, are often the economical choice. Where further nitrogen removal is required, one or more of the alternatives for nitrogen removal (Sec. 13-5) may be considered. The types of disposal/reuse systems used for individual onsite systems are presented in Table 13-1. Typical examples of onsite treatment and disposal systems are shown in Fig. 13-2.

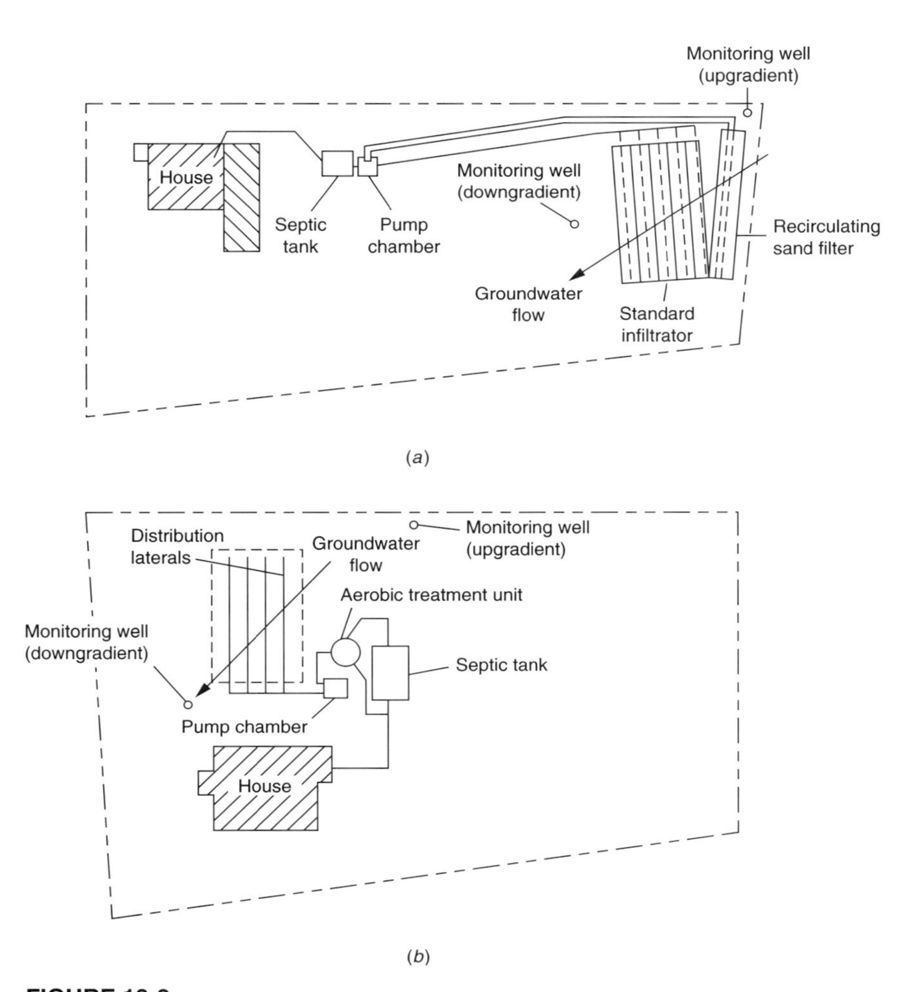

FIGURE 13-2
Typical onsite demonstration systems at Gloucester, Massachusetts, with septic tanks followed by: (*a*) recirculating sand filter and infiltration trenches, (*b*) aerobic treatment unit and pressure-dosed leachfield.

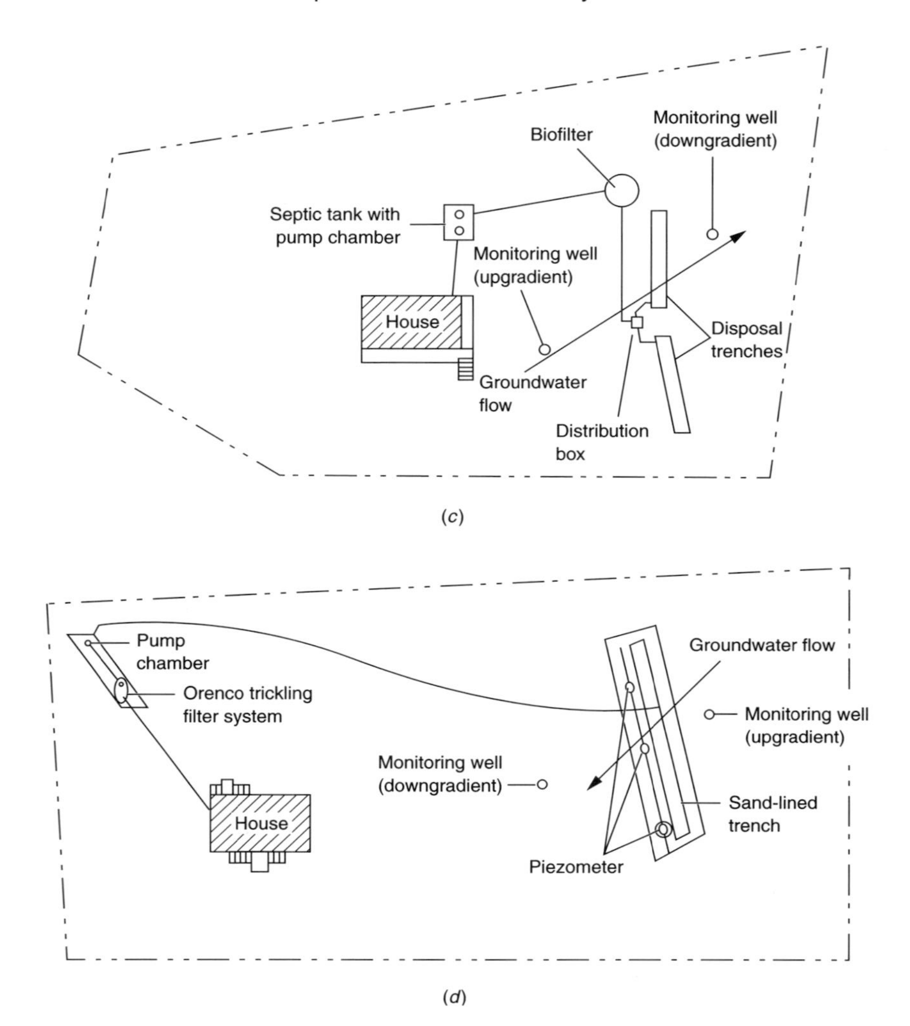

FIGURE 13-2 (*continued*)
Typical onsite demonstration systems at Gloucester, Massachusetts, with septic tanks followed by: (*c*) biofilter followed by conventional gravity leachfield, and (*d*) attached-growth reactor in septic tank followed by sand-lined leachfield.

13-2 EFFLUENT DISPOSAL/REUSE OPTIONS

Alternative effluent disposal systems (presented in Table 13-2) have been developed to overcome restrictive conditions such as:

- Very rapidly permeable soils
- Very slowly permeable soils
- Shallow soil over bedrock

TABLE 13-2

Selection of alternative effluent disposal methods under various site constraints[a]

Method	Site constraints											
	Soil permeability			Depth to bedrock			Depth to water table		Slope			Small lot size
	Very rapid	Rapid–moderate	Slow–very slow	Shallow and porous	Shallow and nonporous	Deep	Shallow	Deep	0–5%	5–15%	> 15%	
Leachfields (trenches)		×	×[b]			×		×	×	×	×	×[c]
Beds		×				×		×	×			×
Mounds	×	×	×	×	×	×	×	×	×	×		
Fill systems	×	×[d]	×[d]	×	×	×	×	×	×	×	×	×[c]
Sand-lined trenches or beds	×	×	×[b]			×		×	×	×[e]	×[e]	×[c]
Artificially drained systems		×				×	×		×	×	×[e]	
Evaporation infiltration lagoons		×	×[e]			×		×	×			
Evaporation lagoons (lined)[c,f]	×	×	×	×	×	×	×	×	×			
ET beds or trenches (lined)[c,f]	×	×	×	×	×	×	×	×	×	×[g]		
ETA beds or trenches[c]		×	×			×		×	×	×	×	×

[a]Adapted from U.S. EPA (1980).
[b]Construct only during dry soil conditions. Use trench configuration only.
[c]Flow reduction suggested.
[d]Only where surface soil can be stripped to expose sand or sandy loam material.
[e]Trenches only.
[f]High evaporation potential required.
[g]Recommended for south-facing slopes only.
× means that system can function effectively with that constraint.

- Shallow groundwater
- Steep slopes
- Groundwater quality restrictions
- Limited space

The alternatives for reuse of onsite system effluent include drip irrigation, spray irrigation, groundwater recharge, and toilet flushing. Drip irrigation is becoming more popular for water reuse and is described in this chapter. Spray irrigation is more suited to larger flows (commercial, industrial, and small community flows) and is described in detail in Chap. 10. Groundwater recharge is used in areas of deep permeable soils, as described in Chap. 10. Toilet flushing and other nonpotable water reuse are presented in Chap. 12.

13-3 SITE EVALUATION AND ASSESSMENT

The process of selecting a suitable onsite location for onsite disposal involves multiple steps of identification, reconnaissance, and assessment. The process begins with a thorough examination of the soil characteristics which include permeability, depth, texture, structure, and pore sizes. The nature of the soil profile and the soil permeability are of paramount concern in the evaluation and assessment of the site. Other important aspects of the site are depth to groundwater, site slope, existing landscape and vegetation, and surface drainage features. After a potential site has been located, the site evaluation and assessment proceeds, generally in two phases: preliminary site evaluation, and detailed site assessment (Burks and Minnis, 1994).

Preliminary Site Evaluation

The initial step in conducting a preliminary site evaluation is to determine the current and proposed land use and the expected flow and characteristics of the wastewater, and to observe the site characteristics. The next step is to gather information on the following characteristics:

- Soil depth
- Soil permeability (general or qualitative)
- Site slope
- Site drainage
- Existence of streams, drainage courses, or wetlands
- Existing and proposed structures
- Water wells
- Zoning
- Vegetation and landscape

Applicable Regulations

Once the pertinent data have been collected, the local regulatory agency should be contacted to determine the regulatory requirements. The tests needed for the Phase

TABLE 13-3
Typical regulatory factors in onsite systems

Factor	Unit	Typical value
Setback distances (horizontal, separation from wells, springs, surface waters, escarpments, site boundaries, buildings)	ft	See Table 13-8
Maximum slope for onsite disposal field	%	25-30
Soil characteristics:		
Depth	ft	2
Percolation rate	min/in	>1 to <120
Minimum depth to groundwater	ft	3
Septic tank (minimum size)	gal	750
Maximum hydraulic loading rates for leachfields	gal/ft^2·d	1.5
Maximum loading rates for sand filters	gal/ft^2·d	1.2

2 investigation, which can include depth to groundwater during the wettest period of the year, and permeability tests to determine water absorption rates, can also be determined at this time. A list of typical regulatory factors for onsite disposal is presented in Table 13-3.

Detailed Site Assessment

The important parameters that require field investigation are soil type, structure, permeability, and depth, and depth to groundwater. The use of backhoe pits, soil augers, piezometers, and percolation tests may be required to characterize the soil (see Fig. 13-3). Backhoe pits are useful to allow a detailed examination of the soil profile for soil texture, color, degree of saturation, horizons, discontinuities, and restrictions to water movement. An example of a backhoe pit for soil examination is shown in Fig. 13-4. Soil augers are useful in determining the soil depth, soil type, and soil moisture and many hand borings can be made across a site prior to the

FIGURE 13-3
Typical backhoe used to excavate soil examination pits.

FIGURE 13-4
Examination of soil profile by soil scientist in excavated backhoe pit.

siting of a backhoe pit location. Piezometers are occasionally required by regulatory agencies to determine the level and fluctuation of groundwater (NOWRA, 1996).

In most parts of the country, the results of percolation tests are used to determine the required size of the soil absorption area. The allowable hydraulic loading rate for the soil absorption system is determined from a curve or table that relates allowable loading rates to the measured percolation rate. A typical curve relating percolation rate to hydraulic loading rate for subsurface soil absorption systems is shown in Fig. 13-5.

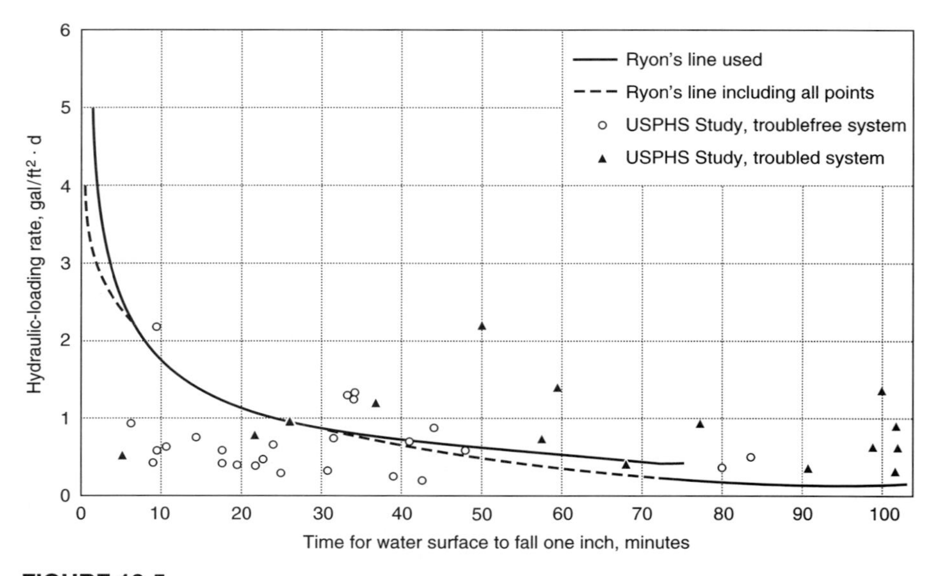

FIGURE 13-5
Permissible hydraulic loading rates for onsite soil absorption systems with various percolation rates (Winneberger, 1984).

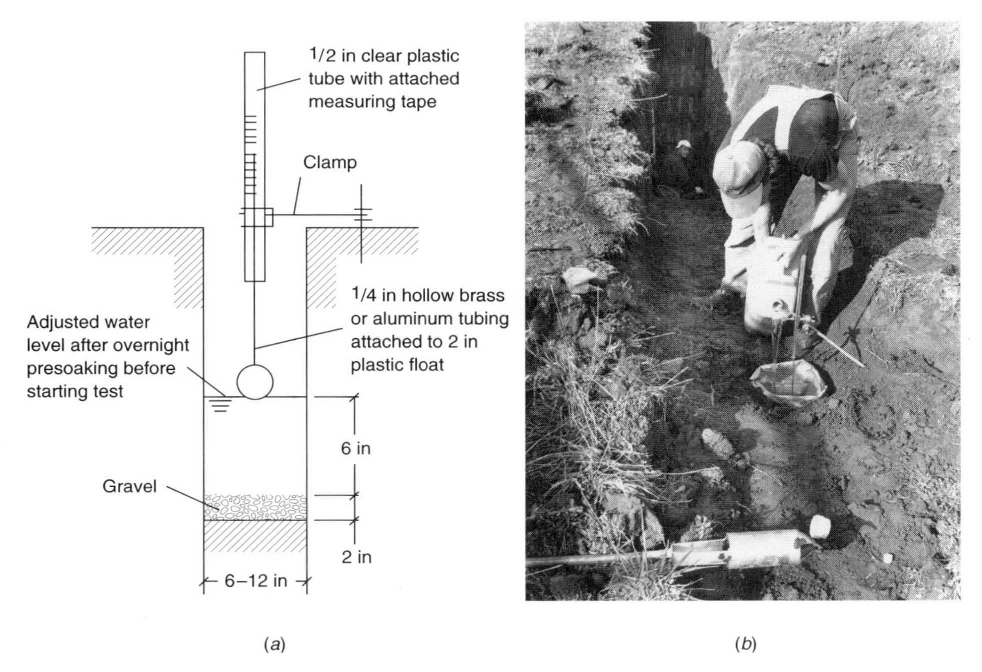

FIGURE 13-6
Definition sketch for the percolation test: (*a*) typical details of test and (*b*) field test setup in which water is being poured into a perforated paper bag to avoid splashing and unnecessary clogging of the sidewall surface area. The test technique depicted in *b* was developed by Winneberger (1984).

In the percolation test, test holes that vary in diameter from 4 to 12 in (100 to 300 mm) are bored in the location of the proposed soil absorption area. The bottom of the test hole is placed at the same depth as the proposed bottom of the absorption area (see Fig. 13-6). Prior to measuring the percolation rate, the hole should be soaked for a period of 24 h. Tests and acceptable procedures used by local regulatory agencies should be checked prior to site investigations.

Although used commonly, the percolation test results, because of the nature of the test, are not related to the performance of the actual leachfields. Many agencies and states are abandoning the test in favor of detailed soil profile evaluations. The percolation test is useful only in identifying soil permeabilities that are very rapid or very slow. Percolation tests should not be used as the sole basis for design of soil absorption systems because of the inherent inaccuracies.

Hydraulic Assimilative Capacity

For facilities that are designed for larger flows than those generated by individual households, or for sites where the hydraulic capacity is borderline within the local regulations, a shallow trench pump-in test or a basin infiltration test can be used. The test, known as an absorption test, has been developed for wastewater disposal

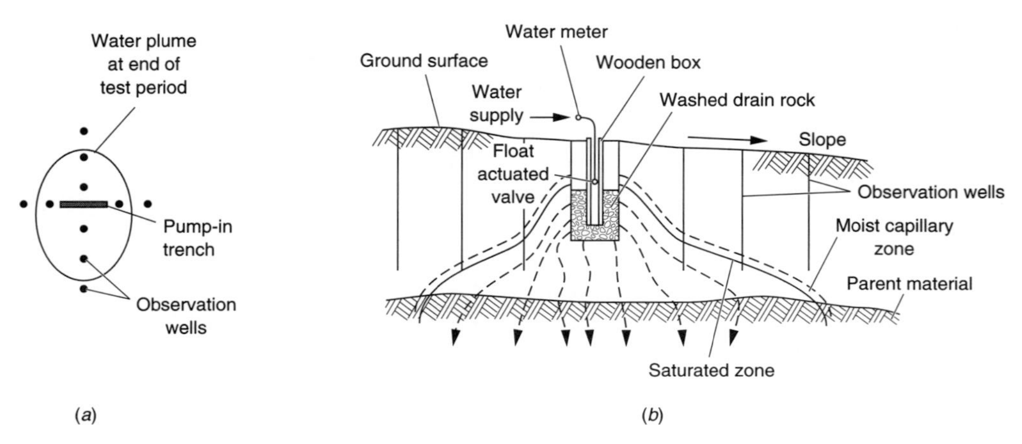

FIGURE 13-7
Definition sketch for hydraulic assimilation test to assess the suitability of a site for wastewater disposal: (*a*) plan view of the pump-in trench and observation wells and (*b*) typical section through long axis.

(Wert, 1997). The procedure allows an experienced person to determine the site absorption capacity. In the shallow trench pump-in test, a trench 6 to 10 ft (2 to 3 m) long is excavated to the depth of the proposed disposal trenches. Gravel is placed in a wooden box in the trench to simulate a leachfield condition. A constant head is maintained by using a pump, water meter, and float. The soil acceptance rate is then calculated by measuring the amount of water that is pumped into the soil over a period of 2 to 8 d. The method is illustrated in Example 13-1, and Fig. 13-7.

EXAMPLE 13-1. HYDRAULIC ASSIMILATION CAPACITY. Determine the soil acceptance rate for an onsite disposal system using the results of a pump-in test as defined in the sketch and the data given below.

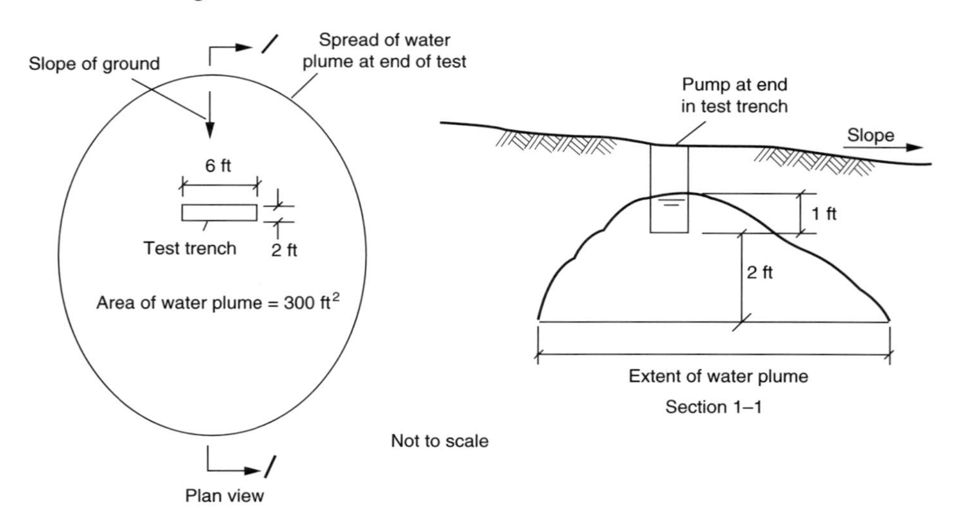

1. Trench dimensions = 6 ft long, 2 ft wide, and 3 ft deep.
2. Total extent of the water plume = 300 ft^2.
3. Water depth in the bottom of the trench is 1 ft.
4. Depth of water below the bottom of the trench at the periphery of the plume = 2 ft.
5. Height of the capillary zone = 1.5 ft.
6. Degree of saturation in the capillary zone = 33%.
7. Soil porosity = 0.35.
8. Water applied during the test = 3600 gal.
9. Total elapsed time = 96 h (4 d).

Solution

1. Account for the water remaining in the trench, the water in the saturated zone, and the water in the capillary fringe.
   a. Water in the trench = 6 ft × 2 ft × 1 ft = 12 ft^3. Volume in gallons = 12 ft^3 × 7.48 gal/ft^3 = 90 gal.
   b. Determine water in the saturated zone (refer to the definition sketch, Fig. 13-7). The volume of soil that is above the area defined by the plume can be considered as a truncated pyramid (less the volume of the trench). The volume of water in the truncated pyramid is determined by multiplying the volume of the soil by the soil porosity:

$$\text{Saturated zone water} = [((300\ \text{ft}^2 + 12\ \text{ft}^2)/2) \times 3\ \text{ft} - 12\ \text{ft}^2] \times 0.35$$
$$= 159.6\ \text{ft}^3 \times 7.48\ \text{gal/ft}^3 = 1194\ \text{gal}$$

   c. Determine water in the capillary zone (equal to area of plume × depth × porosity × degree of saturation):

$$\text{Water in the capillary zone} = 300\ \text{ft}^2 \times 1.5\ \text{ft} \times 0.35 \times 0.33$$
$$= 52\ \text{ft}^3 \times 7.48\ \text{gal/ft}^3 = 389\ \text{gal}$$

   d. Total water remaining (sum of $a + b + c$) = 90 + 1194 + 389 = 1673 gal.

2. Determine the water absorbed by the underlying soil:

$$\text{Water applied} - \text{water remaining} = \text{water absorbed}$$
$$3600\ \text{gal} - 1673\ \text{gal} = 1927\ \text{gal}$$

3. Determine the acceptance rate:

$$\text{Soil acceptance rate} = \frac{\text{total absorbed water}}{\text{area} \times \text{time}}$$
$$= \frac{1927\ \text{gal}}{(300\ \text{ft}^2)(4\ \text{d})}$$
$$= 1.6\ \text{gal/ft}^2\cdot\text{d}$$
$$= (1.6\ \text{gal/ft}^2\cdot\text{d})/(7.48\ \text{gal/ft}^3) = 0.21\ \text{ft/d}$$

Comment. Assuming that the long-term acceptance rate of the biomat that forms with septic tank effluent application is about 0.25 gal/ft^2·d, the measured acceptance rate of 1.6 gal/ft^2·d will not be the limiting factor in the design of a conventional leachfield.

Groundwater Mounding

For multiple-home, commercial, and industrial applications, groundwater mounding evaluations should be conducted. A procedure developed by Finnemore and Hantzsche (1983) can be used. Equation (13-1) can be used along with the definition sketch in Fig. 13-8:

$$h = H + Z_m/2 \tag{13-1}$$

where h = distance from boundary to mid-point of the long-term mound, ft
H = height of stable groundwater table above impermeable boundary, ft
Z_m = long-term maximum rise of the mound, ft

$$Z_m = \left(\frac{QC}{A}\right)\left(\frac{L}{4}\right)^n\left(\frac{1}{Kh}\right)^{0.5n}\left(\frac{t}{S_y}\right)^{1-0.5n} \tag{13-2}$$

where Q = average flow, ft³/d
A = area of disposal field, ft²
C = constant, see Table 13-4
L = length of disposal field, ft
K = horizontal permeability of soil, ft/d
n = exponent, see Table 13-4
S_y = specific yield of receiving soil, see Table 13-5
t = time since beginning of wastewater application, d

The specific yield of a receiving soil or aquifer is the volume of the drainable voids divided by the total volume of the soil or aquifer. The value of K can be determined in the field with a standard hydrogeologic (slug) test. In general, the horizontal con-

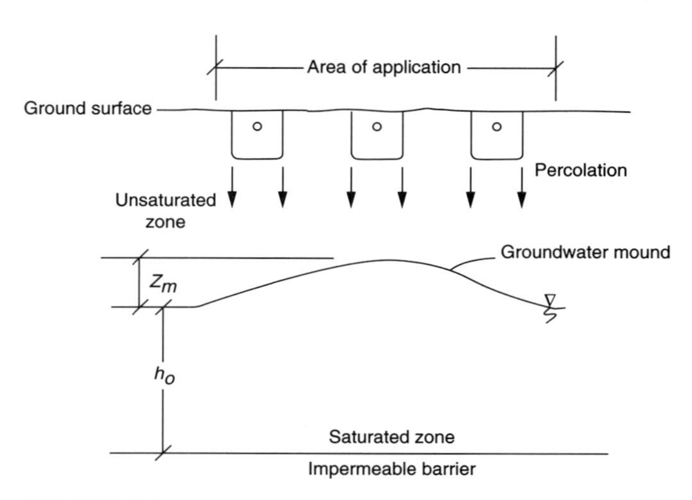

FIGURE 13-8
Definition sketch for groundwater mounding equation to estimate mounding below leachfields and seepage beds.

TABLE 13-4
Groundwater mounding equation constants for Eq. 13-2*

Length-to-width ratio of disposal field (*L/W*)	*C*	*n*
1	3.4179	1.7193
2	2.0748	1.7552
4	1.1348	1.7716
8	0.5922	1.7793

*From Finnemore and Hantzsche (1983).

TABLE 13-5
Representative values of specific yield*

Material	Specific yield, %
Gravel, coarse	23
Gravel, medium	24
Gravel, fine	25
Sand, coarse	27
Sand, medium	28
Sand, fine	23
Silt	8
Clay	3
Sandstone, fine-grained	21
Sandstone, medium-grained	27
Limestone	14
Dune sand	38
Loess	18
Peat	44
Schist	26
Siltstone	12
Till, predominantly silt	6
Till, predominantly sand	16
Till, predominantly gravel	16
Tuff	21

*After Johnson (1967).

ductivity of most soils is significantly higher than the vertical conductivity (Reed and Crites, 1984). As a consequence, a conservative estimate of the mound rise will result if the vertical conductivity is used in Eq. (13-2). At very short time periods Eq. (13-2) will predict high, yet conservative, levels of mound rise. A period of 10 years is recommended (Finnemore and Hantzsche, 1983). An iterative approach, illustrated in Example 13-2, is necessary because a value of Z_m for Eq. (13-1) must be assumed to determine h so that Eq. (13-2) can be solved for Z_m.

EXAMPLE 13-2. GROUNDWATER MOUND HEIGHT ANALYSIS. Determine the mound height rise for a commercial onsite disposal system. The flow is 2000 gal/d and the design for the disposal bed is a square with length and width of 100 ft. Use a value of H of 40 ft, a value of $K = 2$ ft/d, and a Z_0 of 5 ft. The underlying soil is a fine sand. Use a time value of 3650 d (10 yr).

Solution

1. From Table 13-4 for a length-to-width ratio of 1, select a value of $C = 3.4179$ and a value of $n = 1.7193$.

2. Convert flow units.

$$Q = (2000 \text{ gal/d})/(7.48 \text{ gal/ft}^3) = 267 \text{ ft}^3/\text{d}$$

3. Make an initial estimate of Z_m and calculate h. Try $Z_m = 6$ ft. Using Eq. (13-1),

$$h = H + Z_m/2$$
$$= 40 + 6/2 = 43 \text{ ft}$$

4. Select a value of 0.23 for the specific yield S_y from Table 13-5 for fine sand.

5. Use Eq. (13-2) to calculate Z_m:

$$Z_m = \left(\frac{QC}{A}\right)\left(\frac{L}{4}\right)^n\left(\frac{1}{Kh}\right)^{0.5n}\left(\frac{t}{S_y}\right)^{1-0.5n}$$

$$= \left(\frac{267 \times 3.4179}{100^2}\right)\left(\frac{100}{4}\right)^{1.7193}\left(\frac{1}{2 \times 43}\right)^{0.8597}\left(\frac{3650}{0.23}\right)^{0.1404}$$

$$= (0.09125)(253.2)(0.02172)(3.888) = 1.95 \text{ ft}$$

6. Use the value of 1.95 ft for Z_m in Eq. (13-1) to conduct the second iteration.
   a. Calculate h:

$$h = 40 + 1.95/2 = 40.98$$

   b. Recalculate Z_m using $h = 40.98$ in Eq. (13-2):

$$Z_m = (0.0913)(253.2)(0.0226)(3.89) = 2.03 \text{ ft}$$

Comment. After 10 years, the groundwater mound is estimated to rise 2.03 ft and will be 2.97 ft below the surface. With this expected rise, it would be prudent to examine options such as a rectangular disposal site with the long axis perpendicular to the groundwater gradient, or increase the bed area.

In using Eq. (13-1) it is acceptable for the first iteration to assume that $h = H$, especially when H is relatively large (more than 30 ft or so). The calculation in Eq. (13-2) is sensitive to the groundwater depth H when the depth is less than 20 ft or so.

13-4 CUMULATIVE AREAL NITROGEN LOADINGS

As described in Chaps. 2 and 3, nitrogen forms can be transformed when released to the environment. Because the oxidized form of nitrogen, nitrate nitrogen, is a public health concern in drinking water supplies, the areal loading of nitrogen is important.

Nitrogen Loading from Conventional Effluent Leachfields

The nitrogen loading from conventional leachfields depends on the density of housing and the nitrogen in the applied effluent. The impact of the nitrate nitrogen on groundwater quality depends on the nitrogen loading, the water balance, and the background concentration of nitrate nitrogen.

To determine the nitrogen loading, the following procedure is suggested:

1. Determine the wastewater loading rate. From Chap. 4, the unit generation factor is multiplied by the density of the housing units per acre. For example, 150 gal/household times 4 houses per acre yields 600 gal/d·ac.
2. Determine the nitrogen concentration in the applied effluent. Use 60 mg/L.
3. Calculate the nitrogen loading. Multiply the nitrogen concentration by the wastewater loading:

$$L_N = L_W C_N F \tag{13-3}$$

where N loading = lb/ac·d (kg/ha·d)
L_W = wastewater loading, gal/ac·d (m^3/ha·d)
C_N = nitrogen concentration, mg/L (g/m^3)
F = conversion factor, 8.34 lb/[10^6 gal·(mg/L)] (kg/10^3 g)

4. In this example the nitrogen loading is

$$L_N = (600 \text{ gal/ac·d})(60 \text{ mg/L})(8.34)(10^{-6})$$

$$= 0.30 \text{ lb/acre·d}(135 \text{ gal/ac·d})$$

Cumulative Nitrogen Loadings

The loadings of nitrate nitrogen to the groundwater are reduced by denitrification in the soil column. As indicated in Chap. 10, denitrification depends on the carbon available in the soil or the percolating wastewater, and on the soil percolation rate. For sandy, well-drained soils, the denitrification fraction is 15 percent (Hantzsche and Finnemore, 1992). For heavier soils or where high groundwater or slowly permeable subsoils reduce the rate of percolation, the denitrification fraction can be estimated at 25 percent. The percolate nitrate concentration can be calculated from

$$C_{N_p} = C_{N_e}(1 - f) \tag{13-4}$$

where C_{N_p} = nitrate nitrogen in the leachfield percolate, mg/L
C_{N_e} = nitrogen concentration in the applied effluent, mg/L
f = denitrification decimal fraction (0.15 to 0.25)

Calculation of the nitrogen loading from onsite systems is illustrated in Example 13-3.

EXAMPLE 13-3. NITROGEN LOADING RATE IN ONSITE SYSTEMS. A local environmental health ordinance limits the application of septic tank effluent on an areal basis to 45 g/ac·d.

Determine the housing density with conventional septic tank effluent soil absorption systems that will comply with the ordinance. Assume a total nitrogen content in the septic tank effluent of 70 mg/L (see Table 4-15) and a household wastewater generation of 150 gal/d.

Solution

1. Determine the acceptable nitrogen loading rate in lb/ac·d:

$$L_N = (45 \text{ g/ac·d})/(454 \text{ g/lb}) = 0.099 \text{ lb/ac·d}$$

2. Calculate the corresponding wastewater application rate using Eq. (13-3):

$$\text{Wastewater loading} = \frac{\text{nitrogen loading}}{\text{nitrogen conc.} \times 8.34}$$

$$L_W = \frac{0.099 \text{ lb/ac·d}}{(70 \text{ mg/L})(8.34 \text{ lb/gal})(10^{-6})}$$

$$= 169.9 \text{ gal/ac·d}$$

3. Determine the number of households per acre:

$$\text{Households/ac} = (169.9 \text{ gal/ac·d})/(150 \text{ gal/d}) = 1.13$$

4. Calculate the minimum lot size for compliance:

$$\text{Lot size} = 1/1.13 = 0.88 \text{ ac}$$

Comment. This ordinance would be very conservative. If a 25 percent denitrification fraction were recognized in the ordinance, the nitrogen loading rate would be increased to 60 g/ac·d.

13-5 ALTERNATIVE NUTRIENT REMOVAL PROCESSES

Alternative nutrient removal processes have been and continue to be developed for the cost-effective control of nutrients from onsite systems. Nitrogen removal is the most critical of the nutrients because nitrogen can have public health effects as well as eutrophication and toxicological impacts.

Nitrogen Removal

Removal of nitrogen is a critical issue in most onsite disposal systems. Onsite nitrogen removal processes include intermittent sand filters and recirculating granular medium filters (Piluk and Hao, 1994) (Chap. 11), septic tank with attached-growth reactor (internal trickling filters in septic tanks), the RUCK system, and ion exchange.

Septic tank with attached-growth reactor. This system involves a small trickling filter unit placed above the septic tank. Septic tank effluent, which is pumped over the filter, is nitrified as it passes through and over the plastic medium. The system is shown schematically in Fig. 13-9. A number of experimental units

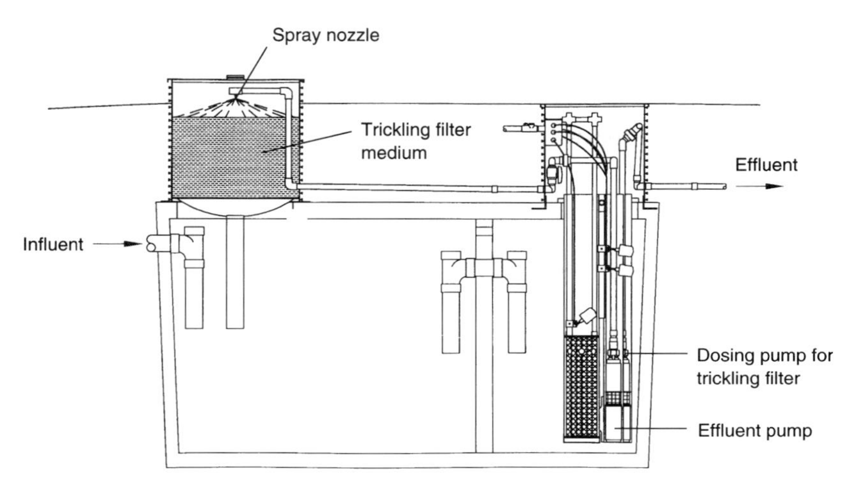

FIGURE 13-9
Septic tank with attached-growth reactor for the removal of nitrogen (courtesy Orenco Systems, Inc.).

have been installed in septic tanks. The best performance with plastic trickling filter medium has been achieved with a hydraulic loading rate of 2.5 gal/min (9.5 L/min) over a 3-ft (0.9-m) -deep unit containing hexagonally corrugated plastic with a surface area of 67 ft^2/ft^3 (226 m^2/m^3). A total nitrogen removal of 78 percent has been reported with an effluent nitrogen concentration of less than 15 mg/L (Ball, 1995). The performance of these systems is summarized in Table 13-6. Alternative filter media that have been tested include the foam medium used in the Waterloo filter and the textile chips used in the textile bioreactor (see Chap. 11).

RUCK system. The RUCK system is a proprietary system that is a variation on the ISF system involving the separation of graywater from blackwater. The process is intended to provide significant (80 percent) nitrogen removal (Laak, 1986). The flow diagram for the RUCK system is shown in Fig. 13-10. Blackwater from toilets, sinks, and showers is treated in one septic tank and then passed through an ISF. The graywater from the kitchen and laundry passes through the second septic tank and

TABLE 13-6
Performance of onsite attached-growth reactors*

	Applied septic tank effluent, mg/L reactor		Attached-growth reactor effluent, mg/L	
Constituent	Range	Typical	Range	Typical
BOD	90–214	125	5–90	25
TSS	30–40	40	5–40	30
Total N	37–109	68	8–30	15

*From Ball (1996).

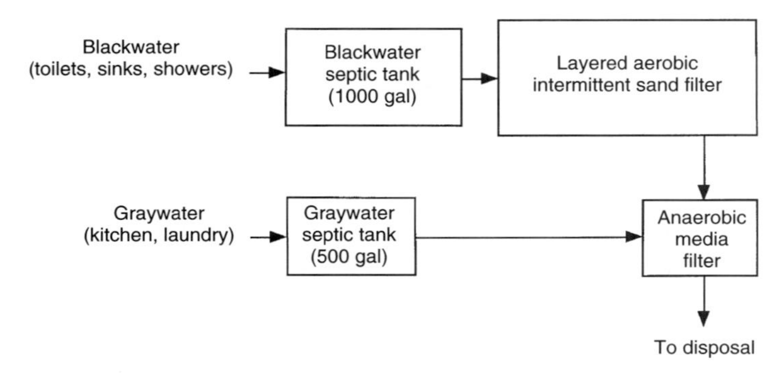

FIGURE 13-10
Flow diagram for RUCK system for the removal of nitrogen (adapted from Laak, 1986).

then mixes with the sand filter effluent. The ammonia nitrogen that is nitrified in the ISF is then denitrified in the anaerobic media filter. If the BOD-to-nitrogen ratio is too low (less than 2:1), supplemental organics, such as soap or methanol, can be added.

RSF-2 system. In the RSF-2 system, a recirculating sand filter is used for nitrification and is combined with an anaerobic filter for denitrification (Sandy et al., 1988). A flow diagram for the RSF-2 system is presented in Fig. 13-11. Septic tank effluent is discharged to one end of a rock storage filter which is directly below and in the same compartment as the RSF. Septic tank effluent flows horizontally through the rock and enters a pump chamber at the other end. The septic tank effluent is pumped over the RSF where it is nitrified. Filtrate is collected from near the top of the rock storage filter into a second pump chamber and is returned to the anaerobic environment of the septic tank where raw wastewater can serve as a carbon source for denitrification. A portion of effluent from the second pump chamber is discharged for disposal.

Experiments with the RSF-2 system produced nitrogen removals of 80 to 90 percent. Total nitrogen concentrations in the effluent ranged from 7.2 to 9.6 mg/L (Sandy et al., 1988). The rock storage zone, filled with 1.5-in (38-mm) rock, was effective in promoting denitrification.

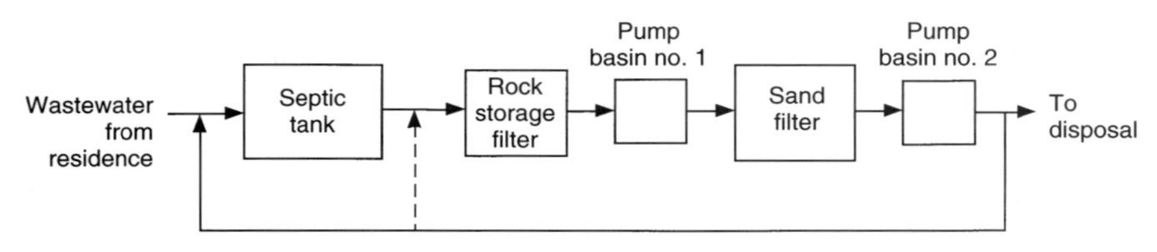

FIGURE 13-11
Flow diagram for RSF-2 system for the removal of nitrogen.

An alternative modification is to add the fixed medium (plastic, textile sheets) for biomass growth into the recirculation tank. Nitrified effluent from the recirculating sand filter is mixed with the incoming septic tank effluent and flows past the attached biomass, where any residual dissolved oxygen is consumed rapidly and the nitrate is denitrified using the organic matter in the septic tank effluent as the carbon source.

Other nitrogen removal methods. Other nitrogen methods that have been conceptualized include ammonia removal by ion exchange and nitrogen removal by denitrification in soil trenches. Biological nitrogen removal processes are described in Chap. 7. Attempts have been made to remove ammonia by ion exchange using zeolite at Los Osos, California, and other locations (Nolte and Associates, 1994). The attempts have been generally unsuccessful to date because of inadequate volumes of zeolite used and the high cost of frequent regeneration or replacement of the ion exchange medium.

Phosphorus Removal

Phosphorus removal is seldom required for onsite systems; however, when it is required, the soil mantle is the most cost-effective place to remove and retain phosphorus (see Chap. 10). Attempts to remove phosphorus in peat beds have usually been unsuccessful unless iron or limestone is present or added to the bed. In Maryland, the use of iron filings plowed into the peat bed was successful in removing phosphorus.

13-6 DISPOSAL OF VARIOUSLY TREATED EFFLUENTS IN SOILS

The disposal of partially treated wastewater into soils involves two major considerations: (1) treatment of the effluent so that it does not contaminate surface water or groundwater and (2) hydraulic flow of the effluent through the soil and away from the site. Pretreatment of the raw wastewater affects the degree of treatment that the soil-aquifer must achieve after the pretreated effluent is applied to the soil absorption system. Treatment of wastewater in soil has been long recognized (Reed and Crites, 1984). The soil is a combined biological, chemical, and physical filter. Wastewater flowing through soil is purified of organic and biological constituents, as described in Chap. 10.

Septic tank effluent has sufficient solids and organic matter to form a biological mat ("biomat") in the subsurface, particularly if gravity-flow application is used. More highly treated effluent and pressure-dosed application results in little, if any, biomat formation and the flow through the soil is inhibited only by the hydraulic conductivity of the soil. Allowable hydraulic loading rates for variously treated effluents are presented in Table 13-7.

TABLE 13-7
Allowable hydraulic loading rates for variously treated effluent*

	Allowable hydraulic loading rates			Mass loading rate, $g/m^2 \cdot d$		
Type of effluent	**in/d**	**$gal/ft^2 \cdot d$**	**mm/d**	**BOD_5**	**TSS**	**TKN**
Restaurant septic tank effluent†	0.12	0.07	3	2.4	0.9	0.24
Domestic septic tank effluent	0.4	0.25	10	1.5	0.8	0.55
Graywater septic tank effluent	0.6	0.37	15	1.8	0.6	0.22
Domestic aerobic unit effluent	0.8	0.50	20	0.7	0.8	0.30
Domestic sand filter effluent	3.0	1.87	76	0.3	0.75	0.75

*Adapted from Siegrist (1988).
†Increased from Siegrist's values for BOD (800 mg/L), TSS (300 mg/L), and TKN (80 mg/L) and derated HLR from 4 mm/d to 3 mm/d.

13-7 DESIGN CRITERIA FOR ONSITE DISPOSAL ALTERNATIVES

Gravity-flow leachfields are the most common type of onsite wastewater disposal. This type of onsite disposal functions well for sites with deep, relatively permeable soils, where groundwater is deep, and the site is relatively level.

Gravity Leachfields

Septic tank effluent flows by gravity into a series of trenches or beds for subsurface disposal. Trenches are usually shallow, level excavations that range in depth from 1 to 5 ft (0.3 to 1.5 m) and in width from 1 to 3 ft (0.3 to 0.9 m). The bottom of the trench is filled with 6 in (150 mm) of washed drainrock. The 4-in (100-mm) perforated distribution pipe is next placed in the center of the trench. Additional drainrock is placed over the top of the distribution pipe, followed by a layer of barrier material, typically building paper or fabric. The purpose of the barrier material is to prevent migration of fines from the backfill into the drainrock and avoid clogging of the drainrock by the clay or silt particles. A typical cross section of a leachfield trench is shown in Fig. 13-12.

The infiltrative surfaces in a leachfield trench are the bottom and the sidewalls. However, as a clogging layer of biological solids, or biomat, develops, the infiltration through the bottom of the trench decreases and the sidewalls become more effective. The long-term route for water passage is through the bottom and sidewalls of the trench.

Bed systems consist of an excavated area or bed with perforated distribution pipes that are 3 to 6 ft (0.9 to 1.8 m) apart. The route for water passage out of the bed is through the bottom. Bed systems can also use infiltration chambers, which create underground caverns over the soil's infiltrative surface and therefore do not need the gravel or barrier material.

Leaching chambers constructed out of concrete are open-bottomed shells that replace perforated pipe and gravel for distribution and storage of the wastewater.

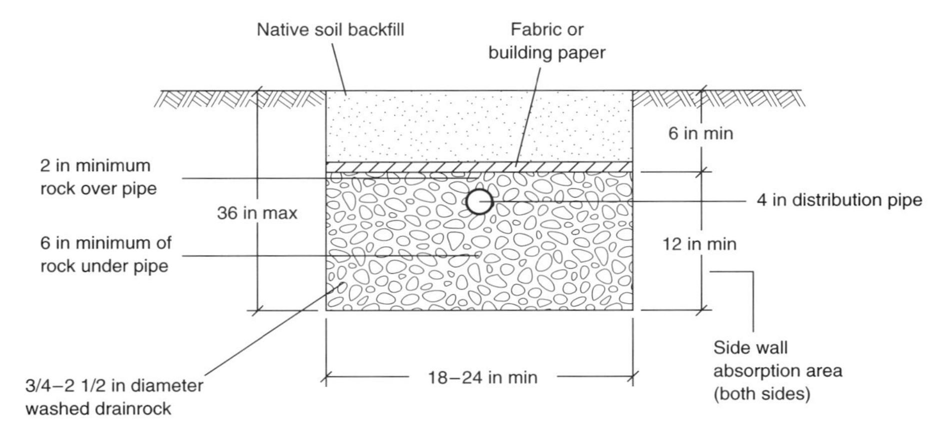

FIGURE 13-12
Typical cross section through conventional leachfield trench.

The chambers interlock to form an underground cavern over the soil. Wastewater is discharged into the cavern through a central weir, trough, or splash plate and is allowed to flow over the infiltrative surface in any direction. Access holes in the top of the chambers allow the surface to be inspected and maintained as necessary. A large number of leaching chamber systems have been installed in the northeastern United States. A typical leaching infiltrator system is shown in Fig. 13-13.

Typical criteria for siting of leachfield systems are presented in Table 13-8. Loading rates for trench and bed systems can be based on percolation test results and regulatory tables, on soil characteristics, or a combination of both. Disposal field

FIGURE 13-13
Infiltrator used for gravelless infiltration (unit here used to cover a recirculating sand filter).

TABLE 13-8
Design considerations in siting leachfields*

Item	Criteria
Landscape form†	Level, well-drained areas, crests of slopes, convex slopes are most desirable. Avoid depressions, bases of slopes, and concave slopes unless suitable surface drainage is provided.
Slope†	0–25 %. Slopes in excess of 25 % can be used, but construction equipment selection is limited.
Typical horizontal setbacks‡	
Water supply wells	50–100 ft
Surface waters, springs	50–100 ft
Escarpments, cuts	10–20 ft
Boundary of property	5–10 ft
Building foundations	10–20 ft
Soil	
Unsaturated depth	2–4 ft (0.6–1.2 m) of unsaturated soil should exist between the bottom of the disposal field and the seasonally high water table or bedrock.
Texture	Soils with sandy or loamy textures are best suited. Gravelly and cobbley soils with open pores and slowly permeable clay soils are less desirable.
Structure	Strong granular, blocky, or prismatic structures are desirable. Platey or unstructured massive soils should be avoided.
Color	Bright, uniform colors indicate well-drained, well-aerated soils; dull, gray, or mottled soils indicate continuous or seasonal saturation and are unsuitable.
Layering	Soils exhibiting layers with distinct textural or structural changes should be evaluated carefully to ensure that water movement will not be severely restricted.
Swelling clays	Presence of swelling clays requires special consideration in construction, and may be unsuitable if extensive.

*Adapted from U.S. EPA (1980).
†Landscape position and slope are more restrictive for seepage beds because of the depth of cut on the upslope side.
‡Intended only as a guide. Safe distance varies from site to site, depending on local codes, topography, soil permeability, groundwater gradients, geology, etc.

loading rates recommended by the U.S. EPA for design, based on bottom area, for various types of soils and observed percolation rates are shown in Table 13-9.

The most conservative criterion on which to base the loading rate is to assume that the percolation rate through the soil will eventually be reduced to coincide with the percolation rate through the biomat. On this assumption, the hydraulic loading rate is 0.125 $gal/ft^2 \cdot d$ (5 $L/m^2 \cdot d$), based on the trench bottom and sidewall area (Winneberger, 1984).

TABLE 13-9
Recommended rates of wastewater application for trench and bed bottom areas[a]

Soil texture	Percolation rate, min/in	Application rate, gal/ft^2·d[b,c]
Gravel, coarse sand	<1	Not suitable[d]
Coarse to medium sand	1–5	1.2
Fine sand, loamy sand	6–15	0.8
Sand loam, loam	16–30	0.6
Loam, porous silt loam	31–60	0.45
Silty clay loam, clay loam[e,f]	61–120	0.2
Clays, colloidal clays	>120	Not suitable[g]

[a] From U.S. EPA (1980).
[b] Rates based on septic tank effluent from a domestic waste source. A factor of safety may be desirable for wastewaters of significantly different strength or character.
[c] May be suitable for sidewall infiltration rates.
[d] Soils with percolation rates less than 1 min/in may be suitable for septic tank effluent if a 2-ft layer of loamy sand or other suitable soil is placed above or in place of the native topsoil.
[e] These soils are suitable if they are without significant amounts of expandable clays.
[f] Soil easily damaged during construction.
[g] Alternative pretreatment may be needed and alternative disposal (wetlands or evapotranspiration systems) may be required.

Where the site soils contain significant amounts of clay, it is suggested that the disposal field be divided into two fields and that the two fields be used alternately every 6 months. When two fields are used, the actual hydraulic loading rate for the field in operation is 0.25 gal/ft^2·d (10 L/m^2·d).

Shallow Gravity Distribution

Shallow leachfields (see Fig. 13-14) offer the benefits of lower cost and higher biological treatment potential because the upper soil layers have the most bacteria and fungi for wastewater renovation (Reed and Crites, 1984). The State of Oregon recently allowed the use of gravelless leachfield trenches that are 10 in (250 mm) deep and 12 in (300 mm) wide (Ball, 1994).

Pressure-Dosed Distribution

Pressure dosing can be done with either a dosing siphon or a pump. A pressure distribution system has the advantages over gravity distribution of providing a uniform small dose to the entire absorption area, promoting unsaturated flow, and providing a consistent drying/reaeration period between doses.

Pressure-dosed distribution can allow the absorption site to be at a higher elevation from the septic tank and will also allow a shallow (6- to 12-in) distribution network. With screened septic tank effluent or sand filter effluent, the distribution

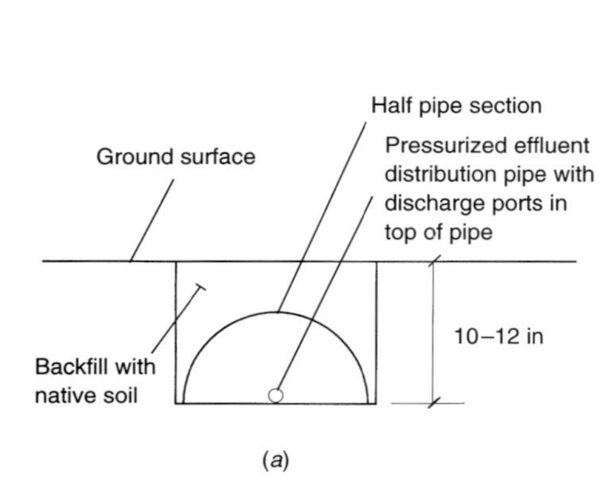

(a)

(b)

FIGURE 13-14
Shallow gravelless leachfield trench (a) definition sketch and (b) View of an excavated shallow gravelless leachfield trench with a 4-in (100-mm) half pipe section used to cover the effluent distribution pipe.

system can use 0.125-in (3-mm) orifices, typically spaced 1 to 2 ft (0.3 to 0.6 m) apart. For unscreened septic tank effluent the orifice size is typically 0.25 in (6 mm).

The spacing and sizing of orifices should be uniform because the objective of pressure dosing is to provide uniform distribution with unsaturated flow beneath the pipe. Pressure dosing is less effective in soils with very low permeabilities. Procedures for hydraulic analysis of pressure distribution systems are presented in Chap. 11.

Imported Fill Systems

Fill systems involve importing suitable off-site soils and placing them over the soil absorption area to overcome limited depth of soil or limited depth to groundwater. Care must be taken in selecting suitable soil to use in a fill system, and in the timing and conditions of importing the soil. Several conditions must be satisfied to construct a successful fill system:

1. Native soil should be scarified prior to import of fill.
2. The fill should be placed when the soil is dry.
3. The fill material should also be dry to prevent compaction.
4. The first 6 in (150 mm) of fill should be mixed thoroughly with the native soil.

At-Grade Systems

The concept of the at-grade system was developed in Wisconsin as an intermediate system between conventional in-ground distribution and the mound system. The aggregate or drainrock is placed on the soil surface (at grade) and a soil cap is added over the top.

Typically, the area for the at-grade system is tilled, the drainrock is placed on the tilled area, distribution pipe is positioned within the drainrock, synthetic fabric is spread over the drainrock, and final soil cover (12 in or 300 mm) is placed over the system. At-grade systems do not need the 24 in (600 mm) of sand that mounds have and are, therefore, less expensive.

Mound Systems

Mound systems are, in effect, bottomless intermittent sand filters. Components of a typical mound, as shown in Fig. 13-15, include a 24-in (600-mm) layer of sand, clean drainrock, distribution laterals, barrier material, and the soil cap. Mounds are pressure-dosed, usually 4 to 6 times per day. Mounds were first developed by the North Dakota Agricultural College in the late 1940s. They were known as NODAK systems and were designed to overcome problems with slowly permeable soils and areas that had high groundwater tables (Ingham, 1980; WPCF, 1990). Mounds may be used on sites that have slopes up to 12 percent, provided the soils are permeable. If the native soils are slowly permeable, the use of mounds should be restricted to slopes of less than 6 percent.

The design of mound systems is a two-step process. Percolation tests are conducted on the native soils on the site at the depth at which the mound base will exist. The values of the measured percolation rate are correlated to the design infiltration rate in Table 13-10. The infiltration rate is then used to calculate the base area of the mound.

The second step is to design the mound section. On the basis of the type of material used to construct the mound, the area of the application bed in the mound is determined. Mound fill materials are listed in Table 13-11 along with the corresponding design infiltration rate for determining the bed area (Otis, 1982).

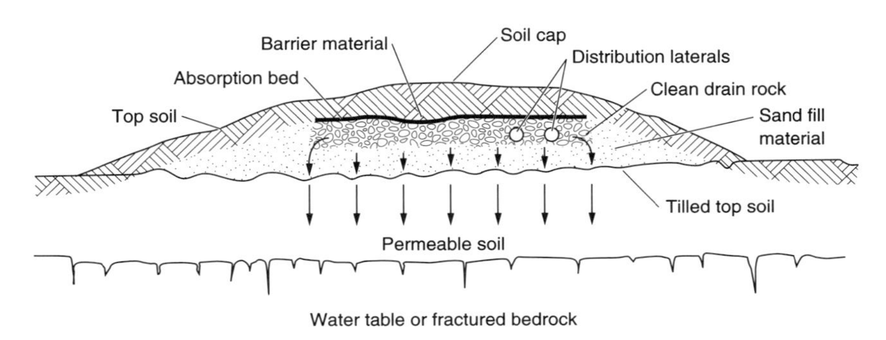

FIGURE 13-15
Typical cross section through mound effluent disposal system.

TABLE 13-10
Infiltration rates for determining base area of mound*

Native onsite soil	Percolation rate, min/in	Infiltration rate, $gal/ft^2 \cdot d$
Sand, sandy loam	0–30	1.2
Loam, silt loams	31–45	0.75
Silt loams, silty clay loams	46–60	0.50
Clay loams, clay	61–120	0.25

*From U.S. EPA (1980).

TABLE 13-11
Mound fill materials and infiltration rates*

Material	Characteristics, % by weight	Infiltration rate, $gal/ft^2 \cdot d$
Medium sand	>25%, 0.25–0.2 mm <30–35%, 0.05–0.25 mm <5–10%, 0.002–0.05 mm	1.2
Sandy loam	5–15% clay	0.6
Sand/sandy loam	88–93% sand	1.2

*From U.S. EPA (1980).

Artificially Drained Systems

Sometimes a high groundwater condition can be overcome by draining the groundwater away from the site. High groundwater tables in the area of the soil absorption fields may be artificially lowered by vertical drains or underdrains. Underdrains can be perimeter drains, used for level sites and sites up to 12 percent in slope, or curtain drains (upslope side only) for sites with slopes greater than 12 percent (see Fig. 13-16) (Nolte and Associates, 1992).

Constructed Wetlands

Constructed wetlands can be used for onsite treatment as well as onsite disposal/reuse. As described in Chap. 9, constructed wetlands can be either the free-water-surface type or the subsurface-flow type. For onsite systems in close proximity to children, the subsurface flow wetlands are most appropriate. A large number of subsurface wetlands have been constructed and placed in operation in Louisiana, Arkansas, Kentucky, Mississippi, Tennessee, Colorado, and New Mexico. These systems serve single-family dwellings, public facilities and parks, apartments, and commercial developments (Burgan and Sievers, 1994 and Reed, 1993).

There are two methods that are recommended for onsite wetlands: the TVA method and the plug-flow method (Reed, 1993). The TVA method (Steiner and

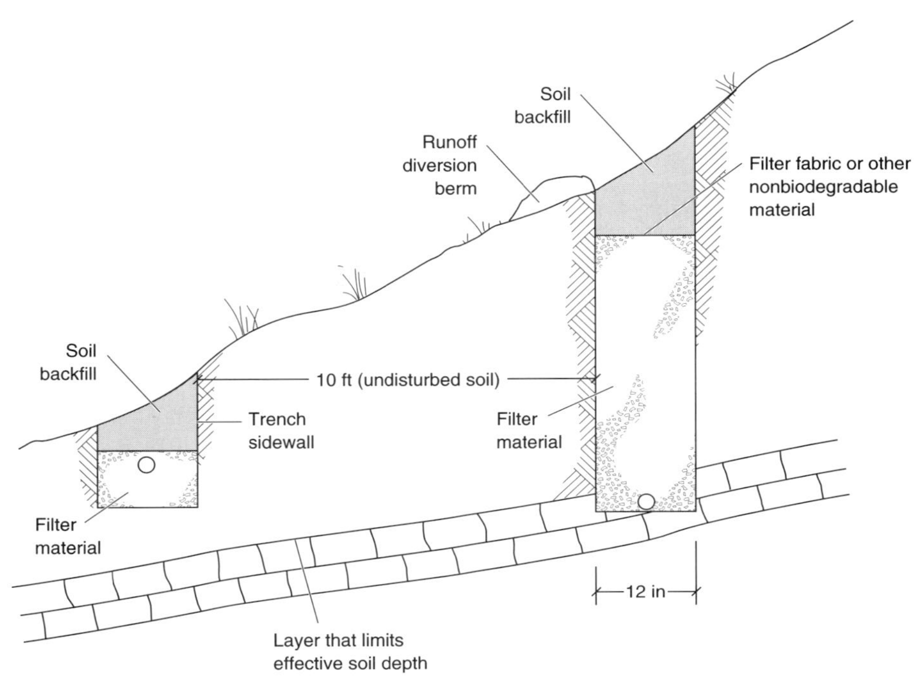

FIGURE 13-16
Artificially drained system for groundwater control.

Watson, 1993) uses a design approach in which the factors controlling the hydraulic performance of the bed and the BOD loading on the entry zone into the wetland are considered. The wetlands consist of two equal-sized wetland cells, operated in series. The first cell is lined with a 20- to 30-mil polyethylene or PVC liner. Washed gravel with an average size of 0.5 in (13 mm) is placed 1 ft (0.3 m) deep. The second cell is not lined and can therefore serve as a disposal cell for treated wastewater. The detention time in a TVA wetland is 3 to 4 days, which is adequate for BOD removal but may not be adequate for ammonia removal.

The plug-flow method, described by Reed (Reed, 1993; Reed et al., 1995), results in a 2.8-d detention time for a 90 percent removal of BOD. When nitrification is necessary, a 6-d detention time is recommended. If total nitrogen removal is required, a 10-d detention time is recommended for cold climates. The plug-flow method assumes similar sized gravel as the TVA method; however, a 2-ft (0.6-m) depth is recommended. An example of an onsite constructed wetland is shown in Fig. 13-17. Design of an onsite constructed wetland is illustrated in Example 13-4.

EXAMPLE 13-4. ONSITE CONSTRUCTED WETLAND. Design an onsite constructed wetland for a single residence with a flow of 300 gal/d. Use a rock depth of 18 in, a flow depth of 15 in, and a rock porosity of 35 percent. Assume no nitrogen removal is required and that a drip irrigation system will be designed to follow the wetlands.

Solution

1. Compute the area needed using the TVA method:

$$A = 1.3 \text{ ft}^2/\text{gal·d } (300 \text{ gal/d}) = 390 \text{ ft}^2$$

2. Compute the area needed using the plug-flow method. Use a detention time of 2.8 d.
   a. Calculate the volume of water in the rock that is needed:

$$\text{Volume} = 300 \text{ gal/d } (2.8 \text{ d}) = 840 \text{ gal}$$
$$= 840 \text{ gal}/(7.48 \text{ gal/ft}^3) = 112.3 \text{ ft}^3$$

   b. Calculate the volume of rock:

$$\text{Water volume/rockporosity} = 112.3/0.35 = 320 \text{ ft}^3$$

   c. Calculate the area of the rock:

$$A = \text{volume/depth} = 320 \text{ ft}^3/1.25 \text{ ft} = 256 \text{ ft}^2$$

3. Compare the two areas. The TVA method, in this case, requires the largest area. Because there is no requirement for nitrogen reduction, the area calculated with the plug-flow method (256 ft^2) should be selected.

Comment. Use a single-cell SF wetland with a 16-ft by 16-ft configuration.

Evapotranspiration Systems

In arid climates, evapotranspiration (ET) systems can be used for effluent disposal. Effluent from the septic tank is applied through perforated pipes to a sand bed underlaid by a liner. The sand depth is typically 24 to 30 in (0.6 to 0.75 m). Bernhart recommends a sand depth of 18 in (0.45 m) (Bernhart, 1973). The surface of the sand bed is covered with a shallow layer of topsoil, which can be planted to water tolerant vegetation. Treated wastewater is drawn up through the sand by capillary forces and

(*a*)

(*b*)

FIGURE 13-17
Examples of onsite constructed wetlands: (*a*) established reed vegetation for single home and (*b*) newly constructed and planted system for 14 homes near Burlington, Vermont (Crites).

by the plant roots and it is evaporated or transpired to the atmosphere. A fine sand (0.1 mm) is recommended to maximize the capillary rise. Observation wells are used to monitor the depth of water in the sand beds. The design of an evapotranspiration system is illustrated in Example 13-5.

The ET system can also be designed without a liner, and the resultant system is referred to as an ETA (evapotranspiration-absorption) system. The ETA approach can be used where percolation is acceptable and possible. An ETA system is similar to an at-grade system, except for the addition of surface vegetation.

Both ET and ETA systems are designed for the hydraulic loading rate. For ET systems, the hydraulic loading rate is the minimum monthly net evapotranspiration rate for at least 10 years of record. For ETA systems, the minimum monthly percolation rate is added to the minimum ET rate to determine the design hydraulic loading rate. The bed area for ET and ETA systems can be determined by

$$A = \frac{Q}{\mathrm{ET} - \mathrm{Pr} + P} \tag{13-5}$$

where A = bed area, ft^2 (m^2)
Q = annual flow, $\mathrm{ft}^3/\mathrm{yr}$ (m^3/yr)
ET = annual potential evapotranspiration rate, ft/yr (m/yr)
Pr = annual precipitation rate, ft/yr (m/yr)
P = annual percolation rate, ft/yr (m/yr)

For ET systems, the percolation rate is zero. For ETA systems, the percolation rate should be determined based on long-term saturated flow conditions.

EXAMPLE 13-5. DESIGN OF AN EVAPOTRANSPIRATION SYSTEM. Design an evapotranspiration system for a commercial establishment with a design flow of 900 gal/d. The annual lake evaporation rate is 45 in/yr and the precipitation rate for the wettest year in ten is 19 in/yr.

Solution

1. Convert the daily flow to an annual flow:

$$Q = 900 \text{ gal/d } (365 \text{ d/yr}) = 328,500 \text{ gal/yr } (1/7.48 \text{ gal/ft}^3) = 43{,}917 \text{ ft}^3/\text{yr}$$

2. Calculate the wastewater hydraulic loading rate:

$$\begin{aligned} L_W &= \mathrm{ET} - \mathrm{Pr} \\ &= 45 - 19 = 26 \text{ in/yr} = 2.17 \text{ ft/yr} \end{aligned}$$

3. Calculate the required bed area:

$$\begin{aligned} A &= \frac{Q}{L_W} \\ &= \frac{43{,}917 \text{ ft}^3/\text{yr}}{2.17 \text{ ft/yr}} \\ &= 20{,}238 \text{ ft}^2 \end{aligned}$$

Comment. A factor of safety, typically 15 to 20 percent, should be added to the bed area to account for variations in precipitation and flowrate.

13-8 DESIGN CRITERIA FOR ONSITE REUSE ALTERNATIVES

Reuse alternatives for onsite systems include drip irrigation and spray irrigation. Other reuse alternatives that include toilet flushing and other urban nonpotable uses are described in Chap. 12.

Drip Irrigation

Drip irrigation technology has advanced over the years to where nonclog emitters are available for both surface and subsurface uses. Sand filter and other high-quality effluent can be used in drip irrigation of landscape and other crops. Periodic chlorination of the drip tubing has been found to be necessary to avoid clogging growths in the distribution lines and emitters (Fig. 13-18).

Modern drip emitters have been designed so that they cannot be clogged by roots. For example, the Geoflow emitters have been treated with a herbicide to protect them from root intrusion. The emitters are designed with a turbulent flow path to minimize clogging from suspended solids. These emitters operate at a flowrate of 1 to 2 gal/h with openings 0.06 to 0.07 in (1.5 to 1.8 mm) in diameter. The drip irrigation system usually requires 15 to 25 lb/in^2 pressure. It may be necessary to flush the lines and to apply periodic doses of chlorine for control of clogging from bacterial growth.

A typical onsite drip irrigation system consists of emitter lines placed on 2-ft (0.6-m) centers with a 2-ft (0.6-m) emitter spacing. This spacing is typical for sandy and loamy soils. Closer spacings of 15 to 18 in (0.4 to 0.45 m) are used on clay soils where lateral movement of water is restricted. The emitter lines are placed at depths

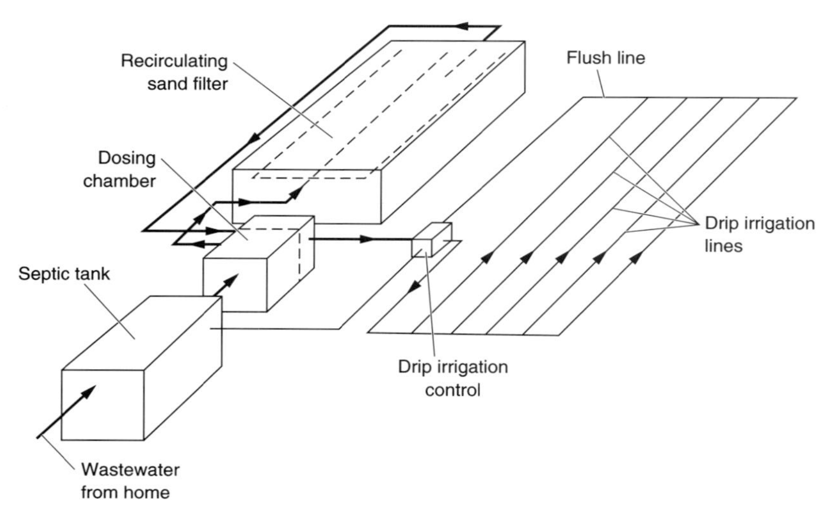

FIGURE 13-18
Example of onsite drip irrigation system.

of 6 to 10 in (150 to 250 mm). The design of a drip irrigation system is illustrated in Example 13-6.

EXAMPLE 13-6. DESIGN OF A DRIP IRRIGATION SYSTEM. Design a drip irrigation system for the reuse of treated effluent from Example 13-5. Use a design infiltration rate of 0.25 gal/ft^2·d.

Solution

1. Determine the area needed for irrigation:

$$A = (300 \text{ gal/d})/(0.25 \text{ gal/ft}^2\cdot\text{d}) = 1200 \text{ ft}^2$$

2. Lay out the 1200 ft^2 as a 40- by 30-ft rectangle.
3. Select a spacing of the drip emitter lines of 2 ft. Use 20 emitter lines that are 30 ft long.
4. Use 1-gal/h emitters, spaced at 3-ft intervals. Calculate the number of emitters:

$$30 \text{ ft per line/3 ft spacing} = 10 \text{ emitters per line}$$

$$20 \text{ lines} \times 10 \text{ emitters per line} = 200 \text{ emitters}$$

5. Calculate the flow discharged from 200 emitters:

$$\text{Flow} = 200 \text{ emitters} \times 1 \text{ gal/h} = 200 \text{ gal/h}$$

6. Calculate the time of operation per day:

$$(300 \text{ gal/d})/(200 \text{ gal/h}) = 1.5 \text{ h/d}$$

7. Select a pump for the application. The pump must be able to supply (300 gal/d) × (1.5 h × 60 min/h) = 3.3 gal/min at a minimum pressure of 20 lb/ft^2.

Comment. The emitters should be buried at a depth of 10 in.

Spray Irrigation

The use of spray irrigation for onsite disposal is relatively limited except in areas where housing density is low and other less expensive alternatives are not appropriate. Flows need to exceed 3 to 5 gal/min (11 to 19 L/min) to operate most single sprinklers. This relatively high flow generally means that spray irrigation is better suited to flows from an industrial, commercial, or institutional facility. In addition, for residential onsite systems the additional treatment may need to include sand filtration and disinfection. The details of spray irrigation site assessment and design are presented in Chap. 10 in the discussion of slow rate land treatment.

Graywater Systems

In older homes and in areas where water conservation/reuse is practiced because of water shortages or lack of wastewater disposal capacity, the laundry water and other nontoilet wastewater is often reused or disposed of separately from the "black" water that goes into the septic tank. The graywater includes organics, nutrients, and pathogens; however, it is perceived as being a benign source of wastewater that can be reused directly for landscape irrigation. Local health departments have allowed graywater reuse in rural areas, but have often denied graywater reuse in urban areas. Recent regulations specify safe and acceptable methods of onsite reuse of graywater (California Resources Agency, 1994). California's graywater standards are now part of the state plumbing code, making it legal to use graywater everywhere in California.

13-9 CORRECTION OF FAILED SYSTEMS

The failure of a subsurface onsite disposal system is defined as the inability of the system to accept and absorb the design flow of effluent at the expected rate. When failure occurs soon after the system is put into operation, the failure may be the result of poor construction (Winneberger, 1984), poor design, or unanticipated high groundwater, or a combination of the three. If high groundwater is the problem, a curtain drain or other drainage improvements may be necessary.

Use of Effluent Screens

When failure occurs after several years of successful operation, the reasons may be unanticipated flow increases, solids carryover from the septic tank, or biological clogging at the infiltration surfaces. A comparison of current flows to design flows should be made to ensure that hydraulic overloading is not the problem. The septic tank should be checked for scum and sludge layers (see Chap. 5) and pumped, if necessary. The use of Orenco or Zabel (or equivalent) effluent screens will minimize the discharge of suspended solids to the disposal system. If biological clogging is occurring, drying for a period of months, rehabilitation using oxidants such as hydrogen peroxide, or upgrading the septic tank effluent using a sand filter or equivalent treatment may be tried.

Use of Hydrogen Peroxide

For rehabilitation, a procedure developed at the University of Wisconsin uses hydrogen peroxide, a very strong oxidizing agent, to destroy the organic deposition and restore the infiltration capacity (Harkin and Jawson, 1977). Lysimeter work at the University of New Hampshire was successful in rehabilitating sandy and loamy sand soil infiltration systems that had clogged because of the buildup of organic material. A 30 percent solution of hydrogen peroxide and water was successful in all cases (Bishop and Logsdon, 1981). A weaker solution of 7.5 percent hydrogen peroxide was successful only for sandy soils. The loading rates for hydrogen peroxide were 0.25 lb/ft^2 of surface for sands, and at least 0.5 lb/ft^2 for silty soils. In subsequent research it was found that one or two applications of hydrogen peroxide may be required to renovate clean sands (Mickelson et al., 1989), and that the infiltration rates may be reduced significantly by peroxide treatment of some soils (Hargett et al., 1984).

Use of Upgraded Pretreatment

A study was conducted in Wisconsin in which failing soil absorption units were rehabilitated by reducing the organic and solids loadings by upgrading the pretreatment. In 1994, 15 failing systems were upgraded by aerobically treated effluent. Of the 15 systems, 12 were able to resume successful operation accepting the higher-quality effluent. Of the three systems that continued to have problems, two continued to need frequent pumping to operate and one system could not be rehabilitated (Converse and Tyler, 1995).

Retrofitting Failed Systems

Other methods of retrofitting failed systems include upgrading the treatment of the septic tank effluent with intermittent sand filters, plastic medium, or textile bioreactors (see Chap. 11), followed by disposal in shallow trenches or drip irrigation. In some cases, the new shallow trench can be above or between the existing failed leachfields.

Long-Term Effects of Sodium on Clay Soils

If clay soil slaking is occurring (see Chap. 10 for effects of sodium on soil permeability), calcium can be added to the system to reverse the effects of the sodium (Patterson, 1997). Calcium thiosulfate (CaS_2O_3) has been used to improve the permeability of slaked soils. Changing to a low-sodium laundry detergent may also reduce the sodium adsorption ratio and alleviate the problem.

13-10 ROLE OF ONSITE MANAGEMENT

There are a number of roles that onsite management can have for individual and decentralized onsite disposal/reuse. The functions of onsite management districts, described in Chap. 15, include monitoring for system failures as well as monitoring for environmental and public health protection. Onsite management can reduce the risk of using innovative, cost-effective technologies and can increase the opportunities for local water reuse.

Reduced Risk of Using Innovative Technologies

Onsite systems without any management must be designed very conservatively because failure could mean abandonment of a residence or business. With management and oversight, innovative technologies can be tried, with the assurance that, should the technology fail, corrective measures or replacement technology can be used. As an example, the town of Paradise, California, has had an onsite management district since 1992, and has encouraged the use of sand filters, aerobic pretreatment of restaurant wastewater, and pressure-dosed distribution. Paradise and Gloucester, Massachusetts, are participating in the EPA-sponsored National Onsite Demonstration Program (NDOP) that pays for installation and monitoring of onsite treatment and disposal/reuse technologies.

Water Reuse in Decentralized Systems

As described in Chap. 12, water reuse near the point of wastewater generation is a major benefit of decentralized wastewater management systems. The concept of water reuse on the periphery of a community has been practiced in a number of areas including the "satellite" reclamation plants used by the Los Angeles County Sanitation Districts. At Stinson Beach, California, the continued use of onsite soil absorption systems and the resultant support of trees and vegetation has resulted in a better allocation of the water resource than if sewers had been installed and the collected water had been discharged to the ocean.

PROBLEMS AND DISCUSSION TOPICS

13-1. Obtain the design criteria used for onsite systems for your county and compare them to the criteria given in this chapter with respect to the sizing of septic tanks and the design of absorption fields. Identify and discuss any major differences and the impact of any differences with respect to the design of onsite systems.

13-2. For an individual home with an expected flow of 150 gal/d, lay out and design a conventional septic tank and leachfield. List your assumptions.

13-3. Using the following data obtained from a trench pump-in test, determine the soil acceptance rate and the saturated coefficient of permeability.

1. Trench size = 10 ft long, 4 ft wide, and 2 ft deep
2. Total extent of the water plume = 500 ft^2
3. Water depth in the bottom of trench = 1 ft
4. Depth of water below trench bottom at periphery of plume = 2 ft
5. Height of capillary zone = 1.5 ft
6. Degree of saturation = 35%
7. Soil porosity = 0.45
8. Total water used in the test = 5000 gal
9. Total time elapsed = 4.5 d

13-4. If the amount of septic tank effluent discharged to a soil absorption system is limited to 900 gal/acre-d, determine the concentration of nitrate in the groundwater, assuming the depth of the receiving aquifer is 50 ft and that the rate of groundwater flow is 1 ft/d and that mixing of the percolate with the groundwater is complete. List the assumptions you have made in solving this problem.

13-5. Compare the amount of nitrogen added, expressed in lb N/ac·yr, from a residential area with 0.5-ac lots, to the amount of nitrogen added as a result of landscape fertilization and to the amount used in conventional agricultural practice, assuming the crop grown is corn. Assume four residents per house. Refer to Chap. 10 for typical fertilization values.

13-6. Compare the amount of nitrogen added in Prob. 5 from residential areas to that added by cattle grazing on irrigated pasture. Assume four head of cattle per acre.

13-7. Size and lay out an onsite system with alternating absorption fields to serve a home having an average occupancy of 3.5. Assume the vertical permeability is 0.5 gal/ft^2·d.

13-8. Size and lay out a shallow pressure-dosed distribution system serving a hotel with 65 units. Assume 80 gal/unit-d and a vertical permeability of 0.8 gal/ft^2·d.

13-9. Design and lay out a mound system to serve an RV park with a wastewater flow of 1000 gal/d.

13-10. Size the pressure pipe needed to dose a pressure distribution system used in conjunction with a shallow trench. The $\frac{1}{8}$-in-diameter orifices are spaced every 2 ft and the length of the trench is 100 ft. Assume the system is dosed 6 times/day with a total flow of 360 gal/d.

13-11. What features need to be incorporated in the design of a soil absorption system in locations where soil freezing is a common occurrence?

13-12. Design a mound system for a cluster of four homes. The flow is 900 gal/d. Assume that the native soil is a silt loam.

13-13. A community leachfield is being designed to treat and dispose of 10,000 gal/d. Design the leachfield using the application rate for a 2-ft-wide trench, based on a fine sand. Determine the potential groundwater mounding for a rectangular field with a length to width ratio of 2:1, a depth to groundwater of 8 ft, a height of stable groundwater table above the impermeable barrier of 20 ft, and a value of K of 1.2 ft/d. Use 10 years as the time frame.

13-14. Design an evapotranspiration bed for an individual home with a flow of 250 gal/d. The ET rate is 45 in/yr and the precipitation rate is 22 in/yr.

REFERENCES

Ball, H. L. (1994) Nitrogen Reduction in an Onsite Trickling Filter/Upflow Filter Wastewater Treatment System, *Proceedings of the 7th International Symposium on Individual and Small Community Sewage Systems,* pp. 499–503, American Society of Agricultural Engineers (ASAE), Atlanta.

Ball, H. L. (1995) Nitrogen Reduction in an Onsite Trickling Filter/Upflow Filter System, *Proceedings of the 8th Northwest On-Site Wastewater Treatment Short Course and Equipment Exhibition,* University of Washington, Seattle.

Ball, H. L. (1996) Personal communication, Roseburg, OR.

Bernhart, A. P. (1973) Treatment and Disposal of Waste Water from Homes by Soil Infiltration and Evapotranspiration, University of Toronto Press, Toronto.

Bishop, P. L., and H. S. Logsdon (1981) Rejuvenation of Failed Soil Absorption Systems, *Journal of Environmental Engineering Division,* ASCE, Vol. 107, No. EE1, p. 47.

Burgan, M. A., and D. M. Sievers (1994) On-site Treatment of Household Sewage via Septic Tank and Two-stage Submerged Bed Wetland, *Proceedings of the 7th International Symposium on Individual and Small Community Sewage Systems,* pp. 77–84, American Society of Agricultural Engineers (ASAE), Atlanta.

Burks, B. D., and M. M. Minnis (1994) Onsite Wastewater Treatment Systems, Hogarth House, Madison, WI.

California Resources Agency (1994) *Graywater Guide*, The Resources Agency, Sacramento.

Converse, J. C., and E. J. Tyler (1995) Aerobically Treated Domestic Wastewater to Renovate Failing Septic Tank-Soil Absorption Fields, *Proceedings of the 8th Northwest On-Site Wastewater Treatment Short Course and Equipment Exhibition,* University of Washington, Seattle.

Finnemore, E. J., and N. N. Hantzsche (1983) Groundwater Mounding Due to Onsite Sewage Disposal, *Journal American Society of Civil Engineers,* IDE, Vol. 109, No. 2, p. 199.

Hantzsche, N. N., and E. J. Finnemore (1992) Predicting Groundwater Nitrate-Nitrogen Impacts, *Ground Water,* Vol. 30, No. 4, pp. 490–499.

Hargett, D. L., E. J. Tyler, J. C. Converse, and R. A. Apfel (1985) Effects of Hydrogen Peroxide as a Chemical Treatment for Clogged Wastewater Absorption Systems, *Onsite Wastewater Treatment,* American Society of Agricultural Engineers, St. Joseph, Michigan.

Harkin, J.N., and M. D. Jawson (1977) Clogging and Unclogging of Septic System Seepage Beds, *Proceedings 2nd Illinois Symposium on Private Sewage Disposal Systems,* Illinois Dept. of Public Health, Springfield.

Ingham, A. T. (1980) Guidelines for Mound Systems, California State Water Resources Control Board, Sacramento.

Johnson, A. I. (1967) *Specific Yield—Compilation of Specific Yields for Various Materials,* U.S. Geological Survey Water-Supply Paper 1662-D.

Laak, R. (1986) *Wastewater Engineering Design for Unsewered Areas,* 2nd ed.,Technomic Publishing, Lancaster, PA.

Mickelson, M., J. C. Converse, and E. J. Tyler (1989) Hydrogen Peroxide Renovation of Clogged Wastewater Soil Absorption Systems in Sands, *Transactions of American Society of Agricultural Engineers,* Vol. 32, No. 5, pp. 1662–1668.

Nolte and Associates (1992) Manual for the Onsite Treatment of Wastewater, Town of Paradise, CA, Sacramento.

Nolte and Associates (1994) Fatal Flaw Analysis of Onsite Alternatives for Los Osos, California, Prepared for Metcalf & Eddy and San Luis Obispo County, Sacramento, CA.

NOWRA (1996) Soil and Site Evaluation for Onsite Wastewater Infiltration System Selection and Design, National Onsite Wastewater Recycling Association, Inc., with Tyler & Associates, Inc., Milwaukee, Wisconsin.

Otis, R. J. (1982) Pressure Distribution Design for Septic Tank Systems, *Journal of Environmental Engineering Division,* ASCE, Vol. 108, No. EE1, p. 123.

Patterson, R. A. (1997) Domestic Wastewater and Sodium Factor, in M. S. Bedinger et al., (eds.), *Site Characterization and Design of Onsite Septic Systems,* ASTM STP 1324, American Society for Testing and Materials, Philadelphia.

Piluk, R. J., and O. J. Hao (1994) Evaluation of Onsite Waste Disposal System for Nitrogen Reduction, *Journal of Environmental Engineering Division,* ASCE, Vol. 115, No. 4, pp. 725–740.

Reed, S. C. (1993) *Subsurface Flow Constructed Wetlands for Wastewater Treatment: A Technology Assessment,* EPA 832-R-93-001, U.S. Environmental Protection Agency, Washington, DC.

Reed, S. C., and R. W. Crites (1984) *Handbook of Land Treatment Systems for Industrial and Municipal Wastes,* Noyes Publications, Park Ridge, NJ.

Reed, S. C., R. W. Crites, and E. J. Middlebrooks (1995) *Natural Systems for Waste Management and Treatment,* 2nd ed., McGraw-Hill, New York.

Sandy, A. T., W. A. Sack, and S. P. Dix (1988) Enhanced Nitrogen Removal Using a Modified Recirculating Sand Filter (RSF2), *Proceedings of the 5th National Symposium on Individual and Small Community Sewage Systems,* American Society of Agricultural Engineers, Chicago.

Siegrist, R. L. (1988) Hydraulic Loading Rates for Soil Absorption Systems Based on Wastewater Quality, in Onsite Wastewater Treatment, *Proceedings of the 5th National Symposium on Individual and Small Community Sewage Systems,* American Society of Agricultural Engineers, Chicago.

Steiner, G. R., and J. T. Watson (1993) *General Design, Construction, and Operational Guidelines for Constructed Wetlands Wastewater Treatment Systems for Small Users Including Individual Residences,* 2nd edition, TVA/WM-93/10, Tennessee Valley Authority, Chattanooga, TN.

U.S. EPA (1980) *Onsite Wastewater Treatment and Disposal Systems—Design Manual,* Municipal Environmental Research Laboratory, U.S. Environmental Protection Agency, Cincinnati, OH.

WPCF (1990) *Natural Systems for Wastewater Treatment,* Manual of Practice FD-16, Water Pollution Control Federation (now Water Environment Federation), Alexandria, VA.

Wert, S. (1997) Personal communication, Roseburg, OR.

Winneberger, J. H. T. (1984) *Septic-Tank Systems, A Consultant's Toolkit,* Vol. 1, Subsurface Disposal of Septic-Tank Effluents, Ann Arbor Science, Stoneham, MA.

CHAPTER 14

Biosolids and Septage Management

Septage is the semiliquid material that is pumped out of septic (or interceptor) tanks. It consists of the sludge that has settled to the bottom of the septic tank over a period of years, and the liquid and surface scum layer (see Fig. 5-44). Septage is characterized typically by significant quantities of grit, oil and grease, solids, and organic material. *Sludge* is the material that settles out of wastewater in Imhoff tanks, clarifiers, ponds, and aquatic and land treatment systems. In raw wastewater the settleable solids that become sludge in pretreatment (Chap. 5) are both inorganic (grit) and organic. In suspended-growth biological processes (Chap. 7) the sludge produced has a significant amount of biological solids. *Biosolids* is the material that remains after sludge stabilization. Biosolids have characteristics that can provide environmental benefits through reuse, distribution, and land application. Despite the differences between septage, sludge, and biosolids, the processes used to treat, dewater, and reuse or dispose of these materials are similar. Important subjects in the management of septage, sludge, and biosolids are: (1) characteristics of the material, (2) treatment options, (3) dewatering options, (4) composting, (5) land application, and (6) landfilling.

14-1 SEPTAGE: CHARACTERISTICS AND QUANTITIES

Important characteristics of septage include expected quantities, chemical and nutrient content, and heavy metal content. Characteristics of septage will vary with the frequency of pumping, whether the source is residential or commercial, and with the use of kitchen food waste grinders (U.S. EPA, 1984).

Septage Quantities

Quantities of septage vary depending on the frequency with which septic tanks are pumped. A typical value for planning purposes is 60 gal/capita·yr (0.227 m^3/cap·yr). If the frequency of septic tank pumping is known, the annual volume of septage can be calculated by (Eq. 14-1).

$$\text{Annual volume} = \frac{\text{(number of septic tanks)(typical volume)}}{\text{(pumpout interval)}} \tag{14-1}$$

where annual volume = septage quantity, gal/yr
typical volume = typical septic tank volume, gal
pumpout interval = time between septic tank pumpings, yr

To determine the maximum and minimum periods of the year, the local septage haulers should be contacted. The major factor that affects septic tank cleaning frequency in cold regions is the weather. The frequency of pumping can be based on the number of people per house and the equations in Chap. 5.

Septage Characteristics

The physical and chemical characteristics of septage are summarized in Table 14-1. As shown in Table 14-1 septage contains high concentrations of solids, grease, BOD and nutrients. Metals are also present in septage, depending on the use of household chemicals and leaching of metal from household piping and joints (Table 14-2).

TABLE 14-1
Physical and chemical characteristics of septage*
Units: mg/L except for pH

Constituent	U.S. mean	EPA mean	EPA suggested design value
BOD	6480	5000	7000
COD	31,900	42,850	15,000
Total solids	34,106	38,800	40,000
TVS	23,100	25,260	25,000
TSS	12,862	13,000	15,000
VSS	19,027	8720	10,000
TKN	588	677	700
NH_3-N	97	157	150
Total P	210	253	250
Alkalinity	970		1000
Grease	5600	9090	8000
pH	1.5–12.6	6.9	6.0
LAS	110–200	157	150

*From U.S. EPA (1984).

TABLE 14-2
Heavy metal concentrations in septage*
Units: mg/L

Constituent	U.S. mean	EPA mean	EPA suggested design value
Aluminum	48	48	50
Arsenic	0.16	0.16	0.2
Cadmium	0.27	0.71	0.7
Chromium	0.92	1.1	1.0
Copper	8.27	6.4	8.0
Iron	191	200	200
Mercury	0.23	0.28	0.25
Manganese	3.97	5	5
Nickel	0.75	0.9	1
Lead	5.2	8.4	10
Selenium	0.076	0.1	0.1
Zinc	27.4	49	40

*From U.S. EPA (1984).

14-2 SLUDGE: CHARACTERISTICS AND QUANTITIES

Sludge characteristics of importance include the expected quantity, chemical and nutrient content, and heavy metal content. Characteristics of sludge vary with the type of wastewater operation or process that produces the sludge as well as the wastewater strength.

Sludge Quantities

Sludge quantities vary with the type of wastewater treatment system as shown in Table 14-3. For primary sedimentation or Imhoff tanks the sludge produced is typically 0.6 tons/Mgal (144 $kg/10^3 m^3$) of wastewater treated. For an activated-sludge

TABLE 14-3
Sludge production from wastewater treatment

	Dry sludge, tons/10^6 gal	
Operation or process	Range	Typical
Primary sedimentation or Imhoff tanks	0.45–0.7	0.60
Activated sludge	0.3–0.4	0.35
Trickling filters	0.25–0.4	0.30
Aerated lagoon	0.35–0.5	0.40*
Extended aeration	0.35–0.5	0.40*
Filtration	0.05–0.1	0.07

*Includes primary sludge.

TABLE 14-4
Typical composition of sludges*

Constituent	Unit	Untreated primary	Waste activated
Total solids (TS)	%	5	0.83–1.16
Volatile solids	% of TS	65	59–88
Nitrogen as N	% of TS	2.5	2.4–5.0
Phosphorus as P_2O_5	% of TS	1.6	2.8–11.0
Potassium as K_2O	% of TS	0.4	0.5–0.7
pH		5.0–8.0	6.5–8.0
Iron	%	2.5	

*From Tchobanoglous and Burton (1991).

system the combined waste activated and primary sludge would amount daily to 0.95 dry tons/Mgal. In contrast aerated lagoons produce a combined 0.4 dry tons/Mgal.

Sludge Characteristics

The sludge from conventional primary sedimentation is relatively dilute, with a typical solids concentration of 5 percent. Data on typical sludge quality for primary and waste-activated sludges are presented in Table 14-4. The fertilizer values of a typical sludge are relatively low with an NPK (nitrogen, phosphorus, potassium) ratio (percent by weight) of 2.5:1.6:0.4. A typical mixed fertilizer will have an NPK ratio of 10:10:10.

The metals in sludge are also of importance in the selection of a sludge utilization or disposal option. Typical concentrations of metals in sludge are presented in Table 14-5.

TABLE 14-5
Heavy metal concentrations in sludge*
Units: mg/kg

Constituent	Typical	EPA median
Arsenic	10	10
Boron	33	
Cadmium	16	10
Cobalt	4	30
Chromium	890	500
Copper	850	800
Mercury	5	6
Manganese	260	260
Molybdenum	30	4
Nickel	82	80
Lead	500	500
Zinc	1740	1700

*From U.S. EPA (1987) and Sommers (1980).

TABLE 14-6
Sludge production in pond systems*

		Facultative lagoons, Utah		Aerated lagoons, Alaska	
Parameter	**Unit**	**A**	**B**	**C**	**D**
Wastewater suspended solids	mg/L	62	69	185	170
Flow rate	Mgal/d	10	0.18	0.18	0.07
Sludge depth	ft	0.29	0.25	1.10	0.91
Solids content of sludge	%	5.9	7.7	8.6	0.9
Time between removal	yr	13	9	5	8
Average sludge production	lb/d	834	44	280	3.1
Sludge production	ton/Mgal	0.04	0.12	0.77	0.02

*From Reed et al. (1995).

Sludge Production from Natural Systems

Natural systems such as land treatment, ponds, aquatic treatment, or constructed wetlands produce significantly less sludge than conventional treatment processes. In land treatment and constructed wetlands, the solids are incorporated into the soil matrix and the vegetation/debris. In ponds and aquatic treatment systems the solids must be removed, but over a 5- to 10-year period typically.

Ponds in Alaska and Utah were studied for solids accumulation and the results are presented in Table 14-6. The sludge production in the facultative ponds in Table 14-6 is about 10 percent of the production of a typical primary treatment system. In warmer climates the solids production would be expected to be lower than in cold climates because of the biological stabilization that occurs more rapidly in warmer climates. The long detention time in ponds allows for the stabilization of organic matter, consolidation of the sludge, and for a significant die-off of fecal coliforms.

Volume-Weight Relationships for Sludge and Septage

The volume of sludge depends mainly on its water content and only slightly on the character of the solid matter. A 5 percent sludge, for example, contains 95 percent water by weight. If the solid matter is composed of fixed (mineral) solids and volatile (organic) solids, the specific gravity of all the solid matter can be computed by

$$\frac{W_s}{S_s \rho_w} = \frac{W_f}{S_f \rho_w} + \frac{W_v}{S_v \rho_w} \tag{14-2}$$

where W_s = weight of solids
S_s = specific gravity of solids
ρ_w = density of water
W_f = weight of fixed solids (mineral matter)
S_f = specific gravity of fixed solids
W_v = weight of volatile solids
S_v = specific gravity of volatile solids

Therefore, if one-third of the solid matter in a sludge containing 95 percent water is composed of fixed mineral solids with a specific gravity of 2.5, and two-thirds is composed of volatile solids with a specific gravity of 1.0, then the specific gravity of all solids, S_s, would be equal to 1.25, as follows:

$$\frac{1}{S_s} = \frac{0.33}{2.5} + \frac{0.67}{1.0} = 0.802$$

$$S_s = \frac{1.0}{0.802} = 1.25$$

If the specific gravity of the water is taken to be 1.00, the specific gravity of the sludge is 1.01, as follows:

$$\frac{1}{S_{sl}} = \frac{0.05}{1.25} + \frac{0.95}{1.00} = 0.99$$

$$S_{sl} = \frac{1.0}{0.99} = 1.01$$

The volume of a sludge may be computed with the following expression:

$$V = \frac{W_s}{\rho_w S_{sl} P_s} \tag{14-3}$$

where W_s = weight of dry solids, lb
ρ_w = density of water, lb/ft^3
S_{sl} = specific gravity of sludge
P_s = percent solids expressed as a decimal

For approximate calculations for a given solids content, it is simple to remember that the volume varies inversely with the percent of solid matter contained in the sludge as given by

$$\frac{V_1}{V_2} = \frac{P_2}{P_1} \qquad \text{(approximate)}$$

where V_1, V_2 = sludge volumes and P_1, P_2 = percent solid matter. The application of these volume and weight relationships is illustrated in Example 14-1.

EXAMPLE 14-1. VOLUME OF UNTREATED AND DIGESTED SLUDGE. Determine the liquid volume before and after digestion and the percent reduction of 1000 lb (dry basis) of primary sludge with the following characteristics:

Item	Primary	Digested
Solids, percent	4	5
Volatile matter, percent	60	60 (destroyed)
Specific gravity of fixed solids	2.5	2.5
Specific gravity of volatile solids	1.0	1.0

Solution

1. Compute the specific gravity of all the solids in the primary sludge using Eq. (14-2):

$$\frac{1}{S_s} = \frac{0.4}{2.5} + \frac{0.6}{1.0} = 0.76$$

$$S_s = \frac{1.0}{0.76} = 1.32 \qquad \text{(primary solids)}$$

2. Compute the specific gravity of the primary sludge:

$$\frac{1}{S_{sl}} = \frac{0.04}{1.32} + \frac{0.96}{1.00} = 0.99$$

$$= \frac{1.0}{0.99} = 1.01$$

3. Compute the volume of the primary sludge using Eq. (14-3):

$$V = \frac{1000\text{ lb}}{(62.4\text{ lb/ft}^3)(1.01)(0.04)} = 397\text{ ft}^3$$

4. Compute the percentage of volatile matter after digestion:

$$\text{Percent volatile matter} = \frac{\text{total volatile solids after digestion}}{\text{total solids after digestion}} \times 100$$

$$= \frac{0.4(0.6 \times 1000)}{400 + 0.4(600)} \times 100 = 37.5\%$$

5. Compute the average specific gravity of all the solids in the digested sludge using Eq. (14-2):

$$\frac{1}{S_s} = \frac{0.625}{2.5} + \frac{0.375}{1.0} = 0.625$$

$$S_s = \frac{1.0}{0.625} = 1.6 \qquad \text{(digested solids)}$$

6. Compute the specific gravity of the digested sludge:

$$\frac{1}{S_{ds}} = \frac{0.05}{1.6} + \frac{0.95}{1.0} = 0.98$$

$$S_{ds} = \frac{1.0}{0.98} = 1.02$$

7. Compute the volume of digested sludge using Eq. (14-3):

$$V = \frac{400 + 0.4(600)}{(62.4)(1.02)(0.05)} = 201\text{ ft}^3$$

8. Determine the percentage reduction in the sludge volume after digestion:

$$\text{Percent reduction} = \frac{397 - 201}{397} \times 100 = 49.3\%$$

14-3 TREATMENT OPTIONS

Septage and sludge generally will require some level of treatment prior to final disposal or reuse. In many cases septage is discharged to a municipal wastewater treatment plant and treated as a wastewater source. In separate treatment situations, the options for treatment include the conventional wastewater treatment processes, the conventional sludge treatment processes, land treatment, or a combination of aquatic treatment and constructed wetlands. Sludge can be treated in digesters or sludge lagoons as a liquid or it can be dewatered and treated by composting or land application.

Septage Treatment

Five alternatives for septage treatment are considered in the following discussion:

- Treatment in a municipal treatment plant
- Treatment as a wastewater source
- Treatment as a sludge
- Land treatment
- Aquatic treatment

Septage treatment in a municipal treatment plant. If septage is to be cotreated with wastewater, it will be necessary to construct a septage receiving station. A typical septage receiving station is shown in Fig. 14-1. As shown in Fig. 14-1, such a station will consist of an unloading area (sloped to allow gravity draining of septage

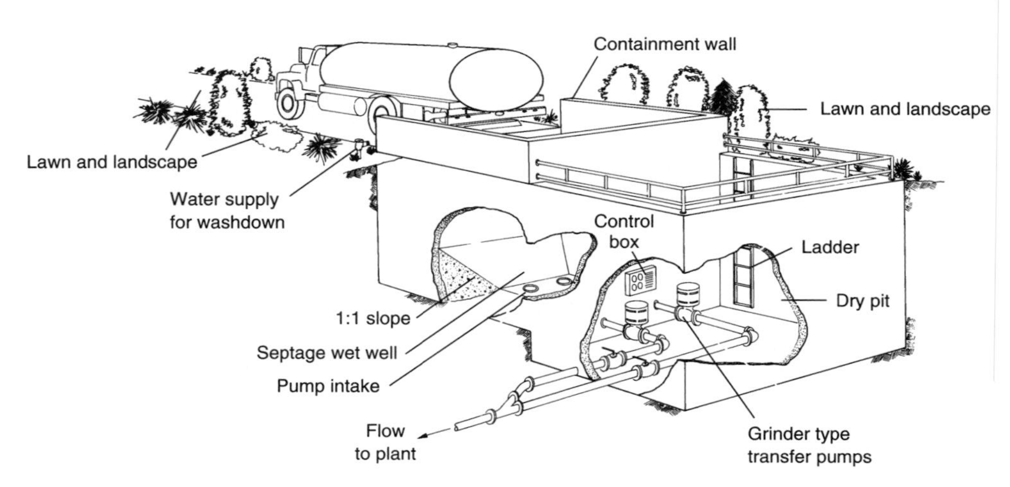

FIGURE 14-1
Typical septage receiving station at a municipal wastewater treatment plant. It is important that the approach pad be sloped so that the contents of the pumper truck can be emptied completely by gravity (German and Tchobanoglous, 1982).

hauling trucks), a septage storage tank, and one or more grinder pumps. The storage tank is used to store the septage so that it can be discharged to the treatment plant. The storage tank should be covered for odor control. Discharge of septage to a headworks is usually preferred for the removal of grit and screenings. If there are no screening or comminution facilities ahead of the septage discharge facility, the septage should be transferred from the storage tank to the treatment plant with grinder pumps. In some cases, this transfer can be accomplished by gravity flow. If the septage is especially strong, it can be diluted with treated wastewater. Chemicals such as lime or chlorine can also be added to the septage in the storage tank to neutralize it, to render it more treatable, or to reduce odors. If the capacity limitations do not exist, the co-treatment and disposal of septage with wastewater is one of the most cost-effective and environmentally sound methods that can be used for the management of septage.

Septage treatment as a wastewater. Because septage is typically 98.5 percent water, it is often treated as a wastewater source. The heavier suspended solids are settled out from the waste stream to allow more effective biological treatment of the remaining organics. Two treatment process flow diagrams for treatment of septage as a wastewater are shown in Fig. 14-2.

In the flow diagram shown in Fig. 14-2*a* the septage is screened for large solids and rags and is treated biologically in a series of aerated pond cells. The effluent is further treated by land treatment. An alternative to pond treatment would be direct land application, either by the slow-rate method (irrigation) or by rapid infiltration.

In the flow diagram shown in Fig. 14-2*b* the septage is treated in a conventional approach that includes primary sedimentation, biological treatment, effluent filtration, and disinfection. The biological treatment can be accomplished with an oxidation ditch, a conventional activated-sludge system, or a sequencing batch reactor (SBR). Sludge produced can be dewatered and composted for reuse or disposed of in a landfill or by land application (WEF, 1997).

Septage treatment as a sludge. Options for treatment of septage as a sludge include composting, land application, aerobic and anaerobic digestion, chemical oxidation, and lime stabilization (U.S. EPA, 1984). The processes used typically for small systems are composting, land application (see next section) and aerobic digestion.

Composting is the stabilization of organic waste through aerobic biological decomposition. As described in more detail later in this chapter, the process can be accomplished in various configurations. The different types of composting include two open-area methods: windrow and static pile composting and in-vessel mechanical composting. Septage must be dewatered prior to composting. Operational parameters for septage composting are presented in Table 14-7. Compost products can be sold or given away.

Aerobic digestion is normally practiced in open tanks or basins. Solids retention times of 30 to 40 d are needed for stabilization at temperatures of 18 to 29°C (U.S. EPA, 1984). Surface aeration or submerged diffusers are used to maintain a minimum dissolved oxygen level of 1 mg/L. Without adequate aeration, odors can develop. With excess aeration, foaming can be a problem. Because of these

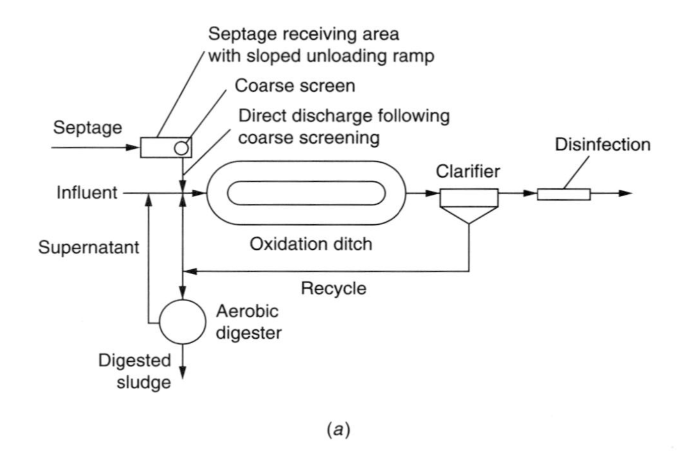

(a)

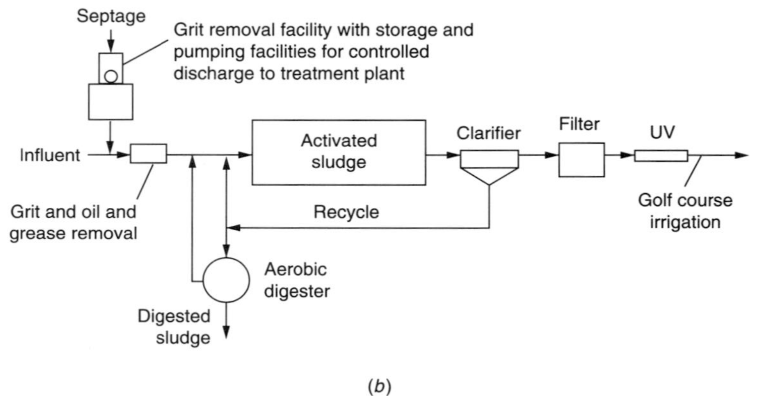

(b)

FIGURE 14-2
Flow diagrams for treatment of septage in biological wastewater treatment facilities: (*a*) direct discharge of septage after screening and (*b*) controlled discharge of septage after screening and grit removal.

limitations, long detention times, and high capital and operating costs (compared to land treatment or lagooning), it is unlikely that aerobic digestion will be economically attractive except for large or land-limited independent septage treatment facilities (U.S. EPA, 1984).

Land treatment. Land treatment or land application is the primary method of septage treatment and disposal in the U.S. (U.S. EPA, 1984). The types of land treatment, described in Chap. 10, include slow rate, overland flow, and rapid infiltration. Land treatment of septage is achieved primarily by the slow-rate method, with many variations on the application technique. A summary of the surface land application techniques is presented in Table 14-8. The subsurface techniques for land application are summarized in Table 14-9.

TABLE 14-7
Operational parameters for septage composting*

Parameter	Optimum range	Control mechanisms
Moisture content of compost mixture	40–60%	Dewatering of septage to 10 to 20% solids followed by addition of bulking material (amendments such as sawdust and woodchips), 3:1 by volume amendment: dewatered septage.
Oxygen	5–15%	Periodic turning (windrow), forced aeration (static pile), mechanical agitation with compressed air (mechanical).
Temperature (compost must reach)	55–65°C	Natural result of biological activity in piles. Too much aeration will reduce temperature.
pH	5–8	Septage is generally within this pH range, adjustments not normally necessary.
Carbon/nitrogen ratio	20:1 to 30:1	Addition of bulking material.

*From U.S. EPA (1984).

TABLE 14-8
Surface land application techniques for septage*

Technique	Liquid or sludge	Characteristic	Advantage	Disadvantage
Hauler truck spreading	Liquid	500- to 2000-gal trucks	Some trucks can be used for collection, transport, and application	Some odor on spreading. Storage needed. Slopes limited to 8%.
Farm tractor and wagon spreading	Liquid or sludge	300- to 3000-gal capacity	Frees the hauler truck during high-usage periods	Some odor on spreading. Storage needed. Slopes limited to 8%. Requires additional equipment.
Ridge and furrow irrigation	Liquid	Surface leveling required	Lower energy requirements than sprinkler application	Storage pond needed. Some odor potential. Slopes between 0.5 and 1.5%.
Overland flow	Liquid	Use on sloping ground with vegetation	Can be applied from ridge roads	Difficult to get uniform solids application. Storage needed. Runoff control. Slopes between 2 and 8%. Extensive site preparation.
Sprinkler irrigation	Liquid	Large orifice nozzles	Use on steep or unleveled land. Uniform solids application.	Odor potential. Higher energy needs. Storage needed. Winterization needed.

*From U.S. EPA (1984).

TABLE 14-9
Subsurface land application techniques for septage*

Technique	Liquid or sludge	Characteristic	Advantage	Disadvantage
Tank truck with plow and furrow cover	Liquid	Single-furrow plow mounted on truck	Minimal odor	Slopes limited to 8%. Storage pond needed in winter. Longer time needed for application than for surface techniques.
Farm tractor with plow and furrow	Liquid	Septage applied to furrow ahead of single plow. Septage spread in narrow swath and immediately covered with plow.	Minimal odor	Slopes limited to 8%. Storage pond needed in winter. Longer time needed for application than for surface techniques.
Subsurface injection	Liquid	Septage placed in opening created by tillage tool. Keep vehicles off the area for 1 to 2 weeks after injection.	Injector can mount on rear of some trucks. No odor.	Slopes limited to 8%. Storage needed in winter.

*From U.S. EPA (1984).

Surface spreading is the most common surface application technique. The hauler truck that pumps out the septic tank is typically the vehicle that applies septage to the land. During wet weather or when the land application site is frozen the septage must be stored. In some instances the soil conditions may require the use of high-flotation-type tires for the application vehicle. In these cases a separate application vehicle is required because flotation-type tires are not suitable for long-distance highway use.

As shown in Table 14-8, other forms of surface application include fixed sprinklers, ridge and furrow irrigation, and overland flow. The principal concerns associated with surface application are runoff control, odors, and potential health risks.

Subsurface application avoids the odor and nuisance problems and limits the potential health risks. Groundwater contamination concerns must be met by proper process design for both surface and subsurface application.

The elements of a land treatment project for septage management include site availability and selection analysis, process design, facility design, construction management, and operation. All of these elements are the same as those for land application of sludge and are discussed under that section.

Aquatic treatment. A relatively new process has been used experimentally to treat septage by a combination of aerated aquatic treatment and constructed wetlands. Settled septage was treated by this solar aquatic system at Harwich, Massachusetts, in 1988 (Teal and Peterson, 1991). The treatment performance for the experimental operation which had a hydraulic detention time of 15 d is reported

TABLE 14-10
Performance of solar aquatic system at Harwich, Massachusetts*

Constituent	Influent, mg/L	10th aquatic cell, mg/L	Final effluent, mg/L
BOD_5	1386	57	4.9
TSS	406	146	5.7
TN	152	69	9.5
NH_4-N	72	29	0.15
NO_3-N	2.1	23	5.4
TP	21	19	3.9

*From Teal and Peterson (1991).

(*a*)

(*b*)

FIGURE 14-3
Aquatic system for septage treatment in a greenhouse environment: (*a*) external view, (*b*) internal view of tanks with vegetation.

in Table 14-10. The system requires a greenhouse for year-round operation in cold climates (see Fig. 14-3).

Sludge Treatment

Sludge from sedimentation and biological treatment processes must be stabilized or treated prior to disposal or reuse. The need for stabilization or treatment depends on the type of disposal or reuse and on the nuisance potential for odor at the site. Sludge is treated to: (1) eliminate offensive odors, (2) reduce or inhibit the potential for putrefaction, and (3) reduce the pathogen content. The means of treatment include: (1) biological reduction of biodegradable volatile solids, (2) chemical oxidation of volatile solids, (3) chemical addition to render the sludge unsuitable for biodegradation, and (4) heating to disinfect or sterilize the sludge. The most common methods of sludge treatment for small wastewater facilities are aerobic digestion and sludge lagoons. Methods that are used less frequently are anaerobic digestion, chemical oxidation, and lime stabilization.

Aerobic digestion. Aerobic digestion is used most commonly to stabilize waste activated sludge from package aerobic treatment plants or small activated-

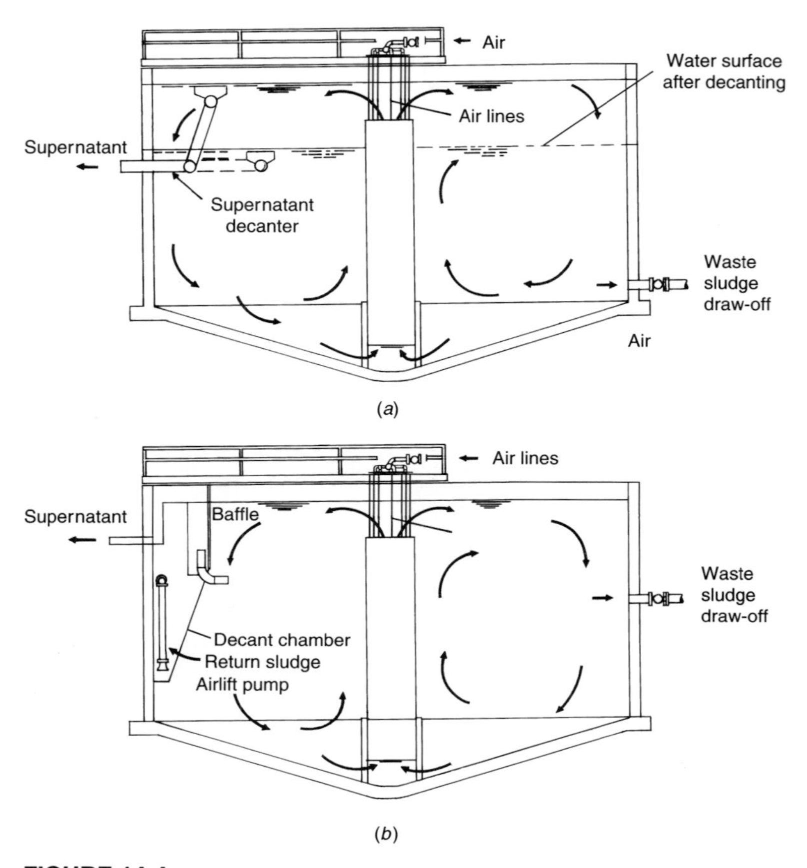

FIGURE 14-4
Typical complete-mix aerobic digester: (*a*) batch operation and (*b*) continuous operation (adapted from Tchobanoglous and Burton, 1991).

sludge plants (see Fig. 14-4). For small plants aerobic digestion is used more often than anaerobic digestion because: (1) operation is relatively easy; (2) capital costs are lower; (3) it produces an odorless, humuslike stable end product; and (4) lower BOD concentrations are produced in the supernatant liquor.

Aerobic digestion is usually conducted in open, unheated concrete tanks. Factors that need consideration in design are temperature, hydraulic detention time, solids loading, oxygen requirement, air requirement, and energy requirement for mixing. Typical values for design criteria are presented in Table 14-11.

Temperature affects the rate of biological treatment and the oxygen transfer rate. In extremely cold climates, consideration should be given to retaining the heat content of the sludge by covering the tanks, providing insulation to aboveground tanks, and providing submerged aeration instead of surface aeration. The BOD removal or volatile suspended solids (VSS) stabilization criteria should be based on the coldest average monthly temperature of the wastewater. Oxygen transfer calculations should be made for the warmest average monthly temperature.

TABLE 14-11
Design criteria for aerobic digestion of sludge*

Parameter	Unit	Value
Hydraulic detention time at 68°F (20°C)		
Waste activated sludge	d	12-16
Waste activated sludge from plants without primary sedimentation	d	16-18
Primary plus waste activated or trickling filter sludge	d	18-22
To meet PSRP	d	40
Solids loading	lb volatile solids/$ft^3 \cdot d$	0.1-0.2
Oxygen requirement	lb O_2/lb of VSS destroyed	2.0
Tank volume	ft^3/capita	3-4
Air requirement	$ft^3/min \cdot 10^3\ ft^3$	20-60
Energy requirement for mixing	$ft^3/min \cdot 10^3\ ft^3$	20-30

*From U.S. EPA (1985).

Aerobic digesters are usually designed for about 40 percent reduction in VSS. The tank volume is usually determined by the flowrate of sludge and the hydraulic detention time:

$$V = Qt \tag{14-4}$$

where V = volume of aerobic digester, ft^3 (m^3)
Q = flowrate of sludge to the digester, ft^3/d (m^3/d)
t = hydraulic detention time, d

Depending on the type of sludge, select a hydraulic detention time from Table 14-11. If the minimum monthly sludge temperature is less than 68°F (20°C) increase the hydraulic detention time by the ratio of the k factors according to

$$k_T = k_{20}(1.06)^{T-20}$$

where k_{20} = 0.10 d^{-1} at temperature of 20°C
k_T = reaction rate constant at temperature T
1.06 = θ value
T = temperature of sludge, °C

Aerobic digestion can also occur at high temperatures (thermophilic) and low temperatures (cryophilic). European experience with autothermal thermophilic aerobic digestion (ATAD) indicates that detention times of 5 to 6 d can be used instead of the 15 to 20 d for mesophilic digestion. Typical ATAD systems operate at 55°C and reach 60°C to 65°C in the second stage (U.S. EPA, 1990). They rely on the heat released during digestion to attain and sustain the desired operating temperatures.

The cryophilic temperature range can go down to 5°C. Researchers in British Columbia have found that sludge age should be maintained between 50 and 60 d when temperatures are in the 5°C range (WPCF, 1985).

Anaerobic digestion. Anaerobic digesters are generally not associated with small wastewater systems because of the complexity. Anaerobic digesters must be mixed and heated, and the sludge pH must be controlled. In addition gas recovery and reuse or burning (flaring) in the open air must be provided. For these reasons anaerobic digestion is usually confined to larger conventional wastewater treatment plants (Tchobanoglous and Burton, 1991; U.S. EPA, 1985).

Lime stabilization. In the lime stabilization process, lime is added to raw sludge in sufficient quantities to raise the pH of the sludge to 12 for at least 2 hours. The high pH kills the microorganisms in the sludge and thereby stabilizes the organics.

The drawbacks of lime stabilization include an increase in the mass of the lime/sludge mixture, which means more sludge for dewatering and disposal; the relatively high cost; and the high pH, which can be a disadvantage in land application to agricultural soil that already has a high pH. Advantages of lime stabilization include the short detention time required, the simplicity of the process, and, where acid soil conditions exist, the high pH of the sludge (a benefit in land application).

Two forms of lime are available commercially: quicklime (CaO) and hydrated lime [$Ca(OH)_2$]. Smaller treatment plants use hydrated lime because it can be mixed with water and applied directly. Lime dosage depends on the type of sludge and the solids content and may range from 200 to 800 lb/ton of dry solids (U.S. EPA, 1985). Postlime treatment can be used to further stabilize digested sludge and aid in dewatering the sludge. Fly ash, cement kiln dust, and carbide lime have also been used as a substitute for lime.

14-4 DEWATERING OPTIONS

Dewatering is a physical unit operation in which the moisture content is reduced (solids content increases). A wide variety of operations and equipment can be used to dewater sludge; however, for small plants the principal methods are

- Drying beds
- Mechanical dewatering
- Sludge freezing
- Reed beds
- Lagoons

Dewatering may be necessary for (1) more efficient hauling of sludge (less water), (2) reduced handling and storage, (3) meeting minimum solids content for landfilling, and (4) meeting minimum solids content for composting or burning.

Drying Beds

Air drying of sludge on sand beds is the most common method of sludge dewatering for small wastewater systems. Sludge drying beds are simple to operate, produce a

FIGURE 14-5
Typical sand drying bed for dewatering digested sludge, showing cracking of sludge on the right and the beginning of cracking in the center.

high solids content, are low cost, and require minimal operator attention. Types of drying beds include:

- Sand
- Paved
- Artificial medium
- Vacuum assisted

Sand drying beds. A typical sand drying bed is shown in Fig. 14-5. The bed consists of 12 in (300 mm) of fine sand underlaid by 8 to 18 in (200 to 460 mm) of gravel. The sand should have an effective size of 0.01 to 0.03 in (0.3 to 0.75 mm), be clean of any fines, and have a uniformity coefficient less than 3.5. The gravel size is typically 0.1 to 1 in (2.5 to 25 mm).

Underdrains are usually plastic perforated pipe or clay tile laid with open joints. Underdrains are 4 in (100 mm) in diameter and are sloped at 1 percent. Beds are divided every 20 ft (6 m) by 2-ft-high (0.6-m) partitions.

Loading rates for conventional sand drying beds range from 10 to 40 lb dry solids/$ft^2 \cdot yr$ (49 to 195 kg/$m^2 \cdot yr$) or 1 to 3 ft^2/capita. Lower loading rates are used with cooler climates and slowly drainable sludges (such as waste activated sludge). For primary plus waste activated sludge, values of loading rates range from 12 to 20 lb/$ft^2 \cdot yr$ (60 to 100 kg/$m^2 \cdot yr$) (U.S. EPA, 1987).

Paved drying beds. Recent improvements in the design of paved drying beds changed the concept from drainage to decanting. The older asphalt- or concrete-paved drying beds were found to inhibit the subsurface drainage and therefore required larger areas than the comparably loaded sand beds. The new design features no subsurface drainage and a decant draw-off piping system. A tractor-mounted horizontal auger is used to regularly mix and aerate the sludge. Because 70 to 80 percent of the water is lost to evaporation, the new type of paved bed is best suited to warm and arid climates.

Loadings depend on the evaporation rate, annual precipitation, and solids content. Final solids contents in the range of 40 to 50 percent can be achieved in 30 to 40 d in arid climate for a 12-in layer of sludge (U.S. EPA, 1987).

To determine the area of the paved beds use

$$A = \frac{1.04S\left(\dfrac{1 - s_d}{s_d} - \dfrac{1 - s_e}{s_e}\right) + PAF}{k_e E_p F} \tag{14-5}$$

where A = bed area, ft^2 (m^2)
S = annual sludge production, dry solids, lb/yr (kg/yr)
s_d = percent dry solids of sludge after decanting, decimal
s_e = percent dry solids of sludge for final disposal, decimal
P = annual precipitation, ft/yr (m/yr)
F = conversion factor, 62.4 lb/ft^3 (1000 kg/m^3)
k_e = reduction factor for sludge versus a free water surface, decimal (use 0.6 for preliminary estimate, pilot test for final design)
E_p = free water pan evaporation rate, ft/yr (m/yr)

The total design area should be divided into at least three beds. For small systems it may be advantageous to plan the orientation of the beds to receive maximum solar radiation (U.S. EPA, 1987). The process design for paved drying beds is illustrated in Example 14-2.

EXAMPLE 14-2. DETERMINE THE AREA NEEDED FOR PAVED DRYING BEDS. The quantity of sludge to be dried is 10,000 lb/yr. The sludge to be dewatered has a solids content of 7 percent and the needed solids content for disposal is 40 percent. Annual precipitation is 1.5 ft and annual free-water pan evaporation is 3.5 ft/yr. Use $k_e = 0.6$.

Solution

1. Solve for the required area using Eq. (14-5):

$$A = \frac{1.04(10{,}000)\left(\dfrac{1 - 0.07}{0.07} - \dfrac{1 - 0.40}{0.40}\right) + (62.4)(1.5)(A)}{(62.4)(0.6)(3.5)}$$

$$= \frac{10{,}400(13.28 - 1.5) + 93.6A}{131}$$

$$= \frac{122{,}512 + 93.6A}{131}$$

$$131A - 93.6A = 122{,}512$$

$$A = 3275\ ft^2$$

2. Determine the solids loading by dividing the solids production by the drying bed area:

$$\text{Solids loading} = \frac{10{,}000\ lb/yr}{3275\ ft^2} = 3.05\ lb/ft^2 \cdot yr$$

Artificial-medium beds. The medium in artificial-medium drying beds was originally stainless steel "wedgewire" but is now made predominantly out of polyurethane. The wedgewire creates a septum for the underdrainage from the drying sludge. The medium consists of wedge-shaped slots that are 0.01 in (0.25 mm) wide.

According to manufacturers of wedgewire systems, polymer-treated aerobically digested sludges can be dewatered to 8 to 12 percent solids within 24 h. Polymer addition is desirable for all sludges, as it rapidly speeds up the dewatering process. As a result of the polymer conditioning and the false bottom for rapid drainage, annual loading rates exceed those loadings for sand beds by an order of magnitude.

Vacuum-assisted drying beds. In vacuum-assisted drying beds, a vacuum is applied to the underside of a rigid, porous medium plate on which polymer-conditioned sludge has been placed (see Fig. 14-6). The vacuum pulls the water through the plate, and the dewatered sludge accumulates on the surface. A solids loading of 2 lb/ft^2·cycle (10 kg/m^2·cycle) has been found acceptable (U.S. EPA, 1987).

For a 24-hour cycle time, a typical system would have three beds. Each bed would be sized for 70 percent of the average daily sludge production. Under this design the system would operate 5 d/wk and provide for the dewatering of 7 d worth of sludge. After each day's dewatering the sludge cake is removed by a front end

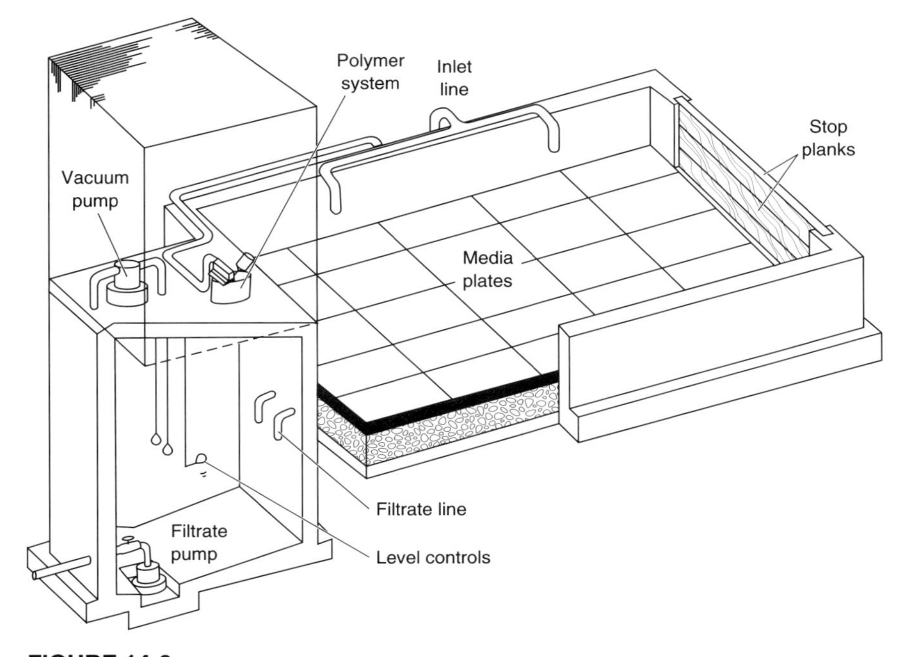

FIGURE 14-6
Typical vacuum-assisted drying bed for dewatering digested sludge (adapted from U.S. EPA, 1987).

loader, and the plates are washed down with a high-pressure hose. Typical polymer dosages are 20 lb/ton (9 kg/Mg) of dry solids (U.S. EPA, 1987).

Mechanical Dewatering

Mechanical dewatering is associated normally with larger treatment plants. The principal methods of mechanical dewatering are belt filter presses, centrifuges, and plate filter presses (Tchobanoglous and Burton, 1991; U.S. EPA, 1987). These types of dewatering are rarely cost-effective for small systems.

Sludge Freezing

Freezing of sludge greatly improves gravity dewatering upon the thawing of the frozen sludge. A waste activated sludge with a solids content of 0.6 percent will increase to about 20 percent solids after freezing. Natural freezing, the only cost-effective method of sludge freezing, can be accomplished only in the northern states and only in discrete layers of 3 in (80 mm) maximum. If the maximum depth of frost penetration is less than 24 in (600 mm), sludge freezing should not be considered as a feasible sludge dewatering option.

Where frost penetration is deep enough to be cost-effective (40 in or 1000 mm), natural sludge freezing can be considered (Reed et al., 1986). Sludge should be liquid (less than 8 percent solids) and applied to beds in 3-in (80-mm) layers. At an air temperature of 23°F (−5°C), a 3-in layer should freeze in 3 d, while at 30°F (−1°C), the freezing would take 2 wk (Reed et al, 1995). Layers deeper than 3 in should be used only in very cold climates. After the sludge layer is frozen, another layer can be applied and the process continued until the thaw period. The total depth of sludge that can be frozen can be calculated from

$$x = 1.76F_p - 40 \qquad \text{(U.S. customary units)} \qquad (14\text{-}6a)$$

$$x = 1.76F_p - 101 \qquad \text{(SI units)} \qquad (14\text{-}6b)$$

where x = total depth of sludge that can be frozen, in (mm), and F_p = maximum depth of frost penetration, in (mm). The depth of frost penetration is published for particular locations in the literature (Whiting, 1975).

A sludge freezing program can be conducted in conjunction with a warm weather system, such as land application, drying beds, or reed beds. The thawed sludge can be landfilled, land applied or composted.

Reed Beds

The reed beds used for sludge dewatering are similar in appearance to subsurface-flow constructed wetlands. The difference is that the liquid sludge is applied to the surface and the filtrate flows through the gravel to the underdrains. Typical beds are constructed of washed river-run gravel 0.8 in (20 mm) in diameter and 10 in

(250 mm) deep, overlaid by a 4-in layer of filter sand (0.3 to 0.6 mm). At least 3 ft (1 m) of freeboard above the gravel layer is provided for 10 yr accumulation of sludge. Phragmites (reeds) are planted on 12-in (300-mm) centers in the gravel layer. Other wetland vegetation can be used, although reeds are the most popular. The first sludge application is made after the reeds are well established. Harvesting of the reeds is practiced typically in the winter by cutting the tops back to a level above the sludge blanket. Harvesting is necessary whenever the plant growth becomes too thick and restricts the even flow of sludge. The harvested material can be composted, burned, or landfilled.

The purpose of the plants is to provide a pathway for continuous drainage of water from the sludge layer. Movement and growth of the plants creates pathways for water to drain from the biosolids into the underdrains. The plants also absorb water from the sludge. Oxygen transfer to the plant roots assists in the biological stabilization and mineralization of the sludge. The reed bed system is a form of passive composting. A typical reed bed for sludge dewatering treatment and storage is shown in Fig. 14-7.

The design loading rates for reed beds range from 12 to 20 $lb/ft^2 \cdot yr$ (60 to 100 $kg/m^2 \cdot yr$). The liquid sludge is applied intermittently, as in sand drying beds. The typical sludge depth applied is 3 to 4 in (75 to 100 mm) every week to 10 d.

Vermistabilization

The use of earthworms to stabilize and dewater sludge is known as *vermistabilization.* In cold climates, the beds need to be covered and heated. Worms are placed on a bed at a single application of 0.4 lb/ft^2 (2 kg/m^2) live weight. A sludge loading rate of 0.2 lb/ft^2 (1 kg/m^2) of volatile solids per week is recommended for liquid primary and liquid waste-activated sludge (Loehr et al., 1984).

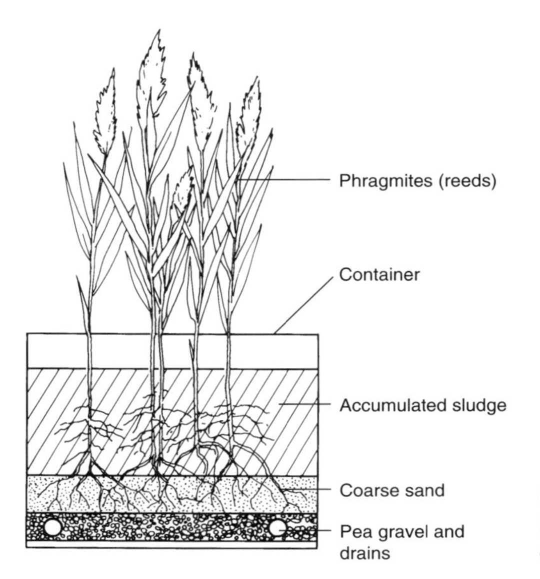

FIGURE 14-7
Reed bed for dewatering and storage of digested sludge.

At Lufkin, Texas, a thickened combination of primary and waste activated sludge with a solids content of 3.5 to 4 percent was sprayed at a rate of 0.05 $lb/ft^2 \cdot d$ (0.24 $kg/m^2 \cdot d$) over beds of earthworms and sawdust. The sawdust depth was 8 in (200 mm) initially and another 1 to 2 in (25 to 50 mm) of sawdust was added to the bed after 2 months of operation. The mixture of earthworms, castings, and sawdust is removed every 6 to 12 months (Donovan, 1981).

The earthworms are harvested by driving a small front-end loader into the bed to move the bed contents into windrows. A new source of food is spread adjacent to the windrow, and within 2 d the earthworms will migrate to the new material. The concentrated earthworms are collected and used to inoculate a new bed. The castings and sawdust residue are removed and the bed is prepared for the next cycle (Reed et al., 1995).

Although the worms stabilize and dewater the sludge, they also accumulate heavy metals. The results of numerous studies are that cadmium, copper, and zinc will accumulate in the earthworms (Pietz et al., 1984). As a result, worms from a sludge stabilization operation may not be suitable for sale as a commercial fish bait operation unless the metal concentrations in the sludge are very low.

Sludge Lagoons

Sludge lagoons can be designed for a number of purposes including sludge drying, intermediate storage in conjunction with land application, and long-term storage. The depths and design criteria of each type of sludge lagoon are discussed below.

Sludge drying lagoons. Drying lagoons can be used to dewater stabilized sludge by combining sedimentation (gravity thickening) and evaporation. Sludge depths of 2 to 4 ft (0.62 to 1.25 m) are typical. After gravity thickening, the supernatant is decanted and returned to the treatment facility. Solids loadings are typically 2.2 to 2.4 $lb/ft^2 \cdot yr$ (36 to 39 $kg/m^2 \cdot yr$) (U.S. EPA, 1987).

When sludge drying lagoons are full, the input of sludge is discontinued and the drying phase begins. As the surface dries, a crust is formed that is broken up mechanically. When the sludge solids content reaches 20 to 30 percent the sludge is removed with front end loaders.

A complete cycle for a single lagoon may take 1 to 3 yr, depending on the climate, final solids content desired, depth of sludge applied, and frequency of turning the drying sludge. A minimum of two cells is needed. Occasional odors, flies, and mosquitoes may be a problem, so a remote site location is important (U.S. EPA, 1987).

Sludge storage basins. Sludge storage basins (SSBs) are often used with seasonal land application systems. A typical example of an SSB is shown in Fig. 14-8. Design criteria for an SSB located near Crescent City, California, which accepts sludge from an extended aeration system that treats 0.75 Mgal/d (2840 m^3/d) of domestic wastewater are presented in Table 14-12.

FIGURE 14-8
Typical sludge storage basin with floating surface aerators.

TABLE 14-12
Design criteria for sludge storage basins at Pelican Bay, California

Item	Unit	Value
Number		2
Operational depth	ft	14
Freeboard, total	ft	4
Rainfall storage freeboard	ft	1
Operational volume	Mgal	8.6
Sludge loading	gal/d	34,000
Volatile solids loading	lb/d	1300
Volatile solids loading rate	$lb/10^3\ ft^2 \cdot d$	12.1
Volatile solids destruction	%	40
Sludge concentration at bottom	%	4 to 7
Sludge retention time	months	2 to 12
Aeration system type		Floating aerators
Number		14
Total horsepower	hp	42
Total aeration capacity	lb O_2/d	1510

Sludge storage basins range in depth from 10 to 16 ft (3 to 5 m). Floating mechanical aerators are usually provided to maintain aerobic conditions in the upper layers. Typical solids loadings for SSBs equipped with surface aerators range from 20 to 50 lb VSS/10^3 $ft^2 \cdot d$ (0.1 to 0.25 kg VSS/$m^2 \cdot d$) of surface area. If the SSBs are not aerated, the volatile solids loads should be less than 20 lb/10^3 $ft^2 \cdot d$.

Multiple basins are used to ensure that one basin can be out of service for up to 6 months for maintenance. Sludge is removed by floating sludge pumps.

Long-term storage. Stabilized sludge can be stored in surface impoundments for many years. Long-term storage is effective in reducing bacteria and virus levels in sludge and also reducing concentrations of volatile solids and nitrogen.

14-5 COMPOSTING

Composting is a biological stabilization process that disinfects sludge and produces a humuslike product that can be used beneficially as a soil amendment. Aerobic composting minimizes the potential for nuisance odors. Composting requires a sludge amendment mixture with an initial solids content of 40 percent.

Compost Process

The process of composting, regardless of the type of reactor or aeration technique, can be described as follows:

1. Dewatered sludge is mixed with an amendment or bulking agent to increase the solids content, provide supplemental carbon, and increase porosity. Amendments, such as sawdust, finished compost, leaves, rice hulls, and peanut hulls become part of the finished product and are selected on the basis of cost and availability for their degradable carbon content. Bulking agents, such as wood chips, bark, or shredded tires, are chosen to increase the solids content and provide porosity. Bulking agents such as these need to be screened out of the finished compost.
2. The mixed sludge and amendment/bulking agent is heated by bacterial action to the point at which pathogens are destroyed.
3. The mixture is aerated for 15 to 30 d by blowers, periodic remixing or a combination of both. Aeration supplies oxygen for the aerobic microbes, controls the temperature, and removes water vapor.
4. Bulking agents are screened out of the finished compost.
5. Compost is cured for an additional 30 to 60 d to complete the stabilization process.

Although composting is an old and relatively simple process, the simultaneous need for temperature rise, aeration, and drying makes the operation somewhat complex. The process has been compared to placing a large log on a fire, "slow to start but once started hard to control" (Haug, 1994).

The three main types of composting systems are the windrow, aerated static pile, and in-vessel mechanical systems. For small systems the windrow and aerated static pile systems are usually most appropriate and affordable. In a 1995 survey of 228 operational systems, 31 percent of the sludge composting facilities were windrow; 48 percent were aerated static pile, and 21 percent were in-vessel mechanical (Goldstein and Steuteville, 1995). A list of operational composting facilities is shown in Table 14-13. Additonal details on the compositing process may be found in Epstein (1997), Haug (1994), and Tchobanoglous et al. (1993).

Windrow Composting

In the windrow process, the sludge/amendment mixture to be composted is placed in long piles (see Fig. 14-9*a*). The windrows are 3 to 6 ft high (1 to 2 m) and 6 to 15 ft wide (2 to 5 m) at the base. The windrow process is conducted normally in uncovered pads and relies on natural ventilation with frequent mechanical mixing of the piles to maintain aerobic conditions. Under typical operating conditions the windrows are turned every other day. The turning is accomplished with specialized equipment (see Fig. 14-9*b*) and serves to aerate the pile and allow moisture to escape. To meet

TABLE 14-13
Operational composting systems*

Location	Type	Sludge volume, dry tons/d
Hot Springs, Arkansas	Windrow	10.0
Chino Basin Municipal Water District, California	Windrow	26.0
Las Virgenes, California	In-vessel	4.7
Ft. Collins, Colorado	Aerated windrow	0.75
Fairfield, Connecticut	In-vessel	3.3
Cooper City Utilities, Florida	Windrow	1.0
Maui County, Hawaii	ASP†	10.0
Lewiston, Idaho	ASP	4.0
Lewiston/Auburn, Maine	In-vessel	9.0
Elkton, Maryland	ASP	2.0
Kalispell, Montana	ASP	1.6
Albuquerque, New Mexico	Windrow	4.5
Yakima, Washington	Windrow	1.5

*From Goldstein and Steuteville (1995).
† ASP = aerated static pile.

the U.S. EPA pathogen reduction requirements the windrows have to be turned five times in 15 d, maintaing a temperature of 55°C (WEF, 1992).

Because anaerobic conditions can develop within the windrow between turnings, putrescible compounds can be formed which can cause offensive odors, especially as the windrows are turned. In many locations negative aeration is provided to limit the formation of odorous compounds. Where air is provided mechanically, the process is known as aerated windrow composting (Epstein, 1997).

(*a*)

(*b*)

FIGURE 14-9
Typical windrow composting system for digested sludge and greenwaste: (*a*) material to be composted is placed in windrows and (*b*) mechanized windrow turning device used to mix and aerate the material being composted.

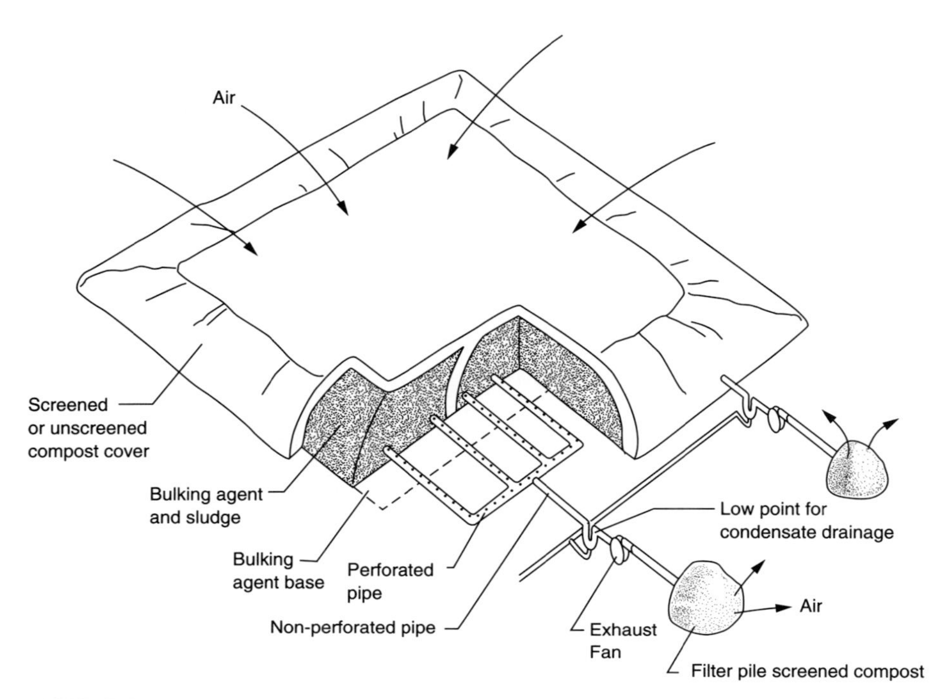

FIGURE 14-10
Aerated static pile composting of digested sludge.

TABLE 14-14
Design considerations for sludge composting

Factor	Discussion
Sludge type	Digested and undigested sludge can be composted. Undigested sludge needs more aeration and odor control considerations in siting and design.
Solids content	Optimum solids content is 40 to 50%. Dewatered sludge will usually require amendments or bulking agent to adjust the solids content.
Temperature	Most efficient operation is when temperature is between 130 and 150°F (55 and 60°C). High temperature inactivates pathogens.
Carbon to nitrogen ratio	20 to 35:1 by weight. Low or high C/N ratio limits the reaction rate. Biodegradable carbon is important to check because lignin is not readily degradable.
Oxygen supply	10 to 15%. Less than 5% can lead to odors.
Detention time for active composting	14 to 21 d for aerated static pile. 30 to 45 d for windrow.

Aerated Static Pile Composting

In the aerated static pile process the material to be composted is placed in a pile and oxygen is provided by mechanical aeration systems. Most states require paved surfaces for the pile construction areas to permit capture and control of runoff and allow operation during wet weather. The most common aeration system involves the use of a grid of subsurface piping (see Fig. 14-10). Aeration piping often consists of flexible plastic drainage tubing assembled on the composting pad. Because the drainage-type aeration piping is inexpensive it is often used only once. Before constructing the static pile, a layer of wood chips is placed over the aeration pipes or grid to provide unifrom air distribution. The static pile is then built up to 8 to 12 ft (2.6 to 3.9 m) using a front end loader. A cover layer of screened or unscreened compost is placed over the sludge to be composted. Typically, oxygen is provided by pulling air through the pile with an exhaust fan. Air that has passed through the compost pile is vented to the atmosphere through a compost filter for odor control.

In-Vessel Mechanical Composting

In-vessel reactors are enclosed in a building or a closed reactor to control temperature and moisture and odors. The in-vessel systems are either plug-flow or agitated-bed type (U.S. EPA, 1989). Because in-vessel systems are more expensive they are best suited to large systems.

Design Considerations

Design considerations for sludge composting are presented in Table 14-14. Most sludges can be composted provided the volatile solids are over 30 percent of the total solids content. Degradability of the volatile solids and the amendment solids is important because some wood materials, such as yard waste may be slowly degradable and may take many months to stabilize.

The pH of the mixture should be between 6 and 11. If "significant" pathogen reduction is acceptable for the use intended, the time and temperature requirement is a minimum of 5 d at 105°F (40°C) with 4 h at 130°F (55°C) or higher for windrow composting. If "further" reduction is necessary, then 3 d at 130°F (55°C) is needed for aerated static piles or 15 d at 130°F (55°C) with five turnings for the windrow method. Temperatures and times of exposure that are required for the destruction of common pathogens are presented in Table 14-15.

A design for sludge composting requires that a mass balance be conducted using the solids content of the sludge and the amendment/bulking agent. The area needed for composting ranges from 0.15 to 0.9 ac/ton (0.05 to 0.33 ha/tonne) of dry solids composted per day, depending on storage requirements and buffer area. The pad area depends on the detention time, amount of bulking agent, and height of the pile:

$$A = \frac{1.1S(R + 1)}{H} \tag{14-7}$$

TABLE 14-15
Temperature and time of exposure required for destruction of some common pathogens and parasites*

Organism	Observations
Salmonella typhosa	No growth beyond 46°C; death within 30 min at 55–60°C and within 20 min at 60°C; destroyed in a short time in compost environment.
Salmonella sp.	Death within 1 h at 55°C and within 15–20 min at 60°C.
Shigella sp.	Death within 1 h at 55°C.
Escherichia coli	Most die within 1 h at 55°C and within 15–20 min at 60°C.
Entamoeba histolytica cysts	Death within a few min at 45°C and within a few seconds at 55° C.
Taenia saginata	Death within a few min at 55°C.
Trichinella spiralis larvae	Quickly killed at 55°C; instantly killed at 60°C.
Brucella abortus or *Br. suis*	Death within 3 min at 62–63°C and within 1 h at 55°C.
Micrococcus pyogenes var. *aureus*	Death within 10 min at 50°C.
Streptococcus pyogenes	Death within 10 min at 54°C.
Mycobacterium tuberculosis var. *hominis*	Death within 15–20 min at 66°C or after momentary heating at 67°C.
Corynebacterium diptheria	Death within 45 min at 55°C.
Necator americanus	Death within 50 min at 45 C.
Ascaris lumbricoides eggs	Death in less than 1 h at temperatures over 50°C.

*From Tchobanoglous et al. (1993).
Note: 1.8 × (°C) + 32 =° F.

where A = pad area for active compost piles, ft^2 (m^2)
S = total volume of sludge produced in 4 weeks, ft^3 (m^3)
R = ratio of bulking agent to sludge, ft^3/ft^3 (m^3/m^3)
H = height of pile, not including cover or base material, ft (m)

Distribution and Marketing

Sludge compost is a humuslike material that can be used beneficially as a soil amendment. It can improve the structure and water-holding capacity of soil, in addition to improving soil aeration and drainage. Users of biosolids from composting include landscape contractors, nurseries, public parks and recreation departments, sod farms, golf courses, and homeowners. Soil blenders may also purchase or take compost and use it in their products. The distribution and marketing of compost are regulated federally under 40 CFR Part 503.

14-6 LAND APPLICATION OF BIOSOLIDS

Applying biosolids to agricultural land is beneficial from a number of standpoints including (NRC, 1996):

- Organic matter improves soil structure, tilth, water-holding capacity, water infiltration, and soil aeration.
- Macronutrients (nitrogen, phosphorus, potassium) aid plant growth.
- Micronutrients (iron, manganese, copper, chromium, selenium, and zinc) aid plant growth.

Organic matter also contributes to the cation exchange capacity (CEC) of the soil, which allows the soil to retain potassium, calcium and magnesium. Organic matter presence improves the biological diversity in soil and makes nutrients more available to the plants (Wegner, 1992).

Land application can also be of great value in silviculture and site reclamation. Forest utilization has been practiced extensively in the northwest (Leonard et al., 1992). Reclamation of disturbed land such as strip-mined land has also been very successful (WEF, 1989 and Sopper and Kerr, 1979). In this book a distinction is made between agricultural or silvicultural land application and high-rate land application or dedicated land disposal. Loading rates of up to 50 tons/ac (112 tonnes/ha) are considered in the agricultural or silvicultural land application category. Dedicated land disposal is addressed in the next section.

Site Selection

Suitable sites for land application of biosolids are similar to those for slow-rate land treatment. Ideal sites have deep silty loam to sandy loam soils; groundwater deeper than 10 ft (3 m); slopes at 0 to 3 percent; no wells, wetlands, or streams; and few neighbors. Site characteristics of importance are topography, site drainage, soil type and depth, soil permeability, soil chemistry, depth to groundwater, proximity to critical areas (requiring setbacks or buffer zones) and accessibility for sludge delivery.

Topography. Suitability of site topographies depends on the type of sludge and the method of application. Liquid sludge can be spread, sprayed, or injected onto sites with rolling terrain up to 15 percent in slope. Forested sites can accommodate slopes up to 30 percent provided that adequate setbacks are observed from streams.

Dewatered sludge is usually spread on agricultural land which requires a tractor and spreader. Slopes up to 15 percent have been used in agriculture and up to 30 percent in silviculture. The presence of any drainage courses on the site should be observed and documented.

Soil characteristics. Desirable soil characteristics include (1) loamy soil, (2) slow to moderate permeability, (3) soil depth of 2 ft (0.6 m) or more, (4) neutral to alkaline soil pH, and (5) well drained to moderately well drained soil. Practically any soil can be adapted to a well-designed and -operated system.

TABLE 14-16
Typical setback distances for land application sites

Setback from	Distance, ft
Residences	200–300
Property boundaries	50–200
Roads	50–100
Flood plains	50–100
Streams	50–200
Private wells	100–200
Public supply wells	500–2000

Depth to groundwater. At least 3 ft (1 m) to groundwater is preferred for land application sites. Seasonal water table fluctuations to within 1.5 ft (0.5 m) of the surface can be tolerated. If the shallow groundwater is excluded as a drinking water aquifer, the groundwater depth can be as shallow as 1.5 ft before problems with trafficability of the soil arise.

Setback distances. Application areas are often reduced by setbacks or buffer zones to separate the active application area from sensitive areas such as residences, wells, roads, surface waters, and property boundaries. Local and state regulations often include minimum distances for setbacks depending on the method of application. State regulations and contact names to check for regulatory updates are found in WEF (1989) and U.S. EPA (1995). Some typical setback distances are presented in Table 14-16. Minimal setbacks are needed when sludge is injected into the subsurface.

EPA Regulations

The U.S. Environmental Protection Agency (U.S. EPA) has published regulations on biosolids (sewage sludge is the term used in the regulations) use and disposal under the Code of Federal Regulations (CFR), 40 CFR Part 503. For land application the regulations provide numerical limits on 10 metals, management practice guidance, and requirements for monitoring, record keeping and reporting. The regulations are summarized in Table 14-17. Calculation of metal loadings is illustrated in Example 14-3.

Management practices. As described in Table 14-18, the Part 503 rule specifies management practices that must be followed when biosolids are land-applied. The practices vary depending on whether the material that is applied is hauled in bulk or in individual bags.

Pathogen reduction alternatives. The Part 503 pathogen reduction requirements for biosolids are divided into categories of Class A and Class B as described

TABLE 14-17

EPA sludge regulations for land application

Pathogen classifications	Class A: no restrictions other than bag labeling (like a fertilizer) Class B: site restrictions
Metal limits	See Table 14-22
Management practices	See Table 14-18
Pathogen reduction alternatives	See Table 14-19
Vector attraction reduction	See Table 14-20
Site restrictions for Class B	See Table 14-21

TABLE 14-18

Land application management practices under EPA Part 503 rule*

For bulk sewage sludge†
Bulk sewage sludge cannot be applied to flooded, frozen, or snow-covered agricultural land, forests, public contact sites, or reclamation sites in such a way that the sewage sludge enters a wetland or other waters of the United States (as defined in 40 CFR Part 122.2), except as provided in a permit issued pursuant to Section 402 (NPDES permit) or Section 404 (Dredge and Fill Permit) of the Clean Water Act, as amended.
Bulk sewage sludge cannot be applied to agricultural land, forests, or reclamation sites that are 10 meters or less from U.S. waters, unless otherwise specified by the permitting authority.
If applied to agricultural lands, forests, or public contact sites, bulk sewage sludge must be applied at a rate that is equal to or less than the agronomic rate for the site. Sewage sludge applied to reclamation sites may exceed the agronomic rate if allowed by the permitting authority.
Bulk sewage sludge must not harm or contribute to the harm of a threatened or endangered species or result in the destruction or adverse modification of the species' critical habitat when applied to the land. Threatened or endangered species and their critical habitats are listed in Section 4 of the Endangered Species Act. Critical habitat is defined as any place where a threatened or endangered species lives and grows during any stage of its life cycle. Any direct or indirect action (or the result of any direct or indirect action) in a critical habitat that diminishes the likelihood of survival and recovery of a listed species is considered destruction or adverse modification of a critical habitat.
For sewage sludge sold or given away in a bag or other container for application to the land*
A label must be affixed to the bag or other container, or an information sheet must be provided to the person who receives this type of sewage sludge in another container. At a minimum, the label or information sheet must contain the following information:
The name and address of the person who prepared the sewage sludge for sale or give-away in a bag or other container.
A statement that prohibits application of the sewage sludge to the land except in accordance with the instructions on the label or information sheet.
An AWSAR (annual whole sludge application rate) for the sewage sludge that does not cause the annual pollutant loading rate limits to be exceeded.

*From U.S. EPA (1995).

†These management practices do not apply if the sewage sludge is of "exceptional quality."

TABLE 14-19
Pathogen reduction alternatives*

Class A

In addition to meeting the requirements in one of the six alternatives listed below, fecal coliform or *Salmonella* sp. bacterial levels must meet specific densities at the time of sewage sludge use or disposal, when prepared for sale or give-away in a bag or other container for application to the land, or when prepared to meet the requirements in 503.10(b), (c), (e), or (f).

Alternative 1 Thermally treated sewage sludge: Use one of four time-temperature regimes.

Alternative 2 Sewage sludge treated in a high-pH–high-temperature process: Specifies pH, temperature, and air-drying requirements.

Alternative 3 For sewage sludge treated in other processes: Demonstrate that the process can reduce enteric viruses and viable helminth ova. Maintain operating conditions used in the demonstration.

Alternative 4 Sewage sludge treated in unknown processes: Demonstration of the process is unnecessary. Instead, test for pathogens—*Salmonella* sp. bacteria, enteric viruses, and viable helminth ova—at the time the sewage sludge is used or disposed, or is prepared for sale or give-away in a bag or other container for application to the land, or when prepared to meet the requirements in 503.10(b), (c), (e), or (f).

Alternative 5 Use of PFRP: Sewage sludge is treated in one of the processes to further reduce pathogens (PFRP).

Alternative 6 Use of a process equivalent to PFRP: Sewage sludge is treated in a process equivalent to one of the PFRPs, as determined by the permitting authority.

Class B

The requirements in one of the three alternatives below must be met in addition to Class B site restrictions.

Alternative 1 Monitoring of indicator organisms: Test for fecal coliform density as an indicator for all pathogens at the time of sewage sludge use or disposal.

Alternative 2 Use of PSRP: Sewage sludge is treated in one of the processes to significantly reduce pathogens (PSRP).

Alternative 3 Use of processes equivalent to PSRP: Sewage sludge is treated in a process equivalent to one of the PSRPs, as determined by the permitting authority.

*From U.S. EPA (1995).

in Table 14-19. The goal of the Class A requirements is to reduce the pathogens in the sludge (including *Salmonella* sp. bacteria, enteric viruses, and viable helminth ova) to below detectable levels. When this goal is achieved, Class A biosolids can be land-applied without any pathogen-related restrictions on the site (U.S. EPA, 1995).

The goal of the Class B requirements is to ensure that pathogens have been reduced to levels that are unlikely to pose a threat to public health and the environment under specific use conditions. Site restrictions on land application of Class B biosolids minimize the potential for human and animal contact with the biosolids until environmental factors have reduced pathogens to below detectable levels.

Vector attraction reduction. There are 10 potential vector attraction reduction measures that can be combined with pathogen reduction alternatives for an acceptable land application project using Class B biosolids. The list, presented in Table 14-20, includes some stabilization processes that also reduce pathogens.

Site restrictions for Class B. The restrictions, listed in Table 14-21, depend on the crops to be used and the contact control for animals and the public. Food crops and turf grass are given the longest time restrictions because of the potential for public exposure (U.S. EPA, 1995).

Design Loading Rates

Design loading rates for biosolids land application can be limited by certain constituents (heavy metals) or by nitrogen. The long-term loadings of heavy metals are based on EPA 503 regulations. The annual loading rate is usually limited by the nitrogen loading rate.

Nitrogen loading rates. Nitrogen loading rates are set typically to match the available nitrogen provided by commercial fertilizers (Chang et al., 1995). Because municipal biosolids represent a slow release organic fertilizer, a combination of ammonia and organic nitrogen must be made according to this equation:

$$L_N = [(\mathrm{NO_3}) + k_v(\mathrm{NH_4}) + f_n(N_o)]F \tag{14-8}$$

where L_N = plant available nitrogen in the application year, lb N/ton (kg N/tonne)
NO_3 = percent nitrate nitrogen in sludge, decimal
k_v = volatilization factor for ammonia loss
= 0.5 for surface-applied liquid sludge
= 0.75 for surface-applied dewatered sludge
= 1.0 for injected liquid or dewatered sludge
NH_4 = percent ammonia nitrogen in sludge, decimal
f_n = mineralization factor for organic nitrogen
= 0.5 for warm climates and digested sludge
= 0.4 for cool climates and digested sludge
= 0.3 for cold climates or composted sludge
N_o = percent organic nitrogen in sludge, decimal
F = conversion factor, 2000 lb/ton dry solids (1000 kg/tonne)

Using Eq. (14-8) properly requires knowledge of the method of application, the nitrogen content of the sludge (nitrate, ammonia, and organic), the type of stabilization, and the type of climate. The use of the mineralization factors simplifies the previously used method of calculating the amount of organic nitrogen mineralized each year and adding up the total for an annual equivalent. The use of Eq. (14-8) is also appropriate if sludge is applied to a single site once every 2 to 3 yr.

TABLE 14-20

Vector attraction reduction*

Requirement	What is required?	Most appropriate for
Option 1 503.33(b)(1)	At least 38% reduction in volatile solids during sewage sludge treatment	Sewage sludge processed by: Anaerobic biological treatment Aerobic biological treatment Chemical oxidation
Option 2 503.33(b)(2)	Less than 17% additional volatile solids loss during bench-scale anaerobic batch digestion of the sewage sludge for 40 additional days at 30°C to 37°C (86°F to 99°F)	Only for anaerobically digested sewage sludge.
Option 3 503.33(b)(3)	Less than 15% additional volatile solids reduction during bench-scale aerobic batch digestion for 30 additional days at 20°C (68°F)	Only for aerobically digested sewage sludge with 2% or less solids—e.g., sewage sludge treated in extended aeration plants.
Option 4 503.33(b)(4)	SOUR at 20°C (68°F) is 1.5 mg O_2 per hour per gram total sewage sludge solids	Sewage sludge from aerobic processes (should not be used for composted sludges). Also for sewage sludge that has been deprived of oxygen for longer than 1–2 hours.
Option 5 503.33(b)(5)	Aerobic treatment of the sewage sludge for at least 14 d at over 40°C (104°F) with an average temperature of over 45°C (113°F)	Composted sewage sludge (options 3 and 4 are likely to be easier to meet for sewage sludge from other aerobic processes).

Option 6 503.33(b)(6)	Addition of sufficient alkali to raise the pH to at least 12 at 25°C (77°F) and maintain a pH of 12 for 2 hours and a pH of 11.5 for 22 more hours	Alkali-treated sewage sludge (alkalies include lime, fly ash, kiln dust, and wood ash).
Option 7 503.33(b)(7)	Percent solids 75% prior to mixing with other materials	Sewage sludges treated by an aerobic or anaerobic process (i.e., sewage sludges that do not contain unstabilized solids generated in primary wastewater treatment).
Option 8 503.33(b)(8)	Percent solids 90% prior to mixing with other materials	Sewage sludges that contain unstabilized solids generated in primary wastewater treatment (e.g., any heat-dried sewage sludges).
Option 9 503.33(b)(9)	Sewage sludge is injected into soil so that no significant amount of sewage sludge is present on the land surface 1 hour after injection, except Class A sewage sludge which must be injected within 8 hours after the pathogen reduction process	Liquid sewage sludge applied to the land. Domestic septage applied to agricultural land, a forest, or a reclamation site.
Option 10 503.33(b)(10)	Sewage sludge is incorporated into the soil within 6 hours after application to land. Class A sewage sludge must be applied to the land surface within 8 hours after the pathogen reduction process, and must be incorporated within 6 hours after application.	Sewage sludge applied to the land. Domestic septage applied to agricultural land, forest, or a reclamation site.

*From U.S. EPA (1995).

TABLE 14-21
Site restrictions for Class B biosolids*

Restrictions for the harvesting of crops and turf
1. Food crops with harvested parts that touch the sewage sludge/soil mixture and are totally above ground shall not be harvested for *14 months* after application of sewage sludge.
2. Food crops with harvested parts below the land surface where sewage sludge remains on the land surface for 4 months or longer prior to incorporation into the soil shall not be harvested for *20 months* after sewage sludge application.
3. Food crops with harvested parts below the land surface where sewage sludge remains on the land surface for less than 4 months prior to incorporation shall not be harvested for *38 months* after sewage sludge application.
4. Food crops, feed crops, and fiber crops, whose edible parts do not touch the surface of the soil, shall not be harvested for *30 d* after sewage sludge application.
5. Turf grown on land where sewage sludge is applied shall not be harvested for *1 yr* after application of the sewage sludge when the harvested turf is placed on either land with a high potential for public exposure or a lawn, unless otherwise specified by the permitting authority.
Restriction for the grazing of animals
1. Animals shall not be grazed on land for 30 d after application of sewage sludge to the land.
Restrictions for public contact
1. Access to land with a high potential for public exposure, such as a park or ballfield, is restricted for 1 yr after sewage sludge application. Examples of restricted access include posting with no trespassing signs or fencing.
2. Access to land with a low potential for public exposure (e.g., private farmland) is restricted for 30 d after sewage sludge application. An example of restricted access is remoteness.

*From U.S. EPA (1995).

The sludge loading rate based on nitrogen loadings is then calculated by

$$L_{SN} = \frac{U}{N_p} \tag{14-9}$$

where L_{SN} = annual sludge loading rate based on N, ton/ac (tonne/ha)
U = crop uptake of nitrogen, lb/ac (see Chap. 10, Table 10-17) (kg/ha)
N_p = plant-available nitrogen in sludge, lb/ton (kg/tonne)

Loading rates based on constituent loading. The constituents of concern are those listed in Table 14-22. To calculate the sludge loading rate based on constituent loading, use

$$L_S = \frac{L_C}{CF} \tag{14-10}$$

where L_S = maximum amount of sludge that can be applied per year, ton/ac·yr (tonne/ha·yr)
L_C = maximum amount of constituent that can be applied per year, lb/ac·yr (kg/ha·yr)
C = constituent concentration in biosolids, decimal (mg/kg)
F = conversion factor, 2000 lb/ton [(0.001 kg/tonne)/mg/kg]

TABLE 14-22
Metals concentrations and loading rates from EPA 503 regulation

(1) Constituent	(2) Ceiling concentration, mg/kg*	(3) Cumulative consituent loading rate, kg/ha†	(4) Consituent concentration for exceptional quality, mg/kg‡	(5) Annual consituent loading rate, kg/ha·yr§
Arsenic	75	41	41	2.0
Cadmium	85	39	39	1.9
Chromium¶				
Copper	4300	1500	1500	75
Lead	840	300	300	15
Mercury	57	17	17	0.85
Molybdenum	75			
Nickel	420	420	420	21
Selenium	100	100	100	5.0
Zinc	7500	2800	2800	140

*Dry weight basis, Table 1 from 503 regulations.
† Dry weight basis, Table 2 from 503 regulations.
‡ Dry weight basis, Table 3 from 503 regulations.
§ Table 4 from 503 regulations.
¶ A February 25, 1994, Federal Register Notice deleted chromium; deleted the molybdenum values for Tables 2, 3, and 4; and raised the selenium value in Table 3 from 36 to 100.

EXAMPLE 14-3. METAL LOADINGS IN LAND APPLICATION. A community has a treatment pond that has been in operation over 20 yr. The sludge level in the primary pond is 2 ft deep and a removal program has been recommended. The metal concentrations (mg/kg) are as follows:

As	50
Cd	20
Cu	2000
Pb	400
Hg	15
Mb	15
Ni	300
Se	70
Zn	3500

Determine if the biosolids are acceptable for land application.

Solution

1. Compare the constituent concentrations for the above metals to the ceiling concentration (column 2) and the constituent concentration for exceptional quality (column 4) in Table 14-22.
   a. All metals concentrations are under the ceiling limits in column 2. The biosolids are suitable for land application.
   b. Arsenic, copper, lead, and zinc exceed the values for exceptional quality. Calculations of annual loadings are necessary.

2. Calculate the allowable annual biosolids loading rates, using Eq. (14-10), for the four metals for the annual constituent loading rates in Table 14-22, column 5.
   a. Arsenic-based loading rate:

$$L_{SAs} = \frac{L_{As}}{C_p(0.001)} = \frac{2 \text{ kg/ha·yr}}{50 \text{ mg/kg}[(0.001 \text{ kg/tonne})/\text{mg/kg}]} = 40 \text{ tonne/ha·yr}$$

   b. Copper-based loading rate:

$$L_{SCu} = \frac{75}{2000(0.001)} = 37.5 \text{ tonne/ha·yr}$$

   c. Lead-based loading rate:

$$L_{SPb} = \frac{15}{400(0.001)} = 37.5 \text{ tonne/ha·yr}$$

   d. Zinc-based loading rate:

$$L_{SZn} = \frac{140}{3500(0.001)} = 40 \text{ tonne/ha·yr}$$

3. Compare the whole biosolids loading rates to determine the limiting rate. The 37.5 tonne/ha biosolids loadings based on lead and copper loadings are limiting.

Comment. Nitrogen loadings typically are more limiting than metal loadings. If the nitrogen loading rate exceeds 37.5 tonne/ha, then the lead and copper loading rates will determine the whole biosolids loading rate.

Land requirements. Once the minimum sludge loading rate is determined [by comparing the values from Eqs. (14-9) and (14-10)], the field area can be calculated by Eq. (14-11):

$$A = \frac{B}{L_S} \tag{14-11}$$

where A = application area required, ac (ha)
B = biosolids production, tons of dry solids/yr (tonne/yr)
L_S = design sludge loading rate, tons of dry solids/ac·yr (tonne/ha·yr)

Application Methods

Application methods for biosolids range from direct injection of liquid biosolids to surface spreading of dewatered biosolids.

Liquid application methods. If the application site is nearby, liquid application is cost-effective. Liquid biosolids can be piped directly from storage basins into injection hoses attached to tractors. Injection vehicles can also hold liquid biosolids and inject the liquid into the root zone as shown in Fig. 14-11. An injection vehicle used in Norway is shown in Fig. 14-12.

FIGURE 14-11
Subsurface injection of biosolids.

FIGURE 14-12
Vehicle used for liquid injection of biosolids in land application (As, Norway).

Surface application methods include spreading by tank wagons, special application vehicles equipped with splash plates and flotation tires, tank trucks, portable or fixed spray irrigation systems, and ridge and furrow irrigation. Sprinkler application from application vehicles has been adapted to forests.

Dewatered biosolids application methods. The most common method of spreading dewatered solids is through the use of manure spreaders. Typical solids concentrations of dewatered biosolids applied to land are between 15 and 20 percent.

FIGURE 14-13
Vehicle used for land application of dewatered biosolids.

TABLE 14-23
Long-term effects of land application*

Location	Years of application	Crops grown	Site characteristics	Application rate, ft/ac·yr	Comments
Albuquerque, New Mexico	5	Forage grass	Dedicated Public	12–14	Dewatered cake
Chicago, Illinois	20	Corn, wheat, soybeans	Public	20–40	Disturbed land reclamation
Denver, Colorado	12	Corn, wheat	Private/Public	1–10	5000 ac corn 18,000 ac wheat
Madison, Wisconsin	13	Corn, soybeans, alfalfa	Private	3–5	250 to 300 farms permitted
Port Huron, Michigan	9	Corn, soybeans, wheat	Private	3–5	3000 ac under contract
Raleigh, North Carolina	11	Corn, soybeans sorghum, hay	Public	2–6	High nitrates in some hay
Reno/Sparks, Nevada	14	Seed garlic, wheat, alfalfa	Private	10–20	Dewatered cake
Springfield, Illinois	19	Corn	Dedicated	20–40	Operational since 1974
Virginia Beach, Virginia	9	Corn, wheat soybeans	Public/Private	2–4	Liquid biosolids on district land, dewatered on private land
Yuma, Arizona	12	Barley, wheat, cotton, alfalfa	Private	3–7	Contract operator, Los Angeles biosolids

*Adapted from Stukenberg et al. (1993).

The advantage of using manure spreaders is that they are widely used in agriculture (see Fig. 14-13). For forest application a side-slinging vehicle has been tested that can apply dewatered biosolids up to 200 ft (61 m) (Leonard et al., 1992).

Long-term effects. Land application of biosolids is a long-standing practice in the United States. In 1993 a study was conducted of 10 land application programs to determine the long-term experience (Stukenberg et al., 1993). Monitoring experience with soil, biosolids, crops, and groundwater was reported. The 10 sites are summarized in Table 14-23. In general the long-term effects were not significant, with crops, soils, and groundwater being comparable to those in the area where commercial fertilizer was used.

14-7 DEDICATED LAND DISPOSAL

Two types of high-rate land application exist—disturbed land reclamation and dedicated land disposal. Disturbed land reclamation consists of a one-time application of 50 to 100 dry tons/ac (112 to 224 tonnes/ha) to correct adverse soil conditions of lack of soil fertility and poor physical properties and to allow revegetation programs to proceed. For disturbed land reclamation to be the sole avenue for sludge reuse a large area of disturbed land must be available on an ongoing basis. The details of successful programs in Pennsylvania and West Virginia should be consulted (Sopper, 1993).

A dedicated land disposal project requires a site where high rates of sludge application on a continuing basis are acceptable environmentally. Sludge for a dedicated land disposal (DLD) operation should be stabilized to minimize odors, insect breeding, and pathogen transmission.

Site Selection

A list of siting criteria for a dedicated land disposal site is presented in Table 14-24. A major issue in DLD siting is nitrogen control. Groundwater contamination can be avoided by

1. Avoiding useful groundwater aquifers
2. Intercepting leachate
3. Seeking impervious geological barrier location protecting the groundwater
4. Ensuring low percolation and deep aquifer, and making calculations to show that the impact will be minimal

Loading Rates

Annual sludge loading rates have ranged from 5 to 1000 tons/ac (12 to 2250 tonnes/ha). The higher rates have been associated with sites that:

- Receive dewatered sludge
- Mechanically incorporate the sludge into the soil
- Have relatively low precipitation
- Have no leachate problems because of site conditions or project design

Design loading rates for DLDs can be estimated by

$$L_S = \frac{E(\text{TS})F}{100} - \text{TS} \tag{14-12}$$

where L_S = annual sludge loading rate, ton/ac (tonne/ha)
E = net soil evaporation rate, in/yr (mm/yr)
TS = total solids content, percent by weight
F = conversion factor, 113.3 ton/in (10 tonne/mm)

The net soil evaporation can be estimated from

$$E = fE_L - P \tag{14-13}$$

where E = net soil evaporation rate, in/yr (mm/yr)
f = 0.7
E_L = lake evaporation rate, in/yr (mm/yr)
P = annual precipitation, in/yr (mm/yr)

Equation (14-12) assumes no infiltration into the soil. If infiltration is allowed, the term E should be increased by the annual infiltration rate in inches per year.

Once the annual loading rate is calculated, the field area can be determined from Eq. (14-11) (dividing the sludge production by the loading rate). Other area requirements include buffer zones, surface runoff control, roads, and supporting facilities.

TABLE 14-24
Siting criteria for dedicated land disposal*

Parameter	Unacceptable condition	Ideal condition
Slope	Deep gullies, slope >12%	<3%
Soil permeability	$> 1 \times 10^5$ cm/s[†]	$\leq 10^{-7}$ cm/s[‡]
Soil depth	<2 ft (0.6 m)	>10 ft (3 m)
Distance to surface water	<300 ft (92 m) to any pond or lake used for recreational or livestock purposes, or any surface water body officially classified under state law	>1000 ft (305 m) from any surface water
Depth to groundwater	<10 ft (3.1 m) to groundwater table (wells tapping shallow aquifers)[§]	>50 ft (15.3 m)
Supply wells	Within 1000 ft (305 m) radius	No wells within 2000 ft (610 m)

*From U.S. EPA (1983).
[†] Permeable soil can be used for DLD if appropriate engineering design preventing DLD leachate from reaching the groundwater is feasible.
[‡] When low-permeability soils are at or too close to the surface, liquid disposal operations can be hindered because of water ponding.
[§] If an exempted aquifer underlies the site, poor-quality leachate may be permitted to enter groundwater.

14-8 LANDFILLING

Landfilling of sludge in a sanitary landfill with municipal solid waste is regulated by the EPA under 40 CFR 258. Landfilling of sludge alone in a monofill is covered under 40 CFR 503. If an acceptable site is convenient, landfilling can be used for disposal of sludge, grit, screenings, and other solids.

Most states have regulations on the use of sludge in landfills. Some states, such as New Jersey, have banned sludge from municipal solid waste (MSW) landfills. Other states restrict the solids content of the sludge to 15 to 20 percent or more and require a minimum ratio of 5:1 (municipal solid waste to sludge) in the landfill.

A concern with landfilling sludge is the increase in leachate from the landfill after sludge is introduced. All new MSW landfills and lateral expansions that receive sludge must be constructed with composite liners and a leachate collection system or equivalent.

14-9 COMPARISON OF BIOSOLIDS ALTERNATIVES

The reuse or disposal of biosolids is an economic choice that must be made within the local, state, and federal regulatory requirements. The alternatives are compared in Example 14-4.

EXAMPLE 14-4. COMPARISON OF BIOSOLIDS MANAGEMENT ALTERNATIVES. A community is comparing sludge management options for the sludge from an activated-sludge process. The alternatives for the waste activated sludge are

1. Aerobic digestion followed by sand drying beds and composting.
2. Aerobic digestion followed by sand drying beds and land application.

The sludge characteristics are

Quantity = 1900 lb/d
Concentration = 2%
Nitrogen content = 3%
Ammonia content = 1%

Climatic characteristics are a warm growing season, 1.5 ft/yr of precipitation, and 4.0 ft/yr of lake evaporation.

Solution

1. Select the detention time for the aerobic digester. From Table 14-11, select 20 d as the midpoint in the range for primary plus waste activated sludge.
2. Calculate the daily volume of sludge:

$$\text{Volume} = \frac{1900 \text{ lb/d}}{8.34 \times 20{,}000 \text{ mg/L}} = 0.0114 \text{ mil gal/d} = 11{,}400 \text{ gal/d}$$

3. Calculate the volume of the aerobic digester:

$$\text{Volume} = Qt = (11{,}400 \text{ gal/d})(20 \text{ d}) = 230{,}000 \text{ gal} = 30{,}750 \text{ ft}^3$$

4. Estimate the air required for the aerobic digester. Select 40 ft^3/min per 1000 ft^3 of volume.

$$\text{Air required} = \left(\frac{40\ \text{ft}^3/\text{min}}{1000\ \text{ft}^3}\right)\left(\frac{30{,}750\ \text{ft}^3}{1000\ \text{ft}^3}\right) = 1230\ \text{ft}^3/\text{min}$$

5. Design the sand drying beds. Estimate the area required using 16 lb/ft^2·yr as the loading rate.

$$\text{Area} = \frac{1900\ \text{lb/d} \times 365\ \text{d/yr}}{16\ \text{lb/ft}^2\cdot\text{yr}} = 43{,}340\ \text{ft}^2 = 1.0\ \text{ac}$$

6. Select windrow composting using yard waste in a 2:1 mixture with dried sludge. Set the windrow height at 4 ft. Use an overall detention time of 6 wk (4 wk composting and 2 wk curing). Assume dried sludge specific weight 1200 lb/yd^3.

$$\text{Area} = \frac{1.1S(R+1)}{H} = \frac{1.1\left(\dfrac{1900\ \text{lb/d}}{1200\ \text{lb/yd}^3}\right)\left(\dfrac{27\ \text{ft}^3}{\text{yd}^3}\right)(6\ \text{wk})\left(\dfrac{7\ \text{d}}{\text{wk}}\right)(2+1)}{4\ \text{ft}} = 1480\ \text{ft}^2$$

7. For land application of dewatered sludge, calculate the available nitrogen for surface application. Use $k_v = 0.75$. Use $f_n = 0.5$ for warm climate and digested sludge.

$$L_N = [0 + k_v(0.01) + f_n(0.03)](2000\ \text{lb/ton})$$
$$= [(0.75)(0.01) + 0.5(0.03)](2000) = 45\ \text{lb/ton}$$

8. Calculate the loading rate and the area of pasture needed for land application. Use a crop uptake value of 200 lb/ac·yr from Table 10-16 (midrange for ryegrass).

$$L_{SN} = \frac{U}{N_p} = \frac{200\ \text{lb/ac}\cdot\text{yr}}{45\ \text{lb/ton}} = 4.4\ \text{tons/ac}$$

$$\text{Area} = \frac{B}{L_{SN}} = \left(\frac{1900\ \text{lb}}{\text{d}}\right)\left(\frac{365\ \text{d}}{\text{yr}}\right)\left(\frac{\text{ton}}{2000\ \text{lb}}\right)\left(\frac{\text{ac}}{4.4\ \text{ton}}\right) = 79\ \text{ac/yr}$$

PROBLEMS AND DISCUSSION TOPICS

14-1. Compare the mass of dry solids produced in a watertight septic tank from a residence with four occupants, assuming a pumping frequency of 7 yr, to the mass of dry solids produced in a conventional activated-sludge package plant.

14-2. Compare the total mass of solids produced in the following two systems. System 1 consists of a septic tank with effluent screen, as described in Prob. 1, with the septic tank effluent discharged to a constructed wetland through a small-diameter gravity sewer. System 2 consists of the same residence with four occupants; however, the wastewater flow is collected via a conventional gravity sewer to an activated sludge treatment plant.

14-3. Compare and contrast the septage from older homes with metal plumbing to newer homes with plastic plumbing.

14-4. Compare the quantity of sludge produced in a constructed wetland to that produced in a conventional activated-sludge plant.

14-5. What are the advantages of dewatering sludge?

14-6. What are some of the potential problems associated with the development of vectors and odors from land application of septage or biosolids?

14-7. What should be done to septage to allow it to be composted?

14-8. Design a land application system for 1000 dry tons per year of biosolids. Assume a crop uptake of 200 lb/ac·yr and a nitrogen content of the dewatered biosolids of 4 percent, of which 10 percent is ammonia.

14-9. What factors must be considered in the use of sludge drying beds for septage?

14-10. Determine the area of a paved bed for sludge drying for sludge with the following characteristics. Stabilized sludge amounts to 200 tons/yr at 5% solids content. The final solids content is 45%, the evaporation rate is 4 ft/yr, and the rainfall is 2 ft/yr.

14-11. Design a sludge storage basin to treat the septage produced from a 1000-person community.

REFERENCES

Chang, A. C., A. L. Page, and T. Asano (1995) Developing Human Health-Related Chemicals Guidelines for Reclaimed Wastewater and Sewage Sludge Applications in Agriculture, World Health Organization, Geneva, Switzerland.

Donovan, J. (1981) Engineering Assessment of Vermicomposting Municipal Wastewater Sludges, EPA-600/2-81-75, available as PB 81-196933 from National Technical Information Service, Springfield, VA.

Epstein, E. (1997) *The Science of Composting*, Technomic Publishing Co. Inc., Lancaster, PA.

German, J., and G. Tchobanoglous (1982) Tulare County Septage Study, prepared for the Board of Supervisors, Tulare County, CA.

Goldstein, N., and R. Steuteville (1995) Biosolids Composting Maintains Steady Growth, *Biocycle* Vol. 36, No. 12, pp. 49–60.

Haug, R. T. (1994) *Compost Engineering, Principles and Practice*, 2nd ed., Ann Arbor Science, Ann Arbor, MI.

Leonard, P., R. King, and M. Lucas (1992) Fertilizing Forests with Biosolids: How to Plan, Operate, and Maintain a Long-Term Program. *Proceedings, The Future Direction of Municipal Sludge (Biosolids) Management*, WEF Specialty Conference, pp. 233–250, Portland, OR.

Loehr, R. C., J. H. Martin, E. F. Neuhauser, and M. R. Malecki (1984) Waste Management Using Earthworms—Engineering and Scientific Relationships, National Science Foundation ISP-8016764, Cornell University, Ithaca, NY.

NRC (1996) *Use of Reclaimed Water and Sludge in Food Crop Production*, National Research Council, National Academy Press, Washington, D.C.

Pietz, R. I., J. R. Peterson, J. E. Prater, and D. R. Zenz (1984) Metal Concentrations in Earthworms from Sewage Sludge Amended Soils at a Strip Mine Reclamation Site, *Journal of Environmental Quality*, Vol. 13, No. 4, pp. 651–654.

Reed, S. C., J. Bouzoun, and W. Medding (1986) A Rational Method for Sludge Dewatering Via Freezing, *Journal of Water Pollution Control Federation,* Vol. 58, pp. 911–916.

Reed, S. C., R. W. Crites, and E. J. Middlebrooks (1995) *Natural Systems for Waste Management and Treatment,* 2nd ed., McGraw-Hill, New York.

Sommers, L. E. (1980) Toxic Metals in Agricultural Crops, in G. Bitton et al. (eds.), *Sludge—Health Risks of Land Application*, Ann Arbor Science, Ann Arbor, MI.

Sopper, W. E. (1993) *Municipal Sludge Use in Land Reclamation*, Lewis Publishers, Boca Raton, FL.

Sopper, W. E., and S. N. Kerr (eds.) (1979) *Utilization of Municipal Sewage Effluent and Sludge on Forest and Disturbed Lands*, Pennsylvania State University Press, University Park, PA.

Stukenberg, J. R., S. Carr, L. W. Jacobs, and S. Bohm (1993) Document Long-Term Experience of Biosolids Land Application Programs, WERF Project 91-ISP-4, Water Environment Research Federation, Alexandria, VA.

Tchobanoglous, G., and F. L. Burton (1991) *Wastewater Engineering: Treatment, Disposal, and Reuse*, 3rd ed., McGraw-Hill, New York.

Tchobanoglous, G., H. Theisen, and S. Vigil (1993) *Integrated Solid Waste Management*, McGraw-Hill, New York.

Teal, J. M., and S. B. Peterson (1991) The Next Generation of Septage Treatment, *Research Journal of Water Pollution Control Federation*, Vol. 63, pp. 84–89.

U.S. EPA (1983) *Process Design Manual for Land Application of Municipal Sludge*, EPA-625/1-83-016, Center for Environmental Research Information, U.S. Environmental Protection Agency, Cincinnati, OH.

U.S. EPA (1984) *Handbook of Septage Treatment and Disposal*, EPA-625/6-84-009, Center for Environmental Research Information, U.S. Environmental Protection Agency, Cincinnati, OH.

U.S. EPA (1985) *Handbook of Estimating Sludge Management Costs,* EPA/625/6-85/010, Center for Environmental Research Information, U.S. Environmental Protection Agency, Cincinnati, OH.

U.S. EPA (1987) *Design Manual—Dewatering Municipal Wastewater Sludges,* EPA/625/1-87/014, Center for Environmental Research Information, U.S. Environmental Protection Agency, Cincinnati, OH.

U.S. EPA (1989) *In-Vessel Composting of Municipal Wastewater Sludge,* EPA/625/8-89/016, Center for Environmental Research Information, U.S. Environmental Protection Agency, Cincinnati, OH.

U.S. EPA (1990) *Autothermal Thermophilic Aerobic Digestion of Municipal Wastewater Sludge*, EPA/625/10-90/007, Center for Environmental Research Information, U.S. Environmental Protection Agency, Cincinnati, OH.

U.S. EPA (1994) *Guide to Septage Treatment and Disposal.*, EPA/625/R-94/002, U.S. Environmental Protection Agency, Cincinnati, OH.

U.S. EPA (1995) *Process Design Manual—Land Application of Sewage Sludge and Domestic Septage,* EPA/625/R-95/001, Center for Environmental Research Information, U.S. Environmental Protection Agency, Cincinnati, OH.

WEF (1989) *Beneficial Use of Waste Solids,* Manual of Practice FD-5, Water Environment Federation, (formerly Water Pollution Control Federation), Alexandria, VA.

WEF (1992) *Design of Municipal Wastewater Treatment Plants*, Vol. II, Chap. 18, WEF Manual of Practice No. 8, Water Environment Federation, Alexandria, VA.

WEF (1997) *Septage Handling*, Manual of Practice No. 24, Water Environment Federation, Alexandria, VA.

Wegner, G. (1992) The Benefits of Biosolids from a Farmer's Perspective. *Proceedings, The Future Direction of Municipal Sludge (Biosolids) Management*, WEF Specialty Conference, pp. 39-44, Portland, OR.

Whiting, D. M. (1975) *Use of Climatic Data in Design of Soil Treatment Systems,* EPA 660/2-75-018, U.S. Environmental Protection Agency, Corvallis, OR.

CHAPTER 15

Management of Decentralized Wastewater Systems

Decentralized wastewater management (DWM), as defined previously, involves the collection, treatment, disposal, and/or reuse of wastewater from individual homes, clusters of homes, and isolated community and commercial facilities at or near the point of generation. The purpose of this chapter is to examine the management of decentralized systems. Topics to be examined include: (1) the need for management of decentralized systems, (2) the types of wastewater management agencies and districts, (3) the functions of wastewater management districts and agencies, (4) the requirements for a successful DWM district or agency, (5) the financing of DWM districts and agencies, (6) examples of DWM districts, (7) centralized monitoring and control, and (8) the future for DWM systems.

15-1 NEED FOR MANAGEMENT OF DECENTRALIZED SYSTEMS

Although most of the treatment units used in decentralized wastewater management systems require very little maintenance, they rarely receive any. As a result, many system failures have occurred. Also, because design standards vary so widely, many DWM systems are designed and constructed inappropriately. In many instances there is no design at all. To allow for the rational development and use of DWM systems some management oversight is required. The purposeful management of DWM systems must be undertaken (1) to overcome the stigma of failed onsite systems, (2) to obtain cost savings by using many recently developed technologies, (3) to allow for the development and testing of new technologies, and (4) to encourage the orderly development of unsewered areas in the context of a sustainable environment.

Overcoming the Stigma of Failed Onsite Systems

Over the years, every imaginable problem has been experienced in the implementation of DWM systems, especially individual onsite systems. Perhaps the most common problem is the premature failure of both new and older wastewater disposal fields, caused by undersizing of the disposal fields (design issue), inappropriate estimate of assimilative capacity (design issue), solids carryover from septic tank (design issue), discharge of grease from septic tank (design and homeowner issue), and the loss of permeability due to the sodium in septic tank effluent (site soil constraint and homeowner issue).

Because so many onsite systems have failed, local regulatory personnel and officials want onsite systems to be designed to function forever without any problems. As a result, a number of arbitrary design and regulatory constraints are now imposed on the implementation of onsite systems. Typically these constraints have evolved anecdotally, in most cases with little or no scientific underpinning. The principal problem caused by arbitrary design and regulatory constraints is that extremely conservative designs have evolved. Acceptance and use of innovative designs is thus an extremely slow process. With a management district, the benefits of operating existing technologies at significantly higher loading rates and testing of new technologies could be realized, as discussed below.

Cost Savings with Many Recently Developed Technologies

Many of the new treatment technologies that have been developed can function effectively at very high rates for a period of a year or two, but not for 20 years. If a management agency were available, high-rate units could be serviced once every 6 or 12 months at minimum cost. For example, the upper layer of an absorbent plastic-medium biofilter could be replaced in a manner similar to that of an oil filter in an automobile. If such a design option were available, the cost of onsite wastewater management could be reduced significantly. It is estimated that the size of many treatment processes could be reduced by 50 to 75 percent, with a concomitant capital cost savings of up to 40 or 50 percent.

Development and Testing of New Technologies

The availability of a management agency will also allow for the development and testing of new technologies. Such testing could be undertaken as part of the agency's ongoing duties and responsibilities. On the basis of the developments of the past 5 years, it is anticipated that many new innovative designs and technologies will be developed in accordance with the concept of routine maintenance.

Orderly Development of Unsewered Areas

As the pressure increases to develop unsewered areas—and locations where the provision of sewerage facilities is unlikely for a variety of economic, social, and environmental reasons—the development of entities for the management of DWM systems is critical if development is to occur in an environmentally sustainable manner. For example, to ensure that individual decentralized systems will function properly, especially in densely developed areas, it is imperative to organize a maintenance district or to contract with a public or private operating agency to conduct periodic inspections and any necessary maintenance. In some cases, it may be necessary to form a management agency to assist in the design of appropriate DWM systems. Large-scale decentralized wastewater management systems should be allowed only if a responsible management agency has been designated.

15-2 TYPES OF DWM DISTRICTS

Both private and public onsite wastewater management agencies and districts have been formed for the management of DWM systems. The specific type of agency or district used depends on local circumstances. For example, a developer of a residential housing development may form a private management district. In small rural communities the formation of public districts is most common. In each case, the district is the responsible legal entity for the continued long-term performance of the onsite systems. Properly constituted and staffed DWM districts and agencies have proven to be an effective means of assuring the long-term performance of onsite systems (Prince and Davis, 1986).

15-3 THE FUNCTIONS OF DWM DISTRICTS

The functions of DWM districts and agencies will vary, depending on the legal authority and enabling legislation under which the district or agency was formed. Typical functions include the following:

- Inventory
- System design and installation
- Plan review and construction inspection
- Inspections
- Notification
- Certification
- Water quality monitoring
- Reporting
- Education

Inventory

As the name implies, inventory involves the collection of information on each system within the jurisdiction of the DWM district or agency. The inventory step is especially important when existing onsite systems are to be incorporated into a district. The inventory will also help to define which systems must be replaced or upgraded

FIGURE 15-1
Plastic risers installed on septic tank. Risers are required for inspection and cleaning.

to conform to adopted rules and regulations. As part of the inventory, a plot plan showing the location of any wastewater management facilities should be developed for each lot. The information on the plot plan will also help with the inspection program. In locations with older septic tanks, one of the first requirements will be to equip all of the septic tanks within the management district with risers to allow for easy inspection (see Fig. 15-1).

System Design and Installation

Depending on whether the DWM district or agency is public or private, and on the legal authority, some districts and agencies have taken on the responsibility of designing and constructing all of the individual systems within the jurisdiction of the management entity. In some cases, the DWM district or agency will design and conduct the construction inspection functions, but contract out the construction of the physical facilities. In most cases, the homeowner retains ownership of the onsite system with an access easement granted to the management entity. Although design of onsite systems by the management agency has been done, most management agencies do not undertake the design and/or construction of the systems.

Plan Review and Construction Inspection

Plans prepared by accredited designers are reviewed for completeness and conformance with the district's or agency's rules and regulations. In some parts of the country, systems requiring variances must also be reviewed by the state agency responsible for environmental protection. Once the plans have been approved, district personnel will inspect and approve the construction at critical junctures. For example, the sand to be used for an intermittent sand filter must be checked for the amount of dirt and other fine materials, which may affect the performance of the filter over the long run (see Fig. 11-10 in Chap.11).

Inspections

Once the existing systems have been inventoried or new systems have been installed, a routine monitoring schedule must be developed to ensure the long-term performance and reliability of these systems. For new or repaired systems, the inspection frequency will typically be annually or biannually. Systems with problems would be inspected more frequently. In some cases, demonstration systems may be installed that might require a weekly, biweekly, or monthly inspection frequency. The purpose of the inspections is to identify any obvious problems, such as a broken tight line leading from the septic tank to the leachfield disposal system, and to identify the need for pumping the contents of the septic tank to remove accumulated scum and sludge.

Notification

Depending on the results of the inspection, the homeowner will be notified if there is any problem that must be corrected or if the contents of the tank need to be pumped.

Certification/Permitting

Depending on the results of the system inspection and the legal structure of the DWM district or agency, the homeowner will be issued a certificate or permit to operate. If the system is functioning properly, a 1- or 2-year permit to operate may be issued. If deficiencies are identified during the inspection, the homeowner may be issued a temporary permit to operate, with the issuance of a regular permit, subject to a follow-up inspection.

Water Quality Monitoring

Because most DWM districts or agencies must report to a county or state environmental protection agency on the performance of the systems within their jurisdiction, a water quality monitoring program must be developed (see Figs. 15-2 and 15-3). The extent of the program will depend on the potential environmental impacts that may result if the systems are not operating properly. Contamination of a groundwater supply source by nitrates is a typical example.

Reporting

Typically, an annual report must be prepared in which the operation of the district is documented, including the results of the monitoring programs. The specific

FIGURE 15-2
Routine inspection of water levels in monitoring wells in Anne Arundel County, Maryland.

FIGURE 15-3
Collection of water sample from groundwater monitoring well in Stinson Beach, California.

topics that must be included in annual reports will usually be worked out with the responsible environmental protection agency.

Education

An important role for any DWM district or agency is public education. When homeowners and tenants do not know how to care for their systems, expensive failures can result. When a septic tank fails, plumbing backs up, wastewater in the leachfield can be forced to the ground surface where its presence may cause a potential health hazard, and natural water bodies may become polluted. Ultimately, the homeowner may have to install a new leachfield. Therefore, it makes good economic and environmental sense to take care of one's septic system. One of the best ways to inform the public is through the development and issuance of a homeowner's guide, in which the operation and maintenance of septic systems are discussed. A typical topic outline for a homeowner's guide is presented in Table 15-1.

TABLE 15-1
Typical outline for a homeowners' and users' guide for onsite wastewater disposal systems

WHY THIS GUIDE?
HOW A SEPTIC SYSTEM WORKS
The Septic Tank
Additional Treatment Facilities
The Soil Absorption System
WHAT CAN GO WRONG?
PREPARING YOUR SEPTIC TANK FOR INSPECTION
Locating Your Septic Tank
Installation of Access Risers and Covers
Record the Location of Your Septic Tank
CLEANING YOUR SEPTIC TANK
Pumping Out Your Septic Tank
Frequency of Cleaning
OTHER MAINTENANCE TIPS AND SUGGESTIONS
Minimize the Liquid Load
Minimize the Solids Load
Don't Use Septic Tank Additives
Keep Toxic Chemicals Out of Your System
Keep Heavy Vehicles Off Your System
If Your Leachfield System Fails
Operation and Inspection of Leachfield
INSPECTING YOUR OWN TANK
RESOURCE GUIDE

15-4 REQUIREMENTS FOR A SUCCESSFUL DWM DISTRICT OR AGENCY

The elements required for a successful DWM district or agency include:

- Regulatory authority
- Well-developed rules and regulations
- Authority to correct a failed system
- Trained personnel
- Economic feasibility
- Well-designed and -constructed onsite systems
- Well-informed public

Regulatory Authority

The regulatory authority for the operation of a DWM district or agency is of great importance, because it will define how the district must function, and ultimately what legal authority the district or agency will have.

Well-Developed Rules and Regulations

A well-developed set of rules and regulations, by which the district or agency will operate, is of utmost importance to assure all of the participants that they will be treated fairly and equitably. Chaos will result if the rules and regulations are developed on an ad hoc or as-needed basis. An outline of a typical set of regulations is shown in Table 15-2. It should be noted that rules and regulations need to be updated every few years to account for regulatory, technological, and community changes.

Authority to Correct a Failed System

Although all of the above elements are important, one of the most important is the authority and the legal means to get a homeowner to correct a failed system. If the DWM district or agency has no hold over the homeowner, getting a failed system corrected has proven, in many cases, to be an impossible task. In Stinson Beach, California, the Onsite Wastewater Management District (OSWMD) is part of the Stinson Beach County Water District, which also has responsibility for providing water service. Ultimately, after all other means have been exhausted, the district has the legal authority to turn off the water supply, which it has done (Stinson Beach Rules and Regulations, 1996). This method has proved to be effective in getting homeowners to correct failed systems.

TABLE 15-2
Typical outline for rules and regulations for wastewater management entity*

Chapter	Title
1	Administrative provisions
2	Definitions
3	Organization and enforcement
4	Permits and inspection
5	Abatement
6	Design and construction standards: General
7	Variances to design standards
8	Waivers to design standards: Repair and replacement
9	Design standards: Conventional septic system
10	Design standards: Special systems
11	Design standards: Special systems—intermittent sand filter design
12	Design standards: Repair system—Bottomless sand filter
13	Subsurface soils and groundwater protection: General
14	Excavation permits and standards
15	Alteration of parcel to meet site criteria
16	Wells and water systems
17	Swimming pools, spas, and hot tub permits and standards

*Adapted from Stinson Beach Water District Wastewater Management Code.

Trained Personnel

In the development of a DWM district, the hiring of properly trained personnel early on is important, especially with respect to plan review and construction inspection of new systems or existing systems that are being repaired. In fact, it may be necessary to hire more personnel for the first one or two years than will be required once the program is established.

Economic Feasibility

Economic feasibility usually refers to the residents' or communities' ability to pay. To be successful, the organization of the district or agency must be based on a sound economic assessment of the community and its economic circumstances.

Well-Designed and -Constructed Onsite Systems

The importance of having well-designed and -constructed onsite systems cannot be overstressed. If newly installed systems start to fail, the chances of having a successful district or agency are diminished. It is for this reason that a well-developed set of rules and regulations, as discussed above, is of great importance.

Well-Informed Public

For a DWM district or agency to be successful, the public it serves must be well informed. The situation with a DWM district or agency is different than that in a city with a centralized wastewater management system, because residents are responsible for their own onsite system. Thus, the better the public is informed about the legal and management responsibilities of the district or agency and the responsibilities of the homeowner, the better will be the cooperation of the residents in meeting the goals of the DWM district or agency.

15-5 FINANCING OF DWM PROGRAMS AND DISTRICTS

Over the years, a number of methods have been developed for financing the implementation of DWM programs, districts, and agencies. Because the specific details will vary for each state, the purpose here is to introduce some of the methods that have been used to finance DWM programs, districts, and agencies. The two methods used most commonly are: (1) revenue-based financing and (2) benefit assessment financing.

Revenue-Based Financing

Revenue-based financing includes any method that involves borrowed funds repaid from the revenue of the wastewater management enterprise. These financing methods include wastewater collection system revenue bonds, certificates of participation, state or federal loan programs, nonprofit corporation bonds, and any combination of the above, typically with a joint-powers authority. Whichever funding technique is used, the money (debt service) is repaid by net system revenues (gross revenues less operations and maintenance costs). For purposes of most financial plans, all of these techniques are essentially the same. The interest rates may vary, but they are all revenue-based.

Other revenue-based financing methods are: (1) revenue bonds and (2) installment sale agreement-backed securities. Revenue bonds typically require a 50 percent plus 1 vote of all residents of the management district or town. An installment sale agreement financing is a method whereby the management district or town enters into an installment sale agreement with another entity (such as a nonprofit corporation or a joint-powers authority). The management district or town promises to make semiannual payments of principal and interest as the purchase price for the financed improvements. The other entity then uses that promise to secure an issue of bonds or certificates of participation. In each case described above (state loan, revenue bonds, or installment sale), the management district or town pledges and covenants to levy and collect wastewater management system charges sufficient to pay operations and maintenance and debt service/installment payments on the loan or bond amount.

Benefit Assessment Financing

Benefit assessment financing is performed under special acts enacted by most state legislatures. The costs of the facilities or needed improvements are levied against each property within the area of benefit, according to direct benefit afforded to each property. Onsite wastewater management system costs are typically levied based on an equivalent dwelling unit (EDU) formulation. Each project component is allocated across the EDUs according to benefit.

The amount of the total levy is equal to the component cost, plus financing costs and a reserve fund. Bonds are issued in the amount of the total levy. Each property owner makes semiannual payments of principal and interest. The payments are due with regular property tax payments and do not constitute debts of the owner of the property, but are debts secured only by the property. The lien amount is not prepaid with a sale of the property unless the buyer/seller desires to prepay.

Repayment of state loans. Benefit assessment financing can also be used as a mechanism for repayment of the state revolving fund loans through assessments rather than through monthly service charges. This option would permit maximizing the amount borrowed at subsidized interest rates, and decrease the amount borrowed by the sale of assessment bonds at market rates, yet keep monthly service charges at a minimum.

Integrated financing district. An integrated financing district (IFD) is a type of benefit assessment financing whereby the management district or town can levy a "contingent assessment" against a property that will benefit from the improvements at some future time. For example, a proposed new development might be given a contingent lien as long as the property remains undivided and undeveloped. When a subdivision map or development plan is approved, the property begins to make annual assessment payments, reducing the payments due from other properties. Another example would be defining a certain class of property or property use that would be exempt from payment until the land use or ownership changes.

An IFD may be useful to ensure that any future developments using the management program will pay a fair share. Further, an IFD should be considered to allow for small or nil initial assessments against certain single-family residences within the formal system. Properties currently developed with a single-family home within what is generally a commercial or multifamily area (i.e., a "nonconforming" land use) that (1) are not part of the degradation problem and (2) do not hook up to the collection/treatment system could be given a contingent assessment.

At such time as a development plan is approved by the DWM district or agency for the parcel, the owner would be required to pay in full the accumulated principal and interest and assume all future annual installments of principal and interest with respect to the original contingent assessments. If the development plan includes a subdivision of the parcel, the owner would be required to prepay all the remaining principal amount of assessment. If the property's loading changes sufficiently to require connection to the system, the owner would begin to make annual installment payments on an assessment amount equal to the original contingent assessment plus all interest accrued to date.

TABLE 15-3
Comparison of onsite wastewater management districts*

Location	Activity							Year started
	Inventory	Education	Inspections	Notification	Reporting	Certification	Water quality monitoring	
Allen Co., Ohio		✓	✓	✓	✓			1972
Cayuga Co., New York	✓	✓	✓	✓	✓	✓	✓	1993
Santa Cruz Co., California	✓	✓	✓		✓		✓	1985
Stinson Beach, California	✓	✓	✓	✓	✓		✓	1978
Sea Ranch, California		✓	✓	✓	✓		✓	1987
Paradise, California	✓	✓	✓	✓	✓	✓	✓	1992
Island Co., Washington	✓	✓		✓	✓			1989
Jefferson Co., Washington		✓	✓	✓	✓			1987
Clark Co., Washington†	✓	✓	✓	✓	✓	✓		1992
Thurston Co., Washington		✓	✓	✓	✓	✓		1990
Kitsap Co., Washington		✓	✓		✓	✓		1995
Mason Co., Washington		✓	✓	✓	✓	✓		1995
Georgetown, California	✓	✓	✓	✓	✓		✓	1971/1985

*Adapted from Washington State DOH (1996).
† Southwest Washington Health District includes Klickitat and Skamania Counties.

15-6 EXAMPLES OF DWM DISTRICTS

There are a number of DWM districts around the country, especially in California and Washington. Examples of onsite wastewater management districts along with their array of activities are presented in Table 15-3. Two examples of onsite districts are presented in more detail. The first, at Stinson Beach, California, resulted from the need to manage onsite systems and water quality in a developed area without resorting to sewers. The second, at Georgetown, California, resulted from a new development with serious problems involving multiple-site restrictions to conventional onsite soil absorption systems.

Stinson Beach Onsite Wastewater Management District

Since its inception as a summer community in the late 1800s, onsite systems of various types have been used for the disposal of liquid wastes in Stinson Beach (see Fig. 15-4). In 1961, a survey was conducted by Marin County, California, to determine the adequacy of wastewater disposal in the Stinson Beach area. From the results of the survey, it was concluded that the existing disposal method involving the use of septic tanks constituted a public health hazard and that a public district should be formed to deal with the problem. The Stinson Beach County Water District (SBCWD) was formed in 1962 to provide sewerage services to Stinson Beach and those surrounding areas not already within the Bolinas Public Utility District.

FIGURE 15-4
View of Stinson Beach, California, taken from Highway 1 looking north.

Between 1961 and 1972, the Marin County Health Services and the State Health Department continued to collect water samples from Easkoot Creek for bacterial analysis. The principal finding derived from the results of bacteriological testing was that the coliform counts usually exceeded the water quality standards established by the State Water Resources Control Board. These findings led the California Regional Water Quality Control Board (RWQCB), San Francisco Bay Region, to adopt Resolutions 73-13 and 73-18 (September 1973); the resolutions required that all onsite septic systems within the Stinson Beach area be eliminated by October 1977. In addition, a ban on new buildings with onsite disposal systems was included in Resolution 73-13.

During this same period of time (1961–1972), nine engineering studies were conducted to determine what should be done to correct the situation. In each case, the proposed plans were rejected because of local opposition or failure of the plans to meet county or state water quality regulations. Prompted by the objections of the residents and a recognized need for further study, the SBCWD, in conjunction with the State Water Resources Control Board, decided to initiate an objective investigation of the engineering and environmental factors associated with all alternatives.

After a number of trials and tribulations, the engineering report was completed in April 1977; the final environmental impact report was completed in October 1977. The principal finding documented in these two reports was that of all the alternatives evaluated, the continued use of onsite systems was the most cost-effective and the most environmentally acceptable. The recommended program for the continued use of onsite systems involved: (1) the establishment of an onsite wastewater management district and (2) the establishment of a sampling and inspection program to monitor onsite system performance.

Even before the engineering study was approved in July 1975, it was apparent that special legislation would be required if the continued use of onsite systems was to be a viable alternative. As conceived at that time, the legislation would make it possible under California law to form a management district for onsite systems within the SBCWD. The SBCWD would then be empowered to work with the RWQCB. Following the development of appropriate language, Senate Bill 1902 was passed by the legislature on September 13, 1976, which made it possible to form a management district for the operation and maintenance of onsite systems. Working together with the board and staff of the RWQCB, the staff of the SBCWD began putting together a set of rules and regulations that could be used as a basis for the management of the proposed program. It took until December 10, 1977, to develop an acceptable set of rules and regulations and to work out the administrative details between the agencies. With the availability of an acceptable set of rules and regulations and the enabling legislation that made the District a legal entity with which the RWQCB could deal, the staff of the RWQCB recommended that the implementation of the program be approved on a trial basis. On January 17, 1978, the RWQCB passed Resolution 78-1 allowing for the continued use of the onsite systems for the disposal of wastewater in the community of Stinson Beach under the management of the SBCWD. After a 3-year test program, if the RWQCB was satisfied with the performance of the management program conducted by the SBCWD including the results of the water quality monitoring program, the RWQCB would consider amending the basin plan to allow for the continued long-term use of onsite systems.

Condition of onsite systems. In the house-to-house onsite survey conducted in 1975–76, some 75 onsite systems were found or assumed to be failing. The observed causes of these failures are summarized in Table 15-4. With respect to the failed systems, it was found that most had received little or no maintenance or were installed improperly, poorly constructed, or overused. Also, many people were hesitant to spend money to maintain or repair/replace their systems with the impending possibility of a proposed sewer.

Program implementation. Implementation of the Onsite Wastewater Management District (OSWMD) by the SBCWD involved: (1) adoption of the program Rules and Regulations, (2) employment of staff, (3) development of office

TABLE 15-4
Observed causes of onsite system failures in Stinson Beach, California, as recorded in 1978

Observed cause	Total number	Percent of total
Clogged leachfield	17	44
Poor leachfield design	10	26
Inadequate septic tank maintenance	6	15
High groundwater/poor drainage	4	10
Undersized septic tank	2	5
Total	39*	100

*Excludes seven systems assumed to be failing, because of refusal of residents to allow system to be surveyed.

procedures, (4) issuance of permits and citations, (5) initiation of the inspection and monitoring program, (6) continuation of the water quality monitoring program, (7) submission of monthly reports to the RWQCB, and (8) cooperative programs.

Program rules and regulations. Perhaps the most important step in the establishment of the OSWMD was the adoption of the program Rules and Regulations, in which the operation of the program was delineated. The Rules and Regulations were adopted formally on December 10, 1977. Utilizing operating experience, the systematic process of reviewing the original rules and regulations for needed additions and corrections was begun in 1980 and continues today. Typically, the rules and regulations are revised to reflect changes in procedures, fees, and recent legislation and court rulings. Revisions have also been made to standardize format and to improve readability. To date, changes to the original Rules and Regulations are usually in the form of new wastewater ordinances and resolutions (Stinson Beach Rules and Regulations, 1996).

Employment of staff. To carry out the work associated with the OSWMD, a full-time supervisor and half-time secretary were hired. The supervisor is responsible for: (1) routine inspections and water quality sampling, (2) inspecting new homes constructed under Marin County Code 18.06, (3) issuing permits to operate or failed-system citations, (4) inspecting corrected systems, and (5) general record keeping. The half-time secretary is responsible for coordinating district work including the work on the OSWMD, accounting, and the maintenance of files.

Office procedures. Development of specific office procedures to handle the paperwork associated with the OSWMD was a significant part of the program implementation. Although this work had, in part, begun with the preparation of the District's Rules and Regulations, in which various reporting forms had been developed, a separate bookkeeping and accounting system had to be established for the OSWMD. In addition, separate files had to be set up for the large amount of paperwork involved, even for such a small district. Work on the standardization of office procedures and practices continued throughout the first and second years.

Failed system citations. One of the first activities of the OSWMD was to issue Failed System Citations and Interim Permits to Operate for all onsite systems in the community. On January 19, 1978, 61 Failed System Citations were issued to homes with failing systems, based on information obtained during the house-to-house onsite survey conducted during 1975–76. Included in this category were homes with cesspools, deteriorated septic tanks and/or drainfields, and clogged drainfields. Systems where the homeowner refused either the initial survey or a wet weather resurvey were also included in this category. On January 24, 1978, 423 Interim Permits to Operate were issued to homes where the onsite systems were found to be operating satisfactorily during the onsite survey conducted during 1975–76. Failed systems that had been corrected since the onsite survey was completed were also issued Interim Permits to Operate. Interim Permits were issued for one year, at which time each system was to be reinspected.

Inspection and monitoring program. The inspection and monitoring program for onsite systems within the community of Stinson Beach was initiated on January 19, 1978. When the program was initiated, each homeowner was requested to provide permanent access to the septic tank on the homeowner's property for the purposes of inspection and routine maintenance. The initial inspection included documenting the following information for each onsite system: age of the septic system components, past maintenance, number of residents, number of bedrooms and bathrooms, tank dimensions and volume, scum and sludge thickness, and tank construction and condition. According to the plan after these details are noted, water is run into the tank from an outside tap. The outlet to the drainfield is observed to record any increase in the liquid level as the tank is being surcharged. Recommendations are made to the homeowner on the basis of the levels of material in the septic tank and the hydraulic detention time calculated from annual water usage. If the system passes the inspection, a 1- or 2-year Permit to Operate is issued. If a system is found to be operating only marginally, it may require special monitoring to determine if it should be cited as a failed system. If a failed system is found, a Failed System Citation is issued.

Inspection of onsite systems that had been issued temporary permits to operate was also initiated on January 19, 1978. Those systems found to be working properly were issued 1- or 2-year Permits to Operate. Onsite systems found to be operating marginally (requiring minor alterations, maintenance, or water conservation) during the onsite survey conducted in 1975–76 and those systems found to be operating marginally as part of the District's routine inspection program were placed in a special monitoring category. High groundwater demonstration systems, alternative waste disposal systems, composting toilet-graywater systems, and other nonconventional onsite systems were also placed in this category.

Water quality monitoring program. The water quality monitoring program was initiated in January 1978. Seven surface water and six groundwater stations are involved. The sampling stations were selected from those installed during the conduct of the onsite survey in 1975–76.

Submission of reports to RWQCB. As agreed with the RWQCB, the SBCWD is to submit monthly Self-Monitoring Reports in which the results of the water quality sampling program and the monthly inspections of special systems are presented and discussed. In addition, the Failed System Status Reports, in which the status of the failed systems is delineated, are submitted every 2 months. An annual report must also be submitted. Included in a summary report prepared for the RWQCB on the operation of the District for the first 3 years was information on the distribution of the different types of onsite systems in use as of December 31, 1981 (see Table 15-5).

Current developments. In 1988, SBCWD was given authority to approve onsite wastewater disposal systems to serve new construction. Unfortunately, staffing levels were not increased and by 1992 the RWQCB imposed a building ban while District staff reviewed and reorganized the OSWMD. During this period of

TABLE 15-5
Summary of onsite sytems used in Stinson Beach, California, for wastewater management as of December 31, 1981*

Type of system following septic tank	Total number
1. Single leachfied or other disposal means	394
2. Dual valved leachfieds	151
3. Mound disposal system	1
4. Composting toilets with graywater systems	6
5. Incinerating toilets with graywater systems	1
Total	553

*Excludes seven systems assumed to be failing, because of refusal of residents to allow system to be surveyed.

review, SBCWD staff audited all files, drafted a new wastewater code (SBCWD Ord. 1994-01 and revised as 1996-01) and reorganized the inspection schedule. The new wastewater code eliminates the relaxed setback criteria and design specifications associated with the previous repair code and requires owners to bring failed systems up to new construction standards. The new wastewater code also formalizes design criteria for sand filter installations.

Currently, District staff reports to the RWQCB staff quarterly and submits an annual report. These reports detail the inspection program and water quality monitoring and update all system reviews. The water quality monitoring program has also been modified; nine surface water sites are sampled bimonthly and the man-made Seadrift Lagoon is sampled weekly for total and fecal coliform densities during summer months. Groundwater is sampled quarterly.

There are now 672 discrete onsite wastewater disposal systems in Stinson Beach. The types of onsite systems used are summarized in Table 15-6. As reported in Table 15-6, 76 percent of the systems are conventional gravity systems, 16 percent of the systems incorporate sand filter pretreatment prior to disposal, and the remaining 8 percent consist of a variety of different systems. All systems are inspected every 1 to 3 years and, upon the first inspection, a nonexpiring discharge permit is issued and filed with the County of Marin on the title of the property. Following each subsequent inspection a notice of inspection/repair order is sent to the property owner detailing requisite system maintenance and repairs. Each system has been entered into a custom database (Microsoft Access) and all system information (i.e., inspection date, type of system, and outstanding repair items) is cataloged. If a system fails inspection or if a property owner fails to comply with a repair order in a timely manner, the District will issue a failed system citation and proceed with enforcement proceedings (Jones, 1997).

Georgetown Divide Public Utilities District

In 1971, the Georgetown Divide Public Utilities District (District) was formed in response to concerns about potential environmental degradation from individual

TABLE 15-6
Summary of onsite systems used in Stinson Beach, California, for wastewater management as of May 1, 1997

Type of system following septic tank number	Total number
Gravity effluent disposal to soil absorption system:	
1. Single leachfield	274
2. Single leachfield with graywater system	5
3. Single leachfield with lift pump to leachfield	2
4. Single seepage bed	10
5. Dual leachfields	172
6. Dual leachfields with graywater system	3
7. Dual leachfields with lift pump to leachfield	44
8. Dual leachfields with lift pump and graywater system	1
Pressure-dosed disposal to soil absorption system:	
9. Dual leachfields	8
10. Single sand-filled leachfield trenches	1
11. Dual sand-filled leachfield trenches	3
Sand filter system:	
12. Intermittent (lined) with leachfield	98
13. Intermittent single bottomless (unlined)	4
14. Intermittent dual bottomless (unlined)	6
15. Recirculating with leachfield	3
Other systems:	
16. Holding tank only	19
17. Holding tank with leachfield	9
18. High groundwater demonstration systems	9
19. Dual mound	1
Total	672

onsite wastewater disposal systems. The District was originally conceived as a temporary entity, necessary only until rising residential densities made the installation of a community-wide wastewater collection system necessary. However, because of the increasing costs of large-scale disposal systems, improvements in onsite disposal system technology, and changing philosophies about wastewater disposal, the District continues to provide initial site inspection, design, management, periodic system inspection, and educational and environmental surveillance services to onsite wastewater disposal system customers.

The District serves the Auburn Lake Trails subdivision, a 2500-ac development located in the Sierra Nevada foothills in El Dorado County, California. When onsite wastewater disposal systems were proposed for the development, field inspections revealed that many of the 1800 lots were only marginally suitable for receiving wastewater, because of limited soil depth, high seasonal groundwater, barriers to conductivity, and unfavorable topographic conditions. To mitigate these adverse factors, public management of the design, installation, and operation of individual onsite systems was proposed. An initial soils investigation, involving the analysis of over 4000 soil profiles and 6000 percolation tests, was conducted to match the performance of onsite wastewater disposal technologies to individual sites. As lots were developed, District inspectors met onsite with contractors to resolve issues of grading, landscaping, and secondary features of the wastewater disposal system predetermined (on the basis of geological data) for the site.

The District operates on the conviction that intimate involvement in all phases ("cradle to grave") of onsite wastewater disposal system operation is necessary to achieve long-term satisfactory performance. District inspectors work with individual site contractors and homeowners during the processes of disposal system design, location, and construction to verify proper installation. Homeowners are instructed about onsite disposal system functions, to ensure continued successful operation. As new residents are added to the District, site information based on field inspections and interviews is added to the District computer database. The database not only records geologic, hydrologic, and topographic conditions for each lot, but it also maintains a historical record of visits and inspections, maintenance issues, and correspondence regarding each homesite. Based on this information, the computer determines the inspection schedule for each system, and generates visitation notices for mailing when system inspections are forthcoming. The computer is also used to track environmental quality, as groundwater and watershed quality data are entered into the system and compared to baseline values established by analyzing samples collected before development.

In 1985, the District reorganized the onsite program under enabling legislation of the California Health and Safety Code. The District is operated by a five-member elected board of directors. The continued viability of the District is due to its success in preventing onsite disposal system failures in an area of environmental sensitivity and complex geology, in spite of the nonideal environment in which these systems were installed. Due to District scrutiny, and communication and cooperation between District, contractor, and homeowner, comprehensive public management of individual onsite wastewater disposal systems has been cost-effective and successful for this community (Prince and Davis, 1986, 1988).

15-7 CENTRALIZED MONITORING AND CONTROL

The next logical step for onsite wastewater management is the centralization of information collection and control of onsite wastewater systems. In centralized wastewater treatment some components of the collection system are decentralized (e.g., lift stations). Telemetry units, programmable logic controllers (PLCs), and supervisory control and data acquisition (SCADA) are all common tools used in the municipal water and wastewater industries. To a limited degree, telemetry has been used in the onsite management field. Recently, a system has been developed for the remote sensing and control of onsite systems by a microprocessor-based control panel (see Fig. 15-5) (Cagle, 1997; ISAC, 1997). The microprocessor-based controller integrated with standard control panel components (motor contactors, etc.), and combined with analog and digital sensor inputs, is the next best thing to having a full-time wastewater treatment plant operator at the site.

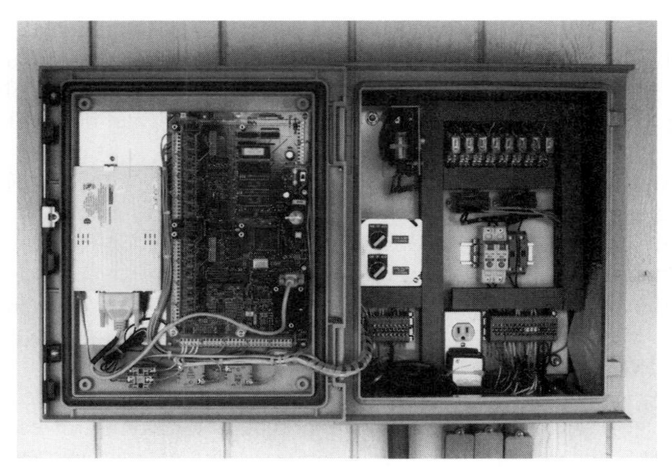

FIGURE 15-5
View of telemetry system for remote monitoring and control of onsite systems (courtesy B. Cagle).

The controller can operate independently because the microprocessor logs, trends, and analyzes the events to adjust process treatment exclusive of human input. Communications are two-way via modem or other communications interface (radio, cellular telephone, etc.). This feature gives the operator the advantage of receiving remote alarm callouts for early preventive maintenance. Pumps, compressors, fans, etc., can all be turned on and off or adjusted from a remote personal computer. Standard logging and trending of data are possible, including pump run times, high-level alarms, and low-level alarms. Any type of event log can be captured by the processor and stored as a report (e.g., flows in gallons per day, percent orifice clogging, etc.) by preprogramming the controller. Data can be stored without downloading for 2 to 3 years, depending on input volume and report requirements.

Reports can be viewed or retrieved remotely and transferred into a familiar database for data management. This aids in remote troubleshooting for better maintenance and operation efficiency. Programming of the system is done by writing rules (in plain English) which define the setpoints (parameters the specific system should run within). Depending on the sequence of operation for a specific process treatment application, the power of processor logic analysis can detect trends in individual site-specific wastewater characteristics that may either cause an automatic adjustment to be made in process treatment by the microprocessor, or signal an alarm callout, or both. Basically any type of sensor can be loaded on the control board that puts out 4 to 20 mA or 0 to 5 V dc. Currently, common sensors being used are liquid level (detect incoming flows and dosing volumes) and pressure (orifice clogging in low-pressure dosing applications). However, sensors for dissolved oxygen, pH, oxidation reduction potential, conductivity, turbidity, temperature, and many other parameters can be read without other peripheral devices such as a datalogger.

15-8 THE FUTURE OF DECENTRALIZED WASTEWATER MANAGEMENT

It must be recognized that the proper collection, treatment, reuse, and disposal of wastewater cannot be accomplished without some expense. As the expense of conventional centralized wastewater management systems continues to increase, and the availability of water supply sources decreases, the role of decentralized systems in wastewater management will become more important. Given the fact that one day, in the not-so-distant future, conventional gravity sewers will become obsolete, movement away from the concept and reality of large regional centralized facilities to the acceptance of decentralized wastewater management systems represents a step into the future.

PROBLEMS AND DISCUSSION TOPICS

15-1. Why has the failure of onsite wastewater systems led to more conservative rather than more innovative designs?

15-2. What does the county where you grew up do with respect to the management of onsite systems?

15-3. What are the advantages and disadvantages of public versus private ownership of onsite septic tanks?

15-4. What regulatory mechanism exists in your state for repairing a failed onsite system?

15-5. Estimate the cost of an onsite management district for 1000 homes if each septic tank is to be inspected once per year and monthly samples must be collected from six monitoring wells and analyzed for nitrate concentrations. Assume the time for inspection is 3 hours per site and the monitoring takes 1 day per month.

15-6. In your state would it be easier to implement private or public entities for the management of onsite systems?

15-7. With whom should the authority to correct failed onsite systems reside?

15-8. In your county, what criteria are used to define a failed onsite system? Is a cesspool considered to be a failed onsite system?

15-9. What are available alternative ways for financing an onsite wastewater management agency?

15-10. In your state, what would be the easiest way of dealing with failed systems in the development of an onsite wastewater management district for both new and older homes?

REFERENCES

Cagle, W. (1997) Personal communication, Auburn, CA.

ISAC (1997) Integrated Systems and Controls, Newcastle, CA.

Jones, B. (1997) Personal communication, Stinson Beach, CA.

Prince, R. N., and M. E. Davis (1986) Evolution of an Onsite Wastewater Management Program, presented at the Summer Meeting of the American Society of Agricultural Engineers, California Polytechnic Institute, San Luis Obispo, CA.

Prince, R. N., and M. E. Davis (1988) On-Site System Management, presented at the Third Annual Midyear Conference of the National Environmental Health Association, Mobile, AL.

Stinson Beach Rules and Regulations (1996) *Wastewater Management Program Rules and Regulations,* Stinson Beach County Water District, Stinson Beach, CA.

Washington State DOH (1996) *On-Site Sewage System Monitoring Programs in Washington State,* Community Environmental Health Programs, Washington State Department of Health, Olympia, WA.

APPENDIX A

Metric Conversion Factors

Factors for the conversion of U.S. Customary Units to the International System (SI) of units

Multiply the U.S. customary unit			To obtain the corresponding SI unit	
Name	**Abbreviation**	**By**	**Name**	**Symbol**
acre	ac	4047	square meter	m^2
acre	ac	0.4047	hectare	ha*
British thermal unit	Btu	1.055	kilojoule	kJ
British thermal units per cubic foot	Btu/ft^3	37.259	kilojoules per cubic meter	kJ/m^3
British thermal units per kilowatt-hour	Btu/kWh	1.055	kilojoules per kilowatt-hour	kJ/kWh
British thermal units per pound	Btu/lb	2.326	kilojoules per kilogram	KJ/Kg
British thermal units per ton	Btu/ton	0.00116	kilojoules per kilogram	kJ/Kg
degree Celsius	°C	plus 273	Kelvin	K
cubic foot	ft^3	0.0283	cubic meter	m^3
cubic foot	ft^3	28.32	liter	L*
cubic feet per minute	ft^3/min	0.0004719	cubic meters per second	m^3/s
cubic feet per minute	ft^3/min	0.4719	liters per second	L*/s
cubic feet per second	ft^3/s	0.0283	cubic meters per second	m^3/s
cubic yard	yd^3	0.7646	cubic meter	m^3
day	d	86.4	kilosecond	ks
degree Fahrenheit	°F	0.555 (°F − 32)	degree Celsius	°C
foot	ft	0.3048	meter	m
feet per minute	ft/min	0.00508	meters per second	m/s
feet per second	ft/s	0.3048	meters per second	m/s

(Continued)

Multiply the U.S. customary unit			To obtain the corresponding SI unit	
Name	**Abbreviation**	**By**	**Name**	**Symbol**
gallon	gal	0.003785	cubic meter	m^3
gallon	gal	3.785	liter	L*
gallons per minute	gal/min	0.0631	liters per second	L*/s
grain	gr	0.0648	gram	g
horsepower	hp	0.746	kilowatt	kW
horsepower-hour	hp-h	2.684	megajoule	MJ
inch	in	2.54	centimeter	cm
inch	in	0.0254	meter	m
kilowatt-hour	kWh	3.600	megajoule	MJ
pound (force)	lb_f	4.448	newton	N
pound (mass)	lb_m	0.4536	kilogram	kg
pounds per acre	lb/ac	0.1122	grams per square meter	g/m^2
pounds per acre	lb/ac	1.122	kilograms per hectare	kg/ha*
pounds per capita per day	lb/capita·d	0.4536	kilograms per capita per day	kg/capita·d
pounds per cubic foot	lb/ft^3	16.019	kilograms per cubic meter	kg/m^3
pounds per cubic yard	lb/yd^3	0.5933	kilograms per cubic meter	kg/m^3
million gallons per day	Mgal/d	0.04381	cubic meters per second	m^3/s
miles	mi	1.609	kilometer	km
miles per hour	mi/h	1.609	kilometers per hour	km/h
miles per hour	mi/h	0.447	meters per second	m/s
miles per gallon	mpg	0.425	kilometers per liter	km/L*
ounce	oz	28.35	gram	g
parts per million	ppm	is approximately	milligrams per liter	mg/L*
pounds per square foot	lb/ft^2	47.88	newtons per square meter	N/m^2
pounds per square inch	lb/in^2	6.895	kilonewtons per square meter	kN/m^2
square foot	ft^2	0.0929	square meter	m^2
square mile	mi^2	2.590	square kilometer	km^2
square yard	yd^2	0.8361	square meter	m^2
ton (2000 pounds mass)	ton (2000 lb_m)	907.2	kilogram	kg
watt-hour	Wh	3.60	kilojoule	kJ
yard	yd	0.9144	meter	m

*Not an SI unit, but commonly used term.

Commonly Used Conversion Factors for Wastewater Treatment Plant Design Parameters

To convert, multiply in direction shown by arrows			
U.S. units	→	←	SI units
Ac	0.4047	2.4711	ha
Btu	1.0551	0.9478	kJ
Btu/lb	2.3241	0.4303	kJ/kg
Btu/ft^3	37.2603	0.0269	kJ/m^3
Btu/ft^2·°F·h	5.6735	0.1763	W/m^2·°C
bu/ac·yr	2.4711	0.4047	bu/ha·yr
ft/h	0.3048	3.2808	m/h
ft/min	18.2880	0.0547	m/h
ft^3/capita	0.0283	35.3147	m^3/capita
ft^3/gal	7.4805	0.1337	m^3/m^3
ft^3/ft·min	0.0929	10.7639	m^3/m·min
ft^3/lb	0.0624	16.0185	m^3/kg
ft^3/Mgal	7.04805×10^{-3}	133.6805	m^3/10^{3}m^3
ft^2/Mgal·d	407.4611	0.0025	m^2/10^{3}m^3·d
ft^3/ft^2·h	0.3048	3.2808	m^3/m^2·h
ft^3/10^{3}gal·min	7.04805×10^{-3}	133.6805	m^3/m^3·min
ft^3/min	1.6990	0.5886	m^3/h
ft^3/10^{3}ft^3·min	0.001	1000.0	m^3/m^3·min
gal	3.7854	0.2642	L
gal/ac·d	0.0094	106.9064	m^3/ha·d
gal/ft·d	0.0124	80.5196	m^3/m·d
gal/ft^2·d	40.7458	2.4542×10^{-2}	mm/d
gal/ft^2·d	0.0407	24.5424	m^3/m^2·d
gal/ft^2·d	0.0017	589.0173	m^3/m^2·h
gal/ft^2·d	0.0283	35.3420	L/m^2·min
gal/ft^2·d	40.7458	2.4542×10^{-2}	L/m^2·d
gal/ft^2·min	2.4448	0.4090	m/h
gal/ft^2·min	40.7458	0.0245	L/m^2·min
gal/ft^2·min	58.6740	0.0170	m^3/m^2·d

(*continued*)

(Continued)

To convert, multiply in direction shown by arrows			
U.S. units	→	←	**SI units**
$hp/10^3 gal$	0.197	5.0763	kW/m^3
$hp/10^3 ft^3$	26.3342	0.0380	$kW/10^3 m^3$
in	25.4	3.9370×10^{-2}	mm
in Hg (60°F)	3.3768	0.2961	kPa Hg (60°F)
lb	0.4536	2.2046	kg
lb/ac	1.1209	0.8922	kg/ha
lb/hp·h	0.6083	1.6440	kg/kW·h
lb/Mgal	0.1198	8.3454	g/m^3
lb/Mgal	1.1983×10^{-4}	8345.4	kg/m^3
lb/ft^2	4.8824	0.2048	kg/m^2
lb_f/in^2 (gage)	6.8948	0.1450	kPa (gage)
$lb/ft^3 \cdot h$	16.0185	0.0624	$kg/m^3 \cdot h$
$lb/10^3 ft^3 \cdot d$	0.0160	62.4280	$kg/m^3 \cdot d$
Mgal/ac·d	0.9354	1.0691	$m^3/m^2 \cdot d$
Mgal/d	3.7854×10^3	0.264×10^{-3}	m^3/d
Mgal/d	4.3813×10^{-2}	22.8245	m^3/s
min/in	3.9370	0.2540	$min/10^2 mm$
ton/ac	2.2417	0.4461	Mg/ha
yd^3	0.7646	1.3079	m^3

Physical Properties of Water

The principal physical properties of water are summarized in Table C-1 in U.S. customary units and in Table C-2 in SI units. They are described briefly below (Webber, 1971).

DENSITY

The density ρ of a fluid is its mass per unit volume. In U.S. customary units it is expressed in slugs per cubic feet. For water, ρ is 1.940 slugs/ft^3 at 32°F. There is a slight decrease in density with increasing temperature.

SPECIFIC WEIGHT

The specific weight γ of a fluid is its weight per unit volume. In U.S. customary units, it is expressed in pounds per cubic foot. The relationship between γ, ρ, and the acceleration due to gravity g is $\gamma = \rho g$. At normal temperatures, the value of γ for water is 62.4 lb_f/ft^3 (9.81 kN/m^3).

MODULUS OF ELASTICITY

For most practical purposes, liquids may be regarded as incompressible. The bulk modulus of elasticity E is given by

$$E = \frac{\Delta p}{\Delta V/V}$$

where Δp is the increase in pressure which, when applied to a volume V, results in a decrease in volume ΔV.

TABLE C-1

Physical properties of water (U.S. customary units)*

Temperature, °F	Specific weight, γ, lb/ft^3	Density,† ρ, slug/ft^3	Modulus of elasticity,† $E/10^3$, lb$_f$/in^2	Dynamic viscosity, $\mu \times 10^5$, lb·s/ft^2	Kinematic viscosity, $\nu \times 10^5$, ft^2/s	Surface tension,‡ σ, lb/ft	Vapor pressure, p_v, lb$_f$/in^2
32	62.42	1.940	287	3.746	1.931	0.00518	0.09
40	62.43	1.940	296	3.229	1.664	0.00614	0.12
50	62.41	1.940	305	2.735	1.410	0.00509	0.18
60	62.37	1.938	313	2.359	1.217	0.00504	0.26
70	62.30	1.936	319	2.050	1.059	0.00498	0.36
80	62.22	1.934	324	1.799	0.930	0.00492	0.51
90	62.11	1.931	328	1.595	0.826	0.00486	0.70
100	62.00	1.927	331	1.424	0.739	0.00480	0.95
110	61.86	1.923	332	1.284	0.667	0.00473	1.27
120	61.71	1.918	332	1.168	0.609	0.00467	1.69
130	61.55	1.913	331	1.069	0.558	0.00460	2.22
140	61.38	1.908	330	0.981	0.514	0.00454	2.89
150	61.20	1.902	328	0.905	0.476	0.00447	3.72
160	61.00	1.896	326	0.838	0.442	0.00441	4.74
170	60.80	1.890	322	0.780	0.413	0.00434	5.99
180	60.58	1.883	318	0.726	0.385	0.00427	7.51
190	60.36	1.876	313	0.678	0.362	0.00420	9.34
200	60.12	1.868	308	0.637	0.341	0.00413	11.52
212	59.83	1.860	300	0.593	0.319	0.00404	14.70

*Adapted from Vennard and Street (1975).
† At atmospheric pressure.
‡ In contact with the air.

TABLE C-2

Physical properties of water (SI units)*

Temperature, °C	Specific weight, γ, kN/m^3	Density,† ρ, kg/m^3	Modulus of elasticity,† $E/10^6$, kN/m^2	Dynamic viscosity, $\mu \times 10^3$, N·s/m^2	Kinematic viscosity, $\nu \times 10^6$, m^2/s	Surface tension,‡ σ, N/m	Vapor pressure, p_v, kN/m^2
0	9.805	999.8	1.98	1.781	1.785	0.0765	0.61
5	9.807	1000.0	2.05	1.518	1.519	0.0749	0.87
10	9.804	999.7	2.10	1.307	1.306	0.0742	1.23
15	9.798	999.1	2.15	1.139	1.139	0.0735	1.70
20	9.789	998.2	2.17	1.002	1.003	0.0728	2.34
25	9.777	997.0	2.22	0.890	0.893	0.0720	3.17
30	9.764	995.7	2.25	0.798	0.800	0.0712	4.24
40	9.730	992.2	2.28	0.653	0.658	0.0696	7.38
50	9.689	988.0	2.29	0.547	0.553	0.0679	12.33
60	9.642	983.2	2.28	0.466	0.474	0.0662	19.92
70	9.589	977.8	2.25	0.404	0.413	0.0644	31.16
80	9.530	971.8	2.20	0.354	0.364	0.0626	47.34
90	9.466	965.3	2.14	0.315	0.326	0.0608	70.10
100	9.399	958.4	2.07	0.282	0.294	0.0589	101.33

*Adapted from Vennard and Street (1975).
† At atmospheric pressure.
‡ In contact with air.

DYNAMIC VISCOSITY

The viscosity of a fluid μ is a measure of its resistance to tangential or shear stress. Viscosity in U.S. customary units is expressed in pound-seconds per square foot.

KINEMATIC VISCOSITY

In many problems concerning fluid motion, the viscosity appears with the density in the form μ/ρ, and it is convenient to use a single term ν, known as the kinematic viscosity and expressed in square feet per second, or stokes, in U.S. customary units. The kinematic viscosity of a liquid diminishes with increasing temperature.

SURFACE TENSION

Surface tension σ is the physical property that enables a drop of water to hold in suspension at a tap, a glass to be filled with liquid slightly above the brim and yet not spill, or a needle to float on the surface of a liquid. The surface-tension force across any imaginary line at a free surface is proportional to the length of the line and acts in a direction perpendicular to it. The surface tension per unit length σ is expressed in pounds per foot. There is a slight decrease in surface tension with increasing temperature.

VAPOR PRESSURE

Liquid molecules that possess sufficient kinetic energy are projected out of the main body of a liquid at its free surface and pass into the vapor. The pressure exerted by this vapor is known as the vapor pressure p_v. The vapor pressure of water at 32°F is 0.09 lb_f/in^2.

REFERENCES

Webber, N. B. (1971) *Fluid Mechanics for Civil Engineers,* SI ed., Chapman and Hall, London.

Vennard, J. K., and R. L. Street (1975) *Elementary Fluid Mechanics,* 5th ed., Wiley, New York.

APPENDIX D

Physical Properties of Selected Gases and the Composition of Air

TABLE D-1
Molecular weight, specific weight, and density of gases found in wastewater at standard conditions (0°C, 1 atm)*

Gas	Formula	Molecular weight	Specific weight, lb/ft^3	Density, g/L
Air		28.97	0.0808	1.2928
Ammonia	NH_3	17.03	0.0482	0.7708
Carbon dioxide	CO_2	44.00	0.1235	1.9768
Carbon monoxide	CO	28.00	0.0781	1.2501
Hydrogen	H_2	2.016	0.0056	0.0898
Hydrogen sulfide	H_2S	34.08	0.0961	1.5392
Methane	CH_4	16.03	0.0448	0.7167
Nitrogen	N_2	28.02	0.0782	1.2507
Oxygen	O_2	32.00	0.0892	1.4289

*Adapted from Perry, R. H., D. W. Green, and J. O. Maloney: (eds.) (1984) *Perry's Chemical Engineers' Handbook*, 6th ed., McGraw-Hill, New York.

Note: To compute the specific weight of air, γ_a, at other temperatures at atmospheric pressure, the following relationship can be used:

$$\gamma_a = \frac{p(144 \text{ in}^2/\text{ft}^2)}{RT}$$

where p = atmospheric pressure = 14.7 lb/in ^{2}
R = universal gas constant = 53.3 ft·lb/(lb air)·°R
T = temperature, °R (Rankine) = (460+°F)

For example, at 68°F, the specific weight of air is

$$\gamma_a = \frac{(14.7 \text{ lb/in}^2)(144 \text{ in}^2/\text{ft}^2)}{[53.3 \text{ ft·lb/(lb air)·°R}][(460 + 68)\text{°R}]} = 0.0752 \text{ lb/ft}^3$$

Note: To compute the specific weight of air, γ_a, at other temperatures and pressures use

$$\frac{p_1 v_1}{T_1} = \frac{p_2 v_2}{T_2}$$

The general formula for the change in atmospheric pressure with elevation is

$$\frac{p_b}{p_a} = \exp\left[-\frac{gM(z_b - z_a)}{g_c RT}\right]$$

where p = pressure, lb/in^2
g = 32.2 ft/s^2
g_c = 32.2 ft·lb$_m$/lb·s^2
M = mole of air (see Table D-1) = 28.97 lb$_m$/lb$_{mol}$
z = elevation, ft
$R = \dfrac{1545 \text{ ft·lb}}{\text{lb}_{mol}\text{·T}}$
T = °R (460 +°F)

TABLE D-2

Composition of dry air at 32°F and 1.0 atmosphere*

Gas	Formula	Percent by volume †,‡	Percent by weight
Nitrogen	N_2	78.03	75.47
Oxygen	O_2	20.99	23.18
Argon	Ar	0.94	1.30
Carbon dioxide	CO_2	0.03	0.05
Other§		0.01	

* *Note:* Values reported in the literature vary depending on the standard conditions.

† Adapted from *North American Combustion Handbook,* 2nd ed., North American Mfg. Co., Cleveland, OH.

‡ For ordinary purposes air is assumed to be composed of 79% N_2 and 21% O_2 by volume.

§ Hydrogen, neon, helium, krypton, xenon.

Note: (0.7803 × 28.02) + (0.2099 × 32.00) + (0.0094 × 39.95) + (0.0003 × 44.00) = 28.97 (see Table D-1 above)

APPENDIX E

Dissolved-Oxygen Concentration in Water as a Function of Temperature, Salinity, and Barometric Pressure

TABLE E-1

Dissolved-oxygen concentration in water as a function of temperature and salinity (barometric pressure = 760 mm Hg)*

Temp, °C	Dissolved-oxygen concentration, mg/L									
	Salinity, parts per thousand									
	0	5	10	15	20	25	30	35	40	45
0	14.60	14.11	13.64	13.18	12.74	12.31	11.90	11.50	11.11	10.74
1	14.20	13.73	13.27	12.83	12.40	11.98	11.58	11.20	10.83	10.46
2	13.81	13.36	12.91	12.49	12.07	11.67	11.29	10.91	10.55	10.20
3	13.45	13.00	12.58	12.16	11.76	11.38	11.00	10.64	10.29	9.95
4	13.09	12.67	12.25	11.85	11.47	11.09	10.73	10.38	10.04	9.71
5	12.76	12.34	11.94	11.56	11.18	10.82	10.47	10.13	9.80	9.48
6	12.44	12.04	11.65	11.27	10.91	10.56	10.22	9.89	9.57	9.27
7	12.13	11.74	11.37	11.00	10.65	10.31	9.98	9.66	9.35	9.06
8	11.83	11.46	11.09	10.74	10.40	10.07	9.75	9.44	9.14	8.85
9	11.55	11.19	10.83	10.49	10.16	9.84	9.53	9.23	8.94	8.66
10	11.28	10.92	10.58	10.25	9.93	9.62	9.32	9.03	8.75	8.47
11	11.02	10.67	10.34	10.02	9.71	9.41	9.12	8.83	8.56	8.30
12	10.77	10.43	10.11	9.80	9.50	9.21	8.92	8.65	8.38	8.12
13	10.53	10.20	9.89	9.59	9.30	9.01	8.74	8.47	8.21	7.96
14	10.29	9.98	9.68	9.38	9.10	8.82	8.55	8.30	8.04	7.80
15	10.07	9.77	9.47	9.19	8.91	8.64	8.38	8.13	7.88	7.65
16	9.86	9.56	9.28	9.00	8.73	8.47	8.21	7.97	7.73	7.50
17	9.65	9.36	9.09	8.82	8.55	8.30	8.05	7.81	7.58	7.36
18	9.45	9.17	8.90	8.64	8.39	8.14	7.90	7.66	7.44	7.22
19	9.26	8.99	8.73	8.47	8.22	7.98	7.75	7.52	7.30	7.09
20	9.08	8.81	8.56	8.31	8.07	7.83	7.60	7.38	7.17	6.96
21	8.90	8.64	8.39	8.15	7.91	7.69	7.46	7.25	7.04	6.84
22	8.73	8.48	8.23	8.00	7.77	7.54	7.33	7.12	6.91	6.72
23	8.56	8.32	8.08	7.85	7.63	7.41	7.20	6.99	6.79	6.60
24	8.40	8.16	7.93	7.71	7.49	7.28	7.07	6.87	6.68	6.49
25	8.24	8.01	7.79	7.57	7.36	7.15	6.95	6.75	6.56	6.38
26	8.09	7.87	7.65	7.44	7.23	7.03	6.83	6.64	6.46	6.28
27	7.95	7.73	7.51	7.31	7.10	6.91	6.72	6.53	6.35	6.17
28	7.81	7.59	7.38	7.18	6.98	6.79	6.61	6.42	6.25	6.08
29	7.67	7.46	7.26	7.06	6.87	6.68	6.50	6.32	6.15	5.98
30	7.54	7.33	7.14	6.94	6.75	6.57	6.39	6.22	6.05	5.89
31	7.41	7.21	7.02	6.83	6.65	6.47	6.29	6.12	5.96	5.80
32	7.29	7.09	6.90	6.72	6.54	6.36	6.19	6.03	5.87	5.71
33	7.17	6.98	6.79	6.61	6.44	6.26	6.10	5.94	5.78	5.63
34	7.05	6.86	6.68	6.51	6.33	6.17	6.01	5.85	5.69	5.54
35	6.93	6.75	6.58	6.40	6.24	6.07	5.92	5.76	5.61	5.46
36	6.82	6.65	6.47	6.31	6.14	5.98	5.83	5.68	5.53	5.39
37	6.72	6.54	6.37	6.21	6.05	5.89	5.74	5.59	5.45	5.31
38	6.61	6.44	6.28	6.12	5.96	5.81	5.66	5.51	5.37	5.24
39	6.51	6.34	6.18	6.03	5.87	5.72	5.58	5.44	5.30	5.16
40	6.41	6.25	6.09	5.94	5.79	5.64	5.50	5.36	5.22	5.09

*From Colt, J.: "Computation of Dissolved Gas Concentrations in Water as Functions of Temperature, Salinity, and Pressure," *American Fisheries Society Special Society Publication 14,* Bethesda, Maryland, 1984.

TABLE E-2

Dissolved-oxygen concentration in water as a function of temperature and barometric pressure (salinity = 0 ppt)*

	Dissolved-oxygen concentration, mg/L									
	Barometric pressure, millimeters of mercury									
Temp, °C	735	740	745	750	755	760	765	770	775	780
0	14.12	14.22	14.31	14.41	14.51	14.60	14.70	14.80	14.89	14.99
1	13.73	13.82	13.92	14.01	14.10	14.20	14.29	14.39	14.48	14.57
2	13.36	13.45	13.54	13.63	13.72	13.81	13.90	14.00	14.09	14.18
3	13.00	13.09	13.18	13.27	13.36	13.45	13.53	13.62	13.71	13.80
4	12.66	12.75	12.83	12.92	13.01	13.09	13.18	13.27	13.35	13.44
5	12.33	12.42	12.50	12.59	12.67	12.76	12.84	12.93	13.01	13.10
6	12.02	12.11	12.19	12.27	12.35	12.44	12.52	12.60	12.68	12.77
7	11.72	11.80	11.89	11.97	12.05	12.13	12.21	12.29	12.37	12.45
8	11.44	11.52	11.60	11.67	11.75	11.83	11.91	11.99	12.07	12.15
9	11.16	11.24	11.32	11.40	11.47	11.55	11.63	11.70	11.78	11.86
10	10.90	10.98	11.05	11.13	11.20	11.28	11.35	11.43	11.50	11.58
11	10.65	10.72	10.80	10.87	10.94	11.02	11.09	11.16	11.24	11.31
12	10.41	10.48	10.55	10.62	10.69	10.77	10.84	10.91	10.98	11.05
13	10.17	10.24	10.31	10.38	10.46	10.53	10.60	10.67	10.74	10.81
14	9.95	10.02	10.09	10.16	10.23	10.29	10.36	10.43	10.50	10.57
15	9.73	9.80	9.87	9.94	10.00	10.07	10.14	10.21	10.27	10.34
16	9.53	9.59	9.66	9.73	9.79	9.86	9.92	9.99	10.06	10.12
17	9.33	9.39	9.46	9.52	9.59	9.65	9.72	9.78	9.85	9.91
18	9.14	9.20	9.26	9.33	9.39	9.45	9.52	9.58	9.64	9.71
19	8.95	9.01	9.07	9.14	9.20	9.26	9.32	9.39	9.45	9.51
20	8.77	8.83	8.89	8.95	9.02	9.08	9.14	9.20	9.26	9.32
21	8.60	8.66	8.72	8.78	8.84	8.90	8.96	9.02	9.08	9.14
22	8.43	8.49	8.55	8.61	8.67	8.73	8.79	8.84	8.90	8.96
23	8.27	8.33	8.39	8.44	8.50	8.56	8.62	8.68	8.73	8.79
24	8.11	8.17	8.23	8.29	8.34	8.40	8.46	8.51	8.57	8.63
25	7.96	8.02	8.08	8.13	8.19	8.24	8.30	8.36	8.41	8.47
26	7.82	7.87	7.93	7.98	8.04	8.09	8.15	8.20	8.26	8.31
27	7.68	7.73	7.79	7.84	7.89	7.95	8.00	8.06	8.11	8.17
28	7.54	7.59	7.65	7.70	7.75	7.81	7.86	7.91	7.97	8.02
29	7.41	7.46	7.51	7.57	7.62	7.67	7.72	7.78	7.83	7.88
30	7.28	7.33	7.38	7.44	7.49	7.54	7.59	7.64	7.69	7.75
31	7.16	7.21	7.26	7.31	7.36	7.41	7.46	7.51	7.46	7.62
32	7.04	7.09	7.14	7.19	7.24	7.29	7.34	7.39	7.44	7.49
33	6.92	6.97	7.02	7.07	7.12	7.17	7.22	7.27	7.31	7.36
34	6.80	6.85	6.90	6.95	7.00	7.05	7.10	7.15	7.20	7.24
35	6.69	6.74	6.79	6.84	6.89	6.93	6.98	7.03	7.08	7.13
36	6.59	6.63	6.68	6.73	6.78	6.82	6.87	6.92	6.97	7.01
37	6.48	6.53	6.57	6.62	6.67	6.72	6.76	6.81	6.86	6.90
38	6.38	6.43	6.47	6.52	6.56	6.61	6.66	6.70	6.75	6.80
39	6.28	6.33	6.37	6.42	6.46	6.51	6.56	6.60	6.65	6.69
40	6.18	6.23	6.27	6.32	6.36	6.41	6.46	6.50	6.55	6.59

*From Colt, J.: "Computation of Dissolved Gas Concentrations in Water as Functions of Temperature, Salinity, and Pressure," *American Fisheries Society Special Publication 14,* Bethesda, Maryland, 1984.

Note: ppt = parts per thousand.

APPENDIX F

Carbonate Equilibrium

The chemical species that make up the carbonate system include gaseous carbon dioxide $[(CO_2)_g]$, aqueous carbon dioxide $[(CO_2)_{aq}]$, carbonic acid $[H_2CO_3]$, bicarbonate $[HCO_3^-]$, carbonate $[CO_2^{-2}]$, and solids containing carbonates. In waters exposed to the atmosphere, the equilibrium concentration of dissolved CO_2 is a function of the liquid phase CO_2 mole fraction and the partial pressure of CO_2 in the atmosphere. Henry's law (see Chap. 2) is applicable to the CO_2 equilibrium between air and water; thus

$$P_g = Hx_g \qquad \text{(F-1)}$$

where P_g = partial pressure of gas, atm
H = Henry's law constant, atm/mole fraction
x_g = equilibrium mole fraction of dissolved gas

$$= \frac{\text{mol gas } (n_g)}{\text{mol gas } (n_n) + \text{mol water } (n_w)}$$

The value of H as a function of temperature is given in Table F-1. Carbon dioxide comprises approximately 0.03 percent of the atmosphere at sea level, where the average atmospheric pressure is 1 atm, or 101.4 kPa. The concentration of aqueous carbon dioxide is determined using Eq. (F-1).

Aqueous carbon dioxide $[(CO_2)_{aq}]$ reacts reversibly with water to form carbonic acid:

$$(CO_2)_{aq} + H_2O \Leftrightarrow H_2CO_3 \qquad \text{(F-2)}$$

The corresponding equilibrium expression is

$$\frac{[H_2CO_3]}{[CO_2]_{aq}} = K_m \qquad \text{(F-3)}$$

TABLE F-1
Henry's law constants for CO_2 as function of temperature

T, °C	H_{CO2}, atm
0	0.728×10^3
10	0.104×10^4
20	0.142×10^4
30	0.186×10^4
40	0.233×10^4
50	0.283×10^4
60	0.341×10^4

TABLE F-2
Carbonate equilibrium constants as function of temperature

T, °C	K_1, mol/L	K_2, mol/L
5	3.02×10^{-7}	2.75×10^{-11}
10	3.46×10^{-7}	3.24×10^{-11}
15	3.80×10^{-7}	3.72×10^{-11}
20	4.17×10^{-7}	4.17×10^{-11}
25	4.47×10^{-7}	4.68×10^{-11}
40	5.07×10^{-7}	6.03×10^{-11}
60	5.07×10^{-7}	7.24×10^{-11}

The value of K_m at 25°C is 1.58×10^{-3}. Note that K_m is unitless. The difficulty of differentiating between $(CO_2)_{aq}$ and H_2CO_3 in solution and the fact that very little H_2CO_3 is ever present in natural waters have led to the use of an effective carbonic-acid value ($H_2CO_3^*$) defined as

$$H_2CO_3^* \Leftrightarrow (CO_2)_{aq} + H_2CO_3 \tag{F-4}$$

Because carbonic acid is a diprotic acid it will dissociate in two steps: first to bicarbonate and then to carbonate. The first dissociation of carbonic acid to bicarbonate can be represented as

$$H_2CO_3^* \Leftrightarrow H^+ + HCO_3^- \tag{F-5}$$

The corresponding equilibrium relationship is defined as

$$\frac{[H^-][HCO_3^-]}{[H_2CO_3^*]} = K_1 \tag{F-6}$$

The value of K_1 at 25°C is 4.47×10^{-7} mol/L. Values of K_1 at other temperatures are given in Table F-2.

The second dissociation of carbonic acid is from bicarbonate to carbonate:

$$HCO_3^- \Leftrightarrow H^+ + CO_3^{-2} \tag{F-7}$$

The corresponding equilibrium relationship is defined as:

$$\frac{[H^-][CO_3^{-2}]}{[HCO_3^*]} = K_2 \tag{F-8}$$

The value of K_2 at 25°C is 4.68×10^{-11} mol/L. Values of K_2 at other temperatures are given in Table F-2.

APPENDIX G

MPN Tables and Their Use

When three serial sample volumes (e.g., dilutions) are used in the bacteriological testing of water, the resulting MPN (most probable number) values per 100 mL can be determined from Tables G-1 and G-2. Table G-2 also contains the corresponding 95 percent confidence limits for the MPN values. The MPN values given in Tables G-1 and G-2 are based on serial sample volumes of 10, 1, and 0.1 mL. If lower or higher serial sample volumes are used, the MPN values given in Tables G-1 and G-2 must be adjusted by using Eq. (G-1). For example, if sample volumes used are 100, 10, and 1 mL, the MPN values from the table are multiplied by 0.1. Similarly, if the sample volumes 1, 0.1, and 0.01 mL were used, the MPN values from the table are multiplied by 10. It should be noted that the MPN values given in Table G-1 are the actual values, in contrast to the MPN values given in Table G-2, which have been rounded.

$$\text{MPN/100 mL} = \text{MPN value (from table)} \times \frac{10}{\begin{array}{c}\text{largest volume tested}\\ \text{in dilution series used}\\ \text{for MPN test}\end{array}} \qquad \text{(G-1)}$$

In situations where more than three test dilutions have been run, the following rule is applied to select the three dilutions to be used in determining the MPN value (Standard Methods, 1995). Choose the highest dilution that gives positive results in all five portions tested (no lower dilution giving any negative results) and the two next higher dilutions. Use the results at these three volumes in computing the MPN value. In the examples given in the table found on page 1031, the significant dilution results are shown in boldface. The number in the numerator represents positive tubes, that in the denominator, the total tubes planted.

TABLE G-1

Most probable number (MPN) of coliforms per 100 mL of sample

Number of positive tubes				Number of positive tubes				Number of positive tubes			
10 mL	1 mL	0.1 mL	MPN	10 mL	1 mL	0.1 mL	MPN	10 mL	1 mL	0.1 mL	MPN
0	0	0		1	0	0	2.0	2	0	0	4.5
0	0	1	1.8	1	0	1	4.0	2	0	1	6.8
0	0	2	3.6	1	0	2	6.0	2	0	2	9.1
0	0	3	5.4	1	0	3	8.0	2	0	3	12
0	0	4	7.2	1	0	4	10	2	0	4	14
0	0	5	9.0	1	0	5	12	2	0	5	16
0	1	0	1.8	1	1	0	4.0	2	1	0	6.8
0	1	1	3.6	1	1	1	6.1	2	1	1	9.2
0	1	2	5.5	1	1	2	8.1	2	1	2	12
0	1	3	7.3	1	1	3	10	2	1	3	14
0	1	4	9.1	1	1	4	12	2	1	4	17
0	1	5	11	1	1	5	14	2	1	5	19
0	2	0	3.7	1	2	0	6.1	2	2	0	9.3
0	2	1	5.5	1	2	1	8.2	2	2	1	12
0	2	2	7.4	1	2	2	10	2	2	2	14
0	2	3	9.2	1	2	3	12	2	2	3	17
0	2	4	11	1	2	4	15	2	2	4	19
0	2	5	13	1	2	5	17	2	2	5	22
0	3	0	5.6	1	3	0	8.3	2	3	0	12
0	3	1	7.4	1	3	1	10	2	3	1	14
0	3	2	9.3	1	3	2	13	2	3	2	17
0	3	3	11	1	3	3	15	2	3	3	20
0	3	4	13	1	3	4	17	2	3	4	22
0	3	5	15	1	3	5	19	2	3	5	25
0	4	0	7.5	1	4	0	11	2	4	0	15
0	4	1	9.4	1	4	1	13	2	4	1	17
0	4	2	11	1	4	2	15	2	4	2	20
0	4	3	13	1	4	3	17	2	4	3	23
0	4	4	15	1	4	4	19	2	4	4	25
0	4	5	17	1	4	5	22	2	4	5	28
0	5	0	9.4	1	5	0	13	2	5	0	17
0	5	1	11	1	5	1	15	2	5	1	20
0	5	2	13	1	5	2	17	2	5	2	23
0	5	3	15	1	5	3	19	2	5	3	26
0	5	4	17	1	5	4	22	2	5	4	29
0	5	5	19	1	5	5	24	2	5	5	32

Number of positive tubes				Number of positive tubes				Number of positive tubes			
10 mL	1 mL	0.1 mL	MPN	10 mL	1 mL	0.1 mL	MPN	10 mL	1 mL	0.1 mL	MPN
3	0	0	7.8	4	0	0	13	5	0	0	23
3	0	1	11	4	0	1	17	5	0	1	31
3	0	2	13	4	0	2	21	5	0	2	43
3	0	3	16	4	0	3	25	5	0	3	58
3	0	4	20	4	0	4	30	5	0	4	76
3	0	5	23	4	0	5	36	5	0	5	95
3	1	0	11	4	1	0	17	5	1	0	33
3	1	1	14	4	1	1	21	5	1	1	46
3	1	2	17	4	1	2	26	5	1	2	64
3	1	3	20	4	1	3	31	5	1	3	84
3	1	4	23	4	1	4	36	5	1	4	110
3	1	5	27	4	1	5	42	5	1	5	130
3	2	0	14	4	2	0	22	5	2	0	49
3	2	1	17	4	2	1	26	5	2	1	70
3	2	2	20	4	2	2	32	5	2	2	95
3	2	3	24	4	2	3	38	5	2	3	120
3	2	4	27	4	2	4	44	5	2	4	150
3	2	5	31	4	2	5	50	5	2	5	180
3	3	0	17	4	3	0	27	5	3	0	79
3	3	1	21	4	3	1	33	5	3	1	110
3	3	2	24	4	3	2	39	5	3	2	140
3	3	3	28	4	3	3	45	5	3	3	180
3	3	4	31	4	3	4	52	5	3	4	210
3	3	5	35	4	3	5	59	5	3	5	250
3	4	0	21	4	4	0	34	5	4	0	130
3	4	1	24	4	4	1	40	5	4	1	170
3	4	2	28	4	4	2	47	5	4	2	220
3	4	3	32	4	4	3	54	5	4	3	280
3	4	4	36	4	4	4	62	5	4	4	350
3	4	5	40	4	4	5	69	5	4	5	430
3	5	0	25	4	5	0	41	5	5	0	240
3	5	1	29	4	5	1	48	5	5	1	350
3	5	2	32	4	5	2	56	5	5	2	540
3	5	3	37	4	5	3	64	5	5	3	920
3	5	4	41	4	5	4	72	5	5	4	1600
3	5	5	45	4	5	5	81				

TABLE G-2
Most probable number (MPN) and 95 percent confidence limits when five tubes are used per dilution (10 mL, 1.0 mL, and 0.1 mL)*

Combination of positive tubes	MPN/ 100 mL	95% Confidence limits		Combination of positive tubes	MPN/ 100 mL	95% Confidence limits	
		Lower	Upper			Lower	Upper
0-0-0	< 2	-	-	4-2-0	22	9.0	56
0-0-1	2	1.0	10	4-2-1	26	12	65
0-1-0	2	1.0	10	4-3-0	27	12	67
0-2-0	4	1.0	13	4-3-1	33	15	77
1-0-0	2	1.0	11	4-4-0	34	16	80
1-0-1	4	1.0	15	5-0-0	23	9.0	86
1-1-0	4	1.0	15	5-0-1	30	10	110
1-1-1	6	2.0	18	5-0-2	40	20	140
1-2-0	6	2.0	18	5-1-0	30	10	120
2-0-0	4	1.0	17	5-1-1	50	20	150
2-0-1	7	2.0	20	5-1-2	60	30	180
2-1-0	7	2.0	21	5-2-0	50	20	170
2-1-1	9	3.0	24	5-2-1	70	30	210
2-2-0	9	3.0	25	5-2-2	90	40	250
2-3-0	12	5.0	29	5-3-0	80	30	250
3-0-0	8	3.0	24	5-3-1	110	40	300
3-0-1	11	4.0	29	5-3-2	140	60	360
3-1-0	11	4.0	29	5-3-3	170	80	410
3-1-1	14	6.0	35	5-4-0	130	50	390
3-2-0	14	6.0	35	5-4-1	170	70	480
3-2-1	17	7.0	40	5-4-2	220	100	580
4-0-0	13	5.0	38	5-4-3	280	120	690
4-0-1	17	7.0	45	5-4-4	350	160	820
4-1-0	17	7.0	46	5-5-0	240	100	940
4-1-1	21	9.0	55	5-5-1	300	100	1300
4-1-2	26	12	63	5-5-2	500	200	2000
				5-5-3	900	300	2900
				5-5-4	1600	600	5300
				5-5-5	≥ 1600		

*From Standard Methods (1995).

In example *c*, the first three dilutions are used so as to throw the positive result in the middle dilution. Where positive results occur in dilutions higher than the three chosen according to the above rule, they are incorporated into the result of the highest chosen dilution up to a total of five. The results of applying this procedure to the data are illustrated in examples *d* and *e*.

Where the quality of the treated effluent is high, ten 10-mL tubes are often used to determine the MPN value. The use of larger volumes of sample increases the detection limits. The MPN values and the 95 percent confidence limits when ten 10-mL tubes are used are given in Table G-3.

TABLE G-3

Most probable number (MPN) and 95 percent confidence limits when ten 10-mL tubes are used*

Combination of positive tubes	MPN/ 100 mL	95% Confidence limits	
		Lower	Upper
0	<1.1	0	3.0
1	1.1	0.03	5.9
2	2.2	0.26	8.1
3	3.6	0.69	10.6
4	5.1	1.3	13.4
5	6.9	2.1	16.8
6	9.2	3.1	21.1
7	12.0	4.3	27.1
8	16.1	5.9	36.8
9	23.0	8.1	59.5
10	>23.0	13.5	Infinite

*From Standard Methods (1995).

Example	1.0 mL	0.1 mL	0.01 mL	0.001 mL	0.0001 mL	Combination of positives	MPN/ 100 mL
a	**5/5**	**5/5**	**2/5**	**0/5**		**5-2-0**	4900
b	**5/5**	5/5	**4/5**	**2/5**	**0/5**	**5-4-2**	2200
c	**5/5**	**0/5**	**1/5**	**0/5**	0/5	**0-1-0**	18
d	**5/5**	5/5	3/5	1/5	1/5		
*d**	**5/5**	**5/5**	**3/5**	**2/5**	0/5	**5-3-2**	1400
e	**5/5**	4/5	1/5	1/5	0/5		
*e**	**5/5**	**4/5**	**2/5**	0/5	0/5	**5-4-2**	2200

*Adjusted values used to determine the MPN from Tables G-1 and G-2.

REFERENCE

Standard Methods (1995) *Standard Methods for the Examination of Water and Waste Water* (1995) 19th ed., American Public Health Association, Washington, DC.

NAME INDEX

SUBJECT INDEX

Where a topic is listed both as an abbreviation and as a spelled out entry under one letter of the alphabet [e.g., UC (uniformity coefficient) and Uniformity coefficient (UC)], appropriate page numbers are given for both citation entries if there are no subheadings under the spelled out entry. If a number of subheadings are included under the spelled out entry, then the *see also* notation is used with the abbreviation [e.g., UC {*See also* Uniformity coefficient (UC)}]. Where a topic is listed under several alphabetical headings, (e.g., Filter, multipass (recirculating) packed-bed, Multipass (recirculating) packed-bed filter, Packed-bed multipass (recirculating) filter), appropriate inclusive page numbers are given for each citation entry. The *see also* notation is used to direct the reader to the alphabetical citation where the complete set of subheadings is presented. For older or unused terms, the *see* notation is used to direct the reader to the appropriate term used in this text.